Lecture Notes in Computer Science 12361

More information about this series at http://www.springer.com/series/7412

Andrea Vedaldi · Horst Bischof ·
Thomas Brox · Jan-Michael Frahm (Eds.)

Computer Vision – ECCV 2020

16th European Conference
Glasgow, UK, August 23–28, 2020
Proceedings, Part XVI

Editors
Andrea Vedaldi
University of Oxford
Oxford, UK

Horst Bischof
Graz University of Technology
Graz, Austria

Thomas Brox
University of Freiburg
Freiburg im Breisgau, Germany

Jan-Michael Frahm
University of North Carolina at Chapel Hill
Chapel Hill, NC, USA

ISSN 0302-9743 ISSN 1611-3349 (electronic)
Lecture Notes in Computer Science
ISBN 978-3-030-58516-7 ISBN 978-3-030-58517-4 (eBook)
https://doi.org/10.1007/978-3-030-58517-4

LNCS Sublibrary: SL6 – Image Processing, Computer Vision, Pattern Recognition, and Graphics

This Springer imprint is published by the registered company Springer Nature Switzerland AG
The registered company address is: Gewerbestrasse 11, 6330 Cham, Switzerland

Foreword

Hosting the European Conference on Computer Vision (ECCV 2020) was certainly an exciting journey. From the 2016 plan to hold it at the Edinburgh International Conference Centre (hosting 1,800 delegates) to the 2018 plan to hold it at Glasgow's Scottish Exhibition Centre (up to 6,000 delegates), we finally ended with moving online because of the COVID-19 outbreak. While possibly having fewer delegates than expected because of the online format, ECCV 2020 still had over 3,100 registered participants.

Although online, the conference delivered most of the activities expected at a face-to-face conference: peer-reviewed papers, industrial exhibitors, demonstrations, and messaging between delegates. In addition to the main technical sessions, the conference included a strong program of satellite events with 16 tutorials and 44 workshops.

Furthermore, the online conference format enabled new conference features. Every paper had an associated teaser video and a longer full presentation video. Along with the papers and slides from the videos, all these materials were available the week before the conference. This allowed delegates to become familiar with the paper content and be ready for the live interaction with the authors during the conference week. The live event consisted of brief presentations by the oral and spotlight authors and industrial sponsors. Question and answer sessions for all papers were timed to occur twice so delegates from around the world had convenient access to the authors.

As with ECCV 2018, authors' draft versions of the papers appeared online with open access, now on both the Computer Vision Foundation (CVF) and the European Computer Vision Association (ECVA) websites. An archival publication arrangement was put in place with the cooperation of Springer. SpringerLink hosts the final version of the papers with further improvements, such as activating reference links and supplementary materials. These two approaches benefit all potential readers: a version available freely for all researchers, and an authoritative and citable version with additional benefits for SpringerLink subscribers. We thank Alfred Hofmann and Aliaksandr Birukou from Springer for helping to negotiate this agreement, which we expect will continue for future versions of ECCV.

August 2020

Vittorio Ferrari
Bob Fisher
Cordelia Schmid
Emanuele Trucco

Preface

Welcome to the proceedings of the European Conference on Computer Vision (ECCV 2020). This is a unique edition of ECCV in many ways. Due to the COVID-19 pandemic, this is the first time the conference was held online, in a virtual format. This was also the first time the conference relied exclusively on the Open Review platform to manage the review process. Despite these challenges ECCV is thriving. The conference received 5,150 valid paper submissions, of which 1,360 were accepted for publication (27%) and, of those, 160 were presented as spotlights (3%) and 104 as orals (2%). This amounts to more than twice the number of submissions to ECCV 2018 (2,439). Furthermore, CVPR, the largest conference on computer vision, received 5,850 submissions this year, meaning that ECCV is now 87% the size of CVPR in terms of submissions. By comparison, in 2018 the size of ECCV was only 73% of CVPR.

The review model was similar to previous editions of ECCV; in particular, it was double blind in the sense that the authors did not know the name of the reviewers and vice versa. Furthermore, each conference submission was held confidentially, and was only publicly revealed if and once accepted for publication. Each paper received at least three reviews, totalling more than 15,000 reviews. Handling the review process at this scale was a significant challenge. In order to ensure that each submission received as fair and high-quality reviews as possible, we recruited 2,830 reviewers (a 130% increase with reference to 2018) and 207 area chairs (a 60% increase). The area chairs were selected based on their technical expertise and reputation, largely among people that served as area chair in previous top computer vision and machine learning conferences (ECCV, ICCV, CVPR, NeurIPS, etc.). Reviewers were similarly invited from previous conferences. We also encouraged experienced area chairs to suggest additional chairs and reviewers in the initial phase of recruiting.

Despite doubling the number of submissions, the reviewer load was slightly reduced from 2018, from a maximum of 8 papers down to 7 (with some reviewers offering to handle 6 papers plus an emergency review). The area chair load increased slightly, from 18 papers on average to 22 papers on average.

Conflicts of interest between authors, area chairs, and reviewers were handled largely automatically by the Open Review platform via their curated list of user profiles. Many authors submitting to ECCV already had a profile in Open Review. We set a paper registration deadline one week before the paper submission deadline in order to encourage all missing authors to register and create their Open Review profiles well on time (in practice, we allowed authors to create/change papers arbitrarily until the submission deadline). Except for minor issues with users creating duplicate profiles, this allowed us to easily and quickly identify institutional conflicts, and avoid them, while matching papers to area chairs and reviewers.

Papers were matched to area chairs based on: an affinity score computed by the Open Review platform, which is based on paper titles and abstracts, and an affinity score computed by the Toronto Paper Matching System (TPMS), which is based on the paper's full text, the area chair bids for individual papers, load balancing, and conflict avoidance. Open Review provides the program chairs a convenient web interface to experiment with different configurations of the matching algorithm. The chosen configuration resulted in about 50% of the assigned papers to be highly ranked by the area chair bids, and 50% to be ranked in the middle, with very few low bids assigned.

Assignments to reviewers were similar, with two differences. First, there was a maximum of 7 papers assigned to each reviewer. Second, area chairs recommended up to seven reviewers per paper, providing another highly-weighed term to the affinity scores used for matching.

The assignment of papers to area chairs was smooth. However, it was more difficult to find suitable reviewers for all papers. Having a ratio of 5.6 papers per reviewer with a maximum load of 7 (due to emergency reviewer commitment), which did not allow for much wiggle room in order to also satisfy conflict and expertise constraints. We received some complaints from reviewers who did not feel qualified to review specific papers and we reassigned them wherever possible. However, the large scale of the conference, the many constraints, and the fact that a large fraction of such complaints arrived very late in the review process made this process very difficult and not all complaints could be addressed.

Reviewers had six weeks to complete their assignments. Possibly due to COVID-19 or the fact that the NeurIPS deadline was moved closer to the review deadline, a record 30% of the reviews were still missing after the deadline. By comparison, ECCV 2018 experienced only 10% missing reviews at this stage of the process. In the subsequent week, area chairs chased the missing reviews intensely, found replacement reviewers in their own team, and managed to reach 10% missing reviews. Eventually, we could provide almost all reviews (more than 99.9%) with a delay of only a couple of days on the initial schedule by a significant use of emergency reviews. If this trend is confirmed, it might be a major challenge to run a smooth review process in future editions of ECCV. The community must reconsider prioritization of the time spent on paper writing (the number of submissions increased a lot despite COVID-19) and time spent on paper reviewing (the number of reviews delivered in time decreased a lot presumably due to COVID-19 or NeurIPS deadline). With this imbalance the peer-review system that ensures the quality of our top conferences may break soon.

Reviewers submitted their reviews independently. In the reviews, they had the opportunity to ask questions to the authors to be addressed in the rebuttal. However, reviewers were told not to request any significant new experiment. Using the Open Review interface, authors could provide an answer to each individual review, but were also allowed to cross-reference reviews and responses in their answers. Rather than PDF files, we allowed the use of formatted text for the rebuttal. The rebuttal and initial reviews were then made visible to all reviewers and the primary area chair for a given paper. The area chair encouraged and moderated the reviewer discussion. During the discussions, reviewers were invited to reach a consensus and possibly adjust their ratings as a result of the discussion and of the evidence in the rebuttal.

After the discussion period ended, most reviewers entered a final rating and recommendation, although in many cases this did not differ from their initial recommendation. Based on the updated reviews and discussion, the primary area chair then made a preliminary decision to accept or reject the paper and wrote a justification for it (meta-review). Except for cases where the outcome of this process was absolutely clear (as indicated by the three reviewers and primary area chairs all recommending clear rejection), the decision was then examined and potentially challenged by a secondary area chair. This led to further discussion and overturning a small number of preliminary decisions. Needless to say, there was no in-person area chair meeting, which would have been impossible due to COVID-19.

Area chairs were invited to observe the consensus of the reviewers whenever possible and use extreme caution in overturning a clear consensus to accept or reject a paper. If an area chair still decided to do so, she/he was asked to clearly justify it in the meta-review and to explicitly obtain the agreement of the secondary area chair. In practice, very few papers were rejected after being confidently accepted by the reviewers.

This was the first time Open Review was used as the main platform to run ECCV. In 2018, the program chairs used CMT3 for the user-facing interface and Open Review internally, for matching and conflict resolution. Since it is clearly preferable to only use a single platform, this year we switched to using Open Review in full. The experience was largely positive. The platform is highly-configurable, scalable, and open source. Being written in Python, it is easy to write scripts to extract data programmatically. The paper matching and conflict resolution algorithms and interfaces are top-notch, also due to the excellent author profiles in the platform. Naturally, there were a few kinks along the way due to the fact that the ECCV Open Review configuration was created from scratch for this event and it differs in substantial ways from many other Open Review conferences. However, the Open Review development and support team did a fantastic job in helping us to get the configuration right and to address issues in a timely manner as they unavoidably occurred. We cannot thank them enough for the tremendous effort they put into this project.

Finally, we would like to thank everyone involved in making ECCV 2020 possible in these very strange and difficult times. This starts with our authors, followed by the area chairs and reviewers, who ran the review process at an unprecedented scale. The whole Open Review team (and in particular Melisa Bok, Mohit Unyal, Carlos Mondragon Chapa, and Celeste Martinez Gomez) worked incredibly hard for the entire duration of the process. We would also like to thank René Vidal for contributing to the adoption of Open Review. Our thanks also go to Laurent Charling for TPMS and to the program chairs of ICML, ICLR, and NeurIPS for cross checking double submissions. We thank the website chair, Giovanni Farinella, and the CPI team (in particular Ashley Cook, Miriam Verdon, Nicola McGrane, and Sharon Kerr) for promptly adding material to the website as needed in the various phases of the process. Finally, we thank the publication chairs, Albert Ali Salah, Hamdi Dibeklioglu, Metehan Doyran, Henry Howard-Jenkins, Victor Prisacariu, Siyu Tang, and Gul Varol, who managed to compile these substantial proceedings in an exceedingly compressed schedule. We express our thanks to the ECVA team, in particular Kristina Scherbaum for allowing open access of the proceedings. We thank Alfred Hofmann from Springer who again serve as the publisher. Finally, we thank the other chairs of ECCV 2020, including in particular the general chairs for very useful feedback with the handling of the program.

August 2020

Andrea Vedaldi
Horst Bischof
Thomas Brox
Jan-Michael Frahm

Organization

General Chairs

Vittorio Ferrari	Google Research, Switzerland
Bob Fisher	University of Edinburgh, UK
Cordelia Schmid	Google and Inria, France
Emanuele Trucco	University of Dundee, UK

Program Chairs

Andrea Vedaldi	University of Oxford, UK
Horst Bischof	Graz University of Technology, Austria
Thomas Brox	University of Freiburg, Germany
Jan-Michael Frahm	University of North Carolina, USA

Industrial Liaison Chairs

Jim Ashe	University of Edinburgh, UK
Helmut Grabner	Zurich University of Applied Sciences, Switzerland
Diane Larlus	NAVER LABS Europe, France
Cristian Novotny	University of Edinburgh, UK

Local Arrangement Chairs

Yvan Petillot	Heriot-Watt University, UK
Paul Siebert	University of Glasgow, UK

Academic Demonstration Chair

Thomas Mensink	Google Research and University of Amsterdam, The Netherlands

Poster Chair

Stephen Mckenna	University of Dundee, UK

Technology Chair

Gerardo Aragon Camarasa	University of Glasgow, UK

Tutorial Chairs

Carlo Colombo	University of Florence, Italy
Sotirios Tsaftaris	University of Edinburgh, UK

Publication Chairs

Albert Ali Salah	Utrecht University, The Netherlands
Hamdi Dibeklioglu	Bilkent University, Turkey
Metehan Doyran	Utrecht University, The Netherlands
Henry Howard-Jenkins	University of Oxford, UK
Victor Adrian Prisacariu	University of Oxford, UK
Siyu Tang	ETH Zurich, Switzerland
Gul Varol	University of Oxford, UK

Website Chair

Giovanni Maria Farinella	University of Catania, Italy

Workshops Chairs

Adrien Bartoli	University of Clermont Auvergne, France
Andrea Fusiello	University of Udine, Italy

Area Chairs

Lourdes Agapito	University College London, UK
Zeynep Akata	University of Tübingen, Germany
Karteek Alahari	Inria, France
Antonis Argyros	University of Crete, Greece
Hossein Azizpour	KTH Royal Institute of Technology, Sweden
Joao P. Barreto	Universidade de Coimbra, Portugal
Alexander C. Berg	University of North Carolina at Chapel Hill, USA
Matthew B. Blaschko	KU Leuven, Belgium
Lubomir D. Bourdev	WaveOne, Inc., USA
Edmond Boyer	Inria, France
Yuri Boykov	University of Waterloo, Canada
Gabriel Brostow	University College London, UK
Michael S. Brown	National University of Singapore, Singapore
Jianfei Cai	Monash University, Australia
Barbara Caputo	Politecnico di Torino, Italy
Ayan Chakrabarti	Washington University, St. Louis, USA
Tat-Jen Cham	Nanyang Technological University, Singapore
Manmohan Chandraker	University of California, San Diego, USA
Rama Chellappa	Johns Hopkins University, USA
Liang-Chieh Chen	Google, USA

Timothy Hospedales	University of Edinburgh and Samsung, UK
Gang Hua	Wormpex AI Research, USA
Slobodan Ilic	Siemens AG, Germany
Hiroshi Ishikawa	Waseda University, Japan
Jiaya Jia	The Chinese University of Hong Kong, SAR China
Hailin Jin	Adobe Research, USA
Justin Johnson	University of Michigan, USA
Frederic Jurie	University of Caen Normandie, France
Fredrik Kahl	Chalmers University, Sweden
Sing Bing Kang	Zillow, USA
Gunhee Kim	Seoul National University, South Korea
Junmo Kim	Korea Advanced Institute of Science and Technology, South Korea
Tae-Kyun Kim	Imperial College London, UK
Ron Kimmel	Technion-Israel Institute of Technology, Israel
Alexander Kirillov	Facebook AI Research, USA
Kris Kitani	Carnegie Mellon University, USA
Iasonas Kokkinos	Ariel AI, UK
Vladlen Koltun	Intel Labs, USA
Nikos Komodakis	Ecole des Ponts ParisTech, France
Piotr Koniusz	Australian National University, Australia
M. Pawan Kumar	University of Oxford, UK
Kyros Kutulakos	University of Toronto, Canada
Christoph Lampert	IST Austria, Austria
Ivan Laptev	Inria, France
Diane Larlus	NAVER LABS Europe, France
Laura Leal-Taixe	Technical University Munich, Germany
Honglak Lee	Google and University of Michigan, USA
Joon-Young Lee	Adobe Research, USA
Kyoung Mu Lee	Seoul National University, South Korea
Seungyong Lee	POSTECH, South Korea
Yong Jae Lee	University of California, Davis, USA
Bastian Leibe	RWTH Aachen University, Germany
Victor Lempitsky	Samsung, Russia
Ales Leonardis	University of Birmingham, UK
Marius Leordeanu	Institute of Mathematics of the Romanian Academy, Romania
Vincent Lepetit	ENPC ParisTech, France
Hongdong Li	The Australian National University, Australia
Xi Li	Zhejiang University, China
Yin Li	University of Wisconsin-Madison, USA
Zicheng Liao	Zhejiang University, China
Jongwoo Lim	Hanyang University, South Korea
Stephen Lin	Microsoft Research Asia, China
Yen-Yu Lin	National Chiao Tung University, Taiwan, China
Zhe Lin	Adobe Research, USA

Haibin Ling	Stony Brooks, State University of New York, USA
Jiaying Liu	Peking University, China
Ming-Yu Liu	NVIDIA, USA
Si Liu	Beihang University, China
Xiaoming Liu	Michigan State University, USA
Huchuan Lu	Dalian University of Technology, China
Simon Lucey	Carnegie Mellon University, USA
Jiebo Luo	University of Rochester, USA
Julien Mairal	Inria, France
Michael Maire	University of Chicago, USA
Subhransu Maji	University of Massachusetts, Amherst, USA
Yasushi Makihara	Osaka University, Japan
Jiri Matas	Czech Technical University in Prague, Czech Republic
Yasuyuki Matsushita	Osaka University, Japan
Philippos Mordohai	Stevens Institute of Technology, USA
Vittorio Murino	University of Verona, Italy
Naila Murray	NAVER LABS Europe, France
Hajime Nagahara	Osaka University, Japan
P. J. Narayanan	International Institute of Information Technology (IIIT), Hyderabad, India
Nassir Navab	Technical University of Munich, Germany
Natalia Neverova	Facebook AI Research, France
Matthias Niessner	Technical University of Munich, Germany
Jean-Marc Odobez	Idiap Research Institute and Swiss Federal Institute of Technology Lausanne, Switzerland
Francesca Odone	Universita di Genova, Italy
Takeshi Oishi	The University of Tokyo, Tokyo Institute of Technology, Japan
Vicente Ordonez	University of Virginia, USA
Manohar Paluri	Facebook AI Research, USA
Maja Pantic	Imperial College London, UK
In Kyu Park	Inha University, South Korea
Ioannis Patras	Queen Mary University of London, UK
Patrick Perez	Valeo, France
Bryan A. Plummer	Boston University, USA
Thomas Pock	Graz University of Technology, Austria
Marc Pollefeys	ETH Zurich and Microsoft MR & AI Zurich Lab, Switzerland
Jean Ponce	Inria, France
Gerard Pons-Moll	MPII, Saarland Informatics Campus, Germany
Jordi Pont-Tuset	Google, Switzerland
James Matthew Rehg	Georgia Institute of Technology, USA
Ian Reid	University of Adelaide, Australia
Olaf Ronneberger	DeepMind London, UK
Stefan Roth	TU Darmstadt, Germany
Bryan Russell	Adobe Research, USA

Mathieu Salzmann	EPFL, Switzerland
Dimitris Samaras	Stony Brook University, USA
Imari Sato	National Institute of Informatics (NII), Japan
Yoichi Sato	The University of Tokyo, Japan
Torsten Sattler	Czech Technical University in Prague, Czech Republic
Daniel Scharstein	Middlebury College, USA
Bernt Schiele	MPII, Saarland Informatics Campus, Germany
Julia A. Schnabel	King's College London, UK
Nicu Sebe	University of Trento, Italy
Greg Shakhnarovich	Toyota Technological Institute at Chicago, USA
Humphrey Shi	University of Oregon, USA
Jianbo Shi	University of Pennsylvania, USA
Jianping Shi	SenseTime, China
Leonid Sigal	University of British Columbia, Canada
Cees Snoek	University of Amsterdam, The Netherlands
Richard Souvenir	Temple University, USA
Hao Su	University of California, San Diego, USA
Akihiro Sugimoto	National Institute of Informatics (NII), Japan
Jian Sun	Megvii Technology, China
Jian Sun	Xi'an Jiaotong University, China
Chris Sweeney	Facebook Reality Labs, USA
Yu-wing Tai	Kuaishou Technology, China
Chi-Keung Tang	The Hong Kong University of Science and Technology, SAR China
Radu Timofte	ETH Zurich, Switzerland
Sinisa Todorovic	Oregon State University, USA
Giorgos Tolias	Czech Technical University in Prague, Czech Republic
Carlo Tomasi	Duke University, USA
Tatiana Tommasi	Politecnico di Torino, Italy
Lorenzo Torresani	Facebook AI Research and Dartmouth College, USA
Alexander Toshev	Google, USA
Zhuowen Tu	University of California, San Diego, USA
Tinne Tuytelaars	KU Leuven, Belgium
Jasper Uijlings	Google, Switzerland
Nuno Vasconcelos	University of California, San Diego, USA
Olga Veksler	University of Waterloo, Canada
Rene Vidal	Johns Hopkins University, USA
Gang Wang	Alibaba Group, China
Jingdong Wang	Microsoft Research Asia, China
Yizhou Wang	Peking University, China
Lior Wolf	Facebook AI Research and Tel Aviv University, Israel
Jianxin Wu	Nanjing University, China
Tao Xiang	University of Surrey, UK
Saining Xie	Facebook AI Research, USA
Ming-Hsuan Yang	University of California at Merced and Google, USA
Ruigang Yang	University of Kentucky, USA

Kwang Moo Yi	University of Victoria, Canada
Zhaozheng Yin	Stony Brook, State University of New York, USA
Chang D. Yoo	Korea Advanced Institute of Science and Technology, South Korea
Shaodi You	University of Amsterdam, The Netherlands
Jingyi Yu	ShanghaiTech University, China
Stella Yu	University of California, Berkeley, and ICSI, USA
Stefanos Zafeiriou	Imperial College London, UK
Hongbin Zha	Peking University, China
Tianzhu Zhang	University of Science and Technology of China, China
Liang Zheng	Australian National University, Australia
Todd E. Zickler	Harvard University, USA
Andrew Zisserman	University of Oxford, UK

Technical Program Committee

Sathyanarayanan N. Aakur
Wael Abd Almgaeed
Abdelrahman Abdelhamed
Abdullah Abuolaim
Supreeth Achar
Hanno Ackermann
Ehsan Adeli
Triantafyllos Afouras
Sameer Agarwal
Aishwarya Agrawal
Harsh Agrawal
Pulkit Agrawal
Antonio Agudo
Eirikur Agustsson
Karim Ahmed
Byeongjoo Ahn
Unaiza Ahsan
Thalaiyasingam Ajanthan
Kenan E. Ak
Emre Akbas
Naveed Akhtar
Derya Akkaynak
Yagiz Aksoy
Ziad Al-Halah
Xavier Alameda-Pineda
Jean-Baptiste Alayrac
Samuel Albanie
Shadi Albarqouni
Cenek Albl
Hassan Abu Alhaija
Daniel Aliaga
Mohammad S. Aliakbarian
Rahaf Aljundi
Thiemo Alldieck
Jon Almazan
Jose M. Alvarez
Senjian An
Saket Anand
Codruta Ancuti
Cosmin Ancuti
Peter Anderson
Juan Andrade-Cetto
Alexander Andreopoulos
Misha Andriluka
Dragomir Anguelov
Rushil Anirudh
Michel Antunes
Oisin Mac Aodha
Srikar Appalaraju
Relja Arandjelovic
Nikita Araslanov
Andre Araujo
Helder Araujo
Pablo Arbelaez
Shervin Ardeshir
Sercan O. Arik
Anil Armagan
Anurag Arnab
Chetan Arora
Federica Arrigoni
Mathieu Aubry
Shai Avidan
Angelica I. Aviles-Rivero
Yannis Avrithis
Ismail Ben Ayed
Shekoofeh Azizi
Ioan Andrei Bârsan
Artem Babenko
Deepak Babu Sam
Seung-Hwan Baek
Seungryul Baek
Andrew D. Bagdanov
Shai Bagon
Yuval Bahat
Junjie Bai
Song Bai
Xiang Bai
Yalong Bai
Yancheng Bai
Peter Bajcsy
Slawomir Bak

Mahsa Baktashmotlagh
Kavita Bala
Yogesh Balaji
Guha Balakrishnan
V. N. Balasubramanian
Federico Baldassarre
Vassileios Balntas
Shurjo Banerjee
Aayush Bansal
Ankan Bansal
Jianmin Bao
Linchao Bao
Wenbo Bao
Yingze Bao
Akash Bapat
Md Jawadul Hasan Bappy
Fabien Baradel
Lorenzo Baraldi
Daniel Barath
Adrian Barbu
Kobus Barnard
Nick Barnes
Francisco Barranco
Jonathan T. Barron
Arslan Basharat
Chaim Baskin
Anil S. Baslamisli
Jorge Batista
Kayhan Batmanghelich
Konstantinos Batsos
David Bau
Luis Baumela
Christoph Baur
Eduardo Bayro-Corrochano
Paul Beardsley
Jan Bednavr'ik
Oscar Beijbom
Philippe Bekaert
Esube Bekele
Vasileios Belagiannis
Ohad Ben-Shahar
Abhijit Bendale
Róger Bermúdez-Chacón
Maxim Berman
Jesus Bermudez-cameo
Florian Bernard
Stefano Berretti
Marcelo Bertalmio
Gedas Bertasius
Cigdem Beyan
Lucas Beyer
Vijayakumar Bhagavatula
Arjun Nitin Bhagoji
Apratim Bhattacharyya
Binod Bhattarai
Sai Bi
Jia-Wang Bian
Simone Bianco
Adel Bibi
Tolga Birdal
Tom Bishop
Soma Biswas
Mårten Björkman
Volker Blanz
Vishnu Boddeti
Navaneeth Bodla
Simion-Vlad Bogolin
Xavier Boix
Piotr Bojanowski
Timo Bolkart
Guido Borghi
Larbi Boubchir
Guillaume Bourmaud
Adrien Bousseau
Thierry Bouwmans
Richard Bowden
Hakan Boyraz
Mathieu Brédif
Samarth Brahmbhatt
Steve Branson
Nikolas Brasch
Biagio Brattoli
Ernesto Brau
Toby P. Breckon
Francois Bremond
Jesus Briales
Sofia Broomé
Marcus A. Brubaker
Luc Brun
Silvia Bucci
Shyamal Buch
Pradeep Buddharaju
Uta Buechler
Mai Bui
Tu Bui
Adrian Bulat
Giedrius T. Burachas
Elena Burceanu
Xavier P. Burgos-Artizzu
Kaylee Burns
Andrei Bursuc
Benjamin Busam
Wonmin Byeon
Zoya Bylinskii
Sergi Caelles
Jianrui Cai
Minjie Cai
Yujun Cai
Zhaowei Cai
Zhipeng Cai
Juan C. Caicedo
Simone Calderara
Necati Cihan Camgoz
Dylan Campbell
Octavia Camps
Jiale Cao
Kaidi Cao
Liangliang Cao
Xiangyong Cao
Xiaochun Cao
Yang Cao
Yu Cao
Yue Cao
Zhangjie Cao
Luca Carlone
Mathilde Caron
Dan Casas
Thomas J. Cashman
Umberto Castellani
Lluis Castrejon
Jacopo Cavazza
Fabio Cermelli
Hakan Cevikalp
Menglei Chai
Ishani Chakraborty
Rudrasis Chakraborty
Antoni B. Chan

Kwok-Ping Chan
Siddhartha Chandra
Sharat Chandran
Arjun Chandrasekaran
Angel X. Chang
Che-Han Chang
Hong Chang
Hyun Sung Chang
Hyung Jin Chang
Jianlong Chang
Ju Yong Chang
Ming-Ching Chang
Simyung Chang
Xiaojun Chang
Yu-Wei Chao
Devendra S. Chaplot
Arslan Chaudhry
Rizwan A. Chaudhry
Can Chen
Chang Chen
Chao Chen
Chen Chen
Chu-Song Chen
Dapeng Chen
Dong Chen
Dongdong Chen
Guanying Chen
Hongge Chen
Hsin-yi Chen
Huaijin Chen
Hwann-Tzong Chen
Jianbo Chen
Jianhui Chen
Jiansheng Chen
Jiaxin Chen
Jie Chen
Jun-Cheng Chen
Kan Chen
Kevin Chen
Lin Chen
Long Chen
Min-Hung Chen
Qifeng Chen
Shi Chen
Shixing Chen
Tianshui Chen
Weifeng Chen
Weikai Chen
Xi Chen
Xiaohan Chen
Xiaozhi Chen
Xilin Chen
Xingyu Chen
Xinlei Chen
Xinyun Chen
Yi-Ting Chen
Yilun Chen
Ying-Cong Chen
Yinpeng Chen
Yiran Chen
Yu Chen
Yu-Sheng Chen
Yuhua Chen
Yun-Chun Chen
Yunpeng Chen
Yuntao Chen
Zhuoyuan Chen
Zitian Chen
Anchieh Cheng
Bowen Cheng
Erkang Cheng
Gong Cheng
Guangliang Cheng
Jingchun Cheng
Jun Cheng
Li cheng
Ming-Ming Cheng
Yu Cheng
Ziang Cheng
Anoop Cherian
Dmitry Chetverikov
Ngai-man Cheung
William Cheung
Ajad Chhatkuli
Naoki Chiba
Benjamin Chidester
Han-pang Chiu
Mang Tik Chiu
Wei-Chen Chiu
Donghyeon Cho
Hojin Cho
Minsu Cho
Nam Ik Cho
Tim Cho
Tae Eun Choe
Chiho Choi
Edward Choi
Inchang Choi
Jinsoo Choi
Jonghyun Choi
Jongwon Choi
Yukyung Choi
Hisham Cholakkal
Eunji Chong
Jaegul Choo
Christopher Choy
Hang Chu
Peng Chu
Wen-Sheng Chu
Albert Chung
Joon Son Chung
Hai Ci
Safa Cicek
Ramazan G. Cinbis
Arridhana Ciptadi
Javier Civera
James J. Clark
Ronald Clark
Felipe Codevilla
Michael Cogswell
Andrea Cohen
Maxwell D. Collins
Carlo Colombo
Yang Cong
Adria R. Continente
Marcella Cornia
John Richard Corring
Darren Cosker
Dragos Costea
Garrison W. Cottrell
Florent Couzinie-Devy
Marco Cristani
Ioana Croitoru
James L. Crowley
Jiequan Cui
Zhaopeng Cui
Ross Cutler
Antonio D'Innocente

Rozenn Dahyot
Bo Dai
Dengxin Dai
Hang Dai
Longquan Dai
Shuyang Dai
Xiyang Dai
Yuchao Dai
Adrian V. Dalca
Dima Damen
Bharath B. Damodaran
Kristin Dana
Martin Danelljan
Zheng Dang
Zachary Alan Daniels
Donald G. Dansereau
Abhishek Das
Samyak Datta
Achal Dave
Titas De
Rodrigo de Bem
Teo de Campos
Raoul de Charette
Shalini De Mello
Joseph DeGol
Herve Delingette
Haowen Deng
Jiankang Deng
Weijian Deng
Zhiwei Deng
Joachim Denzler
Konstantinos G. Derpanis
Aditya Deshpande
Frederic Devernay
Somdip Dey
Arturo Deza
Abhinav Dhall
Helisa Dhamo
Vikas Dhiman
Fillipe Dias Moreira de Souza
Ali Diba
Ferran Diego
Guiguang Ding
Henghui Ding
Jian Ding
Mingyu Ding
Xinghao Ding
Zhengming Ding
Robert DiPietro
Cosimo Distante
Ajay Divakaran
Mandar Dixit
Abdelaziz Djelouah
Thanh-Toan Do
Jose Dolz
Bo Dong
Chao Dong
Jiangxin Dong
Weiming Dong
Weisheng Dong
Xingping Dong
Xuanyi Dong
Yinpeng Dong
Gianfranco Doretto
Hazel Doughty
Hassen Drira
Bertram Drost
Dawei Du
Ye Duan
Yueqi Duan
Abhimanyu Dubey
Anastasia Dubrovina
Stefan Duffner
Chi Nhan Duong
Thibaut Durand
Zoran Duric
Iulia Duta
Debidatta Dwibedi
Benjamin Eckart
Marc Eder
Marzieh Edraki
Alexei A. Efros
Kiana Ehsani
Hazm Kemal Ekenel
James H. Elder
Mohamed Elgharib
Shireen Elhabian
Ehsan Elhamifar
Mohamed Elhoseiny
Ian Endres
N. Benjamin Erichson
Jan Ernst
Sergio Escalera
Francisco Escolano
Victor Escorcia
Carlos Esteves
Francisco J. Estrada
Bin Fan
Chenyou Fan
Deng-Ping Fan
Haoqi Fan
Hehe Fan
Heng Fan
Kai Fan
Lijie Fan
Linxi Fan
Quanfu Fan
Shaojing Fan
Xiaochuan Fan
Xin Fan
Yuchen Fan
Sean Fanello
Hao-Shu Fang
Haoyang Fang
Kuan Fang
Yi Fang
Yuming Fang
Azade Farshad
Alireza Fathi
Raanan Fattal
Joao Fayad
Xiaohan Fei
Christoph Feichtenhofer
Michael Felsberg
Chen Feng
Jiashi Feng
Junyi Feng
Mengyang Feng
Qianli Feng
Zhenhua Feng
Michele Fenzi
Andras Ferencz
Martin Fergie
Basura Fernando
Ethan Fetaya
Michael Firman
John W. Fisher

Matthew Fisher
Boris Flach
Corneliu Florea
Wolfgang Foerstner
David Fofi
Gian Luca Foresti
Per-Erik Forssen
David Fouhey
Katerina Fragkiadaki
Victor Fragoso
Jean-Sébastien Franco
Ohad Fried
Iuri Frosio
Cheng-Yang Fu
Huazhu Fu
Jianlong Fu
Jingjing Fu
Xueyang Fu
Yanwei Fu
Ying Fu
Yun Fu
Olac Fuentes
Kent Fujiwara
Takuya Funatomi
Christopher Funk
Thomas Funkhouser
Antonino Furnari
Ryo Furukawa
Erik Gärtner
Raghudeep Gadde
Matheus Gadelha
Vandit Gajjar
Trevor Gale
Juergen Gall
Mathias Gallardo
Guillermo Gallego
Orazio Gallo
Chuang Gan
Zhe Gan
Madan Ravi Ganesh
Aditya Ganeshan
Siddha Ganju
Bin-Bin Gao
Changxin Gao
Feng Gao
Hongchang Gao
Jin Gao
Jiyang Gao
Junbin Gao
Katelyn Gao
Lin Gao
Mingfei Gao
Ruiqi Gao
Ruohan Gao
Shenghua Gao
Yuan Gao
Yue Gao
Noa Garcia
Alberto Garcia-Garcia
Guillermo Garcia-Hernando
Jacob R. Gardner
Animesh Garg
Kshitiz Garg
Rahul Garg
Ravi Garg
Philip N. Garner
Kirill Gavrilyuk
Paul Gay
Shiming Ge
Weifeng Ge
Baris Gecer
Xin Geng
Kyle Genova
Stamatios Georgoulis
Bernard Ghanem
Michael Gharbi
Kamran Ghasedi
Golnaz Ghiasi
Arnab Ghosh
Partha Ghosh
Silvio Giancola
Andrew Gilbert
Rohit Girdhar
Xavier Giro-i-Nieto
Thomas Gittings
Ioannis Gkioulekas
Clement Godard
Vaibhava Goel
Bastian Goldluecke
Lluis Gomez
Nuno Gonçalves
Dong Gong
Ke Gong
Mingming Gong
Abel Gonzalez-Garcia
Ariel Gordon
Daniel Gordon
Paulo Gotardo
Venu Madhav Govindu
Ankit Goyal
Priya Goyal
Raghav Goyal
Benjamin Graham
Douglas Gray
Brent A. Griffin
Etienne Grossmann
David Gu
Jiayuan Gu
Jiuxiang Gu
Lin Gu
Qiao Gu
Shuhang Gu
Jose J. Guerrero
Paul Guerrero
Jie Gui
Jean-Yves Guillemaut
Riza Alp Guler
Erhan Gundogdu
Fatma Guney
Guodong Guo
Kaiwen Guo
Qi Guo
Sheng Guo
Shi Guo
Tiantong Guo
Xiaojie Guo
Yijie Guo
Yiluan Guo
Yuanfang Guo
Yulan Guo
Agrim Gupta
Ankush Gupta
Mohit Gupta
Saurabh Gupta
Tanmay Gupta
Danna Gurari
Abner Guzman-Rivera

JunYoung Gwak
Michael Gygli
Jung-Woo Ha
Simon Hadfield
Isma Hadji
Bjoern Haefner
Taeyoung Hahn
Levente Hajder
Peter Hall
Emanuela Haller
Stefan Haller
Bumsub Ham
Abdullah Hamdi
Dongyoon Han
Hu Han
Jungong Han
Junwei Han
Kai Han
Tian Han
Xiaoguang Han
Xintong Han
Yahong Han
Ankur Handa
Zekun Hao
Albert Haque
Tatsuya Harada
Mehrtash Harandi
Adam W. Harley
Mahmudul Hasan
Atsushi Hashimoto
Ali Hatamizadeh
Munawar Hayat
Dongliang He
Jingrui He
Junfeng He
Kaiming He
Kun He
Lei He
Pan He
Ran He
Shengfeng He
Tong He
Weipeng He
Xuming He
Yang He
Yihui He
Zhihai He
Chinmay Hegde
Janne Heikkila
Mattias P. Heinrich
Stéphane Herbin
Alexander Hermans
Luis Herranz
John R. Hershey
Aaron Hertzmann
Roei Herzig
Anders Heyden
Steven Hickson
Otmar Hilliges
Tomas Hodan
Judy Hoffman
Michael Hofmann
Yannick Hold-Geoffroy
Namdar Homayounfar
Sina Honari
Richang Hong
Seunghoon Hong
Xiaopeng Hong
Yi Hong
Hidekata Hontani
Anthony Hoogs
Yedid Hoshen
Mir Rayat Imtiaz Hossain
Junhui Hou
Le Hou
Lu Hou
Tingbo Hou
Wei-Lin Hsiao
Cheng-Chun Hsu
Gee-Sern Jison Hsu
Kuang-jui Hsu
Changbo Hu
Di Hu
Guosheng Hu
Han Hu
Hao Hu
Hexiang Hu
Hou-Ning Hu
Jie Hu
Junlin Hu
Nan Hu
Ping Hu
Ronghang Hu
Xiaowei Hu
Yinlin Hu
Yuan-Ting Hu
Zhe Hu
Binh-Son Hua
Yang Hua
Bingyao Huang
Di Huang
Dong Huang
Fay Huang
Haibin Huang
Haozhi Huang
Heng Huang
Huaibo Huang
Jia-Bin Huang
Jing Huang
Jingwei Huang
Kaizhu Huang
Lei Huang
Qiangui Huang
Qiaoying Huang
Qingqiu Huang
Qixing Huang
Shaoli Huang
Sheng Huang
Siyuan Huang
Weilin Huang
Wenbing Huang
Xiangru Huang
Xun Huang
Yan Huang
Yifei Huang
Yue Huang
Zhiwu Huang
Zilong Huang
Minyoung Huh
Zhuo Hui
Matthias B. Hullin
Martin Humenberger
Wei-Chih Hung
Zhouyuan Huo
Junhwa Hur
Noureldien Hussein
Jyh-Jing Hwang
Seong Jae Hwang

Sung Ju Hwang
Ichiro Ide
Ivo Ihrke
Daiki Ikami
Satoshi Ikehata
Nazli Ikizler-Cinbis
Sunghoon Im
Yani Ioannou
Radu Tudor Ionescu
Umar Iqbal
Go Irie
Ahmet Iscen
Md Amirul Islam
Vamsi Ithapu
Nathan Jacobs
Arpit Jain
Himalaya Jain
Suyog Jain
Stuart James
Won-Dong Jang
Yunseok Jang
Ronnachai Jaroensri
Dinesh Jayaraman
Sadeep Jayasumana
Suren Jayasuriya
Herve Jegou
Simon Jenni
Hae-Gon Jeon
Yunho Jeon
Koteswar R. Jerripothula
Hueihan Jhuang
I-hong Jhuo
Dinghuang Ji
Hui Ji
Jingwei Ji
Pan Ji
Yanli Ji
Baoxiong Jia
Kui Jia
Xu Jia
Chiyu Max Jiang
Haiyong Jiang
Hao Jiang
Huaizu Jiang
Huajie Jiang
Ke Jiang
Lai Jiang
Li Jiang
Lu Jiang
Ming Jiang
Peng Jiang
Shuqiang Jiang
Wei Jiang
Xudong Jiang
Zhuolin Jiang
Jianbo Jiao
Zequn Jie
Dakai Jin
Kyong Hwan Jin
Lianwen Jin
SouYoung Jin
Xiaojie Jin
Xin Jin
Nebojsa Jojic
Alexis Joly
Michael Jeffrey Jones
Hanbyul Joo
Jungseock Joo
Kyungdon Joo
Ajjen Joshi
Shantanu H. Joshi
Da-Cheng Juan
Marco Körner
Kevin Köser
Asim Kadav
Christine Kaeser-Chen
Kushal Kafle
Dagmar Kainmueller
Ioannis A. Kakadiaris
Zdenek Kalal
Nima Kalantari
Yannis Kalantidis
Mahdi M. Kalayeh
Anmol Kalia
Sinan Kalkan
Vicky Kalogeiton
Ashwin Kalyan
Joni-kristian Kamarainen
Gerda Kamberova
Chandra Kambhamettu
Martin Kampel
Meina Kan
Christopher Kanan
Kenichi Kanatani
Angjoo Kanazawa
Atsushi Kanehira
Takuhiro Kaneko
Asako Kanezaki
Bingyi Kang
Di Kang
Sunghun Kang
Zhao Kang
Vadim Kantorov
Abhishek Kar
Amlan Kar
Theofanis Karaletsos
Leonid Karlinsky
Kevin Karsch
Angelos Katharopoulos
Isinsu Katircioglu
Hiroharu Kato
Zoltan Kato
Dotan Kaufman
Jan Kautz
Rei Kawakami
Qiuhong Ke
Wadim Kehl
Petr Kellnhofer
Aniruddha Kembhavi
Cem Keskin
Margret Keuper
Daniel Keysers
Ashkan Khakzar
Fahad Khan
Naeemullah Khan
Salman Khan
Siddhesh Khandelwal
Rawal Khirodkar
Anna Khoreva
Tejas Khot
Parmeshwar Khurd
Hadi Kiapour
Joe Kileel
Chanho Kim
Dahun Kim
Edward Kim
Eunwoo Kim
Han-ul Kim

Hansung Kim
Heewon Kim
Hyo Jin Kim
Hyunwoo J. Kim
Jinkyu Kim
Jiwon Kim
Jongmin Kim
Junsik Kim
Junyeong Kim
Min H. Kim
Namil Kim
Pyojin Kim
Seon Joo Kim
Seong Tae Kim
Seungryong Kim
Sungwoong Kim
Tae Hyun Kim
Vladimir Kim
Won Hwa Kim
Yonghyun Kim
Benjamin Kimia
Akisato Kimura
Pieter-Jan Kindermans
Zsolt Kira
Itaru Kitahara
Hedvig Kjellstrom
Jan Knopp
Takumi Kobayashi
Erich Kobler
Parker Koch
Reinhard Koch
Elyor Kodirov
Amir Kolaman
Nicholas Kolkin
Dimitrios Kollias
Stefanos Kollias
Soheil Kolouri
Adams Wai-Kin Kong
Naejin Kong
Shu Kong
Tao Kong
Yu Kong
Yoshinori Konishi
Daniil Kononenko
Theodora Kontogianni
Simon Korman
Adam Kortylewski
Jana Kosecka
Jean Kossaifi
Satwik Kottur
Rigas Kouskouridas
Adriana Kovashka
Rama Kovvuri
Adarsh Kowdle
Jedrzej Kozerawski
Mateusz Kozinski
Philipp Kraehenbuehl
Gregory Kramida
Josip Krapac
Dmitry Kravchenko
Ranjay Krishna
Pavel Krsek
Alexander Krull
Jakob Kruse
Hiroyuki Kubo
Hilde Kuehne
Jason Kuen
Andreas Kuhn
Arjan Kuijper
Zuzana Kukelova
Ajay Kumar
Amit Kumar
Avinash Kumar
Suryansh Kumar
Vijay Kumar
Kaustav Kundu
Weicheng Kuo
Nojun Kwak
Suha Kwak
Junseok Kwon
Nikolaos Kyriazis
Zorah Lähner
Ankit Laddha
Florent Lafarge
Jean Lahoud
Kevin Lai
Shang-Hong Lai
Wei-Sheng Lai
Yu-Kun Lai
Iro Laina
Antony Lam
John Wheatley Lambert
Xiangyuan lan
Xu Lan
Charis Lanaras
Georg Langs
Oswald Lanz
Dong Lao
Yizhen Lao
Agata Lapedriza
Gustav Larsson
Viktor Larsson
Katrin Lasinger
Christoph Lassner
Longin Jan Latecki
Stéphane Lathuilière
Rynson Lau
Hei Law
Justin Lazarow
Svetlana Lazebnik
Hieu Le
Huu Le
Ngan Hoang Le
Trung-Nghia Le
Vuong Le
Colin Lea
Erik Learned-Miller
Chen-Yu Lee
Gim Hee Lee
Hsin-Ying Lee
Hyungtae Lee
Jae-Han Lee
Jimmy Addison Lee
Joonseok Lee
Kibok Lee
Kuang-Huei Lee
Kwonjoon Lee
Minsik Lee
Sang-chul Lee
Seungkyu Lee
Soochan Lee
Stefan Lee
Taehee Lee
Andreas Lehrmann
Jie Lei
Peng Lei
Matthew Joseph Leotta
Wee Kheng Leow

Gil Levi
Evgeny Levinkov
Aviad Levis
Jose Lezama
Ang Li
Bin Li
Bing Li
Boyi Li
Changsheng Li
Chao Li
Chen Li
Cheng Li
Chenglong Li
Chi Li
Chun-Guang Li
Chun-Liang Li
Chunyuan Li
Dong Li
Guanbin Li
Hao Li
Haoxiang Li
Hongsheng Li
Hongyang Li
Houqiang Li
Huibin Li
Jia Li
Jianan Li
Jianguo Li
Junnan Li
Junxuan Li
Kai Li
Ke Li
Kejie Li
Kunpeng Li
Lerenhan Li
Li Erran Li
Mengtian Li
Mu Li
Peihua Li
Peiyi Li
Ping Li
Qi Li
Qing Li
Ruiyu Li
Ruoteng Li
Shaozi Li
Sheng Li
Shiwei Li
Shuang Li
Siyang Li
Stan Z. Li
Tianye Li
Wei Li
Weixin Li
Wen Li
Wenbo Li
Xiaomeng Li
Xin Li
Xiu Li
Xuelong Li
Xueting Li
Yan Li
Yandong Li
Yanghao Li
Yehao Li
Yi Li
Yijun Li
Yikang LI
Yining Li
Yongjie Li
Yu Li
Yu-Jhe Li
Yunpeng Li
Yunsheng Li
Yunzhu Li
Zhe Li
Zhen Li
Zhengqi Li
Zhenyang Li
Zhuwen Li
Dongze Lian
Xiaochen Lian
Zhouhui Lian
Chen Liang
Jie Liang
Ming Liang
Paul Pu Liang
Pengpeng Liang
Shu Liang
Wei Liang
Jing Liao
Minghui Liao
Renjie Liao
Shengcai Liao
Shuai Liao
Yiyi Liao
Ser-Nam Lim
Chen-Hsuan Lin
Chung-Ching Lin
Dahua Lin
Ji Lin
Kevin Lin
Tianwei Lin
Tsung-Yi Lin
Tsung-Yu Lin
Wei-An Lin
Weiyao Lin
Yen-Chen Lin
Yuewei Lin
David B. Lindell
Drew Linsley
Krzysztof Lis
Roee Litman
Jim Little
An-An Liu
Bo Liu
Buyu Liu
Chao Liu
Chen Liu
Cheng-lin Liu
Chenxi Liu
Dong Liu
Feng Liu
Guilin Liu
Haomiao Liu
Heshan Liu
Hong Liu
Ji Liu
Jingen Liu
Jun Liu
Lanlan Liu
Li Liu
Liu Liu
Mengyuan Liu
Miaomiao Liu
Nian Liu
Ping Liu
Risheng Liu

Sheng Liu
Shu Liu
Shuaicheng Liu
Sifei Liu
Siqi Liu
Siying Liu
Songtao Liu
Ting Liu
Tongliang Liu
Tyng-Luh Liu
Wanquan Liu
Wei Liu
Weiyang Liu
Weizhe Liu
Wenyu Liu
Wu Liu
Xialei Liu
Xianglong Liu
Xiaodong Liu
Xiaofeng Liu
Xihui Liu
Xingyu Liu
Xinwang Liu
Xuanqing Liu
Xuebo Liu
Yang Liu
Yaojie Liu
Yebin Liu
Yen-Cheng Liu
Yiming Liu
Yu Liu
Yu-Shen Liu
Yufan Liu
Yun Liu
Zheng Liu
Zhijian Liu
Zhuang Liu
Zichuan Liu
Ziwei Liu
Zongyi Liu
Stephan Liwicki
Liliana Lo Presti
Chengjiang Long
Fuchen Long
Mingsheng Long
Xiang Long
Yang Long
Charles T. Loop
Antonio Lopez
Roberto J. Lopez-Sastre
Javier Lorenzo-Navarro
Manolis Lourakis
Boyu Lu
Canyi Lu
Feng Lu
Guoyu Lu
Hongtao Lu
Jiajun Lu
Jiasen Lu
Jiwen Lu
Kaiyue Lu
Le Lu
Shao-Ping Lu
Shijian Lu
Xiankai Lu
Xin Lu
Yao Lu
Yiping Lu
Yongxi Lu
Yongyi Lu
Zhiwu Lu
Fujun Luan
Benjamin E. Lundell
Hao Luo
Jian-Hao Luo
Ruotian Luo
Weixin Luo
Wenhan Luo
Wenjie Luo
Yan Luo
Zelun Luo
Zixin Luo
Khoa Luu
Zhaoyang Lv
Pengyuan Lyu
Thomas Möllenhoff
Matthias Müller
Bingpeng Ma
Chih-Yao Ma
Chongyang Ma
Huimin Ma
Jiayi Ma
K. T. Ma
Ke Ma
Lin Ma
Liqian Ma
Shugao Ma
Wei-Chiu Ma
Xiaojian Ma
Xingjun Ma
Zhanyu Ma
Zheng Ma
Radek Jakob Mackowiak
Ludovic Magerand
Shweta Mahajan
Siddharth Mahendran
Long Mai
Ameesh Makadia
Oscar Mendez Maldonado
Mateusz Malinowski
Yury Malkov
Arun Mallya
Dipu Manandhar
Massimiliano Mancini
Fabian Manhardt
Kevis-kokitsi Maninis
Varun Manjunatha
Junhua Mao
Xudong Mao
Alina Marcu
Edgar Margffoy-Tuay
Dmitrii Marin
Manuel J. Marin-Jimenez
Kenneth Marino
Niki Martinel
Julieta Martinez
Jonathan Masci
Tomohiro Mashita
Iacopo Masi
David Masip
Daniela Massiceti
Stefan Mathe
Yusuke Matsui
Tetsu Matsukawa
Iain A. Matthews
Kevin James Matzen
Bruce Allen Maxwell
Stephen Maybank

Helmut Mayer
Amir Mazaheri
David McAllester
Steven McDonagh
Stephen J. Mckenna
Roey Mechrez
Prakhar Mehrotra
Christopher Mei
Xue Mei
Paulo R. S. Mendonca
Lili Meng
Zibo Meng
Thomas Mensink
Bjoern Menze
Michele Merler
Kourosh Meshgi
Pascal Mettes
Christopher Metzler
Liang Mi
Qiguang Miao
Xin Miao
Tomer Michaeli
Frank Michel
Antoine Miech
Krystian Mikolajczyk
Peyman Milanfar
Ben Mildenhall
Gregor Miller
Fausto Milletari
Dongbo Min
Kyle Min
Pedro Miraldo
Dmytro Mishkin
Anand Mishra
Ashish Mishra
Ishan Misra
Niluthpol C. Mithun
Kaushik Mitra
Niloy Mitra
Anton Mitrokhin
Ikuhisa Mitsugami
Anurag Mittal
Kaichun Mo
Zhipeng Mo
Davide Modolo
Michael Moeller
Pritish Mohapatra
Pavlo Molchanov
Davide Moltisanti
Pascal Monasse
Mathew Monfort
Aron Monszpart
Sean Moran
Vlad I. Morariu
Francesc Moreno-Noguer
Pietro Morerio
Stylianos Moschoglou
Yael Moses
Roozbeh Mottaghi
Pierre Moulon
Arsalan Mousavian
Yadong Mu
Yasuhiro Mukaigawa
Lopamudra Mukherjee
Yusuke Mukuta
Ravi Teja Mullapudi
Mario Enrique Munich
Zachary Murez
Ana C. Murillo
J. Krishna Murthy
Damien Muselet
Armin Mustafa
Siva Karthik Mustikovela
Carlo Dal Mutto
Moin Nabi
Varun K. Nagaraja
Tushar Nagarajan
Arsha Nagrani
Seungjun Nah
Nikhil Naik
Yoshikatsu Nakajima
Yuta Nakashima
Atsushi Nakazawa
Seonghyeon Nam
Vinay P. Namboodiri
Medhini Narasimhan
Srinivasa Narasimhan
Sanath Narayan
Erickson Rangel Nascimento
Jacinto Nascimento
Tayyab Naseer
Lakshmanan Nataraj
Neda Nategh
Nelson Isao Nauata
Fernando Navarro
Shah Nawaz
Lukas Neumann
Ram Nevatia
Alejandro Newell
Shawn Newsam
Joe Yue-Hei Ng
Trung Thanh Ngo
Duc Thanh Nguyen
Lam M. Nguyen
Phuc Xuan Nguyen
Thuong Nguyen Canh
Mihalis Nicolaou
Andrei Liviu Nicolicioiu
Xuecheng Nie
Michael Niemeyer
Simon Niklaus
Christophoros Nikou
David Nilsson
Jifeng Ning
Yuval Nirkin
Li Niu
Yuzhen Niu
Zhenxing Niu
Shohei Nobuhara
Nicoletta Noceti
Hyeonwoo Noh
Junhyug Noh
Mehdi Noroozi
Sotiris Nousias
Valsamis Ntouskos
Matthew O'Toole
Peter Ochs
Ferda Ofli
Seong Joon Oh
Seoung Wug Oh
Iason Oikonomidis
Utkarsh Ojha
Takahiro Okabe
Takayuki Okatani
Fumio Okura
Aude Oliva
Kyle Olszewski

Björn Ommer
Mohamed Omran
Elisabeta Oneata
Michael Opitz
Jose Oramas
Tribhuvanesh Orekondy
Shaul Oron
Sergio Orts-Escolano
Ivan Oseledets
Aljosa Osep
Magnus Oskarsson
Anton Osokin
Martin R. Oswald
Wanli Ouyang
Andrew Owens
Mete Ozay
Mustafa Ozuysal
Eduardo Pérez-Pellitero
Gautam Pai
Dipan Kumar Pal
P. H. Pamplona Savarese
Jinshan Pan
Junting Pan
Xingang Pan
Yingwei Pan
Yannis Panagakis
Rameswar Panda
Guan Pang
Jiahao Pang
Jiangmiao Pang
Tianyu Pang
Sharath Pankanti
Nicolas Papadakis
Dim Papadopoulos
George Papandreou
Toufiq Parag
Shaifali Parashar
Sarah Parisot
Eunhyeok Park
Hyun Soo Park
Jaesik Park
Min-Gyu Park
Taesung Park
Alvaro Parra
C. Alejandro Parraga
Despoina Paschalidou
Nikolaos Passalis
Vishal Patel
Viorica Patraucean
Badri Narayana Patro
Danda Pani Paudel
Sujoy Paul
Georgios Pavlakos
Ioannis Pavlidis
Vladimir Pavlovic
Nick Pears
Kim Steenstrup Pedersen
Selen Pehlivan
Shmuel Peleg
Chao Peng
Houwen Peng
Wen-Hsiao Peng
Xi Peng
Xiaojiang Peng
Xingchao Peng
Yuxin Peng
Federico Perazzi
Juan Camilo Perez
Vishwanath Peri
Federico Pernici
Luca Del Pero
Florent Perronnin
Stavros Petridis
Henning Petzka
Patrick Peursum
Michael Pfeiffer
Hanspeter Pfister
Roman Pflugfelder
Minh Tri Pham
Yongri Piao
David Picard
Tomasz Pieciak
A. J. Piergiovanni
Andrea Pilzer
Pedro O. Pinheiro
Silvia Laura Pintea
Lerrel Pinto
Axel Pinz
Robinson Piramuthu
Fiora Pirri
Leonid Pishchulin
Francesco Pittaluga
Daniel Pizarro
Tobias Plötz
Mirco Planamente
Matteo Poggi
Moacir A. Ponti
Parita Pooj
Fatih Porikli
Horst Possegger
Omid Poursaeed
Ameya Prabhu
Viraj Uday Prabhu
Dilip Prasad
Brian L. Price
True Price
Maria Priisalu
Veronique Prinet
Victor Adrian Prisacariu
Jan Prokaj
Sergey Prokudin
Nicolas Pugeault
Xavier Puig
Albert Pumarola
Pulak Purkait
Senthil Purushwalkam
Charles R. Qi
Hang Qi
Haozhi Qi
Lu Qi
Mengshi Qi
Siyuan Qi
Xiaojuan Qi
Yuankai Qi
Shengju Qian
Xuelin Qian
Siyuan Qiao
Yu Qiao
Jie Qin
Qiang Qiu
Weichao Qiu
Zhaofan Qiu
Kha Gia Quach
Yuhui Quan
Yvain Queau
Julian Quiroga
Faisal Qureshi
Mahdi Rad

Filip Radenovic
Petia Radeva
Venkatesh B. Radhakrishnan
Ilija Radosavovic
Noha Radwan
Rahul Raguram
Tanzila Rahman
Amit Raj
Ajit Rajwade
Kandan Ramakrishnan
Santhosh K. Ramakrishnan
Srikumar Ramalingam
Ravi Ramamoorthi
Vasili Ramanishka
Ramprasaath R. Selvaraju
Francois Rameau
Visvanathan Ramesh
Santu Rana
Rene Ranftl
Anand Rangarajan
Anurag Ranjan
Viresh Ranjan
Yongming Rao
Carolina Raposo
Vivek Rathod
Sathya N. Ravi
Avinash Ravichandran
Tammy Riklin Raviv
Daniel Rebain
Sylvestre-Alvise Rebuffi
N. Dinesh Reddy
Timo Rehfeld
Paolo Remagnino
Konstantinos Rematas
Edoardo Remelli
Dongwei Ren
Haibing Ren
Jian Ren
Jimmy Ren
Mengye Ren
Weihong Ren
Wenqi Ren
Zhile Ren
Zhongzheng Ren
Zhou Ren
Vijay Rengarajan
Md A. Reza
Farzaneh Rezaeianaran
Hamed R. Tavakoli
Nicholas Rhinehart
Helge Rhodin
Elisa Ricci
Alexander Richard
Eitan Richardson
Elad Richardson
Christian Richardt
Stephan Richter
Gernot Riegler
Daniel Ritchie
Tobias Ritschel
Samuel Rivera
Yong Man Ro
Richard Roberts
Joseph Robinson
Ignacio Rocco
Mrigank Rochan
Emanuele Rodolà
Mikel D. Rodriguez
Giorgio Roffo
Grégory Rogez
Gemma Roig
Javier Romero
Xuejian Rong
Yu Rong
Amir Rosenfeld
Bodo Rosenhahn
Guy Rosman
Arun Ross
Paolo Rota
Peter M. Roth
Anastasios Roussos
Anirban Roy
Sebastien Roy
Aruni RoyChowdhury
Artem Rozantsev
Ognjen Rudovic
Daniel Rueckert
Adria Ruiz
Javier Ruiz-del-solar
Christian Rupprecht
Chris Russell
Dan Ruta
Jongbin Ryu
Ömer Sümer
Alexandre Sablayrolles
Faraz Saeedan
Ryusuke Sagawa
Christos Sagonas
Tonmoy Saikia
Hideo Saito
Kuniaki Saito
Shunsuke Saito
Shunta Saito
Ken Sakurada
Joaquin Salas
Fatemeh Sadat Saleh
Mahdi Saleh
Pouya Samangouei
Leo Sampaio Ferraz Ribeiro
Artsiom Olegovich Sanakoyeu
Enrique Sanchez
Patsorn Sangkloy
Anush Sankaran
Aswin Sankaranarayanan
Swami Sankaranarayanan
Rodrigo Santa Cruz
Amartya Sanyal
Archana Sapkota
Nikolaos Sarafianos
Jun Sato
Shin'ichi Satoh
Hosnieh Sattar
Arman Savran
Manolis Savva
Alexander Sax
Hanno Scharr
Simone Schaub-Meyer
Konrad Schindler
Dmitrij Schlesinger
Uwe Schmidt
Dirk Schnieders
Björn Schuller
Samuel Schulter
Idan Schwartz

William Robson Schwartz
Alex Schwing
Sinisa Segvic
Lorenzo Seidenari
Pradeep Sen
Ozan Sener
Soumyadip Sengupta
Arda Senocak
Mojtaba Seyedhosseini
Shishir Shah
Shital Shah
Sohil Atul Shah
Tamar Rott Shaham
Huasong Shan
Qi Shan
Shiguang Shan
Jing Shao
Roman Shapovalov
Gaurav Sharma
Vivek Sharma
Viktoriia Sharmanska
Dongyu She
Sumit Shekhar
Evan Shelhamer
Chengyao Shen
Chunhua Shen
Falong Shen
Jie Shen
Li Shen
Liyue Shen
Shuhan Shen
Tianwei Shen
Wei Shen
William B. Shen
Yantao Shen
Ying Shen
Yiru Shen
Yujun Shen
Yuming Shen
Zhiqiang Shen
Ziyi Shen
Lu Sheng
Yu Sheng
Rakshith Shetty
Baoguang Shi
Guangming Shi
Hailin Shi
Miaojing Shi
Yemin Shi
Zhenmei Shi
Zhiyuan Shi
Kevin Jonathan Shih
Shiliang Shiliang
Hyunjung Shim
Atsushi Shimada
Nobutaka Shimada
Daeyun Shin
Young Min Shin
Koichi Shinoda
Konstantin Shmelkov
Michael Zheng Shou
Abhinav Shrivastava
Tianmin Shu
Zhixin Shu
Hong-Han Shuai
Pushkar Shukla
Christian Siagian
Mennatullah M. Siam
Kaleem Siddiqi
Karan Sikka
Jae-Young Sim
Christian Simon
Martin Simonovsky
Dheeraj Singaraju
Bharat Singh
Gurkirt Singh
Krishna Kumar Singh
Maneesh Kumar Singh
Richa Singh
Saurabh Singh
Suriya Singh
Vikas Singh
Sudipta N. Sinha
Vincent Sitzmann
Josef Sivic
Gregory Slabaugh
Miroslava Slavcheva
Ron Slossberg
Brandon Smith
Kevin Smith
Vladimir Smutny
Noah Snavely
Roger D. Soberanis-Mukul
Kihyuk Sohn
Francesco Solera
Eric Sommerlade
Sanghyun Son
Byung Cheol Song
Chunfeng Song
Dongjin Song
Jiaming Song
Jie Song
Jifei Song
Jingkuan Song
Mingli Song
Shiyu Song
Shuran Song
Xiao Song
Yafei Song
Yale Song
Yang Song
Yi-Zhe Song
Yibing Song
Humberto Sossa
Cesar de Souza
Adrian Spurr
Srinath Sridhar
Suraj Srinivas
Pratul P. Srinivasan
Anuj Srivastava
Tania Stathaki
Christopher Stauffer
Simon Stent
Rainer Stiefelhagen
Pierre Stock
Julian Straub
Jonathan C. Stroud
Joerg Stueckler
Jan Stuehmer
David Stutz
Chi Su
Hang Su
Jong-Chyi Su
Shuochen Su
Yu-Chuan Su
Ramanathan Subramanian
Yusuke Sugano

Masanori Suganuma
Yumin Suh
Mohammed Suhail
Yao Sui
Heung-Il Suk
Josephine Sullivan
Baochen Sun
Chen Sun
Chong Sun
Deqing Sun
Jin Sun
Liang Sun
Lin Sun
Qianru Sun
Shao-Hua Sun
Shuyang Sun
Weiwei Sun
Wenxiu Sun
Xiaoshuai Sun
Xiaoxiao Sun
Xingyuan Sun
Yifan Sun
Zhun Sun
Sabine Susstrunk
David Suter
Supasorn Suwajanakorn
Tomas Svoboda
Eran Swears
Paul Swoboda
Attila Szabo
Richard Szeliski
Duy-Nguyen Ta
Andrea Tagliasacchi
Yuichi Taguchi
Ying Tai
Keita Takahashi
Kouske Takahashi
Jun Takamatsu
Hugues Talbot
Toru Tamaki
Chaowei Tan
Fuwen Tan
Mingkui Tan
Mingxing Tan
Qingyang Tan
Robby T. Tan
Xiaoyang Tan
Kenichiro Tanaka
Masayuki Tanaka
Chang Tang
Chengzhou Tang
Danhang Tang
Ming Tang
Peng Tang
Qingming Tang
Wei Tang
Xu Tang
Yansong Tang
Youbao Tang
Yuxing Tang
Zhiqiang Tang
Tatsunori Taniai
Junli Tao
Xin Tao
Makarand Tapaswi
Jean-Philippe Tarel
Lyne Tchapmi
Zachary Teed
Bugra Tekin
Damien Teney
Ayush Tewari
Christian Theobalt
Christopher Thomas
Diego Thomas
Jim Thomas
Rajat Mani Thomas
Xinmei Tian
Yapeng Tian
Yingli Tian
Yonglong Tian
Zhi Tian
Zhuotao Tian
Kinh Tieu
Joseph Tighe
Massimo Tistarelli
Matthew Toews
Carl Toft
Pavel Tokmakov
Federico Tombari
Chetan Tonde
Yan Tong
Alessio Tonioni
Andrea Torsello
Fabio Tosi
Du Tran
Luan Tran
Ngoc-Trung Tran
Quan Hung Tran
Truyen Tran
Rudolph Triebel
Martin Trimmel
Shashank Tripathi
Subarna Tripathi
Leonardo Trujillo
Eduard Trulls
Tomasz Trzcinski
Sam Tsai
Yi-Hsuan Tsai
Hung-Yu Tseng
Stavros Tsogkas
Aggeliki Tsoli
Devis Tuia
Shubham Tulsiani
Sergey Tulyakov
Frederick Tung
Tony Tung
Daniyar Turmukhambetov
Ambrish Tyagi
Radim Tylecek
Christos Tzelepis
Georgios Tzimiropoulos
Dimitrios Tzionas
Seiichi Uchida
Norimichi Ukita
Dmitry Ulyanov
Martin Urschler
Yoshitaka Ushiku
Ben Usman
Alexander Vakhitov
Julien P. C. Valentin
Jack Valmadre
Ernest Valveny
Joost van de Weijer
Jan van Gemert
Koen Van Leemput
Gul Varol
Sebastiano Vascon
M. Alex O. Vasilescu

Subeesh Vasu
Mayank Vatsa
David Vazquez
Javier Vazquez-Corral
Ashok Veeraraghavan
Erik Velasco-Salido
Raviteja Vemulapalli
Jonathan Ventura
Manisha Verma
Roberto Vezzani
Ruben Villegas
Minh Vo
MinhDuc Vo
Nam Vo
Michele Volpi
Riccardo Volpi
Carl Vondrick
Konstantinos Vougioukas
Tuan-Hung Vu
Sven Wachsmuth
Neal Wadhwa
Catherine Wah
Jacob C. Walker
Thomas S. A. Wallis
Chengde Wan
Jun Wan
Liang Wan
Renjie Wan
Baoyuan Wang
Boyu Wang
Cheng Wang
Chu Wang
Chuan Wang
Chunyu Wang
Dequan Wang
Di Wang
Dilin Wang
Dong Wang
Fang Wang
Guanzhi Wang
Guoyin Wang
Hanzi Wang
Hao Wang
He Wang
Heng Wang
Hongcheng Wang
Hongxing Wang
Hua Wang
Jian Wang
Jingbo Wang
Jinglu Wang
Jingya Wang
Jinjun Wang
Jinqiao Wang
Jue Wang
Ke Wang
Keze Wang
Le Wang
Lei Wang
Lezi Wang
Li Wang
Liang Wang
Lijun Wang
Limin Wang
Linwei Wang
Lizhi Wang
Mengjiao Wang
Mingzhe Wang
Minsi Wang
Naiyan Wang
Nannan Wang
Ning Wang
Oliver Wang
Pei Wang
Peng Wang
Pichao Wang
Qi Wang
Qian Wang
Qiaosong Wang
Qifei Wang
Qilong Wang
Qing Wang
Qingzhong Wang
Quan Wang
Rui Wang
Ruiping Wang
Ruixing Wang
Shangfei Wang
Shenlong Wang
Shiyao Wang
Shuhui Wang
Song Wang
Tao Wang
Tianlu Wang
Tiantian Wang
Ting-chun Wang
Tingwu Wang
Wei Wang
Weiyue Wang
Wenguan Wang
Wenlin Wang
Wenqi Wang
Xiang Wang
Xiaobo Wang
Xiaofang Wang
Xiaoling Wang
Xiaolong Wang
Xiaosong Wang
Xiaoyu Wang
Xin Eric Wang
Xinchao Wang
Xinggang Wang
Xintao Wang
Yali Wang
Yan Wang
Yang Wang
Yangang Wang
Yaxing Wang
Yi Wang
Yida Wang
Yilin Wang
Yiming Wang
Yisen Wang
Yongtao Wang
Yu-Xiong Wang
Yue Wang
Yujiang Wang
Yunbo Wang
Yunhe Wang
Zengmao Wang
Zhangyang Wang
Zhaowen Wang
Zhe Wang
Zhecan Wang
Zheng Wang
Zhixiang Wang
Zilei Wang
Jianqiao Wangni

Anne S. Wannenwetsch
Jan Dirk Wegner
Scott Wehrwein
Donglai Wei
Kaixuan Wei
Longhui Wei
Pengxu Wei
Ping Wei
Qi Wei
Shih-En Wei
Xing Wei
Yunchao Wei
Zijun Wei
Jerod Weinman
Michael Weinmann
Philippe Weinzaepfel
Yair Weiss
Bihan Wen
Longyin Wen
Wei Wen
Junwu Weng
Tsui-Wei Weng
Xinshuo Weng
Eric Wengrowski
Tomas Werner
Gordon Wetzstein
Tobias Weyand
Patrick Wieschollek
Maggie Wigness
Erik Wijmans
Richard Wildes
Olivia Wiles
Chris Williams
Williem Williem
Kyle Wilson
Calden Wloka
Nicolai Wojke
Christian Wolf
Yongkang Wong
Sanghyun Woo
Scott Workman
Baoyuan Wu
Bichen Wu
Chao-Yuan Wu
Huikai Wu
Jiajun Wu
Jialin Wu
Jiaxiang Wu
Jiqing Wu
Jonathan Wu
Lifang Wu
Qi Wu
Qiang Wu
Ruizheng Wu
Shangzhe Wu
Shun-Cheng Wu
Tianfu Wu
Wayne Wu
Wenxuan Wu
Xiao Wu
Xiaohe Wu
Xinxiao Wu
Yang Wu
Yi Wu
Yiming Wu
Ying Nian Wu
Yue Wu
Zheng Wu
Zhenyu Wu
Zhirong Wu
Zuxuan Wu
Stefanie Wuhrer
Jonas Wulff
Changqun Xia
Fangting Xia
Fei Xia
Gui-Song Xia
Lu Xia
Xide Xia
Yin Xia
Yingce Xia
Yongqin Xian
Lei Xiang
Shiming Xiang
Bin Xiao
Fanyi Xiao
Guobao Xiao
Huaxin Xiao
Taihong Xiao
Tete Xiao
Tong Xiao
Wang Xiao
Yang Xiao
Cihang Xie
Guosen Xie
Jianwen Xie
Lingxi Xie
Sirui Xie
Weidi Xie
Wenxuan Xie
Xiaohua Xie
Fuyong Xing
Jun Xing
Junliang Xing
Bo Xiong
Peixi Xiong
Yu Xiong
Yuanjun Xiong
Zhiwei Xiong
Chang Xu
Chenliang Xu
Dan Xu
Danfei Xu
Hang Xu
Hongteng Xu
Huijuan Xu
Jingwei Xu
Jun Xu
Kai Xu
Mengmeng Xu
Mingze Xu
Qianqian Xu
Ran Xu
Weijian Xu
Xiangyu Xu
Xiaogang Xu
Xing Xu
Xun Xu
Yanyu Xu
Yichao Xu
Yong Xu
Yongchao Xu
Yuanlu Xu
Zenglin Xu
Zheng Xu
Chuhui Xue
Jia Xue
Nan Xue

Tianfan Xue
Xiangyang Xue
Abhay Yadav
Yasushi Yagi
I. Zeki Yalniz
Kota Yamaguchi
Toshihiko Yamasaki
Takayoshi Yamashita
Junchi Yan
Ke Yan
Qingan Yan
Sijie Yan
Xinchen Yan
Yan Yan
Yichao Yan
Zhicheng Yan
Keiji Yanai
Bin Yang
Ceyuan Yang
Dawei Yang
Dong Yang
Fan Yang
Guandao Yang
Guorun Yang
Haichuan Yang
Hao Yang
Jianwei Yang
Jiaolong Yang
Jie Yang
Jing Yang
Kaiyu Yang
Linjie Yang
Meng Yang
Michael Ying Yang
Nan Yang
Shuai Yang
Shuo Yang
Tianyu Yang
Tien-Ju Yang
Tsun-Yi Yang
Wei Yang
Wenhan Yang
Xiao Yang
Xiaodong Yang
Xin Yang
Yan Yang
Yanchao Yang
Yee Hong Yang
Yezhou Yang
Zhenheng Yang
Anbang Yao
Angela Yao
Cong Yao
Jian Yao
Li Yao
Ting Yao
Yao Yao
Zhewei Yao
Chengxi Ye
Jianbo Ye
Keren Ye
Linwei Ye
Mang Ye
Mao Ye
Qi Ye
Qixiang Ye
Mei-Chen Yeh
Raymond Yeh
Yu-Ying Yeh
Sai-Kit Yeung
Serena Yeung
Kwang Moo Yi
Li Yi
Renjiao Yi
Alper Yilmaz
Junho Yim
Lijun Yin
Weidong Yin
Xi Yin
Zhichao Yin
Tatsuya Yokota
Ryo Yonetani
Donggeun Yoo
Jae Shin Yoon
Ju Hong Yoon
Sung-eui Yoon
Laurent Younes
Changqian Yu
Fisher Yu
Gang Yu
Jiahui Yu
Kaicheng Yu
Ke Yu
Lequan Yu
Ning Yu
Qian Yu
Ronald Yu
Ruichi Yu
Shoou-I Yu
Tao Yu
Tianshu Yu
Xiang Yu
Xin Yu
Xiyu Yu
Youngjae Yu
Yu Yu
Zhiding Yu
Chunfeng Yuan
Ganzhao Yuan
Jinwei Yuan
Lu Yuan
Quan Yuan
Shanxin Yuan
Tongtong Yuan
Wenjia Yuan
Ye Yuan
Yuan Yuan
Yuhui Yuan
Huanjing Yue
Xiangyu Yue
Ersin Yumer
Sergey Zagoruyko
Egor Zakharov
Amir Zamir
Andrei Zanfir
Mihai Zanfir
Pablo Zegers
Bernhard Zeisl
John S. Zelek
Niclas Zeller
Huayi Zeng
Jiabei Zeng
Wenjun Zeng
Yu Zeng
Xiaohua Zhai
Fangneng Zhan
Huangying Zhan
Kun Zhan

Xiaohang Zhan
Baochang Zhang
Bowen Zhang
Cecilia Zhang
Changqing Zhang
Chao Zhang
Chengquan Zhang
Chi Zhang
Chongyang Zhang
Dingwen Zhang
Dong Zhang
Feihu Zhang
Hang Zhang
Hanwang Zhang
Hao Zhang
He Zhang
Hongguang Zhang
Hua Zhang
Ji Zhang
Jianguo Zhang
Jianming Zhang
Jiawei Zhang
Jie Zhang
Jing Zhang
Juyong Zhang
Kai Zhang
Kaipeng Zhang
Ke Zhang
Le Zhang
Lei Zhang
Li Zhang
Lihe Zhang
Linguang Zhang
Lu Zhang
Mi Zhang
Mingda Zhang
Peng Zhang
Pingping Zhang
Qian Zhang
Qilin Zhang
Quanshi Zhang
Richard Zhang
Rui Zhang
Runze Zhang
Shengping Zhang
Shifeng Zhang
Shuai Zhang
Songyang Zhang
Tao Zhang
Ting Zhang
Tong Zhang
Wayne Zhang
Wei Zhang
Weizhong Zhang
Wenwei Zhang
Xiangyu Zhang
Xiaolin Zhang
Xiaopeng Zhang
Xiaoqin Zhang
Xiuming Zhang
Ya Zhang
Yang Zhang
Yimin Zhang
Yinda Zhang
Ying Zhang
Yongfei Zhang
Yu Zhang
Yulun Zhang
Yunhua Zhang
Yuting Zhang
Zhanpeng Zhang
Zhao Zhang
Zhaoxiang Zhang
Zhen Zhang
Zheng Zhang
Zhifei Zhang
Zhijin Zhang
Zhishuai Zhang
Ziming Zhang
Bo Zhao
Chen Zhao
Fang Zhao
Haiyu Zhao
Han Zhao
Hang Zhao
Hengshuang Zhao
Jian Zhao
Kai Zhao
Liang Zhao
Long Zhao
Qian Zhao
Qibin Zhao
Qijun Zhao
Rui Zhao
Shenglin Zhao
Sicheng Zhao
Tianyi Zhao
Wenda Zhao
Xiangyun Zhao
Xin Zhao
Yang Zhao
Yue Zhao
Zhichen Zhao
Zijing Zhao
Xiantong Zhen
Chuanxia Zheng
Feng Zheng
Haiyong Zheng
Jia Zheng
Kang Zheng
Shuai Kyle Zheng
Wei-Shi Zheng
Yinqiang Zheng
Zerong Zheng
Zhedong Zheng
Zilong Zheng
Bineng Zhong
Fangwei Zhong
Guangyu Zhong
Yiran Zhong
Yujie Zhong
Zhun Zhong
Chunluan Zhou
Huiyu Zhou
Jiahuan Zhou
Jun Zhou
Lei Zhou
Luowei Zhou
Luping Zhou
Mo Zhou
Ning Zhou
Pan Zhou
Peng Zhou
Qianyi Zhou
S. Kevin Zhou
Sanping Zhou
Wengang Zhou
Xingyi Zhou

Yanzhao Zhou
Yi Zhou
Yin Zhou
Yipin Zhou
Yuyin Zhou
Zihan Zhou
Alex Zihao Zhu
Chenchen Zhu
Feng Zhu
Guangming Zhu
Ji Zhu
Jun-Yan Zhu
Lei Zhu
Linchao Zhu
Rui Zhu
Shizhan Zhu
Tyler Lixuan Zhu
Wei Zhu
Xiangyu Zhu
Xinge Zhu
Xizhou Zhu
Yanjun Zhu
Yi Zhu
Yixin Zhu
Yizhe Zhu
Yousong Zhu
Zhe Zhu
Zhen Zhu
Zheng Zhu
Zhenyao Zhu
Zhihui Zhu
Zhuotun Zhu
Bingbing Zhuang
Wei Zhuo
Christian Zimmermann
Karel Zimmermann
Larry Zitnick
Mohammadreza Zolfaghari
Maria Zontak
Daniel Zoran
Changqing Zou
Chuhang Zou
Danping Zou
Qi Zou
Yang Zou
Yuliang Zou
Georgios Zoumpourlis
Wangmeng Zuo
Xinxin Zuo

Additional Reviewers

Victoria Fernandez Abrevaya
Maya Aghaei
Allam Allam
Christine Allen-Blanchette
Nicolas Aziere
Assia Benbihi
Neha Bhargava
Bharat Lal Bhatnagar
Joanna Bitton
Judy Borowski
Amine Bourki
Romain Brégier
Tali Brayer
Sebastian Bujwid
Andrea Burns
Yun-Hao Cao
Yuning Chai
Xiaojun Chang
Bo Chen
Shuo Chen
Zhixiang Chen
Junsuk Choe
Hung-Kuo Chu
Jonathan P. Crall
Kenan Dai
Lucas Deecke
Karan Desai
Prithviraj Dhar
Jing Dong
Wei Dong
Turan Kaan Elgin
Francis Engelmann
Erik Englesson
Fartash Faghri
Zicong Fan
Yang Fu
Risheek Garrepalli
Yifan Ge
Marco Godi
Helmut Grabner
Shuxuan Guo
Jianfeng He
Zhezhi He
Samitha Herath
Chih-Hui Ho
Yicong Hong
Vincent Tao Hu
Julio Hurtado
Jaedong Hwang
Andrey Ignatov
Muhammad Abdullah Jamal
Saumya Jetley
Meiguang Jin
Jeff Johnson
Minsoo Kang
Saeed Khorram
Mohammad Rami Koujan
Nilesh Kulkarni
Sudhakar Kumawat
Abdelhak Lemkhenter
Alexander Levine
Jiachen Li
Jing Li
Jun Li
Yi Li
Liang Liao
Ruochen Liao
Tzu-Heng Lin
Phillip Lippe
Bao-di Liu
Bo Liu
Fangchen Liu

Hanxiao Liu
Hongyu Liu
Huidong Liu
Miao Liu
Xinxin Liu
Yongfei Liu
Yu-Lun Liu
Amir Livne
Tiange Luo
Wei Ma
Xiaoxuan Ma
Ioannis Marras
Georg Martius
Effrosyni Mavroudi
Tim Meinhardt
Givi Meishvili
Meng Meng
Zihang Meng
Zhongqi Miao
Gyeongsik Moon
Khoi Nguyen
Yung-Kyun Noh
Antonio Norelli
Jaeyoo Park
Alexander Pashevich
Mandela Patrick
Mary Phuong
Bingqiao Qian
Yu Qiao
Zhen Qiao
Sai Saketh Rambhatla
Aniket Roy
Amelie Royer
Parikshit Vishwas Sakurikar
Mark Sandler
Mert Bülent Sarıyıldız
Tanner Schmidt
Anshul B. Shah
Ketul Shah
Rajvi Shah
Hengcan Shi
Xiangxi Shi
Yujiao Shi
William A. P. Smith
Guoxian Song
Robin Strudel
Abby Stylianou
Xinwei Sun
Reuben Tan
Qingyi Tao
Kedar S. Tatwawadi
Anh Tuan Tran
Son Dinh Tran
Eleni Triantafillou
Aristeidis Tsitiridis
Md Zasim Uddin
Andrea Vedaldi
Evangelos Ververas
Vidit Vidit
Paul Voigtlaender
Bo Wan
Huanyu Wang
Huiyu Wang
Junqiu Wang
Pengxiao Wang
Tai Wang
Xinyao Wang
Tomoki Watanabe
Mark Weber
Xi Wei
Botong Wu
James Wu
Jiamin Wu
Rujie Wu
Yu Wu
Rongchang Xie
Wei Xiong
Yunyang Xiong
An Xu
Chi Xu
Yinghao Xu
Fei Xue
Tingyun Yan
Zike Yan
Chao Yang
Heran Yang
Ren Yang
Wenfei Yang
Xu Yang
Rajeev Yasarla
Shaokai Ye
Yufei Ye
Kun Yi
Haichao Yu
Hanchao Yu
Ruixuan Yu
Liangzhe Yuan
Chen-Lin Zhang
Fandong Zhang
Tianyi Zhang
Yang Zhang
Yiyi Zhang
Yongshun Zhang
Yu Zhang
Zhiwei Zhang
Jiaojiao Zhao
Yipu Zhao
Xingjian Zhen
Haizhong Zheng
Tiancheng Zhi
Chengju Zhou
Hao Zhou
Hao Zhu
Alexander Zimin

Contents – Part XVI

Partially-Shared Variational Auto-encoders for Unsupervised Domain Adaptation with Target Shift

Ryuhei Takahashi[1], Atsushi Hashimoto[2(✉)], Motoharu Sonogashira[1], and Masaaki Iiyama[1]

[1] Kyoto University, Kyoto, Japan
{sonogashira,iiyama}@mm.media.kyoto-u.ac.jp
[2] OMRON SINIC X Corp., Tokyo, Japan
atsushi.hashimoto@sinicx.com

Abstract. Target shift, the different label distributions of source and target domains, is an important problem for practical use of unsupervised domain adaptation (UDA); as we do not know labels in target domain datasets, we cannot ensure an identical label distribution between the two domains. Despite this inaccessibility, modern UDA methods commonly try to match the shape of the feature distributions over the domains while projecting the features to labels by a common classifier. This implicitly assumes the identical label distribution. To overcome this problem, we propose a method that generates a pseudo pair by domain conversion where the label is preserved identically even trained with target-shifted datasets. A pair-wise metric learning enables to align feature over the domains without matching the shape of distributions. We conducted two experiments: one is a regression of pose-estimation, where label distribution is continuous and the target shift problem can seriously degrade the quality of UDA. The other is digit classification task where we can systematically control the distribution difference. The code and dataset are available at https://github.com/iiyama-lab/PS-VAEs.

1 Introduction

Unsupervised domain adaptation (UDA) is one of the most studied topics in recent years. One attractive application of UDA is adaptation from computer graphic (CG) data to sensor-observed data. By constructing a CG-rendering system, we can easily obtain a large amount of supervised data with diversity for training. Because a model straightforwardly trained on CG-rendered dataset hardly works with real observation, training a model with both CG-rendered dataset (source domain) and unsupervised real observation dataset (target domain) by UDA is a necessary but promised approach.

Electronic supplementary material The online version of this chapter (https://doi.org/10.1007/978-3-030-58517-4_1) contains supplementary material, which is available to authorized users.

A. Vedaldi et al. (Eds.): ECCV 2020, LNCS 12361, pp. 1–17, 2020.
https://doi.org/10.1007/978-3-030-58517-4_1

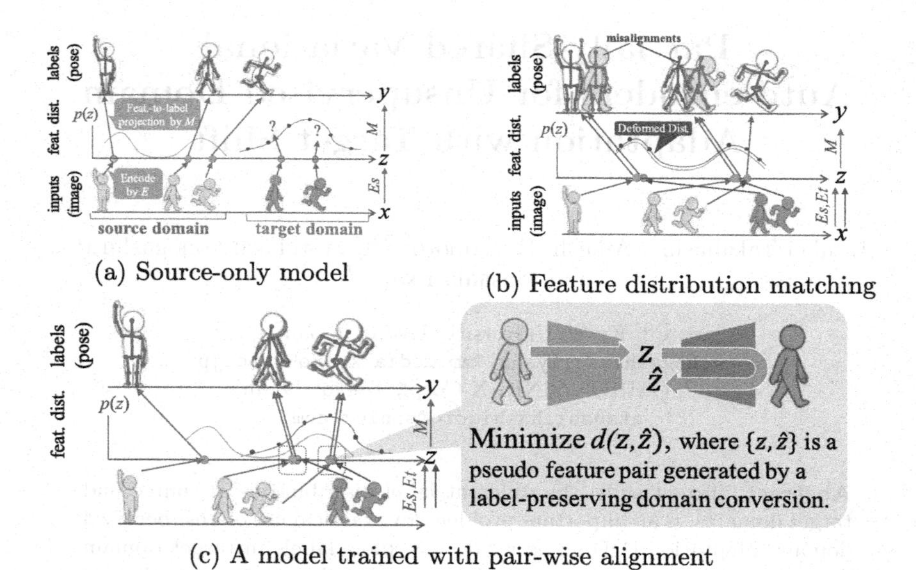

(a) Source-only model

(b) Feature distribution matching

(c) A model trained with pair-wise alignment

Fig. 1. (best viewed in color) Overview of the proposed approach on the problem of (2D) human pose estimation. Note that the feature-to-label projection M is trained only with source domain dataset, where E_s and E_t are domain specific image-to-feature encoders. (a) The naive approach fails in the target domain due to the differences in the feature distributions between the two domains: location difference that illustrates the affection by domain shift and shape difference caused by *target shift* (non-identical label distributions of the two domains). (b) While feature distribution matching attempts to adjust the shape of two feature distributions, it suffers from misalignment in label estimation due to the deformed target domain feature distribution. (c) The proposed method avoid this deformation problem by sample-wise feature distance minimization, where pseudo sample pairs with an identical label are generated via a CycleGAN-based architecture.

As in ADDA [38], the typical approach for UDA is to match feature distributions between the source and target domains [8,19,22]. This approach works impressively with identically-balanced datasets, such as those for digits (MNIST, USPS, and SVHN) and traffic-scene semantic-segmentation (GTA5 [30] to Cityscapes [5]). When the prior label distributions of the source and target domains are mismatched, however, such approaches hardly work without a countermeasure for the mismatch (see Fig. 1). Cluster finding [6,33,34] is another approach for UDA by a class-boundary adjustment; they are not, however, applicable to regression problems due to the absence of class-boundaries in the distribution.

In this paper, we propose a novel UDA method applicable especially to the regression problem with mismatched label distributions. The problem of mismatched label distributions is also known as *target shift* [10,42] or *prior probability shift* [28]. The typical example of this problem is UDA from a balanced source domain dataset to an imbalanced target domain dataset. Some recent studies have tried to overcome this problem by estimating category-wise importance labels [1–3,39,41] or sample-wise importance labels [15]. The former approach is only applicable to classification tasks. The latter is applicable to regression but under-samples the source domain data. In addition, it requires a reliable similarity metric over the domain shift (a domain-shift-free metric) to select important samples; this suffers from the chicken-and-egg situation. Namely, if we have a measure that is hardly affected by domain shift, we can safely apply pair-wise metric learning (e.g., with a Siamese loss), but such metric is not given in general.

In contrast, our method resolves this problem by oversampling with label-preserving data augmentation. This is applicable even to regression with UDA and does not require any preliminary knowledge of the domain-shift-free metrics. Figure 1 shows the overview. Traditional methods [11,19,32,38] matches feature distributions of the two domains (Fig. 1(b)). Since feature distributions are forced to be identical and the feature-to-label projection function is shared by the two domains, the estimated labels in the target domain must distribute identically with that of the source domain. Under the target shift condition, this clearly competes with the assumption of non-identical label distributions.

Our method addresses the problem of target shift by tolerating feature distribution mismatches and instead requiring the sample-wise matches of the labels (Fig. 1(c)). To this end, our method called partially-shared variational auto-encoders (PS-VAEs) organizes a CycleGAN architecture [44] with two VAE branches that share weights as much as possible to realize the label-preserving conversions.

The contribution of this paper is three-fold.

- We propose a novel UDA method that overcomes the target shift problem by oversampling with label-preserving data augmentation, which is applicable to regression. This is the first algorithm that solves regression with UDA under a target shift condition without relying on any prior knowledge of domain-shift-free metrics.
- We tackled the problem of human-pose estimation by UDA with target shift for the first time and outperformed the baselines with a large margin.
- The proposed method showed the versatility under various levels of target shift intensities and different combinations of datasets in the task of digit classification with UDA.

2 Related Work

UDA by a Feature Space Discriminator

The most popular approach in modern UDA methods is to match the feature distributions of the source and target domains so that a classifier trained with the source domain dataset is applicable to target domain samples. There are various options to match the distributions, such as minimizing MMD [22,39], using a gradient-reversal layer with domain discriminators [8], and using alternative adversarial training with domain discriminators [2,16,19,38]. Adversarial

Table 1. Representative UDA methods and their supporting situations. The symbol "(✓)" indicates that the method theoretically supports the situation but this was not experimentally confirmed in the original paper.

	Balance		Imbalance	
	Classification	Regression	Classification	Regression
ADDA [38], UFDN [19], CyCADA [11]	✓	(✓)		
MCD [34]	✓		(✓)	
PADA [39], UAN [40], CDAN-E [23]	✓		✓	
SimGAN [35]		✓		(✓)
Ours	✓	(✓)	✓	✓

training removes domain bias from the feature representation. To preserve information in domain invariant features as much as possible, UFDN [19] involves a VAE module [7] with the discriminator. Another approach is feature whitening [31], which whitens features from each domain at domain-specific alignment layers. This approach does not use adversarial training, but it tries to analytically fit a feature distribution from each domain to a common spherical distribution. As shown in Table 1, all these methods are theoretically applicable to both classification and regression, but it is limited to the situations without target shift.

Cluster Finding Approaches
MCD was proposed by Saito *et al.* [33,34], which does not use distribution matching. Instead, the classifier discrepancy is measured based on the difference of decision boundaries between multiple classifiers. DIRT-T [36] and CLAN [24] are additional approaches focusing on the boundary adjustment. These approaches are potentially be robust against target shift, because they focus only on the boundaries and do not try to match the distributions. CAT [6] is a plug-and-play method that aligns clusters found by other backbone methods. Since these approaches assume an existence of boundaries between clusters, they are not applicable to regression, which have continuous sample distributions (see the second row in Table 1).

UDA with Target Shift
Traditional UDA benchmarks barely discuss the problem of target shift. Most classification datasets, such as MNIST, USPS, and SVHN are balanced. Even the class-imbalance problem is known with semantic segmentation, target shift does not come to the surface as long as source and target domains are similarly imbalanced (i.e., their label distributions can be considered as identical). GTA5→Cityscapes is in the case. CDAN-E [23] is one of the few methods that has potential to deal with target shift although the original paper does not clearly discuss the target shift problem. Partial domain adaptation (PDA) is a variant of UDA with several papers on it [1–3,39,41] (see the third row in Table 1). This problem assumes a situation in which some categories in the source domain do not appear in the target domain. This problem is a special case of target shift in two senses: it assumes the absence of a category and it assumes

only classification tasks. The principle approach for this problem is to estimate the importance weight for each category, and ignore those judged as unimportant (under-sampling). UAN [40] is another recent method that solves PDA and the open-set problem simultaneously. It estimates sample-wise importance weight based on the entropy at the classification output for each target sample. PADACO [15] is designed for a regression problem of head pose estimation under a target shift situation. To obtain sample-wise importance weight with a regression problem, it uses head-pose similarity between source and target samples, where target domain head-pose is estimated by a pretrained backbone, which is source-only model in the paper. Then, the similarity values are converted into fixed sampling weights (under-sampling). Finally, it performs UDA training with a weighted sampling strategy. To obtain better results with this method, it is important to obtain good sampling weights with the backbone, just as CAT [6].

UDA by Domain Conversion
Label-preserving domain conversion is another important approach and includes the proposed method (see fourth and fifth rows in Table 1). Shrivastava *et al.* proposed SimGAN [35], which converts CG images to nearly real images by adversarial training. This method tries to preserve labels by minimizing the self-regularization loss, the pixel-value difference between images before and after conversion. This method can be seen as an approach based on over-sampling with data augmentation in the sense that it generates source-domain-like samples using GAN under the self-regularization constraint. Note that SimGAN is the first deep-learning-based UDA method for regression that is theoretically applicable to the task with target shift. On the other hand, this method still assumes a domain-shift-free metric of the self-regularization loss, which is not always domain-shift-free.

CyCADA [11] combines CycleGAN, ADDA and SimGAN for a better performance. It first generates fake target domain images via CycleGAN. The label-consistency of generated samples are preserved by SimGAN's self-regularization loss; however it has a discriminator that matches the feature distributions. Hence, this methods principally has the same weakness against target shift. SBADA-GAN [32] is yet another CycleGAN-based method with discriminator for feature distribution matching.

In addition to the above methods, there is a recent attempt to solve human-pose estimation by domain adaptation [43]. This method tried to regularize domain difference by sharing a discrete space of body-parts segmentation as an intermediate representation, but the reported score shows that the method dose not work effectively under the UDA setting.

3 Method

3.1 Problem Statement

Let $\{x_s, y_s\} \in X_s \times Y_s$ be samples and their labels in the source domain dataset (Y_s is the label space), and let $x_t \in X_t$ be samples in the target domain dataset.

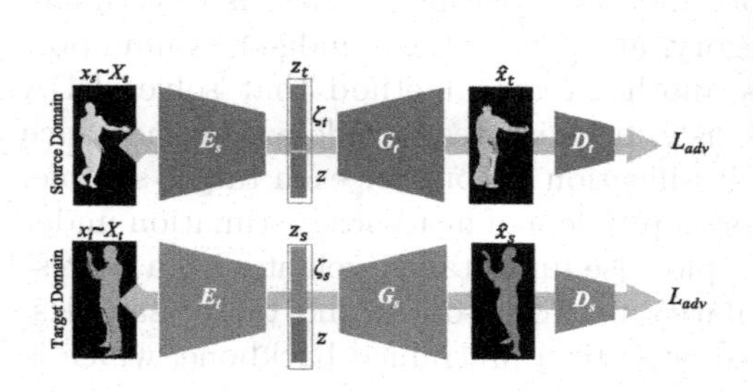

Fig. 2. The basic CycleGAN architecture. We abbreviated the identity loss (L_{id}) and cycle consistency loss (L_{cycle}) for the simplicity.

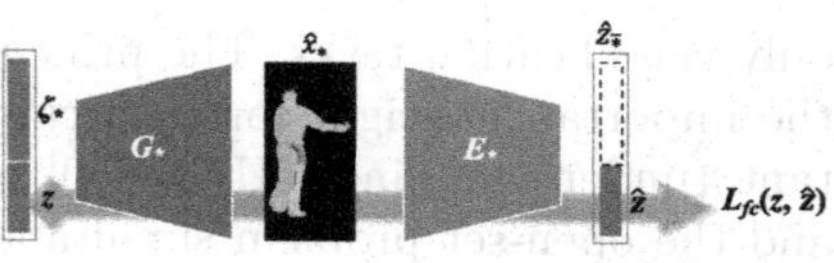

Fig. 3. Feature consistency loss (for both directions $s \rightarrow t$ and $t \rightarrow s$).

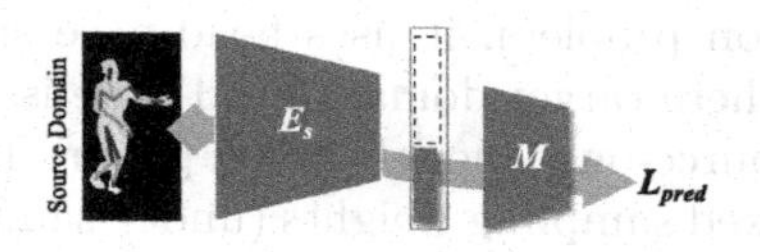

Fig. 4. The prediction loss, which is calculated only with source domain samples.

The target labels Y_t and their distribution $\Pr(Y_t)$ are unknown (i.e., possibly $\Pr(Y_t) \neq \Pr(Y_s)$) in the problem of UDA with target shift. The goal of this problem is to obtain a high-accuracy model for predicting the labels of samples obtained in the target domain.

3.2 Overview of the Proposed Method

The main strategy of the proposed method is to replace the feature distribution matching process with pair-wise feature alignment. To achieve this, starting from the standard CycleGAN as the base architecture (Fig. 2), we add two new losses to generate pseudo pairs $(x_s, \hat{x}_t)$ and $(x_t, \hat{x}_s)$, each of which are expected to have an identical label: L_{fc} for feature alignment (Fig. 3) and L_{pred} for label prediction (Fig. 4). The both losses are calculated only on the domain-invariant component z of the disentangled feature representation z_s (or z_t). After the training, prediction in target domain is done by the path, encoder→predictor ($M \circ E_t$). Section 3.3 describes this modification in detail.

To preserve the label-related content at domain conversion, we further modify the network by sharing weights and introducing VAE's mechanism (Fig. 5). Section 3.4 describes this modification in detail.

3.3 Disentangled CycleGAN with Feature Consistency Loss

The model in Fig. 2 has pairs of encoders E_*, generators G_*, and discriminators D_*, where $* \in \{s, t\}$. $\hat{x}_t$ is generated by $G_t(E_s(x_s))$, and $\hat{x}_s$ by $G_s(E_t(x_t))$. The original CycleGAN [44] is trained by minimizing the cycle consistency loss L_{cyc}, the identity loss L_{id}, and the adversarial loss L_{adv} defined in LSGAN [26]:

$$\min_{E_s, E_t, G_s, G_t} L_{cyc}(X_s, X_t) = \sum_{* \in \{s,t\}} \mathbb{E}_{x \in X_*}[d(x, \hat{\hat{x}})], \tag{1}$$

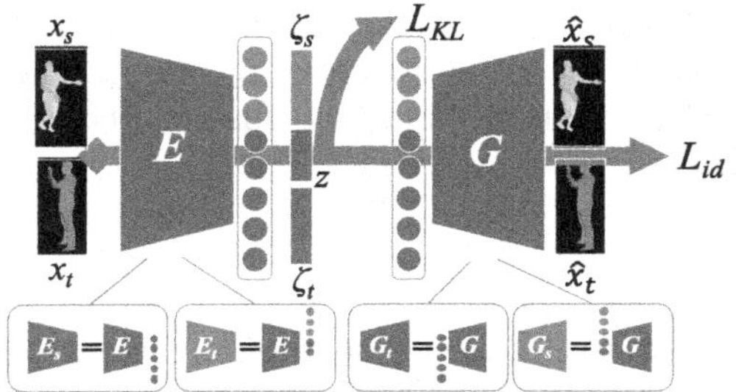

Fig. 5. (best viewed in color) Architecture of partially-shared variational auto-encoders. We note that x_s/x_t is input to E_t/E_s to calculate L_{id}. For a label-preserving domain conversion, the encoders and decoders share parameters other than the connection to ζ_*.

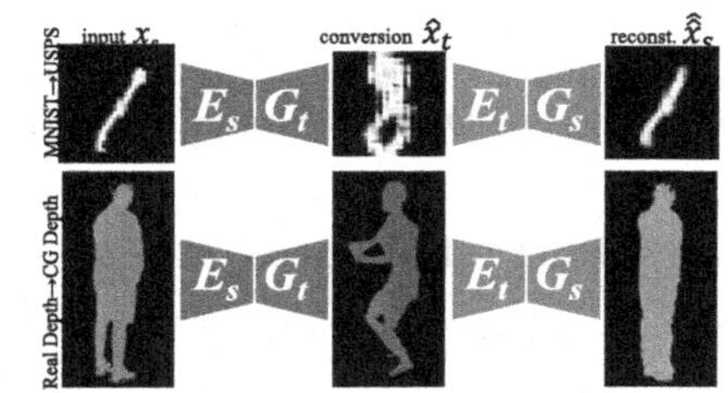

Fig. 6. Misalignment caused by CycleGAN's two image-space discriminators. This is typically seen with a model that does not share encoder weights.

where d is a distance function and $\hat{\hat{x}} = G_*(E_{\bar{*}}(\hat{x}_{\bar{*}}))$. Here, $\bar{*}$ is the opposite domain of $*$.

$$\min_{E_s,E_t,G_s,G_t} L_{id}(X_s, X_t) = \sum_{*\in\{s,t\}} \mathbb{E}_{x\in X_*}[d(x, G_*(E_*(x)))] \tag{2}$$

$$\min_{E_s,E_t,G_s,G_t} \max_{D_s,D_t} L_{adv}(X_s, X_t) = \mathbb{E}_{\{x_s,x_t\}\in X_s\times X_t}[\|D_s(x_s) - 1\|_2 + \|D_s(G_s(E_t(x_t))) + 1\|_2 + \|D_t(x_t) - 1\|_2 + \|D_t(G_t(E_s(x_s))) + 1\|_2] \tag{3}$$

Note that we used spectral normalization [27] in D_s and D_t for a stable adversarial training.

To successfully achieve pair-wise feature alignment, the model divides the output of E_* into $z_{\bar{*}} = \{z, \zeta_{\bar{*}}\}$. Then, it performs feature alignment by using the domain-invariant feature consistency loss L_{fc} (Fig. 3), defined as

$$\min_{E_s,E_t,G_s,G_t} L_{fc}(Z_s, Z_t) = \sum_{*\in\{s,t\}} \mathbb{E}_{z_*\in Z_*}[d(select(z_*), select(E_{\bar{*}}(G_{\bar{*}}(z_*))))], \tag{4}$$

where $Z_* = E_*(X_*)$ and *select* is a function to select z in z_*. Note that gradients are not further back-propagated to $E_{\bar{*}}$ over z_* because updating both z_* and $\hat{z}_*$ in one step leads to bad convergence.

In addition, z obtained from x_s is fed into M to train the classifier/regressor $M : z \rightarrow \hat{y}$ by minimizing the prediction loss $L_{pred}(X_s, Y_s)$ (Fig. 4). The concrete implementation of L_{pred} is task-dependent.

We avoid applying L_{fc} to the whole feature components z_t, as it can hardly reach good local minima because of the competition between the pair-wise feature alignment (by L_{fc}) and CycleGAN (by L_{cyc} and L_{id}). Specifically, training G_t to generate $\hat{x}_t$ must yield a dependency of $\Pr(z_t|x_t)$. This means that $\hat{z}_t$ is trained to have in-domain variation information for x_t. The situation is the same with x_s and z_s. Hence, $\hat{z}_t$ and $\hat{z}_s$ have dependencies on different factors, x_t and x_s, respectively, and it is difficult to match the whole features, $\hat{z}_t$ and $\hat{z}_s$. The disentanglement into z and ζ_* resolves this situation. Note that this architecture is similar to DRIT [18] and MUNIT [12].

3.4 Partially Shared VAEs

Next, we expect E_s and E_t to output a domain-invariant feature z. Even with this implementation, however, CycleGAN can misalign an image's label-related content in domain conversions under a severe target shift, because it has discriminators that match not feature- but image-distributions. Figure 6 shows actual examples of misalignment caused by the image space discriminators. This happens because the decoders G_s and G_t can convert identical zs into different digits (or poses) to better minimize L_{adv} with mismatched label distributions. In such cases, the corresponding encoders also extract identical zs from images with totally different appearance.

To prevent such misalignment and get more stable results, we make the decoders share weights to generate similar content from z, and we make the encoders extract z only from similar content. Figure 5 shows the details of the parameter-sharing architecture, which consists of units called *partially-shared auto-encoders* (PS-AEs). Formally, the partially shared encoders are described as a function $E : x \rightarrow \{z, \zeta_s, \zeta_t\}$. In our implementation, only the last layer is divided into three parts, which outputs z, ζ_s, and ζ_t. E can obviously be substituted for E_s and E_t by discarding ζ_t and ζ_s from the output, respectively. Similarly, the generator $G : \{z, \zeta_s, \zeta_t\} \rightarrow \hat{x}$ shares weights other than for the first layer. The first layer consists of three parts, which output z, ζ_s, and ζ_t. G can be substituted for G_s and G_t by inputting $\{z, \zeta_s, \mathbf{0}\}$ and $\{z, \mathbf{0}, \zeta_t\}$, respectively. Note that the reparameterization trick and L_{kl} minimization are applied only at L_{id} calculation, but not at L_{cycle} calculation.

This implementation brings another advantage for UDA tasks: it can disentangle the feature space by consisting of two variational auto-encoders (VAEs), $G_s \circ E_t$ and $G_t \circ E_s$ (Fig. 5). Putting VAE in a model to obtain a domain-invariant feature is reported as an effective option in recent domain adaptation studies [19,21]. To make PS-AEs a pair of VAEs, we put VAE's resampling process at calculation of L_{id} and add the KL loss defined as

$$\min_{E_s, E_t} L_{KL}(X_s, X_t) = \sum_{* \in \{s,t\}} \mathbb{E}_{z, \zeta_* \in E_*(X_*)}[KL(p_z || q_z) + KL(p_{\zeta_*} || q_{\zeta_*})], \tag{5}$$

where $KL(p||q)$ is the KL divergence between two distributions p and q, p_{ζ_*} is the distribution of ζ_* sampled from X_*, and q_z and q_{ζ_*} are standard normal distributions with the same sizes as z and ζ_*, respectively.

Our full model, *partially-shared variational auto-encoders* (PS-VAEs), is trained by optimizing the weighted sum of the all the above loss functions:

$$L_{total} = L_{adv} + \alpha L_{cyc} + \beta L_{id} + \gamma L_{KL} + \delta L_{fc} + \epsilon L_{pred}, \tag{6}$$

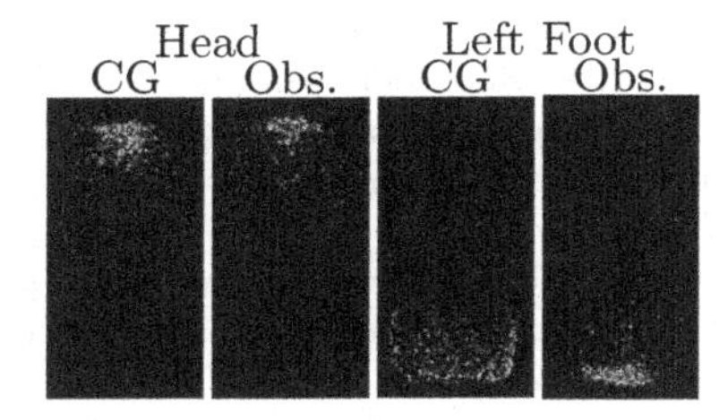

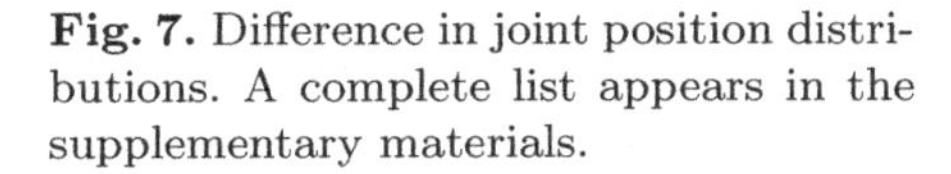

Fig. 7. Difference in joint position distributions. A complete list appears in the supplementary materials.

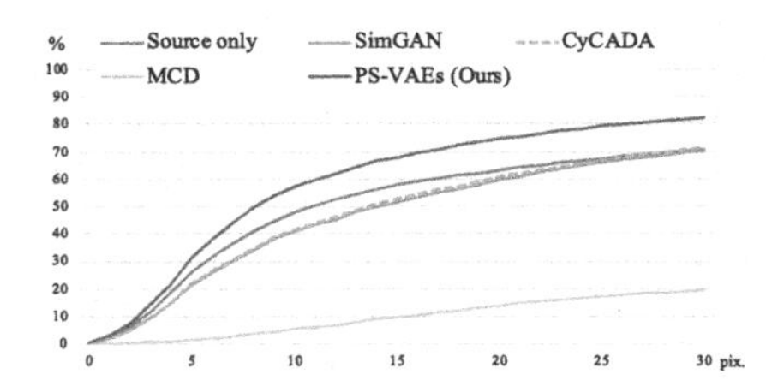

Fig. 8. (best viewed in color) Averaged percentage of joints detected with errors less than N pixels. (Higher is better.)

where α, β, γ, δ, and ϵ are hyper-parameters that should be tuned for each task. For the distance function d, we use the smooth L1 distance [9], which is defined as

$$d(a,b) = \begin{cases} \|a-b\|_2 & \text{if } |a-b| < 1 \\ |a-b| - 0.5 & \text{otherwise} \end{cases} \tag{7}$$

4 Evaluation

4.1 Evaluation on Human-Pose Dataset

We firstly evaluated the proposed method with a regression task on human-pose estimation.

Dataset. For this task, we prepared a synthesized depth image dataset whose poses were sampled with CMU Mocap [4] and rendered with PoserPro2014 [37], as the source domain dataset. Each image had 18 joint positions. In the sampling, we avoided pose duplication by confirming that at least one joint had a position more than 50 mm away from its position in any other samples. The total number of source domain samples was 15000. These were rendered with a choice of two human models (male and female), whose heights were sampled from a normal distribution with respective means of 1.707 and 1.579 m and standard deviations of 56.0 mm and 53.3 mm). For the target dataset, we used depth images from the CMU Panoptic Dataset [14], which were observed with a Microsoft Kinect. We automatically eliminated the background in the target domain data by preprocessing[1]. Finally, we used 15,000 images for training and 500 images (with manually annotated joint positions) for the test.

[1] The details appear in the supplementary material.

Table 2. Accuracy in human-pose estimation by UDA (higher is better). Results were averaged for joints with left and right entries (e.g., the "Shoulder" column lists the average scores for the left and right shoulders). The "Avg." column lists the average scores over all samples, rather than only the joints appearing in this table.

Error less than 10px	Head	Neck	Chest	Waist	Shldr.	Elbow	Wrists	Hands	Knees	Ankles	Foots	Avg.
Source only	69.6	78.6	31.6	34.2	47.3	44.5	38.4	31.5	38.5	54.1	66.2	47.4
MCD	4.6	7.0	0.2	0.6	1.4	0.2	0.3	0.9	0.4	21.0	16.6	5.3
SimGAN	90.2	68.0	10.8	22.6	38.8	26.3	28.5	33.6	35.9	52.5	52.8	40.4
CyCADA	90.0	69.0	15.4	28.2	39.5	27.3	31.3	32.5	35.4	54.4	53.2	41.0
Ours												
CycleGAN+L_{fc}	82.8	79.0	33.8	17.0	40.0	16.4	15.8	28.4	13.8	51.0	51.5	35.5
D-CycleGAN	**93.0**	**85.8**	21.4	**47.8**	42.5	42.5	35.8	39.2	42.5	66.9	64.1	50.8
D-CycleGAN+VAE	40.6	34.2	17.6	41.2	10.1	10.2	7.5	6.4	20.0	28.0	20.2	18.6
PS-AEs	80.6	72.4	**40.8**	28.0	46.5	28.4	25.2	29.4	25.3	58.9	53.9	42.1
PS-VAEs(full)	89.4	84.6	21.4	43.4	**51.7**	**54.4**	**49.4**	**43.9**	**45.6**	**74.5**	**74.0**	**57.0**

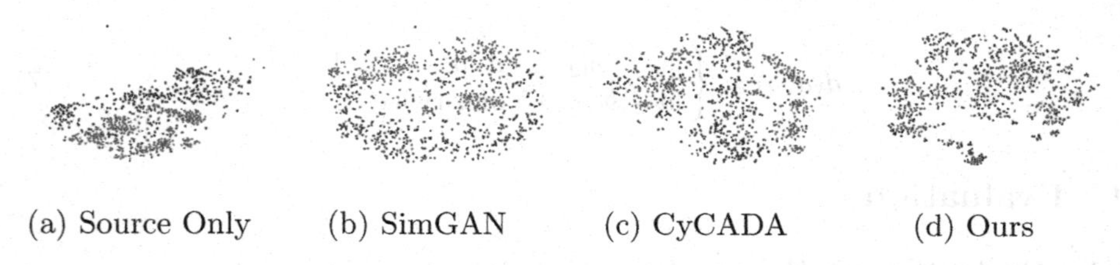

(a) Source Only (b) SimGAN (c) CyCADA (d) Ours

Fig. 9. (best viewed in color) Feature distribution visualized by t-SNE [25]: source domain CG data (blue points) and target domain observed data (red points). (Color figure online)

Experimental Settings. Figure 7 shows the target shift between the source and target domains via the differences in joint positions at the head and foot. L_{pred} was defined as

$$\min_{E_s,M} L_{pred}(Y_s, X_s) = \mathbb{E}_{x_s,y_s \in \{X_s,Y_s\}}(d(M(E_s(x_s)), y_s)). \tag{8}$$

Comparative Methods. We compared the proposed method with the following three baselines.

SimGAN [35] is a method based on image-to-image conversion. To prevent misalignment during conversion, it also minimizes changes in the pixel-values before and after conversion by using a self-regularization loss. The code is borrowed from the implementation of CyCADA.

CyCADA [11] is a CycleGAN-based UDA method. The self-regularization loss is used in this method, too. In addition, it matches feature distributions, like ADDA.

MCD [34] is a method that minimizes a discrepancy defined by the boundary differences obtained from multiple classifiers. This method is expected to be

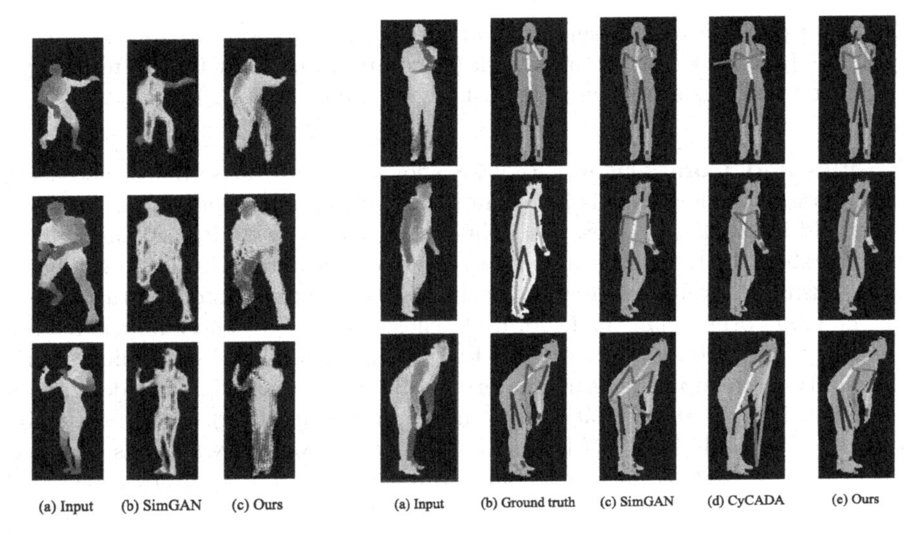

Fig. 10. (best viewed in color) Qualitative comparison on the domain conversion. Detailed structure in body region is lost with SimGAN, but reproduced with our model.

Fig. 11. (best viewed in color) Qualitative results of human-pose estimation. Due to the lack of detailed depth structure as seen in Fig. 10, SimGAN and CyCADA often fail to estimate joints with self-occlusion.

more robust against target shift than methods based on distribution matching, because it does not focus on the entire distribution shape. On the other hand, this kind of approach is theoretically applicable only to classification but not to regression.

All the above methods were implemented with a common network structure, which appears in the supplementary materials.

In addition to the above methods, on the purpose of an ablation study, we compared our full model with the following four different variations.

CycleGAN+L_{fc} does not divide z_s and z_t into the two components, but applied L_{fc} to z_s and z_t directly.

D-CycleGAN stands for disentangled CycleGAN, which divides z_s and z_t into the two components, but weights of encoders and decoders are not shared and not using VAE at the calculation of L_{id}.

D-CycleGAN+VAE is a D-CycleGAN with L_{KL} and the resampling trick of VAE at the calculation of L_{id}.

PS-AEs stands for Partially-shared Auto-Encoders, whose encoders and decoders partially shares weights as described in Sect. 3.4, but not using VAE.

PS-VAEs stands for Partially-shared Variational Auto-Encoders and this is the full model of the proposed method.

All hyper-parameters of baselines and the proposed methods are manually tuned with our best effort for this new task (A comparison under the optimal hyper-parameter settings are given in Sect. 4.2 with the other task.).

Results and Comparison. Figure 8 shows the rate of samples whose estimated joint position error is less than thresholds (the horizontal axis shows the threshold in pixels). To view the joint-wise tendency, we trimmed the figure at the threshold of ten pixels and listed joint-wise accuracy in Table 2. The full model using the proposed methods achieved the best scores on average and for all the joints other than the head, neck, chest, and waist. These four joints have less target shift than others do (see Fig. 7 or the complete list of joint position distributions in supplementary materials). SimGAN was originally designed for a similar task (gaze estimation and hand-pose estimation) and achieved better scores than MCD. CyCADA is an extension of SimGAN and has additional losses for distribution matching, but it did not boost the accuracy in the tasks of UDA with target shift. MCD was originally designed for classification tasks and did not work for this regression task, as expected.

Discussion. Figure 9 shows the feature distributions obtained from four different methods. Because SimGAN does not have any mechanisms to align features in the feature space, the distributions did not merge well. CyCADA better mixes the distributions, but still the components are separated. In contrast, the proposed method merged features quite well despite no discriminators or discrepancy minimization was performed. This indicates that the proposed pair-wise feature alignment by L_{fc} worked well with this UDA task.

A qualitative difference in domain conversion is shown in Fig. 10. SimGAN's self-regularization loss worked to keep the silhouette of generated samples, but subtle depth differences in the body regions were not reproduced well. In addition, the silhouettes were forced to be similar to those of the two human models used to render the CG dataset. This insists that the prior assumption of domain-shift-free metric (i.e., self-regularization loss) could rather reduce the accuracy from source only model. In contrast, the proposed method seems to be able to reproduce such subtle depth differences with a more realistic silhouette. This difference contributed to the prediction quality difference shown in Fig. 11.

D-CycleGAN had actually performed the second best result and D-CycleGAN+VAE and PS-AEs did not work well. First, as UNIT [20] does[2], it seems to be difficult to use VAE with CycleGAN without sharing weights between the encoder-decoder models. After combining all these modifications, the full model of the proposed method outperformed any other methods with a large margin.

[2] Another neural network model that combines CycleGAN and VAE as the proposed model, but for image-to-image translation.

Table 3. Accuracy in the three UDA tasks (Bold & Underline: the best and second best scores. Δ: the degradation from 10% to 50%).

Task		Src. Only	ADDA	UFDN	CDAN-E	PADA	UAN	SimGAN	CyCADA	MCD	Ours
MNIST→USPS	Ref.	-	89.4	97.1	95.6	-	-	-	95.6	94.2	-
	10%	71.0	89.8	**94.0**	91.0	75.3	78.2	72.4	91.8	91.2	93.9
	20%	-	86.9	90.4	90.7	77.7	77.3	86.5	91.0	90.4	**94.8**
	30%	-	79.3	83.2	91.1	79.3	74.9	84.0	80.3	79.0	**93.4**
	40%	-	81.8	82.3	84.2	77.8	75.0	84.3	86.4	78.5	**94.6**
	50%	-	78.5	83.8	79.8	80.2	76.0	76.3	87.6	80.3	**92.6**
	Δ	-	-11.3	-10.2	-11.2	4.9	-2.2	3.9	-4.2	-10.9	-1.3
USPS→MNIST	Ref.	-	90.1	93.7	98.0	-	-	-	96.5	94.1	-
	10%	55.6	**96.0**	93.6	95.8	47.9	83.2	68.3	75.3	**96.0**	94.8
	20%	-	89.0	81.9	**95.8**	39.2	83.4	50.2	75.3	81.5	94.4
	30%	-	81.5	79.2	**95.4**	36.0	79.4	49.9	75.2	79.1	90.8
	40%	-	78.9	72.0	**90.9**	29.8	78.4	63.8	76.7	78.1	82.6
	50%	-	80.5	69.1	**90.7**	25.2	77.7	49.3	70.7	77.4	82.4
	Δ	-	-15.5	-24.6	-4.1	-22.7	-5.5	-19.0	-4.6	-18.4	-12.4
SVHN→MNIST	Ref.	-	76.0	95.0	89.2	-	-	-	90.4	96.2	-
	10%	46.6	75.5	91.1	78.7	30.5	68.0	61.4	**91.4**	90.3	73.7
	20%	-	65.0	70.9	79.4	39.5	64.8	52.5	75.4	**89.7**	72.9
	30%	-	65.2	58.7	73.2	37.3	63.2	57.7	69.7	**80.2**	73.8
	40%	-	50.8	52.6	56.9	36.8	64.9	51.8	70.7	**72.0**	64.4
	50%	-	54.3	43.6	56.1	36.7	66.8	49.3	68.3	65.3	**68.4**
	Δ	-	-21.2	-47.5	-22.6	6.2	-1.2	-12.1	-23.1	-25.0	-5.3

4.2 Evaluation on Digit Classification

To show the versatility of the proposed method with classification task and to systematically analyze the performance of the methods against target shift, we conducted an experiment by a simple UDA task with digit datasets (MNIST [17]↔USPS [13], and SVHN [29]→MNIST), with which the optimal hyper-parameters are provided in many methods.

Controlling the Intensity of Target Shift. To evaluate the performance under a controlled situation with an easy-to-reproduce and high-contrast class-imbalances in the target domain, we adjusted the rate of samples of class '1' from 10% to 50%. Note that the operation is done only to the training samples in the target domain. Those in the source domain and test data are both maintained to be balanced.

When the rate was 10%, the number of samples was exactly the same among the categories. When it was 50%, half the data belonged to category '1,' which was the largest target shift in this experiment. Note that the reference scores reported in the original papers and those at 10% are slightly different due to the controlled numbers of training data. A more detailed explanation of this operation appears in the supplementary materials.

Experimental Settings and Comparative Methods. In this task, L_{pred} is simply given as the following categorical cross-entropy loss:

$$\min_{E_s,M} L_{pred}(Y_s, X_s) = \mathbb{E}_{x_s,y_s \in X_s \times L}[-y_s \log M(E(x_s))] \tag{9}$$

In addition to the comparative methods shown in Sect. 4.1, we prepared the following three additional baselines as the methods that resolve domain shift purely by distribution matching:

ADDA [38] **and UFDN** [19] are methods based on feature distribution matching.

PADA [2] **and UAN** [40] also match feature distributions but while estimating a category- and sample-wise weights, respectively.

CDAN-E [23] uses category-wise distribution matching and thus potentially valid under target shift as long as the target domain samples are assigned to the right category. To ensure the right assignment, the method estimates sample-wise weights.

Note that some of recent state-of-the-art methods for the digit UDA task without target shift was not listed in the experiment due to their reproducibility problem[3]. The detailed implementations (network architecture, hyperparameters, and so on) of the proposed method and the above methods appear in the supplementary material.

Results and Discussion. Table 3 lists the results. The methods based on distribution matching (ADDA and UFDN) were critically affected by target shift. CyCADA was more robust than ADDA and UFDN for the MNIST↔USPS tasks, owing to the self-regularization loss; however it did not work for the SVHN→MNIST task due to the large pixel-value differences. A similar tendency was observed by SimGAN.

MCD stably performed well among all the three tasks; however, it was largely affected by target shifts (≥30%). Similar tendency was observed with CDAN-E (≥40%). From the perspective of the performance drop by target shift, PADA behaved differently from any other methods; it typically works better with a heavier target shift but not so good without target shift. UAN, which was not evaluated on this dataset in the original paper, achieved a poor absolute performance although it was least degraded by the target shift. Overall, our method comparably performed under the various level of target-shifted conditions even in the classification task. This shows the versatility of the method against various UDA tasks.

[3] The authors of [31] provide no implementation and there are currently no other authorized implementations. Two SBADAGAN [32] implementations were available but it was difficult to customize them for this test and the reported accuracy was not reproducible.

5 Conclusion

In this paper, we have proposed a novel approach of partially-shared variational auto-encoders for the problem of UDA with target shift. The traditional approach of feature distribution matching implicitly assumes the identical distribution and will fail with target shift. The proposed method resolves this problem by label-preserving domain conversion; pseudo pair with an identical label is generated with domain conversion and used to resolve domain shift by sample-wise metric learning rather than a distribution matching. The model is specially designed to preserve the labels by sharing weights between two domain conversion branches as much as possible. The experimental results showed its versatile performance on pose estimation and digit classification tasks.

References

1. Cao, Z., Long, M., Wang, J., Jordan, M.I.: Partial transfer learning with selective adversarial networks. In: The IEEE Conference on Computer Vision and Pattern Recognition (2018)
2. Cao, Z., Ma, L., Long, M., Wang, J.: Partial adversarial domain adaptation. In: Ferrari, V., Hebert, M., Sminchisescu, C., Weiss, Y. (eds.) ECCV 2018. LNCS, vol. 11212, pp. 139–155. Springer, Cham (2018). https://doi.org/10.1007/978-3-030-01237-3_9
3. Cao, Z., You, K., Long, M., Wang, J., Yang, Q.: Learning to transfer examples for partial domain adaptation. In: The IEEE Conference on Computer Vision and Pattern Recognition (2019)
4. CMU Graphics Lab.: CMU graphics lab motion capture database. http://mocap.cs.cmu.edu/. Accessed 11 Nov 2019
5. Cordts, M., et al.: The cityscapes dataset for semantic urban scene understanding. In: Proceedings of the IEEE Conference on Computer Vision and Pattern Recognition, pp. 3213–3223 (2016)
6. Deng, Z., Luo, Y., Zhu, J.: Cluster alignment with a teacher for unsupervised domain adaptation. In: The IEEE International Conference on Computer Vision (2019)
7. Doersch, C.: Tutorial on variational autoencoders. In: CoRR (2016)
8. Ghifary, M., Kleijn, W.B., Zhang, M.: Domain adaptive neural networks for object recognition. In: Pham, D.-N., Park, S.-B. (eds.) PRICAI 2014. LNCS (LNAI), vol. 8862, pp. 898–904. Springer, Cham (2014). https://doi.org/10.1007/978-3-319-13560-1_76
9. Girshick, R.: Fast R-CNN. In: The IEEE International Conference on Computer Vision (2015)
10. Gong, M., Zhang, K., Liu, T., Tao, D., Glymour, C., Schölkopf, B.: Domain adaptation with conditional transferable components. In: International Conference on Machine Learning, pp. 2839–2848 (2016)
11. Hoffman, J., et al.: CyCADA: cycle-consistent adversarial domain adaptation. In: Proceedings of the 35th International Conference on Machine Learning (2018)
12. Huang, X., Liu, M.-Y., Belongie, S., Kautz, J.: Multimodal unsupervised image-to-image translation. In: Ferrari, V., Hebert, M., Sminchisescu, C., Weiss, Y. (eds.) ECCV 2018. LNCS, vol. 11207, pp. 179–196. Springer, Cham (2018). https://doi.org/10.1007/978-3-030-01219-9_11

13. Hull, J.J.: A database for handwritten text recognition research. IEEE Trans. Pattern Anal. Mach. Intell. **16**(5), 550–554 (1994)
14. Joo, H., et al.: Panoptic studio: a massively multiview system for social motion capture. In: The IEEE International Conference on Computer Vision (2015)
15. Kuhnke, F., Ostermann, J.: Deep head pose estimation using synthetic images and partial adversarial domain adaption for continuous label spaces. In: The IEEE International Conference on Computer Vision (2019)
16. Laradji, I., Babanezhad, R.: M-ADDA: unsupervised domain adaptation with deep metric learning. In: Proceedings of the 36th International Conference on Machine Learning Workshop (2018)
17. LeCun, Y., Cortes, C.: MNIST handwritten digit database. http://yann.lecun.com/exdb/mnist/ (2010)
18. Lee, H.-Y., Tseng, H.-Y., Huang, J.-B., Singh, M., Yang, M.-H.: Diverse image-to-image translation via disentangled representations. In: Ferrari, V., Hebert, M., Sminchisescu, C., Weiss, Y. (eds.) ECCV 2018. LNCS, vol. 11205, pp. 36–52. Springer, Cham (2018). https://doi.org/10.1007/978-3-030-01246-5_3
19. Liu, A.H., Liu, Y.C., Yeh, Y.Y., Wang, Y.C.F.: A unified feature disentangler for multi-domain image translation and manipulation. In: Advances in Neural Information Processing Systems, vol. 31, pp. 2590–2599 (2018)
20. Liu, M.Y., Breuel, T., Kautz, J.: Unsupervised image-to-image translation networks. In: Guyon, I., et al. (eds.) Advances in Neural Information Processing Systems, vol. 30, pp. 700–708 (2017)
21. Liu, Y., Wang, Z., Jin, H., Wassell, I.: Multi-task adversarial network for disentangled feature learning. In: Proceedings of the IEEE Conference on Computer Vision and Pattern Recognition, pp. 3743–3751 (2018)
22. Long, M., Cao, Y., Wang, J., Jordan, M.I.: Learning transferable features with deep adaptation networks. In: Proceedings of the 32nd International Conference on International Conference on Machine Learning, pp. 97–105 (2015)
23. Long, M., Cao, Z., Wang, J., Jordan, M.I.: Conditional adversarial domain adaptation. In: Advances in Neural Information Processing Systems, pp. 1645–1655 (2018)
24. Luo, Y., Zheng, L., Guan, T., Yu, J., Yang, Y.: Taking a closer look at domain shift: category-level adversaries for semantics consistent domain adaptation. In: The IEEE Conference on Computer Vision and Pattern Recognition (2019)
25. Maaten, L.V.D., Hinton, G.: Visualizing data using t-SNE. J. Mach. Learn. Res. **9**(Nov), 2579–2605 (2008)
26. Mao, X., Li, Q., Xie, H., Lau, R.Y., Wang, Z., Paul Smolley, S.: Least squares generative adversarial networks. In: Proceedings of the IEEE International Conference on Computer Vision, pp. 2794–2802 (2017)
27. Miyato, T., Kataoka, T., Koyama, M., Yoshida, Y.: Spectral normalization for generative adversarial networks. In: International Conference on Learning Representations (2018)
28. Moreno-Torres, J.G., Raeder, T., Alaiz-RodríGuez, R., Chawla, N.V., Herrera, F.: A unifying view on dataset shift in classification. Pattern Recogn. **45**(1), 521–530 (2012)
29. Netzer, Y., Wang, T., Coates, A., Bissacco, A., Wu, B., Ng, A.Y.: Reading digits in natural images with unsupervised feature learning. In: Advances in Neural Information Processing Systems Workshop (2011)

30. Richter, S.R., Vineet, V., Roth, S., Koltun, V.: Playing for data: ground truth from computer games. In: Leibe, B., Matas, J., Sebe, N., Welling, M. (eds.) ECCV 2016. LNCS, vol. 9906, pp. 102–118. Springer, Cham (2016). https://doi.org/10.1007/978-3-319-46475-6_7
31. Roy, S., Siarohin, A., Sangineto, E., Bulo, S.R., Sebe, N., Ricci, E.: Unsupervised domain adaptation using feature-whitening and consensus loss. In: Proceedings of the IEEE Conference on Computer Vision and Pattern Recognition, pp. 9471–9480 (2019)
32. Russo, P., Carlucci, F.M., Tommasi, T., Caputo, B.: From source to target and back: symmetric bi-directional adaptive GAN. In: Proceedings of the IEEE Conference on Computer Vision and Pattern Recognition, pp. 8099–8108 (2018)
33. Saito, K., Ushiku, Y., Harada, T., Saenko, K.: Adversarial dropout regularization. In: The International Conference on Learning Representations (2018)
34. Saito, K., Watanabe, K., Ushiku, Y., Harada, T.: Maximum classifier discrepancy for unsupervised domain adaptation. In: Proceedings of the IEEE Conference on Computer Vision and Pattern Recognition, pp. 3723–3732 (2018)
35. Shrivastava, A., Pfister, T., Tuzel, O., Susskind, J., Wang, W., Webb, R.: Learning from simulated and unsupervised images through adversarial training. In: Proceedings of the IEEE Conference on Computer Vision and Pattern Recognition, pp. 2107–2116 (2017)
36. Shu, R., Bui, H., Narui, H., Ermon, S.: A DIRT-T approach to unsupervised domain adaptation. In: International Conference on Learning Representations (2018)
37. Software, P.: Poser pro 2014. https://www.renderosity.com/mod/bcs/poser-pro-2014/102000. Accessed 10 Nov 2019
38. Tzeng, E., Hoffman, J., Saenko, K., Darrell, T.: Adversarial discriminative domain adaptation. In: The IEEE Conference on Computer Vision and Pattern Recognition (2017)
39. Yan, H., Ding, Y., Li, P., Wang, Q., Xu, Y., Zuo, W.: Mind the class weight bias: weighted maximum mean discrepancy for unsupervised domain adaptation. In: The IEEE Conference on Computer Vision and Pattern Recognition (2017)
40. You, K., Long, M., Cao, Z., Wang, J., Jordan, M.I.: Universal domain adaptation. In: The IEEE Conference on Computer Vision and Pattern Recognition (2019)
41. Zhang, J., Ding, Z., Li, W., Ogunbona, P.: Importance weighted adversarial nets for partial domain adaptation. In: The IEEE Conference on Computer Vision and Pattern Recognition (2018)
42. Zhang, K., Schölkopf, B., Muandet, K., Wang, Z.: Domain adaptation under target and conditional shift. In: International Conference on Machine Learning, pp. 819–827 (2013)
43. Zhang, X., Wong, Y., Kankanhalli, M.S., Geng, W.: Unsupervised domain adaptation for 3D human pose estimation. In: Proceedings of the 27th ACM International Conference on Multimedia (2019)
44. Zhu, J.Y., Park, T., Isola, P., Efros, A.A.: Unpaired image-to-image translation using cycle-consistent adversarial networks. In: The IEEE International Conference on Computer Vision (2017)

Learning Where to Focus for Efficient Video Object Detection

Zhengkai Jiang[1,2](✉), Yu Liu[3](✉), Ceyuan Yang[3], Jihao Liu[3], Peng Gao[3], Qian Zhang[4], Shiming Xiang[1,2], and Chunhong Pan[1,2]

[1] National Laboratory of Pattern Recognition, Institute of Automation, Chinese Academy of Sciences, Beijing, China
{zhengkai.jiang,smxiang}@nlpr.ia.ac.cn

[2] School of Artificial Intelligence, University of Chinese Academy of Sciences, Beijing, China

[3] The Chinese University of Hong Kong, Hong Kong, People's Republic of China
liuyuisanai@gmail.com

[4] Horizon Robotics, Beijing, China

Abstract. Transferring existing image-based detectors to the video is non-trivial since the quality of frames is always deteriorated by part occlusion, rare pose, and motion blur. Previous approaches exploit to propagate and aggregate features across video frames by using optical flow-warping. However, directly applying image-level optical flow onto the high-level features might not establish accurate spatial correspondences. Therefore, a novel module called Learnable Spatio-Temporal Sampling (LSTS) has been proposed to learn semantic-level correspondences among adjacent frame features accurately. The sampled locations are first randomly initialized, then updated iteratively to find better spatial correspondences guided by detection supervision progressively. Besides, Sparsely Recursive Feature Updating (SRFU) module and Dense Feature Aggregation (DFA) module are also introduced to model temporal relations and enhance per-frame features, respectively. Without bells and whistles, the proposed method achieves state-of-the-art performance on the ImageNet VID dataset with less computational complexity and real-time speed. Code will be made available at LSTS.

Keywords: Flow-warping · Learnable Spatio-Temporal Sampling · Spatial correspondences · Temporal relations

1 Introduction

Object detection is a fundamental problem in computer vision and enables various applications, *e.g.*, robot vision and autonomous driving. Recently, deep convolution neural networks have achieved significant process on object

Electronic supplementary material The online version of this chapter (https://doi.org/10.1007/978-3-030-58517-4_2) contains supplementary material, which is available to authorized users.

A. Vedaldi et al. (Eds.): ECCV 2020, LNCS 12361, pp. 18–34, 2020.
https://doi.org/10.1007/978-3-030-58517-4_2

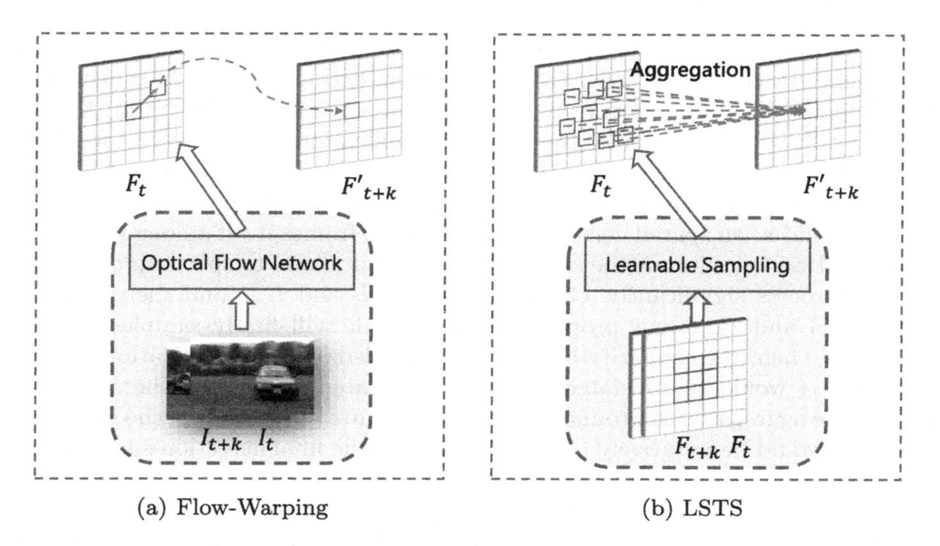

(a) Flow-Warping

(b) LSTS

Fig. 1. Comparison between flow-warping with LSTS for feature propagation. F_t and F_{t+k} denote features extracted from two adjacent frames I_t and I_{t+k}, respectively. Previous work directly applies optical flow to represent the feature-level shift while our LSTS could learn more accurate correspondences from data

detection [12,14,22,24,27]. However, directly applying image object detectors frame-by-frame cannot obtain satisfactory results since frames in a video are always deteriorated by part occlusion, rare pose, and motion blur. The inherent temporal information in the video, as the rich cues of motion, can boost the performance of video object detection (Fig. 1).

Previous efforts on leveraging the temporal information can mainly be divided into two categories. The first one relies on temporal information for post-processing to ensure the object detection results more coherent [13,20,21]. These methods usually apply image-detector to obtain detection results, then associate the results via the box-level matching, e.g., tracker or off-the-shelf optical flow. Such post-processing tends to slow down the processing of detection. Another category [17–19,31,35,37,42–44] exploits the temporal information on feature representation. Specifically, they mainly improve features by aggregating that of adjacent frames to boost the detection accuracy, or propagating features to avoid dense feature extraction to improve the speed.

Nevertheless, when propagating the features across frames, optical flow based warping operation is always required. Such operation would introduce the following disadvantages: (1) Optical flow tends to increase the number of model parameters by a large margin, which makes the applications on embedded devices unfriendly. Take the image detector ResNet101+RFCN [2,16] as an example, the parameter size would increase from 60.0 M to 96.7 M even with the pruned FlowNet [44]. (2) Optical flow cannot represent the correspondences among high-level features accurately. Due to the increase of the receptive field of networks, the small motion drift in the high-level feature always corresponds to large motion movements in the image-level. (3) Optical flow extraction is time-consuming. For example, FlowNet [7] runs at only 10 frames per second (FPS) on KITTI dataset [11], which will hinder the practical application of video detectors.

To address the above issues, *Learnable Spatio-Temporal Sampling* (**LSTS**) module has been proposed to propagate the high-level feature across frames. Such module could learn spatial correspondences across frames itself among the whole datasets. Besides, without the extra optical flow, it allows to speed up the propagation process significantly. Given two frames I_t and I_{t+k} and the extracted features F_t and F_{t+k}, our proposed LSTS module will firstly samples specific locations. Then, the similarity between the sampled locations on feature F_t and feature F_{t+k} would be calculated. Next, the calculated weights together with feature F_t are aggregated to produce propagated feature $F_{t+k}^{'}$. At last, the sampling locations would be iteratively updated guided by the final detection supervision, which allows to propagate and align the high-level features across frames more accurately. Based on LSTS, an efficient video object detection framework is also introduced. The features of keyframes and non-keyframes would be extracted by expensive and light-weight network, respectively. To further leverage the temporal relation across whole videos, *Sparsely Recursive Feature Updating* (**SRFU**) and *Dense Feature Aggregation* (**DFA**) are then proposed to boost the dense low-level features, and enhance the feature representation separately.

Experiments are conducted on the public video object detection benchmark i.e., ImageNet VID datasets. Without bells and whistles, the proposed framework achieves state-of-the-art performance at the real-time speed and brings much fewer model parameters simultaneously. In addition, elaborate ablative studies show the advance of learnable sampling locations over the hand-crafted design.

We summarize the major contributions as follows: 1) LSTS module is proposed to propagate the high-level feature across frames, which could calculate the spatial-temporal correspondences accurately. Different from previous approaches, LSTS treat the offsets of sampling locations as parameters and the optimal offsets would be learned through back-propagation guided by bounding box and category supervision. 2) SRFU and DFA module are proposed to model temporal relation and enhance feature representation, respectively. 3) Experiments on VID dataset demonstrate that the proposed framework achieves state-of-the-art trade-off performance between speed and model parameters.

2 Related Work

Image Object Detection. Recently, state-of-the-art methods for image-based detectors are mainly based on the deep convolutional neural networks [22,24,27]. Generally, the image object detectors can be divided into two paradigms, i.e., the single-stage and the two-stage detectors. Two-stage detectors usually first generate region proposals, which are then refined by classification and regression process through the RCNN stage. ROI pooling [15] was proposed to speed up R-CNN [12]. Faster RCNN [27] utilized anchor mechanism to replace Selective

Search [33] proposal generation process, achieving great performance promotion as well as faster speed. FPN [22] introduced an inherent multi-scale, pyramidal hierarchy of deep convolution networks to build feature pyramids with marginal extra cost and significant improvements. The single-stage detector pipeline is more efficient but achieves less accurate performance. SSD [24] directly generates results from anchor boxes on a pyramid of feature maps. RetinaNet [23] handled extreme foreground and background imbalance issue by a novel loss named focal loss. Usually, the image object detector provides the baseline results for video object detection through frame-by-frame detection.

Video Object Detection. Compared with image object detection, temporal information provides the cue for video object detection, which can be utilized to boost accuracy or efficiency. To improve detection efficiency, a few works [17,40,44] exploited to propagate features across frames to avoid dense expensive feature extraction, which mainly relied on the flow-warping [44] operation. DFF [44] was proposed with an efficient framework which only runs expensive CNN feature extraction on sparse and regularly selected keyframes, achieving more than 10x speedup than using an image detector for per-frame detection. Towards High Performance [42] proposed spare recursive feature aggregation and spatially-adaptive feature updating strategies, which helps run real-time speed with significant performance. On the one hand, the slow flow extraction process is still the bottleneck for higher speed. On the other hand, the image-level flow which is used to propagate high-level feature may hinder the propagation accuracy, resulting in inferior accuracy.

To improve detection accuracy, different methods [4,31,37,43] have been proposed to aggregate features across frames. They either rely on optical flow to propagate the neighbouring frames' features, or establish spatial-temporal relation to propagate the adjacent frames' features. Then the propagated features from the adjacent frames and current frame feature are aggregated to improve the feature. FGFA [43] was proposed to aggregate nearby features for each frame. It achieves better accuracy at the cost of slow speed, which only runs on 3 FPS due to dense prediction and heavy flow extraction process. EDN [6] was proposed to aggregate and propagate object relation to augment object features. SELSA [37] and LRTR [31] were proposed to aggregate feature by modeling temporal proposals. Besides, OGEM [4] utilized object guided external memory network to model the relationship among temporal proposals. However, these methods cannot run real-time due to the multi-frames feature aggregation. Compared with the above works, our proposed method can be much efficient and run at real-time speed.

Flow-Warping for Feature Propagation. Optical flow [7] has been widely used to model motion relation across frames in many video-based applications, such as video action recognition [32] and video object detection [28]. DFF [44] is the first work to propagate deep keyframe feature to non-keyframe using flow-warping operation based on tailored optical-flow extraction network, resulting

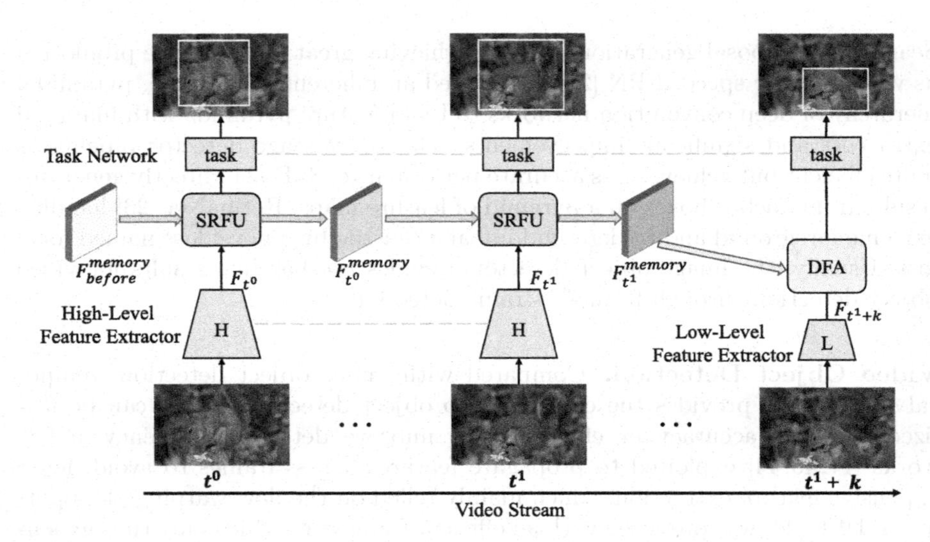

Fig. 2. Framework for inference. For simplicity, frames at t^0, t^1 (keyframes) and $t^1 + k$ (non-keyframe) would be fed into a high-level and a low-level feature extractor respectively. Based on the high-level features, the memory feature F^{memory} is maintained by SRFU to capture the temporal relation, and updated iteratively at keyframes time step. Meanwhile, DFA propagates the memory feature F^{memory} of keyframes to enhance and enrich the low-level features of non-keyframes. LSTS is embedded in SRFU and DFA to propagate and align features across frames accurately. Both the output of SFRU and DFA is produced by the task network to make the final prediction

in 10x speedup but inferior performance. However, optical flow extraction is time-consuming, which means that we are also expected to manually design lightweight optical flow extraction network for higher speed, which can be in the price of losing precision. What's more, it is less robust for feature-level warping using image-level optical flow. Compared with flow-warping based feature propagate, our proposed method is much lightweight and can model the correspondences across frames in the feature-level accurately.

Self-attention for Feature Propagation. Attention mechanism [9,10,26,29, 34,36,39,41] is widely studied in computer vision and natural language processing. Self-attention [34] and non-local [36] are proposed to model the relation of language sequences and to capture long-range dependencies, respectively. An attention function can be described as mapping a query and a set of key-value pairs to an output, where the query, keys, values, and output are all feature maps. Due to the formulation of such attention operation, it can naively be used to model the relation of the features across frames. However, motion across frames is usually limited in a near window, not the whole feature size. Thus MatchTrans [38] was proposed to propagate the features across frames as a local Non-Local [36] manner. Even so, the neighbourhood size is still needed to be carefully designed to match the motion distribution of whole datasets. Compared with MatchTrans [38] module, our proposed LSTS module can adaptively learn the sampling locations, which allows to estimate spatial correspondences across frames more accurately. At the same time, Liu [25] proposed to learn sampling locations for convolution kernel, which shares core idea with us.

3 Methodology

3.1 Framework Overview

In terms of the frame-by-frame detector, frames are divided into two categories i.e., keyframe and non-keyframe. In order to decrease the computational complexity, the feature extractors vary from different types of frames. Specifically, the features of the keyframe and the non-keyframe would derive from the heavy (H) and light (L) feature extraction networks, respectively. In Sect. 3.2, LSTS is proposed to align and propagate the features across frames. In order to leverage the relation among frames, a memory feature F^{memory} is maintained on keyframes, which is gradually updated by the proposed SRFU module (in Sect. 3.3). Besides, with the lightweight feature extractor network, the low-level features on the non-keyframes are usually not capable to obtain good detection results. Thus, DFA module (in Sect. 3.4) is proposed to improve the low-level features on the non-keyframes by utilizing the memory feature F^{memory} on the keyframes. The pipeline of our framework is illustrated in Fig. 2.

3.2 Learnable Spatio-Temporal Sampling

After collecting the features F_t, F_{t+k} of corresponding frames I_t, I_{t+k}, LSTS module allows to calculate the similarity weights of correspondences. As Fig. 3 shows, the procedure of our LSTS module consists of four steps: 1) It first samples some locations on the feature F_t. 2) The correspondence similarity weights are then calculated on the embedded features $f(F_t)$ and $g(F_{t+k})$ by using the sampled locations, where $f(\cdot)$ and $g(\cdot)$ are embedding functions, which aims to reduce the channel of features F_t and F_{t+k} to save computational cost. 3) Next, the calculated weights together with feature F_t are aggregated to obtain propagated feature F'_{t+k}. 4) At last, the sampled locations can be iteratively updated according to final detection loss during training process.

Sampling. To propagate features from F_t to F_{t+k} accurately, we need accurate spatial correspondences across two frames. Motivated by Fig. 5, the correspondences can be limited to the neighbourhood. Thus we first initialize some sampled locations on the neighborhood, which provides coarse correspondences. Besides, with the ability of learning, LSTS can shift and scale the distribution of sampled locations progressively to establish spatial correspondences more accurately. Uniform and Gaussian distribution are applied as two kinds of initialization methods, which will be discussed in detail in the experimental section.

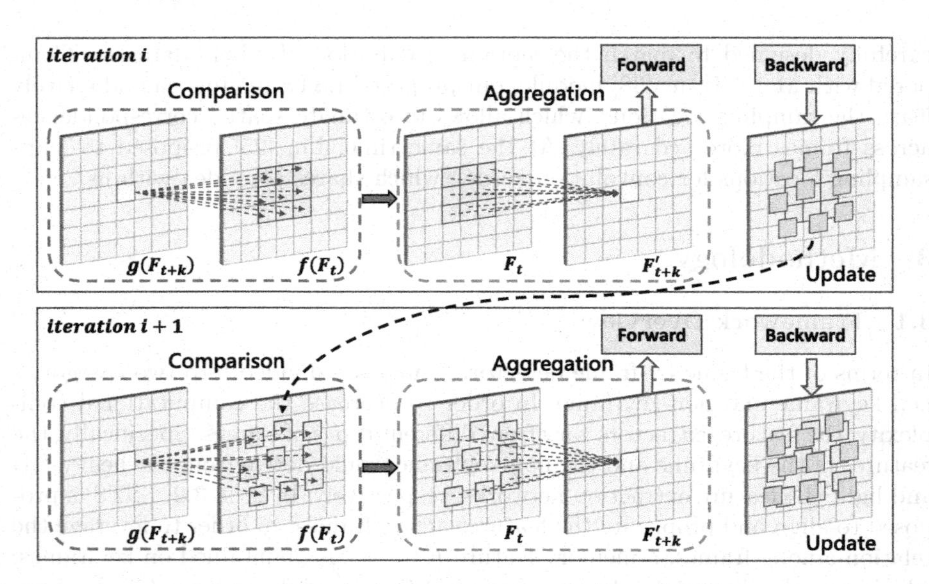

Fig. 3. Illustration of LSTS module. LSTS basically consists of 4 steps: 1) some locations on the feature are randomly sampled. 2) The affinity weight is calculated by similarity comparison. 3) Next, the features F_t together with weights will be aggregated to obtain features F'_{t+k}. 4) the locations would be updated iteratively by back-propagation during training. After training, the final learned sampling locations would be used for inference

Comparison. With the correspondence locations, the similarity of them would be calculated. To save the computational cost, features F_t and F_{t+k} are embedded to $f(F_t)$ and $g(F_{t+k})$, respectively, where f and g are the embedding function. Then the correspondence similarity weights are calculated based on the embedded features $f(F_t)$ and $g(F_{t+k})$. As Fig. 3 shows, $\mathbf{p}_0$ denotes the specific grid location (yellow square) on the feature map $g(F_{t+k})$. The sampled correspondence locations (blue square) on $f(F_t)$ are denoted as $\mathbf{p}_n$, where $n = 1, ..., N$, and N is the number of the sampled locations. Let $f(F_t)_{\mathbf{p}_n}$ and $g(F_{t+k})_{\mathbf{p}_0}$ denote the features at the location $\mathbf{p}_n$ from $f(F_t)$ and at location $\mathbf{p}_0$ from $g(F_{t+k})$, respectively. We aims to calculate the similarity weight between each $f(F_t)_{\mathbf{p}_n}$ and $g(F_{t+k})_{\mathbf{p}_0}$.

Considering $\mathbf{p}_n$ may be in the arbitrary location on the feature map, $f(F_t)_{\mathbf{p}_n}$ firstly requires the bilinear interpolation operation following

$$f(F_t)_{\mathbf{p}_n} = \sum_{\mathbf{q}} G(\mathbf{p}_n, \mathbf{q}) \cdot f(F_t)_{\mathbf{q}}. \tag{1}$$

Here, $\mathbf{q}$ enumerates all integral spatial locations on the feature map $f(F_t)$, and $G(\cdot, \cdot)$ is the bilinear interpolation kernel function as in [3]. After obtaining the value of $f(F_t)_{\mathbf{p}_n}$, we use similarity function **Sim** to measure the distance between

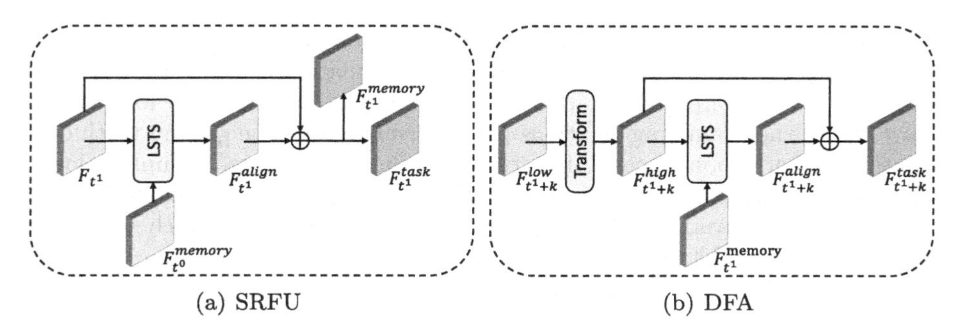

(a) SRFU (b) DFA

Fig. 4. Architecture of SRFU and DFA. $\bigoplus$ is the Aggregation Unit. Transform unit only consists of several convolutions, which is used to improve low-level features on the non-keyframes. For SRFU, LSTS module is utilized to aggregate last keyframe $F_{t^0}^{memory}$ to current keyframe t^1. While for DFA, LSTS module aims to propagate the keyframe memory feature $F_{t^1}^{memory}$ to non-keyframe t^1+k to boost the feature quality to obtain better detection results

the vector $f(F_t)_{\mathbf{p}_n}$ and the vector $g(F_{t+k})_{\mathbf{p}_0}$. Suppose that both $f(F_t)_{\mathbf{p}_n}$ and $f(g_{t+k})_{\mathbf{p}_0}$ are c dimensional vectors, then we have the similarity value $s(\mathbf{p}_n)$:

$$s(\mathbf{p}_n) = \mathbf{Sim}(f(F_t)_{\mathbf{p}_n}, g(F_{t+k})_{\mathbf{p}_0}). \tag{2}$$

A very common function **Sim** can be dot-product. After getting all similarity value $s(\mathbf{p}_n)$ on the location $\mathbf{p}_n$, then the normalized similarity weights can be calculated by:

$$S(\mathbf{p}_n) = \frac{s(\mathbf{p}_n)}{\sum_{n=1}^{N} s(\mathbf{p}_n)}. \tag{3}$$

Aggregation. After obtaining each of the calculated correspondence similarity weights $S(\mathbf{p}_n)$ on location $\mathbf{p}_n$, then the estimated value on location $\mathbf{p}_0$ for F'_{t+k} can be calculated as:

$$F'_{t+k}(\mathbf{p}_0) = \sum_{n=1}^{N} S(\mathbf{p}_n) \cdot F_t(\mathbf{p}_n). \tag{4}$$

Updating. In order to learn the ground truth distribution of correspondences, the sampled locations are also updated by the back-propagation during training. We use the dot-product for function **Sim** for simplicity. Then we have:

$$s(\mathbf{p}_n) = \sum_{\mathbf{q}} G(\mathbf{p}_n, \mathbf{q}) \cdot f(F_t)_{\mathbf{q}} \cdot g(F_{t+k})_{\mathbf{p}_0}. \tag{5}$$

Thus, the gradients for location $\mathbf{p}_n$ can be calculated by:

$$\frac{\partial s(\mathbf{p}_n)}{\partial \mathbf{p}_n} = \sum_{\mathbf{q}} \frac{\partial G(\mathbf{p}_n, \mathbf{q})}{\partial \mathbf{p}_n} \cdot f(F_t)_{\mathbf{q}} \cdot g(F_{t+k})_{\mathbf{p}_0}. \tag{6}$$

$\frac{\partial G(\mathbf{p}_n,\mathbf{q})}{\partial p_l}$ can be easily calculated due to the function $G(.,.)$ is bilinear interpolation kernel. According to the above gradient calculation in Eq. 6, the sampled location $\mathbf{p}_n$ will be iteratively updated according to final detection loss, which allows the learned sampling locations progressively match the ground truth distribution of correspondences. After training, the final learned sampling locations could be applied to propagate and align features across frames during the inference process. And LSTA is the core of SRFU and DFA, which would be introduced next in detail.

3.3 Sparsely Recursive Feature Updating

SRFU module allows to leverage the inherent temporal cues insides videos to propagate and aggregate high-level features of sparse keyframes over the whole video. Specifically, SRFU module maintains and recursively updates a temporal feature F^{memory} over the whole video. As shown in Fig. 4(a), during this procedure, directly updating the memory feature by the new keyframes feature F_{t^1} is sub-optimal due to the motion misalignment during keyframes t^0 and t^1. Thus, our LSTS module could estimate the motion and generate the aligned feature $F_{t^1}^{align}$. After that, an aggregation unit is proposed to generate the updated memory feature $F_{t^1}^{memory}$. Specially, the concatenation of F_{t^1} and $F_{t^1}^{align}$ would be fed into a several layers of convolutions with a softmax operation to produce the corresponding aggregation weights W_{t^1} and $W_{t^1}^{align}$, where $W_{t^1} + W_{t^1}^{align} = 1$.

$$F_{t^1}^{memory} = W_{t^1} \odot F_{t^1} + W_{t^1}^{align} \odot F_{t^1}^{align}. \tag{7}$$

Then the memory feature on the keyframes t_1 can be updated by Eq. 7, where $\odot$ is the Hadamard product (i.e. element-wise multiplication) after broadcasting the weight maps. And the memory feature $F_{t^1}^{memory}$ together with F_{t^1} would be aggregated to generate the task feature for the keyframes. To validate the effectiveness of proposed SRFU, we divide SRFU into Sparse Deep Feature Extraction, Keyframe Memory Update and Quality-Aware Memory Update. Each component of SRFU module will be explained and discussed in detail in the experimental section.

3.4 Dense Feature Aggregation

Considering the computational complexity, lightweight feature extractor networks are utilized for the non-keyframes, which would extract the low-level features. Thus, DFA module allows to reuse the sparse high-level features of keyframe to improve the quality of that of the non-keyframes. Specifically, as shown in Fig. 4(b), the non-keyframes feature $F_{t^1+k}^{low}$ would be fed into a **Transform** unit which only brings few computation cost to predict the semantic-level feature $F_{t^1+k}^{high}$. Due to the motion misalignment between the time step of t^1 and t^1+k, the proposed LSTS module is applied on the keyframe memory feature $F_{t^1}^{memory}$ to generate the aligned feature $F_{t^1+k}^{align}$. After obtaining $F_{t^1+k}^{align}$, an aggregation unit is utilized to predict the aggregation weights $W^{align}_{t^1+k}$ and $W^{high}_{t^1+k}$, where $W^{align}_{t^1+k} + W^{high}_{t^1+k} = 1$.

$$F^{task}_{t^1+k} = W^{align}_{t^1+k} \odot F^{align}_{t^1+k} + W^{high}_{t^1+k} \odot F^{high}_{t^1+k}. \tag{8}$$

Finally, the task feature $F^{task}_{t^1+k}$ on the non-keyframe $t^1 + k$ can be calculated in Eq. 8, where $\odot$ is the Hadamard product (i.e. element-wise multiplication) after broadcasting the weight maps. Comparing with low-level feature $F^{low}_{t^1+k}$, $F^{task}_{t^1+k}$ contains more semantic-level information and allows to obtain good detection results. To validate the effectiveness of proposed DFA, we divide DFA into Non-Keyframe Transform, Non-Keyframe Aggregation and Quality-Aware Non-Keyframe Aggregation. Each component of DFA module will be explained and discussed in detail in the experimental section.

4 Experiments

4.1 Datasets and Evaluation Metrics

We evaluate our method on the ImageNet VID dataset, which is the benchmark for video object detection [28]. And the ImageNet VID dataset is composed of 3862 training videos and 555 validation videos containing 30 object categories. All videos are fully annotated with bounding boxes and tracking IDs. And mean average precision (mAP) is used as the evaluation metric following the previous methods [44].

The ImageNet VID dataset has extreme redundancy among video frames, which prevents efficient training. At the same time, video frames of the ImageNet VID dataset have poorer quality than images in the ImageNet DET [28] dataset. So, we follow the previous method [42] to use both ImageNet VID and DET dataset for training. For the ImageNet DET set, only the same 30 class categories of ImageNet VID are used.

4.2 Implementation Detail

For the training process, each mini-batch contains three images. In both the training and testing stage, the shorter side of the images will be resized to 600 pixels [27]. Feature before `conv4_3` will be treated as Low-Level Feature Extractor. The whole ResNet will be used for High-Level Feature Extractor. Following the setting of most previous methods, the R-FCN detector [2] pretrained on ImageNet [5] with ResNet-101 [16] serves as the single-frame detector. During the training stage, we adopt OHEM strategy [30] and horizontal flip data augmentation. In our experiment, each GPU will hold one sample, namely three images sampled from one video or repetition of the static image. We train our network on an 8-GPUs machine for 4 epochs with SGD optimization, starting with a learning rate of 2.5e-4 and reducing it by a factor of 10 at every 2.33 epochs. The keyframe interval is 10 frames in default, as in [43,44].

Table 1. Performance comparison with the state-of-the-arts on ImageNet VID validation set. In terms of both accuracy and speed, Our method outperforms the most of them and has fewer parameters than the existing optical flow-based models. V means that the speed is tested on TITAN V GPU

Model	Online	mAP (%)	Runtime (FPS)	#Params (M)	Backbone
TCN [21]	✗	47.5	-	-	GoogLeNet
TPN+LSTM [20]	✗	68.4	2.1	-	GoogLeNet
R-FCN [2]	✓	73.9	4.05	60.0	ResNet-101
DFF [44]	✓	73.1	20.25	97.8	ResNet-101
D (& T loss) [8]	✓	75.8	7.8	-	ResNet-101
LWDN [18]	✓	76.3	20	77.5	ResNet-101
FGFA [43]	✗	76.3	1.4	100.5	ResNet-101
ST-lattice [1]	✗	79.5	20	-	ResNet-101
MANet [35]	✗	78.1	5.0	-	ResNet-101
OGEMNet [4]	✓	76.8	14.9	-	ResNet-101
Towards [42]	✓	78.6	13.0	-	ResNet-101 + DCN
RDN [6]	✗	81.8	10.6(V)	-	ResNet-101
SELSA [37]	✗	80.3	-	-	ResNet-101
LRTR [31]	✗	80.6	10	-	ResNet-101
Ours	✓	77.2	**23.0**	**64.5**	ResNet-101
Ours	✓	80.1	21.2	65.5	ResNet-101 + DCN
MANet [35] + [13]	✗	80.3	-	-	ResNet-101
STMN [38] + [13]	✗	80.5	1.2	-	ResNet-101
Ours + [13]	✓	**82.1**	4.6	65.5	ResNet-101 + DCN

Aggregation Unit. The aggregation weights of the features are generated by a quality estimation network. It has three randomly initialized layers: a $3 \times 3 \times 256$ convolution, a $1 \times 1 \times 16$ convolution and a $1 \times 1 \times 1$ convolutions. The output is position-wise raw score map which will be applied on each channel of corresponding features. The normalized weights and the features are fused to obtain the aggregated result.

Transform. To reduce the computational complexity, we only extract low-level features for the non-keyframes, which is a lack of high-level semantic information. A lightweight neural convolution unit containing 3 randomly initialized layers: a $3 \times 3 \times 256$ convolution, a $3 \times 3 \times 512$ convolution and a $3 \times 3 \times 1024$ convolutions has been utilized to compensate the semantic information.

4.3 Results

We compare our method with existing state-of-the-art image and video object detectors. The results are shown in Table 1. From the table, we can make the following conclusion. First of all, our method outperforms most previous approaches considering the speed and accuracy trade-off. Secondly, our proposed approach has fewer parameter comparing with flow-warping based method. Without external optical flow network, our approach can significantly simplify the overall

Table 2. Ablation studies on accuracy, runtime and complexity between ours and flow-warping methods. † belong to SRFU and ‡ belong to DFA

Architecture Component	(a)	(b)	(c)	(d)	(e)	(f)	(g)
Sparse Deep Feature Extraction †	✓	✓	✓	✓	✓	✓	✓
Keyframe Memory Update †		✓					
Quality-Aware Memory Update †			✓	✓	✓	✓	✓
Non-Keyframe Aggregation ‡				✓	✓		
Non-Keyframe Transformer ‡					✓		✓
Quality-Aware Non-Keyframe Aggregation ‡						✓	✓
Optical flow							
mAP (%)	73.0	75.2	75.4	75.5	75.7	75.9	76.1
Runtime (FPS)	29.4	29.4	29.4	19.2	18.9	19.2	18.9
#Params (M)	96.7	96.7	97.0	100.3	100.4	100.4	100.5
Ours							
mAP (%)	**73.5**	**75.8**	**75.9**	**76.4**	**76.5**	**76.8**	**77.2**
Runtime (FPS)	23.8	23.5	23.5	23.3	23.0	23.3	23.0
#Params (M)	63.8	63.7	64.0	64.3	64.4	64.4	64.5

detection framework. Lastly, the results indicate that our LSTS module can learn feature correspondences between consecutive video frames more precise than optical flow-warping based methods.

To conclude, our detector surpasses the static image-based R-FCN detector with a large margin (+3.3%) while maintaining high speed (23.0 FPS). Furthermore, the parameter count (64.5 M) is fewer than other video object detectors using an optical flow network (e.g., around 100 M), which also indicates that our method is more friendly for mobile devices.

4.4 Ablation Studies

We conduct ablation studies on ImageNet VID dataset to demonstrate the effectiveness of proposed LSTS module and the proposed framework. We first introduce the configuration of each element in our proposed framework for ablation studies. Then we compare our LSTS with both optical flow and existing non-optical flow alternatives. Finally, we conduct ablation studies of different modules in our framework.

Effectiveness of the Proposed Framework. We first describe each component about proposed SRFU and DFA. Then we compare our method with optical flow-warping based method under different configurations to validate the effectiveness of our proposed LSTS module. Each component of SRFU and DFA is listed following:

Table 3. Comparisons with MatchTrans [38] **and Non-Local** [36]

Method	mAP (%)	Runtime (FPS)	#Params (M)
Non-Local	74.2	25.0	64.5
MatchTrans	75.5	24.1	64.5
Ours	**77.2**	23.0	64.5

Table 4. Comparisons of LSTS with different initialization methods

Method	Learning	mAP (%)	Runtime (FPS)	#Params (M)
Uniform	✗	75.5	21.7	64.5
	✓	76.8	21.7	64.5
Gaussian	✗	75.5	23.0	64.5
	✓	77.2	23.0	64.5

- Sparse Deep Feature Extraction. The entire backbone network is used to extract feature only on keyframes.
- Keyframe Memory Update. The keyframe feature aggregates with the last keyframe memory to generate the task feature and updated memory feature (see Fig. 4(a)). The weights are naively fixed to 0.5.
- Quality-Aware Memory Update. The keyframe feature aggregates with the last keyframe memory to generate the task feature and updated memory feature using a quality-aware aggregation unit.
- Non-Keyframe Transform. We apply a transform unit on the low-level feature to generate a high-level semantic feature on the non-keyframe.
- Non-Keyframe Aggregation. The task feature for the non-keyframe is naively aggregated with an aligned feature from keyframes, and the current low-level feature is obtained by a part of the backbone network. The weights are set to 0.5.
- Quality-Aware Non-Keyframe Aggregation. The task feature for the non-keyframe is aggregated with an aligned feature from the keyframe using a quality-aware aggregation unit, and the current high-level feature is obtained through a transform unit.

Our frame-by-frame baseline achieves 74.1% mAP and runs at 10.1 FPS. After using the sparse deep feature, we have 73.5% mAP and runs at 23.8 FPS. When applying the quality-aware keyframe memory propagation, we have 75.9% mAP and runs at 23.5 FPS with 64.0 M parameters. Last, non-keyframe quality-aware aggregation can also improve performance which achieves 76.4% mAP at 23.3 FPS. By adding quality-aware memory aggregation, non-keyframe transformer unit, and quality-aware non-keyframe aggregation, our approach can achieve 77.2% mAP and run 23.0 FPS with 64.5 M parameters.

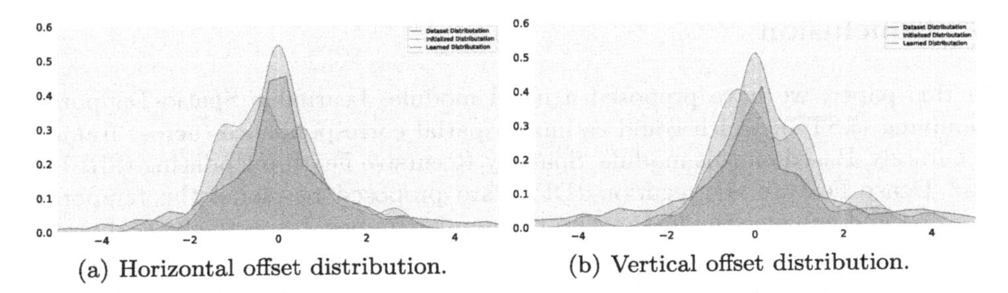

(a) Horizontal offset distribution. (b) Vertical offset distribution.

Fig. 5. The comparison of offset distribution on the horizontal and vertical between ours and the dataset. For the dataset distribution, we random sample 100 videos from the training dataset, then calculate motion across frames by FlowNet [7]. To verify the effectiveness of learnable spatial-temporal sampling, we also compare the learned offset distribution with the initialized Gaussian distribution.

Comparison with Optical Flow-Based Method. Optical flow can predict motion field between consecutive frames. DFF [44] proposed to propagate feature across frames by using flow-warping operation. To validate the effectiveness of LSTS on estimating spatial correspondences, we make a detailed comparison with optical flow. The results can be seen as in Table 2. Our proposed LSTS can outperform optical flow on all settings with fewer model parameters.

Comparison with Non-Optical Flow-Based Method. The results of using different feature propagation methods are listed in Table 3. By attending on the local region, our method outperforms the Non-Local by a large margin. The reason is that the motion distribution is limited to the near center, as shown in Fig. 5. Our method can surpass both the MatchTrans and Non-Local a lot, which show the effectiveness of LSTS.

Learnable Sampled Locations. To demonstrate the effectiveness of learning sampled locations, we perform ablation study on two different initialization methods, Uniform Initialization and Gaussian Initialization.

The first one is just like MatchTrans [38] module with the neighbourhoods are set to 4. While the second is two-dimensional Gaussian Distribution with zero means and one variance The results of different initialization settings can be seen in Table 4. We can figure out, no matter what the initialization methods are, there is a consistent trend that the performance can be significantly boosted by learning sampled locations. To be more specific, Gaussian initialization can achieve 77.2% mAP. Comparing with the fixed Gaussian initialization 75.5%, learnable sampling locations could obtain 1.7% mAP improvement. And Uniform initialization can achieve 76.8% mAP. Comparing with the fixed Uniform initialization 75.5%. learnable sampling locations could obtain 1.3% mAP improvement.

5 Conclusion

In this paper, we have proposed a novel module, Learnable Spatio-Temporal Sampling (LSTS), which could estimate spatial correspondences across frames accurately. Based on this module, Sparsely Recursive Feature Updating (SRFU) and Dense Feature Aggregation (DFA) are proposed to model the temporal relation and enhance the features on the non-keyframes, respectively. Elaborate ablative studies have shown the advancement of our LSTS module and architecture design. Without any whistle and bell, the proposed framework has achieved state-of-the-art performance (82.1% mAP) on ImageNet VID dataset. We hope the proposed differential paradigm could extend to more tasks, such as sampling locations for general convolution operation, sampling locations of aggregating features for semantic segmentation, and so on.

Acknowledgement. This research was supported by the Major Project for New Generation of AI under Grant No. 2018AAA0100400, the National Natural Science Foundation of China under Grants 91646207, 61976208 and 61620106003. We also would like to thank Lin Song for the discussions and suggestions.

References

1. Chen, K., et al.: Optimizing video object detection via a scale-time lattice. In: CVPR, pp. 7814–7823 (2018)
2. Dai, J., Li, Y., He, K., Sun, J.: R-FCN: object detection via region-based fully convolutional networks. In: NeurPIS, pp. 379–387 (2016)
3. Dai, J., et al.: Deformable convolutional networks. In: CVPR, pp. 764–773 (2017)
4. Deng, H., et al.: Object guided external memory network for video object detection. In: ICCV, pp. 6678–6687 (2019)
5. Deng, J., Dong, W., Socher, R., Li, L.J., Li, K., Fei-Fei, L.: ImageNet: a large-scale hierarchical image database. In: CVPR, pp. 248–25 (2009)
6. Deng, J., Pan, Y., Yao, T., Zhou, W., Li, H., Mei, T.: Relation distillation networks for video object detection. In: ICCV, pp. 7023–7032 (2019)
7. Dosovitskiy, A., et al.: FlowNet: learning optical flow with convolutional networks. In: ICCV, pp. 2758–2766 (2015)
8. Feichtenhofer, C., Pinz, A., Zisserman, A.: Detect to track and track to detect. In: ICCV, pp. 3038–3046 (2017)
9. Gao, P., et al.: Dynamic fusion with intra-and inter-modality attention flow for visual question answering. In: CVPR, pp. 6639–6648 (2019)
10. Gao, P., et al.: Question-guided hybrid convolution for visual question answering. In: Ferrari, V., Hebert, M., Sminchisescu, C., Weiss, Y. (eds.) ECCV 2018. LNCS, vol. 11205, pp. 485–501. Springer, Cham (2018). https://doi.org/10.1007/978-3-030-01246-5_29
11. Geiger, A., Lenz, P., Stiller, C., Urtasun, R.: Vision meets robotics: the KITTI dataset. Int. J. Robot. Res. **32**(11), 1231–1237 (2013)
12. Girshick, R., Donahue, J., Darrell, T., Malik, J.: Rich feature hierarchies for accurate object detection and semantic segmentation. In: CVPR, pp. 580–587 (2014)
13. Han, W., et al.: Seq-NMS for video object detection. arXiv preprint arXiv:1602.08465 (2016)

14. He, K., Gkioxari, G., Dollár, P., Girshick, R.: Mask R-CNN. In: ICCV, pp. 2961–2969 (2017)
15. He, K., Zhang, X., Ren, S., Sun, J.: Spatial pyramid pooling in deep convolutional networks for visual recognition. IEEE Trans. Pattern Anal. Mach. Intell. **37**(9), 1904–1916 (2015)
16. He, K., Zhang, X., Ren, S., Sun, J.: Deep residual learning for image recognition. In: CVPR, pp. 770–778 (2016)
17. Hetang, C., Qin, H., Liu, S., Yan, J.: Impression network for video object detection. arXiv preprint arXiv:1712.05896 (2017)
18. Jiang, Z., Gao, P., Guo, C., Zhang, Q., Xiang, S., Pan, C.: Video object detection with locally-weighted deformable neighbors. In: AAAI (2019)
19. Jiang, Z., et al.: Learning motion priors for efficient video object detection. arXiv preprint arXiv:1911.05253 (2019)
20. Kang, K., et al.: Object detection in videos with tubelet proposal networks. In: CVPR, pp. 727–735 (2017)
21. Kang, K., Ouyang, W., Li, H., Wang, X.: Object detection from video tubelets with convolutional neural networks. In: CVPR, pp. 817–825 (2016)
22. Lin, T.Y., Dollár, P., Girshick, R., He, K., Hariharan, B., Belongie, S.: Feature pyramid networks for object detection. In: CVPR, pp. 2117–2125 (2017)
23. Lin, T.Y., Goyal, P., Girshick, R., He, K., Dollár, P.: Focal loss for dense object detection. In: ICCV, pp. 2980–2988 (2017)
24. Liu, W., et al.: SSD: single shot multibox detector. In: Leibe, B., Matas, J., Sebe, N., Welling, M. (eds.) ECCV 2016. LNCS, vol. 9905, pp. 21–37. Springer, Cham (2016). https://doi.org/10.1007/978-3-319-46448-0_2
25. Liu, Y., Liu, J., Zeng, A., Wang, X.: Differentiable kernel evolution. In: CVPR, pp. 1834–1843 (2019)
26. Mnih, V., Heess, N., Graves, A., et al.: Recurrent models of visual attention. In: NeurPIS, pp. 2204–2212 (2014)
27. Ren, S., He, K., Girshick, R., Sun, J.: Faster R-CNN: towards real-time object detection with region proposal networks. In: NeurPIS, pp. 91–99 (2015)
28. Russakovsky, O., et al.: ImageNet large scale visual recognition challenge. Int. J. Comput. Vision **115**(3), 211–252 (2015). https://doi.org/10.1007/s11263-015-0816-y
29. Sharma, S., Kiros, R., Salakhutdinov, R.: Action recognition using visual attention. arXiv preprint arXiv:1511.04119 (2015)
30. Shrivastava, A., Gupta, A., Girshick, R.: Training region-based object detectors with online hard example mining. In: CVPR, pp. 761–769 (2016)
31. Shvets, M., Liu, W., Berg, A.C.: Leveraging long-range temporal relationships between proposals for video object detection. In: ICCV, pp. 9756–9764 (2019)
32. Simonyan, K., Zisserman, A.: Two-stream convolutional networks for action recognition in videos. In: NeurPIS, pp. 568–576 (2014)
33. Uijlings, J.R., Van De Sande, K.E., Gevers, T., Smeulders, A.W.: Selective search for object recognition. Int. J. Comput. Vision **104**(2), 154–171 (2013). https://doi.org/10.1007/s11263-013-0620-5
34. Vaswani, A., et al.: Attention is all you need. In: NeurPIS, pp. 5998–6008 (2017)
35. Wang, S., Zhou, Y., Yan, J., Deng, Z.: Fully motion-aware network for video object detection. In: Ferrari, V., Hebert, M., Sminchisescu, C., Weiss, Y. (eds.) ECCV 2018. LNCS, vol. 11217, pp. 557–573. Springer, Cham (2018). https://doi.org/10.1007/978-3-030-01261-8_33
36. Wang, X., Girshick, R., Gupta, A., He, K.: Non-local neural networks. In: CVPR, pp. 7794–7803 (2018)

37. Wu, H., Chen, Y., Wang, N., Zhang, Z.: Sequence level semantics aggregation for video object detection. In: ICCV, pp. 9217–9225 (2019)
38. Xiao, F., Lee, Y.J.: Video object detection with an aligned spatial-temporal memory. In: Ferrari, V., Hebert, M., Sminchisescu, C., Weiss, Y. (eds.) ECCV 2018. LNCS, vol. 11212, pp. 494–510. Springer, Cham (2018). https://doi.org/10.1007/978-3-030-01237-3_30
39. Xu, K., et al.: Show, attend and tell: neural image caption generation with visual attention. In: ICML, pp. 2048–2057 (2015)
40. Zhang, M., Song, G., Zhou, H., Liu, Y.: Discriminability distillation in group representation learning. In: ECCV (2020)
41. Zhou, H., Liu, J., Liu, Z., Liu, Y., Wang, X.: Rotate-and-render: unsupervised photorealistic face rotation from single-view images. In: CVPR, pp. 5911–5920 (2020)
42. Zhu, X., Dai, J., Yuan, L., Wei, Y.: Towards high performance video object detection. In: CVPR, pp. 7210–7218 (2018)
43. Zhu, X., Wang, Y., Dai, J., Yuan, L., Wei, Y.: Flow-guided feature aggregation for video object detection. In: ICCV, pp. 408–417 (2017)
44. Zhu, X., Xiong, Y., Dai, J., Yuan, L., Wei, Y.: Deep feature flow for video recognition. In: CVPR, pp. 2349–2358 (2017)

Learning Object Permanence from Video

Aviv Shamsian[1(✉)], Ofri Kleinfeld[1(✉)], Amir Globerson[2(✉)], and Gal Chechik[1,3(✉)]

[1] Bar-Ilan University, Ramat-Gan, Israel
mista2311@gmail.com, ofri64@gmail.com, gal.chechik@gmail.com
[2] Tel Aviv University, Tel Aviv, Israel
amir.globerson@gmail.com
[3] NVIDIA Research, Tel-Aviv, Israel

Abstract. *Object Permanence* allows people to reason about the location of non-visible objects, by understanding that they continue to exist even when not perceived directly. Object Permanence is critical for building a model of the world, since objects in natural visual scenes dynamically occlude and contain each-other. Intensive studies in developmental psychology suggest that object permanence is a challenging task that is learned through extensive experience.

Here we introduce the setup of learning Object Permanence from labeled videos. We explain why this learning problem should be dissected into four components, where objects are (1) visible, (2) occluded, (3) contained by another object and (4) carried by a containing object. The fourth subtask, where a target object is carried by a containing object, is particularly challenging because it requires a system to reason about a moving location of an invisible object. We then present a unified deep architecture that learns to predict object location under these four scenarios. We evaluate the architecture and system on a new dataset based on CATER, with per-frame labels, and find that it outperforms previous localization methods and various baselines.

Keywords: Object Permanence · Reasoning · Video Analysis

1 Introduction

Understanding dynamic natural scenes is often challenged by objects that contain or occlude each other. To reason correctly about such visual scenes, systems need to develop a sense of *Object Permanence* (OP) [20]. Namely, the understanding that objects continue to exist and preserve their physical characteristics, even if they are not perceived directly. For example, we want systems

A. Shamsian and O. Kleinfeld—Equal contribution.

Electronic supplementary material The online version of this chapter (https://doi.org/10.1007/978-3-030-58517-4_3) contains supplementary material, which is available to authorized users.

A. Vedaldi et al. (Eds.): ECCV 2020, LNCS 12361, pp. 35–50, 2020.
https://doi.org/10.1007/978-3-030-58517-4_3

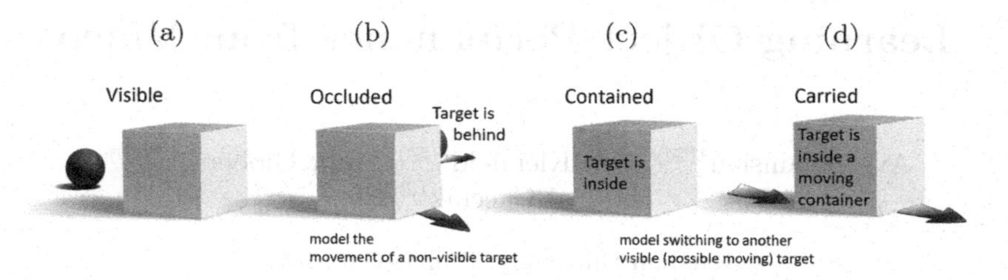

Fig. 1. Inferring object location in rich dynamic scenes involves four different tasks, and two different types of reasoning. (a) The target, a red ball, is fully visible. (b) The target is fully-or partially occluded by the static cube. (c) The target is located inside the cube and fully covered. (d) The non-visible target is located inside another moving object; its location changes even though it is not directly visible.

to learn that a pedestrian occluded by a truck may emerge from its other side, but that a person entering a car would "disappear" from the scene.

The concept of OP received substantial attention in the cognitive development literature. Piaget hypothesized that infants develop OP relatively late (at two years of age), suggesting that it is a challenging task that requires deep modelling of the world based on sensory-motor interaction with objects. Later evidence showed that children learn OP for occluded targets early [1,2]. Still, only at a later age do children develop understanding of objects that are contained by other objects [25]. Based on these experiments we hypothesize that reasoning about the location of non-visible objects may be much harder when they are carried inside other moving objects.

Reasoning about the location of a target object in a video scene involves four different subtasks of increasing complexity (Fig. 1). These four tasks are based on the state of the target object, depending if it is (1) visible, (2) occluded, (3) contained or (4) carried. The *visible* case is perhaps the simplest task, and corresponds to object detection, where one aims to localize an object that is visible. Detection was studied extensively and is viewed as a key component in computer vision systems. The second task, *occlusion*, is to detect a target object which becomes transiently invisible by a moving occluding object (e.g., bicycle behind a truck). Tracking objects under occlusion can be very challenging, especially with long-term occlusions [4,9,11,14,18,30].

Third, in a *containment* scenario, a target object may be located inside another container object and become non-visible [28], for example when person enters a store. Finally, the fourth case of a *carried* object is arguably the most challenging task. It requires inferring the location of a non-visible object located inside a moving containing object (e.g., a person enters a taxi that leaves the scene). The task is particularly challenging because one has to keep a representation of which object should be tracked at every time point and to "switch states" dynamically through time. This task received little attention in the computer vision community so far.

We argue that reasoning about the location of a non-visible object should address two distinct and fundamentally different cases: occlusion and containment. First, to *localize an occluded object*, an agent has to build an internal state that models how the object moves. For example, when we observe a person walking in the street, we can predict her ever-changing location even if occluded by a large bus. In this mode, our reasoning mechanism keeps attending to the person and keeps inferring her location from past data. Second, *localizing contained objects* is fundamentally different. It requires a reasoning mechanism that switches to attend to the containing object, which is visible. Here, even though the object of interest is not-visible, its location can be accurately inferred from the location of the visible containing object. We demonstrate below that incorporating these two reasoning mechanisms leads to more accurate localization in all four subtasks.

Specifically, we develop a unified approach for learning all four object localization subtasks in video. We design a deep architecture that learns to localize objects that may be visible, occluded, contained or carried. Our architecture consists of two reasoning modules designed to reason about (1) carried or contained targets, and (2) occluded or visible targets. The first reasoning component is explicitly designed to answer the question *"Which object should be tracked now?"*. It does so by using an LSTM to weight the perceived locations of the objects in the scene. The second reasoning component leverages the information about which object should be tracked and previous known locations of the target to localize the target, even if it is occluded. Finally, we also introduce a dataset called LA-CATER, based on videos from CATER [8] enriched with new annotations about task type and about ground-truth location of all objects.

Our main novel contributions are: (1) We conceptualize that localizing non-visible objects requires two types of reasoning: about occluded objects and about carried ones. (2) We define four subtypes of localization tasks and introduce annotations for the CATER dataset to facilitate evaluating each of these subtasks. (3) We describe a new unified architecture for all four subtasks, which can capture the two types of reasoning, and we show empirically that it outperforms multiple strong baselines. Our data and code are available for the community at our website[1].

2 Related Work

Relational Reasoning in Synthetic Video Datasets. Recently, several studies provided synthetic datasets to explore object interaction and reasoning. Many of these studies are based on CLEVR [12], a synthetic dataset designed for visual reasoning through visual question answering. CLEVRER [31] extended CLEVR to video, focusing on the causal structures underlying object interactions. It demonstrated that visual reasoning models that thrive on perception based tasks often perform poorly in causal reasoning tasks.

[1] https://chechiklab.biu.ac.il/~avivshamsian/OP/OP_HTML.html.

Most relevant for our paper, CATER [8] is a dataset for reasoning about object action and interactions in video. One of the three tasks defined in CATER, the *snitch localization* task, is closely related to the OP problem studied here. It is defined as localizing a target *at the end of a video*, where the target is usually visible. Our work refines their setup, learning to localize the target through the full video, and breaks down prediction into four types of localization tasks. As a result, we provide a fine-grained insight about the architectures and reasoning that is required for solving the complex localization task.

Architectures for Video Reasoning. Several recent papers studied the effectiveness of CNN-based architectures for video action recognition. Many approaches use 3D convolutions for spatiotemporal feature learning [3,27] and separate the spatial and temporal modalities by adding optical flow as a second stream [6,24]. These models are computationally expensive because 3D convolution kernels may be costly to compute. As a result, they may limited the sequence length to 20–30 frames [3,27]. In [34] it was proposed to sparsely sample video frames to capture temporal relations in action recognition datasets. However, sparse sampling may be insufficient for long occlusion and containment sequences, which is the core of our OP focus.

Another strategy for temporal aggregation is to use recurrent architectures like LSTM [10], connecting the underlying CNN output along the temporal dimension [32]. [7,23,26] combined LSTM with spatial attention, learning to attend to those parts of the video frame that are relevant for the task as the video progresses. In Sect. 6 we experiment with a spatial attention module, which learns to dynamically focus on relevant objects.

Tracking with Object Occlusion. A large body of work has been devoted to tracking objects [18]. For objects under complex occlusion like carrying, early work studied tracking using classical techniques and without deep learning methods. For instance, [11,19] used the idea of object permanence to track objects through long-term occlusions. They located objects using adaptive appearance models, modelling spatial distributions and inter-occlusion relationships. In contrast, the approach presented in this paper focuses on a single deep differentiable model to learn motion reasoning end-to-end. [9] succeeds to track occluded targets by learning how their movement is coupled with the movement of other visible objects. The dataset studied here, CATER [8], has weak object-object motion coupling by design. Specifically, when measuring the correlation between the movement of the target and other object (as in [9]), we found that the correlation in 94% of the videos was not statistically significant.

More recently, models based on Siamese neural network achieved SOTA results in object tracking [5,15,35]. Despite the power of these architectures, tracking highly-occluded objects is still challenging [18]. The tracker of [35], DaSiamRPN, extends the region-proposal sub-network of [15]. It was designed for long-term tracking and handles full occlusion or out-of-view scenarios. DaSiamRPN was used as a baseline for the snitch localization task in CATER [8], and we evaluated its performance for the OP problem in Sect. 6.

Containment. Few recent studies explored the idea of containment relations. [16] recovered incomplete object trajectories by reasoning about containment relations. [28] proposed an unsupervised model to categorize spatial relations, including containment between objects. The containment setup defined in these studies differs from the one defined here in that the contained object is always at least partially visible [28], or the containment does not involve carrying [16,28].

3 The Learning Setup: Reason About Non-visible Objects

We next formally define the OP task and learning setup. We are given a set of videos $v_1, ..., v_N$ where each frame x_t^i in video v_i is accompanied by the bounding box position B_t^i of the target object as its label. The goal is to predict for each frame a bounding box $\hat{B}_t^i$ of the target object that is closest (in terms of L_1 distance) to the ground-truth bounding box B_t^i.

We define four localization tasks: (1) Localizing a visible object, which we define as an object that is at least partially visible. (2) Localizing an occluded object, which we define as an object that is *fully* occluded by another object. (3) Localizing an object contained by another object, thus also completely non visible. (4) Localizing an object that is carried along the surface by a containing object. Thus in this case the target is moving while being completely non-visible. Together, these four tasks form a localization task that we call object-permanence localization task, or OP.

In Sect. 7.2, we also study a semi-supervised learning setup, where at training time the location B_t^i of the target is provided only in frames when it is visible. This would correspond to the case of a child learning object permanence without explicit feedback about where an object is located when it is hidden.

It is instructive to note how the task we address here differs from the tasks of relation or action recognition [13,17,22]. In these tasks, models are trained to output an explicit label that captures the name of the interaction or relation (e.g., "behind", "carry"). In our task, the model aims to predict the location of the target (a regression problem), but it is not trained to name it explicitly (occluded, contained). While it is possible that the model creates some implicit representation describing the visibility type, this is not mandated by the loss or the architecture.

4 Our Approach

We describe a deep network architecture designed to address the four localization subtasks of the OP task. We refer to the architecture as OPNet. It contains three modules, that account for the perception and inference computations which facilitate OP (see Fig. 2).

Perception and detection module (Fig. 2a): A perception module, responsible for detecting and tracking visible objects. We incorporated a Faster R-CNN [21] object detection model, fine-tuned on frames from our dataset, as the perception component of our model. After pre-training, we used the detector to output the bounding boxes together with identifiers of all objects in any given frame. Specifically, we represent a frame using a $K \times 5$ matrix. Each row in the matrix represents an object using 5 values: four values of the bounding box (x_1, y_1, x_2, y_2) and one *visibility bit*, which indicates whether the object is visible or not. As the video progresses, we assign a unique row to each *newly identified* object. If an object is not detected in a given frame, its corresponding information (assigned row) is set to zero. In practice, $K = 15$ was the maximal number of objects in a single video in our dataset. Notably, the videos in the dataset we used do not contain two identical objects, but we found that the detector sometimes mistakes one object for another. Objects in a video form an unordered collection [33]. To increase learning efficiency in this settings, we canonicalize the representation and keep the target as the first item of the set.

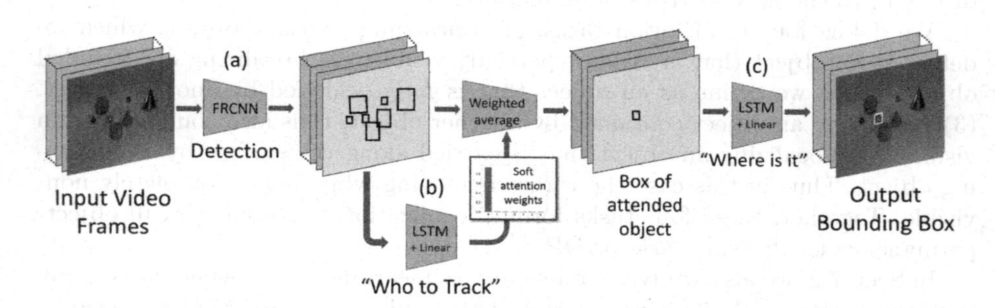

Fig. 2. The architecture of the *Object-Permanence network* (OPNet) consists of three components. (a) A perception module for detection. (b) A reasoning module for inferring which object to track when the target is carried or contained. (c) A second reasoning module for occluded or visible targets, to refine the location of the predicted target.

"Who to track?" module (Fig. 2c): responsible for understanding which object is currently covering the target. This component consists of a single LSTM layer with a hidden dimension of 256 neurons and a linear projection matrix. After applying the LSTM to the object bounding boxes, we project its output to K neurons, each representing a distinct object in the frame. Finally, we apply a softmax layer, resulting in a distribution over the objects in the frame. This distribution can be viewed as an attention mask focusing on the object that covers the target in this frame. Importantly, we do not provide explicit supervision to this attention mask (e.g., by explicitly "telling the model" during training what is the correct attention mask). Rather, our only objective is the location of the

target. The output of this module is 5 numbers per frame. It is computed as the weighted average over the $K \times 5$ outputs of the previous stage, weighted by the attention mask.

"Where is it" module (Fig. 2b): learns to predict the location of occluded targets. This final component consists of a second LSTM and a projection matrix. Using the output of the previous component, this component is responsible for predicting the target localization. It takes the output of the previous step (5 values per frame), feeds it into the LSTM and projects its output to four units, representing the predicted bounding box of the target for each frame.

5 The LA-CATER Dataset

To train models and evaluate their performance on the four OP subtasks defined above, we introduce a new set of annotations to the CATER dataset [8]. We refer to these as *Localization Annotations* (LA-CATER).

The CATER dataset consists of 5,500 videos generated programmatically using the Blender 3D engine. Each video is 10-s long (300 frames) and contains 5 to 10 objects. Each object is characterized by its shape (cube, sphere, cylinder and inverted cone), size (small, medium, large), material (shiny metal and matte rubber) and color (eight colors). Every video contains a golden small sphere referred to as "the snitch", that is used as the target object which needs to be localized.

For the purpose of this study, we generated videos following a similar configuration to the one used by CATER, but we computed additional annotations during video generation. Specifically, we augmented the CATER dataset with ground-truth bounding boxes locations of all objects. These annotations were programmatically extracted from the Blender engine, by projecting the internal 3D coordinates of objects are to the 2D pixel space.

We further annotated videos with detailed frame-level annotations. Each frame was labeled with one of four classes: visible, fully occluded, contained (i.e., covered, static and non-visible) and carried (i.e., covered, moving and non-visible). This classification of frames matches the four localization subtasks of the OP problem.

LA-CATER includes a total number of 14K videos split into train, dev and test datasets. See Table 1 for a classification of video frames to each one of the localization subtasks across the dataset splits. Further details about dataset preparation are provided in the supplementary.

Table 1. Fraction of frames per type in the train, dev and test sets of LA-CATER. Occluded and carried target frames make up less than 8% of the frames, but they present the most challenging prediction tasks.

	Number of Samples	Visible	Occluded	Contained	Carried
Train	9,300	63.00%	3.03%	29.43%	4.54%
Dev	3,327	63.27%	2.89%	29.19%	4.65%
Test	1,371	64.13%	3.07%	28.56%	4.24%

6 Experiments

We describe our experimental setup, compared methods and evaluation metrics. Implementation details are given in the supplementary.

6.1 Baselines and Model Variants

We compare our proposed OPNet with six other architectures designed to solve the OP tasks. Since we are not aware of previously published unified architectures designed to solve all OP tasks at once, we used existing models as components in our baselines. All baseline models receive the predictions of the object detector (perception) component as their input.

(A) Programmed Models. We evaluated two models that are "hard-coded" rather than learned. They are designed to evaluate the power of models that programmatically solve the reasoning task.

- (1) *Detector + Tracker.* Using the detected location of the target, this method initiates a DaSiamRPN tracker [35] to track the target. Whenever the target is no longer visible, the tracker is re-initiated to track the object located in the last known location of the target.
- (2) *Detector + Heuristic.* When the target is not detected, the model switches from tracking the target to tracking the object located closest to last known location of the target. The model also employs an heuristic logic to adjust between the sizes of the current tracked object and the original target.

(B) Learned Models. We evaluated four learned baselines with an increasing level of representation complexity.

- (3) *OPNet.* The proposed model, as presented in Sect. 4.
- (4) *Baseline LSTM.* This model uses a single unidirectional LSTM layer with a hidden state of 512 neurons, operating on the temporal (frames) dimension. The input to the LSTM is the concatenation of the objects input representations. It is the simplest learned baseline as the input representation is not transformed non-linearly before being fed to the LSTM.

- (5) *Non-Linear + LSTM.* This model augments the previous model and increases the complexity of the scene representation. The input representations are upsampled using a linear layer followed by a ReLU activation, resulting in a 256-dimensional vector representation for each object in the frame. These high-dimensional objects representations are concatenated and fed into the LSTM.
- (6) *Transformer + LSTM.* This model augments the previous baselines by introducing a much complex representations for objects in frame. We utilized a transformer encoder [29] after up-sampling the input representations, employing self attention between all the objects in a frame. We used a transformer encoder with 2 layers and 2 attention heads, yielding a single vector containing the target attended values. These attended values, which corresponds to each other object in the frame, are then fed into the LSTM.
- (7) *LSTM + MLP.* This model (*Fig.* 2) ablates the second LSTM module (c) in the model presented in Sect. 4.

6.2 Evaluation Metric

We evaluate model performance at a given frame t by comparing the predicted target localization and the ground truth (GT) target localization. We use two metrics as follows. First, intersection over union (IoU),

$$IoU_t = \frac{B_t^{GT} \cap B_t^p}{B_t^{GT} \cup B_t^p} \quad , \tag{1}$$

where B_t^p denotes the predicted bounding box for frame t and B_t^{GT} denotes the ground truth bounding box for frame t. Second, we evaluate models using their mean average precision (MAP). MAP is computed by employing an indicator function to each frame, determining whether the IoU value is greater than a predefined threshold, then averaging across frames in a single video and all the videos in the dataset.

$$AP = \frac{1}{n}\sum_{t=1}^{n} \mathbf{1_t}\, , \text{ where } \mathbf{1_t} = \begin{cases} 1 & IoU_t > IoU\ threshold \\ 0 & \text{otherwise} \end{cases} \tag{2}$$

$$MAP = \frac{1}{N}\sum_{v=1}^{N} AP_v \quad . \tag{3}$$

These per-frame metrics allow us to quantify the performance on each of the four OP subtasks separately.

7 Results

We start with comparing OPNet with the baselines presented in Sect. 6.1. We then provide more insights into the performance of the models by repeating the evaluations with *"Perfect Perception"* in Sect. 7.1. Section 7.3 describes a semi-supervised setting of training with visible frames only. Finally, in Sect. 7.3 we compare OPNet with the models presented in the CATER paper on the original CATER data. We first compare OPNet and the baselines presented in Sect. 6.1. Table 2 shows IoU for all models in all four sub-tasks and Fig. 3 presents the MAP accuracy of the models across different IoU thresholds.

It can be seen in Table 2 that OPNet performs consistently well across all subtasks and outperforms all other models overall. On the visible and occluded frames performance is similar to other baselines. But on the contained and carried frames, OPNet is significantly better than the other methods. This is likely due to OPNet's explicit modeling of the object to be tracked.

Table 2 also reports results for two variants of OPNet: OPNet (LSTM+MLP) and OPNet (LSTM+LSTM). The former is missing the second module ("Where is it" in Fig. 3) which is meant to handle occlusion, and indeed under-performs for occlusion frames (the "occluded" and "contained" subtasks). This highlights the importance of using the two LSTM modules in Fig. 3.

Table 2. Mean IoU on LA-CATER test data. "±" denotes the standard error of the mean (SEM). OPNet performs consistently well across all subtasks, and is significantly better on *contained* and *carried*

Mean IoU± SEM	Visible	Occluded	Contained	Carried	Overall
Detector + Tracker	90.27 ±0.13	53.62 ±0.58	39.98 ±0.38	34.45 ±0.40	71.23 ±0.51
Detector + Heuristic	90.06 ±0.14	47.03 ±0.73	55.36 ±0.53	55.87 ±0.59	76.91 ±0.43
Baseline LSTM	81.60 ±0.19	59.80 ±0.61	49.18 ±0.64	21.53 ±0.40	67.20 ±0.53
Non-Linear + LSTM	88.25 ±0.14	70.14 ±0.62	55.66 ±0.67	24.58 ±0.44	73.53 ±0.51
Transformer + LSTM	**90.82** ±0.14	**80.40** ±0.61	70.71 ±0.78	28.25 ±0.45	80.27 ±0.50
OPNet (LSTM + MLP)	88.11 ±0.16	55.32 ±0.85	65.18 ±0.89	**57.59** ±0.85	78.85 ±0.52
OPNet (LSTM + LSTM)	88.89 ±0.16	78.83 ±0.56	**76.79** ±0.62	56.04 ±0.77	**81.94** ±0.41

Figure 3 provides interesting insight into the behavior of the programmed models, namely, Detector + Tracker and Detector + Heuristic. These models perform well when the IoU threshold is low. This reflects the fact that they have a good coarse estimate of where the target is, but fail to provide more accurate localization. At the same time, OPNet performs well for accurate localization, presumably due to its learned "Where is it" module.

7.1 Reasoning with Perfect Perception

The OPNet model contains an initial "Perception" module that analyzes the frame pixels to get bounding boxes. Errors in this component will naturally propagate to the rest of the model and adversely affect results. Here we analyze the effect of the perception module by replacing it with ground truth bounding boxes and visibility bits. See supplementary for details on extracting ground-truth annotations. In this setup all errors reflect failure in the reasoning components of the models.

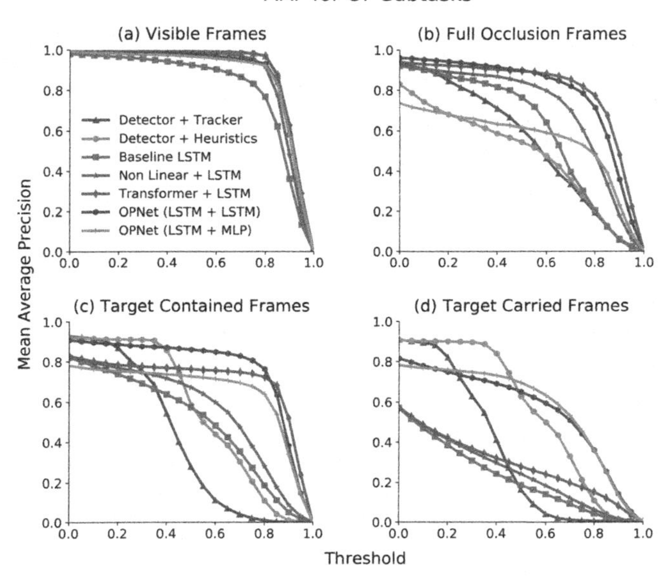

Fig. 3. Mean average precision (MAP) as a function of IoU thresholds. The two programmed models, Detector+Tracker (blue) and Detector+Heuristic (orange) perform well when the IoU threshold is low, providing a good coarse estimate of target location. OPNet performs well on all subtasks. (Color figure online)

Table 3. Mean IoU with *Perfect Perception.* "±" denotes the standard error of the mean (S.E.M.). Results are similar in nature to those with imperfect, detector-based, perception (Table 2). All models improve when using ground-truth perception. The largest improvement due to OPNet is observed in the carried task.

Mean IoU ± SEM	Visible	Occluded	Contained	Carried	Overall
DETECTOR + TRACKER	90.27 ±0.13	53.62 ±0.58	39.98 ±0.38	34.45 ±0.40	71.23 ±0.51
DETECTOR + HEURISTIC	**95.59** ± 0.34	30.40 ± 0.81	59.81 ± 0.47	59.33 ± 0.50	81.24 ± 0.49
BASELINE LSTM	75.22 ± 0.31	50.52 ± 0.75	45.10 ± 0.62	19.12 ± 0.36	61.41 ± 0.53
NON-LINEAR + LSTM	88.63 ± 0.25	65.73 ± 0.82	58.77 ± 0.70	23.89 ± 0.41	74.53 ± 0.54
TRANSFORMER + LSTM	93.99 ± 0.24	**81.31** ± 0.88	75.75 ± 0.85	28.01 ± 0.44	83.78 ± 0.55
OPNET (LSTM + MLP)	88.11 ± 0.16	19.39 ± 0.60	77.40 ± 0.68	**78.25** ± 0.65	83.84 ± 0.48
OPNET (LSTM + LSTM)	88.78 ± 0.25	67.79 ± 0.69	**83.47** ± 0.47	76.42 ± 0.66	**85.44** ± 0.38

Table 3 provides the IoU performance and Fig. 4 the MAP for all compared methods on all four subtasks. The results are similar to the previous results. When compared to the previous section (imperfect, detector-based, perception), the overall trend is the same, but all models improve when using the ground truth perception information. Interestingly, the subtask that improves the most from using ground truth boxes is the carried task. This makes sense, since it is the hardest subtask and the one that most relies on having the correct object locations per frame.

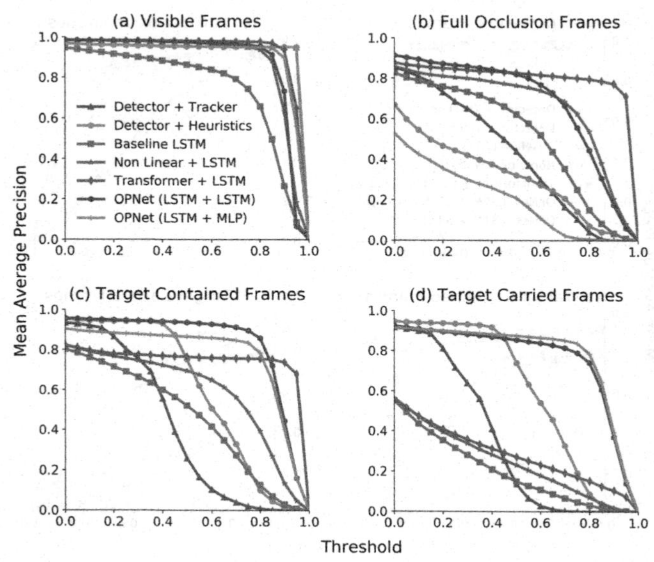

Fig. 4. Mean average precision (MAP) as a function of IoU thresholds for reasoning with Perfect Perception (Sect. 7.1). The most notable performance gain of OPNet (pink and brown curves) was with carried targets (subtask d).

7.2 Learning only from Visible Frames

We now examine a learning setup in which localization supervision is available only for frames where the target object is visible. This setup corresponds more naturally to the process by which people learn object permanence. For instance, imagine a child learning to track a carried (non visible) object for the first time and receiving a surprising feedback only when the object reappears in the scene.

In absence of any supervision when the target is non-visible, incorporating an extra auxiliary loss is needed to account for these frames. Towards this end, we incorporated an auxiliary *consistency loss* that minimizes the change between predictions in consecutive frames. $\mathcal{L}_{consistency} = \frac{1}{n}\sum_{t=1}^{n} \|b_t - b_{t-1}\|^2$. The total

Table 4. IoU for learning with *only visible supervision*. "±" denotes the standard error of the mean (S.E.M.). The models do not perform well when the target is carried.

Mean IoU	Visible	Occluded	Contained	Carried	Overall
BASELINE LSTM	88.61 ± 0.16	80.39 ± 0.54	68.35 ± 0.76	27.39 ± 0.45	78.09 ± 0.49
NON LINEAR + LSTM	89.30 ± 0.15	82.49 ± 0.45	67.25 ± 0.75	27.34 ± 0.45	78.15 ± 0.49
TRANSFORMER + LSTM	88.33 ± 0.15	**83.74** ± 0.44	**69.93** ± 0.77	**27.65** ± 0.54	78.43 ± 0.49
OPNET (LSTM + MLP)	88.45 ± 0.17	48.03 ± 0.82	10.95 ± 0.51	7.28 ± 0.30	61.18 ± 0.69
OPNET (LSTM + LSTM)	**88.95** ± 0.16	81.84 ± 0.48	69.01 ± 0.76	27.50 ± 0.45	**78.50** ± 0.49

loss is defined as an interpolation between the localization loss and the consistency loss, balancing their different scales: $\mathcal{L} = \alpha \cdot \mathcal{L}_{localization} + \beta \cdot \mathcal{L}_{consistency}$ Details on choosing the values of α and β are provided in the supplementary.

Table 4 shows the mean IoU for this setup (compare with Table 2). The baselines perform well when the target is visible, fully occluded or contained without movement. This phenomenon goes hand-in-hand with the inductive bias of the *consistency loss*. Usually, to solve these subtasks, a model only needs to predict the last known target location. This explains why the OPNet (LSTM+MLP) model performs so poorly in this setup.

We note that the performance of non-OPNet models on the carried task is similar to that obtained using full supervision (see Table 2, Sect. 7). This suggests that these models fail to use the supervision for the "carried" task, and further reinforces the observation that localizing carried object is highly challenging.

7.3 Comparison with CATER Data

The original CATER paper [8] considered the "snitch localization" task, aiming to localize the snitch at the last frame of the video, and formalized as a classification problem. The x-y plane was divided with a 6-by-6 grid, and the goal was to predict the correct cell of that grid.

Here we report the performance of OPNet and relevant baselines evaluated with the exact setup of [8], to facilitate comparison with the results reported there. Table 5 shows the accuracy and L_1-distance metrics for this evaluation. OPNet significantly improves over all baselines from [8]. It reduces classification error from 40% to 24%, and the L_1 distance from 1.2 to 0.54.

7.4 Qualitative Examples

To gain insight into the behaviour and limitations of the OPNet model, we provide examples of its successes and failures. The first video[2] provides a "win" example, demonstrating the power of the *"who-to-track"* reasoning component. In this example, the model correctly handles recursive containment that involve "carrying". See Fig. 5 (top row). The second video[3] illustrates a failure, where OPNet fails to switch between tracked objects when the target is "carried". The model accidentally switches to an incorrect cone object (the yellow cone) that already contains another object, not the target. Interestingly, OPNet operates properly when the yellow cone is picked up and switches to track the blue ball that was contained by the yellow cone. It suggests that OPNet learns an implicit representation of the object actions (pick-up, slide, contain etc.) even though it was not explicitly trained to do so. See Fig. 5 (bottom row).

[2] https://youtu.be/FnturB2Blw8.

[3] https://youtu.be/qkdQSHLrGqI.

Table 5. Classification accuracy on CATER using the metrics of [8].

Model	Accuracy (higher is better)	L_1 Distance (lower is better)
DaSiamRPN	33.9	2.4
TSN-RGB + LSTM	25.6	2.6
R3D + LSTM	60.2	1.2
OPNet (Ours)	**74.8**	**0.54**

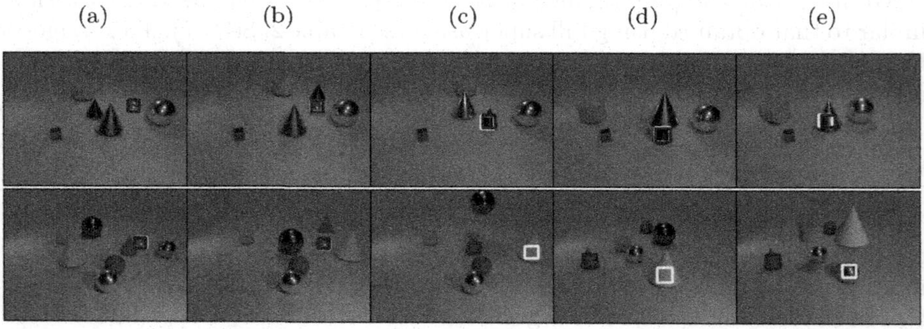

(a) (b) (c) (d) (e)

Fig. 5. Examples of a success case (top row) and a failure case (bottom row) for localizing a carried object. The blue box marks the ground-truth location. The yellow box marks the predicted location. *Top* (a) The target object is visible; (b-c) The target becomes covered and carried by the orange cone; (d-e) The big golden cone covers and carries the orange cone, illustrating recursive containment. The target object is not visible, but OPNet successfully tracks it. *Bottom* (c-d) OPNet accidentally switches to the wrong cone object (the yellow cone instead of the brown cone); (e) OPNet correctly finds when the yellow cone is picked up and switches to track the blue ball underneath. (Color figure online)

8 Conclusion

We considered the problem of localizing one target object in highly dynamic scenes, where the target can be occluded, contained or even carried away, concealed by another object. We name this task *object permanence*, following the cognitive concept of an object that is physically present in a scene but is occluded or carried. We presented an architecture called OPNet, whose components naturally correspond to the perceptual and the reasoning stages of solving OP. Specifically, it has a module that learns to switch attention to an object if it infers that the object contains or carries the target. Our empirical evaluation shows that these components are needed for improving accuracy in this task.

Our results highlight a remaining gap between perfect perception and a pixel-based detector. It is expected that this gap may be even wider when applying OP to more complex natural videos in an open-world setting. It will be interesting to further improve detection architectures to reduce this gap

Acknowledgments. This study was funded by grants to GC from the Israel Science Foundation and Bar-Ilan University (ISF 737/2018, ISF 2332/18). AS is funded by the Israeli innovation authority through the AVATAR consortium. AG received funding from the European Research Council (ERC) under the European Unions Horizon 2020 research and innovation program (grant ERC HOLI 819080).

References

1. Aguiar, A., Baillargeon, R.: 2.5-month-old infants' reasoning about when objects should and should not be occluded. Cogn. Psychol. **39**(2), 116–157 (1999)
2. Baillargeon, R., DeVos, J.: Object permanence in young infants: further evidence. Child Dev. **62**(6), 1227–1246 (1991)
3. Carreira, J., Zisserman, A.: Quo vadis, action recognition? A new model and the kinetics dataset. In: proceedings of the IEEE Conference on Computer Vision and Pattern Recognition, pp. 6299–6308 (2017)
4. Fan, H., et al.: LaSOT: a high-quality benchmark for large-scale single object tracking. In: Proceedings of the IEEE Conference on Computer Vision and Pattern Recognition, pp. 5374–5383 (2019)
5. Fan, H., Ling, H.: Siamese cascaded region proposal networks for real-time visual tracking. In: 2019 IEEE/CVF Conference on Computer Vision and Pattern Recognition (CVPR), June 2019
6. Feichtenhofer, C., Pinz, A., Wildes, R.P.: Spatiotemporal residual networks for video action recognition. Corr abs/1611.02155. arXiv preprint arXiv:1611.02155 (2016)
7. Gao, L., Guo, Z., Zhang, H., Xu, X., Shen, H.T.: Video captioning with attention-based lstm and semantic consistency. IEEE Trans. Multimed. **19**(9), 2045–2055 (2017)
8. Girdhar, R., Ramanan, D.: CATER: a diagnostic dataset for compositional actions and temporal reasoning. arXiv preprint arXiv:1910.04744 (2019)
9. Grabner, H., Matas, J., Van Gool, L., Cattin, P.: Tracking the invisible: learning where the object might be. In: 2010 IEEE Computer Society Conference on Computer Vision and Pattern Recognition, pp. 1285–1292. IEEE (2010)
10. Hochreiter, S., Schmidhuber, J.: Long short-term memory. Neural Comput. **9**(8), 1735–1780 (1997)
11. Huang, Y., Essa, I.: Tracking multiple objects through occlusions. In: 2005 IEEE Computer Society Conference on Computer Vision and Pattern Recognition (CVPR 2005), vol. 2, pp. 1051–1058. IEEE (2005)
12. Johnson, J., Hariharan, B., van der Maaten, L., Fei-Fei, L., Lawrence Zitnick, C., Girshick, R.: CLEVR: a diagnostic dataset for compositional language and elementary visual reasoning. In: Proceedings of the IEEE Conference on Computer Vision and Pattern Recognition, pp. 2901–2910 (2017)
13. Krishna, R., et al.: Visual genome: connecting language and vision using crowdsourced dense image annotations. Int. J. Comput. Vis. **123**(1), 32–73 (2017)
14. Kristan, M., Leonardis, A., et al., J.M.: The sixth visual object tracking vot2018 challenge results. In: ECCV Workshops (2018)
15. Li, B., Yan, J., Wu, W., Zhu, Z., Hu, X.: High performance visual tracking with siamese region proposal network. In: 2018 IEEE/CVF Conference on Computer Vision and Pattern Recognition, pp. 8971–8980 (2018)

16. Liang, W., Zhu, Y., Zhu, S.C.: Tracking occluded objects and recovering incomplete trajectories by reasoning about containment relations and human actions. In: Thirty-Second AAAI Conference on Artificial Intelligence (2018)
17. Lu, C., Krishna, R., Bernstein, M., Fei-Fei, L.: Visual relationship detection with language priors. In: Leibe, B., Matas, J., Sebe, N., Welling, M. (eds.) ECCV 2016. LNCS, vol. 9905, pp. 852–869. Springer, Cham (2016). https://doi.org/10.1007/978-3-319-46448-0_51
18. Mojtaba Marvasti-Zadeh, S., Cheng, L., Ghanei-Yakhdan, H., Kasaei, S.: Deep learning for visual tracking: a comprehensive survey. arXiv-1912 (2019)
19. Papadourakis, V., Argyros, A.: Multiple objects tracking in the presence of long-term occlusions. Comput. Vis. Image Underst. **114**(7), 835–846 (2010)
20. Piaget, J.: The construction of reality in the child (1954)
21. Ren, S., He, K., Girshick, R., Sun, J.: Faster R-CNN: towards real-time object detection with region proposal networks. In: Advances in Neural Information Processing Systems, pp. 91–99 (2015)
22. Sadeghi, M.A., Farhadi, A.: Recognition using visual phrases. In: CVPR 2011, pp. 1745–1752. IEEE (2011)
23. Sharma, S., Kiros, R., Salakhutdinov, R.: Action recognition using visual attention. arXiv preprint arXiv:1511.04119 (2015)
24. Simonyan, K., Zisserman, A.: Two-stream convolutional networks for action recognition in videos. In: Advances in Neural Information Processing Systems, pp. 568–576 (2014)
25. Smitsman, A.W., Dejonckheere, P.J., De Wit, T.C.: The significance of event information for 6-to 16-month-old infants' perception of containment. Dev. Psychol. **45**(1), 207 (2009)
26. Song, S., Lan, C., Xing, J., Zeng, W., Liu, J.: An end-to-end spatio-temporal attention model for human action recognition from skeleton data. In: Thirty-first AAAI Conference on Artificial Intelligence (2017)
27. Tran, D., Bourdev, L., Fergus, R., Torresani, L., Paluri, M.: Learning spatiotemporal features with 3D convolutional networks. In: Proceedings of the IEEE International Conference on Computer Vision, pp. 4489–4497 (2015)
28. Ullman, S., Dorfman, N., Harari, D.: A model for discovering 'containment' relations. Cognition **183**, 67–81 (2019)
29. Vaswani, A., et al.: Attention is all you need (2017)
30. Wu, Y., Lim, J., Yang, M.H.: Object tracking benchmark. IEEE Trans. Pattern Anal. Mach. Intell. **37**(9), 1834–1848 (2015)
31. Yi, K., et al.: CLEVRER: collision events for video representation and reasoning (2019)
32. Yue-Hei Ng, J., Hausknecht, M., Vijayanarasimhan, S., Vinyals, O., Monga, R., Toderici, G.: Beyond short snippets: deep networks for video classification. In: Proceedings of the IEEE Conference on Computer Vision and Pattern Recognition, pp. 4694–4702 (2015)
33. Zaheer, M., Kottur, S., Ravanbakhsh, S., Poczos, B., Salakhutdinov, R.R., Smola, A.J.: Deep sets. In: Advances in Neural Information Processing Systems, pp. 3391–3401 (2017)
34. Zhou, B., Andonian, A., Oliva, A., Torralba, A.: Temporal relational reasoning in videos. European Conference on Computer Vision (2018)
35. Zhu, Z., Wang, Q., Bo, L., Wu, W., Yan, J., Hu, W.: Distractor-aware siamese networks for visual object tracking. In: European Conference on Computer Vision (2018)

Adaptive Text Recognition Through Visual Matching

Chuhan Zhang[1(✉)], Ankush Gupta[2], and Andrew Zisserman[1]

[1] Visual Geometry Group, Department of Engineering Science, University of Oxford, Oxford, UK
{czhang,az}@robots.ox.ac.uk
[2] DeepMind, London, UK
ankushgupta@google.com

Abstract. This work addresses the problems of generalization and flexibility for text recognition in documents. We introduce a new model that exploits the repetitive nature of characters in languages, and decouples the visual decoding and linguistic modelling stages through intermediate representations in the form of *similarity maps*. By doing this, we turn text recognition into a visual matching problem, thereby achieving generalization in appearance and flexibility in classes.

We evaluate the model on both synthetic and real datasets across different languages and alphabets, and show that it can handle challenges that traditional architectures are unable to solve without expensive retraining, including: (i) it can change the number of classes simply by changing the exemplars; and (ii) it can generalize to novel languages and characters (not in the training data) simply by providing a new glyph exemplar set. In essence, it is able to carry out one-shot sequence recognition. We also demonstrate that the model can generalize to unseen fonts without requiring new exemplars from them.

Code, data, and model checkpoints are available at: http://www.robots.ox.ac.uk/~vgg/research/FontAdaptor20/.

Keywords: Text recognition · Sequence recognition · Similarity maps

1 Introduction

Our objective in this work is *generalization* and *flexibility* in text recognition. Modern text recognition methods [2,7,23,32] achieve excellent performance in many cases, but generalization to unseen data, i.e., novel fonts and new languages, either requires large amounts of data for primary training or expensive fine-tuning for each new case.

The text recognition problem is to map an image of a line of text $\boldsymbol{x}$ into the corresponding sequence of characters $\boldsymbol{y} = (y_1, y_2, \ldots, y_k)$, where k is the length of

Electronic supplementary material The online version of this chapter (https://doi.org/10.1007/978-3-030-58517-4_4) contains supplementary material, which is available to authorized users.

A. Vedaldi et al. (Eds.): ECCV 2020, LNCS 12361, pp. 51–67, 2020.
https://doi.org/10.1007/978-3-030-58517-4_4

the string and $y_i \in \mathcal{A}$ are characters in alphabet $\mathcal{A}$ (e.g., {a,b,...,z,<space>}). Current deep learning based methods [7,23,32] cast this in the encoder-decoder framework [8,37], where first the text-line image is encoded through a visual ConvNet [22], followed by a recurrent neural network decoder, with alignment between the visual features and text achieved either through attention [3] or Connectionist Temporal Classification (CTC) [13].

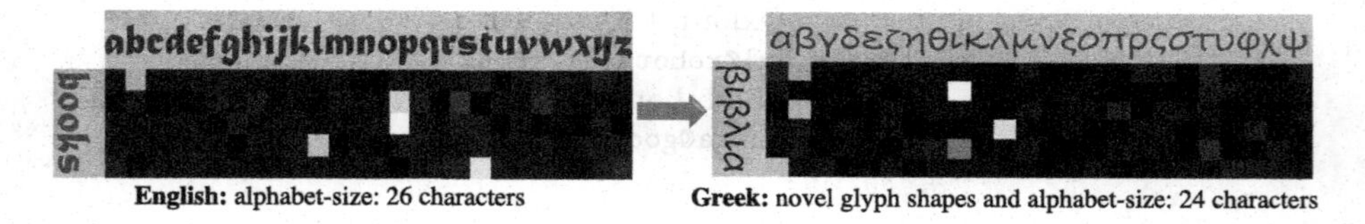

Fig. 1. Visual matching for text recognition. Current text recognition models learn discriminative features specific to character shapes (*glyphs*) from a pre-defined (fixed) alphabet. We train our model instead to establish *visual similarity* between given character glyphs **(top)** and the text-line image to be recognized **(left)**. This makes the model highly adaptable to unseen glyphs, new alphabets/languages, and extensible to novel character classes, e.g., English → Greek, *without* further training. Brighter colors correspond to higher visual similarity.

Impediments to Generalization. The conventional methods for text recognition train the visual encoder and the sequence decoder modules in an end-to-end manner. While this is desirable for optimal co-adaptation, it induces monolithic representations which confound visual and linguistic functions. Consequently, these methods suffer from the following limitations: (1) Discriminative recognition models specialize to fonts and textures in the training set, hence generalize poorly to novel visual styles. (2) The decoder discriminates over a fixed alphabet/number of characters. (3) The encoder and decoder are tied to each other, hence are not inter-operable across encoders for new visual styles or decoders for new languages. Therefore, current text recognition methods generalize poorly and require re-initialization or fine-tuning for new alphabets and languages. Further, fine-tuning typically requires new training data for the target domain and does not overcome these inherent limitations.

Recognition by Matching. Our method is based on a key insight: text is a sequence of repetitions of a finite number of discrete entities. The repeated entities are *characters* in a text string, and *glyphs*, i.e., visual representations of characters/symbols, in a text-line image. We re-formulate the text recognition problem as one of *visual matching*. We assume access to *glyph exemplars* (i.e., cropped images of characters), and task the visual encoder to localize these repeated glyphs in the given text-line image. The output of the visual encoder is a *similarity map* which encodes the visual similarity of each spatial location in the text-line to each glyph in the alphabet as shown in Fig. 1. The decoder ingests this similarity map to infer the most probable string. Figure 2 summarizes the proposed method.

Overcoming Limitations. The proposed model overcomes the above mentioned limitations as follows: (1) Training the encoder for *visual matching* relieves it from learning specific visual styles (fonts, colors etc.) from the training data, improving generalization over novel visual styles. (2) The similarity map is agnostic to the number of different glyphs, hence the model generalizes to novel alphabets (different number of characters). (3) The similarity map is also agnostic to visual styles, and acts as an interpretable interface between the visual encoder and the decoder, thereby disentangling the two.

Contributions. Our main contributions are threefold. First, we propose a novel network design for text recognition aimed at generalization. We exploit the repetition of glyphs in language, and build this similarity between units into our architecture. The model is described in Sects. 3 and 4. Second, we show that the model outperforms state-of-the-art methods in recognition of novel fonts unseen during training (Sect. 5). Third, the model can be applied to novel languages without expensive fine-tuning at test time; it is only necessary to supply glyph exemplars for the new font set. These include languages/alphabets with different number of characters, and novel styles e.g., characters with accents or historical characters "ſ" (also in Sect. 5).

Although we demonstrate our model for *document OCR* where a consistent visual style of glyphs spans the entire document, the method is applicable to scene-text/text-in-the-wild (e.g., SVT [41], ICDAR [18,19] datasets) where each instance has a unique visual style (results in supplementary material).

2 Related Work

Few-shot Recognition. Adapting model behavior based on class exemplars has been explored for few-shot object recognition. Current popular few-shot classification methods, e.g., Prototypical Nets [34], Matching Nets[40], Relation Nets [36], and MAML [11], have been applied only to recognition of *single instances*. Our work addresses the unique challenges associated with one-shot classification of *multiple instances in sequences*. To the best of our knowledge this is the first work to address one-shot *sequence recognition*. We discuss these challenges and the proposed architectural innovations in Sect. 3.4. A relevant work is from Cao et al. [5] which tackles few-shot video classification, but similar to few-shot object recognition methods, they classify the whole video as a *single* instance.

Text Recognition. Recognizing text in images is a classic problem in pattern recognition. Early successful applications were in reading handwritten documents [4,22], and document optical character recognition (OCR) [33]. The

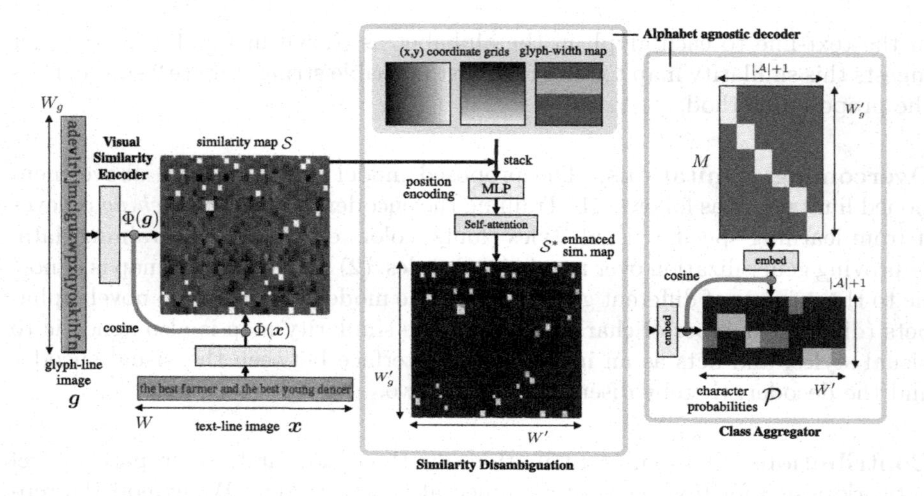

Fig. 2. Visual matching for text recognition. We cast the problem of text recognition as one of visual matching of glyph exemplars in the given text-line image. The visual encoder Φ embeds the glyph-line $\boldsymbol{g}$ and text-line $\boldsymbol{x}$ images and produces a similarity map $\mathcal{S}$, which scores the similarity of each glyph against each position along the text-line. Then, ambiguities in (potentially) imperfect visual matching are resolved to produce the enhanced similarity map $\mathcal{S}^*$. Finally, similarity scores are aggregated to output class probabilities $\mathcal{P}$ using the ground-truth glyph width contained in $\mathcal{M}$.

OCR industry standard—*Tesseract* [33]—employs specialized training data for each supported language/alphabet[1]. Our model enables rapid adaptation to novel visual styles and alphabets and does not require such expensive fine-tuning/specialization. More recently, interest has been focussed towards text in natural images. Current methods either directly classify word-level images [16], or take an encoder-decoder approach [8,37]. The text-image is encoded through a ConvNet, followed by bidirectional-LSTMs for context aggregation. The image features are then aligned with string labels either using Connectionist Temporal Classification (CTC) [13,15,30,35] or through attention [3,6,7,23,31]. Recognizing irregularly shaped text has garnered recent interest which has seen a resurgence of dense character-based segmentation and classification methods [28,42]. Irregular text is rectified before feature extraction either using geometric transformations [24,31,32,44] or by re-generating the text image in canonical fonts and colors [43]. Recently, Baek et al. [2] present a thorough evaluation of text recognition methods, unifying them in a four-stage framework—input transformation, feature extraction, sequence modeling, and string prediction.

[1] Tesseract's specialized training data for 103 languages: https://github.com/tesseract-ocr/tesseract/wiki/Data-Files.

3 Model Architecture

Our model recognizes a given text-line image by localizing glyph exemplars in it through visual matching. It takes both the text-line image and an alphabet image containing a set of exemplars as input, and predicts a sequence of probabilities over N classes as output, where N is equal to the number of exemplars given in the alphabet image. For inference, a glyph-line image is assembled from the individual character glyphs of a reference font simply by concatenating them side-by-side, and text-lines in that font can then be read.

The model has two main components: (1) a visual similarity encoder (Sect. 3.1) which outputs a similarity map encoding the similarity of each glyph in the text-line image, and (2) an alphabet agnostic decoder (Sect. 3.2) which ingests this similarity map to infer the most probable string. In Sect. 3.3 we give details for the training objective. Figure 2 gives a concise schematic of the model.

3.1 Visual Similarity Encoder

The visual similarity encoder is provided with a set of glyphs for the target alphabet, and tasked to localize these glyphs in the input text-line image to be recognized. It first embeds the text-line and glyphs using a shared visual encoder Φ and outputs a *similarity map* $\mathcal{S}$ which computes the visual similarity between all locations in the text-line against all locations in every glyph in the alphabet.

Mathematically, let $\boldsymbol{x} \in \mathbb{R}^{H\times W\times C}$ be the text-line image, with height H, width W and C channels. Let the glyphs be $\{g_i\}_{i=1}^{i=|\mathcal{A}|}$, $g_i \in \mathbb{R}^{H\times W_i \times C}$, where $\mathcal{A}$ is the alphabet, and W_i is the width of the ith glyph. The glyphs are stacked along the width to form a *glyph-line image* $\boldsymbol{g} \in \mathbb{R}^{H\times W_g\times C}$. Embeddings are obtained using the visual encoder Φ for both the text-line $\Phi(\boldsymbol{x}) \in \mathbb{R}^{1\times W'\times D}$ and the glyph-line $\Phi(\boldsymbol{g}) \in \mathbb{R}^{1\times W'_g\times D}$, where D is the embedding dimensionality. The output widths are downsampled by the network stride s (i.e., $W' = \frac{W}{s}$). Finally, each spatial location along the width in the glyph-line image is scored against the every location in the text-line image to obtain the similarity map $\mathcal{S} \in [-1,1]^{W'_g\times W'}$:

$$S_{ij} = \langle \Phi(\boldsymbol{g})_i, \Phi(\boldsymbol{x})_j \rangle = \frac{\Phi(\boldsymbol{g})_i^T \Phi(\boldsymbol{x})_j}{||\Phi(\boldsymbol{g})_i|| \cdot ||\Phi(\boldsymbol{x})_j||} \tag{1}$$

where score is the cosine similarity, and $i \in \{1,\ldots,W'_g\}$, $j \in \{1,\ldots,W'\}$.

3.2 Alphabet Agnostic Decoder

The alphabet agnostic decoder discretizes the similarity maps into probabilities for each character in the alphabet for all spatial locations along the width of the text-line image. Concretely, given the visual similarity map $\mathcal{S} \in \mathbb{R}^{W'_g\times W'}$ it outputs logits over the alphabet for each location in the text-line: $\mathcal{P} \in \mathbb{R}^{|A|\times W'}$, $\mathcal{P}_{ij} = \log p(y_i|\boldsymbol{x}_j)$, where $\boldsymbol{x}_j$ is the jth column in text-line image (modulo encoder stride) and y_i is the ith character in the alphabet $\mathcal{A}$.

A simple implementation would predict the argmax or sum of the similarity scores aggregated over the extent of each glyph in the similarity map. However, this naive strategy does not overcome ambiguities in similarities or produce smooth/consistent character predictions. Hence, we proceed in two steps: first, **similarity disambiguation** resolves ambiguities over the glyphs in the alphabet producing an enhanced similarity map ($\mathcal{S}^*$) by taking into account the glyph widths and position in the line image, and second, **class aggregator** computes character class probabilities by aggregating the scores inside the spatial extent of each glyph in $\mathcal{S}^*$. We detail the two steps next; the significance of each component is established empirically in Sect. 5.4.

Similarity Disambiguation. An ideal similarity map would have square regions of high-similarity. This is because the width of a character in the glyph and text-line images will be the same. Hence, we encode glyph widths along with local x, y coordinates using a small MLP into the similarity map. The input to the MLP at each location is the similarity map value $\mathcal{S}$ stacked with: (1) two channels of x, y coordinates (normalized to $[0, 1]$), and (2) a *glyph width-map* $\mathcal{G}$: $\mathcal{G} = \boldsymbol{w}_g \mathbb{1}^T$, where $\boldsymbol{w}_g \in \mathbb{R}^{W'_g}$ is a vector of glyph widths in pixels; see Fig. 2 for an illustration. For disambiguation over all the glyphs (columns of $\mathcal{S}$), we use a self-attention module [38] which outputs the final enhanced similarity map $\mathcal{S}^*$ of the same size as $\mathcal{S}$.

Class Aggregator. The class aggregator Λ maps the similarity map to logits over the alphabet along the horizontal dimension in the text-line image: $\Lambda : \mathbb{R}^{W'_g \times W'} \mapsto \mathbb{R}^{|\mathcal{A}| \times W'}$, $\mathcal{S}^* \mapsto \mathcal{P}$. This mapping can be achieved by multiplication through a matrix $M \in \mathbb{R}^{|\mathcal{A}| \times W'_g}$ which aggregates (sums) the scores in the span of each glyph: $\mathcal{P} = M\mathcal{S}^*$, such that $M = [m_1, m_2, \ldots, m_{|\mathcal{A}|}]^T$ and $m_i \in \{0, 1\}^{W'_g} = [0, \ldots, 0, 1, \ldots, 1, 0, \ldots, 0]$ where the non-zero values correspond to the span of the ith glyph in the glyph-line image.

In practice, we first embed columns of $\mathcal{S}^*$ and M^T independently using learnt linear embeddings. The embeddings are ℓ_2-normalized before the matrix product (equivalent to cosine similarity). We also expand the alphabet to add an additional "boundary" class (for CTC) using a learnt $m_{|\mathcal{A}|+1}$. Since, the decoder is agnostic to the number of characters in the alphabet, it generalizes to novel alphabets.

3.3 Training Loss

The dense per-pixel decoder logits over the alphabet $\mathcal{P}$ are supervised using the CTC loss [12] ($\mathcal{L}_{CTC}$) to align the predictions with the output label. We also supervise the similarity map output of the visual encoder $\mathcal{S}$ using an auxiliary cross-entropy loss ($\mathcal{L}_{sim}$) at each location. We use ground-truth character bounding-boxes for determining the spatial span of each character. The overall training objective is the following two-part loss,

$$\mathcal{L}_{pred} = \mathcal{L}_{CTC}\left(\text{SoftMax}(\mathcal{P}), \boldsymbol{y}_{gt})\right) \tag{2}$$

$$\mathcal{L}_{sim} = -\sum_{ij} \log(\text{SoftMax}(S_{y_i j})) \tag{3}$$

$$\mathcal{L}_{total} = \mathcal{L}_{pred} + \lambda \mathcal{L}_{sim} \tag{4}$$

where, SoftMax$(\cdot)$ normalization is over the alphabet (rows), $\boldsymbol{y}_{gt}$ is the string label, and y_i is the ground-truth character associated with the ith position in the glyph-line image. The model is insensitive to the value of λ within a reasonbale range (see supplementary), and we use $\lambda = 1$ for a good balance of losses.

3.4 Discussion: One-Shot Sequence Recognition

Our approach can be summarized as a method for one-shot sequence recognition. Note, existing few-shot methods [17,34,36,40] are not directly applicable to this problem of one-shot sequence recognition, as they focus on classification of the whole of the input (e.g. an image) as a single instance. Hence, these cannot address the following unique challenges associated with (text) sequences: (1) segmentation of the imaged text sequence into characters of different widths; (2) respecting language-model/sequence-regularity in the output. We develop a novel neural architectural solutions for the above, namely: (1) A neural architecture with *explicit reasoning over similarity maps* for decoding sequences. The similarity maps are key for *generalization* at both ends—novel fonts/visual styles and new alphabets/languages respectively. (2) Glyph width aware *similarity disambiguation*, which identifies contiguous square blocks in noisy similarity maps from novel data. This is critical for robustness against imprecise visual matching. (3) *Class aggregator*, aggregates similarity scores over the reference width-spans of the glyphs to produce character logit scores over the alphabet. It operates over a variable number of characters/classes and glyph-widths. The importance of each of these components is established in the ablation experiments in Sect. 5.4.

4 Implementation Details

The architectures of the visual similarity encoder and the alphabet agnostic decoder are described in Sect. 4.1 and Sect. 4.2 respectively, followed by training set up in Sect. 4.3.

4.1 Visual Similarity Encoder

The visual similarity encoder (Φ) encodes both the text-line ($\boldsymbol{x}$) and glyph-line ($\boldsymbol{g}$) images into feature maps. The inputs of height 32 pixels, width W and 1 channel (grayscale images) are encoded into a tensor of size $1 \times \frac{W}{2} \times 256$. The glyph-line image's width is held fixed to a constant $W_g = 720$ px: if $\sum_{i=1}^{i=|\mathcal{A}|} W_i < W_g$ the image is padded at the end using the <space> glyph, otherwise the image is downsampled bilinearly to a width of $W_g = 720$ px. The text-line image's input width is free (after resizing to a height of 32 proportionally). The encoder is implemented as a U-Net [29] with two residual blocks [14]; detailed architecture in Table 1. The visual similarity map ($\mathcal{S}$) is obtained by taking the cosine distance between all locations along the width of the encoded features from text-line $\Phi(\boldsymbol{x})$ and glyph-line $\Phi(\boldsymbol{g})$ images.

Table 1. Visual encoder architecture (Sects. 3.1 and 4.1). The input is an image of size $32 \times W \times 1$ (height×width×channels).

layer	kernel	channels in / out	pooling	output size H×W
conv1	3×3	1 / 64	max = (2, 2)	16 × W/2
resBlock1	3×3	64 / 64	max = (1, 2)	8 × W/2
resBlock2	3×3	64 / 128	max = (2, 2)	4 × W/4
upsample	–	–	(2, 2)	8 × W/2
skip	3×3	128+64 / 128	–	8 × W/2
pool	–	–	avg = (2, 1)	4 × W/2
conv2	1×1	128 / 64	–	4 × W/2
reshape	–	64 / 256	–	1 × W/2

4.2 Alphabet Agnostic Decoder

Similarity Disambiguation. We use the self-attention based *Transformer* model [38] with three layers with four attention heads each. The input to this module is the similarity map $\mathcal{S}$ stacked with with local positions (x, y) and glyph widths, which are then encoded through a three-layer (4×16, 16×32, 32×1) MLP with ReLU non-linearity [26].

Class Aggregator. The columns of $\mathcal{S}^*$ and glyph width templates (refer to Sect. 3.2) are embedded independently using linear embeddings of size $W'_g \times W'_g$, where $W'_g = \frac{W_g}{s} = \frac{720}{2} = 360$ (s = encoder stride).

Inference. We decode greedily at inference, as is common after training with CTC loss. No additional language model (LM) is used, except in Experiment VS-3 (Sect. 5.5), where a 6-gram LM learnt from over 10M sentences from the WMT News Crawl (2015) English corpus [1] is combined with the model output with beam-search using the algorithm in [25] (parameters: $\alpha = 1.0$, $\beta = 2.0$, beam-width=15).

4.3 Training and Optimization

The entire model is trained end-to-end by minimizing the training objective Eq. 4. We use online data augmentation on both the text-line and glyph images, specifically random translation, crops, contrast, and blur. All parameters, for both ours and SotA models, are initialized with random weights. We use the Adam optimizer [20] with a constant learning rate of 0.001, a batch size of 12 and train until validation accuracy saturates (typically 100k iterations) on a single Nvidia Tesla P40 GPU. The models are implemented in PyTorch [27].

Fig. 3. Left: FontSynth splits. Randomly selected fonts from each of the five font categories – (1) *regular* (R), (2) *bold* (B), (3) *italic* (I), (4) *light* (L) – used for generating the synthetic training set, and (5) *other* (i.e.none of the first four) – used for the test set. **Right: Synthetic data.** Samples from *FontSynth* **(top)** generated using fonts from MJSynth [16], and *Omniglot-Seq* **(bottom)** generated using glyphs from Omniglot [21] as fonts (Sect. 5.2).

5 Experiments

We compare against state-of-the-art text-recognition models for generalization to novel fonts and languages. We first describe the models used for comparisons (Sect. 5.1), then datasets and evaluation metrics (Sect. 5.2), followed by an overview of the experiments (Sect. 5.3), and a thorough component analysis of the model architecture (Sect. 5.4). Finally, we present the results (Sect. 5.5) of all the experiments.

5.1 State-of-the-art Models in Text Recognition

For comparison to state-of-the-art methods, we use three models: (i) Baek et al. [2] for scene-text recognition; (ii) Tesseract [33], the industry standard for document OCR; and (iii) Chowdhury et al. [9] for handwritten text recognition.

For (i), we use the open-source models provided, but without the transformation module (since documents do not have the scene-text problem of non-rectilinear characters). Note, our visual encoder has similar number of parameters as in the encoder ResNet of [2] (theirs: 6.8M, ours: 4.7M parameters). For (ii) and (iii) we implement the models using the published architecture details. Further details of these networks, and the verifcation of our implementations is provided in the supplementary material.

5.2 Datasets and Metrics

FontSynth. We take 1400 fonts from the MJSynth dataset [16] and split them into five categories by their appearance attributes as determined from their names: (1) regular, (2) bold, (3) italic, (4) light, and (5) others (i.e., all fonts with none of the first four attributes in their name); visualized in Fig. 3 (left). We use the first four splits to create a training set, and (5) for the test set. For training, we select 50 fonts at random from each split and generate 1000 text-line and glyph images for each font. For testing, we use all the 251 fonts in category (5). LRS2 dataset [10] is used as the text source. We call this dataset *FontSynth*; visualization in Fig. 3 (right) and further details in the supplementary.

Omniglot-Seq. Omniglot [21] consists of 50 alphabets with a total of 1623 characters, each drawn by 20 different writers. The original one-shot learning task is defined for *single* characters. To evaluate our sequence prediction network we generate a new *Omniglot-Seq* dataset with *sentence* images as following. We randomly map alphabets in Omniglot to English, and use them as 'fonts' to render text-line images as in FontSynth above. We use the original alphabet splits (30 training, 20 test) and generate data online for training, and 500 lines per alphabet for testing. Figure 3 (right) visualizes a few samples.

Google1000. Google1000 [39] is a standard benchmark for document OCR released as part of ICDAR 2007. It constitutes scans of 1000 public domain historical books in English (EN), French (FR), Italian (IT) and Spanish (ES) languages; Table 2 provides a summary. Figure 4 visualizes a few samples from this dataset. This dataset poses significant challenges due to severe degradation, blur, show-through (from behind), inking, fading, oblique text-lines etc. Type-faces from 18th century are significantly different from modern fonts, containing old ligatures like "ſt,ct,Qi". We use this dataset only for evaluation: further details in supplementary.

Table 2. Google1000 dataset summary. Total number of books, alphabet size and percentage of letters with accent (counting accented characters a new) for various languages in the Google1000.

language →	EN	FR	IT	ES
# books	780	40	40	140
Alphabet size	26	35	29	32
% accented letters	0	2.6	0.7	1.5

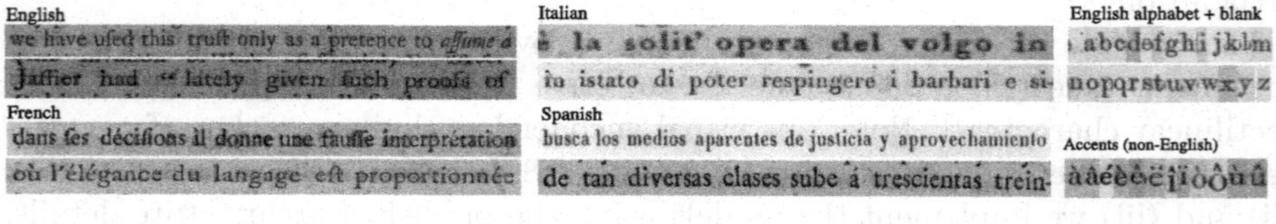

Fig. 4. Google1000 printed books dataset. (left): Text-line image samples from the Google1000 [39] evaluation set for all the languages, namely, English, French, Italian and Spanish. **(right):** *Common* set of glyph exemplars used in our method for *all* books in the evaluation set for English and accents for the other languages.

Evaluation Metrics. We measure the character (CER) and word error rates (WER); definitions in supplementary.

5.3 Overview of Experiments

The goal of our experiments is to evaluate the proposed model against state-of-the-art models for text recognition on their generalization ability to (1) novel visual styles **(VS)** (e.g., novel fonts, background, noise etc.), and (2) novel alphabets/languages **(A)**. Specifically, we conduct the following experiments:

1. **VS-1: Impact of number of training fonts.** We use FontSynth to study the impact of the number of different training fonts on generalization to novel fonts when the exemplars from the testing fonts are provided.
2. **VS-2: Cross glyph matching.** In this experiment, we do *not* assume access to the testing font. Instead of using exemplars from the test font, the most similar font from the training set is selected automatically.
3. **VS-3: Transfer from synthetic to real data.** This evaluates transfer of models trained on synthetic data to real data with historical typeface and degradation.
4. **A-1: Transfer to novel alphabets.** This evaluates transfer of models trained on English to new Latin languages in Google1000 with additional characters in the alphabet (e.g., French with accented characters).
5. **A-2: Transfer to non-Latin glyphs.** The above experiments both train and test on Latin alphabets. Here we evaluate the generalization of the models trained on English fonts to non-Latin scripts in Omniglot-Seq (e.g., from English to Greek).

5.4 Ablation Study

We ablate each major component of the proposed model on the VS-1 experiment to evaluate its significance. Table 3 reports the recognition accuracy on the FontSynth test set when trained on one (R) and all four (R+B+L+I) font attributes. Without the decoder (last row), simply reporting the argmax from the visual similarity map reduces to nearest-neighbors or one-shot Protypical Nets [34] method. This is ineffective for unsegmented text recognition (49% CER vs. 9.4% CER for the full model). Excluding the position encoding in the similarity disambiguation module leads to a moderate drop. The similarity disambiguation *(sim. disamb.)* and linear embedding in class aggregator *(agg. embed.)* are both important, especially when the training data is limited. With more training data, the advantage brought by these modules becomes less significant, while improvement from position encoding does not have such a strong correlation with the amount of training data.

Table 3. Model component analysis. The first row corresponds to the full model; the last row corresponds to reading out characters using the CTC decoder from the output of the visual encoder. *R*, *B*, *L* and *I* correspond to the FontSynth training splits: Regular, Bold, Light and Italic respectively.

sim. enc. $\mathcal{S}$	sim. dis-amb.		agg. embed.	training data			
				R		R+B+L+I	
	pos. enc.	self-attn		CER	WER	CER	WER
✓	✓	✓	✓	**9.4**	**30.1**	**5.6**	**22.3**
✓	✗	✓	✓	11.8	37.9	7.9	22.9
✓	✗	✗	✓	23.9	68.8	13.0	52.0
✓	✓	✓	✗	22.9	65.8	8.5	26.4
✓	✗	✗	✗	25.8	63.1	18.4	45.0
✓	–	–	–	49.0	96.2	38.3	78.9

Table 4. VS-1, VS-2: Generalization to novel fonts with/without known test glyphs and increasing number of training fonts. The mean error rates (in %; ↓ is better) on **FontSynth** test set. For cross matching (*ours-cross*), standard-dev is reported in parenthesis. *R*, *B*, *L* and *I* correspond to the FontSynth training splits; *OS* stands for the Omniglot-Seq dataset (Sect. 5.2).

Training set →			R		R+B		R+B+L		R+B+L+I		R+B+L+I+OS	
model		test glyphs known	CER	WER	CER	WER	CER	WER	CER	WER	CER	WER
CTC Baek et al. [2]		✗	17.5	46.1	11.5	30.3	10.4	28.2	10.4	27.7	—	—
Attn. Baek et al. [2]		✗	16.5	41.0	12.7	34.5	11.1	27.4	10.3	23.6	—	—
Tesseract [33]		✗	19.2	48.6	12.3	37.0	10.8	31.7	9.1	27.8	—	—
Chowdhury et al. [9]		✗	16.2	39.1	12.6	28.6	11.5	29.5	10.5	24.2	—	—
ours-cross	Mean	✗	11.0	33.7	9.3	30.8	9.1	28.6	7.6	22.2	7.0	25.8
	std		(2.9)	(9.8)	(1.4)	(5.9)	(1.1)	(2.2)	(0.2)	(0.9)	(0.9)	(3.7)
ours-cross	Selected	✗	9.8	30.0	8.4	29.4	8.4	27.8	7.2	21.8	5.3	18.3
ours		✓	**9.4**	**30.2**	**8.3**	**28.8**	**8.1**	**27.3**	**5.6**	**22.4**	**3.5**	**12.8**

5.5 Results

VS-1: Impact of Number of Training Fonts. We investigate the impact of the number of training fonts on generalization to unseen fonts. For this systematic evaluation, we train the models on an increasing number of FontSynth splits–regular, regular + bold, regular + bold + light, etc. and evaluate on FontSynth test set. These splits correspond to increments of 50 new fonts with a different appearance attribute. Table 4 summarizes the results. The three baseline SotA models have similar CER when trained on the same amount of data. *Tesseract* [33] has a slightly better performance but generalizes poorly when there is only one attribute in training. Models with an attention-based LSTM (Attn. Baek et al. [2], Chowdhury et al. [9]) achieve lower WER than those without due to better language modelling. Notably, our model achieves the same accuracy with 1 training attribute (CER = 9.4%) as the SotA's with 4 training attributes (CER > 10%), i.e., using 150 (= 3 × 50) less training fonts, proving the strong generalization ability of the proposed method to unseen fonts.

Leveraging Visual Matching. Since, our method does not learn class-specific filters (unlike conventional discriminatively trained models), but instead is trained for visual matching, we can leverage non-English glyphs for training. Hence, we further train on Omniglot-Seq data and drastically reduce the CER from 5.6% (4 attributes) to 3.5%. Being able to leverage language-agnostic data for training is a key strength of our model.

VS-2: Cross Glyph Matching. In VS-1 above, our model assumed privileged access to glyphs from the test image. Here we consider the setting where glyphs exemplars from *training fonts* are used instead. This we term as *cross matching*, denoted 'ours-cross' in Table 4. We randomly select 10 fonts from each font attribute and use those as glyph exemplars. In Table 4 we report the aggregate mean and standard-deviation over all attributes. To automatically find the best font match, we also measure the similarity between the reference and unseen fonts by computing the column-wise entropy in the similarity map $\mathcal{S}$ during inference: Similarity scores within each glyph span are first aggregated to obtain logits $\mathcal{P} \in \mathbb{R}^{|\mathcal{A}| \times W'}$, the averaged entropy of logits over columns $\frac{1}{W'} \sum_i^{W'} -P_i \log(P_i)$ is then used as the criterion to choose the best-matched reference font. Performance from the best-matched exemplar set is reported in 'ours-cross selected' in Table 4. With CER close to the last row where test glyphs are provided, it is shown that the model does not rely on extra information from the new fonts to generalize to different visual styles. Figure 5 details the performance for each attribute separately. The accuracy is largely insensitive to particular font attributes—indicating the strong ability of our model to match glyph shapes. Further, the variation decreases as expected as more training attributes are added.

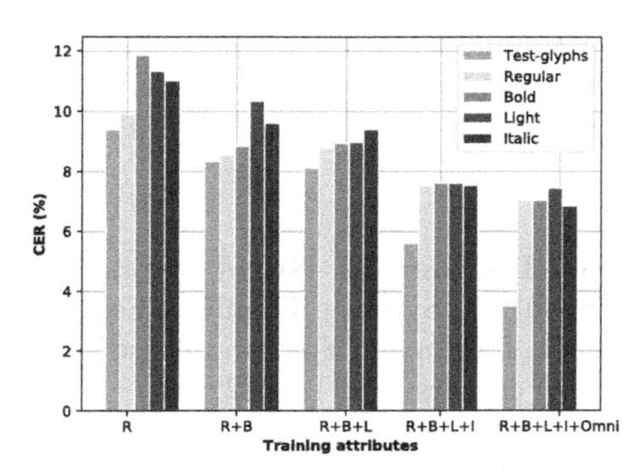

Fig. 5. VS-2: Cross matching on FontSynth. Our model maintains its performance when using training fonts as glyph exemplars instead of test-image glyphs (refer to Sect. 5.5). On the x-axis we show the FontSynth training splits (Fig. 3 left).

VS-3: Transfer from Synthetic to Real Data. We evaluate models trained with synthetic data on the real-world Google1000 test set for generalization to novel visual fonts and robustness against degradation and other nuisance factors in real data. To prevent giving per test sample specific privileged information to our model, we use a common glyph set extracted from Google1000 (visualized in Fig. 4). This glyph set is used for *all* test samples, i.e., is not sample specific. Table 5 compares our model trained on FontSynth+Omniglot-Seq against the SotAs. These models trained on modern fonts are not able to recognize historical ligatures like long s: "ſ" and usually classify it as the character "f". Further, they show worse ability for

Table 5. VS-3: Generalization from synthetic to real data. Mean error rates (in %; ↓ is better) on Google1000 English document for models trained only on synthetic data (Sect. 5.5). LM stands for 6-gram language model.

	CTC Baek [2]		Attn. Baek [2]		Tesseract [33]		Ch. et al. [9]		**ours**	
LM	✗	✓	✗	✓	✗	✓	✗	✓	✗	✓
CER	3.5	3.14	5.4	5.4	4.65	3.8	5.5	5.6	**3.1**	**2.4**
WER	**12.9**	11.4	13.1	13.8	15.9	12.2	14.9	15.6	14.9	**8.0**

handling degradation problems like fading and show-through, and thus are outperformed by our model, especially when supported by a language model (LM) (CER: ours = 2.4% vs. CTC = 3.14%).

A-1: Transfer to Novel Alphabets. We evaluate our model trained on English FontSynth + Omniglot-Seq to other languages in Google1000, namely, French, Italian and Spanish. These languages have more characters than English due to accents (see Table 2). We expand the glyph set from English to include the accented glyphs shown in Fig. 4. For comparsion, we pick the CTC Baek et al.[2] (the SotA with the lowest CER when training data is limited), and adapt it to the new alphabet size by fine-tuning the last linear classifier layer on an increasing number of training samples. Figure 6 summarizes the results. Images for fine-tuning are carefully selected to cover as many new classes as possible. For all three languages, at least 5 images with new classes are required in fine-tuning to match our performance without fine-tuning; Depending on the number of new classes in this language (for French 16 samples are required). Note that for our model we do not need fine-tuning at all, just supplying exemplars of new glyphs gives a good performance.

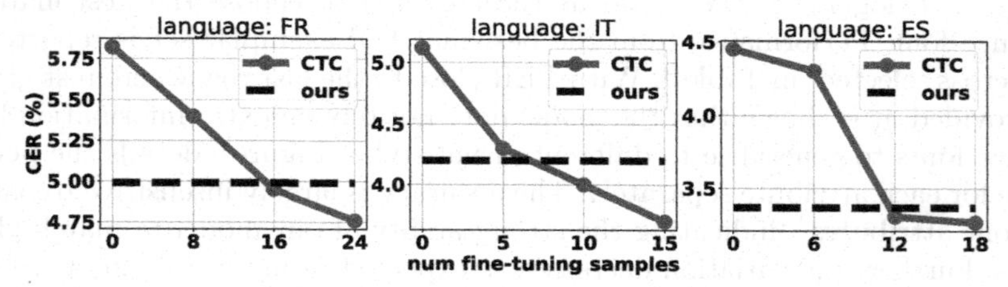

Fig. 6. A-2: Transfer to novel alphabets in Google1000. We evaluate models trained over the English alphabet on novel languages in the Google1000 dataset, namely, French, Italian and Spanish. CER is reported (in %; ↓ is better).

A-2: Transfer to Non-Latin Glyphs. In the above experiments, the models were both trained and tested on English/Latin script and hence, are not tasked to generalize to completely novel glyph shapes. Here we evaluate the generalization ability of our model to new glyph shapes by testing the model trained on FontSynth + Omniglot-Seq on the Omniglot-Seq test set, which consists of novel alphabets/scripts. We provide our model with glyph exemplars from the randomly generated alphabets (Sect. 5.2). Our model achieves CER = 1.8%/7.9%, WER = 7.6%/31.6% (with LM/without LM), which demonstrates strong generalization to novel scripts. Note, the baseline text recognition models trained on FontSynth (English fonts) cannot perform this task, as they cannot process completely new glyph shapes.

6 Conclusion

We have developed a method for text recognition which generalizes to novel visual styles (e.g., fonts, colors, backgrounds etc.), and is not tied to a particular alphabet size/language. It achieves this by recasting the classic text recognition as one of visual matching, and we have demonstrated that the matching can leverage random shapes/glyphs (e.g., Omniglot) for training. Our model is perhaps the first to demonstrate one-shot sequence recognition, and achieves superior generalization ability as compared to conventional text recognition methods without requiring expensive adaptation/fine-tuning. Although the method has been demonstrated for text recognition, it is applicable to other sequence recognition problems like speech and action recognition.

Acknowledgements. This research is funded by a Google-DeepMind Graduate Scholarship and the EPSRC Programme Grant Seebibyte EP/M013774/1. We would like to thank Triantafyllos Afouras, Weidi Xie, Yang Liu and Erika Lu for discussions and proof-reading.

References

1. EMNLP 2015 Tenth Workshop On Statistical Machine Translation. http://www.statmt.org/wmt15/
2. Baek, J., et al.: What is wrong with scene text recognition model comparisons? dataset and model analysis. In: Proceedings of ICCV (2019)
3. Bahdanau, D., Cho, K., Bengio, Y.: Neural machine translation by jointly learning to align and translate (2014). arXiv preprint arXiv:1409.0473
4. Bunke, H., Bengio, S., Vinciarelli, A.: Offline recognition of unconstrained handwritten texts using HMMs and statistical language models. PAMI **26**(6), 709–720 (2004)
5. Cao, K., Ji, J., Cao, Z., Chang, C.Y., Niebles, J.C.: Few-shot video classification via temporal alignment. In: Proceedings of the IEEE/CVF Conference on Computer Vision and Pattern Recognition, pp. 10618–10627 (2020)
6. Cheng, Z., Bai, F., Xu, Y., Zheng, G., Pu, S., Zhou, S.: Focusing attention: towards accurate text recognition in natural images. In: Proceedings of ICCV (2017) 4
7. Cheng, Z., Xu, Y., Bai, F., Niu, Y., Pu, S., Zhou, S.: Aon: towards arbitrarily-oriented text recognition. In: Proceedings of CVPR (2018)
8. Cho, K., et al.: Learning phrase representations using RNN encoder-decoder for statistical machine translation. In: EMNLP (2014)
9. Chowdhury, A., Vig, L.: An efficient end-to-end neural model for handwritten text recognition. In: Proceedings of BMVC (2018)
10. Chung, J.S., Zisserman, A.: Lip reading in the wild. In: Proceedings of ACCV (2016)
11. Finn, C., Abbeel, P., Levine, S.: Model-agnostic meta-learning for fast adaptation of deep networks. In: Proceedings of ICML (2017)
12. Graves, A., Fernández, S., Gomez, F., Schmidhuber, J.: Connectionist temporal classification: labelling unsegmented sequence data with recurrent neural networks. In: Proceedings of ICML. ACM (2006)

13. Graves, A., Schmidhuber, J.: Framewise phoneme classification with bidirectional LSTM and other neural network architectures. Neural Netw. **18**(5–6), 602–610 (2005)
14. He, K., Zhang, X., Ren, S., Sun, J.: Deep residual learning for image recognition. In: Proceedings pf CVPR (2016)
15. He, P., Huang, W., Qiao, Y., Loy, C.C., Tang, X.: Reading scene text in deep convolutional sequences. In: Thirtieth AAAI Conference on Artificial Intelligence (2016)
16. Jaderberg, M., Simonyan, K., Vedaldi, A., Zisserman, A.: Synthetic data and artificial neural networks for natural scene text recognition. In: Workshop on Deep Learning, NIPS (2014)
17. Jia, X., De Brabandere, B., Tuytelaars, T., Gool, L.V.: Dynamic filter networks. In: Proceedings of NIPS (2016)
18. Karatzas, D., et al.: ICDAR 2015 robust reading competition. In: Proceedings of ICDAR, pp. 1156–1160 (2015)
19. Karatzas, D., et al.: ICDAR 2013 robust reading competition. In: Proceedings of ICDAR (2013) 3
20. Kingma, D.P., Ba, J.: Adam: a method for stochastic optimization (2014). arXiv preprint arXiv:1412.6980
21. Lake, B.M., Salakhutdinov, R., Tenenbaum, J.B.: Human-level concept learning through probabilistic program induction. Science **350**, 1332–1338 (2015)
22. LeCun, Y., et al.: Backpropagation applied to handwritten zip code recognition. Neural Comput. **1**(4), 541–551 (1989)
23. Lee, C.Y., Osindero, S.: Recursive recurrent nets with attention modeling for OCR in the wild. In: Proceedings of CVPR (2016)
24. Liu, W., Chen, C., Wong, K.Y.K.: Char-net: a character-aware neural network for distorted scene text recognition. In: Thirty-Second AAAI Conference on Artificial Intelligence (2018)
25. Maas, A., Xie, Z., Jurafsky, D., Ng, A.: Lexicon-free conversational speech recognition with neural networks. In: NAACL-HLT (2015)
26. Nair, V., Hinton, G.E.: Rectified linear units improve restricted Boltzmann machines. In: Proceedings of ICML (2010)
27. Paszke, A., et al.: Automatic differentiation in pytorch (2017)
28. Pengyuan, L., Minghui, L., Cong, Y., Wenhao, W., Xiang, B.: Mask textspotter: an end-to-end trainable neural network for spotting text with arbitrary shapes. In: Proceedings of ECCV (2018)
29. Ronneberger, O., Fischer, P., Brox, T.: U-Net: convolutional networks for biomedical image segmentation. In: Navab, N., Hornegger, J., Wells, W.M., Frangi, A.F. (eds.) MICCAI 2015. LNCS, vol. 9351, pp. 234–241. Springer, Cham (2015). https://doi.org/10.1007/978-3-319-24574-4_28
30. Shi, B., Bai, X., Yao, C.: An end-to-end trainable neural network for image-based sequence recognition and its application to scene text recognition. PAMI **39**, 2298–2304 (2016)
31. Shi, B., Wang, X., Lyu, P., Yao, C., Bai, X.: Robust scene text recognition with automatic rectification. In: Proceedings of CVPR (2016)
32. Shi, B., Yang, M., Wang, X., Lyu, P., Yao, C., Bai, X.: Aster: an attentional scene text recognizer with flexible rectification. PAMI **41**, 2035–2048 (2018)
33. Smith, R.: An overview of the tesseract OCR engine. In: Ninth International Conference on Document Analysis and Recognition (ICDAR 2007), vol. 2, pp. 629–633. IEEE (2007)

34. Snell, J., Swersky, K., Zemel, R.: Prototypical networks for few-shot learning. In: Proceedings of NIPS (2017)
35. Su, B., Lu, S.: Accurate scene text recognition based on recurrent neural network. In: Proceedings of ACCV (2014)
36. Sung, F., et al.: Learning to compare: relation network for few-shot learning. In: Proceedings of the IEEE Conference on Computer Vision and Pattern Recognition, pp. 1199–1208 (2018)
37. Sutskever, I., Vinyals, O., Le, Q.V.: Sequence to sequence learning with neural networks. In: Advances in Neural Information Processing Systems, pp. 3104–3112 (2014)
38. Vaswani, A., et al.: Attention is all you need. In: Proceedings of NIPS (2017)
39. Vincent, L.: Google book search: document understanding on a massive scale. In: Proccedings of Ninth International Conference on Document Analysis and Recognition (ICDAR), Washington, DC, pp. 819–823 (2007)
40. Vinyals, O., Blundell, C., Lillicrap, T., Kavukcuoglu, K., Wierstra, D.: Matching networks for one shot learning. In: Proceedings of NIPS (2016)
41. Wang, K., Belongie, S.: Word spotting in the wild. In: Proceedings of ECCV (2010) 3
42. Wei, F., Wenhao, H., Fei, Y., Xu-Yao, Z., Cheng-Lin, L.: Textdragon: an end-to-end framework for arbitrary shaped text spotting. In: Proceedings of ICCV (2019)
43. Yang, L., Zhaowen, W., Hailin, J., Ian, W.: Synthetically supervised feature learning for scene text recognition. In: Proceedings of ECCV (2018)
44. Zhan, F., Lu, S.: Esir: end-to-end scene text recognition via iterative image rectification. In: Proceedings of CVPR (2019)

Actions as Moving Points

Yixuan Li, Zixu Wang, Limin Wang(✉), and Gangshan Wu

State Key Laboratory for Novel Software Technology,
Nanjing University, Nanjing, China
liyixxxuan@gmail.com,zixuwang1997@gmail.com,{lmwang,gswu}@nju.edu.cn

Abstract. The existing action tubelet detectors often depend on heuristic anchor design and placement, which might be computationally expensive and sub-optimal for precise localization. In this paper, we present a conceptually simple, computationally efficient, and more precise action tubelet detection framework, termed as *MovingCenter Detector* (MOC-detector), by treating an action instance as a trajectory of moving points. Based on the insight that movement information could simplify and assist action tubelet detection, our MOC-detector is composed of three crucial head branches: (1) Center Branch for instance center detection and action recognition, (2) Movement Branch for movement estimation at adjacent frames to form trajectories of moving points, (3) Box Branch for spatial extent detection by directly regressing bounding box size at each estimated center. These three branches work together to generate the tubelet detection results, which could be further linked to yield video-level tubes with a matching strategy. Our MOC-detector outperforms the existing state-of-the-art methods for both metrics of frame-mAP and video-mAP on the JHMDB and UCF101-24 datasets. The performance gap is more evident for higher video IoU, demonstrating that our MOC-detector is particularly effective for more precise action detection. We provide the code at https://github.com/MCG-NJU/MOC-Detector.

Keywords: Spatio-temporal action detection · Anchor-free detection

1 Introduction

Spatio-temporal action detection is an important problem in video understanding, which aims to recognize all action instances present in a video and also localize them in both space and time. It has wide applications in many scenarios, such as video surveillance [12,20], video captioning [31,36] and event detection [5]. Some early approaches [8,21,25,26,32,33] apply an action detector at each frame independently and then generate action tubes by linking these frame-wise detection results [8,21,25,26,32] or tracking one detection result [33]

Y. Li and Z. Wang—Contribute equally to this work. This work is supported by Tencent AI Lab.

Electronic supplementary material The online version of this chapter (https://doi.org/10.1007/978-3-030-58517-4_5) contains supplementary material, which is available to authorized users.

A. Vedaldi et al. (Eds.): ECCV 2020, LNCS 12361, pp. 68–84, 2020.
https://doi.org/10.1007/978-3-030-58517-4_5

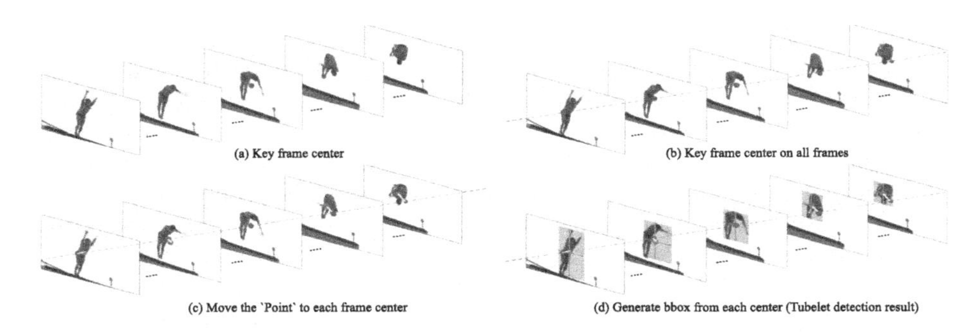

(a) Key frame center
(b) Key frame center on all frames
(c) Move the 'Point' to each frame center
(d) Generate bbox from each center (Tubelet detection result)

Fig. 1. Motivation illustration. We focus on devising an action tubelet detector from a short sequence. Movement information naturally describes human behavior, and each action instance could be viewed as a trajectory of *moving points*. In this view, action tubelet detector could be decomposed into three simple steps: (1) localizing the center point (red dots) at key frame (i.e., center frame), (2) estimating the movement at each frame with respect to the center point (yellow arrows), (3) regressing bounding box size at the calculated center point (green dots) for all frames. Best viewed in color and zoom in. (Color figure online)

across time. These methods fail to well capture temporal information when conducting frame-level detection, and thus are less effective for detecting action tubes in reality. To address this issue, some approaches [11,14,24,27,35,38] try to perform action detection at the clip-level by exploiting short-term temporal information. In this sense, these methods input a sequence of frames and directly output detected tubelets (i.e., a short sequence of bounding boxes). This tubelet detection scheme yields a more principled and effective solution for video-based action detection and has shown promising results on standard benchmarks.

The existing tubelet detection methods [11,14,24,27,35,38] are closely related with the current mainstream object detectors such as Faster R-CNN [23] or SSD [19], which operate on a huge number of pre-defined anchor boxes. Although these anchor-based object detectors have achieved success in image domains, they still suffer from critical issues such as being sensitive to hyper-parameters (e.g., box size, aspect ratio, and box number) and less efficient due to densely placed bounding boxes. These issues are more serious when adapting the anchor-based detection framework from images to videos. First, the number of possible tubelet anchors would grow dramatically when increasing clip duration, which imposes a great challenge for both training and inference. Second, it is generally required to devise more sophisticated anchor box placement and adjustment to consider the variation along the temporal dimension. In addition, these anchor-based methods directly extend 2D anchors along the temporal dimension which predefine each action instance as a cuboid across space and time. This assumption lacks the flexibility to well capture temporal coherence and correlation of adjacent frame-level bounding boxes.

Inspired by the recent advances in anchor-free object detection [4,15,22,30,40], we present a **conceptually simple, computationally efficient, and more precise** action tubelet detector in videos, termed as *MovingCenter detector*

(MOC-detector). As shown in Fig. 1, our detector presents a new tubelet detection scheme by treating each instance as a trajectory of *moving points*. In this sense, an action tubelet is represented by its center point in the key frame and offsets of other frames with respect to this center point. To determine the tubelet shape, we directly regress the bounding box size along the moving point trajectory on each frame. Our MOC-detector yields a fully convolutional one-stage tubelet detection scheme, which not only allows for more efficient training and inference but also could produce more precise detection results (as demonstrated in our experiments).

Specifically, our MOC detector decouples the task of tubelet detection into three sub-tasks: center detection, offset estimation and box regression. First, frames are fed into a 2D efficient backbone network for feature extraction. Then, we devise three separate branches: (1) Center Branch: detecting the action instance center and category; (2) Movement Branch: estimating the offsets of the current frame with respect to its center; (3) Box Branch: predicting bounding box size at the detected center point of each frame. This unique design enables three branches cooperate with each other to generate the tubelet detection results. Finally, we link these detected action tubelets across frames to yield long-range detection results following the common practice [14]. We perform experiments on two challenging action tube detection benchmarks of UCF101-24 [28] and JHMDB [13]. Our MOC-detector outperforms the existing state-of-the-art approaches for both frame-mAP and video-mAP on these two datasets, in particular for higher IoU criteria. Moreover, the fully convolutional nature of MOC detector yields a high detection efficiency of around 25FPS.

2 Related Work

2.1 Object Detection

Anchor-Based Object Detectors. Traditional one-stage [17,19,22] and two-stage object detectors [6,7,10,23] heavily relied on predefined anchor boxes. Two-stage object detectors like Faster-RCNN [23] and Cascade-RCNN [1] devised RPN to generate RoIs from a set of anchors in the first stage and handled classification and regression of each RoI in the second stage. By contrast, typical one-stage detectors utilized class-aware anchors and jointly predicted the categories and relative spatial offsets of objects, such as SSD [19], YOLO [22] and RetinaNet [17].

Anchor-Free Object Detectors. However, some recent works [4,15,30,40,41] have shown that the performance of anchor-free methods could be competitive with anchor-based detectors and such detectors also get rid of computation-intensive anchors and region-based CNN. CornerNet [15] detected object bounding box as a pair of corners, and grouped them to form the final detection. CenterNet [40] modeled an object as the center point of its bounding box and regressed its width and height to build the final result.

2.2 Spatio-Temporal Action Detection

Frame-Level Detector. Many efforts have been made to extend an image object detector to the task of action detection as frame-level action detectors [8,21,25,26,32,33]. After getting the frame detection, linking algorithm is applied to generate final tubes [8,21,25,26,32] and Weinzaepfel et al. [33] utilized a tracking-by-detection method instead. Although flows are used to capture motion information, frame-level detection fails to fully utilize the video's temporal information.

Clip-Level Detector. In order to model temporal information for detection, some clip-level approaches or action tubelet detectors [11,14,16,27,35,38] have been proposed. ACT [14] took a short sequence of frames and output tubelets which were regressed from anchor cuboids. STEP [35] proposed a progressive method to refine the proposals over a few steps to solve the large displacement problem and utilized longer temporal information. Some methods [11,16] linked frame or tubelet proposals first to generate tubes proposal and then did classification.

These approaches are all based on anchor-based object detectors, whose design might be sensitive to anchor design and computationally cost due to large numbers of anchor boxes. Instead, we try to design an anchor-free action tubelet detector by treating each action instance as a trajectory of moving points. Experimental results demonstrate that our proposed action tubelet detector is effective for spatio-temporal action detection, in particular for the high video IoU.

3 Approach

Overview. Action tubelet detection aims at localizing a short sequence of bounding boxes from an input clip and recognizing its action category as well. We present a new tubelet detector, coined as MovingCenter detector (MOC-detector), by viewing an action instance as a trajectory of moving points. As shown in Fig. 2, in our MOC-detector, we take a set of consecutive frames as input and separately feed them into an efficient 2D backbone to extract frame-level features. Then, we design three head branches to perform tubelet detection in an anchor-free manner. The first branch is Center Branch, which is defined on the center (key) frame. This Center Branch localizes the tubelet center and recognizes its action category. The second branch is Movement Branch, which is defined over all frames. This Movement Branch tries to relate adjacent frames to predict the center movement along the temporal dimension. The estimated movement would propagate the center point from key frame to other frames to generate a trajectory. The third branch is Box Branch, which operates on the detected center points of all frames. This branch focuses on determining the spatial extent of the detected action instance at each frame, by directly regressing the height and width of the bounding box. These three branches collaborate together to yield tubelet detection from a short clip, which will be further linked to form action tube detection in a long untrimmed video by following a common

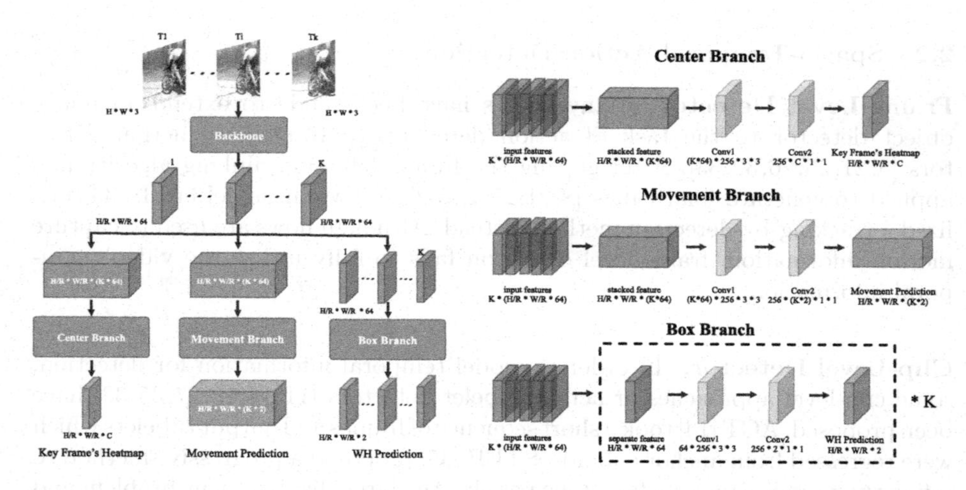

Fig. 2. Pipeline of MOC-detector. In the left, we present the overall MOC-detector framework. The red cuboids represent the extracted features, the blue boxes denote the backbone or detection head, and the gray cuboids are detection results produced by the Center Branch, the Movement Branch, the Box Branch. In the right, we show the detailed design of each branch. Each branch consists of a sequence of one 3*3 conv layer, one ReLu layer and one 1*1 conv layer, which is presented as yellow cuboids. The parameters of convolution are input channel, output channel, convolution kernel height, convolution kernel width.

linking strategy [14]. We will first give a short description of the backbone design, and then provide technical details of three branches and the linking algorithm in the following subsections.

Backbone. In our MOC-detector, we input K frames and each frame is with the resolution of $W \times H$. First K frames are fed into a 2D backbone network sequentially to generate a feature volume $\mathbf{f} \in \mathbb{R}^{K \times \frac{W}{R} \times \frac{H}{R} \times B}$. R is the spatial downsample ratio and B denotes channel number. To keep the full temporal information for subsequent detection, we do not perform any downsampling over the temporal dimension. Specifically, we choose DLA-34 [37] architecture as our MOC-detector feature backbone following CenterNet [40]. This architecture employs an encoder-decoder architecture to extract features for each frame. The spatial downsampling ratio R is 4 and the channel number B is 64. The extracted features are shared by three head branches. Next we will present the technical details of these head branches.

3.1 Center Branch: Detect Center at Key Frame

The Center Branch aims at detecting the action instance center in the key frame (i.e., center frame) and recognizing its category based on the extracted video features. Temporal information is important for action recognition, and thereby we design a temporal module to estimate the action center and recognize its class by concatenating multi-frame feature maps along channel dimension. Specifically, based on the video feature representation $\mathbf{f} \in \mathbb{R}^{\frac{W}{R} \times \frac{H}{R} \times (K \times B)}$, we estimate a center heatmap $\hat{L} \in [0,1]^{\frac{W}{R} \times \frac{H}{R} \times C}$ for the key frame. The C is the number of action classes. The value of $\hat{L}_{x,y,c}$ represents the likelihood of detecting an action instance of class c at location (x, y), and higher value indicates a stronger possibility. Specifically, we employ a standard convolution operation to estimate the center heatmap in a fully convolutional manner.

Training. We train the Center Branch following the common dense prediction setting [15,40]. For i^{th} action instance, we represent its center as key frame's bounding box center and utilize center's position for each action category as the ground truth label (x_{c_i}, y_{c_i}). We generate the ground truth heatmap $L \in [0,1]^{\frac{W}{R} \times \frac{H}{R} \times C}$ using a Gaussian kernel which produces the soft heatmap groundtruth $L_{x,y,c_i} = \exp(-\frac{(x-x_{c_i})^2+(y-y_{c_i})^2}{2\sigma_p^2})$. For other class (i.e., $c \neq c_i$), we set the heatmap $L_{x,y,c} = 0$. The σ_p is adaptive to instance size and we choose the maximum when two Gaussian of the same category overlap. We choose the training objective, which is a variant of focal loss [17], as follows:

$$\ell_{\text{center}} = -\frac{1}{n} \sum_{x,y,c} \begin{cases} (1-\hat{L}_{xyc})^{\alpha} \log(\hat{L}_{xyc}) & \text{if } L_{xyc} = 1 \\ (1-L_{xyc})^{\beta} (\hat{L}_{xyc})^{\alpha} \log(1-\hat{L}_{xyc}) & \text{otherwise} \end{cases} \tag{1}$$

where n is the number of ground truth instances and α and β are hyper-parameters of the focal loss [17]. We set $\alpha = 2$ and $\beta = 4$ following [15,40] in our experiments. It indicates that this focal loss is able to deal with the imbalanced training issue effectively [17].

Inference. After the training, the Center Branch could be deployed in tubelet detection for localizing action instance center and recognizing its category. Specifically, we detect all local peaks which are equal to or greater than their 8-connected neighbors in the estimated heatmap $\hat{L}$ for each class independently. And then keep the top N peaks from all categories as candidate centers with tubelet scores. Following [40], we set N as 100 and detailed ablation studies will be provided in the supplementary material.

3.2 Movement Branch: Move Center Temporally

The Movement Branch tries to relate adjacent frames to predict the movement of the action instance center along the temporal dimension. Similar to Center Branch, Movement Branch also employs temporal information to regress the center offsets of current frame with respect to key frame. Specifically, Movement Branch takes stacked feature representation as input and outputs a movement prediction map $\hat{M} \in \mathbb{R}^{\frac{W}{R} \times \frac{H}{R} \times (K \times 2)}$. $2K$ channels represent center movements from key frame to current frames in X and Y directions. Given the key frame center $(\hat{x}_{key}, \hat{y}_{key})$, $\hat{M}_{\hat{x}_{key},\hat{y}_{key},2j:2j+2}$ encodes center movement at j^{th} frame.

Training. The ground truth tubelet of ith action instance is $[(x_{tl}^1, y_{tl}^1, x_{br}^1, y_{br}^1), ..., (x_{tl}^j, y_{tl}^j, x_{br}^j, y_{br}^j), ..., (x_{tl}^K, y_{tl}^K, x_{br}^K, y_{br}^K)]$, where subscript tl and br represent top-left and bottom-right points of bounding boxes, respectively. Let k be the key frame index, and the ith action instance center at key frame is defined as follows:

$$(x_i^{key}, y_i^{key}) = (\lfloor (x_{tl}^k + x_{br}^k)/2 \rfloor, \lfloor (y_{tl}^k + y_{br}^k)/2 \rfloor). \tag{2}$$

We could compute the bounding box center (x_i^j, y_i^j) of ith instance at j^{th} frame as follows:

$$(x_i^j, y_i^j) = ((x_{tl}^j + x_{br}^j)/2, (y_{tl}^j + y_{br}^j)/2). \tag{3}$$

Then, the ground truth movement of the ith action instance is calculated as follows:

$$m_i = (x_i^1 - x_i^{key}, y_i^1 - y_i^{key}, ..., x_i^K - x_i^{key}, y_i^K - y_i^{key}). \tag{4}$$

For the training of Movement Branch, we optimize the movement map $\hat{M}$ only at the key frame center location and use the ℓ_1 loss as follows:

$$\ell_{\text{movement}} = \frac{1}{n} \sum_{i=1}^{n} |\hat{M}_{x_i^{key}, y_i^{key}} - m_i|. \tag{5}$$

Inference. After the Movement Branch training and given N detected action centers $\{(\hat{x}_i, \hat{y}_i) | i \in \{1, 2, \cdots, N\}\}$ from Center Branch, we obtain a set of movement vector $\{\hat{M}_{\hat{x}_i, \hat{y}_i} | i \in \{1, 2, \cdots, N\}\}$ for all detected action instance. Based on the results of Movement Branch and Center Branch, we could easily generate a trajectory set $T = \{T_i | i \in \{1, 2, \cdots, N\}\}$, and for the detected action center $(\hat{x}_i, \hat{y}_i)$, its trajectory of moving points is calculated as follows:

$$T_i = (\hat{x}_i, \hat{y}_i) + [\hat{M}_{\hat{x}_i, \hat{y}_i, 0:2}, \hat{M}_{\hat{x}_i, \hat{y}_i, 2:4}, \cdots, \hat{M}_{\hat{x}_i, \hat{y}_i, 2K-2:2K}]. \tag{6}$$

3.3 Box Branch: Determine Spatial Extent

The Box Branch is the last step of tubelet detection and focuses on determining the spatial extent of the action instance. Unlike Center Branch and Movement Branch, we assume box detection only depends on the current frame and temporal information will not benefit the class-agnostic bounding box generation. We will provide the ablation study in the supplementary material. In this sense, this branch could be performed in a frame-wise manner. Specifically, Box Branch inputs the single frame's feature $\mathbf{f}^j \in \mathbb{R}^{\frac{W}{R} \times \frac{H}{R} \times B}$ and generates a size prediction map $\hat{S}^j \in \mathbb{R}^{\frac{W}{R} \times \frac{H}{R} \times 2}$ for the jth frame to directly estimate the bounding box size (i.e., width and height). Note that the Box Branch is shared across K frames.

Training. The ground truth bbox size of ith action instance at j^{th} frame can be represented as follows:

$$s_i^j = (x_{br}^j - x_{tl}^j, y_{br}^j - y_{tl}^j). \tag{7}$$

With this ground truth bounding box size, we optimize the Box Branch at the center points of all frames for each tubelet with ℓ_1 Loss as follows:

$$\ell_{\text{box}} = \frac{1}{n} \sum_{i=1}^{n} \sum_{j=1}^{K} |\hat{S}^j_{p_i^j} - s_i^j|. \tag{8}$$

Note that the p_i^j is the ith instance ground truth center at j^{th} frame. So the overall training objective of our MOC-detector is

$$\ell = \ell_{\text{center}} + a\ell_{\text{movement}} + b\ell_{\text{box}}, \tag{9}$$

where we set a=1 and b=0.1 in all our experiments. Detailed ablation studies will be provided in the supplementary material.

Inference. Now, we are ready to generate the tubelet detection results. based on center trajectories T from Movement Branch and size prediction heatmap $\hat{S}$ for each location produced by this branch. For j^{th} point in trajectory T_i, we use (T_x, T_y) to denote its coordinates, and (w,h) to denote Box Branch size output $\hat{S}$ at specific location. Then, the bounding box for this point is calculated as:

$$(T_x - w/2, T_y - h/2, T_x + w/2, T_y + h/2). \tag{10}$$

3.4 Tubelet Linking

After getting the clip-level detection results, we link these tubelets into final tubes across time. As our main goal is to propose a new tubelet detector, we use the same linking algorithm as [14] for fair comparison. Given a video, MOC extracts tubelets and keeps the top 10 as candidates for each sequence of K frames with stride 1 across time, which are linked into the final tubes in a tubelet by tubelet manner. **Initialization:** In the first frame, every candidate starts a new link. At a given frame, candidates which are not assigned to any existing links start new links. **Linking:** one candidate can only be assigned to one existing link when it meets three conditions: (1) the candidate is not selected by other links, (2) the candidate t has the highest score, (3) the overlap between link and candidate is greater than a threshold τ. **Termination:** An existing link stops if it has not been extended in consecutive K frames. We build an action tube for each link, whose score is the average score of tubelets in the link. For each frame in the link, we average the bbox coordinates of tubelets containing that frame. Initialization and termination determine tubes' temporal extents. Tubes with low confidence and short duration are abandoned. As this linking algorithm is online, MOC can be applied for online video stream.

4 Experiments

4.1 Experimental Setup

Datasets and Metrics. We perform experiments on the UCF101-24 [28] and JHMDB [13] datasets. UCF101-24 [28] consists of 3207 temporally untrimmed videos from 24 sports classes. Following the common setting [14,21], we report the action detection performance for the first split only. JHMDB [13] consists of 928 temporally trimmed videos from 21 action classes. We report results averaged over three splits following the common setting [14,21]. AVA [9] is a larger dataset for action detection but only contains a single-frame action instance annotation for each 3s clip, which concentrates on detecting actions on a single key frame. Thus, AVA is not suitable to verify the effectiveness of tubelet action detectors. Following [8,14,33], we utilize frame mAP and video mAP to evaluate detection accuracy.

Implementation Details. We choose the DLA34 [37] as our backbone with COCO [18] pretrain and ImageNet [3] pretrain. We provide MOC results with COCO pretrain without extra explanation. For a fair comparison, we provide two-stream results on two datasets with both COCO pretrain and ImageNet pretrain in Sect. 4.3. The frame is resized to 288×288. The spatial downsample ratio R is set to 4 and the resulted feature map size is 72×72. During training, we use the same data augmentation as [14] to the whole video: photometric transformation, scale jittering, and location jittering. We use Adam with a learning rate 5e-4 to optimize the overall objective. The learning rate adjusts to convergence on the validation set and it decreases by a factor of 10 when performance saturates. The iteration maximum is set to 12 epochs on UCF101-24 [28] and 20 epochs on JHMDB [13].

4.2 Ablation Studies

For efficient exploration, we perform experiments only using RGB input modality, COCO pretrain, and K as 5 without extra explanation. Without special specified, we use exactly the same training strategy in this subsection.

Effectiveness of Movement Branch. In MOC, Movement Branch impacts on both bbox's location and size. Movement Branch moves key frame center to other frames to locate bbox center, named as Move Center strategy. Box Branch estimates bbox size on the current frame center, which is located by Movement Branch not the same with key frame, named as Bbox Align strategy. To explore the effectiveness of Movement Branch, we compare MOC with other two detector designs, called as *No Movement* and *Semi Movement*. We set the tubelet length $K = 5$ in all detection designs with the same training strategy. As shown in Fig. 3, **No Movement** directly removes the Movement Branch and just generates the bounding box for each frame at the same location with

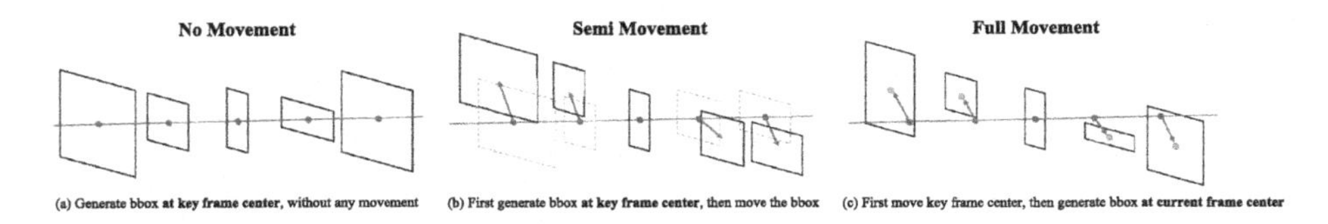

Fig. 3. Illustration of Three Movement Strategies. Note that the arrow represents moving according to Movement Branch prediction, the red dot represents the key frame center and the green dot represents the current frame center, which is localized by moving key frame center according to Movement Branch prediction.

Table 1. Exploration study on MOC detector design with various combinations of movement strategies on UCF101-24.

Method	Strategy		F-mAP@0.5 (%)	Video-mAP (%)			
	Move center	Bbox align		@0.2	@0.5	@0.75	0.5:0.95
No Movement			68.22	68.91	37.77	19.94	19.27
Semi Movement	✓		69.78	76.63	48.82	**27.05**	26.09
Full Movement (MOC)	✓	✓	**71.63**	**77.74**	**49.55**	27.04	**26.09**

key frame center. **Semi Movement** first generates the bounding box for each frame at the same location with key frame center, and then moves the generated box in each frame according to Movement Branch prediction. **Full Movement (MOC)** first moves the key frame center to the current frame center according to Movement Branch prediction, and then Box Branch generates the bounding box for each frame at its own center. The difference between Full Movement and Semi Movement is that they generate the bounding box at different locations: one at the real center, and the other at the fixed key frame center. The results are summarized in Table 1.

First, we observe that the performance gap between No Movement and Semi Movement is 1.56% for frame mAP@0.5 and 11.05% for video mAP@0.5. We find that the Movement Branch has a relatively small influence on frame mAP, but contributes much to improve the video mAP. Frame mAP measures the detection quality in a single frame without tubelet linking while video mAP measures the tube-level detection quality involving tubelet linking. Small movement in short tubelet doesn't harm frame mAP dramatically but accumulating these subtle errors in the linking process will seriously harm the video-level detection. So it demonstrates that the movement information is important for improving video mAP. *Second*, we can see that Full Movement performs slightly better than Semi Movement for both video mAP and frame mAP. Without Bbox Align, Box Branch estimates bbox size at key frame center for all frames, which causes a small performance drop with MOC. This small gap implies that Box Branch is relatively robust to the box center and estimating bbox size at small shifted location only brings a very slight performance difference.

Table 2. Exploration study on the Movement Branch design on UCF101-24 [28]. Note that our MOC-detector adopts the Center Movement.

Method	F-mAP@0.5 (%)	Video-mAP (%)			
		@0.2	@0.5	@0.75	0.5:0.95
Flow guided movement	69.38	75.17	42.28	22.26	21.16
Cost volume movement	69.63	72.56	43.67	21.68	22.46
Accumulated movement	69.40	75.03	46.19	24.67	23.80
Center movement	**71.63**	**77.74**	**49.55**	**27.04**	**26.09**

Table 3. Exploration study on the tubelet duration K on UCF101-24.

Tubelet duration	F-mAP@0.5 (%)	Video-mAP (%)			
		@0.2	@0.5	@0.75	0.5:0.95
$K = 1$	68.33	65.47	31.50	15.12	15.54
$K = 3$	69.94	75.83	45.94	24.94	23.84
$K = 5$	71.63	77.74	49.55	27.04	26.09
$K = 7$	**73.14**	**78.81**	**51.02**	**27.05**	**26.51**
$K = 9$	72.17	77.94	50.16	26.26	26.07

Study on Movement Branch Design. In practice, in order to find an efficient way to capture center movements, we implement Movement Branch in several different ways. The first one is *Flow Guided Movement* strategy which utilizes optical flow between adjacent frames to move action instance center. The second strategy, *Cost Volume Movement*, is to directly compute the movement offset by constructing cost volume between key frame and current frame following [39], but this explicit computing fails to yield better results and is slower due to the constructing of cost volume. The third one is *Accumulated Movement* strategy which predicts center movement between consecutive frames instead of with respect to key frame. The fourth strategy, *Center Movement*, is to employ 3D convolutional operation to directly regress the offsets of the current frame with respect to key frame as illustrated in Sect. 3.2. The results are reported in Table 2.

We notice that the simple Center Movement performs best and choose it as Movement Branch design in our MOC-detector, which directly employs a 3D convolution to regress key frame center movement for all frames as a whole. We will analyze the fail reason for other three designs. For *Flow Guided Movement*, (i) Flow is not accurate and just represents pixel movement, while *Center Movement* is supervised by box movement. (ii) Accumulating adjacent flow to generate trajectory will enlarge error. For the *Cost Volume Movement*, (i) We explicitly calculate the correlation of the current frame with respect to key frame. When regressing the movement of the current frame, it only depends on the current correlation map. However, when directly regressing movement with 3D convolutions, the movement information of each frame will depend on all

Table 4. Comparison with the state of the art on JHMDB (trimmed) and UCF101-24 (untrimmed). Ours (MOC)† is pretrained on ImageNet [3] and **Ours (MOC)** is pretrained on COCO [18].

Method	JHMDB					UCF101-24				
	Frame-mAP@0.5 (%)	Video-mAP (%)				Frame-mAP@0.5 (%)	Video-mAP (%)			
		@0.2	@0.5	@0.75	0.5:0.95		@0.2	@0.5	@0.75	0.5:0.95
2D Backbone										
Saha *et al.* 2016 [25]	–	72.6	71.5	43.3	40.0	–	66.7	35.9	7.9	14.4
Peng *et al.* 2016 [21]	58.5	74.3	73.1	–	–	39.9	42.3	–	–	–
Singh *et al.* 2017 [26]	–	73.8	72.0	44.5	41.6	–	73.5	46.3	15.0	20.4
Kalogeiton *et al.* 2017 [14]	65.7	74.2	73.7	52.1	44.8	69.5	76.5	49.2	19.7	23.4
Yang *et al.* 2019 [35]	–	–	–	–	–	75.0	76.6	–	–	–
Song *et al.* 2019 [27]	65.5	74.1	73.4	52.5	44.8	72.1	77.5	52.9	21.8	24.1
Zhao *et al.* 2019 [38]	–	–	74.7	53.3	45.0	–	78.5	50.3	22.2	24.5
Ours (MOC)†	68.0	76.2	75.4	68.5	54.0	76.9	81.3	**54.4**	29.5	**28.4**
Ours (MOC)	**70.8**	**77.3**	**77.2**	**71.7**	**59.1**	**78.0**	**82.8**	53.8	**29.6**	28.3
3D Backbone										
Hou *et al.* 2017 [11] (C3D)	61.3	**78.4**	76.9	–	–	41.4	47.1	–	–	–
Gu *et al.* 2018 [9] (I3D)	73.3	–	78.6	–	–	**76.3**	–	**59.9**	–	–
Sun *et al.* 2018 [29] (S3D-G)	**77.9**	–	**80.1**	–	–	–	–	–	–	–

frames, which might contribute to more accurate estimation. (ii) As cost volume calculation and offset aggregation involve a correlation without extra parameters, it is observed that the convergence is much harder than *Center Movement*. For *Accumulated Movement*, this strategy also causes the issue of error accumulation and is more sensitive to the training and inference consistency. In this sense, the ground truth movement is calculated at the real bounding box center during training, while for inference, the current frame center is estimated from Movement Branch and might not be so precise, so that *Accumulated Movement* would bring large displacement to the ground truth.

Study on Input Sequence Duration. The temporal length K of the input clip is an important parameter in our MOC-detector. In this study, we report the RGB stream performance of MOC on UCF101-24 [28] by varying K from 1 to 9 and the experiment results are summarized in Table 3. We reduce the training batch size for $K = 7$ and $K = 9$ due to GPU memory limitation.

First, we notice that when $K = 1$, our MOC-detector reduces to the frame-level detector which obtains the worst performance, in particular for video mAP. This confirms the common assumption that frame-level action detector lacks consideration of temporal information for action recognition and thus it is worse than those tubelet detectors, which agrees with our basic motivation of designing an action tubelet detector. *Second*, we see that the detection performance will increase as we vary K from 1 to 7 and the performance gap becomes smaller when comparing $K = 5$ and $K = 7$. From $K = 7$ to $K = 9$, detection performance drops because predicting movement is harder for longer input length. According to the results, we set $K = 7$ in our MOC.

4.3 Comparison with the State of the Art

Finally, we compare our MOC with the existing state-of-the-art methods on the trimmed JHMDB dataset and the untrimmed UCF101-24 dataset in Table 4. For a fair comparison, we also report two-stream results with ImageNet pretrain.

Our MOC gains similar performance on UCF101-24 for ImageNet pretrain and COCO pretrain, while COCO pretrain obviously improves MOC's performance on JHMDB because JHMDB is quite small and sensitive to the pretrain model. Our method significantly outperforms those frame-level action detectors [21,25,26] both for frame-mAP and video-mAP, which perform action detection at each frame independently without capturing temporal information. [14,27,35,38] are all tubelet detectors, our MOC outperforms them for all metrics on both datasets, and the improvement is more evident for high IoU video mAP. This result confirms that our anchor-free MOC detector is more effective for localizing precise tubelets from clips than those anchor-based detectors, which might be ascribed to the flexibility and continuity of MOC detector by directly regressing tubelet shape. Our methods get comparable performance to those 3D backbone based methods [9,11,29]. These methods usually divide action detection into two steps: person detection (ResNet50-based Faster RCNN [23] pretrained on ImageNet), and action classification (I3D [2]/S3D-G [34] pretrained on Kinetics [2] + ROI pooling), and fail to provide a simple unified action detection framework.

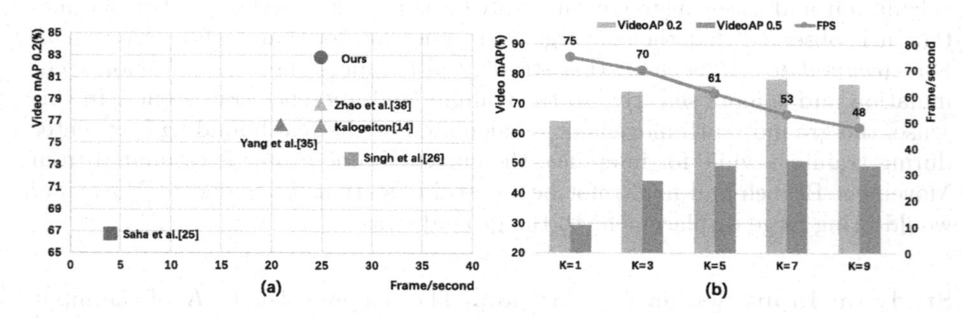

Fig. 4. Runtime Comparison and Analysis. (a) Comparison with other methods. Two-stream results following ACT [14]'s setting. (b) The detection accuracy (green bars) and speeds (red dots) of MOC's online setting.

4.4 Runtime Analysis

Following ACT [14], we evaluate MOC's two-stream offline speed on a single GPU without including flow extraction time and MOC reaches 25 25 fps. In Fig. 4(a), we compare MOC with some existing methods which have reported their speed in the original paper. [14,35,38] are all action tubelet detectors and our MOC gains more accurate detection results with comparable speed. Our MOC can be applied for processing online real-time video stream. To simulate

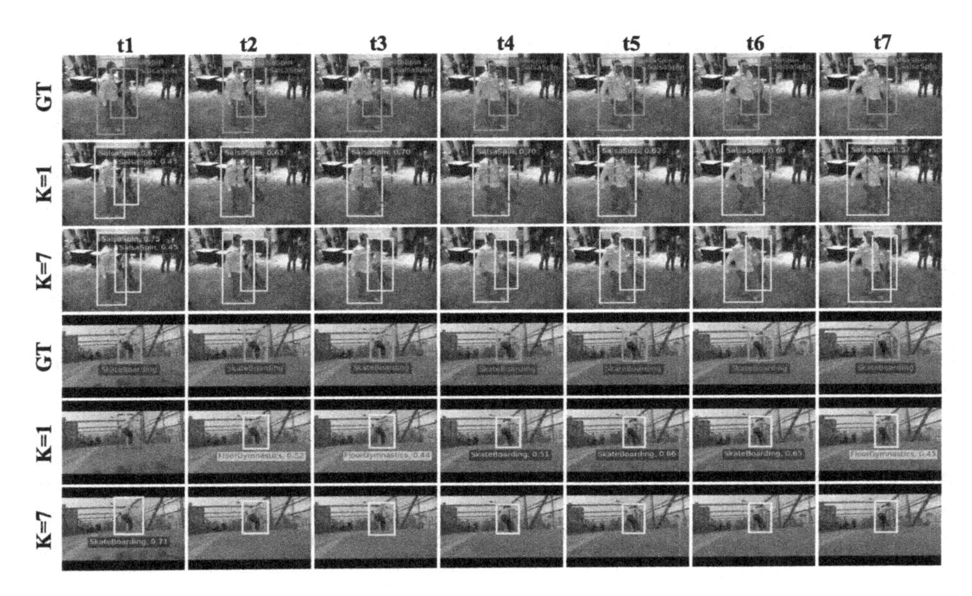

Fig. 5. Examples of Per-frame (K = 1) and Tubelet (K = 7) Detection. The yellow color boxes present detection results, whose categories and scores are provided beside. Yellow categories represent correct and red ones represent wrong. Red dashed boxes represent missed actors. Green boxes and categories are the ground truth. MOC generates one score and category for one tubelet and we mark these in the first frame of the tubelet. Note that we set the visualization threshold as 0.4.

online video stream, we set batch size as 1. Since the backbone feature can be extracted only once, we save previous K–1 frames' features in a buffer. When getting a new frame, MOC's backbone first extracts its feature and combines with the previous K-1 frames' features in the buffer. Then MOC's three branches generate tubelet detections based on these features. After that, update the buffer by adding current frame's feature for subsequent detection. For online testing, we only input RGB as optical flow extraction is quite expensive and the results are reported in Fig. 4(b). We see that our MOC is quite efficient in online testing and it reaches 53 FPS for K = 7.

4.5 Visualization

In Fig. 5, we give some qualitative examples to compare the performance between tubelet duration K = 1 and K = 7. Comparison between the second row and the third row shows that our tubelet detector leads to less missed detection results and localizes action more accurately owing to offset constraint in the same tubelet. What's more, comparison between the fifth and the sixth row presents that our tubelet detector can reduce classification error because some actions can not be discriminated by just looking one frame.

5 Conclusion and Future Work

In this paper, we have presented an action tubelet detector, termed as MOC, by treating each action instance as a trajectory of moving points and directly regressing bounding box size at estimated center points of all frames. As demonstrated on two challenging datasets, the MOC-detector has brought a new state-of-the-art with both metrics of frame mAP and video mAP, while maintaining a reasonable computational cost. The superior performance is largely ascribed to the unique design of three branches and their cooperative modeling ability to perform tubelet detection. In the future, based on the proposed MOC-detector, we try to extend its framework to longer-term modeling and model action boundary in the temporal dimension, thus contributing to spatio-temporal action detection in longer continuous video streams.

Acknowledgements. This work is supported by Tencent AI Lab Rhino-Bird Focused Research Program (No. JR202025), the National Science Foundation of China (No. 61921006), Program for Innovative Talents and Entrepreneur in Jiangsu Province, and Collaborative Innovation Center of Novel Software Technology and Industrialization.

References

1. Cai, Z., Vasconcelos, N.: Cascade r-cnn: delving into high quality object detection. In: Proceedings of the IEEE Conference on Computer Vision and Pattern Recognition, pp. 6154–6162 (2018)
2. Carreira, J., Zisserman, A.: Quo vadis, action recognition? a new model and the kinetics dataset. In: proceedings of the IEEE Conference on Computer Vision and Pattern Recognition, pp. 6299–6308 (2017)
3. Deng, J., Dong, W., Socher, R., Li, L.J., Li, K., Fei-Fei, L.: Imagenet: a large-scale hierarchical image database. In: 2009 IEEE Conference on Computer Vision and Pattern Recognition, pp. 248–255. IEEE (2009)
4. Duan, K., Bai, S., Xie, L., Qi, H., Huang, Q., Tian, Q.: Centernet: keypoint triplets for object detection. In: Proceedings of the IEEE International Conference on Computer Vision, pp. 6569–6578 (2019)
5. Gan, C., Wang, N., Yang, Y., Yeung, D.Y., Hauptmann, A.G.: Devnet: a deep event network for multimedia event detection and evidence recounting. In: Proceedings of the IEEE Conference on Computer Vision and Pattern Recognition, pp. 2568–2577 (2015)
6. Girshick, R.: Fast r-cnn. In: Proceedings of the IEEE International Conference on Computer Vision, pp. 1440–1448 (2015)
7. Girshick, R., Donahue, J., Darrell, T., Malik, J.: Rich feature hierarchies for accurate object detection and semantic segmentation. In: Proceedings of the IEEE Conference on Computer Vision and Pattern Recognition, pp. 580–587 (2014)
8. Gkioxari, G., Malik, J.: Finding action tubes. In: Proceedings of the IEEE Conference on Computer Vision and Pattern Recognition, pp. 759–768 (2015)
9. Gu, C., et al.: Ava: a video dataset of spatio-temporally localized atomic visual actions. In: Proceedings of the IEEE Conference on Computer Vision and Pattern Recognition, pp. 6047–6056 (2018)

10. He, K., Zhang, X., Ren, S., Sun, J.: Spatial pyramid pooling in deep convolutional networks for visual recognition. IEEE Trans. Pattern Anal. Mach. Intell. **37**(9), 1904–1916 (2015)
11. Hou, R., Chen, C., Shah, M.: Tube convolutional neural network (T-CNN) for action detection in videos. In: Proceedings of the IEEE International Conference on Computer Vision, pp. 5822–5831 (2017)
12. Hu, W., Tan, T., Wang, L., Maybank, S.: A survey on visual surveillance of object motion and behaviors. IEEE Trans. Syst. Man Cybern. Part C (Appl. Rev.) **34**(3), 334–352 (2004)
13. Jhuang, H., Gall, J., Zuffi, S., Schmid, C., Black, M.J.: Towards understanding action recognition. In: Proceedings of the IEEE International Conference on Computer Vision, pp. 3192–3199 (2013)
14. Kalogeiton, V., Weinzaepfel, P., Ferrari, V., Schmid, C.: Action tubelet detector for spatio-temporal action localization. In: Proceedings of the IEEE International Conference on Computer Vision, pp. 4405–4413 (2017)
15. Law, H., Deng, J.: Cornernet: detecting objects as paired keypoints. In: Proceedings of the European Conference on Computer Vision (ECCV), pp. 734–750 (2018)
16. Li, D., Qiu, Z., Dai, Q., Yao, T., Mei, T.: Recurrent tubelet proposal and recognition networks for action detection. In: Proceedings of the European conference on computer vision (ECCV), pp. 303–318 (2018)
17. Lin, T.Y., Goyal, P., Girshick, R., He, K., Dollár, P.: Focal loss for dense object detection. In: Proceedings of the IEEE International Conference on Computer Vision, pp. 2980–2988 (2017)
18. Lin, T.Y., et al.: Microsoft COCO: common objects in context. In: Fleet, D., Pajdla, T., Schiele, B., Tuytelaars, T. (eds.) ECCV 2014. LNCS, vol. 8693, pp. 740–755. Springer, Cham (2014). https://doi.org/10.1007/978-3-319-10602-1_48
19. Liu, W., Anguelov, D., Erhan, D., Szegedy, C., Reed, S., Fu, C.-Y., Berg, A.C.: SSD: single shot multiBox detector. In: Leibe, B., Matas, J., Sebe, N., Welling, M. (eds.) ECCV 2016. LNCS, vol. 9905, pp. 21–37. Springer, Cham (2016). https://doi.org/10.1007/978-3-319-46448-0_2
20. Oh, S., et al.: A large-scale benchmark dataset for event recognition in surveillance video. In: CVPR 2011, pp. 3153–3160. IEEE (2011)
21. Peng, X., Schmid, C.: Multi-region two-stream R-CNN for action detection. In: Leibe, B., Matas, J., Sebe, N., Welling, M. (eds.) ECCV 2016. LNCS, vol. 9908, pp. 744–759. Springer, Cham (2016). https://doi.org/10.1007/978-3-319-46493-0_45
22. Redmon, J., Divvala, S., Girshick, R., Farhadi, A.: You only look once: unified, real-time object detection. In: Proceedings of the IEEE Conference on Computer Vision and Pattern Recognition, pp. 779–788 (2016)
23. Ren, S., He, K., Girshick, R., Sun, J.: Faster r-cnn: towards real-time object detection with region proposal networks. In: Advances in Neural Information Processing Systems, pp. 91–99 (2015)
24. Saha, S., Singh, G., Cuzzolin, F.: Amtnet: action-micro-tube regression by end-to-end trainable deep architecture. In: Proceedings of the IEEE International Conference on Computer Vision, pp. 4414–4423 (2017)
25. Saha, S., Singh, G., Sapienza, M., Torr, P.H., Cuzzolin, F.: Deep learning for detecting multiple space-time action tubes in videos (2016). arXiv preprint arXiv:1608.01529
26. Singh, G., Saha, S., Sapienza, M., Torr, P.H., Cuzzolin, F.: Online real-time multiple spatiotemporal action localisation and prediction. In: Proceedings of the IEEE International Conference on Computer Vision, pp. 3637–3646 (2017)

27. Song, L., Zhang, S., Yu, G., Sun, H.: Tacnet: transition-aware context network for spatio-temporal action detection. In: Proceedings of the IEEE Conference on Computer Vision and Pattern Recognition, pp. 11987–11995 (2019)
28. Soomro, K., Zamir, A.R., Shah, M.: Ucf101: a dataset of 101 human actions classes from videos in the wild (2012). arXiv preprint arXiv:1212.0402
29. Sun, C., Shrivastava, A., Vondrick, C., Murphy, K., Sukthankar, R., Schmid, C.: Actor-centric relation network. In: ECCV, pp. 335–351 (2018)
30. Tian, Z., Shen, C., Chen, H., He, T.: Fcos: fully convolutional one-stage object detection. In: The IEEE International Conference on Computer Vision (ICCV) (2019)
31. Venugopalan, S., Rohrbach, M., Donahue, J., Mooney, R., Darrell, T., Saenko, K.: Sequence to sequence-video to text. In: Proceedings of the IEEE International Conference on Computer Vision, pp. 4534–4542 (2015)
32. Wang, L., Qiao, Y., Tang, X., Gool, L.V.: Actionness estimation using hybrid fully convolutional networks. In: CVPR, pp. 2708–2717 (2016)
33. Weinzaepfel, P., Harchaoui, Z., Schmid, C.: Learning to track for spatio-temporal action localization. In: Proceedings of the IEEE International Conference on Computer Vision, pp. 3164–3172 (2015)
34. Xie, S., Sun, C., Huang, J., Tu, Z., Murphy, K.: Rethinking spatiotemporal feature learning: Speed-accuracy trade-offs in video classification. In: Proceedings of the European Conference on Computer Vision (ECCV), pp. 305–321 (2018)
35. Yang, X., Yang, X., Liu, M.Y., Xiao, F., Davis, L.S., Kautz, J.: Step: spatio-temporal progressive learning for video action detection. In: Proceedings of the IEEE Conference on Computer Vision and Pattern Recognition, pp. 264–272 (2019)
36. Yao, L., et al.: Describing videos by exploiting temporal structure. In: Proceedings of the IEEE International Conference on Computer Vision, pp. 4507–4515 (2015)
37. Yu, F., Wang, D., Shelhamer, E., Darrell, T.: Deep layer aggregation. In: Proceedings of the IEEE Conference on Computer Vision and Pattern Recognition, pp. 2403–2412 (2018)
38. Zhao, J., Snoek, C.G.: Dance with flow: two-in-one stream action detection. In: Proceedings of the IEEE Conference on Computer Vision and Pattern Recognition, pp. 9935–9944 (2019)
39. Zhao, Y., Xiong, Y., Lin, D.: Recognize actions by disentangling components of dynamics. In: Proceedings of the IEEE Conference on Computer Vision and Pattern Recognition, pp. 6566–6575 (2018)
40. Zhou, X., Wang, D., Krähenbühl, P.: Objects as points (2019). arXiv preprint arXiv:1904.07850
41. Zhou, X., Zhuo, J., Krahenbuhl, P.: Bottom-up object detection by grouping extreme and center points. In: Proceedings of the IEEE Conference on Computer Vision and Pattern Recognition, pp. 850–859 (2019)

Learning to Exploit Multiple Vision Modalities by Using Grafted Networks

Yuhuang Hu(✉), Tobi Delbruck, and Shih-Chii Liu

Institute of Neuroinformatics, University of Zürich and ETH Zürich, Zürich, Switzerland
{yuhuang.hu,tobi,shih}@ini.uzh.ch

Abstract. Novel vision sensors such as thermal, hyperspectral, polarization, and event cameras provide information that is not available from conventional intensity cameras. An obstacle to using these sensors with current powerful deep neural networks is the lack of large labeled training datasets. This paper proposes a Network Grafting Algorithm (NGA), where a new front end network driven by unconventional visual inputs replaces the front end network of a pretrained deep network that processes intensity frames. The self-supervised training uses only synchronously-recorded intensity frames and novel sensor data to maximize feature similarity between the pretrained network and the grafted network. We show that the enhanced grafted network reaches competitive average precision (AP_{50}) scores to the pretrained network on an object detection task using thermal and event camera datasets, with no increase in inference costs. Particularly, the grafted network driven by thermal frames showed a relative improvement of 49.11% over the use of intensity frames. The grafted front end has only 5–8% of the total parameters and can be trained in a few hours on a single GPU equivalent to 5% of the time that would be needed to train the entire object detector from labeled data. NGA allows new vision sensors to capitalize on previously pretrained powerful deep models, saving on training cost and widening a range of applications for novel sensors.

Keywords: Network Grafting Algorithm · Self-supervised learning · Thermal camera · Event-based vision · Object detection

1 Introduction

Novel vision sensors like thermal, hyperspectral, polarization, and event cameras provide new ways of sensing the visual world and enable new or improved vision system applications. So-called *event cameras*, for example, sense normal visible light, but dramatically sparsify it to pure brightness change events, which provide sub-ms timing and HDR to offer fast vision under challenging illumination

Electronic supplementary material The online version of this chapter (https://doi.org/10.1007/978-3-030-58517-4_6) contains supplementary material, which is available to authorized users.

A. Vedaldi et al. (Eds.): ECCV 2020, LNCS 12361, pp. 85–101, 2020.
https://doi.org/10.1007/978-3-030-58517-4_6

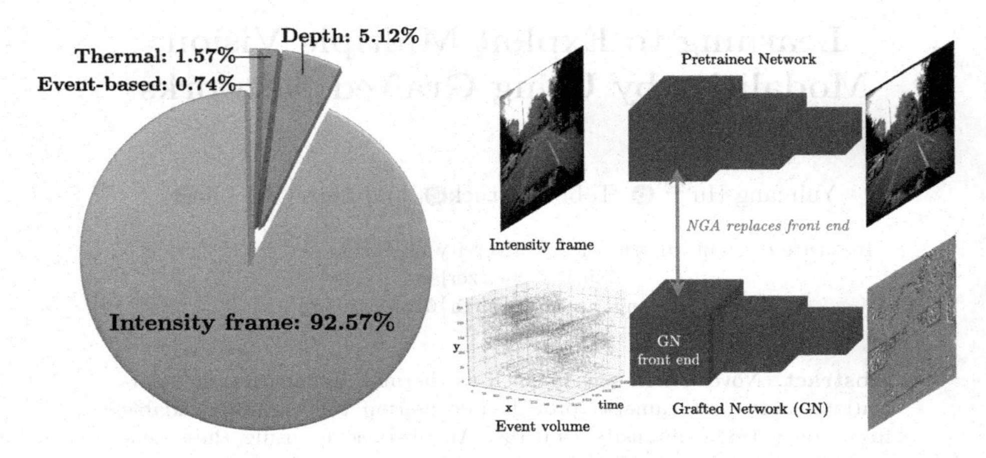

Fig. 1. Types of computer vision datasets. Data from [9].

Fig. 2. A network (blue) trained on intensity frames outputs bounding boxes of detected objects. NGA trains a new GN front end (red) using a small unlabeled dataset of recordings from a DAVIS [4] event camera that concurrently outputs intensity frames and asynchronous brightness change events. The grafted network is obtained by *replacing* the original front end with the GN front end, and is used for inference with the novel camera input data. (Color figure online)

conditions [11,21]. These novel sensors are becoming practical alternatives that complement standard cameras to improve vision systems.

Deep Learning (**DL**) with labeled data has revolutionized vision systems using conventional intensity frame-based cameras. But exploiting DL for vision systems based on novel cameras has been held back by the lack of large labeled datasets for these sensors. Prior work to solve high-level vision problems using inputs other than intensity frames has followed the principles of supervised Deep Neural Network (**DNN**) training algorithms, where the task-specific datasets must be labeled with a tremendous amount of manual effort [2,3,24,31]. Although the community has collected many useful small datasets for novel sensors, the size, variety, and labeling quality of these datasets is far from rivaling intensity frame datasets [2,3,10,15,18,26]. As shown in Fig. 1, among 1,212 surveyed computer vision datasets in [9], 93% are intensity frame datasets. Notably, there are only 28 event-based and thermal datasets.

Particularly for event cameras, another line of DL research employs unsupervised methods to train networks that predict pixel-level quantities such as optical flow [41], depth [40]; and that reconstruct intensity frames [28]. The information generated by these networks can be further processed by a downstream DNN trained to solve tasks such as object classification. This information is exceptionally useful in challenging scenarios such as high-speed motion under difficult lighting conditions. The additional latency introduced by running these networks might be undesirable for fast online applications. For instance, the DNNs used for intensity reconstruction at low QVGA resolution take ~30 ms on a dedicated GPU [28, 33].

This paper introduces a simple yet effective algorithm called the *Network Grafting Algorithm* (**NGA**) to obtain a *Grafted Network* (**GN**) that addresses both issues: 1. the lack of large labeled datasets for training a DNN from scratch, and 2. additional inference cost and latency that comes from running networks that compute pixel-level quantities. With this algorithm, we train a *GN front end* for processing unconventional visual inputs (red block in Fig. 2) to drive a network originally trained on intensity frames. We demonstrate GNs for thermal and event cameras in this paper.

The NGA training encourages the GN front end to produce features that are similar to the features at several early layers of the pretrained network. Since the algorithm only requires pretrained hidden features as the target, the training is self-supervised, that is, no labels are needed from the novel camera data. The training method is described in Sect. 3.1. Furthermore, the newly trained GN has a similar inference cost to the pretrained network and does not introduce additional preprocessing latency. Because the training of a GN front end relies on the pretrained network, the NGA has similarities to Knowledge Distillation (KD) [14], Transfer Learning [27], and Domain Adaptation (DA) [12, 35, 37]. In addition, our proposed algorithm utilizes loss terms proposed for super-resolution image reconstruction and image style transfer [13, 16]. Section 2 elaborates on the similarities and differences between NGA and these related domains.

To evaluate NGA, we start with a pretrained object detection network and obtain a GN for a thermal object detection dataset (Sect. 4.1) to solve the same task. Then, we further demonstrate the training method on car detection using an event camera driving dataset (Sect. 4.2). We show that the GN achieves similar detection precision compared to the original pretrained network. We also evaluate the accuracy gap between supervised and NGA self-supervised with MNIST for event cameras (Sect. 4.3). Finally, we do representation analysis and ablation studies in Sect. 5. Our contributions are as follows:

1. We propose a novel algorithm called NGA that allows the use of networks already trained to solve a high-level vision problem but adapted to work with a new GN front end that processes inputs from thermal/event cameras.
2. The NGA algorithm does not need a labeled thermal/event dataset because the training is self-supervised.
3. The newly trained GN has an inference cost similar to the pretrained network because it directly processes the thermal/event data. Hence, the computation latency brought by *e.g.*, intensity reconstruction from events is eliminated.
4. The algorithm allows the output of these novel cameras to be exploited in situations that are difficult for standard cameras.

2 Related Work

The NGA trains a GN front end such that the hidden features at different layers of the GN are similar to respective pretrained network features on intensity frames. From this aspect, the NGA is similar to Knowledge Distillation [14,32,36] where the knowledge of a teacher network is gradually distilled into a student network (usually smaller than the teacher network) via the soft labels provided by the teacher network. In KD, the teacher and student networks use the same dataset. In contrast, the NGA assumes that the inputs for the pretrained front end and the GN front end come from two *different* modalities that see the same scene concurrently, but this dataset can simply be raw unlabeled recordings. The NGA is also a form of Transfer Learning [27] and Domain Adaptation [12,35,37] that study how to fine-tune the knowledge of a pretrained network on a new dataset. *Our method trains a GN front end from scratch since the network has to process the data from a different sensory modality.*

Another interpretation of maximizing hidden feature similarity can be understood from the algorithms used for super-resolution (SR) image reconstruction and image style transfer. SR image reconstruction requires a network that upsamples a low-resolution image into a high-resolution image. The perceptual loss [16,38] was used to increase the sharpness and maintain the natural image statistics of the reconstruction. Image style transfer networks often aim to transfer an image into a target artistic style where Gram loss [13] is often employed. While these networks learn to match either a high-resolution image ground truth or an artistic style, we train the GN front end to output features that match the hidden features of the pretrained network. For training the front end, we draw inspiration from these studies and propose the use of combinations of training loss metrics including perceptual loss and Gram loss.

3 Methods

We first describe the details of NGA in Sect. 3.1, then the the event camera and its data representation in Sect. 3.2. Finally in Sect. 3.3, we discuss the details of the thermal and event datasets.

3.1 Network Grafting Algorithm

The NGA uses a pretrained network `N` that takes an intensity frame I_t at time t, and produces a grafted network `GN` whose input is a thermal frame or an event volume V_t. I_t and V_t are synchronized during the training. The `GN` should perform with similar accuracy on the same network task, such as object detection. During inference with the thermal or event camera, I_t is not needed. The rest of this section sets up the constructions of `N` and `GN`, then the NGA is described.

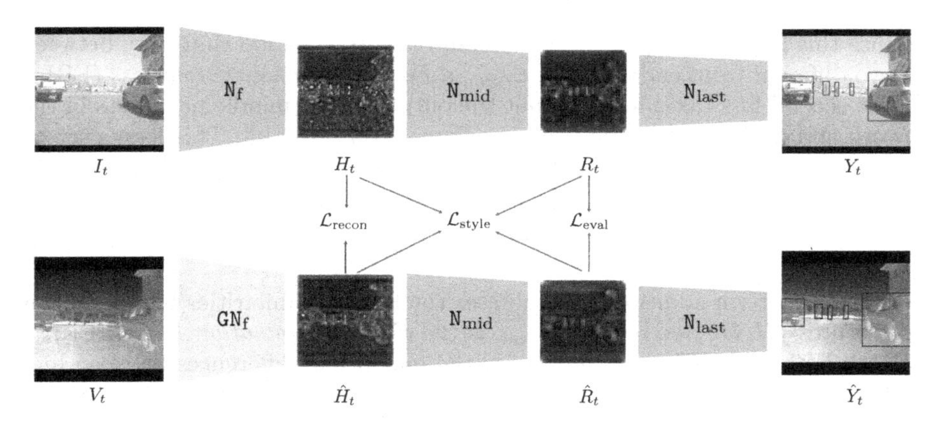

Fig. 3. NGA. (**top**) Pretrained Network. (**bottom**) Grafted Network. Arrows point from variables to relevant loss terms. I_t and V_t here are an intensity frame and a thermal frame, respectively. The intermediate features $\hat{H}_t$, H_t, $\hat{R}_t$, R_t are shown as heat maps averaged across channels. The object bounding boxes predicted by the original and the grafted network are outlined in red and blue correspondingly. (Color figure online)

Pretrained Network Setup. The pretrained network $\texttt{N}$ consists of three blocks: $\{\texttt{N}_\text{f}$ (Front end), $\texttt{N}_\text{mid}$ (Middle net), $\texttt{N}_\text{last}$ (Remaining layers)$\}$. Each block is made up of several layers and the outputs of each of the three blocks are defined as

$$H_t = \texttt{N}_\text{f}(I_t), \qquad R_t = \texttt{N}_\text{mid}(H_t), \qquad Y_t = \texttt{N}_\text{last}(R_t) \tag{1}$$

where H_t is the front end features, R_t is the middle net features, and Y_t is the network prediction. The separation of the network blocks is studied in Sect. 5.2. The top row in Fig. 3 illustrates the three blocks of the pretrained network.

Grafted Network Setup. We define a *GN front end* $\texttt{GN}_\text{f}$ that takes V_t as the input and outputs grafted front end features, $\hat{H}_t$, of the same dimension as H_t. $\texttt{GN}_\text{f}$ combined with $\texttt{N}_\text{mid}$ and $\texttt{N}_\text{last}$ produces the predictions $\hat{Y}$:

$$\hat{H}_t = \texttt{GN}_\text{f}(V_t), \qquad \hat{Y}_t = \texttt{N}_\text{last}(\texttt{N}_\text{mid}(\hat{H}_t)) \tag{2}$$

We define $\texttt{GN} = \{\texttt{GN}_\text{f}, \texttt{N}_\text{mid}, \texttt{N}_\text{last}\}$ as the *Grafted Network* (bottom row of Fig. 3).

Network Grafting Algorithm. The NGA trains the grafted network $\texttt{GN}$ to reach a similar performance to that of the pretrained network $\texttt{N}$ by increasing the representation similarity between features $H = \{H_t|\forall t\}$ and $\hat{H} = \{\hat{H}_t|\forall t\}$.

The loss function for the training of the $\texttt{GN}_\text{f}$ consists of a combination of three losses. The first loss is the Mean-Squared-Error (MSE) between H and $\hat{H}$:

$$\mathcal{L}_\text{recon} = \text{MSE}(H, \hat{H}) \tag{3}$$

Because this loss term captures the amount of representation similarity between the two different front ends, we call $\mathcal{L}_{\text{recon}}$ a *Feature Reconstruction Loss* (FRL).

The second loss takes into account the output of the middle net layers in the network and draws inspiration from the Perception Loss [16]. This loss is set by the MSE between the middle net frame features $R = \{R_t | \forall t\}$ and the grafted middle net features $\hat{R} = \{\mathtt{N}_{\text{mid}}(\hat{H}_t) | \forall t\}$:

$$\mathcal{L}_{\text{eval}} = \text{MSE}(R, \hat{R}) \tag{4}$$

Since this loss term additionally evaluates the feature similarities between front end features $\{H, \hat{H}\}$, we refer to $\mathcal{L}_{\text{eval}}$ as the *Feature Evaluation Loss* (FEL).

Both FRL and FEL terms minimize the magnitude differences between hidden features. To further encourage the GN front end to generate intensity frame-like textures, we introduce the *Feature Style Loss* (FSL) based on the mean-subtracted Gram loss [13] that computes a Gram matrix using feature columns across channels (indexed using i, j). The Gram matrix represents image texture rather than spatial structure. This loss is defined as:

$$\text{Gram}(F)^{(i,j)} = \sum_{\forall t} \tilde{F}_t^{(i)\top} \tilde{F}_t^{(j)}, \quad \text{where } \tilde{F}_t = F_t - \text{mean}(F_t) \tag{5}$$

$$\mathcal{L}_{\text{style}} = \gamma_h \text{MSE}(\text{Gram}(H), \text{Gram}(\hat{H})) + \gamma_r \text{MSE}(\text{Gram}(R), \text{Gram}(\hat{R})) \tag{6}$$

The final loss function is a weighted sum of the three loss terms:

$$\mathcal{L}_{\text{tot}} = \alpha \mathcal{L}_{\text{recon}} + \beta \mathcal{L}_{\text{eval}} + \mathcal{L}_{\text{style}} \tag{7}$$

For all experiments in the paper, we set $\alpha = \beta = 1$, $\gamma_h \in \{10^5, 10^6, 10^7\}$, $\gamma_r = 10^7$. The loss terms and their associated variables are shown in Fig. 3. The importance of each loss term is studied in Sect. 5.3.

3.2 Event Camera and Feature Volume Representation

Event cameras such as the DAVIS camera [4,21] produce a stream of asynchronous "events" triggered by local brightness (log intensity) changes at individual pixels. Each output event of the event camera is a four-element tuple $\{t, x, y, p\}$ where t is the timestamp, (x, y) is the location of the event, and p is the event polarity. The polarity is either positive (brightness increasing) or negative (brightness decreasing). To preserve both spatial and temporal information captured by the polarity events, we use the event voxel grid [28,41]. Assuming a volume of N events $\{(t_i, x_i, y_i, p_i)\}_{i=1}^N$ where i is the event index, we divide this volume into D event slices of equal temporal intervals such that the d-th slice S_d is defined as follows:

$$\forall x, y; \quad S_d(x, y) = \sum_{x_i = x, y_i = y} p_i \max(0, 1 - |d - \tilde{t}_i|) \tag{8}$$

and $\tilde{t}_i = (D-1)\frac{t_i - t_1}{t_N - t_1}$ is the normalized event timestamp. The event volume is then defined as $V_t = \{S_d\}_{d=0}^{D-1}$. In Sect. 4, $D = 3, 10$ and $N = 25,000$. Prior work has shown that this spatio-temporal view of the input scene activity covering a constant number of brightness change events is simple but effective for optical flow computation [41] and video reconstruction [28].

3.3 Datasets

Two different vision datasets were used in the experiments in this paper and are presented in the subsections.

Thermal Dataset for Object Detection. The FLIR Thermal Dataset [10] includes labeled recordings from a thermal camera for driving on Santa Barbara, CA area streets and highways for both day and night. The thermal frames were captured using a FLIR IR Tau2 thermal camera with an image resolution of 640×512. The dataset has parallel RGB intensity frames and thermal frames in an 8-bit JPEG format with AGC. Since the standard camera is placed alongside the thermal camera, a constant spatial displacement is expected, and this shift is corrected for the training samples. The dataset has 4,855 training intensity-thermal pairs, and 1,256 testing pairs, of which 60% are daytime and 40% are nighttime driving samples. We excluded samples where the intensity frames are corrupted. The annotated object classes are `car`, `person`, and `bicycle`.

Event Camera Dataset. The Multi Vehicle Stereo Event Camera Dataset (MVSEC) [39] is a collection of event camera recordings for studying 3D perception and optical flow estimation. The `outdoor_day2` recording is carried out in an urban area of West Philadelphia. This recording was selected for the car detection experiment because of its better quality compared to other recordings, and it has a large number of cars in the scenes distributed throughout the entire recording. We generated in total 7,000 intensity frames and event volume pairs from this recording. Each event volume contains $N = 25,000$ events. The first 5,000 pairs are used as the training dataset, and the last 2,000 pairs are used as the testing dataset. There are no temporally overlapping pairs between the training and testing datasets.

Because MVSEC does not provide ground truth bounding boxes for cars, we pseudo-labeled data pairs of the testing dataset for intensity frames that contain at least one car detected by the Hybrid Task Cascade (HTC) Network [6], which provides state-of-the-art results in object detection. We only use the bounding boxes with 80% or higher confidence to obtain high-quality bounding boxes. To compare the effect of using different numbers of event slices D in an event volume on the detection results, we additionally created two versions of this dataset: DVS-3 where $D = 3$ and DVS-10 where $D = 10$.

4 Experiments

We use the NGA to train a GN front end for a pretrained object detection network. In this case, we use the YOLOv3 network [29] that was trained using the COCO dataset [22] with 80 objects. This network was chosen because it still provides good detection accuracy and could be deployed on a low-cost embedded real-time platform. The pretrained network is referred to as YOLOv3-N and the grafted thermal/event-driven networks as YOLOv3-GN in the rest of the paper. The training inputs consist of 224×224 image patches randomly cropped from the training pairs. No other data augmentation is performed. All networks are trained for 100 epochs with the Adam optimizer [17], a learning rate of 10^{-4}, and a mini-batch size of 8. Each model training takes ~1.5 h using an NVIDIA RTX 2080 Ti, which is only 5% of the 2 days it typically requires to train one of the object detectors used in this paper on standard labeled datasets. More results from the experiments on the different vision datasets are presented in the supplementary material.

4.1 Object Detection on Thermal Driving Dataset

This section presents the experimental results of using the NGA to train an object detector for the thermal driving dataset.

Fig. 4. Examples of six testing pairs from the thermal driving dataset. The red boxes are objects detected by the original intensity-driven YOLOv3 network and the blue boxes show the objects detected by the thermal-driven network. The magenta box shows cars detected by the thermal-driven GN that are missed by the intensity-driven network when the intensity frame is underexposed. Best viewed in color. (Color figure online)

Figure 4 shows six examples of object detection results from the original intensity-driven YOLOv3 network and the thermal-driven network. These examples show that when the intensity frame is well-exposed, the prediction difference between YOLOv3-N and YOLOv3-GN appears to be small. However, when the intensity frame is either underexposed or noisy, the thermal-driven network detects many more objects than the pretrained network. For instance, in the magenta box of Fig. 4, most cars are not detected by the intensity-driven network but they are detected by the thermal-driven network.

The detection precision (AP_{50}) results over the entire test set (Table 1) show that the accuracy of our pretrained YOLOv3-N on the intensity frames (30.36) is worse than on thermal frames (39.92) because 40% of the intensity night frames look noisy and are underexposed. The YOLOv3-GN thermal-driven network achieved the highest AP_{50} detection precision (45.27) among all our YOLOv3 variants while requiring training of only 5.17% (3.2M) parameters with NGA. A baseline Faster R-CNN which was trained on the same labeled thermal dataset [10] achieved a higher precision of 53.97. However, it required training of 47M parameters which is 15X more than the YOLOv3-GN. Overall, the results show that the self-supervised GN front end significantly improves the accuracy results of the original network on the thermal dataset.

For comparison with other object detectors, we also use the `mmdetection` framework [7] to process the intensity frames using pretrained SSD [23], Faster R-CNN [30] and Cascade R-CNN [5] detectors. All have worse AP_{50} scores than any of the YOLOv3 networks, so YOLOv3 was a good choice for evaluating the effectiveness of NGA.

Table 1. Object detection AP_{50} scores on the FLIR driving dataset. The training of YOLOv3-GN repeats five times.

Network	Modality	AP_{50}	# Trained params
This work			
YOLOv3-GN	Thermal	**45.27 ± 1.14**	**3.2M**
YOLOv3-N	Intensity	30.36	62M
YOLOv3-N	Thermal	39.92	62M
SSD	Intensity	8.00	36M
Faster R-CNN	Intensity	23.82	42M
Cascade R-CNN	Intensity	27.10	127M
Baseline supervised thermal object detector			
Faster R-CNN [8]	Thermal	53.97	47M

4.2 Car Detection on Event Camera Driving Dataset

To study if the NGA is also effective for exploiting another visual sensor, *e.g.*, an event camera, we evaluated car detection results using the pretrained network YOLOv3-N and a grafted network YOLOv3-GN using the MVSEC dataset.

The event camera operates over a larger dynamic range of lighting than an intensity frame camera and therefore will detect moving objects even in poorly lighted scenes. From the six different data pairs in the MVSEC testing dataset (Fig. 5), we see that the event-driven YOLOv3-GN network detects most of the cars found in the intensity frames and additional cars not detected in the intensity frame (see the magenta box in the figure). These examples help illustrate how event cameras and the event-driven network can complement the pretrained network in challenging situations.

Fig. 5. Examples of testing pairs from the MVSEC dataset. The event volume is visualized after averaging across slices. The predicted bounding boxes (in red) from the intensity-driven network can be compared with the predicted bounding boxes (in blue) from the event-driven network. The magenta box shows cars detected by the event-driven network that are missed by the intensity-driven network. Best viewed in color. (Color figure online)

Table 2 compares the accuracy of the intensity and event camera detection networks on the testing set. As might be expected for these well-exposed and sharp daytime intensity frames, the YOLOv3-`N` produces the highest average precision (AP). Surprisingly, the YOLOv3-`GN` with DVS-3 input achieves close to the same accuracy, although it was never explicitly trained to detect objects on this type of data. We also tested if the pretrained network would perform poorly on the DVS-3 event dataset. The AP_{50} is almost 0 (not reported in the table) and confirms that the intensity-driven front end fails at processing the event volume and that using a GN front end is essential for acceptable accuracy.

We also compare the performances of the event-driven networks that receive as input, the two datasets with different numbers of event slices for the event volume, *i.e.*, DVS-3, and DVS-10. The network trained on DVS-10 shows a better score of $AP_{50} = 70.35$, which is only 3.18 lower than the original YOLOv3 network accuracy. Table 2 also shows the effect on accuracy when varying the number of training samples. Even when trained using only 40% of training data (2k samples), the YOLOv3-`GN` still shows strong detection precision at 66.75. But when the NGA has access to only 10% of the data (500 samples) during training, the detection performance drops by 22.47% compared to the best event-driven network. Although the NGA requires far less data compared to standard supervised training, training with only a few hundreds of samples remains challenging and could benefit from data augmentation to improve performance.

To study the benefit of using the event camera brightness change events to complement its intensity frame output, we combined the detection results from both the pretrained network and event-driven network (Row **Combined** in Table 2). After removing duplicated bounding boxes through non-maximum suppression, the AP_{50} score of the combined prediction is higher by 1.92 than the prediction of the pretrained network using intensity frames.

Reference AP_{50} scores from three additional intensity frame detectors implemented using the `mmdetection` toolbox are also reported in the table for comparison.

Table 2. AP_{50} scores for car detection on the MVSEC driving dataset (five runs).

Network	Modality	AP_{50}	# Trained params
YOLOv3-N	Intensity	73.53	62M
YOLOv3-GN	DVS-3	70.14 ± 0.36	3.2M
YOLOv3-GN	DVS-10	70.35 ± 0.51	3.2M
YOLOv3-GN	DVS-10 (40% samples)	66.75 ± 0.30	3.2M
YOLOv3-GN	DVS-10 (10% samples)	47.88 ± 1.86	3.2M
Combined	Intensity + DVS-10	**75.45**	N/A
SSD	Intensity	36.17	36M
Faster R-CNN	Intensity	71.89	42M
Cascade R-CNN	Intensity	85.16	127M

4.3 Comparing NGA and Standard Supervised Learning

Intuitively, a network trained in a supervised manner should perform better than a network trained through self-supervision. To study this, we evaluate the accuracy gap between classification networks trained with supervised learning, and the NGA using event recordings of the MNIST handwritten digit recognition dataset, also called N-MNIST dataset [26]. Each event volume is prepared by setting $D = 3$. The training uses the Adam optimizer, a learning rate of 10^{-3} and a batch size of 256.

First, we train the LeNet-N network with the standard LeNet-5 architecture [20] using the intensity samples in the MNIST dataset. Next, we train LeNet-GN with the NGA by using parallel MNIST and N-MNIST sample pairs. We also train an event-driven LeNet-supervised network from scratch on N-MNIST using standard supervised learning with the labeled digits. The results in Table 3 show that the accuracy of the LeNet-GN network is only 0.36% lower than that of the event-driven LeNet-supervised network even with the training of a front end which has only 8% of the total network parameters, and without the availability of labeled training data. The LeNet-GN also performed better or on par with other models that have been tested on the N-MNIST dataset [19,25,34].

Table 3. Classification results on MNIST and N-MNIST datasets.

Network	Dataset	Error Rate (%)	# Trained params
LeNet-N	MNIST	0.92	64k
LeNet-GN	N-MNIST	1.47 ± 0.05	5k
LeNet-supervised	N-MNIST	1.11 ± 0.06	64k
HFirst [25]	N-MNIST	28.80	–
HOTS [19]	N-MNIST	19.20	–
HATS [34]	N-MNIST	0.90	–

5 Network Analysis

To understand the representational power of the GN features, Sect. 5.1 presents a qualitative study that shows how the grafted front end features represent useful visual input under difficult lighting conditions. To design an effective GN, it is important to select what parts of the network to graft. Sections 5.2 and 5.3 describe studies on the network variants and the importance of the loss terms.

5.1 Decoding Grafted Front End Features

Previous experiments show that the grafted front end features provide useful information for the GN in the object detection tasks. In this section, we provide qualitative evidence that the grafted features often faithfully represent the input scene. Specifically, we decode the grafted features by optimizing a decoded intensity frame I_t^{d} that produces features through the intensity-driven network best matching the grafted features $\hat{H}_t$, by minimizing:

$$\arg\min_{I_t^{\mathrm{d}}} \mathrm{MSE}(\mathtt{N_f}(I_t^{\mathrm{d}}), \hat{H}_t) + 5 \times \mathrm{TV}(I_t^{\mathrm{d}}) \tag{9}$$

Fig. 6. Decoded frames of image pairs taken from both the thermal and event datasets. Each column represents an example image from either the thermal dataset (the leftmost two columns) or the event dataset (the rightmost two columns). The top panel of each column shows either the thermal frame or the event volume. The middle panel shows the raw intensity frames. The bottom panel shows the decoded intensity frames (see main text). Labeled regions in the decoded frames show details that are not visible in the four original intensity frames. The figure is best viewed in color. (Color figure online)

where TV(·) is a total variation regularizer for encouraging spatial smoothness [1]. The decoded intensity frame I_t^{d} is initialized randomly and has the same spatial dimension as the intensity frame, then the pixel values of I_t^{d} are optimized for 1k iterations using an Adam optimizer with learning rate of 10^{-2}.

Figure 6 shows four examples from the thermal dataset and the event dataset. Under extreme lighting conditions, the intensity frames are often under/over-exposed while the decoded intensity frames show that the thermal/event front end features can represent the same scene better (see the labeled regions).

5.2 Design of Grafted Network

The backbone network of YOLOv3 is called Darknet-53, and consists of five residual blocks (Fig. 7). Selecting the correct set of residual blocks used for the NGA front end is important. Six combinations of the front end and middle net by using different numbers of residual blocks: {S1, S4}, {S1, S5}, {S2, S4}, {S2, S5}, {S3, S4} and {S3, S5} are tested. S1, S2, S3 indicate front end variants with different number of residual blocks that uses 0.06% (40k), 0.45% (279k), and 5.17% (3.2M) of total parameters (62M) respectively. The number of blocks for S4 and S5 vary depending on the chosen variant. Figure 8 shows the AP_{50} scores for different combinations of front end and middle net variants. The best separation of the network blocks is {S3, S4}. In the YOLOv3 network, the detection results improve sharply when the front end includes more layers. On the other hand, the difference in AP_{50} between using S4 or S5 for the middle net is not significant. These results suggest that using a deeper front end is better than a shallow front end, especially when training resources are not a constraint.

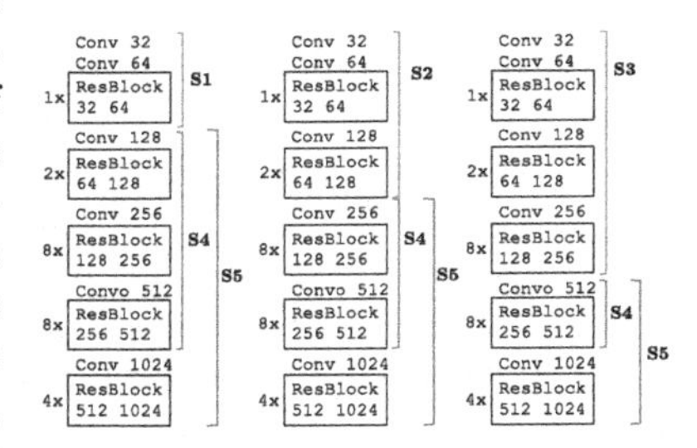

Fig. 7. YOLOv3 backbone: Darknet-53 [29]. The front end variants are S1, S2 and S3. The middle net variants are S4 and S5. `Conv` represents a convolution layer, `ResBlock` represents a residual block.

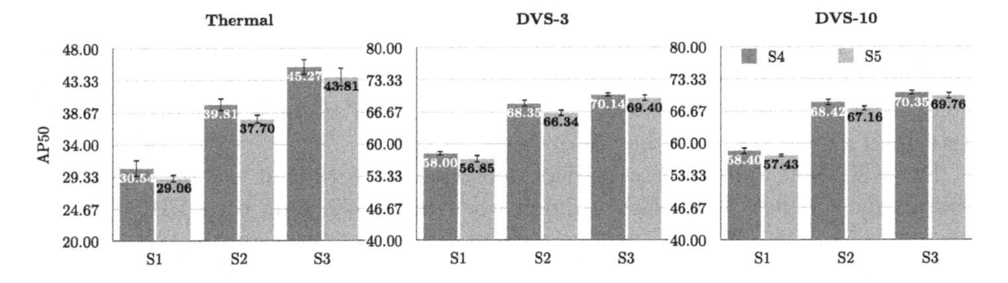

Fig. 8. Results of different front end and middle net variants in Fig. 7 for both thermal and event datasets in AP_{50}. Experiments for each variant are repeated five times.

5.3 Ablation Study on Loss Terms

The NGA training includes three loss terms: FRL, FEL, and FSL. We studied the importance of these loss terms by performing an ablation study using both the thermal dataset and the event dataset. These experiments are done on the network configuration {S3, S4} that gave the best accuracy (see Fig. 8). The detection precision scores are shown in Fig. 9 for different loss configurations. The FRL and the FEL are the most critical loss terms, while the role of the FSL is less significant. The effectiveness of different loss combinations seems task-dependent and sometimes fluctuates, *e.g.*, FRL+FEL for thermal and FEL+FSL for DVS-10. The trend lines indicate that using a combination of loss terms is most likely to produce better detection scores.

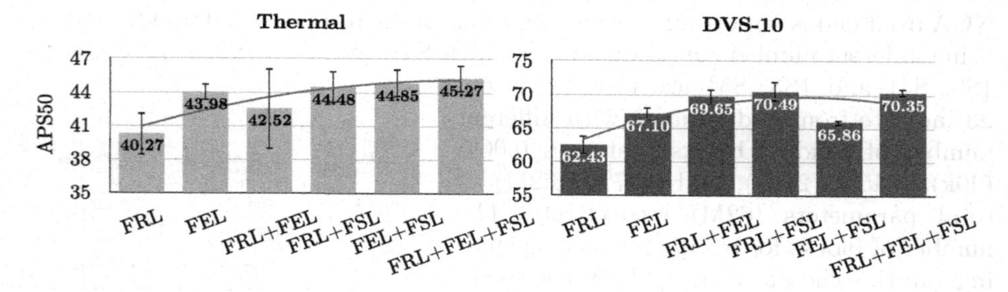

Fig. 9. GN performance (AP_{50}) trained with different loss configurations. Results are from five repeats of each loss configuration.

6 Conclusion

This paper proposes the Network Grafting Algorithm (NGA) that replaces the front end of a network that is pretrained on a large labelled dataset so that the new grafted network (GN) also works well with a different sensor modality. Training the GN front end for a different modality, in this case, a thermal camera or an event camera, requires only a reasonably small unlabeled dataset ($\sim$5k samples) that has spatio-temporally synchronized data from both modalities. By comparison, the COCO dataset on which many object detection networks are trained has 330k images. Ordinarily, training a network with a new sensor type and limited labeled data requires a lot of careful data augmentation. NGA avoids this by exploiting the new sensor data even if unlabeled because the pretrained network already has informative features.

The NGA was applied on an object detection network that was pretrained on a big image dataset. The NGA training was conducted using the FLIR thermal dataset [10] and the MVSEC driving dataset [39]. After training, the GN reached a similar or higher average precision (AP_{50}) score compared to the precision achieved by the original network. Furthermore, the inference cost of the GN is similar to that of the pretrained network, which eliminates the latency cost for computing low-level quantities, particularly for event cameras. This newly proposed NGA widens the use of these unconventional cameras to a broader range of computer vision applications.

Acknowledgements. This work was funded by the Swiss National Competence Center in Robotics (NCCR Robotics).

References

1. Aly, H.A., Dubois, E.: Image up-sampling using total-variation regularization with a new observation model. IEEE Trans. Image Process. **14**(10), 1647–1659 (2005)
2. Anumula, J., Neil, D., Delbruck, T., Liu, S.C.: Feature representations for neuromorphic audio spike streams. Front. Neurosci. **12**, 23 (2018). https://doi.org/10.3389/fnins.2018.00023
3. Bahnsen, C.H., Moeslund, T.B.: Rain removal in traffic surveillance: does it matter? IEEE Trans. Intell. Transp. Syst. **20**(8), 1–18 (2018)
4. Brandli, C., Berner, R., Yang, M., Liu, S.C., Delbruck, T.: A 240 × 180 130 dB 3 μs latency global shutter spatiotemporal vision sensor. IEEE J. Solid-State Circ. **49**(10), 2333–2341 (2014)
5. Cai, Z., Vasconcelos, N.: Cascade R-CNN: delving into high quality object detection. In: The IEEE Conference on Computer Vision and Pattern Recognition (CVPR) (2018)
6. Chen, K., et al.: Hybrid task cascade for instance segmentation. In: 2019 IEEE Conference on Computer Vision and Pattern Recognition (CVPR) (2019)
7. Chen, K., et al.: MMDetection: open MMLab detection toolbox and benchmark (2019). CoRR abs/1906.07155
8. Devaguptapu, C., Akolekar, N., Sharma, M.M., Balasubramanian, V.N.: Borrow from anywhere: pseudo multi-modal object detection in thermal imagery. In: The IEEE Conference on Computer Vision and Pattern Recognition (CVPR) Workshops (2019)
9. Fisher, R.: CVonline: Image Databases (2020). http://homepages.inf.ed.ac.uk/rbf/CVonline/Imagedbase.htm
10. FLIR: Free FLIR thermal dataset for algorithm training (2018). https://www.flir.com/oem/adas/adas-dataset-form/
11. Gallego, G., et al.: Event-based vision: a survey (2019). CoRR abs/1904.08405
12. Ganin, Y., Lempitsky, V.: Unsupervised domain adaptation by backpropagation. In: Bach, F., Blei, D. (eds.) Proceedings of the 32nd International Conference on Machine Learning. Proceedings of Machine Learning Research, 07–09 July 2015, vol. 37, pp. 1180–1189. PMLR, Lille (2015)
13. Gatys, L.A., Ecker, A.S., Bethge, M.: Image style transfer using convolutional neural networks. In: 2016 IEEE Conference on Computer Vision and Pattern Recognition (CVPR), pp. 2414–2423 (2016)
14. Hinton, G., Vinyals, O., Dean, J.: Distilling the knowledge in a neural network. In: NIPS Deep Learning and Representation Learning Workshop (2015)
15. Hu, Y., Liu, H., Pfeiffer, M., Delbruck, T.: DVS benchmark datasets for object tracking, action recognition, and object recognition. Front. Neurosci. **10**, 405 (2016)
16. Johnson, J., Alahi, A., Fei-Fei, L.: Perceptual losses for real-time style transfer and super-resolution. In: Leibe, B., Matas, J., Sebe, N., Welling, M. (eds.) ECCV 2016. LNCS, vol. 9906, pp. 694–711. Springer, Cham (2016). https://doi.org/10.1007/978-3-319-46475-6_43

17. Kingma, D.P., Ba, J.: Adam: a method for stochastic optimization. In: Proceedings of the 3rd International Conference on Learning Representations (ICLR) (2014)
18. Krišto, M., Ivašić-Kos, M.: Thermal imaging dataset for person detection. In: 2019 42nd International Convention on Information and Communication Technology, Electronics and Microelectronics (MIPRO), pp. 1126–1131 (2019)
19. Lagorce, X., Orchard, G., Galluppi, F., Shi, B.E., Benosman, R.B.: Hots: s hierarchy of event-based time-surfaces for pattern recognition. IEEE Trans. Pattern Anal. Mach. Intell. **39**(7), 1346–1359 (2017)
20. Lecun, Y., Bottou, L., Bengio, Y., Haffner, P.: Gradient-based learning applied to document recognition. Proc. IEEE **86**(11), 2278–2324 (1998)
21. Lichtsteiner, P., Posch, C., Delbruck, T.: A 128 $\times$ 128 120 dB 15 μs latency asynchronous temporal contrast vision sensor. IEEE J. Solid-State Circ. **43**(2), 566–576 (2008)
22. Lin, T.Y., et al.: Microsoft COCO: common objects in context. In: Fleet, D., Pajdla, T., Schiele, B., Tuytelaars, T. (eds.) ECCV 2014. LNCS, vol. 8693, pp. 740–755. Springer, Cham (2014). https://doi.org/10.1007/978-3-319-10602-1_48
23. Liu, W., et al.: SSD: single shot multibox detector. In: Leibe, B., Matas, J., Sebe, N., Welling, M. (eds.) ECCV 2016. LNCS, vol. 9905, pp. 21–37. Springer, Cham (2016). https://doi.org/10.1007/978-3-319-46448-0_2
24. Moeys, D.P., et al.: Steering a predator robot using a mixed frame/event-driven convolutional neural network. In: 2016 Second International Conference on Event-based Control, Communication, and Signal Processing (EBCCSP), pp. 1–8 (2016)
25. Orchard, G., Meyer, C., Etienne-Cummings, R., Posch, C., Thakor, N., Benosman, R.: Hfirst: a temporal approach to object recognition. IEEE Trans. Pattern Anal. Mach. Intell. **37**(10), 2028–2040 (2015)
26. Orchard, G., Jayawant, A., Cohen, G.K., Thakor, N.: Converting static image datasets to spiking neuromorphic datasets using saccades. Front. Neurosci. **9**, 437 (2015)
27. Pan, S.J., Yang, Q.: A survey on transfer learning. IEEE Trans. Knowl. Data Eng. **22**(10), 1345–1359 (2010)
28. Rebecq, H., Ranftl, R., Koltun, V., Scaramuzza, D.: Events-To-Video: bringing modern computer vision to event cameras. In: IEEE Conference on Computer Vision and Pattern Recognition (CVPR) (2019)
29. Redmon, J., Farhadi, A.: YOLOv3: An incremental improvement (2018). arXiv
30. Ren, S., He, K., Girshick, R., Sun, J.: Faster R-CNN: towards real-time object detection with region proposal networks. In: Cortes, C., Lawrence, N.D., Lee, D.D., Sugiyama, M., Garnett, R. (eds.) Advances in Neural Information Processing Systems, vol. 28, pp. 91–99. Curran Associates Inc., New York (2015)
31. Rodin, C.D., de Lima, L.N., de Alcantara Andrade, F.A., Haddad, D.B., Johansen, T.A., Storvold, R.: Object classification in thermal images using convolutional neural networks for search and rescue missions with unmanned aerial systems. In: 2018 International Joint Conference on Neural Networks (IJCNN), pp. 1–8 (2018)
32. Romero, A., Ballas, N., Kahou, S.E., Chassang, A., Gatta, C., Bengio, Y.: FitNets: hints for thin deep nets. In: International Conference on Laerning Representations (ICLR) (2015)
33. Scheerlinck, C., Rebecq, H., Gehrig, D., Barnes, N., Mahony, R.E., Scaramuzza, D.: Fast image reconstruction with an event camera. In: 2020 IEEE Winter Conference on Applications of Computer Vision (WACV), pp. 156–163 (2020)
34. Sironi, A., Brambilla, M., Bourdis, N., Lagorce, X., Benosman, R.: Hats: histograms of averaged time surfaces for robust event-based object classification. In: The IEEE Conference on Computer Vision and Pattern Recognition (CVPR) (2018)

35. Sun, Y., Tzeng, E., Darrell, T., Efros, A.A.: Unsupervised domain adaptation through self-supervision (2019)
36. Yim, J., Joo, D., Bae, J., Kim, J.: A gift from knowledge distillation: fast optimization, network minimization and transfer learning. In: IEEE Conference on Computer Vision and Pattern Recognition (CVPR) (2017)
37. You, K., Long, M., Cao, Z., Wang, J., Jordan, M.I.: Universal domain adaptation. In: The IEEE Conference on Computer Vision and Pattern Recognition (CVPR) (2019)
38. Zhang, R., Isola, P., Efros, A.A., Shechtman, E., Wang, O.: The unreasonable effectiveness of deep features as a perceptual metric. In: IEEE Conference on Computer Vision and Pattern Recognition (CVPR) (2018)
39. Zhu, A.Z., Thakur, D., Özaslan, T., Pfrommer, B., Kumar, V., Daniilidis, K.: The multivehicle stereo event camera dataset: an event camera dataset for 3d perception. IEEE Robot. Autom. Lett. **3**(3), 2032–2039 (2018)
40. Zhu, A.Z., Yuan, L., Chaney, K., Daniilidis, K.: Unsupervised event-based learning of optical flow, depth, and egomotion. In: IEEE Conference on Computer Vision and Pattern Recognition (CVPR) (2019)
41. Zhu, A.Z., Yuan, L., Chaney, K., Daniilidis, K.: Unsupervised event-based optical flow using motion compensation. In: Leal-Taixé, L., Roth, S. (eds.) ECCV 2018. LNCS, vol. 11134, pp. 711–714. Springer, Cham (2019). https://doi.org/10.1007/978-3-030-11024-6_54

Geometric Correspondence Fields: Learned Differentiable Rendering for 3D Pose Refinement in the Wild

Alexander Grabner[1,2(✉)], Yaming Wang[2], Peizhao Zhang[2], Peihong Guo[2], Tong Xiao[2], Peter Vajda[2], Peter M. Roth[1], and Vincent Lepetit[1]

[1] Graz University of Technology, Graz, Austria
{alexander.grabner,pmroth,lepetit}@icg.tugraz.at
[2] Facebook Inc., Menlo Park, USA
{wym,stzpz,phg,xiaot,vajdap}@fb.com

Abstract. We present a novel 3D pose refinement approach based on differentiable rendering for objects of arbitrary categories in the wild. In contrast to previous methods, we make two main contributions: First, instead of comparing real-world images and synthetic renderings in the RGB or mask space, we compare them in a feature space optimized for 3D pose refinement. Second, we introduce a novel differentiable renderer that learns to approximate the rasterization backward pass from data instead of relying on a hand-crafted algorithm. For this purpose, we predict deep cross-domain correspondences between RGB images and 3D model renderings in the form of what we call geometric correspondence fields. These correspondence fields serve as pixel-level gradients which are analytically propagated backward through the rendering pipeline to perform a gradient-based optimization directly on the 3D pose. In this way, we precisely align 3D models to objects in RGB images which results in significantly improved 3D pose estimates. We evaluate our approach on the challenging Pix3D dataset and achieve up to 55% relative improvement compared to state-of-the-art refinement methods in multiple metrics.

1 Introduction

Recently, there have been significant advances in single image 3D object pose estimation thanks to deep learning [7,32,42]. However, the accuracy achieved by today's feed-forward networks is not sufficient for many applications like augmented reality or robotics [8,44]. As shown in Fig. 1, feed-forward networks can robustly estimate the coarse high-level 3D rotation and 3D translation of objects in the wild (*left*) but fail to predict fine-grained 3D poses (*right*) [48].

Electronic supplementary material The online version of this chapter (https://doi.org/10.1007/978-3-030-58517-4_7) contains supplementary material, which is available to authorized users.

A. Vedaldi et al. (Eds.): ECCV 2020, LNCS 12361, pp. 102–119, 2020.
https://doi.org/10.1007/978-3-030-58517-4_7

To improve the accuracy of predicted 3D poses, refinement methods aim at aligning 3D models to objects in RGB images. In this context, many methods train a feed-forward network that directly predicts 3D pose updates given the input image and a 3D model rendering under the current 3D pose estimate [23, 38, 52]. In contrast, more recent methods use differentiable rendering [27] to explicitly optimize an objective function conditioned on the input image and renderer inputs like the 3D pose [17, 34]. These methods yield more accurate 3D pose updates because they exploit prior knowledge about the rendering pipeline.

Fig. 1. Given an initial 3D pose predicted by a feed-forward network (*left*), we predict deep cross-domain correspondences between real-world RGB images and synthetic 3D model renderings in the form of geometric correspondence fields (*middle*) that enable us to refine the 3D pose in a differentiable rendering framework (*right*). (Color figure online)

However, existing approaches based on differentiable rendering have significant shortcomings because they rely on comparisons in the RGB or mask space. First, methods which compare real-world images and synthetic renderings in the RGB space require photo-realistic 3D model renderings [27]. Generating such renderings is difficult because objects in the real world are subject to complex scene lighting, unknown reflection properties, and cluttered backgrounds. Moreover, many 3D models only provide geometry but no textures or materials which makes photo-realistic rendering impossible [39]. Second, methods which rely on comparisons in the mask space need to predict accurate masks from real-world RGB images [17, 34]. Generating such masks is difficult even using state-of-the-art approaches like Mask R-CNN [11]. Additionally, masks discard valuable object shape information which makes 2D-3D alignment ambiguous. As a consequence, the methods described above are not robust in practice. Finally, computing gradients for the non-differentiable rasterization operation in rendering is still an open research problem and existing approaches rely on hand-crafted approximations for this task [14, 17, 27].

To overcome these limitations, we compare RGB images and 3D model renderings in a feature space optimized for 3D pose refinement and learn to approximate the rasterization backward pass in differentiable rendering from data. In particular, we introduce a novel network architecture that jointly performs both tasks. Our network maps real-world images and synthetic renderings to a common feature space and predicts deep cross-domain correspondences in the form of *geometric correspondence fields* (see Fig. 1, *middle*). Geometric correspondence fields hold 2D displacement vectors between corresponding 2D object points in RGB images and 3D model renderings similar to optical flow [5]. These predicted 2D displacement vectors serve as pixel-level gradients that enable us to approximate the rasterization backward pass and compute accurate gradients for renderer inputs like the 3D pose, the 3D model, or the camera intrinsics that minimize an ideal geometric reprojection loss.

Our approach has three main advantages: First, we can leverage depth, normal, and object coordinate [2] renderings which provide 3D pose information more explicitly than RGB and mask renderings [23]. Second, we avoid task-irrelevant appearance variations in the RGB space and 3D pose ambiguities in the mask space [9]. Third, we learn to approximate the rasterization backward pass from data instead of relying on a hand-crafted algorithm [14,17,27].

To demonstrate the benefits of our novel 3D pose refinement approach, we evaluate it on the challenging Pix3D [39] dataset. We present quantitative as well as qualitative results and significantly outperform state-of-the-art refinement methods in multiple metrics by up 55% relative. Finally, we combine our refinement approach with feed-forward 3D pose estimation [8] and 3D model retrieval [9] methods to predict fine-grained 3D poses for objects in the wild without providing initial 3D poses or ground truth 3D models at runtime **given only a single RGB image**. To summarize, our main contributions are:

- We introduce the first refinement method based on differentiable rendering that does not compare real-world images and synthetic renderings in the RGB or mask space but in a feature space optimized for the task at hand.
- We present a novel differentiable renderer that learns to approximate the rasterization backward pass instead of relying on a hand-crafted algorithm.

2 Related Work

In the following, we discuss prior work on differentiable rendering, 3D pose estimation, and 3D pose refinement.

2.1 Differentiable Rendering

Differentiable rendering [27] is a powerful concept that provides inverse graphics capabilities by computing gradients for 3D scene parameters from 2D image observations. This novel technique recently gained popularity for 3D reconstruction [20,33], scene lighting estimation [1,22], and texture prediction [16,49].

However, rendering is a non-differentiable process due to the rasterization operation [17]. Thus, differentiable rendering approaches either try to mimic rasterization with differentiable operations [26,34] or use conventional rasterization and approximate its backward pass [6,14,45].

In this work, we also approximate the rasterization backward pass but, in contrast to existing methods, do not rely on hand-crafted approximations. Instead, we train a network that performs the approximation. This idea is not only applicable for 3D pose estimation but also for other tasks like 3D reconstruction, human pose estimation, or the prediction of camera intrinsics in the future.

2.2 3D Pose Estimation

Modern 3D pose estimation approaches build on deep feed-forward networks and can be divided into two groups: Direct and correspondence-based methods.

Direct methods predict 3D pose parameters as raw network outputs. They use classification [41,42], regression [30,46], or hybrid variants of both [28,48] to estimate 3D rotation and 3D translation [21,31,32] in an end-to-end manner. Recent approaches additionally integrate these techniques into detection pipelines to deal with multiple objects in a single image [18,20,44,47].

In contrast, correspondence-based methods predict keypoint locations and recover 3D poses from 2D-3D correspondences using PnP algorithms [36,38] or trained shape models [35]. In this context, different methods predict sparse object-specific keypoints [35–37], sparse virtual control points [7,38,40], or dense unsupervised 2D-3D correspondences [2,3,8,15,43].

In this work, we use the correspondence-based feed-forward approach presented in [8] to predict initial 3D poses for refinement.

2.3 3D Pose Refinement

3D pose refinement methods are based on the assumption that the projection of an object's 3D model aligns with the object's appearance in the image given the correct 3D pose. Thus, they compare renderings under the current 3D pose to the input image to get feedback on the prediction.

A simple approach to refine 3D poses is to generate many small perturbations and evaluate their accuracy using a scoring function [25,50]. However, this is computationally expensive and the design of the scoring function is unclear. Therefore, other approaches try to predict iterative 3D pose updates with deep networks instead [23,29,38,52]. In practice, though, the performance of these methods is limited because they cannot generalize to 3D poses or 3D models that have not been seen during training [38].

Recent approaches based on differentiable rendering overcome these limitations [17,27,34]. Compared to the methods described above, they analytically propagate error signals backward through the rendering pipeline to compute more accurate 3D pose updates. In this way, they exploit knowledge about the 3D scene geometry and the projection pipeline for the optimization.

In contrast to existing differentiable rendering approaches that rely on comparisons in the RGB [27] or mask [17,34] space, we compare RGB images and 3D model renderings in a feature space that is optimized for 3D pose refinement.

3 Learned 3D Pose Refinement

Given a single RGB image, a 3D model, and an initial 3D pose, we compute iterative updates to refine the 3D pose, as shown in Fig. 2. For this purpose, we first introduce the objective function that we optimize at runtime (see Sect. 3.1). We then explain how we compare the input RGB image to renderings under the current 3D pose in a feature space optimized for refinement (see Sect. 3.2), predict pixel-level gradients that minimize an ideal geometric reprojection loss in the form of geometric correspondence fields (see Sect. 3.3), and propagate gradients backward through the rendering pipeline to perform a gradient-based optimization directly on the 3D pose (see Sect. 3.4).

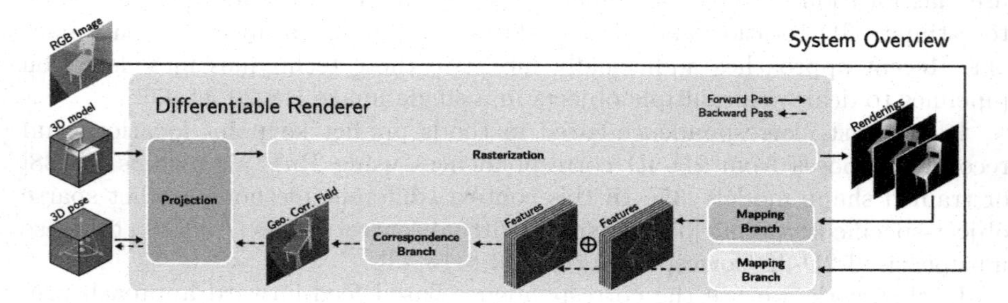

Fig. 2. Overview of our system. In the forward pass (⟶), we generate 3D model renderings under the current 3D pose. In the backward pass (← - -), we map the RGB image and our renderings to a common feature space and predict a geometric correspondence field that enables us to approximate the rasterization backward pass and compute gradients for the 3D pose that minimize an ideal geometric reprojection loss. (Color figure online)

3.1 Runtime Object Function

Our approach to refine the 3D pose of an object is based on the numeric optimization of an objective function at runtime. In particular, we seek to minimize an ideal geometric reprojection loss

$$e(\mathcal{P}) = \frac{1}{2} \sum_i \|\text{proj}(\mathbf{M}_i, \mathcal{P}_{\text{gt}}) - \text{proj}(\mathbf{M}_i, \mathcal{P})\|_2^2 \quad (1)$$

for all provided 3D model vertices $\mathbf{M}_i$. In this case, $\text{proj}(\cdot)$ performs the projection from 3D space to the 2D image plane, $\mathcal{P}$ denotes the 3D pose parameters, and $\mathcal{P}_{\text{gt}}$ is the ground truth 3D pose. Hence, it is clear that $\text{argmin} e(\mathcal{P}) = \mathcal{P}_{\text{gt}}$.

To efficiently minimize $e(\mathcal{P})$ using a gradient-based optimization starting from an initial 3D pose, we compute gradients for the 3D pose using the Jacobian of $e(\mathcal{P})$ with respect to $\mathcal{P}$. Applying the chain rule yields the expression

$$\left(\frac{\partial e(\mathcal{P})}{\partial \mathcal{P}}\right)(\mathcal{P}_{\text{curr}}) = \sum_i \left[\frac{\partial \text{proj}(\mathbf{M}_i, \mathcal{P})}{\partial \mathcal{P}}\right]^T \left[\text{proj}(\mathbf{M}_i, \mathcal{P}_{\text{gt}}) - \text{proj}(\mathbf{M}_i, \mathcal{P}_{\text{curr}})\right], \quad (2)$$

where $\mathcal{P}_{\text{curr}}$ is the current 3D pose estimate and the point where the Jacobian is evaluated. In this case, the term $\left[\frac{\partial \text{proj}(\mathbf{M}_i, \mathcal{P})}{\partial \mathcal{P}}\right]^T$ can be computed analytically because it is simply a sequence of differentiable operations. In contrast, the term $\left[\text{proj}(\mathbf{M}_i, \mathcal{P}_{\text{gt}}) - \text{proj}(\mathbf{M}_i, \mathcal{P}_{\text{curr}})\right]$ cannot be computed analytically because the 3D model vertices projected under the ground truth 3D pose, *i.e.*, $\text{proj}(\mathbf{M}_i, \mathcal{P}_{\text{gt}})$, are unknown at runtime and can only be observed indirectly via the input image.

However, for visible vertices, this term can be calculated given a geometric correspondence field (see Sect. 3.4). Thus, we introduce a novel network architecture that learns to predict geometric correspondence fields given an RGB image and 3D model renderings under the current 3D pose estimate in the following. Moreover, we embed this network in a differentiable rendering framework to approximate the rasterization backward pass and compute gradients for renderer inputs like the 3D pose of an object in an end-to-end manner (see Fig. 2).

3.2 Refinement Feature Space

The first step in our approach is to render the provided 3D model under the current 3D pose using the forward pass of our differentiable renderer (see Fig. 2). In particular, we generate depth, normal, and object coordinate [2] renderings. These representations provide 3D pose and 3D shape information more explicitly than RGB or mask renderings which makes them particularly useful for 3D pose refinement [9]. By concatenating the different renderings along the channel dimension, we leverage complementary information from different representations in the backward pass rather than relying on a single type of rendering [27].

Next, we begin the backward pass of our differentiable renderer by mapping the input RGB image and our multi-representation renderings to a common feature space. For this task, we use two different network branches that bridge the domain gap between the real and rendered images (see Fig. 2). Our mapping branches use a custom architecture based on task-specific design choices:

First, we want to predict local cross-domain correspondences under the assumption that the initial 3D pose is close to the ground truth 3D pose. Thus, we do not require features with global context but features with local discriminability. For this reason, we use small fully convolutional networks which are fast, memory-efficient, and learn low-level features that generalize well across different objects. Because the low-level structures appearing across different objects are similar, we do not require a different network for each object [38] but address objects of all categories with a single class-agnostic network for each domain.

Second, we want to predict correspondences with maximum spatial accuracy. Thus, we do not use pooling or downsampling but maintain the spatial resolution throughout the network. In this configuration, consecutive convolutions provide sufficient receptive field to learn advanced shape features which are superior to simple edge intensities [18], while higher layers benefit from full spatial parameter sharing during training which increases generalization. As a consequence, the effective minibatch size during training is higher than the number of images per minibatch because only a subset of all image pixels contributes to each computed feature. In addition, the resulting high spatial resolution feature space provides an optimal foundation for computing spatially accurate correspondences.

For the implementation of our mapping branches, we use fully convolutional networks consisting of an initial 7×7 Conv-BN-ReLU block, followed by three residual blocks [12,13], and a 1×1 Conv-BN-ReLU block for dimensionality reduction. This architecture enforces local discriminability and high spatial resolution and maps RGB images and multi-representation renderings to $W \times H \times 64$ feature maps, where W and H are the spatial dimensions of the input image.

3.3 Geometric Correspondence Fields

After mapping RGB images and 3D model renderings to a common feature space, we compare their feature maps and predict cross-domain correspondences. For this purpose, we concatenate their feature maps and use another fully convolutional network branch to predict correspondences, as shown in Fig. 2.

In particular, we regress per-pixel correspondence vectors in the form of geometric correspondence fields (see Fig. 1, *middle*). Geometric correspondence fields hold 2D displacement vectors between corresponding 2D object points in real-world RGB images and synthetic 3D model renderings similar to optical flow [5]. These displacement vectors represent the projected relative 2D motion of individual 2D object points that is required to minimize the reprojection error and refine the 3D pose. A geometric correspondence field has the same spatial resolution as the respective input RGB image and two channels, *i.e.*, $W \times H \times 2$.

If an object's 3D model and 3D pose are known, we can render the ground truth geometric correspondence field for an arbitrary 3D pose. For this task, we first compute the 2D displacement $\nabla \mathbf{m}_i = \text{proj}(\mathbf{M}_i, \mathcal{P}_{\text{gt}}) - \text{proj}(\mathbf{M}_i, \mathcal{P}_{\text{curr}})$ between the projection under the ground truth 3D pose $\mathcal{P}_{\text{gt}}$ and the current 3D pose $\mathcal{P}_{\text{curr}}$ for each 3D model vertex $\mathbf{M}_i$. We then generate a ground truth geometric correspondence field $G(\mathcal{P}_{\text{curr}}, \mathcal{P}_{\text{gt}})$ by rendering the 3D model using a shader that interpolates the per-vertex 2D displacements $\nabla \mathbf{m}_i$ across the projected triangle surfaces using barycentric coordinates.

In our scenario, predicting correspondences using a network has two advantages compared to traditional correspondence matching [10]. First, predicting correspondences with convolutional kernels is significantly faster than exhaustive feature matching during both training and testing [51]. This is especially important in the case of dense correspondences. Second, training explicit correspondences can easily result in degenerated feature spaces and requires tedious regularization and hard negative sample mining [4].

For the implementation of our correspondence branch, we use three consecutive 7×7 Conv-BN-ReLU blocks followed by a final 7×7 convolution which reduces the channel dimensionality to two. For this network, a large receptive field is crucial to cover correspondences with high spatial displacement.

However, in many cases local correspondence prediction is ambiguous. For example, many objects are untextured and have homogeneous surfaces, *e.g.*, the backrest and the seating surface of the chair in Fig. 1, which cause unreliable correspondence predictions. To address this problem, we additionally employ a geometric attention module which restricts the correspondence prediction to visible object regions with significant geometric discontinuities, as outlined in white underneath the 2D displacement vectors in Fig. 1. We identify these regions by finding local variations in our renderings.

In particular, we detect rendering-specific intensity changes larger than a certain threshold within a local 5×5 window to construct a geometric attention mask w^{att}. For each pixel of w^{att}, we compute the geometric attention weight

$$w^{att}_{x,y} = \max_{u,v \in W} \left(\delta^R \Big(R(\mathcal{P}_{\text{curr}})_{x,y}, R(\mathcal{P}_{\text{curr}})_{x-u,y-v} \Big) \right) > t^R . \tag{3}$$

In this case, $R(\mathcal{P}_{\text{curr}})$ is a concatenation of depth, normal, and object coordinate renderings under the current 3D pose $\mathcal{P}_{\text{curr}}$, (x, y) is a pixel location, and (u, v) are pixel offsets within the window W. The comparison function $\delta^R(\cdot)$ and the threshold t^R are different for each type of rendering. For depth renderings, we compute the absolute difference between normalized depth values and use a threshold of 0.1. For normal renderings, we compute the angle between normals and use a threshold of 15°. For object coordinate renderings, we compute the Euclidean distance between 3D points and use a threshold of 0.1. If any of these thresholds applies, the corresponding pixel (x, y) in our geometric attention mask w^{att} becomes active. Because we already generated these renderings before, our geometric attention mechanism requires almost no additional computations and is available during both training and testing.

Training. During training of our system, we optimize the learnable part of our differentiable renderer, *i.e.*, a joint network $f(\cdot)$ consisting of our two mapping branches and our correspondence branch with parameters θ (see Fig. 2). Formally, we minimize the error between predicted $f(\cdot)$ and ground truth $G(\cdot)$ geometric correspondence fields as

$$\min_{\theta} \sum_{x,y} w^{att}_{x,y} \| f(I, R(\mathcal{P}_{\text{curr}}); \theta)_{x,y} - G(\mathcal{P}_{\text{curr}}, \mathcal{P}_{\text{gt}})_{x,y} \|_2^2 . \tag{4}$$

In this case, w^{att} is a geometric attention mask, I is an RGB image, $R(\mathcal{P}_{\text{curr}})$ is a concatenation of depth, normal, and object coordinate renderings generated under a random 3D pose $\mathcal{P}_{\text{curr}}$ produced by perturbing the ground truth 3D pose $\mathcal{P}_{\text{gt}}$, $G(\mathcal{P}_{\text{curr}}, \mathcal{P}_{\text{gt}})$ is the ground truth geometric correspondence field, and (x, y) is a pixel location. In particular, we first generate a random 3D pose $\mathcal{P}_{\text{curr}}$ around the ground truth 3D pose $\mathcal{P}_{\text{gt}}$ for each training sample in each iteration. For this purpose, we sample 3D pose perturbations from normal distributions and apply them to $\mathcal{P}_{\text{gt}}$ to generate $\mathcal{P}_{\text{curr}}$. For 3D rotations, we use absolute perturbations with $\sigma = 5°$. For 3D translations, we use relative perturbations with $\sigma = 0.1$. We then render the ground truth geometric correspondence field $G(\mathcal{P}_{\text{curr}}, \mathcal{P}_{\text{gt}})$ between the perturbed 3D pose $\mathcal{P}_{\text{curr}}$ and the ground truth 3D pose $\mathcal{P}_{\text{gt}}$ as described above, generate concatenated depth, normal, and object coordinate renderings $R(\mathcal{P}_{\text{curr}})$ under the perturbed 3D pose $\mathcal{P}_{\text{curr}}$, and compute the geometric attention mask w^{att}. Finally, we predict a geometric correspondence field using our network $f(I, R(\mathcal{P}_{\text{curr}}); \theta)$ given the RGB image I and the renderings $R(\mathcal{P}_{\text{curr}})$, and optimize for the network parameters θ.

In this way, we train a network that performs three tasks: First, it maps RGB images and multi-representation 3D model renderings to a common feature space. Second, it compares features in this space. Third, it predicts geometric correspondence fields which serve as pixel-level gradients that enable us to approximate the rasterization backward pass of our differentiable renderer.

3.4 Learned Differentiable Rendering

In the classic rendering pipeline, the only non-differentiable operation is the rasterization [27] that determines which pixels of a rendering have to be filled, solves the visibility of projected triangles, and fills the pixels using a shading computation. This discrete operation raises one main challenge: Its gradient is zero, which prevents gradient flow [17]. However, we must flow non-zero gradients from pixels to projected 3D model vertices to perform differentiable rendering.

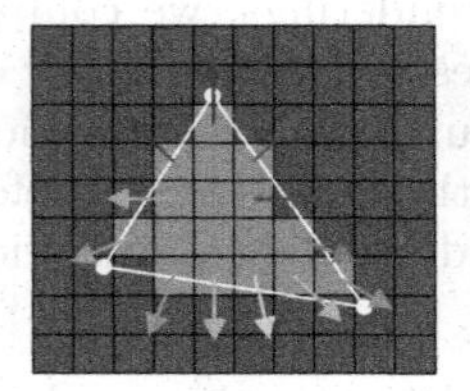
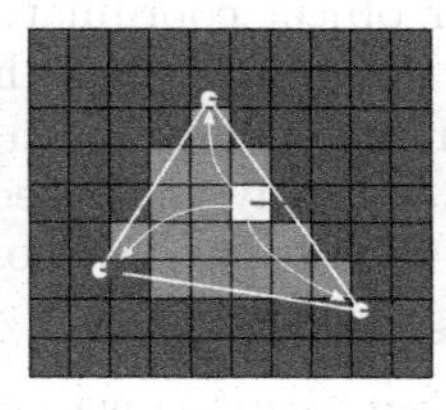
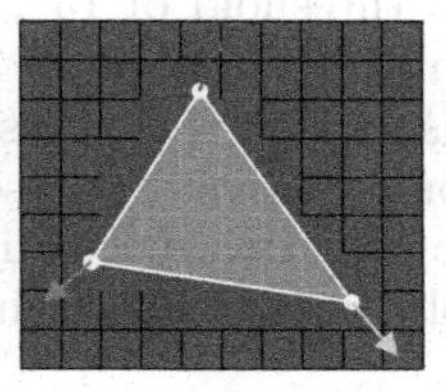

Fig. 3. To approximate the rasterization backward pass, we predict a geometric correspondence field (*left*), disperse the predicted 2D displacement of each pixel among the vertices of its corresponding visible triangle (*middle*), and normalize the contributions of all pixels. In this way, we obtain gradients for projected 3D model vertices (*right*).

We solve this problem using geometric correspondence fields. Instead of actually differentiating a loss in the image space and relying on hand-crafted comparisons between pixel intensities to approximate the gradient flow from pixels to projected 3D model vertices [14,17], we first use a network to predict per-pixel 2D displacement vectors in the form of a geometric correspondence field, as shown in Fig. 2. We then compute gradients for projected 3D model vertices by simply accumulating the predicted 2D displacement vectors using our knowledge of the projected 3D model geometry, as illustrated in Fig. 3.

Formally, we compute the gradient of a projected 3D model vertex $\mathbf{m}_i$ as

$$\nabla \mathbf{m}_i = \frac{1}{\sum_{x,y} w_{x,y}^{att}\, w_{x,y}^{bar,i}} \sum_{x,y} w_{x,y}^{att}\, w_{x,y}^{bar,i}\, f(I, R(\mathcal{P}_{\text{curr}}); \theta)_{x,y} \qquad (5)$$

$$\forall x, y : \mathbf{m}_i \in \triangle_{\texttt{IndexMap}_{x,y}} .$$

In this case, $f(I, R(\mathcal{P}_{\text{curr}}); \theta)$ is a geometric correspondence field predicted by our network $f(\cdot)$ with frozen parameters θ given an RGB image I and concatenated

3D model renderings $R(\mathcal{P}_{\text{curr}})$ under the current 3D pose estimate $\mathcal{P}_{\text{curr}}$, $w_{x,y}^{att}$ is a geometric attention weight, $w_{x,y}^{bar,i}$ is a barycentric weight for $\mathbf{m}_i$, and (x, y) is a pixel position. We accumulate predicted 2D displacement vectors for all positions (x, y) for which $\mathbf{m}_i$ is a vertex of the triangle $\triangle_{\texttt{IndexMap}_{x,y}}$ visible at (x, y). For this task, we generate an `IndexMap` which stores the index of the visible triangle for each pixel during the forward pass of our differentiable renderer.

Inference. Our computed $\nabla \mathbf{m}_i$ approximate the second term in Eq. (2) that cannot be computed analytically. In this way, our approach combines local per-pixel 2D displacement vectors into per-vertex gradients and further computes accurate global 3D pose gradients considering the 3D model geometry and the rendering pipeline. Our experiments show that this approach generalizes better to unseen data than predicting 3D pose updates with a network [23,52].

During inference of our system, we perform iterative updates to refine $\mathcal{P}_{\text{curr}}$. In each iteration, we compute a 3D pose gradient by evaluating our refinement loop presented in Fig. 2. For our implementation, we use the Adam optimizer [19] with a small learning rate and perform multiple updates to account for noisy correspondences and achieve the best accuracy.

4 Experimental Results

To demonstrate the benefits of our 3D pose refinement approach, we evaluate it on the challenging Pix3D [39] dataset which provides in-the-wild images for objects of different categories. In particular, we quantitatively and qualitatively compare our approach to state-of-the-art refinement methods in Sect. 4.1, perform an ablation study in Sect. 4.2, and combine our refinement approach with feed-forward 3D pose estimation [8] and 3D model retrieval [9] methods to predict fine-grained 3D poses without providing initial 3D poses or ground truth 3D models in Sect. 4.3. We follow the evaluation protocol of [8] and report the median (*MedErr*) of rotation, translation, pose, and projection distances. Details on evaluation setup, datasets, and metrics as well as extensive results and further experiments are provided in our **supplementary material**.

4.1 Comparison to the State of the Art

We first quantitatively compare our approach to state-of-the-art refinement methods. For this purpose, we perform 3D pose refinement on top of an initial 3D pose estimation baseline. In particular, we predict initial 3D poses using the feed-forward approach presented in [8] which is the state of the art for single image 3D pose estimation on the Pix3D dataset (*Baseline*). We compare our refinement approach to traditional image-based refinement without differentiable rendering [52] (*Image Refinement*) and mask-based refinement with differentiable rendering [17] (*Mask Refinement*).

RGB-based refinement with differentiable rendering [27] is not possible in our setup because all available 3D models lack textures and materials.

This approach even fails if we compare grey-scale images and renderings because the image intensities at corresponding locations do not match. As a consequence, 2D-3D alignment using a photo-metric loss is impossible.

For *Image Refinement*, we use grey-scale instead of RGB renderings because all available 3D models lack textures and materials. In addition, we do not perform a single full update [52] but perform up to 1000 iterative updates with a small learning rate of $\eta = 0.05$ using the Adam optimizer [19] for all methods.

For *Mask Refinement*, we predict instance masks from the input RGB image using Mask R-CNN [11]. To achieve maximum accuracy, we internally predict masks at four times the original spatial resolution proposed in Mask R-CNN and fine-tune a model pre-trained on COCO [24] on Pix3D.

Table 1 (upper part) summarizes our results. In this experiment, we provide the ground truth 3D model of the object in the image for refinement. Compared to the baseline, *Image Refinement* only achieves a small improvement in the rotation, translation, and pose metrics. There is almost no improvement in the projection metric ($MedErr_P$), as this method does not minimize the reprojection error. Traditional refinement methods are not aware of the rendering pipeline and the underlying 3D scene geometry and can only provide coarse 3D pose updates [52]. In our in-the-wild scenario, the number of 3D models, possible 3D pose perturbations, and category-level appearance variations is too large to simulate all permutations during training. As a consequence, this method cannot generalize to examples which are far from the ones seen during training and only achieves small improvements.

Table 1. Quantitative 3D pose refinement results on the Pix3D dataset. In the case of provided ground truth 3D models, our refinement significantly outperforms previous refinement methods across all metrics by **up to 55%** relative. In the case of automatically retrieved 3D models (+ Retrieval [9]), we reduce the 3D pose error ($MedErr_{R,t}$) compared to the state of the art for single image 3D pose estimation on Pix3D (*Baseline*) **by 55%** relative **without using additional inputs.**

Method	Input	Rotation $MedErr_R$ $\cdot 1$	Translation $MedErr_t$ $\cdot 10^2$	Pose $MedErr_{R,t}$ $\cdot 10^2$	Projection $MedErr_P$ $\cdot 10^2$
Baseline [8]	RGB Image	6.75	6.21	4.76	3.71
Image Refinement [52]	RGB Image + 3D Model	6.46	5.43	4.31	3.67
Mask Refinement [17]	RGB Image + 3D Model	3.56	4.06	2.96	1.90
Our Refinement	RGB Image + 3D Model	**2.56**	**1.74**	**1.34**	**1.27**
Image Refinement [52] + Retrieval [9]	RGB Image	6.47	5.51	4.33	3.74
Mask Refinement [17] + Retrieval [9]	RGB Image	5.47	5.25	4.15	3.12
Our Refinement + Retrieval [9]	RGB Image	**3.79**	**2.65**	**2.14**	**2.18**

Additionally, we observe that after the first couple of refinement steps, the predicted updates are not accurate enough to refine the 3D pose but start to jitter without further improving the 3D pose. Moreover, for many objects, the

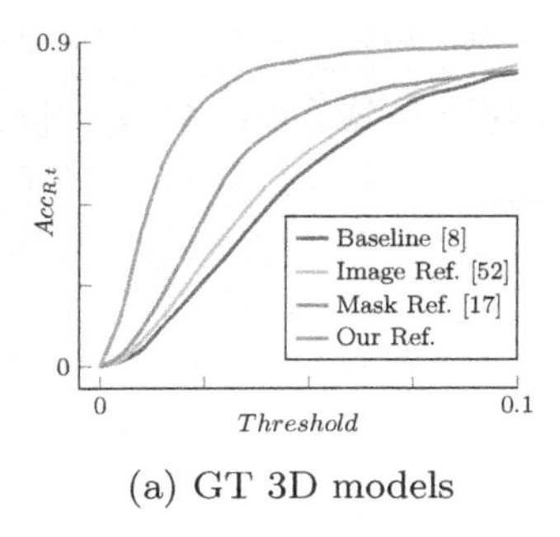

(a) GT 3D models

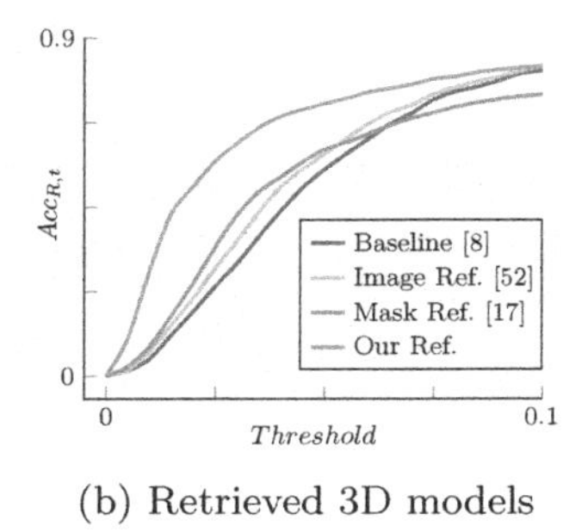

(b) Retrieved 3D models

Fig. 4. Evaluation on 3D pose accuracy at different thresholds. We significantly outperform other methods on strict thresholds using both GT and retrieved 3D models.

prediction fails and the iterative updates cause the 3D pose to drift off. Empirically, we obtain the best overall results for this method using only 20 iterations. For all other methods based on differentiable rendering, we achieve the best accuracy after the full 1000 iterations. A detailed analysis on this issue is presented in our supplementary material.

Next, *Mask Refinement* improves upon *Image Refinement* by a large margin across all metrics. Due to the 2D-3D alignment with differentiable rendering, this method computes more accurate 3D pose updates and additionally reduces the reprojection error ($MedErr_P$). However, we observe that *Mask Refinement* fails in two common situations: First, when the object has holes and the mask is not a single blob the refinement fails (see Fig. 5, *e.g.*, 1st row - right example). In the presence of holes, the hand-crafted approximation for the rasterization backward pass accumulates gradients with alternating signs while traversing the image. This results in unreliable per-vertex motion gradients. Second, simply aligning the silhouette of objects is ambiguous as renderings under different 3D poses can have similar masks. The interior structure of the object is completely ignored. As a consequence, the refinement gets stuck in poor local minima. Finally, the performance of *Mask Refinement* is limited by the quality of the target mask predicted from the RGB input image [11].

In contrast, our refinement overcomes these limitations and significantly outperforms the baseline as well as competing refinement methods across all metrics by **up to 70%** and **55%** relative. Using our geometric correspondence fields, we bridge the domain gap between real-world images and synthetic renderings and align both the object outline as well as interior structures with high accuracy.

Our approach performs especially well in the fine-grained regime, as shown in Fig. 4a. In this experiment, we plot the 3D pose accuracy $Acc_{R,t}$ which gives the percentage of samples for which the 3D pose distance $e_{R,t}$ is smaller than a varying threshold. For strict thresholds close to zero, our approach outperforms other refinement methods by a large margin. For example, at the threshold 0.015, we achieve more than 55% accuracy while the runner-up *Mask Refinement* achieves only 19% accuracy.

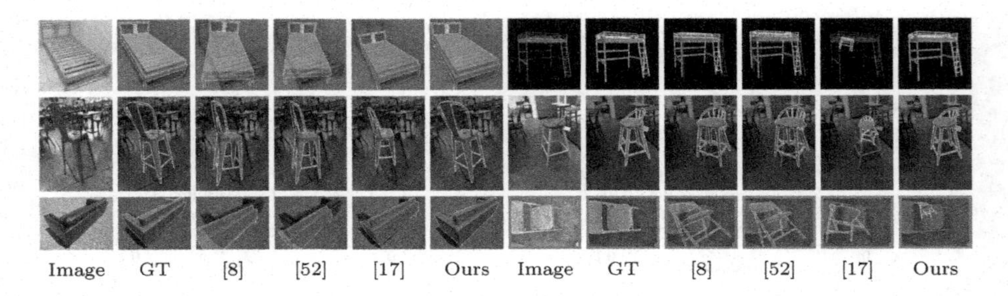

Fig. 5. Qualitative 3D pose refinement results for objects of different categories. We project the ground truth 3D model on the image using the 3D pose estimated by different methods. Our approach overcomes the limitations of previous methods and predicts fine-grained 3D poses for objects in the wild. The last example shows a failure case (indicated by the red frame) where the initial 3D pose is too far from the ground truth 3D pose and no refinement method can converge. More qualitative results are presented in our **supplementary material**. Best viewed in **digital zoom**. (Color figure online)

This significant performance improvement is also reflected in our qualitative examples presented in Fig. 5. Our approach precisely aligns 3D models to objects in RGB images and computes 3D poses which are in many cases visually indistinguishable from the ground truth. Even if the initial 3D pose estimate (*Baseline*) is significantly off, our method can converge towards the correct 3D pose (see Fig. 5, *e.g.*, 1st row - left example). Finally, Fig. 6 illustrates the high quality of our predicted geometric correspondence fields.

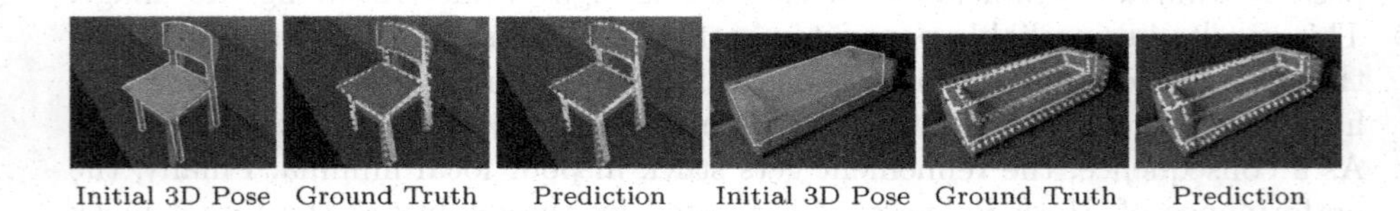

Fig. 6. Qualitative examples of our predicted geometric correspondence fields. Our predicted 2D displacement vectors are highly accurate. Best viewed in **digital zoom**.

4.2 Ablation Study

To understand the importance of individual components in our system, we conduct an ablation study in Table 2. For this purpose, we modify a specific system component, retrain our approach, and evaluate the performance impact.

If we use smaller kernels with less receptive field (3×3 vs 7×7) or fewer layers (2 vs 4) in our correspondence branch, the performance drops significantly. Also, using shallow mapping branches which only employ a single Conv-BN-ReLU block to simulate simple edge and ridge features results in low accuracy because the computed features are not discriminative enough. If we perform refinement without our geometric attention mechanism, the accuracy degrades due to unreliable correspondence predictions in homogeneous regions.

Next, the choice of the rendered representation is important for the performance of our approach. While using masks only performs poorly, depth, normal, and object coordinate renderings increase the accuracy. Finally, we achieve the best accuracy by exploiting complementary information from multiple different renderings by concatenating depth, normal, and object coordinate renderings.

By inspecting failure cases, we observe that our method does not converge if the initial 3D pose is too far from the ground truth 3D pose (see Fig. 5, last example). In this case, we cannot predict accurate correspondences because our computed features are not robust to large viewpoint changes and the receptive field of our correspondence branch is limited. In addition, occlusions cause our refinement to fail because there are no explicit mechanisms to address them. We plan to resolve this issue in the future by predicting occlusion masks and correspondence confidences. However, other refinement methods also fail in these scenarios (see Fig. 5, last example).

4.3 3D Model Retrieval

So far, we assumed that the ground truth 3D model required for 3D pose refinement is given at runtime. However, we can overcome this limitation by automatically retrieving 3D models from single RGB images. For this purpose, we combine all refinement approaches with the retrieval method presented in [9], where the 3D model database essentially becomes a part of the trained model. In this way, we perform initial 3D pose estimation, 3D model retrieval, and 3D pose refinement given only a single RGB image. This setting allows us to benchmark refinement methods against feed-forward baselines in a fair comparison.

Table 2. Ablation study of our method. Using components which increase the discriminability of learned features is important for the performance of our approach. Also, our geometric attention mechanism and the chosen type of rendering effect the accuracy.

Method	Rotation $MedErr_R$ $\cdot 1$	Translation $MedErr_t$ $\cdot 10^2$	Pose $MedErr_{R,t}$ $\cdot 10^2$	Projection $MedErr_P$ $\cdot 10^2$
Ours less receptive	3.01	2.12	1.75	1.56
Ours fewer layers	2.84	1.98	1.58	1.41
Ours shallow features	2.76	2.00	1.55	1.42
Ours without attention	2.70	1.92	1.45	1.37
Ours only MASK	2.98	1.85	1.44	1.41
Ours only OBJ COORD	2.57	1.80	1.39	1.31
Ours only DEPTH	2.60	1.77	1.38	1.29
Ours only NORMAL	2.58	1.76	1.36	1.30
Ours	**2.56**	**1.74**	**1.34**	**1.27**

The corresponding results are presented in Table 1 (lower part) and Fig. 4b. Because the retrieved 3D models often differ from the ground truth 3D models, the refinement performance decreases compared to given ground truth 3D models. Differentiable rendering methods lose more accuracy than traditional refinement methods because they require 3D models with accurate geometry.

Still, all refinement approaches perform remarkably well with retrieved 3D models. As long as the retrieved 3D model is reasonably close to the ground truth 3D model in terms of geometry, our refinement succeeds. Our method achieves even lower 3D pose error ($MedErr_{R,t}$) with retrieved 3D models than *Mask Refinement* with ground truth 3D models. Finally, using our joint 3D pose estimation-retrieval-refinement pipeline, we reduce the 3D pose error ($MedErr_{R,t}$) compared to the state of the art for single image 3D pose estimation on Pix3D (*Baseline*) **by 55%** relative **without using additional inputs**.

5 Conclusion

Aligning 3D models to objects in RGB images is the most accurate way to predict 3D poses. However, there is a domain gap between real-world images and synthetic renderings which makes this alignment challenging in practice. To address this problem, we predict deep cross-domain correspondences in a feature space optimized for 3D pose refinement and combine local 2D displacement vectors into global 3D pose updates using our novel differentiable renderer. Our method outperforms existing refinement approaches by up to 55% relative and can be combined with feed-forward 3D pose estimation and 3D model retrieval to predict fine-grained 3D poses for objects in the wild given only a single RGB image. Finally, our novel learned differentiable rendering framework can be used for other tasks in the future.

References

1. Azinovic, D., Li, T.M., Kaplanyan, A., Niessner, M.: Inverse path tracing for joint material and lighting estimation. In: Conference on Computer Vision and Pattern Recognition, pp. 2447–2456 (2019)
2. Brachmann, E., Krull, A., Michel, F., Gumhold, S., Shotton, J., Rother, C.: Learning 6D object pose estimation using 3D object coordinates. In: Fleet, D., Pajdla, T., Schiele, B., Tuytelaars, T. (eds.) ECCV 2014. LNCS, vol. 8690, pp. 536–551. Springer, Heidelberg (2014). https://doi.org/10.1007/978-3-319-10605-2_35
3. Brachmann, E., Michel, F., Krull, A., Ying Yang, M., Gumhold, S., Rother, C.: Uncertainty-driven 6D pose estimation of objects and scenes from a single RGB image. In: Conference on Computer Vision and Pattern Recognition, pp. 3364–3372 (2016)
4. Choy, C.B., Gwak, J., Savarese, S., Chandraker, M.: Universal correspondence network. In: Advances in Neural Information Processing Systems, pp. 2414–2422 (2016)
5. Dosovitskiy, A., et al.: FlowNet: learning optical flow with convolutional networks. In: Conference on Computer Vision and Pattern Recognition, pp. 2758–2766 (2015)

6. Genova, K., Cole, F., Maschinot, A., Sarna, A., Vlasic, D., Freeman, W.T.: Unsupervised training for 3D morphable model regression. In: Conference on Computer Vision and Pattern Recognition, pp. 8377–8386 (2018)
7. Grabner, A., Roth, P.M., Lepetit, V.: 3D pose estimation and 3D model retrieval for objects in the wild. In: Conference on Computer Vision and Pattern Recognition, pp. 3022–3031 (2018)
8. Grabner, A., Roth, P.M., Lepetit, V.: GP^2C: geometric projection parameter consensus for joint 3D pose and focal length estimation in the wild. In: International Conference on Computer Vision, pp. 2222–2231 (2019)
9. Grabner, A., Roth, P.M., Lepetit, V.: Location field descriptors: single image 3D model retrieval in the wild. In: International Conference on 3D Vision, pp. 583–593 (2019)
10. Hartley, R., Zisserman, A.: Multiple View Geometry in Computer Vision. Cambridge University Press, Cambridge (2003)
11. He, K., Gkioxari, G., Dollár, P., Girshick, R.: Mask R-CNN. In: International Conference on Computer Vision, pp. 2980–2988 (2017)
12. He, K., Zhang, X., Ren, S., Sun, J.: Deep residual learning for image recognition. In: Conference on Computer Vision and Pattern Recognition, pp. 770–778 (2016)
13. He, K., Zhang, X., Ren, S., Sun, J.: Identity mappings in deep residual networks. In: Leibe, B., Matas, J., Sebe, N., Welling, M. (eds.) ECCV 2016. LNCS, vol. 9908, pp. 630–645. Springer, Heidelberg (2016). https://doi.org/10.1007/978-3-319-46493-0_38
14. Henderson, P., Ferrari, V.: Learning to generate and reconstruct 3D meshes with only 2D supervision. In: British Machine Vision Conference, pp. 139:1–139:13 (2018)
15. Jafari, O.H., Mustikovela, S.K., Pertsch, K., Brachmann, E., Rother, C.: iPose: instance-aware 6D pose estimation of partly occluded objects. In: Jawahar, C., Li, H., Mori, G., Schindler, K. (eds.) ACCV 2018. LNCS, vol. 11363, pp. 477–492. SPringer, Heidelberg (2018). https://doi.org/10.1007/978-3-030-20893-6_30
16. Kanazawa, A., Tulsiani, S., Efros, A.A., Malik, J.: Learning category-specific mesh reconstruction from image collections. In: European Conference on Computer Vision, pp. 371–386 (2018)
17. Kato, H., Ushiku, Y., Harada, T.: Neural 3D mesh renderer. In: Conference on Computer Vision and Pattern Recognition, pp. 3907–3916 (2018)
18. Kehl, W., Manhardt, F., Tombari, F., Ilic, S., Navab, N.: SSD-6D: making RGB-based 3D detection and 6D pose estimation great again. in: International Conference on Computer Vision, pp. 1530–1538 (2017)
19. Kingma, D.P., Ba, J.: Adam: a method for stochastic optimization. arXiv:1412.6980 (2014)
20. Kundu, A., Li, Y., Rehg, J.M.: 3D-RCNN: instance-level 3D object reconstruction via render-and-compare. In: Conference on Computer Vision and Pattern Recognition, pp. 3559–3568 (2018)
21. Li, C., Bai, J., Hager, G.D.: A unified framework for multi-view multi-class object pose estimation. In: European Conference on Computer Vision, pp. 1–16 (2018)
22. Li, T.M., Aittala, M., Durand, F., Lehtinen, J.: Differentiable Monte Carlo ray tracing through edge sampling. In: ACM SIGGRAPH Asia, pp. 222:1–222:11 (2018)
23. Li, Y., Wang, G., Ji, X., Xiang, Y., Fox, D.: DeepIM: deep iterative matching for 6D pose estimation. In: European Conference on Computer Vision, pp. 683–698 (2018)

24. Lin, T.Y., et al.: Microsoft COCO: common objects in context. In: Fleet, D., Pajdla, T., Schiele, B., Tuytelaars, T. (eds.) ECCV 2014. LNCS, vol. 8693, pp. 740–755. Springer, Heidelberg (2014). https://doi.org/10.1007/978-3-319-10602-1_48
25. Liu, L., Lu, J., Xu, C., Tian, Q., Zhou, J.: Deep fitting degree scoring network for monocular 3D object detection. In: Conference on Computer Vision and Pattern Recognition, pp. 1057–1066 (2019)
26. Liu, S., Li, T., Chen, W., Li, H.: Soft rasterizer: a differentiable renderer for image-based 3D reasoning. In: International Conference on Computer Vision, pp. 7708–7717 (2019)
27. Loper, M.M., Black, M.J.: OpenDR: an approximate differentiable renderer. In: Fleet, D., Pajdla, T., Schiele, B., Tuytelaars, T. (eds.) ECCV 2014. LNCS, vol. 8695, pp. 154–169. Springer, Heidelberg (2014). https://doi.org/10.1007/978-3-319-10584-0_11
28. Mahendran, S., Ali, H., Vidal, R.: A mixed classification-regression framework for 3D pose estimation from 2D images. In: British Machine Vision Conference, pp. 238:1–238:12 (2018)
29. Manhardt, F., Kehl, W., Navab, N., Tombari, F.: Deep model-based 6D pose refinement in RGB. In: European Conference on Computer Vision, pp. 800–815 (2018)
30. Massa, F., Marlet, R., Aubry, M.: Crafting a multi-task CNN for viewpoint estimation. In: British Machine Vision Conference, pp. 91:1–91:12 (2016)
31. Mottaghi, R., Xiang, Y., Savarese, S.: A coarse-to-fine model for 3D pose estimation and sub-category recognition. In: Conference on Computer Vision and Pattern Recognition, pp. 418–426 (2015)
32. Mousavian, A., Anguelov, D., Flynn, J., Kosecka, J.: 3D bounding box estimation using deep learning and geometry. In: Conference on Computer Vision and Pattern Recognition, pp. 7074–7082 (2017)
33. Nguyen-Phuoc, T.H., Li, C., Balaban, S., Yang, Y.: RenderNet: a deep convolutional network for differentiable rendering from 3D shapes. In: Advances in Neural Information Processing Systems, pp. 7891–7901 (2018)
34. Palazzi, A., Bergamini, L., Calderara, S., Cucchiara, R.: End-to-end 6-DoF object pose estimation through differentiable rasterization. In: European Conference on Computer Vision Workshops, pp. 1–14 (2018)
35. Pavlakos, G., Zhou, X., Chan, A., Derpanis, K., Daniilidis, K.: 6-DoF object pose from semantic keypoints. In: International Conference on Robotics and Automation, pp. 2011–2018 (2017)
36. Peng, S., Liu, Y., Huang, Q., Zhou, X., Bao, H.: 3D object class detection in the wild. In: Conference on Computer Vision and Pattern Recognition, pp. 4561–4570 (2019)
37. Pepik, B., Stark, M., Gehler, P., Ritschel, T., Schiele, B.: 3D object class detection in the wild. In: Conference on Computer Vision and Pattern Recognition Workshops, pp. 1–10 (2015)
38. Rad, M., Lepetit, V.: BB8: a scalable, accurate, robust to partial occlusion method for predicting the 3D poses of challenging objects without using depth. In: International Conference on Computer Vision, pp. 3828–3836 (2017)
39. Sun, X., et al.: Pix3D: dataset and methods for single-image 3D shape modeling. In: Conference on Computer Vision and Pattern Recognition, pp. 2974–2983 (2018)
40. Tekin, B., Sinha, S.N., Fua, P.: Real-time seamless single shot 6D object pose prediction. In: Conference on Computer Vision and Pattern Recognition, pp. 292–301 (2018)

41. Tulsiani, S., Carreira, J., Malik, J.: Pose induction for novel object categories. In: International Conference on Computer Vision, pp. 64–72 (2015)
42. Tulsiani, S., Malik, J.: Viewpoints and keypoints. In: Conference on Computer Vision and Pattern Recognition, pp. 1510–1519 (2015)
43. Wang, H., Sridhar, S., Huang, J., Valentin, J., Song, S., Guibas, L.: Normalized object coordinate space for category-level 6D object pose and size estimation. In: Conference on Computer Vision and Pattern Recognition, pp. 2642–2651 (2019)
44. Wang, Y., et al.: 3D pose estimation for fine-grained object categories. In: European Conference on Computer Vision Workshops (2018)
45. Wu, J., Zhang, C., Xue, T., Freeman, W.T., Tenenbaum, J.: Learning a probabilistic latent space of object shapes via 3D generative-adversarial modeling. In: Advances in Neural Information Processing Systems, pp. 82–90 (2016)
46. Xiang, Y., et al.: ObjectNet3D: a large scale database for 3D object recognition. In: Leibe, B., Matas, J., Sebe, N., Welling, M. (eds.) ECCV 2016. LNCS, vol. 9912, pp. 160–176. Springer, Heidelberg (2016). https://doi.org/10.1007/978-3-319-46484-8_10
47. Xiang, Y., Schmidt, T., Narayanan, V., Fox, D.: PoseCNN: a convolutional neural network for 6D object pose estimation in cluttered scenes. In: Robotics: Science and Systems Conference, pp. 1–10 (2018)
48. Xiao, Y., Qiu, X., Langlois, P.A., Aubry, M., Marlet, R.: Pose from shape: deep pose estimation for arbitrary 3D objects. In: British Machine Vision Conference, pp. 120:1–120:14 (2019)
49. Yao, S., Hsu, T.M., Zhu, J.Y., Wu, J., Torralba, A., Freeman, W.T., Tenenbaum, J.: 3D-Aware Scene Manipulation via Inverse Graphics. In: Advances in Neural Information Processing Systems. pp. 1887–1898 (2018)
50. Zabulis, X., Lourakis, M.I.A., Stefanou, S.S.: 3D pose refinement using rendering and texture-based matching. In: International Conference on Computer Vision and Graphics, pp. 672–679 (2014)
51. Zagoruyko, S., Komodakis, N.: Learning to compare image patches via convolutional neural networks. In: Conference on Computer Vision and Pattern Recognition, pp. 4353–4361 (2015)
52. Zakharov, S., Shugurov, I., Ilic, S.: DPOD: dense 6D pose object detector in RGB images. In: International Conference on Computer Vision, pp. 1941–1950 (2019)

3D Fluid Flow Reconstruction Using Compact Light Field PIV

Zhong Li[1], Yu Ji[2], Jingyi Yu[2,3], and Jinwei Ye[4(✉)]

[1] University of Delaware, Newark, DE, USA
[2] DGene, Baton Rouge, LA, USA
[3] ShanghaiTech University, Shanghai, China
[4] Louisiana State University, Baton Rouge, LA, USA
jye@csc.lsu.edu

Abstract. Particle Imaging Velocimetry (PIV) estimates the fluid flow by analyzing the motion of injected particles. The problem is challenging as the particles lie at different depths but have similar appearances. Tracking a large number of moving particles is particularly difficult due to the heavy occlusion. In this paper, we present a PIV solution that uses a compact lenslet-based light field camera to track dense particles floating in the fluid and reconstruct the 3D fluid flow. We exploit the focal symmetry property in the light field focal stacks for recovering the depths of similar-looking particles. We further develop a motion-constrained optical flow estimation algorithm by enforcing the local motion rigidity and the Navier-Stoke fluid constraint. Finally, the estimated particle motion trajectory is used to visualize the 3D fluid flow. Comprehensive experiments on both synthetic and real data show that using a compact light field camera, our technique can recover dense and accurate 3D fluid flow.

Keywords: Volumetric flow reconstruction · Particle Imaging Velocimetry (PIV) · Light field imaging · Focal stack

1 Introduction

Recovering time-varying volumetric 3D fluid flow is a challenging problem. Successful solutions can benefit applications in many science and engineering fields, including oceanology, geophysics, biology, mechanical and environmental engineering. In experimental fluid dynamics, a standard methodology for measuring fluid flow is called Particle Imaging Velocimetry (PIV) [1]: the fluid is seeded with tracer particles, whose motions are assumed to follow the fluid dynamics

This work was performed when Zhong Li was a visiting student at LSU.

Electronic supplementary material The online version of this chapter (https://doi.org/10.1007/978-3-030-58517-4_8) contains supplementary material, which is available to authorized users.

A. Vedaldi et al. (Eds.): ECCV 2020, LNCS 12361, pp. 120–136, 2020.
https://doi.org/10.1007/978-3-030-58517-4_8

faithfully, then the particles are tracked over time and their motion trajectories in 3D are used to represent the fluid flows.

Although being highly accurate, existing PIV solutions usually require complex and expensive equipment and the setups end up bulky. For example, standard laser-based PIV methods [6,16] use ultra high speed laser beam to illuminate particles in order to track their motions. One limitation of these method is that the measured motion field only contains 2D in-plane movement restricted on the 2D fluid slice being scanned, as the laser beam can only scan one depth layer at a time. To fully characterize the fluid, it is necessary to recover the 3D flow motion within the entire fluid volume. Three-dimensional PIV such as tomographic PIV (Tomo-PIV) [9] use multiple cameras to capture the particles and resolve their depths in 3D using multi-view stereo. But such multi-camera systems need to be well calibrated and fully synchronized. More recently, the Rainbow PIV solutions [46,47] use color to encode particles at different depths in order to recover the 3D fluid flow. However, this setup requires specialized illumination source with diffractive optics for color-encoding and the optical system needs to be precisely aligned.

In this paper, we present a flexible and low-cost 3D PIV solution that only uses one compact lenslet-based light field camera as the acquisition device. A light field camera, in essence, is a single-shot, multi-view imaging device [33]. The captured light field records 4D spatial and angular light rays scattered from the tracer particles. As commercial light field cameras (*e.g.* Lytro Illum and Raytrix R42) can capture high resolution light field, we are able to resolve dense particles in 3D fluid volume. Small baseline of the lenslet array further helps recover subtle particle motions at sub-pixel level. In particular, our method benefits from the post-capture refocusing capability of light field. We use the focal stack to establish correspondences among particles at different depths. To resolve heavily occluded particles, we exploit the focal stack symmetry (*i.e.*, intensities are symmetric in the focal stack around the ground truth disparity [25,41]) for accurate particle 3D reconstruction.

Given the 3D locations of particles at each time frame, we develop a physics-based optical flow estimation algorithm to recover the particle's 3D velocity field, which represents the 3D fluid flows. In particular, we introduce two new regularization terms to refine the classic variational optical flow [17]: 1) one-to-one particle correspondence term to maintain smooth and consistent flow motions across different time frames; and 2) divergence-free regularization term derived from the Navier-Stoke Equations to enforce the physical properties of incompressible fluid. These terms help resolve ambiguities in particle matching caused by similar appearances while enforcing the reconstruction to obey physical laws. Through synthetic and real experiments, we show that using a simple single camera setup, our approach outperforms state-of-the-art PIV solutions on recovering volumetric 3D fluid flows of various types.

2 Related Work

In computer vision and graphics, much effort has been made in modeling and recovering transparent objects or phenomena directly from images (*e.g.*, fluid [32,49], gas flows [4,20,28,48], smoke [12,14], and flames [13,19], *etc.*). As these objects do not have their own appearances, often a known pattern is assumed and the light paths traveled through the transparent medium are estimated for 3D reconstruction. A comprehensive survey can be found in [18]. However, many of these imaging techniques are designed to recover the 3D density field, which does not explicitly reveal the internal flow motion.

Our method, instead, aims at estimating the 3D flow motion in terms of velocity field. The measurement procedure is similar to the Particle Imaging Velocimetry (PIV) method that estimates flow motion from movement by injecting tracer particles. Traditional PIV [6,16] recovers 2D velocity fields on thin fluid slices using high speed laser scanning. As 3D volumetric flow is critical to fully characterize the fluid behavior, recovering a 3D velocity field within the entire volume is of great interest.

To recover 3D velocity field of a dense set of particles, stereoscopic cameras [3,35] are used to estimate the particle depth. Tomographic PIV (Tomo-PIV) [9,22,36] use multiple (usually three to six) cameras to determine 3D particle locations by space carving. Aguirre-Pablo *et al.* [2] perform Tomo-PIV using mobile devices. However, the accuracy of reconstruction is compromised due to the low resolution of mobile cameras. Other notable 3D PIV approaches include defocusing PIV [21,45], Holographic PIV [39,50], and synthetic aperture PIV [5,31]. All these systems use an array of cameras for acquisition and each measurement requires elaborate calibration and synchronization. In contrast, our setup is more flexible by using a single compact light field camera. Recently proposed rainbow PIV [46,47] use color-coded illumination to recover depth from a single camera. However, both the light source and camera are customized with special optical elements and only sparse set of particles can be resolved. Proof-of-concept simulations [27] and experiments [10] using compact light field or plenoptic cameras for PIV have been performed and showed efficacy. However, the depth estimation and particle tracking algorithms used in these early works are rather primitive and are not optimized according to light field properties. As result, the recovered particles are relatively sparse and the reconstruction accuracy is lower than traditional PIV. Shi *et al.* [37,38] use ray tracing to estimate particle velocity with a light field camera, and conduct comparison with Tomo-PIV. In our approach, we exploit the focal stack symmetry [25] of light fields for more accurate depth reconstruction in presence of heavily occluded dense particles.

To recover the flow motion, standard PIV uses 3D cross-correlation to match local windows between neighboring time frames [9,44]. Although many improvements (for instance, matching with adaptive window sizes [22]) have been made, the window-based solutions suffer problems at regions with few visible particles. Another class of methods directly track the path of individual particles over time [29,36]. However, with increased particle density, tracking is challenging under occlusions. Heitz *et al.* [15] propose the application of variational optical flow to fluid flow estimation. Vedula *et al.* [43] extend the optical flow to dynamic environment and introduce the scene flow. Lv et al. [26] use a neural network to recover 3D scene flow. Unlike natural scenes that have diverse features, our PIV scenes only contain similar-looking particles. Therefore, existing optical flow or scene flow algorithms are not directly applicable to our problem. Some methods [23, 47] incorporate physical constraints such as the Stokes equation into optical flow framework to recover fluid flows that obey physical laws. However, these physics-based regularizations are in high-orders and are difficult to solve. In our approach, we introduce two novel regularization terms: 1) rigidity-enforced particle correspondence term and 2) divergence-free term to refine the basic variational optical flow framework for estimating the motion of dense particles.

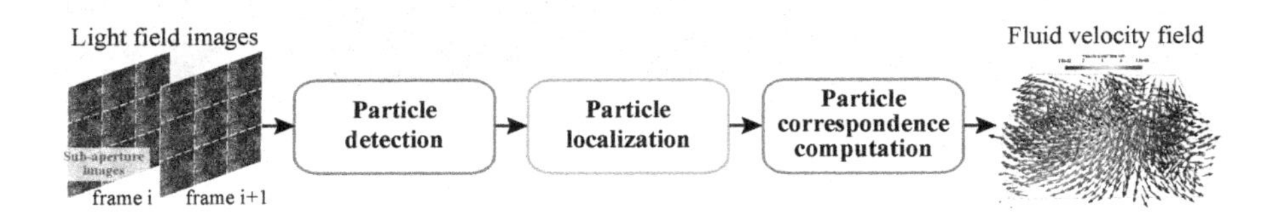

Fig. 1. Overall pipeline of our light field PIV 3D fluid flow reconstruction algorithm.

3 Our Approach

Figure 1 shows the algorithmic pipeline of volumetric 3D fluid flow reconstruction using light field PIV. For each time frame, we first detect particles in the light field sub-aperture images using the IDL particle detector [7]. We then estimate particle depths through a joint optimization that exploits light field properties. After we obtain 3D particle locations, we compare two consecutive frames to establish one-to-one particle correspondences and finally solve the 3D velocity field using a constrained optical flow.

3.1 3D Particle Reconstruction

We first describe our 3D particle reconstruction algorithm that exploits various properties of light field.

Focal Stack Symmetry. A focal stack is a sequence of images focused at different depth layers. Due to the post-capture refocusing capability, a focal stack can be synthesized from a light field by integrating captured light rays. Lin *et al.* [25] conduct symmetry analysis on focal stacks and show that *non-occluding* pixels in a focal stack exhibit symmetry along the focal dimension centered at the in-focus slice. In contrast, occluding boundary pixels exhibit local asymmetry as the outgoing rays are not originated from the same surface. Such property is called focal stack symmetry. As shown in Fig. 2, in a focal stack, a particle exhibits symmetric defocus effect centered at the in-focus slice. It's also worth noting that occluded particles could be seen in the focal stack as the occluder becomes extremely out-of-focus. Utilizing the focal stack symmetry helps resolve heavily occluded particles and hence enhances the accuracy and robustness of particle depth estimation.

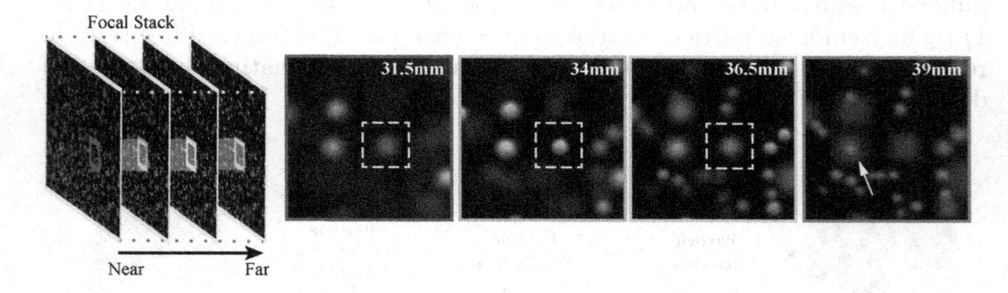

Fig. 2. Focal stack symmetry. We show zoom-in views of four focal slices on the right. A particle exhibits symmetric defocus effect (e.g., 31.5 mm and 36.5 mm slices) centered at the in-focus slice (34 mm). In the 39 mm slice, an occluded particle could be seen as the occluder becomes extremely out-of-focus.

Given a particle light field, we synthesize a focal stack from the sub-aperture images by integrating rays from the same focal slice. Each focus slice f has a corresponding disparity d that indicates the in-focus depth layer. Let $I(p, f)$ be the intensity of a pixel p at focal slice f. For symmetry analysis, we define an in-focus score $\kappa(p, f)$ a pixel p at focal slice f as:

$$\kappa(p, f) = \int_0^{\delta_{max}} \rho(I(p, f + \delta) - I(p, f - \delta))d\delta \tag{1}$$

where δ represents tiny disparity/focal shift and δ_{max} is maximum shift amount; $\rho(\nu) = 1 - e^{-|\nu|_2/(2\sigma^2)}$ is a robust distance function with σ controlling its sensitivity to noises. According to the focal stack symmetry, the intensity profile $I(p, f)$ is locally symmetric around the true surface depth. Therefore, if the pixel p is in focus at its true depth sparity $\hat{d}$, $\kappa(p, \hat{d})$ should be 0. Hence given an estimated disparity d at p, the closer distance between d and $\hat{d}$, the smaller the $\kappa(p, \hat{d})$. We then formulate the focal stack symmetry term β_{fs} for particle depth estimation by summing up $\kappa(p, d)$ for all pixels in a focal slice f with disparity d:

$$\beta_{fs}(d) = \sum_p \kappa(p, d) \tag{2}$$

Color and Gradient Consistency. Besides the focal stack symmetry, we also consider the color and gradient data consistency across sub-aperture images for depth estimation using data terms similar to [25]. Specifically, by comparing each sub-aperture image with the center view, we define a cost metric $C(i, p, d)$ as:

$$C(i, p, d) = |I_c(\omega(p)) - I_i(\omega(p + d(p)\chi(i)))| \tag{3}$$

where i is the sub-aperture image index; I_c and I_i refers to the center view and sub-aperture image respectively; $\omega(p)$ refers to a small local window centered around pixel p; $d(p)$ is an estimate disparity at pixel p; and $\chi(i)$ is a scalar that scale the disparity $d(p)$ according to the relative position between I_c and I_i as $d(p)$ is the pixel-shift between neighboring sub-aperture images.

The cost metric C measures the intensity similarity between shifted pixels in sub-aperture images given an estimated disparity. By summing up C for all pixels, we obtain the sum of absolute differences (SAD) term for color consistency measurement:

$$\beta_{sad}(d) = \frac{1}{N} \sum_{i \in N} \sum_{p} C \tag{4}$$

where N is the total number of sub-aperture images (excluding the center view).

Besides the color consistency, we also consider the consistency in gradient domain. We first take partial derivates of cost metric C (Eq. 3) in both x and y directions: $D_x = \partial C / \partial x$ and $D_y = \partial C / \partial y$ and then formulate the following weighted sum of gradient differences (GRAD) for gradient consistency measurement:

$$\beta_{grad}(d) = \frac{1}{N} \sum_{i \in N} \sum_{p} \mathcal{W}(i)\mathcal{D}_x + (1 - \mathcal{W}(i))\mathcal{D}_y \tag{5}$$

In Eq. 5, $\mathcal{W}(i)$ is a weighing factor that determines the contribution of horizontal gradient cost ($\mathcal{D}_x$) according to the relative positions of the two sub-aperture images being compared. It is defined as $\mathcal{W}(i) = \frac{\Delta i_x}{\Delta i_x + \Delta i_y}$, where Δi_x and Δi_y are the position differences between sub-aperture images along x and y directions. For example, $\mathcal{W}(i) = 1$ if the target view is located at the horizontal extent of the reference view. In this case, only the gradient costs in the x direction are aggregated.

Particle Depth Estimation. Finally, combining Eq. 2, 4, and 5, we form the following energy function for optimizing the particle disparity d:

$$\beta(d) = \beta_{fs}(d) + \lambda_{sad}\beta_{sad}(d) + \lambda_{grad}\beta_{grad}(d) \tag{6}$$

In our experiments, the two weighting factors are set as $\lambda_{sad} = 0.8$ and $\lambda_{grad} = 0.9$. We use the Levenberg-Marquardt (LM) optimization to solve Eq. 6. Finally, using the calibrated light field camera intrinsic parameters, we are able to convert the particle disparity map to 3D particle location. The pipeline of our 3D particle reconstruction algorithm is shown in Fig. 3.

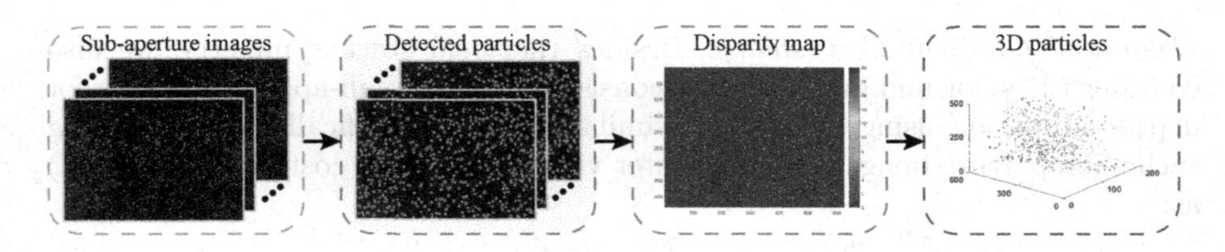

Fig. 3. Our 3D particle reconstruction algorithm pipeline.

3.2 Fluid Flow Reconstruction

After we reconstruct 3D particles in each frame, we compare two consecutive frames to estimate the volumetric 3D fluid flow.

Given two sets of particle locations S_1 and S_2 recovered from consecutive frames, we first convert S_1 and S_2 into voxelized 3D volumes as occupancy probabilities Θ_1 and Θ_2 through linear interpolation. Our goal is to solve per-voxel 3D velocity vector $\mathbf{u} = [u, v, w]$ for the whole volume.

In particular, we solve this problem under the variational optical flow framework [17] and propose two novel regularization terms, the correspondence term and the divergence-free term, for improved accuracy and efficiency. Our overall energy function E_{total} is combination of regularization terms and is written as:

$$E_{total} = E_{data} + \lambda_1 E_{smooth} + \lambda_2 E_{corres} + \lambda_3 E_{div} \tag{7}$$

where λ_1, λ_2, and λ_3 are term balancing factors. Please see our supplementary material for mathematical details of solving this energy function. In the following, we describe the algorithmic details of each regularization term.

Basic Optical Flow. The data term E_{data} and smooth term E_{smooth} are adopted from basic optical flow. They are derived from the brightness constancy assumption. E_{data} enforces consistency between occupancy possibilities Θ_1 and Θ_2 at corresponding voxels and E_{smooth} constrain the fluid motion to be piece-wise smooth. In our case, E_{data} and E_{smooth} can be written as:

$$E_{data}(\mathbf{u}) = \int ||\Theta_2(\mathbf{p} + \mathbf{u}) - \Theta_1(\mathbf{p})||_2^2 d\mathbf{p} \tag{8}$$

.

$$E_{smooth}(\mathbf{u}) = ||\nabla \cdot \mathbf{u}||_2^2 \tag{9}$$

where $\mathbf{p}$ refers to a voxel in fluid volume and ∇ is the gradient operator.

Correspondence Term. We propose a novel correspondence term for more accurate flow estimation. Notice that E_{data} in the basic optical flow only enforces voxel-level consistency while particle-to-particle correspondences are not guaranteed. We therefore develop a correspondence term E_{corres} to enforce one-to-one particle matching. E_{corres} helps improve matching accuracy especially in regions with high particle density.

Let's consider two sets of particles: $S_1 = \{s_1 | s_1 \in \mathbb{R}^3\}$ as reference and $S_2 = \{s_2 | s_2 \in \mathbb{R}^3\}$ as target. E_{corres} enforces the one-to-one particle matching between the target and reference sets. To formulate E_{corres}, we first estimate correspondences between particles in S_1 and S_2. We solve this problem by estimating transformations that map particles in S_1 to S_2.

In particular, we employ a deformable graph similar to [42] that considers local geometric similarity and rigidity. To build the graph, we uniformly sample a set of particles in S_1 and use them as graph nodes $\mathbf{G} = \{g_1, g_2, g_3, ..., g_m\}$. We then aim to estimate a set of affine transformations $\mathbf{A} = \{A_i\}_{i=1}^m$ and $\mathbf{b} = \{b_i\}_{i=1}^m$ for each graph node. We then use these graph nodes as control points to deform particles in S_1 instead of computing transformations for individual particles. Given the graph node transformations $\mathbf{A}$ and $\mathbf{b}$, we can transform every particle $s_1 \in S_1$ to it new location s_1' using a weighted linear combination of graph nodes transformations:

$$s_1' = f(s_1, \mathbf{A}, \mathbf{b}) = \sum_{i=1}^{m} \varpi_i(s_1)(\mathbf{A}(s_1 - g_i) + g_i + b_i) \tag{10}$$

where the weight $\varpi_i(s_1) = max(0, (1 - ||s_1 - g_i||^2/R^2)^3)$ models a graph node g_i influence on a particle $s_1 \in S_1$ according to their Euclidean distance. This restricts the particle transformation to be only affected by nearby graph nodes. In our experiment, we consider the nearest four graph nodes and R is the particle's distance to its nearest graph node.

To obtain the graph node transformations $\mathbf{A}$ and $\mathbf{b}$, we solve an optimization problem with energy function:

$$\Psi_{total} = \Psi_{data} + \alpha_1 \Psi_{rigid} + \alpha_2 \Psi_{smooth} \tag{11}$$

Ψ_{data} is the data term aims to minimize particle-to-particle distances after transformation and is thus formulated as:

$$\Psi_{data} = \sum_{s_1 \in S_1} ||s_1' - c_i||^2 \tag{12}$$

where c_i is the closest point to s_1' in S_2.

Ψ_{rigid} is a rigidity regularization term that enforces the local rigidity of affine transformation. Ψ_{rigid} can be written as:

$$\Psi_{rigid} = \sum_{\mathbf{G}} ||A_i^T A_i - \mathbb{I}||_F^2 + (det(A_i) - 1)^2 \tag{13}$$

where $\mathbb{I}$ is an identity matrix.

The last term Ψ_{smooth} enforces the spatial smoothness of nearby nodes and is written as:

$$\Psi_{smooth} = \sum_{\mathbf{G}} \sum_{k \in \Omega(i)} ||A_i(g_k - g_i) + g_i + b_i - (g_k + b_k)||^2 \tag{14}$$

where $\Omega(i)$ refers to the set of nearest four neighbors of g_i.

The overall energy function Ψ_{total} can be optimized with an iterative Gauss-Newton algorithm and the affine transformations $\mathbf{A}$ and $\mathbf{b}$ are thus solved. In our experiment, we use $\alpha_1 = 50$ and $\alpha_2 = 10$ for Eq. 11.

By applying Eq. 11, we can transform every particle $s_1 \in S_1$ to it new location s_1' using the graph nodes' transformations. We then find S_1's corresponding set S_2^c in the target S_2 using a nearest neighbor search (ie, $s_2^c = \text{nnsearch}(s_1', s_2)$). After we establish the one-to-one correspondences between S_1 and S_2, our correspondence term can be formulated based on the color consistency assumption as follow:

$$E_{corres}(\mathbf{u}, S_1, S_2^c) = \sum_{s_1 \in S_1, s_2^c \in S_2^c} ||s_2^c - (s_1 + \mathbf{u}(s_1))||_2^2 \tag{15}$$

We show the effectiveness of the correspondence term by comparing the velocity field obtained with vs. without E_{corres}. The results are shown in Fig. 4. This comparison demonstrates that our correspondence term greatly improves matching accuracy and hence benefits flow reconstruction.

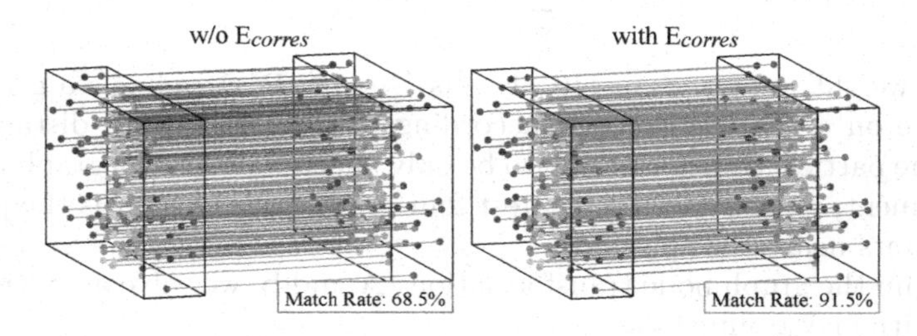

Fig. 4. Particle matching between source and target volumes with vs. without using the correspondence term E_{corres}. In our plots, green lines indicate correct correspondences and red lines indicate incorrect ones. (Color figure online)

Divergence-Free Term. To enforce the physical properties of incompressible fluid, we add a divergence-free regularization term E_{div} to the optical flow framework. Based on the Navier-Stoke equations, fluid velocity $\mathbf{u}$ can be split into into two distinct components: irrotational component ∇P and solenoidal component $\mathbf{u}_{sol} = [u_{sol}, v_{sol}, w_{sol}]$ with the Helmholtz decomposition. The Irrotational component ∇P is curl-free and is determined by the gradient of a scalar function P (eg, pressure). The solenoidal component $\mathbf{u}_{sol}$ is divergence-free and models an incompressible flow. From the divergence-free property, we have:

$$\nabla \cdot \mathbf{u}_{sol} = 0 \tag{16}$$

where $\nabla = [\frac{\partial}{\partial x}, \frac{\partial}{\partial y}, \frac{\partial}{\partial z}]^T$ is the divergence operator. Since $\mathbf{u} = \mathbf{u}_{sol} + \nabla P$, taking divergence on both sides, we have:

$$\nabla \cdot \mathbf{u} = \nabla^2 P \tag{17}$$

We solve Eq. 17 by Poisson integration and compute the scalar field as $P = (\nabla^2)^{-1}(\nabla \cdot \mathbf{u})$. We then project $\mathbf{u}$ into the divergence-free vector field: $\mathbf{u}_{sol} = \mathbf{u} - \nabla P$. Similar to [11], we formulate a divergence-free term E_{div} that enforces the flow velocity field $\mathbf{u}$ close to its divergence-free component $\mathbf{u}_{sol}$:

$$E_{div}(\mathbf{u}) = ||\mathbf{u} - \mathbf{u}_{sol}||_2^2 \tag{18}$$

4 Experimental Results

To evaluate our fluid flow reconstruction algorithm, we perform experiments on both synthetic and real data under the light field PIV setting. We also evaluate our method on the John Hopkins Turbulence Database (JHUTDB) [24,34] that has the ground truth fluid flow. All experiments are performed on a PC with Intel i7-4700K CPU with 16G of memory. On the computational time, the entire process takes about 2 min: 30 s for particle location estimation and 40 s for correspondence matching, and 50 s for velocity field reconstruction.

4.1 Synthetic Data

We first evaluate our proposed approach on simulated flows: a vortex flow and a drop flow. The flows are simulated within a volume of $100 \times 100 \times 20$ voxels. We randomly sample tracer particles within the fluid volume. The particle density is 0.02 per voxel. We render light fields images with angular resolution 7×7 and spatial resolution 434×625. We simulate the advection of particles over time following the method in [40]. We apply our algorithms on the rendered light fields to recover 3D fluid flows. In Fig. 5, we show our recovered velocity fields in comparison with the ground truth ones. Qualitatively, our reconstructed vector fields are highly consistent with the ground truth ones.

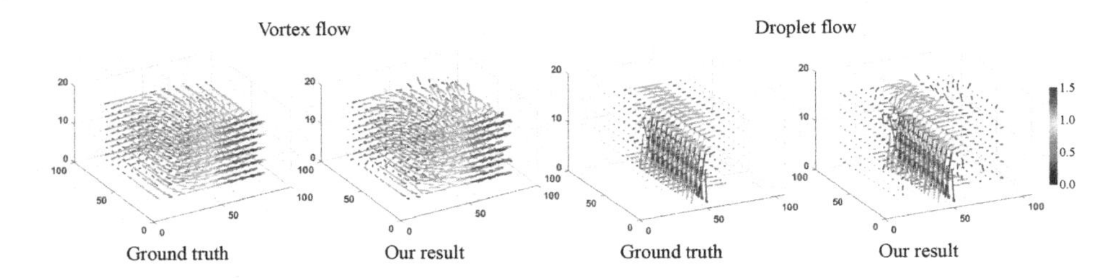

Fig. 5. Synthetic results in comparison with the ground truth.

We perform quantitative evaluations using two error metrics: the average end-point error (AEE) and the average angular error (AAE). AEE is computed as the averaged Euclidean distance between the estimated particle positions and ground truth ones. AAE is computed with the average difference of vector in the velocity field. We compare our method with the multi-scale Horn-Schunck (H & S) [30] and the rainbow PIV [47]. Specifically, we apply H & S on our recovered 3D particles and use it as the baseline algorithm for flow estimation. With this comparison, we hope to demonstrate the effectiveness of our regularization terms in flow estimation. For rainbow PIV, we have implemented a renderer to generate depth-dependent spectral images of virtual particles. To ensure fairness, the rendered images have the same spatial resolution as our input light field (ie, 434×625).

We also perform ablation study by testing two variants of our method: "w/o E_{corres}" that takes out the correspondence term and "w/o E_{div}" that takes out the divergence-free term. The experiments are performed on the vortex flow with particle density 0.02. Quantitative evaluations are shown in Fig. 6. The error maps of recovered velocity fields for our ablation study are shown Fig. 7. We can see that our method achieves the best performance when both regularization terms are imposed. Our outperforms both H & S and the rainbow PIV at various particle density levels. Further, our accumulated error over time grows much slower than the other two state-of-the-arts.

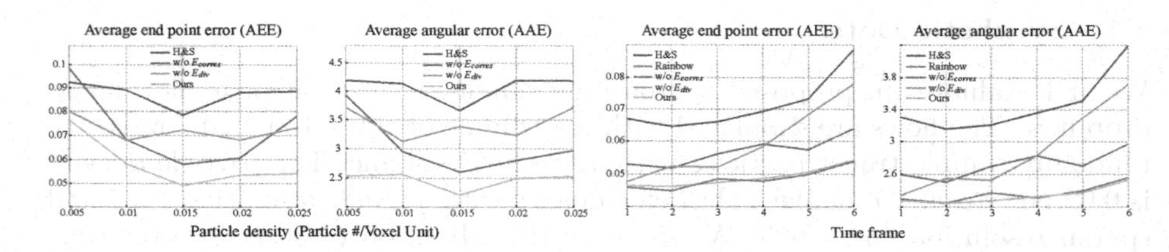

Fig. 6. Quantitative evaluation. The left two plots show errors with respect to different particle densities. The right two plots show accumulated errors over time.

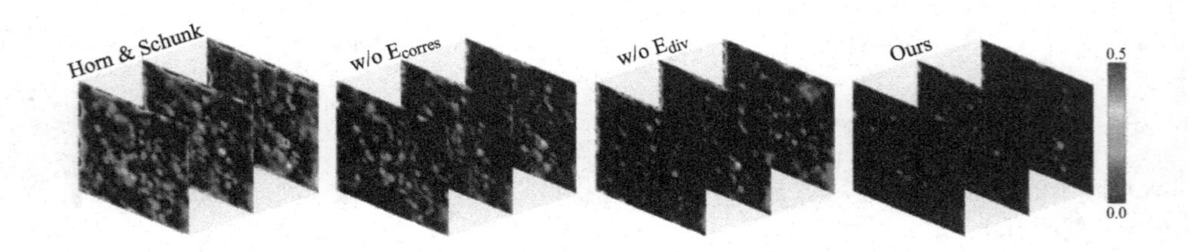

Fig. 7. Ablation study. We show the error maps of estimated velocity field at three fluid volume slices.

4.2 John Hopkins Turbulence Database (JHUTDB)

Next we conduct experiments on data generated from the Johns Hopkins Turbulence Database (JHUTDB) [24]. To reduce processing time, we crop out a volume of $256 \times 128 \times 80$ voxels for each turbulence in the dataset. The norm of the velocity field at each location ranges from 0 to 2.7 voxels per time step. We generate random tracer particles with density 0.025 per voxels and advect the particles according to the turbulence velocity field. In our evaluation, we render two light field images at two consecutive frames to estimate the particle locations and reconstruct the velocity field. Our reconstruction results in comparison with the ground truth is shown in Fig. 8. We show our reconstructed velocity volume in x, y, z directions. We also show the error map of magnitudes to illustrate that our method is highly accurate.

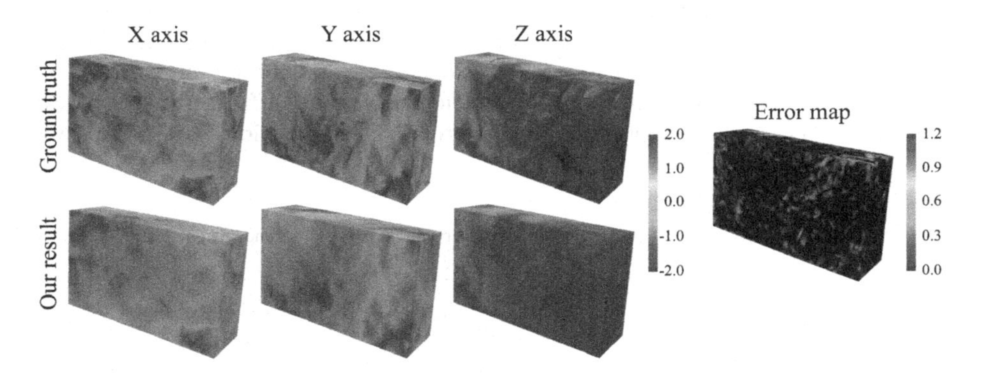

Fig. 8. JHUTDB velocity field reconstruction results.

4.3 Real Data

We finally test our method on real captured flow data. Figure 9 shows our acquisition system for capturing real 3D flows. We use a Lytro Illum light field camera with 30 mm focal length to capture the tracer particles in fluid. As Illum does not have video mode, we use an external control board to trigger the camera at high frequency to capture consecutive time frames. Due to the limitation of on-chip image buffer size, our acquisition cannot achieve very high frame rate. In our experiment, we set the trigger frequency to be 10 Hz. The capture light field has angular resolution 15×15 and spatial resolution 625 × 434. We use the light field calibration toolbox [8] to process and decode raw light field data into sub-aperture images. We

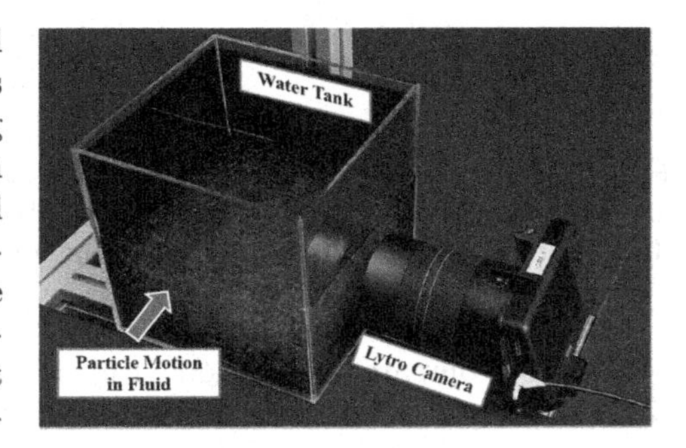

Fig. 9. Our real experiment setup. We use a compact light field camera in PIV setting.

use the center view as reference for depth estimation and the effective depth volume that we are able to reconstruct is around $600 \times 500 \times 200$ (mm), slightly lower than the capture image because we enforce rectangular volumes inside the perspective view frustum.

We use green polyethylene microspheres with density 1g/cc and size 1000–1180 µm as tracer particles. Before dispersing the particles, we mix some surfactant with the particles to reduce surface tension caused by water in order to minimize agglomeration between particles. We test on three types of flows: vortex, double vortex, and random complex flows.

Figure 10 shows our recovered fluid flow velocity field and path line visualization (please refer to the supplemental material for more reconstruction results). We show three flow types, vortex, double vortex, and random complex flows. The left column shows the velocity field between first and the second frame. The right column shows the path line visualization through 1–4 frames. We can see that our reconstructions well depicts the intended fluid motions and are highly reliable.

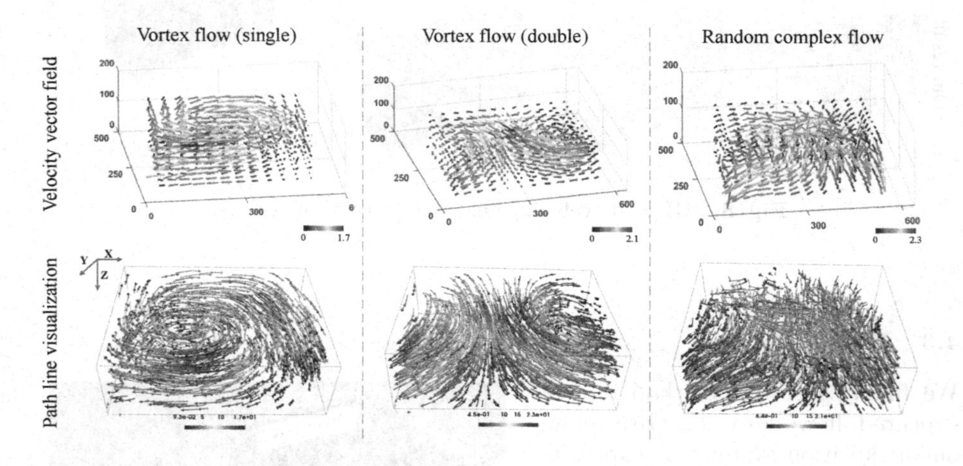

Fig. 10. Real experiment results. We show our recovered velocity fields (upper row) and path line visualizations on four consecutive frames (lower row) for three types of flows: vortex, double vortex and a random complex flow.

We also compare our method with a recent state-of-the-art scene flow method [26] on the real data. The scene flow method takes two consecutive RGB-D images as inputs and use rigidity transform network and flow network for motion estimation. Since the method also needs depth map as input, we first calculate a depth map for the center view of light field and then combine the depth map with the sub-aperture color image and use them as input for [26]. The flow estimation results are shown in Fig. 11. We show the projected scene flows and the flow vector fields for three types of flows (single vortex, double vortex, and random flow). The scene flow method fails to recover the flow structures,

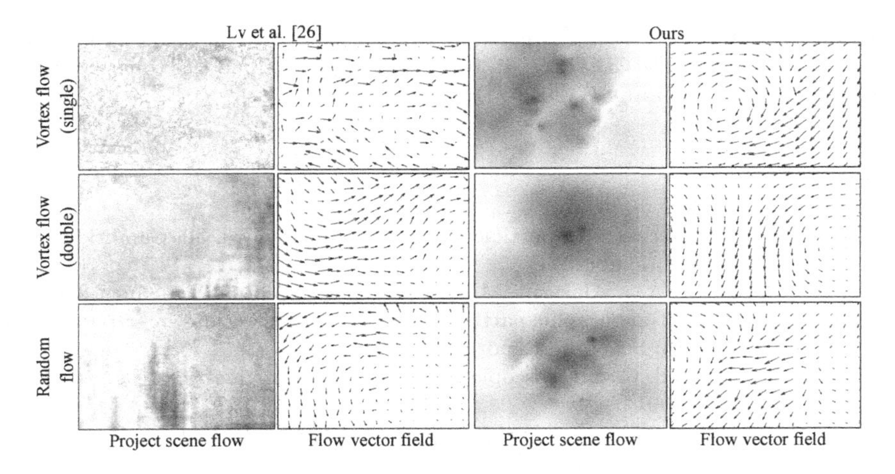

Fig. 11. Comparison result with scene flow (Lv et al. [26]) on the real data. We compare the project scene flow and the flow vector field on three types of flows.

especially for vortex flows. This is because our particles are heavily occluded and have very similar appearances. Further, the scene flow algorithm does not take the physical properties of fluid into consideration.

5 Conclusions

In this paper, we have presented a light field PIV solution that uses a commercial compact light field camera to recover volumetric 3D fluid motion from tracer particles. We have developed a 3D particle reconstruction algorithm by exploiting the light field focal stack symmetry in order to handle heavily occluded particles. To recover the fluid flow, we have refined the classical optical flow framework by introducing two novel regularization terms: 1) the correspondence term to enforce one-to-one particle matching; and 2) the divergence-free term to enforce the physical properties of incompressible fluid. Comprehensive synthetic and real experiments as well as comparisons with the state-of-the-arts have demonstrated the effectiveness of our method.

Although our method can faithfully recover fluid flows in a small to medium volume, our method still has several limitations. First of all, due to the small baseline of compact light field camera, the resolvable depth range is rather limited. As a result, our volumetric velocity field's resolution along the z-axis is much smaller than its x- or y-resolutions. One way to enhance the z-resolution is using a second light field camera capturing the fluid volume from an orthogonal angle. Second, in our fluid flow reconstruction step, only two consecutive frames are considered. Hence motion continuity might not always be satisfied. Adding temporal constraint to our optimization framework can be further improved.

Acknowledgments. This work is partially supported by the National Science Foundation (NSF) under Grant CBET-1706130 and CRII-1948524, and the Louisiana Board of Regent under Grant LEQSF (2018-21)-RD-A-10.

References

1. Adrian, R.J., Westerweel, J.: Particle Image Velocimetry, vol. 30. Cambridge University Press, Cambridge (2011)
2. Aguirre-Pablo, A.A., Alarfaj, M.K., Li, E.Q., Hernández-Sánchez, J.F., Thoroddsen, S.T.: Tomographic particle image velocimetry using smartphones and colored shadows. In: Scientific Reports (2017)
3. Arroyo, M., Greated, C.: Stereoscopic particle image velocimetry. Meas. Sci. Technol. **2**(12), 1181 (1991)
4. Atcheson, B., et al.: Time-resolved 3D capture of non-stationary gas flows. ACM Trans. Graph. (TOG) **27**, 132 (2008)
5. Belden, J., Truscott, T.T., Axiak, M.C., Techet, A.H.: Three-dimensional synthetic aperture particle image velocimetry. Meas. Sci. Technol. **21**(12), 125403 (2010)
6. Brücker, C.: 3D scanning piv applied to an air flow in a motored engine using digital high-speed video. Meas. Sci. Technol. **8**(12), 1480 (1997)
7. Crocker, J.C., Grier, D.G.: Methods of digital video microscopy for colloidal studies. J. Colloid Interface Sci. **179**(1), 298–310 (1996)
8. Dansereau, D.G., Pizarro, O., Williams, S.B.: Decoding, calibration and rectification for lenselet-based plenoptic cameras. In: Proceedings of the IEEE Conference on Computer Vision and Pattern Recognition, pp. 1027–1034 (2013)
9. Elsinga, G.E., Scarano, F., Wieneke, B., van Oudheusden, B.W.: Tomographic particle image velocimetry. Exp. Fluids **41**(6), 933–947 (2006)
10. Fahringer, T., Thurow, B.: Tomographic reconstruction of a 3-D flow field using a plenoptic camera. In: 42nd AIAA Fluid Dynamics Conference and Exhibit, p. 2826 (2012)
11. Gregson, J., Ihrke, I., Thuerey, N., Heidrich, W.: From capture to simulation: connecting forward and inverse problems in fluids. ACM Trans. Graph. (TOG) **33**(4), 139 (2014)
12. Gu, J., Nayar, S.K., Grinspun, E., Belhumeur, P.N., Ramamoorthi, R.: Compressive structured light for recovering inhomogeneous participating media. IEEE Trans. Pattern Anal. Mach. Intell. **35**, 1 (2013)
13. Hasinoff, S.W., Kutulakos, K.N.: Photo-consistent reconstruction of semitransparent scenes by density-sheet decomposition. IEEE Trans. Pattern Anal. Mach. Intell. **29**, 870–885 (2007)
14. Hawkins, T., Einarsson, P., Debevec, P.: Acquisition of time-varying participating media. ACM Trans. Graph. (ToG) **24**, 812–815 (2005)
15. Heitz, D., Mémin, E., Schnörr, C.: Variational fluid flow measurements from image sequences: synopsis and perspectives. Exp. Fluids **48**(3), 369–393 (2010)
16. Hori, T., Sakakibara, J.: High-speed scanning stereoscopic piv for 3D vorticity measurement in liquids. Meas. Sci. Technol. **15**(6), 1067 (2004)
17. Horn, B.K., Schunck, B.G.: Determining optical flow. Artif. Intell. **17**(1–3), 185–203 (1981)
18. Ihrke, I., Kutulakos, K.N., Lensch, H.P., Magnor, M., Heidrich, W.: Transparent and specular object reconstruction. In: Computer Graphics Forum, vol. 29, pp. 2400–2426. Wiley Online Library (2010)

19. Ihrke, I., Magnor, M.A.: Image-based tomographic reconstruction of flames. In: Symposium on Computer Animation (2004)
20. Ji, Y., Ye, J., Yu, J.: Reconstructing gas flows using light-path approximation. In: 2013 IEEE Conference on Computer Vision and Pattern Recognition, pp. 2507–2514 (2013)
21. Kajitani, L., Dabiri, D.: A full three-dimensional characterization of defocusing digital particle image velocimetry. Meas. Sci. Technol. **16**(3), 790 (2005)
22. Lasinger, K., Vogel, C., Schindler, K.: Volumetric flow estimation for incompressible fluids using the stationary stokes equations. In: 2017 IEEE International Conference on Computer Vision (ICCV), pp. 2584–2592. IEEE (2017)
23. Lasinger, K., Vogel, C., Schindler, K.: Volumetric flow estimation for incompressible fluids using the stationary stokes equations. In: 2017 IEEE International Conference on Computer Vision (ICCV), pp. 2584–2592 (2017)
24. Li, Y., et al.: A public turbulence database cluster and applications to study lagrangian evolution of velocity increments in turbulence. J. Turbul. (9), N31 (2008)
25. Lin, H., Chen, C., Bing Kang, S., Yu, J.: Depth recovery from light field using focal stack symmetry. In: Proceedings of the IEEE International Conference on Computer Vision, pp. 3451–3459 (2015)
26. Lv, Z., Kim, K., Troccoli, A., Sun, D., Rehg, J.M., Kautz, J.: Learning rigidity in dynamic scenes with a moving camera for 3D motion field estimation. In: Proceedings of the European Conference on Computer Vision (ECCV), pp. 468–484 (2018)
27. Lynch, K., Fahringer, T., Thurow, B.: Three-dimensional particle image velocimetry using a plenoptic camera. In: 50th AIAA Aerospace Sciences Meeting including the New Horizons Forum and Aerospace Exposition, p. 1056 (2012)
28. Ma, C., Lin, X., Suo, J., Dai, Q., Wetzstein, G.: Transparent object reconstruction via coded transport of intensity. In: Proceedings of the IEEE Conference on Computer Vision and Pattern Recognition, pp. 3238–3245 (2014)
29. Maas, H., Gruen, A., Papantoniou, D.: Particle tracking velocimetry in three-dimensional flows. Exp. Fluids **15**(2), 133–146 (1993)
30. Meinhardt, E., Pérez, J.S., Kondermann, D.: Horn-schunck optical flow with a multi-scale strategy. IPOL J. **3**, 151–172 (2013)
31. Mendelson, L., Techet, A.H.: Quantitative wake analysis of a freely swimming fish using 3D synthetic aperture PIV. Exp. Fluids **56**(7), 135 (2015)
32. Morris, N.J., Kutulakos, K.N.: Dynamic refraction stereo. IEEE Trans. Pattern Anal. Mach. Intell. **33**(8), 1518–1531 (2011)
33. Ng, R., Levoy, M., Brédif, M., Duval, G., Horowitz, M., Hanrahan, P., et al.: Light field photography with a hand-held plenoptic camera. Comput. Sci. Tech. Rep. CSTR **2**(11), 1–11 (2005)
34. Perlman, E., Burns, R., Li, Y., Meneveau, C.: Data exploration of turbulence simulations using a database cluster. In: Proceedings of the 2007 ACM/IEEE Conference on Supercomputing, p. 23. ACM (2007)
35. Pick, S., Lehmann, F.O.: Stereoscopic PIV on multiple color-coded light sheets and its application to axial flow in flapping robotic insect wings. Exp. Fluids **47**(6), 1009 (2009)
36. Schanz, D., Gesemann, S., Schröder, A.: Shake-the-box: lagrangian particle tracking at high particle image densities. Exp. Fluids **57**(5), 70 (2016)
37. Shi, S., Ding, J., Atkinson, C., Soria, J., New, T.H.: A detailed comparison of single-camera light-field PIV and tomographic PIV. Exp. Fluids **59**, 1–13 (2018)

38. Shi, S., Ding, J., New, T.H., Soria, J.: Light-field camera-based 3D volumetric particle image velocimetry with dense ray tracing reconstruction technique. Exp. Fluids **58**, 1–16 (2017)
39. Soria, J., Atkinson, C.: Towards 3C–3D digital holographic fluid velocity vector field measurement? Tomographic digital holographic PIV (tomo-HPIV). Meas. Sci. Technol. **19**(7), 074002 (2008)
40. Stam, J.: Stable fluids. In: Proceedings of the 26th Annual Conference on Computer Graphics and Interactive Techniques, pp. 121–128. ACM Press/Addison-Wesley Publishing Co. (1999)
41. Strecke, M., Alperovich, A., Goldluecke, B.: Accurate depth and normal maps from occlusion-aware focal stack symmetry. In: 2017 IEEE Conference on Computer Vision and Pattern Recognition (CVPR), pp. 2529–2537. IEEE (2017)
42. Sumner, R.W., Schmid, J., Pauly, M.: Embedded deformation for shape manipulation. ACM Trans. Graph. (TOG) **26**, 80 (2007)
43. Vedula, S., Baker, S., Rander, P., Collins, R.T., Kanade, T.: Three-dimensional scene flow. In: Proceedings of the Seventh IEEE International Conference on Computer Vision, vol. 2, pp. 722–729 (1999)
44. Wieneke, B.: Volume self-calibration for 3D particle image velocimetry. Exp. Fluids **45**(4), 549–556 (2008)
45. Willert, C., Gharib, M.: Three-dimensional particle imaging with a single camera. Exp. Fluids **12**(6), 353–358 (1992)
46. Xiong, J., Fu, Q., Idoughi, R., Heidrich, W.: Reconfigurable rainbow PIV for 3D flow measurement. In: 2018 IEEE International Conference on Computational Photography (ICCP), pp. 1–9. IEEE (2018)
47. Xiong, J., et al.: Rainbow particle imaging velocimetry for dense 3D fluid velocity imaging. ACM Trans. Graph. (TOG) **36**(4), 36 (2017)
48. Xue, T., Rubinstein, M., Wadhwa, N., Levin, A., Durand, F., Freeman, W.T.: Refraction wiggles for measuring fluid depth and velocity from video. In: Fleet, D., Pajdla, T., Schiele, B., Tuytelaars, T. (eds.) ECCV 2014. LNCS, vol. 8691, pp. 767–782. Springer, Cham (2014). https://doi.org/10.1007/978-3-319-10578-9_50
49. Ye, J., Ji, Y., Li, F., Yu, J.: Angular domain reconstruction of dynamic 3D fluid surfaces. In: 2012 IEEE Conference on Computer Vision and Pattern Recognition, pp. 310–317. IEEE (2012)
50. Zhang, J., Tao, B., Katz, J.: Turbulent flow measurement in a square duct with hybrid holographic PIV. Exp. Fluids **23**(5), 373–381 (1997)

Contextual Diversity for Active Learning

Sharat Agarwal[1], Himanshu Arora[2], Saket Anand[1](✉), and Chetan Arora[3]

[1] IIIT-Delhi, New Delhi, India
{sharata,anands}@iiitd.ac.in
[2] Flixstock Inc., New Delhi, India
himanshu@flixstock.com
[3] Indian Institute of Technology Delhi, New Delhi, India
chetan@cse.iitd.ac.in

Abstract. Requirement of large annotated datasets restrict the use of deep convolutional neural networks (CNNs) for many practical applications. The problem can be mitigated by using active learning (AL) techniques which, under a given annotation budget, allow to select a subset of data that yields maximum accuracy upon fine tuning. State of the art AL approaches typically rely on measures of visual diversity or prediction uncertainty, which are unable to effectively capture the variations in spatial context. On the other hand, modern CNN architectures make heavy use of spatial context for achieving highly accurate predictions. Since the context is difficult to evaluate in the absence of ground-truth labels, we introduce the notion of contextual diversity that captures the confusion associated with spatially co-occurring classes. Contextual Diversity (CD) hinges on a crucial observation that the probability vector predicted by a CNN for a region of interest typically contains information from a larger receptive field. Exploiting this observation, we use the proposed CD measure within two AL frameworks: (1) a core-set based strategy and (2) a reinforcement learning based policy, for active frame selection. Our extensive empirical evaluation establish state of the art results for active learning on benchmark datasets of Semantic Segmentation, Object Detection and Image classification. Our ablation studies show clear advantages of using contextual diversity for active learning. The source code and additional results are available at https://github.com/sharat29ag/CDAL.

1 Introduction

Deep convolutional neural networks (CNNs) have acheived state of the art (SOTA) performance on various computer vision tasks. One of the key driving factors for this success has been the effort gone in preparing large amounts

Sharat Agarwal and Himanshu Arora—Equal contribution.
Himanshu Arora—Work done while the author was at IIIT-Delhi.

Electronic supplementary material The online version of this chapter (https://doi.org/10.1007/978-3-030-58517-4_9) contains supplementary material, which is available to authorized users.

A. Vedaldi et al. (Eds.): ECCV 2020, LNCS 12361, pp. 137–153, 2020.
https://doi.org/10.1007/978-3-030-58517-4_9

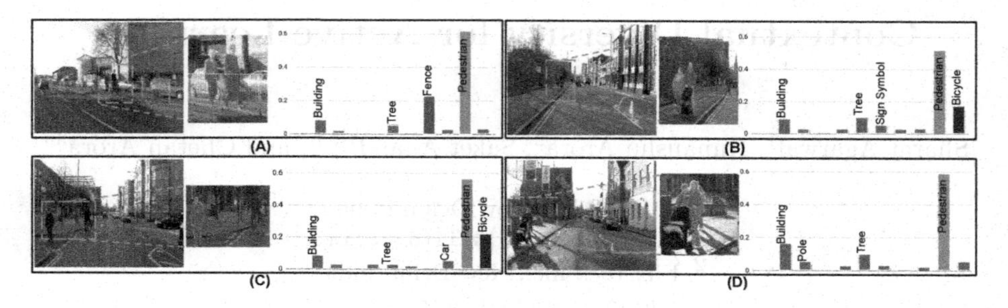

Fig. 1. Illustration showing 4 frames from Camvid. Each subfigure shows the full RGB image, region of interest with ground truth overlaid, and the average probability for the 'pedestrian' class with bars color coded by class. We observe that the confusion reflected by the average probability vector corresponding to a class in a frame is also influenced by the object's background. Notice the confusion of pedestrian class with fence in (A) and with bicycle in (C), each of which appear in the neighborhood of a pedestrian instance. We propose a novel *contextual diversity* based measure that exploits the above structure in probability vectors to help select images containing objects in diverse backgrounds. Including this set of images for training helps improving accuracy of CNN-based classifiers, which rely on the local spatial neighborhoods for prediction. For the above example our contextual diversity based selection picks $\{(A),(C),(D)\}$ as opposed to the set $\{(B),(C),(D)\}$ picked by a maximum entropy based strategy (best viewed in color). (Color figure online)

of labeled training data. As CNNs become more popular, they are applied to diverse tasks from disparate domains, each of which may incur annotation costs that are task as well as domain specific. For instance, the annotation effort in image classification is substantially lower than that of object detection or semantic segmentation in images or videos. Similarly, annotations of RGB images may be cheaper than MRI/CT images or Thermal IR images, which may require annotators with specialized training.

The core idea of Active Learning (AL) is to leverage the current knowledge of a machine learning model to select most *informative* samples for labeling, which would be more beneficial to model improvement compared to a randomly chosen data point [32]. With the effectiveness of deep learning (DL) based models in recent years, AL strategies have been investigated for these models as well. Here, it has been shown that DL models trained with a fraction of available training samples selected by active learning can achieve nearly the same performance as when trained with all available data [31,35,40]. Since DL models are expensive to train, AL strategies for DL typically operate in a batch selection setting, where a set of images are selected and annotated followed by retraining or fine-tuning of the model using the selected set.

Traditional AL techniques [13,19,20,24,33] have mostly been based on *uncertainty* and have exploited the ambiguity in the predicted output of a model. As most measures of uncertainty employed are based on predictions of individual samples, such approaches often result in highly correlated selections in the batch

AL setting. Consequently, more recent AL techniques attempt to reduce this correlation by following a strategy based on the *diversity* and *representativeness* of the selected samples [17,36,39]. Existing approaches that leverage these cues are still insufficient in adequately capturing the spatial and semantic context within an image and across the selected set. Uncertainty, typically measured through entropy, is also unable to capture the class(es) responsible for the resulting uncertainty. On the other hand, visual diversity and representativeness are able to capture the semantic context across image samples, but are typically measured using global cues in a feature space that do not preserve information about the spatial location or relative placement of the image's constituent objects.

Spatial context is an important aspect of modern CNNs, which are able to learn discriminative semantic representations due to their large receptive fields. There is sufficient evidence that points to the brittleness of CNNs as object locations, or the spatial context, in an image are perturbed [30]. In other words, a CNN based classifier's misclassification is not simply attributed to the objects from the true class, but also to other classes that may appear in the object's spatial neighborhood. This crucial observation also points to an important gap in the AL literature, where existing measures are unable to capture uncertainty arising from the diversity in spatial and semantic context in an image. Such a measure would help select a training set that is diverse enough to cover a *variety of object classes* and their *spatial co-occurrence* and thus improve generalization of CNNs. The objective of this paper is to achieve this goal by designing a novel measure for active learning which helps select frames having objects in diverse contexts and background. Figure 1 describes an illustrative comparison of some of the samples selected by our approach with the entropy based one.

In this paper, we introduce the notion of contextual diversity, which permits us to unify the model prediction uncertainty with the diversity among samples based upon spatial and semantic context in the data. We summarize our contributions below:

- We introduce a novel information-theoretic distance measure, Contextual Diversity (CD), to capture the diversity in spatial and semantic context of various object categories in a dataset.
- We demonstrate that using CD with core-set based active learning [31] almost always beats the state of the art across three visual recognition tasks: semantic segmentation, object detection and image classification. We show an improvement of 1.1, 1.1, and 1.2 units on the three tasks, over the state of the art performance achieving 57.2, 73.3, and 80.9 respectively.
- Using CD as a reward function in an RL framework further improves the AL performance and achieves an improvement of 2.7, 2.1, and 2.3 units on the respective visual recognition tasks over state of the art (57.2, 73.3, and 80.9 respectively).
- Through a series of ablation experiments, we show that CD complements existing cues like visual diversity.

2 Related Work

Active learning techniques can be broadly categorized into the following categories. Query by committee methods operate on consensus by several models [2,10]. However, these approaches in general are too computationally expensive to be used with deep neural networks and big datasets. Diversity-based approaches identify a subset of a dataset that is sufficiently representative of the entire dataset. Most approaches in this category leverage techniques like clustering [29], matrix partitioning [12], or linkage based similarity [3]. Uncertainty based approaches exploit the ambiguity in the predicted output of a model. Some representative techniques in this class include [13,19,20,24,33]. Some approaches attempt to combine both uncertainty and diversity cues for active sample selection. Some notable works in this category include [17,21,36,39]. Recently, generative models have also been used to synthesize informative samples for Active Learning [27,28,44]. In the following, we give a detailed review of three recent state of the art AL approaches applied to vision related tasks. We compare with these methods later in the experiment sections over the three visual recognition tasks.

Core-Set. Sener and Savarese [31] have modeled active learning as a *core-set* selection problem in the feature space learned by convolutional neural networks (CNNs). Here, the core-set is defined as a selected subset of points such that the union of $\mathbb{R}^n$-balls of radius δ around these points contain *all* the remaining unlabeled points. The main advantage of the method is in its theoretical guarantees, which claim that the difference between the loss averaged over all the samples and that averaged over the selected subset does not depend on the *number of samples* in the selected subset, but only on the radius δ. Following this result, Sener and Savarese used approximation algorithms to solve a facility location problem using a Euclidean distance measure in the feature space. However, as was noted by [35], reliance on Euclidean distance in a high-dimensional feature space is ineffective. Our proposed contextual diversity measure relies on KL divergence, which is known to be an effective surrogate for distance in the probability space [6]. Due to distance like properties of our measure, the proposed approach, named *contextual diversity based active learning using core-sets* (CDAL-CS), respects the theoretical guarantees of core-set, yet does not suffer from curse of dimensionality.

Learning Loss. Yoo and Kweon [40] have proposed a novel measure of uncertainty by learning to predict the loss value of a data sample. They sampled data based on the ranking obtained on the basis of predicted loss value. However, it is not clear if the sample yielding the largest loss, is also the one that leads to most performance gain. The samples with the largest loss, could potentially be outliers or label noise, and including them in the training set may be misleading to the network. The other disadvantage of the technique is that, there is no obvious way to choose the diverse samples based upon the predicted loss values.

Variational Adversarial Active Learning (VAAL). Sinha et al. [35] have proposed to use a VAE to map both the labeled and unlabeled data into a latent

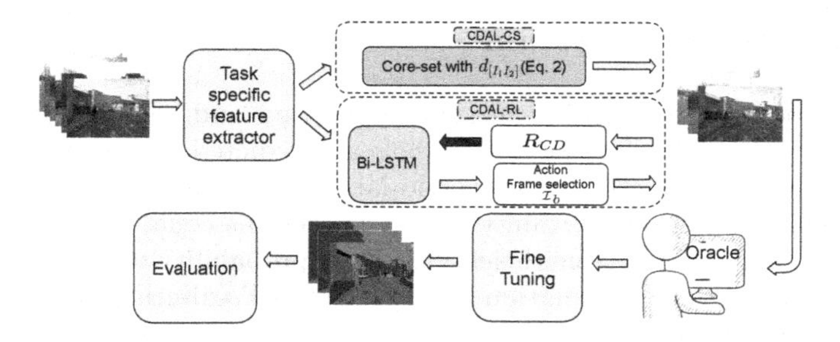

Fig. 2. The architecture for the proposed frame selection technique. Two of the strengths of our technique are its unsupervised nature and its generalizability to variety of tasks. The frame selection can be performed in either way by CDAL-CS or CDAL-RL modules. Based on the visual task, a pre-trained model can be readily integrated. The top scoring frames are selected to be annotated and are used to fine tune the model to be evaluated over the main task.

space, followed by a discriminator to distinguish between the two based upon their latent space representation. The sample selection is simply based on the output probability of the discriminator. Similar to [40], there seem to be no obvious way to choose diverse samples in their technique based on the discriminator score only. Further, there is no guarantee that the representation learnt by their VAE is closer to the one used by the actual model for the task. Therefore, the most informative frame for the discriminator need not be the same for the target model as well. Nonetheless, in the empirical analysis, VAAL demonstrates state of the art performance among the other active learning techniques for image classification and semantic segmentation tasks.

Reinforcement Learning for Active Learning. Recently, there has been an increasing interest in application of RL based methods to active learning. RALF [7] takes a meta-learning approach where the policy determines a weighted combination of pre-defined uncertainty and diversity measure, which is then used for sample selection. Both [38] and [23] train the RL agents using ground truth based rewards for one-shot learning and person re-identification separately. This requires their method to have a large, annotated training set to learn the policy, and therefore is hard to generalize to more annotation heavy tasks like object detection and semantic segmentation. In [15], an RL framework minimizes time taken for object-level annotation by choosing between bounding box verification and drawing tasks. Fang et al. [9] design a special state space representation to capture uncertainty and diversity for active learning for text data. This design makes it harder to generalize their model to other tasks. Contrary to most of these approaches, our RL based formulation, CDAL-RL, takes a task specific state representation and uses the contextual diversity based reward that combines uncertainty and diversity in an unsupervised manner.

3 Active Frame Selection

One of the popular approaches in semi-supervised and self-supervised learning is to use *pseudo-labels*, which are labels as predicted by the current model [4,18,36]. However, directly using pseudo-labels in training, without appropriately accounting for the uncertainty in the predictions could lead to overfitting and confirmation bias [1]. Nonetheless, the class probability vector predicted by a model contains useful information about the model's discriminative ability. In this section, we present the proposed *contextual diversity* (CD), an information-theoretic measure that forms the basis for Contextual Diversity based Active Learning (CDAL). At the heart of CD is our quantification of the model's predictive uncertainty defined as a mixture of softmax posterior probabilities of pseudo-labeled samples. This mixture distribution effectively captures the spatial and semantic context over a set of images. We then derive the CD measure, which allows us to select diverse and uncertain samples from the unlabeled pool for annotation and finally suggest two strategies for active frame selection. First (CDAL-CS), inspired by the core-set [31] approach and the second (CDAL-RL) using a reinforcement learning framework. An overview of our approach to Active Learning is illustrated in Fig. 2.

3.1 Contextual Diversity

Deep CNNs have large receptive fields to capture sufficient spatial context for learning discriminative semantic features, however, it also leads to feature interference making the output predictions more ambiguous [30]. This spatial pooling of features adds to confusion between classes, especially when a model is not fully trained and has noisy feature representations. We quantify this ambiguity by defining the *class-specific confusion*.

Let $C = \{1, \ldots, n_C\}$ be the set of classes to be predicted by a Deep CNN based model. Given a region r within an input image $\mathbf{I}$, let $\boldsymbol{P}_r(\widehat{y} \mid \mathbf{I}; \boldsymbol{\theta})$ be the *softmax probability vector* as predicted by the model $\boldsymbol{\theta}$. For convenience of notation, we will use $\boldsymbol{P}_r$ instead of $\boldsymbol{P}_r(\widehat{y}|\mathbf{I}; \boldsymbol{\theta})$ as the subscript r implies the conditioning on its constituent image $\mathbf{I}$ and the model $\boldsymbol{\theta}$ is fixed in one step of sample selection. These regions could be pixels, bounding boxes or the entire image itself depending on the task at hand. The pseudo-label for the region $r \subseteq \mathbf{I}$ is defined as $\widehat{y}_r = \arg\max_{j \in C} \boldsymbol{P}_r[j]$, where the notation $\boldsymbol{P}_r[j]$ denotes the jth element of the vector. We emphasize that this abstraction of regions is important as it permits us to define overlapping regions within an image and yet have different predictions, thereby catering to tasks like object detection. Let $\mathcal{I} = \cup_{c \in C} \mathcal{I}^c$ be the total pool of *unlabeled* images, where $\mathcal{I}^c$ is the set of images, each of which have at least one region classified by the model into class c. Further, let $\mathcal{R}^c_{\mathbf{I}}$ be the set of regions within image $\mathbf{I} \in \mathcal{I}^c$ that are assigned a pseudo-label c. The collection of *all* the regions that the model believes belong to class c is contained within the set $\mathcal{R}^c_{\mathcal{I}} = \cup_{\mathbf{I} \in \mathcal{I}^c} \mathcal{R}^c_{\mathbf{I}}$. We assume that for a sufficiently large unlabeled pool $\mathcal{I}$, there will be a non-empty set $\mathcal{R}^c_{\mathcal{I}}$. For a given model $\boldsymbol{\theta}$ over the unlabeled pool $\mathcal{I}$, we now define the class-specific confusion for class c by the following mixture distribution $\boldsymbol{P}_{\mathcal{I}}^{c}$

$$\boldsymbol{P}_{\mathcal{I}}^{c} = \frac{1}{|\mathcal{I}^{c}|} \sum_{\mathbf{I} \in \mathcal{I}^{c}} \left[\frac{\sum_{r \in \mathcal{R}_{\mathbf{I}}^{c}} w_r \boldsymbol{P}_r(\widehat{y} \mid \mathbf{I}; \boldsymbol{\theta})}{\sum_{r \in \mathcal{R}_{\mathbf{I}}^{c}} w_r} \right] \tag{1}$$

with $w_r \geq 0$ as the mixing weights. While the weights could take any non-negative values, we are interested in capturing the predictive uncertainty of the model. Therefore, we choose the weights to be the Shannon entropy of $w_r = \mathbb{H}(\boldsymbol{P}_r) = -\sum_{j \in C} \boldsymbol{P}_r[j] \log_2 \boldsymbol{P}_r[j] + \epsilon$, where $\epsilon > 0$ is a small constant and avoid any numerical instabilities. If the model were perfect, $\boldsymbol{P}_{\mathcal{I}}^{c}$ would be a one-hot encoded vector[1], but for an insufficiently trained model $\boldsymbol{P}_{\mathcal{I}}^{c}$ will have a higher entropy indicating the confusion between class c and all other classes ($c' \in C, c' \neq c$). We use $\boldsymbol{P}_{\mathbf{I}}^{c}$ to denote the mixture computed from a single image $\mathbf{I} \in \mathcal{I}^{c}$.

As discussed in Sect. 1, in CNN based classifiers, this uncertainty stems from spatial and semantic context in the images. For instance, in the semantic segmentation task shown in Fig. 1, the model may predict many pixels as of class 'pedestrian' (c = pedestrian) with the highest probability, yet it would have a sufficiently high probability of another class like 'fence' or 'bicycle'. In such a case, $\boldsymbol{P}_{\mathcal{I}}^{c}[j]$ will have high values at $j = \{\text{fence, bicycle}\}$, reflecting the chance of confusion between these classes across the unlabeled pool $\mathcal{I}$. As the predictive ability of the model increases, we expect the probability mass to get more concentrated at $j = c$ and therefore reduce the overall entropy of $\boldsymbol{P}_{\mathcal{I}}^{c}$. It is easy to see that the total Shannon's entropy $h_{\mathcal{I}} = \sum_{c \in C} \mathbb{H}(\boldsymbol{P}_{\mathcal{I}}^{c})$ reduces with the cross-entropy loss.

Annotating an image and using it to train a model would help resolve the confusion constituent in that image. Based on this intuition, we argue that the annotation effort for a new image is justified only if its inclusion increases the informativeness of the selected subset, i.e., when an image captures a *different kind* of confusion than the rest of the subset. Therefore, for a given pair of images $\mathbf{I}_1$ and $\mathbf{I}_2$, we quantify the disparity between their constituent class-specific confusion by the *pairwise contextual diversity* defined using a symmetric KL-divergence as

$$d_{[\mathbf{I}_1, \mathbf{I}_2]} = \sum_{c \in C} \mathbb{1}^{c}(\mathbf{I}_1, \mathbf{I}_2) \left(0.5 * \mathrm{KL}(\boldsymbol{P}_{\mathbf{I}_1}^{c} \parallel \boldsymbol{P}_{\mathbf{I}_2}^{c}) + 0.5 * \mathrm{KL}(\boldsymbol{P}_{\mathbf{I}_2}^{c} \parallel \boldsymbol{P}_{\mathbf{I}_1}^{c}) \right). \tag{2}$$

In Eq. (2), $\mathrm{KL}(\cdot \parallel \cdot)$ denotes the KL-divergence between the two mixture distributions. We use the indicator variable denoted by $\mathbb{1}^{c}(\cdot)$ that takes a value of one only if both $\mathbf{I}_1, \mathbf{I}_2 \in \mathcal{I}^{c}$, otherwise zero. This variable ensures that the disparity in class-specific confusion is considered only when both images have at least one region pseudo-labeled as class c, i.e., when both images have a somewhat reliable measure of confusion w.r.t. class c. This confusion disparity accumulated

[1] We ignore the unlikely event where the predictions are perfectly consistent over the large unlabeled pool $\mathcal{I}$, yet different from the *true* label.

over all classes is the pairwise contextual diversity measure between two images. Given that the KL-divergence captures a distance between two distributions, $d_{[\mathbf{I}_1,\mathbf{I}_2]}$ can be used as a distance measure between two images in the probability space. Thus, using pairwise distances, we can take a core-set [31] style approach for sample selection. Additionally, we can readily aggregate $d_{[\mathbf{I}_m,\mathbf{I}_n]}$ over the selected batch of images, $\mathcal{I}_b \subseteq \mathcal{I}$ to compute the aggregate contextual diversity

$$d_{\mathcal{I}_b} = \sum_{\mathbf{I}_m,\mathbf{I}_n \in \mathcal{I}_b} d_{[\mathbf{I}_m,\mathbf{I}_n]}. \tag{3}$$

We use this term as the primary reward component in our RL framework. In addition to the intuitive motivation of using the contextual diversity, we show extensive comparisons in Sect. 4 and ablative analysis in Sect. 5.

3.2 Frame Selection Strategy

CDAL-CS. Our first frame selection strategy is contextual diversity based active learning using core-set (CDAL-CS), which is inspired by the theoretically grounded core-set approach [31]. To use core-set with Contextual Diversity, we simply replace the Euclidean distance with the pairwise contextual diversity (Eq. 2) and use it in the K-Center-Greedy algorithm [31, Algo. 1], which is reproduced in the supplementary material for completeness.

CDAL-RL. Reinforcement Learning has been used for frame selection [38,43] for tasks like active one-shot learning and video summarization. We use contextual diversity as part of the reward function to learn a Bi-LSTM-based policy for frame selection. Our reward function comprises of the following three components.

Contextual Diversity (R_{cd}). This is simply the aggregated contextual diversity, as given in Eq. (3), over the selected subset of images $\mathcal{I}_b$.

Visual Representation (R_{vr}). We use this reward to incorporate the visual representativeness over the whole unlabeled set using the image's feature representation. Let $\mathbf{x}_i$ and $\mathbf{x}_j$ be the feature representations of an image $\mathbf{I}_i \in \mathcal{I}$ and of $\mathbf{I}_j \in \mathcal{I}_b$ respectively, then

$$R_{vr} = \exp\left(-\frac{1}{|\mathcal{I}|}\sum_{i=1}^{|\mathcal{I}|} \min_{j\in\mathcal{I}_b}\left(\|\mathbf{x}_i - \mathbf{x}_j\|_2\right)\right) \tag{4}$$

This reward prefers to pick images that are spread out across the feature space, akin to k-medoid approaches.

Semantic Representation (R_{sr}). We introduce this component to ensure that the selected subset of images are reasonably balanced across all the classes and define it as

$$R_{sr} = \sum_{c\in C} \log\left(|\mathcal{R}^c_{\mathcal{I}_b}|/\lambda\right) \tag{5}$$

Here, λ is a hyper-parameter that is set to a value such that a selection that has substantially small representation of a class ($|\mathcal{R}^c_{\mathcal{I}_b}| \ll \lambda$) gets penalized. We use this reward component only for the semantic segmentation application where certain classes (e.g., 'pole') may occupy a relatively small number of regions (pixels).

We define the total reward as $R = \alpha R_{cd} + (1 - \alpha)(R_{vr} + R_{sr})$ and use it to train our LSTM based policy network. To emphasize the CD component in the reward function we set α to 0.75 across all tasks and experiments. The precise value of α does not influence results significantly as shown by the ablation experiments reported in the supplementary.

3.3 Network Architecture and Training

The contextual diversity measure is agnostic to the underlying *task network* and is computed using the predicted softmax probability. Therefore in Sect. 4, our task network choice is driven by reporting a fair comparison with the state-of-the-art approaches on the respective applications. In the core-set approach [31], images are represented using the feature embeddings and pairwise distances are Euclidean. Contrarily, our representation is the mixture distribution computed in Eq. (1) over a single image and the corresponding distances are computed using pairwise contextual diversity in Eq. (2).

For CDAL-RL, we follow a policy gradient based approach using the REINFORCE algorithm [37] and learn a Bi-LSTM *policy network*, where the reward used is as described in the previous section. The input to the policy network at a given time step is a representation of each image extracted using the task network. This representation is the vectorized form of an $n_C \times n_C$ matrix, where the columns of the matrix are set to $\boldsymbol{P}^c_{\mathbf{I}}$ for all $c \in C$ such that $\mathbf{I} \in \mathcal{I}^c$, and zero vectors otherwise. The binary action (select or not) for each frame is modeled as a Bernoulli random variable using a sigmoid activation at the output of the Bi-LSTM policy network. The LSTM cell size is fixed to 256 across all experiments with the exception of image classification, where we also show results with a cell size of 1024 to accommodate for a larger set of 100 classes. For REINFORCE, we use learning rate $= 10^{-5}$, weight decay $= 10^{-5}$, max epoch $= 60$ and #episodes $= 5$. We achieve the best performance when we train the policy network from scratch in each iteration of AL, however, in Sect. 5 we also analyze and compare other alternatives. It is worth noting that in the AL setting, the redundancy within a large unlabeled pool may lead to multiple subsets that are equally good selections. CDAL-RL is no exception and multiple subsets may achieve the same reward, thus rendering the specific input image sequence to our Bi-LSTM policy network, irrelevant.

4 Results and Comparison

We now present empirical evaluation of our approach on three visual recognition tasks of semantic segmentation, object detection and image classification[2].

[2] Additional results and ablative analysis is presented in the supplementary.

Datasets. For semantic segmentation, we use Cityscapes [5] and BDD100K [42]. Both these datasets have 19 object categories, with pixel-level annotation for 10 K and 3475 frames for BDD100K and Cityscapes respectively. We report our comparisons using the mIoU metric. For direct comparisons with [40] over the object detection task, we combine the training and validation sets from PASCAL VOC 2007 and 2012 [8] to obtain 16,551 unlabeled pool and evaluate the model performance on the test set of PASCAL VOC 2007 using the mAP metric. We evaluate the image classification task using classification accuracy as the metric over the CIFAR-10 and CIFAR-100 [16] datasets, each of which have 60 K images evenly divided over 10 and 100 categories respectively.

Compared Approaches. The two recent works [35,40] showed state of the art AL performance on various visual recognition tasks and presented a comprehensive empirical comparison with prior work. We follow their experimental protocol for a fair comparison and present our results over all the three tasks. For the semantic segmentation task, we use the reported results for VAAL and its other competitors from [35], which are core-set [31], Query-by-Committee (QBC) [17], MC-Dropout [11] and Suggestive Annotation (SA) [39]. We refer to our contextual diversity based approaches as CDAL-CS for its core-set variant and CDAL-RL for the RL variant, which uses the combined reward R as defined in Sect. 3.2. The object detection experiments are compared with learn loss [40] and its competitors – core-set, entropy based and random sampling – using results reported in [40]. For the image classification task, we again compare with VAAL, core-set, DBAL [10] and MC-Dropout. All the CDAL-RL results are reported after averaging over three independent runs. In Sect. 5 we demonstrate the strengths of *CD* through various ablative analysis on the Cityscapes dataset. Finally, in the supplementary material, we show further comparisons with region based approaches [14,26], following their experimental protocol on the Cityscapes dataset.

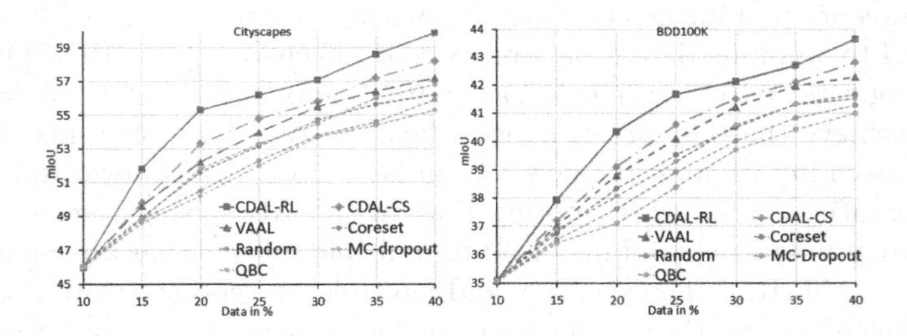

Fig. 3. Quantitative comparison for the Semantic Segmentation problem over Cityscapes **(left)** and BDD100K **(right)**. Note: DRN results 62.95% and 44.95% mIoU on 100% data for Cityscapes and BDD respectively (Color figure online).

4.1 Semantic Segmentation

Despite the tediousness associated with semantic segmentation, there are limited number of works for frame-level selection using AL. A recent approach applied to this task is VAAL [35], which achieves state-of-the-art performance while presenting a comprehensive comparison with previously proposed approaches. We follow the experimental protocol of VAAL [35], and also use the same backbone model of dilated residual network (DRN) [41] for a fair comparison. As in their case, the annotation budget is set to 150 and 400 for Cityscapes and BDD100K respectively. The evaluation metric is mIoU. For each dataset, we evaluate the performance of an AL technique at each step, as the number of samples to be selected are increased from 10% to 40% of the unlabeled pool size, in steps of 5%. Figure 3 shows the comparison over the two datasets.

We observe that for both challenging benchmarks, the two variants of CDAL comprehensively outperform all other approaches by a significant margin. Our CDAL-RL approach can acheive current SOTA 57.2 and 42.8 mIoU by reducing the labeling effort by 300 and 800(10%) frames on cityscapes and BDD100k respectively. A network's performance on this task is the most affected by the spatial context, due to the fine-grained spatial labeling necessary for improving the mIoU metric. We conclude that the CD measure effectively captures the spatial and semantic context and simultaneously helps select the most informative samples. There exist region-level AL approaches to semantic segmentation, where only certain regions are annotated in each frame [14,26]. Our empirical analysis in the supplementary material shows that our CDAL based frame selection strategy is complementary to the region-based approaches.

4.2 Object Detection

For the object detection task, we compare with the learning loss approach [40] and the competing methods therein. For a fair comparison, we use the same base detector network as SSD [22] with a VGG-16 [34] backbone and use the same hyperparameter settings as described in [40].

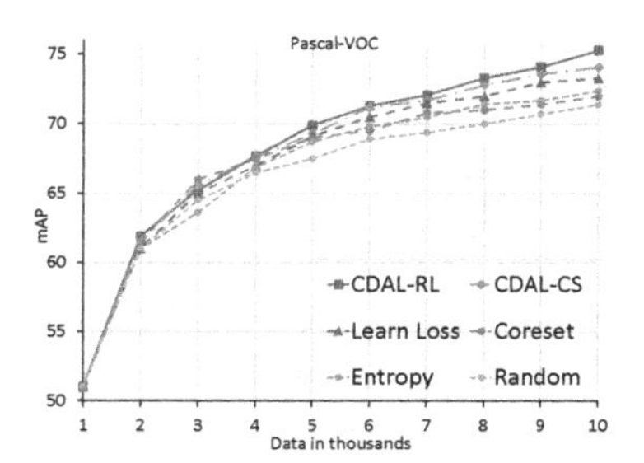

Fig. 4. Quantitative comparison for Object Detection over PASCAL-VOC dataset. We follow the experimental protocol of the learning loss method [40]. Note: SSD results 77.43% mAP on 100% data of PASCAL-VOC (07+12).

Figure 4 shows the comparisons, where we see in most cases, both variants of CDAL perform better than the other approaches. During the first few cycles of active learning, i.e., until about 5 K training samples are selected for annotation, CDAL performs nearly as well as core-set, which outperforms all the other approaches. In the later half of the active learning cycles with 5 K to 10K selected samples, CDAL variants outperform all the other approaches including core-set [31]. CDAL-RL achieved 73.3 mAP

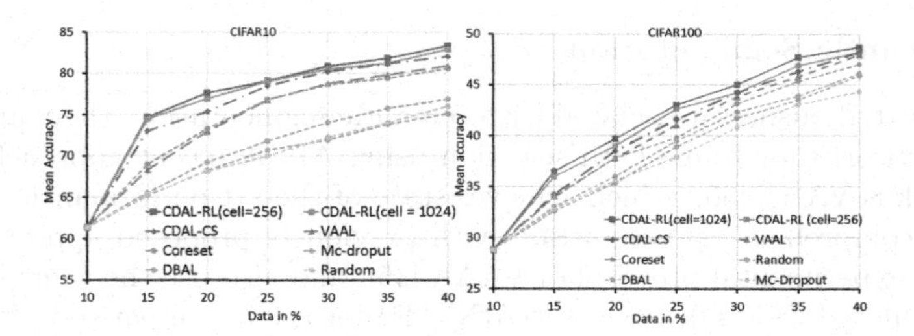

Fig. 5. Quantitative comparison for Image Classification over CIFAR-10 **(left)** and CIFAR-100 **(right)**. CDAL-RL (cell = n) indicates that the LSTM policy network has a cell size n. Note: VGG16 results 90.2% and 63.14% accuracy on 100% data for CIFAR10 and CIFAR100 respectively (Color figure online).

using 8k data where learning loss [40] achieved it by 10k hence reducing 2k labeled samples.

4.3 Image Classification

One of the criticisms often made about the active learning techniques is their relative difficulty in scaling with the number of classes in a given task. For example it has been reported in [35], that core-set [31] does not scale well for large number of classes. To demonstrate the strength of contextual diversity cues when the number of classes is large, we present the evaluations on the image classification task using CIFAR-10 and CIFAR-100. Figure 5 shows the comparison. It is clear that CDAL convincingly outperforms the state of the art technique, VAAL [35] on both the datasets. We can see that CDAL-RL can achieve ~81% accuracy on CIFAR10 by using 5000 (10%) less samples than VAAL [35] and similarly 2500 less samples are required in CIFAR100 to beat SOTA of 47.95% accuracy. These results indicate that CDAL can gracefully scale with the number of classes, which is not surprising as CD is a measure computed by accumulating KL-divergence, which scales well with high-dimensions unlike the Euclidean distance. It is worth noting that an increase in the LSTM cell size to 1024, helps improve the performance on CIFAR-100, without any significant effect on the CIFAR-10 performance. A higher dimension of the LSTM cell has higher capacity which better accommodates a larger number of classes. For completeness, we include more ablations of CDAL for image classification in the supplementary.

We also point out that in image classification, the entire image qualifies as a *region* (as defined in Sect. 3.1), and the resulting mixture $\boldsymbol{P}_{\mathbf{I}}^{c}$ comprising of a single component still captures confusion arising from the spatial context. Therefore, when a batch $\mathcal{I}_b$ is selected using the contextual diversity measure, the selection is diverse in terms of classes and their confusion.

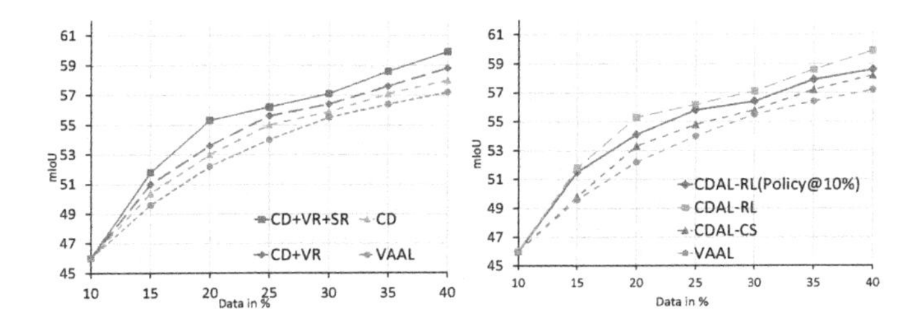

Fig. 6. **(left)** Ablation with individual reward components on Cityscapes. **(right)** Cityscapes results when the CDAL-RL policy was learned only once in the first iteration with 10% randomly selected frames (Color figure online).

5 Analysis and Ablation Experiments

In the previous section, we showed that contextual diversity consistently outperforms state of the art active learning approaches over three different visual recognition tasks. We will now show a series of ablation experiments to demonstrate the value of contextual diversity in an active learning problem. Since active learning is expected to be the most useful for the semantic segmentation task with highest amount of annotation time per image, we have chosen the task for our ablation experiments. We have designed all our ablation experiments on the Cityscapes dataset using the DRN model in the same settings as in Sect. 4.1.

Reward Component Ablation. We first investigate the performance of various components of the reward used in our approach. Figure 6(left) shows the performance of CDAL in three different reward settings: only contextual diversity ($R = R_{cd}$), contextual diversity and visual representation (CD+VR, i.e., $R = \alpha R_{cd} + (1-\alpha)R_{vr}$) and all the three components including semantic representations (CD+VR+SR, i.e., $R = \alpha R_{cd} + (1-\alpha)(R_{vr} + R_{sr})$). It is clear that contextual diversity alone outperforms the state of the art, VAAL [35], and improves further when the other two components are added to the reward function. As mentioned in Sect. 3.2, the value of $\alpha = 0.75$ was not picked carefully, but only to emphasize the CD component, and remains fixed in all experiments.

Policy Training Analysis. Our next experiment analyzes the effect of learning the Bi-LSTM-based policy only once, in the first AL iteration. We train the policy network using the *randomly selected* 10% and use it in each of the AL iterations for frame selection without further fine-tuning. The results are shown in Fig. 6(right), where we can see that this policy denoted by CDAL-RL(policy@10%), still outperforms VAAL and CDAL-CS in *all* iterations of AL. Here CDAL-RL is the policy learned under the setting in Sect. 4.1, where the policy network is trained from scratch in each AL iteration. An interesting observation is the suitability of the contextual diversity measure as a reward, and that it led to learning a meaningful policy even with randomly selected data.

Visualization of CDAL-Based Selection. In Fig. 7(a), we show t-SNE plots [25] to visually compare the distribution of the points selected by the CDAL

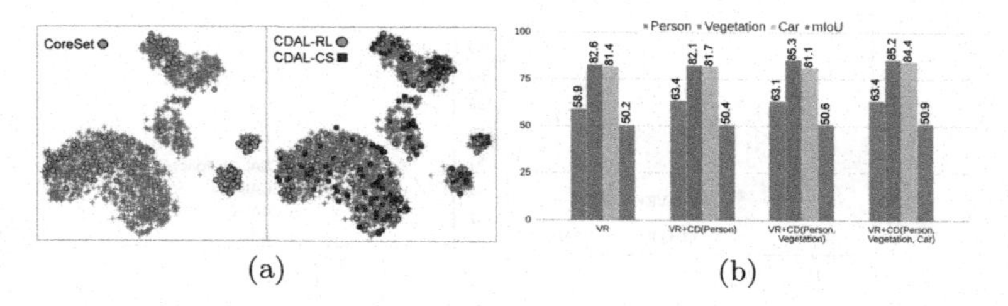

Fig. 7. (a) t-sne plots comparison with [31] on Cityscapes: CoreSet (left), and CDAL (right) (b) Performance analysis when CD reward is computed for an increasing number of classes. (Color figure online)

variants and that of core-set. We use the Cityscapes training samples projected into the feature space of the DRN model. The red points in the plots show the unlabeled samples. The left plot shows green points as samples selected by core-set and the right plot shows green and blue points are selected by CDAL-RL and CDAL-CS respectively. It is clear that both variants of the contextual diversity based selection have better spread across the feature space when compared with the core-set approach, which is possibly due to the distance concentration effect as pointed by [35]. This confirms the limitation of the Euclidean distance in a high-dimensional feature space corresponding to the DRN model. On the other hand, CDAL selects points that are more uniformly distributed across the entire feature space, reflecting better representativeness capability.

Class-wise Contextual Diversity Reward. The CD is computed by accumulating the symmetric KL-divergence (cf. Eq. (2)) over all classes. Therefore, it is possible to use the R_{cd} reward only for a few, and not all, the classes. Figure 7(b) shows the segmentation performance as we incorporate the contextual diversity (CD) from zero to three classes. The initial model is trained using only the visual representation reward (R_{vr}) and is shown as the leftmost group of color-coded bars. As we include the R_{cd} term in the reward with the CD only being computed for the *Person* class, we see a substantial rise in the IoU score corresponding to *Person*, as well as a marginal overall improvement. As expected, when we include both, the *Person* and *Vegetation* classes in the CD reward, we see substantial improvements in both the classes. The analysis indicates that the contextual diversity reward indeed helps mitigating class-specific confusion.

Limitations of CDAL. While we show competitive performance of the policy without retraining (Fig. 6(right)), for best performance retraining at each AL iteration is preferred. For large datasets, this requires larger unrolling of the LSTM network incurring more computational and memory costs. Another limitation of CDAL in general is in the case of image classification, where the entire image is treated as a single region and thus is unable to fully leverage spatial context.

6 Conclusion

We have introduced a novel contextual diversity based measure for active frame selection. Through experiments for three visual recognition tasks, we showed that contextual diversity when used as a distance measure with core-set, or as a reward function for RL, is a befitting choice. It is designed using an information-theoretic distance measure, computed over the mixture of softmax distributions of pseudo-labeled data points, which allows it to capture the model's predictive uncertainty as well as class-specific confusion. We have only elaborated the promising empirical results in this paper, and plan to investigate deeper theoretical interpretations of contextual diversity that may exist.

Acknowledgement. The authors acknowledge the partial support received from the Infosys Center for Artificial Intelligence at IIIT-Delhi. This work has also been partly supported by the funding received from DST through the IMPRINT program (IMP/2019/000250).

References

1. Arazo, E., Ortego, D., Albert, P., O'Connor, N.E., McGuinness, K.: Pseudo-labeling and confirmation bias in deep semi-supervised learning (2019)
2. Beluch, W.H., Genewein, T., Nürnberger, A., Köhler, J.M.: The power of ensembles for active learning in image classification. In: Proceedings of the IEEE Conference on Computer Vision and Pattern Recognition, pp. 9368–9377 (2018)
3. Bilgic, M., Getoor, L.: Link-based active learning. In: NIPS Workshop on Analyzing Networks and Learning with Graphs (2009)
4. Caron, M., Bojanowski, P., Joulin, A., Douze, M.: Deep clustering for unsupervised learning of visual features. In: European Conference on Computer Vision (2018)
5. Cordts, M., et al.: The cityscapes dataset for semantic urban scene understanding. In: Proceedings of the IEEE Conference on Computer Vision and Pattern Recognition, pp. 3213–3223 (2016)
6. Dabak, A.G.: A geometry for detection theory. Ph.D. thesis, Rice Unviersity (1992)
7. Ebert, S., Fritz, M., Schiele, B.: RALF: a reinforced active learning formulation for object class recognition. In: 2012 IEEE Conference on Computer Vision and Pattern Recognition, pp. 3626–3633 (2012)
8. Everingham, M., Van Gool, L., Williams, C.K., Winn, J., Zisserman, A.: The pascal visual object classes (VOC) challenge. Int. J. Comput. Vis. **88**(2), 303–338 (2010)
9. Fang, M., Li, Y., Cohn, T.: Learning how to active learn: a deep reinforcement learning approach. In: Proceedings of the 2017 Conference on Empirical Methods in Natural Language Processing, pp. 595–605 (2017)
10. Gal, Y., Islam, R., Ghahramani, Z.: Deep Bayesian active learning with image data. In: Proceedings of the 34th International Conference on Machine Learning, vol. 70, pp. 1183–1192. JMLR. org (2017)
11. Gorriz, M., Carlier, A., Faure, E., Giro-i Nieto, X.: Cost-effective active learning for melanoma segmentation. arXiv preprint arXiv:1711.09168 (2017)
12. Guo, Y.: Active instance sampling via matrix partition. In: Advances in Neural Information Processing Systems, pp. 802–810 (2010)

13. Joshi, A.J., Porikli, F., Papanikolopoulos, N.: Multi-class active learning for image classification. In: 2009 IEEE Conference on Computer Vision and Pattern Recognition, pp. 2372–2379. IEEE (2009)
14. Kasarla, T., Nagendar, G., Hegde, G., Balasubramanian, V., Jawahar, C.: Region-based active learning for efficient labeling in semantic segmentation. In: 2019 IEEE Winter Conference on Applications of Computer Vision (WACV), pp. 1109–1118, January 2019
15. Konyushkova, K., Uijlings, J., Lampert, C.H., Ferrari, V.: Learning intelligent dialogs for bounding box annotation. In: The IEEE Conference on Computer Vision and Pattern Recognition (CVPR), June 2018
16. Krizhevsky, A., Hinton, G., et al.: Learning multiple layers of features from tiny images. Technical report, Citeseer (2009)
17. Kuo, W., Häne, C., Yuh, E., Mukherjee, P., Malik, J.: Cost-sensitive active learning for intracranial hemorrhage detection. In: Frangi, A.F., Schnabel, J.A., Davatzikos, C., Alberola-López, C., Fichtinger, G. (eds.) MICCAI 2018, Part III. LNCS, vol. 11072, pp. 715–723. Springer, Cham (2018). https://doi.org/10.1007/978-3-030-00931-1_82
18. Lee, D.H.: Pseudo-label: the simple and efficient semi-supervised learning method for deep neural networks. In: ICML Workshop on Challenges in Representation Learning (WREPL) (2013)
19. Lewis, D.D., Catlett, J.: Heterogeneous uncertainty sampling for supervised learning. In: Machine Learning Proceedings 1994, pp. 148–156. Elsevier (1994)
20. Lewis, D.D., Gale, W.A.: A sequential algorithm for training text classifiers. In: Croft, B.W., van Rijsbergen, C.J. (eds.) SIGIR 1994, pp. 3–12. Springer, Heidelberg (1994). https://doi.org/10.1007/978-1-4471-2099-5_1
21. Li, X., Guo, Y.: Adaptive active learning for image classification. In: 2013 IEEE Conference on Computer Vision and Pattern Recognition, pp. 859–866 (2013)
22. Liu, W., et al.: SSD: single shot multibox detector. In: Leibe, B., Matas, J., Sebe, N., Welling, M. (eds.) ECCV 2016. LNCS, vol. 9905, pp. 21–37. Springer, Heidelberg (2016). https://doi.org/10.1007/978-3-319-46448-0_2
23. Liu, Z., Wang, J., Gong, S., Lu, H., Tao, D.: Deep reinforcement active learning for human-in-the-loop person re-identification. In: The IEEE International Conference on Computer Vision (ICCV), October 2019
24. Luo, W., Schwing, A., Urtasun, R.: Latent structured active learning. In: Advances in Neural Information Processing Systems, pp. 728–736 (2013)
25. van der Maaten, L., Hinton, G.: Visualizing data using t-SNE. J. Mach. Learn. Res. **9**, 2579–2605 (2008)
26. Mackowiak, R., Lenz, P., Ghori, O., Diego, F., Lange, O., Rother, C.: CEREALS - cost-effective region-based active learning for semantic segmentation. In: British Machine Vision Conference 2018, BMVC 2018, 3–6 September 2018. Northumbria University, Newcastle (2018)
27. Mahapatra, D., Bozorgtabar, B., Thiran, J.P., Reyes, M.: Efficient active learning for image classification and segmentation using a sample selection and conditional generative adversarial network. In: Frangi, A., Schnabel, J., Davatzikos, C., Alberola-López, C., Fichtinger, G. (eds.) MICCAI 2018. LNCS, pp. 580–588. Springer, Heidelberg (2018). https://doi.org/10.1007/978-3-030-00934-2_65
28. Mayer, C., Timofte, R.: Adversarial sampling for active learning. arXiv preprint arXiv:1808.06671 (2018)
29. Nguyen, H.T., Smeulders, A.: Active learning using pre-clustering. In: Proceedings of the Twenty-First International Conference on Machine Learning, p. 79. ACM (2004)

30. Rosenfeld, A., Zemel, R.S., Tsotsos, J.K.: The elephant in the room. CoRR abs/1808.03305 (2018)
31. Sener, O., Savarese, S.: Active learning for convolutional neural networks: a core-set approach. In: International Conference on Learning Representations (2018)
32. Settles, B.: Active learning. Synthesis Lect. Arti. Intell. Mach. Learn. **6**(1), 1–114 (2012)
33. Settles, B., Craven, M.: An analysis of active learning strategies for sequence labeling tasks. In: Proceedings of the Conference on Empirical Methods in Natural Language Processing, pp. 1070–1079. Association for Computational Linguistics (2008)
34. Simonyan, K., Zisserman, A.: Very deep convolutional networks for large-scale image recognition. In: International Conference on Learning Representations (2015)
35. Sinha, S., Ebrahimi, S., Darrell, T.: Variational adversarial active learning. In: The IEEE International Conference on Computer Vision (ICCV), October 2019
36. Wang, K., Zhang, D., Li, Y., Zhang, R., Lin, L.: Cost-effective active learning for deep image classification. IEEE Trans. Circ. Syst. Video Technol. **27**(12), 2591–2600 (2017)
37. Williams, R.J.: Simple statistical gradient-following algorithms for connectionist reinforcement learning. Mach. Learn. **8**(3–4), 229–256 (1992)
38. Woodward, M., Finn, C.: Active one-shot learning. In: NIPS Deep RL Workshop (2017)
39. Yang, L., Zhang, Y., Chen, J., Zhang, S., Chen, D.Z.: Suggestive annotation: a deep active learning framework for biomedical image segmentation. In: Descoteaux, M., Maier-Hein, L., Franz, A., Jannin, P., Collins, D., Duchesne, S. (eds.) MICCAI 2017. LNCS, vol. 10435, pp. 399–407. Springer, Heidelberg (2017). https://doi.org/10.1007/978-3-319-66179-7_46
40. Yoo, D., Kweon, I.S.: Learning loss for active learning. In: The IEEE Conference on Computer Vision and Pattern Recognition (CVPR), June 2019
41. Yu, F., Koltun, V., Funkhouser, T.: Dilated residual networks. In: Proceedings of the IEEE Conference on Computer Vision and Pattern Recognition, pp. 472–480 (2017)
42. Yu, F., Xian, W., Chen, Y., Liu, F., Liao, M., Madhavan, V., Darrell, T.: Bdd100k: a diverse driving video database with scalable annotation tooling. arXiv preprint arXiv:1805.04687 (2018)
43. Zhou, K., Qiao, Y., Xiang, T.: Deep reinforcement learning for unsupervised video summarization with diversity-representativeness reward (2018)
44. Zhu, J.J., Bento, J.: Generative adversarial active learning. arXiv preprint arXiv:1702.07956 (2017)

Temporal Aggregate Representations for Long-Range Video Understanding

Fadime Sener[1,2(✉)], Dipika Singhania[2], and Angela Yao[2]

[1] University of Bonn, Bonn, Germany
sener@cs.uni-bonn.de
[2] National University of Singapore, Singapore, Singapore
{dipika16,ayao}@comp.nus.edu.sg

Abstract. Future prediction, especially in long-range videos, requires reasoning from current and past observations. In this work, we address questions of temporal extent, scaling, and level of semantic abstraction with a flexible multi-granular temporal aggregation framework. We show that it is possible to achieve state of the art in both next action and dense anticipation with simple techniques such as max-pooling and attention. To demonstrate the anticipation capabilities of our model, we conduct experiments on Breakfast, 50Salads, and EPIC-Kitchens datasets, where we achieve state-of-the-art results. With minimal modifications, our model can also be extended for video segmentation and action recognition.

Keywords: Action anticipation · Temporal aggregation

1 Introduction

We tackle long-range video understanding, specifically anticipating not-yet observed but upcoming actions. When developing intelligent systems, one needs not only to recognize what is *currently* taking place – but also predict what will happen *next*. Anticipating human actions is essential for applications such as smart surveillance, autonomous driving, and assistive robotics.

While action anticipation is a niche (albeit rapidly growing) area, the key issues that arise are germane to long-range video understanding as a whole. How should temporal or sequential relationships be modelled? What temporal extent of information and context needs to be processed? At what temporal scale should they be derived, and how much semantic abstraction is required? The answers to these questions are not only entangled with each other but also depend very much on the videos being analyzed. Here, one needs to distinguish between clipped actions, e.g. of UCF101 [32], versus the multiple actions in long video streams, e.g. of the Breakfast [18]. In the former, the actions and

Electronic supplementary material The online version of this chapter (https://doi.org/10.1007/978-3-030-58517-4_10) contains supplementary material, which is available to authorized users.

A. Vedaldi et al. (Eds.): ECCV 2020, LNCS 12361, pp. 154–171, 2020.
https://doi.org/10.1007/978-3-030-58517-4_10

video clips are on the order of a few seconds, while in the latter, it is several minutes. As such, temporal modelling is usually not necessary for simple action recognition [13], but more relevant for understanding complex activities [28,30].

Temporal models that are built into the architecture [6,8,12,29] are generally favoured because they allow frameworks to be learned end-to-end. However, this means that the architecture also dictates the temporal extent that can be accounted for. This tends to be short, either due to difficulties in memory retention or model size. As a result, the context for anticipation can only be drawn from a limited extent of recent observations, usually on the order of seconds [19,26,38]. This, in turn, limits the temporal horizon and granularity of the prediction.

One way to ease the computational burden, especially under longer temporal extents, is to use higher-level but more compact feature abstractions, e.g. by using detected objects, people [42] or sub-activity labels [1,15] based on the outputs of video segmentation algorithms [29]. Such an approach places a heavy load on the initial task of segmentation and is often difficult to train end-to-end. Furthermore, since labelling and segmenting actions from video are difficult tasks, their errors may propagate onwards when anticipating future actions.

Motivated by these questions of temporal modelling, extent, scaling, and level of semantic abstraction, we propose a general framework for encoding long-range video. We aim for flexibility in frame input, i.e. ranging from low-level visual features to high-level semantic labels, and the ability to meaningfully integrate recent observations with long-range context in a computationally efficient way. To do so, we split video streams into snippets of equal length and max-pool the frame features within the snippets. We then create ensembles of multi-scale feature representations that are aggregated bottom-up based on scaling and temporal extent. Temporal aggregation [16] is a form of summarization used in database systems. Our framework is loosely analogous as it summarizes the past observations through aggregation, so we name it "temporal aggregates". We summarize our main contributions as follows:

- We propose a simple and flexible single-stage framework of multi-scale temporal aggregates for videos by relating recent to long-range observations.
- Our representations can be applied to several video understanding tasks; in addition to action anticipation, it can be used for recognition and segmentation with minimal modifications and is able to achieve competitive results.
- Our model has minimal constraints regarding the type of anticipation (dense or next action), type of the dataset (instructional or daily activities), and type of input features (visual features or frame-level labels).
- We conduct experiments on Breakfast [18], 50Salads [34] and EPIC-Kitchens [5].

2 Related Works

Action recognition has advanced significantly with deep networks in recent years. Notable works include two steam networks [31,40], 3D convolutional networks [3,36], and RNNs [7,44]. These methods have been designed to encode clips of a few seconds and are typically applied to the classification of *trimmed* videos containing a single action [14,32]. In our paper, we work with long *untrimmed* sequences of complex activities. Such long videos are not simply a composition of independent short actions, as the actions are related to each other with sequence dynamics. Various models for complex activity understanding have been addressed before [6,8,30]; these approaches are designed to work on instructional videos by explicitly modelling their sequence dynamics. These models are not flexible enough to be extended to daily activities with loose orderings. Also, when only partial observations are provided, e.g. for anticipation, these models cannot be trained in a single stage.

Action anticipation aims to forecast actions before they occur. Prior works in immediate anticipation were initially limited to movement primitives like *reaching* [17] or interactions such as *hugging* [38]. [24] presents a model for predicting both the next action and its starting position. [5] presents a large daily activities dataset, along with a challenge for anticipating the next action one second before occurrence. [26] proposes next action anticipation from recent observations. Recently, [10] proposed using an LSTM to summarize the past and another LSTM for future prediction. These works assume near-past information, whereas we make use of long-range past.

Dense anticipation predicts actions multiple steps into the future. Previous methods [1,15] to date require having already segmented temporal observations. Different than these, our model can perform dense anticipation in a single stage without any pre-segmented nor labelled inputs.

The **role of motion and temporal dynamics** has been well-explored for video understanding, though the focus has been on short clips [3,13,22]. Some works use longer-term temporal contexts, though still in short videos [21,35]. Recently, Wu *et al.* [42] proposed integrating long-term features with 3D CNNs in short videos and showed the importance of temporal context for action recognition. Our model is similar in spirit to [42] in that we couple the recent with the long-range past using attention. One key difference is that we work with ensembles of multiple scalings and granularities, whereas [42] work at a single frame-level granularity. As such, we can handle long videos up to tens of minutes, while they are only able to work on short videos. Recently, Feichtenhofer *et al.* [9] proposed SlowFast networks, which, similar to our model, encode time-wise multi-scale representations. These approaches are limited to short videos and cannot be extended to minutes-long videos due to computational constraints.

3 Representations

We begin by introducing the representations, which are inputs to the building blocks of our framework, see Fig. 1. We had two rationales when designing our network. First, we relate recent to long-range observations, since some past actions directly determine future actions. Second, to represent recent and long-range past at various granularities, we pool snippets over multiple scales.

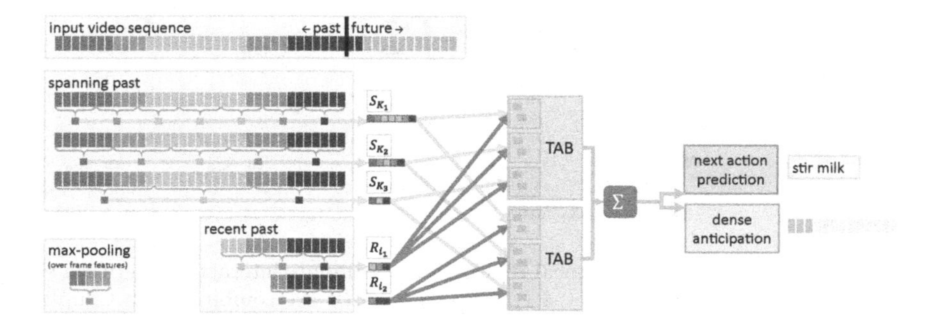

Fig. 1. Model overview: In this example we use 3 scales for computing the "spanning past" features $\mathbf{S}_{K_1}, \mathbf{S}_{K_2}, \mathbf{S}_{K_3}$, and 2 starting points to compute the "recent past" features, $\mathbf{R}_{i_1}, \mathbf{R}_{i_2}$, by max-pooling over the frame features in each snippet. Each recent snippet is coupled with all the spanning snippets in our Temporal Aggregation Block (TAB). An ensemble of TAB outputs is used for dense or next action anticipation.

3.1 Pooling

For a video of length T, we denote the feature representation of a single video frame indexed at time t as $f_t \in \mathbb{R}^D, 1 \leq t \leq T$. f_t can be derived from low-level features, such as IDT [39] or I3D [3], or high-level abstractions, such as sub-activity labels derived from segmentation algorithms. To reduce computational load, we work at a snippet-level and define a snippet feature $\mathbf{F}_{ij;K}$ as the concatenation of max-pooled features from K snippets, where snippets are partitioned consecutively from frames i to j:

$$\begin{aligned} \mathbf{F}_{ij;K} &= [F_{i,i+k}, F_{i+k+1,i+2k}, ..., F_{j-k+1,j}], \text{ where} \\ (F_{p,q})_d &= \max_{p \leq t \leq q} \{f_t\}_d, \; 1 \leq d \leq D, \;\; k = {(j-i)}/{K}. \end{aligned} \tag{1}$$

Here, $F_{p,q}$ indicates the maximum over each dimension d of the frame features in a given snippet between frames p and q, though it can be substituted with other alternatives. In the literature, methods representing snippets or segments of frames range from simple sampling and pooling strategies to more complex representations such as learned pooling [23] and LSTMs [27]. Especially for long snippets, it is often assumed that a learned representation is necessary [11,20], though their effectiveness over simple pooling is still controversial [40]. The learning of novel temporal pooling approaches goes beyond the scope of this work and is an orthogonal line of development. We verify established methods (see Sect. 5.2) and find that a simple max-pooling is surprisingly effective and sufficient.

3.2 Recent vs. Spanning Representations

Based on different start and end frames i and j and number of snippets K, we define two types of snippet features: *"recent"* features $\{\mathcal{R}\}$ from recent observations, and *"spanning"* features $\{\mathcal{S}\}$ drawn from the entire video. The recent

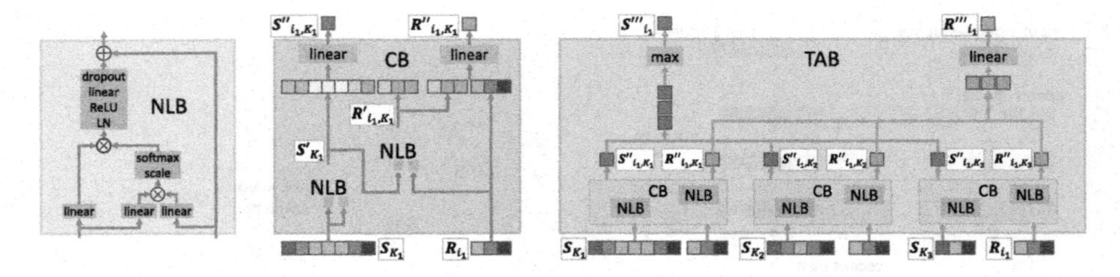

Fig. 2. Model components: Non-Local Blocks (NLB) compute interactions between two representations via attention (Sect. 4.1). Two such NLBs are combined in a Coupling Block (CB), which calculates the attention-reweighted spanning and recent representations (Sect. 4.2). We couple each recent with all spanning representations via individual CBs and combine their outputs in a Temporal Aggregation Block (TAB) (Sect. 4.3). The outputs of multiple such TABs are combined to perform anticipation, Fig. 1.

snippets cover a couple of seconds (or up to a minute, depending on the temporal granularity) before the current time point while spanning snippets refer to the long-range past and may last up to ten minutes. For *"recent"* snippets, the end frame j is fixed to the current time point t, and the number of snippets is fixed to K_R. The recent snippet features $\mathcal{R}$ can be defined as a feature bank of snippet features with different start frames i, i.e.

$$\mathcal{R} = \{\mathbf{F}_{i_1 t;K_R}, \mathbf{F}_{i_2 t;K_R}, ..., \mathbf{F}_{i_R t;K_R}\} = \{\mathbf{R}_{i_1}, \mathbf{R}_{i_2}, ..., \mathbf{R}_{i_R}\}, \tag{2}$$

where $\mathbf{R}_i \in \mathbb{R}^{D \times K_R}$ is a shorthand to denote $\mathbf{F}_{i\,t;K_R}$, since endpoint t and number of snippets K_R are fixed. In Fig. 1 we use two starting points to compute the "recent" features and represent each with $K_R = 3$ snippets (▬ & ▬).

For *"spanning"* snippets, i and j are fixed to the start of the video and current time point, i.e. $i = 0, j = t$. Spanning snippet features $\mathcal{S}$ are defined as a feature bank of snippet features with varying number of snippets K, i.e.

$$\mathcal{S} = \{\mathbf{F}_{0\,t;K_1}, \mathbf{F}_{0\,t;K_2}, ..., \mathbf{F}_{0\,t;K_S}\} = \{\mathbf{S}_{K_1}, \mathbf{S}_{K_2}, ..., \mathbf{S}_{K_S}\}, \tag{3}$$

where $\mathbf{S}_K \in \mathbb{R}^{D \times K}$ is a shorthand for $\mathbf{F}_{0\,t;K}$. In Fig. 1 we use three scales to compute the "spanning" features with $K = \{7, 5, 3\}$ (▬, ▬ & ▬).

Key to both types of representations is the ensemble of snippet features from multiple scales. We achieve this by varying the number of snippets K for the spanning past. For the recent past, it is sufficient to keep the number of snippets K_R fixed, and vary only the start point i, due to redundancy between $\mathcal{R}$ and $\mathcal{S}$ for the snippets that overlap. For our experiments, we work with snippets ranging from seconds to several minutes.

4 Framework

In Fig. 2 we present an overview of the components used in our framework, which we build in a bottom-up manner, starting with the recent and spanning features $\mathcal{R}$ and $\mathcal{S}$, which are coupled with non-local blocks (NLB) (Sect. 4.1) within coupling blocks (CB) (Sect. 4.2). The outputs of the CBs from different scales are then aggregated inside temporal aggregation blocks (TAB) (Sect. 4.3). Outputs of different TABs can then be chained together for either next action anticipation or dense anticipation (Sects. 5.3, 5.5).

4.1 Non-Local Blocks (NLB)

We apply non-local operations to capture relationships amongst the spanning snippets and between spanning and recent snippets. Non-local blocks [41] are a flexible way to relate features independently from their temporal distance and thus capture long-range dependencies. We use the modified non-local block from [42], which adds layer normalization [2] and dropout [33] to the original one in [40]. Figure 2 (left) visualizes the architecture of the block, the operation of which we denote as NLB$(\cdot, \cdot)$.

4.2 Coupling Block (CB)

Based on the NLB, we define attention-reweighted spanning and recent outputs:

$$\mathbf{S}'_K = NLB(\mathbf{S}_K, \mathbf{S}_K) \quad \text{and} \quad \mathbf{R}'_{i,K} = NLB(\mathbf{S}'_K, \mathbf{R}_i). \tag{4}$$

$\mathbf{R}'_{i,K}$ is coupled with either $\mathbf{R}_i$ or $\mathbf{S}'_K$ via concatenation and a linear layer. This results in the fixed-length representations $\mathbf{R}''_{i,K}$ and $\mathbf{S}''_{i,K}$, where i is the starting point of the recent snippet and K is the scale of the spanning snippet.

4.3 Temporal Aggregation Block (TAB)

The final representation for recent and spanning past is computed by aggregating outputs from multiple CBs. For the same recent starting point i, we concatenate $\mathbf{R}''_{i,K_1}, ..., \mathbf{R}''_{i,K_S}$ for all spanning scales and pass the concatenation through a linear layer to compute $\mathbf{R}'''_i$. The final spanning representation $\mathbf{S}'''_i$ is a max over all $\mathbf{S}''_{i,K_1}, ..., \mathbf{S}''_{i,K_S}$. We empirically find that taking the max outperforms other alternatives like linear layers and/or concatenation for the spanning past (Sect. 5.2). TAB outputs, by varying recent starting points $\{i\}$ and scales of spanning snippets $\{K\}$, are multi-granular video representations that aggregate and encode both the recent and long-range past. We name these **temporal aggregate representations**. Figure 1 shows an example with 2 recent starting points and 3 spanning scales. These representations are generic and can be applied in various video understanding tasks (see Sect. 4.4) from long streams of video.

4.4 Prediction Model

Classification: For single-label classification tasks such as next action anticipation, temporal aggregate representations can be used directly with a classification

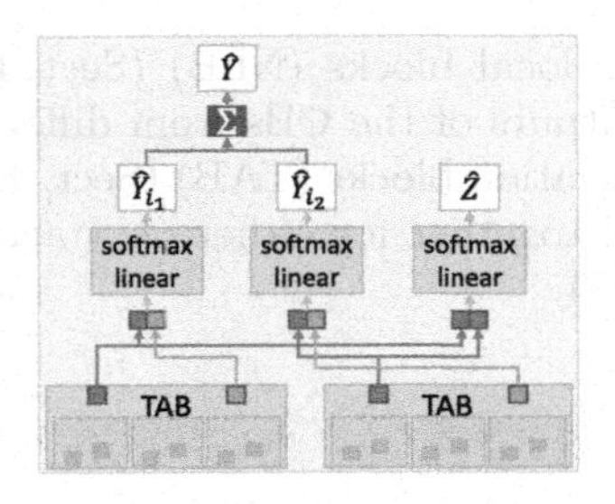

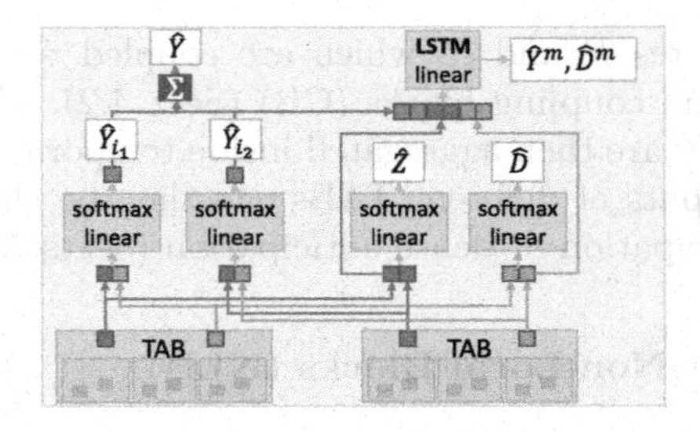

Fig. 3. Prediction models for classification (left) and sequence prediction (right).

layer (linear + softmax). A cross-entropy loss based on ground truth labels Y can be applied to the predictions $\hat{Y}_i$, where Y is either the current action label for recognition, or the next action label for next action prediction (see Fig. 3).

When the individual actions compose a complex activity (e.g. "take bowl" and "pour milk" as part of "making cereal" in Breakfast [18]), we can add an additional loss based on the complex activity label Z. Predicting Z as an auxiliary task helps with anticipation. For this we concatenate $\mathbf{S}'''_{i_1}, \ldots, \mathbf{S}'''_{i_R}$ from all TABs and pass them through a classification layer to obtain $\hat{Z}$. The overall loss is a sum of the cross entropies over the action and complex activity labels:

$$\mathcal{L}_{\text{cl}} = \mathcal{L}_{\text{comp}} + \mathcal{L}_{\text{action}} = -\sum_{n=1}^{N_Z} Z_n \log(\hat{Z})_n - \sum_{r=1}^{R}\sum_{n=1}^{N_Y} Y_n \log(\hat{Y}_{i_r})_n, \tag{5}$$

where i_r is one of the R recent starting points, and N_Y and N_Z are the total number of actions and complex activity classes respectively. During inference, the predicted scores are summed for a final prediction, i.e. $\hat{Y} = \max_n(\sum_{r=1}^{R} \hat{Y}_{i_r})_n$.

We frame sequence segmentation as a classification task and predict frame-level action labels of complex activities. Multiple sliding windows with fixed start and end times are generated and then classified into actions using Eq. 5.

Sequence Prediction: The dense anticipation task predicts frame-wise actions of the entire future sequence. Previously, [1] predicted future segment labels via classification and regressed the durations. We opt to estimate both via classification. The sequence duration is discretized into N_D intervals and represented as one-hot encodings $D \in \{0,1\}^{N_D}$. For dense predictions, we perform multi-step estimates. We first estimate the current action and complex activity label, as per Eq. 6. The current duration D is then estimated via a classification layer applied to the concatenation of recent temporal aggregates $\mathbf{R}'''_{i_1}, ..., \mathbf{R}'''_{i_R}$.

For future actions, we concatenate all recent and spanning temporal aggregates $\mathbf{R}'''_{i_1}, ..., \mathbf{R}'''_{i_R}$ and $\mathbf{S}'''_{i_1}, ..., \mathbf{S}'''_{i_R}$ and the classification layer outputs $\hat{Y}_{i_1}, ..., \hat{Y}_{i_R}$, and pass the concatenation through a linear layer before feeding the output to a one-layer LSTM. The LSTM predicts at each step m an action vector $\hat{Y}^m$ and a duration vector $\hat{D}^m$ (see Fig. 3). The dense anticipation loss is a sum of the cross-entropies over the current action, its duration, future actions and durations, and task labels respectively:

Table 1. Dataset details and our respective model parameters.

Dataset	Video duration median, mean ±std	# classes	# segments	$\{i\}$ (in seconds)	K_R	$\{K\}$
Breakfast(@15fps)	15.1 s, 26.6 s ±36.8	48	11.3K	$\{t-10, t-20, t-30\}$	5	$\{10, 15, 20\}$
50Salads(@30fps)	29.7 s, 38.4 s ±31.5	17	0.9K	$\{t-5, t-10, t-15\}$	5	$\{5, 10, 15\}$
EPIC(@60fps)	1.9 s, 3.7 s ±5.6	2513	36.6K	$\{t-1.6, t-1.2, t-0.8, t-0.4\}$	2	$\{2, 3, 5\}$

$$\mathcal{L}_{\text{dense}} = \mathcal{L}_{\text{cl}} - \sum_{n=1}^{N_D} D_n \log(\hat{D})_n - \frac{1}{M} \sum_{m=1}^{M} \left(\sum_{n=1}^{N_Y} Y_n^m \log(\hat{Y}^m)_n + \sum_{n=1}^{N_D} D_n^m \log(\hat{D}^m)_n \right) \tag{6}$$

During inference we sum the predicted scores (post soft-max) for all starting points i_r to predict the current action as $\max_n(\sum_{r=1}^{R} \hat{Y}_{i_r})_n$. The LSTM is then applied recurrently to predict subsequent actions and durations.

5 Experiments

5.1 Datasets and Features

We experiment on Breakfast [18], 50Salads [34] and EPIC-Kitchens [5]. The sequences in each dataset reflect realistic durations and orderings of actions, which is crucial for real-world deployment of anticipation models. Relevant datasets statistics are given in Table 1. One notable difference between these datasets is the label granularity; it is very fine-grained for EPIC, hence their 2513 action classes, versus the coarser 48 and 17 actions of Breakfast and 50Salads. As a result, the median action segment duration is 8-16x shorter.

Feature-wise, we use pre-computed Fisher vectors [1] and I3D [3] for Breakfast, Fisher vectors for 50Salads, and appearance, optical flow and object-based features provided by [10] for EPIC. Results for Breakfast and 50Salads are averaged over the predefined 4 and 5 splits respectively. Since 50Salads has only a single complex activity (making salad) we omit complex activity prediction for it. For EPIC, we report results on the test set. Evaluation measures are class accuracy (Acc.) for next action prediction and mean over classes [1] for dense prediction. We report Top-1 and Top-5 accuracies to be consistent with [10,26].

Hyper-parameters for spanning $\{K\}$, recent scales K_R and recent starting points $\{i\}$ are given in Table 1. We cross validated the parameters on different splits of 50Salads and Breakfast; on EPIC, we select parameters with the validation set [10].

5.2 Component Validation

We verify each component's utility via a series of ablation studies summarized in Table 2. As our main motivation was to develop a representation for anticipation

Table 2. Ablations on the influence of different model components.

Pooling type	frame sampling	GRU	BiLSTM	mean-pooling	max-pooling
Acc.	32.1	37.9	38.7	36.6	**40.1**

Influence of	Changes in components	Acc. (Drop)
Non-Local Blocks (NLB)	replace all NLBs with concatenation + linear layer	33.7 (6.4%)
Coupling Blocks (CB)	only couple the $\mathbf{S}_K$ and $\mathbf{S}_K$ in CBs	35.1 (5.0%)
	only couple the $\mathbf{R}_i$ and $\mathbf{R}_i$ in CBs	34.2 (5.9%)
	replace CBs with concatenation + linear layer	33.4 (6.7%)
Temporal Aggregation Blocks (TAB)	a single CB is used in TABs	38.0 (2.1%)
	three CBs are used in a single TAB	37.7 (2.4%)
	a single a CB is used without any TABs	32.1 (8.0%)

Recent & Spanning Repr.								
(a)	starting points i	$i_1 = t - 10$	$i_2 = t - 20$	$i_3 = t - 30$	$i_4 = 0$	$\{\mathbf{i_1}, \mathbf{i_2}, \mathbf{i_3}\}$		
	Acc.	36.9	37.7	37.2	35.1	40.1		
(b)	spanning scales K	{5}	{10}	{15}	{20}	{10, 15}	{10,15,20}	{5,10,15,20}
	Acc.	37.4	38.0	37.5	37.4	39.0	**40.1**	40.2
(c)	recent scales K_R	1	3	**5**	10			
	Acc.	38.7	39.5	**40.1**	38.6			

in long video streams, we validate on Breakfast for next action anticipation. Our full model gets a performance of 40.1% accuracy averaged over actions.

Video Representation: Several short-term feature representations have been proposed for video, e.g. 3D convolutions [36], or combining CNNs and RNNs for sequences [7,44]. For long video streams, it becomes difficult to work with all the raw features. Selecting representative features can be as simple as sub-sampling the frames [9,43], or pooling [40], to more complex RNNs [44]. Current findings in the literature are not in agreement. Some propose learned strategies [20,25], while others advocate pooling [40]. Our experiments align with the latter, showing that max-pooling is superior to both sampling (+8%) and the GRU (+2.2%) or bi-directional LSTM [4] (+1.4%). The performance of GRU and BiLSTM are comparable to average-pooling, but require much longer training and inference time. For us, max-pooling works better than average pooling; this contradicts the findings of [40]. We attribute this to the fact that we pool over minutes-long snippets and it is likely that mean- smooths away salient features that are otherwise preserved by max-pooling. We conducted a similar ablations on EPIC, where we observed a 1.3% increase with max- over mean-pooling.

Recent and Spanning Representations: In our ablations, unless otherwise stated, an ensemble of 3 spanning scales $K = \{10, 15, 20\}$ and 3 recent starting points $i = \{t - 10, t - 20, t - 30\}$ are used. Table 2(**a**) compares single starting points for the recent snippet features versus an ensemble. With a single starting point, points too near to and too far from the current time decrease the performance. The worst individual result is with $i_4 = 0$, i.e. using the entire sequence; the peak is at $i_2 = t - 20$, though an ensemble is still best. In Table 2(**b**), we show the influence of spanning snippet scales. These scales determine the temporal snippet granularity; individually, results are not significantly different across the scales, but as we begin to aggregate an ensemble, the results improve. The ensemble with 4 scales is best but only marginally better than 3, at the expense of a larger network, so we choose $K = \{10, 15, 20\}$. In Table 2(**c**), we show the influence of recent snippet scales, we find $K_R = 5$ performs best.

Coupling Blocks: Previous studies on simple video understanding have shown the benefits of using features from both the recent and long-range past [21,42]. A naïve way to use both is to simply concatenate, though combining the two in a learned way, e.g. via attention, is superior (+6.4%). To incorporate attention, we apply NLBs [41], which is an adaptation of the attention mechanism that is popularly used in machine translation. When we replace our CBs with concatenation and a linear layer, there is a drop of 6.7%. When we do not use coupling but separately pass the $\mathbf{R}_i$ and $\mathbf{S}_K$ through concatenation and a linear layer, there is a drop of 7.5%. We find also that coupling the recent $\mathbf{R}_i$ and long range $\mathbf{S}_K$ information is critical. Coupling only recent information (−5.9%) does not keep sufficient context, whereas coupling only long-range past (−5%) does not leave sufficient representation for the more relevant recent aspects.

Temporal Aggregation Blocks (TAB) are the most critical component. Omitting them and classifying a single CB's outputs significantly decreases accuracy (−8%). The strength of the TAB comes from using ensembles of coupling blocks as input (single, −2.1%) and using the TABs in an ensemble (single, −2.4%).

Additional Ablations: When we omit the auxiliary complex activity prediction, i.e. removing the Z term from Eq. 6 ("no Z"), we observe a slight performance drop of 1.1%. In our model we max pool over all $\mathbf{S}''_{i,K_1}, ..., \mathbf{S}''_{i,K_S}$ in our TABs. When we replace the max-pooling with concatenation + linear, we reach an accuracy of 37.4. We also try to disentangle the ensemble effect from the use of multi-granular representations. When we fix the spanning past scales K to $\{15, 15, 15\}$ and all the starting points to $i = t - 20$, we observe a drop of 1.2% in accuracy which indicates the importance of our multi-scale representation.

5.3 Anticipation on Procedural Activities - Breakfast & 50 Salads

Next Action Anticipation predicts the action that occurs 1 s from the current time t. We compare to the state of the art in Table 3 with two types of frame inputs: spatio-temporal features (Fisher vectors or I3D) and frame-wise action labels (either from ground truth or via a separate segmentation algorithm) on Breakfast. Compared to previous methods using only visual features as input, we outperform CNN (FC7) features [38] and spatio-temporal features R(2+1)D [26] by a large margin (+32.3% and +8.1%). While the inputs are different, R(2+1)D features were shown to be comparable to I3D features [37]. Since [26] uses only recent observations, we conclude that incorporating the spanning past into the prediction model is essential.

Our method degrades when we replace I3D with the weaker Fisher vectors (40.1% vs 29.7%). Nevertheless, this result is competitive with methods using action labels [1] (30.1% with RNN) derived from segmentation algorithms [29] using Fisher vectors as input. For fair comparison, we report a variant without the complex activity prediction ("no Z"), which has a slight performance

Table 3. Next action anticipation comparisons on Breakfast and 50Salads, given different frame inputs (GT action labels , Fisher vector , I3D features).

Method	Input	Segmentation Method and Feature	Breakfast	50Salads
[38]	FC7 features	-	8.1	6.2
[26]	R(2+1)D	-	32.3	
RNN [1]	segmentation	[29], Fisher	**30.1**	30.1
CNN [1]	segmentation	[29], Fisher	27.0	29.8
ours no Z	Fisher	-	29.2	**31.6**
ours	Fisher	-	29.7	
ours	I3D	-	40.1	40.7
ours	segmentation	ours, I3D	43.1	
ours	segmentation + I3D	ours, I3D	47.0	
ours	frame GT	-	64.7	63.8
ours	frame GT + I3D	-	63.1	

drop (−0.5%). If we use action labels as inputs instead of visual features, our performance improves from 40.1% to 43.1%; merging labels and visual features gives another 4% boost to 47%. In this experiment we use segmentation results from our own framework, (see Sect. 5.6). However, if we substitute ground truth instead of segmentation labels, there is still a 17% gap. This suggests that the quality of the segmentation matters. When the segmentation is very accurate, adding additional features does not help and actually slightly deteriorates results (see Table 3 "frame GT" vs. "frame GT + I3D").

In Table 3, we also report results for 50Salads. Using Fisher vectors we both outperform the state of the art by 1.8% and the baseline with CNN features [38] by 25.4%. Using I3D features improves the accuracy by 9.1% over Fisher vectors.

Dense Anticipation predicts frame-wise actions; accuracies are given for specific portions of the remaining video (Pred.) after observing a given percentage of the past (Obs.). We refer the reader to the supplementary for visual results. Competing methods [1] and [15] have two stages; they first apply temporal video segmentation and then use outputs [29], i.e. frame-wise action labels, as inputs for anticipation. We experiment with both action labels and visual features.

For Breakfast (Table 4, left), when using GT frame labels, we outperform the others, for shorter prediction horizons. For 50Salads (Table 4, right), we outperform the state of the art for the observed 20%, and our predictions are more accurate on long-range anticipation (Pred. 50%). We outperform [1] when we use visual features as input (B Features (Fisher)). When using the segmentations (from [29], which has a frame-wise temporal segmentation accuracy of 36.8% and 42.9% for the observed 20% and 30% of video respectively), we are comparable to state of the art [15]. We further merge visual features with action labels for dense anticipation. With Fisher vectors and the frame labels obtained from [29], we observe a huge performance increase in performance compared to only using the frame labels (up to +7%) in Breakfast. In 50Salads, this increase is not significant nor consistent. This may be due to the better performing segmentation algorithm on 50Salads (frame-wise accuracy of 66.8% and 66.7% for 20% and 30% observed respectively). We observe further improvements on Breakfast once

Table 4. Dense anticipation mean over classes on Breakfast and 50salads, given different frame inputs (GT action labels , Fisher vectors , I3D features).

	Breakfast								50salads							
Obs.	20%				30%				20%				30%			
Pred.	10%	20%	30%	50%	10%	20%	30%	50%	10%	20%	30%	50%	10%	20%	30%	50%
A	**Labels** (GT)								**Labels** (GT)							
RNN[1]	60.4	50.4	45.3	40.4	61.5	50.3	45.0	41.8	42.3	31.2	25.2	16.8	44.2	29.5	20.0	10.4
CNN[1]	58.0	49.1	44.0	39.3	60.3	50.1	45.2	40.5	36.1	27.6	21.4	15.5	37.4	24.8	20.8	14.1
Ke[15]	64.5	**56.3**	**50.2**	**44.0**	66.0	55.9	**49.1**	**44.2**	45.1	33.2	27.6	17.3	**46.4**	**34.8**	**25.2**	13.8
ours	**65.5**	55.5	46.8	40.1	**67.4**	**56.1**	47.4	41.5	**47.2**	**34.6**	**30.5**	**19.1**	44.8	32.7	23.5	**15.3**
B	**Features** (Fisher)								**Features** (Fisher)							
CNN[1]	12.8	11.6	11.2	10.3	17.7	16.9	15.5	14.1								
ours	**15.6**	**13.1**	**12.1**	**11.1**	**19.5**	**17.0**	**15.6**	**15.1**	25.5	19.9	18.2	15.1	30.6	22.5	19.1	11.2
C	**Labels** (Fisher + [29] (Acc. 36.8/42.9))								**Labels** (Fisher + [29] (Acc. 66.8/66.7))							
RNN[1]	18.1	**17.2**	15.9	15.8	21.6	20.0	19.7	19.2	30.1	25.4	18.7	13.5	30.8	17.2	14.8	9.8
CNN[1]	17.9	16.4	15.4	14.5	22.4	20.1	19.7	18.8	21.2	19.0	16.0	9.9	29.1	20.1	17.5	10.9
Ke[15]	18.4	**17.2**	16.4	**15.8**	22.8	**20.4**	19.6	**19.8**	32.5	**27.6**	21.3	**16.0**	**35.1**	**27.1**	22.1	15.6
ours	**18.8**	16.9	**16.5**	15.4	**23.0**	20.0	**19.9**	18.6	**32.7**	26.3	**21.9**	15.6	32.3	25.5	**22.7**	**17.1**
	Concatenate B and C								Concatenate B and C							
ours	**25.0**	**21.9**	**20.5**	**18.1**	**23.0**	**20.5**	19.8	**19.8**	**34.7**	25.9	**23.7**	15.7	34.5	26.1	19.0	15.5
D	**Features** (I3D)															
ours	24.2	21.1	20.0	18.1	30.4	26.3	23.8	21.2								
E	**Labels** (I3D + our seg. (Acc. 54.7/57.8))															
ours	**37.4**	31.2	30.0	26.1	39.5	34.1	31.0	**27.9**								
	Concatenate D and E															
ours	37.1	**31.8**	**30.1**	**27.1**	**39.8**	**34.2**	**31.9**	27.8								

we substitute Fisher vectors with I3D features and segmentations from our own framework (I3D + ours seg.). Similar to next action anticipation, performance drops when using only visual features as input (I3D is better than Fisher vectors). When using I3D features and the frame label outputs of our segmentation method, we obtain our model's best performance, with a slight increase over using only frame label outputs.

5.4 How Much Spanning Past Is Necessary?

We vary the duration of spanning snippets (Eq. 3) with start time i as fractions of the current time t; $i = 0$ corresponds to the full sequence, i.e. 100% of the spanning past, while $i = t$ corresponds to none, i.e. using only recent snippets since the end points j remain fixed at t. Using the entire past is best for Breakfast (Fig. 4 left). Interestingly, this effect is not observed on EPIC (Fig. 4 right). Though we see a small gain by 1.2% until 40% past for the appearance features (rgb), beyond this, performance saturates. We believe this has to do with the fine granularity of labels in EPIC; given that the median action duration is only 1.9 s, one could observe as many as 16 actions in 30 s. Given that the dataset has only 28.5K samples split over 2513 action classes, we speculate that the model cannot learn all the variants of long-range relationships beyond 30 s. Therefore, increasing the scope of the spanning past does not further increase the performance. Based on experiments on the validation set, we set the spanning scope to 6 s for EPIC for the rest of the paper.

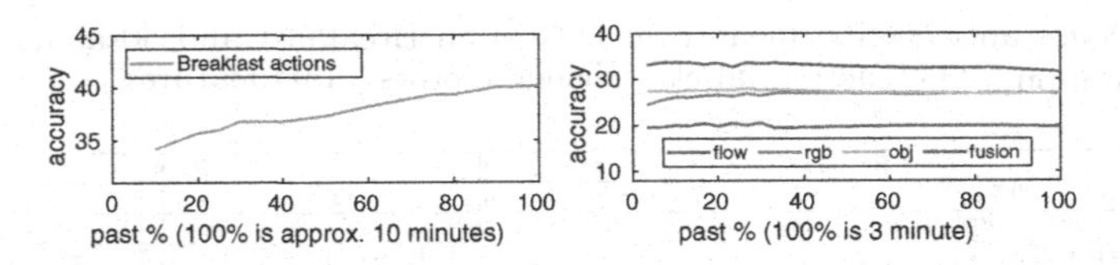

Fig. 4. Effect of spanning scope on instructional vs. daily activities. For EPIC we report Top-5 Acc. on the validation set with rgb, flow and object features and late fusion. (Color figure online)

5.5 Recognition and Anticipation on Daily Activities - EPIC

The **anticipation** task of EPIC requires anticipating the future action $\tau_\alpha = 1$ s before it starts. For fair comparison to the state of the art [10] (denoted by "RU"), we directly use features (appearance, motion and object) provided by the authors. We train our model separately for each feature modality with the same hyper-parameters and fuse predictions from the different modalities by voting. Note that for experiments on this dataset we do not use the entire past for computing our spanning snippet features (see Sect. 5.4). Results on hold-out test data of EPIC are given in Table 5 for seen kitchens (S1) with the same environments as the training data and unseen kitchens (S2) of held out environments. We outperform state of the art, RU [10], in the Top-1 and Top-5 action accuracies by 2% and 2.7% on S1 and by 1.8% and 2.3% on S2 using the same features suggesting superior temporal reasoning abilities of our model. When we add verb and noun classification to our model as auxiliary tasks to help with anticipation, "ours v+n", our performance improves for action and especially for noun and verb scores. For challenge results see supplementary.

For **recognition**, we classify pre-trimmed action segments. We adjust the scope of our spanning and recent snippets according to the action start and end times t_s and t_e. Spanning features are computed on a range of $[t_s-6, t_e+6]$; the first recent snippet scope is fixed to $[t_s, t_e]$ and the rest to $[t_s-1, t_e+1]$, $[t_s-2, t_e+2]$ and $[t_s-3, t_e+3]$. Remaining hyper-parameters are kept the same. In Table 5, we compare to state of the art; we outperform all other methods including SlowFast networks with audio data [43] (+5.4% on S1, +2.2% on S2 for Top-1) and LFB [42], which also uses non-local blocks (+8.6% on S1, +5% on S2 for Top-1) and RU [10] by approximately +7% on both S1 and S2. Together with the anticipation results we conclude that our method generalizes to both anticipation and recognition tasks and is able to achieve state-of-the-art results on both, while [10] performs very well on anticipation but poorly on recognition.

5.6 Temporal Video Segmentation

We compare our performance against the state of the art, MS-TCN (I3D) [8], in Table 6 on Breakfast. We test our model with 2s and 5s windows. We report the frame-wise accuracy (Acc), segment-wise edit distance (Edit) and F1 scores at overlapping thresholds of 10%, 25% and 50%. In the example sequences, in the

Table 5. Action anticipation and recognition on EPIC tests sets S1 and S2

		Action Anticipation						Action Recognition					
		Top-1 Accuracy%			Top-5 Accuracy%			Top-1 Accuracy%			Top-5 Accuracy%		
		Verb	Noun	Action	Verb	Noun	Action	Verb	Noun	Action	Verb	Noun	Action
S1	[26]	30.7	16.5	9.7	76.2	42.7	25.4	-	-	-	-	-	-
	TSN [5]	31.8	16.2	6.0	76.6	42.2	28.2	48.2	36.7	20.5	84.1	62.3	39.8
	RU [10]	33.0	22.8	14.4	79.6	50.9	33.7	56.9	43.1	33.1	85.7	67.1	55.3
	LFB [42]	-	-	-	-	-	-	60.0	45.0	32.7	88.4	71.8	55.3
	[43]	-	-	-	-	-	-	65.7	46.4	35.9	89.5	71.7	57.8
	ours	31.4	22.6	16.4	75.2	47.2	**36.4**	63.2	49.5	41.3	87.3	70.0	63.5
	ours v+n	**37.9**	**24.1**	**16.6**	**79.7**	**54.0**	36.1	**66.7**	**49.6**	**41.6**	**90.1**	**77.0**	**64.1**
S2	[26]	28.4	12.4	7.2	69.8	32.2	19.3	-	-	-	-	-	-
	TSN [5]	25.3	10.4	2.4	68.3	29.5	6.6	39.4	22.7	10.9	74.3	45.7	25.3
	RU [10]	27.0	15.2	8.2	69.6	34.4	21.1	43.7	26.8	19.5	73.3	48.3	37.2
	LFB [42]	-	-	-	-	-	-	50.9	31.5	21.2	77.6	57.8	39.4
	[43]	-	-	-	-	-	-	**55.8**	32.7	24.0	**81.7**	58.9	43.2
	ours	27.5	**16.6**	10.0	66.8	32.8	23.4	52.0	31.5	26.2	76.8	52.7	45.7
	ours v+n	**29.5**	16.5	**10.1**	**70.1**	**37.8**	**23.4**	54.6	**33.5**	**27.0**	80.4	**61.0**	**46.4**

Table 6. Exemplary segmentation and comparisons on Breakfast.

	F1@{10, 25, 50}			Edit	Acc.
MS-TCN (I3D) [8]	52.6	48.1	37.9	**61.7**	**66.3**
ours I3D 2s	52.3	46.5	34.8	51.3	65.3
ours I3D 5s	**59.2**	**53.9**	**39.5**	54.5	64.5
ours I3D GT.seg.	-	-	-	-	**75.9**

F1 scores and edit distances in Table 6, we observe more fragmentation in our segmentation for 2 s than for 5 s. However, for 2 s, our model produces better accuracies, as the 5 s windows are smoothing the predictions at action boundaries. Additionally we provide our model's upper bound, "ours I3D GT.seg.", for which we classify GT action segments instead of sliding windows. The results indicate that there is room for improvement, which we leave as future work. We show that we are able to easily adjust our method from its main application and already get close to the state of the art with slight modifications.

6 Discussion and Conclusion

This paper presented a temporal aggregate model for long-range video understanding. Our method computes recent and spanning representations pooled from snippets of video that are related via coupled attention mechanisms. Validating on three complex activity datasets, we show that temporal aggregates are either comparable or outperform the state of the art on three video understanding tasks: action anticipation, recognition and temporal video segmentation.

In developing our framework, we faced questions regarding temporal extent, scaling, and level of semantic abstraction. Our experiments show that max-pooling is a simple and efficient yet effective way of representing video snippets; this is the case even for snippets as long as two minutes. For learning temporal relationships in long video, attention mechanisms relating the present to long range context can successfully model and anticipate upcoming actions. The extent of context that is beneficial, however, may depend on the nature of activity (instructional vs. daily) and label granularity (coarse vs. fine) of the dataset.

We found significant advantages to using ensembles of multiple scales, both in recent and spanning snippets. Our aggregates model is flexible and can take as input either visual features or frame-wise action labels. We achieve competitive performance with either form of input, though our experiments confirm that higher levels of abstraction such as labels are more preferable for anticipation. Nevertheless, there is still a large gap between what can be anticipated with inputs from current segmentation algorithms in comparison to ground truth labels, leaving room for segmentation algorithms to improve.

Acknowledgments. This work was funded partly by the German Research Foundation (DFG) YA 447/2-1 and GA 1927/4-1 (FOR2535 Anticipating Human Behavior) and partly by National Research Foundation Singapore under its NRF Fellowship Programme [NRF-NRFFAI1-2019-0001] and Singapore Ministry of Education (MOE) Academic Research Fund Tier 1 T1251RES1819.

References

1. Abu Farha, Y., Richard, A., Gall, J.: When will you do what? - Anticipating temporal occurrences of activities. In: IEEE Conference on Computer Vision and Pattern Recognition (CVPR), pp. 5343–5352 (2018)
2. Ba, J.L., Kiros, J.R., Hinton, G.E.: Layer normalization. arXiv preprint arXiv:1607.06450 (2016)
3. Carreira, J., Zisserman, A.: Quo vadis, action recognition? A new model and the kinetics dataset. In: Proceedings of the IEEE Conference on Computer Vision and Pattern Recognition (CVPR), pp. 6299–6308 (2017)
4. Conneau, A., Kiela, D., Schwenk, H., Barrault, L., Bordes, A.: Supervised learning of universal sentence representations from natural language inference data. In: Proceedings of the Conference on Empirical Methods in Natural Language Processing (EMNLP), pp. 670–680 (2017)
5. Damen, D., et al.: Scaling egocentric vision: the dataset. In: Ferrari, V., Hebert, M., Sminchisescu, C., Weiss, Y. (eds.) ECCV 2018. LNCS, vol. 11208, pp. 753–771. Springer, Cham (2018). https://doi.org/10.1007/978-3-030-01225-0_44
6. Ding, L., Xu, C.: Weakly-supervised action segmentation with iterative soft boundary assignment. In: IEEE Conference on Computer Vision and Pattern Recognition (CVPR), pp. 6508–6516 (2018)
7. Donahue, J., et al.: Long-term recurrent convolutional networks for visual recognition and description. In: Proceedings of the IEEE Conference on Computer Vision and Pattern Recognition (CVPR), pp. 2625–2634 (2015)
8. Farha, Y.A., Gall, J.: MS-TCN: multi-stage temporal convolutional network for action segmentation. In: Proceedings of the IEEE Conference on Computer Vision and Pattern Recognition (CVPR), pp. 3575–3584 (2019)
9. Feichtenhofer, C., Fan, H., Malik, J., He, K.: SlowFast networks for video recognition. In: Proceedings of the IEEE International Conference on Computer Vision, pp. 6202–6211 (2019)

10. Furnari, A., Farinella, G.M.: What would you expect? Anticipating egocentric actions with rolling-unrolling LSTMs and modality attention. In: International Conference on Computer Vision (ICCV) (2019)
11. Girdhar, R., Ramanan, D., Gupta, A., Sivic, J., Russell, B.: ActionVLAD: learning spatio-temporal aggregation for action classification. In: Proceedings of the IEEE Conference on Computer Vision and Pattern Recognition, pp. 971–980 (2017)
12. Huang, D.-A., Fei-Fei, L., Niebles, J.C.: Connectionist temporal modeling for weakly supervised action labeling. In: Leibe, B., Matas, J., Sebe, N., Welling, M. (eds.) ECCV 2016. LNCS, vol. 9908, pp. 137–153. Springer, Cham (2016). https://doi.org/10.1007/978-3-319-46493-0_9
13. Huang, D.A., et al.: What makes a video a video: analyzing temporal information in video understanding models and datasets. In: Proceedings of the IEEE Conference on Computer Vision and Pattern Recognition (CVPR), pp. 7366–7375 (2018)
14. Kay, W., et al.: The kinetics human action video dataset. arXiv preprint arXiv:1705.06950 (2017)
15. Ke, Q., Fritz, M., Schiele, B.: Time-conditioned action anticipation in one shot. In: IEEE Conference on Computer Vision and Pattern Recognition (CVPR), June 2019
16. Kline, N., Snodgrass, R.T.: Computing temporal aggregates. In: Proceedings of the Eleventh International Conference on Data Engineering, pp. 222–231. IEEE (1995)
17. Koppula, H.S., Saxena, A.: Anticipating human activities using object affordances for reactive robotic response. IEEE Trans. Pattern Anal. Mach. Intell. (PAMI) **38**(1), 14–29 (2015)
18. Kuehne, H., Arslan, A., Serre, T.: The language of actions: recovering the syntax and semantics of goal-directed human activities. In: Proceedings of the IEEE Conference on Computer Vision and Pattern Recognition (CVPR), pp. 780–787 (2014)
19. Lan, T., Chen, T.-C., Savarese, S.: A hierarchical representation for future action prediction. In: Fleet, D., Pajdla, T., Schiele, B., Tuytelaars, T. (eds.) ECCV 2014. LNCS, vol. 8691, pp. 689–704. Springer, Cham (2014). https://doi.org/10.1007/978-3-319-10578-9_45
20. Lee, J., Natsev, A.P., Reade, W., Sukthankar, R., Toderici, G.: The 2nd YouTube-8M large-scale video understanding challenge. In: Leal-Taixé, L., Roth, S. (eds.) ECCV 2018. LNCS, vol. 11132, pp. 193–205. Springer, Cham (2019). https://doi.org/10.1007/978-3-030-11018-5_18
21. Li, F., et al.: Temporal modeling approaches for large-scale Youtube-8M video understanding. arXiv preprint arXiv:1707.04555 (2017)
22. Lin, J., Gan, C., Han, S.: TSM: temporal shift module for efficient video understanding. In: Proceedings of the IEEE International Conference on Computer Vision (ICCV), pp. 7083–7093 (2019)
23. Lin, R., Xiao, J., Fan, J.: NeXtVLAD: an efficient neural network to aggregate frame-level features for large-scale video classification. In: Leal-Taixé, L., Roth, S. (eds.) ECCV 2018. LNCS, vol. 11132, pp. 206–218. Springer, Cham (2019). https://doi.org/10.1007/978-3-030-11018-5_19
24. Mahmud, T., Hasan, M., Roy-Chowdhury, A.K.: Joint prediction of activity labels and starting times in untrimmed videos. In: Proceedings of the IEEE International Conference on Computer Vision, pp. 5773–5782 (2017)
25. Miech, A., Laptev, I., Sivic, J.: Learnable pooling with context gating for video classification. arXiv preprint arXiv:1706.06905 (2017)

26. Miech, A., Laptev, I., Sivic, J., Wang, H., Torresani, L., Tran, D.: Leveraging the present to anticipate the future in videos. In: IEEE Conference on Computer Vision and Pattern Recognition (CVPR) Workshops, p. 0 (2019)
27. Ostyakov, P., et al.: Label denoising with large ensembles of heterogeneous neural networks. In: Leal-Taixé, L., Roth, S. (eds.) ECCV 2018. LNCS, vol. 11132, pp. 250–261. Springer, Cham (2019). https://doi.org/10.1007/978-3-030-11018-5_23
28. Richard, A., Gall, J.: Temporal action detection using a statistical language model. In: Proceedings of the IEEE Conference on Computer Vision and Pattern Recognition (CVPR), pp. 3131–3140 (2016)
29. Richard, A., Kuehne, H., Gall, J.: Weakly supervised action learning with RNN based fine-to-coarse modeling. In: IEEE Conference on Computer Vision and Pattern Recognition (CVPR), pp. 754–763 (2017)
30. Sener, F., Yao, A.: Unsupervised learning and segmentation of complex activities from video. In: Proceedings of the IEEE Conference on Computer Vision and Pattern Recognition (CVPR) (2018)
31. Simonyan, K., Zisserman, A.: Two-stream convolutional networks for action recognition in videos. In: Advances in Neural Information Processing Systems, pp. 568–576 (2014)
32. Soomro, K., Zamir, A.R., Shah, M.: UCF101: a dataset of 101 human actions classes from videos in the wild. CoRR abs/1212.0402 (2012). http://arxiv.org/abs/1212.0402
33. Srivastava, N., Hinton, G., Krizhevsky, A., Sutskever, I., Salakhutdinov, R.: Dropout: a simple way to prevent neural networks from overfitting. J. Mach. Learn. Res. **15**(1), 1929–1958 (2014)
34. Stein, S., McKenna, S.J.: Combining embedded accelerometers with computer vision for recognizing food preparation activities. In: Proceedings of the 2013 ACM International Joint Conference on Pervasive and Ubiquitous Computing, pp. 729–738. ACM (2013)
35. Tang, Y., Zhang, X., Wang, J., Chen, S., Ma, L., Jiang, Y.-G.: Non-local NetVLAD encoding for video classification. In: Leal-Taixé, L., Roth, S. (eds.) ECCV 2018. LNCS, vol. 11132, pp. 219–228. Springer, Cham (2019). https://doi.org/10.1007/978-3-030-11018-5_20
36. Tran, D., Bourdev, L., Fergus, R., Torresani, L., Paluri, M.: Learning spatiotemporal features with 3D convolutional networks. In: Proceedings of the IEEE International Conference on Computer Vision (ICCV), pp. 4489–4497 (2015)
37. Tran, D., Wang, H., Torresani, L., Ray, J., LeCun, Y., Paluri, M.: A closer look at spatiotemporal convolutions for action recognition. In: IEEE Conference on Computer Vision and Pattern Recognition (CVPR), pp. 6450–6459 (2018)
38. Vondrick, C., Pirsiavash, H., Torralba, A.: Anticipating visual representations from unlabeled video. In: IEEE Conference on Computer Vision and Pattern Recognition (CVPR), pp. 98–106 (2016)
39. Wang, H., Schmid, C.: Action recognition with improved trajectories. In: Proceedings of the IEEE International Conference on Computer Vision (ICCV), pp. 3551–3558 (2013)
40. Wang, L., et al.: Temporal segment networks: towards good practices for deep action recognition. In: Leibe, B., Matas, J., Sebe, N., Welling, M. (eds.) ECCV 2016. LNCS, vol. 9912, pp. 20–36. Springer, Cham (2016). https://doi.org/10.1007/978-3-319-46484-8_2
41. Wang, X., Girshick, R., Gupta, A., He, K.: Non-local neural networks. In: IEEE Conference on Computer Vision and Pattern Recognition (CVPR), pp. 7794–7803 (2018)

42. Wu, C.Y., Feichtenhofer, C., Fan, H., He, K., Krähenbühl, P., Girshick, R.: Long-term feature banks for detailed video understanding. In: IEEE Conference on Computer Vision and Pattern Recognition (CVPR) (2019)
43. Xiao, F., Lee, Y.J., Grauman, K., Malik, J., Feichtenhofer, C.: Audiovisual SlowFast networks for video recognition. arXiv preprint arXiv:2001.08740 (2020)
44. Yue-Hei Ng, J., Hausknecht, M., Vijayanarasimhan, S., Vinyals, O., Monga, R., Toderici, G.: Beyond short snippets: deep networks for video classification. In: Proceedings of the IEEE Conference on Computer Vision and Pattern Recognition (CVPR), pp. 4694–4702 (2015)

Stochastic Fine-Grained Labeling of Multi-state Sign Glosses for Continuous Sign Language Recognition

Zhe Niu(✉) and Brian Mak(✉)

Department of Computer Science and Engineering, The Hong Kong University of Science and Technology, Clear Water Bay, Kowloon, Hong Kong
{zniu,mak}@cse.ust.hk

Abstract. In this paper, we propose novel stochastic modeling of various components of a continuous sign language recognition (CSLR) system that is based on the transformer encoder and connectionist temporal classification (CTC). Most importantly, We model each sign gloss with multiple states, and the number of states is a categorical random variable that follows a learned probability distribution, providing stochastic fine-grained labels for training the CTC decoder. We further propose a stochastic frame dropping mechanism and a gradient stopping method to deal with the severe overfitting problem in training the transformer model with CTC loss. These two methods also help reduce the training computation, both in terms of time and space, significantly. We evaluated our model on popular CSLR datasets, and show its effectiveness compared to the state-of-the-art methods.

1 Introduction

Sign language is the primary communication medium among the deaf. It conveys meaning using gestures, facial expressions and upper body posture, etc., and has linguistic rules that are different from those of spoken languages. Sign language recognition (SLR) is the task of converting a sign language video to the corresponding sequence of (sign) glosses (i.e., "words" in a sign language), which are the basic units of the sign language semantics. Both isolated sign language recognition (ISLR) [15] and continuous sign language recognition (CSLR) have been attempted. ISLR classifies a gloss-wise segmented video into its corresponding gloss, whereas CSLR classifies a sentence-level sign video into its corresponding sequence of glosses. The latter task is more difficult and is the focus of this paper.

Most of the modern CSLR architectures contain three components: visual model, contextual model and alignment model. The visual model first extracts the visual features from the input video frames, based on which the contextual model further mines the correlation between the glosses. Convolutional neural

Electronic supplementary material The online version of this chapter (https://doi.org/10.1007/978-3-030-58517-4_11) contains supplementary material, which is available to authorized users.

A. Vedaldi et al. (Eds.): ECCV 2020, LNCS 12361, pp. 172–186, 2020.
https://doi.org/10.1007/978-3-030-58517-4_11

networks (CNNs) and recurrent neural networks (RNNs) are commonly used architectures for the visual and contextual model, respectively. In CSLR, sign glosses occur (time-wise) monotonically with the corresponding events in the video. Thus, an alignment model is required to find the proper mapping between the video frames and glosses such that the model can be trained. Methods like [11,12,14] align the video frames to glosses by applying Viterbi search on the hidden Markov models (HMMs). While others [2,5,17,25,27] adopt the connectionist temporal classification (CTC) method, where a soft full-sum alignment is calculated as the final training objective. For both HMM-based and CTC-based CSLR models, it is usually necessary to fine-tune the lower-level visual feature extractor during model training, as it has been shown that the visual network cannot learn effective features in end-to-end training [14,17].

To address this problem, we propose to use the transformer encoder [22] as the contextual model for CSLR, which has been shown effective in tasks such as machine translation [6,22] and speech recognition [16]. The residual connections between layers in the transformer encoder help backpropagate the errors better to the visual model. Moreover, to improve model robustness and to alleviate the overfitting problem, we propose dropping video frames stochastically and randomly stopping the gradients of some frames during training. We call these two procedures *stochastic frame dropping* (SFD) and *stochastic gradient stopping* (SGS), respectively. More importantly, we perform detailed modeling and allow each gloss model to have multiple states, but the number of states for each gloss model is variable and is modeled by a probability distribution that is trained jointly with the rest of the system. We named this method *stochastic fine-grained labeling* (SFL). SFL provides stochastic finer-grained labels for the CTC loss, thus provides more supervision in the temporal domain.

Overall, the main contributions of our work are:

1. We propose stochastic frame dropping (SFD) and stochastic gradient stopping (SGS) to reduce video memory footprint, improve model robustness and alleviate the overfitting problem during model training.
2. We introduce stochastic fine-grained labeling (SFL) to model glosses with multiple states. The number of states of any gloss is variable and follows a probability distribution. As a result, the performance of our SLR model is further improved.

The rest of this paper is organized as follows: In Sect. 2, we review related works. Section 3 introduces the use of stochastic modeling in three components of our model. Section 4 presents the experimental evaluation of our proposed methods and discussions on the findings. Finally we conclude in Sect. 5.

2 Related Works

In the past, many works [12] tackle the CSLR problem using the Gaussian mixture model-hidden Markov model (GMM-HMM) with hand-crafted visual features. Since the features are hand-crafted, they are usually not robust and not optimal as they are not optimized jointly with the rest of their CSLR systems.

To leverage the power of deep learning for automatic feature extraction, hybrid models [11,13,14] are proposed, which combine deep neural network models with HMM. As HMM requires the computation of priors, these methods are usually trained with epoch-wise re-alignment. To avoid prior estimation, some other methods [17,25,27] try to replace HMMs with the connectionist temporal classification (CTC) method, which provides a soft full-sum alignment and re-aligns after each mini-batch. However, as the visual model fails to learn representative features, epoch-wise iterative fine-tuning of the visual model is required.

Sequence-to-sequence (Seq2Seq) architecture [1] has also been attempted for CSLR, in which, the alignments are calculated by a global weighted summation of the input sequence. As there is no monotonicity constraint, the model tends to misalign. Moreover, since the decoder takes the ground truth labels as inputs during training, the model suffers from the exposure bias problem, where errors accumulate with each decoding step during testing. [26] utilizes a transformer-based Seq2Seq model and applies reinforcement learning to alleviate exposure bias. On the other hand, [17] uses CTC in additional to a Seq2Seq decoder to make up for the performance degradation. Work [3] proposes to jointly train the SLR and SLT tasks in a CNN-Transformer framework by combining the CTC loss on sign glosses and cross-entropy loss on spoken words, which relies on pretraining the visual model via a CNN+LSTM+HMM setup [11].

In this work, we choose CTC as the alignment model so that we do not need to do re-alignment frequently for the prior estimation in HMM-based models and for alleviating the exposure bias problem in the Seq2Seq architecture. Compared with other CTC works, our method avoid the necessity of having to fine-tune the visual model iteratively after each epoch.

3 Methodology

In this section, we introduce our stochastic multi-states (SMS) framework. We will first give an overview of the framework, then describe stochastic frame dropping (SFD), stochastic gradient stopping (SGS) and stochastic fine-grained labeling (SFL), respectively in detail.

3.1 Framework Overview

The overall framework of our model is presented in Fig. 1. The design of the network follows the visual-contextual-alignment model scheme. For the visual model, we choose 2D convolutional neural network (2D-CNN) to extract visual features from individual frames. For the contextual model, we select the transformer encoder with relative positional encoding. The connectionist temporal classification (CTC) is adopted as the alignment model.

Given an RGB video with T frames $\mathbf{x} = (\mathbf{x}_1, \ldots, \mathbf{x}_T)$, the visual model (CNN) first extracts visual features $\mathbf{z} = (\mathbf{z}_1, \ldots, \mathbf{z}_T)$ from individual frames. After this, the contextual model extracts the temporal correlation between the visual vectors. Then, the posterior probabilities of sub-gloss states are calculated based on

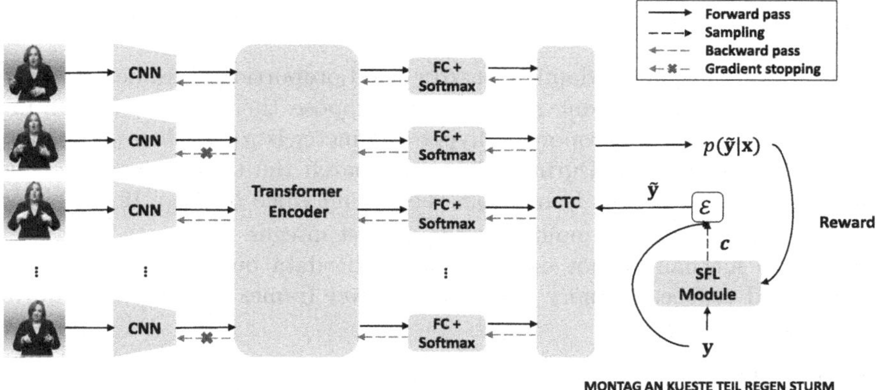

Fig. 1. Overview of our framework. Video frames are first processed by CNN to produce a visual feature sequence which is then fed into a transformer encoder. The output of the transformer encoder is further sent to a fully connected (FC) + softmax layer to produce the probability of sub-gloss states at each time step conditioned on the input video. In the meantime, the target sequence $\mathbf{y}$ is fed into the stochastic fine-grained labeling (SFL) module, and a sequence of sub-gloss state numbers $\mathbf{c}$ is sampled. The extension function $\mathcal{E}(\mathbf{y}, \mathbf{c})$ extends the input gloss sequence $\mathbf{y}$ according to the state number sequence $\mathbf{c}$ to produce a sequence of sub-gloss states $\tilde{\mathbf{y}}$. Based on the probability the sequence produces and the fine-grained label $\tilde{\mathbf{y}}$, CTC calculates the probability of the fine-grained labels $\tilde{\mathbf{y}}$ given the video $p(\tilde{\mathbf{y}}|\mathbf{x})$, which is further used to reward the SFL module for producing state number sequence that leads to higher $p(\tilde{\mathbf{y}}|\mathbf{x})$.

the features output by the contextual model. At the same time, the SFL module takes the corresponding target gloss sequence $\mathbf{y} = (y_1, \ldots, y_L)$ with length L as input and generates a probability distribution of the number of states for each gloss. Based on this distribution, a sequence of state numbers $\mathbf{c} = (c_1, \ldots, c_L)$ is sampled and the original gloss sequence is extended to a sub-gloss state sequence $\tilde{\mathbf{y}} = (\tilde{y}_1, \ldots, \tilde{y}_S)$, where $S = \sum_{l=1}^{L} c_l$. Finally, CTC is applied to calculate the posterior probability of the sub-gloss state sequence $p(\tilde{\mathbf{y}}|\mathbf{x})$.

3.2 Visual Model

We use ResNet [8] as the visual model as they are powerful enough but relatively lightweight compared to models like GoogLeNet [21] or VGG [20]. The ResNet is pre-trained on ImageNet [18]. The final fully connected layer is replaced by a linear layer that suits the dimension of our contextual model.

Stochastic Frame Dropping
Due to the limited amount of sign language data, overfitting is a main issue during training. To avoid the network overfitting salient frames and ignoring the less representative ones, we introduce a stochastic frame dropping (SFD) technique that stochastically drops out some frames during training. The frame

dropping also changes the rate of signs and introduces signing speed variations into the training data.

During training, we randomly discard a fixed proportion of frames in a video by uniform sampling without replacement. Suppose there are T' frames in a video originally, and the proportion hyper-parameter is p_{drop}, then $\lceil T' \times p_{\text{drop}} \rceil$ frames will be discarded. During testing, to match the training condition, we evenly select every $\frac{1}{p_{\text{drop}}}$-th frame from the testing video to drop. If $\frac{n}{p_{\text{drop}}}$ is not a whole integer, it will be rounded to the nearest integer.

The SFD mechanism not only augments the data but also improves time efficiency and reduces memory footprint as fewer frames are processed during training and testing.

Stochastic Gradient Stopping
To further prevent overfitting, reduce memory consumption and speed up training, we propose the stochastic gradient stopping (SGS) training technique. This method avoids to compute back-propagation for a part of the input frames during visual feature extraction. Similar to SFD, a hyper-parameter p_{stop} is set for the proportion of frames whose gradient will be stopped. Again, the SGS frames are sampled stochastically and uniformly without replacement. Denote the output of CNN, i.e., the sequence of visual features as $\mathbf{z} = (\mathbf{z}_1, \ldots, \mathbf{z}_T)$, the gradient of the CTC objective with respect to any CNN parameter ϕ can be written as follows:

$$\begin{aligned} \nabla_\phi \log p(\mathbf{y}|\mathbf{x}) &= \sum_{i=t}^{T} \langle \nabla_{\mathbf{z}_t} \log p(\mathbf{y}|\mathbf{z}), \nabla_\phi \mathbf{z}_t \rangle \\ &\approx T \cdot \mathbb{E}_t \left[\langle \nabla_{\mathbf{z}_t} \log p(\mathbf{y}|\mathbf{z}), \nabla_\phi \mathbf{z}_t \rangle \right] \\ &\approx T \cdot \frac{1}{K} \sum_{k=1}^{K} \langle \nabla_{\mathbf{z}_{t_k}} \log p(\mathbf{y}|\mathbf{z}), \nabla_\phi \mathbf{z}_{t_k} \rangle \end{aligned} \tag{1}$$

where $t_k \sim \mathcal{U}\{1, \ldots, T\}$ and $K = \lfloor T \times p_{stop} \rfloor$.

In SGS, the mean gradient of CNN parameter ϕ is approximated by its sample mean, which introduces noise to the gradient to prevent CNN from overfitting. Since a part of the frames are detached from the computation graph and the intermediate outputs of those frames are no longer required to be held for back-propagation, SGS reduces memory footprint during training. Moreover, the training procedure is sped up as less computation of back-propagation is needed.

3.3 Contextual Model

The transformer encoder [22] is adopted as the contextual model to further extract the temporal information between frames. Relative positional encoding [19] is used instead of the absolute position encoding [22]. For each transformer encoder layer, the relative positional encoding holds two sets of learnable vectors $\{\mathbf{a}_m \in \mathbb{R}^d\}_{m=-M}^{M}$ and $\{\mathbf{b}_m \in \mathbb{R}^d\}_{m=-M}^{M}$. For each attention head, denote the query at the i-th time step as $\mathbf{q}_i \in \mathbb{R}^d$ and the key, value at the j-th time step as $\mathbf{k}_j, \mathbf{v}_j \in \mathbb{R}^d$. Instead of calculating the score before the softmax function as

$$s_{i,j} = \frac{\mathbf{q}_i^\top \mathbf{k}_j}{\sqrt{d}}, \tag{2}$$

the relative positional encoding calculates this score by injecting the relative positional information $\mathbf{a}_{\mathrm{clip}(i-j)}$ to give

$$\begin{aligned} s_{i,j} &= \frac{\mathbf{q}_i^\top \mathbf{k}_j + \mathbf{q}_i^\top \mathbf{a}_{\mathrm{clip}(i-j)}}{\sqrt{d}} \\ \mathrm{clip}(m) &= \max(-M, \min(M, m)). \end{aligned} \tag{3}$$

And for the context vector $\mathbf{c}_i$, instead of calculating it as the weighted average of the value vectors:

$$\mathbf{c}_i = \sum_j \alpha_{i,j} \mathbf{v}_j \tag{4}$$

where $\boldsymbol{\alpha}_i = \mathrm{softmax}(\mathbf{s}_i)$, the relative positional encoding injects the positional vector $\mathbf{b}_{\mathrm{clip}(i-j)}$ into the calculation of $\mathbf{c}_i$ to give

$$\mathbf{c}_i = \sum_j \alpha_{i,j} (\mathbf{v}_j + \mathbf{b}_{\mathrm{clip}(i-j)}). \tag{5}$$

The relative positional encoding is more suitable for video tasks as the input sequence is continuous in time and consecutive frames are more correlated.

3.4 Alignment Model

The contextual model produces a spatio-temporal feature vector sequence with T time steps. To align the feature vector sequence to the target label sequence, we propose the stochastic fine-grained labeling (SFL) mechanism to enhance the supervision along the temporal domain based on the CTC method.

Connectionist Temporal Classification (CTC)
CTC introduces a sequence of hidden variables $\boldsymbol{\pi} = (\pi_1, \ldots, \pi_T), \pi_t \in \mathcal{V} \cup \{blank\}$, where $\mathcal{V}$ is the vocabulary and *blank* is a special token for representing silent time steps and separating consecutive repeating glosses. The hidden state π_t indicates the alignment between the input time step t and the corresponding gloss in the target sentence. Given the input sequence $\mathbf{x} = (\mathbf{x}_1, \ldots, \mathbf{x}_T)$ and the target sequence $\mathbf{y} = (y_1, \ldots, y_L)$, consider the conditional probability:

$$\begin{aligned} p(\mathbf{y}|\mathbf{x}) &= \sum_{\boldsymbol{\pi}} p(\mathbf{y}|\boldsymbol{\pi}, \mathbf{x}) p(\boldsymbol{\pi}|\mathbf{x}) \\ &= \sum_{\pi \in \mathcal{B}^{-1}(\mathbf{y})} p(\boldsymbol{\pi}|\mathbf{x}) \\ &\approx \sum_{\pi \in \mathcal{B}^{-1}(\mathbf{y})} \prod_{t=1}^{T} p(\pi_t|\mathbf{x}), \end{aligned} \tag{6}$$

where $\mathcal{B} : (\mathcal{V} \cup \{blank\})^T \rightarrow \mathcal{V}^L$ is a function that maps a hidden sequence to its corresponding sequence of glosses. More specifically, $\mathcal{B}$ converts the hidden sequence to the gloss sequence by first removing the consecutive repeating words and then the *blank* symbol in the hidden sequence. In CTC, the term $p(\boldsymbol{\pi}|\mathbf{x})$ is approximated by $\prod_{t=1}^{T} p(\pi_t|\mathbf{x})$ as the hidden variables are assumed to be independent given the input $\mathbf{x}$. The CTC loss is defined as

$$\mathcal{L}_{\text{CTC}}(\mathbf{x}, \mathbf{y}) = -\log p(\mathbf{y}|\mathbf{x}) \ . \tag{7}$$

During training, the CTC loss is minimized so that $p(\mathbf{y}|\mathbf{x})$ is maximized. In testing, the prefix beam search algorithm [7] is used to decode the conditional probability sequence. It retains only the k most probable prefixes at each decoding time step so as to reduces the search space and speed up decoding.

Stochastic Fine-Grained Labeling
HMM-based CSLR methods have exploited multiple hidden states to represent each gloss to increase the label granularity and improve recognition performance [11,14]. As one sign gloss usually consists of multiple motion primitives, using multiple states instead of one single state for one gloss helps the network learn more discriminative features at different time steps of one gloss.

Given an input video $\mathbf{x} = (\mathbf{x}_1, \ldots, \mathbf{x}_T)$, the corresponding gloss sequence: $\mathbf{y} = (y_1, \ldots, y_L)$, a sub-gloss state number sequence $\mathbf{c} = (c_1, \ldots, c_L)$ and the maximal state number $c_{\max}$, the fine-grained label $\tilde{\mathbf{y}}$ is defined as an extension of the original gloss sequence:

$$\tilde{\mathbf{y}} = \mathcal{E}(\mathbf{y}, \mathbf{c}) = (y_1^1, \ldots, y_1^{c_1}, y_2^1 \ldots, y_2^{c_2}, \ldots, y_L^1, \ldots, y_L^{c_L}) \tag{8}$$

where $y_i^j \in \mathcal{V} \times \{1, \ldots, c_{\max}\}$ and $j \in \{1, \ldots, c_i\}$. The extension function $\mathcal{E}(\mathbf{y}, \mathbf{c})$ takes the gloss sequence $\mathbf{y}$ and the sub-gloss state number sequence $\mathbf{c}$ as inputs, and extends the gloss sequence to a sub-gloss state sequence $\tilde{\mathbf{y}}$. In HMM-based methods, the number of states is usually fixed for each gloss as there is no prior knowledge about the optimal number of states. Skip transitions are used to allow exiting a gloss earlier. To avoid extra effort of manually fine-tuning the skip/exit penalty, we propose the stochastic fine-grained labeling (SFL) method that allows the model to automatically learn the number of states for each gloss. SFL utilizes the REINFORCE algorithm [23] to estimate the distribution of the number of states during training. By sampling different state number sequence $\mathbf{c}$ based on the target sequence, we reinforce the probability of the state number sequences that produce lower CTC loss.

The log conditional probability $\log p(\mathbf{y}|\mathbf{x})$ in CTC loss (Eq. 7) can be rewritten by introducing the state number sequence $\mathbf{c}$ as latent variables as follows:

$$\begin{aligned} \log p(\mathbf{y}|\mathbf{x}) &= \log \sum_{\mathbf{c}} p(\mathbf{c}|\mathbf{x}) \frac{p(\mathbf{y}, \mathbf{c}|\mathbf{x})}{p(\mathbf{c}|\mathbf{x})} \\ &\geq \sum_{\mathbf{c}} p(\mathbf{c}|\mathbf{x}) \log \frac{p(\mathbf{y}, \mathbf{c}|\mathbf{x})}{p(\mathbf{c}|\mathbf{x})} \quad \text{(Jensen's inequality)} \\ &= \mathbb{E}_{\mathbf{c} \sim p(\mathbf{c}|\mathbf{x})} [\log p(\mathbf{y}, \mathbf{c}|\mathbf{x}) - \log p(\mathbf{c}|\mathbf{x})] \ . \end{aligned} \tag{9}$$

Instead of directly maximizing $\log p(\mathbf{y}|\mathbf{x})$, the lower bound given by Eq. 9 is maximized, which can be further written as follows by taking the extended sub-gloss sequence $\tilde{\mathbf{y}} = \mathcal{E}(\mathbf{y}, \mathbf{c})$ into account:

$$\mathbb{E}_{\mathbf{c}\sim p(\mathbf{c}|\mathbf{x})}[\log p(\mathbf{y}, \mathbf{c}|\mathbf{x}) - \log p(\mathbf{c}|\mathbf{x})] = \mathbb{E}_{\mathbf{c}\sim p(\mathbf{c}|\mathbf{x})}[\log p(\tilde{\mathbf{y}}|\mathbf{x})] + H[\mathbf{c}]. \tag{10}$$

Here $H[\mathbf{c}]$ is the entropy of the state number distribution whose gradient can be calculated explicitly given the model. The gradient of the term $\mathbb{E}_{\mathbf{c}\sim p(\mathbf{c}|\mathbf{x})}[\log p(\tilde{\mathbf{y}}|\mathbf{x})]$ with respect to the network parameter θ can be written as follows after approximating it by the Monte Carlo method with the log derivative trick:

$$\begin{aligned} &\nabla_\theta \mathbb{E}_{\mathbf{c}\sim p(\mathbf{c}|\mathbf{x})}\left[\log p(\tilde{\mathbf{y}}|\mathbf{x})\right] \\ =&\mathbb{E}_{\mathbf{c}\sim p(\mathbf{c}|\mathbf{x})}\left[\nabla_\theta \log p(\tilde{\mathbf{y}}|\mathbf{x}) + \log p(\tilde{\mathbf{y}}|\mathbf{x})\nabla_\theta \log p(\mathbf{c}|\mathbf{x})\right] \\ \approx&\frac{1}{N}\sum_{i=1}^{N}[\nabla_\theta \log p(\mathcal{E}(\mathbf{y}, \mathbf{c}^{(i)})|\mathbf{x}) + \mathcal{R}(\mathbf{c}^{(i)})\nabla_\theta \log p(\mathbf{c}^{(i)}|\mathbf{x})] \end{aligned} \tag{11}$$

where $\mathcal{R}(\mathbf{c}^{(i)}) = \log p(\mathcal{E}(\mathbf{y}, \mathbf{c}^{(i)})|\mathbf{x})$.

Given $\tilde{\mathbf{y}}$ and $\mathbf{x}$, the term $\log p(\tilde{\mathbf{y}}|\mathbf{x})$ and its gradient can be directly calculated using the existing CTC method mentioned above. To reduce the variance of the Monte Carlo estimator of the gradient, similar to [24], we introduce a reward baseline term $b(\mathbf{x})$ into the gradient estimator as follows:

$$\hat{\mathcal{R}}(\mathbf{c}^{(i)}) = \log p(\mathcal{E}(\mathbf{y}, \mathbf{c}^{(i)})|\mathbf{x}) - b(\mathbf{x}). \tag{12}$$

Approximation Trick

Directly sampling the state number sequence $\mathbf{c}$ from $p(\mathbf{c}|\mathbf{x})$ is impractical as its length may not match the length of the gloss sequence $\mathbf{y}$, and causes a large proportion of invalid samples at the beginning of training (as $p(\mathbf{y}, \mathbf{c}|\mathbf{x}) = 0$ for such cases). To tackle this problem, we further simplify the model by assuming (1) the most probable gloss sequence dominates the others given a video, and (2) the numbers of states is only dependent on the corresponding gloss. Thus, we have

$$\begin{aligned} p(\mathbf{c}|\mathbf{x}) &= \sum_{\mathbf{y}} p(\mathbf{y}|\mathbf{x})p(\mathbf{c}|\mathbf{x}, \mathbf{y}) \\ &\approx \max_{\mathbf{y}} p(\mathbf{y}|\mathbf{x}) \prod_{l=1}^{L} p(c_l|y_l)\ . \end{aligned} \tag{13}$$

In our implementation, $p(c_l|y_l)$ is a categorical distribution produced by the SFL module, which is a simple two-layer feed-forward neural network. It takes a gloss y_l as input and outputs the corresponding distribution state number distribution $p(c_l|y_l)$. In training, we follow the empirical distribution and set $p(\mathbf{y}|\mathbf{x}) = 1$ for the data sample $(\mathbf{x}, \mathbf{y})$ in the training dataset. As the distribution of $\mathbf{c}$ has been changed from being conditioned on $\mathbf{x}$ to $\mathbf{y}$, we also update our baseline from $b(\mathbf{x})$ to $b(\mathbf{y})$. For the baseline estimation, we choose another two-layer feed-forward network and train it with the MSE loss between $b(\mathbf{y}) = \frac{1}{L}\sum_{l=1}^{L} b(y_l)$ and the uncalibrated reward $\mathcal{R}(\mathbf{c}^{(i)})$.

4 Experiments

4.1 Dataset and Metrics

PHOENIX-2014. PHOENIX-2014 [12] is a popular German sign language dataset collected from weather forecast broadcast. The dataset contains a total of 963k frames captured by an RGB camera in 25 frames per second. It has a vocabulary size of 1081. There are 5672, 540 and 629 data samples in the training, development, and testing sets, respectively. The dataset contains 9 signers who appear in all three splits.

PHOENIX-2014-T. PHOENIX-2014-T [4] is an extension to the PHOENIX-2014 dataset with different sentence boundaries. It is designed for sign language translation but can also be used to evaluate the CSLR task. The dataset has a vocabulary size of 1085. There are 7096, 519 and 642 samples in the training, development, and testing sets, respectively. Similar to the PHOENIX-2014 dataset, there are 9 signers who appear in all three splits.

Metrics. For evaluation, we use word error rate (WER) as the metric, which is defined as the minimal summation of the substitution, insertion and deletion operations to convert the recognized sentence to the corresponding reference sentence:

$$\text{WER} = \frac{\#\text{ substitutions } + \#\text{ insertions } + \#\text{ deletions}}{\#\text{ glosses in reference}} \tag{14}$$

For both datasets, the evaluation script comes along with the dataset is used for computing the WER.

4.2 Basic Settings

Data Processing and Augmentation. For PHOENIX-2014 and PHOENIX-2014-T dataset, we follow the commonly used setting to adopt the full-frame videos with a resolution of 210×260. The frames are first resized to 256×256 and then cropped to 224×224. During training, random cropping is used and in testing, center cropping is adopted.

Model Hyperparameters. For the visual model, we adopt a 18-layer 2D ResNet pre-trained on ImageNet [18] with the fully connected layer removed. For the contextual model, we use a 2-layer transformer encoder with 4 heads ($h = 4$), model dimension $d = 512$ and position-wise feed-forward layer dimension $d_{ff} = 2048$.

Training and Decoding. The model is trained with a batch size of 8 using the Adam optimizer [10] with $\beta_1 = 0.9$ and $\beta_2 = 0.999$. We schedule the learning rate for the i-th epoch as $\eta_i = \eta_0 \cdot 0.95^{\lfloor \frac{i}{2} \rfloor}$ with initial learning rate $\eta_0 = 1 \times 10^{-4}$. Each model is trained for 30 epochs. For stochastic fine-grained labeling (SFL) based model, the number of Monte Carlo samples N is set to 32.

During training, if the length of the target sequence exceeds the number of input frames, the CTC loss of the corresponding sample will be zeroed out. If it is related to Monte Carlo sampling of $p(\mathbf{c}^{(i)}|\mathbf{y})$, the calibrated reward $\hat{\mathcal{R}}(\mathbf{c}^{(i)})$ will be set to zero. For decoding, we adopt the prefix beam search [7] algorithm. All testing results are generated using a beam width of 10. In multiple-state models, we adopt a pseudo-language model which follows the rules below:

1. Within one gloss, the transition is strictly left to right without skipping.
2. For transition between two glosses, only the transition to the first state of the destination glosses from the last state of the source gloss is allowed.

4.3 Results and Analysis

In this section, we discuss the effectiveness of our proposed methods based on the experimental results on PHOENIX-2014 dataset.

Stochastic Frame Dropping. We conduct experiments with 0%, 25%, 50% and 75% frame dropping rates p_{drop}. To keep the window size M of relative position encoding consistent, we use $M = 16, 12, 8$ and 4 for p_{drop} = 0%, 25%, 50% and 75%, respectively. In this stage, the number of states for all glosses is set to 1 and no gradient is stopped. Figure 2 (Left) shows the performance with different dropping rates. The results show that if the dropping rate is low (i.e., 0% and 25%), the model tends to give worse results as there are fewer variants in the random dropping and the model tends to overfit the retained salient frames. On the other hand, if the dropping rate is too high (i.e, 75%), the model may lose too much information resulting in a slight increase of WER. We select $p_{\text{drop}} = 50\%$ as the default setting for the following experiments.

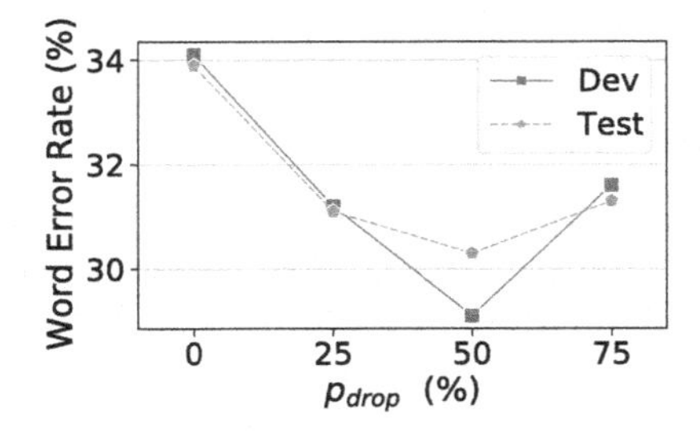

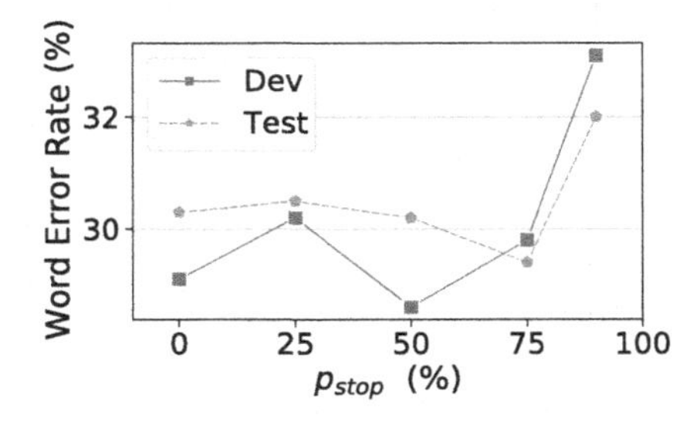

Fig. 2. WER of models trained with p_{drop} = 0%, 25%, 50% and 75% (Left), and WER of models trained with p_{stop} = 0%, 25%, 50%, 75% and 90% given p_{drop} = 50% (Right).

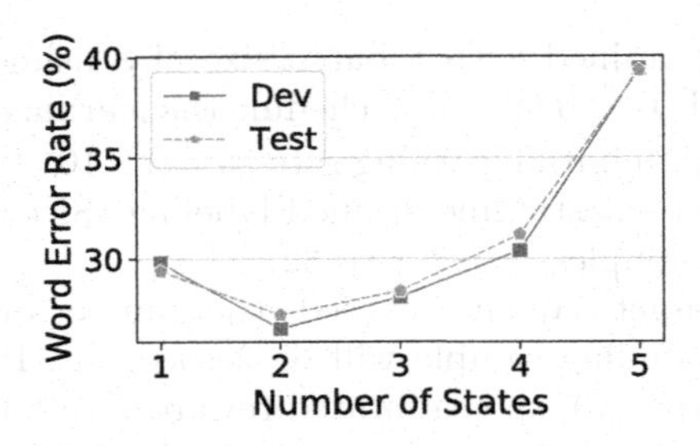

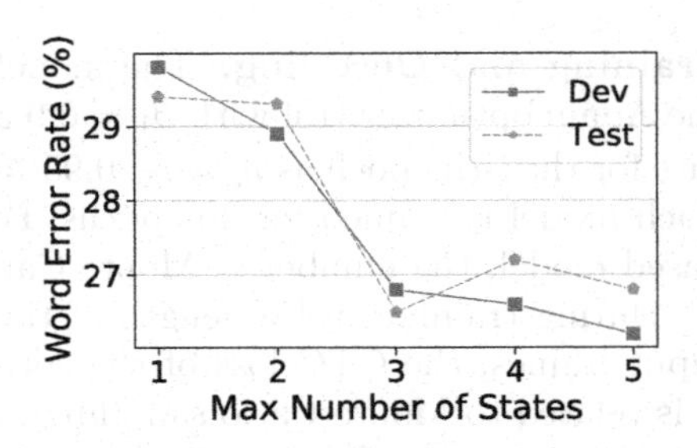

Fig. 3. Comparison between different numbers of states under deterministic fine-grained labeling (Left) and different maximal numbers of states under stochastic fine-grained labeling (Right).

Stochastic Gradient Stopping. For stochastic gradient stopping, we train models with different p_{stop} based on $p_{\text{drop}} = 50\%$ setting. As shown in Fig. 2, the performance is not effected too much by gradient dropping probably because the visual model has already been pre-trained on ImageNet. Among the selected p_{stop}, the optimal p_{stop} is 75% for the Dev set and 50% for the Test set. In the following experiments, $p_{\text{stop}} = 75\%$ is selected as the default setting.

Table 1. The comparison between the REINFORCE method with a simple uniform distribution over the number of states. The maximal numbers of states C are set to 5.

	Dev (%)		Test (%)	
	del/ins	WER	del/ins	WER
Uniform	18.5/2.3	30.2	19.3/2.0	30.3
REINFORCE	7.9/6.5	26.2	7.5/6.3	26.8

Deterministic Fine-Grained Labeling. We first test our model on a deterministic fine-grained labeling (DFL) setting where the numbers of states for all glosses are fixed to a constant C. The distribution of the number of states c_l for the l-th gloss y_l is set as:

$$p(c_l|y_l) = \begin{cases} 1, \ c_l = C \\ 0, \ \text{otherwise} \end{cases} \tag{15}$$

As shown in Fig. 3 (Left), when the number of states is set to $C = 2$, the WER drops to 26.5% on the Dev set and 27.2% on the Test set. The performance improves from the baseline ($C = 1$) when the number of states is 2 or 3, but worsens as the number of states further goes up. This is reasonable as the average number of frames per gloss for the PHOENIX-2014 training dataset is 12.2. With $p_{\text{drop}} = 50\%$, half of the frames are stochastically dropped and the average number of frames per gloss becomes 6.1. When the number of states is large (e.g., $C = 4, 5$), the shorter videos that contain more glosses will violate the CTC length constraint and hence discarded during training.

Stochastic Fine-Grained Labeling. We test different maximal numbers of states for the stochastic fine-grained labeling (SFL) method in Fig. 3 (Right). The WER starts to decrease as the maximal number of states exceeds 2. To further show the necessity of the REINFORCE algorithm, we compare the the REINFORCE method with a model trained with a uniform distribution over the number of states, i.e., $p(c_l|y_l) = \frac{1}{C}$. As shown in Table 1, the performance of the model trained with uniformly distributed number of states is much worse than that of the model trained with the REINFORCE algorithm.

Qualitative Results. Figure 4 shows the posteriors of sub-gloss states produced by our model. It can be seen that our proposed SFL method prefers to assign glosses with 2 or 4 states, and each state lasts for 1–3 frames under the 50% frame dropping rate.

4.4 Comparison with State-of-the-Arts

In this section, we compare our method with other state-of-the-art (SOTA) CSLR methods on the two datasets mentioned in Sect. 4.1.

PHONIEX-2014. Table 2 shows that our final result with the use of language model outperforms the best SOTA result on the PHOENIX-2014 dataset by 1.1% and 0.7% on the Dev and Test set, respectively.

PHONIEX-2014-T. Table 3 shows our result compared to other SOTA results on the PHOENIX-2014-T dataset. Our model achieves comparable results among the models trained with gloss annotation only.

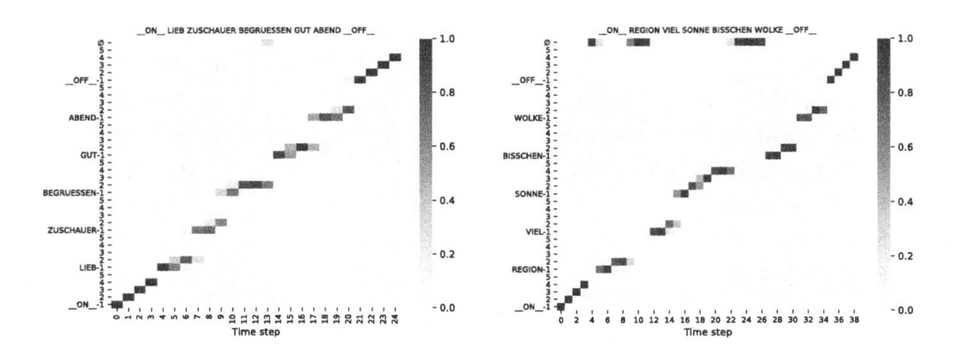

Fig. 4. Examples of sub-glosses posteriors. Left: `__ON__ LIEB ZUSCHAUER BEGRUESSEN GUT ABEND __OFF__`. Right: `__ON__ REGION VIEL SONNE BISSCHEN WOLKE __OFF__`. The number on the y-axis indicates the i-th state for particular glosses and $\varnothing$ indicates blank frame.

Table 2. The comparison with state-of-the-art works on the PHOENIX-2014 dataset. Baseline stands for our network architecture without SFD, SGS, and SFD. LM stands for the use of the language model provided by the PHOENIX-2014 dataset during decoding. Some notes on experimental setting: For SFD, $p_{\text{drop}} = 0.5$; for SGS, $p_{\text{stop}} = 0.75$; for DFL, $C = 2$; for SFL, $C_{max} = 5$.

Method	Dev (%)		Test (%)	
	del/ins	WER	del/ins	WER
Deep Sign [13]	–	38.3	–	38.8
Re-sign [14]	–	27.1	–	26.8
SubUNets [2]	–	40.8	–	40.7
Staged-Opt [5]	–	39.4	–	38.7
LS-HAN [9]	–	–	–	38.3
Align-iOpt [17]	12.9/2.6	37.1	13.0/2.5	36.7
SF-Net [25]	–	35.6	–	34.9
DPD+TEM [27]	9.5/3.2	35.6	9.3/3.1	34.5
CNN-LSTM-HMM [11]	–	26.0	–	26.0
Baseline	9.0/9.3	34.1	8.9/9.2	33.9
SFD	10.1/6.4	29.1	9.9/6.6	30.3
SFD+SGS	9.9/6.9	29.8	9.3/6.6	29.4
SFD+SGS+DFL	8.0/6.5	26.5	8.1/6.3	27.2
SFD+SGS+SFL	7.9/6.5	26.2	7.5/6.3	26.8
SFD+SGS+SFL+LM	10.3/4.1	**24.9**	10.4/3.6	**25.3**

Table 3. The comparison with state-of-the-art works on the PHOENIX-2014-T dataset. Our SFD+SGS+SFL model is trained with only the full-frame stream and gloss annotation.

Method	Annotation				WER (%)	
	Gloss	Mouth	Hand	Text	Dev	Test
CNN-LSTM-HMM (1-Stream) [11]	✓				24.5	26.5
CNN-LSTM-HMM (2-Stream) [11]	✓	✓			24.5	25.4
CNN-LSTM-HMM (3-Stream) [11]	✓	✓	✓		22.1	24.1
SLT (Gloss) [3]	✓				24.9	24.6
SLT (Gloss+Text) [3]	✓			✓	24.6	24.5
SFD+SGS+SFL	✓				25.1	26.1

5 Conclusions

In this paper, we propose stochastic modeling of various components of a continuous sign language recognition architecture. We use ResNet18 as the visual model, transformer encoder as the contextual model and a stochastic fine-grained labeling version of connectionist temporal classification (CTC) as the alignment model. We addressed the issue of unsatisfactory performance when training the CNN-transformer model with CTC loss in an end-to-end manner by introducing the SFD, SGS, and further improve the model performance by introducing SFL. Our model outperforms the state-of-the-art results by 1.1%/0.7% on the dev set and the test set of the PHOENIX-2014 dataset, and achieves competitive results on the PHOENIX-2014-T dataset.

Acknowledgements. This work was supported by the Research Grants Council of the Hong Kong Special Administrative Region, China (Project Nos. HKUST16200118 and T45-407/19N-1).

References

1. Bahdanau, D., Cho, K., Bengio, Y.: Neural machine translation by jointly learning to align and translate. arXiv preprint arXiv:1409.0473 (2014)
2. Camgoz, N.C., Hadfield, S., Koller, O., Bowden, R.: SubUNets: end-to-end hand shape and continuous sign language recognition. In: 2017 IEEE International Conference on Computer Vision (ICCV), pp. 3075–3084. IEEE (2017)
3. Camgoz, N.C., Koller, O., Hadfield, S., Bowden, R.: Sign language transformers: joint end-to-end sign language recognition and translation. In: Proceedings of the IEEE/CVF Conference on Computer Vision and Pattern Recognition, pp. 10023–10033 (2020)
4. Cihan Camgoz, N., Hadfield, S., Koller, O., Ney, H., Bowden, R.: Neural sign language translation. In: Proceedings of the IEEE Conference on Computer Vision and Pattern Recognition, pp. 7784–7793 (2018)
5. Cui, R., Liu, H., Zhang, C.: Recurrent convolutional neural networks for continuous sign language recognition by staged optimization. In: Proceedings of the IEEE Conference on Computer Vision and Pattern Recognition, pp. 7361–7369 (2017)
6. Devlin, J., Chang, M.W., Lee, K., Toutanova, K.: BERT: pre-training of deep bidirectional transformers for language understanding. arXiv preprint arXiv:1810.04805 (2018)
7. Hannun, A.Y., Maas, A.L., Jurafsky, D., Ng, A.Y.: First-pass large vocabulary continuous speech recognition using bi-directional recurrent DNNs. arXiv preprint arXiv:1408.2873 (2014)
8. He, K., Zhang, X., Ren, S., Sun, J.: Deep residual learning for image recognition. In: Proceedings of the IEEE Conference on Computer Vision and Pattern Recognition, pp. 770–778 (2016)
9. Huang, J., Zhou, W., Zhang, Q., Li, H., Li, W.: Video-based sign language recognition without temporal segmentation. In: Thirty-Second AAAI Conference on Artificial Intelligence (2018)
10. Kingma, D.P., Ba, J.: Adam: a method for stochastic optimization. arXiv preprint arXiv:1412.6980 (2014)
11. Koller, O., Camgoz, C., Ney, H., Bowden, R.: Weakly supervised learning with multi-stream CNN-LSTM-HMMs to discover sequential parallelism in sign language videos. IEEE Trans. Pattern Anal. Mach. Intell. (2019)
12. Koller, O., Forster, J., Ney, H.: Continuous sign language recognition: towards large vocabulary statistical recognition systems handling multiple signers. Comput. Vis. Image Underst. **141**, 108–125 (2015)

13. Koller, O., Zargaran, O., Ney, H., Bowden, R.: Deep sign: hybrid CNN-HMM for continuous sign language recognition. In: Proceedings of the British Machine Vision Conference (2016)
14. Koller, O., Zargaran, S., Ney, H.: Re-Sign: Re-aligned end-to-end sequence modelling with deep recurrent CNN-HMMs. In: Proceedings of the IEEE Conference on Computer Vision and Pattern Recognition, pp. 4297–4305 (2017)
15. Liu, Z., Qi, X., Pang, L.: Self-boosted gesture interactive system with ST-Net. In: 2018 ACM Multimedia Conference on Multimedia Conference, pp. 145–153. ACM (2018)
16. Pham, N.Q., Nguyen, T.S., Niehues, J., Muller, M., Waibel, A.: Very deep self-attention networks for end-to-end speech recognition. arXiv preprint arXiv:1904.13377 (2019)
17. Pu, J., Zhou, W., Li, H.: Iterative alignment network for continuous sign language recognition. In: Proceedings of the IEEE Conference on Computer Vision and Pattern Recognition, pp. 4165–4174 (2019)
18. Russakovsky, O., Deng, J., Su, H., Krause, J., Satheesh, S., Ma, S., Huang, Z., Karpathy, A., Khosla, A., Bernstein, M., et al.: ImageNet large scale visual recognition challenge. Int. J. Comput. Vis. **115**(3), 211–252 (2015)
19. Shaw, P., Uszkoreit, J., Vaswani, A.: Self-attention with relative position representations. arXiv preprint arXiv:1803.02155 (2018)
20. Simonyan, K., Zisserman, A.: Very deep convolutional networks for large-scale image recognition. arXiv preprint arXiv:1409.1556 (2014)
21. Szegedy, C., et al.: Going deeper with convolutions. In: Proceedings of the IEEE Conference on Computer Vision and Pattern Recognition, pp. 1–9 (2015)
22. Vaswani, A., et al.: Attention is all you need. In: Advances in Neural Information Processing Systems, pp. 5998–6008 (2017)
23. Williams, R.J.: Simple statistical gradient-following algorithms for connectionist reinforcement learning. Mach. Learn. **8**(3–4), 229–256 (1992)
24. Xu, K., et al.: Show, attend and tell: neural image caption generation with visual attention. In: International Conference on Machine Learning, pp. 2048–2057 (2015)
25. Yang, Z., Shi, Z., Shen, X., Tai, Y.W.: SF-Net: structured feature network for continuous sign language recognition. arXiv preprint arXiv:1908.01341 (2019)
26. Zhang, Z., Pu, J., Zhuang, L., Zhou, W., Li, H.: Continuous sign language recognition via reinforcement learning. In: International Conference on Image Processing (ICIP), pp. 285–289 (2019)
27. Zhou, H., Zhou, W., Li, H.: Dynamic pseudo label decoding for continuous sign language recognition. In: 2019 IEEE International Conference on Multimedia and Expo (ICME), pp. 1282–1287. IEEE (2019)

General 3D Room Layout from a Single View by Render-and-Compare

Sinisa Stekovic[1](✉), Shreyas Hampali[1], Mahdi Rad[1], Sayan Deb Sarkar[1], Friedrich Fraundorfer[1], and Vincent Lepetit[1,2]

[1] Institute for Computer Graphics and Vision, Graz University of Technology, Graz, Austria
{sinisa.stekovic,hampali,rad,sayan.sarkar,fraundorfer, lepetit}@icg.tugraz.at

[2] Université Paris-Est, École des Ponts ParisTech, Paris, France
https://www.tugraz.at/index.php?id=40222

Abstract. We present a novel method to reconstruct the 3D layout of a room—walls, floors, ceilings—from a single perspective view in challenging conditions, by contrast with previous single-view methods restricted to cuboid-shaped layouts. This input view can consist of a color image only, but considering a depth map results in a more accurate reconstruction. Our approach is formalized as solving a constrained discrete optimization problem to find the set of 3D polygons that constitute the layout. In order to deal with occlusions between components of the layout, which is a problem ignored by previous works, we introduce an analysis-by-synthesis method to iteratively refine the 3D layout estimate. As no dataset was available to evaluate our method quantitatively, we created one together with several appropriate metrics. Our dataset consists of 293 images from ScanNet, which we annotated with precise 3D layouts. It offers three times more samples than the popular NYUv2 303 benchmark, and a much larger variety of layouts.

Keywords: Room layout · 3D geometry · Analysis-by-synthesis

1 Introduction

The goal of layout estimation is to identify the layout components—floors, ceilings, walls—and their 3D geometry from one or multiple views, despite the presence of clutter such as furniture, as illustrated in Fig. 1. This is a fundamental problem in scene understanding from images, with potential applications in many domains, including robotics and augmented reality. When enough images from different views are available, it is possible to recover complex 3D layouts by first building a dense point cloud [1,19]. Single view scenarios are far more challenging even when depth information is available, since layout components may

Electronic supplementary material The online version of this chapter (https://doi.org/10.1007/978-3-030-58517-4_12) contains supplementary material, which is available to authorized users.

A. Vedaldi et al. (Eds.): ECCV 2020, LNCS 12361, pp. 187–203, 2020.
https://doi.org/10.1007/978-3-030-58517-4_12

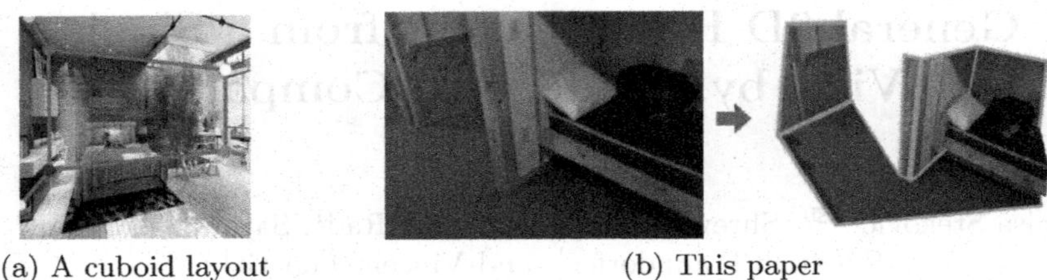

(a) A cuboid layout (b) This paper

Fig. 1. (a) Most current methods for single view layout estimation make the assumption that the view contains a single room with a cuboid shape. This makes the problem significantly simpler as the structure and number of corners remain fixed, but can only handle a fraction of indoor scenes. (b) By contrast, our method is able to estimate general 3D layouts from a single view, even in case of self-occlusions. Its input is either an RGBD image, or an RGB image from which a depth map is predicted.

occlude each other, entirely or partially, and large chunks of the layout are then missing from the point cloud. Moreover, typical scenes contain furniture and the walls, the floors, and the ceilings might be obstructed. Important features such as corners or edges might be only partially observable or even not visible at all.

As shown in Fig. 1(a), many recent methods for single view scenarios avoid these challenges by assuming that the room is a simple 3D cuboid [4,9,11,16,20, 27,32] or that the image contains at most 3 walls, a floor, and a ceiling [37]. This is a very strong assumption, which is not valid for many rooms or scenes, such as the ones depicted in Fig. 1(b) and Fig. 4. In addition, most of these methods only provide the *2D projection* of the layout [9,16,27,32], which is not sufficient for many applications. Other methods rely on panoramic images from viewpoints that do not create occlusions [29,35,36], which is not always feasible.

The very recent method by Howard-Jenkins *et al.* [10] is probably the only method to be able to recover general layouts from a single perspective view. However, it does not provide quantitative evaluation for this task, but only for cuboid layouts in the case of single views. In fact, it often does not estimate well the extents of the layout components, and how they are connected together.

In this paper we introduce a formalization of the problem and an algorithm to solve it. Our algorithm takes as input a single view which can be an RGBD image, or even only a color image: When a depth map is not directly available, it is robust enough to rely on a predicted one from the color image [18,25]. As shown on the right of Fig. 1(b), its output is a 3D model that is "structured", in the sense that the layout components connected in the scene are also connected in the 3D model in the same way, similarly to what a human designer would do. Moreover, we introduce a novel dataset to quantitatively evaluate our method.

More exactly, we formalize the problem of recovering a 3D polygonal model of the layout as a constrained discrete optimization problem. This optimization selects the polygons constituting the layout from a large set of potential polygons. To generate these polygons, like [10] and earlier work [22] for point clouds, we rely on 3D planes rather than edges and/or corners to keep the approach simple in

Table 1. Comparison between test sets of different datasets for single-view layout estimation on real images. ScanNet-Layout does not provide a training set.

Dataset	Layout	Mode	Cam. Param.	Eval. Metrics	#TestSamples
Hedau *et al.* [8]	Cuboid	RGB	Varying	2D	105
LSUN [33]	Cuboid	RGB	Varying	2D	1000
NYUv2 303 [31]	Cuboid	RGBD	Constant	2D	100
ScanNet-Layout (ours)	General	RGBD	Constant	3D	293

terms of perception and model creation. However, as mentioned above, not all 3D planes required in the construction of the layout are visible in the image. Hence, we rely on an analysis-by-synthesis approach, sometimes referred to as 'render-and-compare'. Such approaches do not always require a realistic rendering, in terms of texture or lighting, as in [15,30] for example: We render a depth map for our current layout estimate, and compare it to the measured or predicted depth map. From the differences, we introduce some of the missing polygons to improve our layout estimate. We iterate this process until convergence.

Our approach therefore combines machine learning and geometric reasoning. Applying "pure" machine learning to these types of problems is an appealing direction but it is challenging to rely only on machine learning to obtain structured 3D models as we do. Under the assumption that the room is box-shaped [4,11], this is possible because of the strong prior on feasible 2D layouts [9,16]. In the case of general layouts, this is difficult, as the variability of the layouts are almost infinite (see Fig. 4 for examples). Moreover, only very limited annotated data is available for the general problem. Thus, we use machine learning only to extract image cues on the 3D layout from the perspective view, and geometric reasoning to adapt to general configurations based on these image cues.

To evaluate our method, we manually annotated 293 perspective views from the ScanNet test dataset [5] together with 5 novel 2D and 3D metrics, as there was no existing benchmark for the general problem. This is three times more images than NYUv2 303 [28,31], a popular benchmark for evaluating cuboid layouts. Other single-view layout estimation benchmarks are Hedau *et al.* [8] and LSUN [33], that are cuboid datasets with only 2D annotations, and Structured3D [34], a dataset containing a large number of synthetic scenes generated under the Manhattan world assumption. Our ScanNet-Layout dataset is therefore more general, and is publicly available. Table 1 summarizes the difference between benchmarks. We also compare our method to cuboid-specific methods on NYUv2 303, which contains only cuboids rooms, to show that our method performs comparably to these specialized methods while being more general.

Main Contributions. First, we introduce a formalization of the general layout estimation from single views into a constrained discrete optimization problem. Second, We propose an algorithm based on this formalization and are able to generate a simplistic 3D model for general layouts from single perspective views (RGB or RGBD). Finally, we provide a novel benchmark with a dataset and new metrics for the evaluation of methods for general layout estimations from single perspective views.

2 Related Work

We divide here previous works on layout estimation into two categories. Methods from the first category start by identifying features such as room corners or edges from the image. Like our own approach, methods from the second category rely on 3D planes and their intersections to build the layout. We discuss them below.

2.1 Layout Generation from Image Features

Some approaches to layout estimation, mostly for single-view scenarios, attempt to identify features in the image such as room corners or edges, before connecting them into a 2D layout or lifting them in 3D to generate a 3D room layout. Extracting such features, and lifting them in 3D are, however, very challenging. A common assumption is the Manhattan constraint that enforces orthogonality and parallelism between the layout components of the scene, often done by estimating vanishing points [8,20,24,27], a process that can be very sensitive to noise.

Another assumption used by most of the current methods is that only one box-shaped room is visible in the image [9,16,20,27,31,32]. This is a strong prior that achieves good results, but only if the assumption is correct. For example, in [20], 3D cuboid models are fitted to an edge map extracted from the image, starting from an initial hypothesis obtained from vanishing points. From this 3D cuboid assumption, RoomNet [16] defines a limited number of 11 possible 2D room layouts, and trains a CNN to detect the 2D corners of the box-shaped room. [32] relies on segmentation to identify these corners more robustly. These last approaches [16,32] are limited to the recovery of a 2D cuboid layout.

Yet another approach is to directly predict the 3D layout from the image: [4,11] not only predict the layout but also the objects and humans present in the image, using them as additional constraints. Such constraints are very interesting, however, this approach also requires the '3D cuboid assumption', as it predicts the camera pose with respect to a 3D cuboid.

[29,35,36] relax the cuboid assumption and can recover more general layouts. However, in addition to the Manhattan assumption, this line of work requires panoramic images captured so that they do not exhibit occlusions from walls. This requirement can be difficult to fulfill or even impossible for some scenes. By contrast, our method does not require the cuboid or the Manhattan assumptions, and handles occlusions in the input view to handle variety of general scenes.

2.2 Layout Generation from 3D Planes

An alternative to inferring room layouts from image features like room corners is to identify planes and infer the room layout from these plane structures. If complete point clouds are available, for example, from multiple RGB or RGBD images, identifying such planes is straightforward, and has been successfully applied for this task [2,12,21,26]. Recently, successes in single RGB image based depth estimation [6,17,25] and 3D plane detection [18] opened up the possibility of layout generation from single RGB images. For example, Zou *et al.* [37] finds the layout planes in dominant room directions and then reasons on the depth map to estimate the extents of the layout components. However, even though this method does not assume strict cuboid layouts, it assumes the presence of only five layout components—floor, ceiling, left wall, front wall and right wall.

Like our method, the work of Howard-Jenkins *et al.* [10] uses plane detected in images by a CNN to infer the non-cuboid 3D room layouts. The main contribution of their work is in the design of a network architecture to detect planar regions of the layout from a single image and to infer the 3D plane parameters from it. For this task, we use PlaneRCNN [18], which has very similar functionalities. By intersecting these planes, they can be first delineated, and thanks to a clustering and voting scheme, with help of predicted bounding boxes for the planes, the parts of the planes relevant to the layout can be identified.

However, [10] heavily relies on predicted proposal regions to estimate the extents of layout components. As their qualitative results show, it sometimes struggles to find the correct extents as the proposal regions can be very noisy, and the layout components can be disconnected. It also does not provide any quantitative evaluation for the general room layout estimation problem for single views and it is limited to cuboid rooms from the NYUv2 303 dataset [31].

In contrast, we formalize the problem as a discrete optimization problem, allowing us to reason about occlusions between layout components in the camera view, which often happens in practice, and retrieve structured 3D models.

3 Approach

We describe our approach in this section. We formalize the general layout estimation problem as a constrained discrete optimization problem (Sect. 3.1), explain how we generate a first set of candidate polygons from plane intersections (Sect. 3.2), detail our cost function involved in our formalization (Sect. 3.3), and how we optimize it (Sect. 3.4). When one or more walls are hidden in the image, this results in an imperfect layout, and we show how to augment the set of candidate polygons to include these hidden walls, and iterate until we obtain the final layout (Sect. 3.5). Finally, we describe how we can output a structured 3D model for the layout (Sect. 3.6).

3.1 Formalization

We formalize the problem of estimating a 3D polygonal layout $\hat{\mathcal{R}}$ for a given input image I as solving the following constrained discrete optimization problem:

$$\hat{\mathcal{R}} = \arg \min_{\mathcal{X} \subset \mathcal{R}_0(I)} K(\mathcal{X}, I) \text{ such that } p(\mathcal{X}) \text{ is a partition of } I \,, \tag{1}$$

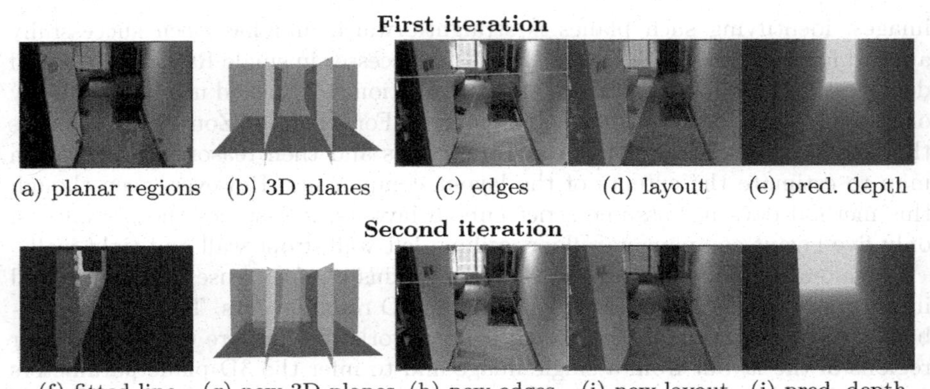

Fig. 2. Approach overview. We detect planar regions (a) for the layout components using PlaneRCNN and a semantic segmentation, and obtain equations of the corresponding 3D planes (b). The planes intersections give a set of candidate edges for the layout (c). From these edges, we find a first layout estimate in 2D (d) and 3D (e) as a set of polygons that minimizes the cost. From the depth discrepancy (f) for the layout estimate and the input view, we find missing planes (g), and extend the set of candidate edges (h). We iterate until we find a layout consistent with the color image (i) and the depth map (j).

where $K(\mathcal{X}, I)$ is a cost function defined below, $\mathcal{R}_0(I)$ is a set of 3D polygons for image I, and $p(\mathcal{X})$ is the set of projections in the input view of the polygons in $\mathcal{X}$. In words, we look for the subset of polygons in $\mathcal{R}_0(I)$, whose projections partition the input image I, and that minimizes $K(\cdot)$.

There are two options when it comes to defining precisely $K(\mathcal{X}, I)$ and $\mathcal{R}_0(I)$: Either $\mathcal{R}_0(I)$ is defined as the set of all possible 3D polygons, and $K(\mathcal{X}, I)$ includes constraints to ensure that the polygons in $\mathcal{X}$ reproject on image cues for the edges and corners of the rooms, or $\mathcal{R}_0(I)$ contains only polygons with edges that correspond to edges of the room. As discussed in the introduction, extracting wall edges and corners from images is difficult in general, mostly because of lack of training data. We therefore chose the second option. We describe below first how we create the set $\mathcal{R}_0(I)$ of candidate 3D polygons, which includes the polygons constituting the 3D layout, and then the cost function $K(\mathcal{X}, I)$.

3.2 Set of Candidate 3D Polygons $\mathcal{R}_0(I)$

As discussed in the introduction, we rely on the intersection of planes to identify good edge candidates to constitute the polygons of the layout. We then group these edges into polygons to create $\mathcal{R}_0(I)$.

Set of 3D Planes $\mathcal{P}_0$. First, we run on the RGB image a) PlaneRCNN [18] to detect planar regions and b) DeepLabv3+ [3] to obtain a semantic segmentation. We keep only the planar regions that correspond to wall, ceiling, or floor segments (the supplementary material provides more details). We denote by $\mathcal{S}(I)$ the set of such regions. An example is shown in Fig. 2(a). PlaneRCNN provides the equations of the 3D planes it detects, or, if a depth map of the image is available, we fit a 3D plane to each detected region to obtain more accurate parameters. The depth map can be measured or predicted from the input image I [18,25].

As can be seen in Fig. 2(a), the regions provided by PlaneRCNN typically do not extend to the full polygonal regions that constitute the layout. To find these polygons, we rely on the intersections of the planes in $\mathcal{P}_0$ as detailed below. In order to limit the extent of the polygons to the borders of the input image, we also include in $\mathcal{P}_0$ the four 3D planes of the camera frustum, which pass through two neighbouring image corners and the camera center.

Some planes required to create some edges of the layout may not be in this first set $\mathcal{P}_0$. This is the case for example for the plane of the hidden wall on the left of the scene in Fig. 2. Through an analysis-by-synthesis approach, we can detect the absence of such planes, and add plausible planes to recover the missing edges and obtain the correct layout. This will be detailed in Sect. 3.5.

Set of 3D Corners $\mathcal{C}_0$. By computing the intersections of each triplet of planes in $\mathcal{P}_0$, we get a set $\mathcal{C}_0$ of candidate 3D corners for the layout. To build a structured layout, it is important to keep track of the planes that generated the corners and, thus, we define each corner $C_j \in \mathcal{C}_0$ as a set of 3 planes:

$$C_j = \{P_j^1, P_j^2, P_j^3\}, \tag{2}$$

where $P_j^1 \in \mathcal{P}_0$, $P_j^2 \in \mathcal{P}_0$, $P_j^3 \in \mathcal{P}_0$, and $P_j^1 \neq P_j^2$, $P_j^1 \neq P_j^3$, and $P_j^2 \neq P_j^3$. For numerical stability, we do not consider the cases where at least two planes are almost parallel, or when the 3 planes almost intersect on a line. Furthermore, we discard the corners that reproject outside the image. We also discard those corners that have negative depth values.

Set of 3D Edges $\mathcal{E}_0$. We then obtain a set $\mathcal{E}_0$ of candidate 3D edges by pairing the corners in $\mathcal{C}_0$ that share exactly 2 planes:

$$E_k = \{C_{\sigma(k)}, C_{\sigma'(k)}\}, \tag{3}$$

where $\sigma(k)$ and $\sigma'(k)$ are 2 functions giving the indices of the corners that are the extremities of edge E_k. Figure 2(c) gives an example of set $\mathcal{E}_0$.

Set of 3D Polygons $\mathcal{R}_0(I)$. We finally create the set $\mathcal{R}_0(I)$ of candidate polygons as the set of all closed loops of edges in $\mathcal{E}_0$ that lie on the same plane and do not intersect each other.

3.3 Cost Function $K(\mathcal{X}, I)$

Our cost function is split into a 3D and a 2D part:

$$K(\mathcal{X}, I) = K_{3D}(\mathcal{X}, I) + \lambda K_{2D}(\mathcal{X}, I). \tag{4}$$

For all our experiments, we used $\lambda = 1$.

Cost function $K_{3D}(\cdot)$ measures the dissimilarity with the depth map $D(I)$ for the input view, and the depth map $D'(\mathcal{X})$ created from the polygons in $\mathcal{X}$, as illustrated in Fig. 2(e). It is based on the observation that the layout should always be located behind the objects of the scene:

$$K_{3D}(\mathcal{X}, I) = \frac{1}{|I|} \sum_{\mathbf{x}} \max(D(I)[\mathbf{x}] - D'(X)[\mathbf{x}], 0), \quad (5)$$

where the sum is over all the image locations $\mathbf{x}$ and $|I|$ denotes the total number of image locations. Since the projections of the polygons in $\mathcal{X}$ are constrained to form a partition of I, $K_{3D}(\cdot)$ can be rewritten as

$$K_{3D}(\mathcal{X}, I) = \frac{1}{|I|} \sum_{R \in \mathcal{X}} \sum_{\mathbf{x} \in p(R)} \max(D(I)[\mathbf{x}] - D'(\mathcal{X})[\mathbf{x}], 0) = \frac{1}{|I|} \sum_{R \in \mathcal{X}} k_{3D}(R, I), \quad (6)$$

where $p(R)$ is the projection of polygon R in the image. $K_{3D}(\cdot)$ is computed as a sum of terms, each term depending on a single polygon in $\mathcal{X}$. These terms are precomputed for each polygon in $\mathcal{R}_0(I)$, to speed up the computation of $K_{3D}(\cdot)$.

Cost function $K_{2D}(\cdot)$ measures the dissimilarity between the polygons in $\mathcal{X}$ and the image segmentation into planar regions $\mathcal{S}(I)$:

$$\begin{aligned} K_{2D}(\mathcal{X}, I) &= \sum_{R \in \mathcal{X}} \Big((1 - \text{IoU}(p(R), S(I, R))) + \text{IoU}(p(R), \mathcal{S}(I) \setminus S(I, R)) \Big) \\ &= \sum_{R \in \mathcal{X}} k_{2D}(R, I), \end{aligned} \quad (7)$$

where IoU is the Intersection over Union score, $S(I, R)$ is the planar region detected by Plane-RCNN and corresponding to the plane of polygon R. Like $K_{3D}(\cdot)$, $K_{2D}(\cdot)$ can be computed as a sum of terms that can be precomputed before optimization. Computing cost function $K(\cdot)$ is therefore very fast.

3.4 Optimization

To find the solution to our constrained discrete optimization problem introduced in Eq. (1), we simply consider all the possible subsets $\mathcal{X}$ in $\mathcal{R}_0(I)$ that pass the partition constraint, and keep the one that minimizes $K(\mathcal{X}, I)$.

The number N of polygons in $\mathcal{R}_0(I)$ varies with the scene, but is typically of a few tens. For example, we obtain 12 candidate polygons in total for the example of Fig. 2. The number of non-empty subsets to evaluate is theoretically $2^N - 1$, which is slightly higher than 2000 for the same example. However, most of these subsets can be trivially discarded: Associating polygons with corresponding planes and considering that only one polygon per plane is possible significantly reduces the number of possibilities, to 36 in this example. The number can be further reduced by removing the polygons that do not have a plausible shape to be part of a room layout. Such shapes can be easily recognized by considering the distance between the non-touching edges of the polygon. Finally, this reduces the number to merely 20 plausible subsets of polygons in the case of the

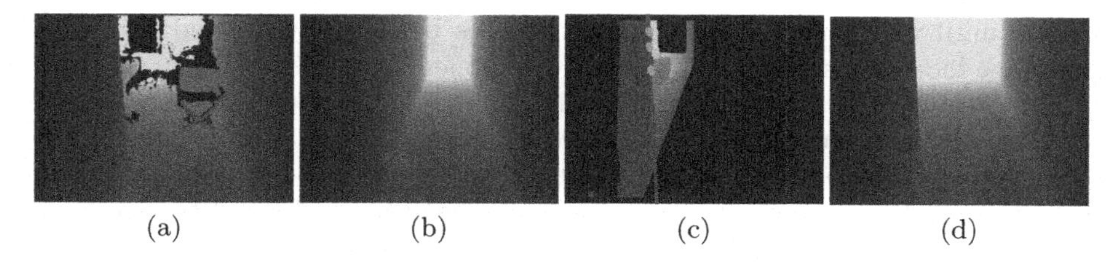

Fig. 3. Layout refinement. We identify planes which are occluded by other layout planes but necessary for the computation of the layout. First, we compare the depth map for the input view (a) to the rendered layout depth (b). (c) If the discrepancy is large, we fit a line (shown in red) through the points with the largest discrepancy change (orange). By computing the plane passing through the line and the camera center, we obtain a layout (d) consistent with the depth map for the input view. (Color figure online)

example. Precomputing the k_{3D} and k_{2D} terms takes about 1s, and the optimization itself takes about 400 ms—most of this time is spent to guarantee the partition constraint in our implementation, which we believe can be significantly improved. The supplementary material details the computation time more.

3.5 Iterative Layout Refinement

As mentioned above in Sect. 3.2, we often encounter cases where some of the planes required to create the layout are not in $\mathcal{P}_0$ because they are hidden by another layout plane. Fortunately, we can detect such mistakes, and fix them by adding a plane to $\mathcal{P}_0$ before running the layout creation described above again.

To detect missing planes, we render the depth map $D'(\hat{\mathcal{R}})$ for the current layout estimate $\hat{\mathcal{R}}$ and measure the discrepancy with the depth map $D(I)$ for the image as illustrated in Fig. 3. As the depth maps $D(I)$ acquired by RGBD cameras typically contain holes around edges, we use the depth completion method by [13] before measuring the discrepancy. If discrepancy is large, *i.e.* there are many pixel locations where the rendered map has smaller values than the original depth map, this indicates a mistake in the layout estimate that can be fixed by adding a plane. This is because the layout cannot be in front of objects.

There is a range of planes that can improve the layout estimate. We chose the conservative option that does not introduce parts not visible in the input image. For a polygon R in $\hat{\mathcal{R}}$ with a large difference between $D'(\hat{\mathcal{R}})$ and $D(I)$, we first identify the image locations with the largest discrepancy changes, and fit a line to these points using RANSAC, as shown in Fig. 2(f). We then add the plane P that passes through this line and the camera center to $\mathcal{P}_0$ to obtain a new set of planes $\mathcal{P}_1$. This is illustrated in Fig. 2(g): the intersection between P and R will create the edge missing from the layout, which is visible in Fig. 2(h). From $\mathcal{P}_1$, we obtain successively the new sets $\mathcal{C}_1$ (corners), $\mathcal{E}_1$ (edges), and $\mathcal{R}_1$ (polygons), and solve again the problem of Eq. (1) after replacing $\mathcal{R}_0$ by $\mathcal{R}_1$. We repeat this process until we do not improve the differences between $D'(\hat{\mathcal{R}})$ and $D(I)$, for the image locations segmented as layout components.

For about 5% of the test samples, the floor plane is not visible because of occlusions by some furniture. When none of the detected planes belongs to the floor class, we create an additional plane by assuming that the camera is $1.5m$ above the floor. For the plane normal, we take the average of the outer products between the normals of the walls and the $[0, 0, 1]^\top$ vector.

3.6 Structured Output

Once the solution $\hat{\mathcal{R}}$ to Eq. (1) is found, it is straightforward to create a structured 3D model for the layout. Each 3D polygon in $\hat{\mathcal{R}}$ is defined as a set of coplanar 3D edges, each edge is defined as a pair of corners, and each corner is defined from 3 planes. We therefore know which corners and edges the polygons share and how they are connected to each other. For example, the 3D layout of Fig. 1 is made of 14 corners, 18 edges, and 5 polygons.

4 Evaluation

We evaluate our approach in this section. First, we present our new benchmark for evaluating 3D room layouts from single perspective views, and our proposed metrics. Second, we evaluate our approach on our benchmark and include both quantitative and qualitative results on general room layouts. For reference, we show that our approach performs similarly to methods assuming cuboid layouts on the NYUv2 303 benchmark, which only includes cuboid layouts, without making such strong assumptions. More qualitative results, detailed computation times, and implementation details are given in the supplementary material.

We have considered additionally evaluating our approach on the LSUN [33] and the Hedau [8] room layout benchmarks. However, because these datasets do not provide the camera intrinsic parameters, these datasets were unpractical for the evaluation of our approach. We have also considered evaluating our approach on 3D layout annotations of the NYUv2 dataset [28] from [7]. However, as the annotations are not publicly available anymore, we were not able to produce any quantitative results for this dataset. Furthermore, the improved annotations from [37] are not publicly available anymore, and the authors were unfortunately not able to provide these annotations in time for this submission.

4.1 ScanNet-Layout Benchmark

Dataset Creation. For our ScanNet-Layout dataset, we manually labelled 293 views sampled from the 100 ScanNet test scenes [5], for testing purposes only. As shown in Fig. 4 and in the supplementary material, these views span different layout settings, are equally distributed to represent both cuboid and general room layouts, challenging views that are neglected in previous room layout datasets,

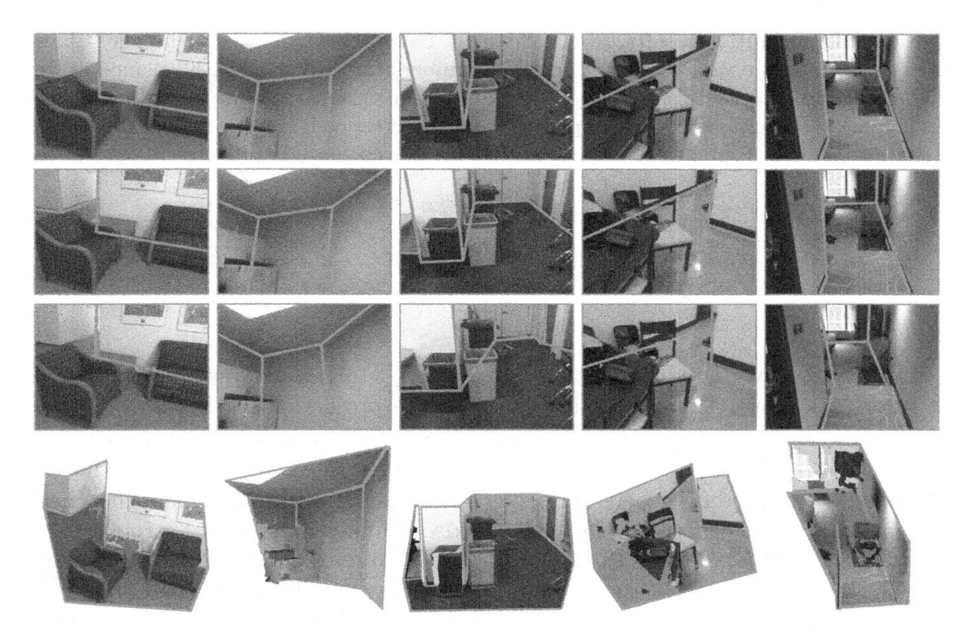

Fig. 4. Results of our method on the ScanNet-Layout. First row: manual annotations; Second row: predictions using an RGBD input; Third row: Predictions using an RGB input. Fourth row: 3D models created using the RGBD mode of our approach. Furniture is shown only to demonstrate consistency of our predictions with the geometry of the scene. Our approach performs well in both RGBD and RGB modes. Our approach in RGB mode fails to detect one component in the third example, due to noisy predictions from PlaneRCNN. The rest of the examples show that, when depth information is not available, predictions from CNN can still be utilized in many different scenarios. More qualitative results, including a video, can be found in the supplementary material.

and in some cases we include similar viewpoints to evaluate effects of noise (*e.g.* motion blur). The ScanNet-Layout dataset is available on our project page.

To manually annotate the 3D layouts, we first drew the layout components as 2D polygons. For each polygon, we then annotated the image region where it is directly visible without any occlusions from objects or other planes. From these regions, we could compute the 3D plane equations for the polygons. Since we could not recover the plane parameters for completely occluded layout components, we only provide 2D polygons without 3D annotations for them.

Evaluation Metrics. To quantitatively evaluate the fidelity of the recovered layout structures and their 2D and 3D accuracy, we introduce 2D and 3D metrics. For the 2D metrics, we first establish one-to-one correspondences $\mathcal{C}$ between the N predicted polygons $\hat{\mathcal{R}}$ and the M ground truth polygons $\mathcal{R}_{\text{gt}}$. Starting with the largest ground truth polygon, we iteratively find the matching predicted polygon with highest intersection over union. At each iteration, we remove the ground truth polygon and its match from further consideration. The metrics are:

Table 2. Quantitative results on our ScanNet-Layout benchmark. (↑: higher values are better, ↓: lower values are better) The numbers for Hirzer *et al.* [9] demonstrate that the approaches assuming cuboid layouts under-perform on our ScanNet-Layout benchmark. Our approach in RGB performs much better as it is not restricted by these assumptions. Our approach in RGBD mode shows even more improvement.

	Mode	IoU ↑ (%)	PE ↓ (%)	EE ↓	RMSE ↓	RMSE_{uts} ↓
Hirzer [9]	RGB	48.6 ± 22.2	24.1 ± 15.1	29.6 ± 19.4	–	–
Ours	RGB	*63.5 ± 25.2*	*16.3 ± 14.7*	*22.3 ± 14.9*	*0.5 ± 0.5*	*0.4 ± 0.5*
Ours	RGBD	**75.9 ± 23.4**	**9.9 ± 12.9**	**11.9 ± 13.2**	**0.2 ± 0.4**	**0.2 ± 0.3**

- Intersection over Union (IoU): $\frac{2}{M+N}\sum_{(R_{\text{gt}},R)\in\mathcal{C}} \text{IoU}(R_{\text{gt}}, R)$, where IoU is the Intersection-over-Union measure between the projections of the 2 polygons. This metric is very demanding on the global structure and 2D accuracy;
- Pixel Error (PE): $\frac{1}{|I|}\sum_{\mathbf{x}\in I} \text{PE}(\mathbf{x})$, with $\text{PE}(\mathbf{x}) = 0$ if the ground truth polygon and the predicted polygon projected at image location $\mathbf{x}$ were matched together, and 1 otherwise. This metric also evaluates the global structure;
- Edge Error (EE): This is the symmetrical Chamfer distance [23] between the polygons in $\hat{\mathcal{R}}$ and $\mathcal{R}_{\text{gt}}$, and evaluates the accuracy of the layout in 2D;
- Root Mean Square Error (RMSE) between the predicted layout depth $D(\mathcal{R})$ and the ground truth layout depth $D(\mathcal{R}_{\text{gt}})$, excluding the pixels that lie on completely occluded layout components, as we could not recover 3D data for these components. This metric evaluates the accuracy of the 3D layout.
- RMSE_{uts} that computes the RMSE after scaling the predicted layout depth to the range of ground truth layout depth by factor $s = \text{median}(D(\mathcal{R}_{\text{gt}}))/\text{median}(D(\mathcal{R}))$. This metric is used when the depth map is predicted from the image, as the scale of depth prediction methods is not reliable.

We note that the PE and EE metrics are extensions of existing metrics in cuboid layout benchmarks. As the PE metric is forgiving when missing out small components, we introduce the IoU metric that drastically penalizes such errors.

4.2 Evaluation on ScanNet-Layout

We evaluate our method on ScanNet-Layout under two different experimental settings: When depth information is directly measured by a depth camera, and when only the color image is available as input. In this case, we use PlaneR-CNN [18] to estimate both the planes parameters and the depth map.

Table 2 reports the quantitative results. The authors of [9], one of the state-of-the-art methods for cuboid layout estimation, kindly provided us with the results of their method. As this method is specifically designed for cuboid layouts, it fails on more general cases, but also on many cuboid examples for viewpoints not well represented in the training sets of layout benchmarks [8,31,33] (Fig. 5(b)).

The good performance of our method for all metrics shows that the layouts are recovered accurately in 2D and 3D. When measured depth is not used, performance decreases due to noisy estimates of the plane parameters, but in many

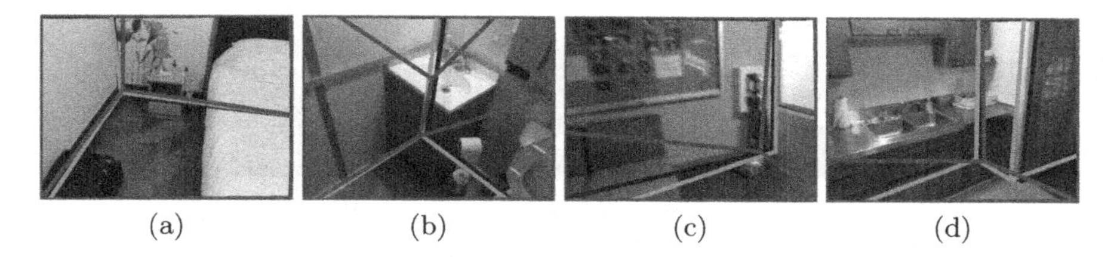

(a) (b) (c) (d)

Fig. 5. Visual comparison to Hirzer *et al.* [9], which assumes only cuboid layouts. The layouts estimated by the Hirzer method are shown in red, the layouts recovered by our approach using RGB information only are shown in green, and ground truth is shown in blue. (a) Both approaches perform similarly. (b) The Hirzer method makes a mistake even for this cuboid layout. (c) and (d) The Hirzer method fails as the cuboid assumption does not hold. Our approach performs well in all of the examples. (Color figure online)

Table 3. Quantitative results on NYUv2 303, a standard benchmark for cuboid room layout estimation. Our method performs similarly to the other methods designed for cuboid rooms without using this assumption. While Hirzer *et al.* [9] performs best on this benchmark, it fails on ScanNet-Layout, even for some of the cuboid rooms (Fig. 5).

	Mode	PE ↓	Median PE ↓
Zhang *et al.* [31]	RGBD	8.04	–
Ours	RGBD	8.9	4.6
Schwing *et al.* [27]	RGB	13.66	–
Zhang *et al.* [31]	RGB	13.94	–
RoomNet [16] (from [9])	RGB	12.31	–
Hirzer *et al.* [9]	RGB	8.49	–
Howard-Jenkins *et al.* [10]	RGB	12.19	–
Ours	RGB	13.0	10.1

cases, the predictions are still accurate. This can be best observed in qualitative comparisons with RGBD and RGB views in Fig. 4. In many cases, RGB information is enough to estimate 3D layouts and are comparable to results with RGBD information. However, the third example clearly demonstrates that small errors in planes parameters can lead to visible errors.

4.3 Evaluation on NYUv2 303

For reference, we evaluate our approach on the NYUv2 303 dataset [28,31]. It is designed to evaluate methods predicting 2D room layouts under the cuboid assumption. We show that our method also performs well on it without exploiting this assumption. This dataset only provides annotations for the room corners under the cuboid assumption. Since the output of our method is more general, we transform it into the format expected by the NYUv2 303 benchmark. For each

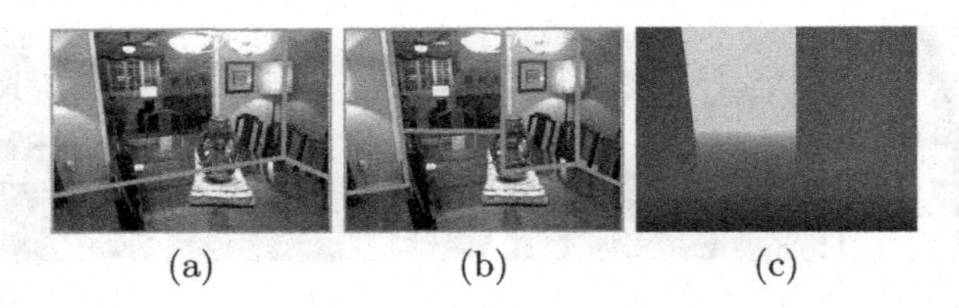

Fig. 6. Qualitative result on NYUv2 303. (a) Layout obtained by enforcing the cuboid assumption, (b) the original layout retrieved by our method, which corresponds better to the scenes. (c) Depth map computed from our estimated layout.

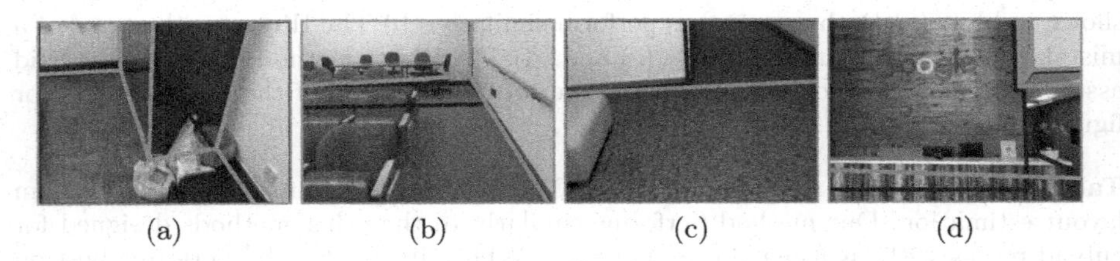

Fig. 7. Failure cases on ScanNet-Layout, our estimations in green, ground truth in blue. (a): Some furniture were segmented as walls. (b): PlaneRCNN did not detect the second floor plane. Even human observers may fail to see this plane. (c): Large areas in the measured depth map were missing along edges. Filling these areas with [13] is not always sufficient to detect discrepancy. (d): As the floor is not visible in the image, it is unclear whether the manual annotation or the estimated floor polygon is correct. (Color figure online)

of the possible cuboid layout components—1 floor, 1 ceiling, 3 walls—we find the planes for which its normal vector best fits the layout component: When fewer than 3 walls are visible, the annotations of walls in the dataset are ambiguous and we apply the Hungarian algorithm [14] to find good correspondences.

Table 3 gives the quantitative results. When depth is available, our method is slightly worse than the Zhang *et al.* [31] method, designed for cuboid rooms. When using only color images, our method performs similarly to the other approaches, specialized for cuboid rooms, even if the recent Hirzer *et al.* [9] method performs best on this dataset. Here, we used the depth maps predicted by [25] to estimate the plane parameters. Figure 6 shows that some layouts we retrieved fit the scene better than the manually annotated layout. To conclude, our method performs closely but is more general than the cuboid-based methods.

4.4 Failure Cases

Figure 7 shows the frequent causes of failures. Most of the failures are due to noisy outputs from PlaneRCNN and DeepLabv3+ that lead to both false-positive and missing layout planes. Our render-and-compare approach is not robust enough to large noise in depth and this should be addressed in future work.

5 Conclusion

We presented a formalization of the general room layout estimation into a constrained discrete optimization problem, and an algorithm to solve this problem. The occasional errors made by our method come from the detection of the planar regions, the semantic segmentation, and the predicted depth maps, pointing to the fact that future progress in these fields will improve our layout estimates.

Acknowledgment. This work was supported by the Christian Doppler Laboratory for Semantic 3D Computer Vision, funded in part by Qualcomm Inc.

References

1. Budroni, A., Boehm, J.: Automated 3D reconstruction of interiors from point clouds. Int. J. Archit. Comput. **8**, 55–73 (2010)
2. Cabral, R., Furukawa, Y.: Piecewise planar and compact floorplan reconstruction from images. In: Conference on Computer Vision and Pattern Recognition (2014)
3. Chen, L.C., Zhu, Y., Papandreou, G., Schroff, F., Adam, H.: Encoder-decoder with atrous separable convolution for semantic image segmentation. In: European Conference on Computer Vision (2018)
4. Chen, Y., Huang, S., Yuan, T., Qi, S., Zhu, Y., Zhu, S.C.: Holistic++ scene understanding: single-view 3D holistic scene parsing and human pose estimation with human-object interaction and physical commonsense. In: International Conference on Computer Vision (2019)
5. Dai, A., Chang, A.X., Savva, M., Halber, M., Funkhouser, T., Niessner, M.: ScanNet: richly-annotated 3D reconstructions of indoor scenes. In: Conference on Computer Vision and Pattern Recognition (2017)
6. Godard, C., Aodha, O.M., Brostow, G.J.: Unsupervised monocular depth estimation with left-right consistency. In: Conference on Computer Vision and Pattern Recognition (2017)
7. Guo, R., Hoiem, D.: Support surface prediction in indoor scenes. In: International Conference on Computer Vision (2013)
8. Hedau, V., Hoiem, D., Forsyth, D.: Recovering the spatial layout of cluttered rooms. In: International Conference on Computer Vision (2009)
9. Hirzer, M., Roth, P.M., Lepetit, V.: Smart hypothesis generation for efficient and robust room layout estimation. In: IEEE Winter Conference on Applications of Computer Vision (2020)
10. Howard-Jenkins, H., Li, S., Prisacariu, V.: Thinking outside the box: generation of unconstrained 3D room layouts. In: Jawahar, C.V., Li, H., Mori, G., Schindler, K. (eds.) ACCV 2018. LNCS, vol. 11361, pp. 432–448. Springer, Cham (2019). https://doi.org/10.1007/978-3-030-20887-5_27
11. Huang, S., Qi, S., Xiao, Y., Zhu, Y., Wu, Y.N., Zhu, S.C.: Cooperative holistic scene understanding: unifying 3D object, layout, and camera pose estimation. In: Advances in Neural Information Processing Systems (2018)
12. Ikehata, S., Yang, H., Furukawa, Y.: Structured indoor modeling. In: International Conference on Computer Vision (2015)
13. Ku, J., Harakeh, A., Waslander, S.L.. In defense of classical image processing: fast depth completion on the CPU. In: CRV (2018)

14. Kuhn, H.W., Yaw, B.: The Hungarian method for the assignment problem. Naval Res. Logist, Quart (1955)
15. Kundu, A., Li, Y., Rehg, J.M.: 3D-RCNN: instance-level 3D object reconstruction via render-and-compare. In: Conference on Computer Vision and Pattern Recognition (2018)
16. Lee, C.Y., Badrinarayanan, V., Malisiewicz, T., Rabinovich, A.: Roomnet: end-to-end room layout estimation. In: International Conference on Computer Vision (2017)
17. Lee, J.H., Han, M.K., Ko, D.W., Suh, I.H.: From big to small: multi-scale local planar guidance for monocular depth estimation. arXiv Preprint (2019)
18. Liu, C., Kim, K., Gu, J., Furukawa, Y., Kautz, J.: PlanerCNN: 3D plane detection and reconstruction from a single image. In: Conference on Computer Vision and Pattern Recognition (2019)
19. Liu, C., Wu, J., Furukawa, Y.: FloorNet: A unified framework for floorplan reconstruction from 3D scans. In: European Conference on Computer Vision (2018)
20. Mallya, A., Lazebnik, S.: Learning informative edge maps for indoor scene layout prediction. In: International Conference on Computer Vision (2015)
21. Murali, S., Speciale, P., Oswald, M.R., Pollefeys, M.: Indoor Scan2BIM: building information models of house interiors. In: International Conference on Intelligent Robots and Systems (2017)
22. Nan, L., Wonka, P.: Polyfit: polygonal surface reconstruction from point clouds. In: International Conference on Computer Vision (2017)
23. Olson, C.F., Huttenlocher, D.P.: Automatic target recognition by matching oriented edge pixels. J. Mach. Learn. Res. **6**, 103–113 (1997)
24. Ramalingam, S., Pillai, J.K., Jain, A., Taguchi, Y.: Manhattan junction catalogue for spatial reasoning of indoor scenes. In: Conference on Computer Vision and Pattern Recognition (2013)
25. Ramamonjisoa, M., Lepetit, V.: SharpNet: fast and accurate recovery of occluding contours in monocular depth estimation. In: International Conference on Computer Vision Workshops (2019)
26. Sanchez, V., Zakhor, A.: Planar 3D modeling of building interiors from point cloud data. In: International Conference on Computer Vision (2012)
27. Schwing, A.G., Hazan, T., Pollefeys, M., Urtasun, R.: Efficient structured prediction for 3D indoor scene understanding. In: Conference on Computer Vision and Pattern Recognition (2012)
28. Silberman, N., Hoiem, D., Kohli, P., Fergus, R.: Indoor segmentation and support inference from RGBD images. In: Fitzgibbon, A., Lazebnik, S., Perona, P., Sato, Y., Schmid, C. (eds.) ECCV 2012. LNCS, vol. 7576, pp. 746–760. Springer, Heidelberg (2012). https://doi.org/10.1007/978-3-642-33715-4_54
29. Sun, C., Hsiao, C.W., Sun, M., Chen, H.T.: HorizonNet: learning room layout with 1D representation and pano stretch data augmentation. In: Conference on Computer Vision and Pattern Recognition (2019)
30. Xu, Y., Zhu, S.C., Tung, T.: DenseRaC: joint 3D pose and shape estimation by dense render-and-compare. In: International Conference on Computer Vision (2019)
31. Zhang, J., Kan, C., Schwing, A.G., Urtasun, R.: Estimating the 3D layout of indoor scenes and its clutter from depth sensors. In: International Conference on Computer Vision (2013)
32. Zhang, W., Zhang, W., Gu, J.: Edge-semantic learning strategy for layout estimation in indoor environment. IEEE Trans. Cybern. **50**, 2730–2739 (2019)

33. Zhang, Y., Yu, F., Song, S., Xu, P., Seff, A., Xiao, J.: Large-scale scene understanding challenge: room layout estimation. In: Conference on Computer Vision and Pattern Recognition Workshops (2015)
34. Zheng, J., Zhang, J., Li, J., Tang, R., Gao, S., Zhou, Z.: Structured3D: a large photo-realistic dataset for structured 3D modeling. In: European Conference on Computer Vision (2020)
35. Zou, C., Colburn, A., Shan, Q., Hoiem, D.: LayoutNet: reconstructing the 3D room layout from a single RGB image. In: Conference on Computer Vision and Pattern Recognition (2018)
36. Zou, C., et al.: 3D manhattan room layout reconstruction from a single 360 image. arXiv Preprint (2019)
37. Zou, C., Guo, R., Li, Z., Hoiem, D.: Complete 3D scene parsing from an RGBD image. Int. J. Comput. Vis. **127**, 143–162 (2019)

Neural Dense Non-Rigid Structure from Motion with Latent Space Constraints

Vikramjit Sidhu[1,2], Edgar Tretschk[1], Vladislav Golyanik[1(✉)], Antonio Agudo[3], and Christian Theobalt[1]

[1] Max Planck Institute for Informatics, SIC, Saarbrücken, Germany
golyanik@mpi-inf.mpg.de
[2] Saarland University, SIC, Saarbrücken, Germany
[3] Institut de Robòtica i Informàtica Industrial, CSIC-UPC, Barcelona, Spain

Abstract. We introduce the first dense neural non-rigid structure from motion (N-NRSfM) approach, which can be trained end-to-end in an unsupervised manner from 2D point tracks. Compared to the competing methods, our combination of loss functions is fully-differentiable and can be readily integrated into deep-learning systems. We formulate the deformation model by an auto-decoder and impose subspace constraints on the recovered latent space function in a frequency domain. Thanks to the state recurrence cue, we classify the reconstructed non-rigid surfaces based on their similarity and recover the period of the input sequence. Our N-NRSfM approach achieves competitive accuracy on widely-used benchmark sequences and high visual quality on various real videos. Apart from being a standalone technique, our method enables multiple applications including shape compression, completion and interpolation, among others. Combined with an encoder trained directly on 2D images, we perform scenario-specific monocular 3D shape reconstruction at interactive frame rates. To facilitate the reproducibility of the results and boost the new research direction, we open-source our code and provide trained models for research purposes (http://gvv.mpi-inf.mpg.de/projects/Neural_NRSfM/).

Keywords: Neural Non-Rigid Structure from Motion · Sequence period detection · Latent space constraints · Deformation auto-decoder

1 Introduction

Non-Rigid Structure from Motion (NRSfM) reconstructs non-rigid surfaces and camera poses from monocular image sequences using multi-frame 2D correspondences calculated across the input views. It relies on motion and deformation cues as well as weak prior assumptions, and is object-class-independent in contrast

Electronic supplementary material The online version of this chapter (https://doi.org/10.1007/978-3-030-58517-4_13) contains supplementary material, which is available to authorized users.

A. Vedaldi et al. (Eds.): ECCV 2020, LNCS 12361, pp. 204–222, 2020.
https://doi.org/10.1007/978-3-030-58517-4_13

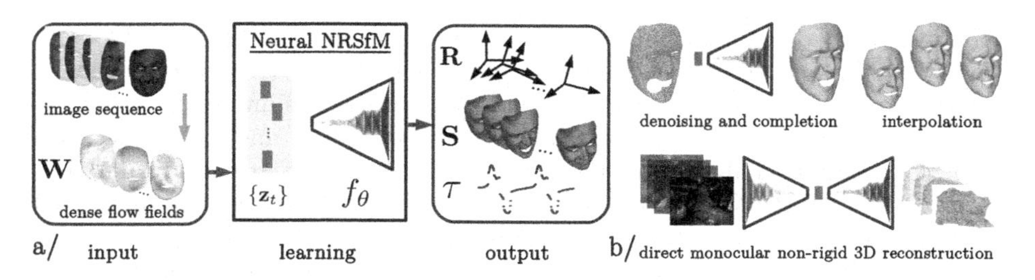

Fig. 1. Neural non-rigid structure from motion (N-NRSfM). Our approach reconstructs monocular image sequences in 3D from dense flow fields (shown using the Middlebury optical flow scheme [9]). In contrast to all other methods, we represent the deformation model with a neural auto-decoder f_θ which decodes latent variables $\mathbf{z}_t$ into 3D shapes (a/). This brings a higher expressivity and flexibility which results in state-of-the-art results and new applications such as shape completion, denoising and interpolation, as well as direct monocular non-rigid 3D reconstruction (b/).

to monocular 3D reconstruction methods which make use of parametric models [59]. Dense NRSfM has achieved remarkable progress during the last several years [1,8,19,37,51]. While the accuracy of dense NRSfM has been recently only marginally improved, learning-based direct methods for monocular rigid and non-rigid 3D reconstruction have become an active research area in computer vision [13,33,47,54,66].

Motivated by these advances, we make the first step towards learning-based dense NRSfM, as it can be seen in Fig. 1. At the same time, we remain in the classical NRSfM setting without strong priors (which restrict to object-specific scenarios) or assuming the availability of training data with 3D geometry. We find that among several algorithmic design choices, replacing an explicit deformation model by an implicit one, *i.e.,* a neural network with latent variables for each shape, brings multiple advantages and enables new applications compared to the previous work such as temporal state segmentation, shape completion, interpolation and direct monocular non-rigid 3D reconstruction (see Fig. 1-b/ for some examples).

By varying the number of parameters in our neural component, we can express our assumption on the complexity of the observed deformations. We observe that most real-world deformations evince state recurrence which can serve as an additional reconstruction constraint. By imposing constraints on the latent space, we can thus detect a period of the sequence, denoted by τ, *i.e.,* the duration in frames after which the underlying non-rigid 3D states repeat, and classify the recovered 3D states based on their similarity. Next, by attaching an image encoder to the learnt neural deformation model (deformation auto-decoder), we can perform in testing direct monocular non-rigid 3D reconstruction at interactive frame rates. Moreover, an auto-decoder represents non-rigid states in a compressed form due to its compactness.

Note that the vast majority of the energy functions proposed in the literature so far is not fully differentiable or cannot be easily used in learning-based systems due to computational or memory requirements [1,8,19,37]. We combine a data loss, along with constraints in the metric and trajectory spaces, a temporal smoothness loss as well as latent space constraints into single energy—with the non-rigid shape parametrised by an auto-decoder—and optimise it with the back-propagation algorithm [49]. The experimental evaluation indicates that the proposed N-NRSfM approach obtains competitive solutions in terms of 3D reconstruction, and outperforms competing methods on several sequences, but also represents a useful tool for non-rigid shape analysis and processing.

Contributions. In summary, the primary contributions of this work are:

⋆ The first, to the best of our belief, fully differentiable dense neural NRSfM approach with a novel auto-decoder-based deformation model (Sects. 3, 4);
⋆ Subspace constraints on the latent space imposed in the Fourier domain. They enhance the reconstruction accuracy and enable temporal classification of the recovered non-rigid 3D states with period detection (Sect. 4.2);
⋆ Several applications of the deformation model including shape compression, interpolation and completion, as well as fast direct non-rigid 3D reconstruction from monocular image sequences (Sect. 4.4);
⋆ An extensive experimental evaluation of the core N-NRSfM technique and its applications with state-of-the-art results (Sect. 5).

2 Related Work

Recovering a non-rigid 3D shape from a single monocular camera has been an active research area in the past two decades. In the literature, two main classes of approaches have proved most effective so far: template-based formulations and NRSfM. On the one hand, template-based approaches relied on establishing correspondences with a reference image in which the 3D shape is already known in advance [42,53]. To avoid ambiguities, additional constraints were included in the optimisation, such as the inextensibility [42,65], as rigid as possible priors [68], providing very robust solutions but limiting its applicability to almost inelastic surfaces. While the results provided by template-based approaches are promising, knowing a 3D template in advance can become a hard requirement. In order to avoid that, NRSfM approaches have reduced these requirements, making their applicability easier. In this context, NRSfM has been addressed in the literature by means of model-based approaches, and more recently, by the use of deep-learning-based methods. We next review the most related work to solve this problem by considering both perspectives.

Non-Rigid Structure from Motion. NRSfM has been proposed to solve the problem from 2D tracking data in a monocular video (in the literature, 2D trajectories are collected in a measurement matrix). The most standard approach to address the inherent ambiguity of the NRSfM problem is by assuming the underlying 3D shape is low-rank. In order to estimate such low-rank model, both factorisation- [11] and optimisation-based approaches [43,61] have been proposed, considering single low-dimensional shape spaces [16,19], or a union of temporal [69] or spatio-temporal subspaces [3]. Low-rank models were also extended to the other domains, by exploiting pre-defined trajectory basis [7], the combination of shape-trajectory vectors [28,29], and the force space that induces the deformations [5]. On top of these models, additional spatial [38] or temporal [2,10,39] smoothness constraints, as well as shape priors [12,21, 35] have also been considered. However, in contrast to their rigid counterparts, NRSfM methods are typically sparse, limiting their application to a small set of salient points. Whereas several methods are adaptations of sparse techniques to dense data [22,51], other techniques were explicitly designed for the dense setting [1,19,37] relying on sophisticated optimisation strategies.

Neural Monocular Non-Rigid 3D Reconstruction. Another possibility to perform monocular non-rigid 3D reconstruction is to use learning-based approaches. Recently, many works have been presented for rigid [13,18,30,40,66] and non-rigid [27,47,54,62] shape reconstruction. These methods exploited a large and annotated dataset to learn the solution space, limiting their applicability to the type of shapes that are observed in the dataset. Unfortunately, this supervision is a hard task to be handled in real applications, where the acquisition of 3D data to train a neural network is not trivial.

While there has been work at the intersection of NRSfM and deep learning, the methods require large training datasets [34,41,52] and address only the sparse case [34,41]. *C3DPO* [41] learns basis shapes from 2D observations and does not require 3D supervision, similar to our approach. Neural methods for monocular non-rigid reconstruction have to be trained for every new object class or shape configuration within the class. In contrast to the latter methods—and similar to the classical NRSfM—we solely rely on motion and deformation cues. Our approach is unsupervised and requires only dense 2D point tracks for the recovery of non-rigid shapes. Thus, we combine the best of both worlds, *i.e.*, the expressivity of neural representations for deformation models and improvements upon weak prior assumptions elaborated in previous works on dense NRSfM. We leverage the latter in the way so that we find an energy function which is fully differentiable and can be optimised with modern machine-learning tools.

3 Revisiting NRSfM

We next review the NRSfM formulation that will be used later to describe our neural approach. Let us consider a set of P points densely tracked across T frames. Let $\mathbf{s}_t^p = [x_t^p, y_t^p, z_t^p]^\top$ be the 3D coordinates of the p-th point in image t, and $\hat{\mathbf{w}}_t^p = [u_t^p, v_t^p]^\top$ its 2D position according to an orthographic projection. In order to simplify subsequent formulation, the camera translation $\mathbf{t}_t = \sum_p \hat{\mathbf{w}}_t^p / P$ can be subtracted from the 2D projections, considering centred measurements as $\mathbf{w}_t^p = \hat{\mathbf{w}}_t^p - \mathbf{t}_t$. We can then build a linear system to map the 3D-to-2D point coordinates as:

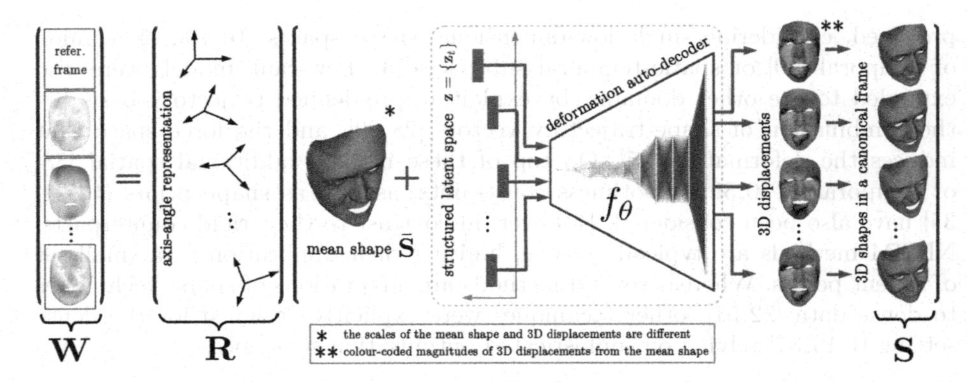

Fig. 2. Overview of our N-NRSfM approach to factorise a measurement matrix W into motion R and shape Sfactors. To enable an end-to-end learning, we formulate a fully-differentiable neural energy function, where each $\mathbf{S}_t$ is mapped by a deformation auto-decoder f_{θ} from a latent space $\mathbf{z}_t$, plus a mean shape $\bar{\mathbf{S}}$. After obtaining optimal network parameters $\boldsymbol{\theta}$, the latent space becomes structured allowing the scene deformation pattern analysis.

$$\underbrace{\begin{bmatrix} \mathbf{w}_1^1 & \dots & \mathbf{w}_1^P \\ \vdots & \ddots & \vdots \\ \mathbf{w}_T^1 & \dots & \mathbf{w}_T^P \end{bmatrix}}_{\mathbf{W}} = \underbrace{\begin{bmatrix} \mathbf{R}_1 & \dots & \mathbf{0} \\ \vdots & \ddots & \vdots \\ \mathbf{0} & \dots & \mathbf{R}_T \end{bmatrix}}_{\mathbf{R}} \underbrace{\begin{bmatrix} \mathbf{s}_1^1 & \dots & \mathbf{s}_1^P \\ \vdots & \ddots & \vdots \\ \mathbf{s}_T^1 & \dots & \mathbf{s}_T^P \end{bmatrix}}_{\mathbf{S}}, \tag{1}$$

where $\mathbf{W}$ is a $2T \times P$ measurement matrix with the 2D measurements arranged in columns, $\mathbf{R}$ is a $2T \times 3T$ block diagonal matrix made of T truncated 2×3 camera rotations $\mathbf{R}_t \equiv \boldsymbol{\Pi}\mathbf{G}_t$ with the full rotation matrix $\mathbf{G}_t$ and $\boldsymbol{\Pi} = \begin{bmatrix} 100 \\ 010 \end{bmatrix}$; and $\mathbf{S}$ is a $3T \times P$ matrix with the non-rigid 3D shapes. Every $\mathbf{G}_t$ lies in the $SO(3)$ group, that we enforce using an axis-angle representation encoding the rotation by a vector $\boldsymbol{\alpha}_t = (\alpha_t^x, \alpha_t^y, \alpha_t^z)$, that can be related to $\mathbf{G}_t$ by the Rodrigues' rotation formula. On balance, the problem consists in estimating the time-varying 3D shape $\mathbf{S}^t$ as well as the camera motion $\mathbf{G}^t$ with $t = \{1, \dots, T\}$, from 2D trajectories $\mathbf{W}$.

4 Deformation Model with Shape Auto-Decoder

In the case of dynamic objects, the 3D shape changes as a function of time. Usually, this function is unknown, and many efforts have been made to model it. The type of deformation model largely determines which observed non-rigid states can be accurately reconstructed, *i.e.,* the goal is to find a simple model with large expressibility. In this context, perhaps the most used model in the literature consists in enforcing the deformation shape to lie in a linear subspace [11]. While this model has been proved to be effective, the form in which the shape bases are estimated can be decisive. For example, it is well known that some constraints cannot be effectively imposed in factorisation methods [11,67], forcing the proposal of more sophisticated optimisation approaches [3,16,69]. In this paper, we propose to depart from the traditional formulations based on linear subspace models and embrace a different formulation that can regress the deformation modes in a unsupervised manner during a neural network training, see Fig. 2 for a method overview. By controlling the architecture and composition of the layers, we can express our assumptions about the complexity and type of the observed deformations. We will use the name of Neural Non-Rigid Structure from Motion (N-NRSfM) to denote our approach.

4.1 Modelling Deformation with Neural Networks

We propose to implement our non-rigid model network as a deformation auto-decoder $f_{\boldsymbol{\theta}}$, as it was done for rigid shape categories [44], where $\boldsymbol{\theta}$ denotes the learned network parameters. Specifically, we construct $f_{\boldsymbol{\theta}}$ as a series of nine fully-connected layers with small hidden dimensions $(2, 8, 8, 8, 16, 32, 32, B, |\mathbf{S}_t|)$, and exponential linear unit (ELU) activations [14] (except after the penultimate and final layers). B—set to 32 by default—can be interpreted as an analogue to the number of basis shapes in linear subspace models. $f_{\boldsymbol{\theta}}$ is a function of the latent space $\mathbf{z}_t$, that is related to the shape space $\mathbf{S}_t$ by means of:

$$\mathbf{S}_t = \bar{\mathbf{S}} + f_{\boldsymbol{\theta}}(\mathbf{z}_t), \tag{2}$$

where $\bar{\mathbf{S}}$ is a $3 \times P$ mean shape matrix. We can also obtain the time-varying shape $\mathbf{S}$ in Eq. (1) by $\mathbf{S} = (\mathbf{1}_T \otimes \bar{\mathbf{S}}) + f_{\boldsymbol{\theta}}(\mathbf{z})$, with $\mathbf{1}_T$ a T-dimensional vector of ones and $\otimes$ a Kronecker product. The fully-connected layers of $f_{\boldsymbol{\theta}}$ are initialised using He initialisation [31], and the bias value of the last layer is set to a rigid shape estimate $\bar{\mathbf{S}}$, which is kept fixed during optimisation. Both $\bar{\mathbf{S}}$ and $\mathbf{R}_t$ with $t = \{1, \ldots, T\}$ are initialised by rigid factorisation [60] from $\mathbf{W}$. Note that we estimate displacements (coded by $f_{\boldsymbol{\theta}}(\mathbf{z}_t)$) from $\bar{\mathbf{S}}$ instead of absolute point positions. Considering that, the weight matrix of the final fully-connected layer of $f_{\boldsymbol{\theta}}$ can be interpreted as a low-rank linear subspace where every vector denotes a 3D displacement from the mean shape. This contributes to the compactness of the recovered space and serves as an additional constraint, similar to the common practice of the principal component analysis [46].

To learn $\boldsymbol{\theta}$, and update it during training, we require gradients with respect to a full energy $\mathbf{E}$ that we will propose later, such that:

$$\frac{\partial \mathbf{E}}{\partial \boldsymbol{\theta}} = \sum_{t=1}^{T} \frac{\partial \mathbf{E}}{\partial \mathbf{S}_t} \frac{\partial \mathbf{S}_t}{\partial \boldsymbol{\theta}}, \tag{3}$$

connecting $f_{\boldsymbol{\theta}}$ into a fully-differentiable loss function, in which $\mathbf{S}_t$, $t = \{1, \ldots, T\}$ are optimised as free variables via gradients. We next describe our novel energy function $\mathbf{E}$, which is compatible with $f_{\boldsymbol{\theta}}$ and supports gradient back-propagation.

4.2 Differentiable Energy Function

To solve the NRSfM problem as it was defined in Sect. 3, we propose to minimise a differentiable energy function with respect to motion parameters $\mathbf{R}$ and shape ones (coded by $\boldsymbol{\theta}$ and $\mathbf{z}$) as:

$$\mathbf{E} = \mathbf{E}_{\text{data}}(\boldsymbol{\theta}, \mathbf{z}, \mathbf{R}) + \beta\, \mathbf{E}_{\text{temp}}(\boldsymbol{\theta}, \mathbf{z}) + \gamma\, \mathbf{E}_{\text{spat}}(\boldsymbol{\theta}, \mathbf{z}) + \eta\, \mathbf{E}_{\text{traj}}(\boldsymbol{\theta}, \mathbf{z}) + \omega\, \mathbf{E}_{\text{latent}}(\mathbf{z}), \quad (4)$$

where $\mathbf{E}_{\text{data}}$ is a data term, and $\{\mathbf{E}_{\text{temp}}, \mathbf{E}_{\text{spat}}, \mathbf{E}_{\text{traj}}, \mathbf{E}_{\text{latent}}\}$ encode the priors that we consider. β, γ, η and ω are weight coefficients to balance the influence of every term. We now describe each of these terms in detail.

The data term $\mathbf{E}_{\text{data}}$ is derived from the projection equation (1), and it is to penalise the image re-projection errors as:

$$\mathbf{E}_{\text{data}}(\boldsymbol{\theta}, \mathbf{z}, \mathbf{R}) = \left\| \mathbf{W} - \mathbf{R}\left((\mathbf{1}_T \otimes \bar{\mathbf{S}}) + f_{\boldsymbol{\theta}}(\mathbf{z})\right) \right\|_{\epsilon}, \quad (5)$$

where $\|\cdot\|_{\epsilon}$ denotes the Huber loss of a matrix.

The temporal smoothness term $\mathbf{E}_{\text{temp}}$ enforces temporal-preserving regularisation of the 3D shape via its latent space as:

$$\mathbf{E}_{\text{temp}}(\boldsymbol{\theta}, \mathbf{z}) = \sum_{t=1}^{T-1} \| f_{\boldsymbol{\theta}}(\mathbf{z}_{t+1}) - f_{\boldsymbol{\theta}}(\mathbf{z}_t) \|_{\epsilon}. \quad (6)$$

Thanks to this soft-constraint prior, our algorithm can generate clean surfaces that also stabilise the camera motion estimation.

The spatial smoothness term $\mathbf{E}_{\text{spat}}$ imposes spatial-preserving regularisation for a neighbourhood. This is especially relevant for dense observations, where most of the points in a local neighbourhood can follow a similar motion pattern. To define this constraint, let $\mathcal{N}(\mathbf{p})$ be a 1-ring neighbourhood of $\mathbf{p} \in \mathbf{S}_t$, that will be used to define a Laplacian term (widely used in computer graphics [55]). For robustness, we complete the spatial smoothness with a depth penalty term. Combining both ideas, we define this term as:

$$\mathbf{E}_{\text{spat}}(\boldsymbol{\theta}, \mathbf{z}) = \underbrace{\sum_{t=0}^{T-1} \sum_{\mathbf{p} \in \mathbf{S}_t} \left\| \mathbf{p} - \frac{1}{|\mathcal{N}(\mathbf{p})|} \sum_{\mathbf{q} \in \mathcal{N}(\mathbf{p})} \mathbf{q} \right\|_1}_{\text{Laplacian smoothing}} \underbrace{- \lambda \sum_{t=1}^{T} \| \mathcal{P}_z(\mathbf{G}_t \mathbf{S}_t) \|_2}_{\text{depth control}}, \quad (7)$$

where $\mathcal{P}_z$ denotes an operator to extract z-coordinates, $\|\cdot\|_1$ and $\|\cdot\|_2$ are the l_1- and l_2-norm, respectively, and $\lambda > 0$ is a weight coefficient. Thanks to the depth term, our N-NRSfM approach automatically achieves more supervision over the z-coordinate of the 3D shapes, since it can lead to an increase in the shape extent along the z-axis.

The point trajectory term $\mathbf{E}_{\text{traj}}$ imposes a subspace constraint on point trajectories throughout the whole sequence, as it was exploited by [6,7]. To this end, the 3D point trajectories are coded by a linear combination of K fixed trajectory vectors by a $T \times K$ matrix $\mathbf{\Phi}$ together with a $3K \times P$ matrix $\mathbf{A}$ of unknown coefficients. The penalty term can be then written as:

$$\mathbf{E}_{\text{traj}}(\boldsymbol{\theta}, \mathbf{z}) = \left\| (\mathbf{1}_T \otimes \bar{\mathbf{S}}) + f_{\boldsymbol{\theta}}(\mathbf{z}) - (\mathbf{\Phi} \otimes \mathbf{I}_3)\mathbf{A} \right\|_{\epsilon}, \quad \mathbf{\Phi} = \begin{pmatrix} \phi_{1,1} & \dots & \phi_{1,K} \\ \vdots & \ddots & \vdots \\ \phi_{T,1} & \dots & \phi_{T,K} \end{pmatrix}, \tag{8}$$

where $\phi_{t,k} = \frac{\sigma_k}{\sqrt{2}} \cos\left(\frac{\pi}{2T}(2t-1)(k-1)\right)$, with $\sigma_k = 1$ for $k = 1$, and $\sigma_k = \sqrt{2}$, otherwise. $\mathbf{I}_3$ is a 3×3 identity matrix. We experimentally find that this term is not redundant with the rest of terms, and it provides a soft regularisation of $f_{\boldsymbol{\theta}}$.

Finally, the latent term $\mathbf{E}_{\text{latent}}$ imposes sparsity constraints over the latent vector $\mathbf{z}$. This type of regularisation is enabled by the new form to express the deformation model with an auto-decoder $f_{\boldsymbol{\theta}}$, and it can be expressed as:

$$\mathbf{E}_{\text{latent}}(\mathbf{z}) = \left\| \mathcal{F}(\mathbf{z}) \right\|_1, \tag{9}$$

where $\mathcal{F}(\cdot)$ denotes the Fourier transform (FT) operator. Thanks to this penalty term, we can impose several effects which were previously not possible. First, $\mathbf{E}_{\text{latent}}$ imposes structure on the latent space by encouraging the sparsity of the Fourier series and removing less relevant frequency components. In other words, this can be interpreted as subspace constraints on the trajectory of the latent space variable, where the basis trajectories are periodic functions. Second, by analysing the structured latent space, we can extract the period of a periodic sequence and temporally segment the shapes according to their similarity. Our motivation for $\mathbf{E}_{\text{latent}}$ is manifold and partially comes from the observation that many real-world scenes evince recurrence, *i.e.*, they repeat their non-rigid states either in periodic or non-periodic manner.

Period Detection and Sequence Segmentation. The period of the sequence can be recovered from the estimated $\mathcal{F}(\mathbf{z})$, by extracting the dominant frequency in terms of energy within the frequency spectrum. If a dominant frequency ω_d is identified, its period can be directly computed as $\tau = \frac{T}{\omega_d}$. Unfortunately, in some real scenarios, the frequency spectrum that we obtain may not be unimodal (two or more relevant peaks can be observed in the spectrum), and therefore we obtain $\tau = T$. Irrespective whether a sequence is periodic or not, the latent space is temporally segmented so that similar values are decoded into similar shapes. This enables applications such as shape interpolation, completing and denoising.

4.3 Implementation Details

The proposed energy in Eq. (4) and the deformation auto-decoder $f_{\boldsymbol{\theta}}$ are fully-differentiable by construction, and therefore the gradients that flow into $\mathbf{S}_t$ can be further back-propagated into $\boldsymbol{\theta}$. Our deformation model is trained to simultaneously recover the motion parameters $\mathbf{R}$, the latent space $\mathbf{z}$ to encode shape deformations, and the model parameters $\boldsymbol{\theta}$. Additionally, the trajectory coefficients in $\mathbf{A}$ are also learned in this manner (see Eq. (8)). For initialisation, we use rigid factorisation to obtain $\mathbf{R}$ and $\bar{\mathbf{S}}$, random values in the interval $[-1, 1]$ for $\mathbf{z}$, and a null matrix for $\mathbf{A}$. The weights $\beta, \gamma, \eta, \omega, \lambda$ are determined empirically and selected from the determined ranges in most experiments we describe in Sect. 5, unless mentioned otherwise. The values we set are 10^2 for $\mathbf{E}_{\text{data}}$, $\beta = 1$, $\gamma \in [10^{-6}; 10^{-4}]$, $\eta \in [1; 10]$, $\omega = 1$, $\lambda \in [0; 10^{-3}]$ and $B = 32$ in f_{θ}. In addition, we use $K = 7$ as default value to define our low-rank trajectory model in Eq. (8).

Our N-NRSfM approach is implemented in pytorch [45]. As all the training data are available at the same time, we use the RProp optimiser [48] with a learning rate of 0.0001, and train for 60,000 epochs. All experiments are performed on NVIDIA Tesla V100 and K80 GPUs with a Debian 9 Operating System. Depending on the size of the dataset, training takes between three (*e.g.*, the *back* sequence [50]) and twelve (the *barn-owl* sequence [26]) hours on our hardware.

4.4 Applications of the Deformation Auto-Decoder f_{θ}

Our deformation auto-decoder f_{θ} can be used for several applications which were not easily possible in the context of NRSfM before, including shape denoising, shape completion and interpolation as well as correspondence-free monocular 3D reconstruction of non-rigid surfaces with reoccurring deformations.

Shape Compression, Interpolation, Denoising and Completion. The trained f_{θ} combined with $\bar{\mathbf{S}}$ represents a compressed version of a 4D reconstruction and requires much less memory compared to the uncompressed shapes in the explicit representation $\mathbf{S}_t$ with $t = \{1, \ldots, T\}$. The number of parameters required to capture all 3D deformations accurately depends on the complexity of the observed deformations, and not on the length of a sequence. Thus, the longer a sequence with repetitive states is, the higher is the compression ratio c. Next, let us suppose we are given a partial and noisy shape $\tilde{\mathbf{S}}$, and we would like to obtain a complete and smooth version of it $\mathbf{S}_{\theta}$ upon the learned deformation model prior. We use our pre-trained auto-decoder and optimise for the latent code $\mathbf{z}$, using the per-vertex error as the loss. In the case of a partial shape, the unknown vertices are assumed to have some dummy values. Moreover, since the learned latent space is smooth and statistically assigns similar variables to similar shapes (displacements), we can interpolate the latent variables which will result in the smooth interpolation of the shapes (displacements).

Direct Monocular Non-rigid 3D Reconstruction with Occlusion Handling. Pre-trained f_{θ} can also be combined with other machine-learning components. We are interested in direct monocular non-rigid 3D reconstruction for endoscopic scenarios (though N-NRSfM is not restricted to those). Therefore, we train an image encoder which relates images to the resulting latent space of shapes (after the N-NRSfM training). Such image-to-mesh encoder-decoder is also robust against moderate partial scene occlusions—which frequently occur is endoscopic scenarios—as the deformations model f_{θ} can also rely on partial observations. We build the image encoder based on ResNet-50 [32] pre-trained on the ImageNet [17] dataset.

At test time, we can reconstruct a surface from a single image, assuming state recurrence. Since the latent space is structured, we are modelling in-between states obtained by interpolation of the observed surfaces. This contrasts to the DSPR method [25], which *de facto* allows only state re-identification. Next, with the gradual degradation of the views, the accuracy of our image-to-surface reconstructor degrades gracefully. We can feed images with occlusions or a constant camera pose bias—such as those observed by changing from the left to the right camera in stereo recordings—and still expect accurate reconstructions.

5 Experiments

In this section, we describe the experimental results. We first compare our N-NRSfM approach to competing approaches on several widely-used benchmarks and real datasets following the established evaluation methodology for NRSfM (Sect. 5.1). We next evaluate how accurately our method detects the periods and how well it segments sequences with non-periodic state recurrence (Sect. 5.2). For the sequences with 3D ground truth geometry $\mathbf{S}^{\text{GT}}$, we report the 3D error e_{3D}—after shape-wise orthogonal Procrustes alignment—defined as $e_{3D} = \frac{1}{T}\sum_t \frac{\|\mathbf{S}_t^{\text{GT}} - \mathbf{S}_t\|_{\mathcal{F}}}{\|\mathbf{S}_t^{\text{GT}}\|_{\mathcal{F}}}$, where $\|\cdot\|_{\mathcal{F}}$ denoted Frobenius norm. Note that e_{3D} also implicitly evaluates the accuracy of $\mathbf{R}_t$ because of the mutual dependence between $\mathbf{R}_t$ and $\mathbf{S}_t$. Finally, for periodic sequences, we compare the estimated pulse τ with the known one τ^{GT}.

5.1 Quantitative Comparisons

We use three benchmark datasets in the quantitative comparison: *synthetic faces* (two sequences with 99 frames and two different camera trajectories denoted by *traj. A* and *traj. B*, with 28,000 points per frame) [19], *expressions* (384 frames with 997 points per frame) [4], and Kinect *t-shirt* (313 frames with 77,000 points) and *paper* (193 frames with 58,000 points) sequences taken from [64]. In the case if 3D ground truth shapes are available, ground truth dense point tracks are obtained by a virtual orthographic camera. Otherwise, dense correspondences are calculated by multi-frame optical flow [20,57].

Synthetic Faces. e_{3D} for the *synthetic faces* are reported in Table 1. We compare our N-NRSfM to Metric Projections (MP) [43], Trajectory Basis (TB) approach [7], Variational Approach (VA) [19], Dense Spatio-Temporal Approach (DSTA) [15], Coherent Depth Fields (CDF) [23], Consolidating Monocular Dynamic Reconstruction (CMDR) [24,25], Grassmannian Manifold (GM) [37], Jum- ping Manifolds (JM) [36], Scalable Monocular Surface Reconstruction (SMSR) [8], Expectation-Maximisation Finite Element Method (EM-FEM) [1] and Probabilistic Point Trajectory Approach (PPTA) [6]. Our N-NRSfM comes close to the most accurate methods on *traj. A* and comes in the middle on *traj. B* among all methods. Note that GM and JM use Procrustes alignment with scaling, which results in the comparison having slightly differing metrics. Still, we

Table 1. e_{3D} for the synthetic face sequence [19]. * denotes methods which use Procrustes analysis for shape alignment, whereas most methods use orthogonal Procrustes. "†" indicates sequential method. Compared to the default settings, the lowest e_{3D} of N-NRSfM is obtained with $B = 10$, $K = 30$, $\lambda = 0$ and $\eta = 10$ for *traj. A* (denoted by "♭") and $K = 40$ for *traj. B* (denoted by "♮").

	TB [7]	**MP** [43]	**VA** [19]	**DSTA** [15]	**CDF** [23]	**CMDR** [24]
traj. A	0.1252	0.0611	0.0346	0.0374	0.0886	0.0324
traj. B	0.1348	0.0762	0.0379	0.0428	0.0905	0.0369
	GM* [37]	**JM*** [36]	**SMSR** [8]	**PPTA** [6]	**EM-FEM** [1]†	**N-NRSfM (ours)**
traj. A	0.0294	0.0280	0.0304	0.0309	0.0389	0.045/0.032♭
traj. B	0.0309	0.0327	0.0319	0.0572	0.0304	0.049/0.0389♮

Table 2. Qualitative comparison on the *expressions* dataset [4].

	EM-LDS [61]	PTA [7]	CSF2 [29]	KSTA [28]	GMLI [4]	N-NRSfM (ours)
Expr.	0.044	0.048	0.03	0.035	**0.026**	**0.026**

Table 3. Quantitative comparison on the Kinect *paper* and *t-shirt* [64].

	TB [7]	MP [43]	DSTA [15]	GM [37]	JM [36]	N-NRSfM (ours)
paper	0.0918	0.0827	0.0612	0.0394	0.0338	**0.0332**
t-shirt	0.0712	0.0741	0.0636	0.0362	0.0386	**0.0309**

include these methods for completeness. *Traj. B* is reportedly more challenging compared to *traj. A* for all tested methods which we also confirm in our runs. We observed that without the depth control term in Eq. (7), the e_{3D} on *traj. B* was higher by ~30%. Figure 4-(a) displays the effect of Eq. (7) on the 3D reconstructions from real images, when the dense point tracks and initialisations can be noisy.

Expressions. The usage of *expressions* allows us to compare N-NRSfM to even more methods from the literature including Expectation-Maximisation Linear Dynamical System (EM-LDS) [61], Column Space Fitting, version 2 (CSF2) [29], Kernel Shape Trajectory Approach (KSTA) [28] and Global Model with Local Interpretation (GMLI) [4]. The results are summarised in Table 2. We achieve $e_{3D} = 0.026$ on par with GMLI, *i.e.*, currently the best method on this sequence. The complexity of facial deformations in the *expressions* is similar to those of the *synthetic faces* [19]. This experiment shows that our novel neural model for NRSfM with constraints in metric and trajectory space is superior to multiple older NRSfM methods.

Kinect Sequences. For a fair evaluation, we pre-process the Kinect *t-shirt* and *paper* sequences along with their respective reference depth measurements as

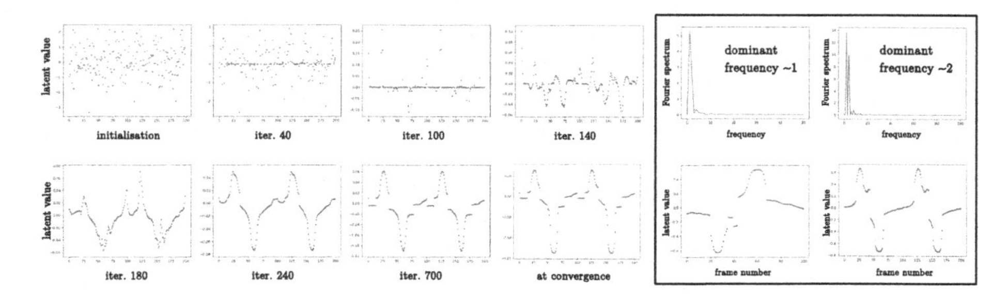

Fig. 3. Latent space evolution during the training of N-NRSfM on *actor mocap* [25]. We show which effect our latent space constraints have on the latent space function. **Left:** The evolution of the latent space function from start until convergence. **Right:** Frequency spectrum for the case with 100 and 200 frames.

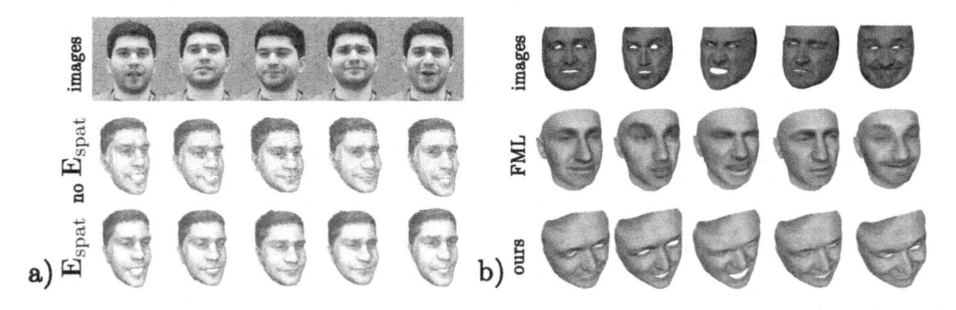

Fig. 4. a): 3D reconstructions of the *real face* with and without $\mathbf{E}_{\text{spat}}$. b): Images of the *actor mocap*; and 3D reconstructions by FML [58] and our approach.

described in Kumar *et al.* [37]. As it is suggested there, we run multi-frame optical flow [20] with default parameters to obtain dense correspondences. e_{3D} for the Kinect sequences are listed in Table 3. Visualisations of selected reconstructions of Kinect sequences can be found in Fig. 6-(top row). On Kinect *paper* and *t-shirt* sequences, we outperform all competing methods, including the current state of the art by significant margins of 1% and 20%, respectively. These sequences evince larger deformations compared to the face sequence, and, on the other hand, a simpler camera trajectory.

5.2 Period Detection and Sequence Segmentation

We evaluate the capability of our N-NRSfM method in period detection and sequence segmentation on the *actor mocap* sequence (100 frames with $3.5 \cdot 10^4$ points per frame) [25,63]. It has highly deformed facial expressions with ground truth shapes, ground truth dense flow fields and rendered images under orthographic projection. We duplicate the sequence and run N-NRSfM on the obtained point tracks. Our approach reconstructs the entire sequence and returns the frequency equal to 2, as can be seen in the Fourier spectrum. Given 200 input frames, it implies a period of 100. The latent space function for this experiment and the evolution of the latent space function are shown in Fig. 3. Note that for the same shapes, the resulting latent variables are also similar. This confirms that our N-NRSfM segments the sequence based on the shape similarity.

Next, we confirm that the period detection works well on real heart bypass surgery sequence [56] with 201 frames and 68,000 point per frame (see Fig. 6-(bottom right) for the exemplary frames and our reconstructions). This sequence evinces natural periodicity, and the flow fields are computed individually for every frame without duplication. We emphasise that images do not repeat as—even though the states are recurrent—they are observed under varying illumination and different occlusions. We recover the dominant frequency of 7.035, whereas the observed number of heartbeats amounts to ∼7.2. Knowing that the video was recorded at 24 frames per second, we obtain the pulse τ of $\tau = 7.035\text{beats} \cdot \frac{24\text{fps}}{201\text{frames}} = 0.84$ beats per second or ∼50 beats per minute—which is in the expected pulse range of a human during bypass surgery.

5.3 Qualitative Results and Applications

The *actor mocap* sequence allows us to qualitatively compare N-NRSfM to a state-of-the-art method for monocular 3D face reconstruction. Thus, we run the Face Model Learning (FML) approach of Tewari *et al.* [58] on it and show qualitative results in Fig. 4-(b). We observe that it is difficult to recognise the person in the FML 3D estimates ($e_{3D} = 0.092$ after Procrustes alignment of the ground truth shapes and FML reconstructions with rescaling of the latter). Since FML runs per-frame, its 3D shapes evince variation going beyond changing facial expressions, *i.e.*, it changes the identity. In contrast, N-NRSfM produces recognizable and consistent shapes at the cost of accurate dense correspondences across an image batch ($e_{3D} = 0.0181$, ∼5 times lower compared to $e_{3D} = 0.092$ of FML).

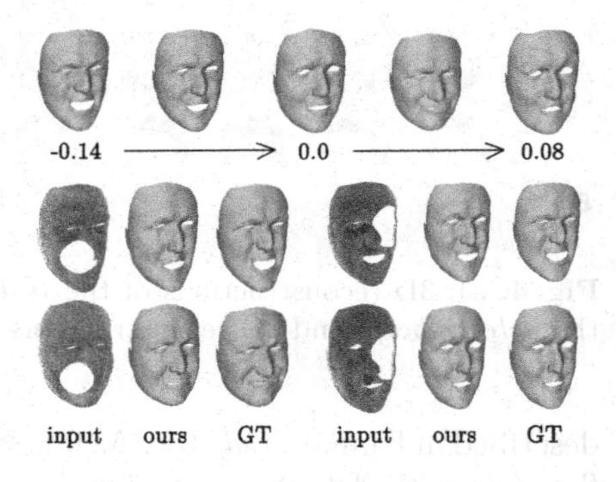

Fig. 5. Shape interpolation and completion. Top: A shape interpolation over the *actor* sequence is performed. **Bottom:** A series of three shapes are displayed, *i.e.*, input data, our estimation after completion, and the ground truth (GT).

Our auto-decoder f_θ is a flexible building block which can be used in multiple applications which were not easily possible with classical NRSfM methods. Those include shape completion, denoising, compression and interpolation, fast direct monocular non-rigid 3D reconstruction as well as sequence segmentation.

Shape Interpolation and Completion. To obtain shape interpolations, we can linearly interpolate the latent variables, see Fig. 5-(top row) for an example with the *actor mocap* reconstructions. Note that the interpolation result depends on the shape order in the latent space. For shape with significantly differing latent variables, it is possible that the resulting interpolations will not be equivalent to linear interpolations between the shapes and include non-linear point trajectories. Results of shape denoising and completion are shown in Fig. 5-(bottom rows). We feed point clouds with missing areas (mouth and the upper head area) and obtain surfaces completed upon our learned f_θ prior.

Direct Monocular Non-rigid 3D Reconstruction. We attach an image encoder to f_θ—as described in Sect. 4.4—and test it in the endoscopic scenario with the *heart* sequence. Our reconstructions follow the cardiac cycle outside of the image sub-sequence, which has been used for the training. Please, see our supplemental material for extra visualisations.

Real Image Sequences. Finally, we reconstruct several real image sequence, *i.e.*, *barn owl* [26], *back* [50] (see Fig. 6) and *real face* [19] (see Fig. 4-(a) which also highlights the influence of the spatial smoothness term). All our reconstructions are of high visual quality and match state of the art. Please, see our supplementary video for time-varying visualisations.

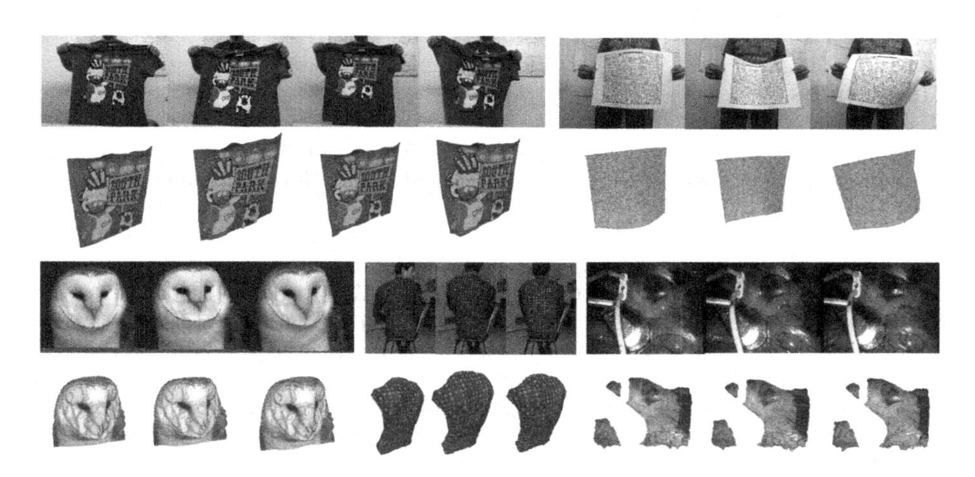

Fig. 6. Qualitative results on real sequences. In all cases, from left to right. **Top:** *T-shirt* and *paper* sequences [64]. **Bottom:** *Barn owl* [26], *back* [50] and *heart* [56] sequences. On both Kinect sequences, we achieve the lowest e_{3D} among all tested methods. The *heart* sequence is also used in the experiment with direct monocular non-rigid 3D reconstruction.

6 Concluding Remarks

This paper introduces the first end-to-end trainable neural dense NRSfM method with a deformation model auto-decoder and learnable latent space function. Our approach operates on dense 2D point tracks without 3D supervision. Structuring the latent space to detect and exploit periodicity is a promising first step towards new regularisation techniques for NRSfM. Period detection and temporal segmentation of the reconstructed sequences, automatically learned deformation model, shape compression, completion and interpolation—all that is obtained with a single neural component in our formulation. Experiments have shown that the new model results in smooth and accurate surfaces while achieving low 3D reconstruction errors in a variety of scenarios. One of the limitations of N-NRSfM is the sensitivity to inaccurate points tracks and the dependence on the mean shape obtained by rigid initialisation. We also found that our method does not cope well with large and sudden changes, even though the mean shape is plausible. Another limitation is the handling of articulated motions.

We believe that our work opens a new perspective on dense NRSfM. In future research, more sophisticated neural components for deformation models can be tested to support stronger non-linear deformations and composite scenes. Moreover, we plan to generalise our model to sequential NRSfM scenarios.

Acknowledgement. This work was supported by the ERC Consolidator Grant 4DReply (770784) and the Spanish Ministry of Science and Innovation under project HuMoUR TIN2017-90086-R. The authors thank Mallikarjun B R for help with running the FML method [58] on our data.

References

1. Agudo, A., Montiel, J.M.M., Agapito, L., Calvo, B.: Online dense non-rigid 3D shape and camera motion recovery. In: British Machine Vision Conference (BMVC) (2014)
2. Agudo, A., Montiel, J.M.M., Calvo, B., Moreno-Noguer, F.: Mode-shape interpretation: re-thinking modal space for recovering deformable shapes. In: Winter Conference on Applications of Computer Vision (WACV) (2016)
3. Agudo, A., Moreno-Noguer, F.: DUST: dual union of spatio-temporal subspaces for monocular multiple object 3D reconstruction. In: Computer Vision and Pattern Recognition (CVPR) (2017)
4. Agudo, A., Moreno-Noguer, F.: Global model with local interpretation for dynamic shape reconstruction. In: Winter Conference on Applications of Computer Vision (WACV) (2017)
5. Agudo, A., Moreno-Noguer, F.: Force-based representation for non-rigid shape and elastic model estimation. Trans. Pattern Anal. Mach. Intell. (TPAMI) **40**(9), 2137–2150 (2018)
6. Agudo, A., Moreno-Noguer, F.: A scalable, efficient, and accurate solution to non-rigid structure from motion. Comput. Vis. Image Underst. (CVIU) **167**, 121–133 (2018)
7. Akhter, I., Sheikh, Y., Khan, S., Kanade, T.: Trajectory space: a dual representation for nonrigid structure from motion. Trans. Pattern Anal. Mach. Intell. (TPAMI) **33**(7), 1442–1456 (2011)
8. Ansari, M., Golyanik, V., Stricker, D.: Scalable dense monocular surface reconstruction. In: International Conference on 3D Vision (3DV) (2017)
9. Baker, S., Scharstein, D., Lewis, J.P., Roth, S., Black, M.J., Szeliski, R.: A database and evaluation methodology for optical flow. Int. J. Comput. Vis. (IJCV) **92**(1), 1–31 (2011)

10. Bartoli, A., Gay-Bellile, V., Castellani, U., Peyras, J., Olsen, S., Sayd, P.: Coarse-to-fine low-rank structure-from-motion. In: Computer Vision and Pattern Recognition (CVPR) (2008)
11. Bregler, C., Hertzmann, A., Biermann, H.: Recovering non-rigid 3D shape from image streams. In: Computer Vision and Pattern Recognition (CVPR) (2000)
12. Bue, A.D.: A factorization approach to structure from motion with shape priors. In: Computer Vision and Pattern Recognition (CVPR) (2008)
13. Choy, C.B., Xu, D., Gwak, J.Y., Chen, K., Savarese, S.: 3D-R2N2: a unified approach for single and multi-view 3D object reconstruction. In: Leibe, B., Matas, J., Sebe, N., Welling, M. (eds.) ECCV 2016. LNCS, vol. 9912, pp. 628–644. Springer, Cham (2016). https://doi.org/10.1007/978-3-319-46484-8_38
14. Clevert, D., Unterthiner, T., Hochreiter, S.: Fast and accurate deep network learning by exponential linear units (elus). In: International Conference on Learning Representations (ICLR) (2016)
15. Dai, Y., Deng, H., He, M.: Dense non-rigid structure-from-motion made easy - a spatial-temporal smoothness based solution. In: International Conference on Image Processing (ICIP), pp. 4532–4536 (2017)
16. Dai, Y., Li, H., He, M.: Simple prior-free method for non-rigid structure-from-motion factorization. Int. J. Comput. Vis. (IJCV) **107**, 101–122 (2014)
17. Deng, J., Dong, W., Socher, R., Li, L.J., Li, K., Fei-Fei, L.: ImageNet: a large-scale hierarchical image database. In: Computer Vision and Pattern Recognition (CVPR) (2009)
18. Fan, H., Su, H., Guibas, L.J.: A point set generation network for 3D object reconstruction from a single image. In: Computer Vision and Pattern Recognition (CVPR) (2017)
19. Garg, R., Roussos, A., Agapito, L.: Dense variational reconstruction of non-rigid surfaces from monocular video. In: Computer Vision and Pattern Recognition (CVPR) (2013)
20. Garg, R., Roussos, A., Agapito, L.: A variational approach to video registration with subspace constraints. Int. J. Comput. Vis. (IJCV) **104**(3), 286–314 (2013)
21. Golyanik, V., Fetzer, T., Stricker, D.: Accurate 3D reconstruction of dynamic scenes from monocular image sequences with severe occlusions. In: Winter Conference on Applications of Computer Vision (WACV), pp. 282–291 (2017)
22. Golyanik, V., Stricker, D.: Dense batch non-rigid structure from motion in a second. In: Winter Conference on Applications of Computer Vision (WACV), pp. 254–263 (2017)
23. Golyanik, V., Fetzer, T., Stricker, D.: Introduction to coherent depth fields for dense monocular surface recovery. In: British Machine Vision Conference (BMVC) (2017)
24. Golyanik, V., Jonas, A., Stricker, D.: Consolidating segmentwise non-rigid structure from motion. In: Machine Vision Applications (MVA) (2019)
25. Golyanik, V., Jonas, A., Stricker, D., Theobalt, C.: Intrinsic Dynamic Shape Prior for Fast, Sequential and Dense Non-Rigid Structure from Motion with Detection of Temporally-Disjoint Rigidity. arXiv e-prints (2019)
26. Golyanik, V., Mathur, A.S., Stricker, D.: NRSfm-Flow: recovering non-rigid scene flow from monocular image sequences. In: British Machine Vision Conference (BMVC) (2016)
27. Golyanik, V., Shimada, S., Varanasi, K., Stricker, D.: HDM-Net: monocular non-rigid 3D reconstruction with learned deformation model. In: Bourdot, P., Cobb, S., Interrante, V., kato, H., Stricker, D. (eds.) EuroVR 2018. LNCS, vol. 11162, pp. 51–72. Springer, Cham (2018). https://doi.org/10.1007/978-3-030-01790-3_4

28. Gotardo, P.F.U., Martinez, A.M.: Kernel non-rigid structure from motion. In: International Conference on Computer Vision (ICCV), pp. 802–809 (2011)
29. Gotardo, P.F.U., Martinez, A.M.: Non-rigid structure from motion with complementary rank-3 spaces. In: Computer Vision and Pattern Recognition (CVPR), pp. 3065–3072 (2011)
30. Groueix, T., Fisher, M., Kim, V.G., Russell, B., Aubry, M.: AtlasNet: a Papier-Mâché approach to learning 3D surface generation. In: Computer Vision and Pattern Recognition (CVPR) (2018)
31. He, K., Zhang, X., Ren, S., Sun, J.: Delving deep into rectifiers: surpassing human-level performance on imagenet classification. In: International Conference on Computer Vision (ICCV), pp. 1026–1034 (2015)
32. He, K., Zhang, X., Ren, S., Sun, J.: Deep residual learning for image recognition. In: Computer Vision and Pattern Recognition (CVPR), pp. 770–778 (2016)
33. Kanazawa, A., Tulsiani, S., Efros, A.A., Malik, J.: Learning category-specific mesh reconstruction from image collections. In: Ferrari, V., Hebert, M., Sminchisescu, C., Weiss, Y. (eds.) ECCV 2018. LNCS, vol. 11219, pp. 386–402. Springer, Cham (2018). https://doi.org/10.1007/978-3-030-01267-0_23
34. Kong, C., Lucey, S.: Deep non-rigid structure from motion. In: International Conference on Computer Vision (ICCV) (2019)
35. Kovalenko, O., Golyanik, V., Malik, J., Elhayek, A., Stricker, D.: Structure from articulated motion: accurate and stable monocular 3D reconstruction without training data. Sensors **19**(20), 4603 (2019)
36. Kumar, S.: Jumping manifolds: geometry aware dense non-rigid structure from motion. In: Computer Vision and Pattern Recognition (CVPR) (2019)
37. Kumar, S., Cherian, A., Dai, Y., Li, H.: Scalable dense non-rigid structure-from-motion: a grassmannian perspective. In: Computer Vision and Pattern Recognition (CVPR) (2018)
38. Lee, M., Cho, J., Choi, C.H., Oh, S.: Procrustean normal distribution for non-rigid structure from motion. In: Computer Vision and Pattern Recognition (CVPR) (2013)
39. Lee, M., Choi, C.H., Oh, S.: A procrustean Markov process for non-rigid structure recovery. In: Computer Vision and Pattern Recognition (CVPR) (2014)
40. Mescheder, L., Oechsle, M., Niemeyer, M., Nowozin, S., Geiger, A.: Occupancy networks: learning 3D reconstruction in function space. In: Computer Vision and Pattern Recognition (CVPR) (2019)
41. Novotny, D., Ravi, N., Graham, B., Neverova, N., Vedaldi, A.: C3DPO: canonical 3D pose networks for non-rigid structure from motion. In: International Conference on Computer Vision (ICCV) (2019)
42. Östlund, J., Varol, A., Ngo, D.T., Fua, P.: Laplacian meshes for monocular 3D shape recovery. In: Fitzgibbon, A., Lazebnik, S., Perona, P., Sato, Y., Schmid, C. (eds.) ECCV 2012. LNCS, vol. 7574, pp. 412–425. Springer, Heidelberg (2012). https://doi.org/10.1007/978-3-642-33712-3_30
43. Paladini, M., Del Bue, A., Xavier, J., Agapito, L., Stosić, M., Dodig, M.: Optimal metric projections for deformable and articulated structure-from-motion. Int. J. Comput. Vis. (IJCV) **96**(2), 252–276 (2012)
44. Park, J.J., Florence, P., Straub, J., Newcombe, R., Lovegrove, S.: Deepsdf: learning continuous signed distance functions for shape representation. In: Computer Vision and Pattern Recognition (CVPR) (2019)
45. Paszke, A., et al.: An imperative style, high-performance deep learning library. In: Advances in Neural Information Processing Systems (NeurIPS) (2019)

46. Pearson, K.: On lines and planes of closest fit to systems of points in space. Philoso. Mag. **2**, 559–572 (1901)
47. Pumarola, A., Agudo, A., Porzi, L., Sanfeliu, A., Lepetit, V., Moreno-Noguer, F.: Geometry-aware network for non-rigid shape prediction from a single view. In: Computer Vision and Pattern Recognition (CVPR) (2018)
48. Riedmiller, M., Braun, H.: A direct adaptive method for faster backpropagation learning: the RPROP algorithm. In: International Conference on Neural Networks (ICNN), pp. 586–591 (1993)
49. Rumelhart, D.E., Hinton, G.E., Williams, R.J.: Learning representations by back-propagating errors. Nature **323**, 533–536 (1986)
50. Russell, C., Fayad, J., Agapito, L.: Energy based multiple model fitting for non-rigid structure from motion. In: Computer Vision and Pattern Recognition (CVPR), pp. 3009–3016 (2011)
51. Russell, C., Fayad, J., Agapito, L.: Dense non-rigid structure from motion. In: 2012 Second International Conference on 3D Imaging, Modeling, Processing, Visualization Transmission (3DIMPVT) (2012)
52. Sahasrabudhe, M., Shu, Z., Bartrum, E., Alp Güler, R., Samaras, D., Kokkinos, I.: Lifting autoencoders: unsupervised learning of a fully-disentangled 3D morphable model using deep non-rigid structure from motion. In: International Conference on Computer Vision Workshops (ICCVW) (2019)
53. Salzmann, M., Fua, P.: Reconstructing sharply folding surfaces: a convex formulation. In: Computer Vision and Pattern Recognition (CVPR), pp. 1054–1061 (2009)
54. Shimada, S., Golyanik, V., Theobalt, C., Stricker, D.: IsMo-GAN: adversarial learning for monocular non-rigid 3D reconstruction. In: Computer Vision and Pattern Recognition Workshops (CVPRW) (2019)
55. Sorkine, O.: Laplacian mesh processing. In: Annual Conference of the European Association for Computer Graphics (Eurographics) (2005)
56. Stoyanov, D.: Stereoscopic scene flow for robotic assisted minimally invasive surgery. In: Ayache, N., Delingette, H., Golland, P., Mori, K. (eds.) MICCAI 2012. LNCS, vol. 7510, pp. 479–486. Springer, Heidelberg (2012). https://doi.org/10.1007/978-3-642-33415-3_59
57. Taetz, B., Bleser, G., Golyanik, V., Stricker, D.: Occlusion-aware video registration for highly non-rigid objects. In: Winter Conference on Applications of Computer Vision (WACV) (2016)
58. Tewari, A., et al.: FML: face model learning from videos. In: Computer Vision and Pattern Recognition (CVPR) (2019)
59. Tewari, A., et al.: MoFA: model-based deep convolutional face autoencoder for unsupervised monocular reconstruction. In: International Conference on Computer Vision (ICCV) (2017)
60. Tomasi, C., Kanade, T.: Shape and motion from image streams under orthography: a factorization method. Int. J. Comput. Vis. (IJCV) **9**(2), 137–154 (1992)
61. Torresani, L., Hertzmann, A., Bregler, C.: Nonrigid structure-from-motion: estimating shape and motion with hierarchical priors. Trans. Pattern Anal. Mach. Intell. (TPAMI) **30**(5), 878–892 (2008)
62. Tsoli, A., Argyros, A.A.: Patch-based reconstruction of a textureless deformable 3D surface from a single RGB image. In: International Conference on Computer Vision Workshops (ICCVW) (2019)
63. Valgaerts, L., Wu, C., Bruhn, A., Seidel, H.P., Theobalt, C.: Lightweight binocular facial performance capture under uncontrolled lighting. ACM Trans. Graph. (TOG) **31**(6), 187:1–187:11 (2012)

64. Varol, A., Salzmann, M., Fua, P., Urtasun, R.: A constrained latent variable model. In: Computer Vision and Pattern Recognition (CVPR) (2012)
65. Vicente, S., Agapito, L.: Soft inextensibility constraints for template-free non-rigid reconstruction. In: Fitzgibbon, A., Lazebnik, S., Perona, P., Sato, Y., Schmid, C. (eds.) ECCV 2012. LNCS, vol. 7574, pp. 426–440. Springer, Heidelberg (2012). https://doi.org/10.1007/978-3-642-33712-3_31
66. Wang, N., Zhang, Y., Li, Z., Fu, Y., Liu, W., Jiang, Y.G.: Pixel2mesh: generating 3D mesh models from single RGB images. In: European Conference on Computer Vision (ECCV) (2018)
67. Xiao, J., Chai, J., Kanade, T.: A closed-form solution to non-rigid shape and motion recovery. In: European Conference on Computer Vision (ECCV) (2004)
68. Yu, R., Russell, C., Campbell, N.D.F., Agapito, L.: Direct, dense, and deformable: template-based non-rigid 3D reconstruction from RGB video. In: International Conference on Computer Vision (ICCV) (2015)
69. Zhu, Y., Huang, D., Torre, F.D.L., Lucey, S.: Complex non-rigid motion 3D reconstruction by union of subspaces. In: Computer Vision and Pattern Recognition (CVPR), pp. 1542–1549 (2014)

Multimodal Memorability: Modeling Effects of Semantics and Decay on Video Memorability

Anelise Newman(✉), Camilo Fosco, Vincent Casser, Allen Lee, Barry McNamara, and Aude Oliva

Massachusetts Institute of Technology, Cambridge, MA, USA
aonewman@mit.edu, anelisenewman@gmail.com

Abstract. A key capability of an intelligent system is deciding when events from past experience must be remembered and when they can be forgotten. Towards this goal, we develop a predictive model of human visual event memory and how those memories decay over time. We introduce *Memento10k*, a new, dynamic video memorability dataset containing human annotations at different viewing delays. Based on our findings we propose a new mathematical formulation of memorability decay, resulting in a model that is able to produce the first quantitative estimation of how a video decays in memory over time. In contrast with previous work, our model can predict the probability that a video will be remembered at an arbitrary delay. Importantly, our approach combines visual and semantic information (in the form of textual captions) to fully represent the meaning of events. Our experiments on two video memorability benchmarks, including Memento10k, show that our model significantly improves upon the best prior approach (by 12% on average).

Keywords: Memorability estimation · Memorability decay · Multimodal video understanding

1 Introduction

Deciding which moments from past experience to remember and which ones to discard is a key capability of an intelligent system. The human brain is optimized to remember what it deems to be important and forget what is uninteresting or redundant. Thus, human memorability is a useful measure of what content is interesting and likely to be retained by a human viewer. If a system can

A. Newman and C. Fosco—Equal contribution.

Electronic supplementary material The online version of this chapter (https://doi.org/10.1007/978-3-030-58517-4_14) contains supplementary material, which is available to authorized users.

A. Vedaldi et al. (Eds.): ECCV 2020, LNCS 12361, pp. 223–240, 2020.
https://doi.org/10.1007/978-3-030-58517-4_14

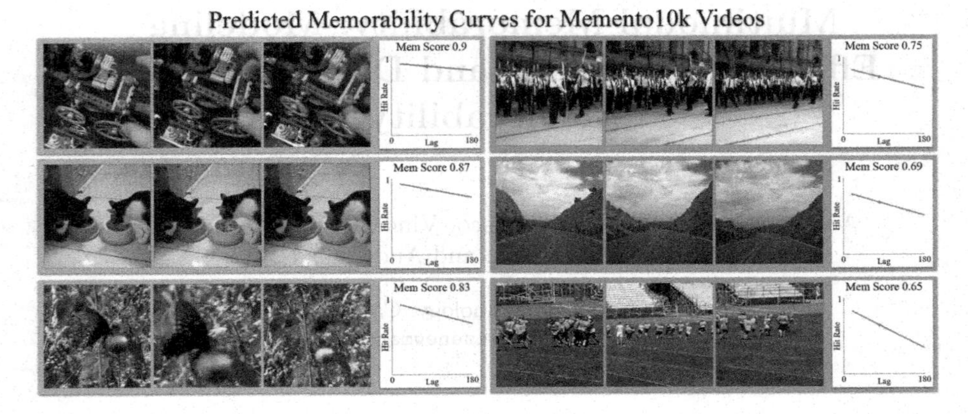

Fig. 1. How do visual and semantic features impact memory decay over time? We introduce a multimodal model, SemanticMemNet, that leverages visual and textual information to predict the memorability decay curve of a video clip. We show predictions on videos from Memento10k, our new video memorability dataset. SemanticMemNet is the first model to predict a decay curve that represents how quickly a video falls off in memory over time.

predict which information will be highly memorable, it can evaluate the utility of incoming data and compress or discard what is deemed to be irrelevant. It can also filter to select the content that will be most memorable to humans, which has potential applications in design and education.

However, memorability of dynamic events is challenging to predict because it depends on many factors. First, different visual representations are forgotten at different rates: while some events persist in memory even over long periods, others are forgotten within minutes [3,22,25,34]. This means that the probability that someone will remember a certain event varies dramatically as a function of time, introducing challenges in terms of how memorability is represented and measured. Second, memorability depends on both visual and semantic factors. In human cognition, language and vision often act in concert for remembering an event. Events described with richer and more distinctive concepts are remembered for longer than events attached to shallower descriptions, and certain semantic categories of objects or places are more memorable than others [25,33].

In this paper, we introduce a new dataset and a model for video memorability prediction that address these challenges[1]. Memento10k, the most dynamic video memorability dataset to date, contains both human annotations at different viewing delays and human-written captions, making it ideal for studying the effects of delay and semantics on memorability. Based on this data, we propose a mathematical formulation of memorability decay that allows us to estimate

[1] Dataset, code, and models can be found at our website: http://memento.csail.mit.edu/.

the probability that an event will be remembered at any point over the first ten minutes after viewing. We introduce SemanticMemNet, a multimodal model that relies on visual and semantic features to predict the decay in memory of a short video clip (Fig. 1). SemanticMemNet is the first model that predicts the entire memorability decay curve, which allows us to estimate the probability that a person will recall a given video after a certain delay. We also enhance our model's features to include information about video semantics by jointly predicting verbal captions. SemanticMemNet achieves state-of-the-art performance on video memorability prediction on two different video memorability baselines.

To summarize, our key contributions are:

- We present a new **multi-temporal, multimodal memory dataset** of 10,000 video clips, *Memento10k*. With over 900,000 human memory annotations at different delay intervals and 50,000 captions describing the events, it is the largest repository of dynamic visual memory data.
- We propose a **new mathematical formulation for memorability decay** which estimates the decay curve of human memory for individual videos.
- We introduce **SemanticMemNet**, a model capitalizing on learning visual and semantic features to predict both the memorability strength and the memory decay curve of a video clip.

2 Related Work

Memorability in Cognitive Science. Four landmark results in cognitive science inspired our current approach. First, memorability is an *intrinsic* property of an image: people are remarkably consistent in which images they remember and forget [3,11,22,25–27,30,36]. Second, semantic features, such as a detailed conceptual representation or a verbal description, boost visual memory and aid in predicting memorability [33,35,44,45]. Third, most stimuli decay in memory, with memory performances falling off predictably over time [22,25,34,43]. Finally, the classical old-new recognition paradigm (i.e. people press a key when they recognize a repeated stimulus in a sequence) allows researchers to collect objective measurements of human memory at a large scale [9,26] and variable time scales, which we draw on to collect our dataset.

Memorability in Computer Vision. The intrinsic nature of visual memorability means that visual stimuli themselves contain visual and semantic features that can be captured by machine vision. For instance, earlier works [24–26] pointed to content of images that were predictive of their memorability (i.e. people, animals and manipulable objects are memorable, but landscapes are often forgettable). Later work replicated the initial findings and extended the memorability prediction task to many photo categories [1,5,17,19,31,38,48], faces [3,29,40], visualizations [7,8] and videos [13,15,39].

The development of large-scale image datasets augmented with memorability scores [30] allowed convolutional neural networks to predict image memorability at near human-level consistency [1,5,19,30,48] and even generate realistic memorable and forgettable photos [21,40]. However, similar large-scale work on video memorability prediction has been limited. In the past year, Cohendet et al. introduced a video-based memorability dataset [14] that is the only other large benchmark comparable to Memento10k, and made progress towards building a predictive model of video memorability [13,15]. Other works on video memorability have largely relied on smaller datasets collected using paradigms that are more challenging to scale. [23] collected memorability and fMRI data on 2400 video clips and aligned audio-visual features with brain data to improve prediction. [39] used a language-based recall task to collect memorability scores as a function of response time for 100 videos. They find that a combination of semantic, spatio-temporal, saliency, and color features can be used to predict memorability scores. [15] collected a long-term memorability dataset using clips from popular movies that people have seen before.

Past work has confirmed the usefulness of semantic features for predicting memorability [13,39,41]. Work at the intersection of Computer Vision and NLP has aimed to bridge the gap between images and text by generating natural language descriptions of images using encoder-decoder architectures (e.g.. [28,42, 47]) or by creating aligned embeddings for visual and textual content [18,20,32]. Here, we experiment with both approaches in order to create a memorability model that jointly learns to understand visual features and textual labels.

3 Memento10k: A Multimodal Memorability Dataset

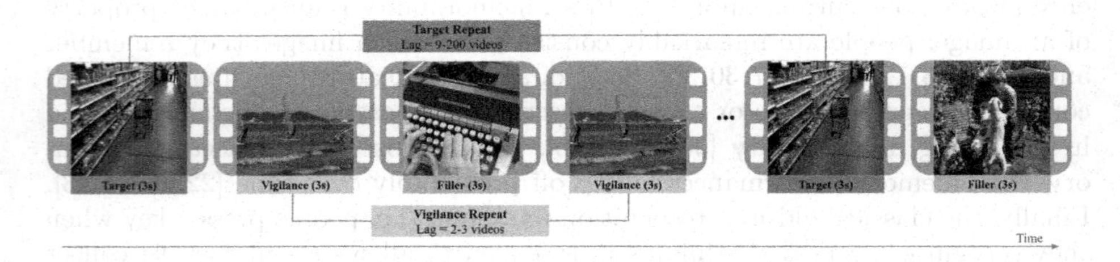

Fig. 2. Task flow diagram of *The Memento Video Memory Game.* Participants see a continuous stream of videos and press the space bar when they see a repeat.

Memento10k focuses on both the visual and semantic underpinnings of video memorability. The dataset contains 10,000 video clips augmented with memory scores, action labels, and textual descriptions (five human-generated captions per video). Importantly, our memorability annotations occur at presentation delays ranging from several seconds to ten minutes, which, for the first time, allows us to model how memorability falls off with time.

The Memento Video Memory Game. In our experiment, crowdworkers from Amazon's Mechanical Turk (AMT) watched a continuous stream of three-second video clips and were asked to press the space bar when they saw a repeated

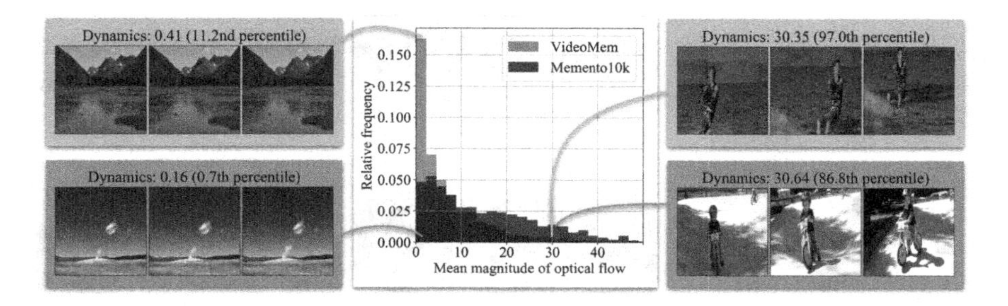

Fig. 3. Video distribution by level of motion for VideoMem and Memento10k. The motion metric for each video is calculated by averaging optical flow magnitude over the entire video. A high percentage of videos in the VideoMem dataset are almost static, while Memento10k is much more balanced. We show select examples with low and high levels of motion.

video (Fig. 2). Importantly, we varied the lag for repeated videos (the number of videos between the first and the second showing of a repeat), which allowed us to study the evolution of memorability over time. Our policies came from [26,30] and a summary of paradigm details can be found in the supplement.

Three seconds, the length of a Memento clip, is about the average duration of human working memory [2,4] and most human visual memory performances plateau at three seconds of exposure [9,10]. This makes three seconds a good "atomic" length for a single item held in memory. Previous work has shown that machine learning models can learn robust features even from short clips [37].

Dynamic, In-The-Wild Videos. Memento10k is composed of natural videos scraped from the Internet[2]. To limit our clips to non-artificial scenes with everyday context, we asked crowdworkers whether each clip was a "home video" and discarded videos that did not meet this criterion. After removing clips that contained undesirable properties (i.e. watermarks), we were left with 10,000 "clean" videos, which we break into train (7000), validation (1500), and test (1500) sets.

The Memento10k dataset is a significant step towards understanding memorability of real-world events. First, it is the most dynamic memorability dataset to date with videos containing a variety of motion patterns including camera motion and moving objects. The mean magnitude of optical flow in Memento10k is nearly double that of VideoMem [13] (approximately 15.476 for Memento vs. 7.296 for VideoMem, see Fig. 3), whose clips tend to be fairly static. Second, Memento10k's diverse, natural content enables the study of memorability in an everyday context: Memento10k was compiled from in-the-wild amateur videos like those found on social media or real-life scenes, while VideoMem is composed of professional footage. Third, Memento10k's greater number of annotations (90 versus 38 per video in VideoMem), spread over lags of 30 30 s to 10 min, leads to higher ground-truth human consistency and allows for a robust estimation

[2] The Memento videos have partial overlap with the Moments in Time [37] dataset.

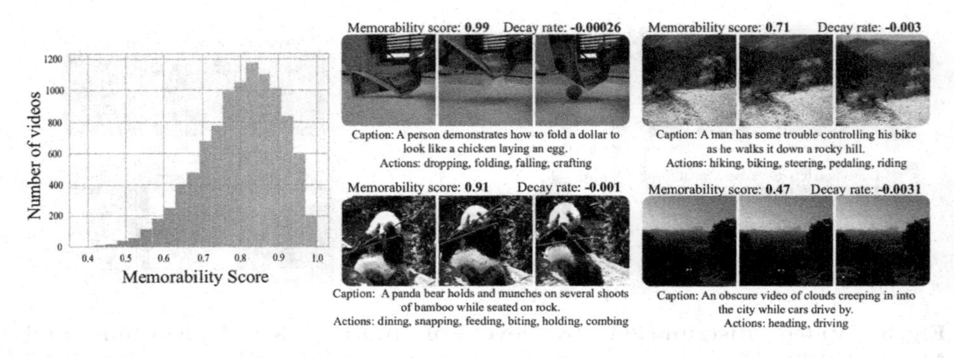

Fig. 4. The Memento10k dataset contains the memorability scores, alpha scores (decay rates), action labels, and five unique captions for 10,000 videos. **Left:** The distribution of memorability scores over the entire dataset. **Right:** Example clips from the Memento10k dataset along with their memorability score, decay rate, actions, and an example caption.

of a video's decay rate. Finally, we provide more semantic information such as action labels as well as 5 detailed captions per video. In this paper, we use both benchmarks to evaluate the generalization of our model.

Semantic Annotations. We augment our dataset with captions, providing a source of rich textual data that we can use to relate memorability to semantic concepts (examples in Fig. 4). We asked crowdworkers to describe the events in the video clip in full sentences and we manually vetted the captions for quality and corrected spelling mistakes. Each video has 5 unique captions from different crowdworkers. More details on caption collection are in the supplement.

Human Results. The Memento10k dataset contains over 900,000 individual annotations, making it the biggest memorability dataset to date. We measured human consistency of these annotations following [13,26]: we randomly split our participant pool into two groups and calculate the Spearman's rank correlation between the memorability rankings produced by each group, where the rankings are generated by sorting videos by raw hit rate. The average rank correlation (ρ) over 25 random splits is 0.73 (compared to 0.68 for images in [30], and 0.616 for videos in [13]). This high consistency between human observers confirms that videos have strong intrinsic visual, dynamic or semantic features that a model can learn from to predict memorability of new videos.

Figure 5 illustrates some qualitative results of our experiment. We see some similar patterns as with image memorability: memorable videos tend to contain saturated colors, people and faces, manipulable objects, and man-made spaces, while less memorable videos are dark, cluttered, or inanimate. Additionally, videos with interesting motion patterns can be highly memorable whereas static videos often have low memorability.

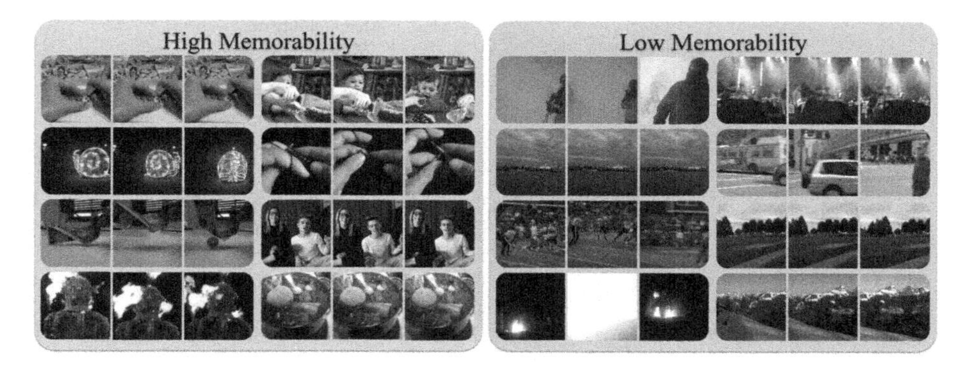

Fig. 5. Examples of high- and low-memorability video clips. Videos involving people, faces, hands, man-made spaces, and moving objects are in general more memorable, while clips containing distant/outdoor landscapes or dark, cluttered, or static content tend to be less memorable.

4 Memory Decay: A Theoretical Formulation

Most memories decay over time. In psychology this is known as the forgetting curve, which estimates how the memory of an item naturally degrades. Because Memento10k's memorability annotations occur at lags of anywhere from 9 videos (less than 30 s) to 200 videos (around 9 min), we have the opportunity to calculate the strength of a given video clip's memory at different lags.

A naive method for calculating a memorability score is to simply take the video's target hit rate, or the fraction of times that the repeated video was correctly detected. However, since we expect a video's hit rate to go down with time, annotations at different lags are not directly comparable. Instead, we derive an equation for how each video's hit rate declines as a function of lag.

First, for the lags tested in our study, we observe that hit rate decays linearly as a function of lag. This is notable because previous work on image memorability has found that images follow a log-linear decay curve [22,25,43]. Figure 6 (left) shows that a linear trend best fits our raw annotations; this holds for videos across the memorability distribution (see the supplement for more details).

Second, in contrast to prior work, we find that different videos decay in memory at different rates. Instead of assuming that all stimuli decay at one universal decay rate, α, as in [30], we assume that each video decays at its own rate, $\alpha^{(v)}$. Following the procedure laid out in [30], we find a memorability score and decay rate for each video that approximates our annotations. We define the memorability of video v as $m_T^{(v)} = \alpha^{(v)}T + c^{(v)}$, where T is the lag (the interval in videos between the first and second presentation) and $c^{(v)}$ is the base memorability of the video. If we know $m_T^{(v)}$ and $\alpha^{(v)}$, we can then calculate the video's memorability at a different lag t with

$$m_t^{(v)} = m_T^{(v)} + \alpha^{(v)}(t - T) \quad (1)$$

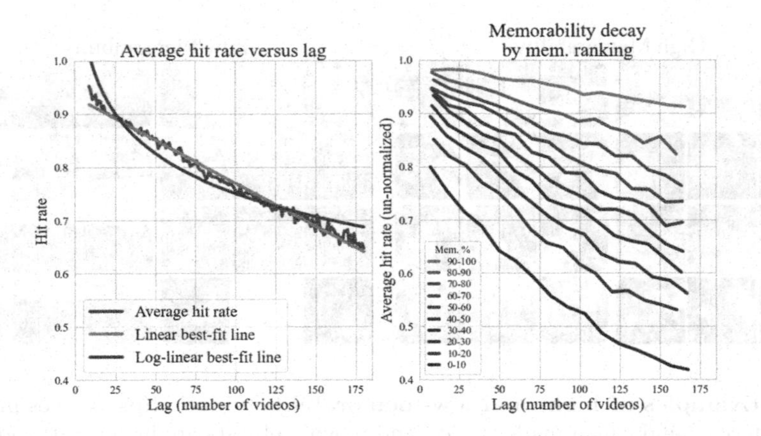

Fig. 6. Our data suggests a **memory model where each video decays linearly** in memory according to an individual decay rate $\alpha^{(v)}$. **Left:** A linear trend is a better approximation for our raw data ($r = -0.991$) than a log-linear trend ($r = -0.953$). **Right:** We confirm our assumption that α varies by video by grouping videos into deciles based on their normalized memorability score and plotting group average hit rate as a function of lag. Videos with lower memorability show a faster rate of decay.

To obtain values for $m_T^{(v)}$ and $\alpha^{(v)}$, we minimize the L2 norm between the raw binary annotations from our experiment $x_j^{(v)}$, $j \in \{0, ..., n^{(v)}\}$ and the predicted memorability score at the corresponding lag, $m_t^{(v)}$. The error equation is:

$$E(\alpha^{(v)}, m_T^{(v)}) = \sum_{j=1}^{n^{(v)}} \left\| x_j^{(v)} - m_t^{(v)} \right\|_2^2 = \sum_{j=1}^{n^{(v)}} \left\| x_j^{(v)} - \left[m_T^{(v)} + \alpha^{(v)}(t_j^{(v)} - T) \right] \right\|_2^2$$

We find update equations for $m_T^{(v)}$ and $\alpha^{(v)}$ by taking the derivative with respect to each and setting it to zero:

$$\alpha^{(v)} \leftarrow \frac{\frac{1}{n^{(v)}} \sum_{j=1}^{n^{(v)}} (t_j^{(v)} - T) \left[x_j^{(v)} - m_T^{(v)} \right]}{\frac{1}{n^{(v)}} \sum_{j=1}^{n^{(v)}} \left[t_j^{(v)} - T \right]^2} \qquad m_T^{(v)} \leftarrow \frac{1}{n^{(v)}} \sum_{j=1}^{n^{(v)}} \left[x_j^{(v)} - \alpha^{(v)}(t_j^{(v)} - T) \right] \tag{2}$$

We initialize $\alpha^{(v)}$ to $-5\text{e}{-4}$ and $m_T^{(v)}$ to each video's mean hit rate. We set our base lag T to 80 and optimize for 10 iterations to produce $\alpha^{(v)}$ and $m_{80}^{(v)}$ for each video. We thus define a video's "memorability score", for the purposes of memorability ranking, as its hit rate at a lag of 80; however, we can use Eq. 1 to calculate its hit rate at an arbitrary lag within the range that we studied.

Next, we validate our hypothesis that videos decay in memory at different rates. We bucket the Memento10k videos into 10 groups based on their normalized memorability scores and plot the raw data (average hit rate as a function of lag) for each group. Figure 6 (right) confirms that different videos decay at different rates.

Table 1. SemanticMemNet ablation study. We experiment with different ways of incorporating visual and semantic features into memorability prediction. We measure performance by calculating the Spearman's rank correlation (RC) of the predicted memorability rankings with ground truth rankings on both Memento10k and VideoMem.

	Approach	RC - Memento10k (test set)	RC - VideoMem (validation set)
	Human consistency	0.730	0.616
(Sect. 5.1)	Flow stream only	0.579	0.425
	Frames stream only	0.595	0.527
	Video stream only	0.596	0.492
	Flow + Frames + Video	0.659	0.555
(Sect. 5.2)	Video stream + captions	0.602	0.512
	Video stream + triplet loss	0.599	-
	SemanticMemNet (ours)	**0.663**	**0.556**

5 Modeling Experiments

In this section, we explore different architecture choices for modeling video memorability that take into consideration both visual and semantic features. We also move away from a conception of memorability as a single value, and instead predict the *decay curve* of a video, resulting in the first model that predicts the raw probability that a video will be remembered at a particular lag.

Metrics. A commonly-used metric for evaluating memorability is the Spearman rank correlation (RC) between the memorability ranking produced by the ground-truth memorability scores versus the predicted scores. This is a popular metric [13,26,30] because memorability rankings are generally robust across experimental designs and choice of lag; however, it does not measure the accuracy of predicted memorability values. As such, when evaluating the quality of the decay curve produced by our model, we will also consider R^2 for our predictions.

5.1 Modeling Visual Features

Baseline: Static Frames. We evaluate the extent to which static visual features contribute to video memorability by training a network to predict a video's memorability from a single frame. We first train an ImageNet-pretrained DenseNet-121 to predict image memorability by training on the LaMem dataset [30], then finetune on the video memorability datasets. At test time, a video's memorability score is calculated by averaging predictions over every 4th frame (about 22 frames total for Memento videos).

3D Architectures: Video and Optical Flow. Training on the RGB videos allows the network to access information on both motion and visual features, while training on optical flow lets us isolate the effects of motion. We train I3D architectures [12] on raw video and optical flow (computed using OpenCV's TV-L1 implementation). Our models were pretrained on the ImageNet and Kinetics datasets. We test the different visual feature architectures on both Memento10k and VideoMem videos. Our results are in the top section of Table 1. Out of the three input representations (frames, flow, and video), optical flow achieves the poorest performance, probably because of the lack of access to explicit visual features like color and shape. Static frames perform remarkably well on their own, outperforming a 3D-video representation on VideoMem (as VideoMem is a fairly static dataset, this result is reasonable). Even with the relatively high level of motion in the Memento10k dataset (see Fig. 3), the video and frames streams perform comparably. For both datasets, combining the three streams maximizes performance, which is consistent with previous work [12] and reinforces that both visual appearance and motion are relevant to predicting video memorability. In fact, our three-stream approach leverages motion information from the optical flow stream to refine its predictions, as shown in Fig. 7.

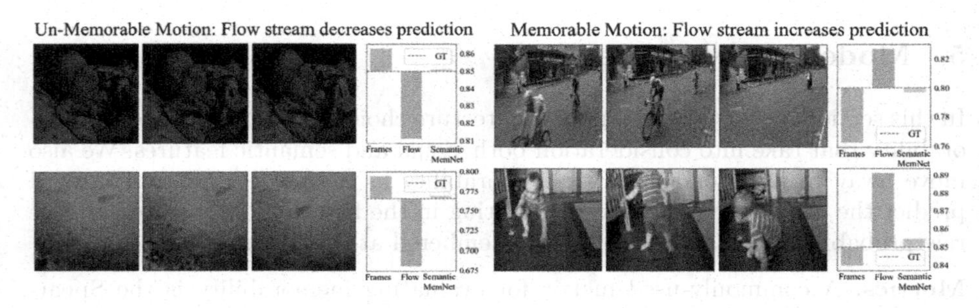

Fig. 7. Our model leverages visual and motion information to produce accurate memorability scores. Here, we compare the contributions of the frames stream (static information only) and the optical flow stream (motion information only) to separate the contributions of visual features and temporal dynamics. **Left:** Flow decreases the frames prediction. The frames stream detects memorable features like a human face or saturated colors, but the flow stream detects a static video and predicts lower memorability. **Right:** Flow increases the frames prediction. The flow stream picks up on dynamic patterns like fast bikers or a baby falling and increases the memorability prediction.

5.2 Modeling Semantic Features

It is well-known that semantics are an important contributor to memorability [13,15,33,39]. To increase our model's ability to extract semantic information, we jointly train it on memorability prediction and a captioning task, which ensures that the underlying representation learned for both problems contains relevant event semantics. To test this approach, we enhance the video stream from the previous section with an additional module that aims to solve one of two tasks:

generating captions or learning a joint text-video embedding. We augment the video stream as opposed to the flow or frames streams because it contains both visual and motion information required to reconstruct a video caption.

Caption Generation. Our first approach is to predict captions directly. This has the benefit of forcing the model to encode a rich semantic representation of the image, taking into account multiple objects, actions, and colors. However, it also involves learning an English language model, which is tangential to the task of predicting memorability. We feed the output features of the video I3D base into an LSTM that learns to predict the ground-truth captions. For Memento10k, we tokenize our 5 ground-truth captions and create a vocabulary of 3,870 words that each appear at least 5 times in our training set. We use pre-trained FastText word embeddings [6] to map our tokens to 300-dimensional feature vectors that we feed to the recurrent module. The VideoMem dataset provides a single brief description with each video, which we process the same way. We train with teacher forcing and at test-time feed the output of the LSTM back into itself.

Mapping Videos into a Semantic Embedding. Our second approach is to learn to map videos into a sentence-level semantic embedding space using a triplet loss. We pre-compute sentence embeddings for the captions using the popular transformer-based network BERT [16].[3] At training time, we stack a fully-connected layer on top of the visual encoder's output features and use a triplet loss with squared distance to ensure that the embedded representation is closer to the matching caption than a randomly selected one from our dataset. This approach has the benefit that our network does not have to learn a language model, but it may not pick up on fine-grained semantic actors in the video.

Captioning Results. The results of our captioning experiments are in the second section of Table 1. Caption generation outperforms the semantic embedding approach on Memento10k. Learning to generate captions provides a boost over only the video stream for both datasets. Figure 9 contains examples of captions generated by our model.

5.3 Modeling Memorability Decay

Up until this point, we have evaluated our memorability predictions by converting them to rankings and comparing them to the ground truth. However, the Memento10k data and our parameterization of the memorability decay curve unlocks a richer representation of memorability, where $m_t^{(v)}$ is the true probability that an arbitrary person remembers video v at lag t. Thus, we also investigate techniques for predicting the ground-truth values of the memorability decay curve. Again, we consider two alternative architectures.

Mem-α Model. "Mem-α" models produce two outputs by regressing to a video's memory score and decay coefficient. To train these models, we define a loss that consists of uniformly sampling 100 values along the true and predicted

[3] Computed using [46].

Table 2. Multi-lag memorability prediction: Rank correlation (RC) and raw predictions (R^2) at 3 different lags (t, representing the number of intervening videos).

Approach	RC	R^2		
		t=40	t=80	t=160
Mem-α	**0.604**	0.146	0.227	0.121
Recur. head	0.599	**0.298**	**0.364**	**0.219**

memorability curves and calculating the Mean Absolute Error on the resulting pairs. Equation 1 can then be used to predict the raw hit rate at a different lag.

Recurrent Decay Model. This model directly outputs multiple probability values corresponding to different points on the decay curve. It works by injecting the feature vector produced by the video encoder into the hidden state of an 8-cell LSTM, where the cells represent evenly spaced lags from $t = 40$ through 180. At each time step, the LSTM modifies the encoded video representation, which is then fed into a multi-layer perceptron to generate the hit rate at that lag. The ground truth values used during training are calculated from $\alpha^{(v)}$ and $m_{80}^{(v)}$ using Eq. 1.

Decay Results. We evaluate the models in two ways. First, we calculate rank correlation with ground truth, based on the memorability scores (defined as $m_{80}^{(v)}$). We also compare their raw predictions for different values of t, for which we report R^2. The results are in Table 2.

We find that the Mem-α has better performance than the recurrent decay model in terms of rank correlation. However, the recurrent decay model outperforms the Mem-α model at predicting the raw memorability values, and exhibits good results for low and high lags as well. It makes sense that the performance of the Mem-α model falls off at lags further away from $T = 80$, since any error in the prediction of alpha (the slope of the decay curve) is amplified as we extrapolate away from the reference lag. These two models present a trade-off between simplicity and ranking accuracy (mem-α) and numerical accuracy along the entire decay curve (recurrent decay). Because of its relative simplicity and strong RC score, we use an Mem-α architecture for our final predictions.

6 Model Results

SemanticMemNet (Fig. 8) combines our findings from the three previous sections. We use a three-stream encoder that operates on three different input representations: 1) the raw frames, 2) the entire video as a 3D unit, and 3) the 3D optical flow. We jointly train the video stream to output memorability scores and captions for the video. Each of our streams predicts both the memorability and the decay rate, which allows us to predict the probability that an observer will recall the video at an arbitrary lag within the range we studied.

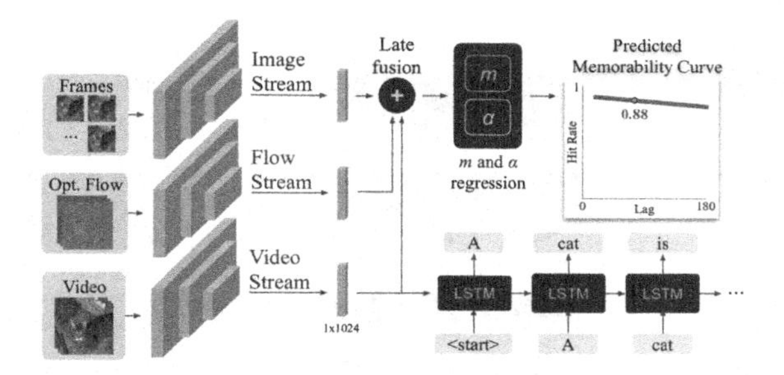

Fig. 8. The architecture of SemanticMemNet. An I3D is jointly trained to predict memorability and semantic captions for an input video. Its memorability predictions are combined with a frames-based and optical flow stream to produce m_{80} and α, the parameters of the memorability decay curve.

To evaluate the effectiveness of our model, we compare against prior work in memorability prediction. MemNet [30] is a strong image memorability baseline; we apply it to our videos by averaging its predictions over 7 frames uniformly sampled from the video. "Feature extraction and regression" is based on the approach from Shekhar et al. [39], where semantic, spatio-temporal, saliency, and color features are first extracted from the video and then a regression is learned to predict the memorability score. The final two baselines are the best-performing models from Cohendet et al. [13]. (Further details about baseline implementation can be found in the supplement.) The results of our evaluations are in Table 3. Our model achieves state-of-the-art performance on both Memento10k and VideoMem. Example predictions generated by our model are in Fig. 9.

Table 3. Comparison to state-of-the-art on Memento10k and VideoMem. Our approach, SemanticMemNet, approaches human consistency and outperforms previous approaches. *Uses ground-truth captions at test-time.

Approach	**RC - Memento10k (test set)**	**RC - VideoMem (validation set)**[4]
Human consistency	0.730	0.616
MemNet Baseline [30]	0.485	0.425
Feature extraction + regression (as in [39])*	0.615	0.427
Cohendet et al. (ResNet3D) [13]	0.574	0.508
Cohendet et al. (Semantic)[13]	0.552	0.503
SemanticMemNet	**0.663**	**0.556**

We use the VideoMem validation set as the test set has not been made public.

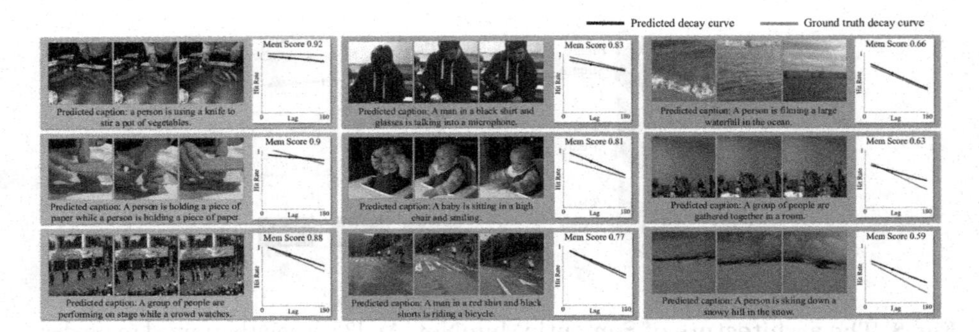

Fig. 9. Memorability and captions predictions from SemanticMemNet. For each example, we plot the predicted memorability decay curve based on SemanticMemNet's values in purple, as well as the ground truth in gray. (Color figure online)

Fig. 10. Under and overpredictions of SemanticMemNet. Our network underestimates the memorability of visually bland scenes with a single distinctive element, like a whale sighting **(a)**. It can fail on out-of-context scenes, like someone surfing on a flooded concrete river **(b)**, or surpising events, like a man getting dragged into a lake by a cow **(c)**. By contrast, it overestimates the memorability of choppy, dynamic scenes without clear semantic content **(d)** and of scenes that contain memorable elements, such as humans and faces, but that are overly cluttered **(e)**, dark **(f)**, or shaky.

7 Conclusion

Our Contributions. We introduce a novel task (memorability decay estimation) and a new dynamic dataset with memorability scores at different delays. We propose a mathematical formulation for memorability decay and a model that takes advantage of it to estimate both memorability and decay.

Limitations and Future Work. Memorability is not a solved problem. Figure 10 analyzes instances where our model fails because of competing visual attributes or complex semantics. Furthermore, there is still room for improvement in modeling memorability decay (Table 2) and extending our understanding of memorability to longer sequences. Our approach makes progress towards continuous memorability prediction for long videos (i.e. first-person live streams, YouTube videos) where memorability models should handle past events *and* their decay rates, to assess memorability of events a t different points in the past. To encourage exploration in this direction, we have released a live demo[4] of SemanticMemNet that extracts memorable segments from longer video clips.

The Utility of Memorability. Video memorability models open the door to many exciting applications in computer vision. They can be used to provide guidance to designers, educators, and models to generate clips that will be durable in memory. They can improve summarization by selecting segments likely to be retained. They can act as a measure of the utility of different segments in space-constrained systems; for instance, a camera in a self-driving car or a pair of virtual assistant glasses could discard data once it has fallen below a certain memorability threshold. Predicting visual memory will lead to systems that make intelligent decisions about what information to delete, enhance, and preserve.

Acknowledgment. We thank Zoya Bylinskii and Phillip Isola for their useful discussions and Alex Lascelles and Mathew Monfort for helping with the dataset.

References

1. Akagunduz, E., Bors, A.G., Evans, K.K.: Defining image memorability using the visual memory schema. IEEE Trans. Pattern Anal. Mach. Intell. **42**, 2165–2178 (2019)
2. Baddeley, A.: Working memory. Science **255**(5044), 556–559 (1992). https://doi.org/10.1126/science.1736359, https://science.sciencemag.org/content/255/5044/556
3. Bainbridge, W., Isola, P., Aude, O.: The intrinsic memorability of face photographs. J. Exp. Psychol. Gen. **142**, 1323–1334 (2013)
4. Barrouillet, P., Bernardin, S., Camos, V.: Time constraints and resource sharing in adults' working memory spans. J. Exp. Psychol. Gen. **133**, 83–100 (2004). https://doi.org/10.1037/0096-3445.133.1.83
5. Baveye, Y., Cohendet, R., Perreira Da Silva, M., Le Callet, P.: Deep learning for image memorability prediction: the emotional bias. In: Proceedings of the 24th ACM International Conference on Multimedia, pp. 491–495. ACM (2016)
6. Bojanowski, P., Grave, E., Joulin, A., Mikolov, T.: Enriching word vectors with subword information. arXiv preprint arXiv:1607.04606 (2016)
7. Borkin, A., et al.: Beyond memorability: visualization recognition and recall. IEEE Trans. Vis. Comput. Graph. **22**(1), 519–528 (2016)
8. Borkin, M., et al.: What makes a visualization memorable? IEEE Trans. Vis. Comput. Graph. **19**(12), 2306–2315 (2013)
9. Brady, T.F., Konkle, T., Alvarez, G.A., Oliva, A.: Visual long-term memory has a massive storage capacity for object details. Proc. Nat. Acad. Sci. **105**(38), 14325–14329 (2008)
10. Brady, T.F., Konkle, T., Gill, J., Oliva, A., Alvarez, G.A.: Visual long-term memory has the same limit on fidelity as visual working memory. Psychol. Sci. **24**(6), 981–990 (2013). pMID: 23630219. https://doi.org/10.1177/0956797612465439
11. Bylinskii, Z., Isola, P., Bainbridge, C., Torralba, A., Oliva, A.: Intrinsic and extrinsic effects on image memorability. Vis. Res. **116**, 165–178 (2015)

[4] http://demo.memento.csail.mit.edu/.

12. Carreira, J., Zisserman, A.: Quo vadis, action recognition? A new model and the kinetics dataset. In: 2017 IEEE Conference on Computer Vision and Pattern Recognition, CVPR 2017, Honolulu, HI, USA, 21–26 July 2017, pp. 4724–4733 (2017). https://doi.org/10.1109/CVPR.2017.502
13. Cohendet, R., Demarty, C., Duong, N.Q.K., Martin, E.: VideoMem: constructing, analyzing, predicting short-term and long-term video memorability. In: Proceedings of the IEEE International Conference on Computer Vision, pp. 2531–2540 (2019)
14. Cohendet, R., et al.: MediaEval 2018: Predicting Media Memorability Task. CoRR abs/1807.01052 (2018). http://arxiv.org/abs/1807.01052
15. Cohendet, R., Yadati, K., Duong, N.Q., Demarty, C.H.: Annotating, understanding, and predicting long-term video memorability. In: Proceedings of the 2018 ACM on International Conference on Multimedia Retrieval, pp. 178–186. ACM (2018)
16. Devlin, J., Chang, M., Lee, K., Toutanova, K.: BERT: pre-training of deep bidirectional transformers for language understanding. CoRR abs/1810.04805 (2018). http://arxiv.org/abs/1810.04805
17. Dubey, R., Peterson, J., Khosla, A., Yang, M.H., Ghanem, B.: What makes an object memorable? In: Proceedings of the IEEE International Conference on Computer Vision, pp. 1089–1097 (2015)
18. Engilberge, M., Chevallier, L., Pérez, P., Cord, M.: Finding beans in burgers: Deep semantic-visual embedding with localization. CoRR abs/1804.01720 (2018). http://arxiv.org/abs/1804.01720
19. Fajtl, J., Argyriou, V., Monekosso, D., Remagnino, P.: AMNet: memorability estimation with attention. In: Proceedings of the IEEE Conference on Computer Vision and Pattern Recognition, pp. 6363–6372 (2018)
20. Frome, A., et al.: DeVise: a deep visual-semantic embedding model. In: Burges, C.J.C., Bottou, L., Welling, M., Ghahramani, Z., Weinberger, K.Q. (eds.) Advances in Neural Information Processing Systems 26, pp. 2121–2129. Curran Associates, Inc. (2013). http://papers.nips.cc/paper/5204-devise-a-deep-visual-semantic-embedding-model.pdf
21. Goetschalckx, L., Andonian, A., Oliva, A., Isola, P.: GANalyze: towards visual definition of cognitive image properties. In: IEEE International Conference on Computer Vision, ICCV 2019, Seoul, Korea, pp. 5744–5753 (2019)
22. Goetschalckx, L., Moors, P., Wagemans, J.: Image memorability across longer time intervals. Memory **26**, 581–588 (2017). https://doi.org/10.1080/09658211.2017.1383435
23. Han, J., Chen, C., Shao, L., Xintao, H., Jungong, H., Tianming, L.: Learning computational models of video memorability from fMRI brain imaging. IEEE Trans. Cybern. **45**(8), 1692–1703 (2015)
24. Isola, P., Parikh, D., Torralba, A., Oliva, A.: Understanding the intrinsic memorability of images. In: Advances in Neural Information Processing Systems 24: 25th Annual Conference on Neural Information Processing Systems 2011, pp. 2429–2437 (2011)
25. Isola, P., Xiao, J., Parikh, D., Torralba, A., Oliva, A.: What makes a photograph memorable? IEEE Trans. Pattern Anal. Mach. Intell. **36**(7), 1469–1482 (2014)
26. Isola, P., Xiao, J., Torralba, A., Oliva, A.: What makes an image memorable? In: The 24th IEEE Conference on Computer Vision and Pattern Recognition, CVPR 2011, Colorado Springs, CO, USA, 20–25 June 2011, pp. 145–152 (2011). https://doi.org/10.1109/CVPR.2011.5995721

27. Jaegle, A., Mehrpour, V., Mohsenzadeh, Y., Meyer, T., Oliva, A., Rust, N.: Population response magnitude variation in inferotemporal cortex predicts image memorability. ELife **8**, e47596 (2019)
28. Karpathy, A., Li, F.: Deep visual-semantic alignments for generating image descriptions. CoRR abs/1412.2306 (2014). http://arxiv.org/abs/1412.2306
29. Khosla, A., Bainbridge, W., Torralba, A., Oliva, A.: Modifying the memorability of face photographs. In: Proceedings of the IEEE International Conference on Computer Vision, pp. 3200–3207 (2013)
30. Khosla, A., Raju, A.S., Torralba, A., Oliva, A.: Understanding and predicting image memorability at a large scale. In: Proceedings of the IEEE International Conference on Computer Vision, pp. 2390–2398 (2015)
31. Khosla, A., Xiao, J., Torralba, A., Oliva, A.: Memorability of image regions. In: Advances in Neural Information Processing Systems, pp. 305–313 (2012)
32. Kiros, R., Salakhutdinov, R., Zemel, R.S.: Unifying visual-semantic embeddings with multimodal neural language models. CoRR abs/1411.2539 (2014). http://arxiv.org/abs/1411.2539
33. Konkle, T., Brady, T., Alvarez, G., Oliva, A.: Conceptual distinctiveness supports detailed visual long-term memory for real-world objects. J. Exp. Psychol. Gen. **139**, 558–578 (2010). https://doi.org/10.1037/a0019165
34. Konkle, T., Brady, T., Alvarez, G., Oliva, A.: Scene memory is more detailed than you think: the role of categories in visual long-term memory. Psychol. Sci. **21**, 1551–6 (2010). https://doi.org/10.1177/0956797610385359
35. Koutstaal, W., Reddy, C., Jackson, E., Prince, S., Cendan, D., Schacter, D.: False recognition of abstract versus common objects in older and younger adults: testing the semantic categorization account. J. Exp. Psychol. Learn. Mem. Cogn. **29**, 499–510 (2003)
36. Mohsenzadeh, Y., Mullin, C., Oliva, A., Pantazis, D.: The perceptual neural trace of memorable unseen scenes. Sci. Rep. **8**, 6033 (2019)
37. Monfort, M., et al.: Moments in time dataset: one million videos for event understanding. IEEE Trans. Pattern Anal. Mach. Intell. **42**, 502–508 (2019)
38. Perera, S., Tal, A., Zelnik-Manor, L.: Is image memorability prediction solved? In: The IEEE Conference on Computer Vision and Pattern Recognition Workshops (2019)
39. Shekhar, S., Singal, D., Singh, H., Kedia, M., Shetty, A.: Show and recall: learning what makes videos memorable. In: Proceedings of the IEEE International Conference on Computer Vision, pp. 2730–2739 (2017)
40. Sidorov, O.: Changing the image memorability: from basic photo editing to GANs. In: The IEEE Conference on Computer Vision and Pattern Recognition Workshops (2019)
41. Squalli-Houssaini, H., Duong, N., Gwenaëlle, M., Demarty, C.H.: Deep learning for predicting image memorability. In: IEEE International Conference on Acoustics, Speech, and Signal Processing (ICASSP) (2018)
42. Venugopalan, S., Rohrbach, M., Donahue, J., Mooney, R., Darrell, T., Saenko, K.: Sequence to sequence - video to text. In: Proceedings of the IEEE International Conference on Computer Vision (ICCV) (2015)
43. Võ, M.L.H., Bylinskii, Z., Oliva, A.: Image memorability in the eye of the beholder: tracking the decay of visual scene representations. bioRxiv, p. 141044 (2017)
44. Vogt, S., Magnussen, S.: Long-term memory for 400 pictures on a common theme. Exp. Psychol. **54**, 298–303 (2007)
45. Wiseman, S., Neisser, U.: Perceptual organization as a determinant of visual recognition memory. Am. J. Psychol. **87**, 675–681 (1974)

46. Xiao, H.: Bert-as-service (2018). https://github.com/hanxiao/bert-as-service
47. Xu, K., et al.: Show, attend and tell: Neural image caption generation with visual attention. CoRR abs/1502.03044 (2015). http://arxiv.org/abs/1502.03044
48. Zarezadeh, S., Rezaeian, M., Sadeghi, M.T.: Image memorability prediction using deep features. In: 2017 Iranian Conference on Electrical Engineering (ICEE), pp. 2176–2181. IEEE (2017)

Yet Another Intermediate-Level Attack

Qizhang Li[1], Yiwen Guo[1(✉)], and Hao Chen[2]

[1] ByteDance AI Lab, Beijing, China
{liqizhang,guoyiwen.ai}@bytedance.com
[2] University of California, Davis, USA
chen@ucdavis.edu

Abstract. The transferability of adversarial examples across deep neural network (DNN) models is the crux of a spectrum of black-box attacks. In this paper, we propose a novel method to enhance the black-box transferability of baseline adversarial examples. By establishing a linear mapping of the intermediate-level discrepancies (between a set of adversarial inputs and their benign counterparts) for predicting the evoked adversarial loss, we aim to take full advantage of the optimization procedure of mulch-step baseline attacks. We conducted extensive experiments to verify the effectiveness of our method on CIFAR-100 and ImageNet. Experimental results demonstrate that it outperforms previous state-of-the-arts considerably. Our code is at https://github.com/qizhangli/ila-plus-plus.

Keywords: Adversarial examples · Transferability · Feature maps

1 Introduction

The adversarial vulnerability of deep neural networks (DNNs) has been extensively studied over the years [1,3,8,10,11,24,25,34]. It has been demonstrated that intentionally crafted perturbations, that are small enough to be imperceptible to human eyes, on a natural image can fool advanced DNNs to make arbitrary (incorrect) predictions. Along with this intriguing phenomenon, it is also pivotal that the adversarial examples crafted on one DNN model can fail another with a non-trivial success rate [8,34]. Such a property, called the transferability (or generalization ability) of adversarial examples, plays a vital role in many black-box adversarial scenarios [27,28], where the architecture and parameters of the victim model is hardly accessible.

Endeavors have been devoted to studying the transferability of adversarial examples. Very recently, intermediate-layer attacks [15,18,41] have been proposed to improve the transferability. It was empirically shown that larger mid-layer disturbance (in feature maps) leads to higher transferability in general. In this paper, we propose a new method for improving the transferability of adversarial examples generated by any baseline attack, just like in [15]. Our method

Q. Li—Work done during an internship at Bytedance AI Lab, under the guidance of Yiwen Guo who is the corresponding author.

A. Vedaldi et al. (Eds.): ECCV 2020, LNCS 12361, pp. 241–257, 2020.
https://doi.org/10.1007/978-3-030-58517-4_15

operates on the mid-layer feature maps of a source model as well. It attempts to take full advantage of the directional guides gathered at each step of the baseline attack, by maximizing the scalar projection on a spectrum of intermediate-level discrepancies. The effectiveness of the method was testified on a variety of image classification models on CIFAR-100 [20] and ImageNet [30], and we show that it outperforms previous state-of-the-arts considerably.

2 Related Work

Adversarial attacks can be categorized into white-box attacks and black-box attacks, according to how much information of a victim model is leaked to the adversary [27]. Initial attempts of performing black-box attacks rely on the transferability of adversarial examples [23,27,28]. Despite the excitement about the possibility of performing attacks under challenging circumstances, early transfer-based methods often suffer from low success rates, and thus an alternative trail of research that estimates gradient from queries also becomes prosperity [2,4,9,16,17,26,36,38]. Nevertheless, there exist applications where queries are difficult and costly to be issued to the victim models, and it is also observed that some stateful patterns can be detected in such methods [5].

Recently, a few methods have been proposed to enhance the transferability of adversarial examples, boosting the transfer-based attacks substantially. They show that maximizing disturbance in intermediate-level feature maps instead of the final cross-entropy loss delivers higher adversarial transferability. To be more specific, Zhou et al. [41] proposed to maximize the discrepancy between an adversarial example and its benign counterpart on DNN intermediate layers and simultaneously reduce spatial variations of the obtained results. Requiring a target example in addition, Inkawhich et al. [18] also advocated performing attacks on the intermediate layers. The most related work to ours comes from Huang et al. [15]. Their method works by maximizing the scalar projection of the adversarial example onto a guided direction (which can be obtained by performing one of many off-the-shelf attacks [7,8,21,24,41]) beforehand, on a specific intermediate layer. Our method is partially motivated by Huang et al. [15]. It is also proposed to enhances the adversarial transferability, yet our method takes the whole optimization procedure of the baseline attacks rather than their final results as guidelines. As will be discussed, we believe temporary results probably provide more informative and more transferable guidance than the final result of the baseline attack. The problem setting will be explained in the following subsection.

2.1 Problem Setting

In this paper, we focus on enhancing the transferability of off-the-shelf attacks, just like Huang et al. intermediate-level attack (ILA) [15]. We mostly consider multi-step attacks which are generally more powerful on the source models. Suppose that a basic iterative FGSM (I-FGSM) is performed a priori as the **baseline attack**, we have

$$\mathbf{x}_{t+1}^{\text{adv}} = \mathbf{\Pi}_{\Psi}(\mathbf{x}_t^{\text{adv}} + \alpha \cdot \text{sgn}(\nabla L(\mathbf{x}_t^{\text{adv}}, y))), \tag{1}$$

in which Ψ is a presumed valid set for the adversarial examples and $\mathbf{\Pi}_{\Psi}$ denotes a projection onto the set, given $\mathbf{x}_0^{\text{adv}} = \mathbf{x}$ and its original prediction y. The typical I-FGSM performs attacks after running Eq. (1) for p times to obtain the final adversarial example $\mathbf{x}_p^{\text{adv}}$. We aim to improve the success rate of the generated adversarial example on some victim models whose architecture and parameters are unknown to the adversary. As depicted in Fig. 1, the whole pipeline consists of two phases. The first phase is to perform the baseline attack just as normal, precursor to **the enhancement phase** where our method or ILA can be applied.

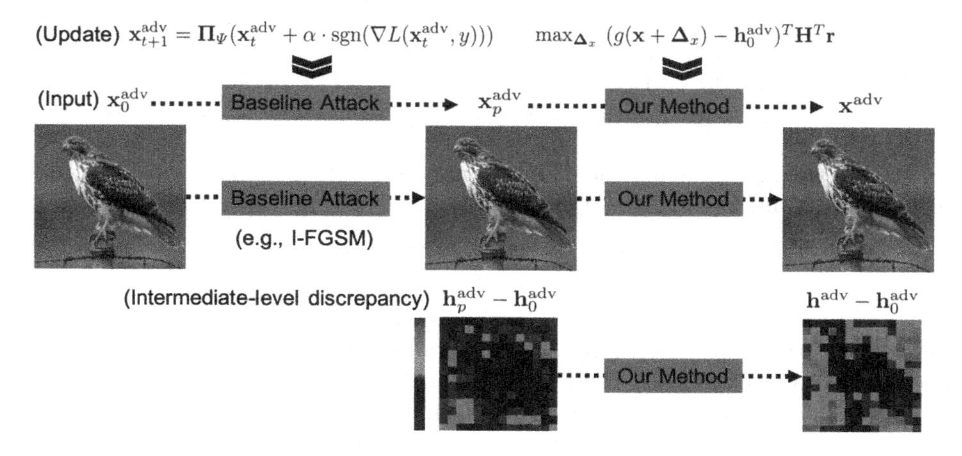

Fig. 1. Pipeline of our method for enhancing the black-box transferability of adversarial examples, which is comprised of two sequential phases, one for performing the baseline attack (e.g., I-FGSM [21], PGD [24], MI-FGSM [7], etc) and the other for enhancing the baseline result x_p^{adv}. In particular, the chartreuse-yellow background in $\mathbf{h}_p^{\text{adv}} - \mathbf{h}_0^{\text{adv}}$ on the left heatmap indicates a much lower disturbance than that in $\mathbf{h}^{\text{adv}} - \mathbf{h}_0^{\text{adv}}$ on the right. The discrepancies of feature maps are illustrated from a spatial size of 14×14.

3 Our Method

As has been mentioned, adversarial attacks are mounted by maximizing some prediction loss, e.g., the cross-entropy loss [8,24] and the hinged logit-difference loss [3]. The applied prediction loss, which is dubbed **adversarial loss** in this paper, describes how likely the input shall be mis-classified by the current model. For introducing our method, we will first propose a new objective function that utilizes the temporary results $\mathbf{x}_0^{\text{adv}} \ldots \mathbf{x}_t^{\text{adv}} \ldots \mathbf{x}_{p-1}^{\text{adv}}$ as well as the final result $\mathbf{x}_p^{\text{adv}}$ of a multi-step attack that takes $p+1$ optimization steps in total, e.g., I-FGSM whose update rule is introduced in Eq. (1).

We also advocate mounting attacks on an intermediate layer of the source model, just like prior arts [15,18,41]. Concretely, given $\mathbf{x}_t^{\text{adv}}$ as a (possibly adversarial) input, we can get the mid-layer output $\mathbf{h}_t^{\text{adv}} = g(\mathbf{x}_t^{\text{adv}}) \in \mathbb{R}^m$ and the adversarial loss $l_t := L(\mathbf{x}_t^{\text{adv}}, y)$ from the source model with $L(\cdot, \cdot)$, at a specific intermediate layer. With a multi-step baseline attack running for a sufficiently long period of time, we can collect a set of **intermediate-level discrepancies** (i.e., "perturbations" of feature maps) and adversarial loss values $\{(\mathbf{h}_t^{\text{adv}} - \mathbf{h}_0^{\text{adv}}, l_t)\}$, and further establish a direct mapping of the intermediate-level discrepancies to predicting the adversarial loss. For instance, a linear (regression) model can be obtained by simply solving a regularized problem.

$$\min_{\mathbf{w}} \sum_{t=0}^{p} (\mathbf{w}^T(\mathbf{h}_t^{\text{adv}} - \mathbf{h}_0^{\text{adv}}) - l_t)^2 + \lambda \|\mathbf{w}\|^2, \tag{2}$$

in which $\mathbf{w} \in \mathbb{R}^m$ is the parameter vector to be learned. The above optimization problems can be written in a matrix/vector form: $\min_{\mathbf{w}} \|\mathbf{r} - \mathbf{H}\mathbf{w}\|^2 + \lambda\|\mathbf{w}\|^2$, in which the t-th row of $\mathbf{H} \in \mathbb{R}^{(p+1)\times m}$ is $(\mathbf{h}_t^{\text{adv}} - \mathbf{h}_0^{\text{adv}})^T$ and the t-th entry of $\mathbf{r} \in \mathbb{R}^{p+1}$ is l_t, and the problem has a closed-form solution: $\mathbf{w}^* = (\mathbf{H}^T\mathbf{H} + \lambda\mathbf{I}_m)^{-1}\mathbf{H}^T\mathbf{r}$.

Rather than maximizing the conventional cross-entropy loss as in FGSM [8], I-FGSM [21], and PGD [24], we opt to optimizing

$$\max_{\mathbf{\Delta}_x} (g(\mathbf{x} + \mathbf{\Delta}_x) - \mathbf{h}_0^{\text{adv}})^T \mathbf{w}^*, \quad \text{s.t. } (\mathbf{x} + \mathbf{\Delta}_x) \in \Psi \tag{3}$$

to generate pixel-level perturbations with maximum *expected* adversarial loss in the sense of the established mapping from the feature space to the loss space. Both one-step (e.g., FGSM) and multi-step algorithms (e.g., I-FGSM and PGD) can be used to naturally solve the optimization problem (3). Here we mostly consider the multi-step algorithms, and as will be explained, our method actually boils down to ILA [15] in a one-step case. Note that the intermediate-level feature maps are extremely high dimensional. The matrix $(\mathbf{H}^T\mathbf{H} + \lambda\mathbf{I}_m) \in \mathbb{R}^{m\times m}$ thus becomes very high dimensional as well, and calculating its inverse is computational demanding, if not infeasible. While on the other hand, multi-step baseline attacks only update for tens or at most hundreds of iterations in general, and we have $p \ll m$. Therefore, we utilize the Woodbury identity

$$\begin{aligned} \mathbf{H}^T\mathbf{H} + \lambda\mathbf{I}_m &= \frac{1}{\lambda}I - \frac{1}{\lambda^2}\mathbf{H}^T(\frac{1}{\lambda}\mathbf{H}\mathbf{H}^T + \mathbf{I}_p)^{-1}\mathbf{H} \\ &= \frac{1}{\lambda}I - \frac{1}{\lambda}\mathbf{H}^T(\mathbf{H}\mathbf{H}^T + \lambda\mathbf{I}_p)^{-1}\mathbf{H} \end{aligned} \tag{4}$$

so as to calculate the matrix inverse of $(\mathbf{H}\mathbf{H}^T + \lambda\mathbf{I}_p)$ instead, for gaining higher computational efficiency. We can then rewrite the derived optimization problem in Eq. (3) as

$$\begin{aligned} &\max_{\mathbf{\Delta}_x} (g(\mathbf{x} + \mathbf{\Delta}_x) - \mathbf{h}_0^{\text{adv}})^T(\mathbf{I}_p - \mathbf{H}^T(\mathbf{H}\mathbf{H}^T + \lambda\mathbf{I}_p)^{-1}\mathbf{H})\mathbf{H}^T\mathbf{r}, \\ &\text{s.t. } (\mathbf{x} + \mathbf{\Delta}_x) \in \Psi. \end{aligned} \tag{5}$$

It is worth mentioning that, with a drastically large "regularizing" parameter λ, we have $\mathbf{H}^T(\mathbf{H}\mathbf{H}^T + \lambda\mathbf{I})^{-1}\mathbf{H} \approx \mathbf{0}$ and, in such a case, the optimization problem in Eq. (5) approximately boils down to: $\max_{\mathbf{\Delta}_x} (g(\mathbf{x}+\mathbf{\Delta}_x)-\mathbf{h}_0^{\text{adv}})^T\mathbf{H}^T\mathbf{r}$. If only the intermediate-level discrepancy evoked by the final result $\mathbf{x}_p^{\text{adv}}$ along with its corresponding adversarial loss is used in the optimization (or a single-step baseline attack is applied), the optimization problem is mathematically equivalent to that considered by Huang et al. [15], making their ILA a special case of our method. In fact, the formulation of our method suggests a maximized projection on a linear combination of the intermediate-level discrepancies, which are derived from the temporary results $\mathbf{x}_0^{\text{adv}} \dots \mathbf{x}_t^{\text{adv}} \dots \mathbf{x}_{p-1}^{\text{adv}}$ and the final result $\mathbf{x}_p^{\text{adv}}$ of the multi-step baseline attack. Since the temporary results possibly provide complementary guidance to the final result, our method can be more effective.

In (3) and (5), we encourage the perturbation $g(\mathbf{x}+\mathbf{\Delta}_x)-\mathbf{h}_0^{\text{adv}}$ on feature maps to align with $\mathbf{w}^*$, to gain more powerful attacks on the source model. In the meanwhile, the magnitude of the intermediate-level discrepancy $\|g(\mathbf{x}+\mathbf{\Delta}_x)-\mathbf{h}_0^{\text{adv}}\|$ is anticipated to be large to improve the transferability of the generated adversarial examples, as also advocated in ILA. Suppose that we are given two directional guides that would lead to similar adversarial loss values on the source model, yet remarkably different intermediate-level disturbance via optimization using for instance ILA. One may anticipate the one that causes larger disturbance in an intermediate layer to show better black-box transferability. Nevertheless, it is not guaranteed that the final result of the baseline attack offers an exciting prospect of achieving satisfactory intermediate-level disturbance in the followup phase. By contrast, our method endows the enhancement phase some capacities of exploring a variety of promising directions and their linear combinations that trade off the adversarial loss on the source model and the black-box transferability. Experimental results in Sect. 4.3 shows that our method indeed achieves more significant intermediate-level disturbance in practice.

3.1 Intermediate-Level Normalization

In practice, the intermediate-level discrepancies at different timestamps t and t' during a multi-step attack have very different magnitude, varying from ~ 0 to ≥ 100 for CIFAR-100. To take full advantage of the intermediate-level discrepancies in Eq. (3), we suggest performing data normalization before solving the linear regression problem. That being said, we suggest $\tilde{\mathbf{w}}^* = (\tilde{\mathbf{H}}^T\tilde{\mathbf{H}}+\lambda\mathbf{I}_m)^{-1}\tilde{\mathbf{H}}^T\mathbf{r}$, in which the t-th row of the matrix $\tilde{\mathbf{H}}$ is the normalized intermediate-level discrepancy $(\mathbf{h}_t^{\text{adv}}-\mathbf{h}_0^{\text{adv}})/\|\mathbf{h}_t^{\text{adv}}-\mathbf{h}_0^{\text{adv}}\|$ obtained at the t-th iteration of the baseline attack. We here optimize a similar problem as in Eq. (3), i.e.,

$$\max_{\mathbf{\Delta}_x} \, (g(\mathbf{x}+\mathbf{\Delta}_x)-\mathbf{h}_0^{\text{adv}})^T\tilde{\mathbf{w}}^*, \quad \text{s.t. } (\mathbf{x}+\mathbf{\Delta}_x)\in\Psi, \tag{6}$$

as both $\frac{(g(\mathbf{x}+\mathbf{\Delta}_x)-\mathbf{h}_0^{\text{adv}})^T\tilde{\mathbf{w}}^*}{\|g(\mathbf{x}+\mathbf{\Delta}_x)-\mathbf{h}_0^{\text{adv}}\|}$ and $\|g(\mathbf{x}+\mathbf{\Delta}_x)-\mathbf{h}_0^{\text{adv}}\|$ are expected to be maximized.

4 Experimental Results

In this section, we show experimental results to verify the efficacy of our method. We will first compare the usefulness of different intermediate-level discrepancies when being applied as the directional guides in our framework and ILA, and then compare plausible settings of our method on CIFAR−100. We will show that our method significantly outperforms its competitors on CIFAR−100 and ImageNet in Sect. 4.3. Our experimental setting are deferred to Sect. 4.4.

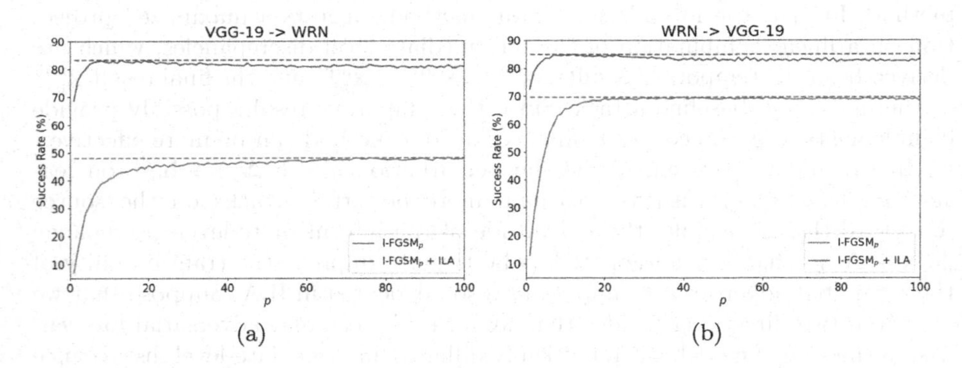

(a) (b)

Fig. 2. How the transferability of the baseline adversarial example (a) crafted on VGG-19 to attack WRN (enhanced by ILA or not) and (b) crafted on WRN to attack VGG-19 (enhanced by ILA or not) varies with p. The dashed lines indicate the performance with the optimal p values. We see that the most transferable I-FGSM$_p$+ILA examples ($\epsilon = 0.03$) are obtained around $p = 10$, and the success rate declines consistently with greater p for $p \geq 10$.

4.1 Delve into the Multi-step Baseline Attacks

We conducted a comprehensive study on the adversarial transferability of contemporary results in multi-step baseline attacks and how competent they are in assisting subsequent methods like ILA [15] and ours. We performed experiments on CIFAR-100 [20], an image classification dataset that consisting of 60000 images from 100 classes. It was officially divided into a training set of 50000 images and a test set of 10000 images. We considered two models in this study: VGG-19 [32] with batch normalization [19] and wide ResNet (WRN) [39] (specifically, WRN-28-10). Their architectures are very different, since the latter is equipped with skip connections and it is much deeper than the former. We collected pre-trained models from Github[1], and they show 28.05% and 18.14% prediction errors respectively on the official test set. We randomly chose 3000 images that could be correctly classified by the two models to initialize the baseline attack, and the success rate over 3000 crafted adversarial examples was considered.

We applied I-FGSM as the baseline attack and utilized adversarial examples crafted on one model (i.e., VGG-19/WRN) to attack the other (i.e., WRN/VGG-19). We tested the success rate when: (1) directly adopting the generated I-FGSM

[1] https://github.com/bearpaw/pytorch-classification.

adversarial examples on the victim models and (2) adopting ILA on the basis of I-FGSM. Untargeted attacks were performed under a constraint of the ℓ_∞ norm with $\epsilon = 0.03$. We denote by I-FGSM$_p$ the results of I-FGSM running for p steps, and denote by I-FGSM$_p$+ILA the ILA outcomes on the basis of I-FGSM$_p$. The success rates of using one model to attack the other are summarized in Fig. 2, with varying p. Apparently, ILA operates better with relatively earlier results from I-FGSM (i.e., I-FGSM$_p$ with a relatively smaller p). The most transferable adversarial examples can be gathered around $p = 10$ when it is equipped with ILA, and further increasing p would lead to declined success rates. While *without ILA*, running more I-FGSM iterations are more beneficial to the transferability.

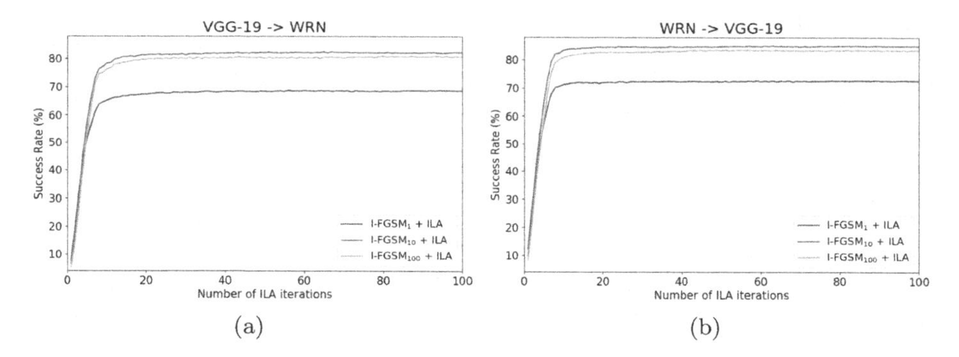

Fig. 3. How the transferability of I-FGSM adversarial examples (a) crafted on VGG-19 to attack WRN and (b) crafted on WRN to attack VGG-19, are enhanced by *ILA*. We let $\epsilon = 0.03$.

In more detail, Fig. 3 shows how much the transferability is improved along with ILA. We see that I-FGSM$_{10}$+ILA consistently outperforms I-FGSM$_{100}$+ILA. We evaluated the performance of our method based on I-FGSM$_p$ examples similarly, and the results are illustrated in Fig. 4 and 5, one with intermediate-level normalization and the other without. We set $\lambda \to \infty$, and how the performance of our method varies with λ will be discussed in Sect. 4.2. Obviously, the same tendency as demonstrated in Fig. 3 can also be observed in Fig. 4 and 5. That being said, earlier results from the multi-step baseline attack I-FGSM are more effective as guide directions for both ILA and our method. As illustrated in Fig. 2, the baseline attack converges faster than expected, making many "training samples" in $\{(\mathbf{h}_t^{\text{adv}} - \mathbf{h}_0^{\text{adv}}, l_t)\}$ highly correlated, with or without intermediate-level normalization. Using relatively early results relieve the problem and is thus beneficial to our method. The performance gain on ILA further suggests that earlier results from I-FGSM overfit less on the source model, and they are more suitable as the directional guides. In what follows, we fix $p = 10$ without any further clarification, which also reduces the computational complexity of our method for calculating $\mathbf{w}^*$ or $\tilde{\mathbf{w}}^*$ (by at least $10\times$), in comparison with $p = 100$.

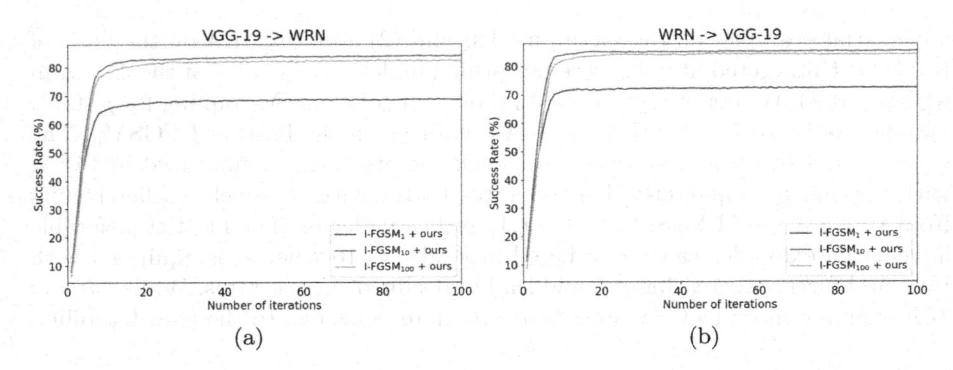

Fig. 4. How the transferability of I-FGSM examples (a) crafted on VGG-19 to attack WRN and (b) crafted on WRN to attack VGG-19, are enhanced by *our method*. The range of the y axes are kept the same as in Fig. 3 for easy comparison. We let $\epsilon = 0.03$.

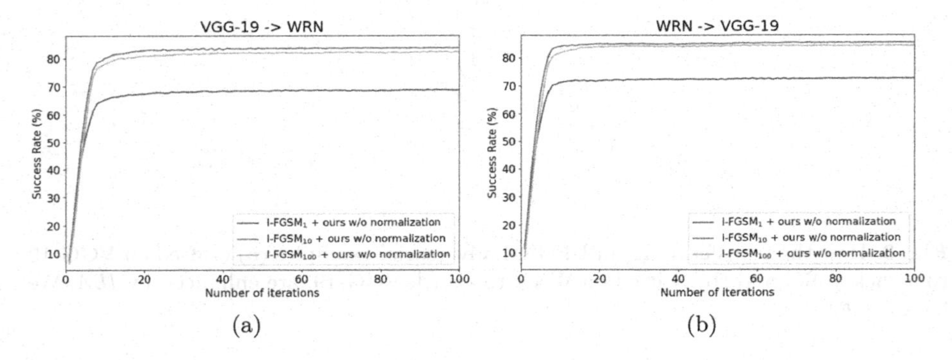

Fig. 5. How the transferability of I-FGSM examples (a) crafted on VGG-19 to attack WRN and (b) crafted on WRN to attack VGG-19, are enhanced by *our method*. Intermediate-level normalization is *NOT* performed. The range of the y axes are kept the same as with Fig. 3 and 4 for easy comparison. We let $\epsilon = 0.03$.

Notice that the ranges of the vertical axes are the same for Fig. 3, 4, and 5. It can easily be observed from the figures that our method, with either $p = 10$ or $p = 100$, achieves superior performance in comparison with ILA in the same setting. With $p = 1$, the two methods demonstrate exactly the same results, as has been discussed in Sect. 3. More comparative studies will be conducted in Sect. 4.3. Based on I-FGSM$_{10}$, our method shows a success rate of 85.53% with normalization and 85.27% without, when attacking VGG-19 using WRN. That being said, the intermediate-level normalization slightly improves our method, and we will keep it for all the experiments in the sequel of the paper.

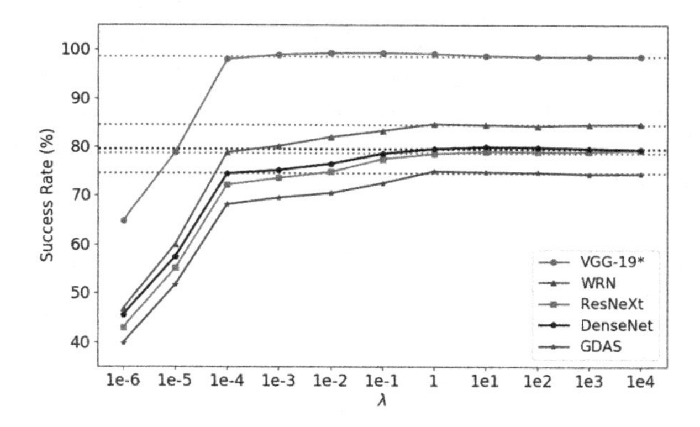

Fig. 6. How the performance of our method varies with λ. The dotted lines indicates the success rates when setting $\lambda \to \infty$. We let $\epsilon = 0.03$.

4.2 Our Method with Varying λ

One seemingly important hyper-parameter in our method is λ, which controls the smoothness of the linear regression model with $\mathbf{w}^*$ or $\tilde{\mathbf{w}}^*$. Here we report the performance with varying λ values and evaluate how different choices for λ affects the final result. The experiment was also performed on CIFAR-100. To make the study more comprehensive, we tested with a few more victim models, including a ResNeXt, a DenseNet, and a convolutional network called GDAS [6][2] whose architecture is learned via neural architecture search [40,42]. We used VGG-19 as the source model and the others (i.e., WRN, ResNeXt, DenseNet, and GDAS) as victim models. Obviously in Fig. 6, small λ values lead to unsatisfactory success rates on the source and victim models, and relatively large λs (even approaching infinity) share similar performance. Specifically, the optimal average success rate 79.55% is obtained with $\lambda = 10$, while we can still get 79.38% with $\lambda \to \infty$. Here we would like to mention that, with the Woodbury identity and a scaling factor $1/\lambda$ in Eq. (4) being eliminated when it is substituted into (5), we have $\mathbf{H}^T(\mathbf{H}\mathbf{H}^T + \lambda\mathbf{I})^{-1}\mathbf{H} \to \mathbf{0}$, *but NOT* $\mathbf{w}^* \to \mathbf{0}$ or $\tilde{\mathbf{w}}^* \to \mathbf{0}$.

As has been discussed in Sect. 3, setting an infinitely large λ leads to a simpler optimization problem and lower computational cost when computing the parameters of the linear mapping. In particular, the calculation of matrix inverse can be omitted with $\lambda \to \infty$. Although not empirically optimal, letting $\lambda \to \infty$ leads to decent performance of our method, hence in the following experiments we fixed the setting and solved the optimization problem in Eq. (5) instead of (3) to keep the introduced computational burden on the intermediate layer at a lower level. As such, the main computational cost shall come from back-propagation, which is inevitable in ILA as well. We compare the run-time of our method with that of ILA on both CPU and GPU in Table 1, where how long it takes to craft 100 adversarial examples on the VGG-19 source model by using the two methods is

[2] https://github.com/D-X-Y/AutoDL-Projects.

reported. The experiment was performed on an Intel Xeon Platinum CPU and an NVIDIA Tesla-V100 GPU. The code was implemented using PyTorch [29]. For fair comparison, the two methods were both executed for 100 iterations for generating a single adversarial example. It can be seen that both methods show similar run-time in practice.

Table 1. Run-time comparison between our method and ILA.

	CPU (s)	GPU (s)
ILA [15]	70.073	2.336
Ours	71.708	2.340

4.3 Compare with the State-of-the-Arts

In this subsection, we show more experimental results to verify the effectiveness of our method. We tried attacking 14 models on CIFAR-100 and ImageNet, 4 for the former and 10 for the latter. For CIFAR-100, the victim models include WRN, ResNeXt, DenseNet, and GDAS, as previously introduced, while for ImageNet, the tested victim models include VGG-19 [32] with batch normalization, ResNet-152 [12], Inception v3 [33], DenseNet [14], MobileNet v2 [31], SENet [13], ResNeXt [37] (precisely ResNeXt-101-32x8d), WRN [39] (precisely WRN-101-2), PNASNet [22], and MNASNet [35]. The source models for the two datasets are VGG-19 and ResNet-50, respectively. Pre-trained models on ImageNet are collected from Github[3] and the torchvision repository[4]. We mostly compare our method with ILA, since the two share the same problem setting as introduced in Sect. 2.1. We first compare their performance on the basis of I-FGSM, which is viewed as the most basic baseline attack in the paper. Table 2 and 3 summarize the results on CIFAR-100 and ImageNet, respectively. It can be seen that our method outperforms ILA remarkably on almost all test cases. As has been explained, both methods work better with relatively earlier I-FGSM results. The results in Table 2 and 3 are obtained with $p = 10$. We also tested with $p = 100$, 30, and 20, and our method is superior to ILA in all these settings. Both methods chose the same intermediate layer according to the procedure introduced in [15].

In addition to ILA, there exist several other methods in favor of black-box transferability, yet most of them are orthogonal to our method and ILA and they can be applied as baseline attacks in a similar spirit to the I-FGSM baseline. We tried adopting the two methods based on PGD [24], TAP [41], and I-FGSM with momentum (MI-FGSM) [7], which are probably more powerful than I-FGSM. Their default setting all choose the cross-entropy loss for optimization with an ℓ_∞ constraint. In fact, I-FGSM can be regarded as a special case of the PGD

[3] https://github.com/Cadene/pretrained-models.pytorch.

[4] https://github.com/pytorch/vision/tree/master/torchvision/models.

Table 2. Performance of transfer-based attacks on CIFAR-100 using I-FGSM with an ℓ_∞ constraint of the adversarial perturbation in the untargeted setting. The symbol * indicates when the source model is the target. The best average results are in bold.

Dataset	Method	ϵ	VGG-19*	WRN	ResNeXt	DenseNet	GDAS	**Average**
CIFAR-100	-	0.1	100.00%	74.90%	69.33%	71.77%	66.93%	70.73%
		0.05	100.00%	64.67%	57.63%	61.13%	56.00%	59.86%
		0.03	100.00%	48.27%	41.20%	43.83%	39.13%	43.11%
	ILA [15]	0.1	99.07%	97.53%	96.90%	97.30%	96.03%	96.94%
		0.05	99.03%	93.90%	90.73%	91.60%	88.73%	91.24%
		0.03	98.77%	82.73%	76.53%	77.87%	72.83%	77.49%
	Ours	0.1	98.83%	97.80%	97.07%	97.50%	96.51%	**97.22%**
		0.05	98.87%	94.03%	91.27%	91.73%	89.37%	**91.60%**
		0.03	98.53%	84.57%	78.70%	79.60%	74.63%	**79.38%**

Table 3. Performance of transfer-based attacks on ImageNet using I-FGSM with ℓ_∞ constraint in the untargeted setting. We use the symbol * to indicate when the source model is used as the target. The lower sub-table is the continuation of the upper sub-table. The best average results are marked in bold.

Dataset	Method	ϵ	ResNet-50*	VGG-19	ResNet-152	Inception v3	DenseNet	MobileNet v2
ImageNet	-	0.1	100.00%	67.70%	61.10%	36.36%	65.00%	65.60%
		0.05	100.00%	54.46%	44.74%	24.68%	49.90%	52.12%
		0.03	100.00%	36.80%	26.56%	13.72%	32.08%	34.56%
	ILA [15]	0.1	99.96%	97.62%	96.96%	87.94%	96.76%	96.54%
		0.05	99.96%	88.74%	86.02%	61.20%	86.42%	85.62%
		0.03	99.96%	69.96%	63.14%	34.86%	64.52%	65.68%
	Ours	0.1	99.92%	97.60%	96.98%	88.46%	97.02%	96.74%
		0.05	99.92%	89.40%	87.12%	64.96%	88.14%	86.98%
		0.03	99.90%	72.88%	67.82%	39.40%	68.38%	69.20%
Dataset	Method	ϵ	SENet	ResNeXt	WRN	PNASNet	MNASNet	**Average**
ImageNet	-	0.1	45.32%	56.36%	56.96%	35.34%	63.68%	55.34%
		0.05	29.92%	41.74%	40.82%	22.76%	49.46%	41.06%
		0.03	15.94%	23.46%	24.32%	11.90%	33.12%	25.25%
	ILA [15]	0.1	93.76%	96.00%	95.62%	91.04%	96.70%	94.89%
		0.05	74.36%	82.54%	81.80%	65.74%	84.32%	79.68%
		0.03	46.50%	59.24%	58.58%	37.22%	64.78%	56.45%
	Ours	0.1	94.00%	96.16%	95.74%	91.22%	96.86%	**95.08%**
		0.05	76.26%	84.00%	83.50%	69.24%	86.26%	**81.59%**
		0.03	50.26%	63.48%	62.72%	42.16%	67.94%	**60.42%**

attack with a random restart radius of zero. TAP and MI-FGSM are specifically designed for transfer-based black-box attacks and the generated adversarial examples generally show better transferability than the I-FGSM examples. Table 4 shows that TAP outperforms the other three (including the basic I-FGSM) multi-step baselines. MI-FGSM and PGD are the second and third

Table 4. Performance of transfer-based attacks on ImageNet. Different baseline attacks are compared in the same setting of $\epsilon = 0.03$. The best average result is marked in bold.

	MI-FGSM [7]			PGD [24]			TAP [41]		
	-	ILA [15]	Ours	-	ILA [15]	Ours	-	ILA [15]	Ours
ResNet-50*	100.00%	99.94%	99.90%	100.00%	99.94%	99.88%	100.00%	99.98%	99.96%
VGG-19	46.46%	67.18%	70.28%	40.80%	70.38%	72.22%	58.34%	78.00%	77.96%
ResNet-152	37.90%	60.76%	63.62%	31.06%	64.32%	68.02%	45.04%	67.42%	68.52%
Inception v3	21.50%	33.98%	37.26%	16.60%	37.76%	41.52%	25.50%	40.70%	42.88%
DenseNet	42.14%	63.02%	65.86%	37.78%	67.14%	69.94%	49.02%	70.56%	71.98%
MobileNet v2	45.78%	63.92%	67.04%	39.02%	66.62%	69.66%	54.98%	72.72%	73.84%
SENet	24.60%	45.26%	48.14%	18.28%	46.32%	49.60%	33.68%	55.30%	56.26%
ResNeXt	34.28%	56.08%	59.64%	27.78%	60.16%	63.72%	41.30%	64.50%	66.20%
WRN	34.20%	56.28%	59.66%	27.92%	60.08%	62.82%	45.08%	66.24%	67.06%
PNASNet	18.36%	34.82%	38.56%	13.82%	38.50%	42.68%	22.20%	42.24%	44.76%
MNASNet	43.26%	62.34%	65.36%	37.08%	65.08%	67.76%	53.64%	71.64%	72.64%
Average	34.85%	54.36%	57.54%	29.01%	57.64%	60.79%	42.88%	62.93%	**64.21%**

best, while the basic I-FGSM performs the worst in the context of adversarial transferability *without further enhancement*. Nevertheless, when further equipped with our method or ILA for transferability enhancement, PGD and I-FGSM become the second and third best, respectively, and TAP is still the winning solution showing 64.21% success rates. The MI-FGSM-related results imply that introducing momentum leads to less severe overfitting on the source model, yet such a benefit diminishes when being used as directional guides for ILA and our method. Whatever baseline attack is applied, our method always outperforms ILA in our experiment, which is conducted on ImageNet with $\epsilon = 0.03$.

It can be observed from all results thus far that our method bears a slightly decreased success rate on the source model, yet it delivers an increased capability of generating transferable adversarial examples. It is discussed in Sect. 3 that our method provides an advantage over the status quo that it is not guaranteed to achieve optimal intermediate-level disturbance. To further analyze the functionality of our method, we illustrate the cross-entropy loss and intermediate-level disturbance on the ImageNet adversarial examples crafted using our method and ILA in Fig. 7. It depicts that our method gives rise to larger intermediate-level disturbance in comparison to ILA, with a little sacrifice of the adversarial loss. As has been explained, more significant intermediate-level disturbance indicates higher transferability in general, which well-explains the superiority of our method in practice. Figure 7 also demonstrates the slightly deteriorating effect on the performance of our method in the white-box setting, which does not really matter under the considered threat model though.

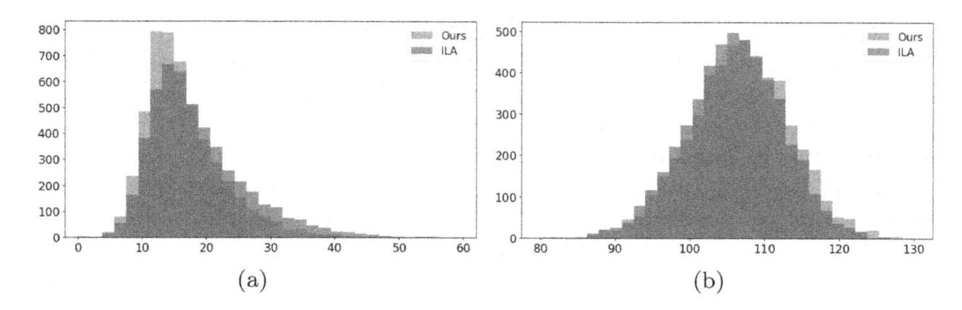

Fig. 7. Comparison of our method and ILA in the sense of (a) the cross-entropy loss and (b) intermediate-level disturbance on the ResNet−50 source model on ImageNet.

The success of our method is ascribed to effective aggregations of diverse directional guides over the whole procedure of the given baseline attack. Obviously, it seems also plausible to ensemble different baseline attacks to gain even better results in practice, since this probably gives rise to a more informative and sufficient "training set" for predicting the adversarial loss linearly. To experimentally testify the conjecture, we directly collected 20 intermediate-level discrepancies and their corresponding adversarial loss from two of the introduced baseline attacks: I-FGSM and PGD, for learning the linear regression model, i.e., 10 from each of them, and we tested our method similarly. We evaluated the performance of such an straightforward ensemble on ImageNet and it shows an average success rate of 62.82% under $\epsilon = 0.03$. Apparently, it is superior to that using the I-FGSM (60.42%) or PGD (60.79%) baseline results solely.

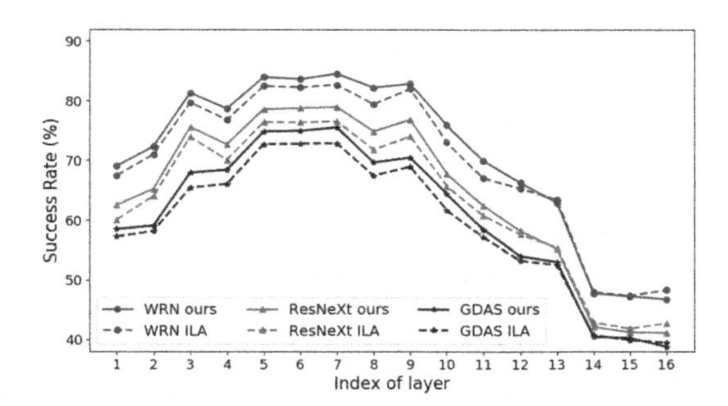

Fig. 8. Comparison of our method and ILA on CIFAR-100 with varying choices of the intermediate layer (on the VGG-19 source model) to calculate the intermediate-level discrepancies. Best viewed in color. We tested under $\epsilon = 0.03$.

Since $\lambda \to \infty$ was set, our method did not fine-tune more hyper-parameters compared to ILA. A crucial common hyper-parameter of the two methods is the location where the intermediate-level discrepancies are calculated. We compare them with various settings of the location on CIFAR-100 in Fig. 8. It can be seen that our method consistently outperforms ILA in almost all test cases from the first to the 13-th layer on the source model VGG-19. Both methods achieve their optimal results *at the same location*, therefore we can use the same procedure for selecting layers as introduced in ILA. The results on DenseNet are very similar to those on ResNeXt, and thus not plotted for clearer illustration. Notice that even the worst results of the intermediate-level methods on these victim models are better than the baseline results. The ImageNet results on three representative victim models are given in Fig. 9, and the same conclusions can be made. For CIFAR-100, the intermediate-level discrepancies were calculated right after each convolutional layer, while for ImageNet, we calculated at the end of each computational block.

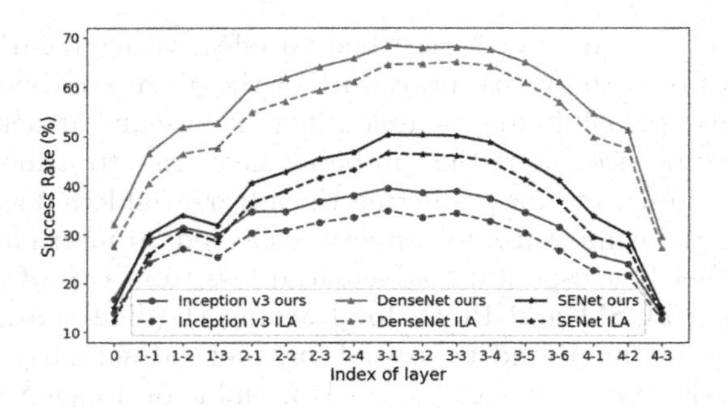

Fig. 9. Comparison of our method and ILA on ImageNet with varying choices of the intermediate layer (on the ResNet-50 source model) to calculate the intermediate-level discrepancies. The index of layer shown as "3–1" indicates the first block of the third meta-block. Best viewed in color. We tested under $\epsilon = 0.03$.

4.4 Experimental Settings and ℓ_2 Attacks

We mostly consider untargeted ℓ_∞ attacks in the black-box setting, just like prior arts of our particular interest [15,18,41]. The element-wise maximum allowed perturbation, i.e., the ℓ_∞ norm, was constrained to be lower than a positive scalar ϵ. We tested with $\epsilon = 0.1, 0.05$, and 0.03 in our experiments. In addition to the ℓ_∞ attacks, ℓ_2 attacks were also tested and the same conclusions could be made, i.e., our method still outperforms ILA and the original baseline considerably. Owing to the space limit of the paper, we only report some representative results here. On CIFAR−100, the I-FGSM baseline achieves an average success rate of 47.23%, based on which ILA and our method achieve 73.73% and 75.78%, respectively. To be more specific, on the victim models including VGG−19*, our method achieves 81.27% (for WRN), 75.4% (for ResNeXt), 75.60% (for DenseNet), 70.83% (for GDAS), and 97.73% (for VGG-19*) success rates. On ImageNet, our method is also remarkably superior to the two competitors in the sense of the average ℓ_2 success rate (ours: 76.73%, ILA: 74.68%, and the baseline: 54.79%). The same victim models as in Table 3 were used. On CIFAR-100 and ImageNet, the ℓ_2 norm of the perturbations was constrained to be lower than 1.0 and 10, respectively.

In ℓ_∞ cases, the step-size for I-FGSM, PGD, TAP, and MI-FGSM were uniformly set as 1/255, on both CIFAR-100 and ImageNet, while in ℓ_2 cases, we used 0.1 and 1.0 on the two datasets respectively. Other hyper-parameters were kept the same for all methods under both the ℓ_∞ and ℓ_2 constraints. We randomly sampled 3000 and 5000 test images that are correctly classified by the victim models from the two datasets respectively to initialize the baseline attacks and generate 3000 and 5000 adversarial examples using each method, as suggested in many previous works in the literature. For CIFAR-100, they were sampled from the official test set consisting of 10000 images, while for ImageNet, they were sampled from the validation set. We run our method and ILA for 100 iterations on the two datasets, such that they *both reached performance plateaux*. Input images to all DNNs were re-scaled to [0, 1] and the default pre-processing pipeline was adopted when feeding images to the DNNs. For ILA and TAP, followed the open-source implementation from Huang et al. [15]. An optional setting for implementing transfer-based attacks is to save the adversarial examples as bitmap images (or not) before feeding them to the victim models. The adversarial examples will be in an 8-bit image format for the former and a 32-bit floating-point format for the latter. We consider the former to be more realistic in practice.

Our learning objective does not employ an *explicit* term for encouraging large norms of the intermediate-level discrepancies. It is possible to further incorporate one such term in (3) and (5). However, an additional hyper-parameter will be introduced inevitably, as discussed by ILA regarding the flexible loss [15]. We shall consider such a formulation in future work. Our code is at https://github.com/qizhangli/ila-plus-plus.

5 Conclusions

In this paper, we have proposed a novel method for improving the transferability of adversarial examples. It operates on baseline attack(s) whose optimization procedures can be analyzed to extract a set of directional guides. By establishing a linear mapping to estimating the adversarial loss using intermediate-layer feature maps, we have developed an adversarial objective function that could take full advantage of the baseline attack(s). The effectiveness of our method has been shown via comprehensive experimental studies on CIFAR-100 and ImageNet.

Acknowledgment. This material is based upon work supported by the National Science Foundation under Grant No. 1801751.

This research was partially sponsored by the Combat Capabilities Development Command Army Research Laboratory and was accomplished under Cooperative Agreement Number W911NF-13-2-0045 (ARL Cyber Security CRA). The views and conclusions contained in this document are those of the authors and should not be interpreted as representing the official policies, either expressed or implied, of the Combat Capabilities Development Command Army Research Laboratory or the U.S. Government.

The U.S. Government is authorized to reproduce and distribute reprints for Government purposes not withstanding any copyright notation here on.

References

1. Athalye, A., Carlini, N., Wagner, D.: Obfuscated gradients give a false sense of security: circumventing defenses to adversarial examples. In: ICML (2018)
2. Brendel, W., Rauber, J., Bethge, M.: Decision-based adversarial attacks: reliable attacks against black-box machine learning models. In: ICLR (2018)
3. Carlini, N., Wagner, D.: Towards evaluating the robustness of neural networks. In: IEEE Symposium on Security and Privacy (SP) (2017)
4. Chen, P.Y., Zhang, H., Sharma, Y., Yi, J., Hsieh, C.J.: Zoo: zeroth order optimization based black-box attacks to deep neural networks without training substitute models. In: Proceedings of the 10th ACM Workshop on Artificial Intelligence and Security, pp. 15–26. ACM (2017)
5. Chen, S., Carlini, N., Wagner, D.: Stateful detection of black-box adversarial attacks. arXiv preprint arXiv:1907.05587 (2019)
6. Dong, X., Yang, Y.: Searching for a robust neural architecture in four gpu hours. In: CVPR (2019)
7. Dong, Y., et al.: Boosting adversarial attacks with momentum. In: CVPR (2018)
8. Goodfellow, I.J., Shlens, J., Szegedy, C.: Explaining and harnessing adversarial examples. In: ICLR (2015)
9. Guo, C., Gardner, J.R., You, Y., Wilson, A.G., Weinberger, K.Q.: Simple black-box adversarial attacks. In: ICML (2019)
10. Guo, Y., Chen, L., Chen, Y., Zhang, C.: On connections between regularizations for improving dnn robustness. IEEE Trans. Pattern Anal. Mach. Intell. (2020)
11. Guo, Y., Zhang, C., Zhang, C., Chen, Y.: Sparse dnns with improved adversarial robustness. In: NeurIPS, pp. 242–251 (2018)
12. He, K., Zhang, X., Ren, S., Sun, J.: Deep residual learning for image recognition. In: CVPR (2016)
13. Hu, J., Shen, L., Sun, G.: Squeeze-and-excitation networks. In: CVPR (2018)
14. Huang, G., Liu, Z., Van Der Maaten, L., Weinberger, K.Q.: Densely connected convolutional networks. In: CVPR (2017)
15. Huang, Q., Katsman, I., He, H., Gu, Z., Belongie, S., Lim, S.N.: Enhancing adversarial example transferability with an intermediate level attack. In: ICCV (2019)
16. Ilyas, A., Engstrom, L., Athalye, A., Lin, J.: Black-box adversarial attacks with limited queries and information. In: ICML (2018)
17. Ilyas, A., Engstrom, L., Madry, A.: Prior convictions: black-box adversarial attacks with bandits and priors. In: ICLR (2019)
18. Inkawhich, N., Wen, W., Li, H.H., Chen, Y.: Feature space perturbations yield more transferable adversarial examples. In: CVPR (2019)
19. Ioffe, S., Szegedy, C.: Batch normalization: accelerating deep network training by reducing internal covariate shift. In: ICML (2015)
20. Krizhevsky, A., Hinton, G.: Learning Multiple Layers of Features From Tiny Images. Technical report, Citeseer (2009)
21. Kurakin, A., Goodfellow, I., Bengio, S.: Adversarial machine learning at scale. In: ICLR (2017)
22. Liu, C., et al.: Progressive neural architecture search. In: ECCV (2018)
23. Liu, Y., Chen, X., Liu, C., Song, D.: Delving into transferable adversarial examples and black-box attacks. In: ICLR (2017)

24. Madry, A., Makelov, A., Schmidt, L., Tsipras, D., Vladu, A.: Towards deep learning models resistant to adversarial attacks. In: ICLR (2018)
25. Moosavi-Dezfooli, S.M., Fawzi, A., Frossard, P.: DeepFool: a simple and accurate method to fool deep neural networks. In: CVPR (2016)
26. Bhagoji, A.N., He, W., Li, B., Song, D.: Practical black-box attacks on deep neural networks using efficient query mechanisms. In: Ferrari, V., Hebert, M., Sminchisescu, C., Weiss, Y. (eds.) ECCV 2018. LNCS, vol. 11216, pp. 158–174. Springer, Cham (2018). https://doi.org/10.1007/978-3-030-01258-8_10
27. Papernot, N., McDaniel, P., Goodfellow, I.: Transferability in machine learning: from phenomena to black-box attacks using adversarial samples. arXiv preprint arXiv:1605.07277 (2016)
28. Papernot, N., McDaniel, P., Goodfellow, I., Jha, S., Celik, Z.B., Swami, A.: Practical black-box attacks against machine learning. In: Asia Conference on Computer and Communications Security (2017)
29. Paszke, A., et al.: Pytorch: an imperative style, high-performance deep learning library. In: NeurIPS (2019)
30. Russakovsky, O., et al.: Imagenet large scale visual recognition challenge. Int. J. Comput. Vis. **115**(3), 211–252 (2015). https://doi.org/10.1007/s11263-015-0816-y
31. Sandler, M., Howard, A., Zhu, M., Zhmoginov, A., Chen, L.C.: Mobilenetv 2: inverted residuals and linear bottlenecks. In: CVPR (2018)
32. Simonyan, K., Zisserman, A.: Very deep convolutional networks for large-scale image recognition. In: ICLR (2015)
33. Szegedy, C., Vanhoucke, V., Ioffe, S., Shlens, J., Wojna, Z.: Rethinking the inception architecture for computer vision. In: CVPR (2016)
34. Szegedy, C., et al.: Intriguing properties of neural networks. In: ICLR (2014)
35. Tan, M., et al.: Mnasnet: platform-aware neural architecture search for mobile. In: CVPR (2019)
36. Tu, C.C., et. al.: Autozoom: autoencoder-based zeroth order optimization method for attacking black-box neural networks. In: AAAI (2019)
37. Xie, S., Girshick, R., Dollár, P., Tu, Z., He, K.: Aggregated residual transformations for deep neural networks. In: CVPR (2017)
38. Yan, Z., Guo, Y., Zhang, C.: Subspace attack: exploiting promising subspaces for query-efficient black-box attacks. In: NeurIPS (2019)
39. Zagoruyko, S., Komodakis, N.: Wide residual networks. In: BMVC (2016)
40. Zela, A., Siems, J., Hutter, F.: Nas-bench-1shot1: benchmarking and dissecting one-shot neural architecture search. In: ICLR (2020)
41. Zhou, W., et al.: Transferable adversarial perturbations. In: ECCV (2018)
42. Zoph, B., Le, Q.V.: Neural architecture search with reinforcement learning. arXiv preprint arXiv:1611.01578 (2016)

Topology-Change-Aware Volumetric Fusion for Dynamic Scene Reconstruction

Chao Li and Xiaohu Guo(✉)

Department of Computer Science, The University of Texas at Dallas, Richardson, USA
{Chao.Li2,xguo}@utdallas.edu

Abstract. Topology change is a challenging problem for 4D reconstruction of dynamic scenes. In the classic volumetric fusion-based framework, a mesh is usually extracted from the TSDF volume as the canonical surface representation to help estimating deformation field. However, the surface and Embedded Deformation Graph (EDG) representations bring conflicts under topology changes since the surface mesh has fixed-connectivity but the deformation field can be discontinuous. In this paper, the classic framework is re-designed to enable 4D reconstruction of dynamic scene under topology changes, by introducing a novel structure of Non-manifold Volumetric Grid to the re-design of both TSDF and EDG, which allows connectivity updates by cell splitting and replication. Experiments show convincing reconstruction results for dynamic scenes of topology changes, as compared to the state-of-the-art methods.

Keywords: Reconstruction · Topology change · Fusion · Dynamic scene

1 Introduction

As the development of Virtual Reality, Augmented Reality and 5G technologies, the demand on 4D reconstruction (space + time) techniques has been raised. Especially with the latest advancements of consumer-level RGB-D cameras, the interest has been growing in developing such 4D reconstruction techniques to capture various dynamic scenes. Volumetric fusion-based techniques [9,28,43] allow the 4D reconstruction of dynamic scenes with a single RGB-D camera, by incrementally fusing the captured depth into a volume encoded by Truncated Signed Distance Fields (TSDF) [7]. The philosophy of such volumetric fusion-based reconstruction is to decompose the 4D information into representations of 3D-space and 1D-time individually. The 3D-space information includes two parts: the geometry of the scene is represented in a canonical volume [28] (or key volumes [10]) encoded by TSDF; the deformation field of the scene is represented

Electronic supplementary material The online version of this chapter (https://doi.org/10.1007/978-3-030-58517-4_16) contains supplementary material, which is available to authorized users.

A. Vedaldi et al. (Eds.): ECCV 2020, LNCS 12361, pp. 258–274, 2020.
https://doi.org/10.1007/978-3-030-58517-4_16

by the transformations on an Embedded Deformation Graph (EDG) [36]. Along the 1D-time, the deformation field varies and the geometry becomes more complete by fusing more coming frames. In order to estimate the deformation field, an intermediate geometry representation, usually a surface mesh, is extracted to solve the model-to-frame registration. However, in the current fusion framework, this intermediate geometry representation and the EDG built on top of it cannot handle topology change cases when the deformation is discontinuous over 3D space, because they have fixed connectivity between vertices or nodes.

Defining a more flexible data structure to handle topology changes is non-trivial. In this paper, our key contribution is the fundamental re-design of the volumetric fusion framework, by revisiting the data structures of geometry and deformation field. We introduce *Non-manifold Volumetric Grids* into the TSDF representation, by allowing the volumetric grids to replicate themselves and break connections, and design the EDG in a similar non-manifold structure. Such a novel design overcomes the issue brought by fixed connectivity of EDG and intermediate mesh (extracted from TSDF grids) and allows their flexible connectivity update throughout the scanning process.

Our second contribution is the proposal of a novel topology-change-aware non-rigid registration method inspired by line process [3]. This approach efficiently and effectively solves the discontinuity issue due to topology changes by adapting weights to loosen the regularization constraints on edges where topology changes happen. Based on such a registration framework, we also propose a topology change event detection approach to guide the connectivity updates of EDG and volumetric grids by fully utilizing line process weights.

2 Related Work

The most popular methods to reconstruct 4D dynamic scene are using a pre-defined template, such as skeleton [42], human body model [43] or pre-scanned geometry [46] as prior knowledge, and reconstruct human body parts [24,31,37,43]. To eliminate the dependency on such priors, some template-less fusion-based methods were proposed to utilize more advanced structure to merge and store geometry information across motion sequences [6,9,10,15,18,19,22,28].

However, there are still two major problems related to 4D dynamic scene reconstruction. Firstly, all of exiting methods are still vulnerable to fast and occluded motion of dynamic scene. Fast motions introduce motion-blur and can severely degrade the tracking accuracy of correspondences between frames which affects geometry fusion. The problem is partially solved in [9] by Spectral Embedding, and by [20] with their high frame rate RGB-D sensors. The second issue is notorious topology change handling problem, which is our focus here. Only a few methods are proposed to handle topology changes. Key volumes were proposed in [10] and [9] to set a new key frame and reinitialize model tracking when a topology change happens. [33] and [34] propose new methods to tackle this issue by aligning TSDF volumes between two frames. However, the resolution of TSDF volume in these methods are lower than that of other mesh-based fusion methods because their fully volumetric registration has scalability limitations.

Furthermore, they cannot provide the segmentation information that we offer in our method: separated objects will be reconstructed as independent meshes.

Currently most of the template-less dynamic 4D reconstruction methods [9,10,18,19,22,28,30,42] use TSDF as the underlying surface representation. However, in dynamic scene reconstruction, the deformation field could be discontinuous, which cannot be represented with a fixed connectivity intermediate mesh and EDG. The approach we propose here will allow the dynamic updates to the TSDF volumetric grids conforming to the discontinuity of the deformation fields. Compared to level set variants [11,29], which support surface splitting and merging in physical simulation and usually have a noise-free complete mesh, our method aims to incrementally reconstruct geometry from noisy partial scans.

Zampogiannis *et al.* [45] proposed a topology-change-aware non-rigid point cloud registration approach by detecting topology change regions based on stretch and compression measurement in both forward and backward motion. However, how to recover the geometry of dynamic scenes under such topology changes is not explored. Inspired by methods in computer animation – virtual node [26] and non-manifold level set [25], we re-design the non-manifold level set and adapt it to the fusion-based 4D reconstruction framework. Tsoli and Argyros [38] presented a method to track topologically changed deformable surfaces with a pre-defined template given input RGB-D images. Compared to their work, our method is template-less, gradually reconstructing the geometry and updating the connectivity of EDG and TSDF volume grids. Bojsen-Hansen *et al.* [4] explored in another direction to solve surface tracking with evolving topology. But our method can detect topology changes in live frames, recover the changed geometry in the canonical space and playback the entire motion sequence on top of the geometry with new topology.

There is also a set of works related to dynamic scene reconstruction but not focused on voxel-based techniques: 1) Other template/mesh-based deformation approaches [5,21,40]; 2) Methods for learning-based schemes that may handle larger changes [1,12–14,21,39]; 3) Methods on point correspondence based interpolation that do not require the prior of a mesh representation and are more flexible with respect to topological changes [2,23,41,44]; 4) Finally, some point distribution based approaches that do not require correspondence search and provide even more flexibility [8,17,35].

3 System Overview

The system takes RGB-D images $\{\mathcal{C}_n, \mathcal{D}_n\}$ of the nth frame, and outputs a reconstructed surface mesh $\overline{\mathcal{M}_n}$ in the canonical space and a per-frame deformation field that transforms that surface into the live frame. The topology changes will be reflected by updating the connectivity of EDG and TSDF volume in the canonical space. In this way, although the topology of $\{\overline{\mathcal{M}_1}, \cdots, \overline{\mathcal{M}_n}\}$ might evolve over time, we can still replicate the topology of the ending frame $\overline{\mathcal{M}_n}$ to all of the earlier frames. Thus we can enable the playback of motions on top of reconstructed meshes with new topology. Figure 1 shows a flowchart of our 4D reconstruction system, composed of two modules: Topology-Change-Aware Registration, and Topology-Change-Aware Geometric Fusion.

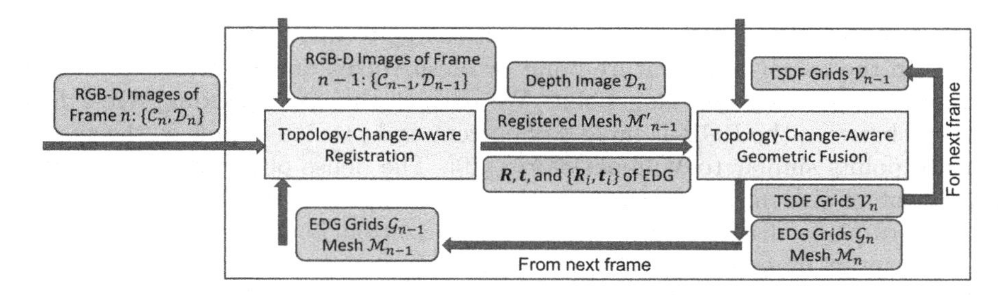

Fig. 1. Computational flowchart of our proposed 4D reconstruction system.

4 Technical Details

Now we describe our reconstruction system in detail. In the first module, the line process based deformation estimation and non-manifold grid based re-design of EDG are the enabler of topology-change-aware registration.

4.1 Topology-Change-Aware Registration

We represent the deformation field through an EDG, of which each node $g^{\mathcal{G}}$ provides a 3DOF displacement $\mathbf{t}_i$ for deformation. For each point (surface vertex or voxel) $\mathbf{x}_c$ in canonical space, $\mathbf{T}(\mathbf{x}_c) = \mathbf{R}\sum_i \alpha_i(\mathbf{x}_c + \mathbf{t}_i) + \mathbf{t}$ transforms this point from canonical space into the live frame via trilinear interpolation, where i is the node index of $\mathbf{x}_c$-belonged EDG cell and α_i is the interpolation weight. When a new n^{th} frame comes in, we update global rotation $\mathbf{R}$, global translation $\mathbf{t}$, and local displacement $\mathbf{t}_i$ on nodes, based on the reconstructed mesh $\mathcal{M}_{n-1}$ from previous frame.

Estimating the Deformation Field. The registration can be decomposed into two steps: rigid alignment, and non-rigid alignment. The rigid alignment is to estimate the global rotation $\mathbf{R}$ and global translation $\mathbf{t}$ by using dense projective ICP [32]. During the non-rigid alignment, we estimate current local deformation field $\{\mathbf{R}_i, \mathbf{t}_i\}$ given the previous reconstructed mesh $\mathcal{M}_{n-1}$ and the RGB-D images $\{\mathcal{C}_n, \mathcal{D}_n\}$ of this frame by minimizing an energy function.

Similar to VolumeDeform [19], we design the energy function as a combination of the following three terms:

$$E_{total}(\mathbf{X}) = \omega_s E_{spr}(\mathbf{X}) + \omega_d E_{dense}(\mathbf{X}) + \omega_r E_{reg}(\mathbf{X}), \quad (1)$$

$$E_{spr}(\mathbf{X}) = \sum_{\mathbf{f}\in\mathcal{F}} \|(\mathbf{T}(\mathbf{f}) - \mathbf{y})\|^2, \quad (2)$$

$$E_{dense}(\mathbf{X}) = \sum_{\mathbf{x}\in\mathcal{M}_{n-1}} [\mathbf{n}_y^\top(\mathbf{T}(\mathbf{x}) - \mathbf{y})]^2. \quad (3)$$

Here E_{spr} is a sparse feature based alignment term. E_{dense} is a dense depth based measurement and E_{reg} is a regularization term. The weights ω_s, ω_d and

ω_r control the relative influence of different energy terms. $\mathbf{y}$ is the corresponding point (in the target) of a feature point or mesh vertex and $\mathbf{n}_y$ is the estimated normal of each corresponding point. We extract the corresponding SIFT features $\mathcal{F}$ between the RGB-D images of current and previous frame as the sparse feature points similar to VolumeDeform [19]. The dense objective enforces the alignment of the surface mesh $\mathcal{M}_{n-1}$ with the captured depth data based on a point-to-plane distance metric. The regularization is an as-rigid-as-possible (ARAP) prior by enforcing the one-ring neighborhood of a node to have similar transformations. However, such ARAP prior is not able to detect potential topology changes, i.e., the breaking of connection between neighboring nodes. In this paper, we propose to use a line process [3] to account for the discontinuity caused by topology changes. The regularization term is:

$$E_{reg} = \sum_i \sum_{j \in \mathcal{N}(i)} [l_{ij} \|\mathbf{R}_i(\mathbf{g}_i - \mathbf{g}_j) - (\tilde{\mathbf{g}_i} - \tilde{\mathbf{g}_j})\|^2 + \Psi(l_{ij})], \tag{4}$$

$$\tilde{\mathbf{g}_i} = \mathbf{g}_i + \mathbf{t}_i, \Psi(l_{ij}) = \mu(\sqrt{l_{ij}} - 1)^2, \tag{5}$$

where $\mathbf{g}_i$ and $\mathbf{g}_j$ are the positions of the two nodes in EDG $\mathcal{G}_{n-1}$ from previous frame. The first term in E_{reg} is exactly the ARAP prior measuring the similarity of transformations between neighboring nodes, except for the multiplication of a line process parameter l_{ij} indicating the presence ($l_{ij} \to 0$) or absence ($l_{ij} \to 1$) of a discontinuity between nodes i and j. The function $\Psi(l_{ij})$ is the "penalty" of introducing a discontinuity between the two nodes. μ is a weight controlling the balance of these two terms. The original strategy of how to set μ is discussed in paper [3]. We will introduce our settings of μ in detail in next part. All unknowns to be solved in the entire energy function are:

$$\mathbf{X} = (\underbrace{\cdots, \mathbf{R}_i^\top, \cdots}_{\text{rotation matrices}} | \underbrace{\cdots, \mathbf{t}_i^\top, \cdots}_{\text{displacements}} | \underbrace{\cdots, l_{ij}, \cdots}_{\text{line process}})^\top. \tag{6}$$

These three groups of unknowns are solved with alternating optimization (see details in *Supplementary Document*). After the optimization, the new warped surface mesh $\mathcal{M}_n$ can be used as the initial surface to estimate the deformation field for the next frame.

Topology Change Event Detection. When detecting topology change events, we run an extra backward registration from the registered mesh to the source RGB-D image based on previous registration result, and find all cutting edges of EDG cells according to line process weights from both forward and backward registration. There are several reasons to add this backward registration. (1) Re-using the EDG instead of resampling a new EDG from the registered mesh will preserve the correct graph node connectivity (edges along the separating boundaries having longer length due to stretching) when there is an open-to-close topology change event while the resampled EDG would not have that correct one. (2) It will help reducing the number of "false positive" cases when only considering the forward registration. "False positive" cases are usually caused by finding bad correspondences with outliers. This can be solved by using bidirectional correspondence search and adding backward registration follows the same way. (3) This backward registration is still computationally light-weight without the need to re-generate a new EDG and all computed line process weights can be directly used to guide the topology change event detection.

The formula to compute l_{ij} is:

$$l_{ij} = \left(\frac{\mu}{\mu + \|\boldsymbol{R}_i(\boldsymbol{g}_i - \boldsymbol{g}_j) - [\boldsymbol{g}_i + \boldsymbol{t}_i - (\boldsymbol{g}_j + \boldsymbol{t}_j)]\|^2}\right)^2. \quad (7)$$

We want to set the threshold of l_{ij} to distinguish between highly stretched (or compressed) edges and normal edges. In our assumption, if the ratio of an edge stretched (or compressed) to the normal length is 20%, there exists a potential topology change event. Then a good approximation of μ is 20% $\times$ *cell length*. In practice, if $l_{ij} < 0.5$ in the forward registration step and $l_{ij} < 0.8$ in the backward registration, it will be classified as a cutting edge, and there is a new topology change event detected.

In order to demonstrate that our topology change detection really works well, we run it on some public datasets used in [45], as shown in Fig. 2. Our approach can also successfully detect all topology change events and update the connectivity of EDG and TSDF grids to reflect such topology changes accordingly in reconstructed geometry. It is worth noting that our method can handle a more complex case like seq "alex (close to open)" (from [33]) – hand moving from contacting with body to no contact, which is not demonstrated in [45]. Besides that, Zampogiannis et al. [45] did not address how to reconstruct the geometry of dynamic scenes under such topology changes, as will be introduced below.

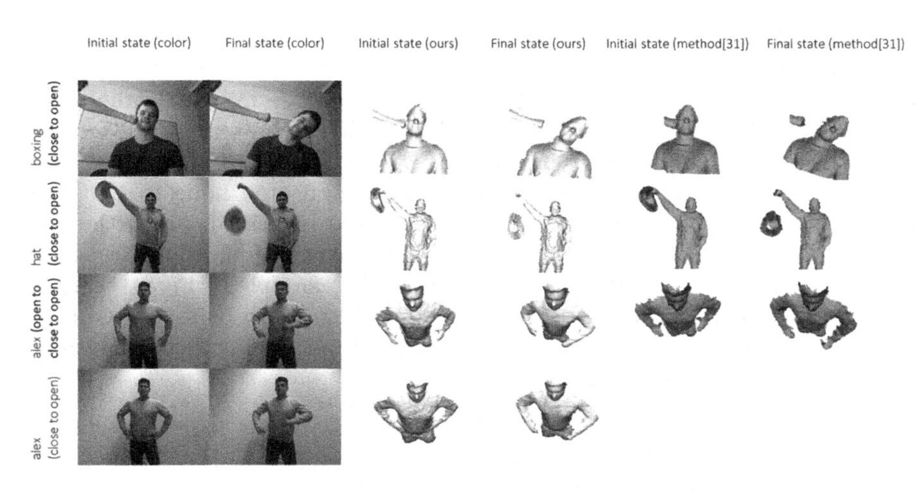

Fig. 2. Effectiveness of our topology change detection on real data. Row 1 to 3 are cases shown in Zampogiannis et al.'s paper [45]. Row 4 is another challenging case from KillingFusion [33].

Updating the Connectivity of EDG. The most fundamental innovation in this work is to allow the cells of volumetric structure to duplicate themselves, and to allow nodes (or grid points) to have non-manifold connectivity. In EDG $\mathcal{G}$, each cell $c^{\mathcal{G}}$ has exactly 8 nodes $\{g^{\mathcal{G}}\}$ located at its corners. Each node $g^{\mathcal{G}}$ can be affiliated with up to 8 cells $\{c^{\mathcal{G}}\}$ in the manifold case. At the beginning of the 4D reconstruction, we assume all connectivity between nodes are manifold, i.e., all nodes are affiliated with 8 cells except for those on the boundary of volume. Figure 3 illustrates the algorithm of our non-manifold EDG connectivity update.

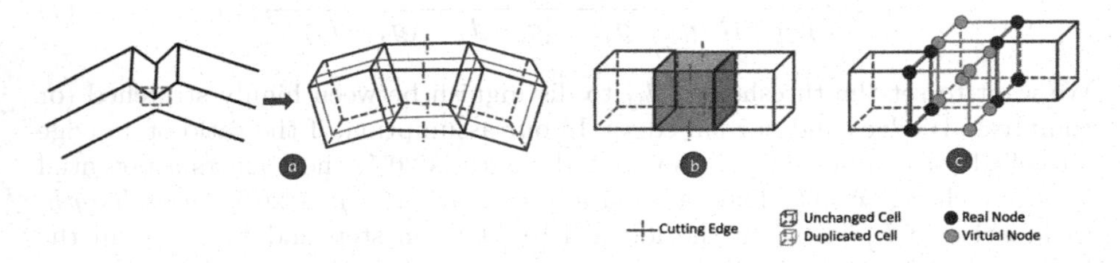

Fig. 3. (a) Cutting edges (in the live frame). (b) "To-be-duplicated" cells found based on edge cutting (in the canonical space). (c) Final non-manifold cells (orange cells are illustrated with a small displacement to distinguish two duplicated cell which are actually at the same location). (Color figure online)

Input: (a) A set of cutting edges detected by the method mentioned above; and (b) a set of candidate cells to be duplicated based on cutting edge detection.

Step 1 [Cell separation]: We separate each candidate cell $c^{\mathcal{G}}$ by removing all cutting edges and computing its connected components (CCs).

Step 2 [Cell duplication based on CCs]: The candidate cells are duplicated depending on its number of CCs. In each duplicated cell $c^{(d)}$ we categorize its nodes into two types: (1) Real Nodes $\{g^{(r)}\}$ being those from the original cell before duplication, and (2) Virtual Nodes $\{g^{(v)}\}$ being those added to make up the duplicated cells. For each virtual node $g^{(v)}$, it will only be affiliated with its duplicated cell. The transformation of each duplicated node in EDG also needs to be determined. For real nodes, they could inherit all properties from the original nodes. For virtual nodes, their displacement could be extrapolated from real nodes belonging to the same cell. In the example of Fig. 3, there are 4 cutting edges on the orange cell $c^{\mathcal{G}}$ causing its 8 nodes to be separated into 2 CCs, thus the original cell $c^{\mathcal{G}}$ is replaced with 2 duplicated cells $\{c^{(d)}\}$ residing at the same location of canonical space.

Step 3 [Restoring connectivity]: For any pair of geometrically adjacent duplicated cells $c^{\mathcal{G}}$ (in the canonical space), given two nodes from them respectively, merge these two nodes if: (1) they are both real nodes and copied from the same original node, or (2) they are both virtual nodes, copied from the same original node and connected with the same real nodes. In the example of Fig. 3(c) all four nodes on the left face of the front orange cell are merged with four nodes of the left cell by the node-merging rules.

The result is shown in Fig. 3(c). After restoring the connectivity, the final EDG has been fully assembled, respecting the topology change of the target RGB-D image. After a few edge cutting and cell duplication operations, the connectivity of nodes will become non-manifolds.

4.2 Topology-Change-Aware Geometric Fusion

Now we describe how to update and fuse the TSDF volume based on the deformation field estimated from the previous step and the depth image $\mathcal{D}_n$ in the nth frame. In order to accelerate the registration running speed and improve the reconstruction quality of geometry, a strategy of multi-level grids is employed in this paper. The resolution of EDG is typically lower than that of TSDF volume, with a ratio of $1:(2k+1)$ in each dimension ($k \in \{1,2,3\}$ in our experiments). Thus, care needs to be taken when updating the connectivity of TSDF volume if the resolution of TSDF volume grid is different from that of EDG.

Updating TSDF Volume. Once the deformation field is estimated, the connectivity of EDG should be propagated to TSDF volume and the depth image should be fused as well. Figure 4 shows key steps on how to propagate the connectivity to TSDF volume.

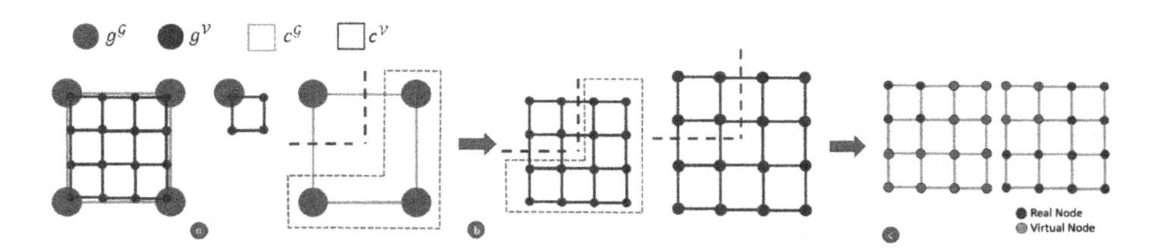

Fig. 4. (a) A cell $c^{\mathcal{G}}$ of EDG, its embedded TSDF volume cells $\{c^{\mathcal{V}}\}$ and a set of voxels $\{g^{\mathcal{V}}\}$ belonging to a node $g^{\mathcal{G}}$ of this cell. (b) Connectivity propagation from an EDG cell to its embedded TSDF volume cells. (c) Connectivity update of TSDF volume.

Input: (a) EDG cells and their embedded TSDF volume cells; and (b) a set of cutting edges in EDG.

Step 1 [Cell separation]: Each EDG cell contains $(2k+1)^3$ TSDF cells and $(2k+2)^3$ TSDF voxels. Each EDG node controls $(k+1)^3$ voxels. Figure 4(a) shows a 2D case when $k=1$. We separate each volume cell $c^{\mathcal{V}}$ by considering the connected components (CCs) of its associated EDG cell – the CCs belonging of each voxel is the same as its associated EDG node. If two vertices of an edge belong to different CCs, this edge is treated as a cutting edge(Fig. 4(b)).

Step 2 [Cell duplication based on CCs]: TSDF volume cells are duplicated depending on the number of CCs of an EDG cell $c^{\mathcal{G}}$, as shown in Fig. 4(c). Therefore, even though the number of CCs of TSDF volume cell on the top left is 1, it will still be duplicated as two copies: one copy containing all real nodes while the other copy containing all virtual nodes.

For those virtual nodes in the TSDF volumetric structure, their TSDF values need to be updated with caution. Here we use the following three updating rules: (1) For all real nodes, since we need to keep the continuity of their TSDF, we directly inherit their TSDF value from the original cell. (2) For all virtual nodes that are connected to real nodes, if their connected real node has negative TSDF value (meaning inside the surface), we set the TSDF of the corresponding virtual node by negating that value, i.e. $-d \rightarrow +d$. (3) For all remaining virtual nodes that have not been assigned TSDF values, we simply set their values as $+1$. Figure 5 shows an illustration of these TSDF updating rules. Note that all these TSDF values might continue to be updated by the depth fusion step that follows.

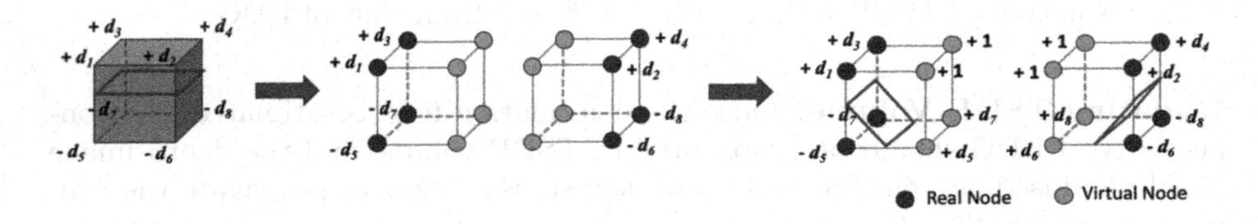

Fig. 5. The updating rule for the signed distances on virtual nodes of TSDF grids. The green surfaces denote the zero crossing surface inside the cells.

Step 3 [Restoring connectivity]: For any pair of geometrically adjacent duplicate cells $c^{\mathcal{V}}$ (in the canonical space), given two nodes $g^{\mathcal{V}}$ from them respectively, the merging rule is a bit different from the one used for EDG cell $c^{\mathcal{G}}$. We merge two nodes $g^{\mathcal{V}}$ if they are copied from the same original node and they are: (1) both real nodes, or (2) both virtual nodes.

Because the connectivity update of EDG is propagated to the TSDF grid, the geometry represented by TSDF could reflect topology changes and each cell $c^{\mathcal{V}}$ in the volume could find its correct EDG cell association. Next, all voxels will be warped to the live frame by the estimated deformation field. Similar to [28], depth information is fused into the volume in the canonical space.

Preparing for the Next Frame. In order to guide the estimation of deformation field for the next coming frame, we need to extract a surface mesh from the TSDF volume in the canonical space. Since the TSDF volumetric grid could become non-manifold, the marching cubes method needs to be modified to make it adapted to the topology changes.

Extended Marching Cubes Method: In the classic fusion framework, each TSDF volume cell is unique. Given the position of the left-front-bottom voxel in the canonical frame, the only corresponding EDG/TSDF grid cell is returned in $O(1)$ time. Now because of cell duplication, this rule will not hold. Therefore, for each voxel, we also store cell information. For each EDG node, we just need to store the id of its belonged EDG cell. For TSDF volume, we do not want to maintain another list of all volume cells. We directly store the list of voxel ids for one specific volume cell – the cell having this voxel as its left-front-bottom voxel. There are two benefits brought by adding this extra information: (1) it will help identifying the corresponding TSDF volume cell for every voxel once cells are duplicated; (2) after extracting the surface mesh by marching cubes method, each vertex also inherits the id of its belonged EDG cell, which makes it convenient to warp the mesh according to the deformation field defined by EDG. Finally, we extract triangle mesh for each TSDF volumetric cell in parallel and merge vertices on shared edges between cells.

Expanding EDG: As the 3D model grows by fusion of new geometry, the support of deformation field – EDG should also be expanded. Because we have a predefined grid structure for EDG and the primitive element of our EDG connectivity update algorithm is EDG cell, different from other fusion-based methods, we directly activate those EDG cells which embed the newly added geometry part to maintain the ability to separate and duplicate cells when there are new topology changes.

5 Experimental Results

There are several specific public datasets on topology change problems. Tsoli and Argyros [38] provided both synthetic and real data, from which the synthetic data is generated through physics-based simulation in Blender and the real data is captured with Kinect v2. Slavcheva *et al.* [33] also published their data. We evaluate qualitatively and quantitatively our method based on those mentioned datasets and the experimental results from the authors. Then ablation study is included to show the effect of different key components in our entire pipeline.

5.1 Evaluation on Synthetic Data

The baseline methods we select for synthetic data evaluation are CPD [27], MFSF [16], Tsoli and Argyros's method [38] and VolumeDeform [19]. The first three methods are template based non-rigid registration methods. Specifically, Tsoli and Argyros's method can deal with deformable surfaces that undergo topology changes. VolumeDeform and our method are both template-less fusion-based reconstruction methods. DynamicFusion [28] has bad performance on this synthetic dataset because it cannot deal well with deformations parallel to camera screen, so we do not compare with it.

We select two metrics proposed in Tsoli and Argyros's paper [38]: (1) Euclidean distance from ground truth; and (2) the number of vertices off the surface. We believe metric 1 can quantitatively evaluate the overall reconstruction quality while metric 2 provides a deeper insight about how the topologically

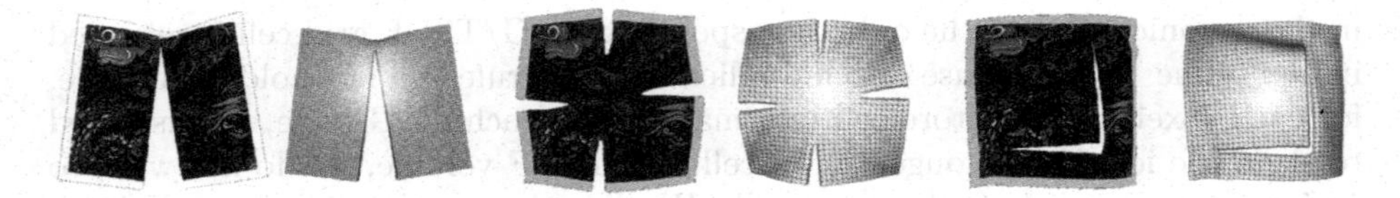

Fig. 6. Our reconstruction results on Tsoli and Argyros's synthetic dataset: seq1, seq2, and seq3, from left to right.

changed parts are reconstructed. There will be lots of vertices "off the surface" if the topologically changed part is not well considered and processed. We refer the readers to Tsoli and Argyros's paper [38] for detailed definition of these metrics. Here, the distance measurement for both metrics are expressed as a percentage of the cell width of the underlying grid. Because VolumeDeform and our method are reconstruction methods without any pre-defined template, to be consistent with Tsoli and Argyros's experiment, we allocate the volume according to the same grid cell width and the resolution of their template in x and y axis directions.

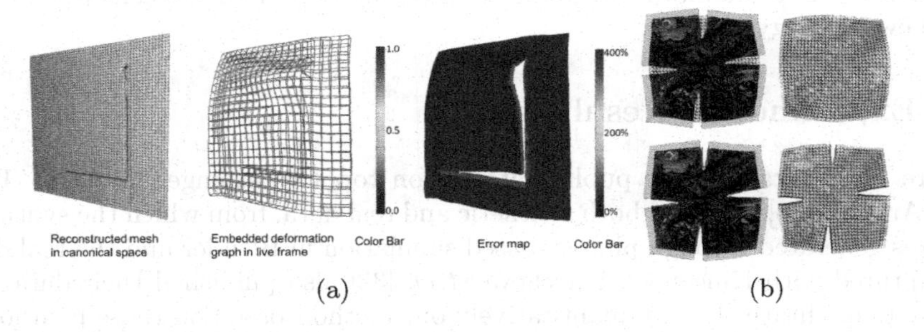

Fig. 7. (a) Qualitative comparison of our reconstructed seq3 data with the ground truth. (b) Reconstruction results on seq2 of Tsoli and Argyros's dataset by VolumeDeform (top row) and our method (bottom row). (Color figure online)

Figure 7(a) shows a reconstruction result on frame #36 of seq3 in Tsoli and Argyros's dataset. The color *Red*-*Green*-*Blue* on the EDG edge represents line process weights l_{ij} from 1 to 0. The error map using the color-bar on the right shows the Euclidean distance from ground truth, expressed as the percentage of the cell width in TSDF volume. We can see that the reconstructed mesh in live frame reflects the topology change in this case and so does the reconstructed mesh in canonical space. The line process weights of edges also represent the presence of deformation discontinuity.

We evaluate all five methods on synthetic dataset: a single cut (seq1), multiple non intersecting cuts (seq2), two intersecting cuts (seq3) (Fig. 6). Figure 8 show the performance of each method based on the two error metrics. Our method outperforms all other methods on seq2 and seq3 in terms of the distance from

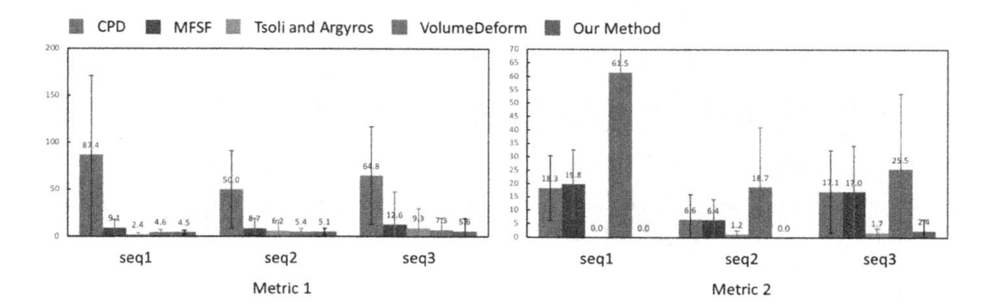

Fig. 8. Quantitative comparison with other methods. Metric 1: Euclidean distance from ground truth. Metric 2: number of vertices off the surface.

ground truth. Only Tsoli and Argyros's method does a better job on seq1 than ours. Under metric 2, our method outperforms all other methods on seq2. On seq1, our method is better than all other methods except Tsoli and Argyros's method. On seq3, our method has a bit higher average error than Tsoli and Argyros's method. Figure 7(b) displays the reason why VolumeDeform performs well under metric 1 but much worse under metric 2. It is because VolumeDeform keeps a fixed-topology grid structure to represent the deformation field and the geometry, and has no mechanism to deal with topology changes.

5.2 Comparison to State-of-the-Art on Real Data

Our method inherits from the classic DynamicFusion [28] framework, so two characteristics of DynamicFusion are kept: (1) open-to-close motions can be solved very well and (2) geometry will grow as more regions are observed during the reconstruction. Figure 9 shows some reconstruction results on VolumeDeform datasets. In the boxing sequence, some key frames reconstruction results illustrate that our method works well on an open-to-close-to-open motion. In the second sequence, the reconstructed geometry of upper body is rendered from a different viewpoint to make it easier to see the geometry growth during fusion.

The methods we compare for real data are VolumeDeform [19] and KillingFusion [33]. Figure 10 shows such comparison, where the first row is a bread breaking sequence and the second row is a paper tearing sequence. The leftmost a couple of images are RGB images for reference: images of starting frame and current live frame. The remaining 3 pairs of images show the reconstruction results by our method, VolumeDeform and KillingFusion. We can see that VolumeDeform could not update geometry correctly while both KillingFusion and our method could handle topology changes. But we can see that KillingFusion produces less smooth reconstructed surfaces compared to ours, even though all three methods use the same resolution of TSDF volume. The entire reconstructed sequences shown in Fig. 2, 6, 9, 10 are in the *Supplementary Video.*

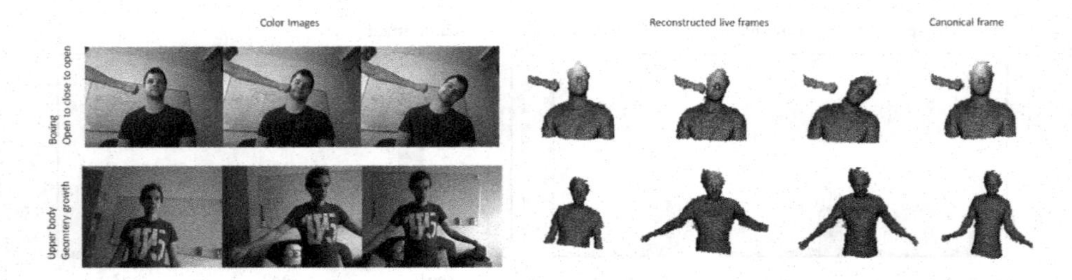

Fig. 9. Reconstruction results on real open-to-close and geometry growth data. Top row: open-to-close case; bottom row: geometry growth on body and arms.

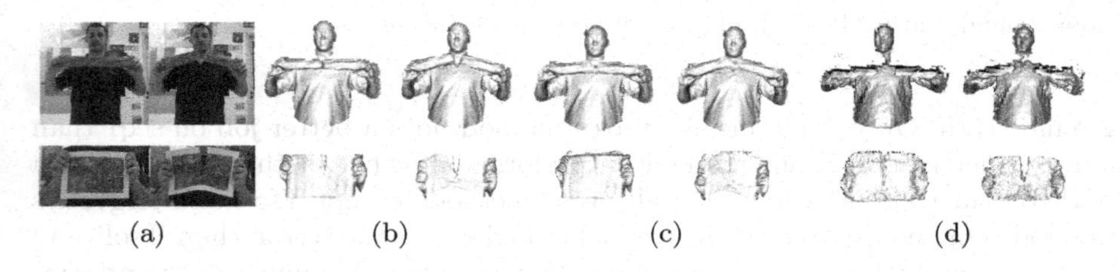

Fig. 10. Results on real data with topology changes. From left to right: (a) starting frame and live frame; (b) reconstructed geometry in canonical frame and live frame by our method; (c) VolumeDeform; (d) KillingFusion.

5.3 Ablation Study

Effect of Line Process Based Registration: Figure 11 shows the comparison of registration results with/without line process in the ARAP regularity term. It could be noted that Fig. 11(b) has better registration result than Fig. 11(c) in the tearing part. The line process weights in Fig. 11(b) also indicate the discontinuity of edges which help identifying cutting edges given a threshold.

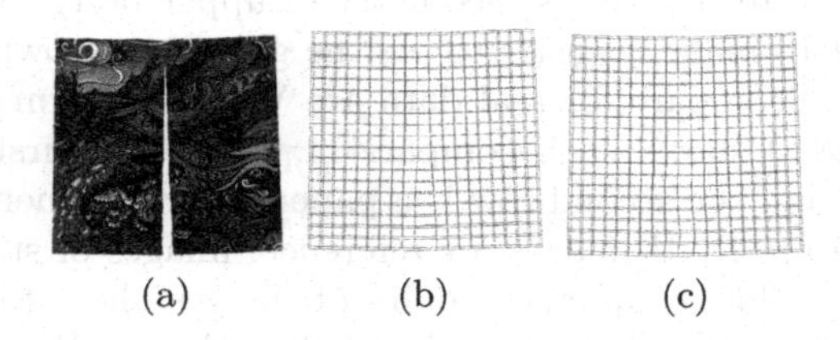

Fig. 11. Effect of line process: (a) target point cloud, (b) with line process, (c) without line process. Color *Red*-*Green*-*Blue* on the edge means l_{ij} from 1 to 0. (Color figure online)

Effect of Connectivity Update: Figure 12 demonstrates the effect of connectivity update. Without the connectivity update, topology changes will not be correctly reconstructed even though our topology-change-aware registration could help aligning surface towards the target point cloud.

Fig. 12. Effect of connectivity update. Left: input point cloud. Middle: result without connectivity update. Right: result with connectivity update.

Effect of Different Resolutions: As previous work points out (Fig. 10 in [19]), higher resolution of TSDF volume results in better reconstructed details and vice versa. This is a common issue of all fusion-based reconstruction, and so is our algorithm. Due to the assumption of all cutting edges being cut in mid-points, lower resolution of EDG may cause inaccurate cutting positions. However, we have two ways to alleviate such an effect: 1) Increasing the resolution of EDG; 2) Our multi-level grids and connectivity propagation algorithm. Moreover, although EDG may have a lower resolution but a higher resolution of TSDF can complement this by reconstructing more detailed geometry. In the bread breaking and paper tearing sequences, the voxel resolution is 6mm while cell resolution is 30 mm.

6 Conclusion and Future Work

In this paper we introduce a new topology-change-aware fusion framework for 4D dynamic scene reconstruction, by proposing the non-manifold volumetric grids for both EDG and TSDF, as well as developing an efficient approach to estimate a topology-change-aware deformation field and detect topology change events. Our method also has some limitations. One failure case caused by mid-point cutting assumption is cloth tearing with complex boundary. A lower resolution EDG tends to make the tearing boundary towards a line. There also exists other topology cases that our method is not designed to handle such as surface merging cases from genus 0 to higher genus, e.g. a ball morphs to a donut. Our system currently runs at around 5 FPS. But our system design is oriented towards parallel computation, as discussed in the *Supplementary Document*. In the future, we would like to perform code optimization and fully implement it in CUDA to achieve real-time performance.

Acknowledgement. This research is partially supported by National Science Foundation (2007661). The opinions expressed are solely those of the authors, and do not necessarily represent those of the National Science Foundation.

References

1. Baran, I., Vlasic, D., Grinspun, E., Popović, J.: Semantic deformation transfer. In: ACM SIGGRAPH 2009 Papers, pp. 1–6 (2009)
2. Bertholet, P., Ichim, A.E., Zwicker, M.: Temporally consistent motion segmentation from RGB-D video. Comput. Graph. Forum **37**, 118–134 (2018)
3. Black, M.J., Rangarajan, A.: On the unification of line processes, outlier rejection, and robust statistics with applications in early vision. Int. J. Comput. Vis. **19**(1), 57–91 (1996). https://doi.org/10.1007/BF00131148
4. Bojsen-Hansen, M., Li, H., Wojtan, C.: Tracking surfaces with evolving topology. ACM Trans. Graph. **31**(4) (2012). Article no. 53–1
5. Chen, X., Feng, J., Bechmann, D.: Mesh sequence morphing. Comput. Graph. Forum **35**, 179–190 (2016)
6. Collet, A., et al.: High-quality streamable free-viewpoint video. ACM Trans. Graph. (ToG) **34**(4), 69 (2015)
7. Curless, B., Levoy, M.: A volumetric method for building complex models from range images. In: Proceedings of the 23rd Annual Conference on Computer Graphics and Interactive Techniques, SIGGRAPH 1996, pp. 303–312. ACM (1996)
8. Digne, J., Cohen-Steiner, D., Alliez, P., de Goes, F., Desbrun, M.: Feature-preserving surface reconstruction and simplification from defect-laden point sets. J. Math. Imaging Vis. **48**(2), 369–382 (2013). https://doi.org/10.1007/s10851-013-0414-y
9. Dou, M., et al.: Motion2fusion: real-time volumetric performance capture. ACM Trans. Graph. (TOG) **36**(6), 246 (2017)
10. Dou, M., et al.: Fusion4D: real-time performance capture of challenging scenes. ACM Trans. Graph. **35**(4), 114 (2016)
11. Enright, D., Marschner, S., Fedkiw, R.: Animation and rendering of complex water surfaces. In: Proceedings of the 29th Annual Conference on Computer Graphics and Interactive Techniques, pp. 736–744 (2002)
12. Fröhlich, S., Botsch, M.: Example-driven deformations based on discrete shells. Comput. Graph. Forum **30**, 2246–2257 (2011)
13. Gao, L., Chen, S.Y., Lai, Y.K., Xia, S.: Data-driven shape interpolation and morphing editing. Comput. Graph. Forum **36**, 19–31 (2017)
14. Gao, L., Lai, Y.K., Huang, Q.X., Hu, S.M.: A data-driven approach to realistic shape morphing. Comput. Graph. Forum **32**, 449–457 (2013)
15. Gao, W., Tedrake, R.: SurfelWarp: efficient non-volumetric single view dynamic reconstruction. arXiv preprint arXiv:1904.13073 (2019)
16. Garg, R., Roussos, A., Agapito, L.: A variational approach to video registration with subspace constraints. Int. J. Comput. Vis. **104**(3), 286–314 (2013). https://doi.org/10.1007/s11263-012-0607-7
17. Golla, T., Kneiphof, T., Kuhlmann, H., Weinmann, M., Klein, R.: Temporal upsampling of point cloud sequences by optimal transport for plant growth visualization. Comput. Graph. Forum (2020)
18. Guo, K., Xu, F., Yu, T., Liu, X., Dai, Q., Liu, Y.: Real-time geometry, albedo, and motion reconstruction using a single RGB-D camera. ACM Trans. Graph. (TOG) **36**(3), 32 (2017)
19. Innmann, M., Zollhöfer, M., Nießner, M., Theobalt, C., Stamminger, M.: VolumeDeform: real-time volumetric non-rigid reconstruction. In: Leibe, B., Matas, J., Sebe, N., Welling, M. (eds.) ECCV 2016. LNCS, vol. 9912, pp. 362–379. Springer, Cham (2016). https://doi.org/10.1007/978-3-319-46484-8_22

20. Kowdle, A., et al.: The need 4 speed in real-time dense visual tracking. ACM Trans. Graph. **37**(6), 220:1–220:14 (2018)
21. Letouzey, A., Boyer, E.: Progressive shape models. In: 2012 IEEE Conference on Computer Vision and Pattern Recognition, pp. 190–197. IEEE (2012)
22. Li, C., Zhao, Z., Guo, X.: ArticulatedFusion: real-time reconstruction of motion, geometry and segmentation using a single depth camera. In: Ferrari, V., Hebert, M., Sminchisescu, C., Weiss, Y. (eds.) ECCV 2018. LNCS, vol. 11212, pp. 324–340. Springer, Cham (2018). https://doi.org/10.1007/978-3-030-01237-3_20
23. Li, H., et al.: Temporally coherent completion of dynamic shapes. ACM Trans. Graph. (TOG) **31**(1), 1–11 (2012)
24. Li, H., Yu, J., Ye, Y., Bregler, C.: Realtime facial animation with on-the-fly correctives. ACM Trans. Graph. **32**(4) (2013). Article no. 42–1
25. Mitchell, N., Aanjaneya, M., Setaluri, R., Sifakis, E.: Non-manifold level sets: a multivalued implicit surface representation with applications to self-collision processing. ACM Trans. Graph. (TOG) **34**(6), 247 (2015)
26. Molino, N., Bao, Z., Fedkiw, R.: A virtual node algorithm for changing mesh topology during simulation. ACM Trans. Graph. (TOG) **23**, 385–392 (2004)
27. Myronenko, A., Song, X.: Point set registration: coherent point drift. IEEE Trans. Pattern Anal. Mach. Intell. **32**(12), 2262–2275 (2010)
28. Newcombe, R.A., Fox, D., Seitz, S.M.: DynamicFusion: reconstruction and tracking of non-rigid scenes in real-time. In: Proceedings of the IEEE Conference on Computer Vision and Pattern Recognition, pp. 343–352 (2015)
29. Osher, S., Fedkiw, R.P.: Level Set Methods and Dynamic Implicit Surfaces, vol. 200. Springer, New York (2005)
30. Oswald, M.R., Stühmer, J., Cremers, D.: Generalized connectivity constraints for spatio-temporal 3D reconstruction. In: Fleet, D., Pajdla, T., Schiele, B., Tuytelaars, T. (eds.) ECCV 2014. LNCS, vol. 8692, pp. 32–46. Springer, Cham (2014). https://doi.org/10.1007/978-3-319-10593-2_3
31. Pons-Moll, G., Baak, A., Helten, T., Müller, M., Seidel, H.P., Rosenhahn, B.: Multisensor-fusion for 3D full-body human motion capture. In: 2010 IEEE Computer Society Conference on Computer Vision and Pattern Recognition, pp. 663–670. IEEE (2010)
32. Rusinkiewicz, S., Levoy, M.: Efficient variants of the ICP algorithm. In: 3DIM, vol. 1, pp. 145–152 (2001)
33. Slavcheva, M., Baust, M., Cremers, D., Ilic, S.: KillingFusion: non-rigid 3D reconstruction without correspondences. In: Proceedings of the IEEE Conference on Computer Vision and Pattern Recognition, pp. 1386–1395 (2017)
34. Slavcheva, M., Baust, M., Ilic, S.: SobolevFusion: 3D reconstruction of scenes undergoing free non-rigid motion. In: Proceedings of the IEEE Conference on Computer Vision and Pattern Recognition, pp. 2646–2655 (2018)
35. Solomon, J., et al.: Convolutional Wasserstein distances: efficient optimal transportation on geometric domains. ACM Trans. Graph. (TOG) **34**(4), 1–11 (2015)
36. Sumner, R.W., Schmid, J., Pauly, M.: Embedded deformation for shape manipulation. ACM Trans. Graph. **26**(3) (2007)
37. Tkach, A., Pauly, M., Tagliasacchi, A.: Sphere-meshes for real-time hand modeling and tracking. ACM Trans. Graph. (TOG) **35**(6), 222 (2016)
38. Tsoli, A., Argyros, A.A.: Tracking deformable surfaces that undergo topological changes using an RGB-D camera. In: 2016 Fourth International Conference on 3D Vision (3DV), pp. 333–341. IEEE (2016)
39. Von-Tycowicz, C., Schulz, C., Seidel, H.P., Hildebrandt, K.: Real-time nonlinear shape interpolation. ACM Trans. Graph. (TOG) **34**(3), 1–10 (2015)

40. Xu, D., Zhang, H., Wang, Q., Bao, H.: Poisson shape interpolation. Graph. Models **68**(3), 268–281 (2006)
41. Xu, W., Salzmann, M., Wang, Y., Liu, Y.: Deformable 3D fusion: from partial dynamic 3D observations to complete 4D models. In: Proceedings of the IEEE International Conference on Computer Vision, pp. 2183–2191 (2015)
42. Yu, T., et al.: BodyFusion: real-time capture of human motion and surface geometry using a single depth camera. In: Proceedings of the IEEE International Conference on Computer Vision, pp. 910–919 (2017)
43. Yu, T., et al.: DoubleFusion: real-time capture of human performances with inner body shapes from a single depth sensor. In: Proceedings of the IEEE Conference on Computer Vision and Pattern Recognition, pp. 7287–7296 (2018)
44. Yuan, Q., Li, G., Xu, K., Chen, X., Huang, H.: Space-time co-segmentation of articulated point cloud sequences. Comput. Graph. Forum **35**, 419–429 (2016)
45. Zampogiannis, K., Fermuller, C., Aloimonos, Y.: Topology-aware non-rigid point cloud registration. IEEE Trans. Pattern Anal. Mach. Intell. (2019)
46. Zollhöfer, M., et al.: Real-time non-rigid reconstruction using an RGB-D camera. ACM Trans. Graph. (ToG) **33**(4), 156 (2014)

Early Exit or Not: Resource-Efficient Blind Quality Enhancement for Compressed Images

Qunliang Xing[1], Mai Xu[1,2(✉)], Tianyi Li[1], and Zhenyu Guan[1]

[1] School of Electronic and Information Engineering, Beihang University, Beijing, China
{xingql,maixu,tianyili,guanzhenyu}@buaa.edu.cn

[2] Hangzhou Innovation Institute of Beihang University, Hangzhou, China

Abstract. Lossy image compression is pervasively conducted to save communication bandwidth, resulting in undesirable compression artifacts. Recently, extensive approaches have been proposed to reduce image compression artifacts at the decoder side; however, they require a series of architecture-identical models to process images with different quality, which are inefficient and resource-consuming. Besides, it is common in practice that compressed images are with unknown quality and it is intractable for existing approaches to select a suitable model for blind quality enhancement. In this paper, we propose a resource-efficient blind quality enhancement (RBQE) approach for compressed images. Specifically, our approach blindly and progressively enhances the quality of compressed images through a dynamic deep neural network (DNN), in which an early-exit strategy is embedded. Then, our approach can automatically decide to terminate or continue enhancement according to the assessed quality of enhanced images. Consequently, slight artifacts can be removed in a simpler and faster process, while the severe artifacts can be further removed in a more elaborate process. Extensive experiments demonstrate that our RBQE approach achieves state-of-the-art performance in terms of both blind quality enhancement and resource efficiency.

Keywords: Blind quality enhancement · Compressed images · Resource-efficient · Early-exit

1 Introduction

We are embracing an era of visual data explosion. According to Cisco mobile traffic forecast [4], the amount of mobile visual data is predicted to grow nearly 10-fold from 2017 to 2022. To overcome the bandwidth-hungry bottleneck caused

Electronic supplementary material The online version of this chapter (https://doi.org/10.1007/978-3-030-58517-4_17) contains supplementary material, which is available to authorized users.

A. Vedaldi et al. (Eds.): ECCV 2020, LNCS 12361, pp. 275–292, 2020.
https://doi.org/10.1007/978-3-030-58517-4_17

by a deluge of visual data, lossy image compression, such as JPEG [40], JPEG 2000 [28] and HEVC-MSP [37], has been pervasively used. However, compressed images inevitably suffer from compression artifacts, such as blocky effects, ringing effects and blurring, which severely degrade the Quality of Experience (QoE) [35,39] and the performance of high-level vision tasks [17,48].

For enhancing the quality of compressed images, many approaches [8,12,13, 16,22,42,46,47] have been proposed. Their basic idea is that one model needs to be trained for enhancing compressed images with similar quality reflected by a particular value of Quantization Parameter (QP) [37], and then a series of architecture-identical models need to be trained for enhancing compressed images with different quality. For example, [12,42,46] train 5 deep models to handle compressed images with QP = 22, 27, 32, 37 and 42. There are three main drawbacks to these approaches. (1) QP cannot faithfully reflect image quality, and thus it is intractable to manually select a suitable model based on QP value. (2) These approaches consume large computational resources during the training stage since many architecture-identical models need to be trained. (3) Compressed images with different quality are enhanced with the same computational complexity, such that these approaches impose excessive computational costs on "easy" samples (high-quality compressed images) but lack sufficient computation on "hard" samples (low-quality compressed images). Intuitively, the quality enhancement of images with different quality can be partly shared in a single framework, such that the joint computational costs can be reduced. More importantly, slight artifacts should be removed in a simpler and faster process, while the severe artifacts need to be further removed through a more elaborate process. Therefore, an ideal framework should automatically conduct a simple or elaborate enhancement process by distinguishing "easy" and "hard" samples, as a blind quality enhancement task.

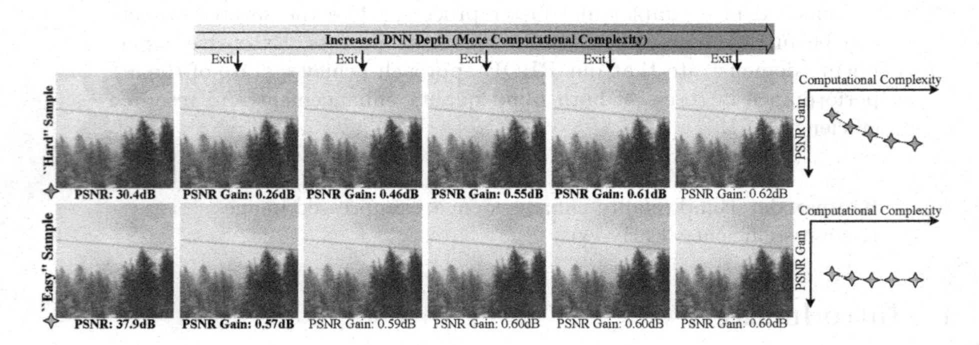

Fig. 1. Examples of quality enhancement on "easy" and "hard" samples, along with increased computational complexity.

In this paper, we propose a resource-efficient blind quality enhancement (RBQE) approach for compressed images. Specifically, we first prove that there exist "easy"/"hard" samples for quality enhancement on compressed images. We demonstrate that "easy" samples are those with slight compression artifacts, while "hard" samples are those with severe artifacts. Then, a novel dynamic deep neural network (DNN) is designed, which progressively enhances the quality of compressed image, assesses the enhanced image quality, and automatically decides whether to terminate (early exit) or continue the enhancement. The quality assessment and early-exit decision are managed by a Tchebichef moments-based Image Quality Assessment Module (IQAM), which is strongly sensitive to compression artifacts. Finally, our RBQE approach can perform "easy to hard" quality enhancement in an end-to-end manner. This way, images with slight compression artifacts can be simply and rapidly enhanced, while those with severe artifacts need to be further enhanced. Some examples are shown in Fig. 1. Also, experimental results verify that our RBQE approach achieves state-of-the-art performance for blind quality enhancement in both efficiency and efficacy.

To the best of knowledge, our approach is a first attempt to manage quality enhancement of compressed images in a resource-efficient manner. To sum up, the contributions are as follows:

(1) We prove that "easy"/"hard" samples exist in quality enhancement, as the theoretical foundation of our approach.
(2) We propose the RBQE approach with a simple yet effective dynamic DNN architecture, which processes "easy to hard" paradigm for blind quality enhancement.
(3) We develop a Tchebichef moments-based IQAM, workable for early-exit determination in our dynamic DNN structure.

2 Related Work

2.1 Quality Enhancement for Compressed Images

Due to the astonishing development of Convolutional Neural Networks (CNNs) [9,34,36] and large-scale image datasets [7], several CNN-based quality enhancement approaches have been successfully applied to JPEG-compressed images. Dong et al. [8] proposed a shallow four-layer Artifacts Reduction Convolutional Neural Network (AR-CNN), which is the pioneer of CNN-based quality enhancement of JPEG-compressed images. Later, Deep Dual-Domain (D3) approach [43] and Deep Dual-domain Convolutional neural Network (DDCN) [13] were proposed for JPEG artifacts removal, which are motivated by dual-domain sparse coding and utilize the quantization prior of JPEG compression. DnCNN [49] is a milestone for reducing both Additive White Gaussian Noise (AWGN) and JPEG artifacts. It is a 20-layer deep network employing residual learning [15] and batch normalization [19], which can yield better results than Block-Matching and 3-D filtering (BM3D) approach [5]. It also achieves blind denoising by mixing and sampling training data randomly with different levels of noise.

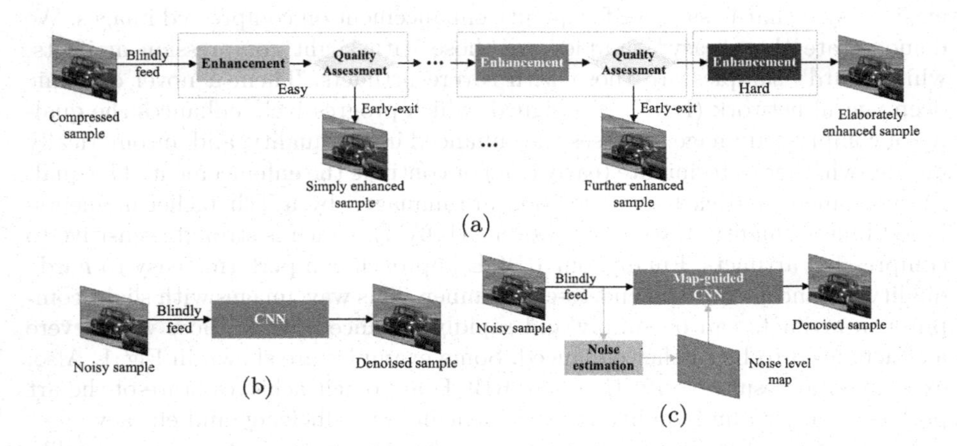

Fig. 2. Proposed resource-efficient blind quality enhancement paradigm (a) vs. two traditional blind denoising paradigms (b) and (c). Our paradigm dynamically processes samples with early exits for "easy" samples, while traditional paradigms (b) and (c) statically process images with equal computational costs on both "easy" and "hard" samples.

Most recently, extensive works have been devoted to the latest video/image coding standard, HEVC/HEVC-MSP [2,25,26,31,32,37,45]. Due to the elaborate coding strategies of HEVC, the approaches for JPEG-compressed images [8, 13,43,49], especially those utilizing the prior of JPEG compression [13,43], cannot be directly used for quality enhancement of HEVC-compressed images. In fact, HEVC [37] codec already incorporates the in-loop filters, which consist of Deblocking Filter (DF) [33] and Sample Adaptive Offset (SAO) filter [10], to suppress blocky effects and ringing effects. However, these handcrafted filters are far from optimum, resulting in still visible artifacts in compressed images. To alleviate this issue, Wang et al. [42] proposed the DCAD approach, which is the first attempt for CNN-based non-blind quality enhancement of HEVC-compressed images. Later, Yang et al. [46] proposed a novel QE-CNN for quality enhancement of images compressed by HEVC-MSP. Unfortunately, they are all non-blind approaches, typically requiring QP information before quality enhancement.

2.2 Blind Denoising for Images

In this section, we briefly review the CNN-based blind denoiser, as the closest field of blind quality enhancement of compressed images. The existing approaches for CNN-based blind denoising can be roughly summarized into two paradigms based on the mechanism of noise level estimation, as shown in Fig. 2(b) and (c). The first paradigm implicitly estimates the noise level. To achieve blind denoising, images with various levels of noise are mixed and randomly sampled during training [38,49]. Unfortunately, the performance is always far from optimum, as stated in [14,50]. It degrades severely when there is a mismatch of noise levels between training and test data. The second paradigm explicitly estimates the noise level. It sets a noise level estimation sub-net before a non-blind denoising sub-net. For example, [14] generates a noise level map to guide the subsequent non-blind denoising. This paradigm can always yield better results than the first paradigm, yet it is not suitable for quality enhancement of compressed artifacts, mainly due to two reasons. (1) The generated noise level map cannot well represent the level of compression artifacts. The compression artifacts are much more complex than generic noise since it is always assumed to be signal-independent and white [43]. (2) Both "easy" and "hard" samples are processed in the same deep architecture consuming equal computational resources, resulting in low efficiency. In this paper, we provide a brand-new paradigm for image reconstruction (as shown in Fig. 2(a)) and exemplify it by our proposed RBQE on quality enhancement of compressed images. It is worth mentioning that our brand-new paradigm also has the potential for the blind denoising task.

3 Proposed Approach

In this section, we propose our RBQE approach for blind quality enhancement. Specifically, we solve three challenging problems that are crucial to the resource-efficient paradigm of our approach. (1) Which samples are "simple"/"hard" in quality enhancement? (to be discussed in Sect. 3.1) (2) How to design a dynamic network for progressive enhancement? (to be discussed in Sect. 3.2) (3) How to measure compression artifacts of enhanced compressed images for early exits? (to be discussed in Sect. 3.3).

3.1 Motivation

Our RBQE approach is motivated by the following two propositions. **Proposition 1**: "Easy" samples (i.e., high-quality compressed images) can be simply enhanced, while "hard" samples (i.e., low-quality compressed images) should be further enhanced. **Proposition 2**: The quality enhancement process with different computational complexity can be jointly optimized in a single network through an "easy to hard" manner, rather than a "hard to easy" manner.

Proof of Proposition 1. We construct a series of vanilla CNNs with different depths and feed them with "easy" and "hard" samples, respectively. Specifically, a series of vanilla CNNs with the layer number from 4 to 11 are constructed. Each layer includes $64\times3\times3$ filters, except for the last layer with $1\times3\times3$ filter. Beside, ReLU [30] activation and global residual learning [15] are adopted. The training, validation and test sets (including 400, 100 and 100 raw images, respectively) are randomly selected from Raw Image Database (RAISE) [6] without overlapping. They are all compressed by HM16.5[1] under intra-coding configuration [37] with

[1] HM16.5 is the latest HEVC reference software.

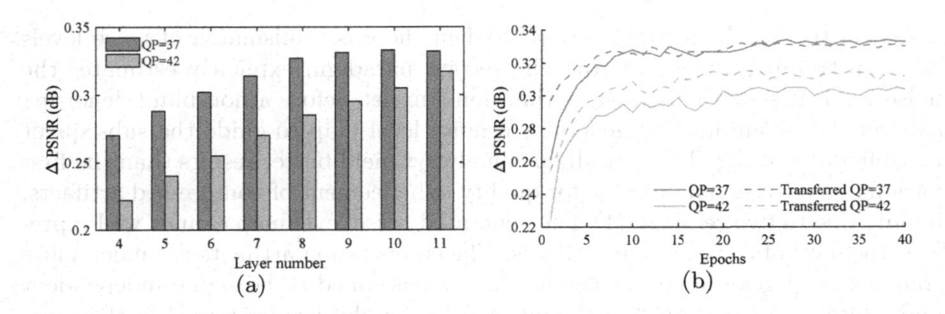

Fig. 3. (a) Average improved peak signal-to-noise ratio (ΔPSNR) of vanilla CNNs over the test set. (b) Average ΔPSNR curves alongside increased epochs, for vanilla CNN models and their transferred models over the validation set during the training stage.

QP = 37 and 42 for obtaining "easy" and "hard" samples, respectively. Then, we train the vanilla CNNs with the "easy" samples, and then obtain converged models "QP = 37". Similarly, we train the CNNs with the "hard" samples and then obtain converged models "QP = 42". As shown in Fig. 3(a), the performance of QP = 42 models improves significantly with the increase of layer numbers, while the performance of QP = 37 models gradually becomes saturated once the layer number excesses 9. Therefore, it is possible to enhance the "easy" samples with a simpler architecture and fewer computational resources, while further enhancing the "hard" samples in a more elaborate process.

Proof of Proposition 2. The advantage of the "easy to hard" strategy has been pointed out in neuro-computation [11]. Here, we investigate its efficacy on image quality enhancement through the experiments of transfer learning. If the filters learned from "easy" samples can be transferred to enhance "hard" samples more successfully than the opposite manner, then our proposition can be proved. Here, we construct 2 identical vanilla CNNs with 10 convolutional layers. The other settings conform to the above. We train these 2 models with the training sets of images compressed at QP = 37 and 42, respectively, and accordingly these 2 models are called "QP = 37" and "QP = 42". After convergence, they exchange their parameters for the first 4 layers and restart training with their own training sets. Note that the exchanged parameters are frozen during the training stage. We name the model transferred from QP = 42 to QP = 37 as "transferred QP = 37" and the model transferred from QP = 37 to QP = 42 as "transferred QP = 42". Figure 3(b) shows the validation-epoch curves of the original 2 models and their transferred models. As shown in this figure, the transferred QP = 42 model improves the performance of the QP = 42 model, while the transferred QP = 37 model slightly degrades the performance of the QP = 37 model. Consequently, the joint simple and elaborate enhancement process should be conducted in an "easy to hard" manner rather than a "hard to easy" manner. Besides, the experimental results of Sect. 4 show that the simple and elaborate enhancement process can be jointly optimized in a single network. In summary, proposition 2 can be proved. The above propositions can be also validated by JPEG-compressed images, as detailed in the supplementary material.

Given the above two propositions, we propose our RBQE approach for resource-efficient quality enhancement of compressed images in an "easy to hard" manner.

3.2 Dynamic DNN Architecture with Early-Exit Strategy

Notations. In this section, we present the DNN architecture of the proposed RBQE approach for resource-efficient quality enhancement. We first introduce the notations for our RBQE approach. The input sample is denoted by $\mathbf{S}_{\text{in}}$. The convolutional layer is denoted by $C_{i,j}$, where i denotes the level and j denotes the index of the convolutional layer on the same level. In addition, I is the total number of levels. Accordingly, the feature maps generated from $C_{i,j}$ are denoted by $\mathbf{F}_{i,j}$. The enhancement residuals are denoted by $\{\mathbf{R}_j\}_{j=2}^{I}$. Accordingly, the output enhanced samples are denoted by $\{\mathbf{S}_{\text{out},j}\}_{j=2}^{I}$.

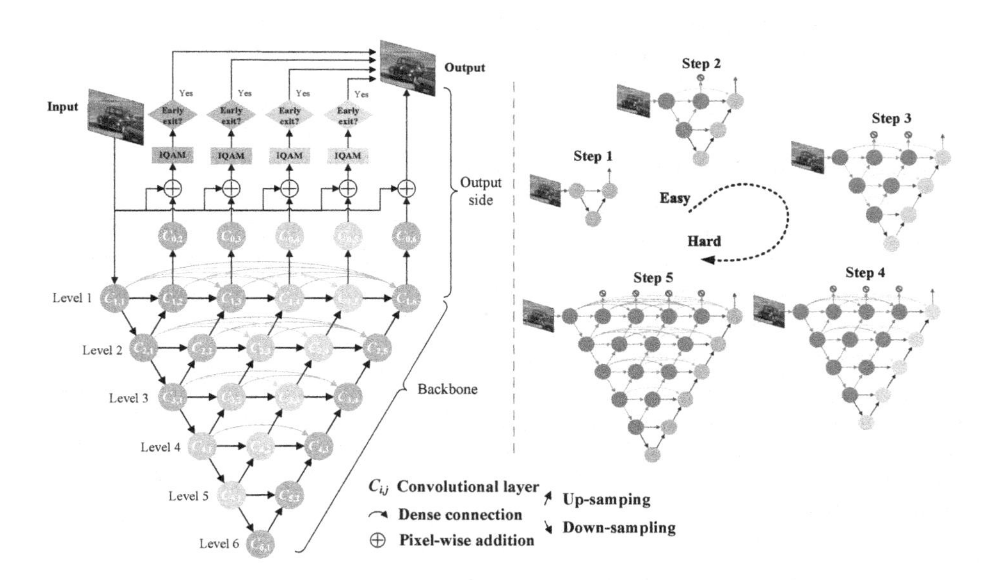

Fig. 4. Dynamic DNN architecture and early-exit strategy of our RBQE approach. The computations of gray objects (arrays and circles) are accomplished in the previous step and inherited in the current step.

Architecture. To better illustrate the architecture of RBQE, we separate the backbone and the output side of RBQE, as shown in the left half of Fig. 4. In this figure, we take RBQE with 6 levels as an example. The backbone of RBQE is a progressive UNet-based structure. Convolutional layers $C_{1,1}$ and $C_{2,1}$ can be seen as the encoding path of the smallest 2-level UNet, while $\{C_{i,1}\}_{i=1}^{6}$ are the encoding path of the largest 6-level UNet. Therefore, the backbone of RBQE can be considered as a compact combination of 5 different-level UNets. In the backbone of RBQE, the input sample is first fed into the convolutional layer $C_{1,1}$. After that, the feature maps generated by $C_{1,1}$ (i.e., $\mathbf{F}_{1,1}$) are progressively down-sampled and convoluted by $\{C_{i,1}, i = 2, 3, ..., 6\}$. This way, we obtain feature maps $\{\mathbf{F}_{i,1}\}_{i=1}^{6}$ at 6 different levels, the size of which progressively becomes smaller from level 1 to 6. In accordance with the encoder-decoder architecture of the UNet approach, $\{\mathbf{F}_{i,1}\}_{i=1}^{6}$ are then progressively up-sampled and convoluted until level 1. Moreover, based on the progressive UNet structure, we adopt dense connections [18] at each level. For example, at level 1, $\mathbf{F}_{1,1}$ are directly fed into the subsequent convolutional layers at the same level: $\{C_{1,j}\}_{j=2}^{6}$. The adoption of dense connection does not only encourage the reuse of encoded low-level fine-grained features by decoders, but also largely decreases the number of parameters, leading to a lightweight structure for RBQE.

At the output side, the obtained feature maps $\{\mathbf{F}_{1,j}\}_{j=2}^{6}$ are further convoluted by independent convolutional layers $\{C_{0,j}\}_{j=2}^{6}$, respectively. In this step, we obtain the enhancement residuals: $\{\mathbf{R}_j\}_{j=2}^{6}$. For each residual $\mathbf{R}_j$, it is then added into the input sample $\mathbf{S}_{\text{in}}$ for calculating the enhanced image $\mathbf{S}_{\text{out},j}$:

$$\mathbf{S}_{\text{out},j} = \mathbf{S}_{\text{in}} + \mathbf{R}_j. \tag{1}$$

To assess the quality of enhanced image $\mathbf{S}_{\text{out},j}$, we feed it into IQAM, which is to be presented in Sect. 3.3.

The backbone of RBQE is motivated by [51], which extends the UNet architecture to a wide UNet for medical image segmentation. Here, we advance the wide UNet in the following aspects: (1) The wide UNet adopts deep supervision [21] directly for the feature maps $\{\mathbf{F}_{1,j}\}_{j=2}^{5}$. Here, we further process the output feature maps $\{\mathbf{F}_{1,j}\}_{j=2}^{I}$ independently through the convolutional layers in the output side $\{C_{0,j}\}_{j=2}^{I}$. This process can alleviate the interference between outputs, while slightly increase the computational costs. (2) The work of [51] manually selects one of the 4 different-level UNet-based structures in the test stage, based on the requirement for speed and accuracy. Here, we incorporate IQAM into RBQE and provide early exits in the test stage. Therefore, all UNet-based structures are progressively and automatically selected to generate the output. The early-exit strategy and proposed IQAM are presented in the following.

Early-Exit Strategy. Now we explain the early-exit strategy of RBQE. Similarly, we take the RBQE structure with 6 levels as an example. The backbone of RBQE can be ablated progressively into 5 different-level UNet-based structures, as depicted in the right half of Fig. 4. For example, the smallest UNet-based structure with 2 levels consists of 3 convolutional layers: $C_{1,1}$, $C_{2,1}$ and $C_{1,2}$. In addition to these 3 layers, the 3-level UNet-based structure includes 3 more convolutional layers: $C_{3,1}$, $C_{2,2}$ and $C_{1,3}$. Similarly, we can identify the layers of the remaining 3 UNet-based structures. Note that the interval activation layers are omitted for simplicity. We denote the parameters of the i-level UNet-based structure by θ_i. This way, the output enhanced samples $\{\mathbf{S}_{\text{out},j}\}_{j=2}^{6}$ can be formulated as:

$$\mathbf{S}_{\text{out},j} = \mathbf{S}_{\text{in}} + \mathbf{R}_j(\theta_j),\ j = 2, 3, ..., 6. \tag{2}$$

In the test stage, $\{\mathbf{S}_{\text{out},j}\}_{j=2}^{6}$ are obtained and assessed progressively. That is, we first obtain $\mathbf{S}_{\text{out},2}$ and send it to IQAM. If $\mathbf{S}_{\text{out},2}$ is assessed to be qualified as the output, the quality enhancement process is terminated. Otherwise, we further obtain $\mathbf{S}_{\text{out},3}$ and assess its quality through IQAM. The same procedure applies to $\mathbf{S}_{\text{out},4}$ and $\mathbf{S}_{\text{out},5}$. If $\{\mathbf{S}_{\text{out},j}\}_{j=2}^{5}$ are all rejected by IQAM, $\mathbf{S}_{\text{out},6}$ is output without assessment. This way, we successfully perform the early-exit strategy for "easy" samples, which are expected to output in the early stage.

3.3 Image Quality Assessment for Enhanced Images

In this section, we introduce IQAM for blind quality assessment and automatic early-exit decision. Most existing blind denoising approaches (e.g.., [14,42,46,49]) ignore the characteristics of compression artifacts; however, these characteristics are important to assess the compression artifacts. Motivated by [23], this paper considers two dominant factors that degrade the quality of enhanced compressed images: (1) blurring in the textured area and (2) blocky effects in the smooth area.

Specifically, the enhanced image is first partitioned into non-overlapping patches. The patches should cover all potential compression block boundaries. Then, these patches are classified into smooth and textured ones according to their sum of squared non-DC Tchebichef moment (SSTM) values that measure the patch energy [23,29]. We take a 4×4 patch as an example, of which Tchebichef moments can be denoted by $\mathbf{M}$:

$$\mathbf{M} = \begin{pmatrix} m_{00} & \cdots & m_{03} \\ \vdots & \ddots & \vdots \\ m_{30} & \cdots & m_{33} \end{pmatrix}. \tag{3}$$

If the patch is classified as a smooth one, we evaluate its score of blocky effects $\mathcal{Q}_{\text{S}}$ by calculating the ratio of the summed absolute 3rd order moments to the SSTM value [24]:

$$e_{\text{h}} = \frac{\sum_{i=0}^{3} |m_{i3}|}{\left(\sum_{i=0}^{3}\sum_{j=0}^{3} |m_{ij}|\right) - |m_{00}| + C}, \tag{4}$$

$$e_{\text{v}} = \frac{\sum_{j=0}^{3} |m_{3j}|}{\left(\sum_{i=0}^{3}\sum_{j=0}^{3} |m_{ij}|\right) - |m_{00}| + C}, \tag{5}$$

$$\mathcal{Q}_{\text{S}} = \log_{(1-T_e)}\left(1 - \frac{e_{\text{v}} + e_{\text{h}}}{2}\right), \tag{6}$$

where e_{v} and e_{h} measure the energy of vertical and horizontal blocky effects, respectively; C is a small constant to ensure numerical stability; T_e is a perception threshold. The average quality score of all smooth patches is denoted by $\bar{\mathcal{Q}}_{\mathrm{S}}$. If the patch is classified as a textured one, we first blur it using a Gaussian filter. Similarly, we obtain the Tchebichef moments of this blurred patch $\mathbf{M}'$. Then, we evaluate its blurring score $\mathcal{Q}_{\mathrm{T}}$ by calculating the similarity between $\mathbf{M}$ and $\mathbf{M}'$:

$$\mathbf{S}(i,j) = \frac{2m_{ij}m'_{ij} + C}{(m_{ij})^2 + (m'_{ij})^2 + C}, \; i,j = 0,1,2,3, \tag{7}$$

$$\mathcal{Q}_{\mathrm{T}} = 1 - \frac{1}{3 \times 3} \sum_{i=0}^{3} \sum_{j=0}^{3} \mathbf{S}(i,j), \tag{8}$$

where $\mathbf{S}(i,j)$ denotes the similarity between two moment matrices. The average quality score of all textured patches is denoted by $\bar{\mathcal{Q}}_{\mathrm{T}}$. The final quality score $\mathcal{Q}$ of the enhanced image is calculated as

$$\mathcal{Q} = (\bar{\mathcal{Q}}_{\mathrm{S}})^{\alpha} \cdot (\bar{\mathcal{Q}}_{\mathrm{T}})^{\beta}, \tag{9}$$

where α and β are the exponents balancing the relative importance between blurring and blocky effects. If $\mathcal{Q}$ exceeds a threshold $T_{\mathcal{Q}}$, the enhanced image is directly output at early exits of the enhancement process. Otherwise, the compressed image needs to be further enhanced by RBQE. Please refer to the supplementary material for additional details.

The advantages of IQAM are as follow: (1) IQAM is constructed based on Tchebichef moments [29], which are highly interpretable for evaluating blurring and blocky effects. (2) The quality score $\mathcal{Q}$ obtained by IQAM is positively and highly correlated to the evaluation metrics of objective image quality, e.g.., PSNR and structural similarity (SSIM) index. See the supplementary material for the validation of such correlation, which is verified over 1,000 pairs of raw/compressed images. (3) With IQAM, we can balance the tradeoff between enhanced quality and efficiency by simply tuning threshold $T_{\mathcal{Q}}$.

3.4 Loss Function

For each output, we minimize the mean-squared error (MSE) between the input compressed image and output enhanced image:

$$\mathcal{L}_j(\theta_j) = \|\mathbf{S}_{\mathrm{out},j}(\theta_j) - \mathbf{S}_{\mathrm{in}}\|_2^2, \; j = 2,3,...,I. \tag{10}$$

Although MSE is known to have limited correlation with the perceptual quality of images [44], it can still yield high accuracy in terms of other metrics, such as PSNR and SSIM [12,14]. The loss function of our RBQE approach (i.e., $\mathcal{L}_{\mathrm{RBQE}}$) can be formulated as the weighted combination of these MSE losses:

$$\mathcal{L}_{\mathrm{RBQE}} = \sum_{j=2}^{I} w_j \cdot \mathcal{L}_j(\theta_j), \tag{11}$$

where w_j denotes the weight of $\mathcal{L}_j(\theta_j)$. By minimizing the loss function, we can obtain the converged RBQE model that simultaneously enhances the quality of input compressed images with different quality in a resource-efficient manner.

4 Experiments

In this section, we present the experimental results to verify the performance of the proposed RBQE approach for resource-efficient blind quality enhancement. Since HEVC-MSP [37] is a state-of-the-art image codec and JPEG [40] is a widely used image codec, our experiments mainly focus on quality enhancement of both HEVC-MSP and JPEG images.

4.1 Dataset

The recent works have adopted large-scale image datasets such as BSDS500 [1] and ImageNet [7], which are widely used for image denoising, segmentation and other vision tasks. However, the images of these datasets are compressed by unknown codecs and compression settings, thus containing various unknown artifacts. To obtain "clean" data without any unknown artifact, we adopt the RAISE dataset, from which 3,000, 1,000 and 1,000 non-overlapping raw images are as the training, validation and test sets, respectively. These images are all center-cropped into 512 × 512 images. Then, we compress the cropped raw images by HEVC-MSP using HM16.5 under intra-coding configuration [37], with QP = 22, 27, 32, 37 and 42. Note that QPs ranging from 22 to 42 can reflect the dramatically varying quality of compressed images, also in accordance with existing works [12,42,46]. For JPEG, we use the JPEG encoder of Python Imaging Library (PIL) [27] to compress the cropped raw images with quality factor (QF) = 10, 20, 30, 40 and 50. Note that these QFs are also used in [49].

4.2 Implementation Details

We set the number of levels $I = 6$ for the DNN architecture of RBQE. Then, $\{C_{i,1}\}_{i=1}^{6}$ are conducted by two successive $32 \times 3 \times 3$ convolutions. The other $C_{i,j}$ are conducted by two successive separable convolutions [3]. Note that each separable convolution consists of a depth-wise $k \times 3 \times 3$ convolution (k is the input channel number) and a point-wise $32 \times 1 \times 1$ convolution. The down-sampling is achieved through a $32 \times 3 \times 3$ convolution with the stride of 2, while the up-sampling is achieved through a transposed $32 \times 2 \times 2$ convolution with the stride of 2. For each group of feature maps $\mathbf{F}_{i,j}$, it is further processed by an efficient channel attention layer [41] before being feeding into other convolutional layers. Additionally, ReLU [30] nonlinearity activation is adopted between neighboring convolutions, except the successive depth-wise and point-wise convolutions within each separable convolution. For IQAM, we set $\alpha = 0.9$, $\beta = 0.1$, $C = 1e{-}8$ and $T_e = 0.05$ through a 1000-image validation. Additionally, as discussed in Sect. 4.3, T_Q is set to 0.89 and 0.74 for HEVC-MSP-compressed and JPEG-compressed images, respectively.

Table 1. Average ΔPSNR (dB) over the HEVC-MSP and JPEG test sets.

HEVC-MSP						JPEG					
QP	CBDNet	DnCNN	DCAD	QE-CNN	RBQE	QF	CBDNet	DnCNN	DCAD	QE-CNN	RBQE
22	0.470	0.264	0.311	0.082	**0.604**	50	1.342	1.078	1.308	1.230	**1.552**
27	0.385	0.414	0.278	0.182	**0.487**	40	1.393	1.362	1.356	1.290	**1.582**
32	0.375	0.405	0.314	0.275	**0.464**	30	1.459	1.550	1.415	1.352	**1.626**
37	0.403	0.314	0.353	0.313	**0.494**	20	1.581	1.572	1.501	1.420	**1.713**
42	0.411	0.186	0.321	0.264	**0.504**	10	1.726	1.121	1.676	1.577	**1.920**
ave	0.409	0.317	0.316	0.223	**0.510**	ave	1.500	1.337	1.451	1.374	**1.678**

In the training stage, batches with QP from 22 to 42 are mixed and randomly sampled. In accordance with the "easy to hard" paradigm, we set $\{w_j\}_{j=2}^6$ to $\{2, 1, 1, 0.5, 0.5\}$ for QP = 22 or QF = 50, to $\{1, 2, 1, 0.5, 0.5\}$ for QP = 27 or QF = 40, to $\{0.5, 1, 2, 1, 0.5\}$ for QP = 32 or QF = 30, to $\{0.5, 0.5, 1, 2, 1\}$ for QP = 37 or QF = 20, and to $\{0.5, 0.5, 1, 1, 2\}$ for QP = 42 or QF = 10. This way, high-quality samples are encouraged to output at early exits, while low-quality samples are encouraged to output at late exits. We apply the Adam optimizer [20] with the initial learning rate lr = $1e{-4}$ to minimize the loss function.

4.3 Evaluation

In this section, we validate the performance of our RBQE approach for the blind quality enhancement of compressed images. In our experiments, we compare our approach with 4 state-of-the-art approaches: DnCNN [49], CBDNet [14], QE-CNN [46] and DCAD [42]. Among them, QE-CNN and DCAD are the latest non-blind quality enhancement approaches for compressed images. For these non-blind approaches, the training batches of different QPs are mixed and randomly sampled in the training stage, such that they can also manage blind quality enhancement. Note that there is no blind approach for quality enhancement of compressed images. Thus, the state-of-the-art blind denoisers (i.e., DnCNN and CBDNet) are used for comparison, which are modified for blind quality enhancement by retraining over compressed images. For fair comparison, all compared approaches are retrained over our training set.

Evaluation on Efficacy. To evaluate the efficacy of our approach, Table 1 presents the ΔPSNR results of our RBQE approach and other compared approaches over the images compressed by HEVC-MSP. As shown in this table, the proposed RBQE approach outperforms all other approaches in terms of ΔPSNR. Specifically, the average ΔPSNR of RBQE is 0.510 dB, which is 24.7% higher than that of the second-best CBDNet (0.409 dB), 60.9% higher than that of DnCNN (0.317 dB), 61.4% higher than that of DCAD (0.316 dB), and 128.7% higher than that of QE-CNN (0.223 dB). Similar results can be found in Table 1 for the quality enhancement of JPEG images.

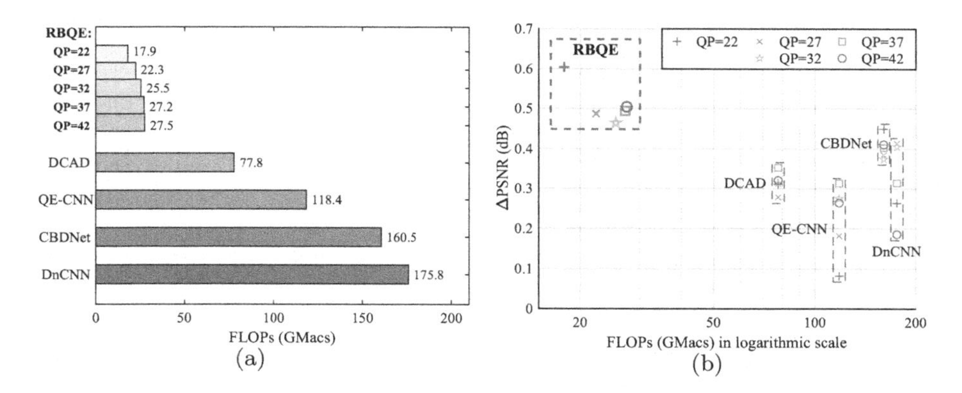

Fig. 5. (a) Average FLOPs (GMacs) over the HEVC-MSP test set. (b) Average FLOPs (GMacs) vs. improved peak signal-to-noise ratio (ΔPSNR), for blind quality enhancement by our RBQE and compared approaches over the HEVC-MSP test set.

Evaluation on Efficiency. More importantly, the proposed RBQE approach is in a resource-efficient manner. To evaluate the efficiency of the RBQE approach, Fig. 5 shows the average consumed floating point operations (FLOPs)[2] by our RBQE and other compared approaches. Note that the results of Fig. 5 are averaged over all images in our test set. As can be seen in this figure, RBQE consumes only 27.5 GMacs for the "hardest" samples, i.e., the images compressed at QP $=$ 42 and 17.9 GMacs for the "easiest" samples, i.e., the images compressed at QP $=$ 22. In contrast, DCAD, QE-CNN, CBDNet and DnCNN consume constantly 77.8, 118.4, 160.5 and 175.8 GMacs for all samples that are either "easy" or "hard" samples compressed at 5 different QPs. Similar results can also be found for the JPEG test set, as reported in the supplementary material. In summary, our RBQE approach achieves the highest ΔPSNR results, while consuming minimal computational resources especially for "easy" samples.

Tradeoff Between Efficacy and Efficiency. As aforementioned, we can simply control the tradeoff between efficacy and efficiency by tuning T_Q. As shown in Fig. 6(a), the average Δ PSNR improves along with the increased consumed FLOPs by enlarging T_Q. In this paper, we choose $T_Q = 0.89$ for HEVC-MSP-compressed images, since the improvement of average Δ PSNR gradually becomes saturated, especially when $T_Q > 0.89$. Due to the similar reason, we choose $T_Q = 0.74$ for JPEG-compressed images. In a word, the tradeoff between efficacy and efficiency of quality enhancement can be easily controlled in our RBQE approach.

Ablation Studies. To verify the effectiveness of the early-exit structure of our RBQE approach, we progressively ablate the 5 outermost decoding paths.

[2] Note that the definition of FLOPs follows [15,18], i.e., the number of multiply-adds.

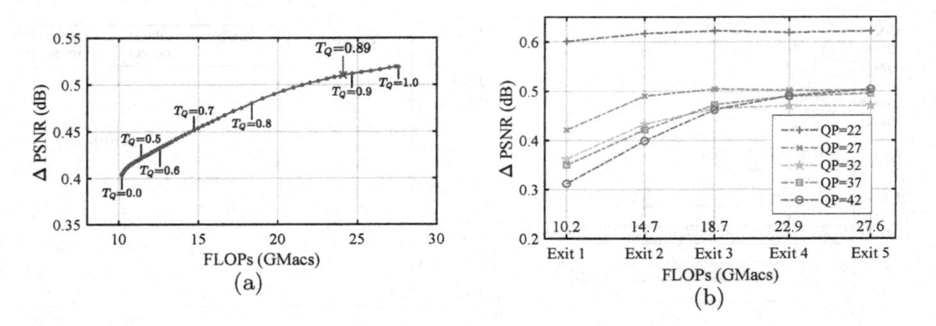

Fig. 6. (a) Average ΔPSNR and FLOPs under a series T_Q on HEVC test set. (b) Ablation results of the early-exit strategy.

Specifically, for the HEVC-MSP images compressed at QP = 22, we force their enhancement process to be terminated at 5 different exits (i.e., ignoring the automatic decision by IQAM), respectively, and then we obtain the brown curve in Fig. 6(b). Similarly, we can obtain the other 4 curves. As shown in this figure, "simplest" (i.e., QP = 22) samples can achieve ΔPSNR = 0.601 dB at the first exit, which is only 0.02 dB lower than that at the last exit. However, the expense is 270% FLOPs when outputting those samples at the last exit instead of the first one. In the opposite, the ΔPSNR of "hardest" (i.e., QP = 42) samples output from the last exit is 0.192 dB higher than that from the first exit. Therefore, "easy" samples can be simply enhanced while slightly sacrificing quality enhancement performance; meanwhile, more resources provided to "hard" samples can result in significantly higher ΔPSNR. This is in accordance with our motivation and also demonstrates the effectiveness of the early exits proposed in our RBQE approach.

5 Conclusions

In this paper, the RBQE approach has been proposed with a simple yet effective DNN structure to blindly enhance the quality of compressed images in a resource-efficient manner. Different from the traditional quality enhancement approaches, the proposed RBQE approach progressively enhances the quality of compressed images, which assesses the enhanced quality and then automatically terminates the enhancement process according to the assessed quality. To achieve this, our RBQE approach incorporates the early-exit strategy into a UNet-based structure, such that compressed images can be enhanced in an "easy to hard" manner. This way, "easy" samples can be simply enhanced and output at the early exits, while "hard" samples can be further enhanced and output at the late exits. Finally, we conducted extensive experiments on enhancing HEVC-compressed and JPEG-compressed images, and the experimental results validated that our proposed RBQE approach consistently outperforms the state-of-the-art quality enhancement approaches, while consuming minimal computational resources.

Acknowledgment. This work was supported by the NSFC under Project 61876013, Project 61922009, and Project 61573037.

References

1. Arbeláez, P., Maire, M., Fowlkes, C., Malik, J.: Contour detection and hierarchical image segmentation. IEEE Trans. Pattern Anal. Mach. Intell. **33**(5), 898–916 (2011). https://doi.org/10.1109/TPAMI.2010.161
2. Cai, Q., Song, L., Li, G., Ling, N.: Lossy and lossless intra coding performance evaluation: HEVC, H. 264/AVC, JPEG 2000 and JPEG LS. In: Asia Pacific Signal and Information Processing Association Annual Summit and Conference, pp. 1–9. IEEE (2012)
3. Chollet, F.: Xception: deep learning with depthwise separable convolutions. In: IEEE Conference on Computer Vision and Pattern Recognition (CVPR), pp. 1251–1258 (2017)
4. Cisco Systems Inc.: Cisco visual networking index: global mobile data traffic forecast update, 2017–2022 white paper. https://www.cisco.com/c/en/us/solutions/collateral/service-provider/visual-networking-index-vni/white-paper-c11-738429.html
5. Dabov, K., Foi, A., Katkovnik, V., Egiazarian, K.: Image denoising by sparse 3-D transform-domain collaborative filtering. IEEE Trans. Image Process. (TIP) **16**(8), 2080–2095 (2007)
6. Dang-Nguyen, D.T., Pasquini, C., Conotter, V., Boato, G.: Raise: a raw images dataset for digital image forensics. In: The 6th ACM Multimedia Systems Conference, pp. 219–224. ACM (2015)
7. Deng, J., Dong, W., Socher, R., Li, L.J., Li, K., Fei-Fei, L.: Imagenet: a large-scale hierarchical image database. In: IEEE Conference on Computer Vision and Pattern Recognition (CVPR), pp. 248–255. IEEE (2009)
8. Dong, C., Deng, Y., Change Loy, C., Tang, X.: Compression artifacts reduction by a deep convolutional network. In: IEEE International Conference on Computer Vision (ICCV), pp. 576–584 (2015)
9. Fan, Z., Wu, H., Fu, X., Huang, Y., Ding, X.: Residual-guide network for single image deraining. In: Proceedings of the 26th ACM International Conference on Multimedia, pp. 1751–1759 (2018)
10. Fu, C.M., et al.: Sample adaptive offset in the HEVC standard. IEEE Trans. Circuits Syst. Video Technol. (TCSVT) **22**(12), 1755–1764 (2012)
11. Gluck, M.A., Myers, C.E.: Hippocampal mediation of stimulus representation: a computational theory. Hippocampus **3**(4), 491–516 (1993)
12. Guan, Z., Xing, Q., Xu, M., Yang, R., Liu, T., Wang, Z.: MFQE 2.0: a new approach for multi-frame quality enhancement on compressed video. IEEE Trans. Pattern Anal. Mach. Intell. (TPAMI), 1 (2019). https://doi.org/10.1109/TPAMI.2019.2944806
13. Guo, J., Chao, H.: Building dual-domain representations for compression artifacts reduction. In: Leibe, B., Matas, J., Sebe, N., Welling, M. (eds.) ECCV 2016. LNCS, vol. 9905, pp. 628–644. Springer, Cham (2016). https://doi.org/10.1007/978-3-319-46448-0_38

14. Guo, S., Yan, Z., Zhang, K., Zuo, W., Zhang, L.: Toward convolutional blind denoising of real photographs. In: IEEE Conference on Computer Vision and Pattern Recognition (CVPR), pp. 1712–1722 (2019)
15. He, K., Zhang, X., Ren, S., Sun, J.: Deep residual learning for image recognition. In: IEEE Conference on Computer Vision and Pattern Recognition (CVPR), pp. 770–778 (2016)
16. He, X., Hu, Q., Zhang, X., Zhang, C., Lin, W., Han, X.: Enhancing HEVC compressed videos with a partition-masked convolutional neural network. In: IEEE International Conference on Image Processing (ICIP), pp. 216–220. IEEE (2018)
17. Hennings-Yeomans, P.H., Baker, S., Kumar, B.V.: Simultaneous super-resolution and feature extraction for recognition of low-resolution faces. In: IEEE Conference on Computer Vision and Pattern Recognition (CVPR), pp. 1–8. IEEE (2008)
18. Huang, G., Liu, Z., Van Der Maaten, L., Weinberger, K.Q.: Densely connected convolutional networks. In: IEEE Conference on Computer Vision and Pattern Recognition (CVPR), pp. 4700–4708 (2017)
19. Ioffe, S., Szegedy, C.: Batch normalization: accelerating deep network training by reducing internal covariate shift. arXiv preprint arXiv:1502.03167 (2015)
20. Kingma, D.P., Ba, J.: Adam: A method for stochastic optimization. arXiv preprint arXiv:1412.6980 (2014)
21. Lee, C.Y., Xie, S., Gallagher, P., Zhang, Z., Tu, Z.: Deeply-supervised nets. In: Artificial Intelligence and Statistics, pp. 562–570 (2015)
22. Li, K., Bare, B., Yan, B.: An efficient deep convolutional neural networks model for compressed image deblocking. In: IEEE International Conference on Multimedia and Expo (ICME), pp. 1320–1325. IEEE (2017)
23. Li, L., Zhou, Y., Lin, W., Wu, J., Zhang, X., Chen, B.: No-reference quality assessment of deblocked images. Neurocomputing **177**, 572–584 (2016)
24. Li, L., Zhu, H., Yang, G., Qian, J.: Referenceless measure of blocking artifacts by Tchebichef kernel analysis. IEEE Signal Process. Lett. **21**(1), 122–125 (2013)
25. Li, S., Xu, M., Ren, Y., Wang, Z.: Closed-form optimization on saliency-guided image compression for HEVC-MSP. IEEE Trans. Multimed. (TMM) **20**(1), 155–170 (2017)
26. Liu, Y., Hamidouche, W., Déforges, O., Lui, Y., Dforges, O.: Intra Coding Performance Comparison of HEVC, H.264/AVC, Motion-JPEG2000 and JPEGXR Encoders. Research report, IETR/INSA Rennes, September 2018. https://hal.archives-ouvertes.fr/hal-01876856
27. Lundh, F.: Python imaging library (PIL). http://www.pythonware.com/products/pil
28. Marcellin, M.W., Gormish, M.J., Bilgin, A., Boliek, M.P.: An overview of JPEG-2000. In: Data Compression Conference (DCC), pp. 523–541. IEEE (2000)
29. Mukundan, R., Ong, S., Lee, P.A.: Image analysis by Tchebichef moments. IEEE Trans. Image Process. (TIP) **10**(9), 1357–1364 (2001)
30. Nair, V., Hinton, G.E.: Rectified linear units improve restricted Boltzmann machines. In: The 27th International Conference on Machine Learning (ICML), pp. 807–814 (2010)
31. Nguyen, T., Marpe, D.: Performance analysis of HEVC-based intra coding for still image compression. In: Picture Coding Symposium (PCS), pp. 233–236. IEEE (2012)
32. Nguyen, T., Marpe, D.: Objective performance evaluation of the HEVC main still picture profile. IEEE Trans. Circuits Syst. Video Technol. (TCSVT) **25**(5), 790–797 (2014)

33. Norkin, A., et al.: HEVC deblocking filter. IEEE Trans. Circuits Syst. Video Technol. (TCSVT) **22**(12), 1746–1754 (2012)
34. Ren, D., Zuo, W., Hu, Q., Zhu, P., Meng, D.: Progressive image deraining networks: a better and simpler baseline. In: Proceedings of the IEEE Conference on Computer Vision and Pattern Recognition (CVPR), pp. 3937–3946 (2019)
35. Seshadrinathan, K., Soundararajan, R., Bovik, A.C., Cormack, L.K.: Study of subjective and objective quality assessment of video. IEEE Trans. Image Process. (TIP) **19**(6), 1427–1441 (2010)
36. Simonyan, K., Zisserman, A.: Very deep convolutional networks for large-scale image recognition. arXiv preprint arXiv:1409.1556 (2014)
37. Sullivan, G.J., Ohm, J.R., Han, W.J., Wiegand, T.: Overview of the high efficiency video coding (HEVC) standard. IEEE Trans. Circuits Syst. Video Technol. (TCSVT) **22**(12), 1649–1668 (2012)
38. Tai, Y., Yang, J., Liu, X., Xu, C.: Memnet: a persistent memory network for image restoration. In: IEEE International Conference on Computer Vision (ICCV), pp. 4539–4547 (2017)
39. Tan, T.K., Weerakkody, R., Mrak, M., Ramzan, N., Baroncini, V., Ohm, J.R., Sullivan, G.J.: Video quality evaluation methodology and verification testing of HEVC compression performance. IEEE Trans. Circuits Syst. Video Technol. (TCSVT) **26**(1), 76–90 (2015)
40. Wallace, G.K.: The JPEG still picture compression standard. IEEE Trans. Consum. Electron. (TCE) **38**(1), xviii–xxxiv (1992). https://doi.org/10.1109/30.125072
41. Wang, Q., Wu, B., Zhu, P., Li, P., Zuo, W., Hu, Q.: ECA-Net: efficient channel attention for deep convolutional neural networks. arXiv preprint arXiv:1910.03151 (2019)
42. Wang, T., Chen, M., Chao, H.: A novel deep learning-based method of improving coding efficiency from the decoder-end for HEVC. In: Data Compression Conference (DCC), pp. 410–419. IEEE (2017)
43. Wang, Z., Liu, D., Chang, S., Ling, Q., Yang, Y., Huang, T.S.: D3: deep dual-domain based fast restoration of JPEG-compressed images. In: IEEE Conference on Computer Vision and Pattern Recognition (CVPR), pp. 2764–2772 (2016)
44. Wang, Z., Bovik, A.C., Sheikh, H.R., Simoncelli, E.P., et al.: Image quality assessment: from error visibility to structural similarity. IEEE Trans. Image Process. (TIP) **13**(4), 600–612 (2004)
45. Xu, M., Li, T., Wang, Z., Deng, X., Yang, R., Guan, Z.: Reducing complexity of HEVC: a deep learning approach. IEEE Trans. Image Process. (TIP) **27**(10), 5044–5059 (2018)
46. Yang, R., Xu, M., Liu, T., Wang, Z., Guan, Z.: Enhancing quality for HEVC compressed videos. IEEE Trans. Circuits Syst. Video Technol. (TCSVT) 2039–2054 (2018)
47. Yang, R., Xu, M., Wang, Z., Li, T.: Multi-frame quality enhancement for compressed video. In: IEEE Conference on Computer Vision and Pattern Recognition (CVPR), pp. 6664–6673 (2018)
48. Zhang, H., Yang, J., Zhang, Y., Nasrabadi, N.M., Huang, T.S.: Close the loop: joint blind image restoration and recognition with sparse representation prior. In: IEEE International Conference on Computer Vision (ICCV), pp. 770–777. IEEE (2011)
49. Zhang, K., Zuo, W., Chen, Y., Meng, D., Zhang, L.: Beyond a Gaussian denoiser: residual learning of deep CNN for image denoising. IEEE Trans. Image Process. (TIP) **26**(7), 3142–3155 (2017)

50. Zhang, K., Zuo, W., Zhang, L.: FFDNet: toward a fast and flexible solution for CNN-based image denoising. IEEE Trans. Image Process. (TIP) **27**(9), 4608–4622 (2018)
51. Zhou, Z., Rahman Siddiquee, M.M., Tajbakhsh, N., Liang, J.: UNet++: a nested U-Net architecture for medical image segmentation. In: Stoyanov, D., et al. (eds.) DLMIA/ML-CDS -2018. LNCS, vol. 11045, pp. 3–11. Springer, Cham (2018). https://doi.org/10.1007/978-3-030-00889-5_1

PatchNets: Patch-Based Generalizable Deep Implicit 3D Shape Representations

Edgar Tretschk[1], Ayush Tewari[1], Vladislav Golyanik[1(✉)], Michael Zollhöfer[2], Carsten Stoll[2], and Christian Theobalt[1]

[1] Max Planck Institute for Informatics, Saarland Informatics Campus, Saarbrücken, Germany
golyanik@mpi-inf.mpg.de
[2] Facebook Reality Labs, Pittsburgh, USA

Abstract. Implicit surface representations, such as signed-distance functions, combined with deep learning have led to impressive models which can represent detailed shapes of objects with arbitrary topology. Since a continuous function is learned, the reconstructions can also be extracted at any arbitrary resolution. However, large datasets such as ShapeNet are required to train such models.

In this paper, we present a new mid-level patch-based surface representation. At the level of patches, objects across different categories share similarities, which leads to more generalizable models. We then introduce a novel method to learn this patch-based representation in a canonical space, such that it is as object-agnostic as possible. We show that our representation trained on one category of objects from ShapeNet can also well represent detailed shapes from any other category. In addition, it can be trained using much fewer shapes, compared to existing approaches. We show several applications of our new representation, including shape interpolation and partial point cloud completion. Due to explicit control over positions, orientations and scales of patches, our representation is also more controllable compared to object-level representations, which enables us to deform encoded shapes non-rigidly.

Keywords: Implicit functions · Patch-based surface representation · Intra-object class generalizability

1 Introduction

Several 3D shape representations exist in the computer vision and computer graphics communities, such as point clouds, meshes, voxel grids and implicit functions. Learning-based approaches have mostly focused on voxel grids due to their regular structure, suited for convolutions. However, voxel grids [5] come

---

Electronic supplementary material The online version of this chapter (https://doi.org/10.1007/978-3-030-58517-4_18) contains supplementary material, which is available to authorized users.

A. Vedaldi et al. (Eds.): ECCV 2020, LNCS 12361, pp. 293–309, 2020.
https://doi.org/10.1007/978-3-030-58517-4_18

with large memory costs, limiting the output resolution of such methods. Point cloud based approaches have also been explored [23]. While most approaches assume a fixed number of points, recent methods also allow for variable resolution outputs [17,28]. Point clouds only offer a sparse representation of the surface. Meshes with fixed topology are commonly used in constrained settings with known object categories [32]. However, they are not suitable for representing objects with varying topology. Very recently, implicit function-based representations were introduced [4,17,21]. DeepSDF [21] learns a network which represents the continuous signed distance functions for a class of objects. The surface is represented as the 0-isosurface. Similar approaches [4,17] use occupancy networks, where only the occupancy values are learned (similar to voxel grid-based approaches), but in a continuous representation. Implicit functions allow for representing (closed) shapes of arbitrary topology. The reconstructed surface can be extracted at any resolution, since a continuous function is learned.

All existing implicit function-based methods rely on large datasets of 3D shapes for training. Our goal is to build a generalizable surface representation which can be trained with much fewer shapes, and can also generalize to different object categories. Instead of learning an object-level representation, our *PatchNet* learns a mid-level representation of surfaces, at the level of patches. At the level of patches, objects across different categories share similarities. We learn these patches in a canonical space to further abstract from object-specific details. Patch extrinsics (position, scale and orientation of a patch) allow each patch to be translated, rotated and scaled. Multiple patches can be combined in order to represent the full surface of an object. We show that our patches can be learned using very few shapes, and can generalize across different object categories, see Fig. 1. Our representation also allows to build object-level models, *ObjectNets*, which is useful for applications which require an object-level prior.

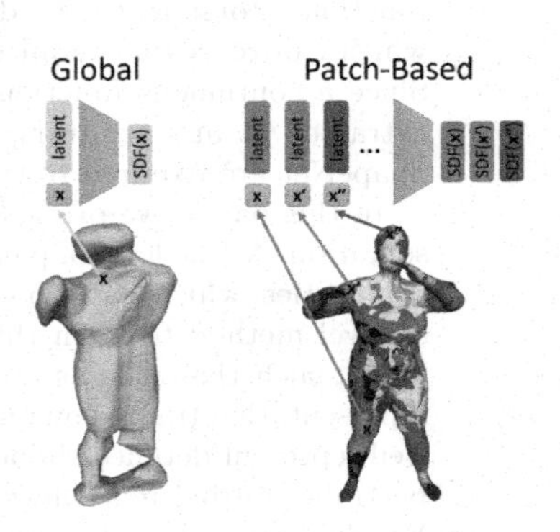

Fig. 1. In contrast to a global approach, our patch-based method generalizes to human shapes after being trained on rigid ShapeNet objects.

We demonstrate several applications of our trained models, including partial point cloud completion from depth maps, shape interpolation, and a generative model for objects. While implicit function-based approaches can reconstruct high-quality and detailed shapes, they lack controllability. We show that our patch-based implicit representation *natively* allows for controllability due to the explicit control over patch extrinsics. By user-guided rigging of the patches to the surface, we allow for articulated deformation of humans without re-encoding the deformed shapes. In addition to the generalization and editing capabilities, our representation includes all advantages of implicit surface modeling. Our patches can represent shapes of any arbitrary topology, and our reconstructions can be extracted at any arbitrary resolution using *Marching Cubes* [16]. Similar to DeepSDF [21], our network uses an auto-decoder architecture, combining classical optimization with learning, resulting in high-quality geometry.

2 Related Work

Our patch-based representation relates to many existing data structures and approaches in classical and learning-based visual computing. In the following, we focus on the most relevant existing representations, methods and applications.

Global Data Structures. There are multiple widely-used data structures for geometric deep learning such as voxel grids [5], point clouds [23], meshes [32] and implicit functions [21]. To alleviate the memory limitations and speed-up training, improved versions of voxel grids with hierarchical space partitioning [24] and tri-linear interpolation [27] were recently proposed. A mesh is an explicit discrete surface representation which can be useful in monocular rigid 3D reconstruction [13,32]. Combined with patched-based policies, this representation can suffer from stitching artefacts [12]. All these data structures enable a limited level of detail given a constant memory size. In contrast, other representations such as sign distance functions (SDF) [6] represent surfaces implicitly as the zero-crossing of a volumetric level set function.

Recently, neural counterparts of implicit representations and approaches operating on them were proposed in the literature [4,17,18,21]. Similarly to SDFs, these methods extract surfaces as zero level sets or decision boundaries, while differing in the type of the learned function. Thus, DeepSDF is a learnable variant of SDFs [21], whereas Mescheder *et al.* [17] train a spatial classifier (indicator function) for regions inside and outside of the scene. In theory, both methods allow for surface extraction at unlimited resolution. Neural implicit functions have already demonstrated their effectiveness and robustness in many follow-up works and applications such as single-view 3D reconstruction [15,25] as well as static [28] and dynamic [19] object representation. While SAL [1] perform shape completion from noisy full raw scans, one of our applications is shape completion from partial data with local refinement. Unlike the aforementioned global approaches, PatchNets generalize much better, for example to new categories.

Patch-Based Representations. Ohtake *et al.* [20] use a combination of implicit functions for versatile shape representation and editing. Several neural techniques use mixtures of geometric primitives as well [7,9,11,31,33]. The latter have been shown as helpful abstractions in such tasks as shape segmentation, interpolation, classification and recognition, as well as 3D reconstruction. Tulsiani *et al.* [31] learn to assemble shapes of various categories from explicit 3D geometric primitives (*e.g.,* cubes and cuboids). Their method discovers a consistent structure and allows to establish semantic correspondences between the samples. Genova *et al.* [11] further develop the idea and learn a general template from data which is composed of implicit functions with local support. Due to the function choice, *i.e.,* scaled axis-aligned anisotropic 3D Gaussians, shapes with sharp edges and thin structures are challenging for their method. In *CVXNets* [7], solid objects are assembled in a piecewise manner from convex elements. This results in a differentiable form which is directly usable in physics and graphics engines. Deprelle *et al.* [9] decompose shapes into learnable combinations of deformable elementary 3D structures. VoronoiNet [33] is a deep generative network which operates on a differentiable version of Voronoi diagrams. The concurrent NASA method [8] focuses on articulated deformations, which is one of our applications. In contrast to other patch-based approaches, our learned patches are not limited to hand-crafted priors but instead are more flexible and expressive.

3 Proposed Approach

We represent the surface of any object as a combination of several surface patches. The patches form a mid-level representation, where each patch represents the surface within a specified radius from its center. This representation is generalizable across object categories, as most objects share similar geometry at the patch level. In the following, we explain how the patches are represented using artificial neural networks, the losses required to train such networks, as well as the algorithm to combine multiple patches for smooth surface reconstruction.

3.1 Implicit Patch Representation

We represent a full object i as a collection of $N_P = 30$ patches. A patch p represents a surface within a sphere of radius $r_{i,p} \in \mathbb{R}$, centered at $\mathbf{c}_{i,p} \in \mathbb{R}^3$. Each patch can be oriented by a rotation about a canonical frame, parametrized by Euler angles $\phi_{i,p} \in \mathbb{R}^3$. Let $\mathbf{e}_{i,p} = (r_{i,p}, \mathbf{c}_{i,p}, \phi_{i,p}) \in \mathbb{R}^7$ denote all extrinsic patch parameters. Representing the patch surface in a canonical frame of reference lets us normalize the query 3D point, leading to more object-agnostic and generalizable patches.

The patch surface is represented as an implicit signed-distance function (SDF), which maps 3D points to their signed distance from the closest surface. This offers several advantages, as these functions are a continuous representation of the surface, unlike point clouds or meshes. In addition, the surface can be extracted at any resolution without large memory requirement, unlike for voxel grids. In contrast to prior work [11,33], which uses simple patch primitives, we parametrize the patch surface as a neural network (PatchNet). Our network architecture is based on the auto-decoder of DeepSDF [21]. The input to the network is a *patch latent code* $\mathbf{z} \in \mathbb{R}^{N_z}$ of length $N_z = 128$, which describes the patch surface, and a 3D query point $\mathbf{x} \in \mathbb{R}^3$. The output is the scalar SDF value of the surface at $\mathbf{x}$. Similar to DeepSDF, we use eight weight-normalized [26] fully-connected layers with 128 output dimensions and ReLU activations, and we also concatenate $\mathbf{z}$ and $\mathbf{x}$ to the input of the fifth layer. The last fully-connected layer outputs a single scalar to which we apply *tanh* to obtain the SDF value.

3.2 Preliminaries

Preprocessing: Given a watertight mesh, we preprocess it to obtain SDF values for 3D point samples. First, we center each mesh and fit it tightly into the unit sphere. We then sample points, mostly close to the surface, and compute their truncated signed distance to the object surface, with truncation at 0.1. For more details on the sampling strategy, please refer to [21].

Auto-Decoding: Unlike the usual setting, we do not use an encoder that regresses patch latent codes and extrinsics. Instead, we follow DeepSDF [21] and auto-decode shapes: we treat the patch latent codes and extrinsics of each object as free variables to be optimized for during training. I.e., instead of back-propagating into an encoder, we employ the gradients to learn these parameters directly during training.

Initialization: Since we perform auto-decoding, we treat the patch latent codes and extrinsics as free variables, similar to classical optimization. Therefore, we can directly initialize them. All patch latent codes are initially set to zero, and the patch positions are initialized by greedy farthest point sampling of point samples of the object surface. We set each patch radius to the minimum such that each surface point sample is covered by its closest patch. The patch orientation aligns the z-axis of the patch coordinate system with the surface normal.

3.3 Loss Functions

We train PatchNet by auto-decoding N full objects. The patch latent codes of an object i are $\mathbf{z}_i = [\mathbf{z}_{i,0}, \mathbf{z}_{i,1}, \ldots, \mathbf{z}_{i,N_P-1}]$, with each patch latent code of length N_z. Patch extrinsics are represented as $\mathbf{e}_i = [\mathbf{e}_{i,0}, \mathbf{e}_{i,1}, \ldots, \mathbf{e}_{i,N_P-1}]$. Let θ denote the trainable weights of PatchNet. We employ the following loss function:

$$\mathcal{L}(\mathbf{z}_i, \mathbf{e}_i, \theta) = \mathcal{L}_{\text{recon}}(\mathbf{z}_i, \mathbf{e}_i, \theta) + \mathcal{L}_{\text{ext}}(\mathbf{e}_i) + \mathcal{L}_{\text{reg}}(\mathbf{z}_i) \ . \tag{1}$$

Here, $\mathcal{L}_{recon}$ is the surface reconstruction loss, $\mathcal{L}_{ext}$ is the extrinsic loss guiding the extrinsics for each patch, and $\mathcal{L}_{reg}$ is a regularizer on the patch latent codes.

Reconstruction Loss: The reconstruction loss minimizes the SDF values between the predictions and the ground truth for each patch:

$$\mathcal{L}_{\text{recon}}(\mathbf{z}_i, \mathbf{e}_i, \theta) = \frac{1}{N_P} \sum_{p=0}^{N_P-1} \frac{1}{|S(\mathbf{e}_{i,p})|} \sum_{\mathbf{x} \in S(\mathbf{e}_{i,p})} \left\| f(\mathbf{x}, \mathbf{z}_{i,p}, \theta) - s(\mathbf{x}) \right\|_1, \tag{2}$$

where $f(\cdot)$ and $s(\mathbf{x})$ denote a forward pass of the network and the ground truth truncated SDF values at point $\mathbf{x}$, respectively; $S(\mathbf{e}_{i,p})$ is the set of all (normalized) point samples that lie within the bounds of patch p with extrinsics $\mathbf{e}_{i,p}$.

Extrinsic Loss: The composite extrinsic loss ensures all patches contribute to the surface and are placed such that the surfaces are learned in a canonical space:

$$\mathcal{L}_{\text{ext}}(\mathbf{e}_i) = \mathcal{L}_{\text{sur}}(\mathbf{e}_i) + \mathcal{L}_{\text{cov}}(\mathbf{e}_i) + \mathcal{L}_{\text{rot}}(\mathbf{e}_i) + \mathcal{L}_{\text{scl}}(\mathbf{e}_i) + \mathcal{L}_{\text{var}}(\mathbf{e}_i) \ . \tag{3}$$

$\mathcal{L}_{\text{sur}}$ ensures that every patch stays close to the surface:

$$\mathcal{L}_{\text{sur}}(\mathbf{e}_i) = \omega_{\text{sur}} \cdot \frac{1}{N_P} \sum_{p=0}^{N_P-1} \max(\min_{\mathbf{x}\in\mathbf{O}_i} \left\|\mathbf{c}_{i,p} - \mathbf{x}\right\|_2^2, t) \ . \tag{4}$$

Here, $\mathbf{O}_i$ is the set of surface points of object i. We use this term only when the distance between a patch and the surface is greater than a threshold $t = 0.06$.

A symmetric coverage loss $\mathcal{L}_{\text{cov}}$ encourages each point on the surface to be covered by a at least one patch:

$$\mathcal{L}_{\text{cov}}(\mathbf{e}_i) = \omega_{\text{cov}} \cdot \frac{1}{|\mathbf{U}_i|} \sum_{\mathbf{x}\in\mathbf{U}_i} \frac{w_{i,p,\mathbf{x}}}{\sum_p w_{i,p,\mathbf{x}}} (\left\|\mathbf{c}_{i,p} - \mathbf{x}\right\|_2 - r_{i,p}) \ , \tag{5}$$

where $\mathbf{U}_i \subseteq \mathbf{O}_i$ are all surface points that are not covered by any patch, *i.e.*, outside the bounds of all patches. $w_{i,p,\mathbf{x}}$ weighs the patches based on their distance from $\mathbf{x}$, with $w_{i,p,\mathbf{x}} = \exp\left(-0.5 \cdot ((\left\|\mathbf{c}_{i,p} - \mathbf{x}\right\|_2 - r_{i,p})/\sigma)^2\right)$ where $\sigma = 0.05$.

We also introduce a loss to align the patches with the surface normals. This encourages the patch surface to be learned in a canonical frame of reference:

$$\mathcal{L}_{\text{rot}}(\mathbf{e}_i) = \omega_{\text{rot}} \cdot \frac{1}{N_P} \sum_{p=0}^{N_P-1} (1 - \langle \phi_{i,p} \cdot [0,0,1]^T, \mathbf{n}_{i,p} \rangle)^2 \ . \tag{6}$$

Here, $\mathbf{n}_{i,p}$ is the surface normal at the point $\mathbf{o}_{i,p}$ closest to the patch center, *i.e.*, $\mathbf{o}_{i,p} = \operatorname{argmin}_{\mathbf{x}\in\mathbf{O}_i} \left\|\mathbf{x} - \mathbf{c}_{i,p}\right\|_2$.

Finally, we introduce two losses for the extent of the patches. The first loss encourages the patches to be reasonably small. This prevents significant overlap between different patches:

$$\mathcal{L}_{\text{scl}}(\mathbf{e}_i) = \omega_{\text{scl}} \cdot \frac{1}{N_P} \sum_{p=0}^{N_P-1} r_{i,p}^2 \ . \tag{7}$$

The second loss encourages all patches to be of similar sizes. This prevents the surface to be reconstructed only using very few large patches:

$$\mathcal{L}_{\text{var}}(\mathbf{e}_i) = \omega_{\text{var}} \cdot \frac{1}{N_P} \sum_{p=0}^{N_P-1} (r_{i,p} - m_i)^2 \ , \tag{8}$$

where m_i is the mean patch radius of object i.

Regularizer: Similar to DeepSDF, we add an ℓ_2-regularizer on the latent codes assuming a Gaussian prior distribution:

$$\mathcal{L}_{\text{reg}}(\mathbf{z}_i) = \omega_{\text{reg}} \cdot \frac{1}{N_P} \sum_{p=0}^{N_P-1} \left\| \mathbf{z}_{i,p} \right\|_2^2 . \tag{9}$$

Optimization: At training time, we optimize the following problem:

$$\underset{\theta, \{\mathbf{z}_i\}_i, \{\mathbf{e}_i\}_i}{\text{argmin}} \sum_{i=0}^{N-1} \mathcal{L}(\mathbf{z}_i, \mathbf{e}_i, \theta) . \tag{10}$$

At test time, we can reconstruct any surface using our learned patch-based representation. Using the same initialization of extrinsics and patch latent codes, and given point samples with their SDF values, we optimize for the patch latent codes and the patches extrinsics with fixed network weights.

3.4 Blended Surface Reconstruction

For a smooth surface reconstruction of object i, *e.g.* for Marching Cubes, we blend between different patches in the overlapping regions to obtain the blended SDF prediction $g_i(\mathbf{x})$. Specifically, $g_i(\mathbf{x})$ is computed as a weighted linear combination of the SDF values $f(\mathbf{x}, \mathbf{z}_{i,p}, \theta)$ of the overlapping patches:

$$g_i(\mathbf{x}) = \sum_{p \in P_{i,\mathbf{x}}} \frac{w_{i,p,\mathbf{x}}}{\sum_{p \in P_{i,\mathbf{x}}} w_{i,p,\mathbf{x}}} f(\mathbf{x}, \mathbf{z}_{i,p}, \theta), \tag{11}$$

with $P_{i,\mathbf{x}}$ denoting the patches which overlap at point $\mathbf{x}$. For empty $P_{i,\mathbf{x}}$, we set $g_i(\mathbf{x}) = 1$. The blending weights are defined as:

$$w_{i,p,\mathbf{x}} = \exp\left(-\frac{1}{2} \left(\frac{\left\| \mathbf{c}_{i,p} - \mathbf{x} \right\|_2}{\sigma} \right)^2 \right) - \exp\left(-\frac{1}{2} \left(\frac{r_{i,p}}{\sigma} \right)^2 \right), \tag{12}$$

with $\sigma = r_{i,p}/3$. The offset ensures that the weight is zero at the patch boundary.

4 Experiments

In the following, we show the effectiveness of our patch-based representation on several different problems. For an ablation study of the loss functions, please refer to the supplemental.

4.1 Settings

Datasets. We employ *ShapeNet* [3] for most experiments. We perform preprocessing with the code of Stutz *et al.* [29], similar to [10,17], to make the meshes watertight and normalize them within a unit cube. For training and test splits, we follow Choy *et al.* [5]. The results in Tables 1 and 2 use the full test set. Other results refer to a reduced test set, where we randomly pick 50 objects from each of the 13 categories. In the supplemental, we show that our results on the reduced test set are representative of the full test set. In addition, we use Dynamic FAUST [2] for testing. We subsample the test set from DEMEA [30] by concatenating all test sequences and taking every 20th mesh. We generate 200k SDF point samples per shape during preprocessing.

Metrics. We use three error metrics. For Intersection-over-Union (IoU), higher is better. For Chamfer distance (Chamfer), lower is better. For F-score, higher is better. The supplementary material contains further details on these metrics.

Training Details. We train our networks using *PyTorch* [22]. The number of epochs is 1000, the learning rate for the network is initially $5 \cdot 10^{-4}$, and for the patch latent codes and extrinsics 10^{-3}. We half both learning rates every 200 epochs. For optimization, we use Adam [14] and a batch size of 64. For each object in the batch, we randomly sample 3k SDF point samples. The weights for the losses are: $\omega_{\text{scl}} = 0.01$, $\omega_{\text{var}} = 0.01$, $\omega_{\text{sur}} = 5$, $\omega_{\text{rot}} = 1$, $\omega_{\text{sur}} = 200$. We linearly increase ω_{reg} from 0 to 10^{-4} for 400 epochs and then keep it constant.

Baseline. We design a "global-patch" baseline similar to DeepSDF, which only uses a single patch without extrinsics. The patch latent size is 4050, matching ours. The learning rate scheme is the same as for our method.

4.2 Surface Reconstruction

We first consider surface reconstruction.

Results. We train our approach on a subset of the training data, where we randomly pick 100 shapes from each category. In addition to comparing with our baseline, we compare with DeepSDF [21] as setup in their paper. Both DeepSDF and our baseline use the subset. Qualitative results are shown in Figs. 2 and 3.

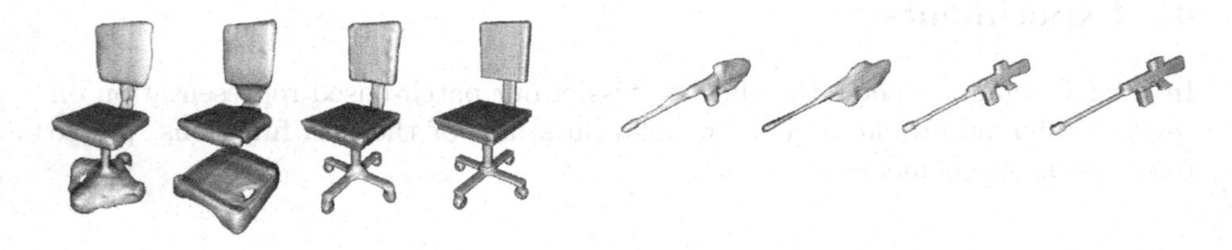

Fig. 2. Surface Reconstruction. From left to right: DeepSDF, baseline, ours, groundtruth.

Table 1. Surface Reconstruction. We significantly outperform DeepSDF [21] and our baseline on all categories of ShapeNet almost everywhere.

Category	IoU			Chamfer			F-score		
	DeepSDF	Baseline	Ours	DeepSDF	Baseline	Ours	DeepSDF	Baseline	Ours
Airplane	84.9	65.3	**91.1**	0.012	0.077	**0.004**	83.0	72.9	**97.8**
Bench	78.3	68.0	**85.4**	0.021	0.065	**0.006**	91.2	80.6	**95.7**
Cabinet	92.2	88.8	**92.9**	**0.033**	0.055	0.110	**91.6**	86.4	91.2
Car	87.9	83.6	**91.7**	**0.049**	0.070	**0.049**	82.2	74.5	**87.7**
Chair	81.8	72.9	**90.0**	0.042	0.110	**0.018**	86.6	75.5	**94.3**
Display	91.6	86.5	**95.2**	**0.030**	0.061	0.039	93.7	87.0	**97.0**
Lamp	74.9	63.0	**89.6**	0.566	0.438	**0.055**	82.5	69.4	**94.9**
Rifle	79.0	68.5	**93.3**	0.013	0.039	**0.002**	90.9	82.3	**99.3**
Sofa	92.5	85.4	**95.0**	0.054	0.226	**0.014**	92.1	84.2	**95.3**
Speaker	91.9	86.7	**92.7**	**0.050**	0.094	0.243	87.6	79.4	**88.5**
Table	84.2	71.9	**89.4**	0.074	0.156	**0.018**	91.1	79.2	**95.0**
Telephone	96.2	95.0	**98.1**	0.008	0.016	**0.003**	97.7	96.2	**99.4**
Watercraft	85.2	79.1	**93.2**	0.026	0.041	**0.009**	87.8	80.2	**96.4**
Mean	77.4	76.5	**92.1**	0.075	0.111	**0.044**	89.9	80.6	**94.8**

Table 1 shows the quantitative results for surface reconstruction. We significantly outperform DeepSDF and our baseline almost everywhere, demonstrating the higher-quality afforded by our patch-based representation.

We also compare with several state-of-the-art approaches on implicit surface reconstruction, OccupancyNetworks [17], Structured Implicit Functions [11] and Deep Structured Implicit Functions [10][1]. While they are trained on the full ShapeNet shapes, we train our model only on a small subset. Even in this disadvantageous and challenging setting, we outperform these approaches on most categories, see Table 2. Note that we compute the metrics consistently with Genova *et al.* [10] and thus can directly compare to numbers reported in their paper.

Generalization. Our patch-based representation is more generalizable compared to existing representations. To demonstrate this, we design several experiments with different training data. We modify the learning rate schemes to equalize the number of network weight updates. For each experiment, we compare our method with the baseline approaches described above. We use a reduced ShapeNet test set, which consists of 50 shapes from each category. Figure 3 shows qualitative results and comparisons. We also show cross-dataset generalization by evaluating on 647 meshes from the Dynamic FAUST [2] test set. In the first experiment, we train the network on shapes from the *Cabinet* category and try to reconstruct shapes from every other category. We significantly outperform the baselines almost everywhere, see Table 3. The improvement is even more noticeable for cross dataset generalization with around 70% improvement in the F-score compared to our global-patch baseline.

[1] DSIF is also known as *Local Deep Implicit Functions for 3D Shape*.

Table 2. Surface Reconstruction. We outperform OccupancyNetworks (OccNet) [17], Structured Implicit Functions (SIF) [11], and Deep Structured Implicit Functions (DSIF) [10] almost everywhere.

Category	IoU				Chamfer				F-score			
	OccNet	SIF	DSIF	Ours	OccNet	SIF	DSIF	Ours	OccNet	SIF	DSIF	Ours
Airplane	77.0	66.2	**91.2**	91.1	0.016	0.044	0.010	**0.004**	87.8	71.4	96.9	**97.8**
Bench	71.3	53.3	**85.6**	85.4	0.024	0.082	0.017	**0.006**	87.5	58.4	94.8	**95.7**
Cabinet	86.2	78.3	**93.2**	92.9	0.041	0.110	**0.033**	0.110	86.0	59.3	**92.0**	91.2
Car	83.9	77.2	90.2	**91.7**	0.061	0.108	**0.028**	0.049	77.5	56.6	87.2	**87.7**
Chair	73.9	57.2	87.5	**90.0**	0.044	0.154	0.034	**0.018**	77.2	42.4	90.9	**94.3**
Display	81.8	69.3	94.2	**95.2**	0.034	0.097	**0.028**	0.039	82.1	56.3	94.8	**97.0**
Lamp	56.5	41.7	77.9	**89.6**	0.167	0.342	0.180	**0.055**	62.7	35.0	83.5	**94.9**
Rifle	69.5	60.4	89.9	**93.3**	0.019	0.042	0.009	**0.002**	86.2	70.0	97.3	**99.3**
Sofa	87.2	76.0	94.1	**95.0**	0.030	0.080	0.035	**0.014**	85.9	55.2	92.8	**95.3**
Speaker	82.4	74.2	90.3	**92.7**	0.101	0.199	**0.068**	0.243	74.7	47.4	84.3	**88.5**
Table	75.6	57.2	88.2	**89.4**	0.044	0.157	0.056	**0.018**	84.9	55.7	92.4	**95.0**
Telephone	90.9	83.1	97.6	**98.1**	0.013	0.039	0.008	**0.003**	94.8	81.8	98.1	**99.4**
Watercraft	74.7	64.3	90.1	**93.2**	0.041	0.078	0.020	**0.009**	77.3	54.2	93.2	**96.4**
Mean	77.8	66.0	90.0	**92.1**	0.049	0.118	**0.040**	0.044	81.9	59.0	92.2	**94.8**

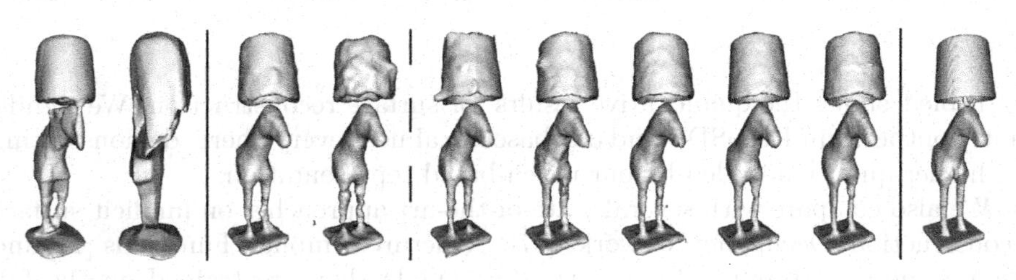

Fig. 3. Generalization. From left to right: DeepSDF, baseline, ours on one category, ours on one shape, ours on 1 shape per category, ours on 3 per category, ours on 10 per category, ours on 30 per category, ours on 100 per category, and groundtruth.

In the second experiment, we evaluate the amount of training data required to train our network. We train both our network as well as the baselines on 30, 10, 3 and 1 shapes per-category of ShapeNet. In addition, we also include an experiment training the networks on a single randomly picked shape from ShapeNet. Figure 4 shows the errors for ShapeNet (mean across categories) and Dynamic FAUST. The performance of our approach degrades only slightly with a decreasing number of training shapes. However, the baseline approach of DeepSDF degrades much more severely. This is even more evident for cross dataset generalization on Dynamic FAUST, where the baseline cannot perform well even with a larger number of training shapes, while we perform similarly across datasets.

Ablation Experiments. We perform several ablative analysis experiments to evaluate our approach. We first evaluate the number of patches required to reconstruct surfaces. Table 4 reports these numbers on the reduced test set. The patch networks here are trained on the reduced training set, consisting of 100

Table 3. Generalization. Networks trained on the *Cabinet* category, but evaluated on every category of ShapeNet, as well as on Dynamic FAUST. We significantly outperform the baseline (BL) and DeepSDF (DSDF) almost everywhere.

Category	IoU			Chamfer			F-score		
	BL	DSDF	Ours	BL	DSDF	Ours	BL	DSDF	Ours
Airplane	33.5	56.9	**88.2**	0.668	0.583	**0.005**	33.5	61.7	**96.3**
Bench	49.1	58.8	**80.4**	0.169	0.093	**0.006**	63.6	76.3	**93.3**
cabinet	86.0	91.1	**91.4**	0.045	**0.025**	0.121	86.4	**92.6**	91.7
Car	78.4	83.7	**92.0**	0.101	0.074	**0.050**	62.7	73.9	**87.2**
Chair	50.7	61.8	**86.9**	0.473	0.287	**0.012**	49.1	65.2	**92.5**
Display	83.2	87.6	**94.4**	0.111	0.065	**0.052**	83.9	89.6	**96.9**
Lamp	49.7	59.3	**86.6**	0.689	2.645	**0.082**	50.4	64.5	**93.4**
Rifle	56.4	56.1	**91.8**	0.114	2.669	**0.002**	71.0	54.7	**99.1**
Sofa	81.1	87.3	**94.8**	0.245	0.193	**0.010**	74.2	84.6	**95.2**
Speaker	83.2	88.3	**90.5**	0.163	**0.080**	0.232	71.8	80.1	**84.9**
Table	55.0	73.6	**88.4**	0.469	0.222	**0.020**	61.8	82.8	**95.0**
Telephone	90.4	94.7	**97.3**	0.051	0.015	**0.004**	90.8	96.1	**99.2**
Watercraft	66.5	73.5	**91.8**	0.115	0.157	**0.006**	63.0	74.2	**96.2**
Mean	66.4	74.8	**90.3**	0.263	0.547	**0.046**	66.3	76.6	**93.9**
DFAUST	57.8	71.2	**94.4**	0.751	0.389	**0.012**	25.0	45.4	**94.0**

Table 4. Ablative Analysis. We evaluate the performance using different numbers of patches, as well as using variable sizes of the patch latent code/hidden dimensions, and the training data. The training time is measured on an Nvidia V100 GPU.

	IoU	Chamfer	F-score	Time
$N_P = 3$	73.8	0.15	72.9	1 h
$N_P = 10$	85.2	0.049	88.0	1.5 h
Size 32	82.8	0.066	84.7	1.5 h
Size 512	95.3	0.048	97.2	8 h
Full dataset	92.2	0.050	94.8	156 h
Ours	91.6	0.045	94.5	2h

shapes per ShapeNet category. As expected, the performance becomes better with a larger number of patches, since this would lead to smaller patches which can capture more details and generalize better. We also evaluate the impact of different sizes of the latent codes and hidden dimensions used for the patch network. Larger latent codes and hidden dimensions lead to higher quality results. Similarly, training on the full training dataset, consisting of $33k$ shapes leads to

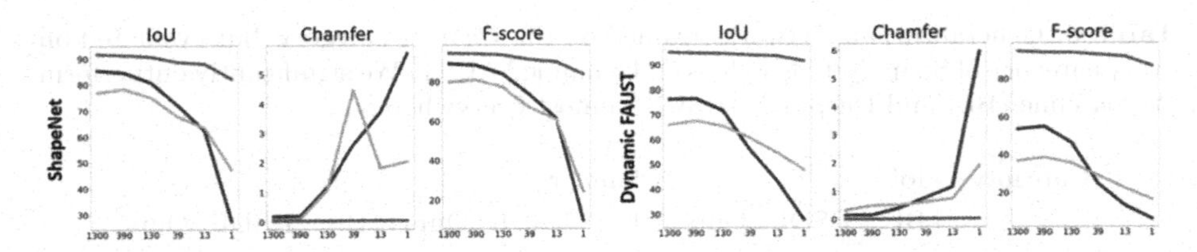

Fig. 4. Generalization. We train our PatchNet (green), the global-patch baseline (orange), and DeepSDF (blue) on different numbers of shapes (x-axis). Results on different metrics on our reduced test sets are shown on the y-axis. For IoU and F-score, higher is better. For Chamfer distance, lower is better.

higher quality. However, all design choices with better performance come at the cost of longer training times, see Table 4.

4.3 Object-Level Priors

We also experiment with *category-specific* object priors. We add ObjectNet (four FC layers with hidden dimension 1024 and ReLU activations) in front of PatchNet and our baselines. From object latent codes of size 256, ObjectNet regresses patch latent codes and extrinsics as an intermediate representation usable with PatchNet. ObjectNet effectively increases the network capacity of our baselines.

Training. We initialize all object latents with zeros and the weights of ObjectNet's last layer with very small numbers. We initialize the bias of ObjectNet's last layer with zeros for patch latent codes and with the extrinsics of an arbitrary object from the category as computed by our initialization in Sect. 3.2. We pretrain PatchNet on ShapeNet. For our method, the PatchNet is kept fixed from this point on. As training set, we use the full training split of the ShapeNet category for which we train. We remove $\mathcal{L}_{\text{rot}}$ completely as it significantly lowers quality. The $L2$ regularization is only applied to the object latent codes. We set $\omega_{\text{var}} = 5$. ObjectNet is trained in three phases, each lasting 1000 epochs. We use the same initial learning rates as when training PatchNet, except in the last phase, where we reduce them by a factor of 5. The batch size is 128.
Phase I: We pretrain ObjectNet to ensure good patch extrinsics. For this, we use the extrinsic loss, $\mathcal{L}_{\text{ext}}$ in Eq. 3, and the regularizer. We set $\omega_{\text{scl}} = 2$.
Phase II: Next, we learn to regress patch latent codes. First, we add a layer that multiplies the regressed scales by 1.3. We then store these extrinsics. Afterwards, we train using $\mathcal{L}_{\text{recon}}$ and two $L2$ losses that keep the regressed position and scale close to the stored extrinsics, with respective weights $1, 3$, and 30.
Phase III: The complete loss $\mathcal{L}$ in Eq. 1, with $\omega_{\text{scl}} = 0.02$, yields final refinements.

Coarse Correspondences. Figure 5 shows that the learned patch distribution is consistent across objects, establishing coarse correspondences between objects.

Fig. 5. Coarse Correspondences. Note the consistent coloring of the patches.

Interpolation. Due to the implicitly learned coarse correspondences, we can encode test objects into object latent codes and then linearly interpolate between them. Figure 6 shows that interpolation of the latent codes leads to a smooth morph between the decoded shapes in 3D space.

Generative Model. We can explore the learned object latent space further by turning ObjectNet into a generative model. Since auto-decoding does not yield an encoder that inputs a known distribution, we have to estimate the unknown input distribution. Therefore, we fit a multivariate Gaussian to the object latent codes obtained at training time. We can then sample new object latent codes from the fitted Gaussian and use them to generate new objects, see Fig. 6.

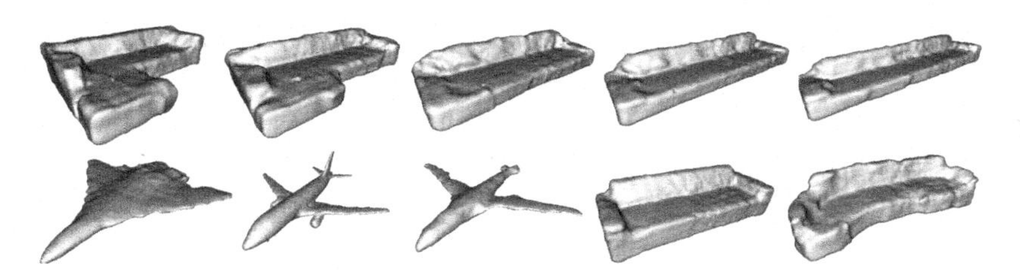

Fig. 6. Interpolation (top). The left and right end points are encoded test objects. Generative Models (bottom). We sample object latents from ObjectNet's fitted prior.

Partial Point Cloud Completion. Given a partial point cloud, we can optimize for the object latent code which best explains the visible region. ObjectNet acts as a prior which completes the missing parts of the shape. For our method, we pretrained our PatchNet on a different object category and keep it fixed, and then train ObjectNet on the target category, which makes this task more challenging for us. We choose the versions of our baselines where the eight final layers are pretrained on all categories and finetuned on the target shape category. We evaluated several other settings, with this one being the most competitive. See the supplemental for more on surface reconstruction with object-level priors.

Optimization: We initialize with the average of the object latent codes obtained at training time. We optimize for 600 iterations, starting with a learning rate of 0.01 and halving it every 200 iterations. Since our method regresses the patch latent codes and extrinsics as an intermediate step, we can further refine the result by treating this intermediate patch-level representation as free variables. Specifically, we refine the patch latent code for the last 100 iterations with a learning rate of 0.001, while keeping the extrinsics fixed. This allows to integrate details not captured by the object-level prior. Figure 7 demonstrates this effect. During optimization, we use the reconstruction loss, the $L2$ regularizer and the coverage loss. The other extrinsics losses have a detrimental effect on patches that are outside the partial point cloud. We use 8k samples per iteration.

We obtain the partial point clouds from depth maps similar to Park *et al.* [21]. We also employ their free-space loss, which encourages the network to regress positive values for samples between the surface and the camera. We use 30% free-space samples. We consider depth maps from a fixed and from a per-scene random viewpoint. For shape completion, we report the F-score between the full groundtruth mesh and the reconstructed mesh. Similar to Park *et al.* [21], we also compute the mesh accuracy for shape completion. It is the 90th percentile of shortest distances from the surface samples of the reconstructed shape to surface samples of the full groundtruth. Table 5 shows how, due to local refinement on the patch level, we outperform the baselines everywhere.

Fig. 7. Shape Completion. (Sofa) from left to right: Baseline, DeepSDF, ours unrefined, ours refined. (Airplane) from left to right: Ours unrefined, ours refined.

Table 5. Partial Point Cloud Completion from Depth Maps. We complete depth maps from a fixed camera viewpoint and from per-scene random viewpoints.

	Sofas fixed		Sofas random		Airplanes fixed		Airplanes Random	
	Acc	F-score	Acc	F-score	Acc	F-score	Acc	F-score
Baseline	0.094	43.0	0.092	42.7	0.069	58.1	0.066	58.7
DeepSDF-based baseline	0.106	33.6	0.101	39.5	0.066	56.9	0.065	55.5
Ours	0.091	48.1	0.077	49.2	0.058	60.5	0.056	59.4
Ours+refined	**0.052**	**53.6**	**0.053**	**52.4**	**0.041**	**67.7**	**0.043**	**65.8**

4.4 Articulated Deformation

Our patch-level representation can model some articulated deformations by *only* modifying the patch extrinsics, without needing to adapt the patch latent codes. Given a template surface and patch extrinsics for this template, we first encode it into patch latent codes. After manipulating the patch extrinsics, we can obtain an articulated surface with our smooth blending from Eq. 11, as Fig. 8 demonstrates.

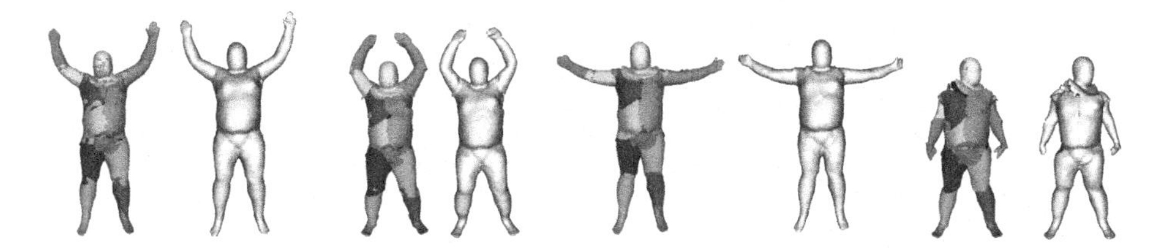

Fig. 8. Articulated Motion. We encode a template shape into patch latent codes (first pair). We then modify the patch extrinsics, while keeping the patch latent codes fixed, leading to non-rigid deformations (middle two pairs). The last pair shows a failure case due to large non-rigid deformations away from the template. Note that the colored patches move rigidly across poses while the mixture deforms non-rigidly.

5 Concluding Remarks

Limitations. We sample the SDF using DeepSDF's sampling strategy, which might limit the level of detail. Generalizability at test time requires optimizing patch latent codes and extrinsics, a problem shared with other auto-decoders. We fit the reduced test set in 71 min due to batching, one object in 10 min.
Conclusion. We have presented a mid-level geometry representation based on patches. This representation leverages the similarities of objects at patch level leading to a highly generalizable neural shape representation. For example, we show that our representation, trained on one object category can also represent other categories. We hope that our representation will enable a large variety of applications that go far beyond shape interpolation and point cloud completion.

Acknowledgements. This work was supported by the ERC Consolidator Grant 4DReply (770784), and an Oculus research grant.

References

1. Atzmon, M., Lipman, Y.: Sal: sign agnostic learning of shapes from raw data. In: Computer Vision and Pattern Recognition (CVPR) (2020)
2. Bogo, F., Romero, J., Pons-Moll, G., Black, M.J.: Dynamic FAUST: registering human bodies in motion. In: Computer Vision and Pattern Recognition (CVPR) (2017)
3. Chang, A.X., et al.: ShapeNet: an information-rich 3D model repository. arXiv preprint arXiv:1512.03012 (2015)
4. Chen, Z., Zhang, H.: Learning implicit fields for generative shape modeling. In: Computer Vision and Pattern Recognition (CVPR) (2019)
5. Choy, C.B., Xu, D., Gwak, J.Y., Chen, K., Savarese, S.: 3D-R2N2: a unified approach for single and multi-view 3D object reconstruction. In: Leibe, B., Matas, J., Sebe, N., Welling, M. (eds.) ECCV 2016. LNCS, vol. 9912, pp. 628–644. Springer, Cham (2016). https://doi.org/10.1007/978-3-319-46484-8_38
6. Curless, B., Levoy, M.: A volumetric method for building complex models from range images. In: SIGGRAPH (1996)

7. Deng, B., Genova, K., Yazdani, S., Bouaziz, S., Hinton, G., Tagliasacchi, A.: CvxNets: learnable convex decomposition. In: Advances in Neural Information Processing Systems Workshops (2019)
8. Deng, B., Lewis, J., Jeruzalski, T., Pons-Moll, G., Hinton, G., Norouzi, M., Tagliasacchi, A.: Nasa: neural articulated shape approximation (2020)
9. Deprelle, T., Groueix, T., Fisher, M., Kim, V., Russell, B., Aubry, M.: Learning elementary structures for 3D shape generation and matching. In: Advances in Neural Information Processing Systems (NeurIPS) (2019)
10. Genova, K., Cole, F., Sud, A., Sarna, A., Funkhouser, T.: Local deep implicit functions for 3D shape. In: Computer Vision and Pattern Recognition (CVPR) (2020)
11. Genova, K., Cole, F., Vlasic, D., Sarna, A., Freeman, W.T., Funkhouser, T.: Learning shape templates with structured implicit functions. In: International Conference on Computer Vision (ICCV) (2019)
12. Groueix, T., Fisher, M., Kim, V., Russell, B., Aubry, M.: A papier-mache approach to learning 3D surface generation. In: Computer Vision and Pattern Recognition (CVPR) (2018)
13. Kato, H., Ushiku, Y., Harada, T.: Neural 3D mesh renderer. In: Computer Vision and Pattern Recognition (CVPR) (2018)
14. Kingma, D.P., Ba, J.: Adam: a method for stochastic optimization. In: International Conference on Learning Representations (ICLR) (2015)
15. Liu, S., Saito, S., Chen, W., Li, H.: Learning to infer implicit surfaces without 3D supervision. In: Advances in Neural Information Processing Systems (NeurIPS) (2019)
16. Lorensen, W.E., Cline, H.E.: Marching cubes: a high resolution 3D surface construction algorithm. In: Conference on Computer Graphics and Interactive Techniques (1987)
17. Mescheder, L., Oechsle, M., Niemeyer, M., Nowozin, S., Geiger, A.: Occupancy networks: learning 3D reconstruction in function space. In: Computer Vision and Pattern Recognition (CVPR) (2019)
18. Michalkiewicz, M., Pontes, J.K., Jack, D., Baktashmotlagh, M., Eriksson, A.: Implicit surface representations as layers in neural networks. In: International Conference on Computer Vision (ICCV) (2019)
19. Niemeyer, M., Mescheder, L., Oechsle, M., Geiger, A.: Occupancy flow: 4D reconstruction by learning particle dynamics. In: International Conference on Computer Vision (CVPR) (2019)
20. Ohtake, Y., Belyaev, A., Alexa, M., Turk, G., Seidel, H.P.: Multi-level partition of unity implicits. In: ACM Transactions on Graphics (TOG) (2003)
21. Park, J.J., Florence, P., Straub, J., Newcombe, R., Lovegrove, S.: DeepSDF: learning continuous signed distance functions for shape representation. In: Computer Vision and Pattern Recognition (CVPR) (2019)
22. Paszke, A., et al.: PyTorch: an imperative style, high-performance deep learning library. In: Advances in Neural Information Processing Systems (NeurIPS) (2019)
23. Qi, C.R., Yi, L., Su, H., Guibas, L.J.: PointNet++: deep hierarchical feature learning on point sets in a metric space. In: Advances in Neural Information Processing Systems (NeurIPS) (2017)
24. Riegler, G., Osman Ulusoy, A., Geiger, A.: OctNet: learning deep 3D representations at high resolutions. In: Computer Vision and Pattern Recognition (CVPR) (2017)

25. Saito, S., Huang, Z., Natsume, R., Morishima, S., Kanazawa, A., Li, H.: PIFu: pixel-aligned implicit function for high-resolution clothed human digitization. In: International Conference on Computer Vision (ICCV) (2019)
26. Salimans, T., Kingma, D.P.: Weight normalization: a simple reparameterization to accelerate training of deep neural networks. In: Advances in Neural Information Processing Systems (NeurIPS) (2016)
27. Shimada, S., Golyanik, V., Tretschk, E., Stricker, D., Theobalt, C.: DispVoxNets: non-rigid point set alignment with supervised learning proxies. In: International Conference on 3D Vision (3DV) (2019)
28. Sitzmann, V., Zollhöfer, M., Wetzstein, G.: Scene representation networks: continuous 3D-structure-aware neural scene representations. In: Advances in Neural Information Processing Systems (NeurIPS) (2019)
29. Stutz, D., Geiger, A.: Learning 3D shape completion under weak supervision. Int. J. Comput. Vision (IJCV) **31**, 1–10 (2018)
30. Tretschk, E., Tewari, A., Zollhöfer, M., Golyanik, V., Theobalt, C.: DEMEA: deep mesh autoencoders for non-rigidly deforming objects. In: European Conference on Computer Vision (ECCV) (2020)
31. Tulsiani, S., Su, H., Guibas, L.J., Efros, A.A., Malik, J.: Learning shape abstractions by assembling volumetric primitives. In: Computer Vision and Pattern Recognition (CVPR) (2017)
32. Wang, N., Zhang, Y., Li, Z., Fu, Y., Liu, W., Jiang, Y.-G.: Pixel2Mesh: generating 3D mesh models from single RGB images. In: Ferrari, V., Hebert, M., Sminchisescu, C., Weiss, Y. (eds.) ECCV 2018. LNCS, vol. 11215, pp. 55–71. Springer, Cham (2018). https://doi.org/10.1007/978-3-030-01252-6_4
33. Williams, F., Parent-Levesque, J., Nowrouzezahrai, D., Panozzo, D., Moo Yi, K., Tagliasacchi, A.: Voronoinet: general functional approximators with local support. In: Computer Vision and Pattern Recognition Workshops (CVPRW) (2020)

How Does Lipschitz Regularization Influence GAN Training?

Yipeng Qin[1,3](✉), Niloy Mitra[2], and Peter Wonka[3]

[1] Cardiff University, Cardiff, UK
qiny16@cardiff.ac.uk
[2] UCL/Adobe Research, London, UK
n.mitra@cs.ucl.ac.uk
[3] KAUST, Thuwal, Saudi Arabia
pwonka@gmail.com

Abstract. Despite the success of Lipschitz regularization in stabilizing GAN training, the exact reason of its effectiveness remains poorly understood. The direct effect of K-Lipschitz regularization is to restrict the $L2$-norm of the neural network gradient to be smaller than a threshold K (e.g., $K = 1$) such that $\|\nabla f\| \leq K$. In this work, we uncover an even more important effect of Lipschitz regularization by examining its impact on the loss function: *It degenerates GAN loss functions to almost linear ones by restricting their domain and interval of attainable gradient values.* Our analysis shows that loss functions are only successful if they are degenerated to almost linear ones. We also show that loss functions perform poorly if they are not degenerated and that a wide range of functions can be used as loss function as long as they are sufficiently degenerated by regularization. Basically, Lipschitz regularization ensures that all loss functions *effectively work in the same way*. Empirically, we verify our proposition on the MNIST, CIFAR10 and CelebA datasets.

Keywords: Generative adversarial network (GAN) · Lipschitz regularization · Loss functions

1 Introduction

Generative Adversarial Networks (GANs) are a class of generative models successfully applied to various applications, e.g., pose-guided image generation [17], image-to-image translation [23,29], texture synthesis [5], high resolution image synthesis [27], 3D model generation [28], urban modeling [13]. Goodfellow et al. [7] proved the convergence of GAN training by assuming that the generator is always updated according to the temporarily optimal discriminator at each

Electronic supplementary material The online version of this chapter (https://doi.org/10.1007/978-3-030-58517-4_19) contains supplementary material, which is available to authorized users.

A. Vedaldi et al. (Eds.): ECCV 2020, LNCS 12361, pp. 310–326, 2020.
https://doi.org/10.1007/978-3-030-58517-4_19

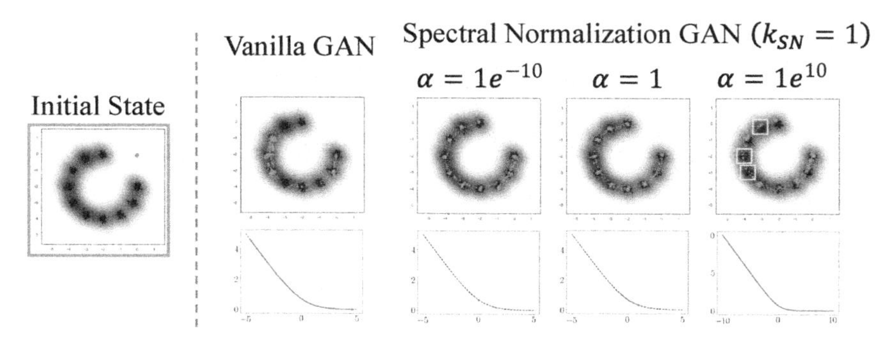

Fig. 1. An illustrative 2D example. First row: model distribution (orange point) vs. data distribution (black points). Second row: domains of the loss function (red curve). It can be observed that the performance of spectral normalized GANs [21] worsen when their domains are enlarged ($\alpha = 1e^{10}$, yellow boxes). However, good performance can always be achieved when their domains are restricted to near-linear ones ($\alpha = 1e^{-10}$ and $\alpha = 1$). Please see Sects. 3.2 and 4.1 for the definitions of α and k_{SN}, respectively.

training step. In practice, this assumption is too difficult to satisfy and GANs remain difficult to train. To stabilize the training of the GANs, various techniques have been proposed regarding the choices of architectures [10,24], loss functions [2,19], regularization and normalization [2,8,20,21]. We refer interested readers to [14,16] for extensive empirical studies.

Among them, the Lipschitz regularization [8,21] has shown great success in stabilizing the training of various GANs. For example, [18] and [4] observed that the gradient penalty Lipschitz regularizer helps to improve the training of the LS-GAN [19] and the NS-GAN [7], respectively; [21] observed that the NS-GAN, with their spectral normalization Lipschitz regularizer, works better than the WGAN [2] regularized by gradient penalty (WGAN-GP) [8].

In this paper, we provide an analysis to better understand the *coupling* of Lipschitz regularization and the choice of loss function. Our main insight is that the rule of thumb of using small Lipschitz constants (e.g., $K = 1$) is degenerates the loss functions to almost linear ones by restricting their domain and interval of attainable gradient values (see Fig. 1). These degenerate losses improve GAN training. Because of this, the exact shapes of the loss functions before degeneration do not seem to matter that much. We demonstrate this by two experiments. First, we show that when K is sufficiently small, even GANs trained with non-standard loss functions (e.g., cosine) give comparable results to all other loss functions. Second, we can directly degenerate loss functions by introducing domain scaling. This enables successful GAN training for a wide range of K for all loss functions, which only worked for the Wasserstein loss before. Our contributions include:

- We discovered an important effect of Lipschitz regularization. It restricts the domain of the loss function (Fig. 2).

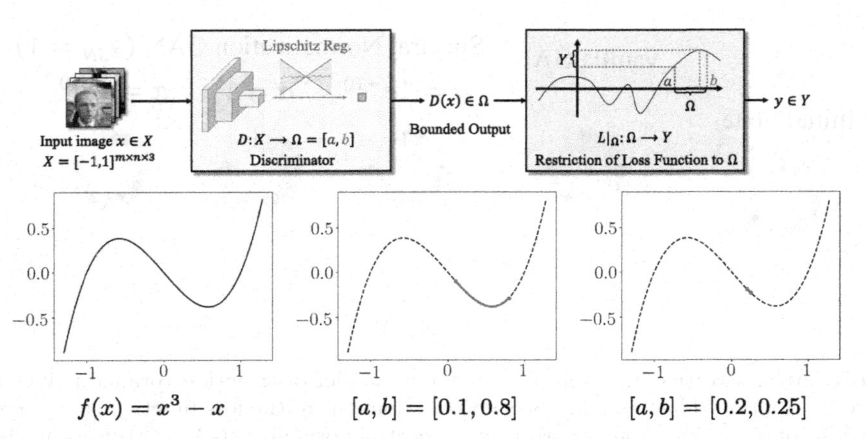

Fig. 2. First row: Applying Lipschitz regularization restricts the domain of the loss function to an interval $\Omega = [a, b]$. Second row: Illustration of the domain restriction. We take a third-order polynomial loss function $f(x) = x^3 - x$ as an example. Its restricted domain $[a, b]$ is shown in red. (a) without restriction $f(x)$ is non-convex. (b) Restricting the domain of $f(x)$ makes it convex. (c) $f(x)$ is almost linear when its domain is restricted to a very small interval.

– Our analysis suggests that although the choice of loss functions matters, the successful ones currently being used are all near-linear within the chosen small domains and actually work in the same way.

2 Related Work

2.1 GAN Loss Functions

A variety of GAN loss functions have been proposed from the idea of understanding the GAN training as the minimization of statistical divergences. Goodfellow et al. [7] first proposed to minimize the Jensen-Shannon (JS) divergence between the model distribution and the target distribution. In their method, the neural network output of the discriminator is first passed through a sigmoid function to be scaled into a probability in [0, 1]. Then, the cross-entropy loss of the probability is measured. Following [4], we refer to such loss as the *minimax* (MM) loss since the GAN training is essentially a minimax game. However, because of the saturation at both ends of the sigmoid function, the MM loss can lead to vanishing gradients and thus fails to update the generator. To compensate for it, Goodfellow et al. [7] proposed a variant of MM loss named the *non-saturating* (NS) loss, which heuristically amplifies the gradients when updating the generator.

Observing that the JS divergence is a special case of the f-divergence, Nowozin et al. [22] extended the idea of Goodfellow et al. [7] and showed that any f-divergence can be used to train GANs. Their work suggested a new direction of improving the performance of the GANs by employing “better” divergence measures.

Following this direction, Arjovsky et al. first pointed out the flaws of the JS divergence used in GANs [1] and then proposed to use the Wasserstein distance instead (WGAN) [2]. In their implementation, the raw neural network output of the discriminator is directly used (i.e. the WGAN loss function is an identity function) instead of being passed through the sigmoid cross-entropy loss function. However, to guarantee that their loss is a valid Wasserstein distance metric, the discriminator is required to be Lipschitz continuous. Such requirement is usually fulfilled by applying an extra Lipschitz regularizer to the discriminator. Meanwhile, Mao et al. [19] proposed the Least-Square GAN (LS-GAN) to minimize the Pearson χ^2 divergence between two distributions. In their implementation, the sigmoid cross-entropy loss is replaced by a quadratic loss.

2.2 Lipschitz Regularization

The first practice of applying the Lipschitz regularization to the discriminator came together with the WGAN [2]. While at that time, it was not employed to improve the GAN training but just a requirement of the Kantorovich-Rubinstein duality applied. In [2], the Lipschitz continuity of the discriminator is enforced by *weight clipping*. Its main idea is to clamp the weights of each neural network layer to a small fixed range $[-c, c]$, where c is a small positive constant. Although weight clipping guarantees the Lipschitz continuity of the discriminator, the choice of parameter c is difficult and prone to invalid gradients.

To this end, Gulrajani et al. [8] proposed the *gradient penalty* (GP) Lipschitz regularizer to stabilize the WGAN training, i.e. WGAN-GP. In their method, an extra regularization term of discriminator's gradient magnitude is weighted by parameter λ and added into the loss function. In [8], the gradient penalty regularizer is one-centered, aiming at enforcing 1-Lipschitz continuity. While Mescheder et al. [20] argued that the zero-centered one should be more reasonable because it makes the GAN training converge. However, one major problem of gradient penalty is that it is computed with finite samples, which makes it intractable to be applied to the entire output space. To sidestep this problem, the authors proposed to heuristically sample from the straight lines connecting model distribution and target distribution. However, this makes their approach heavily dependent on the support of the model distribution [21].

Addressing this issue, Miyato et al. [21] proposed the *spectral normalization* (SN) Lipschitz regularizer which enforces the Lipschitz continuity of a neural network in the operator space. Observing that the Lipschitz constant of the entire neural network is bounded by the product of those of its layers, they break down the problem to enforcing Lipschitz regularization on each neural network layer. These simplified sub-problems can then be solved by normalizing the weight matrix of each layer according to its largest singular value.

3 Restrictions of GAN Loss Functions

In this section, first we derive why a K-Lipschitz regularized discriminator *restricts* the domain and interval of attainable gradient values of the loss function to intervals bounded by K (Sect. 3.1). Second, we propose a scaling method to restrict the domain of the loss function without changing K (Sect. 3.2).

3.1 How Does the Restriction Happen?

Let us consider a simple discriminator $D(x) = L(f(x))$, where x is the input, f is a neural network with scalar output, L is the loss function. During training, the loss function L works by backpropagating the gradient $\nabla L = \partial L(f(x))/\partial f(x)$ to update the neural network weights:

$$\frac{\partial D(x)}{\partial W^n} = \frac{\partial \boldsymbol{L(f(x))}}{\partial \boldsymbol{f(x)}} \frac{\partial f(x)}{\partial W^n} \tag{1}$$

where W^n is the weight matrix of the n-th layer. Let X and Ω be the domain and the range of f respectively (i.e., $f : X \rightarrow \Omega$), it can be easily derived that the attainable values of ∇L is determined by Ω (i.e., $\nabla L : \Omega \rightarrow \Psi$). Without loss of generality, we assume that $x \in X = [-1, 1]^{m \times n \times 3}$ are normalized images and derive the bound of the size of Ω as follows:

Theorem 1. *If the discriminator neural network f satisfies the k-Lipschitz continuity condition, we have $f : X \rightarrow \Omega \subset \mathbb{R}$ satisfying $|\min(\Omega) - \max(\Omega)| \leq k\sqrt{12mn}$.*

Proof. Given a k-Lipschitz continuous neural newtork f, for all $x_1, x_2 \in X$, we have:

$$|f(x_1) - f(x_2)| \leq k\|x_1 - x_2\|. \tag{2}$$

Let $x_b, x_w \in X$ be the pure black and pure white images that maximize the Euclidean distance:

$$\|x_b - x_w\| = \sqrt{(-1-1)^2 \cdot m \cdot n \cdot 3} = \sqrt{12mn}. \tag{3}$$

Thus, we have:

$$\begin{aligned} |f(x_1) - f(x_2)| &\leq k\|x_1 - x_2\| \\ &\leq k\|x_b - x_w\| = k\sqrt{12mn}. \end{aligned} \tag{4}$$

Thus, the range of f is restricted to Ω, which satisfies:

$$|\min(\Omega) - \max(\Omega)| \leq k\sqrt{12mn} \tag{5}$$

Theorem 1 shows that the size of Ω is bounded by k. However, k can be unbounded when Lipschitz regularization is not enforced during training, which results in an unbounded Ω and a large interval of attainable gradient values. On the contrary, when K-Lipschitz regularization is applied (i.e., $k \leq K$), the loss function L is restricted as follows:

Corollary 1 (Restriction of Loss Function). *Assume that f is a Lipschitz regularized neural network whose Lipschitz constant $k \leq K$, the loss function L is C^2-continuous with M as the maximum absolute value of its second derivatives in its domain. Let Ψ be the interval of attainable gradient values that $\nabla L : \Omega \rightarrow \Psi$, we have*

$$\big|\min(\Omega) - \max(\Omega)\big| \leq K\sqrt{12mn} \tag{6}$$

$$\big|\min(\Psi) - \max(\Psi)\big| \leq M \cdot K\sqrt{12mn} \tag{7}$$

Corollary 1 shows that under a mild condition (C^2-continuous), applying K-Lipschitz regularization restricts the domain Ω and thereby the interval of attainable gradient values Ψ of the loss function L to intervals bounded by K. When K is small, e.g., $K = 1$ [4,8,18,21], the interval of attainable gradient values of the loss function is considerably reduced, which prevents the backpropagation of vanishing or exploding gradients and thereby stabilizes the training. Empirically, we will show that these restrictions are indeed significant in practice and strongly influence the training.

Change in Ω_i During Training. So far we analyzed the restriction of the loss function by a static discriminator. However, the discriminator neural network f is dynamically updated during training and thus its range $\Omega^{\cup} = \cup_i \Omega_i$, where Ω_i is the discriminator range at each training step i. Therefore, we need to analyze two questions:

(i) How does the size of Ω_i change during training?

(ii) Does Ω_i shift during training (Fig. 3)?

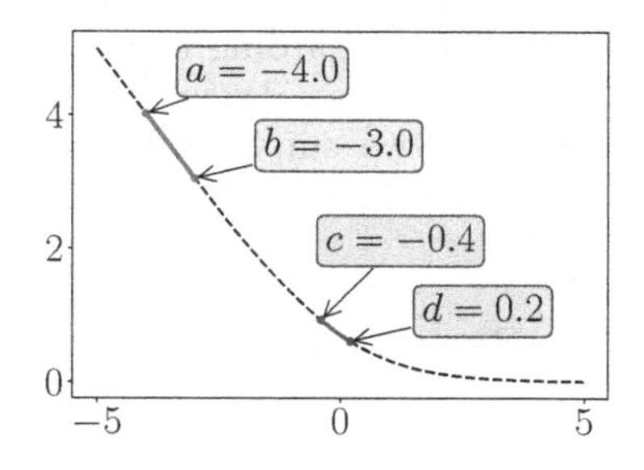

Fig. 3. Domain $[a, b]$ shifts to $[c, d]$.

For question (i), the size of Ω_i is always bounded by the Lipschitz constant K throughout the training (Corollary 1). For question (ii), the answer depends on the discriminator loss function:

- The shifting of Ω_i is prevented if the loss function is strictly convex. For example, the discriminator loss function of NS-GAN [7] (Table 1) is strictly convex and has a unique minimum when $f(x) = f(g(z)) = 0$ at convergence. Thus, minimizing it forces Ω_i to be positioned around 0 and prevents it from shifting. The discriminator loss function of LS-GAN [19] (Table 1) has a similar behavior. Its Ω_i is positioned around 0.5, since its minimum is achieved when $f(x) = f(g(z)) = 0.5$ at convergence. In this scenario, the Ω_i is relatively fixed throughout the training. Thus, $\Omega^{\cup}$ is still roughly bounded by the Lipschitz constant K.
- When the discriminator loss functions is not strictly convex, Ω_i may be allowed to shift. For example, the WGAN [2] discriminator loss function

Table 1. The GAN loss functions used in our experiments. $f(\cdot)$ is the output of the discriminator neural network; $g(\cdot)$ is the output of the generator; x is a sample from the training dataset; z is a sample from the noise distribution. LS-GAN# is the zero-centered version of LS-GAN [18], and for the NS-GAN, $f^*(\cdot) = \text{sigmoid}[f(\cdot)]$. The figure on the right shows the shape of the loss functions at different scales. The dashed lines show non-standard loss functions: cos and exp.

GAN types	Discriminator Loss	Generator Loss
NS-GAN	$L_D = -\mathbb{E}[\log(f^*(x))] - \mathbb{E}[\log(1 - f^*(g(z)))]$	$L_G = -\mathbb{E}[\log(f^*(g(z)))]$
LS-GAN	$L_D = \mathbb{E}[(f(x) - 1)^2] + \mathbb{E}[f(g(z))^2]$	$L_G = \mathbb{E}[(f(g(z)) - 1)^2]$
LS-GAN#	$L_D = \mathbb{E}[(f(x) - 1)^2] + \mathbb{E}[(f(g(z)) + 1)^2]$	$L_G = \mathbb{E}[(f(g(z)) - 1)^2]$
WGAN	$L_D = \mathbb{E}[f(x)] - \mathbb{E}[f(g(z))]$	$L_G = \mathbb{E}[f(g(z))]$
COS-GAN	$L_D = -\mathbb{E}[\cos(f(x) - 1)] - \mathbb{E}[\cos(f(g(z)) + 1)]$	$L_G = -\mathbb{E}[\cos(f(g(z)) - 1)]$
EXP-GAN	$L_D = \mathbb{E}[\exp(f(x))] + \mathbb{E}[\exp(-f(g(z)))]$	$L_G = \mathbb{E}[\exp(f(g(z)))]$

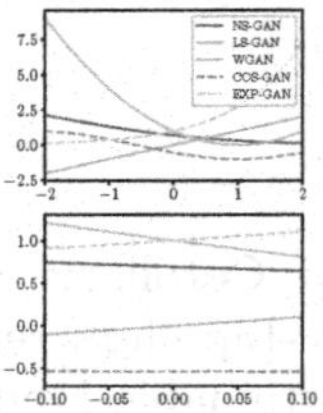

(Table 1) is linear and achieves its minimum when $f(x) = f(g(z))$ at convergence. Thus, it does not enforce the domain Ω_i to be fixed. However, the linear WGAN loss function has a constant gradient that is independent of Ω_i. Thus, regarding to the interval of attainable gradient values (Eq. 7), we can view it as a degenerate loss function that still fits in our discussion. Interestingly, we empirically observed that the domain Ω_i of WGANs also get relatively fixed at late stages of the training (Fig. 4).

3.2 Restricting Loss Functions by Domain Scaling

As discussed above, applying K-Lipschitz regularization not only restricts the gradients of the discriminator, but as a side effect also restricts the domain of the loss function to an interval Ω. However, we would like to investigate these two effects separately. To this end, we propose to decouple the restriction of Ω from the Lipschitz regularization by scaling the domain of loss function L by a positive constant α as follows,

$$L_\alpha(\Omega) = L(\alpha \cdot \Omega)/\alpha. \tag{8}$$

Note that the α in the denominator helps to preserve the gradient scale of the loss function. With this scaling method, we can effectively restrict L to an interval $\alpha \cdot \Omega$ without adjusting K.

Degenerate Loss Functions. To explain why this works, we observe that any loss function degenerates as its domain Ω shrinks to a single value. According to Taylor's expansion, let $\omega, \omega + \Delta\omega \in \Omega$, we have:

$$L(\omega + \Delta\omega) = L(\omega) + \frac{L'(\omega)}{1!}\Delta\omega + \frac{L''(\omega)}{2!}(\Delta\omega)^2 + \cdots. \tag{9}$$

As $|\max(\Omega) - \min(\Omega)|$ shrinks to zero, we have $L(\omega + \Delta\omega) \approx L(\omega) + L'(\omega)\Delta\omega$ showing that we can approximate any loss function by a linear function with constant gradient as its domain Ω shrinks to a single value. Let $\omega \in \Omega$, we implement the degeneration of a loss function by scaling its domain Ω with an extremely small constant α:

$$\lim_{\alpha \to 0} \frac{\partial L_\alpha(\omega)}{\partial \omega} = \frac{1}{\alpha} \cdot \frac{\partial L(\alpha \cdot \omega)}{\partial \omega} = \frac{\partial L(\alpha \cdot \omega)}{\partial(\alpha \cdot \omega)} = \nabla L(0). \tag{10}$$

In our work, we use $\alpha = 1e^{-25}$, smaller values are not used due to numerical errors (NaN).

4 Experiments

To support our proposition, first we empirically verify that applying K-Lipschitz regularization to the discriminator has the side-effect of restricting the domain and interval of attainable gradient values of the loss function. Second, with the proposed scaling method (Sect. 3.2), we investigate how the varying restrictions of loss functions influence the performance of GANs when the discriminator is regularized with a fixed Lipschitz constant. Third, we show that restricting the domain of any loss function (using decreasing α) converges to the same (or very similar) performance as WGAN-SN.

4.1 Experiment Setup

General Setup. In the following experiments, we use two variants of the standard CNN architecture [2,21,24] for the GANs to learn the distributions of the MNIST, CIFAR10 datasets at 32×32 resolution and the CelebA dataset [15] at 64×64 resolution. Details of the architectures are shown in the supplementary material. We use a batch size of 64 to train the GANs. Similar to [2], we observed that the training could be unstable with a momentum-based optimizer such as Adam, when the discriminator is regularized with a very small Lipschitz constant K. Thus, we choose to use an RMSProp optimizer with learning rate 0.00005. To make a fair comparison, we fix the number of discriminator updates in each iteration $n_{dis} = 1$ for all the GANs tested (i.e., we do not use multiple discriminator updates like [1,2]). Unless specified, we stop the training after 10^5 iterations.

Lipschitz Regularizers. In general, there are two state-of-the-art Lipschitz regularizers: the gradient penalty (GP) [8] and the spectral normalization (SN) [21]. In their original settings, both techniques applied only 1-Lipschitz regularization to the discriminator. However, our experiments require altering the Lipschitz constant K of the discriminator. To this end, we propose to control K for both techniques by adding parameters k_{GP} and k_{SN}, respectively.

- For the gradient penalty, we control its strength by adjusting the target gradient norm k_{GP},

$$L = L_{GAN} + \lambda \mathop{\mathbb{E}}_{\hat{x} \in P_{\hat{x}}} [(\| \nabla_{\hat{x}} D(\hat{x}) \| - k_{GP})^2], \tag{11}$$

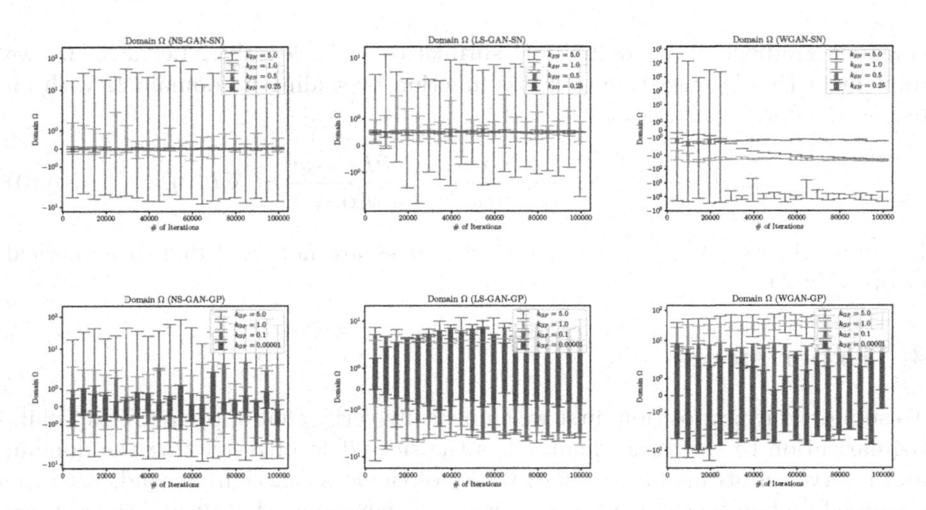

Fig. 4. Relationship between domain Ω and k_{GP}, k_{SN} for different loss functions on CelebA dataset, where k_{GP}, k_{SN} are the parameters controlling the strength of the Lipschitz regularizers. The domain Ω shrinks with decreasing k_{GP} or k_{SN}. Each column shares the same loss function while each row shares the same Lipschitz regularizer. NS-GAN: Non-Saturating GAN [7]; LS-GAN: Least-Square GAN [19]; WGAN: Wasserstein GAN [2]; GP: gradient penalty [8]; SN: spectral normalization [21]. Note that the y-axis is in log scale.

where L_{GAN} is the GAN loss function without gradient penalty, λ is the weight of the gradient penalty term, $P_{\hat{x}}$ is the distribution of linearly interpolated samples between the target distribution and the model distribution [8]. Similar to [4,8], we use $\lambda = 10$.

- For the spectral normalization, we control its strength by adding a weight parameter k_{SN} to the normalization of each neural network layer,

$$\bar{W}_{SN}(W, k_{SN}) := k_{SN} \cdot W/\sigma(W), \tag{12}$$

where W is the weight matrix of each layer, $\sigma(W)$ is its largest singular value. The relationship between k_{SN} and K can be quantitatively approximated as $K \approx k_{SN}^n$ [21], where n is the number of neural network layers in the discriminator. While for k_{GP}, we can only describe its relationship against K qualitatively as: the smaller k_{GP}, the smaller K. The challenge on finding a quantitative approximation resides in that the gradient penalty term $\lambda\, \mathbb{E}_{\hat{x}\in P_{\hat{x}}}[(\|\nabla_{\hat{x}} D(\hat{x})\| - k_{GP})^2]$ has no upper bound during training (Eq. 11). We also verified our claims using Stable Rank Normalization (SRN)+SN [25] as the Lipschitz regularizer, whose results are shown in the supplementary material.

Loss Functions. In Table 1 we compare the three most widely-used GAN loss functions: the Non-Saturating (NS) loss function [7], the Least-Squares (LS) loss function [19] and the Wasserstein loss function [2]. In addition, we also test the performance of the GANs using some non-standard loss functions, $\cos(\cdot)$ and

Table 2. Domain Ω and the interval of attained gradient values $\nabla L(\Omega)$ against k_{SN} on the CelebA dataset.

k_{SN}	Ω	$\nabla L(\Omega)$
5.0	[−8.130, 126.501]	[−1.000, −0.000]
1.0	[−0.683, 2.091]	[−0.664, −0.110]
0.5	[−0.178, 0.128]	[−0.545, −0.468]
0.25	[−0.006, 0.006]	[−0.502, −0.498]

NS-GAN-SN, $L(\cdot) = -\log(\text{sigmoid}(\cdot))$

k_{SN}	Ω	$\nabla L(\Omega)$
5.0	[−2.460, 12.020]	[−4.921, 24.041]
1.0	[−0.414, 1.881]	[−0.828, 3.762]
0.5	[0.312, 0.621]	[0.623, 1.242]
0.25	[0.478, 0.522]	[0.956, 1.045]

LS-GAN-SN, $L(\cdot) = (\cdot)^2$

$\exp(\cdot)$, to support the observation that the restriction of the loss function is the dominating factor of Lipschitz regularization. Note that both the $\cos(\cdot)$ and $\exp(\cdot)$ loss functions are (locally) convex at convergence, which helps to prevent shifting Ω_i (Sect. 3.1).

Quantitative Metrics. To quantitatively measure the performance of the GANs, we follow the best practice and employ the Fréchet Inception Distance (FID) metric [11] in our experiments. The smaller the FID score, the better the performance of the GAN. The results on other metrics, *i.e.* Inception scores [3] and Neural Divergence [9], are shown in the supplementary material.

4.2 Empirical Analysis of Lipschitz Regularization

In this section, we empirically analyze how varying the strength of the Lipschitz regularization impacts the domain, interval of attained gradient values, and performance (FID scores) of different loss functions (Sect. 3.1).

Domain vs. Lipschitz Regularization. In this experiment, we show how the Lipschitz regularization influences the domain of the loss function. As Fig. 4 shows, we plot the domain Ω as intervals for different iterations under different k_{GP} and k_{SN} for the gradient penalty and the spectral normalization regularizers respectively. It can be observed that: (i) For both regularizers, the interval Ω shrinks as k_{GP} and k_{SN} decreases. However, k_{SN} is much more impactful than k_{GP} in restricting Ω. Thus, we use spectral normalization to alter the strength of the Lipschitz regularization in the following experiments. (ii) For NS-GANs and LS-GANs, the domains Ω_i are rather fixed during training. For WGANs, the domains Ω_i typically shift at the beginning of the training, but then get relatively fixed in later stages.

Interval of Attained Gradient Values vs. Lipschitz Regularization. Similar to the domain, the interval of attained gradient values of the loss function also shrinks with the increasing strength of Lipschitz regularization. Table 2 shows the corresponding interval of attained gradient values of the NS-GAN-SN and LS-GAN-SN experiments in Fig. 4. The interval of attained gradient values of

Table 3. FID scores vs. k_{SN} (typically fixed as 1 [21]) on different datasets. When $k_{SN} \leq 1.0$, all GANs have similar performance except the WGANs (slightly worse). For the line plots, x-axis shows k_{SN} (in log scale) and y-axis shows the FID scores. From left to right, the seven points on each line have k_{SN} = 0.2, 0.25, 0.5, 1.0, 5.0, 10.0, 50.0 respectively. Lower FID scores are better.

Dataset	GANs	FID Scores						
		$k_{SN} = 0.2$	0.25	0.5	1.0	5.0	10.0	50.0
MNIST	NS-GAN-SN	5.41	3.99	4.20	3.90	144.28	156.60	155.41
	LS-GAN-SN	5.14	**3.96**	**3.90**	4.42	36.26	59.04	309.35
	WGAN-SN	6.35	6.06	4.44	4.70	**3.58**	**3.50**	**3.71**
	COS-GAN-SN	5.41	4.83	4.05	3.86	291.44	426.62	287.23
	EXP-GAN-SN	**4.38**	4.93	4.25	**3.69**	286.96	286.96	286.96
CIFAR10	NS-GAN-SN	29.23	**24.37**	23.29	**15.81**	41.04	49.67	48.03
	LS-GAN-SN	**28.04**	26.85	23.14	17.30	33.53	39.90	349.35
	WGAN-SN	29.20	25.07	26.61	21.75	**21.63**	**21.45**	**23.36**
	COS-GAN-SN	29.45	25.31	**20.73**	15.88	309.96	327.20	370.13
	EXP-GAN-SN	30.89	24.74	20.90	16.66	401.24	401.24	401.24
CelebA	NS-GAN-SN	18.59	12.71	**8.04**	6.11	18.95	17.04	184.06
	LS-GAN-SN	20.34	**12.14**	8.85	5.69	12.40	13.14	399.39
	WGAN-SN	23.26	17.93	8.48	9.41	**9.03**	**7.37**	**7.82**
	COS-GAN-SN	20.59	13.93	8.88	**5.20**	356.70	265.53	256.44
	EXP-GAN-SN	**18.23**	13.65	9.18	5.88	328.94	328.94	328.94

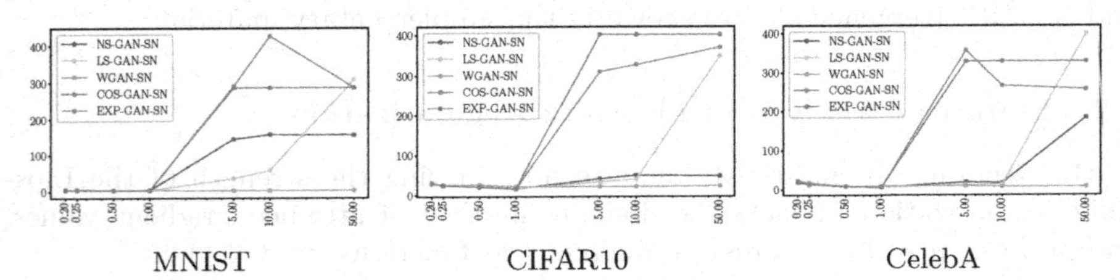

WGAN-SN are not included as they are always zero. It can be observed that the shrinking interval of attained gradient values avoids the saturating and exploding parts of the loss function. For example when $k_{SN} = 5.0$, the gradient of the NS-GAN loss function saturates to a value around 0 while that of the LS-GAN loss function explodes to 24.041. However, such problems do not happen when $k_{SN} \leq 1.0$. Note that we only compute the interval of attained gradient values on one of the two symmetric loss terms used in the discriminator loss function (Table 1). The interval of attained gradient values of the other loss term follows similar patterns.

FID Scores vs. Lipschitz Regularization. Table 3 shows the FID scores of different GAN loss functions with different k_{SN}. It can be observed that:

- When $k_{SN} \leq 1.0$, all the loss functions (including the non-standard ones) can be used to train GANs stably. However, the FID scores of all loss functions slightly worsen as k_{SN} decreases. We believe that the reason for

Table 4. FID scores vs. α. For the line plots, the x-axis shows α (in log scale) and the y-axis shows the FID scores. Results on other datasets are shown in the supplementary material. Lower FID scores are better.

Dataset	GANs	FID Scores $\alpha = 1e^{-11}$	$1e^{-9}$	$1e^{-7}$	$1e^{-5}$	$1e^{-3}$	$1e^{-1}$	Line Plot
CelebA	NS-GAN-SN	9.08	7.05	7.84	18.51	18.41	242.64	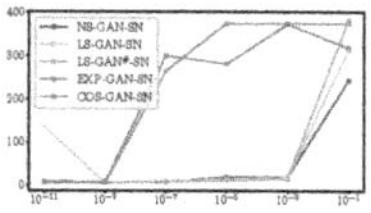
	LS-GAN-SN	135.17	6.57	10.67	13.39	17.42	311.93	
	LS-GAN$^{\#}$-SN	6.66	5.68	8.72	11.13	14.90	383.61	
	COS-GAN-SN	8.00	6.31	300.55	280.84	373.31	318.53	
	EXP-GAN-SN	8.85	6.09	264.49	375.32	375.32	375.32	

$k_{SN} = 50.0$

Dataset	GANs	FID Scores $\alpha = 1e^{1}$	$1e^{3}$	$1e^{5}$	$1e^{7}$	$1e^{9}$	$1e^{11}$	Line Plot
MNIST	NS-GAN-SN	6.55	148.97	134.44	133.82	130.21	131.87	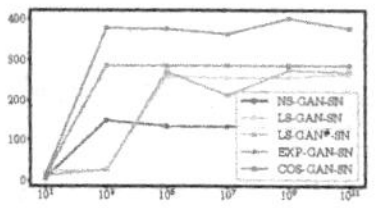
	LS-GAN-SN	23.37	26.96	260.05	255.73	256.96	265.76	
	LS-GAN$^{\#}$-SN	13.43	26.51	271.85	212.74	274.63	269.96	
	COS-GAN-SN	11.79	377.62	375.72	363.45	401.12	376.39	
	EXP-GAN-SN	11.02	286.96	286.96	286.96	286.96	286.96	

$k_{SN} = 1.0$

such performance degradation comes from the trick used by modern optimizers to avoid divisions by zero. For example in RMSProp [26], the moving average of the squared gradients are kept for each weight. In order to stabilize training, gradients are divided by the square roots of their moving averages in each update of the weights, where a small positive constant ϵ is included in the denominator to avoid dividing by zero. When k_{SN} is large, the gradients are also large and the effect of ϵ is negligible. While when k_{SN} is very small, the gradients are also small so that ϵ can significantly slow down the training and worsen the results.

- When $k_{SN} \leq 1.0$, the performance of WGAN is slightly worse than almost all the other GANs. Similar to the observation of [12], we ascribe this problem to the shifting domain of WGANs (Fig. 4 (c)(f)). The reason for the domain shift is that the Wasserstein distance only depends on the difference between $\mathbb{E}[f(x)]$ and $\mathbb{E}[f(g(z)]$ (Table 1). For example, the Wasserstein distances $\mathbb{E}[f(x)] - \mathbb{E}[f(g(z)]$ are the same for i) $\mathbb{E}[f(x)] = 0.5, \mathbb{E}[f(g(z)) = -0.5$ and ii) $\mathbb{E}[f(x)] = 100.5, \mathbb{E}[f(g(z)) = 99.5$.
- When $k_{SN} \geq 5.0$, the WGAN works normally while the performance of all the other GANs worsen and even break (very high FID scores, e.g. ≥ 100). The reasons for the stable performance of WGAN are two-fold: i) due to the KR duality, the Wasserstein distance is insensitive to the Lipschitz constant K. Let $W(\mathbb{P}_r, \mathbb{P}_g)$ be the Wasserstein distance between the data distribution $\mathbb{P}_r$ and the generator distribution $\mathbb{P}_g$. As discussed in [2], applying K-Lipschitz regularization to WGAN is equivalent to estimating $K \cdot W(\mathbb{P}_r, \mathbb{P}_g)$,

which shares the same solution as 1-Lipschitz regularized WGAN. ii) To fight the exploding and vanishing gradient problems, modern neural networks are intentionally designed to be scale-insensitive to the backpropagated gradients (e.g. ReLU [6], RMSProp [26]). This largely eliminates the scaling effect caused by k_{SN}. This observation also supports our claim that the restriction of the loss function is the dominating factor of the Lipschitz regularization.
- The best performance is obtained by GANs with strictly convex (e.g. NS-GAN) and properly restricted (e.g. $k_{SN} = 1$) loss functions that address the shifting domain and exploding/vanishing gradient problems at the same time. However, there is no clear preference and even the non-standard ones (e.g., COS-GAN) can be the best. We believe that this is due to the subtle differences of the convexity among loss functions and propose to leave it to the fine-tuning of loss functions using the proposed domain scaling.

Qualitative results are in the supplementary material.

4.3 Empirical Results on Domain Scaling

In this section, we empirically verify our claim that the restriction of the loss function is the dominating factor of the Lipschitz regularization. To illustrate it, we decouple the restriction of the domain of the loss function from the Lipschitz regularization by the proposed domain scaling method (Sect. 3.2).

Table 4(a) shows that (i) the FID scores of different loss functions generally improve with decreasing α. When $\alpha \leq 10^{-9}$, we can successfully train GANs with extremely large Lipschtiz constant ($K \approx k_{SN}^n = 50^4 = 6.25 \times 10^6$), whose FID scores are comparable to the best ones in Table 3. (ii) The FID scores when $\alpha \leq 10^{-11}$ are slightly worse than those when $\alpha \leq 10^{-9}$. The reason for this phenomenon is that restricting the domain of the loss function converges towards the performance of WGAN, which is slightly worse than the others due to its shifting domain. To further illustrate this point, we scale the domain by $\alpha = 1e^{-25}$ and show the FID scores of WGAN-SN and those of different loss functions in Table 5. It can be observed that all loss functions have similar performance. Since domain scaling does not restrict the neural network gradients, it does not suffer from the above-mentioned numerical problem of division by zero ($k_{SN} \leq 1.0$, Table 3). Thus, it is a better alternative to tuning k_{SN}.

Table 5. FID scores of WGAN-SN and some extremely degenerate loss functions ($\alpha = 1e^{-25}$) on different datasets. We use $k_{SN} = 50$ for all our experiments.

GANs	FID Scores		
	MNIST	CIFAR10	CELEBA
WGAN-SN	3.71	23.36	7.82
NS-GAN-SN	3.74	21.92	8.10
LS-GAN#-SN	3.81	21.47	8.51
COS-GAN-SN	3.96	23.65	8.30
EXP-GAN-SN	3.86	21.91	8.22

Table 4(b) shows that the FID scores of different loss functions generally worsen with less restricted domains. Note that when $\alpha \geq 10^5$, the common practice of 1-Lipschitz regularization fails to stabilize the GAN training. Note that the LS-GAN-SN has some abnormal behavior (e.g. $\alpha = 1e^{-11}$ in Table 4(a) and $\alpha = 1e^{1}$ in Table 4(b)) due to the conflict between its 0.5-centered domain and our zero-centered domain scaling method (Eq. 8). This can be easily fixed by using the zero-centered LS-GAN#-SN (see Table 1).

Bias over Input Samples. When weak Lipschitz regularization (large Lipschitz constant K) is applied, we observed mode collapse for NS-GAN and crashed training for LS-GAN, EXP-GAN and COS-GAN (Fig. 5, more results in supplementary material). We conjecture that this phenomenon is rooted in the inherent bias of neural networks over input samples: neural networks may "prefer" some input (class of) samples over the others by outputting higher/lower values, even though all of them are real samples from the training dataset. When the above-mentioned loss functions are used, such different outputs result in different backpropagated gradients $\nabla L = \partial D(f(x))/\partial f(x)$. The use of weak Lipschitz regularization further enhances the degree of unbalance among backpropagated gradients and causes mode collapse or crashed training. Note that mode collapse happens when ∇L is bounded (e.g. NS-GAN) and crashed training happens when ∇L is unbounded (e.g. LS-GAN, EXP-GAN) or "random" (e.g. COS-GAN). However, when strong Lipschitz regularization is applied, all loss functions degenerate to almost linear ones and balance the backpropagated gradients, thereby improve the training.

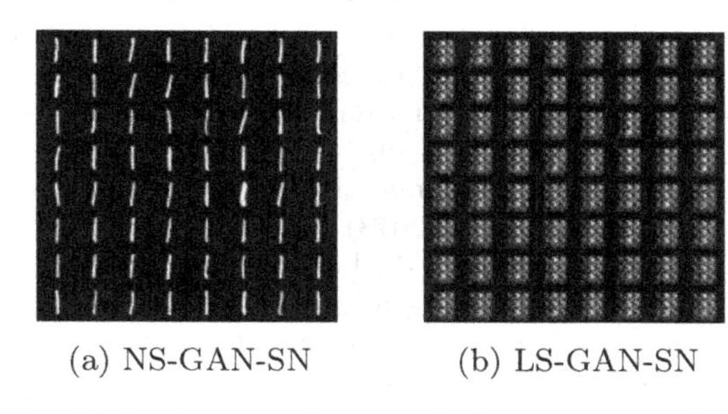

(a) NS-GAN-SN (b) LS-GAN-SN

Fig. 5. (a) Mode collapse and (b) crashed training on MNIST, $k_{SN} = 50.0$, $\alpha = 1e^{-1}$.

5 Conclusion

In this paper, we studied the *coupling* of Lipschitz regularization and the loss function. Our key insight is that instead of keeping the neural network gradients small, the dominating factor of Lipschitz regularization is its restriction on the domain and interval of attainable gradient values of the loss function. Such restrictions stabilize GAN training by avoiding the bias of the loss function over input samples, which is a new step in understanding the exact reason for Lipschitz regularization's effectiveness. Furthermore, our finding suggests that although different loss functions can be used to train GANs successfully, they actually work in the same way because all of them degenerate to near-linear ones within the chosen small domains.

Acknowledgement. This work was supported in part by the KAUST Office of Sponsored Research (OSR) under Award No. OSR-CRG2018-3730.

References

1. Arjovsky, M., Bottou, L.: Towards principled methods for training generative adversarial networks. In: International Conference on Learning Representations (2017)
2. Arjovsky, M., Chintala, S., Bottou, L.: Wasserstein generative adversarial networks. In: Precup, D., Teh, Y.W. (eds.) Proceedings of the 34th International Conference on Machine Learning. Proceedings of Machine Learning Research, vol. 70, pp. 214–223. PMLR, International Convention Centre, Sydney, Australia, 06–11 August 2017. http://proceedings.mlr.press/v70/arjovsky17a.html
3. Barratt, S., Sharma, R.: A note on the inception score. arXiv preprint arXiv:1801.01973 (2018)
4. Fedus, W., Rosca, M., Lakshminarayanan, B., Dai, A.M., Mohamed, S., Goodfellow, I.: Many paths to equilibrium: GANs do not need to decrease a divergence at every step. In: International Conference on Learning Representations (2018). https://openreview.net/forum?id=ByQpn1ZA-
5. Frühstück, A., Alhashim, I., Wonka, P.: Tilegan: synthesis of large-scale non-homogeneous textures. ACM Trans. Graph. **38**(4) (2019). https://doi.org/10.1145/3306346.3322993
6. Glorot, X., Bordes, A., Bengio, Y.: Deep sparse rectifier neural networks. In: Gordon, G., Dunson, D., Dudík, M. (eds.) Proceedings of the Fourteenth International Conference on Artificial Intelligence and Statistics. Proceedings of Machine Learning Research, vol. 15, pp. 315–323. PMLR, Fort Lauderdale, 11–13 April 2011. http://proceedings.mlr.press/v15/glorot11a.html
7. Goodfellow, I., et al.: Generative adversarial nets. In: Ghahramani, Z., Welling, M., Cortes, C., Lawrence, N.D., Weinberger, K.Q. (eds.) Advances in Neural Information Processing Systems 27, pp. 2672–2680. Curran Associates, Inc. (2014). http://papers.nips.cc/paper/5423-generative-adversarial-nets.pdf
8. Gulrajani, I., Ahmed, F., Arjovsky, M., Dumoulin, V., Courville, A.C.: Improved training of wasserstein gans. In: Guyon, I., Luxburg, U.V., Bengio, S., Wallach, H., Fergus, R., Vishwanathan, S., Garnett, R. (eds.) Advances in Neural Information Processing Systems 30, pp. 5767–5777. Curran Associates, Inc. (2017). http://papers.nips.cc/paper/7159-improved-training-of-wasserstein-gans.pdf
9. Gulrajani, I., Raffel, C., Metz, L.: Towards GAN benchmarks which require generalization. In: International Conference on Learning Representations (2019). https://openreview.net/forum?id=HkxKH2AcFm
10. He, K., Zhang, X., Ren, S., Sun, J.: Deep residual learning for image recognition. In: Proceedings of the IEEE Conference on Computer Vision and Pattern Recognition (CVPR), June 2016
11. Heusel, M., Ramsauer, H., Unterthiner, T., Nessler, B., Hochreiter, S.: Gans trained by a two time-scale update rule converge to a local nash equilibrium. In: Guyon, I., Luxburg, U.V., Bengio, S., Wallach, H., Fergus, R., Vishwanathan, S., Garnett, R. (eds.) Advances in Neural Information Processing Systems 30, pp. 6626–6637. Curran Associates, Inc. (2017). http://papers.nips.cc/paper/7240-gans-trained-by-a-two-time-scale-update-rule-converge-to-a-local-nash-equilibrium.pdf
12. Karras, T., Aila, T., Laine, S., Lehtinen, J.: Progressive growing of GANs for improved quality, stability, and variation. In: International Conference on Learning Representations (2018). https://openreview.net/forum?id=Hk99zCeAb
13. Kelly, T., Guerrero, P., Steed, A., Wonka, P., Mitra, N.J.: Frankengan: Guided detail synthesis for building mass models using style-synchonized gans. ACM Trans. Graph. 37(6), December 2018. doi: 10.1145/3272127.3275065

14. Kurach, K., Lucic, M., Zhai, X., Michalski, M., Gelly, S.: The GAN landscape: Losses, architectures, regularization, and normalization (2019). https://openreview.net/forum?id=rkGG6s0qKQ
15. Liu, Z., Luo, P., Wang, X., Tang, X.: Deep learning face attributes in the wild. In: Proceedings of the IEEE International Conference on Computer Vision (ICCV), December 2015
16. Lucic, M., Kurach, K., Michalski, M., Gelly, S., Bousquet, O.: Are gans created equal? a large-scale study. In: Bengio, S., Wallach, H., Larochelle, H., Grauman, K., Cesa-Bianchi, N., Garnett, R. (eds.) Advances in Neural Information Processing Systems 31, pp. 700–709. Curran Associates, Inc. (2018). http://papers.nips.cc/paper/7350-are-gans-created-equal-a-large-scale-study.pdf
17. Ma, L., Jia, X., Sun, Q., Schiele, B., Tuytelaars, T., Van Gool, L.: Pose guided person image generation. In: Guyon, I., Luxburg, U.V., Bengio, S., Wallach, H., Fergus, R., Vishwanathan, S., Garnett, R. (eds.) Advances in Neural Information Processing Systems 30, pp. 406–416. Curran Associates, Inc. (2017). http://papers.nips.cc/paper/6644-pose-guided-person-image-generation.pdf
18. Mao, X., Li, Q., Xie, H., Lau, R.Y.K., Wang, Z., Smolley, S.P.: On the effectiveness of least squares generative adversarial networks. IEEE Trans. Pattern Anal. Mach. Intell. **41**(12), 2947–2960 (2019)
19. Mao, X., Li, Q., Xie, H., Lau, R.Y., Wang, Z., Paul Smolley, S.: Least squares generative adversarial networks. In: Proceedings of the IEEE International Conference on Computer Vision (ICCV), October 2017
20. Mescheder, L., Geiger, A., Nowozin, S.: Which training methods for GANs do actually converge? In: Dy, J., Krause, A. (eds.) Proceedings of the 35th International Conference on Machine Learning. Proceedings of Machine Learning Research, vol. 80, pp. 3481–3490. PMLR, Stockholmsmssan, Stockholm Sweden, 10–15 Jul 2018. http://proceedings.mlr.press/v80/mescheder18a.html
21. Miyato, T., Kataoka, T., Koyama, M., Yoshida, Y.: Spectral normalization for generative adversarial networks. In: International Conference on Learning Representations (2018). https://openreview.net/forum?id=B1QRgziT-
22. Nowozin, S., Cseke, B., Tomioka, R.: f-gan: Training generative neural samplers using variational divergence minimization. In: Lee, D.D., Sugiyama, M., Luxburg, U.V., Guyon, I., Garnett, R. (eds.) Advances in Neural Information Processing Systems 29, pp. 271–279. Curran Associates, Inc. (2016). http://papers.nips.cc/paper/6066-f-gan-training-generative-neural-samplers-using-variational-divergence-minimization.pdf
23. Park, T., Liu, M.Y., Wang, T.C., Zhu, J.Y.: Semantic image synthesis with spatially-adaptive normalization. In: Proceedings of the IEEE/CVF Conference on Computer Vision and Pattern Recognition (CVPR), June 2019
24. Radford, A., Metz, L., Chintala, S.: Unsupervised representation learning with deep convolutional generative adversarial networks. arXiv preprint arXiv:1511.06434 (2015)
25. Sanyal, A., Torr, P.H., Dokania, P.K.: Stable rank normalization for improved generalization in neural networks and gans. In: International Conference on Learning Representations (2020). https://openreview.net/forum?id=H1enKkrFDB
26. Tieleman, T., Hinton, G.: Lecture 6.5-rmsprop: divide the gradient by a running average of its recent magnitude. COURSERA: Neural Networks Mach. Learn. **4**(2), 26–31 (2012)

27. Wang, T.C., Liu, M.Y., Zhu, J.Y., Tao, A., Kautz, J., Catanzaro, B.: High-resolution image synthesis and semantic manipulation with conditional gans. In: Proceedings of the IEEE Conference on Computer Vision and Pattern Recognition (CVPR), June 2018
28. Wu, J., Zhang, C., Xue, T., Freeman, B., Tenenbaum, J.: Learning a probabilistic latent space of object shapes via 3d generative-adversarial modeling. In: Lee, D.D., Sugiyama, M., Luxburg, U.V., Guyon, I., Garnett, R. (eds.) Advances in Neural Information Processing Systems 29, pp. 82–90. Curran Associates, Inc. (2016). http://papers.nips.cc/paper/6096-learning-a-probabilistic-latent-space-of-object-shapes-via-3d-generative-adversarial-modeling.pdf
29. Zhu, J.Y., Park, T., Isola, P., Efros, A.A.: Unpaired image-to-image translation using cycle-consistent adversarial networks. In: Proceedings of the IEEE International Conference on Computer Vision (ICCV), October 2017

Infrastructure-Based Multi-camera Calibration Using Radial Projections

Yukai Lin[1], Viktor Larsson[1(✉)], Marcel Geppert[1], Zuzana Kukelova[2], Marc Pollefeys[1,3], and Torsten Sattler[4]

[1] Department of Computer Science, ETH Zürich, Zürich, Switzerland
vlarsson@inf.ethz.ch
[2] VRG, Faculty of Electrical Engineering, Czech Technical University in Prague, Prague, Czech Republic
[3] Microsoft Mixed Reality & AI Zurich Lab, Zürich, Switzerland
[4] Department of Electrical Engineering, Chalmers University of Technology, Gothenburg, Sweden

Abstract. Multi-camera systems are an important sensor platform for intelligent systems such as self-driving cars. Pattern-based calibration techniques can be used to calibrate the intrinsics of the cameras individually. However, extrinsic calibration of systems with little to no visual overlap between the cameras is a challenge. Given the camera intrinsics, infrastructure-based calibration techniques are able to estimate the extrinsics using 3D maps pre-built via SLAM or Structure-from-Motion. In this paper, we propose to fully calibrate a multi-camera system from scratch using an infrastructure-based approach. Assuming that the distortion is mainly radial, we introduce a two-stage approach. We first estimate the camera-rig extrinsics up to a single unknown translation component per camera. Next, we solve for both the intrinsic parameters and the missing translation components. Extensive experiments on multiple indoor and outdoor scenes with multiple multi-camera systems show that our calibration method achieves high accuracy and robustness. In particular, our approach is more robust than the naive approach of first estimating intrinsic parameters and pose per camera before refining the extrinsic parameters of the system. The implementation is available at https://github.com/youkely/InfrasCal.

1 Introduction

Being able to perceive the surrounding environment is a crucial ability for any type of autonomous intelligent system, including self-driving cars [11,34] and robots [36]. Multi-camera systems (see *e.g.* Fig. 1) are popular sensors for this task: they are cheap to build and maintain, consume little energy, and provide high-resolution data under a wide range of conditions. Enabling 360° perception

Electronic supplementary material The online version of this chapter (https://doi.org/10.1007/978-3-030-58517-4_20) contains supplementary material, which is available to authorized users.

A. Vedaldi et al. (Eds.): ECCV 2020, LNCS 12361, pp. 327–344, 2020.
https://doi.org/10.1007/978-3-030-58517-4_20

around a vehicle [34] using such systems, makes visual localization [2,7] and SLAM [24] more robust.

(a) Pentagonal camera rig

(b) GoPro helmet

Fig. 1. Multi-camera rigs. Our estimated rig calibrations are overlayed in the figure. (a) Pentagonal camera rig with five stereo pairs. (b) Ski helmet with five GoPro Hero7 Black attached covering 360° panoramic view.

Multi-camera systems need to be calibrated before use. This includes calibrating the intrinsic parameters of each camera, *i.e.* focal length, principal point and distortion parameters, as well as the extrinsic parameters between cameras, *i.e.*, the relative poses between them. An accurate and efficient calibration is often crucial for safe and robust performance. A standard approach to this problem, implemented in calibration toolboxes such as Kalibr [27], is to use a calibration pattern to record data which covers the full field-of-view (FoV) of the cameras. Although this method is powerful in achieving high accuracy, it is computationally expensive and recording a calibration dataset with adequate motion/view coverage is cumbersome, especially for wide-FoV cameras. Moreover, it is incapable of calibrating the camera-rigs with little or even no overlapping fields of view, which is often the case for applications in autonomous vehicles.

Another approach to handle such scenarios are sequence- and infrastructure-based calibration [12,13]. In both cases, the methods require prior knowledge of the intrinsics before the extrinsics calibration, which still requires a per camera pre-calibration step using calibration patterns.

In this paper, we introduce an infrastructure-based calibration that calibrates both intrinsics and extrinsics in a single pipeline. Our method uses a pre-build map of sparse feature points as a substitute for the calibration patterns. The map is easily built by a Structure-from-Motion pipeline, *e.g.* COLMAP [33]. We calibrate the camera-rigs in a two-stage process. In the first stage, the camera poses are estimated under a radial camera assumption, where the extrinsics are recovered up to an unknown translation along the principal axis. In the second stage, the intrinsics and the remaining translation parameters are jointly estimated in a robust way. We demonstrate the accuracy and robustness through extensive experiments in indoor and outdoor datasets with different multi-camera systems.

The **main contributions** of this paper are: (**1**) We propose an infrastructure-based calibration method for performing multi-camera rig intrinsics and extrinsics calibration in an user-friendly way as we remove the need for pre-calibration for each camera or tedious recording for calibration pattern data. (**2**) In contrast to current methods, we show that it is possible to first (partially) estimate the camera rig's extrinsic parameters before estimating the internal calibration for each camera. (**3**) Our proposed method is experimentally shown to give high-quality camera calibrations in a variety of environments and hardware setups.

2 Related Work

Pattern-Based Calibration. A pattern-based calibration method estimates camera parameters using special calibration patterns such as AprilTags [28] or checkerboards [27,37,42]. The patterns are precisely designed so that they can be accurately estimated via camera systems. We note that the pattern-based calibration of multi-camera systems usually requires the camera pairs to have overlapping FoV, since the pattern must be visible in multiple images to constrain the rig's extrinsic parameters. Some works [19,23,31] calibrate the cameras without assuming the overlapping FoV. Kumar et al. [19] show that the use of an additional mirror can help to create overlap between cameras. Li et al. [23] only require the neighboring cameras to partially observe the calibration patterns at the same time but the observed parts do not necessarily need to overlap. Robinson et al. [31] calibrate the extrinsic parameters for non-overlapping cameras by temporarily adding an additional camera during calibration with an overlapping FoV with both cameras. We note that the use of a calibration pattern board always introduces a certain viewing constraint or extra effort to calibrate the cameras with non-overlapping FoV. Furthermore, the calibration of wide FoV cameras is especially cumbersome. The pattern needs to be close to the camera to cover any significant part of the image but if it is too close, it leads to problems where the pattern is out of focus. Thus, to get accurate calibration results, it is typically necessary to capture a large number of images.

Infrastructure-Based Calibration. Rather than using calibration patterns, infrastructure-based calibration uses natural scene features to estimate camera parameters. Carrera et al. [5] propose a feature-based extrinsic calibration method through a SLAM-based feature matching among the maps for each camera. Heng et al. [12] simplify that approach to rely on a prior high-accuracy map, removing the need for inter-camera feature correspondences and loop closures. Their pipeline first infers camera poses via the P3P method for calibrated cameras, and subsequently, an initial estimate of the camera-rig transformations and rig poses. A final non-linear refinement step optimizes the camera-rig transformations, rig poses and optionally intrinsics.

Our method is most similar to the work of Heng et al. [12] in that we use a prebuilt sparse feature map for calibration. However, their method relies on a known intrinsics input which still requires calibration patterns for intrinsic calibration.

Our method does not require a prior intrinsics knowledge and performs complete calibration, both intrinsic and extrinsic, using the sparse map.

Compared to checkerboard-style calibration objects, infrastructure-based methods are able to get significantly more constraints per-image since there are typically more feature points observed which acts as a virtual large calibration pattern. In practice, infrastructure-based calibration provides a much wider application range than pattern-based calibration.

Camera Pose Estimation with Unknown Intrinsic Parameters. Given a sparse set of 2D-3D correspondences between an image and a 3D point cloud (a map), it is possible to recover the camera pose. If the cameras' internal calibration is known, i.e. the mapping from image pixels to viewing rays, the absolute pose estimation problem becomes minimal with three correspondences. This problem is usually referred to as the Perspective-Three-Points (P3P) problem [9]. In settings where the intrinsic parameters are unknown, the estimation problem becomes more difficult and more correspondences are necessary. Most modern cameras can be modeled as having square pixels (i.e. zero skew and unit aspect ratio). Due to this, most work on camera pose estimation with unknown/partially known calibration has focused on the case of unknown focal length. The minimal problem for this case was originally solved by Bujnak et al. [3]. Since then, there have been several papers improving on the original solver [18,20,30,41,43]. The case of unknown focal length and principal point was considered by Triggs [39] and later Larsson et al. [21]. When all of the intrinsic parameters are unknown, the Direct-Linear-Transform (DLT) [10] can be applied. Camera pose estimation with unknown radial distortion was first considered by Josephson and Byröd [15]. There have been multiple works improving on this paper in different aspects; faster runtime [17,20], support for other distortion models [22] and even non-parametric distortion models [4].

Radial Alignment Constraint and the 1D Radial Camera Model. Focal length and radial distortion only scales the images points radially outwards from the principal point (assuming this is the center of distortion). This observation was used by Tsai [40] to derive constraints on the camera pose which are independent of the focal length and distortion parameters. For a 2D-3D correspondence, the idea is to only require that the 3D point projects onto the radial line passing through the 2D image point, and to ignore the radial offset. This constraint is called the Radial Alignment Constraint (RAC). This later gave rise to the 1D radial camera model (see [38]) which re-interprets the camera as projecting 3D points onto radial lines instead of 2D points. Since forward motion also moves the projections radially, it is not possible to estimate the forward translation using these constraints. In practice, the 1D radial camera model turns out to be equivalent to only considering the top two rows of the normal projection matrix. Instead of reprojection error, radial reprojection error measures the distance from 2D point to projected radial line, which is invariant to focal length and radial distortion parameters.

These ideas have also been applied to absolute pose estimation with unknown radial distortion. In Kukelova et al. [17], the authors present a two-stage approach which first estimates the camera pose up to an unknown forward translation using the RAC. In a second step the last translation component is jointly estimated with the focal length and distortion parameters. This was later extended in Larsson et al. [22]. Camposeco et al. [4] applied a similar approach to non-parametric distortion models.

In this paper we take a similar approach as [4,17,22]. However, instead of using just one frame, we can leverage multiple frames for the upgrade step since we consider multi-camera systems. We show it is possible to use joint poses of multiple (non-parallel) 1D radial cameras to transform the frames into the camera coordinate system for each single camera.

3 Multi-camera Calibration

Now we present our framework for calibration of a multi-camera system. Our approach is similar to the infrastructure-based calibration method from Heng et al. [12]. We improve on their approach in the following aspects:

- We leverage state-of-the-art absolute pose solvers [17,22] to also perform estimation of the camera intrinsic parameters, thus completely removing any need for pre-calibrating each camera independently.
- We present a new robust estimation scheme to initialize the rig extrinsic parameters. Our experiments show that this greatly improves the robustness of the calibration method, especially on datasets with shorter image sequences.
- Finally we show it is possible to first partially estimate the rig extrinsics and pose before recovering the camera intrinsic parameters. This partial extrinsic knowledge allows us to more easily incorporate information from multiple images into the estimation.

Similarly to Heng et al. [12] we assume that we have a sparse map of the environment. The input to our method is then a synchronized image sequence captured by the multi-camera system as it moves around in the mapped environment. The main steps of our pipeline are presented below and detailed in the next sections.

1. **Independent 1D radial pose estimation.** We independently estimate a 1D radial camera pose (see Sect. 2) for each image using RANSAC [6].
2. **Radial camera rig initialization.** We robustly fit a multi-camera rig with the 1D radial camera model to the estimated individual camera poses.
3. **Radial Bundle Adjustment.** We optimize the partial rig extrinsics and poses by minimizing the radial reprojection error (see Sect. 2). Here we additionally refine the principal point for each camera.
4. **Forward translation and intrinsic estimation.** Using the partially known extrinsic parameters and poses of the rig we can transform all 2D-3D correspondences into the camera coordinate system (up to the unknown forward translation). This allows us to use the entire image sequence when initializing the intrinsic parameters and the forward translations [4].

5. **Final refinement.** Finally, we perform bundle adjustment over rig poses, rig extrinsic and intrinsic parameters, minimizing the reprojection error over the entire sequence.

The entire calibration pipeline is illustrated in Fig. 2.

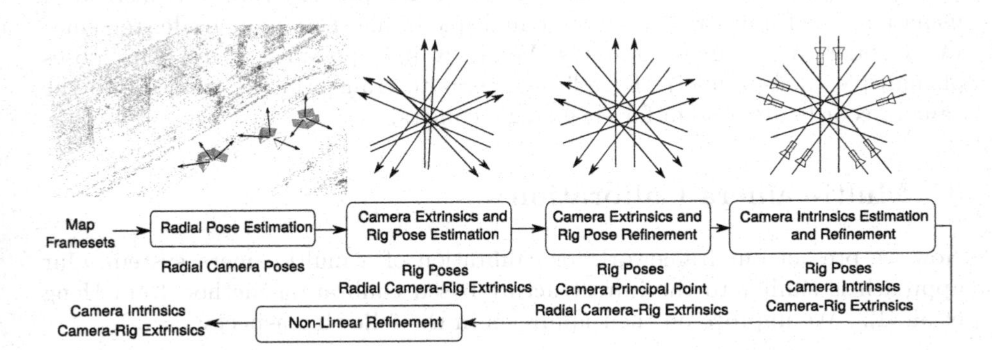

Fig. 2. Illustration of the calibration pipeline. The output of each step is placed below each block. In the first step we independently estimate the 1D radial pose of each camera. Next we robustly fit a rig to the estimated poses. We refine the rig extrinsic parameters and poses by minimizing the radial reprojection errors. Then we upgrade each camera by estimating the last translation component jointly with the internal calibration. Finally we refine all parameters by minimizing the reprojection error.

3.1 The Sparse Map and Input Framesets

One of the inputs to our method is a pre-built sparse feature map, which can be built using a standard Structure-from-Motion pipeline, e.g. COLMAP [33]. It is necessary to build a high-accuracy map in order to produce accurate calibration result. The map can be used as long as there is no large change in the environment. The correct scale of the map can be derived either from a calibrated multi-camera system, e.g. stereo system, or by the user measuring some distances in both the real world and the map and scaling the map accordingly, e.g. by using a checkerboard. In addition, a sequence of synchronized images captured in the map is recorded as the calibration dataset. We define a frameset to be a set of images captured at the same timestamp from all different cameras.

3.2 Initial Camera Pose Estimation

The first step of our pipeline is to independently estimate the pose of each image with respect to the pre-built map. Using the 1D radial camera model allows us to estimate the pose of the camera (up to forward translation) without knowing the camera intrinsic parameters (see Sect. 2). Similarly to Heng et al. [12], to find

2D-3D correspondences between the query image and the map we use a bag-of-words-based image retrieval against the mapping images, followed by 2D-2D image matching. For local features/descriptors we use upright SIFT [25], but any local feature could be used. Once the putative 2D-3D correspondences are found we use the minimal solver from Kukelova et al. [17] (see Sect. 2) in RANSAC to estimate the 1D radial camera pose. The principal point for each camera is initialized to the image center, a valid assumption for common cameras, and could be recovered accurately in later steps. Note that this only estimates the orientation and two components of the camera translation. At this stage we filter out any camera poses with too few inliers.

Alternatively we can also use the solvers from [17,22] which directly solve for the intrinsic parameters. However, estimating the intrinsic parameters from a single image turns out to be significantly less stable. See Sect. 5.4 for a comparison of the errors in the intrinsic calibration when we perform the intrinsic calibration at this stage of the pipeline.

3.3 Camera Extrinsics and Rig Poses Estimation

In the previous step we estimated the absolute poses for each image independently. Since we used the 1D radial camera model we only recovered the pose up to an unknown forward translation, i.e. we estimated

$$T_{ij} = \begin{bmatrix} R & \begin{pmatrix} t_x \\ t_y \\ ? \end{pmatrix} \end{bmatrix}, \tag{1}$$

which transforms from the map coordinate system to the coordinate system of ith camera in the jth frameset. The goal now is to use the initial estimates to recover both the rig extrinsic parameters as well as the rig pose for each frameset in a robust way. In [12], they simplify this problem by assuming that there is at least one frameset where each camera was able to get a pose estimate. In our experiments this assumption was often not satisfied for shorter image sequences, leading to the method completely failing to initialize.

Let P_i be the transform from the rig-centric coordinate system to the ith camera and let Q_j be the transform from the map coordinate system to the rig-centric coordinate system for the jth frameset. A rig-centric coordinate system can be set to any rig-fixed coordinate frame since we only consider the relative extrinsics. In our case, it is set initially to be the same as the first camera with the unknown forward translation being zero. For noise-free measurements we should have

$$T_{ij} = P_i Q_j, \quad (i,j) \in \Omega, \tag{2}$$

where Ω is the set of images that were successfully estimated in the previous step, i.e. $(i,j) \in \Omega$ if camera i in frameset j was successfully registered. Since we did not estimate the third translation component of T_{ij}, we restrict ourselves to finding the first two rows of the camera matrices, i.e.

$$\hat{T}_{ij} = \hat{P}_i Q_j, \quad (i,j) \in \Omega, \tag{3}$$

where $\hat{T}_{ij}$ denotes the first two rows of T_{ij} and similarly for $\hat{P}_i$. As described in Sect. 2 we can interpret $\hat{P}_i$ as 1D radial camera poses. If some $\hat{P}_i$ are known, then the rig poses Q_j can be found by solving

$$\begin{bmatrix} \hat{T}_{1j} \\ \hat{T}_{2j} \\ \vdots \end{bmatrix} = \begin{bmatrix} \hat{P}_1 \\ \hat{P}_2 \\ \vdots \end{bmatrix} Q_j \quad \text{where} \quad Q_j = \begin{bmatrix} R & \boldsymbol{t} \\ \mathbf{0}^T & 1 \end{bmatrix}, \tag{4}$$

which has a closed form solution using SVD [35]. Note that this requires that at least two cameras have non-parallel principal axes. We discuss this limitation more in Sect. 4. In turn, if the rig poses Q_j are known, we can easily recover the rig extrinsic parameters as $\hat{P}_i = \hat{T}_{ij} Q_j^{-1}$.

To robustly fit the rig extrinsics $\hat{P}_i$ and rig poses Q_j to the estimated absolute poses $\hat{T}_{ij}$, we solve the following minimization problem,

$$\min_{\{\hat{P}_i\},\{Q_j\}} \sum_{(i,j)\in\Omega} \rho\left(d\left(\hat{T}_{ij},\, \hat{P}_i Q_j\right)\right), \tag{5}$$

where ρ is a robust loss function and d is a weighted sum of the rotation and translation errors. Since this is a non-convex problem we perform a robust initialization scheme based on a greedy assignment in RANSAC.

In our case we randomly select any frameset with at least two cameras as initialization and assign the corresponding $\hat{P}_i$ using the relative poses from this frameset. Note that this might leave some $\hat{P}_i$ unassigned. We then use these assigned poses to estimate the rig poses Q_j of any other frameset which also contains the already assigned $\hat{P}_i$. We can then iterate between assigning any of the missing $\hat{P}_i$ and estimating new Q_j. This back-and-forth search repeats until all of the rig extrinsics and rig poses are assigned. We repeat the entire process multiple times in a RANSAC-style fashion, keeping track of the best assignment with minimal radial reprojection over all frames. Finally, for the best solution we perform local optimization of (5) using Levenberg-Marquardt.

This approach is similar to the rotation averaging methods in [8,29] which repeatedly build random minimum spanning trees in the pose-graph and assigns the absolute rotations based on these.

3.4 Camera Extrinsics and Rig Poses Refinement

We further refine the camera rig extrinsics and rig poses by performing bundle adjustment to minimize the radial reprojection error. In this step we also optimize over the principal point for each camera which was initialized to the image center. Let $\boldsymbol{X}_p$ be a 3D point and $\boldsymbol{x}_{ijp}$ its observation in the ith camera in frameset j. Then we optimize

$$\min_{\hat{P}_i, Q_j, \boldsymbol{c}_i} \sum_{i,j,p} \rho\left(\left\|\pi_r\left(\hat{P}_i Q_j \boldsymbol{X}_p,\, \boldsymbol{x}_{ijp} - \boldsymbol{c}_i\right) - (\boldsymbol{x}_{ijp} - \boldsymbol{c}_i)\right\|^2\right), \tag{6}$$

where ρ is a robust loss function, $\boldsymbol{c}_i$ is the principal point of the ith camera and $\pi_r : \mathbb{R}^2 \times \mathbb{R}^2 \to \mathbb{R}^2$ is the orthogonal projection of the second argument onto the line generated by the first, i.e. $\pi_r(\boldsymbol{u}, \boldsymbol{v}) = \frac{\boldsymbol{u}^T\boldsymbol{v}}{\boldsymbol{u}^T\boldsymbol{u}}\boldsymbol{u}$.

3.5 Camera Upgrading and Refinement

In this step, we estimate the internal calibration as well as the remaining unknown translation component for each camera. By transforming all 2D-3D correspondences into the rig frame, we can leverage data from all framesets.

From the previous step we have estimated the camera rig extrinsics P_i, except for the third component of the translation vector, i.e. $t_{z,i}$. The 3D points mapped into each camera's coordinate system can then be written as

$$\boldsymbol{Z}_{ijp} + t_{z,i}\boldsymbol{e}_z = P_i Q_j \boldsymbol{X}_p \quad \text{where} \quad \boldsymbol{e}_z = \left(0, 0, 1\right)^T. \tag{7}$$

Now we can use the minimal solvers from Kukelova et al. [17] and Larsson et al. [22] for jointly estimating $t_{z,i}$ and the intrinsic parameters. To further remove outlier correspondences, we again use RANSAC to robustly initialize the parameters. Additionally, we perform non-linear optimization to refine the intrinsics and $t_{z,i}$ by minimizing the reprojection error as

$$\min_{\theta_i, t_z} \sum_{j,p} \rho\left(\left\|\pi_{\theta_i}\left(\boldsymbol{Z}_{ijp} + t_{z,i}\boldsymbol{e}_z\right) - \boldsymbol{x}_{ijp}\right\|^2\right), \tag{8}$$

where θ_i are the intrinsic parameters and π_{θ_i} denotes the projection into image space. Note that here we use full distortion model instead of pure radial distortion. This is done for each camera individually.

3.6 Final Refinement

In the final step, we optimize all the camera intrinsics, extrinsics and rig poses by minimizing the reprojection error. The optimization problem is

$$\min_{P_i, Q_j, \theta_i} \sum_{i,j,p} \rho\left(\left\|\pi_{\theta_i}(P_i Q_j \boldsymbol{X}_p) - \boldsymbol{x}_{ijp}\right\|^2\right). \tag{9}$$

Optionally the 3D scene points can be added into optimization problem, in case the scene points are not accurate enough.

4 Implementation

Our implementation is based on the infrastructure-based calibration from the CamOdoCal library [14]. The sparse map is built by COLMAP [33], which uses upright SIFT [25] features and descriptors. For the camera model, pinhole with radial-tangential distortion and pinhole with equidistant distortion [16] are supported to suit different cameras. The optimization is solved with the Levenberg-Marquardt algorithm using the Ceres Library [1] and we use the Cauchy loss with scale parameter 1 as the robust loss function.

Limitations. Note that to robustly fit the rig extrinsics among different framesets requires that at least two cameras in the rig have non-parallel principal axes, otherwise Eq. 5 fails to determine the rig pose. However, camera rigs with parallel principal axes, usually stereo camera setups, can be easily calibrated through existing calibration methods. Other cases, *e.g.* two cameras with opposite direction, commonly equipped in mobile phones, can be calibrated by our proposed calibration variant **Inf+RD+RA** described in Sect. 5.1, which uses pose solvers that can estimate both the poses and intrinsics per frame.

5 Experimental Evaluation

For the experimental evaluation of our method we first consider two different multi-camera systems, one pentagonal camera rig with ten cameras arranged in five stereo pairs (Fig. 1a) and a ski helmet with five GoPro Hero7 Black cameras attached (Fig. 1b). For the GoPro cameras we record in wide FoV mode, which roughly corresponds to 120° horizontal FoV. The cameras on the pentagonal rig have circa 70° horizontal FoV.

Fig. 3. Sample images of the environment. *Left:* Indoor environment in a lab room. *Right:* Outdoor environment on an urban road.

5.1 Evaluation Datasets and Setup

To validate our method we record datasets in both indoor and outdoor environments. See Fig. 3 for example images. For each dataset we record a mapping sequence with the GoPro Hero Black 7 in linear mode[1], calibration sequences with both the pentagonal rig and the GoPro helmet, and Aprilgrid sequences to allow for offline calibration and validation. We use the calibration toolbox Kalibr [27] on the Aprilgrid datasets to create a *ground-truth* calibration for comparison. As far as we know there is no competing method that performs infrastructure-based multi-camera calibration with unknown intrinsic parameters. We augment the original pipeline from Heng et al. [12] with radial distortion solvers from Larsson et al. [22] as candidates to join the comparison. In particular, we compare the following approaches:

[1] Linear mode provides in-camera undistorted images with a reduced FoV.

- **Inf+K.** The infrastructure-based method from Heng et al. [12].
- **Inf+K+RI.** Same as Inf+K but with refinement of the intrinsic parameters during the final bundle adjustment.
- **Inf+RD.** We replace the P3P solver in [12] with the pose solvers from Larsson et al. [22] which also estimate distortion parameters and focal length.
- **Inf+RD+RA.** We add a robust rig averaging similar to Sect. 3.3.
- **Inf+1DR+RA.** The proposed pipeline as described in Sect. 3.2–3.6 which delays estimation of the intrinsic parameters using 1D radial cameras.

Note that **Inf+K** and **Inf+K+RI** use the intrinsic parameters from running Kalibr on the Aprilgrid images and join the competition as references. To evaluate the resulting calibrations we robustly align the estimated camera rigs with the camera rigs obtained from Kalibr [27] and compute the difference in the rotations (degrees) and camera centers (centimetres). To evaluate the intrinsic parameters we validate the calibration on the Aprilgrid datasets and report the average reprojection error (pixels).

5.2 Calibration Accuracy and Run-Time on Full Image Sequence

First we aim to evaluate the accuracy of the calibrations by running the methods on the entire calibration sequences. See Fig. 4 for a visualization of camera poses recovered in the *Outdoor* dataset. The results can be found in Table 1. We can see that, using infrastructure-based calibration methods, we are able to obtain similar quality results as classical Aprilgrid based methods. In this case, the three methods **Inf+RD**, **Inf+RD+RA**, and **Inf+1DR+RA** all had very similar performance. Note also that the ground truth we are comparing to is not necessarily perfect. In practice, we find that with similar datasets recorded at the same time, the extrinsic results differ up to $0.3°$ and 0.5 cm.

For run-time, we run our method on a DELL Laptop equipped with 16 GB RAM, an i7-9750H CPU and a GTX1050 GPU. A comparison of the processing time of each pipeline is shown in Table 2. Our method **Inf+1DR+RA** takes a similar amount of time while removing the need for pre-calibration required by **Inf+K+RI**, and runs much faster than the pattern-based method **Kalibr**.

5.3 Evaluation of Robustness on Shorter Image Sequences

In the previous section we saw that if we have enough data we are able to achieve high-quality calibration results. In this section we instead evaluate the robustness of the method when input data is more limited. For many applications this is an important scenario since you might want to find the camera calibration as quickly as possible to enable other tasks which depend on knowing the camera calibration. To perform the experiment we select multiple sub-sequences and try to calibrate from these. For each sequence we select 10 framesets which approximately differ by one second (the datasets were captured at normal walking speed). Table 3 shows the percentage of frames where the calibration-methods were able to calibrate the complete rig, as well as the percentage of sequences

Table 1. Evaluation of calibration accuracy. The errors are with respect to the calibration obtained from the Aprilgrid datasets with Kalibr [27]. Note that **Inf+K** and **Inf+K+RI** use the ground-truth intrinsic parameters as input.

	Inf+ K	K+RI	RD	RD+RA	1DR+RA
GoPro Helmet/Indoor					
Reproj. error (px)	0.283	0.270	0.526	0.412	0.270
Rot. error (degree)	0.193	0.320	0.328	0.319	0.321
Trans. error (cm)	0.780	0.418	0.430	0.435	0.426
GoPro Helmet/Outdoor					
Reproj. error (px)	0.339	0.337	0.337	0.336	0.337
Rot. error (degree)	0.141	0.188	0.187	0.187	0.187
Trans. error (cm)	0.642	0.392	0.385	0.390	0.384
Pentagonal/Indoor					
Reproj. error (px)	0.230	0.281	0.280	0.308	0.282
Rot. error (degree)	0.293	0.548	0.545	0.543	0.543
Trans. error (cm)	1.316	0.366	0.372	0.377	0.376
Pentagonal/Outdoor					
Reproj. error (px)	0.198	0.268	0.268	0.263	0.271
Rot. error (degree)	0.295	0.570	0.566	0.568	0.567
Trans. error (cm)	2.217	0.441	0.419	0.417	0.423

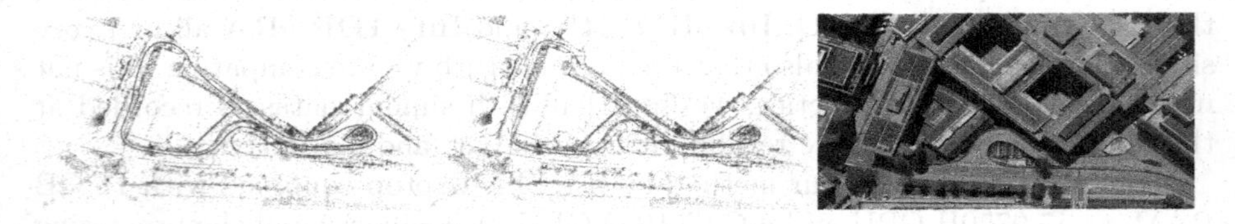

Fig. 4. Experiments in outdoor urban environment. *Left:* The sparse reconstruction from COLMAP [33] with mapping sequence shown in red. *Middle:* The same scene with frames used for calibration in red. *Right:* Aerial view of the scene. (Color figure online)

which gave good calibrations (defined as rotation error below 1° and translation below 1 cm for indoor and 2 cm for outdoor). The total number of sequences were 313 (penta) and 173 (GoPro). Table 3 shows the superior robustness of our approach.

Table 2. Run-Time Comparison. Table lists the average runtime (in minutes) for different methods on calibration sequences with 500 framesets. The runtime for Inf+K+RI and Inf+1DR+RA consists of indoor/outdoor cases.

Runtime (min)	**Inf+K+RI**	**Inf+1DR+RA**	**Kalibr**
GoPro Helmet	9.5/11.3	10.9/12.2	24.5
Pentagonal	7.0/9.6	11.0/15.4	113.0

Table 3. Comparison of robustness for shorter image sequences. Table shows the percentage of sequences which we were able to estimate a complete frameset and the percentage of sequences of sequences that were accurately calibrated. A good calibration is defined in Sect. 5.3.

	Inf+	**RD**	**RD+RA**	**1DR+RA**
GoPro Helmet/Indoor	Complete	54.5	**98.3**	**98.3**
	Good	44.9	75.6	**79.0**
GoPro Helmet/Outdoor	Complete	67.6	97.7	**98.3**
	Good	38.1	45.5	**48.3**
Pentagonal/Indoor	Complete	31.9	68.4	**69.0**
	Good	23.0	43.1	**44.4**
Pentagonal/Outdoor	Complete	28.8	79.2	**80.5**
	Good	21.1	38.3	**41.5**

5.4 Evaluation of Initial Estimates

In this section we evaluate the effect of delaying the estimation of the intrinsic parameters on the quality of the initial estimates, i.e. before running bundle adjustment. Similar to the evaluation for robustness in Sect. 5.3, we run the different methods on multiple sub-sequences and evaluate the extrinsics error of the initial estimates. A qualitative example of the extrinsics is shown in Fig. 5 (Left) and it is obvious that the extrinsics estimate for **Inf+1DR+RA** is much better and almost close to the final result. Figure 5 (Right) shows the distribution of the extrinsics error for both methods, where **Inf+1DR+RA** outperforms **Inf+RD+RA** especially in position error. However, as shown in Table 3 most of these initial errors can be recovered in the final refinement.

5.5 Evaluation on RobotCar Dataset

In addition to the experiments mentioned above, we evaluate our calibration method on the public benchmark RobotCar Dataset [26]. We select a short sequence of 30 s from the 2014/12/16 datasets (frame No.500 to frame No.900) recorded in the morning to calibrate the three Grasshopper2 cameras pointing left, back and right respectively. The map and calibration groundtruth is obtained from the RobotCar Seasons Dataset [32]. The calibration takes only

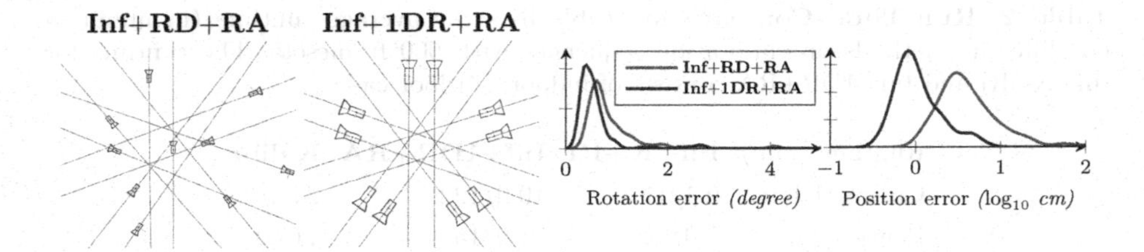

Fig. 5. *Left:* Qualitative example of rig initializations before final refinement. *Right:* Distribution of rotation and translation errors before final refinement.

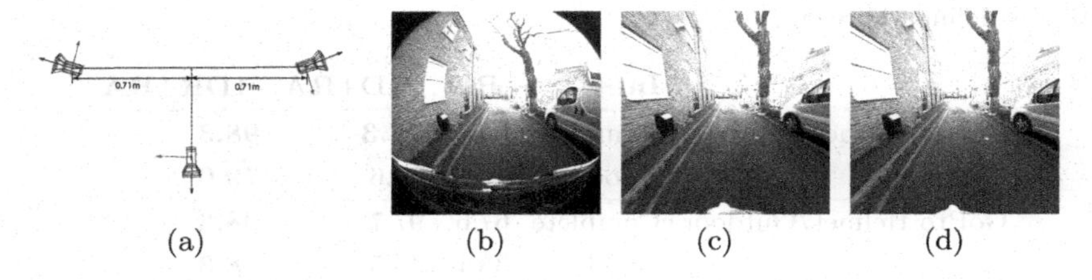

Fig. 6. Results on RobotCar datasets. The extrinsics for out method (blue) and groundtruth (red) are plotted in (a). To validate the intrinsics, the raw image (b) is undistorted using our calibrated results (c) and groundtruth parameters (d). (Color figure online)

3 min on a normal PC and the extrinsic results are shown in Fig. 6(a). The position error is 1.04 cm and rotation error is 0.213°. To validate the intrinsic parameters, we compare the results directly from undistorting the raw image Fig. 6(b). Figure 6(c) and Fig. 6(d) are the undistorted image for our method and the groundtruth respectively. Although this benchmark is designed for visual localization and place recognition algorithms under changing conditions, we show our method robustly and accurately estimates the camera calibration parameters even with real vehicle vision data in urban roads.

5.6 Application: Robot Localization in a Garden

Finally we evaluate our proposed framework in a real robotics application, namely localization in an outdoor environment. We attach the pentagonal rig to a small robot which autonomously navigates in a garden. We record several datasets of the robot driving around in the garden. From one of the recordings we build a map using the calibration obtained from Aprilgrid calibration with Kalibr. We then calibrate the camera rig using one of the other datasets and evaluate localization performance on the rest of the datasets. The position of the robot is tracked with a TopCon laser tracker yielding accurate position used as groundtruth. The plot of Kalibr extrinsics and results from our results shown in Fig. 7(a) confirms the high accuracy of our calibration method. In Fig. 7(b) and Fig. 7(c) we plot the localized trajectory of two different localization datasets

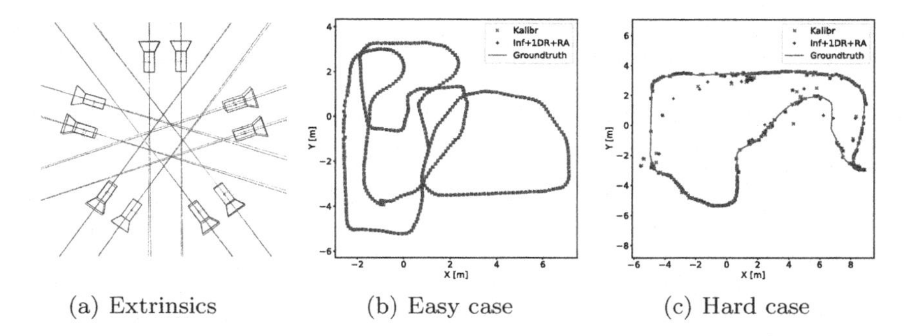

(a) Extrinsics (b) Easy case (c) Hard case

Fig. 7. Results in gardening datasets. The extrinsics are plotted in (a). (b) shows the localization trajectory of an easy dataset and (c) a hard one. The Kalibr results are indicated by red and our method by blue. (Color figure online)

using the calibration results of Kalibr and our method. The median position errors for the two sequences are 3.56 cm and 9.22 cm using results from the proposed method, and 3.77 cm and 9.67 cm using calibration with Kalibr. Using a calibration estimated from the map we are able to achieve slightly higher accuracy for localization compared to the pattern-based approach.

6 Conclusions

We have proposed a method for complete calibration, both intrinsic and extrinsic, of multi-camera systems. Due to the use of natural scene features, our calibration method can be used in any arbitrary indoor and outdoor environments without the aid of other calibration patterns or setups. The extensive experiments and real case application demonstrate the high accuracy, efficiency and robustness of our proposed calibration method. Given the practical usefulness of our approach, we expect it to have large impact in the robotics and autonomous vehicle community.

Acknowledgement. This work was supported by the Swedish Foundation for Strategic Research (Semantic Mapping and Visual Navigation for Smart Robots), the Chalmers AI Research Centre (CHAIR) (VisLocLearn), OP VVV project Research Center for Informatics No. CZ.02.1.01/0.0/0.0/16_019/0000765, and EU Horizon 2020 research and innovation program under grant No. 688007 (TrimBot2020). Viktor Larsson was supported by an ETH Zurich Postdoctoral Fellowship.

References

1. Agarwal, S., Mierle, K., et al.: Ceres solver, 2013 (2018). http://ceres-solver.org
2. Arth, C., Wagner, D., Klopschitz, M., Irschara, A., Schmalstieg, D.: Wide area localization on mobile phones. In: 2009 8th IEEE International Symposium on Mixed and Augmented Reality, pp. 73–82. IEEE (2009)

3. Bujnak, M., Kukelova, Z., Pajdla, T.: A general solution to the P4P problem for camera with unknown focal length. In: Computer Vision and Pattern Recognition (CVPR) (2008)
4. Camposeco, F., Sattler, T., Pollefeys, M.: Non-parametric structure-based calibration of radially symmetric cameras. In: International Conference on Computer Vision (ICCV) (2015)
5. Carrera, G., Angeli, A., Davison, A.J.: SLAM-based automatic extrinsic calibration of a multi-camera rig. In: International Conference on Robotics and Automation (ICRA) (2011)
6. Fischler, M.A., Bolles, R.C.: Random sample consensus: a paradigm for model fitting with applications to image analysis and automated cartography. Commun. ACM **24**(6), 381–395 (1981)
7. Geppert, M., Liu, P., Cui, Z., Pollefeys, M., Sattler, T.: Efficient 2D-3D matching for multi-camera visual localization. In: International Conference on Robotics and Automation (ICRA) (2019)
8. Govindu, V.M.: Robustness in motion averaging. In: Narayanan, P.J., Nayar, S.K., Shum, H.-Y. (eds.) ACCV 2006. LNCS, vol. 3852, pp. 457–466. Springer, Heidelberg (2006). https://doi.org/10.1007/11612704_46
9. Haralick, B.M., Lee, C.N., Ottenberg, K., Nölle, M.: Review and analysis of solutions of the three point perspective pose estimation problem. Int. J. Comput. Vis. **13**(3), 331–356 (1994). https://doi.org/10.1007/BF02028352
10. Hartley, R., Zisserman, A.: Multiple View Geometry in Computer Vision. Cambridge University Press, Cambridge (2003)
11. Heng, L., et al.: Project autovision: localization and 3D scene perception for an autonomous vehicle with a multi-camera system. In: International Conference on Robotics and Automation (ICRA) (2019)
12. Heng, L., Furgale, P., Pollefeys, M.: Leveraging image-based localization for infrastructure-based calibration of a multi-camera rig. J. Field Robot. **32**(5), 775–802 (2015)
13. Heng, L., Lee, G.H., Pollefeys, M.: Self-calibration and visual SLAM with a multi-camera system on a micro aerial vehicle. Auton. Robot. **39**(3), 259–277 (2015). https://doi.org/10.1007/s10514-015-9466-8
14. Heng, L., Li, B., Pollefeys, M.: CamOdoCal: automatic intrinsic and extrinsic calibration of a rig with multiple generic cameras and odometry. In: International Conference on Intelligent Robots and Systems (IROS) (2013)
15. Josephson, K., Byrod, M.: Pose estimation with radial distortion and unknown focal length. In: Computer Vision and Pattern Recognition (CVPR) (2009)
16. Kannala, J., Brandt, S.S.: A generic camera model and calibration method for conventional, wide-angle, and fish-eye lenses. Trans. Pattern Anal. Mach. Intell. (PAMI) **28**(8), 1335–1340 (2006)
17. Kukelova, Z., Bujnak, M., Pajdla, T.: Real-time solution to the absolute pose problem with unknown radial distortion and focal length. In: International Conference on Computer Vision (ICCV) (2013)
18. Kukelova, Z., Heller, J., Fitzgibbon, A.: Efficient intersection of three quadrics and applications in computer vision. In: Computer Vision and Pattern Recognition (CVPR) (2016)
19. Kumar, R.K., Ilie, A., Frahm, J.M., Pollefeys, M.: Simple calibration of non-overlapping cameras with a mirror. In: Computer Vision and Pattern Recognition (CVPR) (2008)

20. Larsson, V., Kukelova, Z., Zheng, Y.: Making minimal solvers for absolute pose estimation compact and robust. In: International Conference on Computer Vision (ICCV) (2017)
21. Larsson, V., Kukelova, Z., Zheng, Y.: Camera pose estimation with unknown principal point. In: Computer Vision and Pattern Recognition (CVPR) (2018)
22. Larsson, V., Sattler, T., Kukelova, Z., Pollefeys, M.: Revisiting radial distortion absolute pose. In: International Conference on Computer Vision (ICCV) (2019)
23. Li, B., Heng, L., Koser, K., Pollefeys, M.: A multiple-camera system calibration toolbox using a feature descriptor-based calibration pattern. In: International Conference on Intelligent Robots and Systems (IROS) (2013)
24. Liu, P., Geppert, M., Heng, L., Sattler, T., Geiger, A., Pollefeys, M.: Towards robust visual odometry with a multi-camera system. In: International Conference on Intelligent Robots and Systems (IROS) (2018)
25. Lowe, D.G.: Distinctive image features from scale-invariant keypoints. Int. J. Comput. Vis. **60**(2), 91–110 (2004). https://doi.org/10.1023/B:VISI.0000029664.99615.94
26. Maddern, W., Pascoe, G., Gadd, M., Barnes, D., Yeomans, B., Newman, P.: Real-time kinematic ground truth for the Oxford RobotCar Dataset. arXiv preprint arXiv:2002.10152 (2020). https://arxiv.org/pdf/2002.10152
27. Maye, J., Furgale, P., Siegwart, R.: Self-supervised calibration for robotic systems. In: 2013 IEEE Intelligent Vehicles Symposium (IV), pp. 473–480. IEEE (2013)
28. Olson, E.: AprilTag: a robust and flexible visual fiducial system. In: International Conference on Robotics and Automation (ICRA) (2011)
29. Olsson, C., Enqvist, O.: Stable structure from motion for unordered image collections. In: Heyden, A., Kahl, F. (eds.) SCIA 2011. LNCS, vol. 6688, pp. 524–535. Springer, Heidelberg (2011). https://doi.org/10.1007/978-3-642-21227-7_49
30. Penate-Sanchez, A., Andrade-Cetto, J., Moreno-Noguer, F.: Exhaustive linearization for robust camera pose and focal length estimation. Trans. Pattern Anal. Mach. Intell. (PAMI) **35**(10), 2387–2400 (2013)
31. Robinson, A., Persson, M., Felsberg, M.: Robust accurate extrinsic calibration of static non-overlapping cameras. In: Felsberg, M., Heyden, A., Krüger, N. (eds.) CAIP 2017. LNCS, vol. 10425, pp. 342–353. Springer, Cham (2017). https://doi.org/10.1007/978-3-319-64698-5_29
32. Sattler, T., et al.: Benchmarking 6DOF outdoor visual localization in changing conditions. In: Computer Vision and Pattern Recognition (CVPR) (2018)
33. Schonberger, J.L., Frahm, J.M.: Structure-from-motion revisited. In: Computer Vision and Pattern Recognition (CVPR) (2016)
34. Schwesinger, U., et al.: Automated valet parking and charging for e-mobility. In: Intelligent Vehicles Symposium (IV). IEEE (2016)
35. Sorkine-Hornung, O., Rabinovich, M.: Least-squares rigid motion using SVD. Computing **1**(1), 1–5 (2017)
36. Strisciuglio, N., et al.: Trimbot 2020: an outdoor robot for automatic gardening. In: ISR 2018; 50th International Symposium on Robotics, pp. 1–6. VDE (2018)
37. Sturm, P.F., Maybank, S.J.: On plane-based camera calibration: a general algorithm, singularities, applications. In: Computer Vision and Pattern Recognition (CVPR) (1999)
38. Thirthala, S., Pollefeys, M.: Radial multi-focal tensors. Int. J. Comput. Vis. **96**(2), 195–211 (2012). https://doi.org/10.1007/s11263-011-0463-x
39. Triggs, B.: Camera pose and calibration from 4 or 5 known 3D points. In: International Conference on Computer Vision (ICCV) (1999)

40. Tsai, R.: A versatile camera calibration technique for high-accuracy 3D machine vision metrology using off-the-shelf TV cameras and lenses. IEEE J. Robot. Autom. **3**(4), 323–344 (1987)
41. Wu, C.: P3.5P: pose estimation with unknown focal length. In: Computer Vision and Pattern Recognition (CVPR) (2015)
42. Zhang, Q., Pless, R.: Extrinsic calibration of a camera and laser range finder (improves camera calibration). In: International Conference on Intelligent Robots and Systems (IROS) (2004)
43. Zheng, Y., Sugimoto, S., Sato, I., Okutomi, M.: A general and simple method for camera pose and focal length determination. In: Computer Vision and Pattern Recognition (CVPR) (2014)

MotionSqueeze: Neural Motion Feature Learning for Video Understanding

Heeseung Kwon[1,2], Manjin Kim[1], Suha Kwak[1], and Minsu Cho[1,2(✉)]

[1] POSTECH, Pohang University of Science and Technology, Pohang, Korea
mscho@postech.ac.kr
[2] NPRC, The Neural Processing Research Center, Seoul, Korea
http://cvlab.postech.ac.kr/research/MotionSqueeze/

Abstract. Motion plays a crucial role in understanding videos and most state-of-the-art neural models for video classification incorporate motion information typically using optical flows extracted by a separate off-the-shelf method. As the frame-by-frame optical flows require heavy computation, incorporating motion information has remained a major computational bottleneck for video understanding. In this work, we replace external and heavy computation of optical flows with internal and light-weight learning of motion features. We propose a trainable neural module, dubbed *MotionSqueeze*, for effective motion feature extraction. Inserted in the middle of any neural network, it learns to establish correspondences across frames and convert them into motion features, which are readily fed to the next downstream layer for better prediction. We demonstrate that the proposed method provides a significant gain on four standard benchmarks for action recognition with only a small amount of additional cost, outperforming the state of the art on Something-Something-V1 & V2 datasets.

Keywords: Video understanding · Action recognition · Motion feature learning · Efficient video processing

1 Introduction

The most distinctive feature of videos, from those of images, is motion. In order to grasp a full understanding of a video, we need to analyze its motion patterns as well as the appearance of objects and scenes in the video [20,27,31,38]. With significant progress of neural networks on the image domain, convolutional neural networks (CNNs) have been widely used to learn appearance features from video frames [5,31,35,40] and recently extended to learn temporal features using spatio-temporal convolution across multiple frames [2,35]. The results, however, have shown that spatio-temporal convolution alone is not enough for learning

Electronic supplementary material The online version of this chapter (https://doi.org/10.1007/978-3-030-58517-4_21) contains supplementary material, which is available to authorized users.

A. Vedaldi et al. (Eds.): ECCV 2020, LNCS 12361, pp. 345–362, 2020.
https://doi.org/10.1007/978-3-030-58517-4_21

motion patterns; convolution is effective in capturing translation-equivariant patterns but not in modeling relative movement of objects [39,46]. As a result, most state-of-the-art methods still incorporate explicit motion features, *i.e.,* dense optical flows, extracted by an external off-the-shelf methods [2,21,31,37,43]. This causes a major computational bottleneck in video-processing models for two reasons. First, calculating optical flows frame-by-frame is a time-consuming process; obtaining optical flows of a video is typically an order of magnitude slower than feed-forwarding the video through a deep neural network. Second, processing optical flows often requires a separate stream in the model to learn motion representations [31], which results in doubling the number of parameters and the computational cost. To address these issues, several methods have attempted to internalize motion modeling [7,20,27,34]. They, however, all either impose a heavy computation on their architectures [7,27] or underperform other methods using external optical flows [20,34].

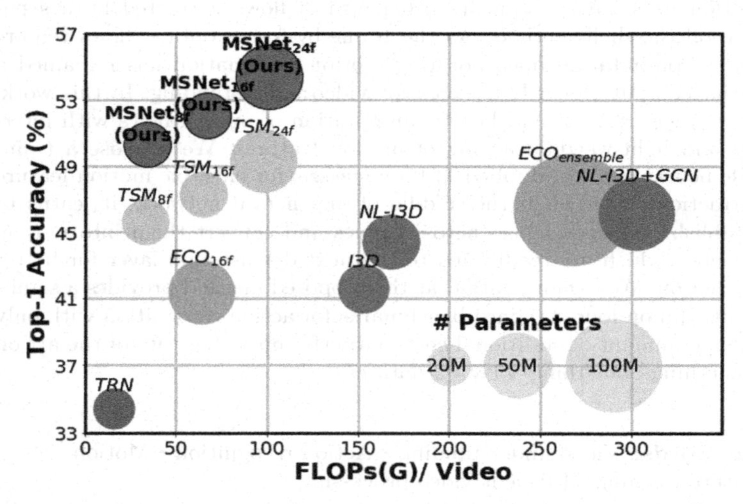

Fig. 1. Video classification performance comparison on Something-Something-V1 [10] in terms of accuracy, computational cost, and model size. The proposed method (MSNet) achieves the best trade-off between accuracy and efficiency compared to state-of-the-art methods of TSM [21], TRN [47], ECO [48], I3D [2], NL-I3D [41], and GCN [42]. (Color figure online)

To tackle the limitation of the existing methods, we propose an end-to-end trainable block, dubbed the *MotionSqueeze* (MS) module, for effective motion estimation. Inserted in the middle of any neural network for video understanding, it learns to establish correspondences across adjacent frames efficiently and convert them into effective motion features. The resultant motion features are readily fed to the next downstream layer and used for final prediction. To validate the proposed MS module, we develop a video classification architecture, dubbed the MotionSqueeze network (MSNet), that is equipped with the MS module. In comparison with recent methods, shown in Fig. 1, the proposed method provides the best trade-off in terms of accuracy, computational cost, and model size in video understanding.

2 Related Work

Video Classification Architectures. One of the main problems in video understanding is to categorize videos given a set of pre-defined target classes. Early methods based on deep neural networks have focused on learning spatio-temporal or motion features. Tran *et al.* [35] propose a 3D CNN (C3D) to learn spatio-temporal features while Simonyan and Zisserman [31] employ an independent temporal stream to learn motion features from precomputed optical flows. Carreira and Zisserman [2] design two-stream 3D CNNs (two-stream I3D) by integrating two former methods, and achieve the state-of-the-art performance at that time. As the two-stream 3D CNNs are powerful but computationally demanding, subsequent work has attempted to improve the efficiency. Tran *et al.*[37] and Xie *et al.*[43] propose to decompose 3D convolutional filters into 2D spatial and 1D temporal filters. Chen *et al.* [3] adopt group convolution techniques while Zolfaghari *et al.* [48] propose to study mixed 2D and 3D networks with the frame sampling method of temporal segment networks (TSN) [40]. Tran *et al.* [36] analyze the effect of 3D group convolutional networks and propose the channel-separated convolutional network (CSN). Lin *et al.* [21] propose the temporal shift module (TSM) that simulates 3D convolution using 2D convolution with a part of input feature channels shifted along the temporal axis. It enables 2D convolution networks to achieve a comparable classification accuracy to 3D CNNs. Unlike these approaches, we focus on efficient learning of motion features.

Learning Motions in a Video. While two-stream-based architectures [8,9,30, 31] have demonstrated the effectiveness of pre-computed optical flows, the use of optical flows typically degrades the efficiency of video processing. To address the issue, Ng *et al.* [26] use a multi-task learning of both optical flow estimation and action classification and Stroud *et al.* [32] propose to distill motion features from pre-trained two-stream 3D CNNs. These methods do not use pre-computed optical flows during inference, but still need them at the training phase. Other methods design network architectures that learn motions internally without external optical flows [7,16,20,27,34]. Sun *et al.* [34] compute spatial and temporal gradients between appearance features to learn motion features. Lee *et al.* [20] and Jiang *et al.* [16] propose a convolutional module to extract motion features by spatial shift and subtraction operation between appearance features. Despite their computational efficiency, they do not reach the classification accuracy of two-stream networks [31]. Fan *et al.* [7] implement the optimization process of TV-L1 [44] as iterative neural layers, and design an end-to-end trainable architecture (TVNet). Piergiovanni and Ryoo [27] extend the idea of TVNet by calculating channel-wise flows of feature maps at the intermediate layers of the CNN. These variational methods achieve a good performance, but require a high computational cost due to iterative neural layers. In contrast, our method learns to extract effective motion features with a marginal increase of computation.

Learning Visual Correspondences. Our work is inspired by recent methods that learn visual correspondences between images using neural networks [6,11, 19,24,28,33]. Fischer *et al.* [6] estimate optical flows using a convolutional neural network, which construct a correlation tensor from feature maps and regresses displacements from it. Sun *et al.* [33] use a stack of correlation layers for coarse-to-fine optical flow estimation. While these methods require dense ground-truth optical flows in training, the structure of correlation computation and subsequent displacement estimation is widely adopted in other correspondence problems with different levels of supervision. For example, recent methods for semantic correspondence, *i.e.*, matching images with intra-class variation, typically follow a similar pipeline to learn geometric transformation between images in a more weakly-supervised regime [11,19,24,28]. In this work, motivated by this line of research, we develop a motion feature module that does not require any correspondence supervision for learning.

Similarly to our work, a few recent methods [22,45] have attempted to incorporate correspondence information for video understanding. Zhao *et al.* [45] use correlation information between feature maps of consecutive frames to replace optical flows. The size of their full model, however, is comparable to the two-stream networks [31]. Liu *et al.* [22] propose the correspondences proposal (CP) module to learn correspondences in a video. Unlike ours, they focus on analyzing spatio-temporal relationship within the whole video, rather than motion, and the model is not fully differentiable and thus less effective in learning. In contrast, we introduce a fully-differentiable motion feature module that can be inserted in the middle of any neural network for video understanding.

The main contribution of this work is three-fold.

- We propose an end-to-end trainable, model-agnostic, and lightweight module for motion feature extraction.
- We develop an efficient video recognition architecture that is equipped with the proposed motion module.
- We demonstrate the effectiveness of our method on four different benchmark datasets and achieve the state-of-the-art on Something-Something-V1&V2.

3 Proposed Approach

The overall architecture for video understanding is illustrated in Fig. 2. Let us assume a neural network that takes a video of T frames as input and predicts the category of the video as output, where convolutional layers are used to transform input frames into frame-wise appearance features. The proposed motion feature module, dubbed *MotionSqueeze (MS) module*, is inserted to produce frame-wise motion features using pairs of adjacent appearance features. The resultant motion features are added to the appearance features for final prediction. In this section, we first explain the MS module, and describe the details of our network architecture for video understanding.

3.1 MotionSqueeze (MS) Module

The MS module is a learnable motion feature extractor, which can replace the use of explicit optical flows for video understanding. As described in Fig. 3, given two feature maps from adjacent frames, it learns to extract effective motion features in three steps: correlation computation, displacement estimation, and feature transformation.

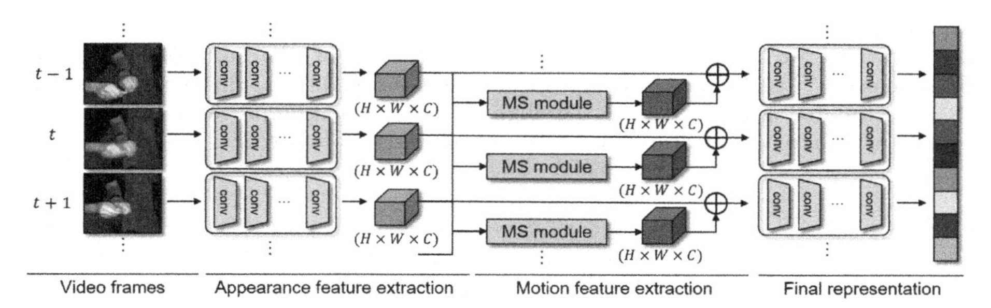

Fig. 2. Overall architecture of the proposed approach. The model first takes T video frames as input and converts them into frame-wise appearance features using convolutional layers. The proposed *MotionSqueeze (MS) module* generates motion features using the frame-wise appearance features, and combines the motion features into the next downstream layer. $\oplus$ denotes element-wise addition.

Correlation Computation. Let us denote the two adjacent input feature maps by $\mathbf{F}^{(t)}$ and $\mathbf{F}^{(t+1)}$, each of which is 3D tensors of size $H \times W \times C$. The spatial resolution is $H \times W$ and the C dimensional features on spatial position $\mathbf{x}$ by $\mathbf{F}_{\mathbf{x}}$. A correlation score of position $\mathbf{x}$ with respect to displacement $\mathbf{p}$ is defined as

$$s(\mathbf{x}, \mathbf{p}, t) = \mathbf{F}_{\mathbf{x}}^{(t)} \cdot \mathbf{F}_{\mathbf{x}+\mathbf{p}}^{(t+1)}, \tag{1}$$

where $\cdot$ denotes dot product. For efficiency, we compute the correlation scores of position $\mathbf{x}$ only in its neighborhood of size $P = 2k+1$ by restricting a maximum displacement: $\mathbf{p} \in [-k, k]^2$. For t_{th} frame, a resultant correlation tensor $\mathbf{S}^{(t)}$ is of size $H \times W \times P^2$. The cost of computing the correlation tensor is equivalent to that of 1×1 convolutions with P^2 kernels; the correlation computation can be implemented as 2D convolutions on t_{th} feature map using $t+1_{\text{th}}$ feature map as P^2 kernels. The total FLOPs in a single video amounts to $THWCP^2$. We apply a convolution layer before computing correlations, which learns to weight informative feature channels for learning visual correspondences. In practice, we set the neighborhood $P = 15$ given the spatial resolution 28×28 and apply an $1 \times 1 \times 1$ layer with $C/2$ channels. For correlation computation, we adopt C++/Cuda implemented version of correlation layer in FlowNet [6].

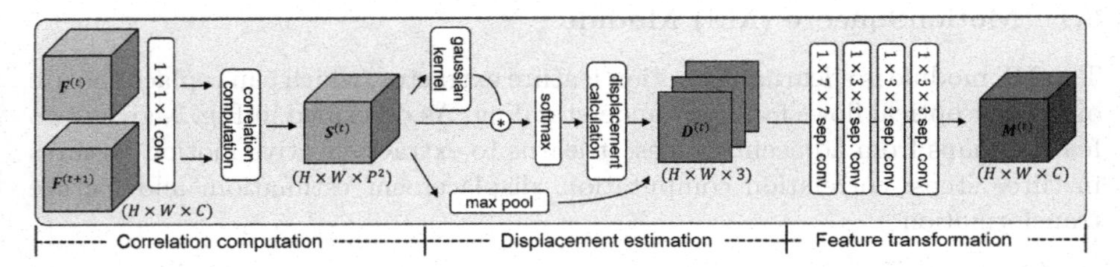

Fig. 3. Overall process of MotionSqueeze (MS) module. The MS module estimates motion across two frame-wise feature maps ($\mathbf{F}^{(t)}, \mathbf{F}^{(t+1)}$) of adjacent frames. A correlation tensor $\mathbf{S}^{(t)}$ is obtained by computing correlations, and then a displacement tensor $\mathbf{D}^{(t)}$ is estimated using the tensor. Through the transformation process of convolution layers, the final motion feature $\mathbf{M}^{(t)}$ is obtained. See text for details.

Displacement Estimation. From the correlation tensor $\mathbf{S}^{(t)}$, we estimate a displacement field for motion information. A straightforward but non-differentiable method would be to take the best matching displacement for position $\mathbf{x}$ by $\mathrm{argmax}_{\mathbf{p}} s(\mathbf{x}, \mathbf{p}, t)$. To make the operation differentiable, we can use a weighted average of displacements using softmax, called *soft-argmax* [13,19], which is defined as

$$d(\mathbf{x}, t) = \sum_{\mathbf{p}} \frac{\exp(s(\mathbf{x}, \mathbf{p}, t))}{\sum_{\mathbf{p}'} \exp(s(\mathbf{x}, \mathbf{p}', t))} \mathbf{p}. \tag{2}$$

This method, however, is sensitive to noisy outliers in the correlation tensor since it is influenced by all correlation values. We thus use the *kernel-soft-argmax* [19] that suppresses such outliers by masking a 2D Gaussian kernel on the correlation values; the kernel is centered on each target position so that the estimation is more influenced by closer neighbors. Our kernel-soft-argmax for displacement estimation is defined as

$$d(\mathbf{x}, t) = \sum_{\mathbf{p}} \frac{\exp(g(\mathbf{x}, \mathbf{p}, t) s(\mathbf{x}, \mathbf{p}, t)/\tau)}{\sum_{\mathbf{p}'} \exp(g(\mathbf{x}, \mathbf{p}', t) s(\mathbf{x}, \mathbf{p}', t)/\tau)} \mathbf{p}, \tag{3}$$

where

$$g(\mathbf{x}, \mathbf{p}, t) = \frac{1}{\sqrt{2\pi}\sigma} \exp\left(\frac{\mathbf{p} - \mathrm{argmax}_{\mathbf{p}} s(\mathbf{x}, \mathbf{p}, t)}{\sigma^2}\right). \tag{4}$$

Note that $g(\mathbf{x}, \mathbf{p}, t)$ is the Gaussian kernel and we empirically set the standard deviation σ to 5. τ is a temperature factor adjusting the softmax distribution; as τ decreases, softmax approaches argmax. We set $\tau = 0.01$ in our experiments.

In addition to the estimated displacement map, we use a confidence map of correlation as auxiliary motion information, which is obtained by pooling the highest correlation on each position $\mathbf{x}$:

$$s^*(\mathbf{x}, t) = \max_{\mathbf{p}} s(\mathbf{x}, \mathbf{p}, t). \tag{5}$$

The confidence map may be useful for identifying displacement outliers and learning informative motion features.

We concatenate the (2-channel) displacement map and the (1-channel) confidence map into a displacement tensor $\mathbf{D}^{(t)}$ of size $H \times W \times 3$ for the next step of motion feature transformation. An example of them is visualized in Fig. 4.

Feature Transformation. We convert the displacement tensor $\mathbf{D}^{(t)}$ to an effective motion feature $\mathbf{M}^{(t)}$ that is readily incorporated into downstream layers. The tensor $\mathbf{D}^{(t)}$ is fed to four *depth-wise separable convolution* [14] layers, one $1 \times 7 \times 7$ layer followed by three $1 \times 3 \times 3$ layers, and transformed into a motion feature $\mathbf{M}^{(t)}$ with the same number of channels C as that of the original input $\mathbf{F}^{(t)}$. The depth-wise separable convolution approximates 2D convolution with a significantly less computational cost [4,29,36]. Note that all depth-wise and point-wise convolution layers are followed by batch normalization [15] and ReLU [25]. As in the temporal stream layers of [31], this feature transformation process is designed to learn task-specific motion features with convolution layers by interpreting the semantics of displacement and confidence. As illustrated in Fig. 2, the MS module generates motion feature $\mathbf{M}^{(t)}$ using two adjacent appearance features $\mathbf{F}^{(t)}$ and $\mathbf{F}^{(t+1)}$ and then add it to the input of the next layer. Given T frames, we simply pad the final motion feature $\mathbf{M}^{(T)}$ with $\mathbf{M}^{(T-1)}$ by setting $\mathbf{M}^{(T)} = \mathbf{M}^{(T-1)}$.

3.2 MotionSqueeze Network (MSNet)

The MS module can be inserted into any video understanding architecture to improve the performance by motion feature modeling. In this work, we introduce standard convolutional neural networks (CNNs) with the MS module, dubbed *MSNet*, for video classification. We adopt the ImageNet-pretrained ResNet [12] as the CNN backbone and insert TSM [21] for each residual block of the ResNet. TSM enables 2D convolution to obtain the effect of 3D convolution by shifting a part of input feature channels along the temporal axis before the convolution operation. Following the default setting in [21], we shift 1/8 of the input features channels forward and another 1/8 of the channels backward in each TSM.

The overall architecture of the proposed model is shown in Fig. 2; a single MS module is inserted after the third stage of the ResNet. We fuse the motion feature into the appearance feature by element-wise addition:

$$\mathbf{F}'^{(t)} = \mathbf{F}^{(t)} + \mathbf{M}^{(t)}. \tag{6}$$

In Sect. 4.5, we extensively evaluate different fusion methods, *e.g.*, concatenation and multiplication, and show that additive fusion is better than the others. After fusing both features, the combined feature is passed through the next downstream layers. The network outputs over T frames are temporally averaged to produce a final output and the cross-entropy with softmax is used as a loss function for training. By default setting, MSNet learns both appearance and motion features jointly in a single network at the cost of only 2.5% and 1.2% increase in FLOPs and the number of parameters, respectively.

4 Experiments

4.1 Datasets

Something-Something V1&V2 [10] are trimmed video datasets for human action classification. Both datasets consist of 174 classes with 108,499 and 220,847 videos in total, respectively. Each video contains one action and the duration spans from 2 to 6 s. Something-Something V1&V2 are motion-oriented datasets where temporal relationships are more salient than in others.

Kinetics [17] is a popular large-scale video dataset, consisting of 400 classes with over 250,000 videos. Each video lasts around 10 s with a single action.
HMDB51 [18] contains 51 classes with 6,766 videos. Kinetics and HMDB-51 focus more on appearance information rather than motion.

4.2 Implementation Details

Clip Sampling. In both training and testing, instead of an entire video, we use a clip of frames that are sampled from the video. We use the segment-based sampling method [40] for the Something-Something V1&V2 while adopting the dense frame sampling method [2] for Kinetics and HMDB-51.

Training. For each video, we sample a clip of 8 or 16 frames, resize them into 240×320 images, and crop 224×224 images from the resized images [48]. The minibatch SGD with Nestrov momentum is used for optimization, and the batch size is set to 48. We use scale jittering for data augmentation. For the Something-Something V1&V2, we set the training epochs to 40 and the initial learning rate to 0.01; the learning rate is decayed by 1/10 after 20_{th} and 30_{th} epochs. For Kinetics, we set the training epochs to 80 and the initial learning rate to 0.01; the learning rate is decayed by 1/10 after 40 and 60 epochs. In training our model on HMDB-51, we fine-tune the Kinetics-pretrained model as in [21, 37]. We set the training epochs to 35 and the initial learning rate to 0.001; the learning rate is decayed by 1/10 after 15_{th} and 30_{th} epochs.

Table 1. Performance comparison on Something-Something V1&V2. The symbol † denotes the reproduced by ours.

Model	Flow	#frame	FLOPs × clips	#param	SomethingV1		SomethingV2	
					top-1	top-5	top-1	top-5
TSN [40]		8	16G × 1	10.7M	19.5	–	33.4	–
TRN [47]		8	16G × N/A	18.3M	34.4	–	48.8	–
TRN Two-stream [47]	✓	8 + 8	16G×N/A	18.3M	42.0	–	55.5	–
MFNet [20]		10	N/A × 10	–	43.9	73.1	–	–
CPNet [22]		24	N/A × 96	–	–	–	57.7	84.0
ECO_{En} *Lite* [48]		92	267 × 1	150M	46.4	–	–	–
ECO Two-stream [48]	✓	92 + 92	N/A × 1	300M	49.5	–	–	–
I3D from [42]		32	153G × 2	28.0M	41.6	72.2	–	–
NL-I3D from [42]		32	168G × 2	35.3M	44.4	76.0	–	–
NL-I3D + GCN [42]		32	303G × 2	62.2M	46.1	76.8	–	–
S3D-G [43]		64	71G × 1	11.6M	48.2	78.7	–	–
DFB-Net [23]		16	N/A × 1	–	50.1	79.5	–	–
STM [16]		16	67G × 30	24.0M	50.7	80.4	64.2	89.8
TSM [21]		8	33G × 1	24.3M	45.6	74.2	58.8	85.4
TSM [21]		16	65G × 1	24.3M	47.3	77.1	61.2	86.9
TSM_{En} [21]		16 + 8	98G × 1	48.6M	49.7	78.5	62.9	88.1
TSM Two-stream [21]	✓	16 + 16	129G × 1	48.6M	52.6	81.9	65.0†	89.4†
TSM Two-stream [21]	✓	16 + 16	129G × 6	48.6M	–	–	66.0	90.5
MSNet-R50 (ours)		8	34G × 1	24.6M	50.9	80.3	63.0	88.4
MSNet-R50 (ours)		16	67G × 1	24.6M	52.1	82.3	64.7	89.4
MSNet-$R50_{En}$ (ours)		16 + 8	101G × 1	49.2M	54.4	83.8	66.6	90.6
MSNet-$R50_{En}$ (ours)		16 + 8	101G × 10	49.2M	**55.1**	**84.0**	**67.1**	**91.0**

Inference. Given a video, we sample a clip and test its center crop. For Something-Something V1&V2, we evaluate both the single clip prediction and the average prediction of 10 randomly-sampled clips. For Kinetics and HMDB-51, we evaluate the average prediction of uniformly-sampled 10 clips from each video.

4.3 Comparison with State-of-the-Art Methods

Table 1 summarizes the results on Something-Something V1&V2. Each section of the table contains results of 2D CNN methods [20, 22, 40, 47], 3D CNN methods [16, 23, 42, 43, 48], ResNet with TSM (TSM ResNet) [21], and the proposed method, respectively. Most of the results are copied from the corresponding papers, except for TSM ResNet; we evaluate the official pre-trained model of TSM ResNet using a single center-cropped clip per video in terms of top-1 and top-5 accuracies. Our method, which uses TSM ResNet as a backbone, achieves 50.9% and 63.0% on Something-Something V1 and V2 at top-1 accuracy, respectively, which outperforms most of 2D CNN and 3D CNN methods, while using a single clip with 8 input frames only. Compared to the TSM ResNet baseline, our method obtains a significant gain of about 5.3% points and 4.2% points at top-1 accuracy at the cost of only 2.5% and 1.2% increase in FLOPs and parameters, respectively. When using 16 frames, our method further improves achieving 52.1% and 64.7% at top-1 accuracy, respectively. Following the evaluation procedure of two-stream networks, we evaluate the ensemble model (MSNet-R50$_{En}$) by averaging prediction scores of the 8-frame and 16-frame models. With the same number of clips for evaluation, it achieves top-1 accuracy 1.8% points and 1.6% points higher than TSM two-stream networks with 22% less computation, even no optical flow needed. Our 10-clip model achieves 55.1% and 67.1% at top-1 accuracy on Something-Something V1 and V2, respectively, which is the state-of-the-art on both of the datasets. As shown in Fig. 1, our model provides the best trade-off in terms of accuracy, FLOPs, and the number of parameters.

Table 2. Performance comparison with motion representation methods. The symbol ‡ denotes that we only report the backbone FLOPs.

Model	Flow	#frame	FLOPs × clips	Speed (V/s)	Kinetics	HMDB51
					Top-1	Top-1
ResNet-50 from [27]		32	132G × 25	22.8	61.3	59.4
R(2 + 1)D [37]		32	152 G× 115	8.7	72.0	74.3
MFNet from [27]		10	80G‡ × 10	–	–	56.8
OFF(RGB) [34]		1	N/A × 25	–	–	57.1
TVNet [7]		18	N/A × 250	–	–	71.0
STM [16]		16	67G × 30	–	73.7	72.2
Rep-flow (ResNet-50) [27]		32	132G‡ ×25	3.7	68.5	76.4
Rep-flow (R(2 + 1)D) [27]		32	152G ‡ × 25	2.0	75.5	77.1
ResNet-50 Two-stream from [27]	✓	32 + 32	264G × 25	0.2	64.5	66.6
R(2+1)D Two-stream [37]	✓	32 + 32	304 G× 115	0.2	73.9	**78.7**
OFF(RGB+Flow+RGB Diff) [34]	✓	1 + 5 + 5	N/A × 25	–	–	74.2
TSM (reproduced)		8	33G × 10	64.1	73.5	71.9
MSNet-R50 (ours)		8	34G × 10	54.2	75.0	75.8
MSNet-R50 (ours)		16	67G × 10	31.2	**76.4**	77.4

4.4 Comparison with Other Motion Representation Methods

Table 2 summarizes comparative results with other motion representation methods [7,16,20,27,34] based on RGB frames. The comparison is done on Kinetics and HMDB51 since the previous methods commonly report their results on them. Each section of the table contains results of conventional 2D and 3D CNNs, motion representation methods [7,16,20,27,34], two-stream CNNs with optical flows [12,37], and the proposed method, respectively. OFF, MFNet, and STM [16,20,34] use a sub-network or lightweight modules to calculate temporal gradients of frame-wise feature maps. TVNet [7] and Rep-flow [27] internalize iterative TV-L1 flow operations in their networks. As shown in Table 2, the proposed model using 16 frames outperforms all the other conventional CNNs and the motion representation methods [2,7,12,16,20,27,34,37], while being competitive with the R(2+1)D two-stream [37] that uses pre-computed optical flows. Furthermore, our model is highly efficient than all the other methods in terms of FLOPs, clips, and the number of frames.

Run-time. We also evaluate in Table 2 the inference speeds of several models to demonstrate the efficiency of our method. All the run-times reported are measured on a single GTX Titan Xp GPU, ignoring the time of data loading. For this experiment, we set the spatial size of the input to 224×224 and the batch size to 1. The official codes are used for ResNet, TSM ResNet, and Rep-flow [12, 21,27] except for R(2+1)D[37] we implemented. In evaluating Rep-flow [27], we use 20 iterations for optimization as in the original paper. The speed of the two-stream networks [31,37] includes computation time for TV-L1 method on the GPU. The run-time results clearly show the cost of iterative optimizations used in two-stream networks and Rep-flow. In contrast, our model using 16 frames is about 160× faster than the two-stream networks. Compared to Rep-flow ResNet-50, our method performs about 4× faster due to the absence of the iterative optimization process in Rep-flow.

Table 3. Performance comparison with different displacement estimations.

Model	FLOPs	top-1	top-5
Baseline	14.6G	41.5	71.8
S	14.8G	43.8	74.9
KS	14.9G	44.6	75.4
KS + CM	14.9G	45.5	76.5
KS + CM + BD	15.1G	46.0	76.7

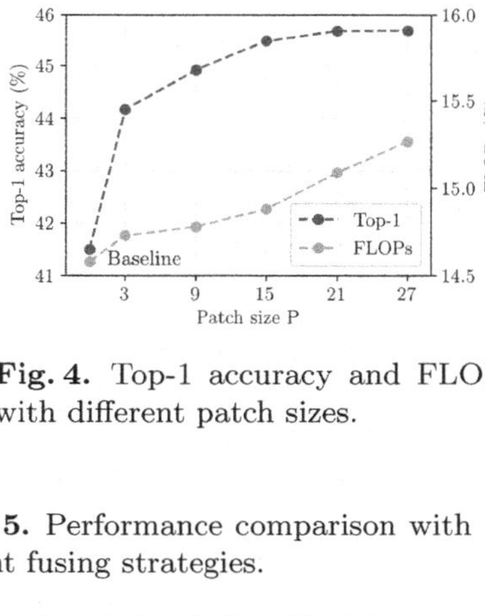

Fig. 4. Top-1 accuracy and FLOPs with different patch sizes.

Table 4. Performance comparison with different positions of the MS module.

Model	FLOPs	top-1	top-5
Baseline	14.6G	41.5	71.8
res_2	15.6G	45.1	76.1
res_3	14.9G	45.5	76.5
res_4	14.7G	42.6	73.2
res_5	14.6G	41.1	71.8
$res_{2,3,4}$	16.0G	45.7	76.8

Table 5. Performance comparison with different fusing strategies.

Model	FLOPs	top-1	top-5
Baseline	14.6G	41.5	71.8
MS only	14.1G	38.8	70.7
Multiply	14.9G	44.5	75.9
Concat	15.7G	45.0	76.1
Add	14.9G	45.5	76.5

4.5 Ablation Studies

We conduct ablation studies of the proposed method on Something-Something V1 [10] dataset. We use ImageNet pre-trained TSM ResNet-18 as a default backbone and use 8 input frames for all experiments in this section.

Displacement Estimation in MS Module. In Table 3, we experiment with different variants of the displacement tensor $\mathbf{D}^{(t)}$ in the MS module. We first compare soft-argmax ('S') and kernel-soft-argmax ('KS') for displacement estimation. As shown in the upper part of Table 3. the kernel-soft-argmax outperforms the soft-argmax, showing the noise reduction effect of Gaussian kernel. In the lower part of Table 3, we evaluate the effect of additional features: confidence maps ('CM') and backward displacement tensor ('BD'). The backward displacement tensor is estimated from $\mathbf{F}^{(t+1)}$ to $\mathbf{F}^{(t)}$. We concatenate the forward and backward displacement tensors, and then pass them to the feature transformation layers. We obtain 0.9% points gain by appending the confidence map to the displacement tensor. Furthermore, by adding backward displacement we obtain another 0.5% points gain at top-1 accuracy, indicating that forward and backward displacement maps complement each other to enrich motion information. We use the kernel-soft-argmax with the confidence map ('KS + CM') as a default method for all other experiments.

Size of Matching Region. In Fig. 4, we evaluate performance varying the spatial size of matching regions of the MS module. Even with a small matching region $P = 3$, it provides a noticeable performance gain of over 2.7% points to the baseline. The performance tends to increase as the matching region becomes larger due to the larger displacement it can handle between frames. The performance is saturated after $P = 15$.

Position of MS Module. In Table 4, we evaluate different positions of the MS module. We denote that res_N by the N-th stage of the ResNet. For each stage, it is inserted right after its final residual block. The result shows that while the MS module is beneficial in most cases, both accuracy and efficiency gains depend on the position of the module. It performs the best at res_3; appearance features from res_2 are not strong enough for accurate feature matching while spatial resolutions of appearance features from res_4 and res_5 are not high enough. The position of the module also affects FLOPs; the computational cost quadratically increases with spatial resolution due to convolution layers of the feature transformation. When inserting multiple MS modules ($res_{2,3,4}$) at the backbone, it marginally improves top-1 accuracy as 0.2% points. Multiple modules appear to generate similar motion information even in different levels of features.

Fusing Strategy of MS Module. In Table 5, we evaluate different fusion strategies for the MS module; 'MS only', 'multiply', 'concat', and 'add'. In the

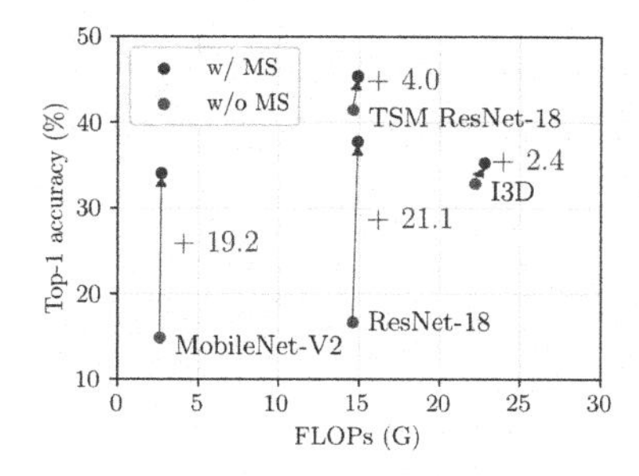

Fig. 5. Top-1 accuracy and FLOPs with MS module on different backbones.

Table 6. Performance comparison with two-stream networks.

Model	Flow	FLOPs	top-1	top-5
Baseline		14.6G	41.5	71.8
Two-stream$_{8+(8\times5)}$	✓	31.4G	46.8	77.3
Two-stream$_{8+(8\times1)}$	✓	28.9G	44.7	75.2
Two-stream$_{8+(8\times1)(low)}$	✓	28.9G	44.1	74.9
MSNet		14.9G	45.5	76.5

case of 'MS only', we only pass $\mathbf{M}^{(t)}$ into downstream layers without $\mathbf{F}^{(t)}$. We apply element-wise multiplication and element-wise addition, respectively, for 'multiply' and 'add'. In the case of 'concat', we concatenate $\mathbf{F}^{(t)}$ and $\mathbf{M}^{(t)}$, whose channel size is transformed to C via an $1 \times 1 \times 1$ convolution layer. 'MS only' is less accurate than the baseline because visual semantic information is discarded. While both 'multiply' and 'concat' clearly improve the accuracy, 'add' achieves the best performance with 45.5% at top-1 accuracy. We find that additive fusion is the most effective and stable in amplifying appearance features of moving objects.

Effect of MS Module on Different Backbones. In Fig. 5, we also evaluate the effect of the MS module on ResNet-18, MobileNet-V2, and I3D. We insert one MS module where the spatial resolution of the feature map remains the same. For ResNet-18 and MobileNet-V2, we finetune models pre-trained on ImageNet. We train I3D from scratch. Our MS module benefits both 2D CNNs and 3D CNNs to obtain higher accuracy. The module significantly improves ResNet-18 and MobileNet-V2 by 21.3% and 19.2% points, respectively, in top-1 accuracy. Since 2D CNNs do not use any spatio-temporal features, it obtains significantly higher gain from the MS module. The MS module also improves I3D and TSM ResNet-18 by 2.4% and 4.0% points, respectively, in top-1 accuracy. The gain on 3D CNNs, although relatively small, verifies that the motion features by the MS module are complementary even to the spatio-temporal features; the MS module learns explicit motion information across adjacent frames whereas TSM covers long-term temporal length using (pseudo-)temporal convolutions.

Comparison with Two-Stream Networks. In Table 6, we compare the proposed method with variants of TSM two-stream networks [31] that use TV-L1 optical flows [44]. We denote the two-stream networks by Two-stream$_{N_r+(N_f\times N_s)}$ where N_r, N_f and N_s indicate the number of frames, optical flows, and their stacking size, respectively. For each frame, the two-stream networks use N_s stacked optical flows, which are extracted using the subsequent frames in the

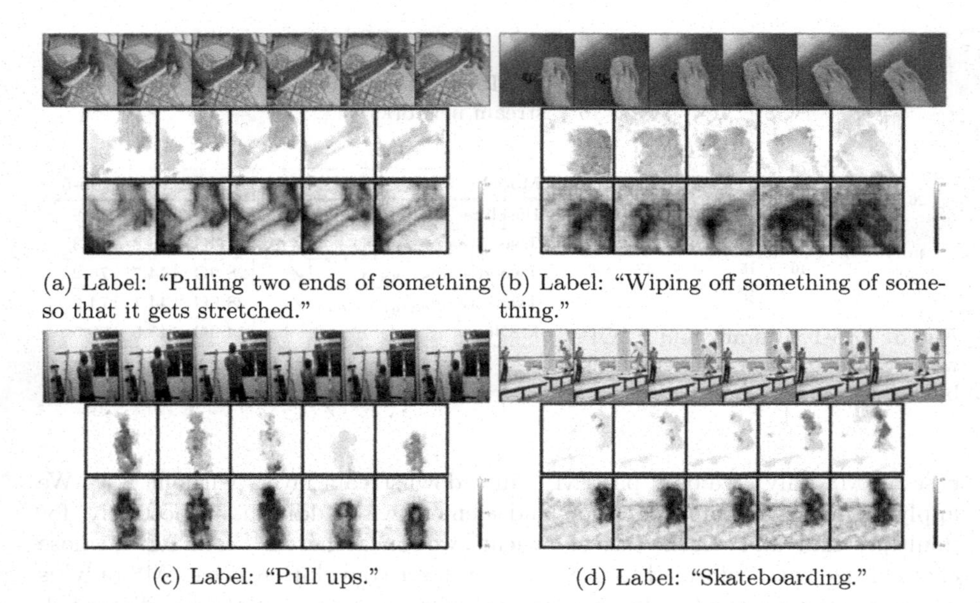

(a) Label: "Pulling two ends of something so that it gets stretched." (b) Label: "Wiping off something of something." (c) Label: "Pull ups." (d) Label: "Skateboarding."

Fig. 6. Visualization on Something-Something-V1 (top) and Kinetics (bottom) datasets. RGB images, displacement maps, and the confidence maps are shown from the top row in each subfigure.

original video. Note that those frames for optical flow extraction are not used in our method (MSNet). The second row of Table 6, Two-stream$_{8+(8\times5)}$, shows the performance of standard TSM two-stream networks that use 5 stacked optical flows for the temporal stream. Using the multiple optical flows for each frame outperforms our model in terms of accuracy but requires substantially larger FLOPs as well as an additional computation for calculating optical flows. For a fair comparison, we report the performance of the two-stream networks, Two-stream$_{8+(8\times1)}$, that do not stack multiple optical flows. Our model outperforms the two-stream networks by 0.8% points at top-1 accuracy, with about two times fewer FLOPs. Note that both Two-stream$_{8+(8\times5)}$ and Two-stream$_{8+(8\times1)}$ use optical flows obtained from the original video with a higher frame rate than the input video clip (sampled frames); our method (MSNet) observes the input video clip only. We thus evaluate other two-stream networks, Two-stream$_{8+(8\times1)(low)}$, that uses low-fps optical flows as input; we sample a sequence of frames 3 fps from the original video and extract TV-L1 optical flows using the sequence. As shown in Table 6, the top-1 accuracy gap between ours and the two-stream network increases to 1.4% points. The result implies that given low-fps videos, our method may further improve over the two-stream networks.

4.6 Visualization

In Fig. 6, we present visualization results on Something-Something V1 and Kinetics datasets. They show that our MS module effectively learns to estimate motion without any direct supervision used in training. The first row of each subfigure shows 6 uniformly sampled frames from a video. The second and third rows show color-coded displacement maps [1] and confidence maps, respectively; we apply min-max normalization on the confidence map. The resolution of all the displacement and confidence maps is set to 56×56 for better visualization. As shown in the figures, the MS module captures reliable displacements in most cases: horizontal and vertical movements (Fig. 6a, 6c, 6d), rotational movements (Fig. 6b), and non-severe deformation (Fig. 6a, 6d). See the supplementary material for additional details and results. We will make our code and data available online.

5 Conclusion

We have presented an efficient yet effective motion feature block, the MS module, that learns to generate motion features on the fly for video understanding. The MS module can be readily inserted into any existing video architectures and trained by backpropagation. The ablation studies on the module demonstrate the effectiveness of the proposed method in terms of accuracy, computational cost, and model size. Our method outperforms existing state-of-the-art methods on Something-Something-V1&V2 for video classification with only a small amount of additional cost.

Acknowledgements. This work is supported by Samsung Advanced Institute of Technology (SAIT), and also by Basic Science Research Program (NRF-2017R1E1A1A010 77999, NRF-2018R1C1B6001223) and Next-Generation Information Computing Development Program (NRF-2017M3C4A7069369) through the National Research Foundation of Korea (NRF) funded by the Ministry of Science, ICT.

References

1. Baker, S., Scharstein, D., Lewis, J., Roth, S., Black, M.J., Szeliski, R.: A database and evaluation methodology for optical flow. Int. J. Comput. Vision (IJCV) **92**(1), 1–31 (2011)
2. Carreira, J., Zisserman, A.: Quo vadis, action recognition? a new model and the kinetics dataset. In: Proceedings of IEEE Conference on Computer Vision and Pattern Recognition (CVPR) (2017)
3. Chen, Y., Kalantidis, Y., Li, J., Yan, S., Feng, J.: Multi-fiber networks for video recognition. In: Proceedings of European Conference on Computer Vision (ECCV) (2018)
4. Chollet, F.: Xception: deep learning with depthwise separable convolutions. In: Proceedings of IEEE Conference on Computer Vision and Pattern Recognition (CVPR) (2017)

5. Donahue, J., et al.: Long-term recurrent convolutional networks for visual recognition and description. In: Proceedings of IEEE Conference on Computer Vision and Pattern Recognition (CVPR) (2015)
6. Dosovitskiy, A., et al.: Flownet: learning optical flow with convolutional networks. In: Proceedings of IEEE International Conference on Computer Vision (ICCV) (2015)
7. Fan, L., Huang, W., Gan, C., Ermon, S., Gong, B., Huang, J.: End-to-end learning of motion representation for video understanding. In: Proceedings of IEEE Conference on Computer Vision and Pattern Recognition (CVPR) (2018)
8. Feichtenhofer, C., Pinz, A., Wildes, R.: Spatiotemporal residual networks for video action recognition. In: Proceedings of Neural Information Processing Systems (NeurIPS) (2016)
9. Feichtenhofer, C., Pinz, A., Zisserman, A.: Convolutional two-stream network fusion for video action recognition. In: Proceedings of IEEE Conference on Computer Vision and Pattern Recognition (CVPR) (2016)
10. Goyal, R., et al.: The "something something" video database for learning and evaluating visual common sense. In: Proceedings of IEEE International Conference on Computer Vision (ICCV) (2017)
11. Han, K., et al.: Scnet: learning semantic correspondence. In: Proceeding of IEEE International Conference on Computer Vision (ICCV) (2017)
12. He, K., Zhang, X., Ren, S., Sun, J.: Deep residual learning for image recognition. In: Proceedings of IEEE Conference on Computer Vision and Pattern Recognition (CVPR) (2016)
13. Honari, S., Molchanov, P., Tyree, S., Vincent, P., Pal, C., Kautz, J.: Improving landmark localization with semi-supervised learning. In: Proceedings of IEEE Conference on Computer Vision and Pattern Recognition (CVPR) (2018)
14. Howard, A.G., et al.: Mobilenets: efficient convolutional neural networks for mobile vision applications (2017). arXiv preprint arXiv:1704.04861
15. Ioffe, S., Szegedy, C.: Batch normalization: accelerating deep network training by reducing internal covariate shift (2015). arXiv preprint arXiv:1502.03167
16. Jiang, B., Wang, M., Gan, W., Wu, W., Yan, J.: Stm: spatiotemporal and motion encoding for action recognition. In: Proceedings of IEEE International Conference on Computer Vision (ICCV) (2019)
17. Kay, W., et al.: The kinetics human action video dataset (2017). arXiv preprint arXiv:1705.06950
18. Kuehne, H., Jhuang, H., Garrote, E., Poggio, T., Serre, T.: HMDB: a large video database for human motion recognition. In: Proceedings of IEEE International Conference on Computer Vision (ICCV) (2011)
19. Lee, J., Kim, D., Ponce, J., Ham, B.: SFNET: learning object-aware semantic correspondence. In: Proceedings of IEEE Conference on Computer Vision and Pattern Recognition (CVPR) (2019)
20. Lee, M., Lee, S., Son, S., Park, G., Kwak, N.: Motion feature network: fixed motion filter for action recognition. In: Proceedings of European Conference on Computer Vision (ECCV) (2018)
21. Lin, J., Gan, C., Han, S.: Tsm: temporal shift module for efficient video understanding. In: Proceedings of IEEE International Conference on Computer Vision (ICCV) (2019)
22. Liu, X., Lee, J.Y., Jin, H.: Learning video representations from correspondence proposals. In: Proceedings of IEEE Conference on Computer Vision and Pattern Recognition (CVPR) (2019)

23. Martinez, B., Modolo, D., Xiong, Y., Tighe, J.: Action recognition with spatial-temporal discriminative filter banks. In: Proceedings of IEEE International Conference on Computer Vision (ICCV) (2019)
24. Min, J., Lee, J., Ponce, J., Cho, M.: Hyperpixel flow: semantic correspondence with multi-layer neural features. In: Proceedings of IEEE International Conference on Computer Vision (ICCV) (2019)
25. Nair, V., Hinton, G.E.: Rectified linear units improve restricted Boltzmann machines. In: Proceedings of International Conference on Machine Learning (ICML) (2010)
26. Ng, J.Y.H., Choi, J., Neumann, J., Davis, L.S.: Actionflownet: learning motion representation for action recognition. In: Proceedings of Winter Conference on Applications of Computer Vision (WACV) (2018)
27. Piergiovanni, A., Ryoo, M.S.: Representation flow for action recognition. In: Proceedings of IEEE Conference on Computer Vision and Pattern Recognition (CVPR) (2019)
28. Rocco, I., Arandjelovic, R., Sivic, J.: Convolutional neural network architecture for geometric matching. In: Proceedings of IEEE Conference on Computer Vision and Pattern Recognition (CVPR) (2017)
29. Sandler, M., Howard, A., Zhu, M., Zhmoginov, A., Chen, L.C.: Mobilenetv 2: inverted residuals and linear bottlenecks. In: Proceedings of IEEE Conference on Computer Vision and Pattern Recognition (CVPR) (2018)
30. Sevilla-Lara, L., Liao, Y., Güney, F., Jampani, V., Geiger, A., Black, M.J.: On the integration of optical flow and action recognition. In: Proceedings of German Conference on Pattern Recognition (GCPR) (2018)
31. Simonyan, K., Zisserman, A.: Two-stream convolutional networks for action recognition in videos. In: Proceedings of Neural Information Processing Systems (NeurIPS) (2014)
32. Stroud, J., Ross, D., Sun, C., Deng, J., Sukthankar, R.: D3D: distilled 3D networks for video action recognition. In: Proceedings of Winter Conference on Applications of Computer Vision (WACV) (2020)
33. Sun, D., Yang, X., Liu, M.Y., Kautz, J.: Pwc-net: CNNs for optical flow using pyramid, warping, and cost volume. In: Proceedings of IEEE Conference on Computer Vision and Pattern Recognition (CVPR) (2018)
34. Sun, S., Kuang, Z., Sheng, L., Ouyang, W., Zhang, W.: Optical flow guided feature: a fast and robust motion representation for video action recognition. In: Proceedings of IEEE Conference on Computer Vision and Pattern Recognition (CVPR) (2018)
35. Tran, D., Bourdev, L., Fergus, R., Torresani, L., Paluri, M.: Learning spatiotemporal features with 3D convolutional networks. In: Proceedings of IEEE International Conference on Computer Vision (ICCV) (2015)
36. Tran, D., Wang, H., Torresani, L., Feiszli, M.: Video classification with channel-separated convolutional networks. In: Proceedings of IEEE International Conference on Computer Vision (ICCV) (2019)
37. Tran, D., Wang, H., Torresani, L., Ray, J., LeCun, Y., Paluri, M.: A closer look at spatiotemporal convolutions for action recognition. In: Proceedings of IEEE Conference on Computer Vision and Pattern Recognition (CVPR) (2018)
38. Wang, H., Kläser, A., Schmid, C., Cheng-Lin, L.: Action recognition by dense trajectories. In: Proceedings of IEEE Conference on Computer Vision and Pattern Recognition (CVPR) (2011)

39. Wang, L., Qiao, Y., Tang, X.: Action recognition with trajectory-pooled deep-convolutional descriptors. In: Proceedings of IEEE Conference on Computer Vision and Pattern Recognition (CVPR) (2015)
40. Wang, L., et al.: Temporal segment networks: towards good practices for deep action recognition. In: Proceedings of European Conference on Computer Vision (ECCV) (2016)
41. Wang, X., Girshick, R., Gupta, A., He, K.: Non-local neural networks. In: Proceedings of IEEE Conference on Computer Vision and Pattern Recognition (CVPR) (2018)
42. Wang, X., Gupta, A.: Videos as space-time region graphs. In: Proceedings of European Conference on Computer Vision (ECCV), pp. 399–417 (2018)
43. Xie, S., Sun, C., Huang, J., Tu, Z., Murphy, K.: rethinking spatiotemporal feature learning: speed-accuracy trade-offs in video classification. In: Proceedings of European Conference on Computer Vision (ECCV) (2018)
44. Zach, C., Pock, T., Bischof, H.: A duality based approach for realtime tv-l 1 optical flow. In: Hamprecht, Fred A., Schnörr, Christoph, Jähne, Bernd (eds.) DAGM 2007. LNCS, vol. 4713, pp. 214–223. Springer, Heidelberg (2007). https://doi.org/10.1007/978-3-540-74936-3_22
45. Zhao, Y., Xiong, Y., Lin, D.: Recognize actions by disentangling components of dynamics. In: Proceedings of IEEE Conference on Computer Vision and Pattern Recognition (CVPR) (2018)
46. Zhao, Y., Xiong, Y., Lin, D.: Trajectory convolution for action recognition. In: Proceedings of Neural Information Processing Systems (NeurIPS) (2018)
47. Zhou, B., Andonian, A., Oliva, A., Torralba, A.: Temporal relational reasoning in videos. In: Proceedings of European Conference on Computer Vision (ECCV) (2018)
48. Zolfaghari, M., Singh, K., Brox, T.: Eco: efficient convolutional network for online video understanding. In: Proceedings of European Conference on Computer Vision (ECCV) (2018)

Polarized Optical-Flow Gyroscope

Masada Tzabari(✉) and Yoav Y. Schechner

Viterbi Faculty of Electrical Engineering, Technion - Israel Institute of Technology, 32000 Haifa, Israel
masada.tz@campus.technion.ac.il, yoav@ee.technion.ac.il
http://www.ee.technion.ac.il/~yoav

Abstract. We merge by generalization two principles of passive optical sensing of motion. One is common spatially resolved imaging, where motion induces temporal readout changes at high-contrast spatial features, as used in traditional optical-flow. The other is the polarization compass, where axial rotation induces temporal readout changes due to the change of incoming polarization angle, relative to the camera frame. The latter has traditionally been modeled for uniform objects. This merger generalizes the brightness constancy assumption and optical-flow, to handle polarization. It also generalizes the polarization compass concept to handle arbitrarily textured objects. This way, scene regions having partial polarization contribute to motion estimation, irrespective of their texture and non-uniformity. As an application, we derive and demonstrate passive sensing of differential ego-rotation around the camera optical axis.

Keywords: Low level vision · Self-calibration · Bio-inspired

1 Introduction

Spatio-temporal image intensity variations indicate motion by low-level vision. This is the *optical-flow* principle. Image-based sensing makes several assumptions. First, the object must have significant spatial contrast of intensity (see Fig. 1). If the scene is nearly uniform, motion estimation is ill-conditioned. Second, spatial resolution should be high, for better conditioning of motion sensing. Third, optical-flow is generally derived assuming *brightness constancy*, i.e, the radiance of observed features is time invariant. Moreover, consider sensing ego-rotation around the camera optical axis. The visual signal useful for this is at the periphery of the field of view. The closer pixels are to the image center (Fig. 1), the less they contribute to image-based sensing of ego-rotation. There, temporal differential variations of intensity tend to null.

Electronic supplementary material The online version of this chapter (https://doi.org/10.1007/978-3-030-58517-4_22) contains supplementary material, which is available to authorized users.

A. Vedaldi et al. (Eds.): ECCV 2020, LNCS 12361, pp. 363–381, 2020.
https://doi.org/10.1007/978-3-030-58517-4_22

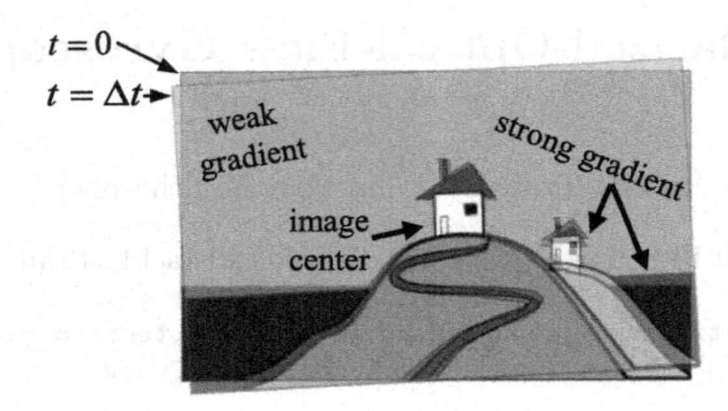

Fig. 1. Types of image regions that are friendly to either optical-flow or polarization-based estimation of rotation. A region of very weak spatial gradient is very suitable for polarization-based estimation. An off-center region of strong gradient is very suitable for optical flow estimation. In the central region of the image, intensity barely changes temporally, despite rotation. Therefore it is not informative in image-based estimation. It is, however, useful in a polarization compass.

There is another principle of passive optical sensing, which can be used for determining orientation and its change (rotation): *polarimetry.* Generally, light coming from a scene towards a sensor has partial linear polarization. The scene polarization is at some angle, set in the object coordinate system. This is the angle of polarization (AOP). Assume a polarization filter is mounted on an optical sensor. The filter has a lateral axis: light is maximally transmitted through the filter if the AOP is aligned with this filter. Multiple measurements by the sensor, in each of which the filter axis is different, can determine the AOP and thus the orientation of the sensing system relative to the object. This principle is known as a *polarization compass.* It is used by many animals [27,82] which exploit celestial or underwater polarization for navigation.

Note that models of the polarization compass *do not use imaging*, i.e, spatial resolution is not required. Rather crude optical sensors having a broad field of view are used. Polarimetry relies on very different - even contradictory - assumptions, relative to image-based methods as optical-flow. Traditional polarimetry has thus far assumed point-wise analysis, oblivious to spatial gradients. *Rotation estimation using polarimetry thrives on spatially uniform objects*, as the sky. On the other hand, image-based methods thrive on high resolution sensing of scenes having high contrast spatial features.

This paper combines the two principles of *optical flow* and *polarization compass.* This enables estimation of motion, and particularly ego-rotation, using both uniform and high-contrast scene regions. Note that in a scene, most image regions have low contrast, and have thus far not been effective for motion estimation by optical-flow. By combining the polarization compass principle with optical flow, partially-polarized areas can help in motion assessments even if they have low contrast, while high-contrast image regions can help as in optical-flow.

On one hand, we generalize *brightness constancy* and *optical-flow* to polarization signals. On the other hand, we generalize the concept of *polarization compass* to spatially-varying signals of arbitrary contrast. Consequently, the paper formulates models and inverse problems that rely on these generalizations. We demonstrate a solution both in simulations and real experiments. We assume polarization-constancy. When does it hold? A major motivation is polarization compass using atmosphere or underwater ambient scattered light [17–19,21,67,79]. In both domains, polarization constancy holds and is invariant to translation. However, when polarization is created by reflection and refraction, it can strongly vary with the direction an object is observed from. There, if translational motion changes the viewpoint significantly, our assumption becomes invalid. The assumption would still hold when the reflective scene is distant relative to lateral translation.

2 Prior Work

Polarization is used by a wide range of animals [13,63–67], particularly for navigation [7,8,20,24,47,82]. Notable animals in this context are bees, ants, squid, spiders, locust, migratory birds, butterflies, crab and octopus. Some of these animals have very crude spatial resolution and most of them have very small brains. In other words, their vision resources are limited. However, observing the sky polarization, even in poor resolution, yields a strong navigational cue which requires minimal computational resources (low level).

Polarization compass is used in bio-inspired robot navigation [17–19,21,69,81], relying mainly on celestial or underwater polarization patterns. These robotic systems therefore can use low resources, and reduce dependency on GPS and inertial measurement units (IMUs). This is our inspiration as well. Low-resource approaches in computer vision [9,12,26,32,53,72,73,75] are significant facilitators of small, distributed agents, using low power, narrow communication bandwidth, reduced computations and crude optical resolution.

Refs. [29,30] use the degree of polarization (DOP), instead of radiance, as input to classic optical-flow. As in our work, [29,30] assume polarization constancy. However, when only the DOP is the input to optical-flow, the formulation is insensitive to the incoming light's AOP, relative to the camera. Hence, the formulation is oblivious to the polarization-compass principle, which provides point-wise sensitivity to the camera orientation and its rotation rate. For example, in a scene having near-uniform DOP as the sky, the DOP does not indicate rotation. Our paper does *not substitute* radiance by DOP. Instead, we *generalize* optical flow to be sensitive to all components of the polarization vector field, and can estimate the rotation rate even from a single pixel.

We estimate the state of an imaging system by observing an unknown scene. This is related to self-calibration. Self-calibration in the context of polarization has been recently suggested [58,85]. In [58], there is no motion of the camera or the object. Self calibration there applies to angles of a polarizer, using redundant images (multi-pixel images in four or more polarizer angles).

More broadly, several scientific communities rely on imaging polarization of light. Specifically, atmospheric sciences and astronomy rely on it [3,16,35,40,59,79], to remotely-sense micro-physical properties of particles or separate scene components. In computational photography, polarization is used in a variety of imaging systems [5,11,14,28,36,37,48,54,56,77,83]. Polarization is helpful for solving a variety of inverse problems [57] in imaging. These include separation of reflection components and surface shape estimation [15,22,31,34,39,41,42,45,49,51,51,80,86], descattering [33,43,44,74,78], and physics-based rendering [11]. Polarization is also used in computational displays [4,46,52] and camera-based communication [84].

Optical-flow is another low-level vision method. As such, it is suitable for low-resource systems, e.g., when computing, power or lag-time constraints prohibit using higher-level processing such as structure-from-motion. It is also independent of external systems [62,70]. Therefore, animals having small brains [71] seem to estimate ego-motion estimation using optical-flow. Similarly, optical-flow is used in airborne and land robots [2,10,23,50,55,68,71,76].

3 Theoretical Background

3.1 Reference Frames and Polarization

An optical axis is perpendicular to a lateral plane. Three frames of reference are in the plane: the object, camera and polarizer frames. Coordinates $\boldsymbol{x}_{\text{cam}} = [x_{\text{cam}}, y_{\text{cam}}]^\top$ in the camera frame express the pixel on the sensor array. Here $\top$ denotes transposition. The object frame is $\boldsymbol{x}_{\text{obj}} = [x_{\text{obj}}, y_{\text{obj}}]^\top$, where x_{obj} and y_{obj} are lateral coordinates of the object. They are respectively referred to here as *horizontal* and *vertical* world coordinates. In the object coordinate frame, radiance from an object location is characterized by intensity $c(\boldsymbol{x}_{\text{obj}})$, DOP $p(\boldsymbol{x}_{\text{obj}})$, and AOP $\theta(\boldsymbol{x}_{\text{obj}})$.

A polarization filter (analyzer) is mounted in front of the camera. The *polarizer axis* is the chief lateral direction in the filter frame. When light is incident at the filter, the polarization component parallel to the polarizer axis is transmitted. The polarization component perpendicular to the polarizer axis is blocked. The angle $\alpha^{\text{in}}_{\text{pol}}$ indicates azimuth of a vector, relative to the polarizer axis.

There are relations between the frames of reference. The polarizer axis is oriented at an angle α relative to the horizon in the object frame. A projection operator $\mathcal{T}$ relates the object and camera coordinates:

$$\boldsymbol{x}_{\text{cam}} = \mathcal{T}\boldsymbol{x}_{\text{obj}}. \tag{1}$$

Light from an object passes through the polarization filter, and then hits the camera sensor array. The measured intensity at the sensor pixel is then

$$I(\boldsymbol{x}_{\text{cam}}) = I(\mathcal{T}\boldsymbol{x}_{\text{obj}}) = \frac{c(\boldsymbol{x}_{\text{obj}})}{2}\left\{1 + p(\boldsymbol{x}_{\text{obj}})\cos 2\left[\alpha - \theta(\boldsymbol{x}_{\text{obj}})\right]\right\}. \tag{2}$$

3.2 Traditional Optical-Flow

Traditional optical-flow assesses a velocity vector field $\boldsymbol{v}(\boldsymbol{x}_{\text{cam}})$ in the camera plane. Traditionally, a polarization filter has not been part of the imaging optics in optical-flow formulations. In unpolarized (radiometric) imagery, the measured intensity at the detector is

$$I(\boldsymbol{x}_{\text{cam}}) = I(\mathcal{T}\boldsymbol{x}_{\text{obj}}) = c(\boldsymbol{x}_{\text{obj}}) \,. \tag{3}$$

Let t denote time. Optical-flow relies on *brightness constancy*: an object patch inherently has static, time invariant radiance:

$$\frac{dc(\boldsymbol{x}_{\text{obj}})}{dt} = 0 \,. \tag{4}$$

Brightness constancy means that images vary in time only due to relative motion between the object coordinates and the camera coordinates. The motion affects the projection operator of Eq. (1). Locally, it is assumed that differential motion is rigid: after differential time dt, a spatial patch which was projected to $\boldsymbol{x}_{\text{cam}}$ is shifted according to

$$\boldsymbol{x}_{\text{cam}} \quad \longrightarrow \quad \boldsymbol{x}_{\text{cam}} + \boldsymbol{v}(\boldsymbol{x}_{\text{cam}})dt. \tag{5}$$

This shift induces a temporal differential change in the measured intensity at $\boldsymbol{x}_{\text{cam}}$, satisfying

$$\frac{\partial I(\boldsymbol{x}_{\text{cam}})}{\partial t} + \nabla I(\boldsymbol{x}_{\text{cam}}) \cdot \boldsymbol{v}(\boldsymbol{x}_{\text{cam}}) = 0. \tag{6}$$

Here ∇I is the spatial gradient field of the measured intensity. Note that Eq. (5) ignores local orientation changes, i.e. traditional methods assume that a local translation-only model may suffice. This assumption cannot hold when polarization is involved, because local orientation changes directly induce changes of measured intensity at the camera, as we discuss.

4 Polarized-Flow

We introduce polarized optical-flow. It generalizes several basic concepts.

- The optical-flow model (Eq. 6) is extended by generalizing brightness constancy (Eq. 4). The essence is extension of Eq. (5) to account for local orientation changes in the projected scene, relative to a camera-mounted polarization filter. This enables quantification of orientation changes even where $\nabla I(\boldsymbol{x}_{\text{cam}}) = 0$.
- Polarization filtering can sense orientation, creating a *polarization compass*. It has so far been formulated for objects having spatially uniform radiance, e.g., zenith view of a clear sky. In polarized-flow, this compass is generalized to arbitrarily nonuniform scenes, where $\nabla I(\boldsymbol{x}_{\text{cam}}) \neq 0$. This is done using a formulation of differential spatio-temporal changes of measured intensities.

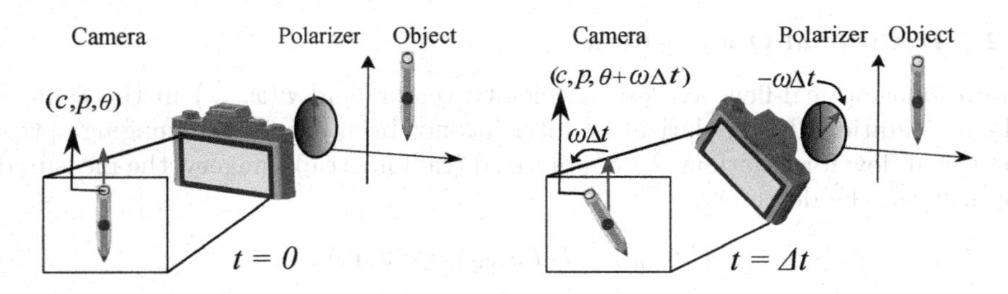

Fig. 2. The camera sensor and the polarizer reference frames are coupled. The coupled camera+polarizer system rotates around the optical axis, which is common to all three reference frames. Angular velocity is positive for counter-clockwise motion. Between shots, the rotation angle is $\omega\Delta t$. The AOP of the object, though static in the object frame, tilts by $\omega\Delta t$ relative to the polarizer axis.

A camera might sense all components of the polarization vector field, by having special pixel-based polarization filters [29,30]. We show that a much simpler sensor can be used: a uniform filter over the whole field of a standard camera. It suffices to obtain enhanced flow estimation and particularly the rotation rate, despite uncertain information about the object's Stokes vector. We use a camera which rigidly rotates in front of a static scene, as illustrated in Fig. 2.

4.1 Polarized Optical-Flow Equation

The transfer of coordinate systems $\mathcal{T}$ (Eq. 1) during projection changes any *orientation* associated with a local patch. Specifically, consider an object for which the AOP $\theta(\boldsymbol{x}_{\text{obj}})$ is uniform. In the camera coordinate system, the projection yields an AOP $\theta(\boldsymbol{x}_{\text{cam}}) = \mathcal{T}\theta(\boldsymbol{x}_{\text{obj}})$.

Now, let $\mathcal{T}$ change in time. The change in $\mathcal{T}$ involves rotation at rate $-\omega$ around the optical axis. Then, during an infinitesimal time of dt

$$\theta(\boldsymbol{x}_{\text{cam}}) \longrightarrow \theta(\boldsymbol{x}_{\text{cam}}) + \omega dt \,. \tag{7}$$

This relation holds when $\theta(\boldsymbol{x}_{\text{cam}})$ is spatially uniform, or when viewing a static object on the optical axis. In the latter case, $\boldsymbol{v}(\boldsymbol{x}_{\text{cam}}) = 0$ in Eq. (5).

Generally, there is change of orientation, lateral motion, and spatial non-uniformity. This compounds Eqs. (5, 7):

$$\theta(\boldsymbol{x}_{\text{cam}}) \longrightarrow \theta[\boldsymbol{x}_{\text{cam}} + \boldsymbol{v}(\boldsymbol{x}_{\text{cam}})dt] + \omega dt \,. \tag{8}$$

During differential time dt, assume that the inherent radiance of the object is static, in the object's own coordinate system. Then Eq. (4) generalizes to

$$\frac{dc(\boldsymbol{x}_{\text{obj}})}{dt} = 0 \qquad \frac{dp(\boldsymbol{x}_{\text{obj}})}{dt} = 0 \qquad \frac{d\theta(\boldsymbol{x}_{\text{obj}})}{dt} = 0 \,. \tag{9}$$

In the camera coordinate system, Eq. (6) generalizes to

$$\begin{aligned}\frac{\partial c(\boldsymbol{x}_{\mathrm{cam}})}{\partial t} + \nabla c(\boldsymbol{x}_{\mathrm{cam}}) \cdot \boldsymbol{v}(\boldsymbol{x}_{\mathrm{cam}}) = 0 \\ \frac{\partial p(\boldsymbol{x}_{\mathrm{cam}})}{\partial t} + \nabla p(\boldsymbol{x}_{\mathrm{cam}}) \cdot \boldsymbol{v}(\boldsymbol{x}_{\mathrm{cam}}) = 0 \\ \frac{\partial \theta(\boldsymbol{x}_{\mathrm{cam}})}{\partial t} + \nabla \theta(\boldsymbol{x}_{\mathrm{cam}}) \cdot \boldsymbol{v}(\boldsymbol{x}_{\mathrm{cam}}) = \omega.\end{aligned} \tag{10}$$

Note specifically how the presence of ω in Eq. (8) affects the orientation (last) relation in Eq. (10). Recall that a camera only measures raw intensity $I(\boldsymbol{x}_{\mathrm{cam}})$, rather than c, p, θ. The polarimetric and intensity variables relate through Eq. (2). Let us transfer Eq. (2) to explicit dependency on camera coordinates, in case the *polarization filter is rigidly tied to the camera.* The camera coordinate system then dictates the polarizer coordinate system. If $\mathcal{T}$ involves rotation of the camera at rate $-\omega$ (in the object coordinate system), then the polarizer axis changes at rate $-\omega$ in the object coordinate system.

Prior to rotation, the angle between the object polarization and the polarizer axis is $[\alpha - \theta(\boldsymbol{x}_{\mathrm{obj}})]$. After time dt, this relative angle changes to $[\alpha + \omega dt - \theta(\boldsymbol{x}_{\mathrm{obj}})]$. Following Eq. (2), the change of the relative angle induces a temporal change in $I(\boldsymbol{x}_{\mathrm{cam}})$, even if c, p, θ are spatially uniform.

We now include both polarization-induced differential changes to $I(\boldsymbol{x}_{\mathrm{cam}})$, as well as changes to $I(\boldsymbol{x}_{\mathrm{cam}})$ created by motion as in common optical-flow. Without loss of generality, prior to rotation, let the polarizer axis be horizontal in the object coordinate system, i.e., $\alpha = 0$. Then,

$$\begin{aligned}&\frac{\partial I(\boldsymbol{x}_{\mathrm{cam}})}{\partial t} + \nabla I(\boldsymbol{x}_{\mathrm{cam}}) \cdot \boldsymbol{v}(\boldsymbol{x}_{\mathrm{cam}}) = \\ &\quad \frac{1}{2}\frac{\partial c}{\partial t} + \frac{1}{2}\left\{p\frac{\partial c}{\partial t} + c\frac{\partial p}{\partial t}\right\}\cos(2\theta) - cp\sin(2\theta)\frac{\partial \theta}{\partial t} + \\ &\quad \frac{1}{2}\frac{\partial c}{\partial x} + \frac{1}{2}\left\{p\frac{\partial c}{\partial x} + c\frac{\partial p}{\partial x}\right\}\cos(2\theta) - cp\sin(2\theta)\frac{\partial \theta}{\partial x} + \\ &\quad \frac{1}{2}\frac{\partial c}{\partial y} + \frac{1}{2}\left\{p\frac{\partial c}{\partial y} + c\frac{\partial p}{\partial y}\right\}\cos(2\theta) - cp\sin(2\theta)\frac{\partial \theta}{\partial y} = \\ &\frac{1 + p\cos 2\theta}{2}\left\{\frac{\partial c}{\partial t} + \nabla c \cdot \boldsymbol{v}\right\} + \frac{c\cos 2\theta}{2}\left\{\frac{\partial p}{\partial t} + \nabla p \cdot \boldsymbol{v}\right\} - cp\sin 2\theta\left\{\frac{\partial \theta}{\partial t} + \nabla \theta \cdot \boldsymbol{v}\right\}.\end{aligned} \tag{11}$$

Using Eq. (10) in Eq. (11) yields the *polarized optical-flow equation*

$$\frac{\partial I(\boldsymbol{x}_{\mathrm{cam}})}{\partial t} + \nabla I(\boldsymbol{x}_{\mathrm{cam}}) \cdot \boldsymbol{v}(\boldsymbol{x}_{\mathrm{cam}}) + \omega c(\boldsymbol{x}_{\mathrm{cam}}) p(\boldsymbol{x}_{\mathrm{cam}}) \sin[2\theta(\boldsymbol{x}_{\mathrm{cam}})] = 0. \tag{12}$$

For the special case $p = 0$, Eq. (12) degenerates to the traditional optical-flow equation, as required. Moreover, because a change in θ is solely due to rotation of the camera+filter relative to the object, the special case $\omega = 0$ also degenerates Eq. (12) to the traditional optical-flow equation, irrespective of p. The model may be applied to ego-rotation sensing.

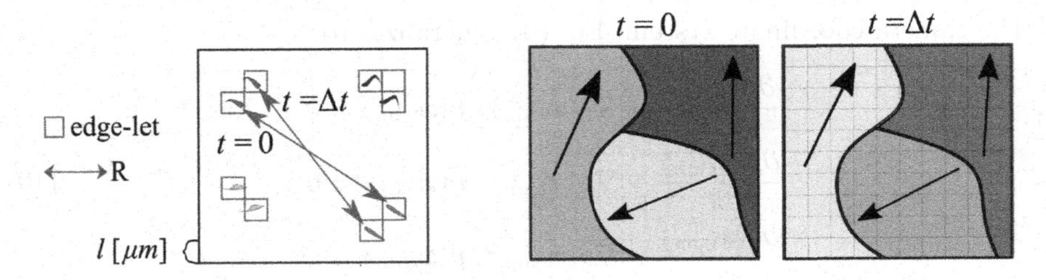

Fig. 3. [Left] The imager has N_{cam} pixels. A typical distance between edge-lets is $R \sim \sqrt{N_{\text{cam}}}/2$ pixels. Between frames, there is time step Δt and the image rotates. If edge-lets move by a pixel, their relative rotation is sensed. Each pixel is of length l microns. [Right] An object is divided to K segments (here K = 3). Each segment has approximately uniform polarization (marked by an arrow). Due to slight rotation of a camera+polarizer rig, the measured intensity per segment changes slightly.

5 Quantifying Sensitivities

We want to assess the significance of the polarization-compass relative to image-based spatially resolving analysis of rotation. The image-based principle is analyzed using pairs of distinct pixels, say edge-lets. The polarimetric principle is analyzed using a broad uniform area in the field of view.

5.1 Image-Based Sensitivity

Let us start with the image-based principle. There are N_{cam} pixels in the camera. Between random pairs of edge-let pixels, a vector is typically of length $R \sim \sqrt{N_{\text{cam}}}/2$ pixels (Fig. 3 [Left]). Rotation means that the inter-pixel vector tilts slightly within a time step of Δt. A single pixel is the basic resolution of this vector tilt. Hence the edge-let pair can resolve a rotation rate $\Delta\omega_{\text{pair}}$ given by

$$\Delta\omega_{\text{pair}} \sim [\Delta t\sqrt{N_{\text{cam}}}/2]^{-1}\,, \tag{13}$$

irrespective of the pixel size. Let there be $N_{\text{edge-lets}}$ independent edge-lets, thus $\approx N^2_{\text{edge-lets}}$ pairs. Each of them contributes an independent random error of standard deviation $\Delta\omega_{\text{pair}}$ to the estimated rotation rate. Averaging the contributions, overall, an image-based method can resolve

$$\Delta\omega_{\text{image}} = \sqrt{\sum_{\text{pair}=1}^{N^2_{\text{edge-lets}}} \left(\frac{\Delta\omega_{\text{pair}}}{N^2_{\text{edge-lets}}}\right)^2} \sim [N_{\text{edge-lets}}\Delta t\sqrt{N_{\text{cam}}}/2]^{-1}\,. \tag{14}$$

5.2 Polarimetric Sensitivity

Now, let us deal with polarimetry in a spatially uniform region (segment) (Fig. 3 [Right]). Consider two rotation rates, whose difference is $\Delta\omega$. From Eq. (12), due to rotation during Δt, the intensity of a pixel changes temporally according to

$$\Delta I(\boldsymbol{x}_{\text{cam}}) = \Delta\omega\Delta t \; c(\boldsymbol{x}_{\text{cam}})p(\boldsymbol{x}_{\text{cam}})|\sin[2\theta(\boldsymbol{x}_{\text{cam}})]| \; . \tag{15}$$

For a random orientation $\theta(\boldsymbol{x}_{\text{cam}})$, let us set roughly $|\sin[2\theta(\boldsymbol{x}_{\text{cam}})]| \sim 1/2$. From Eq. (2), $I(\boldsymbol{x}_{\text{cam}}) \sim c(\boldsymbol{x}_{\text{cam}})/2$. Hence, due to the object segment, a resolvable orientation rate is roughly

$$\Delta\omega_{\text{polar}}^{\text{segment}} \sim 4\Delta I \left[Ip\Delta t\right]^{-1} \; , \tag{16}$$

where we drop the dependency on $\boldsymbol{x}_{\text{cam}}$ due to the uniformity of the segment. Here ΔI is the resolvable intensity, which is determined by radiometric noise.

When a camera is well exposed, noise is dominated by photon (Poisson) statistics. Thus $\Delta I \approx \sqrt{I}$, where I is in units of photo-electrons. The typical full-well depth of a pixel is $\sim 1000l^2$ photo-electrons, where l is the pixel width in microns (Fig. 3 [left]). The object occupies N_{cam}/K pixels, where K is the number of image segments, each being uniform. The intensity over this area is aggregated digitally. This aggregation is equivalent to a full well of $\sim 1000N_{\text{cam}}l^2/K$ photo-electrons per segment. Let the sensor be exposed at medium-level i.e., at half-well. Summing photo-electrons from all pixels in a segment, the signal and its uncertainty are, respectively,

$$I \approx 500N_{\text{cam}}l^2/K \; , \quad \Delta I \approx \sqrt{500N_{\text{cam}}l^2/K}. \tag{17}$$

Combining Eqs. (16, 17)

$$\Delta\omega_{\text{polar}}^{\text{segment}} \sim 4\sqrt{K}\left[pl\Delta t\sqrt{500N_{\text{cam}}}\right]^{-1} \; . \tag{18}$$

Averaging information from all K segments in the image, polarimetry can resolve

$$\Delta\omega_{\text{polar}} = \sqrt{\sum_{\text{segment}=1}^{K}\left(\frac{1}{K}\Delta\omega_{\text{polar}}^{\text{segment}}\right)^2} \sim 4\left[pl\Delta t\sqrt{500N_{\text{cam}}}\right]^{-1} \; . \tag{19}$$

As expected, the uncertainty $\Delta\omega_{\text{polar}}$ becomes smaller, as p or l increase. In other words the higher the object polarization or the better light gathering by the camera detector, the lower is the uncertainty. From Eqs. (14, 19),

$$\Delta\omega_{\text{polar}} \approx \frac{N_{\text{edge-lets}}}{11pl}\Delta\omega_{\text{image}} \; . \tag{20}$$

We examine this relation for a few case studies.

- If $l = 10\mu\text{m}$, $p = 0.5$, then the uncertainty ratio is $\approx N_{\text{edge-lets}}/56$. Hence, if there are about 50 reliable edge-lets in the image, polarization contributes similarly to image-based flow.

- If $l = 3\mu\text{m}$, $p = 0.05$, then the uncertainty ratio is $\approx N_{\text{edge-lets}}/1.7$. If as before there are 50 reliable edge-lets in the image, polarization barely contributes information, relative to image-based flow. Polarization contributes here significantly only if the image is extremely smooth, or comprises of very few pixels.
- For a single pixel sensor (or pair of pixels) the uncertainty ratio is $\approx (11pl)^{-1}$, suggesting high potential to polarization contribution. In this case, spatial non-uniformity would have no impact. The mean intensity of an image may be considered equivalent to a single pixel approximation. A constant rotation rate may be estimated by fitting the change in the mean intensity of the image due to rotation to Eq. (2). This would require a sequence of images from four time steps (see examples in *Supplementary Material*).

The study cases above demonstrate polarization has the best added value for scenes having high polarization, few spatial features, and cameras having low-resolution, or large pixels and hence high SNR.

6 Polarized-Flow Gyro Forward Model

Let us now analyze the case of pure rotational motion around the optical center of the rig (camera + polarizing filter), as in Fig. 2. In the coupled frame, the image rotates with angular velocity of ω, around a center pixel $\overline{\boldsymbol{x}}_{\text{cam}}$. Moreover, let the object be static. Then, in the camera reference frame,

$$\boldsymbol{v} = [-\omega(y_{\text{cam}} - \overline{y}_{\text{cam}}),\ \omega(x_{\text{cam}} - \overline{x}_{\text{cam}})]. \tag{21}$$

Define $\Gamma(\boldsymbol{x}_{\text{cam}}) = c(\boldsymbol{x}_{\text{cam}})p(\boldsymbol{x}_{\text{cam}})\sin 2\theta(\boldsymbol{x}_{\text{cam}})$ and $b(\boldsymbol{x}_{\text{cam}}) = \frac{\partial I(\boldsymbol{x}_{\text{cam}})}{\partial t}$. From Eqs. (12, 21),

$$\omega(y_{\text{cam}} - \overline{y}_{\text{cam}})\frac{\partial I(\boldsymbol{x}_{\text{cam}})}{\partial x_{\text{cam}}} - \omega(x_{\text{cam}} - \overline{x}_{\text{cam}})\frac{\partial I(\boldsymbol{x}_{\text{cam}})}{\partial y_{\text{cam}}} - \omega\Gamma(\boldsymbol{x}_{\text{cam}}) = b(\boldsymbol{x}_{\text{cam}}) \tag{22}$$

Define

$$q(\boldsymbol{x}_{\text{cam}}) = \frac{\partial I(\boldsymbol{x}_{\text{cam}})}{\partial x_{\text{cam}}}(y_{\text{cam}} - \overline{y}_{\text{cam}}) - \frac{\partial I(\boldsymbol{x}_{\text{cam}})}{\partial y_{\text{cam}}}(x_{\text{cam}} - \overline{x}_{\text{cam}}) - \Gamma(\boldsymbol{x}_{\text{cam}}). \tag{23}$$

Then, per pixel $\boldsymbol{x}_{\text{cam}}$, Eqs. (12, 22) can be expressed as $b(\boldsymbol{x}_{\text{cam}}) - q(\boldsymbol{x}_{\text{cam}})\omega = 0$. For N pixels, define the vectors $\boldsymbol{q} = [q(\boldsymbol{x}_{\text{cam}_1}), q(\boldsymbol{x}_{\text{cam}_2}), \ldots, q(\boldsymbol{x}_{\text{cam}_N})]^\top$ and $\boldsymbol{b} = [b(\boldsymbol{x}_{\text{cam}_1}), b(\boldsymbol{x}_{\text{cam}_2}), \ldots, b(\boldsymbol{x}_{\text{cam}_N})]^\top$. Then,

$$\boldsymbol{q}\omega = \boldsymbol{b}. \tag{24}$$

7 Solving a Polarized-Flow Gyro Inverse Problem

We now estimate the angular velocity ω. Assume for the moment that vectors $\boldsymbol{q}$ and $\boldsymbol{b}$ are known. Note that Eq. (24) is a set of linear equations with a single unknown. A least-squares solution is given in closed-form:

$$\hat{\omega} = (\boldsymbol{q}^\top \boldsymbol{q})^{-1} \boldsymbol{q}^\top \boldsymbol{b}. \tag{25}$$

Let ω be known, but c, p and θ be unknown. Equation (22) is sensitive only to the vector $\boldsymbol{\Gamma} = [\Gamma(\boldsymbol{x}_{\text{cam}_1}), \Gamma(\boldsymbol{x}_{\text{cam}_2}), \ldots, \Gamma(\boldsymbol{x}_{\text{cam}_N})]^\top$. Let us estimate it by

$$\hat{\boldsymbol{\Gamma}} = \arg\min_{\boldsymbol{\Gamma}} \left[\sum_{\boldsymbol{x}_{\text{cam}}} |b(\boldsymbol{x}_{\text{cam}}) - q(\boldsymbol{x}_{\text{cam}})\omega|^2 + \mu \|\nabla^2 \boldsymbol{\Gamma}\|_2^2 \right], \tag{26}$$

where μ is a regularization weight. Equation (26) is quadratic in Γ, and thus has a closed-form solution. It can be applied when analyzing small images. For high-resolution images, Eq. (26) is solved iteratively by gradient descent.

Equation (26) assumes that ω is known, while Eq. (25) assumes that $\boldsymbol{\Gamma}$ is known. We alternate between these two processes, starting from an initial $\hat{\omega}$. In our implementation, we ran 50 iterations of Eq. (26) between each calculation of Eq. (25), and repeated to convergence. Note that for the application of ego-rotation estimation, our main interest is estimation of $\hat{\omega}$. The estimation of $\hat{\boldsymbol{\Gamma}}$ is a means to that end, hence we tolerate errors in $\hat{\boldsymbol{\Gamma}}$.

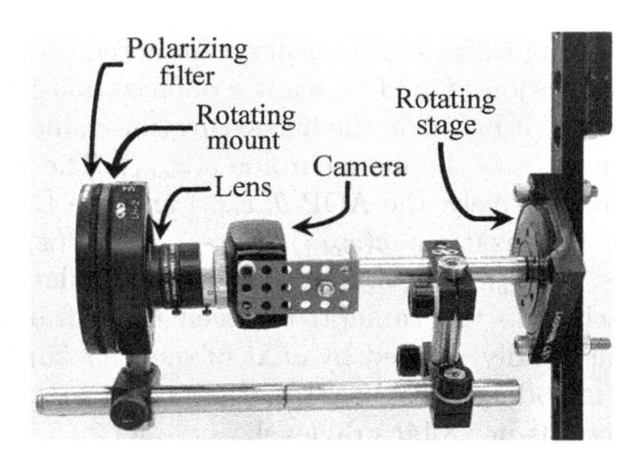

Fig. 4. The coupled camera-polarizer system for simulation and experiments. The polarizer is harnessed to the camera on a rotating mount, allowing measurements of the Stokes vector. The system is assembled on a tripod for stability.

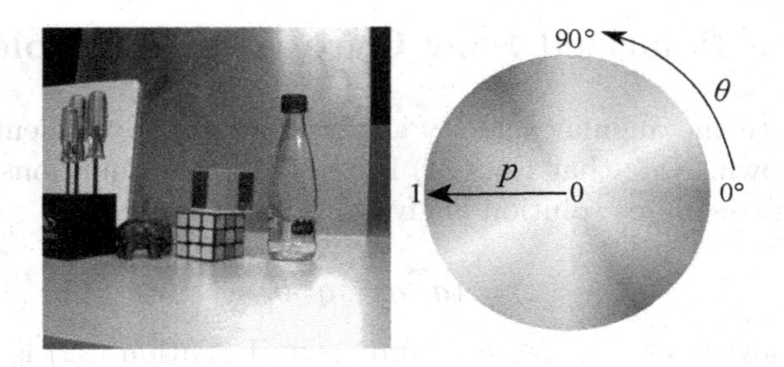

Fig. 5. [Left] Scene used in the simulation, visualized in color using the HSV space. Gamma correction of the intensity was performed for display purposes. [Right] The HSV coordinates are θ (Hue), p (Saturation), and c (Value).

8 Simulation Based on Real Data

Before conducting real experiments, we performed simulations using images $I(\boldsymbol{x}_{\rm cam}, t)$ rendered from real polarimetry. To acquire data, we used a setup shown in Fig. 4. An IDS 4.92 megapixel monochrome camera (UI-3480LE) was fitted with a Fujinon 1/2" 6mm lens (DF6HA-1B). A high transmittance (approximately 98%) Tiffen 72mm linear polarization filter was placed in front of the lens, using a mounted protractor having 2° increments. Both camera and filter were rigidly connected to a rotating stage, having its own protractor. In this section, we held the stage static and rotated only the filter. This way, we performed traditional polarimetric imaging, yielding ground truth $c(\boldsymbol{x}_{\rm obj})$, $p(\boldsymbol{x}_{\rm obj})$, and $\theta(\boldsymbol{x}_{\rm obj})$. Pixels having saturated intensity were noted, in order to be ignored in the subsequent estimation of ω. The scene's polarization field is visualised in Fig. 5. The visualization[1] is based on the hue-saturation-value (HSV) color space. The value channel represents the monochrome $c(\boldsymbol{x}_{\rm obj})$. The hue and saturation channels represent, respectively, the AOP $\theta(\boldsymbol{x}_{\rm obj})$ and the DOP $p(\boldsymbol{x}_{\rm obj})$.

Using the scene polarization, $c(\boldsymbol{x}_{\rm obj}), p(\boldsymbol{x}_{\rm obj})$ and $\theta(\boldsymbol{x}_{\rm obj})$, we simulated two acquired frames $I(\boldsymbol{x}_{\rm cam}, 1), I(\boldsymbol{x}_{\rm cam}, 2)$, where a simulated polarization filter was rigidly attached to the camera. Between these frames, the simulated camera+filter rig was rigidly rotated by $\omega \Delta t$ of our choosing. We define $\Delta t = 1$[frame]. Then, we introduced Poisson (photon) noise to the images, using a scale of 50 photoelectrons per 8-bit graylevel.

In analysis, images were slightly smoothed by a Gaussian filter (STD of 2 pixels). The weight in Eq. (26) was set $\mu = 0.0001$. Ten tests were carried out, each with a different rotation $\omega \in [0.1°/\text{frame}, \dots\ 50°/\text{frame}]$. Each test yielded an estimate of ω. To assess uncertainty, this process was repeated 10 times, using independent Poisson noise samples. Using these samples, per tested rotation angle we obtained an average estimate $\hat{\bar{\omega}}$ and standard deviation $\text{STD}(\hat{\omega})$. We found that typically $\text{STD}(\hat{\omega}) \approx 0.03°/\text{frame}$. The resulting $\text{error}(\omega) \equiv \hat{\bar{\omega}} - \omega$ is plotted in Fig. 6. The approach performed well across a wide range of angles.

[1] The maps of each polarization variable are presented in the *Supplementary Material*.

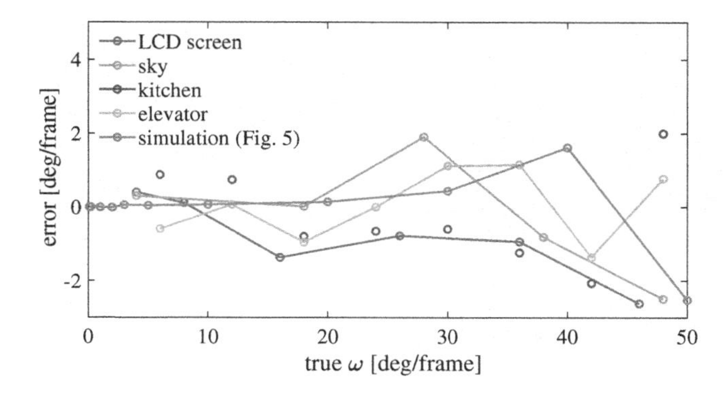

Fig. 6. Plots of error(ω) for the simulated tests and real experiments.

9 Experiments

The experiments use the system described in Sect. 8 and Fig. 4. Here, however, the polarizer protractor was held fixed relative to the camera. The rotating stage that held the camera+polarizer rig was rotated relative to the scene. Only two frames are taken per experiment, each angle in the pair is fixed and known. Rotation between angles is done manually, then a shot is taken. This is in lieu of having two images sampled automatically in a motorized rig. We captured four scenes: *LCD*, *Kitchen*, *Sky*, and *Elevator*. They are visualized in Fig. 7 and described below. Each scene lead to 5–8 experiments, each at a distinct rotation rate $\omega \in [4°/\text{frame}, \ldots\ 48°/\text{frame}]$. This yielded a total of 27 experiments. The resulting error(ω) $= \hat{\hat{\omega}} - \omega$ is plotted in Fig. 6. The approach performed well across a wide range of angles.

- *LCD screen* is simply a computer screen displaying a photo. Its spatial contrast in c is strong. Its DOP is nearly $p = 1$. Slight depolarization is caused by unpolarized light in the lab reflecting off the screen.
- *Kitchen* possesses reflection-induced polarization at a variety of angles. Light thus has a variety of polarization angles. The DOP is generally low. This scene has a lot of intensity edges.
- *Elevator* presents a low light, high DOP scene, with a weak pattern of a rough reflecting metallic surface.
- In *Sky*, the rig views the clear sky through the lab's open window. The viewed sky is highly polarized ($p \approx 0.6$), having nearly uniform radiance. Such a scene can be challenging for traditional image-based ego-rotation estimation.

We imaged a sky-patch using $\omega = 6°/\text{frame}$. Our method erred by 5%. We imaged this sky patch in the same rotation setting, without a polarizer. We ran Horn-Schunck and Lucas-Kanade optical-flow algorithms on the radiance measurements. The resulting flows were fit by least-squares to rotation, yielding respectively, $\hat{\omega} = 0.1°/\text{frame}$ and $\hat{\omega} = 0.03°/\text{frame}$, i.e., far lower than ω.

LCD screen *Kitchen* *Sky* *Elevator*

Fig. 7. *LCD screen*, *Kitchen*, *Sky*, and *Elevator* scenes. The polarization is represented using the HSV color space, as in Fig. 5.

To compare with [29,30], we imaged Stokes vector per pixel of the sky patch, by rotating the polarizer independently of the camera, and then rotating the camera itself. We then ran the Horn-Schunck and Lucas-Kanade algorithms on the DOP field p, as in [29,30]. Using the resulting flow fields for least-squares fit to rotation yielded, respectively, $\hat{\omega} = 10^{-5}$ °/frame and $2 \cdot 10^{-5}$ °/frame. These estimated values are far lower than the true $\omega = 6°$/frame. The spatial uniformity of the radiance and DOP fields yield no reliable information about rotation. The sky's AOP enables proper estimation of rotation.

10 Conclusions

The *polarized optical-flow equation* (12) contains both traditional components of optical-flow and a polarized component. In strong spatial gradients of radiance, flow estimation is similar to optical-flow, while polarization is just an enhancement. In weak spatial gradient, or when spatial resolution is low (low resource sensors) the polarized component may have similar and even higher significance than the optical-flow. The approach can be useful in low-resource outdoor navigation having the sky in view or underwater. Polarization-based navigation aids [17–19,21,67,79] fit low resource systems, but they assume models of the expected polarization pattern. This paper shows that the rotation rate can be derived in unknown scenes. In addition, this rotation cue can assist self-calibration of distant imagers, such as low-resource nano-satellites [61].

The optical-flow model assumes infinitesimal motion. Nevertheless, the estimation results here were found to be reasonable even in rather large angles of rotation. It is worth generalizing analysis for motion having arbitrarily large angles and more complex parametric ego-motion, including translation and 3-axis rotation. We believe that a 3-axis rotation sensor can be achieved by having a rig containing three corresponding cameras, each having a mounted polarizer. Alternatively, this may be achieved using an omnicam mounted with a polarizer having a spatially varying polarization axis [6,60].

There are sensors in which micro-polarizers are attached to individual pixels [1,25]. Adjacent pixels use filters that are 45° or 90° to each other, in analogy to RGB Beyer patterns. It is worth extending the analysis to such cameras. There, a single image yields $\hat{\mathbf{\Gamma}}$, thus significantly simplifying the analysis of Sect. 7.

The work here may shed light on biology. As described in Sect. 2, many animals use a polarization compass. However, many animals have eyes that are both spatially resolving and polarization sensitive. It is common for their eyes to have two interleaved populations of receptor cells, having mutually orthogonal orientations [38]. It is possible to study whether these animals combine these principles, in analogy to our analysis.

We use a particular algorithm for recovery, and as most algorithms, it has pros, cons and parameters. For example, in our implementation, we use an initial optimization step-size which decreases when ω increases. We believe that algorithms that are designed to be efficient and robust in low resource systems would suit the sensing idea of this paper.

Acknowledgements. We thank M. Sheinin, A. Levis, A. Vainiger, T. Loeub, V. Holodovsky, M. Fisher, Y. Gat, and O. Elezra for fruitful discussions. We thank I. Czerninski, O. Shubi, D. Yegudin, and I. Talmon for technical support. Yoav Schechner is the Mark and Diane Seiden Chair in Science at the Technion. He is a Landau Fellow - supported by the Taub Foundation. His work was conducted in the Ollendorff Minerva Center. Minvera is funded through the BMBF. This work is supported by the Israel Science Foundation (ISF fund 542/16).

References

1. Technology Image Sensor: Polarization Products Sony Semiconductor Solutions Group. https://www.sony-semicon.co.jp/e/products/IS/polarization/technology.html. Accessed 16 July 2020
2. Agrawal, P., Ratnoo, A., Ghose, D.: Inverse optical flow based guidance for UAV navigation through urban canyons. Aerosp. Sci. Technol. **68**, 163–178 (2017)
3. Barta, A., Horváth, G.: Underwater binocular imaging of aerial objects versus the position of eyes relative to the flat water surface. JOSA A **20**(12), 2370 (2003)
4. Ben-Ezra, M.: Segmentation with invisible keying signal. In: CVPR, pp. 32–37. IEEE (2000)
5. Berezhnyy, I., Dogariu, A.: Time-resolved Mueller matrix imaging polarimetry. Opt. Express **12**(19), 4635–4649 (2004)
6. Bomzon, Z., Biener, G., Kleiner, V., Hasman, E.: Spatial Fourier-transform polarimetry using space-variant subwavelength metal-stripe polarizers. Opt. Lett. **26**(21), 1711–1713 (2001)
7. Brines, M.L.: Dynamic patterns of skylight polarization as clock and compass. J. Theor. Biol. **86**(3), 507–512 (1980)
8. Brines, M.L., Gould, J.L.: Skylight polarization patterns and animal orientation. J. Exp. Biol. **96**(1), 69–91 (1982)
9. Carey, S.J., Barr, D.R., Dudek, P.: Low power high-performance smart camera system based on SCAMP vision sensor. J. Syst. Architect. **59**(10), 889–899 (2013)
10. Chahl, J.S., Srinivasan, M.V., Zhang, S.W.: Landing strategies in honeybees and applications to uninhabited airborne vehicles. Int. J. Rob. Res. **23**(2), 101–110 (2004)
11. Collin, C., Pattanaik, S., LiKamWa, P., Bouatouch, K.: Computation of polarized subsurface BRDF for rendering. In: Proceedings of Graphics Interface, pp. 201–208. Canadian Information Processing Society (2014)

12. Courroux, S., Chevobbe, S., Darouich, M., Paindavoine, M.: Use of wavelet for image processing in smart cameras with low hardware resources. J. Syst. Architect. **59**(10), 826–832 (2013)
13. Cronin, T.W.: Polarization vision and its role in biological signaling. Integr. Comp. Biol. **43**(4), 549–558 (2003)
14. Dahlberg, A.R., Pust, N.J., Shaw, J.A.: Effects of surface reflectance on skylight polarization measurements at the Mauna Loa Observatory. Opt. Express **19**(17), 16008–16021 (2011)
15. Diamant, Y., Schechner, Y.Y.: Overcoming visual reverberations. In: CVPR, pp. 1–8. IEEE (2008)
16. Diner, D.J., et al.: First results from a dual photoelastic-modulator-based polarimetric camera. Appl. Opt. **49**(15), 2929–2946 (2010)
17. Dupeyroux, J., Diperi, J., Boyron, M., Viollet, S., Serres, J.: A novel insect-inspired optical compass sensor for a hexapod walking robot. In: IROS, pp. 3439–3445. IEEE/RSJ (2017)
18. Dupeyroux, J., Serres, J., Viollet, S.: A hexapod walking robot mimicking navigation strategies of desert ants cataglyphis. In: Vouloutsi, V., et al. (eds.) Living Machines 2018. LNCS (LNAI), vol. 10928, pp. 145–156. Springer, Cham (2018). https://doi.org/10.1007/978-3-319-95972-6_16
19. Dupeyroux, J., Viollet, S., Serres, J.R.: An ant-inspired celestial compass applied to autonomous outdoor robot navigation. Rob. Auton. Syst. **117**, 40–56 (2019)
20. Evangelista, C., Kraft, P., Dacke, M., Labhart, T., Srinivasan, M.: Honeybee navigation: critically examining the role of the polarization compass. Philos. Trans. Roy. Soc. B Biol. Sci. **369**(1636), 20130037 (2014)
21. Fan, C., Hu, X., Lian, J., Zhang, L., He, X.: Design and calibration of a novel camera-based bio-inspired polarization navigation sensor. IEEE Sens. J. **16**(10), 3640–3648 (2016)
22. Farid, H., Adelson, E.H.: Separating reflections from images by use of independent component analysis. JOSA A **16**(9), 2136–2145 (1999)
23. Fernandes, J., Postula, A., Srinivasan, M., Thurrowgood, S.: Insect inspired vision for micro aerial vehicle navigation. In: ACRA, pp. 1–8. ARAA (2011)
24. Foster, J.J., Sharkey, C.R., Gaworska, A.V., Roberts, N.W., Whitney, H.M., Partridge, J.C.: Bumblebees learn polarization patterns. Curr. Biol. **24**(12), 1415–1420 (2014)
25. Garcia, M., Edmiston, C., Marinov, R., Vail, A., Gruev, V.: Bio-inspired color-polarization imager for real-time in situ imaging. Optica **4**(10), 1263 (2017)
26. Gill, P.R., Kellam, M., Tringali, J., Vogelsang, T., Erickson, E., Stork, D.G.: Computational diffractive imager with low-power image change detection. In: Computational Optical Sensing and Imaging, pp. CM3E-2. OSA (2015)
27. Goddard, S.M., Forward, R.B.: The role of the underwater polarized light pattern, in sun compass navigation of the grass shrimp, Palaemonetes Vulgaris. J. Compa. Physiol. A Neuroethol. Sens. Neural Behavi. Physiol. **169**(4), 479–491 (1991)
28. Gruev, V., Van der Spiegel, J., Engheta, N.: Dual-tier thin film polymer polarization imaging sensor. Opt. Express **18**(18), 19292–19303 (2010)
29. Guan, L., Liu, S., Qi Li, S., Lin, W., Yuan Zhai, L., Kui Chu, J.: Study on polarized optical flow algorithm for imaging bionic polarization navigation micro sensor. Optoelectron. Lett. **14**(3), 220–225 (2018)
30. Guan, L., et al.: Study on displacement estimation in low illumination environment through polarized contrast-enhanced optical flow method for polarization navigation applications. Optik **210**, 164513 (2020)

31. Guarnera, G.C., Peers, P., Debevec, P., Ghosh, A.: Estimating surface normals from spherical stokes reflectance fields. In: Fusiello, A., Murino, V., Cucchiara, R. (eds.) ECCV 2012. LNCS, vol. 7584, pp. 340–349. Springer, Heidelberg (2012). https://doi.org/10.1007/978-3-642-33868-7_34
32. Guo, Q., et al.: Compact single-shot metalens depth sensors inspired by eyes of jumping spiders. Proc. Natl. Acad. Sci. **116**(46), 22959–22965 (2019)
33. Gupta, M., Narasimhan, S.G., Schechner, Y.Y.: On controlling light transport in poor visibility environments. In: CVPR, pp. 1–8. IEEE (2008)
34. Gutierrez, D., Narasimhan, S.G., Jensen, H.W., Jarosz, W.: Scattering. In: ACM Siggraph Asia Courses, p. 18 (2008)
35. Hooper, B.A., Baxter, B., Piotrowski, C., Williams, J., Dugan, J.: An airborne imaging multispectral polarimeter. In: Oceans, vol. 2, p. 7. IEEE/MTS (2009)
36. Horisaki, R., Choi, K., Hahn, J., Tanida, J., Brady, D.J.: Generalized sampling using a compound-eye imaging system for multi-dimensional object acquisition. Opt. Express **18**(18), 19367–19378 (2010)
37. Horstmeyer, R., Euliss, G., Athale, R., Levoy, M.: Flexible multimodal camera using a light field architecture. In: ICCP, pp. 1–8. IEEE (2009)
38. Horváth, G.: Polarized Light and Polarization Vision in Animal Sciences, vol. 2. Springer, Heidelberg (2014). https://doi.org/10.1007/978-3-642-54718-8
39. Ihrke, I., Kutulakos, K.N., Lensch, H.P., Magnor, M., Heidrich, W.: State of the art in transparent and specular object reconstruction. In: Eurographics STAR - State of the Art Report (2008)
40. Joos, F., Buenzli, E., Schmid, H.M., Thalmann, C.: Reduction of polarimetric data using Mueller calculus applied to Nasmyth instruments. In: Observatory Operations: Strategies, Processes, and Systems II, vol. 7016, p. 70161I. SPIE (2008)
41. Kadambi, A., Taamazyan, V., Shi, B., Raskar, R.: Polarized 3D: high-quality depth sensing with polarization cues. In: ICCV, pp. 3370–3378. IEEE (2015)
42. Kadambi, A., Taamazyan, V., Shi, B., Raskar, R.: Depth sensing using geometrically constrained polarization normals. Int. J. Comput. Vision **125**(1–3), 34–51 (2017)
43. Kaftory, R., Schechner, Y.Y., Zeevi, Y.Y.: Variational distance-dependent image restoration. In: CVPR, pp. 1–8. IEEE (2007)
44. Kattawar, G.W., Adams, C.N.: Stokes vector calculations of the submarine light field in an atmosphere-ocean with scattering according to a Rayleigh phase matrix: effect of interface refractive index on radiance and polarization. Limnol. Oceanogr. **34**(8), 1453–1472 (1989)
45. Kong, N., Tai, Y.W., Shin, J.S.: A physically-based approach to reflection separation: from physical modeling to constrained optimization. IEEE Trans. Pattern Anal. Mach. Intell. **36**(2), 209–221 (2013)
46. Lanman, D., Wetzstein, G., Hirsch, M., Heidrich, W., Raskar, R.: Polarization fields: dynamic light field display using multi-layer LCDs. ACM Trans. Graph. **30**, 186 (2011)
47. Lerner, A., Sabbah, S., Erlick, C., Shashar, N.: Navigation by light polarization in clear and turbid waters. Philos. Trans. Roy. Soc. B Biol. Sci. **366**(1565), 671–679 (2011)
48. Levis, A., Schechner, Y.Y., Davis, A.B., Loveridge, J.: Multi-view polarimetric scattering cloud tomography and retrieval of droplet size. arXiv preprint arXiv:2005.11423 (2020)
49. Maeda, T., Kadambi, A., Schechner, Y.Y., Raskar, R.: Dynamic heterodyne interferometry. In: ICCP, pp. 1–11. IEEE (2018)

50. McGuire, K., De Croon, G., De Wagter, C., Tuyls, K., Kappen, H.: Efficient optical flow and stereo vision for velocity estimation and obstacle avoidance on an autonomous pocket drone. IEEE Rob. Autom. Lett. **2**(2), 1070–1076 (2017)
51. Miyazaki, D., Saito, M., Sato, Y., Ikeuchi, K.: Determining surface orientations of transparent objects based on polarization degrees in visible and infrared wavelengths. JOSA A **19**(4), 687–694 (2002)
52. Mohan, A., Woo, G., Hiura, S., Smithwick, Q., Raskar, R.: Bokode: imperceptible visual tags for camera based interaction from a distance. ACM Trans. Graph. **28**, 98 (2009)
53. Monjur, M., Spinoulas, L., Gill, P.R., Stork, D.G.: Panchromatic diffraction gratings for miniature computationally efficient visual-bar-position sensing. Sens. Transducers **194**(11), 127 (2015)
54. Mujat, M., Baleine, E., Dogariu, A.: Interferometric imaging polarimeter. JOSA A **21**(11), 2244–2249 (2004)
55. Pijnacker Hordijk, B.J., Scheper, K.Y., De Croon, G.C.: Vertical landing for micro air vehicles using event-based optical-flow. J. Field Rob. **35**(1), 69–90 (2018)
56. Reddy, D., Veeraraghavan, A., Chellappa, R.: P2C2: programmable pixel compressive camera for high speed imaging. In: CVPR, pp. 329–336. IEEE (2011)
57. Schechner, Y.Y.: Inversion by P4: polarization-picture post-processing. Philos. Trans. Roy. Soc. B Biol. Sci. **366**(1565), 638–648 (2011)
58. Schechner, Y.Y.: Self-calibrating imaging polarimetry. In: ICCP, pp. 1–10. IEEE (2015)
59. Schechner, Y.Y., Diner, D.J., Martonchik, J.V.: Spaceborne underwater imaging. In: ICCP, pp. 1–8. IEEE (2011)
60. Schechner, Y.Y., Nayar, S.K.: Generalized mosaicing: polarization panorama. IEEE Trans. Pattern Anal. Mach. Intell. **27**(4), 631–636 (2005)
61. Schilling, K., Schechner, Y., Koren, I.: Cloudct-computed tomography of clouds by a small satellite formation. In: Proceedings of the 12th IAA Symposium on Small Satellites for Earth Observation (2019)
62. Shantaiya, S., Verma, K., Mehta, K.: Multiple object tracking using Kalman filter and optical flow. Eur. J. Adv. Eng. Technol. **2**(2), 34–39 (2015)
63. Shashar, N.: Transmission of linearly polarized light in seawater: implications for polarization signaling. J. Exp. Biol. **207**(20), 3619–3628 (2004)
64. Shashar, N., et al.: Underwater linear polarization: physical limitations to biological functions. Philos. Trans. Roy. Soc. B Biol. Sci. **366**(1565), 649–654 (2011)
65. Shashar, N., Rutledge, P., Cronin, T.: Polarization vision in cuttlefish in a concealed communication channel? J. Exp. Biol. **199**(9), 2077–2084 (1996)
66. Shashar, N., Hagan, R., Boal, J.G., Hanlon, R.T.: Cuttlefish use polarization sensitivity in predation on silvery fish. Vis. Res. **40**, 71–75 (2000)
67. Shashar, N., Milbury, C., Hanlon, R.: Polarization vision in cephalopods: neuroanatomical and behavioral features that illustrate aspects of form and function. Mar. Freshw. Behav. Physiol. **35**, 57–68 (2002)
68. Shen, C., et al.: Optical flow sensor/INS/magnetometer integrated navigation system for MAV in GPS-denied environment. J. Sens. **2016**, 1–10 (2016)
69. Smith, F., Stewart, D.: Robot and insect navigation by polarized skylight. In: BIOSIGNALS, pp. 183–188 (2014)
70. Song, X., Seneviratne, L.D., Althoefer, K.: A Kalman filter-integrated optical flow method for velocity sensing of mobile robots. IEEE/ASME Trans. Mechatron. **16**(3), 551–563 (2010)
71. Srinivasan, M.V.: Honeybees as a model for the study of visually guided flight, navigation, and biologically inspired robotics. Physiol. Rev. **91**(2), 413–460 (2011)

72. Stork, D.G.: Optical elements as computational devices for low-power sensing and imaging. In: Imaging Systems and Applications, pp. ITu4E-4. OSA (2017)
73. Stork, D.G., Gill, P.R.: Special-purpose optics to reduce power dissipation in computational sensing and imaging systems. In: Sensors, pp. 1–3. IEEE (2017)
74. Tanaka, K., Mukaigawa, Y., Matsushita, Y., Yagi, Y.: Descattering of transmissive observation using parallel high-frequency illumination. In: ICCP, pp. 1–8. IEEE (2013)
75. Tessens, L., Morbée, M., Philips, W., Kleihorst, R., Aghajan, H.: Efficient approximate foreground detection for low-resource devices. In: International Conference on Distributed Smart Cameras, pp. 1–8. ACM/IEEE (2009)
76. Thurrowgood, S., Moore, R.J., Soccol, D., Knight, M., Srinivasan, M.V.: A biologically inspired, vision-based guidance system for automatic landing of a fixed-wing aircraft. J. Field Rob. **31**(4), 699–727 (2014)
77. Treeaporn, V., Ashok, A., Neifeld, M.A.: Increased field of view through optical multiplexing. Opt. Express **18**(21), 22432–22445 (2010)
78. Treibitz, T., Schechner, Y.Y.: Polarization: Beneficial for visibility enhancement? In: CVPR, pp. 525–532. IEEE (2009)
79. Tyo, J.S., Goldstein, D.L., Chenault, D.B., Shaw, J.A.: Review of passive imaging polarimetry for remote sensing applications. Appl. Opt. **45**(22), 5453–5469 (2006)
80. Umeyama, S., Godin, G.: Separation of diffuse and specular components of surface reflection by use of polarization and statistical analysis of images. IEEE Trans. Pattern Anal. Mach. Intell. **5**, 639–647 (2004)
81. Wang, D., Liang, H., Zhu, H., Zhang, S.: A bionic camera-based polarization navigation sensor. Sensors **14**(7), 13006–13023 (2014)
82. Wehner, R.D.: Polarized-light navigation by insects. Sci. Am. **235**(1), 106–115 (1976)
83. Xu, Z., Yue, D.K., Shen, L., Voss, K.J.: Patterns and statistics of in-water polarization under conditions of linear and nonlinear ocean surface waves. J. Geophys. Res. Oceans **116**(12), 1–14 (2011)
84. Yuan, W., Dana, K., Varga, M., Ashok, A., Gruteser, M., Mandayam, N.: Computer vision methods for visual MIMO optical system. In: CVPR Workshops, pp. 37–43. IEEE (2011)
85. Zallat, J., Torzynski, M., Lallement, A.: Double-pass self-spectral-calibration of a polarization state analyzer. Opt. Lett. **37**(3), 401–403 (2012)
86. Zhang, L., Hancock, E.R., Atkinson, G.A.: Reflection component separation using statistical analysis and polarisation. In: Vitrià, J., Sanches, J.M., Hernández, M. (eds.) IbPRIA 2011. LNCS, vol. 6669, pp. 476–483. Springer, Heidelberg (2011). https://doi.org/10.1007/978-3-642-21257-4_59

Online Meta-learning for Multi-source and Semi-supervised Domain Adaptation

Da Li[1(✉)] and Timothy Hospedales[1,2]

[1] Samsung AI Center Cambridge, Cambridge, UK
dali.academic@gmail.com, t.hospedales@ed.ac.uk
[2] The University of Edinburgh, Edinburgh, UK

Abstract. Domain adaptation (DA) is the topical problem of adapting models from labelled source datasets so that they perform well on target datasets where only unlabelled or partially labelled data is available. Many methods have been proposed to address this problem through different ways to minimise the domain shift between source and target datasets. In this paper we take an orthogonal perspective and propose a framework to further enhance performance by meta-learning the initial conditions of existing DA algorithms. This is challenging compared to the more widely considered setting of few-shot meta-learning, due to the length of the computation graph involved. Therefore we propose an online shortest-path meta-learning framework that is both computationally tractable and practically effective for improving DA performance. We present variants for both multi-source unsupervised domain adaptation (MSDA), and semi-supervised domain adaptation (SSDA). Importantly, our approach is agnostic to the base adaptation algorithm, and can be applied to improve many techniques. Experimentally, we demonstrate improvements on classic (DANN) and recent (MCD and MME) techniques for MSDA and SSDA, and ultimately achieve state of the art results on several DA benchmarks including the largest scale DomainNet.

Keywords: Meta-learning · Domain adaptation

1 Introduction

Contemporary deep learning methods now provide excellent performance across a variety of computer vision tasks when ample annotated training data is available. However, this performance often degrades rapidly if models are applied to novel domains with very different data statistics from the training data, which is a problem known as domain shift. Meanwhile, data collection and annotation for every possible domain of application is expensive and sometimes impossible.

Electronic supplementary material The online version of this chapter (https://doi.org/10.1007/978-3-030-58517-4_23) contains supplementary material, which is available to authorized users.

A. Vedaldi et al. (Eds.): ECCV 2020, LNCS 12361, pp. 382–403, 2020.
https://doi.org/10.1007/978-3-030-58517-4_23

This challenge has motivated extensive study in the area of domain adaptation (DA), which addresses training models that work well on a target domain using only unlabelled or partially labelled target data from that domain together with labelled data from a source domain.

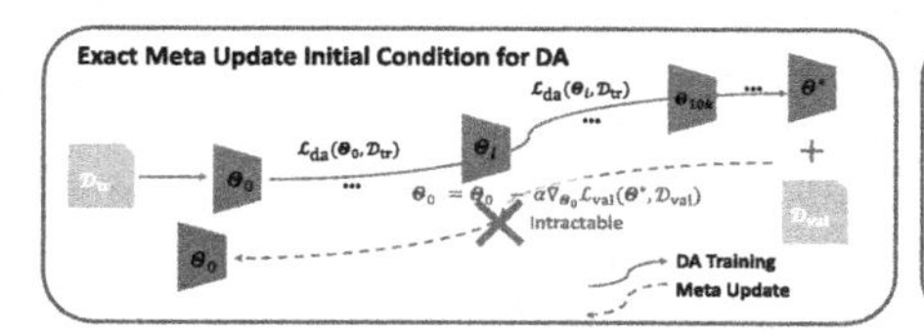

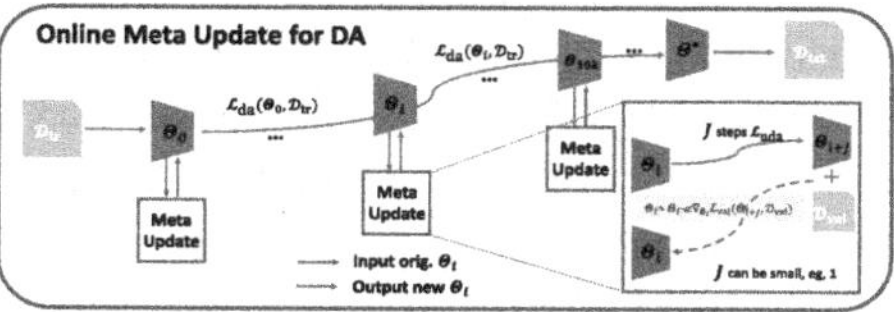

Fig. 1. Left: Exact meta-learning of initial condition with inner loop training DA to convergence is intractable. Right: Online meta-learning alternates between meta-optimization and domain adaptation.

Several variants of the domain adaptation problem have been studied. Single-source domain adaptation (SDA) considers adaptation from a single source domain [5,6], while multi-source domain adaptation (MSDA) considers improving cross-domain generalization by aggregating information across multiple sources [35,47]. Unsupervised domain adaptation (UDA) learns solely from unlabelled data in the target domain [16,43], while semi-supervised domain adaptation (SSDA) learns from a mixture of labelled and unlabelled target domain data [9,10]. The main means of progress has been developing improved methods for aligning representations between source(s) and the target in order to improve generalization. These methods span distribution alignment, for example by maximum mean discrepancy (MMD) [27,48], domain adversarial training [16,43], and cycle consistent image transformation [18,26].

In this paper we adopt a novel research perspective that is complementary to all these existing methods. Rather than proposing a new domain adaptation strategy, we study a meta-learning framework for improving these existing adaptation algorithms. Meta-learning (a.k.a. learning to learn) has a long history [44,45], and has re-surged recently, especially due to its efficacy in improving few-shot deep learning [12,40,50]. Meta-learning pipelines aim to improve learning by training some aspect of a learning algorithm such a comparison metric [50], model optimizer [40] or model initialisation [12], so as to improve outcomes according to some meta-objective such as few-shot learning efficacy [12,40,50] or learning speed [1]. In this paper we provide a first attempt to define a meta-learning framework for improving domain adaptive learning.

We take the perspective of meta-optimizing the *initial condition* [12,33] of domain adaptive learning[1]. While there are several facets of algorithms that can be meta-learned such as hyper-parameters [14] and learning rates [24]; these are somewhat tied to the base learning algorithm (domain adaptive algorithm in our

[1] One may not think of domain adaptation as being sensitive to initial condition, but given the lack of target domain supervision to guide learning, different initialization can lead to a significant 10–15% difference in accuracy (see Supplementary material).

case). In contrast, our framework is *algorithm agnostic* in that it can be used to improve many existing gradient-based domain adaptation algorithms.

Furthermore we develop variants for both unsupervised multi-source adaptation, as well as semi-supervised single source adaptation, thus providing broad potential benefit to existing frameworks and application settings. In particular we demonstrate application of our framework to the classic domain adversarial neural network (DANN) [16] algorithm, as well as the recent maxmium-classifier discrepancy (MCD) [43], and min-max entropy (MME) [41] algorithms.

Meta-learning can often be cleanly formalised as a bi-level optimization problem [14,39]: where an outer loop optimizes the meta-parameter of interest (such as the initial condition in our case) with respect to some meta-loss (such as performance on a validation set); and the inner loop runs the learning algorithm conditioned on the chosen meta-parameter. This is tricky to apply directly in a domain adaptation scenario however, because: (i) The computation graph of the inner loop is typically long (unlike the popular few-shot meta-learning setting [12]), making meta-optimization intractable, and (ii) Especially in unsupervised domain adaptation, there is no labelled data in the target domain to define a supervised learning loss for the outer-loop meta-objective. We surmount these challenges by proposing a simple, fast and efficient meta-learning strategy based on online shortest path gradient descent [36], and defining meta-learning pipelines suited for meta-optimization of domain adaptation problems. Although motivated by initial condition learning, our online algorithm ultimately has the interpretation of intermittently performing meta-update(s) of the parameters in order to achieve the best outcome from the following DA updates (Fig. 1).

Overall, our contributions are: (i) Introducing a meta-learning framework suitable for both multi-source and semi-supervised domain adaptation settings, (ii) We demonstrate the algorithm agnostic nature of our framework through its application to several base domain adaptation methods including MME [41], DANN [16] and MCD [43], (iii) Applying our meta-learner to these base adaptation methods, we achieve state of the art performance on several MSDA and SSDA benchmarks, including the largest-scale DA benchmark, DomainNet [38].

2 Related Work

Single-Source Domain Adaptation. Single-source unsupervised domain adaptation (SDA) is a well established area [5,6,16,19,27,29–31,43,48]. Theoretical results have bound the cross-domain generalization error in terms of domain divergence [4], and numerous algorithms have been proposed to reduce the divergence between source and target features. Representative approaches include minimising MMD distribution shift [27,48] or Wasserstein distance [2,51], adversarial training [16,20,43] or alignment by cycle-consistent image translation [18,26]. Given the difficulty of SDA, studies have considered exploiting semi-supervised or multi-source adaptation where possible.

Semi-supervised Domain Adaptation. This setting assumes that besides the labelled source and unlabelled target domain data, there are a few labelled samples available in the target domain. Exploiting the few target labels allows better domain alignment compared to purely unsupervised approaches. Representative approaches are based on regularization [9], subspace learning [54], label smoothing [10] and entropy minimisation in the target domain [17]. The state of the art method in this area, MME, extends the entropy minimisation idea to adversarial entropy minimisation in a deep network setting [41].

Multi-source Domain Adaptation. This setting assumes there are multiple labelled source domains for training. In deep learning, simply aggregating all source domains data together often already improves performance due to bigger datasets learning a stronger representation. Theoretical results based on $\mathcal{H}$-divergence [4] can still apply after aggregation, and existing SDA methods that attempt to reduce source-target divergence [5,16,43] can be used. Meanwhile, new generalization bounds for MSDA have been derived [38,55], which motivate algorithms that align amongst source domains as well as between source and target. Nevertheless, practical deep network optimization is non-convex, and the degree of alignment achieved depends on the details of the optimization strategy. Therefore our paradigm of meta-learning the initial condition of optimization is compatible with, and complementary to, all this prior work.

Meta-learning for Neural Networks. Meta-Learning (learning to learn) [44, 46] has experienced a recent resurgence. This has largely been driven by its efficacy for few-shot deep learning via initial condition learning [12], optimizer learning [40] and embedding learning [50]. More generally it has been applied to improve optimization efficiency [1], reinforcement learning [37], gradient-based hyperparameter optimization [14] and neural architecture search [25]. We start from the perspective of MAML [12], in terms of focusing on learning initial conditions of neural network optimization. However besides the different application (domain adaptation vs few-shot learning), our resulting algorithm is very different as we end up performing meta-optimization online *while* solving the target task rather than in advance of solving it [12,39]. A few recent studies also attempted online meta-learning [13,53], but these are designed specifically to backprop through RL [53] or few-shot supervised [13] learning. Meta-learning with domain adaptation in the inner loop has not been studied before now.

In terms of learning with multiple domains a few studies [3,11,21] have considered meta-learning for multi-source *domain generalization*, which evaluates the ability of models to generalise directly without adaptation. In practice these methods use supervised learning in their inner optimization. No meta-learning method has been proposed for the domain adaptation problem addressed here.

3 Methodology

3.1 Background

Unsupervised Domain Adaptation. Domain adaptation techniques aim to reduce the domain shift between source domain(s) $\mathcal{D}_S$ and target domain $\mathcal{D}_T$, in order that a model trained on labels from $\mathcal{D}_S$ performs well when deployed on $\mathcal{D}_T$. Commonly such algorithms train a model Θ with a loss $\mathcal{L}_{\text{uda}}$ that breaks

Algorithm 1. Meta-Update Initial Condition

Function UpdateIC(Θ_0, $\mathcal{D}_{\text{tr}}$, $\mathcal{D}_{\text{val}}$, $\mathcal{L}_{\text{inner}}$, $\mathcal{L}_{\text{outer}}$) :
for $j = 1, 2, \ldots, J$ **do** // Inner-level optimization
$\Theta_j = \Theta_{j-1} - \alpha \nabla_{\Theta_{j-1}} \mathcal{L}_{\text{inner}}(\Theta_{j-1}, (\mathcal{D}_{\text{tr}})_j)$
end for
$\Theta_0 = \Theta_0 - \alpha \nabla_{\Theta_0} \mathcal{L}_{\text{outer}}(\Theta_J, \mathcal{D}_{\text{val}})$ //Outer-level step
Output: Θ_0

down into a term for supervised learning on the source domain $\mathcal{L}_{\text{sup}}$ and an adaptation loss $\mathcal{L}_{\text{a}}$ that attempts to align the target and source data

$$\mathcal{L}_{\text{uda}}(\Theta, \mathcal{D}_S, \overline{\mathcal{D}}_T) = \mathcal{L}_{\text{sup}}(\Theta, \mathcal{D}_S) + \lambda \mathcal{L}_{\text{a}}(\Theta, \mathcal{D}_S, \overline{\mathcal{D}}_T). \tag{1}$$

We use notation $\mathcal{D}_S$ and $\overline{\mathcal{D}}_T$ to indicate that the source and target domains contain labelled and unlabelled data respectively. Many existing domain adaptation algorithms [16,38,43,48] fit into this template, differing in their definition of the domain alignment loss $\mathcal{L}_{\text{a}}$. In the case of multi-source adaptation [38], $\mathcal{D}_S$ may contain several source domains $\mathcal{D}_S = \{D_1, \ldots, D_N\}$ and the first supervised learning term $\mathcal{L}_{\text{sup}}$ sums the performance on all of these.

Semi-supervised Domain Adaptation. In the SSDA setting [41], we assume a sparse set of labelled target data $\mathcal{D}_T$ is provided along with a large set of unlabelled target data $\overline{\mathcal{D}}_T$. The goal is to learn a model that fits both the source and few-shot target labels $\mathcal{L}_{\text{sup}}$, while also aligning the unlabelled target data to the source with an adaptation loss $\mathcal{L}_a$.

$$\begin{aligned}\mathcal{L}_{\text{ssda}}(\Theta, \mathcal{D}_S, \overline{\mathcal{D}}_T, \mathcal{D}_T) =& \mathcal{L}_{\text{sup}}(\Theta, \mathcal{D}_S) + \mathcal{L}_{\text{sup}}(\Theta, \mathcal{D}_T) \\ &+ \lambda \mathcal{L}_{\text{a}}(\Theta, \mathcal{D}_S, \overline{\mathcal{D}}_T)\end{aligned} \tag{2}$$

Several existing algorithms [16,28,41] fit this template and hence can potentially be optimized by our framework.

Meta Learning Model Initialisation. The problem of meta-learning the initial condition of an optimization can be seen as a bi-level optimization problem [14,39]. In this view there is a standard task-specific (inner) algorithm of interest whose initial condition we wish to optimize, and an outer-level meta-algorithm that optimizes that initial condition. This setup can be described as

$$\Theta = \underbrace{\underset{\Theta}{\text{argmin}} \;\; \mathcal{L}_{\text{outer}}(\overbrace{\mathcal{L}_{\text{inner}}(\Theta, \mathcal{D}_{\text{tr}})}^{\text{Inner-level}}, \mathcal{D}_{\text{val}})}_{\text{Outer-level}} \tag{3}$$

where $\mathcal{L}_{\text{inner}}(\Theta, \mathcal{D}_{\text{tr}})$ denotes the standard loss of the base task-specific algorithm on its training set. $\mathcal{L}_{\text{outer}}(\Theta^*, \mathcal{D}_{\text{val}})$ denotes the validation set loss *after* $\mathcal{L}_{\text{inner}}$ has been optimized, ($\Theta^* = \text{argmin}\, \mathcal{L}_{\text{inner}}$), when starting from the initial condition set by the outer optimization. The overall goal in Eq. 3 above is thus to set the initial condition of base algorithm $\mathcal{L}_{\text{inner}}$ such that it achieves minimum

Algorithm 2. Meta-Update Initial Condition: SPG

Function UpdateIC(Θ_0, $\mathcal{D}_{\text{tr}}$, $\mathcal{D}_{\text{val}}$, $\mathcal{L}_{\text{inner}}$, $\mathcal{L}_{\text{outer}}$) :
$\tilde{\Theta}_0 = \text{copy}(\Theta_0)$
for $j = 1, 2, \ldots, J$ **do** // Inner-level optimization
$\tilde{\Theta}_j = \tilde{\Theta}_{j-1} - \alpha\nabla_{\tilde{\Theta}_{j-1}}\mathcal{L}_{\text{inner}}(\tilde{\Theta}_{j-1}, (\mathcal{D}_{\text{tr}})_j)$
end for
$\nabla^{\text{short}}_{\Theta_0} = \Theta_0 - \tilde{\Theta}_J$ //Outer-level step
$\Theta_0 = \Theta_0 - \alpha\nabla_{\Theta_0}\mathcal{L}_{\text{outer}}(\Theta_0 - \nabla^{\text{short}}_{\Theta_0}, \mathcal{D}_{\text{val}})$
Output: Θ_0

loss on the validation set. When both losses are differentiable we can in principle solve Eq. 3 by taking gradient steps on $\mathcal{L}_{\text{outer}}$ as shown in Algorithm 1. However, such exact meta-learning requires backpropagating through the path of the inner optimization, which is costly and inaccurate for a long computation graph.

3.2 Meta-learning for Domain Adaptation

Overview. For meta domain adaptation, we would like to instantiate the initial condition learning idea summarised earlier in Eq. 3 in order to initialize popular domain adaptation algorithms such as [16,41,43] that can be represented as problems in the form of Eqs. 1 and 2, so as to maximise the resulting performance in the target domain upon deployment. To this end we will introduce in the following section appropriate definitions of the inner and outer tasks, as well as a tractable optimization strategy.

Multi-source Domain Adaptation. Suppose we have an adequate algorithm to optimize for initial conditions as required in Eq. 3. How could we apply this idea to multi-source unsupervised domain adaptation setting, given that there is no target domain training data to take the role of $\mathcal{D}_{\text{val}}$ in providing the metric for outer loop optimization of the initial condition? Our idea is that in the *multi-source* domain adaptation setting, we can split available source domains into disjoint meta-training and meta-testing domains $\mathcal{D}_S = \mathcal{D}_S^{\text{mtr}} \cup \mathcal{D}_S^{\text{mte}}$, where we actually have labels for both. Now we can let $\mathcal{L}_{\text{inner}}$ be an unsupervised domain method [16,43] $\mathcal{L}_{\text{inner}} := \mathcal{L}_{\text{uda}}$, and ask it to adapt from meta-train to the unlabelled meta-test domain. In the outer loop, we can then use the labels of the meta-test domain as a validation set to evaluate the adaptation performance via a supervised loss $\mathcal{L}_{\text{outer}} := \mathcal{L}_{\text{sup}}$, such as cross-entropy. Thus we aim to find an initial condition for our base domain adaptation method $\mathcal{L}_{\text{uda}}$ that enables it to adapt effectively between source domains

$$\Theta_0 = \underset{\Theta_0}{\text{argmin}} \sum_{\mathcal{D}_S^{\text{mtr}}, \mathcal{D}_S^{\text{mte}} \sim \mathcal{D}_S} \mathcal{L}_{\text{sup}}(\mathcal{L}_{\text{uda}}(\mathcal{D}_S^{\text{mtr}}, \overline{\mathcal{D}}_S^{\text{mte}}; \Theta_0), \mathcal{D}_S^{\text{mte}}) \tag{4}$$

where we use $\mathcal{L}(\cdot; \Theta_0)$ to denote optimizing a loss from starting point Θ_0. This initial condition could be optimized by taking gradient descent steps on the outer supervised loss using UpdateIC from Algorithm 1. The resulting Θ_0 is suited to adapting between all source domains hence should also be good for adapting to the target domain. Thus we would finally instantiate the same UDA algorithm using the learned initial condition, but this time between the full set of source domains, and the true unlabelled target domain $\overline{\mathcal{D}}_T$.

$$\Theta = \underset{\Theta}{\text{argmin}} \;\; \mathcal{L}_{\text{uda}}(\mathcal{D}_S, \overline{\mathcal{D}}_T; \Theta_0) \tag{5}$$

An Online Solution. While conceptually simple, the problem with the direct approach above is that it requires completing domain-adaptive training multiple times in the inner optimization. Instead we propose to perform *online* meta-learning [53,56] by alternating between steps of meta-optimization of Eq. 4 and steps on the final unsupervised domain adaptation problem in Eq. 6. That is, we iterate

$$\begin{aligned} \Theta &= \text{UpdateIC}(\Theta, (\mathcal{D}_S^{\text{mtr}}) \cup (\overline{\mathcal{D}}_S^{\text{mte}}), (\mathcal{D}_S^{\text{mte}}), \mathcal{L}_{\text{uda}}, \mathcal{L}_{\text{sup}}) \\ \Theta &= \Theta - \alpha \nabla_\Theta \mathcal{L}_{\text{uda}}(\Theta, (\mathcal{D}_S), (\overline{\mathcal{D}}_T)) \end{aligned} \tag{6}$$

where $(\mathcal{D})$ denotes minibatch sampling from the corresponding dataset, and we call UpdateIC$(\cdot)$ with a small number of inner-loop optimizations such as $J = 1$.

Our method, summarised in Fig. 1 and Algorithm 3, performs meta-learning online, by simultaneously solving the meta-objective and the target task. It translates to tuning the initial condition between taking optimization steps on the target DA task. This avoids the intractability and instability of backpropagating through the long computational graph in the exact approach that meta-optimizes Θ_0 to completion before doing DA. Online meta-learning is also potentially advantageous in practice due to improving optimization throughout training rather than only at the start – c.f. the vanilla exact method, where the impact of the initial condition on the final outcome is very indirect.

Semi-supervised Domain Adaptation. We next consider how to adapt the ideas introduced previously to the semi-supervised domain adaptation setting. In the MSDA setting above, we divided source domains into meta-train and meta-test, used unlabeled data from meta test to drive adaptation, and then used meta-test labels to validate the adaptation performance. In SSDA we do not have multiple source domains with which to use such a meta-train/meta-test split strategy, but we do have a small amount of labeled data in the target domain that we can use to validate adaptation performance and drive initial condition optimization. By analogy to Eq. 4, we can aim to find the initial condition for the unsupervised component $\mathcal{L}_{\text{uda}}$ of an SSDA method. But now we can use the few labelled examples $\mathcal{D}_T$ to validate the adaptation in the outer loop.

$$\Theta_0 = \underset{\Theta_0}{\text{argmin}} \sum \mathcal{L}_{\text{sup}}(\mathcal{L}_{\text{uda}}(\mathcal{D}_{\text{S}}, \overline{\mathcal{D}}_T; \Theta_0), \mathcal{D}_T) \tag{7}$$

The learned initial condition can then be used to instantiate the final semi-supervised domain adaptive training.

$$\Theta = \underset{\Theta}{\text{argmin}} \;\; \mathcal{L}_{\text{ssda}}(\Theta, \mathcal{D}_S, \overline{\mathcal{D}}_T, \mathcal{D}_T; \Theta_0) \tag{8}$$

Algorithm 3. Online Meta learning: Multi-Source DA

Input: N source domains $\mathcal{D}_S = [D_1, D_2, \ldots, D_N]$ and unlabelled target domain $\overline{\mathcal{D}}_T$.
Initialise: Model parameters Θ, learning rate α, task loss $\mathcal{L}_{\text{sup}}$, UDA method $\mathcal{L}_{\text{uda}}$.
for $i = 1, 2, \ldots, I$ **do**
  $\Theta = \mathbf{UpdateIC}(\Theta, \mathcal{D}_S^{\text{mtr}} \cup \overline{\mathcal{D}}_S^{\text{mte}}, \mathcal{D}_S^{\text{mte}}, \mathcal{L}_{\text{uda}}, \mathcal{L}_{\text{sup}})$
  for $s = 1, 2, \ldots, S$ **do**
    $\Theta = \Theta - \alpha \nabla_\Theta \mathcal{L}_{\text{uda}}(\Theta, (\mathcal{D}_S)_i, (\overline{\mathcal{D}}_T)_i)$ // Domain Adaptation Training
  end for
end for
Output: Θ

An Online Solution. The exact meta SSDA approach above suffers from the same limitations as exact MetaMSDA. So we again apply online meta-learning by iterating between meta-optimization of Eq. 7 and the final supervised domain adaptation problem of Eq. 9.

$$\begin{aligned} \Theta &= \text{UpdateIC}(\Theta, (\mathcal{D}_S) \cup (\overline{\mathcal{D}}_T), (\mathcal{D}_T), \mathcal{L}_{\text{uda}}, \mathcal{L}_{\text{sup}}) \\ \Theta &= \Theta - \alpha \nabla_\Theta \mathcal{L}_{\text{ssda}}(\Theta, \mathcal{D}_S, \overline{\mathcal{D}}_T, \mathcal{D}_T; \Theta) \end{aligned} \tag{9}$$

The final procedure is summarized in Algorithm 4.

3.3 Shortest Path Optimization

Meta-learning Model Initialisation. As described so far, our meta-learning approach to domain adaptation relies on the ability meta-optimize initial conditions using gradient descent steps as described in Algorithm 1. Such steps evaluate a meta-gradient that depends on the parameter Θ^* output by the base domain adaptation algorithm

$$\Theta_0 = \Theta_0 - \alpha \overbrace{\nabla_\Theta \mathcal{L}_{\text{sup}}(\Theta^*, \mathcal{D}_{\text{val}})}^{\text{Meta Gradient}} \tag{10}$$

Evaluating the meta-gradient directly is impractical because: (i) The inner loop that runs the base domain adaptation algorithm may take multiple gradient descent iterations $j = 1 \ldots J$. This will trigger a large chain of higher-order gradients $\nabla_{\Theta_0} \mathcal{L}_{\text{inner}}(\cdot), \ldots, \nabla_{\Theta_{J-1}} \mathcal{L}_{\text{inner}}(\cdot)$. (ii) More fundamentally, several state of the art domain adaptation algorithms [41,43] use multiple optimization steps when making updates on $\mathcal{L}_{\text{inner}}$. For example, to adversarially train the deep feature extractor and classifier modules of the model in Θ. Taking gradient steps on $\mathcal{L}_{\text{outer}}(\Theta^*)$ thus triggers higher-order gradients, even if one only takes a single step $J = 1$ of domain adaptation optimization.

Shortest Path Optimization. To obtain the meta gradient in Eq. 10 efficiently, we use shortest-path gradient (SPG) [36]. Before optimising the innner loop, we copy parameters Θ_0 as $\tilde{\Theta}_0$ and use $\tilde{\Theta}_0$ in the inner-level algorithm.

Algorithm 4. Online Meta Learning: Semi-Supervised DA

Input: N source domains $\mathcal{D}_S = [D_1, D_2, \ldots, D_N]$, labelled and unlabelled target domain data $\mathcal{D}_T$ and $\overline{\mathcal{D}}_T$.
Initialise: Model params Θ, learning rate α, task loss $\mathcal{L}_{\text{sup}}$, UDA method $\mathcal{L}_{\text{uda}}$.
for $i = 1, 2, \ldots, I$ **do**
  $\Theta = \textbf{UpdateIC}(\Theta, \mathcal{D}_S \cup \overline{\mathcal{D}}_T, \mathcal{D}_T, \mathcal{L}_{\text{uda}}, \mathcal{L}_{\text{sup}})$
  for $s = 1, 2, \ldots, S$ **do**
    Θ=Θ-$\alpha\nabla_\Theta(\mathcal{L}_{\text{sup}}(\Theta, (\mathcal{D}_T)_i)$+$\mathcal{L}_{\text{uda}}(\Theta, (\mathcal{D}_S)_i, (\overline{\mathcal{D}}_T)_i))$
// Domain Adaptation Training
  end for
end for
Output: Θ

Then, after finishing the inner loop we get the shortest-path gradient between $\tilde{\Theta}_J$ and Θ_0.

$$\nabla_{\Theta_0}^{\text{short}} = \Theta_0 - \tilde{\Theta}_J \tag{11}$$

Each meta-gradient step (Eq. 10) is then approximated as

$$\Theta_0 = \Theta_0 - \alpha \nabla_{\Theta_0} \mathcal{L}_{\text{sup}}(\Theta_0 - \nabla_{\Theta_0}^{\text{short}}, \mathcal{D}_{\text{val}}) \tag{12}$$

Summary. We now have an efficient implementation of UpdateIC for updating initial conditions as summarised in Algorithm 2. This shortest path approximation has the advantage of allowing efficient initial condition updates both for multiple iterations of inner loop optimization $J > 1$, as well as for inner loop domain adaptation algorithms that use multiple steps [41,43]. We use this implementation for the MSDA and SSDA methods in Algorithms 3 and 4.

4 Experiments

Datasets. We evaluate our method on several multi-source domain adaptation benchmarks including PACS [22], Office-Home [49] and DomainNet [38]; as well as on the semi-supervised setting of Office-Home and DomainNet.

Base Domain Adaptation Algorithms and Ablation. Our Meta-DA framework is designed to complement existing base domain adaptation algorithms. We evaluate it in conjunction with Domain Adversarial Neural Networks (DANN, [16]) – as a representative classic approach to deep domain adaptation; as well as Maximum Classifier Discrepancy (MCD, [43]) and MinMax Entropy (MME, [41]) – as examples of state of the art multi-source and semi-supervised domain adaptation algorithms respectively. Our goal is to evaluate whether our Meta-DA framework can improve these base learners. We note that the MCD algorithm has two variants: (1) A multi-step variant that alternates between updating the classifiers and several steps of updating the feature extractor and

(2) A one-step variant that uses a gradient reversal [16] layer so that classifier and feature extractor can be updated in a single gradient step. We evaluate both of these. Sequential Meta-Learning: As an ablation, we also consider an alternative fast meta-learning approach that performs all meta-updates at the start of learning, before doing DA; rather than performing meta-updates online with DA as in our proposed Meta-DA algorithms.

4.1 Multi-source Domain Adaptation

PACS: Dataset. PACS [22] was initially proposed for domain generalization and had been subsequently been re-purposed [7,34] for multi-source domain adaptation. This dataset has four diverse domains: (A)rt painting, (C)artoon, (P)hoto and (S)ketch with seven object categories 'dog', 'elephant', 'giraffe', 'guitar', 'house', 'horse' and 'person' with 9991 images in total. **Setting.** We follow the setting in [7] and perform leave-one-domain out evaluation, setting each domain as the adaptation target in turn. As per [7], we use the ImageNet pre-trained ResNet-18 as our feature extractor for fair comparison. We train with M-SGD (batch size = 32, learning rate = 2×10^{-3}, momentum = 0.9, weight decay = 10^{-4}). All the models are trained for 5k iterations before testing. **Results.** From the results in Table 1, we can see that: (i) Several recent methods with published results on PACS achieve similar performance, with JiGen [7]

Table 1. Multi-source DA results on PACS. Bold: Best. Red: Second Best.

Method	C, P, S ↦ A	A, P, S ↦ C	A, C, S ↦ P	A, C, P ↦ S	Ave.
Source-only	77.85	74.86	95.73	67.74	79.05
DIAL [34]	87.30	85.50	97.00	66.80	84.15
DDiscovery [34]	**87.70**	**86.90**	97.00	69.60	85.30
JiGen [7]	84.88	81.07	**97.96**	**79.05**	85.74
DANN [16]	84.77	83.83	96.29	69.61	83.62
Meta-DANN (Ours)	87.30	84.90	96.89	73.22	85.58
MCD (n = 4) [43]	86.32	84.51	*97.31*	71.01	84.79
Meta-MCD (n = 4) (Ours)	*87.40*	*86.18*	97.13	*78.26*	**87.24**
MCD (os) [43]	85.99	82.89	97.24	74.49	85.15
Meta-MCD (os) (Ours)	86.67	84.94	96.23	77.70	*86.39*

Table 2. Multi-source domain adaptation on office-home.

Method	C, P, R ↦ A	A, P, R ↦ C	A, C, R ↦ P	A, C, P ↦ R	Ave.
Source-only	67.04	56.04	80.74	82.86	71.67
DSBN [8]	–	–	–	83.00	–
M^3SDA-β [38]	67.20	58.58	79.05	81.18	71.50
DANN [16]	68.23	58.90	79.70	83.08	72.48
Meta-DANN (Ours)	**70.62**	59.13	80.24	82.79	73.20
MCD [43]	69.84	59.84	80.92	82.67	73.32
Meta-MCD (Ours)	70.21	**60.50**	**81.17**	**83.43**	**73.83**

performing best. We additionally evaluate DANN and MCD including one-step MCD (os) and multi-step MCD (n = 4) variants, and find that one-step MCD performs similarly to JiGen. (ii) Applying our Meta-DA framework to DANN and MCD boosts all three base domain adaptation methods by 1.96%, 2.5% and 1.2% respectively. (iii) In particular, our Meta-MCD surpasses the previous state of the art performance set by JiGen. Together these results show the broad applicability and absolute efficacy of our method. Based on these results we focus on the better performing single-step MCD in the following evaluations.

Office-Home: Dataset and Settings. Office-Home was initially proposed for the single-source domain adaptation, containing $\approx 15,500$ images from four domains 'artistic', 'clip art', 'product' and 'real-world' with 65 different categories. We follow the setting in [8] and use ImageNet pretrained ResNet-50 as our backbone. We train all models with M-SGD (batch size = 32, learning rate $= 10^{-3}$, momentum = 0.9 and weight decay $= 10^{-4}$) for $3k$ iterations. **Results.** From Table 2, we see that MCD achieves the best performance among the baselines. Applying our meta-learning framework improves both baselines by a small amount, and Meta-MCD achieves state-of-the-art performance on this benchmark.

Table 3. Multi-source domain adaptation on DomainNet dataset.

	Method	inf,pnt,qdr, rel,skt $\mapsto$ clp	clp,pnt,qdr, rel,skt $\mapsto$ inf	clp,inf,qdr, rel,skt $\mapsto$ pnt	clp,inf,qdr rel,skt $\mapsto$ qdr	clp,inf,qdr, qdr,skt $\mapsto$ rel	clp,inf,qdr, qdr,rel $\mapsto$ skt	Ave.
Various Backbones [38]	Source-only	47.6±0.52	13.0±0.41	38.1±0.45	13.3±0.39	51.9±0.85	33.7±0.54	32.9±0.54
	DAN [27]	45.4±0.49	12.8±0.86	36.2±0.58	15.3±0.37	48.6±0.72	34.0±0.54	32.1±0.59
	RTN [29]	44.2±0.57	12.6±0.73	35.3±0.59	14.6±0.76	48.4±0.67	31.7±0.73	31.1±0.68
	JAN [30]	40.9±0.43	11.1±0.61	35.4±0.50	12.1±0.67	45.8±0.59	32.3±0.63	29.6±0.57
	DANN [16]	45.5±0.59	13.1±0.72	37.0±0.69	13.2±0.77	48.9±0.65	31.8±0.62	32.6±0.68
	ADDA [47]	47.5±0.76	11.4±0.67	36.7±0.53	14.7±0.50	49.1±0.82	33.5±0.49	32.2±0.63
	SE [15]	24.7±0.32	3.9±0.47	12.7±0.35	7.1±0.46	22.8±0.51	9.1±0.49	16.1±0.43
	MCD [43]	54.3±0.64	22.1±0.70	45.7±0.63	7.6±0.49	58.4±0.65	43.5±0.57	38.5±0.61
	DCTN [52]	48.6±0.73	23.5±0.59	48.8±0.63	7.2±0.46	53.5±0.56	47.3±0.47	38.2±0.57
	M^3SDA-β [38]	58.6±0.53	26.0±0.89	52.3±0.55	6.3±0.58	62.7±0.51	49.5±0.76	42.6±0.64
ResNet-18	Source-only	56.58±0.16	18.97±0.10	45.95±0.16	11.52±0.15	60.79±0.17	43.70±0.03	39.58±0.09
	DANN [16]	56.34±0.12	18.66±0.09	47.09±0.08	12.27±0.12	61.34±0.07	45.26±0.34	40.16±0.12
	Meta-DANN (Ours)	57.26±0.17	**19.24±0.09**	47.29±0.16	13.38±0.15	61.21±0.13	45.53±0.17	40.65±0.04
	MCD [43]	57.64±0.28	18.71±0.10	**47.82±0.09**	12.64±0.16	**61.69±0.10**	45.61±0.01	40.69±0.05
	Meta-MCD (Ours)	**58.37±0.21**	19.09±0.08	47.63±0.12	**13.70±0.14**	61.30±0.18	**45.90±0.18**	**41.00±0.05**
ResNet-34	Source-only	61.50±0.06	21.10±0.07	49.13±0.06	13.03±0.18	64.14±0.10	48.19±0.12	42.85±0.05
	DANN [16]	60.95±0.05	20.91±0.11	50.35±0.08	14.53±0.06	64.73±0.02	49.88±0.27	43.56±0.04
	Meta-DANN (Ours)	61.39±0.03	**21.53±0.14**	50.49±0.29	15.31±0.28	64.33±0.09	49.87±0.25	43.82±0.07
	MCD [43]	62.21±0.12	20.49±0.08	**50.87±0.10**	14.66±0.30	**64.78±0.06**	50.10±0.11	43.85±0.05
	Meta-MCD (Ours)	**62.81±0.22**	21.37±0.07	50.53±0.08	**15.47±0.22**	64.58±0.16	**50.40±0.12**	**44.19±0.07**

DomainNet: Dataset. DomainNet is a recently benchmark [38] for multi-source domain adaptation in object recognition. It is the largest-scale DA benchmark so far, with ≈ 0.6 m images across six domains and 345 categories.

Settings. We follow the official train/test split protocol [38][2]. Various feature extraction backbones were used in the original paper [38], making it hard to

[2] Other settings such as optimizer, iterations and data augmentation are not clearly stated in [38], making it hard to replicate their results.

Table 4. Semi-supervised domain adaptation: Office-home.

	Method	$R \mapsto C$	$P \mapsto C$	$P \mapsto A$	$A \mapsto C$	$C \mapsto A$	Ave.
AlexNet	S+T	44.6	44.4	36.1	38.8	37.5	40.3
	DANN [16]	47.2	44.4	36.1	39.8	38.6	41.2
	ADR [42]	45.0	38.9	36.3	40.0	37.3	39.5
	CDAN [28]	41.8	35.8	32.0	34.5	27.9	34.4
	ENT [17]	44.9	41.2	34.6	37.8	31.8	38.1
	MME [41]	**51.2**	47.2	**40.7**	43.8	**44.7**	45.5
	Meta-MME (Ours)	50.3	**48.3**	40.3	**44.5**	44.5	**45.6**
ResNet-34	S+T	57.4	54.5	59.9	56.2	57.6	57.1
	ENT [17]	62.8	61.8	65.4	62.1	65.8	63.6
	MME [41]	64.9	63.8	65.0	63.0	66.6	64.7
	Meta-MME (Ours)	**65.2**	**64.5**	**66.7**	**63.3**	**67.5**	**65.4**

Table 5. Semi-supervised DA on DomainNet.

	Method	$R \mapsto C$	$R \mapsto P$	$P \mapsto C$	$C \mapsto S$	$S \mapsto P$	$R \mapsto S$	$P \mapsto R$	Ave.
AlexNet	S+T	47.1	45.0	44.9	36.4	38.4	33.3	58.7	43.4
	DANN[16]	46.1	43.8	41.0	36.5	38.9	33.4	57.3	42.4
	ADR [42]	46.2	44.4	43.6	36.4	38.9	32.4	57.3	42.7
	CDAN [28]	46.8	45.0	42.3	29.5	33.7	31.3	58.7	41.0
	ENT [17]	45.5	42.6	40.4	31.1	29.6	29.6	60.0	39.8
	MME [41]	55.6	49.0	51.7	39.4	43.0	37.9	**60.7**	48.2
	Meta-MME (Ours)	**56.4**	**50.2**	**51.9**	**39.6**	**43.7**	**38.7**	**60.7**	**48.8**
ResNet-34	S+T	60.0	62.2	59.4	55.0	59.5	50.1	73.9	60.0
	DANN [16]	59.8	62.8	59.6	55.4	59.9	54.9	72.2	60.7
	ADR [42]	60.7	61.9	60.7	54.4	59.9	51.1	74.2	60.4
	CDAN [28]	69.0	67.3	68.4	57.8	65.3	59.0	78.5	66.5
	ENT [17]	71.0	69.2	71.1	60.0	62.1	61.1	78.6	67.6
	MME [41]	72.2	69.7	71.7	61.8	66.8	61.9	78.5	68.9
	Meta-MME (Ours)	**73.5**	**70.3**	**72.8**	**62.8**	**68.0**	**63.8**	**79.2**	**70.1**

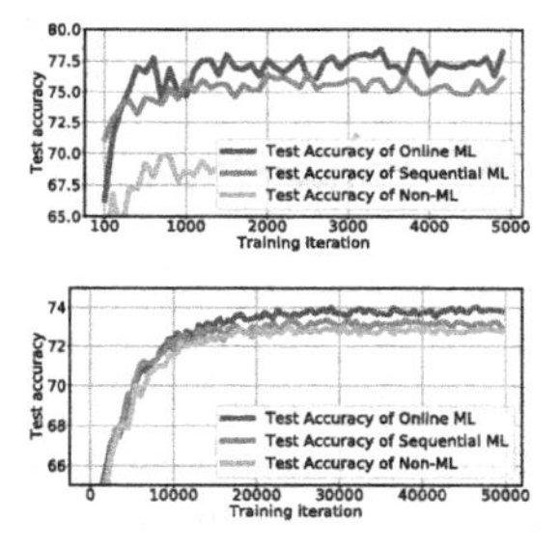

Fig. 2. Vanilla DA vs seq. and online meta (T:MSDA, B:SSDA).

compare results. We use ImageNet pre-trained ResNet-18 and ResNet-34 for our own implementations to facilitate direct comparison. We use M-SGD to train all the competitors (batch size = 32, learning rate = 0.001, momentum = 0.9, weight decay = 0.0001) for $10k$ iterations[3]. We re-train the model three times to generate standard deviations. **Results.** From the results in Table 3, we can see that: (i) The top group of results from [38] show that the dataset is a much more challenging domain adaptation benchmark than previous ones. Most existing domain adaptation methods (typically tuned on small-scale benchmarks) fail to improve over the source-only baseline according to the results in [38]. (ii) The middle

[3] We tried training with up to 50 k, and found it did not lead to clear improvement. So, we train all models for 10 k iterations to minimise cost.

Table 6. Comparison between DG and DA methods on PACS.

Setting	Method	C, P, S→A	A, P, S→C	A, C, S→P	A, C, P→S	Ave.
DG	MetaReg [3]	83.7	77.2	95.5	70.3	81.7
	Epi-FCR [23]	82.1	77.0	93.9	73.0	81.5
	MASF [11]	80.3	77.2	95.0	71.7	81.0
DA	MCD [43]	86.3	84.5	97.3	71.0	84.8
	Meta-MCD	87.4	86.2	97.1	78.3	87.2

group of ResNet-18 results show that our MCD experiment achieves comparable results to those in [38]. (iii) Our Meta-MCD and Meta-DANN methods provide a small but consistent improvement over the corresponding MCD and DANN baselines for both ResNet-18 and ResNet-34 backbones. While the improvement margins are relatively small, this is a significant outcome as the results show that the base DA methods already struggle to make a large improvement over the source-only baseline when using ResNet-18/34; and also the multi-run standard deviation is small compared to the margins. (iv) Overall our Meta-MCD achieves state-of-the-art performance on the benchmark by a small margin.

4.2 Semi-supervised Domain Adaptation

Office-Home: Setting. We follow the setting in [41]. We focus on 3-shot learning in the target domain (three annotated examples only per category), and focus on the five most difficult source-target domain pairs. We use the ImageNet pretrained AlexNet and ResNet-34 as backbone models. We train all the models with M-SGD, with batch size 24 for labelled source and target domains and 48 for the unlabelled target as in [41], learning rate is 10^{-2} and 10^{-3} for the fully-connected and the rest trainable layers. We also use horizontal-flipping and random-cropping data augmentation for training images. **Results.** From the results in Table 4, we can see that our Meta-MME does not impact performance on AlexNet. However, for a modern ResNet-34 architecture, Meta-MME provides a visible ~0.8% accuracy gain over the MME baseline, which results in the state-of-the-art performance of SSDA on this benchmark.

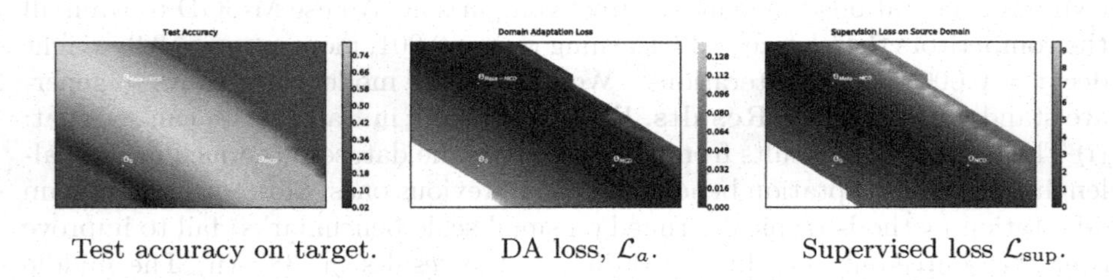

Test accuracy on target. DA loss, $\mathcal{L}_a$. Supervised loss $\mathcal{L}_{\text{sup}}$.

Fig. 3. Performance across weight space slices defined by a common initial condition Θ_0 and MCD and Meta-MCD solutions ($\Theta_{\text{Meta-MCD}}$ and Θ_{MCD} respectively). MSDA PACS benchmark with Sketch target.

DomainNet: Settings. We evaluate DomainNet for 1-1 few-shot domain adaptation as in [41]. We evaluate both AlexNet and modern ResNet-34 backbones, and apply our meta-learning method on MME. As per [41], we train our models using M-SGD where the initial learning rate is 0.01 for the fully-connected layers and 0.001 for the rest of trainable layers. During the training we use the annealing strategy in [16] to decay the learning rate, and use the same batch size as [41]. **Results.** From the results in Table 5, we can see our Meta-MME improves on the accuracy of the base MME algorithm in all pairwise transfer choices, and also for both backbones. These results show the consistent effectiveness of our method, as well as improving state-of-the-art for DomainNet SSDA.

4.3 Further Analysis

Discussion. Our final online algorithm can be understood as performing DA with periodic meta-updates that adjust parameters to optimize the impact of the following DA steps. From the perspective of any given DA step, the role of the preceding meta-update is to tune its initial condition.

Non-meta vs Sequential vs Online Meta. This work is the first to propose meta-learning to improve domain adaptation, and in particular to contribute an efficient and effective online meta-learning algorithm for initial condition training. Exact meta learning is intractable to compare. However, this section we compare our online meta update with the alternative sequential approximation, and non-meta alternatives for both MSDA and SSDA using A, C, P $\rightarrow$ S and R $\rightarrow$ C as examples. For fair comparison, we control the number of meta-updates (UpdateIC) and base DA updates available to both sequential and online meta-learning methods to the same amount. Figure 2 shows that: (1) Sequential meta-learning method already improves the performance on the target domain comparing to vanilla domain adaptation, which confirms the potential for improvement by refining model initialization. (2) The sequential strategy has a slight advantage early in DA training, which makes sense, as all meta-updates occur in advance. But overall our online method that interleaves meta-updates and DA updates leads to higher test accuracy.

Computational Cost. Our Meta-DA imposes only a small computational overhead over the base DA algorithm. For example, comparing Meta-MCD and MCD on ResNet-34 DomainNet, the time per iteration is 2.96s vs 2.49s respectively[4].

Weight-Space Illustration. To investigate our method's mechanism, we train MCD and Meta-MCD from a common initial condition on MSDA PACS when 'Sketch' is the target domain. We use the initial condition Θ_0 and two different solutions ($\Theta_{\text{Meta-MCD}}$ and Θ_{MCD}) to define a plane in weight-space and colour it according to the performance at each point. We can see from Fig. 3(a) that Meta-MCD finds a solution with greater test accuracy. Figures 3(b) and (c) break down the training loss components. We can see that, in this slice, both methods managed to minimize MCD's adaptation (classifier discrepancy) loss

[4] Using GeForce RTX 2080 GPU. Xeon Gold 6130 CPU @ 2.10GHz.

$\mathcal{L}_a$ adequately, but MCD failed to minimize the supervised loss as well as Meta-MCD ($\Theta_{\text{Meta-MCD}}$ is closer to the minima than Θ_{MCD}). Note that both methods were trained to convergence in generating these solutions. This suggests that Meta-MCD's meta-optimization step using meta-train/meta-test splits materially benefits the optimization dynamics of the downstream MSDA task.

Model Agnostic. We emphasize that, although we focused on DANN, MCD and MME, our MetaDA framework can apply to any base DA algorithm. Supplementary C shows some results for JiGen and M^3SDA algorithms.

Comparison Between DA and DG Methods. As a highly related topical problem to domain adaptation, domain generalization assumes no access to the target domain data during the training. DA and DG methods are rarely directly compared. Now we compare our Meta-MCD and MCD with some state of the art DG methods on PACS as shown in Table 6. From the results, we can see that generally DA methods outperform the DG methods with a noticeable margin, which is expected as DA methods 'see' the target domain data at training.

5 Conclusion

We proposed a meta-learning pipeline to improve domain adaptation by initial condition optimization. Our online shortest-path solution is efficient and effective, and provides a consistent boost to several domain adaptation algorithms, improving state of the art in both multi-source and semi-supervised settings. Our approach is agnostic to the base adaptation method, and can potentially be used to improve many DA algorithms that fit a very general template. In future we aim to meta-learn other DA hyper-parameters beyond initial conditions.

A Short-Path Gradient Descent

Optimizing Eq. 3 naively by Algorithm 1 would be costly and ineffective. It is costly because in the case of domain adaptation (unlike for example, few-shot learning [12], the inner loop requires many iterations). So back-propagating through the whole optimization path to update the initial Θ in the outer loop will produce multiple high-order gradients. For example, if the inner loop applies j iterations, we will have

$$\begin{aligned} \Theta^{(1)} &= \Theta - \alpha \nabla_{\Theta^{(0)}} \mathcal{L}_{\text{uda}}(.) \\ &\dots \\ \Theta^{(j)} &= \Theta^{(j-1)} - \alpha \nabla_{\Theta^{(j-1)}} \mathcal{L}_{\text{uda}}(.) \end{aligned} \tag{13}$$

then the outer loop will update the initial condition as

$$\Theta^* = \Theta - \alpha \overbrace{\nabla_{\Theta} \mathcal{L}_{\text{sup}}(\Theta^{(j)}, \mathcal{D}_{\text{val}})}^{\text{Meta Gradient}} \tag{14}$$

where higher-order gradient will be required for all items $\nabla_{\Theta^{(0)}} \mathcal{L}_{\text{uda}}(.)$,..., $\nabla_{\Theta^{(j-1)}} \mathcal{L}_{\text{uda}}(.)$ in the update of Eq. 14.

One intuitive way of eliminating higher-order gradients for computing Eq. 14 is making $\nabla_{\Theta^{(0)}}\mathcal{L}_{\text{uda}}(.), \ldots, \nabla_{\Theta^{(j-1)}}\mathcal{L}_{\text{uda}}(.)$ constant during the optimization. Then, Eq. 14 is equivalent to

$$\Theta^* = \Theta - \alpha \overbrace{\nabla_{\Theta^{(j)}}\mathcal{L}_{\text{sup}}(\Theta^{(j)}, \mathcal{D}_{\text{val}})}^{\text{First-order Meta Gradient}} \tag{15}$$

However, in order to compute Eq. 15, one still needs to store the optimization path of Eq. 13 in memory and back-propagate through it to optimize Θ, which requires high computational load. Therefore, we propose a practical solution an iterative meta-learning algorithm to iteratively optimize the model parameters during training.

Shortest Path Optimization. To obtain the meta gradient in Eq. 15 in a more efficient way, we propose a more scalable and efficient meta-learning method using shortest-path gradient (S-P.G.) [36]. Before the optimization of Eq. 13, we copy the parameters Θ as $\tilde{\Theta}^{(0)}$ and use $\tilde{\Theta}^{(0)}$ in the inner-level algorithm.

$$\tilde{\Theta}^{(j)} = \begin{cases} \tilde{\Theta}^{(0)} - \alpha\nabla_{\Theta^{(0)}}\mathcal{L}_{\text{uda}}(\tilde{\Theta}^{(0)}, \mathcal{D}_{\text{tr}}), \\ \ldots \\ \tilde{\Theta}^{(j-1)} - \alpha\nabla_{\Theta^{(0)}}\mathcal{L}_{\text{uda}}(\tilde{\Theta}^{(j-1)}, \mathcal{D}_{\text{tr}}) \end{cases} \tag{16}$$

then, after finishing the optimization in Eq. 16, we can get the shortest-path gradient between two items $\tilde{\Theta}_i^{(j)}$ and Θ_i.

$$\nabla_{\Theta}^{\text{short}} = \Theta - \tilde{\Theta}^{(j)} \tag{17}$$

Different from Eq. 15, we use this shortest-path gradient $\nabla_{\Theta}^{\text{short}}$ and initial parameter Θ to compute $\mathcal{L}_{\text{sup}}(.)$ as,

$$\mathcal{L}_{\text{sup}}(\Theta_i - \nabla_{\Theta_i}^{\text{short}}, \mathcal{D}_{\text{val}}) \tag{18}$$

Then, one-step meta update of Eq. 18 will be,

$$\begin{aligned} \Theta_i^* &= \Theta_i - \alpha\nabla_{\Theta_i}\mathcal{L}_{\text{sup}}(\Theta_i - \nabla_{\Theta_i}^{\text{short}}, \mathcal{D}_{\text{val}}) \\ &= \Theta_i - \alpha\nabla_{\Theta_i - \nabla_{\Theta_i}^{\text{short}}}\mathcal{L}_{\text{sup}}(\Theta_i - \nabla_{\Theta_i}^{\text{short}}, \mathcal{D}_{\text{val}}) \\ &= \Theta_i - \alpha\nabla_{\tilde{\Theta}_i^{(j)}}\mathcal{L}_{\text{sup}}(\tilde{\Theta}_i^{(j)}, \mathcal{D}_{\text{val}}) \end{aligned} \tag{19}$$

Effectiveness: We can see that one update of Eq. 19 corresponds to that of Eq. 15, which proves that using shortest-path optimization has the equivalent effectiveness to the first-order meta optimization. **Scalability/Efficiency:** The computation memory of the first-order meta-learning increases linearly with the inner-loop update steps, which is constrained by the total GPU memory. However, for the shortest-path optimization, storing the optimization graph is no longer necessary, which makes it scalable and efficient. We also experimentally evaluate that one step shortest-path optimization is 7x faster than one-step first-order meta optimization in our setting. The overall algorithm flow is shown in Algorithm 2.

B Additional Illustrative Schematics

To better explain the contrast between our online meta-learning domain adaptation approach with the sequential meta-learning approach, we add a schematic illustration in Figure 4. The main difference between sequential and online meta-learning approaches is how do we distribute the meta and DA updates. Sequential meta-learning approach performs meta updates and DA updates sequentially. And online meta-learning conducts the alternative meta and DA updates throughout the whole training procedure.

C Additional Experiments

Visualization of the Learned Features. We visualize the learned features of MCD and Meta-MCD on PACS when sketch is the target domain as shown in Fig. 5. We can see that both MCD and Meta-MCD can learn discriminative features. However, the features learned by Meta-MCD is more separable than vanilla MCD. This explains why our Meta-MCD performs better than the vanilla MCD method.

Effect of Varying S. Our online meta-learning method has iteration hyperparameters S and J. We fix $J = 1$ throughout, and analyze the effect of varying S here using the DomainNet MSDA experiment with ResNet-18. The result in Table 7 shows that MetaDA is rather insensitive to this hyperparameter.

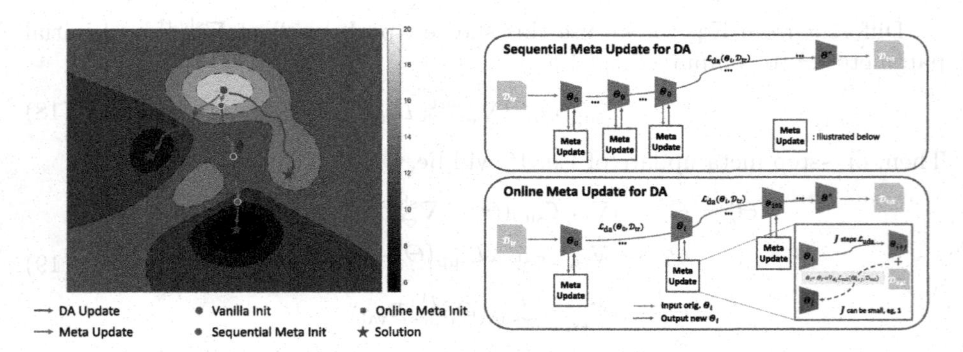

Fig. 4. Illustrative schematics of sequential and online meta domain adaptation. Left: Optimization paths of different approaches on domain adaptation loss (shading). (Solid line) Vanilla gradient descent on a DA objective from a fixed start point. (Multi-segment line) Online meta-learning iterates meta and gradient descent updates. (Two segment line) Sequential meta-learning provides an alternative approximation: update initial condition, then perform gradient descent. Right: (Top) Sequential meta-learning performs meta updates and DA updates sequentially. (Bottom) Online meta-learning alternates between meta-optimization and domain adaptation.

Varying the Number of Source Domains in MSDA. For multi-source DA, the performance of both Meta-DA and the baselines is expected to drop with fewer sources (same for SSDA if fewer labeled target domain points). To disentangle the impact of the number of sources for baseline vs Meta-DA we compare MSDA by Meta-MCD on PACS with 2 vs 3 sources. The results for Meta-MCD vs vanilla MCD are 82.30% vs 80.07% (two source, gap 2.23%) and 87.24% vs 84.79% (three source, gap 2.45%). Meta-DA margin is similar with reduction of domains. Most difference is accounted for by the impact on the base DA algorithm.

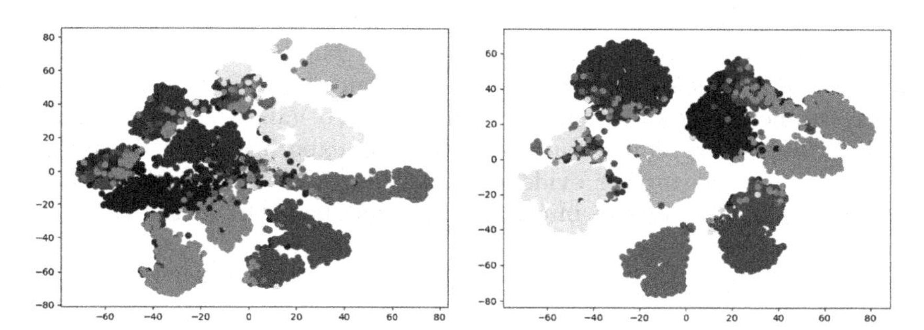

Fig. 5. t-SNE [32] visualization of learned MCD (left) and Meta-MCD (right) features on PACS (sketch as target domain). Different colors indicate different categories.

Table 7. MetaDA is insensitive to the update ratio hyperparameter S – Results for MSDA ResNet-18 performance on DomainNet.

Method	Meta-MCD ($S = 3$)	Meta-MCD ($S = 5$)	Meta-MCD ($S = 10$)
DomainNet (ave.)	41.02	40.98	40.93

Table 8. Test accuracy on PACS. * our run.

Method	C, P, S → A	A, P, S → C	A, C, S → P	A, C, P → S	Ave.
JiGen [7]	84.88	81.07	97.96	79.05	85.74
JiGen*	81.54	85.88	97.25	68.21	83.22
Meta-JiGen	85.21	86.13	97.31	77.91	86.64 (+3.42)

Other Base DA Methods. Besides the base DA methods evaluated in the main paper (DANN, MCD and MME), our method is applicable to any base domain adaptation method. We use the published code of JiGen[5] and M^3SDA[6], and further apply our Meta-DA on the existing code. The results are shown in

[5] https://github.com/fmcarlucci/JigenDG.

[6] https://github.com/VisionLearningGroup/VisionLearningGroup.github.io.

Table 9. Test accuracy on Digit-Five.

Method	mt, up, sv, sy → mm	mm, up, sv, sy → mt	mt, mm, sv, sy → up	mt, mm, up, sy → sv	mt, mm, up, sv →sy	Ave.
M^3SDA-β [38]	72.82	98.43	96.14	81.32	89.58	87.65
Meta-M^3SDA-β	71.73	98.79	97.80	84.81	91.12	88.85 (+1.2)

Table 8 and 9. From the results, we can see that our Meta-JiGen and Meta-M^3SDA-β improves over the base methods by 3.42% and 1.2% accuracy respectively, which confirms our Meta-DA's generality. The reason we excluded these from the main results is that: (i) Re-running JiGen's published code on our compute environment failed to replicate their published numbers. (ii) M^3SDA as a base algorithm is very slow to run comprehensive experiments on. Nevertheless, these results provide further evidence that Meta-DA can be a useful module going forward to plug in and improve future new base DA methods as well as those evaluated here.

Table 10. Test accuracy of MCD on PACS (sketch) with different initialization.

Classifier Init	Kaiming $\mathcal{U}$	Xavier $\mathcal{U}$	Kaiming $\mathcal{N}$	Xavier $\mathcal{N}$
	74.49	73.02	64.27	73.66
Feat. Extr. Init	No perturb	$+\ \epsilon \in \mathcal{N}(0, 0.01)$	$+\ \epsilon \in \mathcal{N}(0, 0.02)$	$+\ \epsilon \in \mathcal{N}(0, 0.03)$
	74.49	71.85	59.99	52.18

Initialization Dependence of Domain Adaptation. One may not think of domain adaptation as being sensitive to initial condition, but given the lack of target domain supervision to guide learning, different initialization can lead to a significant difference in accuracy. To illustrate this we re-ran MCD-based DA on PACS with sketch target using different initializations. From the results in Table 10, we can see that both different classic initialization heuristics, and simple perturbation of a given initial condition with noise can lead to significant differences in final performance. This confirms that studying methods for tuning initialization provide a valid research direction for advancing DA performance.

References

1. Andrychowicz, M., et al.: Learning to learn by gradient descent by gradient descent. In: NeurIPS (2016)
2. Balaji, Y., Chellappa, R., Feizi, S.: Normalized wasserstein distance for mixture distributions with applications in adversarial learning and domain adaptation. In: ICCV (2019)

3. Balaji, Y., Sankaranarayanan, S., Chellappa, R.: Metareg: towards domain generalization using meta-regularization. In: NeurIPS (2018)
4. Ben-David, S., Blitzer, J., Crammer, K., Kulesza, A., Pereira, F., Vaughan, J.W.: A theory of learning from different domains. Mach. Learn. **79**, 151–175 (2010)
5. Ben-David, S., Blitzer, J., Crammer, K., Pereira, F.: Analysis of representations for domain adaptation. In: NeurIPS (2006)
6. Bousmalis, K., Trigeorgis, G., Silberman, N., Krishnan, D., Erhan, D.: Domain separation networks. In: NeurIPS (2016)
7. Carlucci, F.M., D'Innocente, A., Bucci, S., Caputo, B., Tommasi, T.: Domain generalization by solving jigsaw puzzles. In: CVPR (2019)
8. Chang, W.G., You, T., Seo, S., Kwak, S., Han, B.: Domain-specific batch normalization for unsupervised domain adaptation. In: CVPR (2019)
9. Daumé, H.: Frustratingly easy domain adaptation. In: ACL (2007)
10. Donahue, J., Hoffman, J., Rodner, E., Saenko, K., Darrell, T.: Semi-supervised domain adaptation with instance constraints. In: CVPR (2013)
11. Dou, Q., Castro, D.C., Kamnitsas, K., Glocker, B.: Domain generalization via model-agnostic learning of semantic features. In: NeurIPS (2019)
12. Finn, C., Abbeel, P., Levine, S.: Model-agnostic meta-learning for fast adaptation of deep networks. In: ICML (2017)
13. Finn, C., Rajeswaran, A., Kakade, S.M., Levine, S.: Online meta-learning. In: ICML (2019)
14. Franceschi, L., Frasconi, P., Salzo, S., Grazzi, R., Pontil, M.: Bilevel programming for hyperparameter optimization and meta-learning. In: ICML (2018)
15. French, G., Mackiewicz, M., Fisher, M.: Self-ensembling for visual domain adaptation. In: ICLR (2018)
16. Ganin, Y., et al.: Domain-adversarial training of neural networks. In: JMLR (2016)
17. Grandvalet, Y., Bengio, Y.: Semi-supervised learning by entropy minimization. In: NeurIPS (2005)
18. Hoffman, J., et al.: CyCADA: cycle-consistent adversarial domain adaptation. In: ICML (2018)
19. Kim, M., Sahu, P., Gholami, B., Pavlovic, V.: Unsupervised visual domain adaptation: a deep max-margin gaussian process approach. In: CVPR (2019)
20. Lee, C.Y., Batra, T., Baig, M.H., Ulbricht, D.: Sliced wasserstein discrepancy for unsupervised domain adaptation. In: CVPR (2019)
21. Li, D., Yang, Y., Song, Y.Z., Hospedales, T.: Learning to generalize: Meta-learning for domain generalization. In: AAAI (2018)
22. Li, D., Yang, Y., Song, Y.Z., Hospedales, T.M.: Deeper, broader and artier domain generalization. In: ICCV (2017)
23. Li, D., Zhang, J., Yang, Y., Liu, C., Song, Y.Z., Hospedales, T.M.: Episodic training for domain generalization. In: The IEEE International Conference on Computer Vision (ICCV), October 2019
24. Li, Z., Zhou, F., Chen, F., Li, H.: Meta-SGD: learning to learn quickly for few-shot learning. arXiv:1707.09835 (2017)
25. Liu, H., Simonyan, K., Yang, Y.: Darts: differentiable architecture search. In: ICLR (2019)
26. Liu, M.Y., Tuzel, O.: Coupled generative adversarial networks. In: NeurIPS (2016)
27. Long, M., Cao, Y., Wang, J., Jordan, M.I.: Learning transferable features with deep adaptation networks. In: ICML (2015)
28. Long, M., Cao, Z., Wang, J., Jordan, M.I.: Conditional adversarial domain adaptation. In: NeurIPS (2018)

29. Long, M., Zhu, H., Wang, J., Jordan, M.I.: Unsupervised domain adaptation with residual transfer networks. In: NeurIPS (2016)
30. Long, M., Zhu, H., Wang, J., Jordan, M.I.: Deep transfer learning with joint adaptation networks. In: ICML (2017)
31. Luo, Y., Zheng, L., Guan, T., Yu, J., Yang, Y.: Taking a closer look at domain shift: category-level adversaries for semantics consistent domain adaptation. In: CVPR (2019)
32. Maaten, L.V.D., Hinton, G.: Visualizing data using t-SNE. J. Mach. Learn. Res. **9**, 2579–2605 (2008)
33. Maclaurin, D., Duvenaud, D., Adams, R.: Gradient-based hyperparameter optimization through reversible learning. In: ICML (2015)
34. Mancini, M., Porzi, L., Rota Bulò, S., Caputo, B., Ricci, E.: Boosting domain adaptation by discovering latent domains. In: CVPR (2018)
35. Mansour, Y., Mohri, M., Rostamizadeh, A.: Domain adaptation with multiple sources. In: NeurIPS (2009)
36. Nichol, A., Achiam, J., Schulman, J.: On first-order meta-learning algorithms. arXiv:1803.02999 (2018)
37. Parisotto, E., Ghosh, S., Yalamanchi, S.B., Chinnaobireddy, V., Wu, Y., Salakhutdinov, R.: Concurrent meta reinforcement learning. arXiv:1903.02710 (2019)
38. Peng, X., Bai, Q., Xia, X., Huang, Z., Saenko, K., Wang, B.: Moment matching for multi-source domain adaptation. In: CVPR (2019)
39. Rajeswaran, A., Finn, C., Kakade, S., Levine, S.: Meta-learning with implicit gradients. In: NeurIPS (2019)
40. Ravi, S., Larochelle, H.: Optimization as a model for few-shot learning. In: ICLR (2016)
41. Saito, K., Kim, D., Sclaroff, S., Darrell, T., Saenko, K.: Semi-supervised domain adaptation via minimax entropy. In: ICCV (2019)
42. Saito, K., Ushiku, Y., Harada, T., Saenko, K.: Adversarial dropout regularization. In: ICLR (2018)
43. Saito, K., Watanabe, K., Ushiku, Y., Harada, T.: Maximum classifier discrepancy for unsupervised domain adaptation. In: CVPR (2018)
44. Schmidhuber, J.: Learning to control fast-weight memories: an alternative to dynamic recurrent networks. Neural Comput. **4**, 131–139 (1992)
45. Schmidhuber, J., Zhao, J., Wiering, M.: Shifting inductive bias with success-story algorithm, adaptive Levin search, and incremental self-improvement. Mach. Learn. **28**, 105–130 (1997)
46. Thrun, S., Pratt, L. (eds.): Learning to Learn. Kluwer Academic Publishers, Boston (1998)
47. Tzeng, E., Hoffman, J., Saenko, K., Darrell, T.: Adversarial discriminative domain adaptation. In: CVPR (2017)
48. Tzeng, E., Hoffman, J., Zhang, N., Saenko, K., Darrell, T.: Deep domain confusion: Maximizing for domain invariance. arXiv:1412.3474 (2014)
49. Venkateswara, H., Eusebio, J., Chakraborty, S., Panchanathan, S.: Deep hashing network for unsupervised domain adaptation. In: (IEEE) Conference on Computer Vision and Pattern Recognition (CVPR) (2017)
50. Vinyals, O., Blundell, C., Lillicrap, T., Wierstra, D., et al.: Matching networks for one shot learning. In: NeurIPS (2016)
51. Xu, P., Gurram, P., Whipps, G., Chellappa, R.: Wasserstein distance based domain adaptation for object detection. arXiv:1909.08675 (2019)
52. Xu, R., Chen, Z., Zuo, W., Yan, J., Lin, L.: Deep cocktail network: Multi-source unsupervised domain adaptation with category shift. In: CVPR (2018)

53. Xu, Z., van Hasselt, H.P., Silver, D.: Meta-gradient reinforcement learning. In: NeurIPS (2018)
54. Yao, T., Pan, Y., Ngo, C.W., Li, H., Mei, T.: Semi-supervised domain adaptation with subspace learning for visual recognition. In: CVPR (2015)
55. Zhao, H., Zhang, S., Wu, G., Moura, J.M., Costeira, J.P., Gordon, G.J.: Adversarial multiple source domain adaptation. In: NeurIPS (2018)
56. Zheng, Z., Oh, J., Singh, S.: On learning intrinsic rewards for policy gradient methods. In: NeurIPS (2018)

An Ensemble of Epoch-Wise Empirical Bayes for Few-Shot Learning

Yaoyao Liu[1](✉), Bernt Schiele[1], and Qianru Sun[2]

[1] Max Planck Institute for Informatics, Saarland Informatics Campus, Saarbrücken, Germany
{yaoyao.liu,schiele}@mpi-inf.mpg.de
[2] School of Information Systems, Singapore Management University, Singapore, Singapore
qsun@mpi-inf.mpg.de, qianrusun@smu.edu.sg

Abstract. Few-shot learning aims to train efficient predictive models with a few examples. The lack of training data leads to poor models that perform high-variance or low-confidence predictions. In this paper, we propose to meta-learn the ensemble of epoch-wise empirical Bayes models (E^3BM) to achieve robust predictions. "Epoch-wise" means that each training epoch has a Bayes model whose parameters are specifically learned and deployed. "Empirical" means that the hyperparameters, e.g., used for learning and ensembling the epoch-wise models, are generated by hyperprior learners conditional on task-specific data. We introduce four kinds of hyperprior learners by considering inductive *vs.* transductive, and epoch-dependent *vs.* epoch-independent, in the paradigm of meta-learning. We conduct extensive experiments for five-class few-shot tasks on three challenging benchmarks: *mini*ImageNet, *tiered*ImageNet, and FC100, and achieve top performance using the epoch-dependent transductive hyperprior learner, which captures the richest information. Our ablation study shows that both "epoch-wise ensemble" and "empirical" encourage high efficiency and robustness in the model performance (Our code is open-sourced at https://gitlab.mpi-klsb.mpg.de/yaoyaoliu/e3bm).

1 Introduction

The ability of learning new concepts from a handful of examples is well-handled by humans, while in contrast, it remains challenging for machine models whose typical training requires a significant amount of data for good performance [34]. However, in many real-world applications, we have to face the situations of lacking a significant amount of training data, as e.g., in the medical domain. It is thus desirable to improve machine learning models to handle few-shot settings where each new concept has very scarce examples [13, 30, 39, 70].

Electronic supplementary material The online version of this chapter (https://doi.org/10.1007/978-3-030-58517-4_24) contains supplementary material, which is available to authorized users.

A. Vedaldi et al. (Eds.): ECCV 2020, LNCS 12361, pp. 404–421, 2020.
https://doi.org/10.1007/978-3-030-58517-4_24

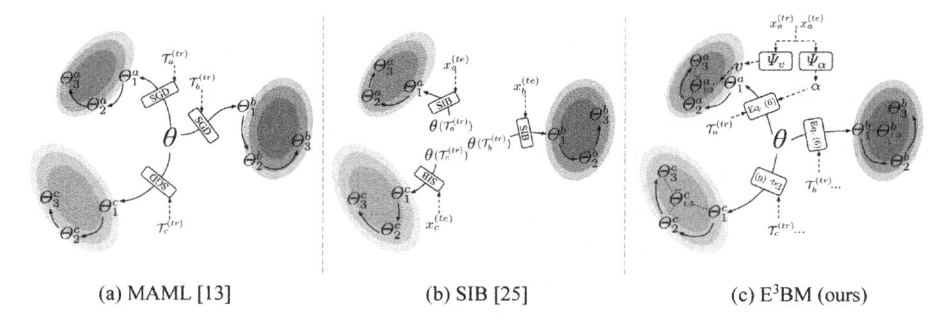

(a) MAML [13] (b) SIB [25] (c) E³BM (ours)

Fig. 1. Conceptual illustrations of the model adaptation on the blue, red and yellow tasks. (a) MAML [13] is the classical inductive method that meta-learns a network initialization θ that is used to learn a single base-learner on each task, e.g., Θ_3^a in the blue task. (b) SIB [25] is a transductive method that formulates a variational posterior as a function of both labeled training data $\mathcal{T}^{(tr)}$ and unlabeled test data $x^{(te)}$. It also uses a single base-learner and optimizes the learner by running several synthetic gradient steps on $x^{(te)}$. (c) Our E^{3}BM is a generic method that learns to combine the epoch-wise base-learners (e.g., Θ_1, Θ_2, and Θ_3), and to generate task-specific learning rates α and combination weights v that encourage robust adaptation. $\bar{\Theta}_{1:3}$ denotes the ensemble result of three base-learners; Ψ_α and Ψ_v denote the hyperprior learners learned to generate α and v, respectively. Note that figure (c) is based on E^{3}BM+MAML, i.e., plug-in our E^{3}BM to MAML baseline. Other plug-in versions are introduced in Sect. 4.4. (Color figure online)

Meta-learning methods aim to tackle the few-shot learning problem by transferring experience from similar few-shot tasks [7]. There are different meta strategies, among which the gradient descent based methods are particularly promising for today's neural networks [1,13–15,20,25,38,70,74,80,82,83,85]. These methods follow a unified meta-learning procedure that contains two loops. The inner loop learns a base-learner for each individual task, and the outer loop uses the validation loss of the base-learner to optimize a meta-learner. In previous works [1,13,14,70], the task of the meta-learner is to initialize the base-learner for the fast and efficient adaptation to the few training samples in the new task.

In this work, we aim to address two shortcomings of the previous works. First, the learning process of a base-learner for few-shot tasks is quite unstable [1], and often results in high-variance or low-confidence predictions. An intuitive solution is to train an ensemble of models and use the combined prediction which should be more robust [6,29,54]. However, it is not obvious how to obtain and combine multiple base-learners given the fact that a very limited number of training examples are available. Rather than learning multiple independent base-learners [79], we propose a novel method of utilizing the sequence of epoch-wise base-learners (while training a single base-learner) as the ensemble. Second, it is well-known that the values of hyperparameters, e.g., for initializing and updating models, are critical for best performance, and are particularly important for few-shot learning. In order to explore the optimal hyperparameters, we propose to employ the empirical Bayes method in the paradigm of meta-learning. In specific, we meta-learn hyperprior learners with meta-training tasks, and use them to generate task-specific hyperparameters, e.g., for updating and ensembling multiple base-learners. We call the resulting novel approach **E^{3}BM**, which learns the **E**nsemble of **E**poch-wise **E**mpirical **B**ayes **M**odels for each few-shot task. Our "epoch-wise models" are *different models* since each one of them is resulted from a specific training epoch and is trained with a specific set of hyperparameter values. During test, E^{3}BM combines the ensemble of models' predictions with soft ensembling weights to produce more robust results. In this paper, we argue that during model adaptation to the few-shot tasks, the most active adapting behaviors actually happen in the early epochs, and then converge to and even overfit to the training data in later epochs. Related works use the single base-learner obtained from the last epoch, so their meta-learners learn only partial adaptation experience [13,14,25,70]. In contrast, our E^{3}BM leverages an ensemble modeling strategy that adapts base-learners at different epochs and each of them has task-specific hyperparameters for updating and ensembling. It thus obtains the optimized combinational adaptation experience. Figure 1 presents the conceptual illustration of E^{3}BM, compared to those of the classical method MAML [13] and the state-of-the-art SIB [25].

Our main contributions are three-fold. (1) A novel few-shot learning approach E^{3}BM that learns to learn and combine an ensemble of epoch-wise Bayes models for more robust few-shot learning. (2) Novel hyperprior learners in E^{3}BM to generate the task-specific hyperparameters for learning and combining epoch-wise Bayes models. In particular, we introduce four kinds of hyperprior learner by considering inductive [13,70] and transductive learning methods [25], and each with either epoch-dependent (e.g., LSTM) or epoch-independent (e.g., epoch-wise FC layer) architectures. (3) Extensive experiments on three challenging few-shot benchmarks, *mini*ImageNet [73], *tiered*ImageNet [58] and Fewshot-CIFAR100 (FC100) [53]. We plug-in our E^{3}BM to the state-of-the-art few-shot learning methods [13,25,70] and obtain consistent performance boosts. We conduct extensive model comparison and observe that our E^{3}BM employing an epoch-dependent transductive hyperprior learner achieves the top performance on all benchmarks.

2 Related Works

Few-Shot Learning & Meta-Learning. Research literature on few-shot learning paradigms exhibits a high diversity from using data augmentation techniques [9,75,77] over sharing feature representation [2,76] to meta-learning [18,72]. In this paper, we focus on the meta-learning paradigm that leverages few-shot learning experiences from similar tasks based on the episodic formulation (see Section 3). Related works can be roughly divided into three categories. (1) *Metric learning methods* [12,24,40,41,64,71,73,78,81] aim to learn a similarity space, in which the learning should be efficient for few-shot examples. The metrics include Euclidean distance [64], cosine distance [8,73], relation module [24,41,71] and graph-based similarity [45,62]. Metric-based task-specific feature representation learning has also been presented in many related works [12,24,41,78]. (2) *Memory network methods* [50,52,53] aim to learn training "experience" from the seen tasks and then aim to generalize to the learning of the unseen ones. A model with external memory storage is designed specifically for fast learning in a few iterations, e.g., Meta Networks [52], Neural Attentive Learner (SNAIL) [50], and Task Dependent Adaptive Metric (TADAM) [53]. (3) *Gradient descent based methods* [1,13,14,20,25,37,38,43,57,70,85] usually employ a meta-learner that learns to fast adapt an NN base-learner to a new task within a few optimization steps. For example, Rusu *et al.* [61] introduced a classifier generator as the meta-learner, which outputs parameters for each specific task. Lee *et al.* [37] presented a meta-learning approach with convex base-learners for few-shot tasks. Finn *et al.* [13] designed a meta-learner called MAML, which learns to effectively initialize the parameters of an NN base-learner for a new task. Sun *et al.* [69,70] introduced an efficient knowledge transfer operator on deeper neural networks and achieved a significant improvement for few-shot learning models. Hu *et al.* [25] proposed to update base-learner with synthetic gradients generated by a variational posterior conditional on unlabeled data. Our approach is closely related to gradient descent based methods [1,13,25,69,70,70]. An important difference is that we learn how to combine an ensemble of epoch-wise base-learners and how to generate efficient hyperparameters for base-learners, while other methods such as MAML [13], MAML++ [1], LEO [61], MTL [69,70], and SIB [25] use a single base-learner.

Hyperparameter Optimization. Building a model for a new task is a process of exploration-exploitation. Exploring suitable architectures and hyperparameters are important before training. Traditional methods are model-free, e.g., based on grid search [4,28,42]. They require multiple full training trials and are thus costly. Model-based hyperparameter optimization methods are adaptive but sophisticated, e.g., using random forests [27], Gaussian processes [65] and input warped Gaussian processes [67] or scalable Bayesian optimization [66]. In our approach, we meta-learn a hyperprior learner to output optimal hyperparameters by gradient descent, without additional manual labor. Related methods using gradient descent mostly work for single model learning in an inductive way [3,10,15,44,46–49]. While, our hyperprior learner generates a sequence of hyperparameters for multiple models, in either the inductive or the transductive learning manner.

Ensemble Modeling. It is a strategy [26,84] to use multiple algorithms to improve machine learning performance, and which is proved to be effective to reduce the problems related to overfitting [35,68]. Mitchell et al. [51] provided a theoretical explanation for it. Boosting is one classical way to build an ensemble, e.g., AdaBoost [16] and Gradient Tree Boosting [17]. Stacking combines multiple models by learning a combiner and it applies to both tasks in supervised learning [6,29,54] and unsupervised learning [63]. Bootstrap aggregating (i.e., Bagging) builds an ensemble of models through parallel training [6], e.g., random forests [22]. The ensemble can also be built on a temporal sequence of models [36]. Some recent works have applied ensemble modeling to few-shot learning. Yoon et al. proposed Bayesian MAML (BMAML) that trains multiple instances of base-model to reduce mete-level overfitting [79]. The most recent work [11] encourages multiple networks to cooperate while keeping predictive diversity. Its networks are trained with carefully-designed penalty functions, different from our automated method using empirical Bayes. Besides, its method needs to train much more network parameters than ours. Detailed comparisons are given in the experiment section.

3 Preliminary

In this section, we introduce the unified episodic formulation of few-shot learning, following [13,57,73]. This formulation was proposed for few-shot classification first in [73]. Its problem definition is different from traditional classification in three aspects: (1) the main phases are not training and test but meta-training and meta-test, each of which includes training and test; (2) the samples in meta-training and meta-testing are not datapoints but episodes, i.e. few-shot classification tasks; and (3) the objective is not classifying unseen datapoints but to fast adapt the meta-learned knowledge to the learning of new tasks.

Given a dataset $\mathcal{D}$ for meta-training, we first sample few-shot episodes (tasks) $\{\mathcal{T}\}$ from a task distribution $p(\mathcal{T})$ such that each episode $\mathcal{T}$ contains a few samples of a few classes, e.g., 5 classes and 1 shot per class. Each episode $\mathcal{T}$ includes a training split $\mathcal{T}^{(tr)}$ to optimize a specific base-learner, and a test split $\mathcal{T}^{(te)}$ to compute a generalization loss to optimize a global meta-learner. For meta-test, given an unseen dataset $\mathcal{D}_{un}$ (i.e., samples are from unseen classes), we sample a test task $\mathcal{T}_{un}$ to have the same-size training/test splits. We first initiate a new model with meta-learned network parameters (output from our hyperprior learner), then train this model on the training split $\mathcal{T}_{un}^{(tr)}$. We finally evaluate the performance on the test split $\mathcal{T}_{un}^{(te)}$. If we have multiple tasks, we report average accuracy as the final result.

4 An Ensemble of Epoch-Wise Empirical Bayes Models

As shown in Fig. 2, E^3BM trains a sequence of epoch-wise base-learners $\{\Theta_m\}$ with training data $\mathcal{T}^{(tr)}$ and learns to combine their predictions $\{z_m^{(te)}\}$ on test data $x^{(te)}$ for the best performance. This ensembling strategy achieves more robustness during prediction. The hyperparameters of each base-learner, i.e., learning rates α and combination weights v, are generated by the hyperprior learners conditional on task-specific data, e.g., $x^{(tr)}$ and $x^{(te)}$. This approach encourages the high diversity and informativeness of the ensembling models.

4.1 Empirical Bayes Method

Our approach can be formulated as an empirical Bayes method that learns two levels of models for a few-shot task. The first level has hyperprior learners that

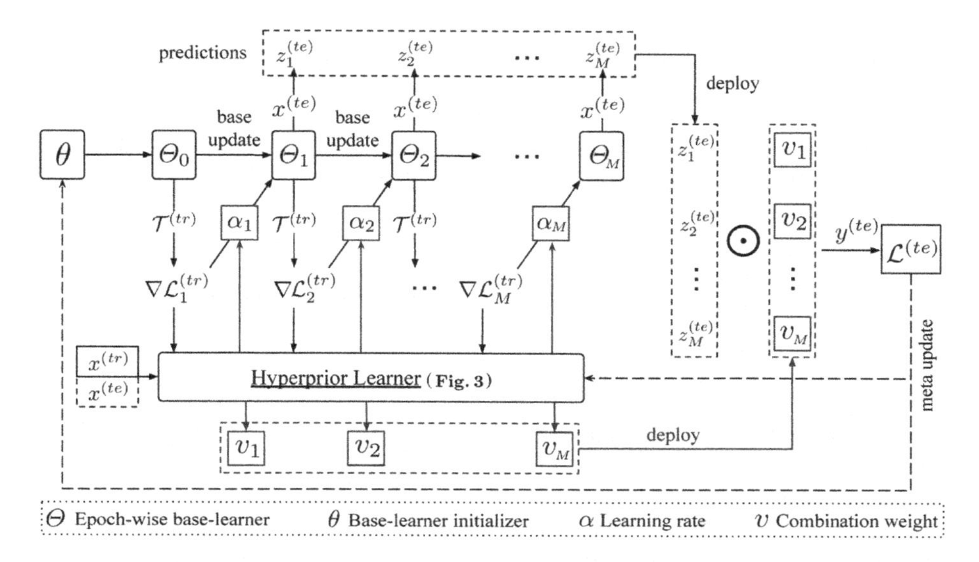

Fig. 2. The computing flow of the proposed E^{3}BM approach in one meta-training episode. For the meta-test task, the computation will be ended with predictions. Hyper-learner predicts task-specific hyperparameters, i.e., learning rates and multi-model combination weights. When its input contains $x^{(te)}$, it is transductive, otherwise inductive. Its detailed architecture is given in Fig. 3.

generate hyperparameters for updating and combining the second-level models. More specifically, these second-level models are trained with the loss derived from the combination of their predictions on training data. After that, their loss of test data are used to optimize the hyperprior learners. This process is also called meta update, see the dashed arrows in Fig. 2.

In specific, we sample K episodes $\{\mathcal{T}_k\}_{k=1}^K$ from the meta-training data $\mathcal{D}$. Let Θ denote base-learner and ψ represent its hyperparameters. An episode $\mathcal{T}_k$ aims to train Θ to recognize different concepts, so we consider to use concepts related (task specific) data for customizing the Θ through a hyperprior $p(\psi_k)$. To achieve this, we first formulate the empirical Bayes method with marginal likelihood according to hierarchical structure among data as follows,

$$p(\mathcal{T}) = \prod_{k=1}^{K} p(\mathcal{T}_k) = \prod_{k=1}^{K} \int_{\psi_k} p(\mathcal{T}_k|\psi_k)p(\psi_k)d\psi_k. \tag{1}$$

Then, we use variational inference [23] to estimate $\{p(\psi_k)\}_{k=1}^K$. We parametrize distribution $q_{\varphi_k}(\psi_k)$ with φ_k for each $p(\psi_k)$, and update φ_k to increase the similarity between $q_{\varphi_k}(\psi_k)$ and $p(\psi_k)$. As in standard probabilistic modeling, we derive an evidence lower bound on the log version of Eq. (1) to update φ_k,

$$\log p(\mathcal{T}) \geqslant \sum_{k=1}^{K} \Big[\mathbb{E}_{\psi_k \sim q_{\varphi_k}}\big[\log p(\mathcal{T}_k|\psi_k)\big] - D_{\mathrm{KL}}(q_{\varphi_k}(\psi_k)||p(\psi_k))\Big]. \tag{2}$$

Therefore, the problem of using $q_{\varphi_k}(\psi_k)$ to approach to the best estimation of $p(\psi_k)$ becomes equivalent to the objective of maximizing the evidence lower bound [5,23,25] in Eq. (2), with respect to $\{\varphi_k\}_{k=1}^K$, as follows,

$$\min_{\{\varphi_k\}_{k=1}^K} \frac{1}{K} \sum_{k=1}^{K} \Big[\mathbb{E}_{\psi_k \sim q_{\varphi_k}}\big[-\log p(\mathcal{T}_k|\psi_k)\big] + D_{\mathrm{KL}}(q_{\varphi_k}(\psi_k)||p(\psi_k))\Big]. \tag{3}$$

To improve the robustness of few-shot models, existing methods sample a significant amount number of episodes during meta-training [13,70]. Each episode employing its own hyperprior $p(\psi_k)$ causes a huge computation burden, making it difficult to solve the aforementioned optimization problem. To tackle this, we leverage a technique called "amortized variational inference" [25,32,59]. We parameterize the KL term in $\{\varphi_k\}_{k=1}^K$ (see Eq. (3)) with a unified deep neural network $\Psi(\cdot)$ taking $x_k^{(tr)}$ (inductive learning) or $\{x_k^{(tr)}, x_k^{(te)}\}$ (transductive learning) as inputs, where $x_k^{(tr)}$ and $x_k^{(te)}$ respectively denote the training and test samples in the k-th episode. In this paper, we call $\Psi(\cdot)$ hyperprior learner. As shown in Fig. 3, we additionally feed the hyperprior learner with the training gradients $\nabla\mathcal{L}_\Theta(\mathcal{T}_k^{(tr)})$ to $\Psi(\cdot)$ to encourage it to "consider" the current state of the training epoch. We mentioned in Sect. 1 that base-learners at different epochs are adapted differently, so we expect the corresponding hyperprior learner to "observe" and "utilize" this information to produce effective hyperparameters. By replacing q_{φ_k} with $q_{\Psi(\cdot)}$, Problem (3) can be rewritten as:

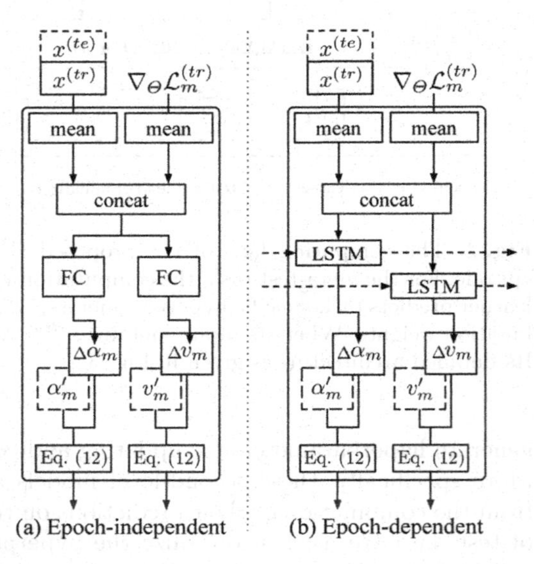

Fig. 3. Two options of hyperprior learner at the m-th base update epoch. In terms of the mapping function, we deploy either FC layers to build epoch-independent hyperprior learners, or LSTM to build an epoch-dependent learner. Values in dashed box were learned from previous tasks.

$$\min_{\Psi} \frac{1}{K} \sum_{k=1}^{K} \Big[\mathbb{E}_{\psi_k \sim q_{\Psi(\cdot)}}\big[-\log p(\mathcal{T}_k|\psi_k)\big] + D_{\mathrm{KL}}(q_{\Psi(\cdot)}(\psi_k)||p(\psi_k))\Big]. \tag{4}$$

Then, we solve Problem (4) by optimizing $\Psi(\cdot)$ with the meta gradient descent method used in classical meta-learning paradigms [13,25,70]. We elaborate the details of learning $\{\Theta_m\}$ and meta-learning $\Psi(\cdot)$ in the following sections.

4.2 Learning the Ensemble of Base-Learners

Previous works have shown that training multiple instances of the base-learner is helpful to achieve robust few-shot learning [12,79]. However, they suffer from the computational burden of optimizing multiple copies of neural networks in parallel, and are not easy to generalize to deeper neural architectures. If include the computation of second-order derivatives in meta gradient descent [13], this burden becomes more unaffordable. In contrast, our approach is free from this problem, because it is built on top of optimization-based meta-learning models, e.g., MAML [13], MTL [70], and SIB [25], which naturally produce a sequence of models along the training epochs in each episode.

Given an episode $\mathcal{T} = \{\mathcal{T}^{(tr)}, \mathcal{T}^{(te)}\} = \{\{x^{(tr)}, y^{(tr)}\}, \{x^{(te)}, y^{(te)}\}\}$, let Θ_m denote the parameters of the base-learner working at epoch m (w.r.t. m-th base-learner or BL-m), with $m \in \{1, ..., M\}$. Basically, we initiate BL-1 with parameters θ (network weights and bias) and hyperparameters (e.g., learning rate α), where θ is meta-optimized as in MAML [13], and α is generated by the proposed hyperprior learner Ψ_α. We then adapt BL-1 with normal gradient descent on the training set $\mathcal{T}^{(tr)}$, and use the adapted weights and bias to initialize BL-2. The general process is thus as follows,

$$\Theta_0 \leftarrow \theta, \tag{5}$$

$$\Theta_m \leftarrow \Theta_{m-1} - \alpha_m \nabla_\Theta \mathcal{L}_m^{(tr)} = \Theta_{m-1} - \Psi_\alpha(\tau, \nabla_\Theta \mathcal{L}_m^{(tr)}) \nabla_\Theta \mathcal{L}_m^{(tr)}, \tag{6}$$

where α_m is the learning rate outputted from Ψ_α, and $\nabla_\Theta \mathcal{L}_m^{(tr)}$ are the derivatives of the training loss, i.e, gradients. τ represents either $x^{(tr)}$ in the inductive setting, or $\{x^{(tr)}, x^{(te)}\}$ in the transductive setting. Note that Θ_0 is introduced to make the notation consistent, and a subscript m is omitted from Ψ_α for conciseness. Let $F(x; \Theta_m)$ denote the prediction scores of input x, so the base-training loss $\mathcal{T}^{(tr)} = \{x^{(tr)}, y^{(tr)}\}$ can be unfolded as,

$$\mathcal{L}_m^{(tr)} = L_{ce}\big(F(x^{(tr)}; \Theta_{m-1}), y^{(tr)}\big), \tag{7}$$

where L_{ce} is the softmax cross entropy loss. During episode test, each base-learner BL-m infers the prediction scores z_m for test samples $x^{(te)}$,

$$z_m = F(x^{(te)}; \Theta_m). \tag{8}$$

Assume the hyperprior learner Ψ_v generates the combination weight v_m for BL-m. The final prediction score is initialized as $\hat{y}_1^{(te)} = v_1 z_1$. For the m-th base epoch, the prediction z_m will be calculated and added to $\hat{y}^{(te)}$ as follows,

$$\hat{y}_m^{(te)} \leftarrow v_m z_m + \hat{y}_{m-1}^{(te)} = \Psi_v(\tau, \nabla_\Theta \mathcal{L}_m^{(tr)}) F(x^{(te)}; \Theta_m) + \hat{y}_{m-1}^{(te)}. \tag{9}$$

In this way, we can update prediction scores without storing base-learners or feature maps in the memory.

4.3 Meta-learning the Hyperprior Learners

As presented in Fig. 3, we introduce two architectures, i.e., LSTM or individual FC layers, for the hyperprior learner. FC layers at different epochs are independent. Using LSTM to "connect" all epochs is expected to "grasp" more task-specific information from the overall training states of the task. In the following, we elaborate the meta-learning details for both designs.

Assume before the k-th episode, we have meta-learned the base learning rates $\{\alpha'_m\}_{m=1}^M$ and combination weights $\{v'_m\}_{m=1}^M$. Next in the k-th episode, specifically at the m-th epoch as shown in Fig. 3, we compute the mean values of τ and $\nabla_{\Theta_m}\mathcal{L}_m^{(tr)}$, respectively, over all samples[1]. We then input the concatenated value to FC or LSTM mapping function as follows,

$$\Delta\alpha_m, \Delta v_m = \text{FC}_m(\text{concat}[\bar{\tau}; \overline{\nabla_{\Theta_m}\mathcal{L}_m^{(tr)}}]), \ \textbf{or} \tag{10}$$

$$[\Delta\alpha_m, \Delta v_m], h_m = \text{LSTM}(\text{concat}[\bar{\tau}; \overline{\nabla_{\Theta_m}\mathcal{L}_m^{(tr)}}], h_{m-1}), \tag{11}$$

where h_m and h_{m-1} are the hidden states at epoch m and epoch $m-1$, respectively. We then use the output values to update hyperparameters as,

$$\alpha_m = \lambda_1\alpha'_m + (1-\lambda_1)\Delta\alpha, \ v_m = \lambda_2 v'_m + (1-\lambda_2)\Delta v, \tag{12}$$

where λ_1 and λ_2 are fixed fractions in $(0,1)$. Using learning rate α_m, we update BL-$(m-1)$ to be BL-m with Eq. (6). After M epochs, we obtain the combination of predictions $\hat{y}_M^{(te)}$ (see Eq. (9)) on test samples. In training tasks, we compute the test loss as,

$$\mathcal{L}^{(te)} = L_{ce}(\hat{y}_M^{(te)}, y^{(te)}). \tag{13}$$

We use this loss to calculate meta gradients to update Ψ as follows,

$$\Psi_\alpha \leftarrow \Psi_\alpha - \beta_1\nabla_{\Psi_\alpha}\mathcal{L}^{(te)}, \ \ \Psi_v \leftarrow \Psi_v - \beta_2\nabla_{\Psi_v}\mathcal{L}^{(te)}, \tag{14}$$

where β_1 and β_2 are meta-learning rates that determine the respective stepsizes for updating Ψ_α and Ψ_v. These updates are to back-propagate the test gradients till the input layer, through unrolling all base training gradients of $\Theta_1 \sim \Theta_M$. The process thus involves a gradient through a gradient [13,14,70]. Computationally, it requires an additional backward pass through $\mathcal{L}^{(tr)}$ to compute Hessian-vector products, which is supported by standard numerical computation libraries such as TensorFlow [19] and PyTorch [55].

4.4 Plugging-In E³BM to Baseline Methods

The optimization of Ψ relies on meta gradient descent method which was first applied to few-shot learning in MAML [13]. Recently, MTL [70] showed more efficiency by implementing that method on deeper pre-trained CNNs (e.g., ResNet-12 [70], and ResNet-25 [69]). SIB [25] was built on even deeper and wider networks (WRN-28-10), and it achieved top performance by synthesizing gradients

[1] In the inductive setting, training images are used to compute $\bar{\tau}$; while in the transductive setting, test images are additionally used.

in transductive learning. These three methods are all optimization-based, and use the single base-learner of the last base-training epoch. In the following, we describe how to learn and combine multiple base-learners in MTL, SIB and MAML, respectively, using our E^3BM approach.

According to [25,70], we pre-train the feature extractor f on a many-shot classification task using the whole set of $\mathcal{D}$. The meta-learner in MTL is called scaling and shifting weights Φ_{SS}, and in SIB is called synthetic information bottleneck network $\phi(\lambda, \xi)$. Besides, there is a common meta-learner called base-learner initializer θ, i.e., the same θ in Fig. 2, in both methods. In MAML, the only base-learner is θ and there is no pre-training for its feature extractor f.

Given an episode $\mathcal{T}$, we feed training images $x^{(tr)}$ and test images $x^{(te)}$ to the feature extractor $f \odot \Phi_{SS}$ in MTL (f in SIB and MAML), and obtain the embedding $e^{(tr)}$ and $e^{(te)}$, respectively. Then in MTL, we use $e^{(tr)}$ with labels to train base-learner Θ for M times to get $\{\Theta_m\}_{m=1}^{M}$ with Eq. (6). In SIB, we use its multilayer perceptron (MLP) net to synthesize gradients conditional on $e^{(te)}$ to indirectly update $\{\Theta_m\}_{m=1}^{M}$. During these updates, our hyperprior learner Ψ_α derives the learning rates for all epochs. In episode test, we feed $e^{(te)}$ to $\{\Theta_m\}_{m=1}^{M}$ and get the combined prediction $\{z_m\}_{m=1}^{M}$ with Eq. (9). Finally, we compute the test loss to meta-update $[\Psi_\alpha; \Psi_v; \Phi_{SS}; \theta]$ in MTL, $[\Psi_\alpha; \Psi_v; \phi(\lambda, \xi); \theta]$ in SIB, and $[f; \theta]$ in MAML. We call the resulting methods MTL+E^3BM, SIB+E^3BM, and MAML+E^3BM, respectively, and demonstrate their improved efficiency over baseline models [13,25,70] in experiments.

5 Experiments

We evaluate our approach in terms of its overall performance and the effects of its two components, i.e. ensembling epoch-wise models and meta-learning hyperprior learners. In the following sections, we introduce the datasets and implementation details, compare our best results to the state-of-the-art, and conduct an ablation study.

5.1 Datasets and Implementation Details

Datasets. We conduct few-shot image classification experiments on three benchmarks: *mini*ImageNet [73], *tiered*ImageNet [58] and FC100 [53]. *mini*ImageNet is the most widely used in related works [13,24,25,25,70,71]. *tiered*ImageNet and FC100 are either with a larger scale or a more challenging setting with lower image resolution, and have stricter training-test splits.
***mini*ImageNet** was proposed in [73] based on ImageNet [60]. There are 100 classes with 600 samples per class. Classes are divided into 64, 16, and 20 classes respectively for sampling tasks for meta-training, meta-validation and meta-test. ***tiered*ImageNet** was proposed in [58]. It contains a larger subset of ImageNet [60] with 608 classes (779, 165 images) grouped into 34 super-class nodes. These nodes are partitioned into 20, 6, and 8 disjoint sets respectively for meta-training, meta-validation and meta-test. Its super-class based training-test split

results in a more challenging and realistic regime with test tasks that are less similar to training tasks. **FC100** is based on the CIFAR100 [33]. The few-shot task splits were proposed in [53]. It contains 100 object classes and each class has 600 samples of 32×32 color images per class. On these datasets, we consider the (5-class, 1-way) and (5-class, 5-way) classification tasks. We use the same task sampling strategy as in related works [1,13,25].

Backbone Architectures. In MAML+E^{3}BM, we use a 4-layer convolution network (4CONV) [1,13]. In MTL+E^{3}BM, we use a 25-layer residual network (ResNet-25) [56,69,78]. Followed by convolution layers, we apply an average pooling layer and a fully-connected layer. In SIB+E^{3}BM, we use a 28-layer wide residual network (WRN-28-10) as SIB [25].

The Configuration of Base-Learners. In MTL [70] and SIB [25], the base-learner is a single fully-connected layer. In MAML [13], the base-learner is the 4-layer convolution network. In MTL and MAML, the base-learner is randomly initialized and updated during meta-learning. In SIB, the base-learner is initialized with the averaged image features of each class. The number of base-learners M in MTL+E^{3}BM and SIB+E^{3}BM are respectively 100 and 3, i.e., the original numbers of training epochs in [25,70].

The Configuration of Hyperprior Learners. In Fig. 3, we show two options for hyperprior learners (i.e., Ψ_α and Ψ_v). Figure 3(a) is the epoch-independent option, where each epoch has two FC layers to produce α and v respectively. Figure 3(b) is the epoch-dependent option which uses an LSTM to generate α and v at all epochs. In terms of the learning hyperprior learners, we have two settings: inductive learning denoted as "Ind.", and transductive learning as "Tra.". "Ind." is the supervised learning in classical few-shot learning methods [13,37,64,70,73]. "Tra." is semi-supervised learning, based on the assumption that all test images of the episode are available. It has been applied to many recent works [24,25,45].

Ablation Settings. We conduct a careful ablative study for two components, i.e., "ensembling multiple base-learners" and "meta-learning hyperprior learners". We show their effects indirectly by comparing our results to those of using arbitrary constant or learned values of v and α. **In terms of** v, we have 5 ablation options: (v1) "E^{3}BM" is our method generating v from Ψ_v; (v2) "learnable" is to set v to be update by meta gradient descent same as θ in [13]; (v3) "optimal" means using the values learned by option (a2) and freezing them during the actual learning; (v4) "equal" is an simple baseline using equal weights; (v5) "last-epoch" uses only the last-epoch base-learner, i.e., v is set to $[0, 0, ..., 1]$. In the experiments of (v1)-(v5), we simply set α as in the following (a4) [13,25,70]. **In terms of** α, we have 4 ablation options: (a1) "E^{3}BM" is our method generating α from Ψ_α; (a2) "learnable" is to set α to be update by meta gradient descent same as θ in [13]; (a3) "optimal" means using the values learned by option (a2) and freezing them during the actual learning; (a4) "fixed" is a simple baseline that uses manually chosen α following [13,25,70]. In the experiments of (a1)-(a4), we simply set v as in (v5), same with the baseline method [70].

Table 1. The 5-class few-shot classification accuracies (%) on *mini*ImageNet, *tiered* ImageNet, and FC100. "(+time, +param)" denote the additional computational time (%) and parameter size (%), respectively, when plugging-in E^{3}BM to baselines (MAML, MTL and SIB). "–" means no reported results in original papers. The **best** and second best results are highlighted.

Methods	Backbone	*mini*ImageNet		*tiered*ImageNet		FC100	
		1-shot	5-shot	1-shot	5-shot	1-shot	5-shot
MatchNets [73]	4CONV	43.44	55.31	–	–	–	–
ProtoNets [64]	4CONV	49.42	68.20	53.31	72.69	–	–
MAML$^\diamond$ [13]	4CONV	48.70	63.11	49.0	66.5	38.1	50.4
MAML++$^\diamond$ [1]	4CONV	52.15	68.32	51.5	70.6	38.7	52.9
TADAM [53]	ResNet-12	58.5	76.7	–	–	40.1	56.1
MetaOptNet [37]	ResNet-12	62.64	78.63	65.99	81.56	41.1	55.5
CAN [24]	ResNet-12	63.85	79.44	69.89	84.23	–	–
CTM [40]	ResNet-18	64.12	80.51	68.41	84.28	–	–
MTL [70]	ResNet-12	61.2	75.5	–	–	**45.1**	57.6
MTL$^\diamond$ [70]	ResNet-25	63.4	80.1	69.1	84.2	43.7	60.1
LEO [61]	WRN-28-10	61.76	77.59	66.33	81.44	–	–
Robust20-dist‡ [12]	WRN-28-10	63.28	**81.17**	–	–	–	–
MAML+E^{3}BM	4CONV	53.2($\uparrow$4.5)	65.1($\uparrow$2.0)	52.1($\uparrow$3.1)	70.2($\uparrow$3.7)	39.9($\uparrow$1.8)	52.6($\uparrow$2.2)
(+time, +param)	–	(8.9, 2.2)	(9.7, 2.2)	(10.6, 2.2)	(9.3, 2.2)	(7.8, 2.2)	(12.1, 2.2)
MTL+E^{3}BM	ResNet-25	**64.3**($\uparrow$0.9)	81.0($\uparrow$0.9)	**70.0**($\uparrow$0.9)	**85.0**($\uparrow$0.8)	45.0($\uparrow$1.3)	**60.5**($\uparrow$0.4)
(+time, +param)	–	(5.9, 0.7)	(10.2, 0.7)	(6.7, 0.7)	(9.5, 0.7)	(5.7, 0.7)	(7.9, 0.7)

(a) Inductive Methods

Methods	Backbone	*mini*ImageNet 1-shot	*mini*ImageNet 5-shot	*tiered*ImageNet 1-shot	*tiered*ImageNet 5-shot	FC100 1-shot	FC100 5-shot
EGNN [31]	ResNet-12	64.02	77.20	65.45	82.52	–	–
CAN+T [24]	ResNet-12	67.19	80.64	73.21	**84.93**	–	–
SIB$^{\diamond\ddagger}$ [25]	WRN-28-10	70.0	79.2	72.9	82.8	45.2	55.9
SIB+E^{3}BM‡	WRN-28-10	**71.4**($\uparrow$1.4)	**81.2**($\uparrow$2.0)	**75.6**($\uparrow$2.7)	84.3($\uparrow$1.5)	**46.0**($\uparrow$0.8)	**57.1**($\uparrow$1.2)
(+time, +param)	–	(2.1, 0.04)	(5.7, 0.04)	(5.2, 0.04)	(4.9, 0.04)	(6.1, 0.04)	(7.3, 0.04)

(b) Transductive Methods

$^\diamond$Our implementation on *tiered*ImageNet and FC100.
‡Input image size: 80 × 80 × 3.

5.2 Results and Analyses

In Table 1, we compare our best results to the state-of-the-arts. In Table 2, we present the results of using different kinds of hyperprior learner, i.e., regarding two architectures (FC and LSTM) and two learning strategies (inductive and transductive). In Fig. 4(a)(b), we show the validation results of our ablative methods, and demonstrate the change during meta-training iterations. In Fig. 4(c)(d), we plot the generated values of v and α during meta-training.

Comparing to the State-of-the-Arts. Table 1 shows that the proposed E^{3}BM achieves the best few-shot classification performance in both 1-shot and 5-shot settings, on three benchmarks. Please note that [12] reports the results of using different backbones and input image sizes. We choose its results under the same setting of ours, i.e., using WRN-28-10 networks and 80 × 80 × 3 images, for fair comparison. In our approach, plugging-in E^{3}BM to the state-of-the-art model SIB achieves 1.6% of improvement on average, based on the identical network architecture. This improvement is significantly larger as 2.9% when taking MAML as baseline. All these show to be more impressive if considering the tiny

Table 2. The 5-class few-shot classification accuracies (%) of using different hyperprior learners, on the *mini*ImageNet, *tiered*ImageNet, and FC100. "Ind." and "Tra." denote the inductive and transductive settings, respectively. The **best** and second best results are highlighted.

No.	Setting			*mini*ImageNet		*tiered*ImageNet		FC100	
	Method	Hyperprior	Learning	1-shot	5-shot	1-shot	5-shot	1-shot	5-shot
1	MTL [70]	–	Ind	63.4	80.1	69.1	84.2	43.7	60.1
2	MTL+E^{3}BM	FC	Ind	64.3	80.9	69.8	84.6	44.8	60.5
3	MTL+E^{3}BM	FC	Tra	64.7	80.7	69.7	84.9	44.7	**60.6**
4	MTL+E^{3}BM	LSTM	Ind	64.3	81.0	70.0	85.0	45.0	60.4
5	MTL+E^{3}BM	LSTM	Tra	64.5	81.1	70.2	**85.3**	45.1	**60.6**
6	SIB [25]	–	Tra	70.0	79.2	72.9	82.8	45.2	55.9
7	SIB+E^{3}BM	FC	Tra	71.3	81.0	75.2	83.8	45.8	56.3
8	SIB+E^{3}BM	LSTM	Tra	**71.4**	**81.2**	**75.6**	84.3	**46.0**	57.1

overheads from pluging-in. For example, using E^{3}BM adds only 0.04% learning parameters to the original SIB model, and it gains only 5.2% average overhead regarding the computational time. It is worth mentioning that the amount of learnable parameters in SIB+E^{3}BM is around 80% less than that of model in [12] which ensembles 5 deep networks in parallel (and later learns a distillation network).

Hyperprior Learners. In Table 2, we can see that using transductive learning clearly outperforms inductive learning, e.g., No. 5 *vs.* No. 4. This is because the "transduction" leverages additional data, i.e., the episode-test images (no labels), during the base-training. In terms of the network architecture, we observe that LSTM-based learners are slightly better than FC-based (e.g., No. 3 *vs.* No. 2). LSTM is a sequential model and is indeed able to "observe" more patterns from the adaptation behaviors of models at adjacent epochs.

Ablation Study. Figure 4(a) shows the comparisons among α related ablation models. Our E^{3}BM (orange) again performs the best, over the models of using any arbitrary α (**red** or light blue), as well as over the model with α optimized by the meta gradient descent (blue) [13]. Figure 4(b) shows that our approach E^{3}BM works consistently better than the ablation models related to v. We should emphasize that E^{3}BM is clearly more efficient than the model trained with meta-learned v (blue) through meta gradient descent [13]. This is because E^{3}BM hyperprior learners generate empirical weights conditional on task-specific data. The LSTM-based learners can leverage even more task-specific information, i.e., the hidden states from previous epochs, to improve the efficiency.

The values of α and v learned by E^{3}BM. Fig. 4(c) (d) shows the values of α and v during the meta-training iterations in our approach. Figure 4(c) show the base-learners working at later training epochs (e.g., BL-100) tend to get smaller values of α. This is actually similar to the common manual schedule, i.e. monotonically decreasing learning rates, of conventional large-scale network

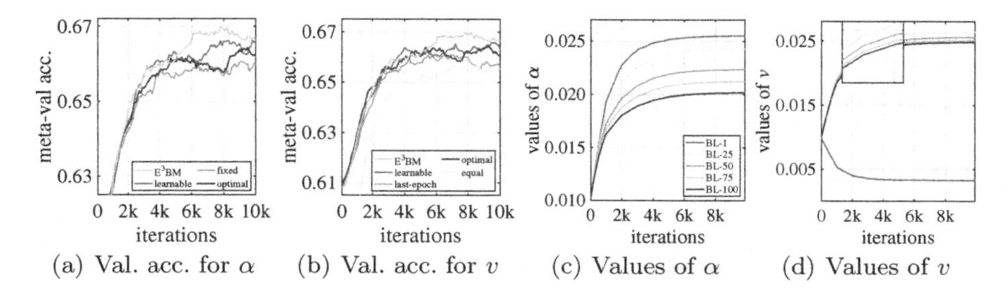

(a) Val. acc. for α (b) Val. acc. for v (c) Values of α (d) Values of v

Fig. 4. (a) (b): The meta-validation accuracies of ablation models. The legends are explained in (a1)–(a4) and (v1)–(v5) in Sect. 5.1 **Ablation settings**. All curves are smoothed with a rate of 0.9 for a better visualization. (c) (d): The values of α and v generated by Ψ_α and Ψ_v, respectively. The setting is using MTL+E^{3}BM, ResNet-25, on *mini*ImageNet, 1-shot. (Color figure online)

training [21]. The difference is that in our approach, this is "scheduled" in a total automated way by hyperprior learners. Another observation is that the highest learning rate is applied to BL-1. This actually encourages BL-1 to make an influence as significant as possible. It is very helpful to reduce meta gradient diminishing when unrolling and back-propagating gradients through many base-learning epochs (e.g., 100 epochs in MTL). Figure 4(d) shows that BL-1 working at the initial epoch has the lowest values of v. In other words, BL-1 is almost disabled in the prediction of episode test. Intriguingly, BL-25 instead of BL-100 gains the highest v values. Our explanation is that during the base-learning, base-learners at latter epochs get more overfitted to the few training samples. Their functionality is thus suppressed. Note that our empirical results revealed that including the overfitted base-learners slightly improves the generalization capability of the approach.

6 Conclusions

We propose a novel E^{3}BM approach that tackles the few-shot problem with an ensemble of epoch-wise base-learners that are trained and combined with task-specific hyperparameters. In specific, E^{3}BM meta-learns the hyperprior learners to generate such hyperparameters conditional on the images as well as the training states for each episode. Its resulting model allows to make use of multiple base-learners for more robust predictions. It does not change the basic training paradigm of episodic few-shot learning, and is thus *generic* and easy to plug-and-play with existing methods. By applying E^{3}BM to multiple baseline methods, e.g., MAML, MTL and SIB, we achieved top performance on three challenging few-shot image classification benchmarks, with little computation or parametrization overhead.

Acknowledgments. This research was supported by the Singapore Ministry of Education (MOE) Academic Research Fund (AcRF) Tier 1 grant. We thank all reviewers and area chairs for their constructive suggestions.

References

1. Antoniou, A., Edwards, H., Storkey, A.: How to train your MAML. In: ICLR (2019)
2. Bart, E., Ullman, S.: Cross-generalization: learning novel classes from a single example by feature replacement. In: CVPR, pp. 672–679 (2005)
3. Bengio, Y.: Gradient-based optimization of hyperparameters. Neural Comput. **12**(8), 1889–1900 (2000)
4. Bergstra, J., Bengio, Y.: Random search for hyper-parameter optimization. J. Mach. Learn. Res. **13**, 281–305 (2012)
5. Blei, D.M., Kucukelbir, A., McAuliffe, J.D.: Variational inference: a review for statisticians. J. Am. Stat. Assoc. **112**(518), 859–877 (2017)
6. Breiman, L.: Stacked regressions. Mach. Learn. **24**(1), 49–64 (1996)
7. Caruana, R.: Learning many related tasks at the same time with backpropagation. In: NIPS, pp. 657–664 (1995)
8. Chen, W.Y., Liu, Y.C., Kira, Z., Wang, Y.C., Huang, J.B.: A closer look at few-shot classification. In: ICLR (2019)
9. Chen, Z., Fu, Y., Zhang, Y., Jiang, Y., Xue, X., Sigal, L.: Multi-level semantic feature augmentation for one-shot learning. IEEE Trans. Image Process. **28**(9), 4594–4605 (2019)
10. Domke, J.: Generic methods for optimization-based modeling. In: AISTATS, pp. 318–326 (2012)
11. Dvornik, N., Schmid, C., Julien, M.: f-VAEGAN-D2: A feature generating framework for any-shot learning. In: ICCV, pp. 10275–10284 (2019)
12. Dvornik, N., Schmid, C., Mairal, J.: Diversity with cooperation: Ensemble methods for few-shot classification. In: ICCV, pp. 3722–3730 (2019)
13. Finn, C., Abbeel, P., Levine, S.: Model-agnostic meta-learning for fast adaptation of deep networks. In: ICML, pp. 1126–1135 (2017)
14. Finn, C., Xu, K., Levine, S.: Probabilistic model-agnostic meta-learning. In: NeurIPS, pp. 9537–9548 (2018)
15. Franceschi, L., Frasconi, P., Salzo, S., Grazzi, R., Pontil, M.: Bilevel programming for hyperparameter optimization and meta-learning. In: ICML, pp. 1563–1572 (2018)
16. Freund, Y., Schapire, R.E.: A decision-theoretic generalization of on-line learning and an application to boosting. J. Comput. Syst. Sci. **55**(1), 119–139 (1997)
17. Friedman, J.H.: Stochastic gradient boosting. Comput. Stat. Data Anal. **38**(4), 367–378 (2002)
18. Geoffrey, H.E., David, P.C.: Using fast weights to deblur old memories. In: CogSci, pp. 177–186 (1987)
19. Girija, S.S.: TensorFlow: large-scale machine learning on heterogeneous distributed systems. **39** (2016). tensorflow.org
20. Grant, E., Finn, C., Levine, S., Darrell, T., Griffiths, T.L.: Recasting gradient-based meta-learning as hierarchical Bayes. In: ICLR (2018)
21. He, T., Zhang, Z., Zhang, H., Zhang, Z., Xie, J., Li, M.: Bag of tricks for image classification with convolutional neural networks. In: CVPR, pp. 558–567 (2019)
22. Ho, T.K.: Random decision forests. In: ICDAR, vol. 1, pp. 278–282 (1995)

23. Hoffman, M.D., Blei, D.M., Wang, C., Paisley, J.: Stochastic variational inference. J. Mach. Learn. Res. **14**(1), 1303–1347 (2013)
24. Hou, R., Chang, H., Bingpeng, M., Shan, S., Chen, X.: Cross attention network for few-shot classification. In: NeurIPS, pp. 4005–4016 (2019)
25. Hu, S.X., et al.: Empirical Bayes meta-learning with synthetic gradients. In: ICLR (2020)
26. Huang, G., Li, Y., Pleiss, G., Liu, Z., Hopcroft, J.E., Weinberger, K.Q.: Snapshot ensembles: Train 1, get M for free. In: ICLR (2017)
27. Hutter, F., Hoos, H.H., Leyton-Brown, K.: Sequential model-based optimization for general algorithm configuration. In: Coello, C.A.C. (ed.) LION 2011. LNCS, vol. 6683, pp. 507–523. Springer, Heidelberg (2011). https://doi.org/10.1007/978-3-642-25566-3_40
28. Jaderberg, M., et al.: Population based training of neural networks. arXiv:1711.09846 (2017)
29. Ju, C., Bibaut, A., van der Laan, M.: The relative performance of ensemble methods with deep convolutional neural networks for image classification. J. Appl. Stat. **45**(15), 2800–2818 (2018)
30. Jung, H.G., Lee, S.W.: Few-shot learning with geometric constraints. IEEE Trans. Neural Netw. Learn. Syst. (2020)
31. Kim, J., Kim, T., Kim, S., Yoo, C.D.: Edge-labeling graph neural network for few-shot learning. In: CVPR, pp. 11–20 (2019)
32. Kingma, D.P., Welling, M.: Auto-encoding variational Bayes. In: ICLR (2014)
33. Krizhevsky, A.: Learning multiple layers of features from tiny images. University of Toronto (2009)
34. Krizhevsky, A., Sutskever, I., Hinton, G.E.: ImageNet classification with deep convolutional neural networks. In: NIPS, pp. 1097–1105 (2012)
35. Kuncheva, L.I., Whitaker, C.J.: Measures of diversity in classifier ensembles and their relationship with the ensemble accuracy. Mach. Learn. **51**(2), 181–207 (2003)
36. Laine, S., Aila, T.: Temporal ensembling for semi-supervised learning. In: ICLR (2017)
37. Lee, K., Maji, S., Ravichandran, A., Soatto, S.: Meta-learning with differentiable convex optimization. In: CVPR, pp. 10657–10665 (2019)
38. Lee, Y., Choi, S.: Gradient-based meta-learning with learned layerwise metric and subspace. In: ICML, pp. 2933–2942 (2018)
39. Li, F., Fergus, R., Perona, P.: One-shot learning of object categories. IEEE Trans. Pattern Anal. Mach. Intell. **28**(4), 594–611 (2006)
40. Li, H., Eigen, D., Dodge, S., Zeiler, M., Wang, X.: Finding task-relevant features for few-shot learning by category traversal. In: CVPR, pp. 1–10 (2019)
41. Li, H., Dong, W., Mei, X., Ma, C., Huang, F., Hu, B.: LGM-Net: learning to generate matching networks for few-shot learning. In: ICML, pp. 3825–3834 (2019)
42. Li, L., Jamieson, K.G., DeSalvo, G., Rostamizadeh, A., Talwalkar, A.: Hyperband: a novel bandit-based approach to hyperparameter optimization. J. Mach. Learn. Res. **18**, 185:1–185:52 (2017)
43. Li, X., et al.: Learning to self-train for semi-supervised few-shot classification. In: NeurIPS, pp. 10276–10286 (2019)
44. Li, Z., Zhou, F., Chen, F., Li, H.: Meta-SGD: learning to learn quickly for few shot learning. arXiv:1707.09835 (2017)
45. Liu, Y., Lee, J., Park, M., Kim, S., Yang, Y.: Learning to propagate labels: transductive propagation network for few-shot learning. In: ICLR (2019)
46. Liu, Y., Su, Y., Liu, A.A., Schiele, B., Sun, Q.: Mnemonics training: multi-class incremental learning without forgetting. In: CVPR, pp. 12245–12254 (2020)

47. Luketina, J., Raiko, T., Berglund, M., Greff, K.: Scalable gradient-based tuning of continuous regularization hyperparameters. In: ICML, pp. 2952–2960 (2016)
48. Maclaurin, D., Duvenaud, D.K., Adams, R.P.: Gradient-based hyperparameter optimization through reversible learning. In: ICML, pp. 2113–2122 (2015)
49. Metz, L., Maheswaranathan, N., Cheung, B., Sohl-Dickstein, J.: Meta-learning update rules for unsupervised representation learning. In: ICLR (2019)
50. Mishra, N., Rohaninejad, M., Chen, X., Abbeel, P.: Snail: a simple neural attentive meta-learner. In: ICLR (2018)
51. Mitchell, T.: Machine Learning. Mcgraw-Hill Higher Education, New York (1997)
52. Munkhdalai, T., Yu, H.: Meta networks. In: ICML, pp. 2554–2563 (2017)
53. Oreshkin, B.N., Rodríguez, P., Lacoste, A.: TADAM: task dependent adaptive metric for improved few-shot learning. In: NeurIPS, pp. 719–729 (2018)
54. Ozay, M., Vural, F.T.Y.: A new fuzzy stacked generalization technique and analysis of its performance. arXiv:1204.0171 (2012)
55. Paszke, A., et al.: PyTorch: an imperative style, high-performance deep learning library. In: NeurIPS, pp. 8024–8035 (2019)
56. Qiao, S., Liu, C., Shen, W., Yuille, A.L.: Few-shot image recognition by predicting parameters from activations. In: CVPR, pp. 7229–7238 (2018)
57. Ravi, S., Larochelle, H.: Optimization as a model for few-shot learning. In: ICLR (2017)
58. Ren, M., et al.: Meta-learning for semi-supervised few-shot classification. In: ICLR (2018)
59. Rezende, D.J., Mohamed, S., Wierstra, D.: Stochastic backpropagation and approximate inference in deep generative models. In: ICML, pp. 1278–1286 (2014)
60. Russakovsky, O., et al.: ImageNet large scale visual recognition challenge. Int. J. Comput. Vision **115**(3), 211–252 (2015)
61. Rusu, A.A., et al.: Meta-learning with latent embedding optimization. In: ICLR (2019)
62. Satorras, V.G., Estrach, J.B.: Few-shot learning with graph neural networks. In: ICLR (2018)
63. Smyth, P., Wolpert, D.: Linearly combining density estimators via stacking. Mach. Learn. **36**(1–2), 59–83 (1999)
64. Snell, J., Swersky, K., Zemel, R.S.: Prototypical networks for few-shot learning. In: NIPS, pp. 4077–4087 (2017)
65. Snoek, J., Larochelle, H., Adams, R.P.: Practical Bayesian optimization of machine learning algorithms. In: NIPS, pp. 2951–2959 (2012)
66. Snoek, J., et al.: Scalable Bayesian optimization using deep neural networks. In: ICML, pp. 2171–2180 (2015)
67. Snoek, J., Swersky, K., Zemel, R.S., Adams, R.P.: Input warping for Bayesian optimization of non-stationary functions. In: ICML, pp. 1674–1682 (2014)
68. Sollich, P., Krogh, A.: Learning with ensembles: how overfitting can be useful. In: NIPS, pp. 190–196 (1996)
69. Sun, Q., Liu, Y., Chen, Z., Chua, T., Schiele, B.: Meta-transfer learning through hard tasks. arXiv:1910.03648 (2019)
70. Sun, Q., Liu, Y., Chua, T.S., Schiele, B.: Meta-transfer learning for few-shot learning. In: CVPR, pp. 403–412 (2019)
71. Sung, F., Yang, Y., Zhang, L., Xiang, T., Torr, P.H.S., Hospedales, T.M.: Learning to compare: relation network for few-shot learning. In: CVPR, pp. 1199–1208 (2018)

72. Thrun, S., Pratt, L.: Learning to learn: introduction and overview. In: Thrun, S., Pratt, L. (eds.) Learning to Learn, pp. 3–17. Springer, Boston (1998). https://doi.org/10.1007/978-1-4615-5529-2_1
73. Vinyals, O., Blundell, C., Lillicrap, T., Kavukcuoglu, K., Wierstra, D.: Matching networks for one shot learning. In: NIPS, pp. 3630–3638 (2016)
74. Wang, X., Huang, T.E., Darrell, T., Gonzalez, J.E., Yu, F.: Frustratingly simple few-shot object detection. In: ICML (2020)
75. Wang, Y., Girshick, R.B., Hebert, M., Hariharan, B.: Low-shot learning from imaginary data. In: CVPR, pp. 7278–7286 (2018)
76. Wang, Y.X., Hebert, M.: Learning from small sample sets by combining unsupervised meta-training with CNNs. In: NIPS, pp. 244–252 (2016)
77. Xian, Y., Sharma, S., Schiele, B., Akata, Z.: f-VAEGAN-D2: a feature generating framework for any-shot learning. In: CVPR, pp. 10275–10284 (2019)
78. Ye, H.J., Hu, H., Zhan, D.C., Sha, F.: Learning embedding adaptation for few-shot learning. arXiv:1812.03664 (2018)
79. Yoon, J., Kim, T., Dia, O., Kim, S., Bengio, Y., Ahn, S.: Bayesian model-agnostic meta-learning. In: NeurIPS, pp. 7343–7353 (2018)
80. Zhang, C., Cai, Y., Lin, G., Shen, C.: DeepEMD: differentiable earth mover's distance for few-shot learning. arXiv:2003.06777 (2020)
81. Zhang, C., Cai, Y., Lin, G., Shen, C.: DeepEMD: few-shot image classification with differentiable earth mover's distance and structured classifiers. In: CVPR, pp. 12203–12213 (2020)
82. Zhang, C., Lin, G., Liu, F., Guo, J., Wu, Q., Yao, R.: Pyramid graph networks with connection attentions for region-based one-shot semantic segmentation. In: ICCV, pp. 9587–9595 (2019)
83. Zhang, C., Lin, G., Liu, F., Yao, R., Shen, C.: CANet: class-agnostic segmentation networks with iterative refinement and attentive few-shot learning. In: CVPR, pp. 5217–5226 (2019)
84. Zhang, L., et al.: Nonlinear regression via deep negative correlation learning. IEEE Trans. Pattern Anal. Mach. Intell. (2019)
85. Zhang, R., Che, T., Grahahramani, Z., Bengio, Y., Song, Y.: MetaGAN: an adversarial approach to few-shot learning. In: NeurIPS, pp. 2371–2380 (2018)

On the Effectiveness of Image Rotation for Open Set Domain Adaptation

Silvia Bucci[1,2](✉), Mohammad Reza Loghmani[3], and Tatiana Tommasi[1,2]

[1] Italian Institute of Technology, Genova, Italy
[2] Politecnico di Torino, Turin, Italy
{silvia.bucci,tatiana.tommasi}@polito.it
[3] Vision for Robotics laboratory, ACIN, TU Wien, 1040 Vienna, Austria
loghmani@acin.tuwien.ac.at

Abstract. Open Set Domain Adaptation (OSDA) bridges the domain gap between a labeled source domain and an unlabeled target domain, while also rejecting target classes that are not present in the source. To avoid negative transfer, OSDA can be tackled by first separating the known/unknown target samples and then aligning known target samples with the source data. We propose a novel method to addresses both these problems using the self-supervised task of rotation recognition. Moreover, we assess the performance with a new open set metric that properly balances the contribution of recognizing the known classes and rejecting the unknown samples. Comparative experiments with existing OSDA methods on the standard Office-31 and Office-Home benchmarks show that: (i) our method outperforms its competitors, (ii) reproducibility for this field is a crucial issue to tackle, (iii) our metric provides a reliable tool to allow fair open set evaluation.

Keywords: Open Set Domain Adaptation · Self-supervised learning

1 Introduction

The current success of deep learning models is showing how modern artificial intelligent systems can manage supervised machine learning tasks with growing accuracy. However, when the level of supervision decreases, all the limitations of the existing data-hungry approaches become evident. For many applications, large amount of supervised data are not readily available, moreover collecting and manually annotating such data may be difficult or very costly. Different sub-fields of computer vision, such as *domain adaptation* [8] and *self-supervised*

S. Bucci, M. R. Loghmani—Equal contributions.

Electronic supplementary material The online version of this chapter (https://doi.org/10.1007/978-3-030-58517-4_25) contains supplementary material, which is available to authorized users.

A. Vedaldi et al. (Eds.): ECCV 2020, LNCS 12361, pp. 422–438, 2020.
https://doi.org/10.1007/978-3-030-58517-4_25

learning [11], aim at designing new learning solutions to compensate for this lack of supervision. Domain adaptation focuses on leveraging a fully supervised data-rich source domain to learn a classification model that performs well on a different but related unlabeled target domain. Traditional domain adaptation methods assume that the target contains exactly the same set of labels of the source (*closed-set* scenario). In recent years, this constraint has been relaxed in favor of the more realistic *open-set* scenario where the target also contains samples drawn from unknown classes. In this case, it becomes important to identify and isolate the unknown class samples before reducing the domain shift to avoid negative transfer. Self-supervised learning focuses on training models on pretext tasks, such as image colorization or rotation prediction, using unlabeled data to then transfer the acquired high-level knowledge to new tasks with scarce supervision. Recent literature has highlighted how self-supervision can be used for domain adaptation: jointly solving a pretext self-supervised task together with the main supervised problem leads to learning robust cross-domain features and supports generalization [5,44]. Other works have also shown that the output of self-supervised models can be used in anomaly detection to discriminate normal and anomalous data [2,17]. However, these works only tackle binary problems (normal and anomalous class) and deal with a single domain.

In this paper, we propose for the first time to use the inherent properties of self-supervision both for cross-domain robustness and for novelty detection to solve *Open-Set Domain Adaptation* (OSDA). To this purpose, we propose a two-stage method called *Rotation-based Open Set* (ROS) that is illustrated in Fig. 1. In the first stage, we separate the known and unknown target samples by training the model on a modified version of the rotation task that consists in predicting the relative rotation between a reference image and the rotated counterpart. In the second stage, we reduce the domain shift between the source domain and the known target domain using, once again, the rotation task. Finally we obtain a classifier that predicts each target sample as either belonging to one of the known classes or rejects it as unknown. While evaluating ROS on the two popular benchmarks *Office-31* [33] and *Office-Home* [41], we expose the reproducibility problem of existing OSDA approaches and assess them with a new evaluation metric that better represents the performance of open set methods. **We can summarize the contributions of our work as following**:

1. we introduce a novel OSDA method that exploits rotation recognition to tackle both known/unknown target separation and domain alignment;
2. we define a new OSDA metric that properly accounts for both known class recognition and unknown rejection;
3. we present an extensive experimental benchmark against existing OSDA methods with two conclusions: (a) we put under the spotlight the urgent need of a rigorous experimental validation to guarantee result reproducibility; (b) our ROS defines the new state-of-the-art on two benchmark datasets.

A Pytorch implementation of our method, together with instructions to replicate our experiments, is available at https://github.com/silvia1993/ROS .

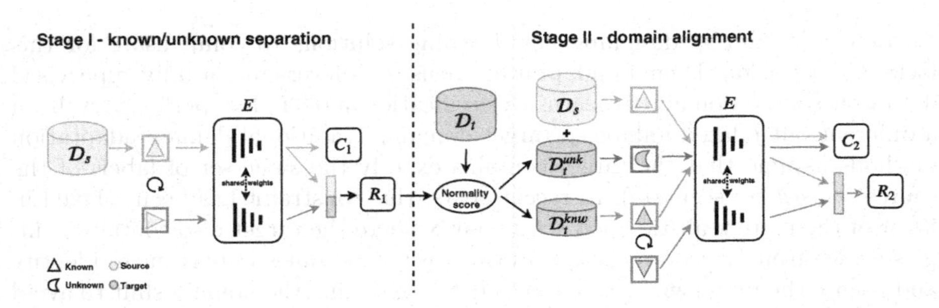

Fig. 1. Schematic illustration of our Rotation-based Open Set (**ROS**). Stage I: the source dataset $\mathcal{D}_s$ is used to train the encoder E, the semantic classifier C_1, and the multi-rotation classifier R_1 to perform known/unknown separation. C_1 is trained using the features of the original image, while R_1 is trained using the concatenated features of the original and rotated image. After convergence, the prediction of R_1 on the target dataset $\mathcal{D}_t$ is used to generate a normality score that defines how the target samples are split into a known target dataset $\mathcal{D}_t^{knw}$ and an unknown target dataset $\mathcal{D}_t^{unk}$. Stage II: E, the semantic+unknown classifier C_2 and the rotation classifier R_2 are trained to align the source and target distributions and to recognize the known classes while rejecting the unknowns. C_2 is trained using the original images from $\mathcal{D}_s$ and $\mathcal{D}_t^{unk}$, while R_2 is trained using the concatenated features of the original and rotated known target samples.

2 Related Work

Self-supervised learning applies the techniques of supervised learning on problems where external supervision is not available. The idea is to manipulate the data to generate the supervision for an artificial task that is helpful to learn useful feature representations. Examples of self-supervised tasks in computer vision include predicting the relative position of image patches [11,28], colorizing a gray-scale image [22,48], and inpainting a removed patch [30]. Arguably, one of the most effective self-supervised tasks is rotation recognition [16] that consists in rotating the input images by multiples of 90° and training the network to predict the rotation angle of each image. This pretext task has been successfully used in a variety of applications including anomaly detection [17] and closed-set domain adaptation [44].

Anomaly detection, also known as outlier or novelty detection, aims at learning a model from a set of *normal* samples to be able to detect out-of-distribution (*anomalous*) instances. The research literature in this area is wide with three main kind of approaches. *Distribution-based* methods [21,47,50,51] model the distribution of the available normal data so that the anomalous samples can be recognized as those with a low likelihood under the learned probability function. *Reconstruction-based* methods [4,12,36,43,49] learn to reconstruct the normal samples from an embedding or a set of basis functions. Anomalous data are then recognized by having a larger reconstruction error with respect to normal samples. *Discriminative* methods [20,23,31,37] train a classifier on the normal data and use its predictions to distinguish between normal and anomalous samples.

Closed-set domain adaptation (CSDA) accounts for the difference between source and target data by considering them as drawn from two different marginal distributions. The literature of DA can be divided into three groups based on the strategy used to reduce the domain shift. *Discrepancy-based* methods [26,39,45] define a metric to measure the distance between source and target data in feature space. This metric is minimized while training the network to reduce the domain shift. *Adversarial* methods [14,32,40] aim at training a domain discriminator and a generator network in an adversarial fashion so that the generator converges to a solution that makes the source and target data indistinguishable for the domain discriminator. *Self-supervised* methods [3,5,15] train a network to solve an auxiliary self-supervised task on the target (and source) data, in addition to the main task, to learn robust cross-domain representations.

Open Set Domain Adaptation (OSDA) is a more realistic version of CSDA, where the source and target distribution do not contain the same categories. The term "OSDA" was first introduced by Busto and Gall [29] that considered the setting where each domain contains, in addition to the shared categories, a set of private categories. The currently accepted definition of OSDA was introduced by Saito *et al.* [34] that considered the target as containing all the source categories and additional set of private categories that should be considered *unknown*. To date, only a handful of papers tackled this problem. *Open Set Back-Propagation* (OSBP) [34] is an adversarial method that consists in training a classifier to obtain a large boundary between source and target samples whereas the feature generator is trained to make the target samples far from the boundary. *Separate To Adapt* (STA) [24] is an approach based on two stages. First, a multi-binary classifier trained on the source is used to estimate the similarity of target samples to the source. Then, target data with extreme high and low similarity are re-used to separate known and unknown classes while the features across domains are aligned through adversarial adaptation. *Attract or Distract* (AoD) [13] starts with a mild alignment with a procedure similar to [34] and refines the decision by using metric learning to reduce the intra-class distance in known classes and push the unknown class away from the known classes. *Universal Adaptation Network* (UAN)[1] [46] uses a pair of domain discriminators to both generate a sample-level transferability weight and to promote the adaptation in the automatically discovered common label set. Differently from all existing OSDA methods, **our approach abandons adversarial training in favor of self-supervision. Indeed, we show that rotation recognition can be used, with tailored adjustments, both to separate known and unknown target samples and to align the known source and target distributions**[2].

[1] UAN is originally proposed for the universal domain adaptation setting that is a superset of OSDA, so it can also be used in the context of this paper.

[2] See the supplementary for a discussion on the use of other self-supervised tasks.

3 Method

3.1 Problem Formulation

Let us denote with $\mathcal{D}_s = \{(\boldsymbol{x}_j^s, y_j^s)\}_{j=1}^{N_s} \sim p_s$ the labeled source dataset drawn from distribution p_s and $\mathcal{D}_t = \{\boldsymbol{x}_j^t\}_{j=1}^{N_t} \sim p_t$ the unlabeled target dataset drawn from distribution p_t. In OSDA, the source domain is associated with a set of *known* classes $y^s \in \{1, \ldots, |\mathcal{C}_s|\}$ that are shared with the target domain $\mathcal{C}_s \subset \mathcal{C}_t$, but the target covers also a set $\mathcal{C}_{t \setminus s}$ of additional classes, which are considered *unknown*. As in CSDA, it holds that $p_s \neq p_t$ and we further have that $p_s \neq p_t^{\mathcal{C}_s}$, where $p_t^{\mathcal{C}_s}$ denotes the distribution of the target domain belonging to the shared label space $\mathcal{C}_s$. Therefore, in OSDA we face both a domain gap ($p_s \neq p_t^{\mathcal{C}_s}$) and a category gap ($\mathcal{C}_s \neq \mathcal{C}_t$). OSDA approaches aim at assigning the target samples to either one of the $|\mathcal{C}_s|$ shared classes or to reject them as *unknown* using only annotated source samples, with the unlabeled target samples available transductively. An important measure characterizing a given OSDA problem is the *openness* that relates the size of the source and target class set. For a dataset pair $(\mathcal{D}_s, \mathcal{D}_t)$, following the definition of [1], the openness $\mathbb{O}$ is measured as $\mathbb{O} = 1 - \frac{|\mathcal{C}_s|}{|\mathcal{C}_t|}$. In CSDA $\mathbb{O} = 0$, while in OSDA $\mathbb{O} > 0$.

3.2 Overview

When designing a method for OSDA, we face two main challenges: *negative transfer* and *known/unknown separation*. Negative transfer occurs when the whole source and target distribution are forcefully matched, thus also the unknown target samples are mistakenly aligned with source data. To avoid this issue, cross-domain adaptation should focus only on the shared $\mathcal{C}_s$ classes, closing the gap between $p_t^{\mathcal{C}_s}$ and p_s. This leads to the challenge of known/unknown separation: recognizing each target sample as either belonging to one of the shared classes $\mathcal{C}_s$ (known) or to one of the target private classes $\mathcal{C}_{t \setminus s}$ (unknown). Following these observations, we structure our approach in two stages: (i) we separate the target samples into known and unknown, and (ii) we align the target samples predicted as known with the source samples (see Fig. 1). The first stage is formulated as an anomaly detection problem where the unknown samples are considered as anomalies. The second stage is formulated as a CSDA problem between source and the known target distribution. Inspired by recent advances in anomaly detection and CSDA [17,44], we solve both stages using the power of self-supervision. More specifically, we use two variations of the rotation classification task to compute a normality score for the known/unknown separation of the target samples and to reduce the domain gap.

3.3 Rotation Recognition for Open Set Domain Adaptation

Let us denote with $rot90(\boldsymbol{x}, i)$ the function that rotates clockwise a 2D image $\boldsymbol{x}$ by $i \times 90°$. Rotation recognition is a self-supervised task that consists in rotating

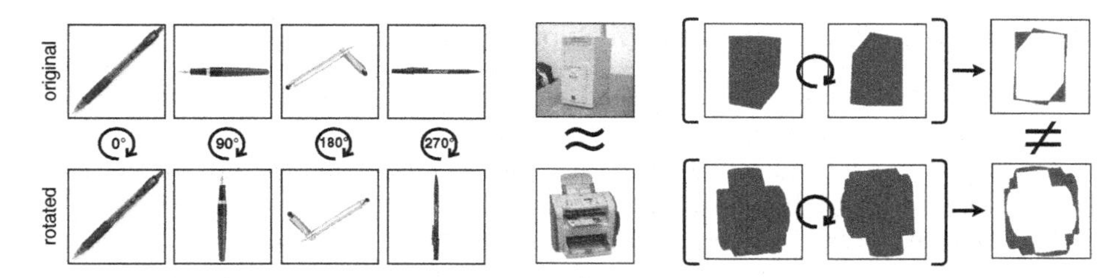

Fig. 2. Are you able to infer the rotation degree of the rotated images without looking at the respective original one?

Fig. 3. The objects on the left may be confused. The relative rotation guides the network to focus on discriminative shape information

a given image x by a random $i \in [1, 4]$ and using a CNN to predict i from the rotated image $\tilde{\boldsymbol{x}} = rot90(\boldsymbol{x}, i)$. We indicate with $|r| = 4$ the cardinality of the label space for this classification task. In order to effectively apply rotation recognition to OSDA, we introduce the following variations.

Relative Rotation: Consider the images in Fig. 2. Inferring by how much each image has been rotated without looking at its original (non-rotated) version is an ill-posed problem since the pens, as all the other object classes, are not presented with a coherent orientation in the dataset. On the other hand, looking at both original and rotated image to infer the relative rotation between them is well-defined. Following this logic, we modify the standard rotation classification task [16] by introducing the original image as an anchor. Finally, we train the rotation classifier to predict the rotation angle given the concatenated features of both original (anchor) and rotated image. As indicated by Fig. 3, the proposed relative rotation has the further effect of boosting the discriminative power of the learned features. It guides the network to focus more on specific shape details rather than on confusing texture information across different object classes.

Multi-rotation Classification: The standard setting of anomaly detection considers samples from one semantic category as the normal class and samples from other semantic categories as anomalies. Rotation recognition has been successfully applied to this setting, but it suffers when including multiple semantic categories in the normal class [17]. This is the case when coping with the known/unknown separation of OSDA, where we have all the $|\mathcal{C}_s|$ semantic categories as known data. To overcome this problem, we propose a simple solution: we extend rotation recognition from a 4-class problem to a $(4 \times |\mathcal{C}_s|)$-class problem, where the set of classes represents the combination of semantic and rotation labels. For example, if we rotate an image of category $y^s = 2$ by $i = 3$, its label for the multi-rotation classification task is $z^s = (y^s \times 4) + i = 11$. In the supplementary material, we discuss the specific merits of the multi-rotation classification task with further experimental evidences. In the following, we indicate with $\boldsymbol{y}, \boldsymbol{z}$ the one-hot vectors respectively for the class and multi-rotation labels.

3.4 Stage I: Known/Unknown Separation

To distinguish between the known and unknown samples of $\mathcal{D}_t$, we train a CNN on the multi-rotation classification task using $\tilde{\mathcal{D}}_s = \{(\boldsymbol{x}^s_j, \tilde{\boldsymbol{x}}^s_j, z^s_j)\}_{j=1}^{4\times N_s}$. The network is composed of an encoder E and two heads: a multi-rotation classifier R_1 and a semantic label classifier C_1. The rotation prediction is computed on the stacked features of the original and rotated image produced by the encoder $\hat{\boldsymbol{z}}^s = \text{softmax}\big(R_1([E(\boldsymbol{x}^s), E(\tilde{\boldsymbol{x}}^s)])\big)$, while the semantic prediction is computed only from the original image features as $\hat{\boldsymbol{y}}^s = \text{softmax}\big(C_1(E(\boldsymbol{x}^s)\big)$. The network is trained to minimize the objective function $\mathcal{L}_1 = \mathcal{L}_{C_1} + \mathcal{L}_{R_1}$, where the semantic loss $\mathcal{L}_{C_1}$ is defined as a cross-entropy and the multi-rotation loss $\mathcal{L}_{R_1}$ combines cross-entropy and center loss [42]. More precisely,

$$\mathcal{L}_{C_1} = -\sum_{j\in\mathcal{D}_s} \boldsymbol{y}^s_j \cdot \log(\hat{\boldsymbol{y}}^s_j), \tag{1}$$

$$\mathcal{L}_{R_1} = \sum_{j\in\tilde{\mathcal{D}}_s} -\lambda_{1,1}\boldsymbol{z}^s_j \cdot \log(\hat{\boldsymbol{z}}^s_j) + \lambda_{1,2}||\boldsymbol{v}^s_j - \gamma(\boldsymbol{z}^s_j)||^2_2, \tag{2}$$

where $||.||_2$ indicates the l_2-norm operator, $\boldsymbol{v}_j$ indicates the output of the penultimate layer of R_1 and $\gamma(\boldsymbol{z}_j)$ indicates the corresponding centroid of the class associated with $\boldsymbol{v}_j$. By using the center loss we further encourage the network to minimize the intra-class variations while keeping far the features of different classes. This supports the following use of the rotation classifier output as a metric to detect unknown category samples.

Once the training is complete, we use E and R_1 to compute the *normality score* $\mathcal{N} \in [0, 1]$ for each target sample, with large $\mathcal{N}$ values indicating normal (known) samples and vice-versa. We start from the network prediction on all the relative rotation variants of a target sample $\hat{\boldsymbol{z}}_i{}^t = \text{softmax}\big(R_1([E(\boldsymbol{x}^t), E(\tilde{\boldsymbol{x}}^t_i)])\big)_i$ and their related entropy $H(\hat{\boldsymbol{z}}^t_i) = \big(\hat{\boldsymbol{z}}^t_i \cdot \log(\hat{\boldsymbol{z}}^t_i)/\log|\mathcal{C}_s|\big)_i$ with $i = 1, \ldots, |r|$. We indicate with $[\hat{\boldsymbol{z}}^t]_m$ the m-th component of the $\hat{\boldsymbol{z}}^t$ vector. The full expression of the normality score is:

$$\mathcal{N}(\boldsymbol{x}^t) = \max\left\{\max_{k=1,\ldots,|\mathcal{C}_s|}\left(\sum_{i=1}^{|r|}[\hat{\boldsymbol{z}}^t_i]_{k\times|r|+i}\right), \left(1 - \frac{1}{|r|}\sum_{i=1}^{|r|}H(\hat{\boldsymbol{z}}^t_i)\right)\right\}. \tag{3}$$

In words, this formula is a function of the ability of the network to correctly predict the semantic class and orientation of a target sample (first term in the braces, *Rotation Score*) as well as of its confidence evaluated on the basis of the prediction entropy (second term, *Entropy Score*). We maximize over these two components with the aim of taking the most reliable metric in each case. Finally, the normality score is used to separate the target dataset into a known target dataset $\mathcal{D}^{knw}_t$ and an unknown target dataset $\mathcal{D}^{unk}_t$. The distinction is made directly through the data statistics using the average of the normality score over the whole target $\bar{\mathcal{N}} = \frac{1}{N_t}\sum_{j=1}^{N_t}\mathcal{N}_j$, without the need to introduce any further parameter:

$$\begin{cases} \boldsymbol{x}^t \in \mathcal{D}_t^{knw} & \text{if} \quad \mathcal{N}(\boldsymbol{x}^t) > \bar{\mathcal{N}} \\ \boldsymbol{x}^t \in \mathcal{D}_t^{unk} & \text{if} \quad \mathcal{N}(\boldsymbol{x}^t) < \bar{\mathcal{N}} \ . \end{cases} \tag{4}$$

It is worth mentioning that only R_1 is directly involved in computing the normality score, while C_1 is only trained for regularization purposes and as a warm up for the following stage. For a detailed pseudo-code on how to compute $\mathcal{N}$ and generate $\mathcal{D}_t^{knw}$ and $\mathcal{D}_t^{unk}$, please refer to the supplementary material.

3.5 Stage II: Domain Alignment

Once the target unknown samples have been identified, the scenario gets closer to that of standard CSDA. On the one hand, we can use $\mathcal{D}_t^{knw}$ to close the domain gap without the risk of negative transfer and, on the other hand, we can exploit $\mathcal{D}_t^{unk}$ to extend the original semantic classifier, making it able to recognize the unknown category. Similarly to Stage I, the network is composed of an encoder E and two heads: a rotation classifier R_2 and a semantic label classifier C_2. The encoder is inherited from the previous stage. The heads also leverage on the previous training phase but have two key differences with respect to Stage I: (1) C_1 has a $|\mathcal{C}_s|$-dimensional output, while C_2 has a $(|\mathcal{C}_s|+1)$-dimensional output because of the addition of the unknown class; (2) R_1 is a multi-rotation classifier with a $(4 \times |\mathcal{C}_s|)$-dimensional output, R_2 is a rotation classifier with a 4-dimensional output. The rotation prediction is computed as $\hat{\boldsymbol{q}} = \text{softmax}\big(R_2([E(\boldsymbol{x}), E(\tilde{\boldsymbol{x}})])\big)$ while the semantic prediction is $\hat{\boldsymbol{g}} = \text{softmax}\big(C_2(E(\boldsymbol{x}))\big)$. The network is trained to minimize the objective function $\mathcal{L}_2 = \mathcal{L}_{C_2} + \mathcal{L}_{R_2}$, where $\mathcal{L}_{C_2}$ combines the supervised cross-entropy and the unsupervised entropy loss for the classification task, while $\mathcal{L}_{R_2}$ is defined as a cross-entropy for the rotation task. The unsupervised entropy loss is used to involve in the semantic classification process also the unlabeled target samples recognized as known. This loss enforces the decision boundary to pass through low-density areas. More precisely,

$$\mathcal{L}_{C_2} = - \sum_{j \in \{\mathcal{D}_s \cup \mathcal{D}_t^{unk}\}} \boldsymbol{g}_j \cdot \log(\hat{\boldsymbol{g}}_j) - \lambda_{2,1} \sum_{j \in \mathcal{D}_t^{knw}} \hat{\boldsymbol{g}}_j \cdot \log(\hat{\boldsymbol{g}}_j), \tag{5}$$

$$\mathcal{L}_{R_2} = -\lambda_{2,2} \sum_{j \in \mathcal{D}_t^{knw}} \boldsymbol{q}_j \cdot \log(\hat{\boldsymbol{q}}_j) \ . \tag{6}$$

Once the training is complete, R_2 is discarded and the target labels are simply predicted as $c_j^t = C_2(E(\boldsymbol{x}_j^t))$ for all $j = 1, \ldots, N_t$.

4 On Reproducibility and Open Set Metrics

OSDA is a young field of research first introduced in 2017. As it is gaining momentum, it is crucial to guarantee the *reproducibility* of the proposed methods and have a valid *metric* to properly evaluate them.

Reproducibility: In recent years, the machine learning community has become painfully aware of a reproducibility crisis [10,19,27]. Replicating the results of state-of-the-art deep learning models is seldom straightforward due to a combination of non-deterministic factors in standard benchmark environments and poor reports from the authors. Although the problem is far from being solved, several efforts have been made to promote reproducibility through checklists [7], challenges [6] and by encouraging authors to submit their code. On our side, we contribute by re-running the state-of-the-art methods for OSDA and compare them with the results reported in the papers (see Sect. 5). Our results are produced using the original public implementation together with the parameters reported in the paper and, in some cases, repeated communications with the authors. We believe that this practice, as opposed to simply copying the results reported in the papers, can be of great value to the community.

Open Set Metrics: The usual metrics adopted to evaluate OSDA are the average class accuracy over the known classes OS^*, and the accuracy of the unknown class *UNK*. They are generally combined in $OS = \frac{|\mathcal{C}_s|}{|\mathcal{C}_s|+1} \times OS^* + \frac{1}{|\mathcal{C}_s|+1} \times UNK$ as a measure of the overall performance. However, we argue (and we already demonstrated in [25]) that treating the unknown as an additional class does not provide an appropriate metric. As an example, let us consider an algorithm that is not designed to deal with unknown classes ($UNK = 0.0\%$) but has perfect accuracy over 10 known classes ($OS^* = 100.0\%$). Although this algorithm is not suitable for open set scenarios because it completely disregards false positives, it presents a high score of $OS = 90.9\%$. With increasing number of known classes, this effect on OS becomes even more acute, making the role of *UNK* negligible. For this reason, we propose a new metric defined as the harmonic mean of OS^* and *UNK*, $HOS = 2\frac{OS^* \times UNK}{OS^* + UNK}$. Differently from *OS*, *HOS* provides a high score only if the algorithm performs well both on known and on unknown samples, independently of $|\mathcal{C}_s|$. Using a harmonic mean instead of a simple average penalizes large gaps between OS^* and *UNK*.

5 Experiments

5.1 Setup: Baselines, Datasets

We validate ROS with a thorough experimental analysis on two widely used benchmark datasets, Office-31 and Office-Home. *Office-31* [33] consists of three domains, Webcam (W), Amazon (A) and Dslr (D), each containing 31 object categories. We follow the setting proposed in [34], where the first 10 classes in alphabetic order are considered known and the last 11 classes are considered unknown. *Office-Home* [41] consists of four domains, Product (Pr), Art (Ar), Real World (Rw) and Clipart (Cl), each containing 65 object categories. Unless otherwise specified, we follow the setting proposed in [24], where the first 25 classes in alphabetic order are considered known classes and the remaining 40 classes are considered unknown. Both the number of categories and the large domain gaps make this dataset much more challenging than Office-31.

We compare ROS against the state-of-the-art methods STA [24], OSBP [34], UAN [46], AoD [13], that we already described in Sect. 2. For each of them, we run experiments using the official code provided by the authors, with the exact parameters declared in the relative paper. The only exception was made for AoD for which the authors have not released the code at the time of writing, thus we report the values presented in their original work. We also highlight that STA presents a practical issue related to the similarity score used to separate known and unknown categories. Its formulation is based on the *max* operator according to the Equation (2) in [24], but appears instead based on *sum* in the implementation code. In our analysis we considered both the two variants (STA_{sum}, STA_{max}) for the sake of completeness. All the results presented in this section, both for ROS and for the baseline methods, are the average over three independent experimental runs. We do not cherry pick the best out of several trials, but only run the three experiments we report.

5.2 Implementation Details

By following standard practice, we evaluate the performances of ROS on Office-31 using two different backbones ResNet-50 [18] and VGGNet [38], both pre-trained on ImageNet [9], and we focus on ResNet-50 for Office-Home. The hyper-parameters values are the same regardless of the backbone and the dataset used. In particular, in both Stage I and Stage II of ROS the batch size is set to 32 with a learning rate of 0.0003 which decreases during the training following an inverse decay scheduling. For all layers trained from scratch, we set the learning rate 10 times higher than the pre-trained ones. We use SGD, setting the weight decay as 0.0005 and momentum as 0.9. In both stages, the weight of the self-supervised task is set three times the one of the semantic classification task, thus $\lambda_{1,1} = \lambda_{2,2} = 3$. In Stage I, the weight of the center loss is $\lambda_{1,2} = 0.1$ and in Stage II the weight of the entropy loss is $\lambda_{2,1} = 0.1$. The network trained in Stage I is used as starting point for Stage II. To take into consideration the extra category, in Stage II we set the learning rate of the new unknown class to twice that of the known classes (already learned in Stage I). More implementation details and a sensitivity analysis of the hyper-parameters are provided in the supplementary material.

5.3 Results

How Does Our Method Compare to the State-of-the-Art? Tables 1 and 2 show the average results over three runs on each of the domain shifts, respectively of Office-31 and Office-Home. To discuss the results, we focus on the HOS metric since it is a synthesis of OS* and UNK, as discussed in Sect. 4. Overall, ROS outperforms the state-of-the-art on a total of 13 out of 18 domain shifts and presents the highest average performance on both Office-31 and Office-Home. The HOS improvement gets up to 2.2% compared to the second best method OSBP. Specifically, ROS has a large gain over STA, regardless of its specific max or sum implementation, while UAN is not a challenging competitor due to its

Table 1. Accuracy (%) averaged over three runs of each method on Office-31 dataset using ResNet-50 and VGGNet as backbones

Office-31	A → W			A → D			D → W			W → D			D → A			W → A			Avg.		
ResNet-50	OS*	UNK	**HOS**	OS*	UNK	**HOS**	OS*	UNK	**HOS**	OS*	UNK	**HOS**	OS*	UNK	**HOS**	OS*	UNK	**HOS**	OS*	UNK	**HOS**
STA_{sum} [24]	92.1	58.0	71.0	95.4	45.5	61.6	97.1	49.7	65.5	96.6	48.5	64.4	94.1	55.0	69.4	92.1	46.2	60.9	94.6	50.5	65.5±0.3
STA_{max}	86.7	67.6	75.9	91.0	63.9	75.0	94.1	55.5	69.8	84.9	67.8	75.2	83.1	65.9	73.2	66.2	68.0	66.1	84.3	64.8	72.5±0.8
OSBP [34]	86.8	79.2	**82.7**	90.5	75.5	**82.4**	97.7	96.7	**97.2**	99.1	84.2	91.1	76.1	72.3	75.1	73.0	74.4	73.7	87.2	80.4	83.7±0.4
UAN [46]	95.5	31.0	46.8	95.6	24.4	38.9	99.8	52.5	68.8	81.5	41.4	53.0	93.5	53.4	68.0	94.1	38.8	54.9	93.4	40.3	55.1±1.4
ROS	88.4	76.7	82.1	87.5	77.8	**82.4**	99.3	93.0	96.0	100.0	99.4	**99.7**	74.8	81.2	**77.9**	69.7	86.6	**77.2**	86.6	85.8	**85.9±0.2**
VGGNet	A → W			A → D			D → W			W → D			D → A			W → A			Avg.		
	OS*	UNK	**HOS**	OS*	UNK	**HOS**	OS*	UNK	**HOS**	OS*	UNK	**HOS**	OS*	UNK	**HOS**	OS*	UNK	**HOS**	OS*	UNK	**HOS**
OSBP [34]	79.4	75.8	77.5	87.9	75.2	81.0	96.8	93.4	**95.0**	98.9	84.2	91.0	74.4	82.4	**78.2**	69.7	76.4	72.9	84.5	81.2	82.6±0.8
ROS	80.3	81.7	**81.0**	81.8	76.5	79.0	99.5	89.9	94.4	99.3	100.0	**99.7**	76.7	79.6	78.1	62.2	91.6	**74.1**	83.3	86.5	**84.4±0.2**
AoD [13]	87.7	73.4	79.9	92.0	71.1	79.3	99.8	78.9	88.1	99.3	87.2	92.9	88.4	13.6	23.6	82.6	57.3	67.7	91.6	63.6	71.9

Table 2. Accuracy (%) averaged over three runs of each method on Office-Home dataset using ResNet-50 as backbone

Office-Home	Pr → Rw			Pr → Cl			Pr → Ar			Ar → Pr			Ar → Rw			Ar → Cl		
	OS*	UNK	**HOS**	OS*	UNK	**HOS**	OS*	UNK	**HOS**	OS*	UNK	**HOS**	OS*	UNK	**HOS**	OS*	UNK	**HOS**
STA_{sum} [24]	78.1	63.3	69.7	44.7	71.5	55.0	55.4	73.7	63.1	68.7	59.7	63.7	81.1	50.5	62.1	50.8	63.4	56.3
STA_{max}	76.2	64.3	69.5	44.2	67.1	53.2	54.2	72.4	61.9	68.0	48.4	54.0	78.6	60.4	68.3	46.0	72.3	55.8
OSBP [34]	76.2	71.7	73.9	44.5	66.3	53.2	59.1	68.1	**63.2**	71.8	59.8	65.2	79.3	67.5	72.9	50.2	61.1	55.1
UAN [46]	84.0	0.1	0.2	59.1	0.0	0.0	73.7	0.0	0.0	81.1	0.0	0.0	88.2	0.1	0.2	62.4	0.0	0.0
ROS	70.8	78.4	**74.4**	46.5	71.2	**56.3**	57.3	64.3	60.6	68.4	70.3	**69.3**	75.8	77.2	**76.5**	50.6	74.1	**60.1**

	Rw → Ar			Rw → Pr			Rw → Cl			Cl → Rw			Cl → Ar			Cl → Pr			Avg.		
	OS*	UNK	**HOS**	OS*	UNK	**HOS**	OS*	UNK	**HOS**	OS*	UNK	**HOS**	OS*	UNK	**HOS**	OS*	UNK	**HOS**	OS*	UNK	**HOS**
STA_{sum}	67.9	62.3	65.0	77.9	58.0	66.4	51.4	57.9	54.2	69.8	63.2	66.3	53.0	63.9	57.9	61.4	63.5	62.5	63.4	62.6	61.9±2.1
STA_{max}	67.5	66.7	67.1	77.1	55.4	64.5	49.9	61.1	54.5	67.0	66.7	66.8	51.4	65.0	57.4	61.8	59.1	60.4	61.8	63.3	61.1±0.3
OSBP	66.1	67.3	66.7	76.3	68.6	72.3	48.0	63.0	54.5	72.0	69.2	**70.6**	59.4	70.3	**64.3**	67.0	62.7	64.7	64.1	66.3	64.7±0.2
UAN	77.5	0.1	0.2	85.0	0.1	0.1	66.2	0.0	0.0	80.6	0.1	0.2	70.5	0.0	0.0	74.0	0.1	0.2	75.2	0.0	0.1±0.0
ROS	67.0	70.8	**68.8**	72.0	80.0	**75.7**	51.5	73.0	**60.4**	65.3	72.2	68.6	53.6	65.5	58.9	59.8	71.6	**65.2**	61.6	72.4	**66.2± 0.3**

low performance on the unknown class. We can compare against AoD only when using VGG for Office-31: we report the original results in gray in Table 2, with the HOS value confirming our advantage.

A more in-depth analysis indicates that the advantage of ROS is largely related in its ability in separating known and unknown samples. Indeed, while our average OS* is similar to that of the competing methods, our average UNK is significantly higher. This characteristic is also visible qualitatively by looking at the t-SNE visualizations in Fig. 4 where we focus on the comparison against the second best method OSBP. Here the features for the known (red) and unknown (blue) target data appear more confused than for ROS.

Is it Possible to Reproduce the Reported Results of the State-of-the-Art? By analyzing the published OSDA papers, we noticed some incoherence in the reported results. For example, some of the results from OSBP are different between the pre-print [35] and the published [34] version, although they present the same description for method and hyper-parameters. Also, AoD [13] compares against the pre-print results of OSBP, while omitting the results of STA. To dissipate these ambiguities and gain a better perspective on the current state-of-the-art methods, in Table 3 we compare the results on Office-31 reported in previous works with the results obtained by running their code. For this analysis we focus on OS since it is the only metric reported for some of the methods. The comparison shows that, despite using the original implementation and the information provided by the authors, the OS obtained by re-running the

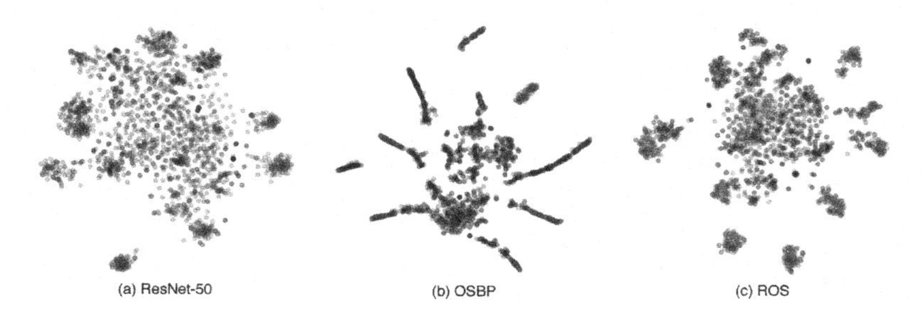

Fig. 4. t-SNE visualization of the target features for the W→A domain shift from Office-31. Red and blue points are respectively features of known and unknown classes (Color figure online)

experiments is between 1.3% and 4.9% lower than the originally published results. The significance of this gap calls for greater attention in providing all the relevant information for reproducing the experimental results. A more extensive reproducibility study is provided in the supplementary material.

Table 3. Reported vs reproduced OS accuracy (%) averaged over three runs

Reproducibility Study								
Office-31 (ResNet-50)						Office-31 (VGGNet)		
STA_{sum}			UAN			OSBP		
$OS_{reported}$	OS_{ours}	gap	$OS_{reported}$	OS_{ours}	gap	$OS_{reported}$	OS_{ours}	gap
92.9	90.6±1.8	**2.3**	89.2	87.9±0.03	**1.3**	89.1	84.2 ±0.4	**4.9**

Why is it Important to Use the HOS Metric? The most glaring example of why OS is not an appropriate metric for OSDA is provided by the results of UAN. In fact, when computing OS from the average (OS*, UNK) in Tables 1 and 2, we can see that UAN has OS = 72.5% for Office-Home and OS = 91.4% for Office-31. This is mostly reflective of the ability of UAN in recognizing the known classes (OS*), but it completely disregards its (in)ability to identify the unknown samples (UNK). For example, for most domain shifts in Office-Home, UAN does not assign (almost) any samples to the unknown class, resulting in UNK = 0.0%. On the other hand, HOS better reflects the open set scenario and assumes a high value only when OS* and UNK are both high.

Is Rotation Recognition Effective for Known/Unknown Separation in OSDA? To better understand the effectiveness of rotation recognition for known/unknown separation, we measure the performance of our Stage I and compare it to the Stage I of STA. Indeed, also STA has a similar two-stage structure, but uses a multi-binary classifier instead of a multi-rotation classifier to separate known and unknown target samples. To assess the performance, we compute the *area under*

Table 4. Ablation analysis on Stage I and Stage II

Ablation Study							
STAGE I (AUC-ROC)	A → W	A → D	D → W	W → D	D → A	W → A	**Avg.**
ROS	90.1	88.1	99.4	99.9	87.5	83.8	**91.5**
Multi-Binary (from STA [24])	83.2	84.1	86.8	72.0	75.7	78.3	79.9
ROS - No Center loss	88.8	83.2	98.8	99.8	84.7	84.5	89.9
ROS - No Anchor	84.5	84.9	99.1	99.9	87.6	86.2	90.4
ROS - No Rot. Score	86.3	82.7	99.5	99.9	86.3	82.9	89.6
ROS - No Ent. Score	80.7	78.7	99.7	99.9	86.6	84.4	88.3
ROS - No Center loss, No Anchor	76.5	79.1	98.3	99.7	85.2	83.5	87.1
ROS - No Rot. Score, No Anchor	83.9	84.6	99.4	99.9	84.7	84.9	89.6
ROS - No Ent. Score, No Anchor	80.1	81.0	99.5	99.7	84.3	83.3	87.9
ROS - No Rot. Score, No Center loss	80.9	81.6	98.9	99.8	85.6	83.3	88.3
ROS - No Ent. Score, No Center loss	76.4	79.8	99.0	98.3	84.4	84.3	87.0
ROS - No Ent. Score, No Center loss, No Anchor	78.6	80.4	99.0	98.9	86.2	83.2	87.7
ROS - No Rot. Score, No Center loss, No Anchor	78.7	82.2	98.3	99.8	85.0	82.6	87.8
STAGE II (HOS)	A → W	A → D	D → W	W → D	D → A	W → A	**Avg.**
ROS	82.1	82.4	96.0	99.7	77.9	77.2	**85.9**
ROS Stage I - GRL [14] Stage II	83.5	80.9	97.1	99.4	77.3	72.6	85.1
ROS Stage I - No Anchor in Stage II	80.0	82.3	94.5	99.2	76.9	76.6	84.9
ROS Stage I - No Anchor, No Entropy in Stage II	80.1	84.4	97.0	99.2	76.5	72.9	85.0

receiver operating characteristic curve (AUC-ROC) over the normality scores $\mathcal{N}$ on Office-31. Table 4 shows that the AUC-ROC of ROS (91.5) is significantly higher than that of the multi-binary used by STA (79.9). Table 4 also shows the performance of Stage I when alternatively removing the center loss (No Center Loss) from Eq. (2) ($\lambda_{1,2} = 0$) and the anchor image (No Anchor) when training R_1, thus passing from relative rotation to the more standard absolute rotation. In both cases, the performance significantly drops compared to our full method, but still outperforms the multi-binary classifier of STA.

Why is the Normality Score Defined the Way It Is? As defined in Eq. (3), our normality score is a function of the rotation score and entropy score. The rotation score is based on the ability of R_1 to predict the rotation of the target samples, while the entropy score is based on the confidence of such predictions. Table 4 shows the results of Stage I when alternatively discarding either the rotation score (No Rot. Score) or the information of the entropy score (No Ent. Score). In both cases the AUC-ROC significantly decreases compared to the full version, justifying our choice.

Is Rotation Recognition Effective for Domain Alignment in OSDA? While rotation classification has already been used for CSDA [44], its application in OSDA, where the shared target distribution could be noisy (*i.e.* contain unknown samples) has not been studied. On the other hand, GRL [14] is used, under different forms, by all existing OSDA methods. We compare rotation recognition and GRL in this context by evaluating the performance of our Stage II when replacing the R_2 with a domain discriminator. Table 4 shows that rotation recognition performs on par with GRL, if not slightly better. Moreover we also evaluate the role of the relative rotation in the Stage II: the results in the last row of Table 4 confirm that it improves over the standard absolute rotation (No Anchor in Stage II) even when the rotation classifier is used as cross-domain adaptation strategy. Finally, the cosine distance between the source and the target domain without adaptation in Stage II (0.188) and with our full method (0.109) confirms that rotation recognition is indeed helpful to reduce the domain gap.

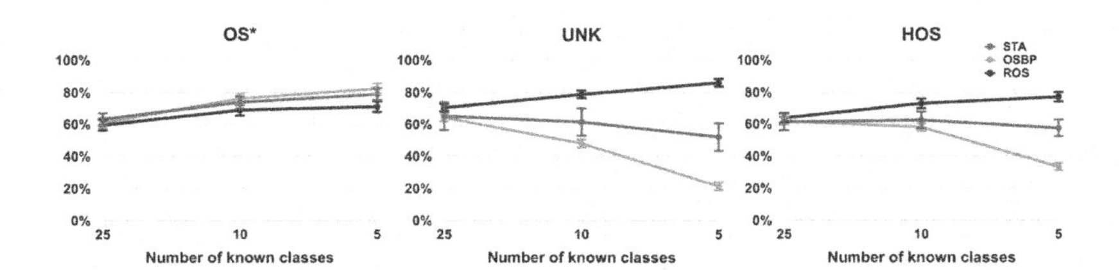

Fig. 5. Accuracy (%) averaged over the three openness configurations.

Is Our Method Effective on Problems with a High Degree of Openness? The standard open set setting adopted in so far, presents a relatively balanced number of shared and private target classes with openness close to 0.5. Specifically it is $\mathbb{O} = 1 - \frac{10}{21} = 0.52$ for Office-31 and $\mathbb{O} = 1 - \frac{25}{65} = 0.62$ for Office-Home. In real-world problems, we can expect the number of unknown target classes to largely exceed the number of known classes, with openness approaching 1. We investigate this setting using Office-Home and, starting from the classes sorted with ID from 0 to 64 in alphabetic order, we define the following settings with increasing openness: **25** known classes $\mathbb{O} = 0.62$, ID: {0-24, 25-49, 40-64}, **10** known classes $\mathbb{O} = 0.85$, ID: {0-9, 10-19, 20-29}, **5** known classes $\mathbb{O} = 0.92$, ID: {0-4, 5-9, 10-14}. Figure 5 shows that the performance of our best competitors, STA and OSBP, deteriorates with larger $\mathbb{O}$ due to their inability to recognize the unknown samples. On the other hand, ROS maintains a consistent performance.

6 Discussion and Conclusions

In this paper, we present ROS: a novel method that tackles OSDA by using the self-supervised task of predicting image rotation. We show that, with simple variations of the rotation prediction task, we can first separate the target samples into known and unknown, and then align the target samples predicted as known with the source samples. Additionally, we propose HOS: a new OSDA metric defined as the harmonic mean between the accuracy of recognizing the known classes and rejecting the unknown samples. HOS overcomes the drawbacks of the current metric OS where the contribution of the unknown classes vanishes with increasing number of known classes.

We evaluate the perfomance of ROS on the standard Office-31 and Office-Home benchmarks, showing that it outperforms the competing methods. In addition, when tested on settings with increasing openness, ROS is the only method that maintains a steady performance. HOS reveals to be crucial in this evaluation to correctly assess the performance of the methods on both known and unknown samples. Finally, the failure in reproducing the reported results of existing methods exposes an important issue in OSDA that echoes the current reproducibility crisis in machine learning. We hope that our contributions can help laying a more solid foundation for the field.

Acknowledgements. This work was partially founded by the ERC grant 637076 RoboExNovo (SB), by the H2020 ACROSSING project grant 676157 (MRL) and took advantage of the NVIDIA GPU Academic Hardware Grant (TT).

References

1. Bendale, A., Boult, T.E.: Towards open set deep networks. In: CVPR (2016)
2. Bergman, L., Hoshen, Y.: Classification-based anomaly detection for general data. In: ICLR (2020)
3. Bousmalis, K., Trigeorgis, G., Silberman, N., Krishnan, D., Erhan, D.: Domain separation networks. In: NeurIPS (2016)
4. Candès, E.J., Li, X., Ma, Y., Wright, J.: Robust principal component analysis? J. ACM **58**(3), 1–37 (2011)
5. Carlucci, F.M., D'Innocente, A., Bucci, S., Caputo, B., Tommasi, T.: Domain generalization by solving jigsaw puzzles. In: CVPR (2019)
6. Reproducibility challenge. https://reproducibility-challenge.github.io/neurips2019/. Accessed 4 Mar 2020
7. The machine learning reproducibility checklist. https://www.cs.mcgill.ca/~jpineau/ReproducibilityChecklist.pdf. Accessed 4 Mar 2020
8. Csurka, G.: Domain Adaptation in Computer Vision Applications, 1st edn. Springer Publishing Company, Incorporated (2017)
9. Deng, J., Dong, W., Socher, R., Li, L.J., Li, K., Fei-Fei, L.: Imagenet: a large-scale hierarchical image database. In: CVPR (2009)

10. Dodge, J., Gururangan, S., Card, D., Schwartz, R., Smith, N.A.: Show your work: improved reporting of experimental results. In: EMNLP (2019)
11. Doersch, C., Gupta, A., Efros, A.A.: Unsupervised visual representation learning by context prediction. In: ICCV (2015)
12. Eskin, E., Arnold, A., Prerau, M., Portnoy, L., Stolfo, S.: A geometric framework for unsupervised anomaly detection. In: Barbará, D., Jajodia, S. (eds.) Applications of Data Mining in Computer Security. Advances in Information Security, vol 6, pp. 77–101. Springer, Boston (2002)
13. Feng, Q., Kang, G., Fan, H., Yang, Y.: Attract or distract: exploit the margin of open set. In: ICCV (2019)
14. Ganin, Y., et al.: Domain-adversarial training of neural networks. J. Mach. Learn. Res. **17**(1), 1–35 (2016)
15. Ghifary, M., Kleijn, W.B., Zhang, M., Balduzzi, D., Li, W.: Deep reconstruction-classification networks for unsupervised domain adaptation. In: Leibe, B., Matas, J., Sebe, N., Welling, M. (eds.) ECCV 2016. LNCS, vol. 9908, pp. 597–613. Springer, Cham (2016). https://doi.org/10.1007/978-3-319-46493-0_36
16. Gidaris, S., Singh, P., Komodakis, N.: Unsupervised representation learning by predicting image rotations. arXiv preprint arXiv:1803.07728 (2018)
17. Golan, I., El-Yaniv, R.: Deep anomaly detection using geometric transformations. In: NeurIPS (2018)
18. He, K., Zhang, X., Ren, S., Sun, J.: Deep residual learning for image recognition. In: CVPR (2016)
19. Henderson, P., Islam, R., Bachman, P., Pineau, J., Precup, D., Meger, D.: Deep reinforcement learning that matters. In: AAAI (2018)
20. Hendrycks, D., Gimpel, K.: A baseline for detecting misclassified and out-of-distribution examples in neural networks. In: ICLR (2017)
21. Kim, J., Scott, C.D.: Robust kernel density estimation. J. Mach. Learn. Res. **13**(1), 2529–2565 (2012)
22. Larsson, G., Maire, M., Shakhnarovich, G.: Colorization as a proxy task for visual understanding. In: CVPR (2017)
23. Liang, S., Li, Y., Srikant, R.: Enhancing the reliability of out-of-distribution image detection in neural networks. In: ICLR (2018)
24. Liu, H., Cao, Z., Long, M., Wang, J., Yang, Q.: Separate to adapt: open set domain adaptation via progressive separation. In: CVPR (2019)
25. Loghmani, M.R., Vincze, M., Tommasi, T.: Positive-unlabeled learning for open set domain adaptation. Pattern Recogn. Lett. **136**, 198–204 (2020)
26. Long, M., Cao, Y., Wang, J., Jordan, M.I.: Learning transferable features with deep adaptation networks. In: ICML (2015)
27. Lucic, M., Kurach, K., Michalski, M., Gelly, S., Bousquet, O.: Are gans created equal? a large-scale study. In: NeurIPS (2018)
28. Noroozi, M., Favaro, P.: Unsupervised learning of visual representations by solving jigsaw puzzles. In: Leibe, B., Matas, J., Sebe, N., Welling, M. (eds.) ECCV 2016. LNCS, vol. 9910, pp. 69–84. Springer, Cham (2016). https://doi.org/10.1007/978-3-319-46466-4_5
29. Panareda Busto, P., Gall, J.: Open set domain adaptation. In: ICCV (2017)
30. Pathak, D., Krähenbühl, P., Donahue, J., Darrell, T., Efros, A.: Context encoders: feature learning by inpainting. In: CVPR (2016)
31. Ruff, L., et al.: Deep one-class classification. In: ICML (2018)
32. Russo, P., Carlucci, F.M., Tommasi, T., Caputo, B.: From source to target and back: symmetric bi-directional adaptive gan. In: CVPR (2018)

33. Saenko, K., Kulis, B., Fritz, M., Darrell, T.: Adapting visual category models to new domains. In: Daniilidis, K., Maragos, P., Paragios, N. (eds.) ECCV 2010. LNCS, vol. 6314, pp. 213–226. Springer, Heidelberg (2010). https://doi.org/10.1007/978-3-642-15561-1_16
34. Saito, K., Yamamoto, S., Ushiku, Y., Harada, T.: Open set domain adaptation by backpropagation. In: Ferrari, V., Hebert, M., Sminchisescu, C., Weiss, Y. (eds.) ECCV 2018. LNCS, vol. 11209, pp. 156–171. Springer, Cham (2018). https://doi.org/10.1007/978-3-030-01228-1_10
35. Saito, K., Yamamoto, S., Ushiku, Y., Harada, T.: Open set domain adaptation by backpropagation. arXiv preprint arXiv:1804.10427 (2018)
36. Schlegl, T., Seeböck, P., Waldstein, S.M., Schmidt-Erfurth, U., Langs, G.: Unsupervised anomaly detection with generative adversarial networks to guide marker discovery. In: IPMI (2017)
37. Schölkopf, B., Williamson, R., Smola, A., Shawe-Taylor, J., Platt, J.: Support vector method for novelty detection. In: NeurIPS (1999)
38. Simonyan, K., Zisserman, A.: Very deep convolutional networks for large-scale image recognition. arXiv preprint arXiv:1409.1556 (2014)
39. Sun, B., Feng, J., Saenko, K.: Return of frustratingly easy domain adaptation. In: AAAI (2016)
40. Tzeng, E., Hoffman, J., Saenko, K., Darrell, T.: Adversarial discriminative domain adaptation. In: CVPR (2017)
41. Venkateswara, H., Eusebio, J., Chakraborty, S., Panchanathan, S.: Deep hashing network for unsupervised domain adaptation. In: CVPR (2017)
42. Wen, Y., Zhang, K., Li, Z., Qiao, Yu.: A discriminative feature learning approach for deep face recognition. In: Leibe, B., Matas, J., Sebe, N., Welling, M. (eds.) ECCV 2016. LNCS, vol. 9911, pp. 499–515. Springer, Cham (2016). https://doi.org/10.1007/978-3-319-46478-7_31
43. Xia, Y., Cao, X., Wen, F., Hua, G., Sun, J.: Learning discriminative reconstructions for unsupervised outlier removal. In: ICCV (2015)
44. Xu, J., Xiao, L., López, A.M.: Self-supervised domain adaptation for computer vision tasks. IEEE Access **7**, 156694–156706 (2019)
45. Xu, R., Li, G., Yang, J., Lin, L.: Larger norm more transferable: an adaptive feature norm approach for unsupervised domain adaptation. In: ICCV (2019)
46. You, K., Long, M., Cao, Z., Wang, J., Jordan, M.I.: Universal domain adaptation. In: CVPR (2019)
47. Zhai, S., Cheng, Y., Lu, W., Zhang, Z.: Deep structured energy based models for anomaly detection. In: ICML (2016)
48. Zhang, R., Isola, P., Efros, A.A.: Colorful image colorization. In: Leibe, B., Matas, J., Sebe, N., Welling, M. (eds.) ECCV 2016. LNCS, vol. 9907, pp. 649–666. Springer, Cham (2016). https://doi.org/10.1007/978-3-319-46487-9_40
49. Zhou, C., Paffenroth, R.C.: Anomaly detection with robust deep autoencoders. In: ACM SIGKDD (2017)
50. Zimek, A., Schubert, E., Kriegel, H.P.: A survey on unsupervised outlier detection in high-dimensional numerical data. Stat. Anal. Data Mining ASA Data Sci. J. **5**, 363–387 (2012)
51. Zong, B., et al.: Deep autoencoding gaussian mixture model for unsupervised anomaly detection. In: ICLR (2018)

Combining Task Predictors via Enhancing Joint Predictability

Kwang In Kim[1(✉)], Christian Richardt[2], and Hyung Jin Chang[3]

[1] UNIST, Ulsan, Korea
kimki@unist.ac.kr
[2] University of Bath, Bath, UK
[3] University of Birmingham, Birmingham, UK

Abstract. Predictor combination aims to improve a (target) predictor of a learning task based on the (reference) predictors of potentially relevant tasks, without having access to the internals of individual predictors. We present a new predictor combination algorithm that improves the target by i) measuring the relevance of references based on their capabilities in predicting the target, and ii) strengthening such estimated relevance. Unlike existing predictor combination approaches that only exploit pairwise relationships between the target and each reference, and thereby ignore potentially useful dependence among references, our algorithm *jointly* assesses the relevance of all references by adopting a Bayesian framework. This also offers a rigorous way to automatically select only relevant references. Based on experiments on seven real-world datasets from visual attribute ranking and multi-class classification scenarios, we demonstrate that our algorithm offers a significant performance gain and broadens the application range of existing predictor combination approaches.

1 Introduction

Many practical visual understanding problems involve learning multiple tasks. When a *target predictor*, e.g. a classification or a ranking function tailored for the task at hand, is not accurate enough, one could benefit from knowledge accumulated in the predictors of other tasks (*references*). The ***predictor combination* problem** studied by Kim et al. [11] aims to improve the target predictor by exploiting the references without requiring access to the internals of any predictors or assuming that all predictors belong to the same class of functions. This is relevant when the forms of predictors are not known (e.g. precompiled binaries) or the best predictor forms differ across tasks. For example, Gaussian process rankers [9] trained on ResNet101 features [7] are identified as the best for the main task, e.g. for image frame retrieval, while convolutional neural networks

Electronic supplementary material The online version of this chapter (https://doi.org/10.1007/978-3-030-58517-4_26) contains supplementary material, which is available to authorized users.

A. Vedaldi et al. (Eds.): ECCV 2020, LNCS 12361, pp. 439–455, 2020.
https://doi.org/10.1007/978-3-030-58517-4_26

are presented as a reference, e.g. classification of objects in images. In this case, existing transfer learning or multi-task learning approaches, such as a parameter or weight sharing, cannot be applied directly.

Kim et al. [11] approached this predictor combination problem for the first time by nonparametrically accessing all predictors based on their evaluations on given datasets, regarding each predictor as a Gaussian process (GP) estimator. Assuming that the target predictor is a noisy observation of an underlying ground-truth predictor, their algorithm projects all predictors onto a Riemannian manifold of GPs and denoises the target by simulating a diffusion process therein. This approach has demonstrated a noticeable performance gain while meeting the challenging requirements of the predictor combination problem. However, it leaves three possibilities to improve. Firstly, this algorithm is inherently (pairwise) metric-based and, therefore, it can model and exploit only pairwise relevance of the target and each reference, while relevant information can lie in the relationship between multiple references. Secondly, this algorithm assumes that all references are noise-free, while in practical applications, the references may also be trained based on limited data points or weak features and thus they can be imperfect. Thirdly, as this algorithm uses the metric defined between GPs, it can only combine one-dimensional target and references.

In this paper, we propose a new predictor combination algorithm that overcomes these three challenges. The proposed algorithm builds on the manifold denoising framework [11] but instead of their metric diffusion process, we newly pose the predictor denoising as an averaging process, which *jointly* exploits *full dependence* of the references. Our algorithm casts the denoising problem into 1) measuring the *joint* capabilities of the references in predicting the target, and 2) optimizing the target as a variable to enhance such prediction capabilities. By adopting Bayesian inference under this setting, identifying relevant references is now addressed by a rigorous Bayesian relevance determination approach. Further, by denoising *all* predictors in a single unified framework, our algorithm becomes applicable even for imperfect references. Lastly, our algorithm can combine multi-dimensional target and reference predictors, e.g. it can improve multi-class classifiers based on one-dimensional rank predictors. Experiments on *relative attribute* ranking and multi-class classification demonstrate that these contributions individually and collectively improve the performance of predictor combination and further extend the application domain of existing predictor combination algorithms.

Related Work. Transfer learning (TL) aims to solve a given learning problem by adapting a source model trained on a different problem [16]. Predictor combination can be regarded as a specific instance of TL. However, unlike predictor combination algorithms, traditional TL approaches improve or newly train predictors of *known* form. Also, most existing algorithms assume that the given source is relevant to the target and, therefore, they do not explicitly consider identifying relevant predictors among many (potentially irrelevant) source predictors.

Another related problem is multi-task learning (MTL), which learns predictors on multiple problems at once [1,2]. State-of-the-art MTL algorithms offer the capability of automatically identifying relevant task groups when not all tasks and the corresponding predictors are mutually relevant. For example, Argyriou et al. [1] and Gong et al. [6], respectively, enforced the sparsity and low-rank constraints in the parameters of predictors to make them aggregate in relevant task groups. Passos et al. [18] performed explicit task clustering ensuring that all tasks (within a cluster) that are fed to the MTL algorithm are relevant. More recently, Zamir et al. [26] proposed to discover a hypergraph that reveals the interdependence of multiple tasks and facilitates transfer of knowledge across relevant tasks.

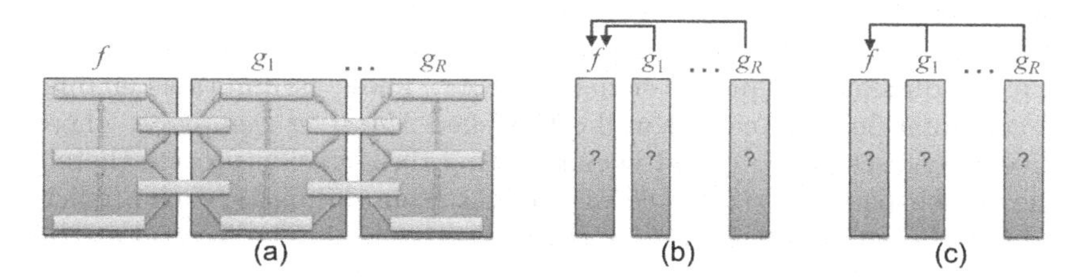

Fig. 1. Illustration of predictor combination algorithms: (a) MTL simultaneously exploits all references $\{g_1, \ldots, g_R\}$ to improve the target predictor f, e.g. by sharing neural network layers. However, they require access to the internals of predictors [2]. (b) Kim et al.'s predictor combination is agnostic to the forms of individual predictors [11] but exploits only pairwise relationships. (c) Our algorithm combines the benefits of both, jointly exploiting all references without requiring their known forms.

While our approach has been motivated by the success of TL and MTL approaches, these approaches are not directly applicable to predictor combination as they share knowledge across tasks via the internal parametric representations [1,6,18] and/or shared deep neural network layers of all predictors (e.g. via shared encoder readouts [26]; see Fig. 1). A closely related approach in this context is Mejjati et al.'s nonparametric MTL approach [13]. Similar to Kim et al. [11], this algorithm assesses predictors based on their sample evaluations, and it (nonparametrically) measures and enforces pairwise statistical dependence among predictors. As this approach is agnostic to the forms of individual predictors, it can be adapted for predictor combination. However, this algorithm shares the same limitations: it can only model pairwise relationships. We demonstrate experimentally that by modeling the joint relevance of all references, our algorithm can significantly outperform both Kim et al.'s original predictor combination algorithm [11] adapted to ranking [10], and Mejjati et al.'s MTL algorithm [13].

2 The Predictor Combination Problem

Suppose we have an *initial predictor* $f^0\colon \mathcal{X} \to \mathcal{Y}$ (e.g. a classification, regression, or ranking function) of a task. The goal of predictor combination is to improve the *target predictor* f^0 based on a set of *reference predictors* $\mathcal{G} = \{g_i\colon \mathcal{X} \to \mathcal{Y}_i\}_{i=1}^R$. The internal structures of the target and reference predictors are unknown and they might have different forms. Crucial to the success of addressing this seriously ill-posed problem is to determine which references (if any) within $\mathcal{G}$ are *relevant* (i.e. useful in improving f^0), and to design a procedure that fully exploits such relevant references without requiring access to the internals of f^0 and $\mathcal{G}$.

Kim et al.'s original predictor combination (*OPC*) [11] approaches this problem by 1) considering the initial predictor f^0 as a noisy estimate of the underlying ground-truth f_{GT}, and 2) assuming f_{GT} and $\mathcal{G}$ are structured such that they all lie on a low-dimensional predictor manifold $\mathcal{M}$. These assumptions enable predictor combination to be cast as well-established *Manifold Denoising*, where one iteratively denoises points on $\mathcal{M}$ via simulating the diffusion process therein [8].

The model space $\mathcal{M}$ of *OPC* consists of Bayesian estimates: each predictor in $\mathcal{M}$ is a GP predictive distribution of the respective task. The natural metric $g_{\mathcal{M}}$ on $\mathcal{M}$, in this case, is induced from the Kullback-Leibler (KL) divergence D_{KL} between probability distributions. Now further assuming that all reference predictors are noise-free, their diffusion process is formulated as a time-discretized evolution of f^t on $\mathcal{M}$: Given the solution f^t at time t and noise-free references $\mathcal{G}$, the new solution f^{t+1} is obtained by minimizing the energy

$$\mathcal{E}_{\mathrm{O}}(f) = D^2_{\mathrm{KL}}(f \mid f^t) + \lambda_{\mathrm{O}} \sum_{i=1}^{R} w_i D^2_{\mathrm{KL}}(f \mid g_i), \tag{1}$$

where $w_i = \exp(-D^2_{\mathrm{KL}}(f^t \mid g_i)/\sigma^2_{\mathrm{O}})$ is inversely proportional to $D_{\mathrm{KL}}(f^t \mid g_i)$, and λ_{O} and σ^2_{O} are hyperparameters. Our supplemental document presents how the iterative minimization of $\mathcal{E}_{\mathrm{O}}$ is obtained by discretizing the diffusion process on $\mathcal{M}$.

In practice, it is infeasible to directly optimize functions, which are infinite-dimensional objects. Instead, *OPC* approximates all predictors $\{f, \mathcal{G}\}$ via their evaluations on a test dataset $X = \{\mathbf{x}_1, \ldots, \mathbf{x}_N\}$, and optimizes the sample f-evaluation $\mathbf{f} = f|_X := [f(\mathbf{x}_1), \ldots, f(\mathbf{x}_N)]^\top$ based on the sample references $\mathcal{G} = \{\mathbf{g}_1, \ldots, \mathbf{g}_R\}$ with $\mathbf{g}_i = g_i|_X$.

At each time step, the relevance of a reference is automatically determined based on its KL-divergence to the current solution: g_i is considered relevant when $D_{\mathrm{KL}}(f^t \mid g_i)$ is small. Then, throughout the iteration, *OPC* robustly denoises f by gradually putting more emphasis on highly relevant references while ignoring outliers. This constitutes the first predictor combination algorithm that improves the target predictor without requiring any known forms of predictors (as the KL-divergences are calculated purely based on predictor evaluations). However, Eq. 1 also highlights the limitations of this approach: it exploits only pairwise relationships between the target predictor and individual references, ignoring the potentially useful information that lies in the dependence between references.

Toy Problem 1. Consider two references, $\{\mathbf{g}_1, \mathbf{g}_2\} \subset \mathbb{R}^{100}$, constructed by uniformly randomly sampling from $\{0, 1\}$. Here, $\{\mathbf{g}_1, \mathbf{g}_2\}$ are regarded as the means of GP predictive distributions with unit variances. We define the ground-truth target as their difference: $\mathbf{f}_{\text{GT}} = \mathbf{g}_1 - \mathbf{g}_2$. By construction, $\mathbf{f}_{\text{GT}}$ is determined by the *relationship* between the references. Now we construct the initial noisy predictor $\mathbf{f}^0$ by adding independent Gaussian noise with standard deviation 1 to $\mathbf{f}_{\text{GT}}$, achieving the rank accuracy of 0.67 (see Sect. 4 for the definition of the visual attribute ranking problem). In this case, applying *OPC* minimizes $\mathcal{E}_{\text{O}}$ (Eq. 1) but shows insignificant performance improvement as no information on $\mathbf{f}_{\text{GT}}$ can be gained by assessing the relevance of the references individually (Table 1). While this problem has been well-studied in existing MTL and TL approaches, the application of these techniques for predictor combination is not straightforward as they require simultaneous training [6,18] and/or shared predictor forms [26]. Another limitation is that *OPC* requires that all predictions are one-dimensional (i.e. $\mathcal{Y}_i \subset \mathbb{R}$). Therefore, it is not capable of, for example, improving the multi-class classification predictor $\mathbf{f}^0$ given the references constructed for ranking tasks.

Table 1. Accuracies of Kim et al.'s original (*OPC*) [11], and our linear (*LPC*) and nonlinear (*NPC*) predictor combination algorithms introduced in Sect. 3, for illustrative toy problems. $\mathbf{g}_1$ and $\mathbf{g}_2$ are random binary vectors while $\mathbf{f}^0$'s are noisy observations of the corresponding ground-truth predictors $\mathbf{f}_{\text{GT}}$'s.

Toy problem	$\mathbf{f}^0$	*OPC* [11] (Eq. 1)	*LPC* (Eq. 7)	*NPC* (Eq. 13)
1: $\mathbf{f}_{\text{GT}} = \mathbf{g}_1 - \mathbf{g}_2$	67.14	67.24	**100**	**100**
2: $\mathbf{f}_{\text{GT}} = \text{XOR}(\mathbf{g}_1, \mathbf{g}_2)$	74.08	74.11	74.24	**100**

3 Joint Predictor Combination Algorithm

Our algorithm takes deterministic predictors instead of Bayesian predictors (i.e. GP predictive distributions) as in *OPC*. When Bayesian predictors are provided as inputs, we simply take their means and discard the predictive variances. This design choice offers a wider range of applications as most predictors—including deep neural networks and support vector machines (SVMs)—are presented as deterministic functions, at the expense of not exploiting potentially useful predictive uncertainties. This assumption has also been adopted by Kim and Chang [10]. Under this setting, our model space is a sub-manifold $\mathcal{M}$ of L^2 space where each predictor has zero mean and unit norm:

$$\forall f \in \mathcal{M}. \quad \int f(\mathbf{x}) \mathrm{d}P(\mathbf{x}) = 0 \quad \text{and} \quad \langle f, f \rangle = 1, \tag{2}$$

where $\langle f, g\rangle := \int f(\mathbf{x})g(\mathbf{x})dP(\mathbf{x})$ and $P(\mathbf{x})$ is the probability distribution of $\mathbf{x}$. This normalization enables scale and shift-invariant assessment of the relevance of references. The Riemannian metric $g_{\mathcal{M}}$ on $\mathcal{M}$ is defined as the *pullback* metric of the ambient L^2 space: when $\mathcal{M}$ is embedded into L^2 via the embedding $\imath$, $g_{\mathcal{M}}(a,b) := \langle \imath(a), \imath(b)\rangle$. *OPC* (Eq. 1) can be adapted for $\mathcal{M}$ by iteratively maximizing the objective $\mathcal{O}_{\mathrm{O}}$ that replaces the KL-divergence D_{KL} with $g_{\mathcal{M}}(\cdot,\cdot)$:

$$\mathcal{O}_{\mathrm{O}}(f) = g_{\mathcal{M}}(f, f^t)^2 + \lambda_{\mathrm{O}} \sum_{i=1}^{R} w_i g_{\mathcal{M}}(f, g_i)^2. \tag{3}$$

For simplicity of exposition, we here assume that the output space is one-dimensional (i.e. $\mathcal{Y}_i = \mathbb{R}$). In Sect. 4, we show how this framework can be extended to multi-dimensional outputs such as for multi-class classification.

The Averaging Process on $\mathcal{M}$. Both *OPC* (Eq. 1) and its adaptation to our model space (Eq. 3) can model only the pairwise relationship between the target f and each reference $g_i \in \mathcal{G}$, while ignoring the dependence present across the references (*joint relevance* of $\mathcal{G}$ on f). We now present a general framework that can capture such joint relevance by iteratively maximizing the objective

$$\mathcal{O}_{\mathrm{J}}(f) = \langle \imath(f), \imath(f^t)\rangle^2 + \lambda_{\mathrm{J}}\langle \imath(f), \mathcal{K}[\imath(f)]\rangle, \tag{4}$$

where $\lambda_{\mathrm{J}} \geq 0$ is a hyperparameter. The linear, non-negative definite averaging operator $\mathcal{K}\colon \imath(\mathcal{M}) \rightarrow \imath(\mathcal{M})$ is responsible to capture the joint relevance of $\mathcal{G}$ on f. Depending on the choice of $\mathcal{K}$, $\mathcal{O}_{\mathrm{J}}$ can accommodate a variety of predictor combination scenarios, including $\mathcal{O}_{\mathrm{O}}$ as a special case for $\mathcal{K}[\imath(f)] = \sum_{i=1}^{R} \imath(g_i) w_i \langle \imath(f), \imath(g_i)\rangle$.

3.1 Linear Predictor Combination (LPC)

Our linear predictability operator $\mathcal{K}_{\mathrm{L}}$ is defined as[1]

$$\mathcal{K}_{\mathrm{L}}[\imath(f)] = \sum_{i,j=1}^{R} \imath(g_i) C^{-1}_{[i,j]} \langle \imath(g_j), \imath(f)\rangle \tag{5}$$

using the *correlation matrix* $C_{[i,j]} = \langle \imath(g_i), \imath(g_j)\rangle$. Interpreting $\mathcal{K}_{\mathrm{L}}$ becomes straightforward when substituting $\mathcal{K}_{\mathrm{L}}$ into the second term of $\mathcal{O}_{\mathrm{J}}$ (Eq. 4):

$$\langle \imath(f), \mathcal{K}_{\mathrm{L}}[\imath(f)]\rangle = \mathbf{c}^\top C^{-1}\mathbf{c}, \tag{6}$$

where $\mathbf{c} = [\langle \imath(f), \imath(g_1)\rangle, \ldots, \langle \imath(f), \imath(g_R)\rangle]^\top$. As each predictor in $\mathcal{M}$ is centered and normalized, all diagonal elements of the correlation matrix C are 1. The off-diagonal elements of C then represent the dependence among the references, making $\langle \imath(f), \mathcal{K}_{\mathrm{L}}[\imath(f)]\rangle$ a measure of *joint correlation* between f and $\mathcal{G} = \{g_i\}_{i=1}^{R}$.

[1] Here, the term 'linear' signifies the capability of $\mathcal{K}_{\mathrm{L}}$ to capture the linear dependence of references, independent of $\mathcal{K}_{\mathrm{L}}$ being a linear operator as well.

In practice, f and $\{g_i\}$ might not be originally presented as embedded elements $\imath(f)$ and $\{\imath(g_i)\}$ of $\mathcal{M}$: i.e. they are not necessarily centered or normalized (Eq. 2). Also, as in the case of *OPC*, it would be infeasible to manipulate infinite-dimensional functions directly. Therefore, we also adopt sample approximations $\{\mathbf{f}, \mathbf{g}_1, \ldots, \mathbf{g}_R\}$ and explicitly project them onto $\mathcal{M}$ via normalization: $\mathbf{f} \rightarrow \overline{\mathbf{f}} := \frac{C_N \mathbf{f}}{\|C_N \mathbf{f}\|}$, where $C_N = \mathbf{1}_{N\times N}/N$, $\mathbf{1}_{N\times N}$ is an $N \times N$ matrix of ones, for the sample size $N = |X|$. For this scenario, we obtain our linear predictor combination (*LPC*) algorithm by substituting Eq. 5 into Eq. 4, and replacing f, f^t, and g_j by $\overline{\mathbf{f}}$, $\overline{\mathbf{f}}^t$, and $\overline{\mathbf{g}}_j$, respectively:

$$\mathcal{O}_{\mathrm{L}}(\mathbf{f}) = \frac{(\mathbf{f}^\top \overline{\mathbf{f}^t})^2}{\mathbf{f}^\top C_N \mathbf{f}} + \lambda_{\mathrm{J}} \mathcal{P}_{\mathrm{L}}, \tag{7}$$

where $\mathcal{P}_{\mathrm{L}} = \frac{\mathbf{f}^\top Q \mathbf{f}}{\mathbf{f}^\top C_N \mathbf{f}}$, $Q = G(G^\top G)^{-1} G$, and $G = [\overline{\mathbf{g}}_1, \ldots, \overline{\mathbf{g}}_R]$. Here, we pre-projected $\mathcal{G}$ and $\mathbf{f}^t$ onto $\mathcal{M}$ while $\mathbf{f}$ is explicitly projected in Eq. 7. Note that our goal is not to simply calculate $\mathcal{P}_{\mathrm{L}}$ for a fixed $\mathbf{f}$, but to optimize $\mathbf{f}$ while enhancing $\mathcal{P}_{\mathrm{L}}$.

Exploiting the joint relevance of references, *LPC* can provide significant accuracy improvements over *OPC*. For example, *LPC* can generate perfect predictions in Toy Problem 1 (Table 1). However, its capability in measuring the joint relevance is limited to linear relationships only. This can be seen by rewriting $\mathcal{P}_{\mathrm{L}}$ explicitly in $\mathbf{f}$ and G:

$$\mathcal{P}_{\mathrm{L}} = \frac{\mathbf{f}^\top Q \mathbf{f}}{\mathbf{f}^\top C_N \mathbf{f}} = 1 - \frac{\sum_{i=1}^{N} (\mathbf{f}_i - q(G_{[i,:]}))^2}{\sum_{i=1}^{N} (\mathbf{f}_i - \sum_{j=1}^{n} \mathbf{f}_j / N)^2}, \tag{8}$$

where $G_{[i,:]}$ represents the i-th row of G, and $q(\mathbf{a}) = \mathbf{w}_q^\top \mathbf{a}$ is the linear function whose weight vector $\mathbf{w}_q = (G^\top G)^{-1} G \mathbf{f}$ is obtained from least-squares regression that takes the reference matrix G as training input and the target predictor variable $\mathbf{f}$ as corresponding labels. Then, $\mathcal{P}_{\mathrm{L}}$ represents the normalized prediction accuracy: the normalizer $\mathbf{f}^\top C_N \mathbf{f}$ is simply the variance of $\mathbf{f}$ elements. For this reason, we call $\mathcal{P}_{\mathrm{L}}$ the (linear) *predictability* of G (and equivalently of $\mathcal{G}$) on $\mathbf{f}$. It takes the maximum value of 1 when the linear prediction (made based on G) perfectly agrees with $\mathbf{f}$ when normalized, and it attains the minimum value 0 when the prediction is no better than taking the mean value of $\mathbf{f}$, in which case the mean squared error becomes the variance. Figure 1 illustrates our algorithm in comparison with MTL and *OPC*.

Toy Problem 2. Under the setting of Toy problem 1, when the target $\mathbf{f}_{\mathrm{GT}}$ is replaced by a variable that is nonlinearly related to the references, e.g. using the logical exclusive OR (XOR) of $\mathbf{g}_1$ and $\mathbf{g}_2$, *LPC* fails to give any noticeable accuracy improvement compared to the baseline $\mathbf{f}^0$.

3.2 Nonlinear Predictor Combination (NPC)

Our final algorithm measures the relevance of $\mathcal{G}$ on $\mathbf{f}$ by predicting $\mathbf{f}$ via Gaussian process (GP) estimation. We use the standard zero-mean Gaussian prior and an i.i.d. Gaussian likelihood with noise variance σ^2 [19]. The resulting prediction is obtained as a Gaussian distribution with mean $\mathbf{m_f}$ and covariance $C_\mathbf{f}$:

$$\mathbf{m_f} = K(K+\sigma^2 I)^{-1}\mathbf{f}, \quad C_\mathbf{f} = K - K(K+\sigma^2 I)^{-1}K, \tag{9}$$

where $K \in \mathbb{R}^{N\times N}$ is defined using the covariance function $k\colon \mathbb{R}^R \times \mathbb{R}^R \to \mathbb{R}$:

$$K_{[i,j]} = k(G_{[i,:]}, G_{[j,:]}) := \exp\left(-\frac{\|G_{[i,:]} - G_{[j,:]}\|^2}{\sigma_k^2}\right). \tag{10}$$

Now we refine the linear predictability $\mathcal{P}_\mathrm{L}$ by replacing $q(G_{[i,:]})$ in Eq. 8 with the corresponding predictive mean $[\mathbf{m_f}]_i$ (where $[\mathbf{a}]_i$ is the i-th element of vector $\mathbf{a}$):

$$\mathcal{P}_\mathrm{N} = \frac{\mathbf{f}^\top Q'\mathbf{f}}{\mathbf{f}^\top C_N\mathbf{f}} = 1 - \frac{\sum_{i=1}^N([\mathbf{f}]_i - [\mathbf{m_f}]_i)^2}{\sum_{i=1}^N([\mathbf{f}]_i - \sum_{j=1}^N[\mathbf{f}]_j/N)^2}, \tag{11}$$

where Q' is a positive definite matrix that replaces Q in Eq. 8:

$$Q' = C_N\left(2K(K+\sigma^2 I)^{-1} - (K+\sigma^2 I)^{-1}KK(K+\sigma^2 I)^{-1}\right)C_N. \tag{12}$$

The matrix Q' becomes Q when the kernel $k(\mathbf{a},\mathbf{b})$ is replaced by the standard dot product $k'(\mathbf{a},\mathbf{b}) = \mathbf{a}^\top\mathbf{b}$. Note that the noise level σ^2 should be strictly positive; otherwise, $\mathbf{f}_i = [\mathbf{m_f}]_i$ for all $i \in \{1,\dots,N\}$, and therefore $\mathcal{P}_\mathrm{N} = 1$ for any $\mathbf{f}$. This means the resulting GP model perfectly overfits to $\mathbf{f}$ and all references are considered perfectly relevant regardless of the actual values of G and $\mathbf{f}$.

Computational Model. Explicitly normalizing $\mathbf{f}$ ($\mathbf{f} \to \overline{\mathbf{f}}$) in the *nonlinear predictability* $\mathcal{P}_\mathrm{N}$ (Eq. 11), substituting Q' into $\mathcal{P}_\mathrm{N}$, and then replacing $\mathcal{P}_\mathrm{L}$ with $\mathcal{P}_\mathrm{N}$ in $\mathcal{O}_\mathrm{L}$ (Eq. 7) yields the following Rayleigh quotient-type objective to maximize:

$$\mathcal{O}_\mathrm{N}(\mathbf{f}) = \frac{\mathbf{f}^\top A\mathbf{f}}{\mathbf{f}^\top C_N\mathbf{f}}, \quad A = (C_N\mathbf{f}^t)(C_N\mathbf{f}^t)^\top + \lambda_\mathrm{J}Q'. \tag{13}$$

For any non-negative definite matrices A and C_N, the maximizer of the Rayleigh quotient $\mathcal{O}_\mathrm{N}$ is the largest eigenvector (the eigenvector corresponding to the maximum eigenvalue) of the generalized eigenvector problem $A\mathbf{f} = \lambda C_N\mathbf{f}$. The computational complexity of solving the generalized eigenvector problem of matrices $\{A, C_N\} \subset \mathbb{R}^{N\times N}$ is $O(N^3)$. As in our case $N = |X|$, solving this problem is infeasible for large-scale problems. To obtain a computationally affordable solution, we first note that A incorporates multiplications by the centering matrix C_N and, therefore, all eigenvectors of A are centered, which implies that they are also eigenvectors of C_N. This effectively renders the generalized eigenvector problem into the standard eigenvector problem of matrix A.

Secondly, we make sparse approximate GP inference by adopting a low-rank approximation of K [20]:

$$K \approx K_{GB}K_{BB}^{-1}K_{GB}^\top, \; K_{GB[i,j]} = k(G_{[i,:]}, B_{[j,:]}), \; K_{BB[i,j]} = k(B_{[i,:]}, B_{[j,:]}), \tag{14}$$

where the i-th row $B_{[i,:]}$ of $B \in \mathbb{R}^{N' \times R}$ represents the i-th *basis vector.* We construct the basis vector matrix B by linearly sampling N' rows from all rows of G. Now substituting the kernel approximation in Eq. 14 into Eq. 12 leads to

$$Q'' = C_N K_{GB} (\lambda_J T) K_{GB}^\top C_N, \quad \text{with} \tag{15}$$

$$T = 2P - P K_{GB}^\top K_{GB} P \quad \text{and} \quad P = (K_{GB}^\top K_{GB} + \lambda K_{BB})^{-1}. \tag{16}$$

Replacing Q' in A with Q'', we obtain $A = YY^\top$, where

$$Y = \left[C_N \mathbf{f}^t, \sqrt{\lambda_J} K_{GB} T^{\frac{1}{2}} \right] \in \mathbb{R}^{N \times (N'+1)} \tag{17}$$

and $T^{\frac{1}{2}}(T^{\frac{1}{2}})^\top = T \in \mathbb{R}^{N' \times N'}$. Note that T is positive definite (PD) for $\sigma^2 > 0$ as Q'' is PD, which can be seen by noting that $0 \leq \frac{\mathbf{f}^\top Q'' \mathbf{f}}{\mathbf{f}^\top C_N \mathbf{f}} \leq 1$: by construction, $\mathbf{f}^\top Q'' \mathbf{f}$ is the prediction accuracy upper bounded by $\mathbf{f}^\top C_N \mathbf{f}$. Therefore, $T^{\frac{1}{2}}$ can be efficiently calculated based on the Cholesky decomposition of T. In the rare case where Cholesky decomposition cannot be calculated, e.g. due to round-off errors, we perform the (computationally more demanding) eigenvalue decomposition $E \Lambda E^\top$ of T, replace all eigenvalues in Λ that are smaller than a threshold $\varepsilon = 10^{-9}$ by ε, and construct $T^{\frac{1}{2}}$ as $E \Lambda^{\frac{1}{2}}$.

Finally, by noting that, when normalized, the largest eigenvector of $YY^\top \in \mathbb{R}^{N \times N}$ is the same as $Y\mathbf{e}$, where $\mathbf{e}$ is the largest eigenvector of $Y^\top Y \in \mathbb{R}^{(N'+1) \times (N'+1)}$, the optimum $\mathbf{f}^*$ of $\mathcal{O}_N$ in Eq. 13 is obtained as $\frac{Y\mathbf{e}}{\|Y\mathbf{e}\|}$ and $\mathbf{e}$ can be efficiently calculated by iterating the power method on $Y^\top Y$. The normalized output $\mathbf{f}^*$ can be directly used in some applications, e.g. ranking. When the absolute values of predictors are important, e.g. in regression and multi-class classification, the standard deviation and the mean of $\mathbf{f}^0$ can be stored before the predictor combination process and $\mathbf{f}^*$ is subsequently inverse normalized.

3.3 Automatic Identification of Relevant Tasks

Our algorithm *NPC* is designed to exploit all references. However, in general, not all references are relevant and therefore, the capability of identifying only relevant references can help. *OPC* does so by defining the weights $\{w_i\}$ (Eq. 1). However, this strategy inherits the limitation of *OPC* in that it does not consider all references jointly. An important advantage of our approach, formulating predictor combination as enhancing the predictability via Bayesian inference, is that the well-established methods of automatic relevance determination can be employed for identifying relevant references. In our GP prediction framework, the contributions of references are controlled by the kernel function k (Eq. 10). The original Gaussian kernel k uses (isotropic) Euclidean distance $\|\cdot\|$ on $\mathcal{X}$ and thus treats all references equally. Now replacing it by an *anisotropic* kernel

$$k_A(\mathbf{a}, \mathbf{b}) = \exp\left(-(\mathbf{a} - \mathbf{b})^\top \Sigma_A (\mathbf{a} - \mathbf{b})\right) \tag{18}$$

with $\Sigma_A = \operatorname{diag}\left[\sigma_A^1, \ldots, \sigma_A^R\right]$ being a diagonal matrix of non-negative entries renders the problem of identifying relevant references into estimating the hyper-parameter matrix Σ_A: when σ_A^i is large, then $\mathbf{g}_i$ is considered relevant and it makes a significant contribution in predicting $\mathbf{f}$, while a small σ_A^i indicates that $\mathbf{g}_i$ makes a minor contribution.

For a fixed target predictor $\mathbf{f}$, identifying the optimal parameter Σ_A^* is a well-studied problem in Bayesian inference: Σ_A^* can be determined by maximizing the *marginal likelihood* [19] $p(\mathbf{f}|G, \Sigma_A)$. This strategy cannot be directly applied to our algorithm as $\mathbf{f}$ is the variable that is optimized depending on the prediction made by GPs. Instead, one could estimate Σ_A^* based on the initial prediction $\mathbf{f}^0$ and G, and fix it throughout the optimization of $\mathbf{f}$. We observed in our preliminary experiments that this strategy indeed led to noticeable performance improvement over using the isotropic kernel k. However, optimizing the GP marginal likelihood $P(\mathbf{f}|G, \Sigma_A)$ for a (nonlinear) Gaussian kernel (Eq. 10) is computationally demanding: this process takes roughly 1,000 times longer than the optimization of $\mathcal{O}_{\mathrm{N}}$ (Eq. 13; for the *AWA2* dataset case; see Sect. 4). Instead, we first efficiently determine surrogate parameters $\Sigma_L = \mathrm{diag}\left[\sigma_L^1, \ldots, \sigma_L^R\right]$ by optimizing the marginal likelihood based on the linear anisotropic kernel $k_L(\mathbf{a}, \mathbf{b}) = \mathbf{a}^\top \Sigma_L \mathbf{b}$. In our preliminary experiments, we observed that once optimized, the relative magnitudes of Σ_L^* elements are similar to these of Σ_A^*, but their global scales differ (see the supplemental document for examples and details of marginal likelihood optimization). In our final algorithm, we determine Σ_A^* by scaling Σ_L^*: $\Sigma_A^* = \Sigma_L^*/\sigma_k^2$ for a hyperparameter $\sigma_k^2 > 0$.

Figure 2 demonstrates the effectiveness of automatic relevance determination: The *OSR* dataset contains 6 target attributes for each data instance, which are defined based on the underlying class labels. The figure shows the average diagonal values of Σ_L^* on this dataset estimated for the first attribute using the remaining 5 attributes, plus 8 additional attributes as references. Two scenarios are considered. In the *random references* scenario, the additional attributes are randomly generated. As indicated by small magnitudes and the corresponding standard deviations of Σ_L^* entries, our algorithm successfully disregarded these irrelevant references. In *class references* scenario, the additional attributes are ground-truth class labels which provide complete information about the target attributes. Our algorithm successfully *picks up* these important references. On average, removing the automatic relevance determination from our algorithm decreases the accuracy improvement (from the initial predictors $\mathbf{f}^0$) by 11.97% (see Table 2).

3.4 Joint Denoising

So far, we assumed that all references in $\mathcal{G}$ are noise-free. However, in practice, they might be noisy estimates of the ground truth. In this case, noise in the references could be propagated to the target predictor during denoising, which would degrade the final output. We account for this by denoising *all* predictors $\mathcal{H} = \{\mathbf{f}, \mathbf{g}_1, \ldots, \mathbf{g}_R\}$ simultaneously. At each iteration t, each predictor $\mathbf{h} \in \mathcal{H}$ is denoised by considering it as the target predictor, and $\mathcal{H} \setminus \{\mathbf{h}\}$ as the references in Eq. 13. In the experiments with the *OSR* dataset, removing this joint denoising process from our final algorithm decreases the average accuracy

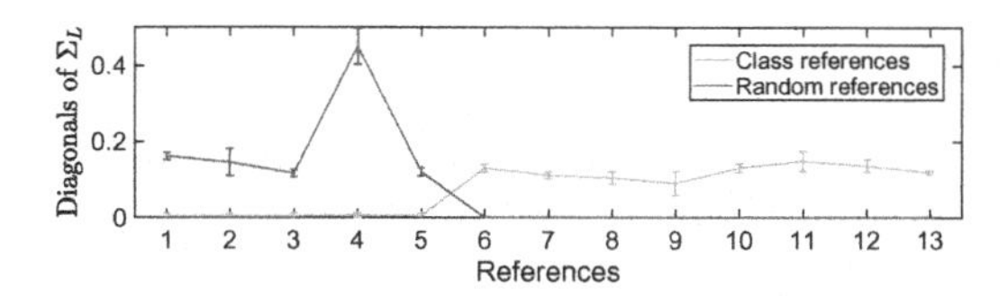

Fig. 2. The average diagonal values of Σ_L^* optimized for the first attribute of the *OSR* dataset as the target with remaining 5 attributes in the same dataset as references 1 to 5, plus 8 additional attributes as references 6 to 13. Σ_L^* values are normalized to sum to one for visualization. The length of each error bar corresponds to twice the standard deviation. *Class references*: References 6–13 are class labels from which attribute labels are generated. *Random references*: References 6–13 are randomly generated. See text.

Table 2. Effect of design choices in our algorithm on the *OSR* dataset. The average rank accuracy improvement over multiple target attributes from the baseline initial predictions $\mathbf{f}^0$ are shown (see Sect. 4 for details). *w/o joint denois.* only denoising the target predictor. *w/o auto. relev.*: without automatic relevance determination. Numbers in parentheses are accuracy ratios w.r.t. *Final NPC*.

Design choices →	*w/o joint denois.*	*w/o auto. relev.*	*Final NPC*
Accuracy improvement	1.96 (91.74%)	1.88 (88.03%)	**2.13 (100%)**

rate by 8.26% (see Table 2). We provide a summary of our complete algorithm in the supplemental document.

Computational Complexity and Discussion. Assuming that $N \gg R$, the computational complexity of our algorithm (Eq. 13) is dominated by calculating the kernel matrix K_{GB} (Eq. 14), which takes $O(NN'R)$ for N data points, N' basis vectors and R references. The second-most demanding part is the calculation of $T^{\frac{1}{2}}$ from T based on Cholesky decomposition (Eq. 15; $O(N'^3)$). As we denoise not only the target predictor but also all references, the overall computational complexity of each denoising step is $O(R \times (NN'R + N'^3))$. On a machine with an Intel Core i7 9700K CPU and an NVIDIA GeForce RTX 2080 Ti GPU, the entire denoising process, including optimization of $\{(\Sigma_A)_i\}_{i=1}^R$ (Eq. 18), took around 10 s for the *AWA2* dataset with 37,322 data points and 79 references for each target attribute. For simplicity, we use the low-rank approximation of K (Eq. 14) for constructing sparse GP predictions, while more advanced methods exist [19]. The number N' of basis vectors is fixed at 300 throughout our experiments. While the accuracy of low-rank approximation (Eq. 14) is in general positively correlated with N', we have not observed any significant performance gain by raising N' to 1,000 in our experiments. GP predictions also generally improve when *optimizing* the basis matrix B, e.g. via the marginal likelihood [21] instead of being selected from datasets as we did. Our efficient eigenvector calculation approach (Eq. 17) can still be applied in these cases.

4 Experiments

We assessed the effectiveness of our predictor combination algorithm in two scenarios: 1) visual attribute ranking [17], and 2) multi-class classification guided by the estimated visual attribute ranks. Given a database of images $X \subset \mathcal{X}$, visual attribute ranking aims to introduce a linear ordering of entries in X based on the strength of semantic attributes present in each image $\mathbf{x} \in X$. For a visual attribute, our goal is to estimate a rank predictor $f\colon \mathcal{X} \to \mathbb{R}$, such that $f(\mathbf{x}_i) > f(\mathbf{x}_j)$ when the attribute is stronger in $\mathbf{x}_i$ than $\mathbf{x}_j$. Parikh and Grauman's original relative attributes algorithm [17] estimates a linear rank predictor $f(\mathbf{x}) = \mathbf{w}^\top \mathbf{x}$ via rank SVMs that use the rank loss $\mathcal{L}$ defined on ground-truth ranked pairs $U \subset X \times X$:

$$\mathcal{E}(f) = \sum_{(\mathbf{x}_i,\mathbf{x}_j)\in U} \mathcal{L}(f, (\mathbf{x}_i, \mathbf{x}_j)) + C\|\mathbf{w}\|^2, \tag{19}$$

$$\mathcal{L}(f, (\mathbf{a}, \mathbf{b})) = \max\left(1 - (f(\mathbf{a}) - f(\mathbf{b})), 0\right)^2. \tag{20}$$

Yang et al. [24] and Meng et al. [14] extended this initial work using deep neural networks (*neural rankers*). Kim and Chang [10] extended the original predictor combination framework of Kim et al. [11] to rank predictor combination.

Experimental Settings. For visual attribute ranking, we use seven datasets, each with annotations for multiple attributes per image. For each attribute, we construct an initial predictor and denoise it via predictor combination using the predictors constructed for the remaining attributes as the reference. The initial predictors are constructed by first training 1) neural rankers, 2) linear and 3) non-linear rank SVMs, and 4) semi-supervised rankers that use the iterated graph Laplacian-based regularizer [27], all using the rank loss $\mathcal{L}$ (Eq. 20). For each attribute, we select the ranker with the highest validation accuracy as *baseline* $\mathbf{f}^0 = f|_X$.

We compare our proposed algorithm to: 1) the baseline predictor $\mathbf{f}^0$, 2) Kim and Chang's adaptation [10] of Kim et al.'s predictor combination approach [11] to visual attribute ranking (*OPC*), and 3) Mejjati et al.'s multi-task learning (*MTL*) algorithm [13]. While the latter was originally designed for MTL problems, it does not require known forms of individual predictors and can be thus adapted for predictor combination. In the supplemental document, we also compare with an adaptation of Evgeniou et al.'s graph Laplacian-based MTL algorithm [4] to the predictor combination setting, which demonstrates that all predictor combination algorithms outperform naïve adaptations of traditional MTL algorithms.

Adopting the experimental settings of Kim et al. [10,11], we tune the hyperparameters of all algorithms on evenly-split training and validation sets. Our algorithm requires tuning the noise level σ^2 (Eq. 12), global kernel scaling σ_k^2, and the regularization parameter λ_J (Eq. 13), which are tuned based on validation accuracy. For the number of iterations S, we use 20 iterations and select the iteration number that achieves the highest validation accuracy. The hyperparameters for other algorithms are tuned similarly (see the supplemental material for details). For each dataset, we repeated experiments 10 times with different training, validation, and test set splits and report the average accuracies.

The *OSR* [17], *Pubfig* [17], and *Shoes* [12] datasets provide 2688, 772 and 14,658 images each and include rank annotations (i.e. strengths of attributes present in images) for 6, 11 and 10 visual attributes, respectively. The attribute annotations in these datasets were obtained from the underlying class labels. For example, each image in *OSR* is also provided with a ground-truth class label out of 8 classes. The attribute ranking is assigned per class-wise comparisons such that all images in a class have stronger (or the same) presence of an attribute than another class. This implies that the class label assigned for each image completely determines its attributes, while attributes themselves might not provide sufficient information to determine classes. Similarly, the attribute annotations for *Pubfig* and *Shoes* are generated from class labels out of 8 and 10 classes, respectively. The input images in *OSR* and *Shoes* are represented as combinations of GIST [15] and color histogram features, while *Pubfig* uses GIST features as provided by the authors [12,17]. In addition, for *OSR*, we extracted 2,048-dimensional features using ResNet101 pre-trained on ImageNet [7] to fairly assess the predictor combination performance when the accuracies of the initial predictors are higher thanks to advanced features (*OSR (ResNet)*).

The *aPascal* dataset is constructed based on the PASCAL VOC 2008 dataset [3] containing 12,695 images with 64 attributes [5]. Each image is represented as a 9,751-dimensional feature vector combining histograms of local texture, HOG, and edge and color descriptors. The Caltech-UCSD Birds-200-2011 (*CUB*) dataset [22] provides 11,788 images with 312 attributes where the images are represented by the ResNet101 features. The Animals With Attributes 2 (*AWA2*) dataset consists of 37,322 images with 85 attributes [23]. We used the ResNet101 features as shared by Xian et al. [23]. For *aPascal*, *CUB*, and *AWA2*, the distributions of attribute values are imbalanced. To ensure that sufficient numbers (300) of training and testing labels exist for each attribute level, we selected 29, 40 and 80 attributes from *aPascal*, *CUB* and *AWA2*, respectively. The ranking accuracy is measured in $100\times$ Kendall's rank correlation coefficient, which is defined as the difference between the numbers of correctly and incorrectly ordered rank pairs, respectively, normalized by the number of total pairs (bounded in $100 \times [-1, 1]$; higher is the better).

The UT Zappos50K (*Zap50K*) contains 50,025 images of shoes with 4 attributes. Each image is represented as a combination of GIST and color histogram features provided by Yu and Grauman [25]. The ground-truth attribute labels are collected by instance-level pairwise comparison collected via *Mechanical Turk* [25].

We also performed multi-class classification experiments on the *OSR*, *Pubfig*, *Shoes*, *aPascal*, and *CUB* datasets based on their respective class labels. The initial predictors $\mathbf{f}^0 : \mathcal{X} \to \mathbb{R}^H$ are obtained as deep neural networks with continuous softmax decisions trained and validated on 20 labels per class. Each prediction is given as an H-dimensional vector with H being the number of classes. Our goal is to improve $\mathbf{f}^0$ using the predictors for visual attribute

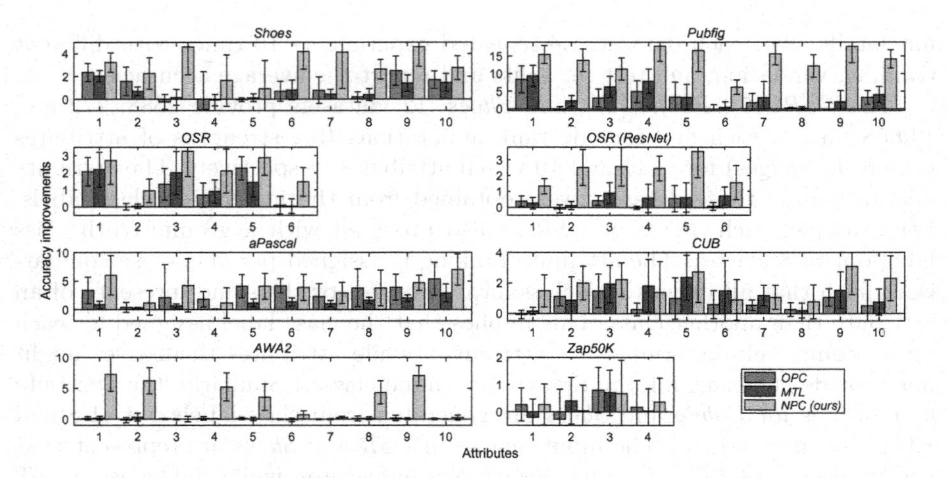

Fig. 3. Average accuracy improvement of different predictor combination algorithms from the *baseline predictors* for up to first 10 attributes. The complete results including statistical significance tests can be found in the supplemental document.

Table 3. Average classification accuracies (%) using rank estimates as references. The numbers in parentheses show the relative accuracy improvement over the baseline $\mathbf{f}^0$.

	Shoes	*Pubfig*	*OSR*	*aPascal*	*CUB*
Baseline $\mathbf{f}^0$	57.90 (0.00)	77.99 (0.00)	76.88 (0.00)	37.86 (0.00)	66.98 (0.00)
OPC [10]	58.52 (1.07)	82.55 (5.85)	77.16 (0.38)	39.77 (5.04)	67.75 (1.15)
MTL [13]	59.51 (2.78)	80.16 (2.78)	77.40 (0.68)	38.38 (1.37)	67.95 (1.45)
NPC (ours)	**62.87** (8.58)	**86.51** (10.9)	**79.71** (3.69)	**40.34** (6.55)	**68.26** (1.92)

ranking as references. It should be noted that our algorithm *jointly improves* all H class-wise predictors as well as ranking references: 1) all class predictors evolve simultaneously, and 2) for improving the predictor of a class, the (evolving) predictors of the remaining classes are used as additional references. For a fair comparison, we denoise class-wise predictors using both the rank predictors and the predictors of the remaining classes as references, also for the other predictor combination algorithms.

Ranking Results. Figure 3 summarizes the results for the relative attributes ranking experiments. Here, we show the results of only the first 10 attributes; the supplemental document contains complete results, which show a similar tendency as presented here. All three predictor combination algorithms frequently achieved significant performance gains over the baseline predictors $\mathbf{f}^0$. Importantly, apart from one case (*Shoes* attribute 4), all predictor combination algorithms did not significantly degrade the performance from the baseline. This demonstrates the utility of predictor combination. However, both *OPC* and *MTL* are limited in that they can only capture pairwise dependence between the target predictor and each reference. By taking into account the dependence present among the references, and thereby *jointly* exploiting them in improving the target predictor, our algorithm further significantly improves the performance: Our algorithm performs best for 87.1% of attributes. In particular, *Ours* showed significant improvement on 6 out of 10 *AWA2* attributes, where the other algorithms achieved no noticeable performance gain. This supports our assumption that multiple attributes indeed can *jointly* supply relevant information for improving target predictors, even if not individually.

Multi-class Classification Results. Table 3 shows the results of improving multi-class classifications. Jointly capturing all rank predictors as well as the multi-dimensional classification predictions as references, our algorithm demonstrates significant performance gains (especially on *Shoes* and *Pubfig*), while other predictor combination algorithms achieved only marginal improvements, confirming the effectiveness of our joint prediction strategy.

5 Discussions and Conclusions

Our algorithm builds upon the assumption that the reference predictors can help improve the target predictor when they can well predict (or explain) the ground-truth $\mathbf{f}_{\mathrm{GT}}$. Since $\mathbf{f}_{\mathrm{GT}}$ is not available during testing, we use the noisy target predictor $\mathbf{f}^t$ at each time step t as a surrogate, which by itself is iteratively denoised. While our experiments demonstrate the effectiveness of this approach in real-world examples, simple failure cases exist. For example, if $\mathbf{f}^0$ (as the initial surrogate to $\mathbf{f}_{\mathrm{GT}}$) is contained in the reference set $\mathcal{G}$, our automatic reference determination approach will pick this up as the single most relevant reference, and therefore, the resulting predictor combination process will simply output $\mathbf{f}^0$ as the final result. We further empirically observed that even when the automatic relevance determination is disabled (i.e. $\Sigma_L = I$), the performance degraded significantly when $\mathbf{f}^0$ is included in $\mathcal{G}$. Also, as shown for the *Zap50K* results, there might be cases where no algorithm shows any significant improvement (indicated by the relatively large error bars). In general, our algorithm may fail when the references do not communicate sufficient *information* for improving the target predictor. Quantifying such utility of references and predicting the failure cases may require a new theoretical analysis framework.

Existing predictor combination algorithms only consider pairwise relationships between the target predictor and each reference. This misses potentially relevant information present in the dependence among the references. We explicitly address this limitation by introducing a new *predictability criterion* that measures how references are *jointly* contributing in predicting the target predictor. Adopting a fully Bayesian framework, our algorithm can automatically select informative references among many potentially irrelevant predictors. Experiments on seven datasets demonstrated the effectiveness of the proposed predictor combination algorithm.

Acknowledgements. This work was supported by UNIST's 2020 Research Fund (1.200033.01), National Research Foundation of Korea (NRF) grant NRF-2019-R1F1A1061603, and Institute of Information & Communications Technology Planning & Evaluation (IITP) grant (No. 2020001336001101, Artificial Intelligence Graduate School support (UNIST)) funded by the Korean government (MSIT).

References

1. Argyriou, A., Evgeniou, T., Pontil, M.: Convex multi-task feature learning. Mach. Learn. **73**(3), 243–272 (2008). https://doi.org/10.1007/s10994-007-5040-8
2. Chen, L., Zhang, Q., Li, B.: Predicting multiple attributes via relative multi-task learning. In: CVPR, pp. 1027–1034 (2014)
3. Everingham, M., Eslami, S.M.A., Van Gool, L., Williams, C.K.I., Winn, J., Zisserman, A.: The PASCAL visual object classes challenge: a retrospective. Int. J. Comput. Vis. **111**(1), 98–136 (2015). https://doi.org/10.1007/s11263-014-0733-5
4. Evgeniou, T., Micchelli, C.A., Pontil, M.: Learning multiple tasks with kernel methods. JMLR **6**, 615–637 (2005)
5. Farhadi, A., Endres, I., Hoiem, D., Forsyth, D.: Describing objects by their attributes. In: CVPR, pp. 1778–1785 (2009)
6. Gong, P., Ye, J., Zhang, C.: Robust multi-task feature learning. In: KDD, pp. 895–903 (2012)
7. He, K., Zhang, X., Ren, S., Sun, J.: Deep residual learning for image recognition. In: CVPR, pp. 770–778 (2016)
8. Hein, M., Maier, M.: Manifold denoising. In: NIPS, pp. 561–568 (2007)
9. Joachims, T.: Optimizing search engines using clickthrough data. In: KDD, pp. 133–142 (2002)
10. Kim, K.I., Chang, H.J.: Joint manifold diffusion for combining predictions on decoupled observations. In: CVPR, pp. 7549–7557 (2019)
11. Kim, K.I., Tompkin, J., Richardt, C.: Predictor combination at test time. In: ICCV, pp. 3553–3561 (2017)
12. Kovashka, A., Parikh, D., Grauman, K.: WhittleSearch: image search with relative attribute feedback. In: CVPR, pp. 2973–2980 (2012)
13. Mejjati, Y.A., Cosker, D., Kim, K.I.: Multi-task learning by maximizing statistical dependence. In: CVPR, pp. 3465–3473 (2018)
14. Meng, Z., Adluru, N., Kim, H.J., Fung, G., Singh, V.: Efficient relative attribute learning using graph neural networks. In: Ferrari, V., Hebert, M., Sminchisescu, C., Weiss, Y. (eds.) Computer Vision – ECCV 2018. LNCS, vol. 11218, pp. 575–590. Springer, Cham (2018). https://doi.org/10.1007/978-3-030-01264-9_34
15. Oliva, A., Torralba, A.: Modeling the shape of the scene: a holistic representation of the spatial envelope. Int. J. Comput. Vis. **42**(3), 145–175 (2001). https://doi.org/10.1023/A:1011139631724
16. Pan, S.J., Yang, Q.: A survey on transfer learning. IEEE Trans. Knowl. Data Eng. **22**(10), 1345–1359 (2010)
17. Parikh, D., Grauman, K.: Relative attributes. In: ICCV, pp. 503–510 (2011)
18. Passos, A., Rai, P., Wainer, J., Daumé III, H.: Flexible modeling of latent task structures in multitask learning. In: ICML, pp. 1103–1110 (2012)
19. Rasmussen, C.E., Williams, C.K.I.: Gaussian Processes for Machine Learning. MIT Press, Cambridge (2006)

20. Seeger, M., Williams, C.K.I., Lawrence, N.D.: Fast forward selection to speed up sparse Gaussian process regression. In: International Workshop on Artificial Intelligence and Statistics (2003)
21. Snelson, E., Ghahramani, Z.: Sparse Gaussian processes using pseudo-inputs. In: NIPS (2006)
22. Wah, C., Branson, S., Welinder, P., Perona, P., Belongie, S.: The Caltech-UCSD Birds-200-2011 Dataset. Technical report CNS-TR-2011-001, California Institute of Technology (2011)
23. Xian, Y., Lampert, C.H., Schiele, B., Akata, Z.: Zero-shot learning - a comprehensive evaluation of the good, the bad and the ugly. IEEE TPAMI **41**(9), 2251–2265 (2019)
24. Yang, X., Zhang, T., Xu, C., Yan, S., Hossain, M.S., Ghoneim, A.: Deep relative attributes. IEEE T-MM **18**(9), 1832–1842 (2016)
25. Yu, A., Grauman, K.: Fine-grained visual comparisons with local learning. In: CVPR, pp. 192–199 (2014)
26. Zamir, A.R., Sax, A., Shen, W., Guibas, L., Malik, J., Savarese, S.: Taskonomy: disentangling task transfer learning. In: CVPR, pp. 3712–3722 (2018)
27. Zhou, X., Belkin, M., Srebro, N.: An iterated graph Laplacian approach for ranking on manifolds. In: KDD, pp. 877–885 (2011)

Multi-scale Positive Sample Refinement for Few-Shot Object Detection

Jiaxi Wu[1,2,3], Songtao Liu[1,2,3], Di Huang[1,2,3](✉), and Yunhong Wang[1,3]

[1] BAIC for BDBC, Beihang University, Beijing 100191, China
[2] SKLSDE, Beihang University, Beijing 100191, China
[3] SCSE, Beihang University, Beijing 100191, China
{wujiaxi,liusongtao,dhuang,yhwang}@buaa.edu.cn

Abstract. Few-shot object detection (FSOD) helps detectors adapt to unseen classes with few training instances, and is useful when manual annotation is time-consuming or data acquisition is limited. Unlike previous attempts that exploit few-shot classification techniques to facilitate FSOD, this work highlights the necessity of handling the problem of scale variations, which is challenging due to the unique sample distribution. To this end, we propose a Multi-scale Positive Sample Refinement (MPSR) approach to enrich object scales in FSOD. It generates multi-scale positive samples as object pyramids and refines the prediction at various scales. We demonstrate its advantage by integrating it as an auxiliary branch to the popular architecture of Faster R-CNN with FPN, delivering a strong FSOD solution. Several experiments are conducted on PASCAL VOC and MS COCO, and the proposed approach achieves state of the art results and significantly outperforms other counterparts, which shows its effectiveness. Code is available at https://github.com/jiaxi-wu/MPSR.

Keywords: Few-shot object detection · Multi-scale refinement

1 Introduction

Object detection makes great progress these years following the success of deep convolutional neural networks (CNN) [3,11,15,31,32]. These CNN based detectors generally require large amounts of annotated data to learn extensive numbers of parameters, and their performance significantly drops when training data are inadequate. Unfortunately, for object detection, labeling data is quite expensive and the samples of some object categories are even hard to collect, such as endangered animals or tumor lesions. This triggers considerable attentions to effective detectors dealing with limited training samples. Few-shot learning is a popular and promising direction to address this issue. However, the overwhelming majority of the existing few-shot investigations focus on object/image

Electronic supplementary material The online version of this chapter (https://doi.org/10.1007/978-3-030-58517-4_27) contains supplementary material, which is available to authorized users.

A. Vedaldi et al. (Eds.): ECCV 2020, LNCS 12361, pp. 456–472, 2020.
https://doi.org/10.1007/978-3-030-58517-4_27

classification, while the efforts on the more challenging *few-shot object detection* (FSOD) task are relatively rare.

With the massive parameters of CNN models, training detectors from scratch with scarce annotations generally incurs a high risk of overfitting. Preliminary research [4] tackles this problem in a transfer learning paradigm. Given a set of base classes with sufficient annotations and some novel classes with only a few samples, the goal is to acquire meta-level knowledge from base classes and then apply it to facilitating few-shot learning in detection of novel classes. Subsequent works [8,16,17,41] strengthen this pipeline by bringing more advanced methods on few-shot image classification, and commonly emphasize to improve classification performance of Region-of-Interest (RoI) in FSOD by using metric learning techniques. With elaborately learned representations, they ameliorate the similarity measurement between RoIs and marginally annotated instances, reporting better detection results. Meanwhile, [16,41] also attempt to deliver more general detectors, which account for all the classes rather than the novel ones only, by jointly using their samples in the training phase.

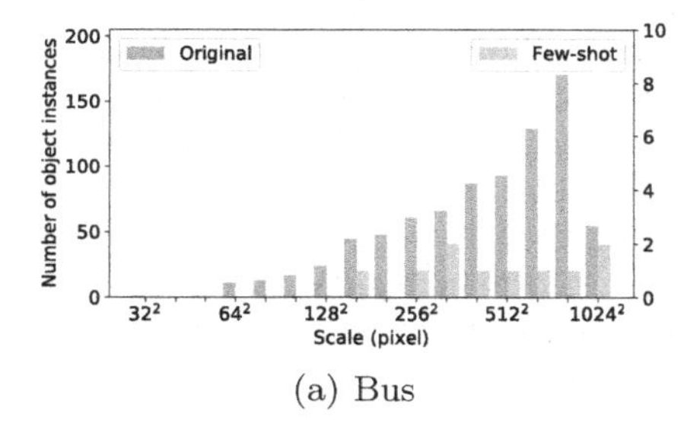

(a) Bus

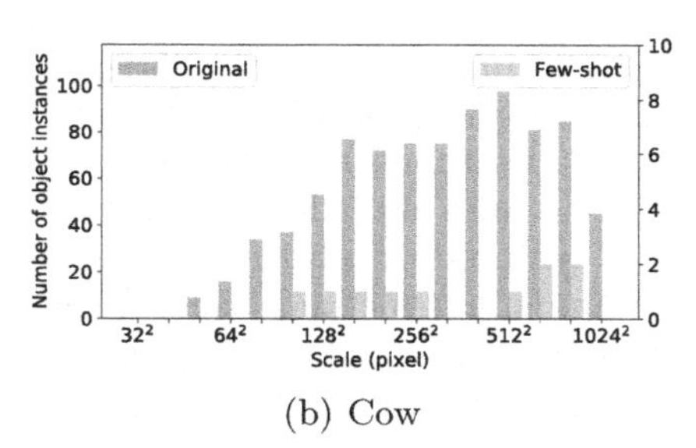

(b) Cow

Fig. 1. Illustration of scale distributions of two specific classes: (a) bus and (b) cow, in PASCAL VOC (Original) and a 10-shot subset (Few-shot). Images are resized with the shorter size at 800 pixels for statistics (Color figure online)

The prior studies demonstrate that the FSOD problem can be alleviated in a similar manner as few-shot image classification. Nevertheless, object detection is much more difficult than image classification, as it involves not only classification but also localization, where the threat of varying scales of objects is particularly evident. The scale invariance has been widely explored in generic supervised detectors [11,18,33,34], while it remains largely intact in FSOD. Moreover, restricted by the quantity of annotations, this scale issue is even more tricky. As shown in Fig. 1, the lack of labels of novel classes leads to a sparse scale space (green bars) which may be totally divergent from the original distribution (yellow bars) of abundant training data. One could assume to make use of current effective solutions from generic object detection to enrich the scale space. For instance, Feature Pyramid Network (FPN), which builds multi-scale feature maps to detect objects at different scales, applies to situations where significant scale variations exist [22]. This universal property does contribute to

FSOD, but it will not mitigate the difference of the scale distribution in the data of novel classes. Regarding image pyramids [11,14], they build multi-scale representations of an image and allow detectors to capture objects in it at different scales. Although they are expected to narrow such a gap between the two scale distributions, the case is not so straightforward. Specifically, multi-scale inputs result in an increase in improper negative samples due to anchor matching. These improper negative samples contain a part of features belonging to the positive samples, which interferes their recognition. With abundant data, the network learns to extract diverse contexts and suppress the improper local patterns. But it is harmful to FSOD where both semantic and scale distributions are sparse and biased.

In this work, we propose a Multi-scale Positive Sample Refinement (MPSR) approach to few-shot object detection, aiming at solving its unique challenge of sparse scale distribution. We take the reputed Faster R-CNN as the basic detection model and employ FPN in the backbone network to improve its tolerance to scale variations. We then exploit an auxiliary refinement branch to generate multi-scale positive samples as object pyramids and further refine the prediction. This additional branch shares the same weights with the original Faster R-CNN. During training, this branch classifies the extracted object pyramids in both the Region Proposal Network (RPN) and the detector head. To keep scale-consistent prediction without introducing more improper negatives, we abandon the anchor matching rules and adaptively assign the FPN stage and spatial locations to the object pyramids as positives. It is worth noting that as we use no extra weights in training, our method achieves remarkable performance gains in an inference cost-free manner and can be conveniently deployed on different detectors.

The contributions of this study are three-fold:

1. To the best of our knowledge, it is the first work to discuss the scale problem in FSOD. We reveal the sparsity of scale distributions in FSOD with both quantitative and qualitative analysis.
2. To address this problem, we propose the MPSR approach to enrich the scale space without largely increasing improper negatives.
3. Comprehensive experiments are carried out, and significant improvements from MPSR demonstrate its advantage.

2 Related Work

Few-Shot Image Classification. There are relatively many historical studies in the area of few-shot image classification that targets recognition of objects with only a handful of images in each class [20,27]. [9] learns to initialize weights that effectively adapt to unseen categories. [1,28] aim to predict network parameters without heavily training on novel images. [19,35,38] employ metric learning to replace linear classifiers with learnable metrics for comparison between query and support samples. Although few-shot image classification techniques are usually used to advance the phase of RoI classification in FSOD, they are different tasks, as FSOD has to consider localization in addition.

Generic Object Detection. Recent object detection architectures are mainly divided into two categories: one-stage detectors and two-stage detectors. One-stage detectors use a single CNN to directly predict bounding boxes [24,25,29,30], and two-stage ones first generate region proposals and then classify them for decision making [11,12,31]. Apart from network design, scale invariance is an important aspect to detectors and many solutions have recently been proposed to handle scale changes [18,22,33,34]. For example, [22] builds multi-scale feature maps to match objects at different scales. [33] performs scale normalization to detect scale-specific objects and adopts image pyramids for multi-scale detection. These studies generally adapt to alleviate large size differences of objects. Few-shot object detection suffers from scale variations in a more serious way where a few samples sparsely distribute in the scale space.

Object Detection with Limited Annotations. To relieve heavy annotation dependence in object detection, there exist two main directions without using external data. One is weakly-supervised object detection, where only image-level labels are provided and spatial supervision is unknown [2]. Research basically concentrates on how to rank and classify region proposals with only coarse labels through multiple instance learning [36,37,39]. Another is semi-supervised object detection that assumes abundant images are available while the number of bounding box annotations is limited [26]. In this case, previous studies confirm the effectiveness of adopting extra images by pseudo label mining [5,10] or multiple instance learning [21]. Both the directions reduce manual annotation demanding to some extent, but they heavily depend on the amount of training images. They have the difficulty in dealing with constrained conditions where data acquisition is inadequate, *i.e.*, few-shot object detection.

Few-Shot Object Detection. Preliminary work [4] on FSOD introduces a general transfer learning framework and presents the Low-Shot Transfer Detector (LSTD), which reduces overfitting by adapting pre-trained detectors to few-shot scenarios with limited training images. Following this framework, RepMet [17] incorporates a distance metric learning classifier into the RoI classification head in the detector. Instead of categorizing objects with fully-connected layers, RepMet extracts representative embedding vectors by clustering and calculates distances between query and annotated instances. [8] is motivated by [19] which scores the similarity in a siamese network and computes pair-wise object relationship in both the RPN and the detection head. [16] is a single-stage detector combined with a meta-model that re-weights the importance of features from the base model. The meta-model encodes class-specific features from annotated images at a proper scale, and the features are viewed as reweighting coefficients and fed to the base model. Similarly, [41] delivers a two-stage detection architecture and re-weights RoI features in the detection head. Unlike previous studies where spatial influence is not considered, we argue that scale invariance is a challenging issue to FSOD, as the samples are few and their scale distribution is sparse. We improve the detector by refining object crops rather than masked images [16,41] or siamese inputs [8] for additional training, which enriches the scale space and ensures the detector being fully trained at all scales.

3 Background

Before introducing MPSR, we briefly review the standard protocols and the basic detector we adopt for completeness. As it is the first work that addresses the challenge of sparse scale distribution in FSOD, we conduct some preliminary attempts with the current effective methods from generic object detection (*i.e.*, FPN and image pyramids) to enrich the scale space and discuss their limitations.

3.1 Baseline Few-Shot Object Detection

Few-Shot Object Detection Protocols. Following the settings in [16,41], object classes are divided into base classes with abundant data and novel classes with only a few training samples. The training process of FSOD generally adopts a two-step paradigm. During base training, the detection network is trained with a large-scale dataset that only contains base classes. Then the detection network is fine-tuned on the few-shot dataset, which only contains a very small number of balanced training samples for both base and novel classes. This two-step training schedule avoids the risk of overfitting with insufficient training samples on novel classes. It also prevents the detector from extremely imbalanced training if all annotations from both base and novel classes are exploited together [41]. To build the balanced few-shot dataset, [16] employs the k-shot sampling strategy, where each object class only has k annotated bounding boxes. Another work [4] collects k images for each class in the few-shot dataset. As k images actually contain an arbitrary number of instances, training and evaluation under this protocol tend to be unstable. We thus use the former strategy following [16].

Basic Detection Model. With the fast development in generic object detection, the base detector in FSOD has many choices. [16] is based on YOLOv2 [30], which is a single-stage detector. [41] is based on a classical two-stage detector, Faster R-CNN [31], and demonstrates that Faster R-CNN provides consistently better results. Therefore, we take the latter as our basic detection model. Faster R-CNN consists of the RPN and the detection head. For a given image, the RPN head generates proposals with objectness scores and bounding-box regression offsets. The RPN loss function is:

$$L_{RPN} = \frac{1}{N_{obj}} \sum_{i=1}^{N_{obj}} L^i_{Bcls} + \frac{1}{N_{obj}} \sum_{i=1}^{N_{obj}} L^i_{Preg}. \tag{1}$$

For the ith anchor in a mini-batch, L^i_{Bcls} is the binary cross-entropy loss over background and foreground and L^i_{Preg} is the smooth L_1 loss defined in [31]. N_{obj} is the total number of chosen anchors. These proposals are used to extract

RoI features and then fed to the detection (RoI) head that outputs class-specific scores and bounding-box regression offsets. The loss function is defined as:

$$L_{RoI} = \frac{1}{N_{RoI}} \sum_{i=1}^{N_{RoI}} L^{i}_{Kcls} + \frac{1}{N_{RoI}} \sum_{i=1}^{N_{RoI}} L^{i}_{Rreg}, \tag{2}$$

where L^{i}_{Kcls} is the log loss over K classes and N_{RoI} is the number of RoIs in a mini-batch. Different from the original implementation in [31], we employ a class-agnostic regression task in the detection head, which is the same as [4]. The total loss is the sum of L_{RPN} and L_{RoI}.

3.2 Preliminary Attempts

FPN for Multi-scale Detection. As FPN is commonly adopted in generic object detection to address the scale variation issue [3,22], we first consider applying it to FSOD in our preliminary experiments. FPN generates several different semantic feature maps at different scales, enriching the scale space in features. Our experiments validate that it is still practically useful under the restricted conditions in FSOD. We thus exploit Faster R-CNN with FPN as our second baseline. However, FPN does not change the distribution in the data of novel classes and the sparsity of scale distribution remains unsolved in FSOD.

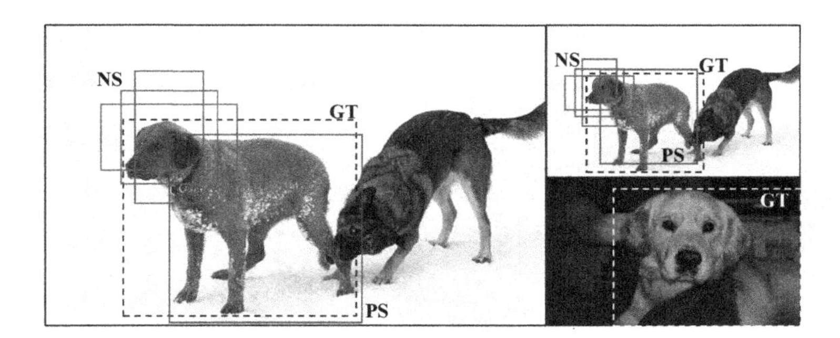

Fig. 2. An example of improper negative samples in FSOD. Negative samples (NS), positive samples (PS) and ground-truth (GT) bounding boxes are annotated. The improper negative samples significantly increase as more scales are involved (top right), while they may even be true positives in other contexts (bottom right) (Color figure online)

Image Pyramids for Multi-scale Training. To enrich object scales, we then consider a multi-scale training strategy which is also widely used in generic object detection for multi-scale feature extraction [11,14] or data augmentation [30]. In few-shot object detection, image pyramids enrich object scales as data augmentation and the sparse scale distribution can be theoretically solved. However, this multi-scale training strategy acts differently in FSOD with the increasing number of improper negative samples. As in Fig. 2, red bounding boxes are negative samples in training while they actually contain part of objects and may even be true positive samples in other contexts (as in bottom right). These improper negative samples require sufficient contexts and clues to suppress, inhibiting being mistaken for potential objects. Such an interference is trivial when abundant annotations are available, but it is quite harmful to the sparse and biased distribution in FSOD. Moreover, with multi-scale training, a large number of extra improper negative samples are introduced, which further hurts the performance.

4 Multi-scale Positive Sample Refinement

4.1 Multi-scale Positive Sample Refinement Branch

Motivated by the above discussion, we employ FPN in the backbone of Faster R-CNN as the advanced version of baseline. To enrich scales of positive samples without largely increasing improper negative samples, we extract each object independently and resize them to various scales, denoted as object pyramids. Specifically, each object is cropped by a square window (whose side is equal to the longer side of the bounding box) with a minor random shift. It is then resized to $\{32^2, 64^2, 128^2, 256^2, 512^2, 800^2\}$ pixels, which is similar to anchor design.

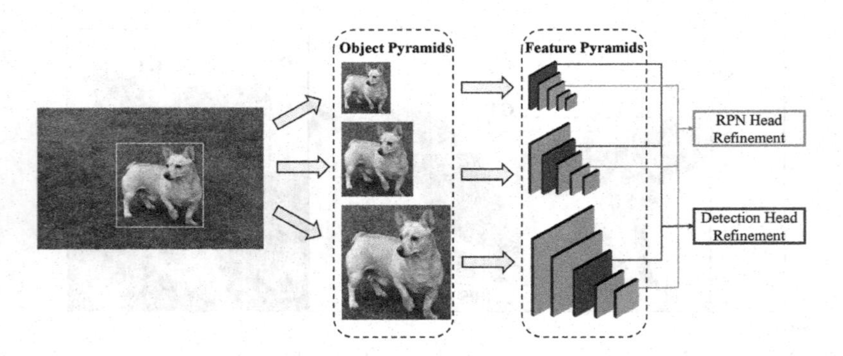

Fig. 3. Multi-scale positive sample feature extraction. The positive sample is extracted and resized to various scales. Specific feature maps from FPN are selected for refinement

In object pyramids, each image only contains a single instance, which is inconsistent to the standard detection pipeline. Therefore, we propose an extra positive sample refinement branch to adaptively project the object pyramids into the standard detection network. For a given object, the standard FPN pipeline samples the certain scale level and the spatial locations as positives for training, operated by anchor matching. However, performing anchor matching on cropped single objects is wasteful and also incurs more improper negatives that hurt the performance for FSOD. As shown in Fig. 3, instead of anchor matching, we manually select the corresponding scale level of feature maps and the fixed center locations as positives for each object, keeping it consistent with the standard FPN assigning rules. After selecting specific features from these feature maps, we feed them directly to the RPN head and the detection head for refinement.

Table 1. FPN feature map selection for different object scales. For each object, two specific feature maps are activated, fed to RPN and detection (RoI) heads respectively

	32^2	64^2	128^2	256^2	512^2	800^2
RPN	P_2	P_3	P_4	P_5	P_6	P_6
RoI	P_2	P_2	P_2	P_3	P_4	P_5

In the RPN head, the multi-scale feature maps of FPN $\{P_2, P_3, P_4, P_5, P_6\}$ represent anchors whose areas are $\{32^2, 64^2, 128^2, 256^2, 512^2\}$ pixels respectively. For a given object, only one feature map with the consistent scale is activated, as shown in Table 1. To simulate that each proposal is predicted by its center location in RPN, we select centric 2^2 features for object refinement. We also put anchors with $\{1:2, 1:1, 2:1\}$ aspect ratios on the sampled locations. These selected anchors are viewed as positives for the RPN classifier.

To extract RoI features for the detection head, only $\{P_2, P_3, P_4, P_5\}$ are used and the original RoI area partitions in the standard FPN pipeline are: $(0^2, 112^2)$, $[112^2, 224^2)$, $[224^2, 448^2)$, $[448^2, \infty)$ [22]. We also select one feature map at a specific scale for each object to keep the scale consistency, as shown in Table 1. As the randomly cropped objects tend to have larger sizes than the original ground truth bounding boxes, we slightly increase the scale range of each FPN stage for better selection. Selected feature maps are adaptively pooled to the same RoI size and fed to the RoI classifier.

4.2 Framework

As shown in Fig. 4, the whole detection framework for training consists of Faster R-CNN with FPN and the refinement branch working in parallel while sharing the same weights. For a given image, it is processed by the backbone network, RPN, RoI Align layer, and the detection head in the standard two-stage detection pipeline [31]. Simultaneously, an independent object extracted from the original image is resized to different scales as object pyramids. The object pyramids are fed into the detection network as described above. The outputs from RPN and detection heads in the MPSR branch include objectness scores and class-specific scores similar to the definitions in Sect. 3.1. The loss function of the RPN head containing Faster R-CNN and the MPSR branch is defined as:

$$L_{RPN} = \frac{1}{N_{obj}+M_{obj}} \sum_{i=1}^{N_{obj}+M_{obj}} L_{Bcls}^i + \frac{1}{N_{obj}} \sum_{i=1}^{N_{obj}} L_{Preg}^i, \tag{3}$$

where M_{obj} is the number of selected positive anchor samples for refinement. The loss function of the detection head is defined as:

$$L_{RoI} = \frac{1}{N_{RoI}} \sum_{i=1}^{N_{RoI}} L^i_{Kcls} + \frac{\lambda}{M_{RoI}} \sum_{i=1}^{M_{RoI}} L^i_{Kcls} + \frac{1}{N_{RoI}} \sum_{i=1}^{N_{RoI}} L^i_{Rreg}, \tag{4}$$

where M_{RoI} is the number of selected RoIs in MPSR. Unlike the RPN head loss where M_{obj} is close to N_{obj}, the number of positives from object pyramids is quite small compared to N_{RoI} in the RoI head. We thus add a weight parameter λ to the RoI classification loss of the positives from MPSR to adjust its magnitude, which is set to 0.1 by default. After the whole network is fully trained, the extra MPSR branch is removed and only Faster R-CNN with FPN is used for inference. Therefore, the MPSR approach that we propose benefits FSOD training without extra time cost at inference.

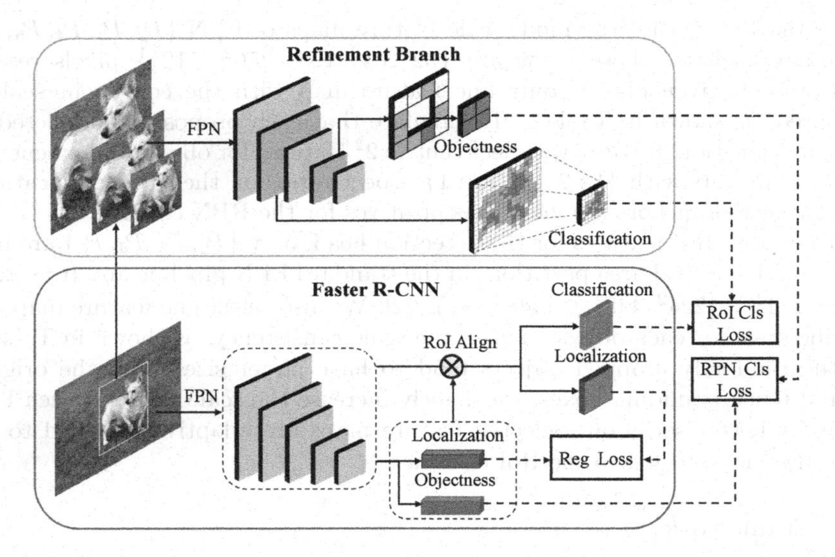

Fig. 4. MPSR architecture. On an input image to Faster R-CNN, the auxiliary branch extracts samples and resizes them to different scales. Each sample is fed to the FPN and specific features are selected to refine RPN and RoI heads in Faster R-CNN

5 Experiments

5.1 Datasets and Settings

We evaluate our method on the PASCAL VOC 2007 [7], 2012 [6] and MS COCO [23] benchmarks. For fair quantitative comparison with state of the art (SOTA) methods, we follow the setups in [16,41] to construct few-shot detection datasets.

PASCAL VOC. Our networks are trained on the modified VOC 2007 trainval and VOC 2012 trainval sets. The standard VOC 2007 test set is used for evaluation. The evaluation metric is the mean Average Precision (mAP). Both the trainval sets are split by object categories, where 5 are randomly chosen as novel classes and the left 15 are base classes. Here we follow [16] to use the same three class splits, where the unseen classes are {"bird", "bus", "cow", "motorbike" ("mbike"), "sofa"}, {"aeroplane" ("aero"), "bottle", "cow", "horse", "sofa"}, {"boat", "cat", "motorbike", "sheep", "sofa"}, respectively. For FSOD experiments, the few-shot dataset consists of images where only k object instances are available for each category and k is set as 1/3/5/10.

MS COCO. COCO has 80 object categories, where the 20 categories overlapped with PASCAL VOC are denoted as novel classes. 5,000 images from the val set, denoted as minival, are used for evaluation while the left images in the train and val sets are used for training. Base and few-shot dataset construction is the same as that in PASCAL VOC except that k is set as 10/30.

Implementation Details. We train and test detection networks on images of a single scale. We resize input images so that their shorter sides are set to 800 pixels and the longer sides are less than 1,333 pixels while maintaining the aspect ratio. Our backbone is ResNet-101 [15] with the RoI Align [13] layer and we use the weights pre-trained on ImageNet [32] in initialization. For efficient training, we randomly sample one object to generate the object pyramid for each image. After training on base classes, only the last fully-connected layer (for classification) of the detection head is replaced. The new classification layer is randomly initialized and none of the network layers is frozen during few-shot fine-tuning. We train our networks with a batchsize of 4 on 2 GPUs, 2 images per GPU. We run the SGD optimizer with the momentum of 0.9 and the parameter decay of 0.0001. For base training on VOC, models are trained for 240k, 8k, and 4k iterations with learning rates of 0.005, 0.0005 and 0.00005 respectively. For few-shot fine-tuning on VOC, we train models for 1,300, 400, 300 iterations and the learning rates are 0.005, 0.0005 and 0.00005, respectively. Models are trained on base COCO classes for 56k, 14k, and 10k iterations. For COCO few-shot fine-tuning, the 10-shot dataset requires 2,800, 700, and 500 iterations, while the 30-shot dataset requires 5,600, 1,400, 1,000 iterations.

5.2 Results

We compare our results with two baseline methods (denoted as Baseline and Baseline-FPN) as well as two SOTA few-shot detection counterparts. Baseline and Baseline-FPN are our implemented Faster R-CNN and Faster R-CNN with FPN described in Sect. 3. YOLO-FS [16] and Meta R-CNN [41] are the SOTA few-shot detectors based on DarkNet-19 and ResNet-101, respectively. It should be noted that due to better implementation and training strategy, our baseline achieves higher performance than SOTA, which is also confirmed by the very recent work [40].

Table 2. Comparison of different methods in terms of mAP (%) of novel classes using the three splits on the VOC 2007 test set

	Class split 1				Class split 2				Class split 3			
Method/Shot	1	3	5	10	1	3	5	10	1	3	5	10
YOLO-FS [16]	14.8	26.7	33.9	47.2	15.7	22.7	30.1	39.2	19.2	25.7	40.6	41.3
Meta R-CNN [41]	19.9	35.0	45.7	51.5	10.4	29.6	34.8	45.4	14.3	27.5	41.2	48.1
Baseline	24.5	40.8	44.6	47.9	16.7	34.9	37.0	40.9	27.3	36.3	41.2	45.2
Baseline-FPN	25.5	41.1	49.6	56.9	15.5	37.7	38.9	43.8	29.9	37.9	46.3	47.8
MPSR (ours)	**41.7**	**51.4**	**55.2**	**61.8**	**24.4**	**39.2**	**39.9**	**47.8**	**35.6**	**42.3**	**48.0**	**49.7**

PASCAL VOC. MPSR achieves 82.1%/82.7%/82.9% on base classes of three splits respectively before few-shot fine-tuning. The main results of few-shot experiments on VOC are summarized in Table 2. It can be seen from this table that the results of the two baselines (*i.e.* Baseline and Baseline-FPN) are close to each other when the number of instances is extremely small (*e.g.* 1 or 3), and Baseline-FPN largely outperforms the other as the number of images increases. This demonstrates that FPN benefits few-shot object detection as in generic object detection. Moreover, our method further improves the performance of Baseline-FPN with any number of training samples in all the three class splits. Specifically, by solving the sparsity of object scales, we achieve a significant increase in mAP compared to the best scores of the two baselines, particularly when training samples are extremely scarce, *e.g.* 16.2% on 1-shot split-1. It clearly highlights the effectiveness of the extra MPSR branch. Regarding other counterparts [16,41], the proposed approach outperforms them by a large margin, reporting the state of the art scores on this dataset.

Table 3. AP (%) of each novel class on the 3-/10-shot VOC dataset of the first class split. mAP (%) of novel classes and base classes are also presented

		Novel classes					Mean	
Shot	Method	Bird	Bus	Cow	Mbike	Sofa	Novel	Base
3	YOLO-FS [16]	26.1	19.1	40.7	20.4	27.1	26.7	64.8
	Meta R-CNN[41]	30.1	44.6	50.8	38.8	10.7	35.0	64.8
	Baseline	34.9	26.9	53.3	50.8	38.2	40.8	45.2
	Baseline-FPN	32.6	29.4	45.5	56.2	41.7	41.1	66.2
	MPSR (ours)	**35.1**	**60.6**	**56.6**	**61.5**	**43.4**	**51.4**	**67.8**
10	YOLO-FS [16]	30.0	62.7	43.2	60.6	39.6	47.2	63.6
	Meta R-CNN [41]	**52.5**	55.9	52.7	54.6	41.6	51.5	67.9
	Baseline	38.6	48.6	51.6	57.2	43.4	47.9	47.8
	Baseline-FPN	41.8	68.4	61.7	66.8	45.8	56.9	70.0
	MPSR (ours)	48.3	**73.7**	**68.2**	**70.8**	**48.2**	**61.8**	**71.8**

Following [16,41], we display the detailed results of 3-/10-shot detection in the first split on VOC in Table 3. Consistently, our Baseline-FPN outperforms the existing methods on both the novel and base classes. This confirms that FPN addresses the scale problem in FSOD to some extent. Furthermore, our method improves the accuracies of Baseline-FPN in all the settings by integrating MPSR, illustrating its advantage.

MS COCO. We evaluate the method using 10-/30-shot setups on MS COCO with the standard COCO metrics. The results on novel classes are provided in Table 4. Although COCO is quite challenging, we still achieve an increase of 0.4% on 30-shot compared with Baseline-FPN while boosting the SOTA mAP from 12.4% (Meta R-CNN) to 14.1%. Specifically, our method improves the recognition of small, medium and large objects simultaneously. This demonstrates that our balanced scales of input objects are effective.

Table 4. AP (%) and AR (%) of 10-/30-shot scores of novel classes on COCO minival

Shot	Method	AP	AP_{50}	AP_{75}	AP_S	AP_M	AP_L	AR_1	AR_{10}	AR_{100}	AR_S	AR_M	AR_L
10	YOLO-FS [16]	5.6	12.3	4.6	0.9	3.5	10.5	10.1	14.3	14.4	1.5	8.4	28.2
	Meta R-CNN [41]	8.7	**19.1**	6.6	2.3	7.7	14.0	12.6	17.8	17.9	**7.8**	15.6	27.2
	Baseline	8.8	18.7	7.1	2.9	8.1	15.0	12.9	17.2	17.2	4.1	14.2	29.1
	Baseline-FPN	9.5	17.3	9.4	2.7	8.4	15.9	14.8	20.6	20.6	4.7	19.3	33.1
	MPSR (ours)	**9.8**	17.9	**9.7**	**3.3**	**9.2**	**16.1**	**15.7**	**21.2**	**21.2**	4.6	**19.6**	**34.3**
30	YOLO-FS [16]	9.1	19.0	7.6	0.8	4.9	16.8	13.2	17.7	17.8	1.5	10.4	33.5
	Meta R-CNN [41]	12.4	25.3	10.8	2.8	11.6	19.0	15.0	21.4	21.7	**8.6**	20.0	32.1
	Baseline	12.6	**25.7**	11.0	3.2	11.8	20.7	15.9	21.8	21.8	5.1	18.0	36.9
	Baseline-FPN	13.7	25.1	13.3	3.6	12.5	**23.3**	**17.8**	**24.7**	**24.7**	5.4	**21.6**	**40.5**
	MPSR (ours)	**14.1**	25.4	**14.2**	**4.0**	**12.9**	23.0	17.7	24.2	24.3	5.5	21.0	39.3

MS COCO to PASCAL VOC. We conduct cross-dataset experiments on the standard VOC 2007 test set. In this setup, all the models are trained on the base COCO dataset and finetured with 10-shot objects in novel classes on VOC. Results of Baseline and Baseline-FPN are 38.5% and 39.3% respectively. They are worse than 10-shot results only trained on PASCAL VOC due to the large domain shift. Cross-dataset results of YOLO-FS and Meta R-CNN are 32.3% and 37.4% respectively. Our MPSR achieves 42.3%, which indicates that our method has better generalization ability in cross-domain situations.

5.3 Analysis of Sparse Scales

We visualize the scale distribution of two categories on the original dataset (Pascal VOC) and 10-shot subset in Fig. 1. It is obvious that the scale distribution in the few-shot dataset is extremely sparse and distinct from the original ones.

Table 5. AP (%) on bus/cow class. Two 10-shot datasets are constructed on VOC split-1, where scales of instances are random or limited. Std over 5 runs are presented

	Bus		Cow	
Method	Random	Limited	Random	Limited
Baseline-FPN	68.4 ± 0.6	39.5 ± 1.3	61.7 ± 0.9	39.9 ± 1.2
MPSR (ours)	73.7 ± 1.6	54.0 ± 1.4	68.2 ± 1.0	52.5 ± 1.6

To quantitatively analyze the negative effect of scale sparsity, we evaluate detectors on two specific 10-shot datasets. We carefully select the bus and cow instances with the scale between 128^2 and 256^2 pixels to construct the "limited" few-shot datasets. As shown in Table 5, such extremely sparse scales lead to a significant drop in performance (*e.g.* for bus, −28.9% on Baseline-FPN). Therefore, it is essential to solve the extremely sparse and biased scale distribution in FSOD. With our MPSR, the reduction of performance is relieved.

Table 6. mAP (%) comparison of novel/base classes on VOC split-1: Baseline-FPN, SNIPER [34], Baseline-FPN with scale augmentation/image pyramids and MPSR

	Novel			Base		
Method/Shot	1	3	5	1	3	5
Baseline-FPN	25.5	41.1	49.6	56.9	66.2	67.9
SNIPER [34]	1.4	21.0	39.7	**67.8**	**74.8**	**76.2**
Scale augmentation	29.8	44.7	49.8	52.7	67.1	68.8
Image pyramids	29.5	48.4	50.4	58.1	67.5	68.3
MPSR (ours)	**41.7**	**51.4**	**55.2**	59.4	67.8	68.4

As in Table 6, we compare MPSR with several methods that are used for scale invariance. SNIPER [34] shows a lower accuracy on novel classes and a higher accuracy on base classes than the baseline. As SNIPER strictly limits the scale range in training, it actually magnifies the sparsity of scales in FSOD. Such low performance also indicates the importance of enriching scales. We also evaluate the scale augmentation and image pyramids with a shorter side of $\{480, 576, 688, 864, 1200\}$ [14]. We can see that our MPSR achieves better results than those two multi-scale training methods on the novel classes. When only one instance is available for each object category, our method exceeds multi-scale training by ∼12%, demonstrating its superiority.

5.4 Ablation Studies

We conduct some ablation studies to verify the effectiveness of the proposed manual selection and refinement method in Table 7.

Table 7. mAP (%) of MPSR with different settings of novel classes on VOC split-1

Baseline FPN	Object pyramids	Manual selection	Refinement		Shot		
			RPN	RoI	1	3	5
✓					25.5	41.1	49.6
✓	✓		✓	✓	30.8	43.6	49.6
✓	✓	✓	✓		36.7	48.0	54.4
✓	✓	✓		✓	33.7	48.2	54.7
✓	✓	✓	✓	✓	41.7	51.4	55.2

Manual Selection. From the first two lines in Table 7, we see that applying anchor matching to object pyramids on both RPN and RoI heads achieves better performance than Baseline-FPN. However, when compared to the last three lines with manual selection rules, anchor matching indeed limits the benefits of object pyramids, as it brings more improper negative samples to interfere few-shot training. It confirms the necessity of the proposed manual refinement rules.

RPN and Detection Refinement. As in the last three lines of Table 7, we individually evaluate RPN refinement and detection (RoI) refinement to analyze their credits in the entire approach. Models with only the RPN and RoI refinement branches exceed Baseline-FPN in all the settings, which proves their effectiveness. Our method combines them and reaches the top score, which indicates that the two branches play complementary roles.

6 Conclusions

This paper targets the scale problem caused by the unique sample distribution in few-shot object detection. To deal with this issue, we propose a novel approach, namely multi-scale positive sample refinement. It generates multi-scale positive samples as object pyramids and refines the detectors at different scales, thus enlarging the scales of positive samples while limiting improper negative samples. We further deliver a strong FSOD solution by integrating MPSR to Faster R-CNN with FPN as an auxiliary branch. Experiments are extensively carried out on PASCAL VOC and MS COCO, and the proposed approach reports better scores compared to current state of the arts, which shows its advantage.

Acknowledgment. This work is funded by the Research Program of State Key Laboratory of Software Development Environment (SKLSDE-2019ZX-03) and the Fundamental Research Funds for the Central Universities.

References

1. Bertinetto, L., Henriques, J.F., Valmadre, J., Torr, P.H.S., Vedaldi, A.: Learning feed-forward one-shot learners. In: Advances in Neural Information Processing Systems (NIPS) (2016)
2. Bilen, H., Vedaldi, A.: Weakly supervised deep detection networks. In: IEEE Conference on Computer Vision and Pattern Recognition (CVPR) (2016)
3. Cai, Z., Vasconcelos, N.: Cascade R-CNN: delving into high quality object detection. In: IEEE Conference on Computer Vision and Pattern Recognition (CVPR) (2018)
4. Chen, H., Wang, Y., Wang, G., Qiao, Y.: LSTD: a low-shot transfer detector for object detection. In: Proceedings of the Thirty-Second AAAI Conference on Artificial Intelligence (2018)
5. Dong, X., Zheng, L., Ma, F., Yang, Y., Meng, D.: Few-example object detection with model communication. IEEE Trans. Pattern Anal. Mach. Intell. (TPAMI) **41**, 1641–1654 (2019)
6. Everingham, M., Eslami, S.M.A., Van Gool, L., Williams, C.K.I., Winn, J., Zisserman, A.: The PASCAL visual object classes challenge: a retrospective. Int. J. Comput. Vision **111**(1), 98–136 (2014). https://doi.org/10.1007/s11263-014-0733-5
7. Everingham, M., Gool, L.V., Williams, C.K.I., Winn, J.M., Zisserman, A.: The pascal visual object classes (VOC) challenge. Int. J. Comput. Vision **88**, 303–338 (2010). https://doi.org/10.1007/s11263-009-0275-4
8. Fan, Q., Zhuo, W., Tang, C.K., Tai, Y.W.: Few-shot object detection with attention-RPN and multi-relation detector. In: IEEE Conference on Computer Vision and Pattern Recognition (CVPR) (2020)
9. Finn, C., Abbeel, P., Levine, S.: Model-agnostic meta-learning for fast adaptation of deep networks. In: International Conference on Machine Learning (ICML) (2017)
10. Gao, J., Wang, J., Dai, S., Li, L.J., Nevatia, R.: NOTE-RCNN: noise tolerant ensemble RCNN for semi-supervised object detection. In: IEEE International Conference on Computer Vision (ICCV) (2019)
11. Girshick, R.B.: Fast R-CNN. In: IEEE International Conference on Computer Vision (ICCV) (2015)
12. Girshick, R.B., Donahue, J., Darrell, T., Malik, J.: Rich feature hierarchies for accurate object detection and semantic segmentation. In: IEEE Conference on Computer Vision and Pattern Recognition (CVPR) (2014)
13. He, K., Gkioxari, G., Dollár, P., Girshick, R.B.: Mask R-CNN. In: IEEE International Conference on Computer Vision (ICCV) (2017)
14. He, K., Zhang, X., Ren, S., Sun, J.: Spatial pyramid pooling in deep convolutional networks for visual recognition. IEEE Trans. Pattern Anal. Mach. Intell. (TPAMI) **37**, 1904–1916 (2015)
15. He, K., Zhang, X., Ren, S., Sun, J.: Deep residual learning for image recognition. In: IEEE Conference on Computer Vision and Pattern Recognition (CVPR) (2016)
16. Kang, B., Liu, Z., Wang, X., Yu, F., Feng, J., Darrell, T.: Few-shot object detection via feature reweighting. In: IEEE International Conference on Computer Vision (ICCV) (2019)
17. Karlinsky, L., et al.: RepMet: representative-based metric learning for classification and few-shot object detection. In: IEEE Conference on Computer Vision and Pattern Recognition (CVPR) (2019)

18. Kim, Y., Kang, B.-N., Kim, D.: SAN: learning relationship between convolutional features for multi-scale object detection. In: Ferrari, V., Hebert, M., Sminchisescu, C., Weiss, Y. (eds.) ECCV 2018. LNCS, vol. 11209, pp. 328–343. Springer, Cham (2018). https://doi.org/10.1007/978-3-030-01228-1_20
19. Koch, G., Zemel, R., Salakhutdinov, R.: Siamese neural networks for one-shot image recognition. In: ICML Deep Learning Workshop (2015)
20. Li, F., Fergus, R., Perona, P.: One-shot learning of object categories. IEEE Trans. Pattern Anal. Mach. Intell. (TPAMI) **28**, 594–611 (2006)
21. Li, Z., et al.: Thoracic disease identification and localization with limited supervision. In: IEEE Conference on Computer Vision and Pattern Recognition (CVPR) (2018)
22. Lin, T., Dollár, P., Girshick, R.B., He, K., Hariharan, B., Belongie, S.J.: Feature pyramid networks for object detection. In: IEEE Conference on Computer Vision and Pattern Recognition (CVPR) (2017)
23. Lin, T.-Y., et al.: Microsoft COCO: common objects in context. In: Fleet, D., Pajdla, T., Schiele, B., Tuytelaars, T. (eds.) ECCV 2014. LNCS, vol. 8693, pp. 740–755. Springer, Cham (2014). https://doi.org/10.1007/978-3-319-10602-1_48
24. Liu, S., Huang, D., Wang, Y.: Receptive field block net for accurate and fast object detection. In: Ferrari, V., Hebert, M., Sminchisescu, C., Weiss, Y. (eds.) ECCV 2018. LNCS, vol. 11215, pp. 404–419. Springer, Cham (2018). https://doi.org/10.1007/978-3-030-01252-6_24
25. Liu, W., et al.: SSD: single shot MultiBox detector. In: Leibe, B., Matas, J., Sebe, N., Welling, M. (eds.) ECCV 2016. LNCS, vol. 9905, pp. 21–37. Springer, Cham (2016). https://doi.org/10.1007/978-3-319-46448-0_2
26. Misra, I., Shrivastava, A., Hebert, M.: Watch and learn: semi-supervised learning of object detectors from videos. In: IEEE Conference on Computer Vision and Pattern Recognition (CVPR) (2015)
27. Munkhdalai, T., Yu, H.: Meta networks. In: International Conference on Machine Learning (ICML) (2017)
28. Qiao, S., Liu, C., Shen, W., Yuille, A.L.: Few-shot image recognition by predicting parameters from activations. In: IEEE Conference on Computer Vision and Pattern Recognition (CVPR) (2018)
29. Redmon, J., Divvala, S., Girshick, R., Farhadi, A.: You only look once: unified, real-time object detection. In: IEEE Conference on Computer Vision and Pattern Recognition (CVPR) (2016)
30. Redmon, J., Farhadi, A.: Yolo9000: better, faster, stronger. In: IEEE Conference on Computer Vision and Pattern Recognition (CVPR) (2017)
31. Ren, S., He, K., Girshick, R.B., Sun, J.: Faster R-CNN: towards real-time object detection with region proposal networks. In: Advances in Neural Information Processing Systems (NIPS) (2015)
32. Russakovsky, O., et al.: ImageNet large scale visual recognition challenge. Int. J. Comput. Vision **115**(3), 211–252 (2015). https://doi.org/10.1007/s11263-015-0816-y
33. Singh, B., Davis, L.S.: An analysis of scale invariance in object detection-SNIP. In: IEEE Conference on Computer Vision and Pattern Recognition (CVPR) (2018)
34. Singh, B., Najibi, M., Davis, L.S.: SNIPER: efficient multi-scale training. In: Advances in Neural Information Processing Systems (NIPS) (2018)
35. Sung, F., Yang, Y., Zhang, L., Xiang, T., Torr, P.H.S., Hospedales, T.M.: Learning to compare: relation network for few-shot learning. In: IEEE Conference on Computer Vision and Pattern Recognition (CVPR) (2018)

36. Tang, P., Wang, X., Bai, X., Liu, W.: Multiple instance detection network with online instance classifier refinement. In: IEEE Conference on Computer Vision and Pattern Recognition (CVPR) (2017)
37. Tang, P., et al.: Weakly supervised region proposal network and object detection. In: Ferrari, V., Hebert, M., Sminchisescu, C., Weiss, Y. (eds.) ECCV 2018. LNCS, vol. 11215, pp. 370–386. Springer, Cham (2018). https://doi.org/10.1007/978-3-030-01252-6_22
38. Vinyals, O., Blundell, C., Lillicrap, T., Kavukcuoglu, K., Wierstra, D.: Matching networks for one shot learning. In: Advances in Neural Information Processing Systems (NIPS) (2016)
39. Wan, F., Liu, C., Ke, W., Ji, X., Jiao, J., Ye, Q.: C-MIL: continuation multiple instance learning for weakly supervised object detection. In: IEEE Conference on Computer Vision and Pattern Recognition (CVPR) (2019)
40. Wang, X., Huang, T.E., Darrell, T., Gonzalez, J.E., Yu, F.: Frustratingly simple few-shot object detection. In: International Conference on Machine Learning (ICML) (2020)
41. Yan, X., Chen, Z., Xu, A., Wang, X., Liang, X., Lin, L.: Meta R-CNN: towards general solver for instance-level low-shot learning. In: IEEE International Conference on Computer Vision (ICCV) (2019)

Single-Image Depth Prediction Makes Feature Matching Easier

Carl Toft[1](✉), Daniyar Turmukhambetov[2], Torsten Sattler[1], Fredrik Kahl[1], and Gabriel J. Brostow[2,3]

[1] Chalmers University of Technology, Gothenburg, Sweden
carltoft@gmail.com
[2] Niantic, San Francisco, USA
[3] University College London, London, UK
https://www.github.com/nianticlabs/rectified-features

Abstract. Good local features improve the robustness of many 3D re-localization and multi-view reconstruction pipelines. The problem is that viewing angle and distance severely impact the recognizability of a local feature. Attempts to improve appearance invariance by choosing better local feature points or by leveraging outside information, have come with pre-requisites that made some of them impractical. In this paper, we propose a surprisingly effective enhancement to local feature extraction, which improves matching. We show that CNN-based depths inferred from single RGB images are quite helpful, despite their flaws. They allow us to pre-warp images and rectify perspective distortions, to significantly enhance SIFT and BRISK features, enabling more good matches, even when cameras are looking at the same scene but in opposite directions.

Keywords: Local feature matching · Image matching

1 Introduction

Matching local features between images is a core research problem in Computer Vision. Feature matching is a crucial step in Simultaneous Localization and Mapping (SLAM) [16,53], Structure-from-Motion (SfM) [64,68], and visual localization [61,62,69]. By extension, good feature matching enables applications such as self-driving cars [28] and other autonomous robots [40] as well as Augmented, Mixed, and Virtual Reality. Handling larger viewpoint changes is often important in practice, *e.g.*, to detect loop closures when revisiting the same place in SLAM [23] or for re-localization under strong viewpoint changes [63].

C. Toft—Work done during an internship at Niantic.

Electronic supplementary material The online version of this chapter (https://doi.org/10.1007/978-3-030-58517-4_28) contains supplementary material, which is available to authorized users.

A. Vedaldi et al. (Eds.): ECCV 2020, LNCS 12361, pp. 473–492, 2020.
https://doi.org/10.1007/978-3-030-58517-4_28

Fig. 1. Left to Right: Two input images, capturing the same section of a sidewalk, but looking in opposite directions. Each RGB input image is shown together with its depth map predicted by a single-image depth prediction network. Based on the predicted depth map, we identify planar regions and remove perspective distortion from them before extracting local features, thus enabling effective feature matching under strong viewpoint changes.

Traditionally, local features are computed in two stages [45]: the feature detection stage determines salient points in an image around which patches are extracted. The feature description stage computes descriptors from these patches. Before extracting a patch, the feature detector typically accounts for certain geometric transformations, thus making the local features robust or even invariant against these transformations. For example, aligning a patch to a dominant direction makes the feature invariant to in-plane rotations [45,51]; detecting salient points at multiple scales introduces robustness to scale changes [41]; removing the effect of affine transformations [47,48] of the image makes the extracted features more robust against viewpoint changes [1,22]. If the 3D geometry of the scene is known, *e.g.*, from 3D reconstruction via SfM, it is possible to undo the effect of perspective projection before feature extraction [33,74,81,82]. The resulting features are, in theory, invariant to viewpoint changes (Fig. 1).

Even without known 3D scene geometry, it is still possible to remove the effect of perspective distortion from a single image [3,56,59]. In principle, vanishing points [9,37,67,85] or repeating structural elements [56,57,75] can be used to rectify planar regions prior to feature detection [14]. However, this process is cumbersome in practice: it is unclear which pixels belong to a plane, so it is necessary to unwarp the full image. This introduces strong distortions for image regions belonging to different planes. As a result, determining a good resolution for the unwarped image is a challenge, since one would like to avoid both too small (resulting in a loss of details) and too high resolutions (which quickly become hard to handle). Figure 2 shows an example of this behaviour on an image from our dataset. This process has to be repeated multiple times to handle multiple planes.

Prior work has shown the advantages of removing perspective distortion prior to feature detection in tasks such as visual localization [59] and image retrieval [3,4,10]. Yet, such methods are not typically used in practice as they are hard to automate. For example, modern SfM [64,73], SLAM [53], and visual

Input image
(1920 x 1080)

Rectified with VP
(1349 x 1188)

Rectified with Ours
(1500 x 1076)

Fig. 2. Perspective rectification of a challenging example. From left to right: input image, perspective rectification with a vanishing point method [9] and Ours. The vanishing point method can heavily distort an image and produce an image where the region of interest only occupies a small part. An output image may become prohibitively large in order to preserve detail. Our method does not have these artifacts as it rectifies planar patches, not the full image.

localization [25,61,62] systems still rely on classical features without any prior removal of perspective effects. This paper shows that convolutional neural networks (CNNs) for single-image depth estimation provide a simple yet effective solution to the practical problems encountered when correcting perspective distortion: although their depth estimates might be noisy (especially for scene geometry far away from the camera), they are typically trained to produce smooth depth gradients [26,38]. This fact can be used to estimate normals, which in turn define planes that can be rectified. Per-pixel normals provide information about which pixels belong to the same plane, thus avoiding the problems of having to unwarp the full image and to repeatedly search for additional planes. Also, depth information can be used to avoid strong distortions by ignoring pixels seen under sharp angles. As a result, our approach is significantly easier to use in practice.

Specifically, this paper makes the following contributions: **1)** We propose a simple and effective method for removing perspective distortion prior to feature extraction based on single-image depth estimation. **2)** We demonstrate the benefits of this approach through detailed experiments on feature matching and visual localization under strong viewpoint changes. In particular, we show that improved performance does not require fine-tuning the depth prediction network per scene. **3)** We propose a new dataset to evaluate the performance of feature matching under varying viewpoints and viewing conditions. We show that our proposed approach can significantly improve matching performance in the presence of dominant planes, without significant degradation if there are no dominant planes.

2 Related Work

Perspective Undistortion via 3D Geometry. If the 3D geometry of the scene is known, *e.g.*, from depth maps recorded by RGB-D sensors, laser scans, or multi-view stereo, it is possible to remove perspective distortion prior to feature extraction [8,33,74,81,82]. More precisely, each plane detected in 3D defines homographies that warp perspective image observations of the plane to orthographic projections. Extracting features from orthographic rather than perspective views makes the features (theoretically) invariant to viewpoint changes[1]. This enables feature matching under strong viewpoint changes [82]. Given known 3D geometry, a single feature match between two images or an image and a 3D model can be sufficient to estimate the full 6-degree-of-freedom (relative) camera pose [3,74]. Following similar ideas, features found on developable surfaces can be made more robust by unrolling the surfaces into a plane prior to feature detection. Known 3D geometry can also be used to remove the need for certain types of invariances [33], *e.g.*, predicting the scale of keypoints from depth [31] removes the need for scale invariance.

Previous work assumed that 3D data is provided together with an image, or is extracted from multiple images. Inspired by this idea, we show that perspective distortion can often be removed effectively using single-image depth predictions made by modern convolutional neural networks (CNNs).

Perspective Undistortion Without 3D Geometry. Known 3D geometry is not strictly necessary to remove the effect of perspective foreshortening on planar structures: either vanishing points [9,15,37,67,85] or repeating geometrical structures [55–58,75] can be used to define a homography for removing perspective distortion for all pixels on the plane [14,39]. In both cases, an orthographic view of the plane can be recovered up to an unknown scale factor and an unknown in-plane rotation, *i.e.*, an unknown similarity transformation. Such approaches have been used to show improved performance for tasks such as image retrieval [3,4,10], visual localization [59], and feature matching [79]. However, they are often brittle and hard to automate for practical use: they do not provide any information about which pixels belong to a given plane. This makes it necessary to warp the full image, which can introduce strong distortion effects for regions that do not belong to the plane. This in turn leads to the problem of selecting a suitable resolution for the unwarped image, to avoid loosing details without creating oversized images that cannot be processed efficiently. As a result, despite their expected benefits, such methods have seen little use in practical applications, *e.g.*, modern SfM or SLAM systems. See Fig. 2 (as well as Sec. 8 in the supplementary material) for an example of this behaviour as seen on an image from our dataset.

In this paper, we show that these problems can easily be avoided by using single-image depth predictions to remove the effect of perspective distortion.

[1] In practice, strong viewpoint changes create strong distortions in the unwarped images, which prevent successful feature matching [33].

Pairwise Image Matching via View Synthesis. An alternative to perspective undistortion for robust feature matching between two images taken from different viewpoints is view synthesis [44,50,52,54]. Such approaches generate multiple affine or projective warps of each of the two images and extract and match features for each warp. Progressive schemes exist, which first evaluate small warps and efficient features to accelerate the process [50]. Still, such approaches are computationally very expensive due to the need to evaluate a large number of potential warps. Methods like ours, based on removing perspective distortion, avoid this computational cost by determining a single warp per region. Such warps can also be estimated locally per region or patch [2,30]. This latter type of approach presupposes that stable keypoints can be detected in perspectively distorted images. Yet, removing perspective effects prior to feature detection can significantly improve performance [8,33].

Datasets. Measuring the performance of local features under strong viewpoint changes, *i.e.*, the scenario where removing perspective distortion could provide the greatest benefit, has a long tradition, so multiple datasets exist for this task [1,2,5,13,48,50,66]. Often, such datasets depict nicely textured scenes, *e.g.*, graffiti, paintings, or photographs, from different viewpoints. Such scenes represent "failure cases" for single-image depth predictions as the networks (not unreasonably) predict the depth of the elements shown in the graffiti *etc.* (*cf.* Fig. 4). This paper thus also contributes a new dataset for measuring performance under strong viewpoint changes that depicts regular street scenes. In contrast to previous datasets, *e.g.*, [5], ours contains both viewpoint and appearance changes occuring at the same time.

Single-Image Depth Prediction. Monocular depth estimation aims at training a neural network to predict a depth map from a single RGB image. Supervised methods directly regress ground-truth depth, acquired with active sensors (LiDAR or Kinect) [20,42,43]; from SfM reconstructions [38]; or manual ordinal annotations [11,38]. However, collecting training data is difficult, costly, and time-consuming. Self-supervised training minimizes a photometric reprojection error between views. These views are either frames of videos [27,32], and/or stereo pairs [24,26,27,72]. Video-only training also needs to estimate the pose between frames (up to scale) and model moving objects [84]. Training with stereo provides metric-accurate depth predictions if the same camera is used at test time. Supervised and self-supervised losses can be combined during training [34,77].

CNNs can be trained to predict normals [18,19,29,71], both depth and normals [36,78,80,83], or 3D plane equations [42,43]. However, normals are either used to regularize depth, or trained exclusively on indoor scenes because of availability of supervised data, which is difficult to collect for outdoor scenes [12].

The approach presented in this paper is not tied to any specific single view depth prediction approach, and simply assumes that approximate depth information is available. Normal estimation networks could also be used in our pipeline, however depth estimation networks are more readily available.

3 Perspective Unwarping

We introduce a method for performing perspective correction of monocular images. The aim is to perform this prior to feature extraction, leading to detection and description of features that are more stable under viewpoint changes. That stability can, for example, establish more numerous correct correspondences between images taken from significantly different viewpoints.

The method is inspired by, and bears close resemblance to, the view-invariant patch descriptor by Wu *et al.* [74]. The main difference is that while Wu's method was designed for alignment of 3D point clouds, our method can be applied to single, monocular images, allowing it to be utilized in applications such as wide-baseline feature matching, single image visual localization, or structure from motion.

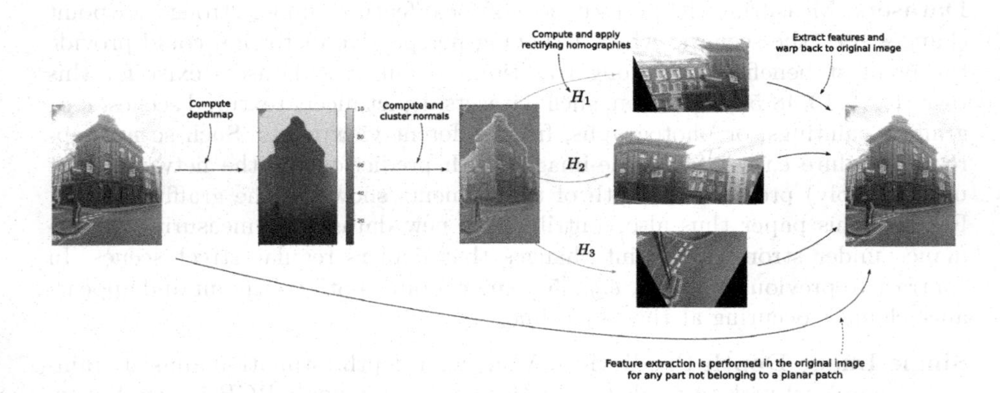

Fig. 3. Proposed pipeline for extracting perspectively corrected features from a single image: Depths are computed using a single-image depth estimation network. The intrinsic parameters of the camera are then used to backproject each pixel into 3D space, and a surface normal is estimated for each pixel. The normals are clustered into three orthogonal directions, and a homography is computed for each cluster to correct for the perspective distortion. Local image features are then extracted from each rectified patch using an off-the-shelf feature extractor and their positions are warped back into the original image. Regular local features are extracted from the parts of the image that do not belong to a planar patch.

A schematic overview of the method is shown in Fig. 3. The central idea is that given a single input image, a network trained for single image depth estimation is used to compute a corresponding dense depth map of the image. Using the camera intrinsics, the depth map gets backprojected into a point cloud, given in the camera's reference frame. From this point cloud, a surface normal vector is estimated for each point. For any given point in the point cloud (and correspondingly, in the image), a rectifying homography H can now be computed. H transforms a patch centered around the point in the image to a corresponding patch, which simulates a virtual camera looking straight down on the patch, *i.e.* a camera whose optical axis coincides with the patch's surface normal. As shown by several experiments in Sect. 5, by performing both feature detection and description in this rectified space, the obtained interest points can sometimes be considerably more robust to viewpoint differences.

In principle this method could be applied to each point independently, but typically a large number of points will share the same normal. Consider for example points lying on a plane, such as points on the ground, points on the same wall, or points on different but parallel planes, such as opposing walls. These points will all be rectified by the same homography. This is utilized in the proposed method by identifying planar regions in the input image, which is done by clustering all normals on the unit sphere. This yields a partitioning of the input image into several connected components, which are then rectified individually. Each input image is thus transformed into a set of rectified patches, consisting of perpendicular views of all dominant planes in the image. Image features can then be extracted, using any off-the-shelf detector and descriptor. In the experiments, results are provided for SIFT [45], SuperPoint [17], ORB [60], and BRISK[35] features.

Note that the rectification process is not dependent on all planes being observed. If only one plane is visible, that may still be rectified on its own. Parts of the image that are not detected to be on a plane are not rectified, but we still extract features from these parts and use them for feature matching. This way, we do not ignore good features just because they are not on a planar surface. For complex, non-planar geometries, large parts of an image may not have planar surfaces. For such images our approach gracefully resorts to standard (non-rectified) feature matching for regions not belonging to the identified planes.

Below, we describe each of the above steps in more detail.

3.1 Depth Estimation

The first step in the perspective correction process is the computation of the depth map. In this paper we use MonoDepth2 [27] which was trained with a Depth Hints loss [72] on several hours of stereo video captured in one European city and three US cities. In addition to stereo, the network was also trained on the MegaDepth dataset [38] and Matterport [7] datasets (see supplementary materials for details). This network takes as input a single image resized to 512×256, and outputs a dense depth map. Under ideal conditions, each pixel in the depth map tells us the calibrated depth in meters. In practice, any method that provides dense depth estimates may be used, and the depths need not be calibrated, *i.e.* depths estimated up to an unknown scale factor may also be used, since the depth map is only used to compute surface normals for each point.

3.2 Normal Computation and Clustering

With the depth map computed, the next step is normal computation. A surface normal is estimated for each pixel in the depth map by considering a 5×5 window centered on the pixel, and fitting a plane to the 25 corresponding back-projected points. The unit normal vector of the plane is taken as an estimate of the surface normal for that pixel.

With the normals computed, they are then clustered to identify regions in the image corresponding to planar surfaces. Since all points on the same plane share the same normals, these normals (and the pixels assigned to them) may be found by performing k-means clustering.

Since the depth map, and by extension the surface normals, are subject to noise, we found that clustering the normals into three clusters or dominant directions, while also enforcing orthogonality between these clusters, gave good results. Each cluster also includes its antipodal point (this means, for example, that two opposing walls would be assigned to the same dominant direction). This assumption seems to correspond to the 3D structure of many scenes: if at least one dominant plane is visible, such as the ground or a building wall, this method will produce satisfactory results. If two are visible, the estimated normals, and thus also the estimated homographies, tend to be more accurate. If no planes are visible, the method gracefully reduces to regular feature extraction. Note also that several different patches in the image can be assigned to the same cluster, but rectified separately as different planes: examples include opposing walls or parallel flat surfaces.

In non-Manhattan world geometries, where several non-perpendicular planes are visible, the estimated normals may not be completely accurate. Thus, our method would apply a homography that would render the planar surface not from a fronto-parallel view, but at a tilt. In most cases, this rectification still removes some effects of perspective distortion.

3.3 Patch Rectification

With the normals clustered into three dominant clusters, each pixel is assigned its normal's respective cluster. Each of these subsets may be further subdivided into their respective connected components. The input image is thus partitioned into a set of patches, each consisting of a connected region of pixels in the image, together with a corresponding estimate of the surface normal for that patch. In Fig. 3, the patches are shown overlaid on the image in different colors.

A rectified view of each patch is now computed, using the estimated patch normal. The patch is warped using a homography, computed as the homography which maps the patch to the patch as it would have been seen in a virtual camera sharing the same camera center as the original camera, but rotated such that its optical axis is parallel to the surface normal (*i.e.* it is facing the patch straight on). The smallest rotation that brings the camera into this position is used.

Lastly, not the entire patch is rectified, since if the plane corresponding to a given patch is seen at a glancing angle in the camera, most of the rectified patch would be occupied by heavily distorted, or stretched, regions. As such, a threshold of 80° is imposed on the maximum angle allowed between the viewing ray from the camera, and the surface normal, and the resulting patch is cropped to only fully contain the region of the patch seen at not too glancing an angle.

3.4 Warping Back

When matching features, the image may now be replaced with its set of rectified patches and patches from non-planar parts of the image. Alternatively, feature extraction may be performed in the non-planar parts of the original image, and in all rectified patches, and the 2D locations of the features in the rectified patches may then be warped back into the original image coordinate system, but with the descriptors unchanged. A perspectively corrected representation of the image has then been computed. The final description thus includes perspectively corrected features for all parts of the image that were deemed as belonging to a plane, and regular features extracted from the original image, from the parts that were deemed non-planar.

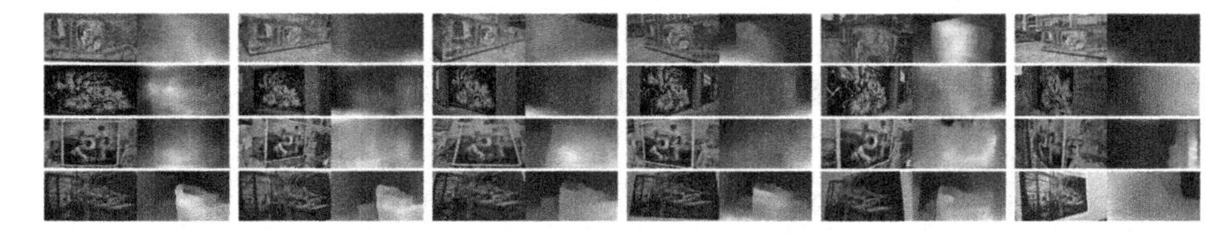

Fig. 4. "Failure" cases of applying a modern single-image depth prediction network [27,72] on images from the HPatches dataset [5] (visualized as inverse depth): the network predicts the depth of the scenes depicted in the graffiti and images rather than understanding that these are drawings/photos attached to a planar surface.

4 Dataset for Strong Viewpoint Changes

A modern, and already well-established, benchmark for evaluating local features is the HPatches dataset [5]. HPatches consists of 116 sequences of 6 images each, where sequences have either illumination or viewpoint changes. Similar to other datasets such as [49], planar scenes that can be modeled as homographies are used for viewpoint changes. Most of the sequences depict paintings or drawings on flat surfaces. Such scenes are ideal for local features as they provide abundant texture. Interestingly, such scenes cause single-image depth prediction to "fail": as shown in Fig. 4, networks predict the depth of the structures shown in the paintings and drawings rather than modeling the fact that the scene is planar. We would argue that this is a rather sensible behavior as the scene's planarity can only be inferred from context, *e.g.*, by observing a larger part of a scene rather than just individual drawings. Still, this behavior implies that standard datasets are not suitable for evaluating the performance of any type of method based on singe-image depth prediction. This motivated us to capture our own dataset that, in contrast to benchmarks such as HPatches, intentionally contains non-planar structures.

In this paper, we thus present a new dataset for evaluating the robustness of local features when matching across large viewpoint variations, and changes in lighting, weather, *etc.*. The dataset consists of 8 separate scenes, where each scene consists of images of one facade or building captured from a wide range of different viewpoints, and in different weather and environmental conditions. Each scene has been revisited up to 5 times, see Table 1.

All images included in the dataset originated from continuous video sequences captured using a consumer smartphone camera. These video sequences were then reconstructed using the Colmap software [64,65], and the poses for a subset of the images were extracted from this reconstruction. Colmap also provides an estimate of the intrinsic parameters for each individual image frame, which are included in the dataset. These are necessary since the focal length of the camera may differ between the images due to the camera's autofocus. Figure 5 shows a set of example images from two of our 8 scenes.

Table 1. Statistics for the scenes in our dataset

scene #	1	2	3	4	5	6	7	8
# img. pairs	3590	3600	3600	2612	3428	3031	2893	3312
# sequences	4	4	3	4	3	5	3	3

Since our scenes are not perfectly planar, measuring feature matching performance by the percentage of matches that are inliers to a homography, as done in [5,49], is not an option for our dataset. Inspired by CVPR 2019 workshops "Image Matching: Local Features and Beyond" and "Long-Term Visual Localization under Changing Conditions", we evaluate feature matching on downstream tasks as opposed to measuring the number of recovered feature matches or repeatability, *etc.*. Hence, we evaluate the performance of local features through the task of accurately estimating the relative pose between pairs of images. This allows us to judge if improvements in feature matching lead to meaningful improvements in practical applications such as localization and structure from motion.

For each scene, a list of image pairs is thus provided. Each image pair has been assigned to one of eighteen different difficulty categories, depending on the distance between the centres of the cameras that captured the images, and the magnitude of their relative rotation. The difficulty categories span the range of almost no difference in rotation up to almost 180° relative rotation. So, the image pairs in the k-th difficulty category have a relative rotation in the range of $[10k, 10(k+1)]$ degrees in one of the axes.

Fig. 5. Six example images, showing two different scenes of the presented benchmark dataset. The dataset contains several scenes, each consisting of over 1000 images of an urban environment captured during different weather conditions, and covering a large range of viewing angles. This dataset permits the evaluation of the degradation of local feature matching methods for increasing viewpoint angle differences. Please see the supplementary material for example images from each of the eight scenes.

The dataset is publicly available on the project webpage www.github.com/nianticlabs/rectified-features.

5 Experiments

This section provides two experiments: Sect. 5.1 shows that perspective unwarping based on the proposed approach can significantly improve feature matching performance on our proposed dataset. Sect. 5.2 shows that our approach can be used to re-localize a car under 180° viewpoint changes, *e.g.*, in the context of loop closure handling for SLAM. We use the SIFT [45] implementation provided by OpenCV [6] for all of our experiments.

5.1 Matching Across Large Viewpoint Changes

First, we evaluate our method on the 8 scenes of our proposed dataset. As a baseline we evaluate the performance of traditional and recently proposed learned local image features. To demonstrate the benefit of perspective rectification, we perform image matching with the same set of local features on the same set of image pairs.

For all image pairs in the dataset, feature matching was performed using SIFT[45], SuperPoint[17], ORB [60] and BRISK[35] features. For each feature type, feature matching was performed between features extracted from the original images, as well as between the perspectively corrected features, as explained in Sect. 3. Our unoptimized implementation performs image rectification in around 0.8 seconds per image. Using the established matches and the known intrinsics of the images, an essential matrix was computed, and the relative camera pose was then retrieved from this. This relative pose was compared to the ground truth relative pose (computed by Colmap as described in Sect. 4). An image pair was considered successfully localized if the difference between the estimated relative rotation and the ground truth relative rotation was smaller than 5°, where the difference between two rotations is taken as the magnitude of the smallest rotation that aligns one rotation with the other. Also included is a curve showing the performance of SIFT features extracted from images rectified using a vanishing point-based rectification method [9].

Figure 6 shows the performance of feature matching directly on the image pairs, *vs.* matching after perspective rectification. The 18 difficulty classes as described in Sect. 4 are listed along the $x-$axis, and the fraction of image pairs successfully localized in that difficulty class is shown on the $y-$ axis.

As can be seen in the figures, extracting perspectively corrected features can improve the pose estimation performance for planar scenes, particularly for SIFT and BRISK features. Overall, SuperPoint features seem to be more robust to viewpoint changes, which is natural since the SuperPoint feature is trained by performing homographic warps of patches. The ORB features show less improvement from using perspectively corrected images. This may have to do with the fact that these are not scale-invariant, and thus only correcting for the projective distortion, but not for the scale, may not be sufficient for obtaining good feature matching performance for these features.

In the supplementary material, localization rate graphs, like the middle and right figures in Fig. 6, can be found for all eight scenes.

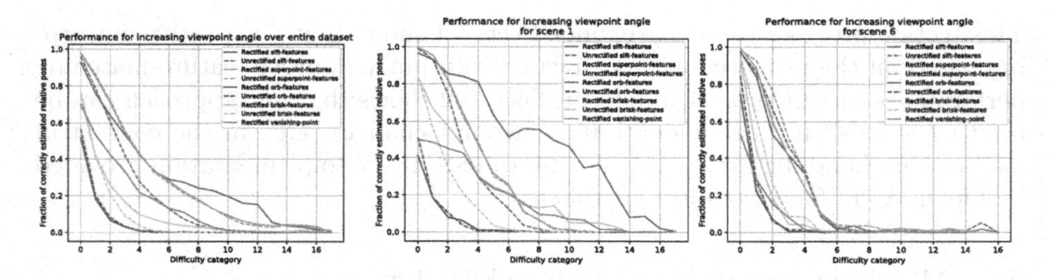

Fig. 6. Performance degradation due to increasing viewpoint difference. For each of the difficulty categories (labelled from 0 to 17 on the x-axis), the y-value shows the fraction of image pairs for which the relative rotation was estimated correctly to within 5°. Feature matching in the original images was compared with our rectification approach, for a variety of local features. Depth-based rectification is helpful overall, particularly for scenes with dominant planes, and more or less reduces to regular feature matching for scenes where no planes can be extracted. *Left:* Results for all images, over all scenes, in the entire dataset. *Middle:* A scene where most image pairs show the same plane, and this plane takes up a large portion of the images. *Right:* A scene containing many small facades, each often occupying a small part of the image, and some non-planar scene structures.

5.2 Re-localization from Opposite Viewpoints

For our next experiment, we consider a re-localization scenario for autonomous driving. More precisely, we consider the problem of re-localizing a car driving

	Seq. 1	Seq. 2
Ours	**98.1 %**	**97.1 %**
Standard	28.2 %	- / -

Fig. 7. Re-localization from Opposite Viewpoints. (a) Satellite imagery of the area covered in the RobotCar "Alternate Route" dataset. We show the trajectory of one of the sequences overlaid in orange. (b) Example of feature matching between rear and frontal images in the RobotCar dataset. Feature matching is performed in the rectified space (bottom), and then visualised in the original images (top). (c) Results of the search for the best front-facing database image for the rear-facing image from (b). (d) Localization results on the two sequences of the RobotCar dataset as the percentage of localized images. We compare our approach based on unwarping the ground plane with matching features in the original images. $*$ Image taken from Google Maps. Imagery ©2020 Google, Map data ©2020

down streets in the opposite direction from its first visit to the scene. Such a problem occurs, for example, during loop closure detection inside SLAM.

We use a subset of the Oxford RobotCar dataset [46], shown in Fig. 7, namely the "Alternate Route" dataset already used in [70]. The dataset consists of two traversals of the scene. We use 3,873 images captured by the front-facing camera of the RobotCar during one traversal as our database representation. 729 images captured by the car's rear-facing camera during the same traversal as well as an additional 717 images captured by the rear camera during a second traversal (captured around 10 min after the first one) are used as query images. As a result, there is a 180° viewpoint change between the query and database images.

We determine the approximate location from which the query images were taken by matching SIFT features between the query and database images. We compare our approach, for which we only unwarp the ground plane and use SIFT features, against a baseline that matches features between the original images[2]. For both approaches, we use a very simple localization approach that exhaustively matches each query image against each database image. For our approach, we select the database image with the largest number of homography inliers, estimated using RANSAC [21]. For the baseline, we select the database image with the largest number of fundamental matrix inliers, since we noticed that most correct matches are found on the buildings on the side of the road, and the corresponding points in the two images are thus not generally related by a homography. Due to the 180° change in viewpoint, the query and its

[2] For the baseline, we only use approximately every 2nd query image.

corresponding database image might be taken multiple meters apart. Thus, it is impossible to use the GPS coordinates provided by the dataset for verification. Instead, we manually verified whether the selected database image showed the same place or not (see also the supp. video).

Table 7(d) shows the percentage of correctly localized queries for our method and the baseline. As can be seen, our approach significantly ourperforms the baseline. A visualization for one query image and its corresponding database image found by our method is shown in Fig. 7(b), while Fig. 7(c) shows the number of homography inliers between this query and all database images. As can be seen, there is a clear peak around the correctly matching database image. This result is representative for most images localized by our approach (*cf.* the supp. video), though the number of inliers in the figure is on the lower end of what is common.

6 Conclusion

The results from Sect. 5 show that our proposed approach can significantly improve feature matching performance in real-world scenes and applications in dominantly planar scenes without a significant degradation in other environments. They further demonstrate that our approach is often easier to use than classical vanishing point-based approaches, which was one of the main motivations for this paper. Yet, our approach has its limitations.

Limitations. Similar to vanishing point-based methods, our approach requires that the planar structures that should be undistorted occupy a large-enough part of an image. If these parts are largely occluded, *e.g.*, by pedestrians, cars, or vegetation, it is unlikely that our approach is able to estimate a stable homography for unwarping. Further, the uncertainty of the depth predictions increases quadratically with the distance of the scene to the camera (as the volume projecting onto a single pixel grows quadratically with the distance). As a result, unwarping planes too far from the camera becomes unreliable. In contrast, it should be possible to relatively accurately undistort faraway scenes based on vanishing points or geometrically repeating elements. This suggests that developing hybrid approaches that adaptivley choose between different cues for perspective undistortion is an interesting avenue for future research.

Another failure case results from the fact that all training images seen by the depth prediction network have been oriented upright. As such, the network fails to produce meaningful estimates for cases where the images are rotated. However, it will be easy to avoid such problems in many practical applications: it is often possible to observe the gravity direction through other sensors or to pre-rotate the image based on geometric cues [76].

Future Work. We have shown that using existing neural networks for single-image depth prediction to remove perspective distortion leads to a simple yet effective approach to improve the performance of existing local features. A natural direction for further work is to integrate the unwarping stage into the learning process for local features. Rather than assuming that perspective distortion is perfectly removed, this would allow the features to compensate for inaccuracies in the undistortion process. Equally interesting is the question whether feature matching under strong viewpoint changes can be used as a self-supervisory signal for training single-image depth predictors: formulating the unwarping stage in a differentiable manner, one could use matching quality as an additional loss when training such networks.

Acknowledgements. The bulk of this work was performed during an internship at Niantic, and the first author would like to thank them for hosting him during the summer of 2019. This work has also been partially supported by the Swedish Foundation for Strategic Research (Semantic Mapping and Visual Navigation for Smart Robots) and the Chalmers AI Research Centre (CHAIR) (VisLocLearn). We would also like to extend our thanks Iaroslav Melekhov, who has captured some of the footage.

References

1. Aanæs, H., Dahl, A., Pedersen, K.S.: Interesting interest points. Int. J. Comput. Vis. **97**, 18–35 (2012). https://doi.org/10.1007/s11263-011-0473-8
2. Altwaijry, H., Trulls, E., Hays, J., Fua, P., Belongie, S.: Learning to match aerial images with deep attentive architectures. In: The IEEE Conference on Computer Vision and Pattern Recognition (CVPR) (2016)
3. Baatz, G., Köser, K., Chen, D., Grzeszczuk, R., Pollefeys, M.: Handling urban location recognition as a 2D homothetic problem. In: Daniilidis, K., Maragos, P., Paragios, N. (eds.) ECCV 2010. LNCS, vol. 6316, pp. 266–279. Springer, Heidelberg (2010). https://doi.org/10.1007/978-3-642-15567-3_20
4. Baatz, G., Köser, K., Chen, D., Grzeszczuk, R., Pollefeys, M.: Leveraging 3D city models for rotation invariant place-of-interest recognition. Int. J. Comput. Vis. (IJCV) **96**(3), 315–334 (2012). https://doi.org/10.1007/s11263-011-0458-7
5. Balntas, V., Lenc, K., Vedaldi, A., Mikolajczyk, K.: HPatches: a benchmark and evaluation of handcrafted and learned local descriptors. In: Proceedings of the IEEE Conference on Computer Vision and Pattern Recognition, pp. 5173–5182 (2017)
6. Bradski, G.: The OpenCV Library. Dr. Dobb's J. Softw. Tools (2000)
7. Chang, A., et al.: Matterport3D: learning from RGB-D data in indoor environments. In: International Conference on 3D Vision (3DV) (2017)
8. Wu, C., Fraundorfer, F., Frahm, J., Pollefeys, M.: 3D model search and pose estimation from single images using VIP features. In: IEEE Computer Society Conference on Computer Vision and Pattern Recognition (CVPR) Workshops (2008)
9. Chaudhury, K., DiVerdi, S., Ioffe, S.: Auto-rectification of user photos. In: 2014 IEEE International Conference on Image Processing (ICIP), pp. 3479–3483. IEEE (2014)
10. Chen, D.M., et al.: City-scale landmark identification on mobile devices. In: IEEE Conference on Computer Vision and Pattern Recognition (CVPR) (2011)
11. Chen, W., Fu, Z., Yang, D., Deng, J.: Single-image depth perception in the wild. In: Advances in Neural Information Processing Systems, pp. 730–738 (2016)
12. Chen, W., Xiang, D., Deng, J.: Surface normals in the wild. In: Proceedings of the IEEE International Conference on Computer Vision, pp. 1557–1566 (2017)

13. Cordes, K., Rosenhahn, B., Ostermann, J.: High-resolution feature evaluation benchmark. In: Wilson, R., Hancock, E., Bors, A., Smith, W. (eds.) CAIP 2013. LNCS, vol. 8047, pp. 327–334. Springer, Heidelberg (2013). https://doi.org/10.1007/978-3-642-40261-6_39
14. Criminisi, A., Reid, I., Zisserman, A.: Single view metrology. Int. J. Comput. Vis. (IJCV) **40**(2), 123–148 (2000). https://doi.org/10.1023/A:1026598000963
15. Criminisi, A.: Single-view metrology: algorithms and applications (invited paper). In: Van Gool, L. (ed.) DAGM 2002. LNCS, vol. 2449, pp. 224–239. Springer, Heidelberg (2002). https://doi.org/10.1007/3-540-45783-6_28
16. Davison, A.J., Reid, I.D., Molton, N.D., Stasse, O.: MonoSLAM: real-time single camera SLAM. PAMI **29**(6), 1052–1067 (2007)
17. DeTone, D., Malisiewicz, T., Rabinovich, A.: SuperPoint: self-supervised interest point detection and description. In: Proceedings of the IEEE Conference on Computer Vision and Pattern Recognition Workshops, pp. 224–236 (2018)
18. Dhamo, H., Navab, N., Tombari, F.: Object-driven multi-layer scene decomposition from a single image. In: Proceedings of the IEEE International Conference on Computer Vision, pp. 5369–5378 (2019)
19. Eigen, D., Fergus, R.: Predicting depth, surface normals and semantic labels with a common multi-scale convolutional architecture. In: Proceedings of the IEEE International Conference on Computer Vision, pp. 2650–2658 (2015)
20. Eigen, D., Puhrsch, C., Fergus, R.: Depth map prediction from a single image using a multi-scale deep network. In: Advances in Neural Information Processing Systems, pp. 2366–2374 (2014)
21. Fischler, M.A., Bolles, R.C.: Random sample consensus: a paradigm for model fitting with applications to image analysis and automated cartography. Commun. ACM **24**(6), 381–395 (1981)
22. Fraundorfer, F., Bischof, H.: A novel performance evaluation method of local detectors on non-planar scenes. In: 2005 IEEE Computer Society Conference on Computer Vision and Pattern Recognition (CVPR 2005) - Workshops (2005)
23. Gálvez-López, D., Tardós, J.D.: Bags of binary words for fast place recognition in image sequences. IEEE Trans. Robot. **28**(5), 1188–1197 (2012). https://doi.org/10.1109/TRO.2012.2197158
24. Garg, R., Bg, V.K., Carneiro, G., Reid, I.: Unsupervised CNN for single view depth estimation: geometry to the rescue. In: Leibe, B., Matas, J., Sebe, N., Welling, M. (eds.) ECCV 2016. LNCS, vol. 9912, pp. 740–756. Springer, Cham (2016). https://doi.org/10.1007/978-3-319-46484-8_45
25. Germain, H., Bourmaud, G., Lepetit, V.: Sparse-to-dense hypercolumn matching for long-term visual localization. In: International Conference on 3D Vision (3DV) (2019)
26. Godard, C., Mac Aodha, O., Brostow, G.J.: Unsupervised monocular depth estimation with left-right consistency. In: Proceedings of the IEEE Conference on Computer Vision and Pattern Recognition, pp. 270–279 (2017)
27. Godard, C., Mac Aodha, O., Firman, M., Brostow, G.J.: Digging into self-supervised monocular depth prediction. In: The International Conference on Computer Vision (ICCV), October 2019
28. Heng, L., et al.: Project autovision: localization and 3D scene perception for an autonomous vehicle with a multi-camera system. In: 2019 International Conference on Robotics and Automation (ICRA) (2019)
29. Hickson, S., Raveendran, K., Fathi, A., Murphy, K., Essa, I.: Floors are flat: leveraging semantics for real-time surface normal prediction. In: Proceedings of the IEEE International Conference on Computer Vision Workshops (2019)

30. Hinterstoisser, S., Lepetit, V., Benhimane, S., Fua, P., Navab, N.: Learning real-time perspective patch rectification. Int. J. Comput. Vis. **91**(1), 107–130 (2011). https://doi.org/10.1007/s11263-010-0379-x
31. Jones, E.S., Soatto, S.: Visual-inertial navigation, mapping and localization: a scalable real-time causal approach. Int. J. Robot. Res. (IJRR) **30**(4), 407–430 (2011)
32. Klodt, M., Vedaldi, A.: Supervising the new with the old: learning SFM from SFM. In: Proceedings of the European Conference on Computer Vision (ECCV), pp. 698–713 (2018)
33. Koser, K., Koch, R.: Perspectively invariant normal features. In: IEEE International Conference on Computer Vision (ICCV) (2007)
34. Kuznietsov, Y., Stuckler, J., Leibe, B.: Semi-supervised deep learning for monocular depth map prediction. In: Proceedings of the IEEE Conference on Computer Vision and Pattern Recognition, pp. 6647–6655 (2017)
35. Leutenegger, S., Chli, M., Siegwart, R.: BRISK: binary robust invariant scalable keypoints. In: 2011 IEEE International Conference on Computer Vision (ICCV), pp. 2548–2555. IEEE (2011)
36. Li, B., Shen, C., Dai, Y., Van Den Hengel, A., He, M.: Depth and surface normal estimation from monocular images using regression on deep features and hierarchical CRFs. In: Proceedings of the IEEE Conference on Computer Vision and Pattern Recognition, pp. 1119–1127 (2015)
37. Li, H., Zhao, J., Bazin, J.C., Chen, W., Liu, Z., Liu, Y.H.: Quasi-globally optimal and efficient vanishing point estimation in Manhattan world. In: The IEEE International Conference on Computer Vision (ICCV) (2019)
38. Li, Z., Snavely, N.: MegaDepth: learning single-view depth prediction from internet photos. In: Computer Vision and Pattern Recognition (CVPR) (2018)
39. Liebowitz, D., Criminisi, A., Zisserman, A.: Creating architectural models from images. Comput. Graph. Forum **18**(3), 39–50 (1999)
40. Lim, H., Sinha, S.N., Cohen, M.F., Uyttendaele, M.: Real-time image-based 6-DOF localization in large-scale environments. In: 2012 IEEE Conference on Computer Vision and Pattern Recognition (2012)
41. Lindeberg, T.: Scale-space theory: a basic tool for analysing structures at different scales. J. Appl. Stat. **21**(2), 224–270 (1994)
42. Liu, C., Kim, K., Gu, J., Furukawa, Y., Kautz, J.: PlaneRCNN: 3D plane detection and reconstruction from a single image. In: Proceedings of the IEEE Conference on Computer Vision and Pattern Recognition, pp. 4450–4459 (2019)
43. Liu, C., Yang, J., Ceylan, D., Yumer, E., Furukawa, Y.: PlaneNet: piece-wise planar reconstruction from a single RGB image. In: Proceedings of the IEEE Conference on Computer Vision and Pattern Recognition, pp. 2579–2588 (2018)
44. Liu, W., Wang, Y., Chen, J., Guo, J., Lu, Y.: A completely affine invariant image-matching method based on perspective projection. Mach. Vis. Appl. **23**(2), 231–242 (2012). https://doi.org/10.1007/s00138-011-0347-7
45. Lowe, D.G.: Distinctive image features from scale-invariant keypoints. Int. J. Comput. Vis. **60**(2), 91–110 (2004). https://doi.org/10.1023/B:VISI.0000029664.99615.94
46. Maddern, W., Pascoe, G., Linegar, C., Newman, P.: 1 year, 1000 km: the Oxford RobotCar dataset. Int. J. Robot. Res. **36**(1), 3–15 (2017)
47. Matas, J., Chum, O., Urban, M., Pajdla, T.: Robust wide-baseline stereo from maximally stable extremal regions. Image Vis. Comput. **22**(10), 761–767 (2004)
48. Mikolajczyk, K., et al.: A comparison of affine region detectors. Int. J. Comput. Vis. **65**(1), 43–72 (2005). https://doi.org/10.1007/s11263-005-3848-x

49. Mikolajczyk, K., Schmid, C.: A performance evaluation of local descriptors. IEEE Trans. Pattern Anal. Mach. Intell. **27**, 1615–1630 (2005)
50. Mishkin, D., Matas, J., Perdoch, M.: MODS: fast and robust method for two-view matching. Comput. Vis. Image Underst. **141**, 81–93 (2015)
51. Yi, K.M., Verdie,Y., Fua, P., Lepetit, V.: Learning to assign orientations to feature points. In: The IEEE Conference on Computer Vision and Pattern Recognition (CVPR) (2016)
52. Morel, J.M., Yu, G.: ASIFT: a new framework for fully affine invariant image comparison. SIAM J. Imaging Sci. **2**(2), 438–469 (2009)
53. Mur-Artal, R., Tardós, J.D.: ORB-SLAM2: an open-source SLAM system for monocular, stereo and RGB-D cameras. IEEE Trans. Robot. **33**(5), 1255–1262 (2017). https://doi.org/10.1109/TRO.2017.2705103
54. Pang, Y., Li, W., Yuan, Y., Pan, J.: Fully affine invariant surf for image matching. Neurocomputing **85**, 6–10 (2012)
55. Pritts, J., Chum, O., Matas, J.: Rectification, and segmentation of coplanar repeated patterns. In: IEEE Conference on Computer Vision and Pattern Recognition (CVPR) (2014)
56. Pritts, J., Kukelova, Z., Larsson, V., Chum, O.: Rectification from radially-distorted scales. In: Jawahar, C.V., Li, H., Mori, G., Schindler, K. (eds.) ACCV 2018. LNCS, vol. 11365, pp. 36–52. Springer, Cham (2019). https://doi.org/10.1007/978-3-030-20873-8_3
57. Pritts, J., Kukelova, Z., Larsson, V., Chum, O.: Radially-distorted conjugate translations. In: The IEEE Conference on Computer Vision and Pattern Recognition (CVPR) (2018)
58. Pritts, J., Rozumnyi, D., Kumar, M.P., Chum, O.: Coplanar repeats by energy minimization. In: Proceedings of the British Machine Vision Conference (BMVC) (2016)
59. Robertson, D.P., Cipolla, R.: An image-based system for urban navigation. In: BMVC (2004)
60. Rublee, E., Rabaud, V., Konolige, K., Bradski, G.R.: ORB: an efficient alternative to sift or surf. In: ICCV, vol. 11, p. 2. Citeseer (2011)
61. Sarlin, P.E., Cadena, C., Siegwart, R., Dymczyk, M.: From coarse to fine: robust hierarchical localization at large scale. In: The IEEE Conference on Computer Vision and Pattern Recognition (CVPR) (2019)
62. Sattler, T., Leibe, B., Kobbelt, L.: Efficient & effective prioritized matching for large-scale image-based localization. IEEE Trans. Pattern Anal. Mach. Intell. **39**(9), 1744–1756 (2017)
63. Schönberger, J.L., Pollefeys, M., Geiger, A., Sattler, T.: Semantic visual localization. In: The IEEE Conference on Computer Vision and Pattern Recognition (CVPR) (2018)
64. Schönberger, J.L., Frahm, J.M.: Structure-from-motion revisited. In: Conference on Computer Vision and Pattern Recognition (CVPR) (2016)
65. Schönberger, J.L., Zheng, E., Frahm, J.-M., Pollefeys, M.: Pixelwise view selection for unstructured multi-view stereo. In: Leibe, B., Matas, J., Sebe, N., Welling, M. (eds.) ECCV 2016. LNCS, vol. 9907, pp. 501–518. Springer, Cham (2016). https://doi.org/10.1007/978-3-319-46487-9_31
66. Shao, H., Svoboda, T., Gool, L.V.: ZuBuD – Zürich buildings database for image based recognition. Technical report 260, Computer Vision Laboratory, Swiss Federal Institute of Technology, April 2003

67. Simon, G., Fond, A., Berger, M.-O.: *A-Contrario* horizon-first vanishing point detection using second-order grouping laws. In: Ferrari, V., Hebert, M., Sminchisescu, C., Weiss, Y. (eds.) ECCV 2018. LNCS, vol. 11214, pp. 323–338. Springer, Cham (2018). https://doi.org/10.1007/978-3-030-01249-6_20
68. Snavely, N., Seitz, S.M., Szeliski, R.: Photo tourism: exploring photo collections in 3D. In: SIGGRAPH (2006)
69. Svärm, L., Enqvist, O., Kahl, F., Oskarsson, M.: City-scale localization for cameras with known vertical direction. IEEE Trans. Pattern Anal. Mach. Intell. **39**(7), 1455–1461 (2017)
70. Toft, C., Olsson, C., Kahl, F.: Long-term 3D localization and pose from semantic labellings. In: Proceedings of the IEEE International Conference on Computer Vision, pp. 650–659 (2017)
71. Wang, X., Fouhey, D., Gupta, A.: Designing deep networks for surface normal estimation. In: Proceedings of the IEEE Conference on Computer Vision and Pattern Recognition, pp. 539–547 (2015)
72. Watson, J., Firman, M., Brostow, G.J., Turmukhambetov, D.: Self-supervised monocular depth hints. In: IEEE International Conference on Computer Vision (ICCV) (2019)
73. Wu, C.: Towards linear-time incremental structure from motion. In: International Conference on 3D Vision (3DV) (2013)
74. Wu, C., Clipp, B., Li, X., Frahm, J.M., Pollefeys, M.: 3D model matching with viewpoint-invariant patches (VIP). In: 2008 IEEE Conference on Computer Vision and Pattern Recognition, pp. 1–8. IEEE (2008)
75. Wu, C., Frahm, J.-M., Pollefeys, M.: Detecting large repetitive structures with salient boundaries. In: Daniilidis, K., Maragos, P., Paragios, N. (eds.) ECCV 2010. LNCS, vol. 6312, pp. 142–155. Springer, Heidelberg (2010). https://doi.org/10.1007/978-3-642-15552-9_11
76. Xian, W., Li, Z., Fisher, M., Eisenmann, J., Shechtman, E., Snavely, N.: UprightNet: geometry-aware camera orientation estimation from single images. In: The IEEE International Conference on Computer Vision (ICCV) (2019)
77. Yang, N., Wang, R., Stuckler, J., Cremers, D.: Deep virtual stereo odometry: leveraging deep depth prediction for monocular direct sparse odometry. In: Proceedings of the European Conference on Computer Vision (ECCV), pp. 817–833 (2018)
78. Yang, Z., Wang, P., Wang, Y., Xu, W., Nevatia, R.: Lego: learning edge with geometry all at once by watching videos. In: Proceedings of the IEEE Conference on Computer Vision and Pattern Recognition, pp. 225–234 (2018)
79. Cao, Y., McDonald, J.: Viewpoint invariant features from single images using 3D geometry. In: Workshop on Applications of Computer Vision (WACV) (2009)
80. Yin, W., Liu, Y., Shen, C., Yan, Y.: Enforcing geometric constraints of virtual normal for depth prediction. In: Proceedings of the IEEE International Conference on Computer Vision, pp. 5684–5693 (2019)
81. Zeisl, B., Köser, K., Pollefeys, M.: Viewpoint invariant matching via developable surfaces. In: Fusiello, A., Murino, V., Cucchiara, R. (eds.) ECCV 2012. LNCS, vol. 7584, pp. 62–71. Springer, Heidelberg (2012). https://doi.org/10.1007/978-3-642-33868-7_7
82. Zeisl, B., Köser, K., Pollefeys, M.: Automatic registration of RGB-D scans via salient directions. In: The IEEE International Conference on Computer Vision (ICCV) (2013)
83. Zhan, H., Weerasekera, C.S., Garg, R., Reid, I.: Self-supervised learning for single view depth and surface normal estimation. In: 2019 International Conference on Robotics and Automation (ICRA) (2019)

84. Zhou, T., Brown, M., Snavely, N., Lowe, D.G.: Unsupervised learning of depth and ego-motion from video. In: The IEEE Conference on Computer Vision and Pattern Recognition (CVPR), July 2017
85. Zhou, Y., Qi, H., Huang, J., Ma, Y.: NeurVPS: neural vanishing point scanning via conic convolution. In: Conference on Neural Information Processing Systems (NeurIPS) (2019)

Deep Reinforced Attention Learning for Quality-Aware Visual Recognition

Duo Li and Qifeng Chen(✉)

The Hong Kong University of Science and Technology, Kowloon, Hong Kong
duo.li@connect.ust.hk, cqf@ust.hk

Abstract. In this paper, we build upon the weakly-supervised generation mechanism of intermediate attention maps in any convolutional neural networks and disclose the effectiveness of attention modules more straightforwardly to fully exploit their potential. Given an existing neural network equipped with arbitrary attention modules, we introduce a meta critic network to evaluate the quality of attention maps in the main network. Due to the discreteness of our designed reward, the proposed learning method is arranged in a reinforcement learning setting, where the attention actors and recurrent critics are alternately optimized to provide instant critique and revision for the temporary attention representation, hence coined as Deep REinforced Attention Learning (DREAL). It could be applied universally to network architectures with different types of attention modules and promotes their expressive ability by maximizing the relative gain of the final recognition performance arising from each individual attention module, as demonstrated by extensive experiments on both category and instance recognition benchmarks.

Keywords: Convolutional Neural Networks · Attention modules · Reinforcement learning · Visual recognition

1 Introduction

Attention is a perception process that aggregates global information and selectively attends to the meaningful parts while neglects other uninformative ones. Mimicking the attention mechanism has allowed deep Convolutional Neural Networks (CNNs) to efficiently extract useful features from redundant information contexts of images, videos, audios, and texts. Consequently, attention modules further push the performance boundary of prevailing CNNs in handling various visual recognition tasks. Recently, popularized attention operators usually follow the modular design which could be seamlessly integrated into feed-forward neural network blocks, such as channel attention [11] and spatial attention [40] modules.

Electronic supplementary material The online version of this chapter (https://doi.org/10.1007/978-3-030-58517-4_29) contains supplementary material, which is available to authorized users.

A. Vedaldi et al. (Eds.): ECCV 2020, LNCS 12361, pp. 493–509, 2020.
https://doi.org/10.1007/978-3-030-58517-4_29

They learn to recalibrate feature maps via inferring corresponding importance factors separately along the spatial or channel dimension.

These attention modules are critical components to capture the most informative features and guide the allocation of network weights to them. Nevertheless, existing attention emerges automatically along with the weak supervision of the topmost classification objective, which is not dedicatedly devised for the intermediate attention generation. Thus, this weakly-supervised optimization scheme may lead to sub-optimal outcomes regarding the attention learning process. In other words, the attention maps learned in such a manner might be opaque in its discrimination ability. Linsley *et al.* propose to supervise the intermediate attention maps with human-derived dense annotations [18], but the annotation procedure could be both labor-intensive and easily affected by subjective biases. To dissolve the above deficiency, we propose Deep REinforced Attention Learning (DREAL) to provide direct supervision for attention modules and fully leverage the representational power of their parameters, thus promoting the final recognition performance. Our method does not require additional annotations and is generic to popular attention modules in CNNs. In addition to the conventional weakly-supervised paradigm, we introduce critic networks in parallel to the main network to evaluate the quality of intermediate attention modules. After investigating the source feature map and the inferred attention map to predict the expected critique[1], the critic network straightforwardly transmits a supervisory signal to the attention module based on the variation of the final recognition performance with or without the effect of this attention module. With this introspective supervision mechanism, the attention module could promptly identify to what degree its behavior benefits the whole model and adapt itself accordingly. If the allocation of attention weights is not favored at the moment, the attention module would correct it instantly according to the feedback from the critic network. In practice, to avoid the unacceptable cost of high-capacity modules, we adopt the recurrent LSTM cell as the critic network, which imposes a negligible parameter and computational burden on the whole network. Furthermore, it implicitly bridges the current layer and the previous layers, enhancing the interactions of features and attention maps at different depths in order to inject more contextual information into the critique.

Considering the supervision for optimizing the critic network, we develop an intuitive criterion that reflects the effect of attention on the amelioration of the final recognition results. This evaluation criterion is non-differentiable so the conventional back-propagation algorithm is hardly applicable. To solve this discrete optimization problem, we encompass the attention-equipped main network and the critic meta network into a reinforcement learning algorithm. Our proposed model can be served as the contextual bandit [15], a primitive instance of reinforcement learning model where all actions are taken in one single shot of the state. Specifically, in a convolutional block, the intermediate feature map is defined as the *state* while the relevant *action* is the attention map conditioned on its current feature map at a training step. The critic network takes the state and

[1] "Critique" refers to the critic value outputted from the critic network in this paper.

action as input and estimates the corresponding critic value. With the joint optimization of the attention actor and the recurrent critic, the quality of attention could be boosted progressively, driven by the signal of reward which measures the relative gain of attention modules in the final recognition accuracy. In a quality-aware style, attention modules would be guided with the direct supervision from critic networks to strengthen the recognition performance by correctly emphasizing meaningful features and suppressing other nuisance factors.

On the ImageNet benchmark, DREAL leads to consistently improved performance for baseline attention neural networks, since attention maps are obtained in a more quality-oriented and reinforced manner. It can be applied to arbitrary attention types in a plug-and-play manner with minimal tunable hyperparameters. To explore its general applicability, the reinforced attention networks are further applied to the person re-identification task, achieving new state-of-the-art results on two popular benchmarks including Market-1501 and DukeMTMC-reID among recent methods which involve the attention mechanism. We also visualize the distribution of some attention maps for a clearer understanding of the improved attention-assisted features, illustrating how the critic network acts on these attention maps. Quantitative and qualitative results provide strong evidence that the learned critic not only improves the overall accuracy but also encodes a meaningful confidence level of the attention maps.

Summarily we make the following contributions to attention-equipped neural network architectures:

- We propose to assess the attention quality of existing modular designs using auxiliary critic networks. To the best of our knowledge, it has never been well studied in the research field to explicitly consider the attention quality of features inside backbone convolutional neural networks before us.
- We further bridge the critic networks and the backbone network with a reinforcement learning algorithm, providing an end-to-end jointly training framework. The formulation of reinforced optimization paves a creative way to solve the visual recognition problem with a quality-aware constraint.
- Our critic networks introduce negligible parameters and computational cost, which could also be completely removed during inference. The critic networks could slot into network models with arbitrary attention types, leading to accuracy improvement validated by comprehensive experiments.

2 Related Work

We revisit attention modules in the backbone network design and reinforcement learning applications associated with attention modeling in previous literature. We clarify the connections and differences of our proposed learning method with these existing works.

Attention Neural Networks. Recently, the attention mechanism is usually introduced to modern neural networks as a generic operation module, augmenting their performance with minimal additional computation. ResAttNet [36] stacks residual attention modules with trunk-and-mask branches. The auxiliary mask branch cascades top-down and bottom-up structure to unfold the feedforward and feedback cognitive process, generating soft weights with mixed attention in an end-to-end fashion. The pioneering SENet [11] builds the foundation of a research area that inserts lightweight modular components to improve the functional form of attention. The proposed SE block adaptively recalibrates channel-wise feature responses by explicitly modeling interdependencies between channels, substantially improving the performance when adapted to any state-of-the-art neural network architectures. The follow-up GENet [10] gathers contextual information spreading over a large spatial extent and redistributes these aggregations to modulate the local features in the spatial domain. To take one step further, MS-SAR [39] collects all responses in the neighborhood regions of multiple scales to compute spatially-asymmetric importance values and reweights the original responses with these recalibration scores. CBAM [40] and BAM [23] come up with to decompose the inference of the three-dimensional attention map along spatial and channel dimensions and arrange them in a sequential or parallel layout for feature refinement. SRM [16] summarizes response statistics of each channel by style pooling and infers recalibration weights through the channel-independent style integration, leveraging the latent capability of style information in the decision making process. ECA-Net [37] applies a local cross-channel interaction strategy that is efficiently implemented by the fast 1D convolution with a kernel of adaptive size. As stated above, most existing methods are dedicated to developing sophisticated feature extraction and recalibration operations, but attention maps are sustained by weakly long-distance supervision. Probably [18] is most related to us regarding the *motivation*, which also attempts to augment the weakly-supervised attention derived from category-level labels. The referred approach first introduces an extra large-scale data set ClickMe with human-annotated salient regions. It then incorporates ClickMe supervision to the intermediate attention learning process of their proposed GALA module (an extension of the seminal SE module). In stark contrast to prior works, we do not propose any new attention modules or leverage external data and annotations for supervision. By employing a shared LSTM to evaluate these attention modules, our approach concentrates on promoting the quality-aware evolution of attention maps via a novel reinforcement learning design. Recently, the non-local modules [38] thrive as a self-attention mechanism. We also elaborate on this sub-area of attention research in the supplementary materials. Generally speaking, our DREAL method could be readily applied to neural networks armed with all aforementioned attention modules regardless of their specific forms.

Deep Reinforcement Learning. Unlike conventional supervised machine learning methods, reinforcement learning has been originated from humans' decision making process [19]. Reinforcement Learning (RL) aims at enabling the agent to make decisions optimally based on rewards it receives from an environment. Recently, the field of RL resurrects with the strong support of deep learning techniques. Deep Reinforcement Learning (DRL), as a principal paradigm, can be roughly divided into two categories: deep Q learning [8,21] and policy gradient [1,30]. In the former class, the goal of deep Q Networks is to fit a Q-value function to capture the expected return for taking a particular action at a given state. In the latter class, policy gradient methods approximate the policy which maximizes the expected future reward using gradient descent.

Deep reinforcement learning has been adopted in the selection procedure of attended parts for computer vision applications. For example, locating the most discriminative ones among a sequence of image patches can be naturally formulated as an MDP process and contributory to a wide array of tasks such as single-label [20] or multi-label [5] image classification, face hallucination [2] and person re-identification [13]. In these exemplars, a policy-guided agent usually traverses the spatial range of a single image to dynamically decide the attended regions via progressively aggregating regional information collected in the past. Distinct from spatially attentive regions in the image space, our research focuses on the attention modules in the backbone networks that are represented with feature-level attention maps instead of image-level saliency maps. In the same spirit, deep reinforcement learning is also utilized in the video space to find appropriate focuses across frames. This kind of attention indicates discarding the misleading and confounding frames within the video for face [25] or action [7] recognition. For comparison, the attention is defined in the spatial domain of an image or the temporal domain of a video segment in the aforementioned works while our formulation is shaped inside the convolutional blocks with attention operators. DRL has also been applied to the field of neural network architecture engineering but mainly focused on network acceleration and automated search, which is depicted in detail in the supplementary materials. Unlike this research line, we propose to measure and boost the quality of attention generation under the reinforcement learning framework. To the best of our knowledge, little progress with reinforcement learning has been made in the fundamental problem of handcrafted attention-equipped CNNs, which is of vital importance in the neural architecture design.

3 Approach

In this section, we first overview the proposed formulation of Deep REinforced Attention Learning (DREAL) and then elaborate on the critic and actor modules within this regime. Finally we describe the optimization procedure in detail.

3.1 Overview

Formally, let $\mathbf{X}$ denote the input image example, the intermediate feature map in a convolutional block is represented as the <u>state</u> $\mathbf{F}(\mathbf{X};\boldsymbol{\mathcal{W}})$, where $\boldsymbol{\mathcal{W}}$ is the weight matrix of the backbone network. The corresponding attention <u>action</u> conditioned on the feature map emerges with an auxiliary operation module, represented as $\mathbf{A}(\mathbf{F};\boldsymbol{\theta})$, where $\boldsymbol{\theta}$ defines the parameters of the attention module.

Given the predefined state-action pair above, a critic network predicts the state-action value (Q-value) function as $Q(\mathbf{A}|\mathbf{F};\phi)$, where ϕ symbolizes the weights of this critic network (deep Q network).

To guide the critic network to predict the actual quality of our attention module, we design a reward R as its direct supervision signal. The reward function reflects the relative gain for the entire network regarding one specific attention module. This reward concerning the lth attention module is defined as

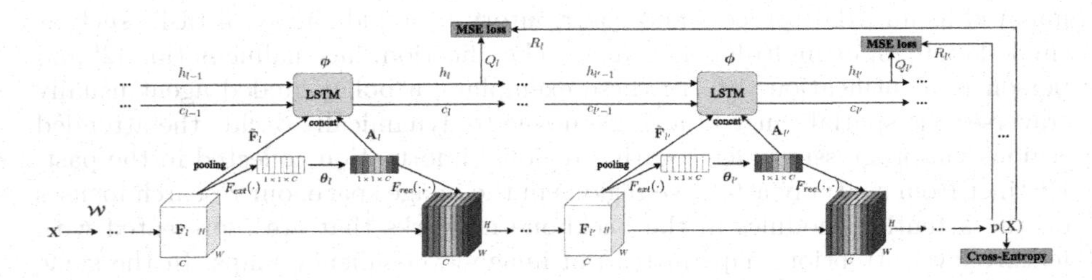

Fig. 1. Schematic illustration of our proposed Deep REinforced Attention Learning, built with the SENet [11] as an instance. Two selected building blocks in the same stage are presented for the purpose of conciseness. Best viewed in color and zoomed in.

$$R_l = \begin{cases} 1 - \frac{\mathbf{p}_c(\mathbf{X}|\mathbf{A}_1,\mathbf{A}_2,\cdots,\mathbf{A}_{l-1},\bar{\mathbf{A}}_l,\mathbf{A}_{l+1},\cdots,\mathbf{A}_L)}{\mathbf{p}_c(\mathbf{X}|\mathbf{A}_1,\mathbf{A}_2,\cdots,\mathbf{A}_L)}, & \\ & \text{if } \mathbf{p}_c(\mathbf{X}|\mathbf{A}) \geq \mathbf{p}_i(\mathbf{X}|\mathbf{A}) \ \forall i = 1, 2, \cdots, K, \\ -\gamma, & \text{otherwise}, \end{cases} \tag{1}$$

where $\mathbf{p}(\mathbf{X}|\mathbf{A})$ or $\mathbf{p}(\mathbf{X}|\mathbf{A}_1, \mathbf{A}_2, \cdots, \mathbf{A}_L)$ denotes the probabilistic prediction of the fully attention-based network with respect to an image sample $\mathbf{X}$, with the subscript i being an arbitrary category label and c being the corresponding ground truth category label drawn from a total of K classes. For further clarification, $\mathbf{p}_c(\mathbf{X}|\mathbf{A}_1, \mathbf{A}_2, \cdots, \mathbf{A}_{l-1}, \bar{\mathbf{A}}_l, \mathbf{A}_{l+1}, \cdots, \mathbf{A}_L)$ defines the prediction output after substituting the attention map from the lth attention module with its mean vector $\bar{\mathbf{A}}_l$ during inference, which helps to bypass the emphasizing or suppressing effect of a specific attention module while retaining all others to isolate its influence on the final prediction. On the first condition of Eq. 1, under the premise that the fully attention-equipped network should have satisfactory recognition ability, we tend to assign large reward value to a certain attention module if the output probability for the true class declines significantly (*i.e.*, the fraction in Eq. 1 becomes small) when this attention module loses its recalibration effect, *i.e.*, substituted by its mean vector. On the second condition of Eq. 1, incorrect prediction of the ground truth label would lead to penalization on all attention modules with a negative reward $-\gamma$, where γ is established as a tunable positive factor. We set the parameter γ as 1 in our main experiments by cross-validation to strike a balance between the positive and negative reward in the above two conditions. Intuitively, this criterion could effectively incentivize attention modules to bring more benefits to the final prediction results.

In the above statement, we have a glance at the general formulation of our proposed DREAL method where the actor generates the attention map and the critic analyzes the gain from the attention actor and guides the actor to maximize this gain. We leave the detailed architectural design of the critic and actor together with the optimization pipeline in the following subsections.

3.2 Recurrent Critic

We take the representative SENet [11] as an exemplar, with the network architecture and computation flow illustrated in Fig. 1. It could be readily extended to other types of attention-equipped networks. The raw feature map $\mathbf{F}_l \in \mathbb{R}^{H\times W\times C}$ in the lth building block is processed with the extraction function $F_{ext}(\cdot)$ to capture non-local context information, which often takes the form of global average pooling in the spatial domain. This processed tensor is fed into the subsequent attention module $\mathbf{A}(\cdot;\boldsymbol{\theta})$ to produce its corresponding attention map $\mathbf{A}_l$, which is then applied to the original feature map $\mathbf{F}_l$ through the recalibration function $F_{rec}(\cdot,\cdot)$. Typically, $F_{rec}(\mathbf{F}_l,\mathbf{A}_l)$ obtained the output tensor through an element-wise multiplication of the state $\mathbf{F}_l$ and action $\mathbf{A}_l$ (broadcast if necessary to match the dimension). With the dynamically selective mechanism, a spectrum of features could be emphasized or suppressed respectively in a channel-wise manner.

Taking consideration of the critic model, even injecting a miniaturized auxiliary network separately into each layer will increase the total amount of parameters as the network depth grows. Furthermore, following this way, critique results of previous layers will be overlooked by subsequent ones. Therefore, we introduce a recurrent critic network design that benefits from parameter sharing and computation re-use to avoid heavily additional overheads. Specifically, an LSTM model is shared by all residual blocks in the same stage, where successive layers have the identical spatial size and similar channel configurations [9]. The dimension of the raw feature map is first reduced to match that of the attention map (usually using global average pooling along the channel or spatial dimension depending on the specific attention types to be evaluated), then they are concatenated and fed into the LSTM cell as the temporary input, together with the hidden and cell state from the previous layer. The LSTM network generates the current hidden state $h_l \in \mathbb{R}$ and cell state $c_l \in \mathbb{R}$ as

$$h_l, c_l = \text{LSTM}(\text{concat}(\tilde{\mathbf{F}}_l, \mathbf{A}_l), h_{l-1}, c_{l-1}; \boldsymbol{\phi}), \tag{2}$$

where $\tilde{\mathbf{F}}_l$ denotes the reduced version of $\mathbf{F}_l$ as stated above. The cell state stores the information from all precedent layers in the same stage, while the new hidden state is a scalar that would be directly extracted to be the output critic value for current attention assessment, written as

$$Q_l(\mathbf{A}_l|\mathbf{F}_l;\boldsymbol{\phi}) = h_l. \tag{3}$$

It is noted that if spatial and channel attention coexist, *e.g.* in the CBAM [40], two individual LSTM models will be employed to process attention maps with different shapes.

The LSTM models not only incorporate the features and attention maps in the current residual block but also recurrently integrate the decisions from previous layers in the same stage, exploring complicated non-linear relationships between them. Thus, the attention-aware features could adjust in a self-adaptive fashion as layers going deeper. The recurrent critic network implicitly captures the inter-layer dependencies to provide a more precise evaluation regarding the influence of the current attention action on the whole network.

Complexity Analysis. The recurrent characteristic permits the critic network to maintain reasonable parameter and computational cost. Both additional parameters and FLOPs approximately amount to $4 \times (2C \times 1 + 1 \times 1)$ for each stage, which is economic and negligible compared to the main network. Specifically, there exist 4 linear transformations that take the concatenated vector with the size of $2C$ and a one-dimensional hidden state as the input to compute two output scalars, *i.e.*, hidden and cell state. Furthermore, since an LSTM is shared throughout the same stage, the number of parameter increments may remain constant with the growing depths, referring to the comparisons between ResNet-50 and ResNet-101 with various attention types in Table 1.

3.3 Attention Actor

We explore various attention types as the actors, including channel, spatial and style modules, which are developed in SENet [11], CBAM [40] and SRM [16] respectively. The detailed forms of these operators are reviewed in the following.

Channel Attention. Different channels in the feature map could contain diverse representations for specific object categories or visual patterns. The channel attention action exploits to emphasize more informative channels and suppress less useful ones. The attention map is represented as

$$\mathbf{A}_c = \sigma(\mathbf{W}_1\delta(\mathbf{W}_0\text{AvgPool}(\mathbf{F}))), \tag{4}$$

where $\mathbf{W}_0 \in \mathbb{R}^{\frac{C}{r} \times C}$ and $\mathbf{W}_1 \in \mathbb{R}^{C \times \frac{C}{r}}$ are weight matrices of two consecutive Fully Connected (FC) layers composing the bottleneck structure, with r being the reduction ratio. σ denotes the sigmoid function and δ refers to the ReLU [22] activation function. AvgPool$(\cdot)$ indicates the global average pooling operation.

Spatial-Channel Attention. Non-local context information is of critical importance on object recognition, which reflects long-range dependence in the spatial domain. The spatial attention action further aggregates such kinds of information and redistribute them to local regions, selecting the most discriminative parts to allocate higher weights. The spatial attention is represented as

$$\mathbf{A}_s = \sigma(\text{conv}_{7\times7}(\text{concat}(AvgPool(\mathbf{F}), MaxPool(\mathbf{F})))), \tag{5}$$

where $\text{conv}_{7\times7}(\cdot)$ defines the convolution operation with the kernel size of 7×7. The concatenation and pooling operations (denoted as *AvgPool* and *MaxPool*) here are along the channel axis, in contrast to the ordinary AvgPool above in the spatial axis. Here, the channel attention map is generated leveraging the clue of highlighted features from global maximum pooling, reformulated as

$$\mathbf{A}_c = \sigma(\mathbf{W}_1\delta(\mathbf{W}_0\text{AvgPool}(\mathbf{F})) + \mathbf{W}_1\delta(\mathbf{W}_0\text{MaxPool}(\mathbf{F}))). \quad (6)$$

The above two attention modules are placed in a sequential manner with the channel-first order.

Style Recalibration. Recently it is revealed that the style information also plays an important role in the decision process of neural networks. The style-based attention action converts channel-wise statistics into style descriptors through a Channel-wise Fully Connected (CFC) layer and re-weight each

Algorithm 1: Deep REinforced Attention Learning

Input: Training dataset $\mathcal{D}$, maximal iterations M, network depth L
Output: Parameters of the backbone network $\boldsymbol{\mathcal{W}}$, attention actors $\boldsymbol{\theta}$ and recurrent critics $\boldsymbol{\phi}$

1 Initialize the model parameters $\boldsymbol{\mathcal{W}}$, $\boldsymbol{\theta}$ and $\boldsymbol{\phi}$
2 **for** $t \leftarrow$ *1 to M* **do**
3   Randomly draw a batch of samples $\mathcal{B}$ from $\mathcal{D}$
4   **foreach** $\mathbf{X}$ *in* $\mathcal{B}$ **do**
5   Compute feature state $\mathbf{F}(\mathbf{X}; \boldsymbol{\mathcal{W}})$
6   Derive attention action $\mathbf{A}(\mathbf{F}; \boldsymbol{\theta})$
7   Estimate critic value $Q(\mathbf{A}|\mathbf{F}; \boldsymbol{\phi})$
8   Bypass the recalibration effect of the attention module and forward to infer the corresponding reward R
9   Calculate loss functions $\mathcal{L}_c$, $\mathcal{L}_q$, $\mathcal{L}_r$
10   Update $\boldsymbol{\mathcal{W}}$ with $\Delta\boldsymbol{\mathcal{W}} \propto \frac{\partial}{\partial\boldsymbol{\mathcal{W}}}\mathcal{L}_c$
11   Update $\boldsymbol{\theta}$ with $\Delta\boldsymbol{\theta} \propto \frac{\partial}{\partial\boldsymbol{\theta}}(\mathcal{L}_c + \mathcal{L}_q)$
12   Update $\boldsymbol{\phi}$ with $\Delta\boldsymbol{\phi} \propto \frac{\partial}{\partial\boldsymbol{\phi}}\mathcal{L}_r$
13   **end**
14 **end**
15 **return** $\boldsymbol{\mathcal{W}}$, $\boldsymbol{\theta}$ and $\boldsymbol{\phi}$

channel with the corresponding importance factor. The style recalibration map is represented as

$$\mathbf{A}_t = \sigma(\text{BN}(\mathbf{W} \cdot \text{concat}(\text{AvgPool}(\mathbf{F}), \text{StdPool}(\mathbf{F})))), \quad (7)$$

where StdPool defines the global standard deviation pooling akin to global average pooling and each row in the weight matrix $\mathbf{W} \in \mathbb{R}^{C\times2}$ of the CFC layer is multiplied individually to each channel representation.

3.4 Reinforced Optimization

Unlike standard reinforcement learning, there does not exist an explicit sequential relationship along the axis of the training step or network depth. The attention action is conditioned on the feature state in a one-shot fashion, which is essentially a one-step Markov Decision Process (MDP). It could be also viewed as a contextual bandit [15] model. Furthermore, the action is a continuous value thus its optimum could be searched through gradient ascent following the solution of continuous Q-value prediction. In order to provide positive guidance for the attention module, the loss function for Q-value prediction is defined as the negative of Eq. 3

$$\mathcal{L}_q = -Q(\mathbf{A}(\mathbf{F};\boldsymbol{\theta})|\mathbf{F};\boldsymbol{\phi}). \tag{8}$$

With the critic network $\boldsymbol{\phi}$ frozen, the attention actor $\boldsymbol{\theta}$ is updated to obtain higher value of critique via this loss function, which implies higher quality of attention.

In the meanwhile, the critic network is optimized via regression to make an accurate quality estimation of the attention action conditioned on the feature state. The Mean Squared Error (MSE) loss is constructed through penalizing the squared Euclidean distance between the predicted Q-value and the actual reward R, represented as

$$\mathcal{L}_r = \|Q(\mathbf{A}(\mathbf{F};\boldsymbol{\theta})|\mathbf{F};\boldsymbol{\phi}) - R\|^2. \tag{9}$$

With the attention actor $\boldsymbol{\theta}$ frozen this time, the critic network $\boldsymbol{\phi}$ is updated to acquire more precise quality-aware evaluation.

The supervised training has been largely in place, which employs the conventional cross-entropy for classification correctness, represented as

$$\mathcal{L}_c = -\frac{1}{|\mathcal{B}|}\sum_{\mathbf{X}\in\mathcal{B}} \log \mathbf{p}_c(\mathbf{X};\mathcal{W},\boldsymbol{\theta}), \tag{10}$$

where $\mathcal{B}$ is a randomly sampled mini-batch within the entire dataset $\mathcal{D}$ and $\mathbf{X}$ denotes an image example with c indicating its corresponding ground truth label.

In this regime, we combine the strength of supervised and reinforcement learning, alternately training the backbone architecture, attention actor models and LSTM-based critic networks. The learning scheme is summarized in Algorithm 1. During inference, recurrent critic networks are all discarded so that the computational cost is exactly identical to that of the original attention-based backbone network.

4 Experiments

In this section, we evaluate the proposed DREAL method on close- and open-set visual recognition tasks: image classification and person re-identification. We make comparisons with extensive baseline attention networks to demonstrate the effectiveness and generality of our method.

4.1 Category Recognition

We employ several attention-based networks as the backbone models, including SENet [11], CBAM [40] and SRM [16], which feature channel, spatial-channel and style attention respectively. We evaluate the reinforced attention networks on the ImageNet [6] dataset, which is one of the most large-scale and challenging object classification benchmarks up to date. It includes over 1.2 million natural images for training as well as 50K images reserved for validation, containing objects spreading across 1,000 predefined categories. Following the common practice of optimization, we adopt the Stochastic Gradient Descent (SGD) optimizer with the momentum of 0.9, the weight decay of 1e-4 and the batch size of 256. We keep in accordance with SENet [11] and train all networks for 100 epochs. The learning rate is initiated from 0.1 and divided by 10 every 30 epochs. For data augmentation, we randomly resize and crop training images to patches of 224×224 size with random horizontal flipping. For evaluation, we resize the shorter sides of validation images to 256 pixels without changing their aspect ratios and crop center regions of the same size as that of training images. As a special note for meta networks of critic, hidden and cell states in the LSTM cells from each stage are initialized as zero scalars. During each training epoch, one building block in each stage is bypassed to measure the corresponding reward, avoiding much additional inference cost. This optimization strategy could guarantee that each LSTM belonging to one stage is optimized all the way along with the main network. All experiments are performed with the PyTorch [24] framework.

Table 1. Recognition error comparisons on the ImageNet validation set. The standard metrics of top-1/top-5 errors are measured using the single-crop evaluation. It is noted that the additional parameters and FLOPs of our proposed reinforced attention networks exist only during the training process, originating from critic networks.

Architecture	Params	GFLOPs	Method	Top-1/Top-5 Err.(%)	Architecture	Params	GFLOPs	Method	Top-1/Top-5 Err.(%)
SE-ResNet-50	28.088M	4.091	*official*	23.29/6.62	SE-ResNet-101	49.326M	7.806	*official*	22.38/6.07
			self impl.	22.616/6.338				*self impl.*	21.488/5.778
	28.119M	4.092	reinforced	**22.152/5.948**		49.358M	7.811	reinforced	**20.732/5.406**
CBAM-ResNet-50	28.089M	4.095	*official*	22.66/6.31	CBAM-ResNet-101	49.330M	7.812	*official*	21.51/5.69
			self impl.	22.386/6.172				*self impl.*	21.518/5.812
	28.154M	4.097	reinforced	**21.802/6.084**		49.394M	7.819	reinforced	**20.682/5.362**
SRM-ResNet-50	25.587M	4.089	*official*	22.87/6.49	SRM-ResNet-101	44.614M	7.801	*official*	21.53/5.80
			self impl.	22.700/6.392				*self impl.*	21.404/5.740
	25.618M	4.090	reinforced	**22.348/6.084**		44.644M	7.806	reinforced	**20.474/5.362**

For the baseline attention networks, we re-implement each network and achieve comparable or even stronger performance compared to those from the original papers. The officially released performance and outcomes of our re-implementation are shown in Table 1, denoted as *official* and *self impl.* respectively. We also report the parameters, computational complexities and validation errors of our reinforced attention networks correspondingly. It is noteworthy that the increment of parameters and computation is completely negligible compared to the baseline counterparts. For SE-ResNet-50, the additionally introduced parameters only occupy 0.11% of the original amount. Thanks to the

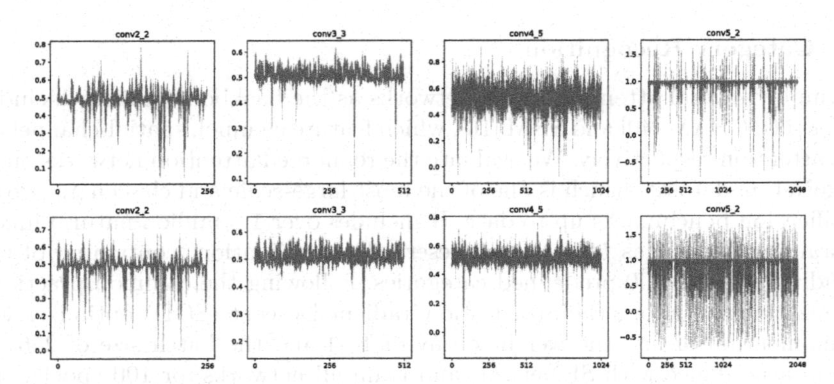

Fig. 2. Distributions of channel-attention vectors on the ImageNet validation set before (*top*) and after (*bottom*) applying DREAL. The x-axis represents channel index.

parameter sharing mechanism of recurrent critics, roughly the same number of network parameters is attached to SE-ResNet-101, which consists of the same number of stages as the 50-layer version. Consequently, the relative increase of parameters is further reduced to 0.06% for this deeper backbone network. Regarding computational cost, the most significant growth among all networks does not exceed 0.1%, which comes from the reinforced CBAM-ResNet-101 with double LSTMs for both spatial and channel attention modeling.

Our reinforced attention networks bring about clear-cut reduction of error rates compared to the strong re-implementation results. The ResNet-101 networks with three types of attention obtain more improvement than their 50-layer versions, which could be attributed to the capability of our method to exploit the potential of more attention representations in these deeper models. While we explore three types of attention networks to demonstrate its wide applicability here, our DREAL method could be integrated into any other types of attention networks conveniently. We also explore more complicated recurrent neural network architectures for critics, but it brings marginally additional benefit with more parameters and computational costs.

Visualization. To provide better intuitive insight of our method, we take SE-ResNet-50 as an example and visualize the distribution of channel-attention vectors before and after applying our method. The attention maps are evaluated on the ImageNet validation set and the distributions of the last residual block in each stage are showcased in Fig. 2, where the solid line indicates the mean values among all validation image examples and the shadow area indicates 3× variance. By comparison, we observe that in certain layers (like `conv3_3` and `conv4_5`), DREAL encourages attention weights to become similar to each other across different channels. It echos the rationale that shallower layers capture fundamental visual patterns, which tend to be **category-agnostic**. In deeper layers (like `conv5_2`), with the guidance of the critic network, attention weights develop a tendence to fluctuate more but within a moderate range, flexibly extracting

category-oriented semantic meaning for the final recognition objective. Visualization results of other layers are provided in the supplementary materials.

4.2 Instance Recognition

We further conduct experiments on the more challenging open-set recognition task to demonstrate the generalization ability of our learning approach. We evaluate the performance of reinforced attention networks on two widely used person re-identification benchmarks, *i.e.* Market-1501 [43] and DukeMTMC-reID [27].

Datasets. Person ReID is an instance recognition task with the target of retrieving gallery images of the same identity as the probe pedestrian image. The Market-1501 dataset is comprised of 32,668 bounding boxes of 1,501 identities generated by a DPM-detector, with original images captured by 6 cameras in front of the supermarket inside the campus of Tsinghua University. The conventional split contains 12,936 training images of 751 identities and 15,913 gallery images of 750 identities as well as 3,368 queries. The DukeMTMC-reID dataset consists of 36,411 images covering 1,812 identities collected by 8 cameras, where only 1,404 identities appear across camera views and the other 408 identities are regarded as distractors. The training split includes 16,522 image examples from 702 persons while the non-overlapping 17,661 gallery samples and 2,228 queries are drawn from the remaining 702 person identities.

Implementation Details. Following the common practice of experimental setup, we adopt the ResNet-50 model as the backbone network due to its strong track record of feature extraction. To achieve fast convergence, the backbone of ReID model is pre-trained on ImageNet for parameter initialization. The last down-sampling operation in the `conv5_x` stage is removed to preserve high resolution for a better output representation. We deploy sequential channel and spatial attention modules on the ResNet model, which resembles the arrangement of CBAM [40]. For data augmentation, input pedestrian images are first randomly cropped to the size of 384×128 for fine-grained representation. Then they are horizontally flipped with the probability of 0.5 and normalized with mean and standard deviation per channel. Finally, the Random Erasing [45] technique is applied to make the model robust to occlusion. In this ranking-based task, we further introduce a triplet loss to encourage inter-class separation and intra-class aggregation with a large margin, which is set as 0.5 in the experiments. To satisfy the demand for triplet loss, we employ the PK sampling strategy [28], randomly selecting P identities and K samples from each identity to form each mini-batch. In our main experiments, we set $\mathrm{P}=16$ and $\mathrm{K}=8$ to generate mini-batches with the size of 128. Furthermore, we apply the label-smoothing regularization [33] to the cross-entropy loss function to alleviate overfitting, where the perturbation probability for original labels is set as 0.1. We also add a Batch Normalization neck after the global average pooling layer to normalize the feature scales. The four losses in total are minimized with the AMSGRAD [26] optimizer ($\beta_1 = 0.9, \beta_2 = 0.999$, weight decay = 5e-4). The learning rate initiates from 3e-4 and is divided by a factor of 10 every 40 epochs within the entire optimization period of 160 epochs. During evaluation, we feed both the original image and its horizontally flipped version into the model and calculate their mean feature representation. The extracted visual features are matched based on the similarities of their cosine distance.

Evaluation Protocols. We conduct evaluation under the single-query mode and adopt Cumulative Matching Characteristics (CMC) and mean Average Precision (mAP) as the evaluation metrics. CMC curve records the hit rate among the top-k ranks and mAP considers both precision and recall to reflect the performance in a more comprehensive manner. Here we choose to report the Rank-1 result in the CMC curve. For the purpose of fairness, we evaluate our method without any post-processing methods, such as re-ranking [44], which is applicable to our method and would significantly boost the performance of mAP especially.

Performance Comparison. As illustrated in the bottom groups of Table 2, we compare our proposed method with the baseline model as well as the attention-based one. We also compare it to other state-of-the-art methods that exploit various types of attention designs, as listed in the top groups of these two sub-tables. It is observed that harnessing the spatial and channel attention mechanism considerably enhances the performance of baseline models, while our proposed reinforced attention networks achieve further improvement over the vanilla attention networks. Specifically, with the proposed method, our model outperforms the vanilla attention network with a margin of 1.4%/0.7% regarding the Rank-1/mAP metric on the Market-1501 dataset. DukeMTMC-reID is a much

Table 2. Comparison to state-of-the-art methods on the Market-1501 (*left*) and DukeMTMC-reID (*right*) benchmarks. Results extracted from the original publications are presented with different decimal points. Bold indicates the best results while italic the runner-up. ResNet-50 is employed as the backbone if no special statement.

Method	Reference	Rank-1(%)	mAP(%)
IDEAL◇	BMVC 2017 [14]	86.7	67.5
MGCAM	CVPR 2018 [31]	83.55	74.25
AACN◇	CVPR 2018 [42]	85.90	66.87
DuATM†	CVPR 2018 [29]	91.42	76.62
HA-CNN‡	CVPR 2018 [17]	91.2	75.7
Mancs	ECCV 2018 [35]	93.1	82.3
AANet	CVPR 2019 [34]	93.89	82.45
ABD-Net	ICCV 2019 [4]	95.60	88.28
MHN-6 (PCB)	ICCV 2019 [3]	95.1	85.0
SONA^{2+3}	ICCV 2019 [41]	*95.58*	*88.83*
baseline	This Paper	93.5	82.8
+ attention		94.7	85.9
+ reinforce		**96.1**	**89.6**

Method	Reference	Rank-1(%)	mAP(%)
AACN◇	CVPR 2018 [42]	76.84	59.25
DuATM†	CVPR 2018 [29]	81.82	64.58
HA-CNN‡	CVPR 2018 [17]	80.5	63.8
Mancs	ECCV 2018 [35]	84.9	71.8
AANet	CVPR 2019 [34]	86.42	72.56
ABD-Net	ICCV 2019 [4]	89.00	*78.59*
MHN-6 (PCB)	ICCV 2019 [3]	89.1	77.2
SONA^{2+3}	ICCV 2019 [41]	*89.38*	78.28
baseline	This Paper	84.8	72.5
+ attention		86.4	76.2
+ reinforce		**89.6**	**79.8**

◇ with the GoogleNet/Inception [32,33] backbone.
† with the DenseNet-121 [12] backbone.
‡ with the dedicate HA-CNN [17] backbone.

more challenging dataset due to the wider camera views and more complex scene variations. In this context, our method could better demonstrate its superiority by leveraging the potential of attention representation. As a result, a more prominent performance gain of 3.2%/2.6% on the Rank-1/mAP metric is achieved. Even horizontally compared with other state-of-the-art methods that utilize dedicatedly designed backbone networks [17] or exploit higher-order attention forms [3,41], our proposed method beats them with consistent margins on both Rank-1 accuracy and mAP results across different datasets. For example, on the Market-1501 benchmark, we surpass the nearest rival method of SONA by 0.5% and 0.8% on the Rank-1 and mAP measurement respectively.

5 Conclusion

In this paper, we have proposed Deep REinforcement Attention Learning (DREAL) to facilitate visual recognition in a quality-aware manner. We employ recurrent critics that assess the attention action according to the performance gain it brings to the whole model. Wrapped up in a reinforcement learning paradigm for joint optimization, critic networks would promote the relevant attention actor to focus on the significant features. Furthermore, the recurrent critic could be used as a plug-and-play module for any pre-existing attention networks with negligible overheads. Extensive experiments on various recognition tasks and benchmarks empirically verify the efficacy and efficiency of our method.

References

1. Ammar, H.B., Eaton, E., Ruvolo, P., Taylor, M.: Online multi-task learning for policy gradient methods. In: ICML (2014)
2. Cao, Q., Lin, L., Shi, Y., Liang, X., Li, G.: Attention-aware face hallucination via deep reinforcement learning. In: CVPR (2017)
3. Chen, B., Deng, W., Hu, J.: Mixed high-order attention network for person re-identification. In: ICCV (2019)
4. Chen, T., et al.: ABD-Net: attentive but diverse person re-identification. In: ICCV (2019)
5. Chen, T., Wang, Z., Li, G., Lin, L.: Recurrent attentional reinforcement learning for multi-label image recognition. In: AAAI (2018)
6. Deng, J., Dong, W., Socher, R., Li, L.J., Li, K., Fei-Fei, L.: ImageNet: a large-scale hierarchical image database. In: CVPR (2009)
7. Dong, W., Zhang, Z., Tan, T.: Attention-aware sampling via deep reinforcement learning for action recognition. In: AAAI (2019)
8. Gu, S., Lillicrap, T., Sutskever, I., Levine, S.: Continuous deep q-learning with model-based acceleration. In: ICML (2016)
9. He, K., Zhang, X., Ren, S., Sun, J.: Deep residual learning for image recognition. In: CVPR (2016)
10. Hu, J., Shen, L., Albanie, S., Sun, G., Vedaldi, A.: Gather-excite: exploiting feature context in convolutional neural networks. In: NeurIPS (2018)

11. Hu, J., Shen, L., Sun, G.: Squeeze-and-excitation networks. In: CVPR (2018)
12. Huang, G., Liu, Z., van der Maaten, L., Weinberger, K.Q.: Densely connected convolutional networks. In: CVPR (2017)
13. Lan, X., Wang, H., Gong, S., Zhu, X.: Deep reinforcement learning attention selection for person re-identification. arXiv e-prints arXiv:1707.02785, July 2017
14. Lan, X., Wang, H., Gong, S., Zhu, X.: Deep reinforcement learning attention selection for person re-identification. In: BMVC (2017)
15. Langford, J., Zhang, T.: The epoch-greedy algorithm for multi-armed bandits with side information. In: NIPS (2008)
16. Lee, H., Kim, H.E., Nam, H.: SRM: a style-based recalibration module for convolutional neural networks. In: ICCV (2019)
17. Li, W., Zhu, X., Gong, S.: Harmonious attention network for person re-identification. In: CVPR (2018)
18. Linsley, D., Shiebler, D., Eberhardt, S., Serre, T.: Learning what and where to attend with humans in the loop. In: ICLR (2019)
19. Littman, M.L.: Reinforcement learning improves behaviour from evaluative feedback. Nature **521**, 445–451 (2015)
20. Mnih, V., Heess, N., Graves, A., Kavukcuoglu, K.: Recurrent models of visual attention. In: NIPS (2014)
21. Mnih, V., et al.: Human-level control through deep reinforcement learning. Nature **518**, 529–533 (2015)
22. Nair, V., Hinton, G.E.: Rectified linear units improve restricted Boltzmann machines. In: ICML (2010)
23. Park, J., Woo, S., Lee, J.Y., Kweon, I.S.: BAM: bottleneck attention module. In: BMVC (2018)
24. Paszke, A., et al.: Pytorch: an imperative style, high-performance deep learning library. In: NeurIPS (2019)
25. Rao, Y., Lu, J., Zhou, J.: Attention-aware deep reinforcement learning for video face recognition. In: ICCV (2017)
26. Reddi, S.J., Kale, S., Kumar, S.: On the convergence of Adam and beyond. In: ICLR (2018)
27. Ristani, E., Solera, F., Zou, R., Cucchiara, R., Tomasi, C.: Performance measures and a data set for multi-target, multi-camera tracking. In: Hua, G., Jégou, H. (eds.) ECCV 2016. LNCS, vol. 9914, pp. 17–35. Springer, Cham (2016). https://doi.org/10.1007/978-3-319-48881-3_2
28. Schroff, F., Kalenichenko, D., Philbin, J.: FaceNet: a unified embedding for face recognition and clustering. In: CVPR (2015)
29. Si, J., et al.: Dual attention matching network for context-aware feature sequence based person re-identification. In: CVPR (2018)
30. Silver, D., Lever, G., Heess, N., Degris, T., Wierstra, D., Riedmiller, M.: Deterministic policy gradient algorithms. In: ICML (2014)
31. Song, C., Huang, Y., Ouyang, W., Wang, L.: Mask-guided contrastive attention model for person re-identification. In: CVPR (2018)
32. Szegedy, C., et al.: Going deeper with convolutions. In: CVPR (2015)
33. Szegedy, C., Vanhoucke, V., Ioffe, S., Shlens, J., Wojna, Z.: Rethinking the inception architecture for computer vision. In: CVPR (2016)
34. Tay, C.P., Roy, S., Yap, K.H.: AANet: attribute attention network for person re-identifications. In: CVPR (2019)

35. Wang, C., Zhang, Q., Huang, C., Liu, W., Wang, X.: Mancs: a multi-task attentional network with curriculum sampling for person re-identification. In: Ferrari, V., Hebert, M., Sminchisescu, C., Weiss, Y. (eds.) ECCV 2018. LNCS, vol. 11208, pp. 384–400. Springer, Cham (2018). https://doi.org/10.1007/978-3-030-01225-0_23
36. Wang, F., et al.: Residual attention network for image classification. In: CVPR (2017)
37. Wang, Q., Wu, B., Zhu, P., Li, P., Zuo, W., Hu, Q.: ECA-Net: efficient channel attention for deep convolutional neural networks. In: CVPR (2020)
38. Wang, X., Girshick, R., Gupta, A., He, K.: Non-local neural networks. In: CVPR (2018)
39. Wang, Y., Xie, L., Qiao, S., Zhang, Y., Zhang, W., Yuille, A.L.: Multi-scale spatially-asymmetric recalibration for image classification. In: Ferrari, V., Hebert, M., Sminchisescu, C., Weiss, Y. (eds.) ECCV 2018. LNCS, vol. 11217, pp. 523–539. Springer, Cham (2018). https://doi.org/10.1007/978-3-030-01261-8_31
40. Woo, S., Park, J., Lee, J.-Y., Kweon, I.S.: CBAM: convolutional block attention module. In: Ferrari, V., Hebert, M., Sminchisescu, C., Weiss, Y. (eds.) ECCV 2018. LNCS, vol. 11211, pp. 3–19. Springer, Cham (2018). https://doi.org/10.1007/978-3-030-01234-2_1
41. Xia, B.N., Gong, Y., Zhang, Y., Poellabauer, C.: Second-order non-local attention networks for person re-identification. In: ICCV (2019)
42. Xu, J., Zhao, R., Zhu, F., Wang, H., Ouyang, W.: Attention-aware compositional network for person re-identification. In: CVPR (2018)
43. Zheng, L., Shen, L., Tian, L., Wang, S., Wang, J., Tian, Q.: Scalable person re-identification: a benchmark. In: ICCV (2015)
44. Zhong, Z., Zheng, L., Cao, D., Li, S.: Re-ranking person re-identification with k-reciprocal encoding. In: CVPR (2017)
45. Zhong, Z., Zheng, L., Kang, G., Li, S., Yang, Y.: Random erasing data augmentation. In: AAAI (2020)

CFAD: Coarse-to-Fine Action Detector for Spatiotemporal Action Localization

Yuxi Li[1], Weiyao Lin[1,2(✉)], John See[3], Ning Xu[4], Shugong Xu[2], Ke Yan[5], and Cong Yang[5]

[1] Department of Electronic Engineering, Shanghai Jiao Tong University, Shanghai, China
wylin@sjtu.edu.cn
[2] Institute for Advanced Communication and Data Science, Shanghai University, Shanghai, China
[3] Faculty of Computing and Informatics, Multimedia University, Cyberjaya, Malaysia
[4] Adobe Research, San Francisco, USA
[5] Clobotics, Shanghai, China

Abstract. Most current pipelines for spatio-temporal action localization connect frame-wise or clip-wise detection results to generate action proposals, where only local information is exploited and the efficiency is hindered by dense per-frame localization. In this paper, we propose Coarse-to-Fine Action Detector (CFAD), an original end-to-end trainable framework for efficient spatio-temporal action localization. The CFAD introduces a new paradigm that first estimates coarse spatio-temporal action tubes from video streams, and then refines the tubes' location based on key timestamps. This concept is implemented by two key components, the Coarse and Refine Modules in our framework. The parameterized modeling of long temporal information in the Coarse Module helps obtain accurate initial tube estimation, while the Refine Module selectively adjusts the tube location under the guidance of key timestamps. Against other methods, the proposed CFAD achieves competitive results on action detection benchmarks of UCF101-24, UCFSports and JHMDB-21 with inference speed that is 3.3× faster than the nearest competitor.

Keywords: Spatiotemporal action detection · Coarse-to-fine paradigm · Parameterized modeling

1 Introduction

Spatial-temporal action detection is the task of recognizing actions from input videos and localizing them in space and time. In contrast to action recognition

Electronic supplementary material The online version of this chapter (https://doi.org/10.1007/978-3-030-58517-4_30) contains supplementary material, which is available to authorized users.

A. Vedaldi et al. (Eds.): ECCV 2020, LNCS 12361, pp. 510–527, 2020.
https://doi.org/10.1007/978-3-030-58517-4_30

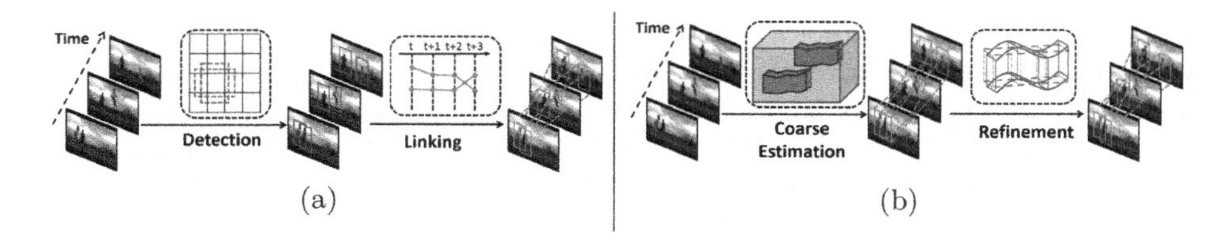

Fig. 1. The comparison between pipelines of detection and linking and our coarse-to-fine framework. (a) workflow of detection and linking method in previous works. (b) Our coarse-to-fine method to detect action tubes. (Best viewed in color.) (Color figure online)

or temporal localization, this task is far more complex, requiring both temporal detection along the time span and spatial detection at each frame when the actions occur.

Most existing methods for spatiotemporal action detection [6,9,19,25,26,29, 38,39] are implemented in two stages (illustrated in Fig. 1(a)). First, a spatial detector is applied to generate dense action box proposals on each frame. Then, these frame-level detections are linked together by a certain heuristic algorithm to generate final output, which is a series of boxes or an *action tube*. Nevertheless, since these approaches take a single or stack of frames as input, the information utilized by the detectors is limited within a fixed time interval, hence limiting the representative capacity of the learned features for classification. The similar problem is encountered in the aspect of localization. During training phase, models could be supervised by only a temporal fragment of the tubes, which can output accurate local proposals but may fail to locate entire tubes in a consistent manner. Additionally, IOU-based linking algorithms may result in accumulative localization error when noisy bounding box proposals are produced. Since the transition within action tubes is usually smooth and gradual, we hypothesize that using lesser number of boxes could be adequate to depict the action tube shape. Current pipelines, in their present state, relies heavily on dense per-frame predictions, which are redundant and a hindrance to efficient action detection.

With these considerations, we depart from classic detect-and-link strategies by proposing a new coarse-to-fine action detector (CFAD) that can generate more accurate action tubes with higher efficiency. Unlike previous approaches that detect dense boxes at first, the CFAD (as illustrated in Fig. 1(b)) goes on a progressive approach of estimating at a rougher level before ironing out the details. This strategy first estimates coarser action tubes, and then selectively refine these tubes at key timestamps. The action tubes are generated via two important components in our pipeline: Coarse Module and Refine Module.

The Coarse Module is designed to address the lack of global information and low efficiency in previous detect-and-link paradigm. In a *global* sense, it supervises the tube regression with the full tube shape information. In addition, within this module, a parameterized modeling scheme is introduced to depict action tubes. Instead of predicting large amount of box location at each frame,

Coarse Module only predict a few trajectory parameters to describe the tube of various endurance. As a result, this module learns a robust representation that accurately and efficiently characterizes action tube changes.

The Refine Module delves into the *local* context of each tube, to find precise temporal locations that are essential to further improve the estimated action tubes, which in turn, improves overall detection performance and efficiency. To properly refine the action tubes, a labelling algorithm is designed to generate labels that guide the learning of key timestamps selection. By a search scheme, the original coarse boxes are replaced by the largest scoring box proposals at these temporal locations, which then interpolate the final tube.

In summary, our contributions are three folds. (1) We propose a novel *coarse-to-fine* framework for the task of spatial-temporal action detection, which differs from the conventional paradigm of detect-and-link. Our new pipeline achieves state-of-the-art results on standard benchmarks with inference speed of 3.3× faster than the nearest competitor. (2) Under this framework, we design a novel action tube estimation method based on parametric modeling to fully exploit global supervision signal and handle time variant box coordinates by predicting limited amount of parameters. (3) We also propose a simple yet effective method of predicting an importance score for each sampled frame which is used to select key timestamps for the refinement of output action tubes.

2 Related Works

2.1 Action Recognition

Deep learning techniques have shown to be effective and powerful in the classification of still images [8,11,28], and some existing works have extended such schemes to the task of human action recognition in video. Direct extensions attempt to model sequential data with serial or parallel networks. [18,33] combined 2D CNN with a RNN structure to model spatial and temporal relations separately. In [27], the authors found that the involvement of optical flow is beneficial for temporal modeling of actions and thus, proposed a two-stream framework that extracts features from RGB and optical flow data using separate parallel networks; the inference result being the combination of both modalities. In [34], the authors designed a 3D convolution architecture to automatically extract a high dimensional representation for input video. The I3D network [3] further improved the 3D convolution technique by inflating convolution kernels of networks pre-trained on ImageNet (2D) [4] into an efficient 3D form for action recognition. Although these methods achieved good results on video classification benchmarks, they can only make video level predictions and are unable to ascertain the position of actors and the duration of action instances.

2.2 Spatio-Temporal Action Detection

The task of spatio-temporal action detection is more complex than direct classification of videos. It requires both correct categorization and accurate localization

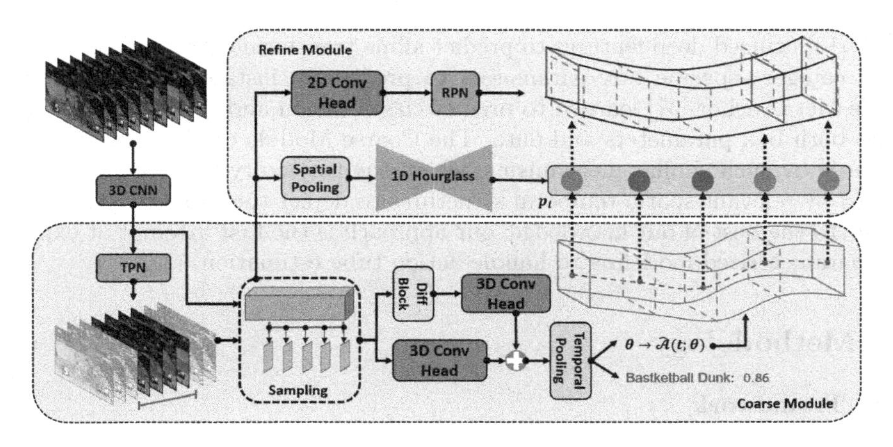

Fig. 2. Overview of the proposed CFAD framework. TPN block denotes the temporal action proposal network. A 3D Conv Head block indicates a cascaded NL-3D Conv structure ("NL" represents the NonLocal Block of [35]). 2D Conv Head block denotes cascaded 2D spatial convolutions. (Best viewed in color) (Color figure online)

of actors during the time interval when the action happens. Gkioxiari *et al.* proposed the first pipeline for this task in [6], where R-CNN [5] was applied on each frame to locate actors and classify actions, the results are then linked by viterbi algorithm. Saha *et al.* [26] designed a potential-based temporal trimming algorithm to extend general detection methods to untrimmed video datasets. Following the workflow of these two works, [9,15,19,21] tried learning more discriminative features of action instances with larger spatial or temporal context, a concept greatly enhanced by [16] through a multi-channel architecture that learns recurrently from tubelet proposals. Some works [12,29] aimed to improve heuristic linking for better localization. Recent works [32,39] proposed innovative two-stream fusion schemes for this task. [38] took a novel route to progressively regressing clip-wise action tubelets and linking them along time. Overall, all these works require temporally dense detections for each video, which is cumbersome. This inefficiency gets worse when optical flow computation is taken into account.

Among the existing works, [38] is the most similar to CFAD with its refinement process. However, our method is different from it from three aspects. Firstly, CFAD estimates coarse level tubes with parametric modeling and global supervision, while [38] relies on per-frame detection. Secondly, our approach does not require further temporal linking or trimming process. Finally, [38] refines the boxes densely for each frame, while our method only refines the box locations at selected key timestamps.

2.3 Weight Prediction

Weight prediction is a meta-learning concept where machine learning models are exploited to predict parameters of other structures [1]. For example, the

STN [13] utilized deep features to predict affine transformation parameters. [10] used category-specific box parameters to predict an instance weighted mask, while MetaAnchor [37] learned to predict classification and regression functions from both box parameters and data. The Coarse Module of our method is also inspired by such similar mechanisms, where the trajectory parameters are predicted by relevant spatio-temporal structures to depict the tube variation along time. To the best of our knowledge, our approach is the first attempt at exploiting parameterized modeling to handle action tube estimation.

3 Methodology

3.1 Framework

In this section, we introduce the proposed Coarse-to-Fine Action Detector (CFAD) in detail. We first formulate the problem and provide an overview of our approach. Then we discuss more elaborately on the two primary components of CFAD – the Coarse Module for tube estimation and Refine Module for final proposal.

One action tube instance in videos can be formulated as a set, $\mathcal{A} = \{(t_i, b_i)|i = 0, \cdots, T_A - 1\}$, where t_i is the timestamp of a certain frame, $b_i = (x_i, y_i, w_i, h_i)$ is the corresponding actor box within this frame, and T_A denotes the total number of bounding boxes in a ground-truth tube. Each tube $\mathcal{A}$ is accompanied with a category label c.

The workflow of CFAD is shown in Fig. 2. Firstly, the input video is resampled to a fixed length T and fed into 3D CNN for spatio-temporal feature extraction. Then the feature is processed a temporal proposal network (TPN) to obtain class-agnostic temporal proposals (t_s, t_e). t_s is the start timestamp and t_e denotes the end timestamp. In this paper, we instantiate the temporal proposal network by implementing one that is similar to [36], readers can refer to Appendix A for architecture detail. Given the temporal proposal, we uniformly sample N 2D features $\{\mathbf{F}_i|i \in [0, N-1]\}$ along the time axis within interval (t_s, t_e), which are sent to Coarse and Refine Module simultaneously. In Coarse Module, these 2D sampled features are used to estimate coarse level action tubes. Next, the estimated tube and the sampled 2D features in the Refine Module are exploited for frame selection and tube refinement at identified key timestamps.

3.2 Coarse Tube Estimation

We design two convolutional brunches in the Coarse Module to process The sampled 2D features, one branch processes the input features directly and the other branch handles the temporal residual component $\{\mathbf{F}_{i+1} - \mathbf{F}_i|i = 0, N-2\}$ of the input features, which is output through the "Diff Block" in Fig. 2. We add residual processing since the temporal residual component can provide more time variant information, which is beneficial to discriminate different actions and predict localization changes along time. For each branch, a Non-Local Block [35]

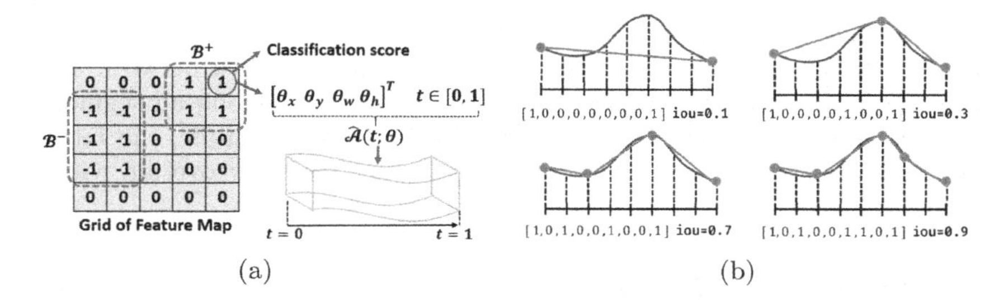

Fig. 3. (a). Illustration of coarse tube estimation, where "1" and "-1" symbols denote the positive and negative samples after matching, and "0" are ignored samples. (b) Key timestamps label selection process in the refine module. For ease of simplification, this figure depicts the case of 1-dimensional linear interpolation. The blue curve is the ground-truth, the orange one is the interpolated curve and green nodes represent selected timestamps. (Best viewed in color) (Color figure online)

is cascaded with a 3D convolution blocks to construct the "3D Conv Head" module in Fig. 2, which aggregates information from both spatial and temporal context. The output of the two branches are fused by element-wise summation and aggregated via temporal average pooling.

To estimate coarse-level action tubes, we adopt a *parameterized modeling* scheme, where we define a coarse-level tube estimation as a parameterized mapping $\hat{\mathcal{A}}(t;\boldsymbol{\theta}) : [0,1] \longrightarrow \mathbb{R}^4$. $\hat{\mathcal{A}}$ tries to predict the coarse spatial location, i.e. [x(t), y(t), w(t), h(t)], given a normalized timestamp t and trajectory parameter $\boldsymbol{\theta}$. The mapping parameters $\boldsymbol{\theta}$ are predicted by the deep features from the temporal pooling block. To this end, we slide predefined anchor boxes of different sizes on the 2D output feature map from temporal pooling block to obtain positive samples $\mathcal{B}^+$ and negative samples $\mathcal{B}^-$ as according to the IOUs between anchors and tubes (illustrated in Fig. 3(a)). For each sample in $\mathcal{B}^+$, the network should predict its corresponding classification score and the tube shape parameter $\boldsymbol{\theta}$ through an additional 1×1 convolution layer.

Segment-Wise Matching. To measure the overlap between an anchor box b_a and ground-truth $\mathcal{A}$, an intuitive idea is to calculate the average value of IOUs between b_a and each boxes belonging to $\mathcal{A}$. However, since tube shapes may include motion and camera shake, such matching strategy might result in small IOU value and induce the imbalance issue of samples. Hence, we design a segment-wise matching scheme to separate positive and negative samples. To be specific, We take the boxes on first K frames in $\mathcal{A}$ as a valid segment for matching positive anchors, where K is a predefined segment length. We take the segment from the beginning of the tube because we found the final model performance is not sensitive to the segment position, readers can refer to Appendix B for detail analysis. If the average overlap between b_a and these K boxes is larger than a threshold, it is taken as a positive sample. Further, if b_a has high overlap with multiple concurrent tubes, we choose the ground-truth with largest segment IOU as the matched tube. On the other hand, to find negative samples, we split the ground-truth tube $\mathcal{A}$ into $\lfloor \frac{T_A}{K} \rfloor$ segments and compute IOUs between b_a and each segments as discussed above, if the maximum IOU among all segments is still less than a threshold, then it is taken as a negative sample, the intuition behind such design is that negative samples should have low overlap with any boxes in $\mathcal{A}$.

Parameterized Modeling. Generally, any parameterized function that takes a single scalar as input and outputs a 4-dimensional vector can be used as the tube mapping. In this paper, we use the family of high order polynomial functions to model action tube variations along the timestamp. This is because action tubes typically change smoothly and gradually, while polynomial functions are capable enough of describing the patterns of tube shape changes. Therefore, the instantiation of parameterized estimation function $\hat{\mathcal{A}}(t;\boldsymbol{\theta})$ can be formulated as:

$$\hat{\mathcal{A}}(t;\boldsymbol{\theta}) = [x(t;\boldsymbol{\theta}_x), y(t;\boldsymbol{\theta}_y), w(t;\boldsymbol{\theta}_w), h(t;\boldsymbol{\theta}_h)] = \left[\boldsymbol{\theta}_x^T \boldsymbol{t}, \boldsymbol{\theta}_y^T \boldsymbol{t}, \boldsymbol{\theta}_w^T \boldsymbol{t}, \boldsymbol{\theta}_h^T \boldsymbol{t}\right] \tag{1}$$

where the trajectory of each coordinate is regarded as a polynomial curve of order k, the predicted parameter matrix $\boldsymbol{\theta} = [\boldsymbol{\theta}_x, \boldsymbol{\theta}_y, \boldsymbol{\theta}_w, \boldsymbol{\theta}_h]$ of size $(k+1) \times 4$ is composed of the polynomial coefficient for each bounding box coordinates. The vector $\boldsymbol{t} = [1, t, t^2, \cdots, t^k]^T$ contains various orders of current timestamp. To learn features that are invariant to anchor transitions, we do not use $\hat{\mathcal{A}}(t;\boldsymbol{\theta})$ to directly estimate the absolute coordinates, but instead perform estimation of relative coordinates w.r.t matched bounding box b_a following the method of encoding in [23].

During training, the model learns to separate positive and negative samples, to predict the correct action classes and relative coordinates of a coarse tube under the supervision of the loss function in Eq. 2,

$$L_{coarse} = \frac{1}{|\mathcal{B}^+ \cup \mathcal{B}^-|} L_c + \frac{1}{|\mathcal{B}^+|} L_r \tag{2}$$

where $|\cdot|$ denotes the size of the set. L_c is the classification loss in [23] while L_r is the regression loss from the supervision of the whole ground-truth tube:

$$L_r = \frac{1}{T_A} \sum_{b_a \in \mathcal{B}^+} \sum_{(t_i, b_i) \in \mathcal{A}} \left|\left| \hat{\mathcal{A}}(\hat{t}_i; \boldsymbol{\theta}_a) - \mathbf{enc}(b_i, b_a) \right|\right|^2 \tag{3}$$

The function $\mathbf{enc}(\cdot,\cdot)$ in Eq. 3 is the same as the encoding function in [23] to encode the 4-dimensional relative offsets from anchor box to ground-truth box. $\boldsymbol{\theta}_a$ is the predicted tube shape parameter associated with anchor b_a. The symbol $\hat{t}_i$ defined in Eq. 4 is the normalized timestamp of ground-truth bounding boxes in tube $\mathcal{A}$. We normalize the input timestamp before calculating the tube shape in order to avoid value explosion when the polynomial order increases.

$$\hat{t}_i = \frac{t_i - t_0}{t_{T_A - 1} - t_0} \quad \forall\ (t_i, b_i) \in \mathcal{A} \tag{4}$$

3.3 Selective Refinement

After the estimated coarse tube $\hat{\mathcal{A}}(t;\boldsymbol{\theta})$ has been generated by the Coarse Module, its location is further refined in the Refine Module. The Refine Module first selects the samples attached with key timestamps for action tube localization, and then refine tube boxes based on these features and guidance of coarse tube.

Key Timestamp Selection. One simple and intuitive refinement scheme is to observe the tube location at each sampled 2D feature map and then refine the box according to the features within that area. However, when the sample number N increases, such a scheme is costly in computation. Since changes in the action tubes are usually smooth, there is only a limited number of sparsely distributed bounding boxes that are decisive to the shape of tubes. Thus, we design a selector network in the Refine Module to dynamically sample key timestamps that are most essential for localization.

In our implementation, we perform importance evaluation by squeezing the input 2D sampled features $\{\mathbf{F}_i | i \in [0, N-1]\}$ with spatial pooling and applying a 1D hourglass network along the time dimension. This outputs an importance score p_i for each sample (shown in Fig. 2). During inference phase, we only take samples that satisfy $p_i \geq \alpha$ as samples of key timestamps and then proceed to refinement.

In the training phase, we heuristically define sets of labels to guide the training of the selector network. Specifically, first the ground-truth action tube $\mathcal{A}$ is uniformly split into $N-1$ segments along temporal axis with N endpoints. The i-th endpoint is associated with the i-th sampled feature $\mathbf{F}_i$, and its normalized timestamp is defined as $s_i = i/(N-1)$. Let the timestamp set be defined as $\mathcal{U} = \{s_i | i = 0, \cdots, N-1\}$ and the key points set as $\mathcal{U}_k$. We start from $\mathcal{U}_k = \{s_0, s_{N-1}\}$ having the start and end points and gradually append other s_i into $\mathcal{U}_k$. The process can be illustrated in Fig. 3(b), whereby for each iteration, we greedily select the timestamps s^* which maximizes the overlap between the interpolated tube and ground-truth $\mathcal{A}$ as in Eq. 5, and append this timestamp into $\mathcal{U}_k$. The process stops when the IOU between interpolated tube and ground-truth tube is larger than a predefined threshold ϵ.

$$s^* = \arg \max_{s_i \in \mathcal{U}/\mathcal{U}_k} IOU\left(\mathbf{Interp}(\mathcal{U}_k \cup \{s_i\}), \mathcal{A}\right) \tag{5}$$

Here, the function $\mathbf{Interp}(\cdot)$ can be any polynomial interpolation over the input timestamp set. To avoid the large numerical oscillation around the endpoint, we choose the piece-wise cubic spline interpolation in this paper as instantiation. We assign feature samples in $\mathcal{U}_k$ with label 1 and samples in $\mathcal{U}/\mathcal{U}_k$ with label 0. We utilize these labels to train the timestamp selector network with binary cross-entropy loss.

Sample-Wise Location Refinement. In the Refine Module, the selected 2D features are first processed by cascaded 2D convolution blocks (shown in Fig. 2), then a class-specific regional proposal network (RPN) [23] is applied over these features to generate bounding box proposals at corresponding timestamps. With the estimated action tube function $\hat{\mathcal{A}}(t;\boldsymbol{\theta})$, we can now obtain the estimated action bounding boxes at i-th sampled timestamps s_i with Eq. 6, where $\mathbf{dec}(\cdot)$ is the inverse operation of $\mathbf{enc}(\cdot,\cdot)$ in Eq. 3.

$$\hat{x}_i, \hat{y}_i, \hat{w}_i, \hat{h}_i = \mathbf{dec}\left(\hat{\mathcal{A}}(s_i;\boldsymbol{\theta})\right) \tag{6}$$

We design a simple local search scheme to refine the estimated bounding box at selected key timestamps. For each selected 2D sample, a searching area Ω is defined as,

$$\Omega = [\hat{x}_i - \sigma\hat{w}_i, \hat{x}_i + \sigma\hat{w}_i] \times [\hat{y}_i - \sigma\hat{h}_i, \hat{y}_i + \sigma\hat{h}_i] \tag{7}$$

where σ is a hyperparameter that controls the size of searching area. We choose the action box proposal (from RPN) with the largest score where its center is located inside Ω, as the replacement of the original coarsely estimated box.

The final output action tube is obtained via interpolation over all refined boxes and unrefined bounding boxes (localized via Eq. 6). The associated action score is the smooth average of classification score and RPN score.

4 Experiment Results

4.1 Experiment Configuration

Datasets. We conduct our experiment on three common datasets for the task of action tube detection – UCF101-24, UCFSports and JHMDB-21 datasets. Although the AVA [7] dataset also includes bounding box annotations, it mainly focuses on the problem of *atomic action classification* on sparse single key frames instead of *spatiotemporal action detection* at the video level, which is the task we are focusing here. Hence, we did not conduct our experiments on the AVA dataset.

The *UCF101-24 dataset* [31] contains 3,207 untrimmed videos with frame level bounding box annotations for 24 sports categories. The dataset is challenging due to frequent camera shake, dynamic actor movements and a large variance in action duration. Following previous works [26], we report results for the first split with 2,275 videos for training and the other videos for validation. We use the corrected annotation [29] for model training and evaluation. The *JHMDB-21* is a subset of HMDB-51 dataset [14], which contains a total of 928 videos with 21 types of actions. All video sequences are temporally trimmed. The results are reported as the average performance over 3 train-val splits. The *UCFSports* dataset [24] contains 150 trimmed videos in total and we report the results on the first split. Note that although videos in JHMDB-21 and UCFSports are trimmed temporally, their samples are still suitable for our framework as they comprised mostly of cases where actions span the whole video.

Table 1. Ablation study on the effectiveness of refinement with different hyperparameter settings.

	UCF101-24				JHMDB-21		
k	2	3	4	5	1	2	3
no refine	46.0	48.4	51.6	50.1	79.7	80.9	80.3
$\sigma = 0.4$	57.5	57.6	58.8	58.0	80.8	82.5	81.3
$\sigma = 0.6$	59.9	60.1	61.7	60.0	81.4	83.2	82.4
$\sigma = 0.8$	60.3	62.0	**62.7**	61.6	82.3	**83.7**	83.2

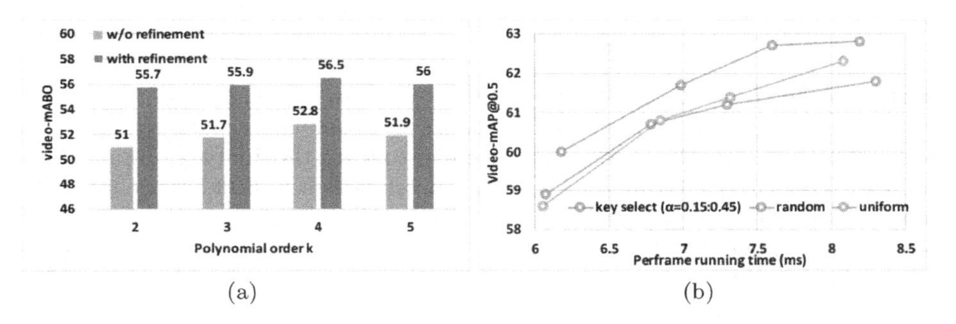

Fig. 4. (a). v-MABO value of action tubes with different polynomial orders on UCF101-24. (b) Time-performance trade-off with different timestamp selection schemes on UCF101-24. (best viewed in color) (Color figure online)

Metrics. We report the video-mAP (v-mAP) [6] with different IOU thresholds as our main evaluation metric for spatial-temporal action localization on all datasets. In addition, frame-level mAP at threshold 0.5 is reported to evaluate per-frame detection performance. A proposal is regarded as positive when its overlap with the ground-truth is larger than threshold δ. We also adopt video-level mean Average Best Overlap (v-MABO) [15] in the ablation study to evaluate the localization performance of our approach. The criterion calculates the mean of largest overlap between ground-truth tubes and action proposals, averaged over all classes.

Implementation Details. We use the I3D network [3] pretrained on Kinetics-600 as our 3D feature extractor, taking the feature from *mixed_5b* layer as our 3D feature. We set the video resampling length T to 96 frames for UCF101-24 and 32 frames for JHMDB-21 and UCFSports. The hyperparameter ϵ in our paper is set to 0.8 and the segment length for matching K is set to 6 frames. The number of sampling points N is set to 16 for UCF101-24, 6 for JHMDB-21 and 8 for UCFSports according to the average length of action instances. For the anchor design, we follow the strategy of [22] by clustering the bounding boxes from training set into 6 centers and taking their respective center coordinates as the

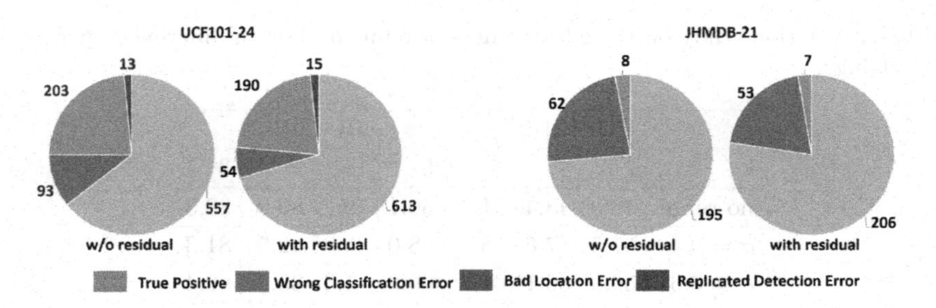

Fig. 5. Statistics of true positive and various false positive proposals of CFAD on UCF101-24 and split-3 of JHMDB-21. (Best viewed in color) (Color figure online)

default anchor boxes. In the training phase, we train the network in two stages. First, we train the temporal proposal network and backbone separately, and then we jointly train the entire network end-to-end to learn the final action tubes. We use the SGD solver to train CFAD with a batch size of 8. In inference stage, to handle concurrent action instances, the Coarse Module outputs at most 3 (the maximum number of instances in a video based on the datasets) estimated tubes followed by a tube-wise non-maximal suppression process with IOU threshold of 0.2 in order to avoid duplicated action tubes.

4.2 Ablation Study

In this section, we report the video-mAP results with $\delta = 0.5$ of various ablation study experiments. The input modality is only RGB data unless specified.

The Effectiveness of Refinement. First, we analyze the effects of the refinement process on the accuracy of coarse tube estimation on UCF101-24 and JHMDB-21. The results are reported in Table 1 where "no refine" denotes the configuration without location refinement. From these results, it is obvious that the Refine Module can bring large improvements in v-mAP regardless of the polynomial order of estimated tubes; the largest performance gain can be up to +14.3% when $k = 2$ on UCF101-24. The improvement is less obvious on JHMDB-21, which we think is owing to the fact that JHMDB-21 is less dynamic and coarse-level estimations may be close to the ground-truth tubes. We also evaluate the v-MABO value on UCF101-24 as shown in Fig. 4(a), where improvements by at most +4.7% are possible by the refinement process. The results show that the Refine component is essential to better detection performance.

Meanwhile, from Table 1, we can also see that as the searching area gets larger, the mAP performance can be improved to some extent, since larger searching area can cover more centers of action proposals. We did not try larger searching area *i.e.* $\sigma > 0.8$ since we find the performance improvement is marginal (less than +0.2%) beyond $\sigma = 0.8$. This is because larger searching area also makes the refinement more vulnerable to noisy proposals.

Polynomial Order Selection. We also report in Table 1 the effect of different polynomial order k which decides the form of estimated tube mapping $\hat{\mathcal{A}}(t;\boldsymbol{\theta})$. Overall, we find that the performance improves along with the increase of k for both with and without refinement, since higher order polynomial functions show stronger ability in characterizing variations of action tubes.

On the other hand, we found that as the order gets larger (than $k = 5$ on UCF101-24 and $k = 3$ on JHMDB-21), the detection performance drops comparatively against the optimal value in both cases. We think the reason behind this is that although higher order polynomial functions are usually more representative, they are more complex requiring more coefficients, and the parameters predicting coefficients of higher orders are more difficult to be trained efficiently since the corresponding gradients are very small. The similar tendency is also reflected in the MABO results on UCF101-24 shown in Fig. 4(a), where the localization did improve (for both refine and no refine cases) from $k = 2$ to $k = 4$, but MABO drops after that with higher orders. Also, we observed during training that configurations with a larger polynomial order tends to slow down the convergence process and possibly result in numerical oscillation of the loss function.

Effectiveness of Residual Processing Branch. Here, we conduct experiments on UCF101-24 and JHMDB-21 to analyze how the temporal residual information impacts the output action tube results. To test this, we remove the branch with differential module ("Diff Block" in Fig. 2) as our baseline. For a detailed comparison, we break down the final proposals into four mutually exclusive types.

- **True Positive**: the proposal classifies an action correctly and has tube-overlap with ground-truth that is larger than δ.
- **Wrong Classification Error**: a proposal with incorrectly classified action although it overlaps more than δ with ground-truth.
- **Bad Localization Error**: a proposal that has correct action class but it overlaps less than δ with ground-truth.
- **Duplicated Detection Error**: a proposal with correct action class and overlaps more than δ with a ground-truth that has been detected.

Figure 5 illustrates the statistics of these proposals with/without the differential module. From Fig. 5, we observe that the residual processing branch is particularly important for accurate action classification. With the help of information from the temporal residual feature, wrongly-classified samples are reduced by 42% on UCF101-24 and 14% on JHMDB-21. Furthermore, models with residual processing also benefit from better tube localization while the overall recall also improves due to the increase in true positive results. These results are evidential of the effectiveness of temporal residual component in Coarse Module.

Key Timestamp Selection. We conduct an experiment on UCF101-24 to analyze the impact of the proposed key timestamp selection mechanism. In the

Table 2. Comparison with state-of-the-art methods (on video-mAP), '–' denotes that the result is not available, '**+OF**' indicates the input is combined with optical flow. All compared methods take both RGB and optical flow as input except [9,25]

Method	JHMDB-21				UCF101-24				UCFSports	
δ	0.2	0.5	0.75	0.5:0.95	0.2	0.3	0.5	0.5:0.95	0.2	0.5
2D backbone										
Saha *et al.* [26]	72.6	71.5	–	–	66.7	54.9	35.9	14.4	–	–
Peng *et al.* [19]	74.3	73.1	–	–	72.8	65.7	30.9	7.1	94.8	94.7
Saha *et al.* [25]	57.8	55.3	–	–	63.1	51.7	33.0	10.7	–	–
Kalogeiton *et al.* [15]	74.2	73.7	52.1	44.8	76.5	–	49.2	23.4	92.7	92.7
Singh *et al.* [29]	73.8	72.0	44.5	41.6	73.5	–	46.3	20.4	–	–
Yang *et al.* [38]	–	–	–	–	76.6	–	–	–	–	–
Zhao *et al.* [39]	–	58.0	42.8	34.6	75.5	-	48.3	23.9	–	92.7
Rizard *et al.* [20]	86.0	84.0	52.8	49.5	82.3	–	51.5	24.1	–	–
Song *et al.* [30]	74.1	73.4	52.5	44.8	77.5	–	52.9	24.1	–	–
Li *et al.* [16]	82.7	81.3	–	–	76.3	71.4	–	–	**97.8**	**97.8**
Li *et al.* [17]	77.3	77.2	**71.7**	**59.1**	82.8	–	53.8	**28.3**	–	–
3D backbone										
Hou *et al.* [9]	78.4	76.9	–	–	73.1	69.4	–	–	95.2	95.2
Gu *et al.* [7]	–	76.3	–	–	–	–	59.9	–	–	–
Su *et al.* [32]	82.6	82.2	63.1	52.8	**84.3**	–	61.0	27.8	–	–
Qiu *et al.* [21]	85.7	84.9	–	–	82.2	75.6	–	–	-	–
CFAD	84.8	83.7	62.4	51.8	79.4	76.7	62.7	25.5	90.2	88.6
CFAD+**OF**	**86.8**	**85.3**	63.8	53.0	81.6	**78.1**	**64.6**	26.7	94.5	92.7

experiment setting, we gradually increase the selection threshold α from 0.15 to 0.45 (in increments of 0.1) and report their respective v-mAP value and per-frame time cost. For comparisons, we design two baseline methods: (1) Random selection of samples from the input N 2D features with their corresponding timestamps taken as key timestamps, denoted as "random". (2) Selection of timestamps across s_i based on a fixed time step, denoted as "uniform".

The time-performance trade-off curves are shown in Fig. 4(b). We can observe that when the per-frame time costs are similar, our dynamic selection scheme is superior to the other two baseline methods. It is also worth noting that when the time cost gets smaller, the performance of "random" and "uniform" deteriorates faster than our scheme. This result indicates that the key timestamp selection process finds the important frames for location refinement and is reasonably robust to the reduction of available 2D features.

4.3 Comparison with State-of-the-Art

In this section, we compare the proposed CFAD with other recent state-of-the-art approaches in the spatio-temporal action localization task on the UCF101-24, JHMDB-21 and UCFSports benchmarks. These results are listed in Table 2. We also evaluate the performance of CFAD with two-stream input, where the

Table 3. Comparison with state-of-the-art methods on frame-level mAP@0.5 on UCF101-24 dataset. '**+OF**' indicates the input is combined with optical flow.

Method	Input modal	Frame-mAP@0.5
Peng *et al.* [19]	RGB+OF	65.7
Kalogeiton *et al.* [15]	RGB+OF	69.5
Yang *et al.* [38]	RGB+OF	75.0
Rizard *et al.* [20]	RGB+OF	73.7
Song *et al.* [30]	RGB+OF	72.1
Gu *et al.* [7]	RGB+OF	**76.3**
CFAD	RGB+OF	72.5
Hou *et al.* [9]	RGB	41.4
Yang *et al.* [38]	RGB	66.7
CFAD	RGB	**69.7**

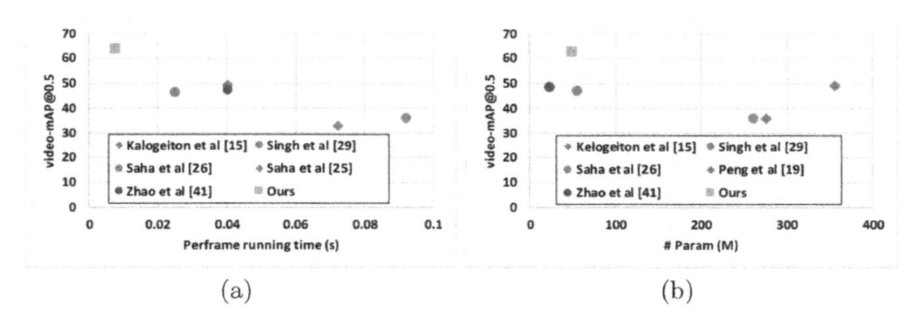

Fig. 6. (a). Comparisons of time-performance trade-off among different state-of-the-art approaches. (b). Comparisons of trade-off between model size and performance among different state-of-the-art approaches. (Best viewed in color) (Color figure online)

optical flow is extracted using the method of [2]. For simplicity, we opt for an early fusion strategy [39] to maintain efficiency of our approach.

It is worth noting that in Table 2, our method with only RGB input outperforms most other approaches that rely on two-stream features on UCF101-24 and JHMDB-21. While it is still worse than the state-of-the-art method on UCFSports, we think the reasons behind this can be that this dataset is relatively simpler and smaller in scale with less dynamic movements, thus it could be more challenging to learn robust tube estimation. For fair benchmarking, we compare our method with other approaches utilizing 3D spatiotemporal features [7,9,21,32]. With RGB as input, CFAD achieves competitive performance on all datasets under different tested threshold criterion. Overall, our method achieves state-of-the-art under small threshold while there is still a margin towards the performance of [17,32] under more strict criterion. Besides, we also observe that the optical flow information is helpful for the overall detection performance.

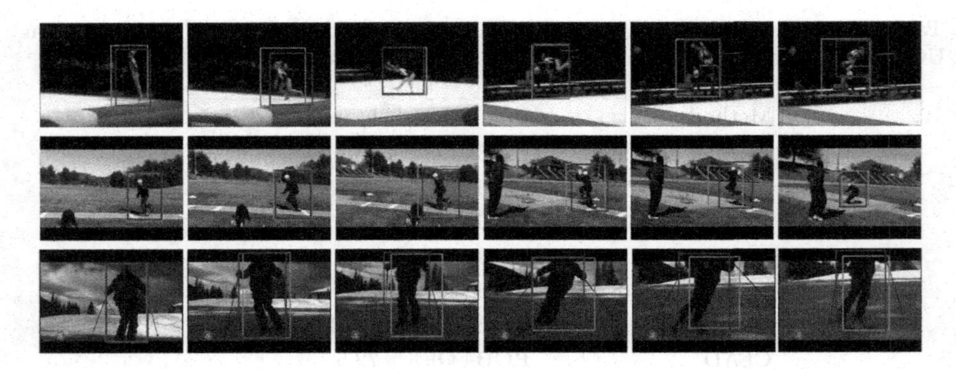

Fig. 7. Visualization of detected action tubes. The green boxes denote the estimated action tubes from the Coarse Module. The red boxes are the final refined action tubes. (Best viewed in color) (Color figure online)

Frame-mAP. In Table 3, we compare CFAD with other approaches on frame-level detection in UCF101-24. In our setting, we assign the video level score of a tube proposal to all boxes included by the tube to generate frame-level proposals. It can be observed that CFAD outperforms three pipelines with two-stream input. While it is still worse than some approaches [7,20,38], we think this is due to the less accurate interpolated boxes between sampled frames, which might result in many false positives with high score (which in turn lowers the overall metric). We argue that although such interpolation sacrifices frame-level accuracy, it enhances the system efficiency and video-level accuracy in return.

Efficiency. We also compare the runtime (inference) and model size of CFAD with RGB input on UCF101-24 against other approaches that also report their runtime. The speed is evaluated based on per-frame processing time, which is obtained by taking the runtime per video and dividing it by input length T. Since some other works only reported per-video time on JHMDB-21 [25,26], we compute their per-frame time in the same manner. The runtime comparison is illustrated in Fig. 6(a) and the model size comparison is reported in Fig. 6(b). We observe that CFAD only requires a small number of parameters (close to [29,39], and much less than others) while achieving superior running speed compared to other state-of-the-art methods. This vast improvement in processing efficiency can be attributed to the coarse-to-fine paradigm of CFAD, which does not require dense per-frame action detection followed by linking, and the RGB input of CFAD avoids the additional computation to process optical flow. Specifically, the proposed CFAD runs $\approx$3.3$\times$ faster than the nearest approach [29] (7.6 ms vs. 25 ms).

4.4 Qualitative Results

Figure 7 shows some qualitative results of detected action tubes from the UCF101-24 dataset. The green boxes denote the estimated action tube output from the Coarse Module while the red boxes are the refined action tubes. We can observe that the selective refinement process has effectively corrected some poorly located action tubes, causing the bounding boxes to wrap tighter and more accurately around the actors. These visuals can evidently explain the robustness of the coarse tube estimation method, and its capability at handling a variety of dynamic actions.

5 Conclusion

In this paper, we propose a novel framework CFAD for spatio-temporal action localization. Its pipeline follows a new coarse-to-fine paradigm, which does away with the need for dense per-frame detections. The action detector comprises of two components (Coarse and Refine Modules) which play vital roles in coarsely estimating and then refining action tubes based on selected timestamps. Our CFAD achieves state-of-the-art results for a good range of thresholds on benchmark datasets and is also an efficient pipeline, running at 3.3× faster than the nearest competitor.

Acknowledgement. The paper is supported in part by the following grants: China Major Project for New Generation of AI Grant (No. 2018AAA0100400), National Natural Science Foundation of China (No. 61971277). The work is also supported by funding from Clobotics under the Joint Research Program of Smart Retail.

References

1. Andrychowicz, M., et al.: Learning to learn by gradient descent by gradient descent. In: NeurIPS, pp. 3981–3989 (2016)
2. Brox, T., Bruhn, A., Papenberg, N., Weickert, J.: High accuracy optical flow estimation based on a theory for warping. In: Pajdla, T., Matas, J. (eds.) ECCV 2004. LNCS, vol. 3024, pp. 25–36. Springer, Heidelberg (2004). https://doi.org/10.1007/978-3-540-24673-2_3
3. Carreira, J., Zisserman, A.: Quo Vadis, action recognition? A new model and the kinetics dataset. In: CVPR, pp. 6299–6308 (2017)
4. Deng, J., Dong, W., Socher, R., Li, L.J., Li, K., Fei-Fei, L.: ImageNet: a large-scale hierarchical image database. In: CVPR, pp. 248–255. IEEE (2009)
5. Girshick, R., Donahue, J., Darrell, T., Malik, J.: Rich feature hierarchies for accurate object detection and semantic segmentation. In: CVPR, pp. 580–587 (2014)
6. Gkioxari, G., Malik, J.: Finding action tubes. In: CVPR, June 2015
7. Gu, C., et al.: AVA: a video dataset of spatio-temporally localized atomic visual actions. In: CVPR, pp. 6047–6056 (2018)
8. He, K., Zhang, X., Ren, S., Sun, J.: Deep residual learning for image recognition. In: CVPR, pp. 770–778 (2016)
9. Hou, R., Chen, C., Shah, M.: An end-to-end 3D convolutional neural network for action detection and segmentation in videos. In: ICCV (2017)
10. Hu, R., Dollár, P., He, K., Darrell, T., Girshick, R.: Learning to segment every thing. In: CVPR, pp. 4233–4241 (2018)

11. Huang, G., Liu, Z., Van Der Maaten, L., Weinberger, K.Q.: Densely connected convolutional networks. In: CVPR, pp. 4700–4708 (2017)
12. Huang, J., Li, N., Zhong, J., Li, T.H., Li, G.: Online action tube detection via resolving the spatio-temporal context pattern. In: ACM MM, pp. 993–1001. ACM (2018)
13. Jaderberg, M., Simonyan, K., Zisserman, A., et al.: Spatial transformer networks. In: NeurIPS, pp. 2017–2025 (2015)
14. Jhuang, H., Gall, J., Zuffi, S., Schmid, C., Black, M.J.: Towards understanding action recognition. In: ICCV, pp. 3192–3199, December 2013
15. Kalogeiton, V., Weinzaepfel, P., Ferrari, V., Schmid, C.: Action tubelet detector for spatio-temporal action localization. In: ICCV, pp. 4405–4413 (2017)
16. Li, D., Qiu, Z., Dai, Q., Yao, T., Mei, T.: Recurrent tubelet proposal and recognition networks for action detection. In: Ferrari, V., Hebert, M., Sminchisescu, C., Weiss, Y. (eds.) ECCV 2018. LNCS, vol. 11210, pp. 306–322. Springer, Cham (2018). https://doi.org/10.1007/978-3-030-01231-1_19
17. Li, Y., Wang, Z., Wang, L., Wu, G.: Actions as moving points. arXiv preprint arXiv:2001.04608 (2020)
18. Li, Z., Gavrilyuk, K., Gavves, E., Jain, M., Snoek, C.G.: VideoLSTM convolves, attends and flows for action recognition. Comput. Vis. Image Underst. **166**, 41–50 (2018)
19. Peng, X., Schmid, C.: Multi-region two-stream R-CNN for action detection. In: Leibe, B., Matas, J., Sebe, N., Welling, M. (eds.) ECCV 2016. LNCS, vol. 9908, pp. 744–759. Springer, Cham (2016). https://doi.org/10.1007/978-3-319-46493-0_45
20. Pramono, R.R.A., Chen, Y.T., Fang, W.H.: Hierarchical self-attention network for action localization in videos. In: ICCV (2019)
21. Qiu, Z., Yao, T., Ngo, C.W., Tian, X., Mei, T.: Learning spatio-temporal representation with local and global diffusion. In: CVPR, pp. 12056–12065 (2019)
22. Redmon, J., Farhadi, A.: Yolo9000: better, faster, stronger. In: CVPR, pp. 7263–7271 (2017)
23. Ren, S., He, K., Girshick, R., Sun, J.: Faster R-CNN: towards real-time object detection with region proposal networks. In: NeurIPS, pp. 91–99 (2015)
24. Rodriguez, M.D., Ahmed, J., Shah, M.: Action MACH a spatio-temporal maximum average correlation height filter for action recognition. In: CVPR, pp. 1–8, June 2008
25. Saha, S., Singh, G., Cuzzolin, F.: AMTNet: action-micro-tube regression by end-to-end trainable deep architecture. In: ICCV, pp. 4414–4423 (2017)
26. Saha, S., Singh, G., Sapienza, M., Torr, P.H., Cuzzolin, F.: Deep learning for detecting multiple space-time action tubes in videos. In: BMVC (2016)
27. Simonyan, K., Zisserman, A.: Two-stream convolutional networks for action recognition in videos. In: NeurIPS, pp. 568–576 (2014)
28. Simonyan, K., Zisserman, A.: Very deep convolutional networks for large-scale image recognition. arXiv preprint arXiv:1409.1556 (2014)
29. Singh, G., Saha, S., Sapienza, M., Torr, P.H., Cuzzolin, F.: Online real-time multiple spatiotemporal action localisation and prediction. In: ICCV, pp. 3637–3646 (2017)
30. Song, L., Zhang, S., Yu, G., Sun, H.: TACNet: transition-aware context network for spatio-temporal action detection. In: CVPR, pp. 11987–11995 (2019)
31. Soomro, K., Zamir, A.R., Shah, M.: UCF101: a dataset of 101 human actions classes from videos in the wild (2012)
32. Su, R., Ouyang, W., Zhou, L., Xu, D.: Improving action localization by progressive cross-stream cooperation. In: CVPR, pp. 12016–12025 (2019)

33. Sun, L., Jia, K., Chen, K., Yeung, D.Y., Shi, B.E., Savarese, S.: Lattice long short-term memory for human action recognition. In: ICCV, pp. 2147–2156 (2017)
34. Tran, D., Bourdev, L., Fergus, R., Torresani, L., Paluri, M.: Learning spatiotemporal features with 3D convolutional networks. In: ICCV, December 2015
35. Wang, X., Girshick, R., Gupta, A., He, K.: Non-local neural networks. In: CVPR, June 2018
36. Xu, H., Das, A., Saenko, K.: R-C3D: region convolutional 3D network for temporal activity detection. In: ICCV, pp. 5783–5792 (2017)
37. Yang, T., Zhang, X., Li, Z., Zhang, W., Sun, J.: MetaAnchor: learning to detect objects with customized anchors. In: NeurIPS, pp. 320–330 (2018)
38. Yang, X., Yang, X., Liu, M.Y., Xiao, F., Davis, L.S., Kautz, J.: STEP: spatio-temporal progressive learning for video action detection. In: CVPR, pp. 264–272 (2019)
39. Zhao, J., Snoek, C.G.: Dance with flow: two-in-one stream action detection. In: CVPR, pp. 9935–9944 (2019)

Learning Joint Spatial-Temporal Transformations for Video Inpainting

Yanhong Zeng[1,2], Jianlong Fu[3(✉)], and Hongyang Chao[1,2(✉)]

[1] School of Data and Computer Science, Sun Yat-sen University, Guangzhou, China
zengyh7@mail2.sysu.edu.cn, isschhy@mail.sysu.edu.cn

[2] Key Laboratory of Machine Intelligence and Advanced Computing, Ministry of Education, Guangzhou, China

[3] Microsoft Research Asia, Beijing, China
jianf@microsoft.com

Abstract. High-quality video inpainting that completes missing regions in video frames is a promising yet challenging task. State-of-the-art approaches adopt attention models to complete a frame by searching missing contents from reference frames, and further complete whole videos frame by frame. However, these approaches can suffer from inconsistent attention results along spatial and temporal dimensions, which often leads to blurriness and temporal artifacts in videos. In this paper, we propose to learn a joint **S**patial-**T**emporal **T**ransformer **N**etwork (**STTN**) for video inpainting. Specifically, we simultaneously fill missing regions in all input frames by self-attention, and propose to optimize STTN by a spatial-temporal adversarial loss. To show the superiority of the proposed model, we conduct both quantitative and qualitative evaluations by using standard stationary masks and more realistic moving object masks. Demo videos are available at https://github.com/researchmm/STTN.

Keywords: Video inpainting · Generative adversarial networks

1 Introduction

Video inpainting is a task that aims at filling missing regions in video frames with plausible contents [2]. An effective video inpainting algorithm has a wide range of practical applications, such as corrupted video restoration [10], unwanted object removal [22,26], video retargeting [16] and under/over-exposed image restoration [18]. Despite of the huge benefits of this technology, high-quality video inpainting still meets grand challenges, such as the lack of high-level understanding of videos [15,29] and high computational complexity [5,33].

Y. Zeng—This work was done when Y. Zeng was an intern at Microsoft Research Asia.

Electronic supplementary material The online version of this chapter (https://doi.org/10.1007/978-3-030-58517-4_31) contains supplementary material, which is available to authorized users.

A. Vedaldi et al. (Eds.): ECCV 2020, LNCS 12361, pp. 528–543, 2020.
https://doi.org/10.1007/978-3-030-58517-4_31

Fig. 1. We propose **S**patial-**T**emporal **T**ransformer **N**etworks for completing missing regions in videos in a spatially and temporally coherent manner. The top row shows sample frames with yellow masks denoting user-selected regions to be removed. The bottom row shows our completion results. [Best viewed with zoom-in] (Color figure online)

Significant progress has been made by using 3D convolutions and recurrent networks for video inpainting [5,16,29]. These approaches usually fill missing regions by aggregating information from nearby frames. However, they suffer from temporal artifacts due to limited temporal receptive fields. To solve the above challenge, state-of-the-art methods apply attention modules to capture long-range correspondences, so that visible contents from distant frames can be used to fill missing regions in a target frame [18,25]. One of these approaches synthesizes missing contents by a weighting sum over the aligned frames with frame-wise attention [18]. The other approach proposes a step-by-step fashion, which gradually fills missing regions with similar pixels from boundary towards the inside by pixel-wise attention [25]. Although promising results have been shown, these methods have two major limitations due to the significant appearance changes caused by complex motions in videos. One limitation is that these methods usually assume global affine transformations or homogeneous motions, which makes them hard to model complex motions and often leads to inconsistent matching in each frame or in each step. Another limitation is that all videos are processed frame by frame without specially-designed optimizations for temporal coherence. Although post-processing is usually used to stabilize generated videos, it is usually time-costing. Moreover, the post-processing may fail in cases with heavy artifacts (Fig. 1).

To relieve the above limitations, we propose to learn a joint **S**patial-**T**emporal **T**ransformer **N**etwork (**STTN**) for video inpainting. We formulate video inpainting as a "multi-to-multi" problem, which takes both neighboring and distant frames as input and simultaneously fills missing regions in all input frames. To fill missing regions in each frame, the transformer searches coherent contents from all the frames along both spatial and temporal dimensions by a proposed multi-scale patch-based attention module. Specifically, patches of different scales are extracted from all the frames to cover different appearance changes caused by complex motions. Different heads of the transformer calculate similarities on spatial patches across different scales. Through such a design, the most relevant patches can be detected and transformed for the missing regions by aggregating attention results from different heads. Moreover, the spatial-temporal transformers can be fully exploited by stacking multiple layers, so that attention results for missing regions can be improved based on updated region features. Last but not least, we further leverage a spatial-temporal adversarial loss for joint optimization [5,6]. Such a loss design can optimize STTN to learn both perceptually pleasing and coherent visual contents for video inpainting.

In summary, our main contribution is to learn joint spatial and temporal transformations for video inpainting, by a deep generative model with adversarial training along spatial-temporal dimensions. Furthermore, the proposed multi-scale patch-based video frame representations can enable fast training and inference, which is important to video understanding tasks. We conduct both quantitative and qualitative evaluations using both stationary masks and moving object masks for simulating real-world applications (e.g., watermark removal and object removal). Experiments show that our model outperforms the state-of-the-arts by a significant margin in terms of PSNR and VFID with relative improvements of 2.4% and 19.7%, respectively. We also show extensive ablation studies to verify the effectiveness of the proposed spatial-temporal transformer.

2 Related Work

To develop high-quality video inpainting technology, many efforts have been made on filling missing regions with spatially and temporally coherent contents in videos [2,13,18,24,29,33]. We discuss representative patch-based methods and deep generative models for video inpainting as below.

Patch-Based Methods: Early video inpainting methods mainly formulate the inpainting process as a patch-based optimization problem [1,7,26,31]. Specifically, these methods synthesize missing contents by sampling similar spatial or spatial-temporal patches from known regions based on a global optimization [24,27,31]. Some approaches try to improve performance by providing foreground and background segments [10,26]. Other works focus on joint estimations for both appearance and optical-flow [13,22]. Although promising results can be achieved, patch-based optimization algorithms typically assume a homogeneous motion field in holes and they are often limited by complex motion in general situations. Moreover, optimization-based inpainting methods often suffer from high computational complexity, which is infeasible for real-time applications [33].

Deep Generative Models: With the development of deep generative models, significant progress has been made by deep video inpainting models. Wang et al. are the first to propose to combine 3D and 2D fully convolution networks for learning temporal information and spatial details for video inpainting [29]. However, the results are blurry in complex scenes. Xu et al. improve the performance by jointly estimating both appearance and optical-flow [33,37]. Kim et al. adopt recurrent networks for ensuring temporal coherence [16]. Chang et al. develop Temporal SN-PatchGAN [35] and temporal shift modules [19] for free-form video

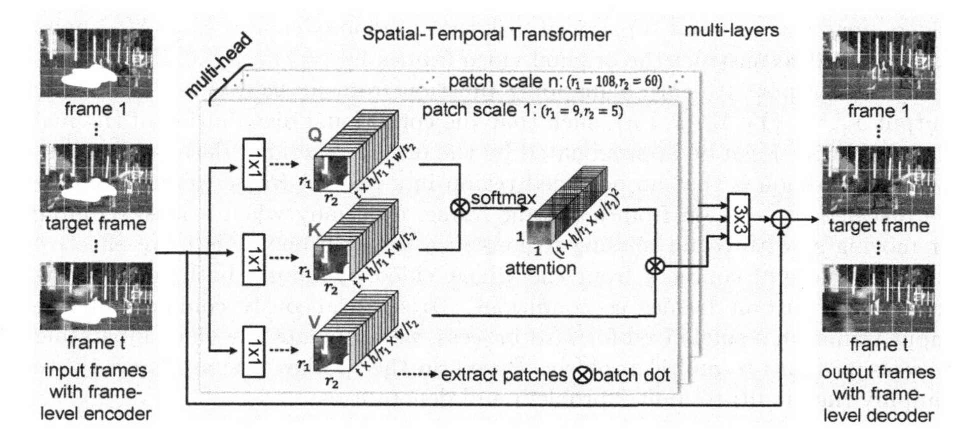

Fig. 2. Overview of the Spatial-Temporal Transformer Networks (STTN). STTN consists of 1) a frame-level encoder, 2) multi-layer multi-head spatial-temporal transformers and 3) a frame-level decoder. The transformers are designed to simultaneously fill holes in all input frames with coherent contents. Specifically, a transformer matches the queries (Q) and keys (K) on spatial patches across different scales in multiple heads, thus the values (V) of relevant regions can be detected and transformed for the holes. Moreover, the transformers can be fully exploited by stacking multiple layers to improve attention results based on updated region features. 1×1 and 3×3 denote the kernel size of 2D convolutions. More details can be found in Sect. 3.

inpainting [5]. Although these methods can aggregate information from nearby frames, they fail to capture visible contents from distant frames.

To effectively model long-range correspondences, recent models have adopted attention modules and show promising results in image and video synthesis [21,34,36]. Specifically, Lee et al. propose to synthesize missing contents by weighted summing aligned frames with frame-wise attention [18]. However, the frame-wise attention relies on global affine transformations between frames, which is hard to handle complex motions. Oh et al. gradually fill holes step by step with pixel-wise attention [25]. Despite promising results, it is hard to ensure consistent attention result in each recursion. Moreover, existing deep video inpainting models that adopt attention modules process videos frame by frame without specially-designed optimization for ensuring temporal coherence.

3 Spatial-Temporal Transformer Networks

3.1 Overall Design

Problem Formulation: Let $X_1^T := \{X_1, X_2, ..., X_T\}$ be a corrupted video sequence of height H, width W and frames length T. $M_1^T := \{M_1, M_2, ..., M_T\}$ denotes the corresponding frame-wise masks. For each mask M_i, value "0" indicates known pixels, and value "1" indicates missing regions. We formulate deep video inpainting as a self-supervised task that randomly creates (X_1^T, M_1^T) pairs as input and reconstruct the original video frames $Y_1^T = \{Y_1, Y_2, ..., Y_T\}$. Specifically, we propose to learn a mapping function from masked video X_1^T to the output $\hat{Y}_1^T := \{\hat{Y}_1, \hat{Y}_2, ..., \hat{Y}_T\}$, such that the conditional distribution of the real data $p(Y_1^T|X_1^T)$ can be approximated by the one of generated data $p(\hat{Y}_1^T|X_1^T)$.

The intuition is that an occluded region in a current frame would probably be revealed in a region from a distant frame, especially when a mask is large or moving slowly. To fill missing regions in a target frame, it is more effective to borrow useful contents from the whole video by taking both neighboring frames and distant frames as conditions. To simultaneously complete all the input frames in a single feed-forward process, we formulate the video inpainting task as a "multi-to-multi" problem. Based on the Markov assumption [11], we simplify the "multi-to-multi" problem and denote it as:

$$p(\hat{Y}_1^T|X_1^T) = \prod_{t=1}^{T} p(\hat{Y}_{t-n}^{t+n}|X_{t-n}^{t+n}, X_{1,s}^T), \tag{1}$$

where X_{t-n}^{t+n} denotes a short clip of neighboring frames with a center moment t and a temporal radius n. $X_{1,s}^T$ denotes distant frames that are uniformly sampled from the videos X_1^T in a sampling rate of s. Since $X_{1,s}^T$ can usually cover most key frames of the video, it is able to describe "the whole story" of the video. Under this formulation, video inpainting models are required to not only preserve temporal consistency in neighboring frames, but also make the completed frames to be coherent with "the whole story" of the video.

Network Design: The overview of the proposed **S**patial-**T**emporal **T**ransformer **N**etworks (**STTN**) is shown in Fig. 2. As indicated in Eq. (1), STTN takes both neighboring frames X_{t-n}^{t+n} and distant frames $X_{1,s}^T$ as conditions, and complete all the input frames simultaneously. Specifically, STTN consists of three components, including a frame-level encoder, multi-layer multi-head spatial-temporal transformers, and a frame-level decoder. The frame-level encoder is built by stacking several 2D convolution layers with strides, which aims at encoding deep features from low-level pixels for each frame. Similarly, the frame-level decoder is designed to decode features back to frames. Spatial-temporal transformers are the core component, which aims at learning joint spatial-temporal transformations for all missing regions in the deep encoding space.

3.2 Spatial-Temporal Transformer

To fill missing regions in each frame, spatial-temporal transformers are designed to search coherent contents from all the input frames. Specifically, we propose to search by a multi-head patch-based attention module along both spatial and temporal dimensions. Different heads of a transformer calculate attentions on spatial patches across different scales. Such a design allows us to handle appearance changes caused by complex motions. For example, on one hand, attentions for patches of large sizes (e.g., frame size $H \times W$) aim at completing stationary backgrounds. On the other hand, attentions for patches of small sizes (e.g., $\frac{H}{10} \times \frac{W}{10}$) encourage capturing deep correspondences in any locations of videos for moving foregrounds.

A multi-head transformer runs multiple "Embedding-Matching-Attending" steps for different patch sizes in parallel. In the Embedding step, features of each frame are mapped into query and memory (i.e., key-value pair) for further retrieval. In the Matching step, region affinities are calculated by matching queries and keys among spatial patches that are extracted from all the frames. Finally, relevant regions are detected and transformed for missing regions in each frame in the Attending step. We introduce more details of each step as below.

Embedding: We use $f_1^T = \{f_1, f_2, ..., f_T\}$, where $f_i \in R^{h \times w \times c}$ to denote the features encoded from the frame-level encoder or former transformers, which is the input of transformers in Fig. 2. Similar to many sequence modeling models, mapping features into key and memory embeddings is an important step in transformers [9,28]. Such a step enables modeling deep correspondences for each region in different semantic spaces:

$$q_i, (k_i, v_i) = M_q(f_i), (M_k(f_i), M_v(f_i)), \tag{2}$$

where $1 \leq i \leq T$, $M_q(\cdot)$, $M_k(\cdot)$ and $M_v(\cdot)$ denote the 1×1 2D convolutions that embed input features into query and memory (i.e., key-value pair) feature spaces while maintaining the spatial size of features.

Matching: We conduct patch-based matching in each head. In practice, we first extract spatial patches of shape $r_1 \times r_2 \times c$ from the query feature of each frame, and we obtain $N = T \times h/r_1 \times w/r_2$ patches. Similar operations are conducted to extract patches in the memory (i.e., key-value pair in the transformer). Such an effective multi-scale patch-based video frame representation can avoid redundant patch matching and enable fast training and inference. Specifically, we reshape the query patches and key patches into 1-dimension vectors separately, so that patch-wise similarities can be calculated by matrix multiplication. The similarity between i-th patch and j-th patch is denoted as:

$$s_{i,j} = \frac{\boldsymbol{p}_i^q \cdot (\boldsymbol{p}_j^k)^T}{\sqrt{r_1 \times r_2 \times c}}, \tag{3}$$

where $1 \leq i, j \leq N$, $\boldsymbol{p}_i^q$ denotes the i-th query patch, $\boldsymbol{p}_j^k$ denotes the j-th key patch. The similarity value is normalized by the dimension of each vector to avoid a small gradient caused by subsequent softmax function [28]. Corresponding attention weights for all patches are calculated by a softmax function:

$$\alpha_{i,j} = \begin{cases} \exp(s_{i,j}) / \sum_{n=1}^{N} \exp(s_{i,n}), & \boldsymbol{p}_j \in \Omega, \\ 0, & \boldsymbol{p}_j \in \bar{\Omega}. \end{cases} \tag{4}$$

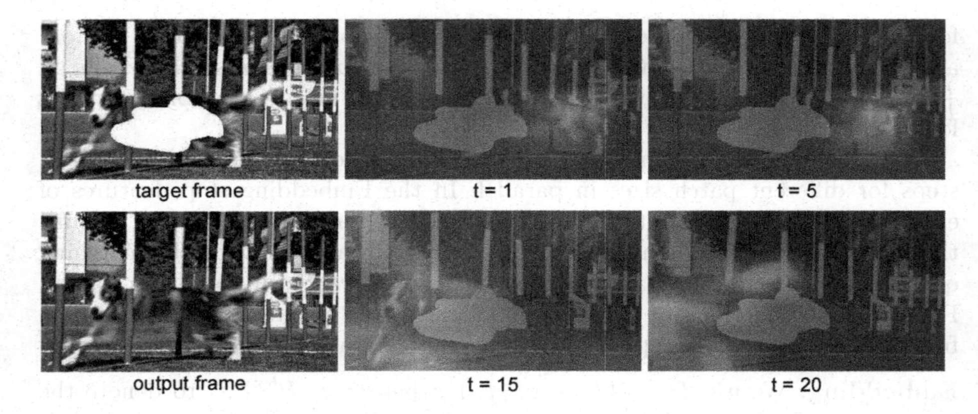

Fig. 3. Illustration of the attention maps for missing regions learned by STTN. For completing the dog corrupted by a random mask in a target frame (e.g., t = 10), our model is able to "track" the moving dog over the video in both spatial and temporal dimensions. Attention regions are highlighted in bright yellow. (Color figure online)

where Ω denotes visible regions outside masks, and $\bar{\Omega}$ denotes missing regions. Naturally, we only borrow features from visible regions for filling holes.

Attending: After modeling the deep correspondences for all spatial patches, the output for the query of each patch can be obtained by weighted summation of values from relevant patches:

$$o_i = \sum_{j=1}^{N} \alpha_{i,j} \boldsymbol{p}_j^v, \tag{5}$$

where $\boldsymbol{p}_j^v$ denotes the j-th value patch. After receiving the output for all patches, we piece all patches together and reshape them into T frames with original spatial size $h \times w \times c$. The resultant features from different heads are concatenated and further passed through a subsequent 2D residual block [12]. This subsequent processing is used to enhance the attention results by looking at the context within the frame itself.

The power of the proposed transformer can be fully exploited by stacking multiple layers, so that attention results for missing regions can be improved based on updated region features in a single feed-forward process. Such a multi-layer design promotes learning coherent spatial-temporal transformations for filling in missing regions. As shown in Fig. 3, we highlight the attention maps learned by STTN in the last layer in bright yellow. For the dog partially occluded by a random mask in a target frame, spatial-temporal transformers are able to "track" the moving dog over the video in both spatial and temporal dimensions and fill missing regions in the dog with coherent contents.

3.3 Optimization Objectives

As outlined in Sect. 3.1, we optimize the proposed STTN in an end-to-end manner by taking the original video frames as ground truths without any other labels. The principle of choosing optimization objectives is to ensure per-pixel reconstruction accuracy, perceptual rationality and spatial-temporal coherence in generated videos [5,8,14,18]. To this end, we select a pixel-wise reconstruction loss and a spatial-temporal adversarial loss as our optimization objectives.

In particular, we include L_1 losses calculated between generated frames and original frames for ensuring per-pixel reconstruction accuracy in results. The L_1 losses for hole regions are denoted as:

$$L_{hole} = \frac{\|M_1^T \odot (Y_1^T - \hat{Y}_1^T)\|_1}{\|M_1^T\|_1}, \tag{6}$$

and corresponding L_1 losses for valid regions are denoted as:

$$L_{valid} = \frac{\|(1 - M_1^T) \odot (Y_1^T - \hat{Y}_1^T)\|_1}{\|1 - M_1^T\|_1}, \tag{7}$$

where $\odot$ indicates element-wise multiplication, and the values are normalized by the size of corresponding regions.

Inspired by the recent studies that adversarial training can help to ensure high-quality content generation results, we propose to use a Temporal PatchGAN (T-PatchGAN) as our discriminator [5,6,34,36]. Such an adversarial loss has shown promising results in enhancing both perceptual quality and spatial-temporal coherence in video inpainting [5,6]. In particular, the T-PatchGAN is composed of six layers of 3D convolution layers. The T-PatchGAN learns to distinguish each spatial-temporal feature as real or fake, so that spatial-temporal coherence and local-global perceptual details of real data can be modeled by STTN. The detailed optimization function for the T-PatchGAN discriminator is shown as follows:

$$L_D = E_{x \sim P_{Y_1^T}(x)}[ReLU(1 - D(x))] + E_{z \sim P_{\hat{Y}_1^T}(z)}[ReLU(1 + D(z))], \tag{8}$$

and the adversarial loss for STTN is denoted as:

$$L_{adv} = -E_{z \sim P_{\hat{Y}_1^T}(z)}[D(z)]. \tag{9}$$

The overall optimization objectives are concluded as below:

$$L = \lambda_{hole} \cdot L_{hole} + \lambda_{valid} \cdot L_{valid} + \lambda_{adv} \cdot L_{adv}. \tag{10}$$

We empirically set the weights for different losses as: $\lambda_{hole} = 1$, $L_{valid} = 1$, $L_{adv} = 0.01$. Since our model simultaneously complete all the input frames in a single feed-forward process, our model runs at 24.3 fps on a single GPU NVIDIA V100. More details are provided in the Section D of our supplementary material.

4 Experiments

4.1 Dataset

To evaluate the proposed model and make fair comparisons with SOTA approaches, we adopt the two most commonly-used datasets in video inpainting, including Youtube-VOS [32] and DAVIS [3]. In particular, **YouTube-VOS** contains 4,453 videos with various scenes, including bedrooms, streets, and so on. The average video length in Youtube-VOS is about 150 frames. We follow the original train/validation/test split (i.e., 3,471/474/508) and report experimental results on the test set for Youtube-VOS. In addition, we also evaluate different approaches on **DAVIS** dataset [3], as this dataset is composed of 150 high-quality videos of challenging camera motions and foreground motions. We follow the setting in previous works [16,33], and set the training/testing split as 60/90 videos. Since the training set of DAVIS is limited (60 videos with at most 90 frames for each), we initialize model weights by a pre-trained model on YouTube-VOS following the settings used in [16,33].

To simulate real-world applications, we evaluate models by using two types of free-form masks, including stationary masks and moving masks [6,16,18]. Because free-form masks are closer to real masks and have been proved to be effective for training and evaluating inpainting models [5,6,20,23]. Specifically, for testing **stationary masks**, we generate stationary random shapes as testing masks to simulate applications like watermark removal. More details of the generation algorithm are provided in the Section B of our supplementary material. Since this type of application targets at reconstructing original videos, we take original videos as ground truths and evaluate models from both quantitative and qualitative aspects. For testing **moving masks**, we use foreground object annotations as testing masks to simulate applications like object removal. Since the ground truths after foreground removal are unavailable, we evaluate the models through qualitative analysis following previous works [16,18,33].

4.2 Baselines and Evaluation Metrics

Recent deep video inpainting approaches have shown state-of-the-art performance with fast computational time [16,18,25,33]. To evaluate our model and make fair comparisons, we select the most recent and the most competitive approaches for comparisons, which are listed as below:

- **VINet** [16] adopts a recurrent network to aggregate temporal features from neighboring frames.
- **DFVI** [33] fills missing regions in videos by pixel propagation algorithm based on completed optical flows.
- **LGTSM** [6] proposes a learnable temporal shift module and a spatial-temporal adversarial loss for ensuring spatial and temporal coherence.
- **CAP** [18] synthesizes missing contents by a deep alignment network and a frame-based attention module.

We fine-tune baselines multiple times on YouTube-VOS [32] and DAVIS [3] by their released models and codes and report their best results in this paper.

We report quantitative results by four numeric metrics, i.e., PSNR [33], SSIM [5], flow warping error [17] and video-based Fréchet Inception Distance (VFID) [5,30]. Specifically, we use PSNR and SSIM as they are the most widely-used metrics for video quality assessment. Besides, the flow warping error is included to measure the temporal stability of generated videos. Moreover, FID has been proved to be an effective perceptual metric and it has been used by many inpainting models [25,30,38]. In practice, we use an I3D [4] pre-trained video recognition model to calculate VFID following the settings in [5,30].

4.3 Comparisons with State-of-the-Arts

Quantitative Evaluation: We report quantitative results for filling stationary masks on Youtube-VOS [32] and DAVIS [3] in Table 1. As stationary masks often involve partially occluded foreground objects, it is challenging to reconstruct a video especially with complex appearances and object motions. Table 1 shows that, compared with SOTA models, our model performs better video reconstruction quality with both per-pixel and overall perceptual measurements. Specifically, our model outperforms the SOTA models by a significant margin, especially in terms of PSNR, flow warp error and VFID. The specific gains are 2.4%, 1.3% and 19.7% relative improvements on Youtube-VOS, respectively. The superior results show the effectiveness of the proposed spatial-temporal transformer and adversarial optimizations in STTN.

Table 1. Quantitative comparisons with state-of-the-art models on Youtube-VOS [32] and DAVIS [3]. Our model outperforms baselines in terms of PSNR [33], SSIM [5], flow warping error (E_{warp}) [17] and VFID [30]. * Higher is better. † Lower is better.

Models		PSNR*	SSIM (%)*	E_{warp} (%)†	VFID†
Youtube-vos	VINet [16]	29.20	94.34	0.1490	0.072
	DFVI [33]	29.16	94.29	0.1509	0.066
	LGTSM [6]	29.74	95.04	0.1859	0.070
	CAP [18]	31.58	96.07	0.1470	0.071
	Ours	**32.34**	**96.55**	**0.1451**	**0.053**
DAVIS	VINet [16]	28.96	94.11	0.1785	0.199
	DFVI [33]	28.81	94.04	0.1880	0.187
	LGTSM [6]	28.57	94.09	0.2566	0.170
	CAP [18]	30.28	95.21	0.1824	0.182
	Ours	**30.67**	**95.60**	**0.1779**	**0.149**

Qualitative Evaluation: For each video from test sets, we take all frames for testing. To compare visual results from different models, we follow the setting

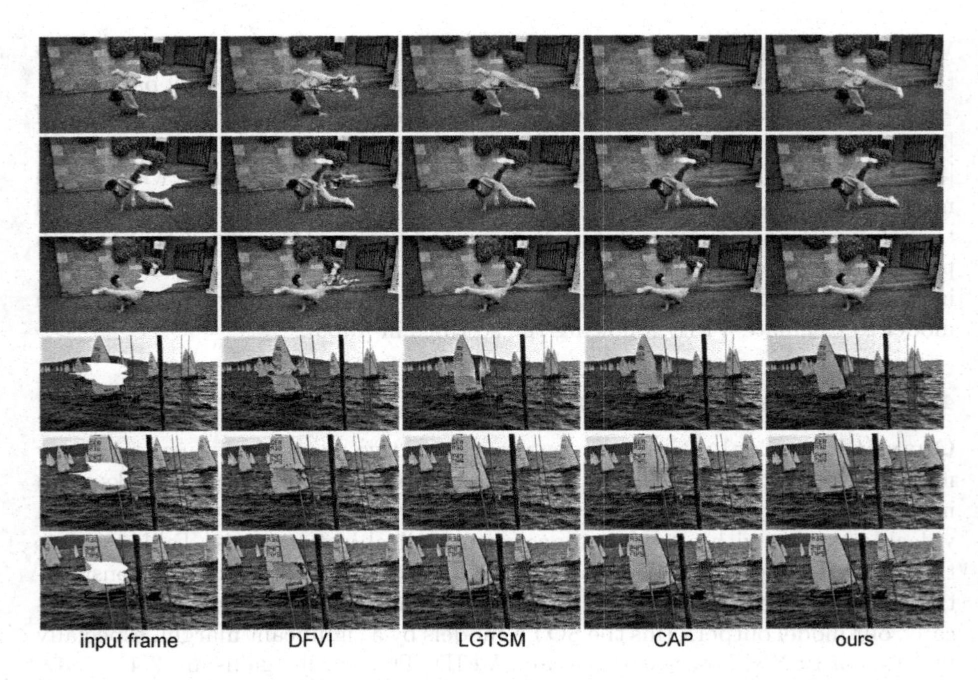

Fig. 4. Visual results for stationary masks. The first column shows input frames from DAVIS [3] (top-3) and YouTube-VOS [32] (bottom-3), followed by results from DFVI [33], LGTSM [6], CAP [18], and our model. Comparing with the SOTAs, our model generates more coherent structures and details of the legs and boats in results.

used by most video inpainting works and randomly sample three frames from the video for case study [18,25,29]. We select the most three competitive models, DFVI [33], LGTSM [6] and CAP [18] for comparing results for stationary masks in Fig. 4. We also show a case for filling in moving masks in Fig. 5. To conduct pair-wise comparisons and analysis in Fig. 5, we select the most competitive model, CAP [18], according to the quantitative comparison results. We can find from the visual results that our model is able to generate perceptually pleasing and coherent contents in results. More video cases are available online[1].

In addition to visual comparisons, we visualize the attention maps learned by STTN in Fig. 6. Specifically, we highlight the top three relevant regions captured by the last transformer in STTN in bright yellow. The relevant regions are selected according to the attention weights calculated by Eq. (4). We can find in Fig. 6 that STTN is able to precisely attend to the objects for filling partially occluded objects in the first and the third cases. For filling the backgrounds in the second and the fourth cases, STTN can correctly attend to the backgrounds.

User Study: We conduct a user study for a more comprehensive comparison. We choose LGTSM [6] and CAP [18] as two strong baselines, since we have

[1] Video demo: https://github.com/researchmm/STTN.

Fig. 5. Visual comparisons for filling moving masks. Comparing with CAP [18], one of the most competitive models for filling moving masks, our model is able to generate visually pleasing results even under complex scenes (e.g., clear faces for the first and the third frames, and better results than CAP for the second frame).

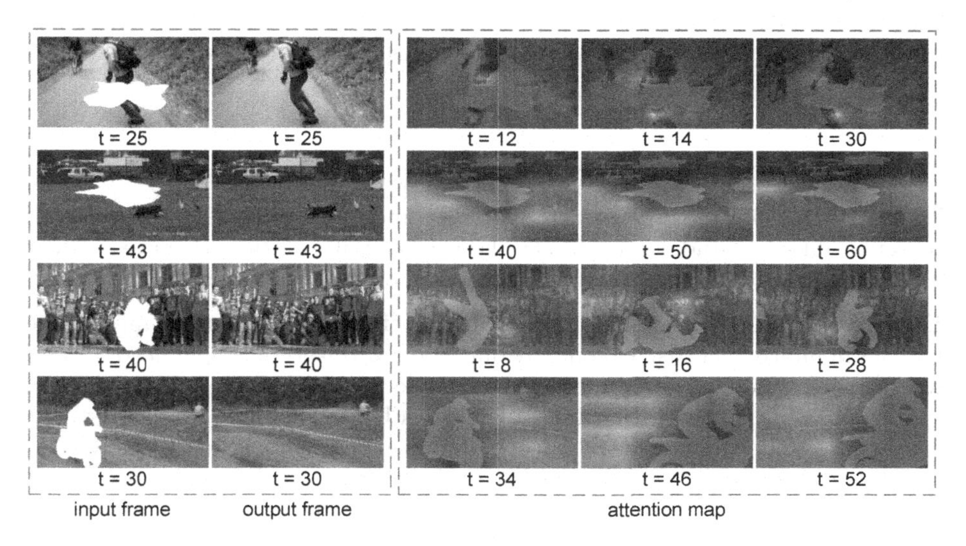

Fig. 6. Illustration of attention maps for missing regions learned by the proposed STTN. We highlight the most relevant patches in yellow according to attention weights. For filling partially occluded objects (the first and the third cases), STTN can precisely attend to the objects. For filling backgrounds (the second and the fourth cases), STTN can correctly attend to the backgrounds. (Color figure online)

observed their significantly better performance than other baselines from both quantitative and qualitative results. We randomly sampled 10 videos (5 from DAVIS and 5 from YouTube-VOS) for stationary masks filling, and 10 videos from DAVIS for moving masks filling. In practice, 28 volunteers are invited to the user study. In each trial, inpainting results from different models are shown to the volunteers, and the volunteers are required to rank the inpainting results. To ensure a reliable subjective evaluation, videos can be replayed multiple times by volunteers. Each participant is required to finish 20 groups of trials without time limit. Most participants can finish the task within 30 min. The results of the user study are concluded in Fig. 7. We can find that our model performs better in most cases for these two types of masks.

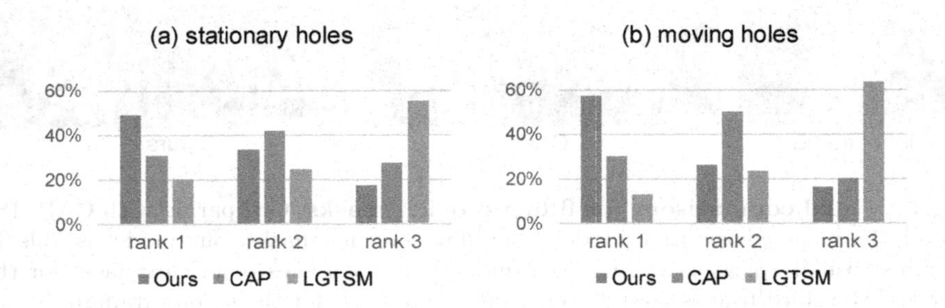

Fig. 7. User study. "Rank x" means the percentage of results from each model being chosen as the x-th best. Our model is ranked in first place in most cases.

4.4 Ablation Study

To verify the effectiveness of the spatial-temporal transformers, this section presents ablation studies on DAVIS dataset [3] with stationary masks. More ablation studies can be found in the Section E of our supplementary material.

Effectiveness of Multi-scale: To verify the effectiveness of using multi-scale patches in multiple heads, we compare our model with several single-head STTNs with different patch sizes. In practice, we select patch sizes according to the spatial size of features, so that the features can be divided into patches without overlapping. The spatial size of features in our experiments is 108×60. Results in Table 2 show that our full model with multi-scale patch-based video frame representation achieves the best performance under this setting.

Table 2. Ablation study by using different patch scales in attention layers. Ours combines the above four scales. * Higher is better. † Lower is better.

Patch size	PSNR*	SSIM (%)*	E_{warp} (%)†	VFID†
108×60	30.16	95.16	0.2243	0.168
36×20	30.11	95.13	0.2051	0.160
18×10	30.17	95.20	0.1961	0.159
9×5	30.43	95.39	0.1808	0.163
Ours	**30.67**	**95.60**	**0.1779**	**0.149**

Table 3. Ablation study by using different stacking number of the proposed spatial-temporal transformers. * Higher is better. † Lower is better.

Stack	PSNR*	SSIM (%)*	E_{warp} (%)†	VFID†
×2	30.17	95.17	0.1843	0.162
×4	30.38	95.37	0.1802	0.159
×6	30.53	95.47	0.1797	0.155
×8 (ours)	**30.67**	**95.60**	**0.1779**	**0.149**

Fig. 8. A failure case. The bottom row shows our results with enlarged patches in the bottom right corner. For reconstructing the dancing woman occluded by a large mask, STTN fails to generate continuous motions and it generates blurs inside the mask.

Effectiveness of Multi-layer: The spatial-temporal transformers can be stacked by multiple layers to repeat the inpainting process based on updated region features. We verify the effectiveness of using multi-layer spatial-temporal transformers in Table 3. We find that stacking more transformers can bring continuous improvements and the best results can be achieved by stacking eight layers. Therefore, we use eight layers in transformers as our full model.

5 Conclusions

In this paper, we propose a novel joint spatial-temporal transformation learning for video inpainting. Extensive experiments have shown the effectiveness of multi-scale patch-based video frame representation in deep video inpainting models. Coupled with a spatial-temporal adversarial loss, our model can be optimized to simultaneously complete all the input frames in an efficient way. The results on YouTube-VOS [32] and DAVIS [3] with challenging free-form masks show the state-of-the-art performance by our model.

We note that STTN may generate blurs in large missing masks if continuous quick motions occur. As shown in Fig. 8, STTN fails to generate continuous dancing motions and it generates blurs when reconstructing the dancing woman in the first frame. We infer that STTN only calculates attention among spatial patches, and the short-term temporal continuity of complex motions are hard to capture without 3D representations. In the future, we plan to extend the proposed transformer by using attention on 3D spatial-temporal patches to improve the short-term coherence. We also plan to investigate other types of temporal losses [17,30] for joint optimization in the future.

Acknowledgments. This project was supported by NSF of China under Grant 61672548, U1611461.

References

1. Barnes, C., Shechtman, E., Finkelstein, A., Goldman, D.B.: PatchMatch: a randomized correspondence algorithm for structural image editing. TOG **28**(3), 24:1–24:11 (2009)
2. Bertalmio, M., Bertozzi, A.L., Sapiro, G.: Navier-stokes, fluid dynamics, and image and video inpainting. In: CVPR, pp. 355–362 (2001)
3. Caelles, S., et al.: The 2018 DAVIS challenge on video object segmentation. arXiv (2018)
4. Carreira, J., Zisserman, A.: Quo vadis, action recognition? A new model and the kinetics dataset. In: CVPR, pp. 6299–6308 (2017)
5. Chang, Y.L., Liu, Z.Y., Lee, K.Y., Hsu, W.: Free-form video inpainting with 3D gated convolution and temporal PatchGAN. In: ICCV, pp. 9066–9075 (2019)
6. Chang, Y.L., Liu, Z.Y., Lee, K.Y., Hsu, W.: Learnable gated temporal shift module for deep video inpainting. In: BMVC (2019)
7. Criminisi, A., Pérez, P., Toyama, K.: Region filling and object removal by exemplar-based image inpainting. TIP **13**(9), 1200–1212 (2004)
8. Gatys, L.A., Ecker, A.S., Bethge, M.: Image style transfer using convolutional neural networks. In: CVPR, pp. 2414–2423 (2016)
9. Girdhar, R., Carreira, J., Doersch, C., Zisserman, A.: Video action transformer network. In: CVPR, pp. 244–253 (2019)
10. Granados, M., Tompkin, J., Kim, K., Grau, O., Kautz, J., Theobalt, C.: How not to be seen-object removal from videos of crowded scenes. Comput. Graph. Forum **31**(21), 219–228 (2012)
11. Hausman, D.M., Woodward, J.: Independence, invariance and the causal Markov condition. Br. J. Philos. Sci. **50**(4), 521–583 (1999)
12. He, K., Zhang, X., Ren, S., Sun, J.: Deep residual learning for image recognition. In: CVPR, pp. 770–778 (2016)
13. Huang, J.B., Kang, S.B., Ahuja, N., Kopf, J.: Temporally coherent completion of dynamic video. TOG **35**(6), 1–11 (2016)
14. Johnson, J., Alahi, A., Fei-Fei, L.: Perceptual losses for real-time style transfer and super-resolution. In: Leibe, B., Matas, J., Sebe, N., Welling, M. (eds.) ECCV 2016. LNCS, vol. 9906, pp. 694–711. Springer, Cham (2016). https://doi.org/10.1007/978-3-319-46475-6_43
15. Kim, D., Woo, S., Lee, J.Y., Kweon, I.S.: Deep blind video decaptioning by temporal aggregation and recurrence. In: CVPR, pp. 4263–4272 (2019)
16. Kim, D., Woo, S., Lee, J.Y., Kweon, I.S.: Deep video inpainting. In: CVPR, pp. 5792–5801 (2019)
17. Lai, W.-S., Huang, J.-B., Wang, O., Shechtman, E., Yumer, E., Yang, M.-H.: Learning blind video temporal consistency. In: Ferrari, V., Hebert, M., Sminchisescu, C., Weiss, Y. (eds.) ECCV 2018. LNCS, vol. 11219, pp. 179–195. Springer, Cham (2018). https://doi.org/10.1007/978-3-030-01267-0_11

18. Lee, S., Oh, S.W., Won, D., Kim, S.J.: Copy-and-paste networks for deep video inpainting. In: ICCV, pp. 4413–4421 (2019)
19. Lin, J., Gan, C., Han, S.: TSM: temporal shift module for efficient video understanding. In: ICCV, pp. 7083–7093 (2019)
20. Liu, G., Reda, F.A., Shih, K.J., Wang, T.-C., Tao, A., Catanzaro, B.: Image inpainting for irregular holes using partial convolutions. In: Ferrari, V., Hebert, M., Sminchisescu, C., Weiss, Y. (eds.) ECCV 2018. LNCS, vol. 11215, pp. 89–105. Springer, Cham (2018). https://doi.org/10.1007/978-3-030-01252-6_6
21. Ma, S., Fu, J., Wen Chen, C., Mei, T.: DA-GAN: instance-level image translation by deep attention generative adversarial networks. In: CVPR, pp. 5657–5666 (2018)
22. Matsushita, Y., Ofek, E., Ge, W., Tang, X., Shum, H.Y.: Full-frame video stabilization with motion inpainting. TPAMI **28**(7), 1150–1163 (2006)
23. Nazeri, K., Ng, E., Joseph, T., Qureshi, F., Ebrahimi, M.: EdgeConnect: generative image inpainting with adversarial edge learning. In: ICCVW (2019)
24. Newson, A., Almansa, A., Fradet, M., Gousseau, Y., Pérez, P.: Video inpainting of complex scenes. SIAM J. Imaging Sci. **7**(4), 1993–2019 (2014)
25. Oh, S.W., Lee, S., Lee, J.Y., Kim, S.J.: Onion-peel networks for deep video completion. In: ICCV, pp. 4403–4412 (2019)
26. Patwardhan, K.A., Sapiro, G., Bertalmio, M.: Video inpainting of occluding and occluded objects. In: ICIP, pp. 11–69 (2005)
27. Patwardhan, K.A., Sapiro, G., Bertalmío, M.: Video inpainting under constrained camera motion. TIP **16**(2), 545–553 (2007)
28. Vaswani, A., et al.: Attention is all you need. In: NeurIPS, pp. 5998–6008 (2017)
29. Wang, C., Huang, H., Han, X., Wang, J.: Video inpainting by jointly learning temporal structure and spatial details. In: AAAI, pp. 5232–5239 (2019)
30. Wang, T.C., et al.: Video-to-video synthesis. In: NeuralPS, pp. 1152–1164 (2018)
31. Wexler, Y., Shechtman, E., Irani, M.: Space-time completion of video. TPAMI **29**(3), 463–476 (2007)
32. Xu, N., et al.: YouTube-VOS: a large-scale video object segmentation benchmark. arXiv (2018)
33. Xu, R., Li, X., Zhou, B., Loy, C.C.: Deep flow-guided video inpainting. In: CVPR, pp. 3723–3732 (2019)
34. Yang, F., Yang, H., Fu, J., Lu, H., Guo, B.: Learning texture transformer network for image super-resolution. In: CVPR, pp. 5791–5800 (2020)
35. Yu, J., Lin, Z., Yang, J., Shen, X., Lu, X., Huang, T.S.: Free-form image inpainting with gated convolution. In: ICCV, pp. 4471–4480 (2019)
36. Zeng, Y., Fu, J., Chao, H., Guo, B.: Learning pyramid-context encoder network for high-quality image inpainting. In: CVPR, pp. 1486–1494 (2019)
37. Zhang, H., Mai, L., Xu, N., Wang, Z., Collomosse, J., Jin, H.: An internal learning approach to video inpainting. In: CVPR, pp. 2720–2729 (2019)
38. Zhang, R., Isola, P., Efros, A.A., Shechtman, E., Wang, O.: The unreasonable effectiveness of deep features as a perceptual metric. In: CVPR, pp. 586–595 (2018)

Single Path One-Shot Neural Architecture Search with Uniform Sampling

Zichao Guo[1(✉)], Xiangyu Zhang[1], Haoyuan Mu[1,2], Wen Heng[1], Zechun Liu[1,3], Yichen Wei[1], and Jian Sun[1]

[1] MEGVII Technology, Beijing, China
{guozichao,zhangxiangyu,hengwen,weiyichen,sunjian}@megvii.com
[2] Tsinghua University, Beijing, China
muhy17@mails.tsinghua.edu.cn
[3] Hong Kong University of Science and Technology, Clear Water Bay, Hong Kong
zliubq@connect.ust.hk

Abstract. We revisit the one-shot Neural Architecture Search (NAS) paradigm and analyze its advantages over existing NAS approaches. Existing one-shot method, however, is hard to train and not yet effective on large scale datasets like ImageNet. This work propose a Single Path One-Shot model to address the challenge in the training. Our central idea is to construct a simplified supernet, where all architectures are single paths so that weight co-adaption problem is alleviated. Training is performed by uniform path sampling. All architectures (and their weights) are trained fully and equally.

Comprehensive experiments verify that our approach is flexible and effective. It is easy to train and fast to search. It effortlessly supports complex search spaces (e.g., building blocks, channel, mixed-precision quantization) and different search constraints (e.g., FLOPs, latency). It is thus convenient to use for various needs. It achieves start-of-the-art performance on the large dataset ImageNet.

1 Introduction

Deep learning automates *feature engineering* and solves the *weight optimization* problem. Neural Architecture Search (NAS) aims to automate *architecture engineering* by solving one more problem, *architecture design*. Early NAS approaches [11,16,21,32,33,36] solves the two problems in a *nested* manner.

Z. Guo and X. Zhang—Equal contribution. This work is done when Haoyuan Mu and Zechun Liu are interns at MEGVII Technology.

Electronic supplementary material The online version of this chapter (https://doi.org/10.1007/978-3-030-58517-4_32) contains supplementary material, which is available to authorized users.

A. Vedaldi et al. (Eds.): ECCV 2020, LNCS 12361, pp. 544–560, 2020.
https://doi.org/10.1007/978-3-030-58517-4_32

A large number of architectures are sampled and trained from scratch. The computation cost is unaffordable on large datasets.

Recent approaches [2–4,12,15,23,26,31] adopt a *weight sharing* strategy to reduce the computation. A supernet subsuming all architectures is trained only once. Each architecture inherits its weights from the supernet. Only fine-tuning is performed. The computation cost is greatly reduced.

Most weight sharing approaches use a continuous relaxation to parameterize the search space [4,12,23,26,31]. The architecture distribution parameters are *jointly* optimized during the supernet training via gradient based methods. The best architecture is sampled from the distribution after optimization. There are two issues in this formulation. *First*, the weights in the supernet are deeply coupled. It is unclear why inherited weights for a specific architecture are still effective. *Second*, joint optimization introduces further coupling between the architecture parameters and supernet weights. The greedy nature of the gradient based methods inevitably introduces bias during optimization and could easily mislead the architecture search.

The one-shot paradigm [2,3] alleviates the second issue. It defines the supernet and performs weight inheritance in a similar way. However, there is no architecture relaxation. The architecture search problem is decoupled from the supernet training and addressed in a separate step. Thus, it is *sequential*. It combines the merits of both *nested* and *joint* optimization approaches above. The architecture search is both efficient and flexible.

The first issue is still problematic. Existing one-shot approaches [2,3] still have coupled weights in the supernet. Their optimization is complicated and involves sensitive hyper parameters. They have not shown competitive results on large datasets.

This work revisits the one-shot paradigm and presents a new approach that further eases the training and enhances architecture search. Based on the observation that the accuracy of an architecture using inherited weights should be predictive for the accuracy using optimized weights, we propose that the supernet training should be *stochastic*. All architectures have their weights optimized simultaneously. This gives rise to a *uniform sampling* strategy. To reduce the weight coupling in the supernet, a simple search space that consists of *single path* architectures is proposed. The training is hyperparameter-free and easy to converge.

This work makes the following contributions.

1. We present a principled analysis and point out drawbacks in existing NAS approaches that use nested and joint optimization. Consequently, we hope this work will renew interest in the one-shot paradigm, which combines the merits of both via sequential optimization.
2. We present a single path one-shot approach with uniform sampling. It overcomes the drawbacks of existing one-shot approaches. Its simplicity enables a rich search space, including novel designs for channel size and bit width, all addressed in a unified manner. Architecture search is efficient and flexible.

Evolutionary algorithm is used to support real world constraints easily, such as low latency.

Comprehensive ablation experiments and comparison to previous works verify that the proposed approach is state-of-the-art in terms of accuracy, memory consumption, training time, architecture search efficiency and flexibility.

2 Review of NAS Approaches

Without loss of generality, the architecture search space $\mathcal{A}$ is represented by a directed acyclic graph (DAG). A network architecture is a subgraph $a \in \mathcal{A}$, denoted as $\mathcal{N}(a, w)$ with weights w.

Neural architecture search aims to solve two related problems. The first is *weight optimization*,

$$w_a = \underset{w}{\text{argmin}}\ \mathcal{L}_{\text{train}}\left(\mathcal{N}(a, w)\right), \tag{1}$$

where $\mathcal{L}_{\text{train}}(\cdot)$ is the loss function on the training set.

The second is *architecture optimization*. It finds the architecture that is trained on the training set and has the best accuracy on the validation set, as

$$a^* = \underset{a\in \mathcal{A}}{\text{argmax}}\text{ACC}_{\text{val}}\left(\mathcal{N}(a, w_a)\right), \tag{2}$$

where $\text{ACC}_{\text{val}}(\cdot)$ is the accuracy on the validation set.

Early NAS approaches perform the two optimization problems in a *nested* manner [1,32,33,35,36]. Numerous architectures are sampled from $\mathcal{A}$ and trained from scratch as in Eq. (1). Each training is expensive. Only small dataset (e.g., CIFAR 10) and small search space (e.g., a single block) are affordable.

Recent NAS approaches adopt a *weight sharing* strategy [2–4,12,15,23,26,31]. The architecture search space $\mathcal{A}$ is encoded in a *supernet*[1], denoted as $\mathcal{N}(\mathcal{A}, W)$, where W is the weights in the supernet. The supernet is trained once. All architectures inherit their weights directly from W. Thus, they share the weights in their common graph nodes. Fine tuning of an architecture is performed in need, but no training from scratch is incurred. Therefore, architecture search is fast and suitable for large datasets like ImageNet.

Most weight sharing approaches convert the discrete architecture search space into a continuous one [4,12,23,26,31]. Formally, space $\mathcal{A}$ is relaxed to $\mathcal{A}(\theta)$, where θ denotes the continuous parameters that represent the *distribution* of the architectures in the space. Note that the new space subsumes the original one, $\mathcal{A} \subseteq \mathcal{A}(\theta)$. An architecture sampled from $\mathcal{A}(\theta)$ could be invalid in $\mathcal{A}$.

An advantage of the continuous search space is that gradient based methods [4,12,22,23,26,31] is feasible. Both weights and architecture distribution parameters are *jointly* optimized, as

$$(\theta^*, W_{\theta^*}) = \underset{\theta, W}{\text{argmin}}\ \mathcal{L}_{train}(\mathcal{N}(\mathcal{A}(\theta), W)). \tag{3}$$

[1] "Supernet" is used as a general concept in this work. It has different names and implementation in previous approaches.

or perform a bi-level optimization, as

$$\begin{aligned} \theta^* &= \underset{\theta}{\operatorname{argmax}} \ \mathrm{ACC}_{\mathrm{val}} \left(\mathcal{N}(\mathcal{A}(\theta), W_\theta^*) \right) \\ s.t.\ W_\theta^* &= \underset{W}{\operatorname{argmin}} \ \mathcal{L}_{train}(\mathcal{N}(\mathcal{A}(\theta), W)) \end{aligned} \tag{4}$$

After optimization, the best architecture a^* is sampled from $\mathcal{A}(\theta^*)$.

Optimization of Eq. (3) is challenging. *First*, the weights of the graph nodes in the supernet depend on each other and become *deeply coupled* during optimization. For a specific architecture, it inherits certain node weights from W. While these weights are decoupled from the others, it is unclear why they are still effective.

Second, joint optimization of architecture parameter θ and weights W introduces further complexity. Solving Eq. (3) inevitably introduces bias to certain areas in θ and certain nodes in W during the progress of optimization. The bias would leave some nodes in the graph well trained and others poorly trained. With different level of maturity in the weights, different architectures are actually non-comparable. However, their prediction accuracy is used as guidance for sampling in $\mathcal{A}(\theta)$ (e.g., used as reward in policy gradient [4]). This would further mislead the architecture sampling. This problem is in analogy to the "dilemma of exploitation and exploration" problem in reinforcement learning. To alleviate such problems, existing approaches adopt complicated optimization techniques (see Table 1 for a summary).

Task Constraints. Real world tasks usually have additional requirements on a network's memory consumption, FLOPs, latency, energy consumption, etc. These requirements only depends on the architecture a, not on the weights w_a. Thus, they are called *architecture constraints* in this work. A typical constraint is that the network's latency is no more than a preset budget, as

$$\mathrm{Latency}(a^*) \leq \mathrm{Lat}_{\mathrm{max}}. \tag{5}$$

Note that it is challenging to satisfy Eq. (2) and Eq. (5) simultaneously for most previous approaches. Some works augment the loss function $\mathcal{L}_{train}$ in Eq. (3) with *soft* loss terms that consider the architecture latency [4,22,23,26]. However, it is hard, if not impossible, to guarantee a hard constraint like Eq. (5).

3 Our Single Path One-Shot Approach

As analyzed above, the coupling between architecture parameters and weights is problematic. This is caused by joint optimization of both. To alleviate the problem, a natural solution is to *decouple* the supernet training and architecture search in two *sequential* steps. This leads to the so called *one-shot* approaches [2,3].

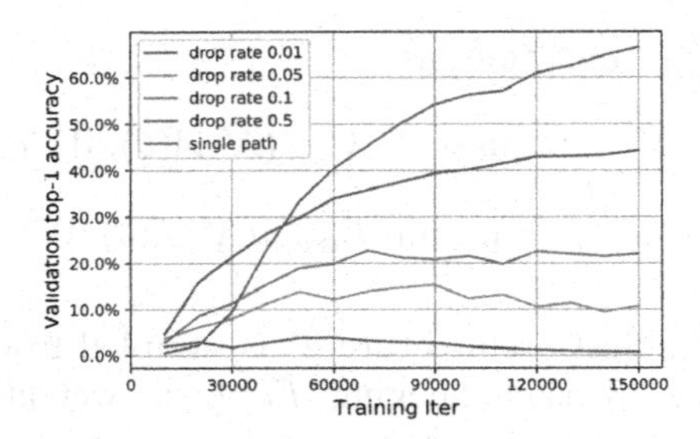

Fig. 1. Comparison of single path strategy and drop path strategy

In general, the two steps are formulated as follows. Firstly, the supernet weight is optimized as

$$W_{\mathcal{A}} = \underset{W}{\operatorname{argmin}}\ \mathcal{L}_{\text{train}}\left(\mathcal{N}(\mathcal{A}, W)\right). \tag{6}$$

Compared to Eq. (3), the continuous parameterization of search space is absent. Only weights are optimized.

Secondly, architecture searched is performed as

$$a^* = \underset{a \in \mathcal{A}}{\operatorname{argmax}}\ \text{ACC}_{\text{val}}\left(\mathcal{N}(a, W_{\mathcal{A}}(a))\right). \tag{7}$$

During search, each sampled architecture a inherits its weights from $W_{\mathcal{A}}$ as $W_{\mathcal{A}}(a)$. The key difference of Eq. (7) from Eq. (1) and (2) is that architecture weights are ready for use. Evaluation of $\text{ACC}_{val}(\cdot)$ only requires inference. Thus, the search is very *efficient*.

The search is also *flexible*. Any adequate search algorithm is feasible. The architecture constraint like Eq. (5) can be exactly satisfied. Search can be repeated many times on the same supernet once trained, using different constraints (e.g., 100 ms latency and 200 ms latency). These properties are absent in previous approaches. These make the one-shot paradigm attractive for real world tasks.

One problem in Sect. 2 still remains. The graph nodes' weights in the supernet training in Eq. (6) are coupled. It is unclear why the inherited weights $W_{\mathcal{A}}(a)$ are still good for an arbitrary architecture a.

The recent one-shot approach [2] attempts to decouple the weights using a "path dropout" strategy. During an SGD step in Eq. (6), each edge in the supernet graph is randomly dropped. The random chance is controlled via a dropout rate parameter. In this way, the co-adaptation of the node weights is reduced during training. Experiments in [2] indicate that the training is very sensitive to the dropout rate parameter. This makes the supernet training hard. A carefully tuned heat-up strategy is used. In our implementation of this work, we also found that the validation accuracy is very sensitive to the dropout rate parameter.

Single Path Supernet and Uniform Sampling. Let us restart to think about the fundamental principle behind the idea of weight sharing. The key to the success of architecture search in Eq. (7) is that, the accuracy of *any* architecture a on a validation set using inherited weight $W_{\mathcal{A}}(a)$ (without extra fine tuning) is highly predictive for the accuracy of a that is fully trained. Ideally, this requires that the weight $W_{\mathcal{A}}(a)$ to approximate the optimal weight w_a as in Eq. (1). The quality of the approximation depends on how well the training loss $\mathcal{L}_{\text{train}}\left(\mathcal{N}(a, W_{\mathcal{A}}(a))\right)$ is minimized. This gives rise to the principle that *the supernet weights $W_{\mathcal{A}}$ should be optimized in a way that all architectures in the search space are optimized simultaneously*. This is expressed as

$$W_{\mathcal{A}} = \underset{W}{\operatorname{argmin}}\ \mathbb{E}_{a \sim \Gamma(\mathcal{A})}\left[\mathcal{L}_{\text{train}}(\mathcal{N}(a, W(a)))\right], \tag{8}$$

where $\Gamma(\mathcal{A})$ is a prior distribution of $a \in \mathcal{A}$. Note that Eq. (8) is an implementation of Eq. (6). In each step of optimization, an architecture a is randomly sampled. Only weights $W(a)$ are activated and updated. So the memory usage is efficient. In this sense, the supernet is no longer a valid network. It behaves as a *stochastic supernet* [22]. This is different from [2].

To reduce the co-adaptation between node weights, we propose a supernet structure that each architecture is a *single path*, as shown in Fig. 3 (a). Compared to the path dropout strategy in [2], the single path strategy is hyperparameter-free. We compared the two strategies within the same search space (as in this work). Note that the original *drop path* in [2] may drop all operations in a block, resulting in a short cut of identity connection. In our implementation, it is forced that one random path is kept in this case since our choice block does not have an identity branch. We randomly select sub network and evaluate its validation accuracy during the training stage. Results in Fig. 1 show that drop rate parameters matters a lot. Different drop rates make supernet achieve different validation accuracies. Our single path strategy corresponds to using drop rate 1. It works the best because our single path strategy can decouple the weights of different operations. The Fig. 1 verifies the benefit of weight decoupling.

The prior distribution $\Gamma(\mathcal{A})$ is important. In this work, we empirically find that *uniform sampling* is good. This is not much of a surprise. A concurrent work [10] also finds that purely random search based on stochastic supernet is competitive on CIFAR-10. We also experimented with a variant that samples the architectures uniformly according to their constraints, named uniform constraint sampling. Specifically, we randomly choose a range, and then sample the architecture repeatedly until the FLOPs of sampled architecture falls in the range. This is because a real task usually expects to find multiple architectures satisfying different constraints. In this work, we find the uniform constraint sampling method is slightly better. So we use it by default in this paper.

We note that sampling a path according to architecture distribution during optimization is already used in previous weight sharing approaches [4,6,20, 22,28,31]. The difference is that, the distribution $\Gamma(\mathcal{A})$ is a *fixed* prior during our training (Eq. (8)), while it is *learnable and updated* (Eq. (3)) in previous

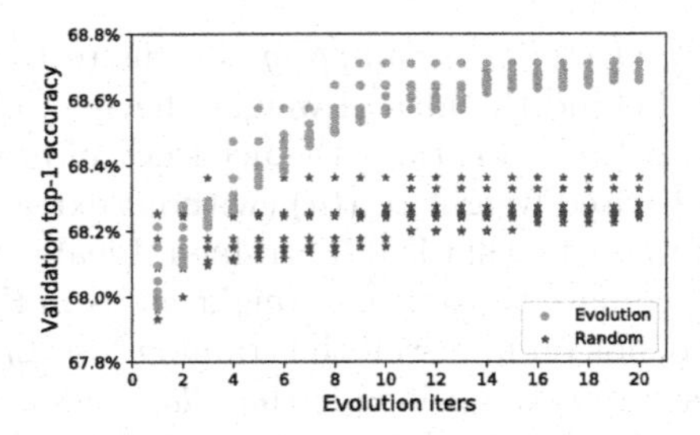

Fig. 2. Evolutionary vs. random architecture search

approaches (e.g. RL [15], policy gradient [4,22], Gumbel Softmax [23,26], APG [31]). As analyzed in Sect. 2, the latter makes the supernet weights and architecture parameters highly correlated and optimization difficult. There is another concurrent work [10] that also proposed to use random sampling of paths in One-Shot model, and performed random search to find the superior architecture. This paper [10] achieved competitive results to several SOTA NAS approaches on CIFAR-10, but didn't verify the method on large dataset ImageNet. It didn't prove the effectiveness of single path sampling compared to the "path dropout" strategy and analyze the correlation of the supernet performance and the final evaluation performance. These questions will be answered in our work, and our experiments also show that random search is not good enough to find superior architecture from the large search space.

Comprehensive experiments in Sect. 4 show that our approach achieves better results than the SOTA methods. Note that there is no such theoretical guarantee that using a fixed prior distribution is *inherently* better than optimizing the distribution during training. Our better result likely indicates that the joint optimization in Eq. (3) is too difficult for the existing optimization techniques.

Supernet Architecture and Novel Choice Block Design. Choice blocks are used to build a *stochastic* architecture. Figure 3 (a) illustrates an example case. A choice block consists of multiple architecture choices. For our single path supernet, each choice block only has one choice invoked at the same time. A path is obtained by sampling all the choice blocks.

The simplicity of our approach enables us to define different types of choice blocks to search various architecture variables. Specifically, we propose two novel choice blocks to support complex search spaces.

Channel Number Search. We propose a new choice block based on weight sharing, as shown in Fig. 3 (b). The main idea is to preallocate a weight tensor with maximum number of channels, and the system randomly selects the channel number and slices out the corresponding subtensor for convolution. With the weight sharing strategy, we found that the supernet can converge quickly.

In detail, assume the dimensions of preallocated weights are (max_c_out, max_c_in, ksize). For each batch in supernet training, the number of current output channels *c_out* is randomly sampled. Then, we slice out the weights for

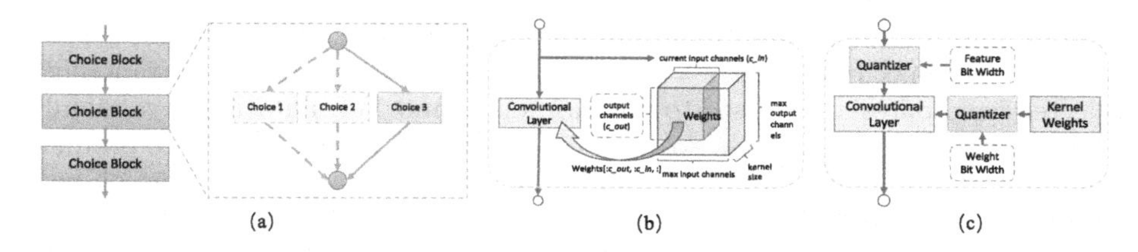

Fig. 3. *Choice blocks* for (a) our single path supernet (b) channel number search (c) mixed-precision quantization search

current batch with the form Weights[: c_out, : c_in, :], which is used to produce the output. The optimal number of channels is determined in the search step.

Mixed-Precision Quantization Search. In this work, We design a novel *choice block* to search the bit widths of the weights and feature maps, as shown in Fig. 3 (c). We also combine the *channel search space* discussed earlier to our *mixed-precision quantization search space*. During supernet training, for each choice block feature bit width and weight bit width are randomly sampled. They are determined in the evolutionary step. See Sect. 4 for details.

Evolutionary Architecture Search. For architecture search in Eq. (7), previous one-shot works [2,3] use random search. This is not effective for a large search space. This work uses an evolutionary algorithm. Note that evolutionary search in NAS is used in [16], but it is costly as each architecture is trained from scratch. In our search, each architecture only performs inference. This is very efficient.

Algorithm 1: Evolutionary Architecture Search

1 **Input**: *supernet weights* $\mathrm{W}_{\mathcal{A}}$*, population size* P*, architecture constraints* $\mathcal{C}$*, max iteration* $\mathcal{T}$*, validation dataset* D_{val}
2 **Output**: *the architecture with highest validation accuracy under architecture constraints*
3 $\mathrm{P}_0 := Initialize\_population(P, \mathcal{C})$; $\mathrm{Topk} := \emptyset$;
4 $n := P/2$; Crossover number
5 $m := P/2$; Mutation number
6 $prob := 0.1$; Mutation probability
7 **for** $i = 1 : \mathcal{T}$ **do**
8 $\quad \mathrm{ACC}_{i-1} := Inference(\mathrm{W}_{\mathcal{A}}, \mathrm{D}_{val}, \mathrm{P}_{i-1})$;
9 $\quad \mathrm{Topk} := Update\_Topk(\mathrm{Topk}, \mathrm{P}_{i-1}, \mathrm{ACC}_{i-1})$;
10 $\quad \mathrm{P}_{crossover} := Crossover(\mathrm{Topk}, n, \mathcal{C})$;
11 $\quad \mathrm{P}_{mutation} := Mutation(\mathrm{Topk}, m, prob, \mathcal{C})$;
12 $\quad \mathrm{P}_i := \mathrm{P}_{crossover} \cup \mathrm{P}_{mutation}$;
13 **end**
14 Return the architecture with highest accuracy in Topk;

The algorithm is elaborated in Algorithm 1. For all experiments, population size $P = 50$, max iterations $\mathcal{T} = 20$ and $k = 10$. For crossover, two randomly selected candidates are crossed to produce a new one. For mutation, a randomly selected candidate mutates its every choice block with probability 0.1 to produce a new candidate. Crossover and mutation are repeated to generate enough new candidates that meet the given architecture constraints. Before the inference of an architecture, the statistics of all the *Batch Normalization (BN)* [9] operations are recalculated on a random subset of training data (20000 images on ImageNet). It takes a few seconds. This is because the BN statistics from the supernet are usually not applicable to the candidate nets. This is also referred in [2].

Figure 2 plots the validation accuracy over generations, using both evolutionary and random search methods. It is clear that evolutionary search is more effective. Experiment details are in Sect. 4.

The evolutionary algorithm is flexible in dealing with different constraints in Eq. (5), because the mutation and crossover processes can be directly controlled to generate proper candidates to satisfy the constraints. Previous RL-based [21] and gradient-based [4,22,23] methods design tricky rewards or loss functions to deal with such constraints. For example, [23] uses a loss function $\text{CE}(a, w_a) \cdot \alpha \log(\text{LAT}(a))^{\beta}$ to balance the accuracy and the latency. It is hard to tune the hyper parameter β to satisfy a hard constraint like Eq. (5).

Summary. The combination of single path supernet, uniform sampling training strategy, evolutionary architecture search, and rich search space design makes our approach simple, efficient and flexible. Table 1 performs a comprehensive comparison of our approach against previous weight sharing approaches on various aspects. Ours is the easiest to train, occupies the smallest memory, best satisfies the architecture (latency) constraint, and easily supports large datasets. Extensive results in Sect. 4 verify that our approach is the state-of-the-art.

4 Experiment Results

Dataset. All experiments are performed on *ImageNet* [17]. We randomly split the original training set into two parts: 50000 images are for validation (50 images for each class exactly) and the rest as the training set. The original validation set is used for testing, on which all the evaluation results are reported, following [4].

Training. We use the same settings (including data augmentation, learning rate schedule, etc.) as [14] for supernet and final architecture training. Batch size is 1024. Supernet is trained for 120 epochs and the best architecture for 240 epochs (300000 iterations) by using 8 *NVIDIA GTX 1080Ti* GPUs.

Search Space: Building Blocks. First, we evaluate our method on the task of *building block selection*, i.e. to find the optimal combination of building blocks under a certain complexity constraint. Our basic building block design is inspired

Table 1. Overview and comparison of SOTA *weight sharing* approaches. Ours is the easiest to train, occupies the smallest memory, best satisfy the architecture (latency) constraint, and easily supports the large dataset. Note that those approaches belonging to the joint optimization category (Eq. (3)) have "Supernet optimization" and "Architecture search" columns merged

Approach	Supernet optimization	Architecture search	Hyper parameters in supernet Training	Memory consumption in supernet training	How to satisfy constraint	Experiment on ImageNet
ENAS[15]	Alternative RL and fine tuning		Short-time fine tuning setting	Single path +RL system	None	No
BSN[22]	Stochastic super networks + policy gradient		Weight of cost penalty	Single path	Soft constraint in training. Not guaranteed	No
DARTS[12]	Gradient-based path dropout		Path dropout rate. Weight of auxiliary loss	Whole supernet	None	Transfer
Proxyless[4]	Stochastic relaxation of the discrete search + policy gradient		Scaling factor of latency loss	Two paths	Soft constraint in training. Not guaranteed.	Yes
FBNet[23]	Stochastic relaxation of the discrete search to differentiable optimization via Gumbel softmax		Temperature parameter in Gumbel softmax. Coefficient in constraint loss	Whole supernet	Soft constraint in training. Not guaranteed.	Yes
SNAS[26]	Same as FBNet		Same as FBNet	Whole supernet	Soft constraint in training. Not guaranteed.	Transfer
SMASH[3]	Hypernet	Random	None	Hypernet+Single Path	None	No
One-Shot[2]	Path dropout	Random	Drop rate	Whole supernet	Not investigated	Yes
Ours	Uniform path sampling	Evolution	None	Single path	Guaranteed in searching. Support multiple constraints.	Yes

by a state-of-the-art manually-designed network – *ShuffleNet v2* [14]. Table 2 shows the overall architecture of the supernet. The "stride" column represents the stride of the first block in each repeated group. There are 20 *choice blocks* in total. Each choice block has 4 candidates, namely "choice_3", "choice_5", "choice_7" and "choice_x" respectively. They differ in kernel sizes and the number of depthwise convolutions. The size of the search space is 4^{20}.

Table 2. Supernet architecture. *CB* - choice block. *GAP* - global average pooling

Input shape	Block	Channels	Repeat	Stride
$224^2 \times 3$	3×3 conv	16	1	2
$112^2 \times 16$	CB	64	4	2
$56^2 \times 64$	CB	160	4	2
$28^2 \times 160$	CB	320	8	2
$14^2 \times 320$	CB	640	4	2
$7^2 \times 640$	1×1 conv	1024	1	1
$7^2 \times 1024$	GAP	-	1	–
1024	fc	1000	1	–

Table 3. Results of building block search. *SPS* – single path supernet

Model	FLOPs	Top-1 acc(%)
All choice_3	324M	73.4
All choice_5	321M	73.5
All choice_7	327M	73.6
All choice_x	326M	73.5
Random select (5 times)	~320M	~73.7
SPS + random search	323M	73.8
Ours (fully-equipped)	319M	**74.3**

We use FLOPs $\leq 330M$ as the complexity constraint, as the FLOPs of a plenty of previous networks lies in [300, 330], including manually-designed networks [8,14,18,30] and those obtained in NAS [4,21,23].

Table 3 shows the results. For comparison, we set up a series of baselines as follows: 1) select a certain block choice only (denoted by "all choice_*" entries); note that different choices have different FLOPs, thus we adjust the channels to meet the constraint. 2) Randomly select some candidates from the search space. 3) Replace our evolutionary architecture optimization with random search used in [2,3]. Results show that random search equipped with our single path supernet finds an architecture only slightly better that random select (73.8 vs. 73.7). It does no mean that our single path supernet is less effective. This is because the random search is too naive to pick good candidates from the large search space. Using evolutionary search, our approach finds out an architecture that achieves superior accuracy (74.3) over all the baselines.

Search Space: Channels. Based on our novel choice block for channel number search, we first evaluate channel search on the baseline structure "all choice_3" (refer to Table 3): for each building block, we search the number of "mid-channels" (output channels of the first 1×1 conv in each building block) varying from 0.2x to 1.6x (with stride 0.2), where "k-x" means k times the number of default channels. Same as building block search, we set the complexity constraint FLOPs $\leq 330M$. Table 4 (first part) shows the result. Our channel search method has higher accuracy (73.9) than the baselines.

Table 4. Results of channel search. * Performances are reported in the form "x (y)", where "x" means the accuracy retrained by us and "y" means accuracy reported by the original paper

Model	FLOPs/Params	Top-1 acc(%)
All choice_3	324M/3.1M	73.4
Rand sel. channels (5 times)	~323M/3.2M	~73.1
Choice_3 + channel search	329M/3.4M	**73.9**
Rand sel. blocks + channels	~325M/3.2M	~73.4
Block search	319M/3.3M	74.3
Block search + channel search	328M/3.4M	**74.7**
MobileNet V1 (0.75x) [8]	325M/2.6M	68.4
MobileNet V2 (1.0x) [18]	300M/3.4M	72.0
ShuffleNet V2 (1.5x) [14]	299M/3.5M	72.6
NASNET-A [36]	564M/5.3M	74.0
PNASNET [11]	588M/5.1M	74.2
MnasNet [21]	317M/4.2M	74.0
DARTS [12]	595M/4.7M	73.1
Proxyless-R (mobile)* [4]	320M/4.0M	74.2 (74.6)
FBNet-B* [23]	295M/4.5M	74.1 (74.1)

To further boost the accuracy, we search building blocks and channels jointly. There are two alternatives: 1) running channel search on the best building block search result; or 2) searching on the combined search space directly. Our experiments show that the first pipeline is slightly better. As shown in Table 4, searching in the joint space achieves the best accuracy (**74.7%** acc.), surpassing the previous state-of-the-art manually-designed [14,18] and automatically-searched models [4,11,12,21,23,36] under complexity of ~300M FLOPs.

Comparison With State-of-the-Arts. Results in Table 4 shows our method is superior. Nevertheless, the comparisons could be unfair because different search spaces and training methods are used in previous works [4]. To make *direct* comparisons, we benchmark our approach to the *same* search space of [4,23]. In addition, we retrain the searched models reported in [4,23] under the same settings to guarantee the fair comparison.

The search space and supernet architecture in *ProxylessNAS* [4] is inspired by *MobileNet v2* [18] and *MnasNet* [21]. It contains 21 *choice blocks*; each choice block has 7 choices (6 different building blocks and one skip layer). The size of the search space is 7^{21}. *FBNet* [23] also uses a similar search space.

Table 5 reports the accuracy and complexities (FLOPs and latency on our device) of 5 models searched by [4,23], as the baselines. Then, for each baseline, our search method runs under the constraints of same FLOPs or same latency,

Table 5. Compared with state-of-the-art *NAS* methods [4,23] *using the same search space.* The latency is evaluated on a single *NVIDIA Titan XP* GPU, with *batchsize* = 32. Accuracy numbers in the brackets are reported by the original papers; others are trained by us. All our architectures are searched from the **same** supernet via evolutionary architecture optimization

Baseline network	FLOPs/Params	Latency	Top-1 acc(%) baseline	Top-1 acc(%) (same FLOPs)	Top-1 acc(%) (same latency)
FBNet-A [23]	249M/4.3M	13 ms	73.0 (73.0)	**73.2**	**73.3**
FBNet-B [23]	295M/4.5M	17 ms	74.1 (74.1)	**74.2**	**74.8**
FBNet-C [23]	375M/5.5M	19 ms	74.9 (74.9)	**75.0**	**75.1**
Proxyless-R(mobile) [4]	320M/4.0M	17 ms	74.2 (74.6)	**74.5**	**74.8**
Proxyless(GPU) [4]	465M/5.3M	22 ms	74.7 (75.1)	**74.8**	**75.3**

respectively. Results shows that for all the cases our method achieves comparable or higher accuracy than the counterpart baselines.

Furthermore, it is worth noting that our architectures under different constraints in Table 5 are searched on the *same* supernet, justifying the flexibility and efficiency of our approach to deal with different complexity constraints: supernet is trained once and searched multiple times. In contrast, previous methods [4,23] have to train multiple supernets under various constraints. According to Table 7, searching is much cheaper than supernet training.

Application: Mixed-Precision Quantization. We evaluate our method on *ResNet-18* and *ResNet-34* as common practice in previous quantization works (e.g. [5, 13,24,29,34]). Following [5,24,34], we only search and quantize the *res-blocks*, excluding the first convolutional layer and the last fully-connected layer. Choices of weight and feature bit widths include $\{(1,2),(2,2),(1,4),(2,4),(3,4),(4,4)\}$ in the search space. As for channel search, we search the number of "bottleneck channels" (i.e. the output channels of the first convolutional layer in each residual block) in $\{0.5x, 1.0x, 1.5x\}$, where "k-x" means k times the number of original channels. The size of the search space is $(3\times 6)^N = 18^N$, where N is the number of choice blocks ($N = 8$ for ResNet-18 and $N = 16$ for ResNet-34). Note that for each building block we use the same bit widths for the two convolutions. We use *PACT* [5] as the quantization algorithm.

Table 6 reports the results. The baselines are denoted as kWkA ($k = 2, 3, 4$), which means uniform quantization of weights and activations with k-bits. Then, our search method runs under the constraints of the corresponding BitOps. We also compare with a recent mixed-precision quantization search approach [24]. Results shows that our method achieves superior accuracy in most cases. Also note that all our results for ResNet-18 and ResNet-34 are searched on the **same** supernet. This is very efficient.

Search Cost Analysis. The search cost is a matter of concern in NAS methods. So we analyze the search cost of our method and previous methods [4,23] (reimplemented by us). We use the search space of our *building blocks* to measure the memory cost of training supernet and overall time cost. All the supernets are

Table 6. Results of mixed-precision quantization search. "kWkA" means k-bit quantization for all the weights and activations

Method	BitOPs	Top1-acc(%)	Method	BitoPs	Top1-acc(%)
ResNet-18	Float point	70.9	ResNet-34	Float point	75.0
2W2A	6.32G	65.6	2W2A	13.21G	70.8
Ours	**6.21G**	**66.4**	Ours	**13.11G**	**71.5**
3W3A	14.21G	68.3	3W3A	29.72G	72.5
DNAS [24]	15.62G	68.7	DNAS [24]	38.64G	73.2
Ours	**13.49G**	**69.4**	Ours	**28.78G**	**73.9**
4W4A	25.27G	69.3	4W4A	52.83G	73.5
DNAS [24]	25.70G	**70.6**	DNAS [24]	57.31G	74.0
Ours	**24.31G**	70.5	Ours	**51.92G**	**74.6**

trained for 150000 iterations with a batch size of 256. All models are trained with 8 GPUs. Table 7 shows that our approach clearly uses less memory than other two methods because of the single path supernet. And our approach is much more efficient overall although we have an extra search step that costs less than 1 GPU day. Note Table 7 only compares a single run. In practice, our approach is more advantageous and more convenient to use when multiple searches are needed. As summarized in Table 1, it guarantees to find out the architecture satisfying constraints within one search. Repeated search is easily supported.

Correlation Analysis. Recently, the effectiveness of many neural architecture search methods based on weight sharing is questioned because of lacking fair comparison on the same search space and adequate analysis on the correlation between the supernet performance and the stand-alone model performance. Some papers [19,25,27] even show that several the state-of-the-art NAS methods perform similarly to the random search. In this work, the fair comparison on the same search space has been showed in Table 5, so we further provider adequate correlation analysis in this part to evaluate the effectiveness of our method.

Table 7. Search Cost. *Gds* - GPU days

Method	Proxyless	FBNet	Ours
Memory cost (8 GPUs in total)	37G	63G	24G
Training time	15 Gds	20 Gds	12 Gds
Search time	0	0	<1 Gds
Retrain time	16 Gds	16 Gds	16 Gds
Total time	31 Gds	36 Gds	29 Gds

Table 8. Correlation in different search spaces

Dataset	Original	Reduce-1	Reduce-2	Reduce-3
CIFAR-10	0.55	0.55	0.58	**0.64**
CIFAR-100	0.56	0.54	0.53	**0.59**
ImageNet-16-120	0.54	0.42	**0.55**	0.53

Correlation analysis requires to achieve the performances of a large number of architectures, but training lots of architectures from scratch is very time-consuming, which also requires a large number of GPU resources, so we use the NAS-Bench-201 [7] to analyze our method. NAS-Bench-201 is a cell-based search space which includes 15,625 architectures in total. It provides the performance of each architecture on CIFAR-10, CIFAR-100, and ImageNet-16-120. So the results on it will be more credible and comparable.

We apply our method on different search spaces and different datasets to verify the effectiveness adequately. The original search space of NAS-Bench-201 consists of 5 possible operations: zeroize, skip connection, 1-by-1 convolution, 3-by-3 convolution, and 3-by-3 average pooling. Based on it, we further design several reduced search spaces, named Reduce-1, Reduce-2, Reduce-3, by deleting some operations. In detail, we delete 1-by-1 convolution and 3-by-3 average pooling respectively from original search space to produce Reduce-1 and Reduce-2 search spaces, and delete both to produce Reduce-3 search space.

As Table 8 shows, our method performs better than random search on different search spaces and different datasets, since the Kendall Tau τ metric of random search should be 0. So the performances of architectures predicted by supernet can reflect the real ranking of architectures to a certain degree. However, the results in Table 8 also reveals a limitation of our method that the predicted ranking of our supernet is partially correlated, but not perfectly correlated to the real ranking. So our method can not guarantee to find the real best architecture in the search space, but is able to find some superior architectures around the best. And we think that the correlation of supernet depends on search space. The simpler search space is, the higher correlation will be achieved.

5 Conclusion

In this paper, we revisit the one-shot NAS paradigm and analyze the drawbacks of previous method. Then we propose a single path one-shot approach which is more simple and effective. Experiments show that our method can achieve better results on several different search spaces. We also analyze the search cost and correlation of our methods. Our method is more efficient can achieve significant correlation on different search spaces. However, there is a limitation in our method that the predicted ranking of our supernet is partially correlated, but not perfectly correlated to the real ranking. And we think that it depends on search space. The simpler search space is, the higher correlation will be achieved.

Acknowledgement. This work is supported by The National Key Research and Development Program of China (No. 2017YFA0700800) and Beijing Academy of Artificial Intelligence (BAAI).

References

1. Baker, B., Gupta, O., Naik, N., Raskar, R.: Designing neural network architectures using reinforcement learning. arXiv preprint arXiv:1611.02167 (2016)
2. Bender, G., Kindermans, P.J., Zoph, B., Vasudevan, V., Le, Q.: Understanding and simplifying one-shot architecture search. In: International Conference on Machine Learning, pp. 549–558 (2018)
3. Brock, A., Lim, T., Ritchie, J.M., Weston, N.: SMASH: one-shot model architecture search through hypernetworks. arXiv preprint arXiv:1708.05344 (2017)
4. Cai, H., Zhu, L., Han, S.: ProxylessNAS: direct neural architecture search on target task and hardware. arXiv preprint arXiv:1812.00332 (2018)
5. Choi, J., Wang, Z., Venkataramani, S., Chuang, P.I.J., Srinivasan, V., Gopalakrishnan, K.: PACT: parameterized clipping activation for quantized neural networks. arXiv preprint arXiv:1805.06085 (2018)
6. Dong, X., Yang, Y.: Searching for a robust neural architecture in four GPU hours. In: Proceedings of the IEEE Conference on Computer Vision and Pattern Recognition, pp. 1761–1770 (2019)
7. Dong, X., Yang, Y.: NAS-Bench-102: extending the scope of reproducible neural architecture search. arXiv preprint arXiv:2001.00326 (2020)
8. Howard, A.G., et al.: MobileNets: efficient convolutional neural networks for mobile vision applications. arXiv preprint arXiv:1704.04861 (2017)
9. Ioffe, S., Szegedy, C.: Batch normalization: accelerating deep network training by reducing internal covariate shift. arXiv preprint arXiv:1502.03167 (2015)
10. Li, L., Talwalkar, A.: Random search and reproducibility for neural architecture search. arXiv preprint arXiv:1902.07638 (2019)
11. Liu, C., et al.: Progressive neural architecture search. In: Proceedings of the European Conference on Computer Vision (ECCV), pp. 19–34 (2018)
12. Liu, H., Simonyan, K., Yang, Y.: Darts: differentiable architecture search. arXiv preprint arXiv:1806.09055 (2018)
13. Liu, Z., Wu, B., Luo, W., Yang, X., Liu, W., Cheng, K.T.: Bi-Real Net: enhancing the performance of 1-bit CNNs with improved representational capability and advanced training algorithm. In: Proceedings of the European Conference on Computer Vision (ECCV), pp. 722–737 (2018)
14. Ma, N., Zhang, X., Zheng, H.T., Sun, J.: ShuffleNet v2: practical guidelines for efficient CNN architecture design. In: Proceedings of the European Conference on Computer Vision (ECCV), pp. 116–131 (2018)
15. Pham, H., Guan, M.Y., Zoph, B., Le, Q.V., Dean, J.: Efficient neural architecture search via parameter sharing. arXiv preprint arXiv:1802.03268 (2018)
16. Real, E., Aggarwal, A., Huang, Y., Le, Q.V.: Regularized evolution for image classifier architecture search. arXiv preprint arXiv:1802.01548 (2018)
17. Russakovsky, O., et al.: Imagenet large scale visual recognition challenge. Int. J. Comput. Vis. **115**(3), 211–252 (2015)
18. Sandler, M., Howard, A., Zhu, M., Zhmoginov, A., Chen, L.C.: MobileNetV 2: inverted residuals and linear bottlenecks. In: Proceedings of the IEEE Conference on Computer Vision and Pattern Recognition, pp. 4510–4520 (2018)

19. Sciuto, C., Yu, K., Jaggi, M., Musat, C., Salzmann, M.: Evaluating the search phase of neural architecture search. arXiv preprint arXiv:1902.08142 (2019)
20. Stamoulis, D., et al.: Single-path NAS: designing hardware-efficient ConvNets in less than 4 hours. arXiv preprint arXiv:1904.02877 (2019)
21. Tan, M., Chen, B., Pang, R., Vasudevan, V., Le, Q.V.: MnasNet: platform-aware neural architecture search for mobile. arXiv preprint arXiv:1807.11626 (2018)
22. Véniat, T., Denoyer, L.: Learning time/memory-efficient deep architectures with budgeted super networks. In: Proceedings of the IEEE Conference on Computer Vision and Pattern Recognition, pp. 3492–3500 (2018)
23. Wu, B., et al.: FBNet: hardware-aware efficient ConvNet design via differentiable neural architecture search. arXiv preprint arXiv:1812.03443 (2018)
24. Wu, B., Wang, Y., Zhang, P., Tian, Y., Vajda, P., Keutzer, K.: Mixed precision quantization of convnets via differentiable neural architecture search. arXiv preprint arXiv:1812.00090 (2018)
25. Xie, S., Kirillov, A., Girshick, R., He, K.: Exploring randomly wired neural networks for image recognition. In: Proceedings of the IEEE International Conference on Computer Vision, pp. 1284–1293 (2019)
26. Xie, S., Zheng, H., Liu, C., Lin, L.: SNAS: stochastic neural architecture search. arXiv preprint arXiv:1812.09926 (2018)
27. Yang, A., Esperança, P.M., Carlucci, F.M.: NAS evaluation is frustratingly hard. arXiv preprint arXiv:1912.12522 (2019)
28. Yao, Q., Xu, J., Tu, W.W., Zhu, Z.: Differentiable neural architecture search via proximal iterations. arXiv preprint arXiv:1905.13577 (2019)
29. Zhang, D., Yang, J., Ye, D., Hua, G.: LQ-Nets: learned quantization for highly accurate and compact deep neural networks. In: Proceedings of the European Conference on Computer Vision (ECCV), pp. 365–382 (2018)
30. Zhang, X., Zhou, X., Lin, M., Sun, J.: ShuffleNet: an extremely efficient convolutional neural network for mobile devices. In: Proceedings of the IEEE Conference on Computer Vision and Pattern Recognition, pp. 6848–6856 (2018)
31. Zhang, X., Huang, Z., Wang, N.: You only search once: single shot neural architecture search via direct sparse optimization. arXiv preprint arXiv:1811.01567 (2018)
32. Zhong, Z., Yan, J., Wu, W., Shao, J., Liu, C.L.: Practical block-wise neural network architecture generation. In: Proceedings of the IEEE Conference on Computer Vision and Pattern Recognition, pp. 2423–2432 (2018)
33. Zhong, Z., et al.: BlockQNN: efficient block-wise neural network architecture generation. arXiv preprint arXiv:1808.05584 (2018)
34. Zhou, S., Wu, Y., Ni, Z., Zhou, X., Wen, H., Zou, Y.: DoReFa-Net: training low bitwidth convolutional neural networks with low bitwidth gradients. arXiv preprint arXiv:1606.06160 (2016)
35. Zoph, B., Le, Q.V.: Neural architecture search with reinforcement learning. arXiv preprint arXiv:1611.01578 (2016)
36. Zoph, B., Vasudevan, V., Shlens, J., Le, Q.V.: Learning transferable architectures for scalable image recognition. In: Proceedings of the IEEE Conference on Computer Vision and Pattern Recognition, pp. 8697–8710 (2018)

Learning to Generate Novel Domains for Domain Generalization

Kaiyang Zhou[1(✉)], Yongxin Yang[1], Timothy Hospedales[2,3], and Tao Xiang[1,3]

[1] University of Surrey, Guildford, UK
{k.zhou,yongxin.yang,t.xiang}@surrey.ac.uk
[2] University of Edinburgh, Edinburgh, UK
t.hospedales@ed.ac.uk
[3] Samsung AI Center, Cambridge, UK

Abstract. This paper focuses on domain generalization (DG), the task of learning from multiple source domains a model that generalizes well to unseen domains. A main challenge for DG is that the available source domains often exhibit limited diversity, hampering the model's ability to learn to generalize. We therefore employ a data generator to synthesize data from pseudo-novel domains to augment the source domains. This explicitly increases the diversity of available training domains and leads to a more generalizable model. To train the generator, we model the distribution divergence between source and synthesized pseudo-novel domains using optimal transport, and maximize the divergence. To ensure that semantics are preserved in the synthesized data, we further impose cycle-consistency and classification losses on the generator. Our method, L2A-OT (Learning to Augment by Optimal Transport) outperforms current state-of-the-art DG methods on four benchmark datasets.

1 Introduction

Humans effortlessly generalize prior knowledge to novel scenarios, a capability that machines still struggle to reproduce. Typically, machine-learning models perform poorly when deployed on test data with a different data distribution than the training data, which is known as the domain shift problem [35]. One line of research towards alleviating the domain shift problem is unsupervised domain adaptation (UDA), which exploits unlabeled target domain data for model adaptation [12,20,33,40,44,53]. Although UDA methods avoid costly data annotation processes from target domains, data collection and per-domain model updates are still required. Meanwhile, UDA's assumption that target data can be collected in advance is not always met in practice [10,37]. This motivates another line of research, namely domain generalization (DG) [2,5,10,15,16,37], which is the main focus in this paper.

Electronic supplementary material The online version of this chapter (https://doi.org/10.1007/978-3-030-58517-4_33) contains supplementary material, which is available to authorized users.

A. Vedaldi et al. (Eds.): ECCV 2020, LNCS 12361, pp. 561–578, 2020.
https://doi.org/10.1007/978-3-030-58517-4_33

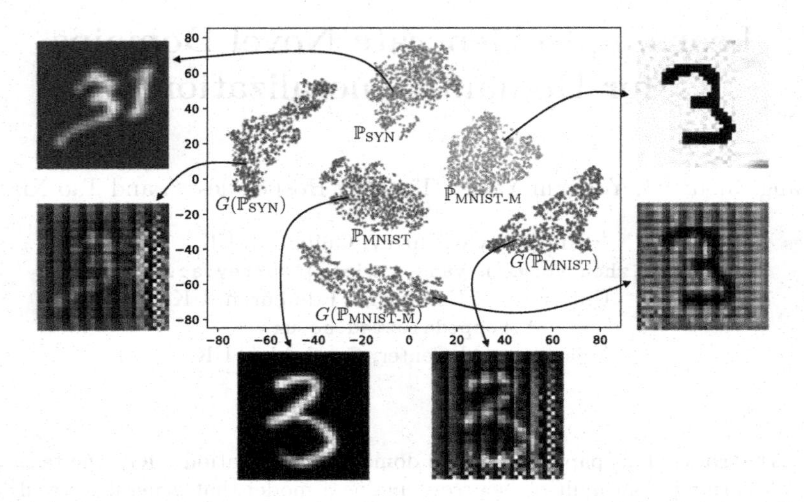

Fig. 1. Motivation of our approach. We improve generalization by increasing the diversity of training domains by learning a generator network G to map images of a source distribution, e.g., $\mathbb{P}_{\text{MNIST}}$, to a novel distribution, i.e. $G(\mathbb{P}_{\text{MNIST}})$. We then combine both source and novel domains for model learning.

DG methods aim to learn models capable of good direct generalization to unseen target domains without data collection or model updating [37]. They usually, but not always [52], leverage multiple source domains to train a generalizable model. Most existing DG methods focus on aligning available source domains [11,15,16,28,29,36], which is mainly inspired by UDA methods that seek to minimize the divergence between source data and unlabeled target data [13,50]. As proved in [4], minimizing the domain divergence can lead to a smaller target error in the UDA setting. However, since DG methods focus on aligning source domains and do not have access to the target data, this theoretical proof does not apply to the DG setting. Recently, meta-learning has been exploited for DG where the key idea is to simulate domain shift by splitting the training data into meta-train and meta-test sets with non-overlapping domains [2,10,26,27,31]. During learning, models are optimized on the meta-train domains in a way that the error is reduced on the meta-test domains. Nevertheless, similar to the alignment-based methods, meta-learning optimizes for reducing the domain gap among source domains, and thus still has the risk of overfitting to seen domains.

In this paper, we address DG from a different perspective, i.e., the most straightforward way to improve model generalization is increasing the diversity of available source domains [49] (see Fig. 1). To this end, we propose *L2A-OT* (*Learning to Augment by Optimal Transport*). The core idea is to learn a conditional generator network that maps source domain images to pseudo-novel domains, and then combine both source and pseudo-novel domain images for training the actual task model. To train the generator, we *maximize* the distance between source domains and the generated pseudo-novel domains, as measured by optimal transport (OT) [41]. This leads to the generated images having a very different distribution from the source domains (Fig. 1). However, this objective alone does not guarantee that the semantic content of the generated images is preserved. Therefore, we further impose two losses on the generator, namely a cycle-consistency loss [64] and a classification loss, for maintaining the structural and semantic consistency respectively.

Our contributions are as follows. **(1)** For the first time, DG is tackled from a perspective of pseudo-novel domain synthesis. **(2)** A novel image generator is formulated which differs from existing generators in the objective (synthesizing pseudo-novel domain images vs. natural photo images). More importantly it has a unique OT-based formulation of objective functions that allow the generator to explore novel domain space and generate diverse data with distributions different from any of the original source domains. We evaluate L2A-OT on three homogeneous DG benchmark datasets[1] including digit recognition [12,24,38], PACS [25] and Office-Home [51] and a heterogeneous DG task in the form of cross-domain person re-identification (re-ID) [22,32,58,59,61]. The results show that L2A-OT surpasses the current state-of-the-art on all datasets.

2 Related Work

Domain Generalization. Many DG methods are based on the idea of domain alignment popularized from the UDA literature [12], with a goal to learn a domain-invariant representation by minimizing the domain discrepancy between sources [11,15,16,28,29,36]. As mentioned earlier, aligning domain distributions is mainly motivated by the theory [4] developed for UDA, which does not apply to DG due to the absence of target data. Therefore, the models learned with domain alignment risk overfitting to source domains and as a result generalize poorly to unseen domains. In recent years, meta-learning [21] has seen increasing interest for DG where the objective is to expose a model to domain shift during training. This can be achieved by dividing source domains into meta-train and meta-test sets without overlapping, and training a model on the meta-train set such that the error on the meta-test set is reduced [2,10,26]. Similar to domain alignment methods, meta-learning methods still risk overfitting since the training data remains unchanged. Moreover, these methods work at feature-level, which is difficult for diagnosis and lacks visual interpretation.

Most related to our work are data augmentation methods, especially those based on adversarial gradients [47,52]. For instance, [47] proposed CrossGrad to perturb input images with adversarial gradients generated by a domain classifier. Different from adversarial gradient-based methods which only produce imperceptible and simple pixel-wise effects (due to the nature of adversarial attack [48]), our approach *learns* a full CNN generator to map source images to

[1] Following [31], homogeneous DG shares the same label space between training and test data while heterogeneous DG has disjoint label space.

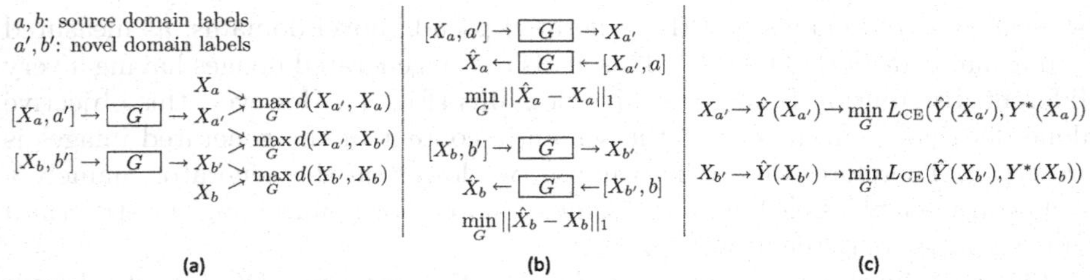

a, b: source domain labels
a', b': novel domain labels

$[X_a, a'] \rightarrow$ G $\rightarrow X_{a'}$, $\max_G d(X_{a'}, X_a)$, $\max_G d(X_{a'}, X_{b'})$
$[X_b, b'] \rightarrow$ G $\rightarrow X_{b'}$, $\max_G d(X_{b'}, X_b)$

(a)

$[X_a, a'] \rightarrow$ G $\rightarrow X_{a'}$
$\hat{X}_a \leftarrow$ G $\leftarrow [X_{a'}, a]$
$\min_G ||\hat{X}_a - X_a||_1$

$[X_b, b'] \rightarrow$ G $\rightarrow X_{b'}$
$\hat{X}_b \leftarrow$ G $\leftarrow [X_{b'}, b]$
$\min_G ||\hat{X}_b - X_b||_1$

(b)

$X_{a'} \rightarrow \hat{Y}(X_{a'}) \rightarrow \min_G L_{\text{CE}}(\hat{Y}(X_{a'}), Y^*(X_a))$

$X_{b'} \rightarrow \hat{Y}(X_{b'}) \rightarrow \min_G L_{\text{CE}}(\hat{Y}(X_{b'}), Y^*(X_b))$

(c)

Fig. 2. Overview of our approach. (a) The conditional generator network G is learned to map input X to novel domains whose distributions are drastically different from the source domains, while keeping the distance between the novel domains as far as possible. (b) A cycle-consistency loss is imposed on G to maintain the structural consistency. (c) The cross-entropy loss is minimized with respect to G, using a pre-trained classifier $\hat{Y}$, for maintaining the semantic consistency.

unseen domains and optimizes it via OT-based distribution divergence to make the new domains as dissimilar as possible to source distributions.

Domain Randomization. Our approach shares a similar high-level intuition with domain randomization (DR) [49], which was originally introduced in the context of robotic learning to improve generalization from simulation to real world. DR aims to diversify the training domains by changing the color and texture of objects, background scenes, lighting conditions, etc. via a computer simulator [49]. Recently, DR has been successfully used in some computer vision applications, such as semantic segmentation [54,55] and vehicle detection for autonomous driving [42]. However, our approach is significantly different from the DR-based methods because we *learn* a CNN generator network from real images rather than using programmatic simulators. Thus our method is more scalable to a wider range of image recognition tasks.

Image-to-Image Translation. Our work is also related to multi-domain image-to-image translation methods such as CycleGAN [64] and StarGAN [6], which use GAN losses [17] to generate realistic images and cycle-consistency losses [64] to achieve translation without using paired training images. Our method is fundamentally different from CycleGAN/StarGAN in that our generator model is learned to map source images to *unseen* domains rather than performing mapping between source domains as did in CycleGAN/StarGAN. We show by experiments that simply doing source-to-source mapping for data augmentation offers little help to DG (see Fig. 5a).

3 Methodology

3.1 Generating Novel-Domain Data

Setup. We are provided with K_s source domains with indices $D_s = \{1, 2, ..., K_s\}$. The goal is to learn a model which can generalize well on an unseen target domain. Without having access to the target data, we propose to improve the model's generalization by synthesizing novel data domains $D_n = \{1, 2, ..., K_n\}$ to augment the original source domains.

Conditional Generator. We learn a conditional generator G (see Sect. 3.4 for detailed architecture design), that maps a source distribution $\mathbb{P}_k$ with $k \in D_s$ to a novel distribution $\mathbb{P}_{\tilde{k}}$ with $\tilde{k} \in D_n$ by conditioning on the novel domain label $\tilde{k}$, i.e. $\mathbb{P}_{\tilde{k}} = G(\mathbb{P}_k, \tilde{k})$. Here $\mathbb{P}$ denotes an empirical distribution rather than the real distribution, which is inaccessible. In practice, we use sampled mini-batches X_k instead of the full empirical distribution $\mathbb{P}_k$. Therefore, the domain translation function is defined as:

$$X_{\tilde{k}} = G(X_k, \tilde{k}). \tag{1}$$

Objective Functions. For each training iteration, we randomly sample for each source domain k a mini-batch X_k, which is transformed to a randomly selected novel domain $\tilde{k} \sim D_n$. The objective is to force the novel distribution to be as dissimilar as possible to any source distribution, thus creating new domains to augment the existing source domains. We have

$$\max_G L_{\text{Novel}} = d(G(X_k, \tilde{k}), X_k), \tag{2}$$

where $d(\cdot, \cdot)$ is a distribution divergence measure (its design will be detailed in Sect. 3.5). Note that Eq. (2) will be summed over all source domains k, and each independently draws a novel domain label $\tilde{k}$.

In addition to maximizing the difference between source and novel distributions, we also maximize the difference between the generated novel distributions, i.e.

$$\max_G L_{\text{Diversity}} = d(X_{\tilde{k}_1}, X_{\tilde{k}_2}), \tag{3}$$

where $\tilde{k}_1, \tilde{k}_2 \in D_n$ and $\tilde{k}_1 \neq \tilde{k}_2$. Equation (3) is summed over all possible pairs of novel distributions generated in one iteration. This diversity constraint diversifies the generated distributions, ensuring that the model benefits from generating $K_n > 1$ novel distributions. It is analogous to the diversity term in some image generation tasks, such as style transfer [30] where the pixel/feature difference between style-transferred instances is maximized. Differently, our formulation focuses on the divergence between data distributions. See Fig. 2a for a graphical illustration.

3.2 Maintaining Semantic Consistency

The model so far is optimizing a powerful CNN generator G for the novelty of the generated distribution (Eq. (2) and (3)). This produces diverse images, but may not preserve their semantic content.

Cycle-Consistency Loss. First, to guarantee structural consistency, we apply a cycle-consistency constraint [64] to the generator,

$$\min_G L_{\text{Cycle}} = ||G(G(X_k, \tilde{k}), k) - X_k||_1, \tag{4}$$

where the outer G aims to reconstruct the original X_k given as input the domain-translated $G(X_k, \tilde{k})$ and the original domain label k. Both G's in the cycle share the same parameters [6]. This is illustrated in Fig. 2b.

Cross-Entropy Loss. Second, to maintain the category label and thus enforce semantic consistency, we further require that the generated data $X_{\tilde{k}}$ is classified into the same category as the original data X_k, i.e.

$$\min_G L_{\text{CE}}(\hat{Y}(X_{\tilde{k}}), Y^*(X_k)), \tag{5}$$

where L_{CE} denotes cross-entropy loss, $\hat{Y}(X_{\tilde{k}})$ the labels of $X_{\tilde{k}}$ predicted by a pretrained classifier and $Y^*(X_k)$ the ground-truth labels of X_k. This is illustrated in Fig. 2c.

3.3 Training

Generator Training. The full objective for G is the weighted combination of Eq. (2), (3), (4) and (5),

$$\begin{aligned} \min_G L_G = & -\lambda_{\text{Domain}}(L_{\text{Novel}} + L_{\text{Diversity}}) \\ & + \lambda_{\text{Cycle}} L_{\text{Cycle}} + \lambda_{\text{CE}} L_{\text{CE}}, \end{aligned} \tag{6}$$

where λ_{Domain}, λ_{Cycle} and λ_{CE} are weighting hyper-parameters.

Task Model Training. The task model F is trained from scratch using both the original data X_k and the synthetic data $X_{\tilde{k}}$ generated as described above. The objective for F is

$$\min_F L_F = (1-\alpha)L_{\text{CE}} + \alpha \tilde{L}_{\text{CE}}, \tag{7}$$

where α is a balancing weight, which is fixed to 0.5 throughout this paper; L_{CE} and $\tilde{L}_{\text{CE}}$ are the cross-entropy losses computed using X_k and $X_{\tilde{k}}$ respectively. The full training algorithm is shown in Alg. 1 (In the Supp.). Note that each source domain $k \in D_s$ will be assigned a unique novel domain $\tilde{k} \in D_n$ as target in each iteration. We set $K_n = K_s$ as default.

3.4 Design of Conditional Generator Network

Our generator model has a conv-deconv structure [6,64] which is shown in Fig. 3. Specifically, the generator model consists of two down-sampling convolution layers with stride 2, two residual blocks [19] and two transposed convolution layers with stride 2 for up-sampling. Following StarGAN [6], the domain indicator is encoded as a one-hot vector with length K_s+K_n (see Fig. 3). During the forward pass, the one-hot vector is first spatially expanded and then concatenated with the image to form the input to G.

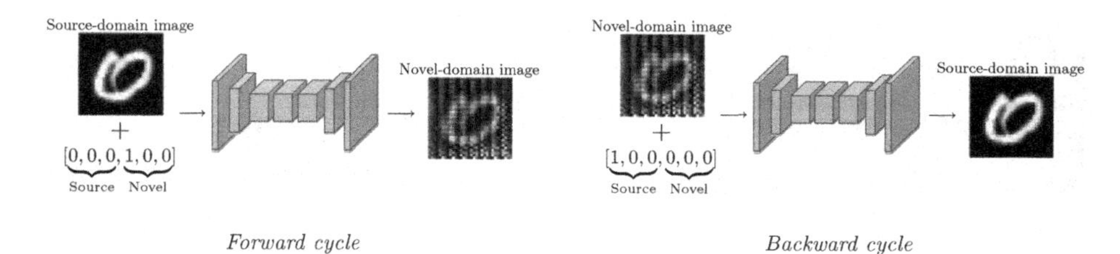

Fig. 3. Architecture of the conditional generator network. Left and right images exemplify the forward cycle and backward cycle respectively in cycle-consistency.

Discussion. Though the design of G is similar to the StarGAN model, their learning objectives are totally different: We aim to generate images that are different from the existing source domain distributions while the StarGAN model is trained to generate images from the existing source domains. In the experiment part we justify that adding novel-domain data is much more effective than adding seen-domain data for DG (see Fig. 5a). Compared with the gradient-based perturbation method in [47], our generator is allowed to model more sophisticated domain shift such as image style changes due to its learnable nature.

3.5 Design of Distribution Divergence Measure

Two common families for estimating the divergence between probability distributions are f-divergence (e.g., KL divergence) and integral probability metrics (e.g., Wasserstein distance). In contrast to most work that minimizes the divergence, we need to maximize it, as shown in Eq. (2) and (3). This strongly suggests to avoid f-divergence because of the near-zero denominators (they tend to generate large but numerically unstable divergence values). Therefore, we choose the second type, specifically the Wasserstein distance, which has been widely used in recent generative modeling methods [1,3,14,45,46].

The Wasserstein distance, also known as optimal transport (OT) distance, is defined as

$$\mathcal{W}_c(\mathbb{P}_a, \mathbb{P}_b) = \inf_{\pi \in \Pi(\mathbb{P}_a, \mathbb{P}_b)} \mathbb{E}_{x_a, x_b \sim \pi}[c(x_a, x_b)], \tag{8}$$

where $\Pi(\mathbb{P}_a, \mathbb{P}_b)$ denotes the set of all joint distributions $\pi(x_a, x_b)$ and $c(\cdot, \cdot)$ the transport cost function. Intuitively, the OT metric computes the minimum cost of transporting masses between distributions in order to turn $\mathbb{P}_b$ into $\mathbb{P}_a$.

As the sampling over $\Pi(\mathbb{P}_a, \mathbb{P}_b)$ is intractable, we resort to using the entropy-regularized Sinkhorn distance [7]. Moreover, to obtain unbiased gradient estimators when using mini-batches, we adopt the generalized (squared) energy distance [45], leading to

$$d(\mathbb{P}_a, \mathbb{P}_b) = 2\mathbb{E}[\mathcal{W}_c(X_a, X_b)] - \mathbb{E}[\mathcal{W}_c(X_a, X'_a)] - \mathbb{E}[\mathcal{W}_c(X_b, X'_b)], \tag{9}$$

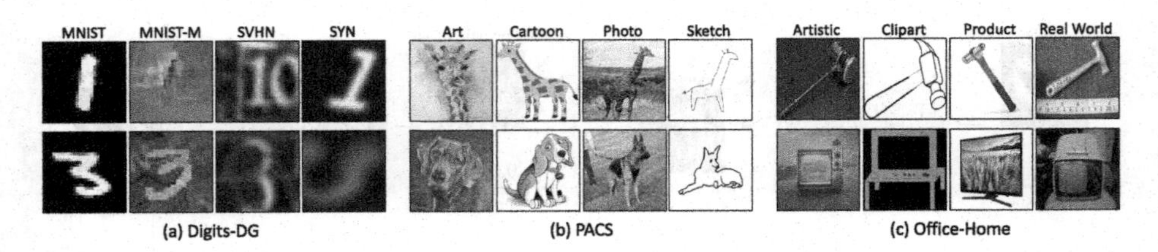

Fig. 4. Example images from different DG datasets.

where X_a and X'_a are independent mini-batches from distribution $\mathbb{P}_a$; X_b and X'_b are independent mini-batches from distribution $\mathbb{P}_b$; $\mathcal{W}_c$ is the Sinkhorn distance defined as

$$\mathcal{W}_c(\cdot,\cdot) = \inf_{M \in \mathcal{M}} \sum_{i,j} [M \odot C]_{i,j}, \tag{10}$$

where the soft-matching matrix M represents the coupling distribution π in Eq. (8) and can be efficiently computed using the Sinkhorn algorithm [14]; C is the pairwise distance matrix computed over two sets of samples.

Following [45], we define the cost function as the cosine distance between instances,

$$c(x_a, x_b) = 1 - \frac{\phi(x_a)^T \phi(x_b)}{||\phi(x_a)||_2 ||\phi(x_b)||_2}, \tag{11}$$

where ϕ is constructed by a CNN (also called critic in [45]), which maps images into a latent space. In practice, ϕ is a fixed CNN that was trained with domain classification loss.

4 Experiments

4.1 Evaluation on Homogeneous DG

Datasets. (1) We use four different digit datasets including MNIST [24], MNIST-M [12], SVHN [38] and SYN [12], which differ drastically in font style, stroke color and background. We call this new dataset **Digits-DG** hereafter. See Fig. 4a for example images. (2) **PACS** [25] is composed of four domains, which are Photo, Art Painting, Cartoon and Sketc.h, with 9,991 images of 7 classes in total. See Fig. 4b for example images. (3) **Office-Home** [51] contains around 15,500 images of 65 classes for object recognition in office and home environments. It has four domains, which are Artistic, Clipart, Product and Real World. See Fig. 4c for example images.

Evaluation Protocol. For fair comparison with prior work, we follow the leave-one-domain-out protocol in [5,25,27]. Specifically, one domain is chosen as the test domain while the remaining domains are used as source domains for model training. The top-1 classification accuracy is used as performance measure. All results are averaged over three runs with different random seeds.

Table 1. Leave-one-domain-out results on Digits-DG.

Method	MNIST	MNIST-M	SVHN	SYN	Avg
Vanilla	95.8	58.8	61.7	78.6	73.7
CCSA [36]	95.2	58.2	65.5	79.1	74.5
MMD-AAE [28]	96.5	58.4	65.0	78.4	74.6
CrossGrad [47]	**96.7**	61.1	65.3	80.2	75.8
JiGen [5]	96.5	61.4	63.7	74.0	73.9
L2A-OT (*ours*)	**96.7**	**63.9**	**68.6**	**83.2**	**78.1**

Baselines. We compare L2A-OT with the recent state-of-the-art DG methods that report results on the same dataset or have code publicly available for reproduction. These include (1) **CrossGrad** [47], the most related work that perturbs input using adversarial gradients from a domain classifier; (2) **CCSA** [36], which learns a domain-invariant representation using a contrastive semantic alignment loss; (3) **MMD-AAE** [28], which imposes a MMD loss on the hidden layers of an autoencoder. (4) **JiGen** [5], which has an auxiliary self-supervision loss to solve the Jigsaw puzzle task [39]; (5) **Epi-FCR** [27], which designs an episodic training strategy; (6) A **vanilla** model trained by aggregating all source domains, which serves as a strong baseline.

Implementation Details. For Digits-DG, the CNN backbone is constructed with four 64-kernel 3×3 convolution layers and a softmax layer. ReLU and 2×2 max-pooling are inserted after each convolution layer. F is trained with SGD, initial learning rate of 0.05 and batch size of 126 (42 images per source) for 50 epochs. The learning rate is decayed by 0.1 every 20 epochs. For all experiments, G is trained with Adam [23] and a constant learning rate of 0.0003. For both PACS and Office-Home, we use ResNet-18 [19] pretrained on ImageNet [8] as the CNN backbone, following [5,9,27]. On PACS, F is trained with SGD, initial learning rate of 0.00065 and batch size of 24 (8 images per source) for 40 epochs. The learning rate is decayed by 0.1 after 30 epochs. On Office-Home, the optimization parameters are similar to those on PACS except that the maximum epoch is 25 and the learning rate decay step is 20. For all datasets, as target data is unavailable during training, the values of hyper-parameters λ_{Domain}, λ_{Cycle} and λ_{CE} are set based on the performance on source validation set,[2] which is a strategy commonly adopted in the DG literature [5,27]. Our implementation is based on `Dassl.pytorch` [63].

Results on Digits-DG. Table 1 shows that L2A-OT achieves the best performance on all domains and consistently outperforms the vanilla baseline by a large margin. Compared with CrossGrad, L2A-OT performs clearly better on MNIST-M, SVHN and SYN, with clear improvements of 2.8%, 3.3% and 3%, respectively. It is worth noting that these three domains are very challenging

[2] The searching space is: $\lambda_{\text{Domain}} \in \{0.5, 1, 2\}$, $\lambda_{\text{Cycle}} \in \{10, 20\}$ and $\lambda_{\text{CE}} \in \{1\}$.

Table 2. Leave-one-domain-out results on PACS dataset.

Method	Art	Cartoon	Photo	Sketc.h	Avg
Vanilla	77.0	75.9	96.0	69.2	79.5
CCSA [36]	80.5	76.9	93.6	66.8	79.4
MMD-AAE [28]	75.2	72.7	96.0	64.2	77.0
CrossGrad [47]	79.8	76.8	96.0	70.2	80.7
JiGen [5]	79.4	75.3	96.0	71.6	80.5
Epi-FCR [27]	82.1	77.0	93.9	73.0	81.5
L2A-OT (*ours*)	**83.3**	**78.2**	**96.2**	**73.6**	**82.8**

Table 3. Leave-one-domain-out results on Office-Home.

Method	Artistic	Clipart	Product	Real World	Avg
Vanilla	58.9	49.4	74.3	76.2	64.7
CCSA [36]	59.9	49.9	74.1	75.7	64.9
MMD-AAE [28]	56.5	47.3	72.1	74.8	62.7
CrossGrad [47]	58.4	49.4	73.9	75.8	64.4
JiGen [5]	53.0	47.5	71.5	72.8	61.2
L2A-OT (*ours*)	**60.6**	**50.1**	**74.8**	**77.0**	**65.6**

with large domain variations compared with their source domains (see Fig. 4a). The huge advantage over CrossGrad can be attributed to L2A-OT's unique generation of unseen-domain data using a fully learnable CNN generator, and using optimal transport to explicitly encourage domain divergence. Compared with the domain alignment methods, L2A-OT surpasses MMD-AAE and CCSA by more than 3.5% on average. The is because L2A-OT enriches the domain diversity of training data, thus reducing overfitting in source domains. L2A-OT clearly beats JiGen because the Jigsaw puzzle transformation does not work well on digit images with sparse pixels [39].

Results on PACS. The results are shown in Table 2. Overall, L2A-OT achieves the best performance on all test domains. L2A-OT clearly beats the latest DG methods, JiGen and Epi-FCR. This is because our classifier benefits from the generated unseen-domain data while JiGen and Epi-FCR, like the domain alignment methods, are prone to overfitting to the source domains. L2A-OT beats CrossGrad on all domains, mostly with a large margin. This again justifies our design of learnable CNN generator over adversarial gradient.

Results on Office-Home. The results are reported in Table 3. Again, L2A-OT achieves the best overall performance, and other conclusions drawn previously also hold. Notably, the simple vanilla model obtains strong results on this benchmark, which are even better than most existing DG methods. This is because

Table 4. Results on cross-domain person re-ID benchmarks.

Method	Market1501 → Duke				Duke → Market1501			
	mAP	R1	R5	R10	mAP	R1	R5	R10
UDA methods								
ATNet [32]	24.9	45.1	59.5	64.2	25.6	55.7	73.2	79.4
CamStyle [59]	25.1	**48.4**	**62.5**	**68.9**	27.4	58.8	78.2	**84.3**
HHL [58]	**27.2**	46.9	61.0	66.7	**31.4**	**62.2**	**78.8**	84.0
DG methods								
Vanilla	26.7	48.5	62.3	67.4	26.1	57.7	73.7	80.0
CrossGrad [47]	27.1	48.5	63.5	69.5	26.3	56.7	73.5	79.5
L2A-OT (*ours*)	**29.2**	**50.1**	**64.5**	**70.1**	**30.2**	**63.8**	**80.2**	**84.6**

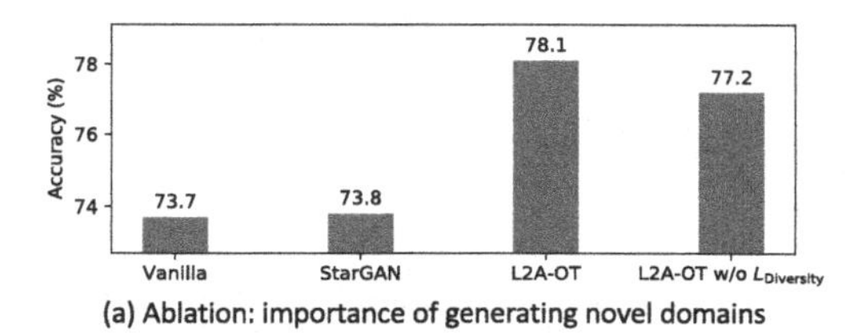

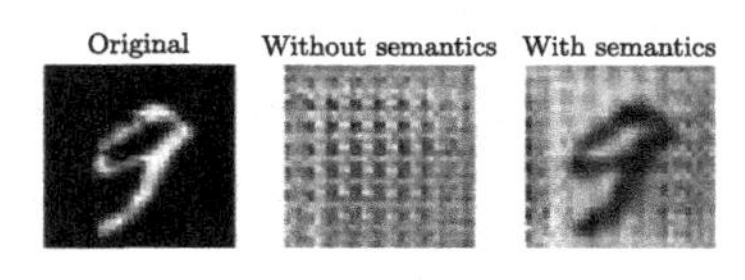

(b) Ablation: importance of semantic constraint

Fig. 5. Ablation study.

the dataset is relatively large, and the domain shift is less severe compared with the style changes on PACS and the font variations on Digits-DG.

4.2 Evaluation on Heterogeneous DG

In this section, we evaluate L2A-OT on a more challenging DG task with disjoint label space between training and test data, namely cross-domain person re-identification (re-ID).

Datasets. We use Market1501 [56] and DukeMTMC-reID (Duke) [43,57]. Market1501 has 32,668 images of 1,501 identities captured by 6 cameras (domains). Duke has 36,411 images of 1,812 identities captured by 8 cameras.

Evaluation Protocol. We follow the recent unsupervised domain adaptation (UDA) methods in the person re-ID literature [32,58,59] and experiment with Market1501 → Duke and Duke → Market1501. Different from the UDA setting, we directly test the source-trained model on the target dataset without adaptation. Note that the cross-domain re-ID evaluation involves training a person classifier on source dataset identities. This is then transferred and used to recognize a disjoint set of people in the target domain of unseen camera views via nearest neighbor. Since the label space is disjoint, this is a *heterogeneous* DG problem. For performance measure, we adopt CMC ranks and mAP [56].

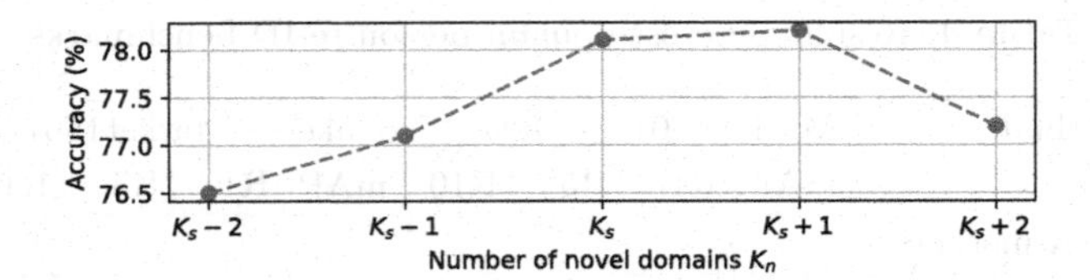

Fig. 6. Results of varying K_n. Here $K_s = 3$.

Table 5. Using two vs. three source domains on Digits-DG where the size of training data is kept identical for all settings for fair comparison.

Source			Target	L2A-OT	Vanilla
MNIST	SVHN	SYN			
✓	✓		MNIST-M	60.9	54.6
✓		✓	MNIST-M	62.1	**59.1**
	✓	✓	MNIST-M	49.7	45.2
✓	✓	✓	MNIST-M	**62.5**	57.1

Implementation Details. For the CNN backbone, we employ the state-of-the-art re-ID model, OSNet-IBN [61,62]. Following [61,62], OSNet-IBN is trained using the standard classification paradigm, i.e. each identity is considered as a class. Therefore, the entire L2A-OT framework remains unchanged. At test time, feature vectors extracted from OSNet-IBN are used to compute ℓ_2 distance for image matching. Our implementation is based on `Torchreid` [60].

Results. In Table 4, we compare L2A-OT with the vanilla model and CrossGrad, as well as state-of-the-art UDA methods for re-ID. As a result, CrossGrad barely improves the vanilla model while L2A-OT achieves clear improvements on both settings. Notably, L2A-OT is highly competitive with the UDA methods, though the latter make the significantly stronger assumption of having access to the target domain data (thus gaining an unfair advantage). In contrast, L2A-OT generates images of unseen styles (domains) for data augmentation, and such more diverse data leads to learning a better generalizable re-ID model.

4.3 Ablation Study

Importance of Generating Novel Domains. To verify that our improvement is brought by the increase in training data distributions by the generated novel domains (i.e. Eq. (2) and (3)), we compare L2A-OT with StarGAN [6], which generates data from the existing source domains by performing source-to-source mapping. The experiment is conducted on Digits-DG and the average performance over all test domains is used for comparison. Figure 5a shows that StarGAN performs only similarly to the vanilla model (StarGAN's 73.8% vs.

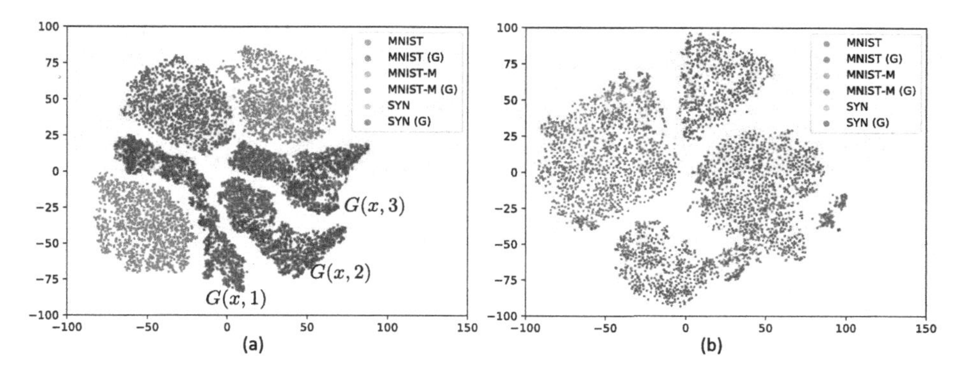

Fig. 7. T-SNE visualization of domain embeddings of (a) L2A-OT and (b) Cross-Grad [47]. X (G) indicates novel data when using the domain X as a source.

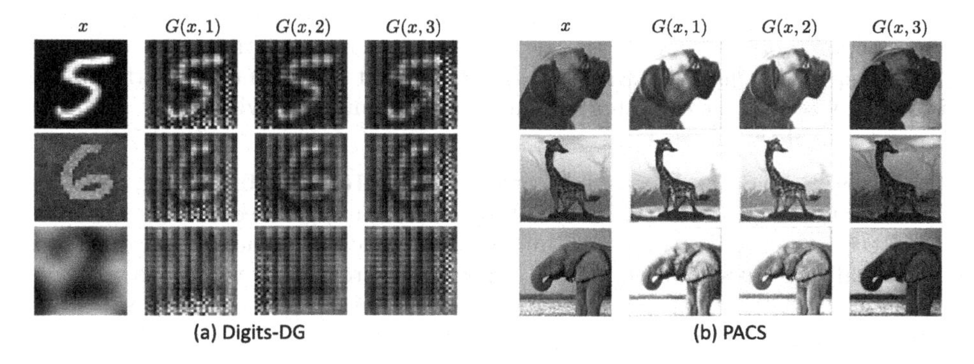

Fig. 8. Visualization of generated images. x: source image. $G(x, i)$: generated image of the i-th novel domain.

vanilla's 73.7%) while L2A-OT obtains a clear improvement of 4.3% over StarGAN. This confirms that increasing domains is far more important than increasing data (of seen domains) for DG. Note that this 4.3% gap is attributed to the combination of the OT-based domain novelty loss (Eq. (2)) and the diversity loss (Eq. (3)). Figure 5a shows that the diversity loss contributes around 1% to the performance, and the rest improvement comes from the diversity loss.

Importance of Semantic Constraint. The cycle-consistency and cross-entropy losses (Eq. (4) and (5)) are essential in the L2A-OT framework for maintaining the semantic content when performing domain translation. Figure 5b shows that without the semantic constraint, the content is completely missing (we found that using these images reduced the result from 78.1% to 73.9%).

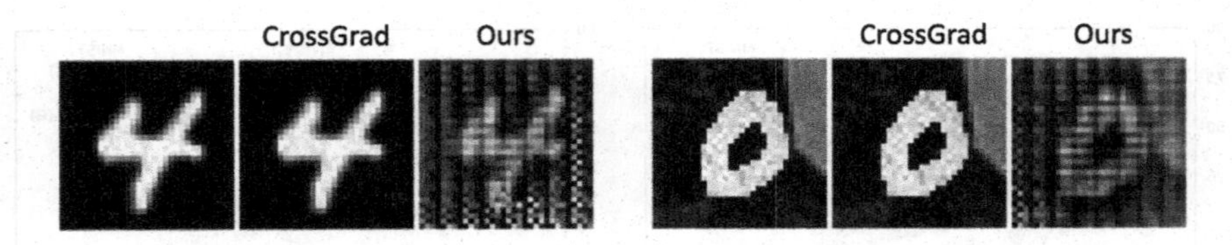

Fig. 9. Comparison between L2A-OT and CrossGrad [47] on image generation.

4.4 Further Analysis

How Many Novel Domains to Generate? Our approach can generate an arbitrary number of novel domains, although we have always doubled the number of domains (set $K_s = K_n$) so far. Figure 6 investigates the significance on the choice of number of novel domains. In principle, synthesizing more domains provides opportunity for more diverse data, but also increases optimization difficulty and is dependent on the source domains. The result shows that the performance is not very sensitive to the choice of novel domain number, with $K_n = K_s$ being a good rule of thumb.

Do More Source Domains Lead to a Better Result? In general, yes. The evidence is shown in Table 5 where the result of using three sources is generally better than using two as we might expect due to the additional diversity. The detailed results show that when using two sources, performance is sensitive to the choice of sources among the available three. This is expected since different sources will vary in transferrability to a given target. However, for both vanilla and L2A-OT the performance of using three sources is better than the performance of using two averaged across the 2-source choices.

Visualizing Domain Distributions. We employ t-SNE [34] to visualize the domain feature embeddings using the validation set of Digits-DG (see Fig. 7a). We have the following observations. (1) The generated distributions are clearly separated from the source domains and evenly fill the unseen domain space. (2) The generated distributions form independent clusters (due to our diversity term in Eq. (3)). (3) G has successfully learned to flexibly transform one source domain to any of the discovered novel domains.

Visualizing Novel-Domain Images. Figure 8 visualizes the output of G. In general, we observe that the generated images from different novel domains manifest different properties and more importantly, are clearly different from the source images. For example, in Digits-DG (Fig. 8a), G tends to generate images with different background patterns/textures and font colors. In PACS (Fig. 8b), G focuses on contrast and color. Figure 8 seems to suggest that the synthesized domains are not drastically different from each other. However, a seemingly limited diversity in the image space to human eyes can be significant to a CNN classifier: both Fig. 1 and Fig. 7a show clearly that the synthesized data points have very different distributions from both the original ones and each other in a feature embedding space, making them useful for learning a domain-generalizable classifier.

L2A-OT vs. CrossGrad. It is clear from Fig. 7b that the new domains generated by CrossGrad largely overlap with the original domains. This is because CrossGrad is based on adversarial attack methods [18], which are designed to make imperceptible changes. This is further verified in Fig. 9 where the images generated by CrossGrad have only subtle differences in contrast to the original images. On the contrary, L2A-OT can model much more complex domain variations that can materially benefit the classifier, thanks to the full CNN image generator and OT-based domain divergence losses.

5 Conclusion

We presented L2A-OT, a novel data augmentation-based DG method that boosts classifier's robustness to domain shift by learning to synthesize images from diverse unseen domains through a conditional generator network. The generator is trained by maximizing the OT distance between source domains and pseudo-novel domains. Cycle-consistency and classification losses are imposed on the generator to further maintain the structural and semantic consistency during domain translation. Extensive experiments on four DG benchmark datasets covering a wide range of visual recognition tasks demonstrate the effectiveness and versatility of L2A-OT.

References

1. Arjovsky, M., Chintala, S., Bottou, L.: Wasserstein generative adversarial networks. In: ICML (2017)
2. Balaji, Y., Sankaranarayanan, S., Chellappa, R.: MetaReg: towards domain generalization using meta-regularization. In: NeurIPS (2018)
3. Bellemare, M.G., et al.: The cramer distance as a solution to biased wasserstein gradients. arXiv preprint arXiv:1705.10743 (2017)
4. Ben-David, Shai., Blitzer, John., Crammer, Koby., Kulesza, Alex., Pereira, Fernando, Vaughan, Jennifer Wortman: A theory of learning from different domains. Mach. Learn. **79**(1), 151–175 (2009). https://doi.org/10.1007/s10994-009-5152-4
5. Carlucci, F.M., D'Innocente, A., Bucci, S., Caputo, B., Tommasi, T.: Domain generalization by solving jigsaw puzzles. In: CVPR (2019)
6. Choi, Y., Choi, M., Kim, M., Ha, J.W., Kim, S., Choo, J.: StarGAN: unified generative adversarial networks for multi-domain image-to-image translation. In: CVPR (2018)
7. Cuturi, M.: Sinkhorn distances: lightspeed computation of optimal transport. In: NeurIPS (2013)
8. Deng, J., Dong, W., Socher, R., Li, L.J., Li, K., Fei-Fei, L.: ImageNet: a large-scale hierarchical image database. In: CVPR (2009)
9. D'Innocente, A., Caputo, B.: Domain generalization with domain-specific aggregation modules. In: GCPR (2018)

10. Dou, Q., Castro, D.C., Kamnitsas, K., Glocker, B.: Domain generalization via model-agnostic learning of semantic features. In: NeurIPS (2019)
11. Gan, C., Yang, T., Gong, B.: Learning attributes equals multi-source domain generalization. In: CVPR (2016)
12. Ganin, Y., Lempitsky, V.S.: Unsupervised domain adaptation by backpropagation. In: ICML (2015)
13. Ganin, Y., et al.: Domain-adversarial training of neural networks. JMLR **17**(1), 2030–2096 (2016)
14. Genevay, A., Peyré, G., Cuturi, M.: Learning generative models with sinkhorn divergences. In: AISTATS (2018)
15. Ghifary, M., Balduzzi, D., Kleijn, W.B., Zhang, M.: Scatter component analysis: a unified framework for domain adaptation and domain generalization. TPAMI **39**(7), 1414–1430 (2017)
16. Ghifary, M., Kleijn, W.B., Zhang, M., Balduzzi, D.: Domain generalization for object recognition with multi-task autoencoders. In: ICCV (2015)
17. Goodfellow, I., et al.: Generative adversarial nets. In: NeurIPS (2014)
18. Goodfellow, I.J., Shlens, J., Szegedy, C.: Explaining and harnessing adversarial examples. In: ICLR (2015)
19. He, K., Zhang, X., Ren, S., Sun, J.: Deep residual learning for image recognition. In: CVPR (2016)
20. Hoffman, J., et al.: CyCADA: cycle-consistent adversarial domain adaptation. In: ICML (2018)
21. Hospedales, T., Antoniou, A., Micaelli, P., Storkey, A.: Meta-learning in neural networks: a survey. arXiv preprint arXiv:2004.05439 (2020)
22. Jin, X., Lan, C., Zeng, W., Chen, Z., Zhang, L.: Style normalization and restitution for generalizable person re-identification. In: CVPR (2020)
23. Kingma, D.P., Ba, J.: Adam: a method for stochastic optimization. In: ICLR (2014)
24. LeCun, Y., Bottou, L., Bengio, Y., Haffner, P.: Gradient-based learning applied to document recognition. Proc. IEEE **86**(11), 2278–2324 (1998)
25. Li, D., Yang, Y., Song, Y.Z., Hospedales, T.M.: Deeper, broader and artier domain generalization. In: ICCV (2017)
26. Li, D., Yang, Y., Song, Y.Z., Hospedales, T.M.: Learning to generalize: meta-learning for domain generalization. In: AAAI (2018)
27. Li, D., Zhang, J., Yang, Y., Liu, C., Song, Y.Z., Hospedales, T.M.: Episodic training for domain generalization. In: ICCV (2019)
28. Li, H., Jialin Pan, S., Wang, S., Kot, A.C.: Domain generalization with adversarial feature learning. In: CVPR (2018)
29. Li, Y., Tiana, X., Gong, M., Liu, Y., Liu, T., Zhang, K., Tao, D.: Deep domain generalization via conditional invariant adversarial networks. In: ECCV (2018)
30. Li, Y., Fang, C., Yang, J., Wang, Z., Lu, X., Yang, M.H.: Diversified texture synthesis with feed-forward networks. In: CVPR (2017)
31. Li, Y., Yang, Y., Zhou, W., Hospedales, T.: Feature-critic networks for heterogeneous domain generalization. In: ICML (2019)
32. Liu, J., Zha, Z.J., Chen, D., Hong, R., Wang, M.: Adaptive transfer network for cross-domain person re-identification. In: CVPR (2019)
33. Long, M., Cao, Y., Wang, J., Jordan, M.I.: Learning transferable features with deep adaptation networks. In: ICML (2015)
34. Maaten, L.V.D., Hinton, G.: Visualizing data using t-SNE. JMLR **9**, 2579–2605 (2008)
35. Moreno-Torres, J.G., Raeder, T., Alaiz-RodríGuez, R., Chawla, N.V., Herrera, F.: A unifying view on dataset shift in classification. PR **45**(1), 521–530 (2012)

36. Motiian, S., Piccirilli, M., Adjeroh, D.A., Doretto, G.: Unified deep supervised domain adaptation and generalization. In: ICCV (2017)
37. Muandet, K., Balduzzi, D., Scholkopf, B.: Domain generalization via invariant feature representation. In: ICML (2013)
38. Netzer, Y., Wang, T., Coates, A., Bissacco, A., Wu, B., Ng, A.Y.: Reading digits in natural images with unsupervised feature learning. In: NeurIPS-W (2011)
39. Noroozi, M., Favaro, P.: Unsupervised learning of visual representations by solving jigsaw puzzles. In: ECCV (2016)
40. Peng, X., Bai, Q., Xia, X., Huang, Z., Saenko, K., Wang, B.: Moment matching for multi-source domain adaptation. In: ICCV (2019)
41. Peyré, G., Cuturi, M., et al.: Computational optimal transport. Found. Trends Mach. Learn. **11**(5–6), 355–607 (2019)
42. Prakash, A., et al.: Structured domain randomization: bridging the reality gap by context-aware synthetic data. In: ICRA (2019)
43. Ristani, E., Solera, F., Zou, R.S., Cucchiara, R., Tomasi, C.: Performance measures and a data set for multi-target, multi-camera tracking. In: ECCV-W (2016)
44. Saito, K., Kim, D., Sclaroff, S., Darrell, T., Saenko, K.: Semi-supervised domain adaptation via minimax entropy. In: ICCV (2019)
45. Salimans, T., Zhang, H., Radford, A., Metaxas, D.: Improving GANs using optimal transport. In: ICLR (2018)
46. Shaham, T.R., Dekel, T., Michaeli, T.: Singan: Learning a generative model from a single natural image. In: ICCV (2019)
47. Shankar, S., Piratla, V., Chakrabarti, S., Chaudhuri, S., Jyothi, P., Sarawagi, S.: Generalizing across domains via cross-gradient training. In: ICLR (2018)
48. Szegedy, C., et al.: Intriguing properties of neural networks. In: ICLR (2014)
49. Tobin, J., Fong, R., Ray, A., Schneider, J., Zaremba, W., Abbeel, P.: Domain randomization for transferring deep neural networks from simulation to the real world. In: IROS (2017)
50. Tzeng, E., Hoffman, J., Saenko, K., Darrell, T.: Adversarial discriminative domain adaptation. In: CVPR (2017)
51. Venkateswara, H., Eusebio, J., Chakraborty, S., Panchanathan, S.: Deep hashing network for unsupervised domain adaptation. In: CVPR (2017)
52. Volpi, R., Namkoong, H., Sener, O., Duchi, J., Murino, V., Savarese, S.: Generalizing to unseen domains via adversarial data augmentation. In: NeurIPS (2018)
53. Xu, R., Li, G., Yang, J., Lin, L.: Larger norm more transferable: an adaptive feature norm approach for unsupervised domain adaptation. In: ICCV (2019)
54. Yue, X., Zhang, Y., Zhao, S., Sangiovanni-Vincentelli, A., Keutzer, K., Gong, B.: Domain randomization and pyramid consistency: simulation-to-real generalization without accessing target domain data. In: ICCV (2019)
55. Zakharov, S., Kehl, W., Ilic, S.: DeceptionNet: network-driven domain randomization. In: ICCV (2019)
56. Zheng, L., Shen, L., Tian, L., Wang, S., Wang, J., Tian, Q.: Scalable person re-identification: a benchmark. In: ICCV (2015)
57. Zheng, Z., Zheng, L., Yang, Y.: Unlabeled samples generated by GAN improve the person re-identification baseline in vitro. In: ICCV (2017)
58. Zhong, Z., Zheng, L., Li, S., Yang, Y.: Generalizing a person retrieval model hetero- and homogeneously. In: ECCV (2018)
59. Zhong, Z., Zheng, L., Zheng, Z., Li, S., Yang, Y.: Camstyle: A novel data augmentation method for person re-identification. TIP **28**(3), 1176–1190 (2019)
60. Zhou, K., Xiang, T.: Torchreid: a library for deep learning person re-identification in pytorch. arXiv preprint arXiv:1910.10093 (2019)

61. Zhou, K., Yang, Y., Cavallaro, A., Xiang, T.: Learning generalisable omni-scale representations for person re-identification. arXiv preprint arXiv:1910.06827 (2019)
62. Zhou, K., Yang, Y., Cavallaro, A., Xiang, T.: Omni-scale feature learning for person re-identification. In: ICCV (2019)
63. Zhou, K., Yang, Y., Qiao, Y., Xiang, T.: Domain adaptive ensemble learning. arXiv preprint arXiv:2003.07325 (2020)
64. Zhu, J.Y., Park, T., Isola, P., Efros, A.A.: Unpaired image-to-image translation using cycle-consistent adversarial networks. In: ICCV (2017)

Continuous Adaptation for Interactive Object Segmentation by Learning from Corrections

Theodora Kontogianni[2(✉)], Michael Gygli[1], Jasper Uijlings[1], and Vittorio Ferrari[1]

[1] Google Research, Zurich, Switzerland
[2] RWTH Aachen University, Aachen, Germany
kontogianni@vision.rwth-aachen.de

Abstract. In interactive object segmentation a user collaborates with a computer vision model to segment an object. Recent works employ convolutional neural networks for this task: Given an image and a set of corrections made by the user as input, they output a segmentation mask. These approaches achieve strong performance by training on large datasets but they keep the model parameters unchanged at test time. Instead, we recognize that user corrections can serve as sparse training examples and we propose a method that capitalizes on that idea to update the model parameters on-the-fly to the data at hand. Our approach enables the adaptation to a particular object and its background, to distributions shifts in a test set, to specific object classes, and even to large domain changes, where the imaging modality changes between training and testing. We perform extensive experiments on 8 diverse datasets and show: Compared to a model with frozen parameters, our method reduces the required corrections (i) by 9%–30% when distribution shifts are small between training and testing; (ii) by 12%–44% when specializing to a specific class; (iii) and by 60% and 77% when we completely change domain between training and testing.

1 Introduction

In interactive object segmentation a human collaborates with a computer vision model to segment an object of interest [11,12,46,53]. The process iteratively alternates between the user providing corrections on the current segmentation and the model refining the segmentation based on these corrections. The objective of the model is to infer an accurate segmentation mask from as few user

T. Kontogianni and M. Gygli—Equal contribution
T. Kontogianni—Work done while interning at Google.

Electronic supplementary material The online version of this chapter (https://doi.org/10.1007/978-3-030-58517-4_34) contains supplementary material, which is available to authorized users.

A. Vedaldi et al. (Eds.): ECCV 2020, LNCS 12361, pp. 579–596, 2020.
https://doi.org/10.1007/978-3-030-58517-4_34

corrections as possible (typically point clicks [8,16] or strokes [22,46] on mislabeled pixels). This enables fast and accurate object segmentation, which is indispensable for image editing [2] and collecting ground-truth segmentation masks at scale [11].

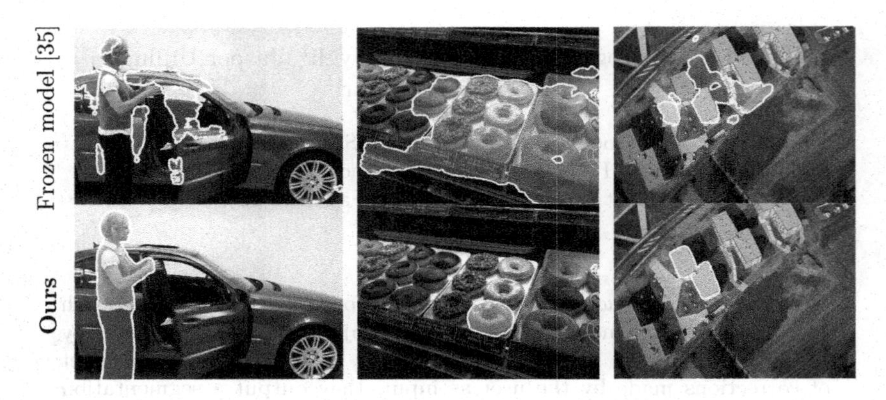

Fig. 1. Example results for a frozen model (top) and our adaptive methods (bottom). A frozen model performs poorly when foreground and background share similar appearance (left), when it is used to segment new object classes absent in the training set (center, donut class), or when the model is tested on a different image domain (aerial) than it is trained on (consumer) (right). By using corrections to adapt the model parameters to a specific test image, or to the test image sequence, our method substantially improves segmentation quality. The input is four corrections in all cases shown.

Current state-of-the-art methods train a convolutional neural network (CNN) which takes an image and user corrections as input and predicts a foreground / background segmentation [10,11,27,30,32,35,53]. At test time, the model parameters are frozen and corrections are only used as additional input to guide the model predictions. But in fact, user corrections directly specify the ground-truth labelling of the corrected pixels. In this paper we capitalize on this observation: we treat user corrections as training examples to adapt our model on-the-fly. We use these user corrections in two ways: (1) in *single image adaptation* we iteratively adapt model parameters to one specific object in an image, given the corrections produced while segmenting that object; (2) in *image sequence adaptation* we adapt model parameters to a sequence of images with an online method, given the set of corrections produced on these images. Each of these leads to distinct advantages over using a frozen model:

During *single image adaptation* our model learns the specific appearance of the current object instance and the surrounding background. This allows the model to adapt even to subtle differences between foreground and background for that specific example. This is necessary when the object to be segmented has similar color to the background (Fig. 1, 1st column), has blurry object boundaries, or low contrast. In addition, a frozen model can sometimes ignore the user corrections and overrule them in its next prediction. We avoid this undesired behavior by updating the model parameters until its predictions respect the user corrections.

During *image sequence adaptation* we continuously adapt the model to a sequence of segmentation tasks. Through this, the model parameters are optimized to the image and class distribution in these tasks, which may consist of different types of images or a set of new classes which are unseen during training. An important case of this is specializing the model for segmenting objects of a single class. This is useful for collecting many examples in high-precision domains, such as *pedestrians* for self-driving car applications. Figure 1, middle column shows an example of specializing to the single, unseen class *donut*. Furthermore, an important property of image sequence adaptation is that it enables us to handle large domain changes, where the imaging modality changes dramatically between training and testing. We demonstrate this by training on consumer photos while testing on medical and aerial images (Fig. 1, right column).

Naturally, single image adaptation and image sequence adaptation can be used jointly, leading to a method that combines their advantages.

In summary: Our innovative idea of treating user corrections as training examples allows to update the parameters of an interactive segmentation model at *test time*. To update the parameters we propose a practical online adaptation method. Our method operates on sparse corrections, balances adaptation *vs.* retaining old knowledge and can be applied to any CNN-based interactive segmentation model. We perform extensive experiments on 8 diverse datasets and show: Compared to a model with frozen parameters, our method reduces the required corrections (i) by 9%–30% when distribution shifts are small between training and testing; (ii) by 12%–44% when specializing to a specific class; (iii) and by 60% and 77% when we completely change domain between training and testing. (iv) Finally, we evaluate on four standard datasets where distribution shifts between training and testing are minimal. Nevertheless, our method did set a new state-of-the-art on all of them, when it was initially released [29].

2 Related Work

Interactive Object Segmentation. Traditional methods approach interactive segmentation via energy minimization on a graph defined over pixels [7,12,22,41,46]. User inputs are used to create an image-specific appearance model based on low-level features (*e.g.* color), which is then used to predict foreground and background probabilities. A pairwise smoothness term between neighboring pixels encourages regular segmentation outputs. Hence these classical methods are based on a weak appearance model which is specialized to one specific image.

Recent methods rely on Convolutional Neural Networks (CNNs) to interactively produce a segmentation mask [3,10,16,26,27,30,32,35,53]. These methods take the image and user corrections (transformed into a guidance map) as input and map them to foreground and background probabilities. This mapping is optimized over a training dataset and remains frozen at test time. Hence these models have a strong appearance model but it is not optimized for the test image or dataset at hand.

Our method combines the advantages of traditional and recent approaches: We use a CNN to learn a strong initial appearance model from a training set. During segmentation of a new test image, we adapt the model to it. It thus learns an appearance model specifically for that image. Furthermore, we also continuously adapt the model to the new image and class distribution of the test set, which may be significantly different from the one the model is originally trained on.

Gradient Descent at Test Time. Several methods iteratively minimize a loss at test time. The concurrent work of [51] uses self-supervision to adapt the feature extractor of a multi-tasking model to the test distribution. Instead, we directly adapt the full model by minimizing the task loss. Others iteratively update the inputs of a model [21,23,27], *e.g.* for style transfer [21]. In the domain of interactive segmentation, [27] updates the guidance map which encodes the user corrections and is input to the model. [49] made this idea more computationally efficient by updating intermediate feature activations, rather than the guidance maps. Instead, our method updates the model parameters, making it more general and allowing it to adapt to individual images as well as sequences.

In-Domain Fine-Tuning. In other applications it is common practice to fine-tune on in-domain data when transferring a model to a new domain [13,39,52, 58]. For example, when supervision for the first frame of a test video is available [13,40,52], or after annotating a subset of an image dataset [39,58]. In interactive segmentation the only existing attempt is [1], which performs polygon annotation [1,15,34]. However, it does not consider adapting to a particular image; their process to fine-tune on a dataset involves 3 different models, so they do it only a few times per dataset; they cannot directly train on user corrections, only on complete masks from previous images; finally, they require a bounding box on the object as input.

Few-Shot and Continual Learning. Our method automatically adapts to distribution shifts and domain changes. It performs domain adaptation from limited supervision, similar to few-shot learning [20,42,43,48]. It also relates to continual learning [19,44], except that the output label space of the classifier is fixed. As in other works, our method needs to balance between preserving existing knowledge and adapting to new data. This is often done by fine-tuning on new tasks while discouraging large changes in the network parameters, either by penalizing changes to important parameters [5,6,28,55] or changing predictions of the model on old tasks [31,38,47]. Alternatively, some training data of the old task is kept and the model is trained on a mixture of the old and new task data [9,44].

3 Method

We adopt a typical interactive object segmentation process [11,12,27,30,35,53]: the model is given an image and makes an initial foreground/background prediction for every pixel. The prediction is then overlaid on the image and presented to the user, who is asked to make a correction. The user clicks on a single pixel to mark that it was incorrectly predicted to be foreground instead of background or vice versa. The model then updates the predicted segmentation based on all corrections received so far. This process iterates until the segmentation reaches a desired quality level.

We start by describing the model we build on (Sect. 3.1). Then, we describe our core contribution: treating user corrections as training examples to adapt the model on-the-fly at test-time (Sect. 3.2). Lastly, we describe how we simulate user corrections to train and test our method (Sect. 3.3).

Fig. 2. Corrections as Training Examples. For learning the initial model parameters, full supervision is available, allowing to compute a loss over all the pixels in the image. At test time, the user provides sparse supervision in the form of corrections. We use these to adapt the model parameters.

3.1 Interactive Segmentation Model

As the basis of our approach, we use a strong re-implementation of [35], an interactive segmentation model based on a convolutional neural network. The model takes an RGB image and the user corrections as input and produces a segmentation mask. As in [11] we encode the position of user corrections by placing binary disks into a *guidance map*. This map has the same resolution as the image and consists of two channels (one channel for foreground and one for background corrections). The guidance map is concatenated with the RGB image to form a 5-channel map $\mathbf{x}$ which is provided as input to the network.

We use DeepLabV3+ [17] as our network architecture, which has demonstrated good performance on semantic segmentation. However, we note that our method does not depend on a specific architecture and can be used with others as well.

For training the model we need a training dataset $\mathcal{D}$ with ground-truth object segmentations, as well as user corrections which we simulate as in [35] (Sect. 3.3). We train the model using the cross-entropy loss over all pixels in an image:

$$\mathcal{L}_{\text{CE}}(\mathbf{x}, \mathbf{y}; \boldsymbol{\theta}) = \frac{1}{|\mathbf{y}|}\{-\mathbf{y} \log \mathbf{f}(\mathbf{x}; \boldsymbol{\theta}) - (1 - \mathbf{y}) \log(1 - \mathbf{f}(\mathbf{x}; \boldsymbol{\theta}))\} \tag{1}$$

where $\mathbf{x}$ is the 5-channel input defined above (image plus guidance maps), $\mathbf{y} \in \{0, 1\}^{H \times W}$ are the pixel labels of the ground-truth object segmentations, and $\mathbf{f}(\mathbf{x}; \boldsymbol{\theta})$ represents the mapping of the convolutional network parameterized by $\boldsymbol{\theta}$. $|\cdot|$ denotes the l_1 norm.

We produce the initial parameters $\boldsymbol{\theta}^*$ of the segmentation model by minimizing $\sum_{(\mathbf{x}_i, \mathbf{y}_i) \in \mathcal{D}} \mathcal{L}_{\text{CE}}(\mathbf{x}_i, \mathbf{y}_i; \boldsymbol{\theta})$ over the training set using stochastic gradient descent.

3.2 Learning from Corrections at Test-Time

Previous interactive object segmentation methods do not treat corrections as training examples. Thus, the model parameters remain unchanged/frozen at test time [10,11,27,30,35,53] and corrections are only used as inputs to guide the predictions. Instead, we treat corrections as ground-truth labels to adapt the model at test time. We achieve this by minimizing the generalized cross-entropy loss over the corrected pixels:

$$\mathcal{L}_{\text{GCE}}(\mathbf{x}, \mathbf{c}; \boldsymbol{\theta}) = \frac{\mathbf{1}[\mathbf{c} \neq -1]^T}{|\mathbf{1}[\mathbf{c} \neq -1]|} \left\{ - \mathbf{c} \log \mathbf{f}(\mathbf{x}; \boldsymbol{\theta}) - (1 - \mathbf{c}) \log \left(1 - \mathbf{f}\left(\mathbf{x}; \boldsymbol{\theta}\right)\right) \right\} \tag{2}$$

where $\mathbf{1}$ is an indicator function and $\mathbf{c}$ is a vector of values $\{1, 0, -1\}$, indicating what pixels were corrected to what label. Pixels that were corrected to be positive are set to 1 and negative pixels to 0. The remaining ones are set to -1, so that they are ignored in the loss. As there are very few corrections available at test time, this loss is computed over a sparse set of pixels. This is in contrast to the initial training which had supervision at every pixel (Sect. 3.1). We illustrate the contrast between the two forms of supervision in Fig. 2.

Dealing with Label Sparsity. In practice, corrections $\mathbf{c}$ are extremely sparse and consist of just a handful of scattered points (Fig. 3). Hence, they offer limited information on the spatial extent of objects and special care needs to be taken to make this form of supervision useful in practice. As our model is initially trained with full supervision, it has learned strong shape priors. Thus, we propose two auxiliary losses to prevent forgetting these priors as the model is adapted. First, we regularize the model by treating the initial mask prediction $\mathbf{p}$ as ground-truth and making it a target in the cross-entropy loss, *i.e.* $\mathcal{L}_{\text{CE}}(\mathbf{x}, \mathbf{p}; \boldsymbol{\theta})$. This prevents the model from focusing only on the user corrections while forgetting the initially good predictions on pixels for which no corrections were given.

Second, inspired by methods for class-incremental learning [5,28,55], we minimize unnecessary changes to the network parameters to prevent it from forgetting crucial patterns learned on the initial training set. Specifically, we add a cost for changing important network parameters:

$$\mathcal{L}_{\mathrm{F}}(\boldsymbol{\theta}) = \boldsymbol{\Omega}^T (\boldsymbol{\theta} - \boldsymbol{\theta}^*)^{\odot 2} \tag{3}$$

where $\boldsymbol{\theta}^*$ are the initial model parameters, $\boldsymbol{\theta}$ are the updated parameters and $\boldsymbol{\Omega}$ is the importance of each parameter. $(\cdot)^{\odot 2}$ is the element-wise square (Hadamard square). Intuitively, this loss penalizes changing the network parameters away from their initial values, where the penalty is higher for important parameters. We compute $\boldsymbol{\Omega}$ using Memory-Aware Synapses (MAS) [5], which estimates importance based on how much changes to the parameters affect the prediction of the model.

Combined Loss. Our full method uses a linear combination of the above losses:

$$\mathcal{L}_{\mathrm{ADAPT}}(\mathbf{x}, \mathbf{p}, \mathbf{c}; \boldsymbol{\theta}) = \lambda \mathcal{L}_{\mathrm{GCE}}(\mathbf{x}, \mathbf{c}; \boldsymbol{\theta}) + (1 - \lambda)\mathcal{L}_{\mathrm{GCE}}(\mathbf{x}, \mathbf{p}; \boldsymbol{\theta}) + \gamma \mathcal{L}_{\mathrm{F}}(\boldsymbol{\theta}) \tag{4}$$

where λ balances the importance of the user corrections *vs.* the predicted mask, and γ defines the strength of parameter regularization. Next, we introduce *single image adaptation* and *image sequence adaptation*, which both minimize Eq. (4). Their difference lies in how the model parameters $\boldsymbol{\theta}$ are updated: individually for each object or over a sequence.

Adapting to a Single Image. We adapt the segmentation model to a particular object in an image by training on the click corrections. We start from the segmentation model with parameters $\boldsymbol{\theta}^*$ fit to the initial training set (Sect. 3.1). Then we update them by running several gradient descent steps to minimize our combined loss Eq. (4) every time the user makes a correction (Algo. in supp. material). We choose the learning rate and the number of update steps such that the updated model adheres to the user corrections. This effectively turns corrections into constraints. This process results in a segmentation mask $\mathbf{p}$, predicted using the updated parameters $\boldsymbol{\theta}$.

Adapting the model to the current test image brings two core advantages. First, it learns about the specific appearance of the object and background in the current image. Hence corrections have a larger impact and can also improve the segmentation of distant image regions which have similar appearance. The model can also adapt to low-level photometric properties of this image, such as overall illumination, blur, and noise, which results in better segmentation in general. Second, our adaptation step makes the corrections effectively hard constraints, so the model will preserve the corrected labeling in later iterations too.

This adaptation is done for each object separately, and the updated $\boldsymbol{\theta}$ is discarded once an object is segmented.

Adapting to an Image Sequence. Here we describe how to continuously adapt the segmentation model to a sequence of test images using an online algorithm. Again, we start from the model parameters $\boldsymbol{\theta}^*$ fit to the initial training set (Sect. 3.1). When the first test image arrives, we perform interactive segmentation using these initial parameters. Then, after segmenting each image $I_t = (\mathbf{x}_t, \mathbf{c}_t)$, the model parameters are updated to $\boldsymbol{\theta}_{t+1}$ by doing a single gradient descent step to minimize Eq. (4) for that image. Thereby we subsample the corrections in the guidance maps to avoid trivial solutions (predict the corrections given the corrections themselves, see supp. material). The updated model parameters are used to segment the next image I_{t+1}.

Through the method described above our model adapts to the whole test image sequence, but does so gradually, as objects are segmented in sequence. As a consequence, this process is fast, does not require storing a growing number of images, and can be used in a online setting. In this fashion it can adapt to changing appearance properties, adapt to unseen classes, and specialize to one particular class. It can even adapt to radically different image domains as we demonstrate in Sect. 4.3.

Combined Adaptation. For a test image I_t, we segment the object using single image adaptation (Algo. in supp. material). After segmenting a test image, we gather all corrections provided for that image and apply a image sequence adaptation step to update the model parameters from $\boldsymbol{\theta}_t$ to $\boldsymbol{\theta}_{t+1}$. At the next image, the image adaptation process will thus start from parameters $\boldsymbol{\theta}_{t+1}$ better suited for the test sequence. This combination allows to leverage the distinct advantages of the two types of adaptation.

3.3 Simulating User Corrections

To train and test our method we rely on simulated user corrections, as is common practice [10,27,30,32,35,53].

Test-Time Corrections. When interactively segmenting an object, the user clicks on a mistake in the predicted segmentation. To simulate this we follow [10, 35,53], which assume that the user clicks on the largest error region. We obtain this error region by comparing the model predictions with the ground-truth and select its center pixel.

Train-Time Corrections. Ideally one wants to train with the same user model that is used at test-time. To make this computationally feasible, we train the model in two stages as in [35]. First, we sample corrections using ground-truth segmentations [10,27,30,32,53]. Positive user corrections are sampled uniformly at random on the object. Negative user corrections are sampled according to three strategies: (1) uniformly at random from pixels around the object, (2) uniformly at random on other objects, and (3) uniformly around the object. We use these corrections to train the model until convergence. Then, we continue training by iteratively sampling corrections following [35]. For each image we keep a set of user corrections $\mathbf{c}$. Given $\mathbf{c}$ we predict a segmentation mask, simulate the next user correction (as done at test time), and add it to $\mathbf{c}$. Based on this additional correction, we predict a new segmentation mask and minimize the loss (Eq. (1)). Initially, **c** corresponds to the corrections simulated in the first stage, and over time more user corrections are added. As we want the model to work well even with few user corrections, we thus periodically reset **c** to the initial clicks [35].

4 Experiments

We extensively evaluate our single image adaptation and image sequence adaptation methods on several standard datasets as well as on aerial and medical images. These correspond to increasingly challenging adaptation scenarios.

Adaptation Scenarios. We first consider *distribution shift*, where the training and test image sets come from the same general domain, consumer photos, but differ in their image and object statistics (Sect. 4.1). This includes differences in image complexity, object size distribution, and when the test set contains object classes absent during training. Then, we consider a *class specialization* scenario, where a sequence of objects of a single class has to be iteratively segmented (Sect. 4.2). Finally we test how our method handles large *domain changes* where the imaging modality changes between training and testing. We demonstrate this by going from consumer photos to aerial and medical images (Sect. 4.3).

Model Details. We use a strong re-implementation of [35] as our interactive segmentation model (Sect. 3.1). We pre-train its parameters on PASCAL VOC12 [18] augmented with SBD [24] (10582 images with 24125 segmented instances of 20 object classes). As a baseline, we use this model as in [35], *i.e.* without updating its parameters at test time. We call this the *frozen model*. This baseline already achieves state-of-the-art results on the PASCAL VOC12 validation set, simply by increasing the encoder resolution compared to [35] (3.44 clicks). This shows that using a fixed set of model parameters

Table 1. Adapting to distribution shifts. Mean number of clicks required to attain a particular IoU score on Berkeley, YouTube-VOS and COCO datasets (Lower is better). Both of our adaptive methods, single image adaptation (IA) and image sequence adaptation (SA) improve over the model that keeps the weights frozen at test time.

	Berkeley [37]	**YouTube-VOS [54]**	**COCO [33]**		
			Seen	Unseen	Unseen 6k
Method	clicks@90%	clicks@85%	clicks@85%		
Frozen model [35]	5.4	7.9	10.0	11.9	13.2
IA	4.9	7.0	9.1	10.7	10.6
SA	5.3	6.9	9.7	10.6	10.0
IA+SA	**4.9**	**6.7**	**9.1**	**9.9**	**9.3**
Δ over frozen model	8.5%	15.2%	9.0%	16.8%	29.5%

works well when the train and test distributions match. We evaluate our proposed method by adapting the parameters of that same model at test time using *single image adaptation* (IA), *image sequence adaptation* (SA), and their combination (IA + SA).

Evaluation Metrics. We use two standard metrics [10,11,27,30,32,35,53]: (1) **IoU@k**, the average intersection-over-union between the ground-truth and predicted segmentation masks, given k corrections per image, and (2) **clicks@q%**, the average number of corrections needed to reach an IoU of $q\%$ on every image (thresholded at 20 clicks). We always report mean performance over 10 runs (standard deviation is negligible at ≈0.01 for clicks@q%).

Hyperparameter Selection. We optimize the hyperparameters for both adaptation methods on a subset of the ADE20k dataset [56,57]. Hence, the hyperparameters are optimized for adapting from PASCAL VOC12 to ADE20k, which is distinct from the distribution shifts and domain changes we evaluate on.

Implementation Details are provided in the supplementary material.

4.1 Adapting to Distribution Shift

We test how well we can adapt the model which is trained on PASCAL VOC12 to other consumer photos datasets.

Datasets. We test on: (1) *Berkeley* [37], 100 images with a single foreground object. (2) *YouTube - VOS* [54], a large video object segmentation dataset. We use the test set of the 2019 challenge, where we take the first frame with ground truth (1169 objects, downscaled to 855 × 480 maximal resolution). (3) *COCO* [33], a large segmentation dataset with 80 object classes. 20 of those overlap with the ones in the PASCAL VOC12 dataset and are thus *seen* during training. The other 60 are *unseen*. We sample 10 objects per class from the validation set and separately report results for seen (200 objects) and unseen classes (600 objects) as in [36,53]. We also study how image sequence adaptation behaves on longer sequences of 100 objects for each unseen class (named *COCO unseen 6k*).

Results. We report our results in Table 1 and Fig. 4. Both types of adaptation improve performance on all tested datasets. On the first few user corrections *single image adaptation* (IA) performs similarly to the frozen model as it is initialized with the same parameters. But as more corrections are provided, it uses these more effectively to adapt its appearance model to a specific image. Thus, it performs particularly well in the high-click regime, which is most useful for objects that are challenging to segment (*e.g.* due to low illumination, Fig. 3), or when very accurate masks are desired.

During *image sequence adaptation* (SA), the model adapts to the test image distribution and thus learns to produce good segmentation masks given just a few clicks (Fig. 4a). As a result, SA outperforms using a frozen model on all datasets with distribution shifts (Table 1). By adapting from images to the video frames of YouTube-VOS, SA reduces the clicks needed to reach 85% IoU by 15%. Importantly, we find that our method adapts fast, making a real difference after just a few images, and then keeps on improving even as the test sequence becomes thousands of images long (Fig. 4b). This translates to a large improvement given a fixed budget of 4 clicks per object: on the COCO unseen 6k split it achieves 69% IoU compared to the 57% of the frozen model (Fig. 4a).

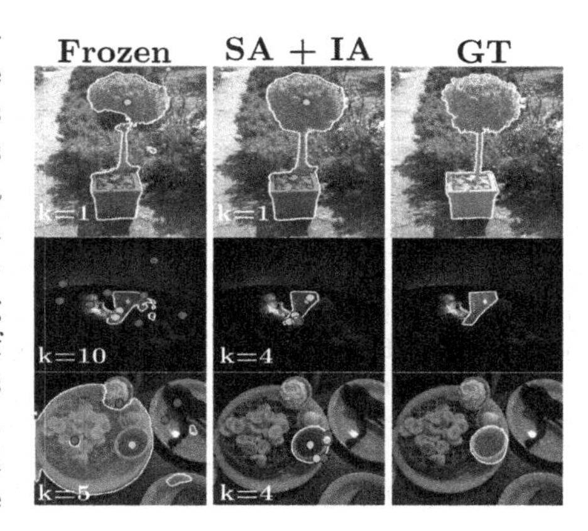

Fig. 3. Qualitative results of the frozen and our combined adaptation model. Red circles are negative clicks and green ones are positive. Green and red areas respectively show the pixels that turned to FG/BG with the latest clicks. Our method produces accurate masks with fewer clicks **k**. (Color figure online)

Generally, the curves for image sequence adaptation grow faster in the low click regime than the single image adaptation ones, but then exhibit stronger diminishing returns in the higher click regime (Fig. 4a). Hence, combining the two compounds their advantages leading to a method that considerably improves over the frozen model on the full range of number of corrections and sequence lengths (Fig. 4a). Compared to the frozen model, our combined method significantly reduces the number of clicks needed to reach the target accuracy on all datasets: from a 9% reduction on Berkeley and COCO seen, to a 30% reduction on COCO unseen 6k.

4.2 Adapting to a Specific Class

When a user segments objects of a single class at test-time, image sequence adaptation naturally specializes its appearance model to that class. We evaluate this phenomenon on 4 COCO classes. We form 4 test image sequences, each focusing on a single class, containing objects of varied appearance. The classes are selected based on how image sequence adaptation performs compared to the frozen model in Sect. 4.1. We selected the following classes, with increasing order of difficulty for image sequence adaptation: (1) donut (2540 objects) (2) bench (3500) (3) umbrella (3979) and (4) bed (1450).

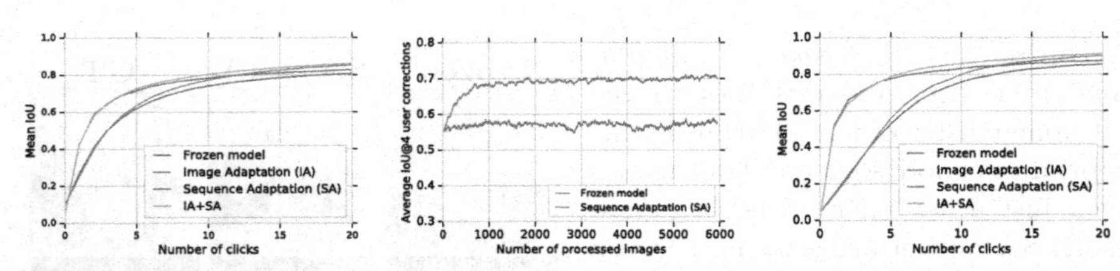

(a) Mean IoU@k for varying k on COCO unseen 6k. Both forms of adaptation significantly improve over a frozen model.

(b) IoU@4 clicks as a function of the number of images processed. SA quickly improves over the model with frozen weights.

(c) IoU@k for varying k when specializing to *donuts*. SA offers large gains by learning a class specific appearance model.

Fig. 4. Results for adapting to dist. shift (a, b) or a specific class (c).

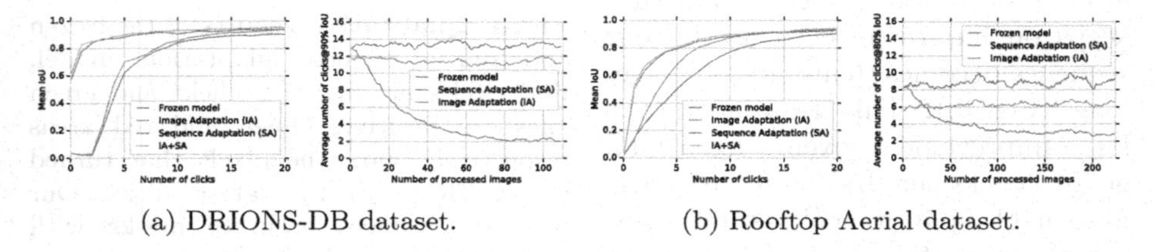

(a) DRIONS-DB dataset.

(b) Rooftop Aerial dataset.

Fig. 5. Results for domain change. For each dataset, we show the mean IoU at k corrections (left in a, b) and the number of clicks to reach the target IoU as a function of the number of images processed (right in a, b). Single image adaptation provides a consistent improvement over the test sequences. Instead, image sequence adaptation adapts its appearance model to the new domain gradually, improving with every image processed (right in a, b).

Results. Table 2, Fig. 4c present results. The class specialization brought by our image sequence adaptation (SA) leads to good masks from very few clicks. For example, on the donut class it reduces clicks@85% by 39% compared to the frozen model and by 44% when combined with single image adaptation (Table 2). Given just 2 clicks, SA reaches 66% IoU for that class, compared to 25% IoU for the frozen model (Fig. 4c). The results for the other classes follow a similar pattern, showing that image sequence adaptation learns an effective appearance model for a single class.

4.3 Adapting to Domain Changes

We test our method's ability of adapting to domain changes by training on consumer photos (PASCAL VOC12) and evaluating on aerial and medical imagery.

Datasets. We explore two test datasets: (1) *Rooftop Aerial* [50], a dataset of 65 aerial images with segmented rooftops and (2) *DRIONS-DB* [14], a dataset of 110 retinal images with a segmentation of the optic disc of the eye fundus. (we use the masks of the first expert). Importantly, the initial model parameters

$\boldsymbol{\theta}^*$ were optimized for the PASCAL VOC12 dataset, which consists of consumer photos. Hence, we explore truly large domain changes here.

Results. Both our forms of adaptation significantly improve over the frozen model (Table 3, Fig. 5). Single image adaptation can only adapt to a limited extent, as it independently adapts to each object instance, always starting from the same initial model parameters $\boldsymbol{\theta}^*$. Nonetheless, it offers a significant improvement, reducing the number of clicks needed to reach the desired IoU by 14%–29%. Image sequence adaptation (SA) shows extremely strong performance, as its adaptation effects accumulate over the duration of the test sequence. It reduces the needed user input by 60% for the Rooftop Aerial dataset and by over 70% for DRIONS-DB. When combining the two types of adaptation, the reduction increases to 77% for the DRIONS-DB dataset (Table 3). Importantly, our method adapts fast: on DRIONS-DB clicks@90% drops quickly and converges to just 2 corrections, as the length of the test sequence increases (Fig. 5a). In contrast, the frozen model performs poorly on both datasets. On the Rooftop Aerial dataset, it needs even more clicks than there are points in the ground truth polygons (8.9 *vs.* 5.1). This shows that even a state-of-the-art model like [35] fails to generalize to truly different domains and highlights the importance of adaptation.

To summarize: We show that our method can bridge large domain changes spanning varied datasets and sequence lengths. With just a single gradient descent step per image, our image sequence adaptation successfully addresses a major shortcoming of neural networks, for the case of interactive segmentation: Their poor generalization to changing distributions [4,45].

4.4 Comparison to Previous Methods

While the main focus of our work is tackling challenging adaptation scenarios, we also compare our method against state-of-the-art interactive segmentation methods on standard datasets. These datasets are typically similar to PASCAL VOC12, hence have a small distribution mismatch between training and testing.

Datasets. (1) Berkeley, introduced in Sect. 4.1 (2) *GrabCut* [46], 49 images with segmentation masks. (3) *DAVIS16* [40], 50 high-resolution videos out of which

Table 2. Class specialization. We test segmenting objects of only one specific class. Our adaptive methods outperforms the frozen model on all tested classes. Naturally, gains are larger for image sequence adaptation, as it can adapt to the class over time.

	clicks @ 85% IoU			
	Donut	**Bench**	**Umbrella**	**Bed**
Frozen model [35]	11.6	15.1	13.1	6.8
IA (Ours)	9.2	14.1	11.9	5.5
SA (Ours)	7.1	14.0	11.1	5.5
IA+SA (Ours)	**6.5**	**13.3**	**10.2**	**5.0**
Δ over frozen model	44.0%	11.9%	22.1%	26.5%

Table 3. Domain change results. We evaluate our model on 2 datasets that belong to different domains: aerial (Rooftop) and medical (DRIONS-DB). Both types of adaptation (IA and SA) outperform the frozen model.

	DRIONS-DB [14]	Rooftop [50]
Method	clicks@90% IoU	clicks@80% IoU
Frozen model [35]	13.3	8.9
IA (Ours)	11.4	6.3
SA (Ours)	3.6	3.6
IA+SA (Ours)	**3.1**	**3.6**
Δ over frozen model	76.7%	59.6%

we sample 10% of the frames uniformly at random as in [27,30] (We note that the standard evaluation protocol of DAVIS16 favors adaptive methods, as the same objects appear repeatedly in the test sequence.) and (4) *PASCAL VOC12 validation*, with 1449 images.

Results. Table 4 shows results. Our adaptation method achieves strong results: At the time of initially releasing our work [29], it outperformed all previous state-of-the-art methods on all datasets (it was later overtaken by [49]). It brings improvements even when the previous methods (which have frozen model parameters) already offers strong performance and need less than 4 clicks on average (PASCAL VOC12, GrabCut). The improvement on PASCAL VOC12 further shows that our method helps even when the training and testing distributions match exactly (the frozen model needs 3.44 clicks).

Importantly, we find that our method outperforms [27,30], even though we use a standard segmentation backbone [17] which predicts at $\frac{1}{4}$ of the input resolution. Instead [27,30] propose specialized network architectures in order to predict at full image resolution, which is crucial for their good performance [27]. We note that our adaptation method is orthogonal to these architectural optimizations and can be combined with them easily.

4.5 Ablation Study

We ablate the benefit of treating corrections as training examples (on COCO unseen 6k). For this, we selectively remove them from the loss (Eq. (4)). For single image adaptation, this leads to a parameter update that makes the model more confident in its current prediction, but this does not improve the segmentation masks. Instead, training on corrections improves clicks@85% from 13.2 to 10.6. For image sequence adaptation, switching off the corrections corresponds to treating the predicted mask as ground-truth and updating the model with it. This approach implicitly contains corrections in the mask and thus improves clicks@85% from 13.2 for the frozen model to 11.9. Explicitly using correction offers an additional gain of almost 2 clicks, down to 10. This shows that treating

Table 4. The focus of our work is handling distribution shifts and domain changes between training and testing (Table 1, 2 and 3). For completeness, we also compare our method against existing methods on standard datasets, where the distribution mismatch between training and testing is small. At the time of initially releasing our work [29], our method outperformed all previous state-of-the-art models on all datasets. Later, F-BRS [49] (CVPR 2020) achieved even better results.

	VOC12 [18] validation	**GrabCut [46]**	**Berkeley [37]**	**DAVIS [40]** 10% of frames
Method	clicks@85%	clicks@90%	clicks@90%	clicks@85%
iFCN w/ GraphCut [53]	6.88	6.04	8.65	–
RIS [32]	5.12	5.00	6.03	–
TSLFN [26]	4.58	3.76	6.49	–
VOS-Wild [10]	5.6	3.8	–	–
ITIS [35]	3.80	5.60	–	–
CAG [36]	3.62	3.58	5.60	–
Latent Diversity [30]	–	4.79	–	5.95
BRS [27]	–	3.60	5.08	5.58
F-BRS [49] (Concurrent Work)	–	2.72	4.57	5.04
IA+SA combined (Ours)	**3.18**	**3.07**	**4.94**	**5.16**

user corrections as training examples is key to our method: They are necessary for single image adaptation and highly beneficial for image sequence adaptation.

4.6 Adaptation Speed

While our method updates the parameters at test time, it remains fast enough for interactive usage. For the model used throughout our paper a parameter update step takes 0.16 s (Nvidia V100 GPU, mixed-precision training, Berkeley dataset). Image sequence adaptation only needs a single update step, done *after* an object is segmented (Sect. 3.2). Thus, the adaptation overhead is negligible here. For single image adaptation we used 10 update steps, for a total time of 1.6 s. We chose this number of steps based on hyperparameter search (see supp. material). In practice, fewer update steps can be used to increase speed, as they quickly show diminishing returns (Fig. 6). We recommend to use 3 update steps, reducing adaptation time to 0.5 s, with a negligible effect on the number of corrections required (average difference of less than 1%, over all datasets).

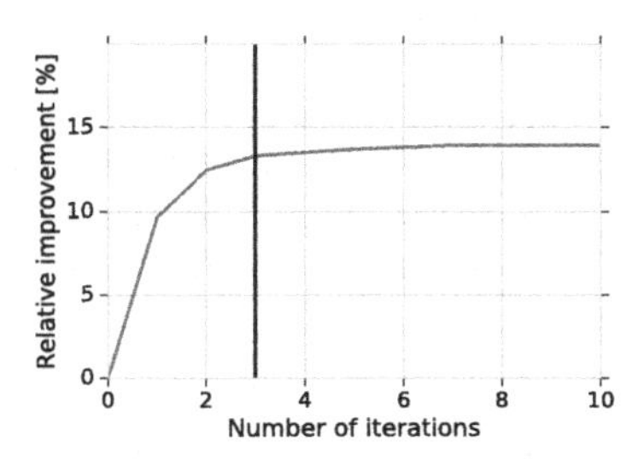

Fig. 6. Iterations *vs.* relative improvement over a frozen model (mean over all datasets).

To increase speed further, the following optimizations are possible: (1) Using a faster backbone, *e.g.* with a ResNet-50 [25], the time for an update step reduces to 0.06 s; (2) Using faster accelerators such as Google Cloud TPUs; (3) Employing a fixed feature extractor and only updating a light-weight segmentation head [30].

5 Conclusion

We propose to treat user corrections as sparse training examples and introduce a novel method that capitalizes on that idea to update the model parameters on-the-fly at test time. Our extensive evaluation on 8 datasets shows the benefits of our method. When distribution shifts between training and testing are small, our methods offers gains of 9%–30%. When specializing to a specific class, our gains are 12%–44%. For large domain changes, where the imaging modality changes between training and testing, it reduces the required number of user corrections by 60% and 77%.

Acknowledgement. We thank Rodrigo Benenson, Jordi Pont-Tuset, Thomas Mensink and Bastian Leibe for their inputs on this work.

References

1. Acuna, D., Ling, H., Kar, A., Fidler, S.: Efficient interactive annotation of segmentation datasets with Polygon-RNN++. In: CVPR (2018)
2. Adobe: Select a subject with just one click (2018). https://helpx.adobe.com/photoshop/how-to/select-subject-one-click.html
3. Agustsson, E., Uijlings, J.R., Ferrari, V.: Interactive full image segmentation by considering all regions jointly. In: CVPR (2019)
4. Alcorn, M.A., et al.: Strike (with) a pose: neural networks are easily fooled by strange poses of familiar objects. In: CVPR (2019)
5. Aljundi, R., Babiloni, F., Elhoseiny, M., Rohrbach, M., Tuytelaars, T.: Memory aware synapses: learning what (not) to forget. In: ECCV (2018)
6. Aljundi, R., Kelchtermans, K., Tuytelaars, T.: Task-free continual learning. In: CVPR (2019)
7. Bai, X., Sapiro, G.: Geodesic matting: a framework for fast interactive image and video segmentation and matting. IJCV **82**(2), 113–132 (2009)
8. Bearman, A., Russakovsky, O., Ferrari, V., Fei-Fei, L.: What's the point: semantic segmentation with point supervision. In: ECCV (2016)
9. Belouadah, E., Popescu, A.: IL2M: class incremental learning with dual memory. In: ICCV (2019)
10. Benard, A., Gygli, M.: Interactive video object segmentation in the wild. arXiv (2017)
11. Benenson, R., Popov, S., Ferrari, V.: Large-scale interactive object segmentation with human annotators. In: CVPR (2019)
12. Boykov, Y., Jolly, M.P.: Interactive graph cuts for optimal boundary and region segmentation of objects in N-D images. In: ICCV (2001)
13. Caelles, S., Maninis, K.K., Pont-Tuset, J., Leal-Taixé, L., Cremers, D., Van Gool, L.: One-shot video object segmentation. In: CVPR (2017)

14. Carmona, E.J., Rincón, M., García-Feijoó, J., Martínez-de-la Casa, J.M.: Identification of the optic nerve head with genetic algorithms. Artif. Intell. Med. **43**(3), 243–259 (2008)
15. Castrejón, L., Kundu, K., Urtasun, R., Fidler, S.: Annotating object instances with a Polygon-RNN. In: CVPR (2017)
16. Chen, D.J., Chien, J.T., Chen, H.T., Chang, L.W.: Tap and shoot segmentation. In: AAAI (2018)
17. Chen, L.C., Zhu, Y., Papandreou, G., Schroff, F., Adam, H.: Encoder-decoder with atrous separable convolution for semantic image segmentation. In: ECCV (2018)
18. Everingham, M., Van Gool, L., Williams, C.K.I., Winn, J., Zisserman, A.: The PASCAL visual object classes challenge 2012 (VOC2012) Results (2012). http://www.pascal-network.org/challenges/VOC/voc2012/workshop/index.html
19. Farquhar, S., Gal, Y.: Towards robust evaluations of continual learning. arXiv (2018)
20. Finn, C., Abbeel, P., Levine, S.: Model-agnostic meta-learning for fast adaptation of deep networks. In: ICML (2017)
21. Gatys, L.A., Ecker, A.S., Bethge, M.: Image style transfer using convolutional neural networks. In: CVPR (2016)
22. Gulshan, V., Rother, C., Criminisi, A., Blake, A., Zisserman, A.: Geodesic star convexity for interactive image segmentation. In: CVPR (2010)
23. Gygli, M., Norouzi, M., Angelova, A.: Deep value networks learn to evaluate and iteratively refine structured outputs. In: ICML (2017)
24. Hariharan, B., Arbeláez, P., Bourdev, L., Maji, S., Malik, J.: Semantic contours from inverse detectors. In: ICCV (2011)
25. He, K., Zhang, X., Ren, S., Sun, J.: Deep residual learning for image recognition. arXiv preprint arXiv:1512.03385 (2015)
26. Hu, Y., Soltoggio, A., Lock, R., Carter, S.: A fully convolutional two-stream fusion network for interactive image segmentation. Neural Netw. **109**, 31–42 (2019)
27. Jang, W.D., Kim, C.S.: Interactive image segmentation via backpropagating refinement scheme. In: CVPR (2019)
28. Kirkpatrick, J., et al.: Overcoming catastrophic forgetting in neural networks. Proc. Nat. Acad. Sci. USA **114**(13), 3521–3526 (2017)
29. Kontogianni, T., Gygli, M., Uijlings, J., Ferrari, V.: Continuous adaptation for interactive object segmentation by learning from corrections. arXiv preprint arXiv:1911.12709v1 (2019)
30. Li, Z., Chen, Q., Koltun, V.: Interactive image segmentation with latent diversity. In: CVPR (2018)
31. Li, Z., Hoiem, D.: Learning without forgetting. IEEE Trans. PAMI **40**(12), 2935–2947 (2017)
32. Liew, J., Wei, Y., Xiong, W., Ong, S.H., Feng, J.: Regional interactive image segmentation networks. In: ICCV (2017)
33. Lin, T.Y., et al.: Microsoft COCO: common objects in context. In: ECCV (2014)
34. Ling, H., Gao, J., Kar, A., Chen, W., Fidler, S.: Fast interactive object annotation with Curve-GCN. In: CVPR (2019)
35. Mahadevan, S., Voigtlaender, P., Leibe, B.: Iteratively trained interactive segmentation. In: BMVC (2018)
36. Majumder, S., Yao, A.: Content-aware multi-level guidance for interactive instance segmentation. In: CVPR (2019)
37. McGuinness, K., O'connor, N.E.: A comparative evaluation of interactive segmentation algorithms. Pattern Recogn. **43**(2), 434–444 (2010)

38. Michieli, U., Zanuttigh, P.: Incremental learning techniques for semantic segmentation. In: ICCV Workshop (2019)
39. Papadopoulos, D.P., Uijlings, J.R.R., Keller, F., Ferrari, V.: We don't need no bounding-boxes: Training object class detectors using only human verification. In: CVPR (2016)
40. Perazzi, F., Pont-Tuset, J., McWilliams, B., Van Gool, L., Gross, M., Sorkine-Hornung, A.: A benchmark dataset and evaluation methodology for video object segmentation. In: CVPR (2016)
41. Price, B.L., Morse, B., Cohen, S.: Geodesic graph cut for interactive image segmentation. In: CVPR (2010)
42. Qi, S., Zhu, Y., Huang, S., Jiang, C., Zhu, S.C.: Human-centric indoor scene synthesis using stochastic grammar. In: CVPR (2018)
43. Ravi, S., Larochelle, H.: Optimization as a model for few-shot learning. In: ICLR (2016)
44. Rebuffi, S., Kolesnikov, A., Sperl, G., Lampert, C.: iCaRL: incremental classifier and representation learning. In: CVPR (2017)
45. Recht, B., Roelofs, R., Schmidt, L., Shankar, V.: Do CIFAR-10 classifiers generalize to CIFAR-10? arXiv (2018)
46. Rother, C., Kolmogorov, V., Blake, A.: GrabCut - interactive foreground extraction using iterated graph cut. SIGGRAPH **23**(3), 309–314 (2004)
47. Shmelkov, K., Schmid, C., Alahari, K.: Incremental learning of object detectors without catastrophic forgetting. In: ICCV (2017)
48. Snell, J., Swersky, K., Zemel, R.: Prototypical networks for few-shot learning. In: NeurIPS (2017)
49. Sofiiuk, K., Petrov, I., Barinova, O., Konushin, A.: F-BRS: rethinking backpropagating refinement for interactive segmentation. In: CVPR (2020)
50. Sun, X., Christoudias, C.M., Fua, P.: Free-shape polygonal object localization. In: ECCV (2014)
51. Sun, Y., Wang, X., Liu, Z., Miller, J., Efros, A.A., Hardt, M.: Test-time training for out-of-distribution generalization. arXiv (2019)
52. Voigtlaender, P., Leibe, B.: Online adaptation of convolutional neural networks for video object segmentation. In: BMVC (2017)
53. Xu, N., Price, B., Cohen, S., Yang, J., Huang, T.: Deep interactive object selection. In: CVPR (2016)
54. Xu, N., et al.: YouTube-VOS: a large-scale video object segmentation benchmark. arXiv (2018)
55. Zenke, F., Poole, B., Ganguli, S.: Continual learning through synaptic intelligence. In: ICML (2017)
56. Zhou, B., Zhao, H., Puig, X., Fidler, S., Barriuso, A., Torralba, A.: Scene parsing through ADE20K dataset. In: CVPR (2017)
57. Zhou, B., Zhao, H., Puig, X., Fidler, S., Barriuso, A., Torralba, A.: Semantic understanding of scenes through the ADE20K dataset. IJCV **127**(3), 302–321 (2018)
58. Zhou, Z., Shin, J., Zhang, L., Gurudu, S., Gotway, M., Liang, J.: Fine-tuning convolutional neural networks for biomedical image analysis: actively and incrementally. In: CVPR (2017)

Impact of Base Dataset Design on Few-Shot Image Classification

Othman Sbai[1,2(✉)], Camille Couprie[1], and Mathieu Aubry[2]

[1] Facebook AI Research, Paris, France
[2] LIGM (UMR 8049) - École des Ponts, UPE, Champs-sur-Marne, France
sbaio@fb.com

Abstract. The quality and generality of deep image features is crucially determined by the data they have been trained on, but little is known about this often overlooked effect. In this paper, we systematically study the effect of variations in the training data by evaluating deep features trained on different image sets in a few-shot classification setting. The experimental protocol we define allows to explore key practical questions. What is the influence of the similarity between base and test classes? Given a fixed annotation budget, what is the optimal trade-off between the number of images per class and the number of classes? Given a fixed dataset, can features be improved by splitting or combining different classes? Should simple or diverse classes be annotated? In a wide range of experiments, we provide clear answers to these questions on the mini-ImageNet, ImageNet and CUB-200 benchmarks. We also show how the base dataset design can improve performance in few-shot classification more drastically than replacing a simple baseline by an advanced state of the art algorithm.

Keywords: Dataset labeling · Few-shot classification · Meta-learning · Weakly-supervised learning

1 Introduction

Deep features can be trained on a base dataset and provide good descriptors on new images [31,39]. The importance of large scale image annotation for the base training is now fully recognized and many efforts are dedicated to creating very large scale datasets. However, little is known on the desirable properties of such dataset, even for standard image classification tasks. To evaluate the impact of the dataset on the quality of learned features, we propose an experimental protocol based on few-shot classification. In this setting, a first model is typically trained to extract features on a base training dataset, and in a second classification stage, features are used to label images of novel classes given only

Electronic supplementary material The online version of this chapter (https://doi.org/10.1007/978-3-030-58517-4_35) contains supplementary material, which is available to authorized users.

A. Vedaldi et al. (Eds.): ECCV 2020, LNCS 12361, pp. 597–613, 2020.
https://doi.org/10.1007/978-3-030-58517-4_35

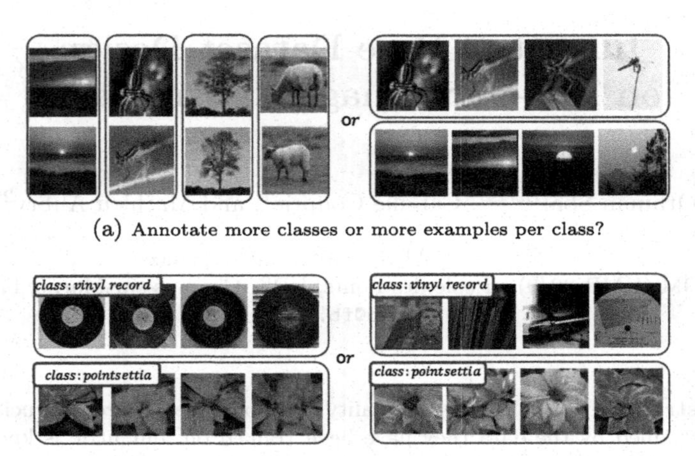

Fig. 1. How should we design the base training dataset and how will it influence the features? a) Many classes with few examples/few classes with many examples; b) Simple or diverse base training images.

few exemplars. Beyond the interest of few-shot classification itself, our protocol is well suited to vary specific parameters in the base training set and answer specific questions about its design, such as the ones presented in Fig. 1.

We believe this work is the first to study, with a consistent approach, the importance of the similarity of training and test data, the suitable trade-off between the number of classes and the number of images per class, the possibility of defining better labels for a given set of images, and the optimal diversity and complexity of the images and classes to annotate. Past studies have mostly focused on feature transfer between datasets and tasks [23,48]. The study most related to ours is likely [23], which asks the question "What makes ImageNet good for transfer learning?". The authors present a variety of experiments on transferring features trained on ImageNet to SUN [47] and Pascal VOC classification and detection [11], as well as a one-shot experiment on ImageNet. However, using AlexNet fc7 features [26], and often relying on the WordNet hierarchy [13], the authors find that variations of the base training dataset do not significantly affect transfer performance, in particular for the balance between image-per-class and classes. This is in strong contrast with our results, which outline the importance of this trade-off in our setup. We believe this might partially be due to the importance of the effect of transfer between datasets, which overshadows the differences in the learned features. Our few-shot learning setting precisely allows to focus on the influence of the training data without considering the complex issues of domain or task transfer.

Our work also aims at outlining data collection strategies and research directions that might lead to new performance boosts. Indeed, several works [6,41] have recently stressed the limitations of performance improvements brought when training on larger datasets, obtained for example by aggregating datasets [41]. On the contrary, [15] shows performance can be improved using a "Selective Joint Fine-Tuning" strategy for transfer learning, selecting only images in the source dataset with low level feature similar to the target dataset and training jointly on both. Our results give insights on why it might happen, showing in particular that a limited number of images per class is often sufficient to obtain good features. Code is available at http://imagine.enpc.fr/~sbaio/fewshot_dataset_design.

Contribution. Our main contribution is an experimental protocol to systematically study the influence of the characteristics of the base training dataset on the resulting deep features for few-shot classification. It leads us to the following key conclusions:

- The similarity of the base training classes and the test classes has a crucial effect and standard datasets for few-shot learning consider only a very specific scenario.
- For a fixed annotation budget, the trade-off between the number of classes and the number of images per class has a major effect on the final performance. The best trade-off usually corresponds to much fewer images per class (~ 60) than collected in most datasets.
- If a dataset with a sub-optimal class number is already available, we demonstrate that a performance boost can be achieved by grouping or splitting classes. While oracle features work best, we show that class grouping can be achieved using self-supervised features.
- Class diversity and difficulty also have an independent influence, easier classes with lower than average diversity leading to better few-shot performances.

While we focus most of our analysis on a single few-shot classification approach and architecture backbone, key experiments for other methods and architectures demonstrate the generality of our results.

2 Related Work and Classical Few-Shot Benchmarks

2.1 Data Selection and Sampling

Training image selection is often tackled through the lens of **active learning** [7]. The goal of active learning is to select a subset of samples to label when training a model, while obtaining similar performance as in the case where the full dataset is annotated. A complete review of classical active learning approaches is beyond the scope of this work and can be found in [38]. A common strategy is to remove redundancy from datasets by designing acquisition functions (entropy, mutual information, and error count) [6,14] to better sample training data. Specifically, [6] introduces an "Adaptive Dataset Subsampling" approach designed to remove redundant samples in datasets. It predicts the uncertainty of ensemble of models to encourage the selection of samples with high "disagreement". Another approach is to select samples close to the boundary decision of the model, which in the case of deep networks can be done using adversarial examples [10]. In [37], the authors adapt active learning strategies to batch training of neural networks and evaluate their method in a transfer learning setting. While these approaches select specific training samples based on their diversity or difficulty, they typically focus on performance on a fixed dataset and classes, and do not analyze performance of learned features on new classes as in our few-shot setting.

Related to active learning is the question of online **sampling strategies** to improve the training with fixed, large datasets [3,12,24,30]. For instance, the study of [3] on class imbalance highlights over-sampling or under-sampling strategies that are privileged in many works. [12] and [24] propose respectively reinforcement learning and importance sampling strategies to select the samples which lead to faster convergence for SGD.

The spirit of our work is more similar to studies that try to understand key properties of good training samples to **remove unnecessary samples** from large datasets. Focusing on the deep training process and inspired by active SVM learning approaches, [43] explored using the gradient magnitude as a measure of the importance of training images. However using this measure to select training examples leads to poor performances on CIFAR and ImageNet. [2] identifies redundancies in datasets such as ImageNet and CIFAR using agglomerative clustering [8]. Similar to us, they use features from a network pre-trained on the full dataset to compute an oracle similarity measure between the samples. However, their focus is to demonstrate that it is possible to slightly reduce the size of datasets (10%) without harming test performance, and they do not explore further the desirable properties of a training dataset.

2.2 Few-Shot Classification

The goal of few-shot image classification is to be able to classify images from novel classes using only a few labeled examples, relying on a large base dataset of annotated images from other classes. Among the many deep learning approaches, the pioneer Matching networks [42] and Prototypical networks [40] tackle the problem from a metric learning perspective. Both methods are meta-learning approaches, i.e. they train a model to learn from sampled classification episodes similar to those of evaluation. MatchingNet considers the cosine similarity to compute an attention over the support set, while ProtoNet employs an ℓ_2 between the query and the class mean of support features.

Recently, [5] revisited few-shot classification and showed that the simple, meta-learning free, Cosine Classifier baseline introduced in [17] performs better or on par with more sophisticated approaches. Notably, its results on the CUB and Mini-ImageNet benchmarks were close to the state-of-the-art [1,27]. Many more approaches have been proposed even more recently in this very active research area (e.g. [28,35]), including approaches relying on other self-supervised tasks (e.g. [16]) and semi-supervised approaches (e.g. [22,25,29]), but a complete review is outside the scope of this work, and exploration of novel methods orthogonal to our goal.

The choice of the base dataset remains indeed largely unexplored in previous studies, whereas we show that it has a huge impact on the performance, and different choices of base datasets might lead to different optimal approaches. The

Meta-dataset [41] study is related to our work from the perspective of analyzing dataset impact on few-shot performance. However, it investigates the effect of meta-training hyper-parameters, while our study focuses on how the base dataset design can improve few-shot classification performance. More recently, [49] investigates the same question of selecting base classes for few-shot learning, leading to a performance better than that of random choice, while highlighting the importance of base dataset selection in few-shot learning.

Since a Cosine Classifier (CC) with a Wide ResNet backbone is widely recognized as a strong baseline [5,16,17,45], we use it as reference, but also report results with two other classical algorithms, namely MatchingNet and ProtoNet.

The classical benchmarks for few-shot evaluation on which we build and evaluate are listed below. Note this is not an exhaustive review, but a selection of diverse datasets which are suited to our goals.

Mini-ImageNet Benchmark. Mini-ImageNet is a common benchmark for few-shot learning of small resolution images [33,42]. It includes 600K images from 100 random classes sampled from the ImageNet-1K [9] dataset and downsampled to 84×84 resolution. It has a standard split of base training, validation and test classes of 64, 16, and 20 classes respectively.

ImageNet Benchmark. For high-resolution images, we consider the few-shot learning benchmark proposed by [19,46]. This benchmark splits the ImageNet-1K dataset into 389 base training, 300 validation and 311 novel classes. The base training set contains 497350 images.

CUB Benchmark. For fine-grained classification, we experiment with the CUB-200-2011 dataset [44]. It contains 11,788 images from 200 classes, each class containing between 40 to 60 images. Following [5,21] we resize the images to 84×84 pixels and use the standard splits in 100 base, 50 validation and 50 novel classes and use exactly the same evaluation protocol as for mini-ImageNet.

3 Base Dataset Design and Evaluation for Few-Shot Classification

In this section, we present the different components of our analysis. First, we explain in detail the main few-shot learning approach that we use to evaluate the influence of training data. Second, we present the large base dataset we use to sample training sets. Third, we discuss the different descriptors of images and classes that we consider, the different splitting and grouping strategies we use for dataset relabeling and the class selection methods we analyze. Finally we give details on architecture and training.

3.1 Dataset Evaluation Using Few-Shot Classification

Few-shot image classification aims at classifying test examples in novel categories using only a few annotated examples per category and typically relying on a

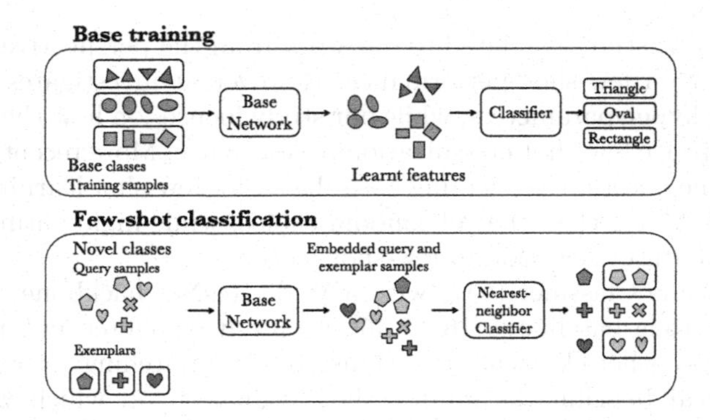

Fig. 2. Illustration of our few-shot learning framework. We train a feature extractor together with a classifier on base training classes. Then, we evaluate the few-shot classification performance of this learned feature extractor to classify novel unseen classes with few annotated examples using a nearest neighbor classifier.

larger base training set with annotated data for training categories. We use the simple but efficient nearest neighbor based approach, visualized in Fig. 2.

More precisely, we start by training a feature extractor f with a cosine classifier on base categories (Fig. 2 top). Then, we define a linear classifier for the novel classes as follows: if z_i for $i = 1...N$ are the labelled examples for a given novel class, we define the classifier weights w for this class as:

$$w = \frac{1}{N} \sum_{i=1}^{N} \frac{f(z_i)}{\|f(z_i)\|}. \tag{1}$$

In other words, we associate each test image to the novel class for which its average cosine similarity with the examples from this novel class is the highest. Previous work on few-shot learning focuses on algorithm design for improving the classifier defined on new labels. Instead, we explore the orthogonal dimension of base training dataset and compare the same baseline classifier using features trained on different base datasets.

3.2 A Large Base Dataset, ImageNet-6K

To investigate a wide variety of base training datasets, we design the ImageNet-6K dataset from which we sample images and classes for our experiments. We require both a large number of classes and a large number of images per class, to allow very diverse image selections, class splittings or groupings. We define ImageNet-6K as the subset from the ImageNet-22K dataset [9,34] containing the largest 6K classes, excluding ImageNet-1K classes. Image duplicates are removed automatically as done in [36]. Each class has more than 900 images. For experiments on mini-ImageNet and CUB, we downsample the images to 84×84, and dub the resulting dataset MiniIN6K. For CUB experiments, to avoid training on classes corresponding to the CUB test set, we additionally look for the most similar images to each of the 2953 images of CUB test set using our oracle features (see Sect. 3.3), and completely remove the 296 classes they belong to. We denote this base dataset MiniIN6K*.

3.3 Class Definition and Sampling Strategies

Image and Class Representation. In most experiments, we represent images by what we call *oracle features*, i.e. features trained on our IN6k or miniIN6K datasets. These features can be expected to provide a good notion of distance between images, but can of course not be used in a practical scenario where no large annotated dataset is available. Each class is represented by its average feature as defined in Eq. 1. This class representation can be used for examples to select training classes close or far from the test classes, or to group similar classes.

We also report results with several alternative representations and metrics. In particular, we experiment with *self-supervised features*, which could be computed on a new type of images from a non-annotated dataset. We tried using features from RotNet [18], DeepCluster [4], and MoCo [20] approaches, and obtained stronger results with MoCo features which we report in the paper. MoCo exploits the self-supervised feature clustering idea and builds a feature dictionary using a contrastive loss. As an additional baseline we report results using deep features with randomly initialized weights and updated batch normalization layers during 1 epoch of miniIN6k. Finally, similar to several prior works, we experiment using the WordNet [13] hierarchy to compute similarity between classes based on the shortest path that relates their synsets and on their respective depths.

Defining New Classes. A natural question is whether for a fixed set of images, different labels could be used to train a better feature extractor.

Given a set of images, we propose to use existing class labels to define new classes by splitting or merging them. Using K-means to cluster images or classes would lead to unbalanced classes, we thus used different strategies for splitting and grouping, which we compare to K-means in the Sup. Mat:

- *Class splitting.* We iteratively split in half every class along the principal component computed over the features of the class images. We refer to this strategy as BPC (Bisection along Principal Component).
- *Class grouping.* To merge classes, we use a simple greedy algorithm which defines meta-classes by merging the two closest classes using their mean features, and repeat the same process for unprocessed classes recursively.

We display examples of resulting grouped and split classes in the Sup. Mat.

Measuring Class Diversity and Difficulty. One of the questions we ask is whether class diversity impacts the trained features' few-shot performance. We therefore analyze results by sampling classes more or less frequently according to their diversity and difficulty:

- *Class diversity.* We use the variance of the normalized features as a measure of class diversity. Classes with low feature variance consist of very similar looking objects or simple visual concepts while the ones with high feature variance represent abstract concepts or include very diverse images.
- *Class difficulty.* To measure the difficulty of a class, we use the validation accuracy of our oracle classifier.

3.4 Architecture and Training Details

We use different architectures and training methods in our experiments. Similar to previous works [5,45], we employ WRN28-10, ResNet10, ResNet18 and Conv4 architectures. The ResNet architectures are adapted to handle 84×84 images by replacing the first convolution with a kernel size of 3 and stride of 1 and removing the first max pooling layer. In addition to the cosine classifier described in Section 3.1, we experiment with the classical Prototypical Networks [40] and Matching Networks [42].

Since we compare different training datasets, we adapt the training schedule depending on the size of the training dataset and the method. For example on MiniIN-6k, we train Prototypical Networks and Matching Networks for 150k episodes, while when training on smaller size datasets we use 40k episodes as in [5]. We use fewer query images per class when training on classes with not enough images per class for Prototypical and Matching Networks.

When training a Cosine Classifier, we train using an SGD optimizer with momentum of 0.9 and weight decay of 5.10^{-4} for 90 epochs starting with an initial learning rate of 0.05 and dividing it by 10 every 30 epochs. We also use a learning rate warmup for the first 6K iterations, that we found beneficial for stabilizing the training and limiting the variance of the results. For large datasets with more than 10^6 images, we use a batch size of 256 and 8 GPUs to speed up the training convergence, while for smaller datasets (most of our experiments are done using datasets of 38400 images, as in MiniIN training set), we use a batch size of 64 images and train on a single GPU. During training, we use a balanced class sampler that ensures sampled images come from a uniform distribution over the classes regardless of their number of images.

On the ImageNet benchmark, we use a ResNet-34 network and trained for 150K dividing the learning rate by 10 after 120K, 135K and 145K iterations using a batch size of 256 on 1 GPU.

Following common practices, during evaluation, we compute the average top-1 accuracy on 15 query examples over 10k episodes sampled from the test set on 5-way tasks for miniIN and CUB benchmarks, while we compute the top-5 accuracy on 6 query examples over 250-way tasks on the ImageNet benchmark.

4 Analysis

4.1 Importance of Base Data and Its Similarity to Test Data

We start by validating the importance of the base training dataset for the few-shot classification, both in terms of size and of the selection of classes. In Table 1,

Table 1. 5-shot, 5-way accuracy on MiniIN and CUB test sets using different base training data, algorithms and backbones. PN: Prototype Networks [40]. MN: Matching Networks [42]. CC: Cosine Classifier. WRN: Wide ResNet28-10. MiniIN6K (resp. MiniIN6K*) Random: 600 images from 64 classes sampled randomly from MiniIN6K (resp. MiniIN6K*). We evaluate the variances over 3 different runs.

		MiniIN test			CUB test		
	Algo.	Base data					
		MiniIN N=38400 C = 64	MiniIN6K Random N=38400 C = 64	MiniIN6K $N\approx$7,1.10^6 C=6000	CUB N=5885 C = 100	MiniIN6K* Random N=38400 C = 64	MiniIN6K* $N\approx$6,8.10^6 C = 5704
WRN	PN [40]	73.64±0.84	70.26±1.30	85.14±0.28	87.84±0.42	52.51±1.57	68.62±0.5
	MN [42]	69.19±0.36	65.45±1.87	82.12±0.27	85.08±0.62	46.32±0.72	59.90±0.45
	CC	**78.95±0.24**	**75.48±1.53**	**96.91**±0.14	**90.32**±0.14	**58.03±1.43**	**90.89**±0.10
Conv4	CC	65.99±0.04	64.05±0.75	74.56±0.12	80.71±0.15	56.44±0.63	66.81±0.30
ResNet10	CC	76.99±0.07	74.17±1.42	91.84±0.06	89.07±0.15	57.01±1.44	82.20±0.44
ResNet18	CC	78.29±0.05	75.14±1.58	93.36±0.19	89.99±0.07	56.64±1.28	88.32±0.23

we report five shot results on the CUB and MiniIN datasets, the one shot results are available in the Sup. Mat. We write N the total number of images in the dataset and C the number of classes. Similar results on ImageNet benchmark can be read in the Sup. Mat. On the miniIN benchmark, we observe that our implementation of the strong CC baseline using a WRN backbone yields slightly better performance using miniIN base classes than the ones reported in [16,27] (76.59). We validate the consistency of our observations by varying algorithms and architectures using the codebase of [5].

Our first finding is that using the whole miniIN-6K dataset for the base training boosts the performance on miniIN by a very large amount, 20% and 18% for 1-shot and 5-shot classification respectively, compared to training on 64 miniIN base classes. Training on IN-6K images also results in a large 10% boost in 5-shot top-5 accuracy on ImageNet benchmark. Another interesting result is that sampling random datasets of 64 classes and 600 images per class leads to a 5-shot performance of 75.48% on MiniIN clearly below the one using the base classes from miniIN 78.95%. A similar observation can be made for different backbones (Conv4, ResNets) and algorithms tested (ProtoNet, MatchingNets), as well as on the ImageNet benchmark. A natural explanation for these differences is that the base training classes from the benchmarks are correlated to the test classes.

To validate this hypothesis, we selected a varying number of base training classes from miniIN-6K closest and farthest to miniIN test classes using either oracle features, MoCo features, or the WordNet hierarchy, and report the results of training using a cosine classifier with WRN architecture in Fig. 3a. A similar experiment on CUB is shown in the Sup. Mat. We use 900 random images for each class. While all features used for class selection yield similarly superior results for closest class selection and worst results for farthest class selection, we observe that using oracle features leads to larger differences than using MoCo features and Wordnet hierarchy. In Fig. 3b, we study the influence of the architecture and training method on the previously observed importance of class similarity to test classes. Similar gaps can be observed in all cases. Note however that for ProtoNet, MatchingNet and smaller backbones with CC, the best performance is not obtained with the largest number of classes.

While these findings themselves are not surprising, the amplitude of performance variations demonstrates the importance of studying the influence of training data and strategies for data selection, especially considering that most advanced few-shot learning strategies only increase performance by a few points compared to strong nearest neighbor based baselines such as CC [5,32].

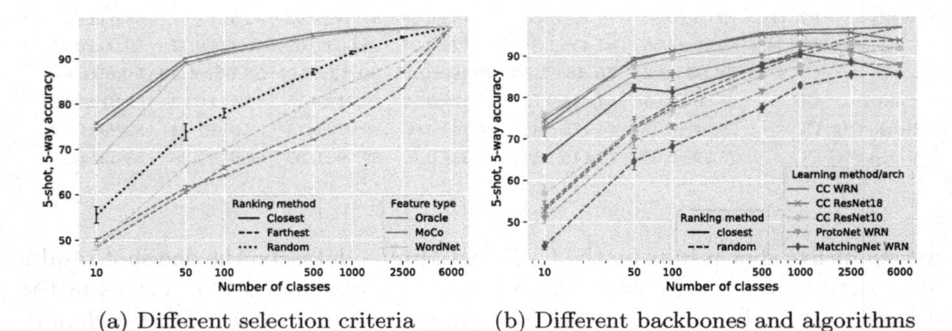

(a) Different selection criteria (b) Different backbones and algorithms

Fig. 3. Five-shot accuracy on miniIN when sampling classes from miniIN-6K randomly or closest/farthest to the miniIN test set using 900 images per class. (a) Comparison between different class selection criteria for selecting classes closest or farthest from the test classes. (b) Comparison of results with different algorithms and backbones using oracle features to select closest classes.

4.2 Effect of the Number of Classes for a Fixed Number of Annotations

An important practical question when building a base training dataset is the number of classes and images to annotate, since the constraint is often the cost of the annotation process. We thus consider a fixed number of annotated images and explore the effect of the trade-off between the number of images per class and the number of classes. In Fig. 4, we visualize the 5-shot performance resulting from this trade-off in the base training classes on the miniIN and CUB benchmarks. In all cases, we select the classes and images randomly from our miniIN6K and miniIN6k* dataset respectively, and plot the variance over 3 runs.

First, in Fig. 4 (a,b) we compare the trade-off for different numbers of annotated images. We sample randomly datasets of 38400 or 3840 images with different number of classes and the same number of image in each class. We also indicate the performance with the standard benchmarks base dataset and the full miniIN6K data. The same graph on ImageNet benchmark can be seen in the Sup. Mat using 50k and 500k images datasets.

As expected, the performance decreases when too few classes or too few images per classes are available. Interestingly, on the miniIN test benchmark (Fig. 4a) the best performance is obtained around 384 classes and 100 images per class with a clear boost (around 5%) over the performance using 600 images for 64 classes which is the trade-off chosen in the miniIN benchmark, we observe that the best trade-off is very different on the CUB benchmark, corresponding to more classes and very few images per class. We believe this is due to the fine-grained nature of the dataset.

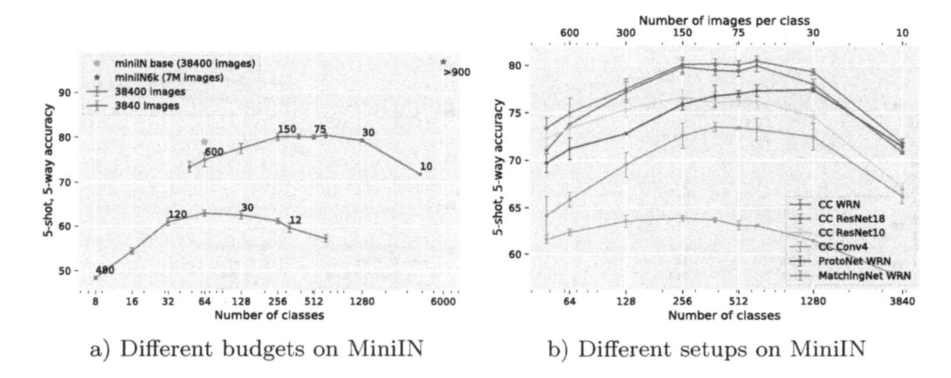

a) Different budgets on MiniIN b) Different setups on MiniIN

Fig. 4. Trade-off between the number of classes and images per class for a fixed image budget. In (a) we show the trade-off for different dataset sizes and points are annotated with the corresponding number of images per class. In (b) we consider a total budget of 38400 annotated images and show the trade-off for different architectures and methods. The top scale shows the number of images per class and the bottom scale the number of classes.

Second, in Fig. 4(b), we study the consistency of these findings for different architectures and few-shot algorithms with a 38400 annotated images budget. While the trade-off depends on the architecture and method, there is always a strong effect, and the optimum tends to correspond to much fewer images per class than in standard benchmarks. For example, the best performance with ProtoNet and MatchingNet on the miniIN benchmark is obtained with as few as 30 images per class. This is interesting since it shows that the ranking of different few-shot approaches may depend on the trade-off between number of base images and classes selected in the benchmark.

The importance of this balance, and the fact that it does not necessarily correspond to the one used in the standard datasets is also important if one wants to pre-train features with limited resources. Indeed, better features can be obtained by using more classes and less images per class compared to using all available images for the classes with the largest number of images as is often done, with the idea to avoid over-fitting. Again, the boost observed for few-shot classification performance is very important compared to the ones provided by many advanced few-shot learning approaches.

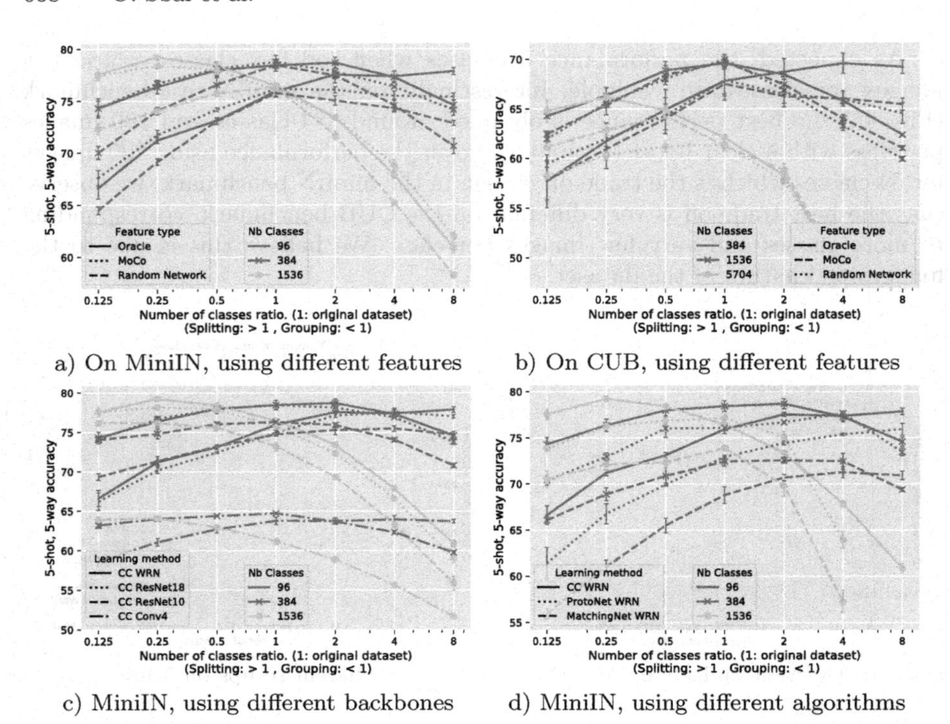

a) On MiniIN, using different features

b) On CUB, using different features

c) MiniIN, using different backbones

d) MiniIN, using different algorithms

Fig. 5. Impact of class grouping or splitting on few-shot accuracy on miniIN and CUB depending on the initial number of classes. Starting from different number of classes C, we group similar classes together into meta-classes or split them into sub-classes to obtain $\alpha \times C$ ones. $\alpha \in \{\frac{1}{8}, \frac{1}{4}, \frac{1}{2}, 1, 2, 4, 8\}$ is the x-axis. Experiments in a) and b) use CC WRN setup.

4.3 Redefining Classes

There are two possible explanations for the improvement provided by the increased number of classes for a fixed number of annotated images discussed in the previous paragraph. The first one is that the images sampled from more random classes cover better the space of natural images, and thus provide images more likely similar to the test images. The second one is that learning a classifier with more classes is itself beneficial to the quality of the features. To investigate whether for fixed data, increasing the number of classes can boost performances, we relabel images inside each class as described in Sect. 3.3.

In Fig. 5, we compare the effect of grouping and splitting classes on three dataset configurations sampled from miniIN-6K and miniIN6K*, with a total number of images of 38400 with different number of classes $C \in \{96, 384, 1536\}$ for miniIN and $C \in \{384, 1536, 5704\}$ for CUB. Given images originally labeled with C classes, we relabel images of each class to obtain $\alpha \times C$ sub-classes. The x-axes represent the class ratio $\alpha \in \{\frac{1}{8}, \frac{1}{4}, \frac{1}{2}, 1, 2, 4, 8\}$. For class ratios lower than 1, we group classes using our greedy iterative grouping, while for ratios α

greater than 1, we split classes using our BCP method. In Fig. 5(a,b), we show three possible behaviors on miniIN and CUB when using our oracle features: (i) if the number of initial classes is higher than the optimal trade-off, grouping is beneficial and splitting hurts performances (yellow curves); (ii) if the number of initial classes is the optimal one, both splitting and grouping decrease performances (blue curves); (iii) if the number of initial classes is smaller than the optimal tradeoff, splitting is beneficial and grouping hurts performance (red curves). This is a strong result, since it shows there is potential to improve performances with a fixed training dataset by redefining new classes. This can be done for grouping using the self-supervised MoCo features. However, we found it is not sufficient to split classes in a way that improves performances. Using random features on the contrary does not lead to any significant improvements. Figure 5c confirms the consistency of results with various architecture on miniIN benchmark. Figure 5d compares these results to the ones obtained with ProtoNet and MatchingNet. Interestingly, we see that since the trade-off for these methods is with much fewer images per class, class splitting increases performances in all the scenarios we considered.

These results outline the need to adapt not only the base training images but also the base training granularity to the target few-shot task and algorithm. They also clearly demonstrate that the performance improvements we observe compared to standard trade-offs by using more classes and less images per class is not only due to the fact that the training data is more diverse, but also to the fact that training a classifier with more classes leads to improved features for few-shot classification.

4.4 Selecting Classes Based on Their Diversity or Difficulty

After observing in Sect. 4.1 the importance of the similarity between base training classes and the test classes, we now study whether the diversity of the base classes or their difficulty is also an important factor. To this end, we compute the measures described in Sect. 3.3 for every miniIN-6K class and rank them by increasing order. Then, we split the ranked classes into 10 bins of similar diversity or validation accuracy. The classes in the obtained bins are correlated to the test classes and thus introduces a bias in the performance due to this similarity instead of the diversity or difficulty we want to study (see the Sup. Mat, showing the similarity of classes in each bin to the test classes). To avoid this sampling bias, we associate to each class its distance to test classes, and sample base classes in each bin only in a small range of similarities, so that the average distance to the test classes is constant over all bins. In Fig. 6 we show the performances obtained by sampling using this strategy 64 classes and 600 images per class for a total of 38400 images in each bin. The performances obtained are shown on miniIN and CUB in Fig. 6a, 6b both using random sampling from the bin and using sampling with distance filtering as explained before. It can be seen that the effect of distance filtering is very strong, decreasing significantly the range of performance variation especially on the CUB dataset, however the difference in performance is still significant, around 5% in all experiments.

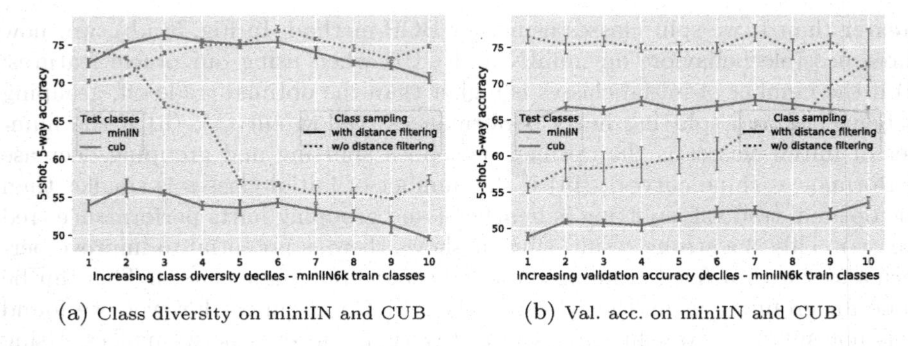

(a) Class diversity on miniIN and CUB

(b) Val. acc. on miniIN and CUB

Fig. 6. Impact of **class selection using class diversity and validation accuracy** on few-shot accuracy on miniIN and CUB. For training, we rank the classes of miniIN-6K in increasing feature variance or validation accuracy and split them into 10 bins from which we sample $C = 64$ classes that we use for base training. Figure 6a, 6b show the importance of selecting classes in each bin while considering their distance to test classes to disentangle both selection effects.

Both for CUB and miniIN, moderate class diversity - avoiding both the most and least diverse classes - seems beneficial, while using the most difficult classes seems to harm performances.

5 Conclusion

Our empirical study outlines the key importance of the base training data in few-shot learning scenarios, with seemingly minor modifications of the base data resulting in large changes in performance, and carefully selected data leading to much better accuracy. We also show that few-shot performance can be improved by automatically relabelling an intial dataset by merging or splitting classes. We hope the analysis and insights that we present will:

1. impact dataset design for practical applications, e.g. given a fixed number of images to label, one should prioritize a large number of different classes and potentially use class grouping strategies using self-supervised features. In addition to base classes similar to test data, one should also prioritize simple classes, with moderate diversity.
2. lead to new evaluations of few-shot learning algorithm, considering explicitly the influence of the base data training in the results: the current miniIN setting of 64 classes and 600 images per class is far from optimal for several approaches. Furthermore, the optimal trade-off between number of classes and number of images per class is different for different few-shot algorithms, suggesting taking into account different base data distributions in future few-shot evaluation benchmarks.
3. inspire advances in few-shot learning, e.g. the design of practical approaches to adapt base training data automatically and efficiently to target few-shot tasks.

Acknowledgements. This work was supported in part by ANR project EnHerit ANR-17-CE23-0008, project Rapid Tabasco. We thank Maxime Oquab, Diane Bouchacourt and Alexei Efros for helpful discussions and feedback.

References

1. Antoniou, A., Storkey, A.J.: Learning to learn via self-critique. In: NeurIPS (2019)
2. Birodkar, V., Mobahi, H., Bengio, S.: Semantic redundancies in image-classification datasets: The 10% you don't need. arXiv preprint arXiv:1901.11409 (2019)
3. Buda, M., Maki, A., Mazurowski, M.A.: A systematic study of the class imbalance problem in convolutional neural networks. arXiv preprint arXiv:1710.05381 (2017)
4. Caron, M., Bojanowski, P., Joulin, A., Douze, M.: Deep clustering for unsupervised learning of visual features. In: ECCV (2018)
5. Chen, W., Liu, Y., Kira, Z., Wang, Y.F., Huang, J.: A closer look at few-shot classification. In: ICLR (2019)
6. Chitta, K., Alvarez, J.M., Haussmann, E., Farabet, C.: Less is more: an exploration of data redundancy with active dataset subsampling. arXiv preprint arXiv:1905.12737 (2019)
7. Cohn, D., Ladner, R., Waibel, A.: Improving generalization with active learning. Mach. Learn. **15**, 201–221 (1994)
8. Defays, D.: An efficient algorithm for a complete link method. Comput. J. **20**(4), 364–366 (1977)
9. Deng, J., Dong, W., Socher, R., Li, L.J., Li, K., Fei-Fei, L.: ImageNet: a large-scale hierarchical image database. In: CVPR (2009)
10. Ducoffe, M., Precioso, F.: Adversarial active learning for deep networks: a margin based approach. arXiv preprint arXiv:1802.09841 (2018)
11. Everingham, M., Van Gool, L., Williams, C.K., Winn, J., Zisserman, A.: The pascal visual object classes (VOC) challenge. Int. J. Comput. Vis. **88**(2), 303–338 (2010)
12. Fan, Y., Tian, F., Qin, T., Liu, T.Y.: Neural data filter for bootstrapping stochastic gradient descent (2016)
13. Fellbaum, C.: WordNet: an electronic lexical database and some of its applications (1998)
14. Gal, Y., Islam, R., Ghahramani, Z.: Deep Bayesian active learning with image data. In: ICML (2017)
15. Ge, W., Yu, Y.: Borrowing treasures from the wealthy: deep transfer learning through selective joint fine-tuning. In: CVPR (2017)
16. Gidaris, S., Bursuc, A., Komodakis, N., Pérez, P., Cord, M.: Boosting few-shot visual learning with self-supervision. In: ICCV (2019)
17. Gidaris, S., Komodakis, N.: Dynamic few-shot visual learning without forgetting. In: CVPR (2018)
18. Gidaris, S., Singh, P., Komodakis, N.: Unsupervised representation learning by predicting image rotations. In: CVPR (2019)
19. Hariharan, B., Girshick, R.: Low-shot visual recognition by shrinking and hallucinating features. In: Proceedings of the IEEE International Conference on Computer Vision, pp. 3018–3027 (2017)
20. He, K., Fan, H., Wu, Y., Xie, S., Girshick, R.: Momentum contrast for unsupervised visual representation learning. arXiv preprint arXiv:1911.05722 (2019)
21. Hilliard, N., Phillips, L., Howland, S., Yankov, A., Corley, C.D., Hodas, N.O.: Few-shot learning with metric-agnostic conditional embeddings. arXiv preprint arXiv:1802.04376 (2018)

22. Hu, S.X., et al.: Empirical bayes transductive meta-learning with synthetic gradients. In: ICLR (2019)
23. Huh, M., Agrawal, P., Efros, A.A.: What makes imagenet good for transfer learning? In: NeurIPS LSCVS 2016 Workshop (2016)
24. Katharopoulos, A., Fleuret, F.: Not all samples are created equal: deep learning with importance sampling. arXiv preprint arXiv:1803.00942 (2018)
25. Kim, J., Kim, T., Kim, S., Yoo, C.D.: Edge-labeling graph neural network for few-shot learning. In: CVPR (2019)
26. Krizhevsky, A., Sutskever, I., Hinton, G.E.: ImageNet classification with deep convolutional neural networks. In: NeurIPS (2012)
27. Lee, K., Maji, S., Ravichandran, A., Soatto, S.: Meta-learning with differentiable convex optimization. In: CVPR (2019)
28. Li, H., Eigen, D., Dodge, S., Zeiler, M., Wang, X.: Finding task-relevant features for few-shot learning by category traversal. In: CVPR (2019)
29. Liu, Y., et al.: Learning to propagate labels: transductive propagation network for few-shot learning. In: ICLR (2019)
30. London, B.: A PAC-Bayesian analysis of randomized learning with application to stochastic gradient descent. In: NeurIPS (2017)
31. Oquab, M., Bottou, L., Laptev, I., Sivic, J.: Learning and transferring mid-level image representations using convolutional neural networks. In: CVPR (2014)
32. Qiao, S., Liu, C., Shen, W., Yuille, A.L.: Few-shot image recognition by predicting parameters from activations. In: CVPR (2018)
33. Ravi, S., Larochelle, H.: Optimization as a model for few-shot learning. In: ICLR (2017)
34. Russakovsky, O., et al.: ImageNet large scale visual recognition challenge. Int. J. Comput. Vis. **115**(3), 211–252 (2015). https://doi.org/10.1007/s11263-015-0816-y
35. Rusu, A.A., et al.: Meta-learning with latent embedding optimization. In: ICLR (2019)
36. Sablayrolles, A., Douze, M., Schmid, C., Jégou, H.: Déja vu: an empirical evaluation of the memorization properties of convnets. arXiv preprint arXiv:1809.06396 (2018)
37. Sener, O., Savarese, S.: Active learning for convolutional neural networks: a core-set approach. In: ICLR (2017)
38. Settles, B.: Active learning literature survey. Tech. rep. University of Wisconsin-Madison Department of Computer Sciences (2009)
39. Sharif Razavian, A., Azizpour, H., Sullivan, J., Carlsson, S.: CNN features off-the-shelf: an astounding baseline for recognition. In: CVPR Workshops (2014)
40. Snell, J., Swersky, K., Zemel, R.: Prototypical networks for few-shot learning. In: NeurIPS (2017)
41. Triantafillou, E., et al.: Meta-dataset: a dataset of datasets for learning to learn from few examples. In: ICLR (2020)
42. Vinyals, O., Blundell, C., Lillicrap, T., Wierstra, D., et al.: Matching networks for one shot learning. In: NeurIPS (2016)
43. Vodrahalli, K., Li, K., Malik, J.: Are all training examples created equal? an empirical study. arXiv preprint arXiv:1811.12569 (2018)
44. Wah, C., Branson, S., Welinder, P., Perona, P., Belongie, S.: The Caltech-UCSD birds-200-2011 dataset (2011)
45. Wang, Y., Chao, W.L., Weinberger, K.Q., van der Maaten, L.: SimpleShot: revisiting nearest-neighbor classification for few-shot learning. arXiv preprint arXiv:1911.04623
46. Wang, Y.X., Girshick, R., Hebert, M., Hariharan, B.: Low-shot learning from imaginary data. In: CVPR (2018)

47. Xiao, J., Hays, J., Ehinger, K.A., Oliva, A., Torralba, A.: Sun database: large-scale scene recognition from abbey to zoo. In: CVPR (2010)
48. Zamir, A.R., Sax, A., Shen, W., Guibas, L.J., Malik, J., Savarese, S.: Taskonomy: disentangling task transfer learning. In: CVPR (2018)
49. Zhou, L., Cui, P., Jia, X., Yang, S., Tian, Q.: Learning to select base classes for few-shot classification. In: Proceedings of the IEEE/CVF Conference on Computer Vision and Pattern Recognition, pp. 4624–4633 (2020)

Invertible Zero-Shot Recognition Flows

Yuming Shen[1], Jie Qin[2(✉)], Lei Huang[2], Li Liu[2], Fan Zhu[2], and Ling Shao[2,3]

[1] EBay, San Jose, USA
ymcidence@gmail.com
[2] Inception Institute of Artificial Intelligence, Abu Dhabi, UAE
qinjiebuaa@gmail.com
[3] Mohamed bin Zayed University of Artificial Intelligence, Abu Dhabi, UAE

Abstract. Deep generative models have been successfully applied to Zero-Shot Learning (ZSL) recently. However, the underlying drawbacks of GANs and VAEs (*e.g.*, the hardness of training with ZSL-oriented regularizers and the limited generation quality) hinder the existing generative ZSL models from fully bypassing the seen-unseen bias. To tackle the above limitations, for the first time, this work incorporates a new family of generative models (*i.e.*, flow-based models) into ZSL. The proposed Invertible Zero-shot Flow (IZF) learns factorized data embeddings (*i.e.*, the semantic factors and the non-semantic ones) with the forward pass of an invertible flow network, while the reverse pass generates data samples. This procedure theoretically extends conventional generative flows to a factorized conditional scheme. To explicitly solve the bias problem, our model enlarges the seen-unseen distributional discrepancy based on a negative sample-based distance measurement. Notably, IZF works flexibly with either a naive Bayesian classifier or a held-out trainable one for zero-shot recognition. Experiments on widely-adopted ZSL benchmarks demonstrate the significant performance gain of IZF over existing methods, in both classic and generalized settings.

Keywords: Zero-Shot Learning · Generative flows · Invertible networks

1 Introduction

With the explosive growth of image classes, there is an ever-increasing need for computer vision systems to recognize images from never-before-seen classes, a task which is known as Zero-Shot Learning (ZSL) [25]. Generally, ZSL aims at recognizing *unseen* images by exploiting relationships between *seen* and *unseen* images. Equipped with prior semantic knowledge (*e.g.*, attributes [26],

---

Electronic supplementary material The online version of this chapter (https://doi.org/10.1007/978-3-030-58517-4_36) contains supplementary material, which is available to authorized users.

A. Vedaldi et al. (Eds.): ECCV 2020, LNCS 12361, pp. 614–631, 2020.
https://doi.org/10.1007/978-3-030-58517-4_36

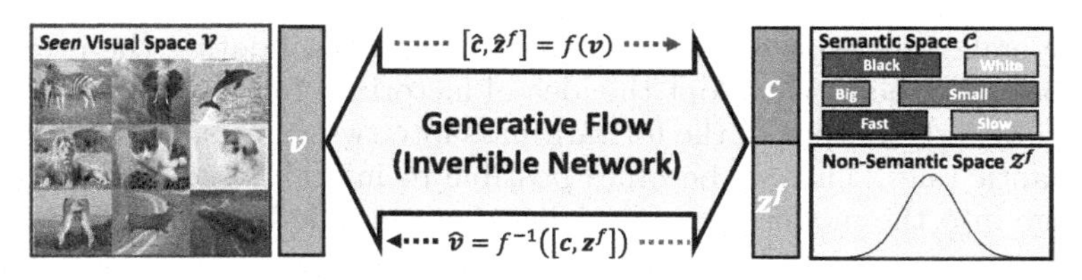

Fig. 1. A brief illustration of IZF for ZSL. We propose a novel factorized conditional **generative flow** with invertible networks.

word embeddings [37]), traditional ZSL models typically mitigate the *seen-unseen* domain gap by learning a visual-semantic projection between images and their semantics. In the context of deep learning [48,49], the recent emergence of generative models has slightly changed this schema by converting ZSL into supervised learning, where a held-out classifier is trained for zero-shot recognition based on the generated *unseen* images. As both *seen* and synthesized *unseen* images are observable to the model, generative ZSL methods largely favor Generalized ZSL (GZSL) [45] and yet perform well in Classic ZSL (CZSL) [25,36,59]. In practice, Generative Adversarial Networks (GANs) [13], Variational Auto-Encoders (VAEs) [22] and Conditional VAEs (CVAEs) [51] are widely employed for ZSL. Despite the considerable success current generative models [27,38,62,64,71] have achieved, their underlying limitations are still inevitable in the context of ZSL.

First, GANs [13] suffer from mode collapse [5] and instability during training with complex learning objectives. It is usually hard to impose additional ZSL-oriented regularizers to the generative side of GANs other than the real/fake game [46]. Second, the Evidence Lower BOund (ELBO) of VAEs/CVAEs [22,51] requires stochastic approximate optimization, preventing them from generating high-quality *unseen* samples for robust ZSL [64]. Third, as only seen data are involved during training, most generative models are not well-addressing the **seen-unseen bias** problem, *i.e.*, generated *unseen* data tend to have the same distribution as *seen* ones. Though these concerns are as well partially noticed by the recent ZSL research [46,64], they either simply bypass the drawback of GAN in ZSL by resorting to VAE or *vice versa*, which can be yet suboptimal.

Therefore, we ought to seek a novel generative model that can bypass the above limitations to further boost the performance of ZSL. Inspired by the recently proposed Invertible Neural Networks (INNs) [2], we find that another branch of generative models, *i.e.*, flow-based generative models [6,7], align well with our insights into generative ZSL models. Particularly, generative flows adopt an identical set of parameters and built-in network for encoding (*forward pass*) and decoding (*reverse pass*). Compared with GANs/VAEs, the forward pass in flows acts as an additional 'encoder' to fully utilize the semantic knowledge.

In this paper, we fully exploit the advantages of generative flows [6,7], based on which a novel ZSL model is proposed, namely Invertible Zero-shot Flow (IZF). In particular, the forward pass of IZF projects visual features to the semantic embedding space, with the reverse pass consolidating the inverse projection between them. We adopt the idea of factorized representations in [54,57] to disentangle the output of the forward pass into two factors, *i.e.*, semantic and non-semantic ones. Thus, it becomes possible to inject category-wise similarity knowledge into the model by regularizing the semantic factors. Meanwhile, the respective reverse pass of IZF performs conditional data generation with factorized embeddings for both *seen* and *unseen* data. We visualize this pipeline in Fig. 1. To further accommodate IZF to ZSL, we propose novel bidirectional training strategies to **1)** centralize the *seen* prototypes for stable classification, and **2)** diverge the distribution of synthesized *unseen* data and real *seen* data to explicitly address the bias problem. Our main contributions include:

1. IZF shapes a novel factorized conditional flow structure that supports exact density estimation. This differs from the existing approximated [2] and the non-factorized [3] approach. To the best of our knowledge, IZF is the first generative flow model for ZSL.
2. A novel mechanism tackling the bias problem is proposed with the merits of the generative nature of IZF, *i.e.*, measuring and diversifying the sample-based *seen-unseen* data distributional discrepancy.

2 Related Work

Zero-Shot Learning. ZSL [25] has been extensively studied in recent years [11,12,42,65]. The evaluation of ZSL can be either classic (CZSL) or generalized (GZSL) [45], while recent research also explores the potential in retrieval [32,47]. CZSL excludes *seen* classes during test, while GZSL considers both *seen* and *unseen* classes, being more popular among recent articles [4,8,19,28]. To tackle the problem of *seen-unseen* domain shift, there propose three typical ways to inject semantic knowledge for ZSL, *i.e.*, **(1)** learning visual→semantic projections [1,10,23,26,44], **(2)** learning semantic→visual projections [43,67,69], and **(3)** learning shared features or multi-modal functions [70]. Recently, deep generative models have been adapted to ZSL, subverting the traditional ZSL paradigm to some extent. The majority of existing generative methods employ GANs [27,35,62], CVAEs [24,38,46] or a mixture of the two [18,64] to synthesize *unseen* data points for a successive classification stage. However, as mentioned in Sect. 1, these models suffer from their underlying drawbacks in ZSL.

Generative Flows. Compared with GANs/VAEs, flow-based generative models [6,7,21] have attracted less research attention in the past few years, probably because this family of models require special neural structures that are in principle invertible for encoding and generation. It was not until the first appearance of the coupling layer in NICE [6] and RealNVP [7] that generative flows with deep INNs became practical and efficient. In [29], flows are extended to a conditional scheme, but the density estimation is not deterministic. The Glow architecture [21] is further introduced with invertible 1×1 convolution for realistic image generation. In [3], conditions are injected into the coupling layers.

IDF [17] and BipartiteFlow [55] define a discrete case of flows. Flows can be combined with adversarial training strategies [14]. In [41], generative flows have also been successfully applied to speech synthesis.

Literally Invertible ZSL. We also notice that some existing ZSL models involve literally *invertible* projections [23,68]. However, these methods are unable to generate samples, failing to benefit GZSL with the held-out classifier schema [62] and our inverse training objectives. In addition, [23,68] are linear models and cannot be paralleled as deep neural networks during training. This limits their model capacity and training efficiency on large-scale data.

3 Preliminaries: Generative Flows and INNs

Density Estimation with Flows. Generative flows are theoretically based on the *change of variables formula.* Given a d-dimensional datum $\mathbf{x} \in \mathcal{X} \subseteq \mathbb{R}^d$ and a pre-defined prior $p_{\mathcal{Z}}$ supporting a set of latents $\mathbf{z} \in \mathcal{Z} \subseteq \mathbb{R}^d$, the *change of variables formula* defines the estimated density of $p_\theta(\mathbf{x})$ using an invertible (also called *bijective*) transformation $f : \mathcal{X} \rightarrow \mathcal{Z}$ as follows:

$$p_\theta(\mathbf{x}) = p_{\mathcal{Z}}\left(f\left(\mathbf{x}\right)\right)\left|\det\frac{\partial f}{\partial \mathbf{x}}\right|. \tag{1}$$

Here θ indicates the set of model parameters and the scalar $|\det\left(\partial f / \partial \mathbf{x}\right)|$ is the absolute value of the determinant of the Jacobian matrix $(\partial f / \partial \mathbf{x})$. One can refer to [6,7] and our **supplementary material** for more details. The choice of the prior $p_{\mathcal{Z}}$ is arbitrary and a zero-mean unite-variance Gaussian is usually adequate, *i.e.*, $p_{\mathcal{Z}}(\mathbf{z}) = \mathcal{N}(\mathbf{z}|\mathbf{0}, \mathbf{I})$. The respective generative process can be written as $\hat{\mathbf{x}} = f^{-1}\left(\mathbf{z}\right)$, where $\mathbf{z} \sim p_{\mathcal{Z}}$. f is usually called the *forward pass*, with f^{-1} being the *reverse pass.*[1] Stacking a series of invertible functions $f = f_1 \circ f_2 \circ \cdots \circ f_k$ literally complies with the name of *flows.*

INNs with Coupling Layers. Generative flows admit networks with **(1)** exactly invertible structure and **(2)** efficiently computed Jacobian determinant. We adopt a typical type of INNs, called the coupling layers [6], which split network inputs/outputs into two respective partitions: $\mathbf{x} = [\mathbf{x}_a, \mathbf{x}_b]$, $\mathbf{z} = [\mathbf{z}_a, \mathbf{z}_b]$. The computation of the layer is defined as:

$$\begin{aligned} f(\mathbf{x}) &= \left[\mathbf{x}_a, \mathbf{x}_b \odot \exp\left(\mathbf{s}(\mathbf{x}_a)\right) + \mathbf{t}(\mathbf{x}_a)\right], \\ f^{-1}(\mathbf{z}) &= \left[\mathbf{z}_a, \left(\mathbf{z}_b - \mathbf{t}(\mathbf{z}_a)\right) \oslash \exp\left(\mathbf{s}(\mathbf{z}_a)\right)\right], \end{aligned} \tag{2}$$

where $\odot$ and $\oslash$ denote element-wise multiplication and division respectively. $\mathbf{s}(\cdot)$ and $\mathbf{t}(\cdot)$ are two arbitrary neural networks with input and output lengths of $d/2$. We show this structure in Fig. 2(b). Its corresponding log-determinant of Jacobian can be conveniently computed by $\sum|\mathbf{s}|$. Coupling layers usually come together with element-wise permutation to build compact transformation.

[1] Note that reverse pass and back-propagation are different concepts.

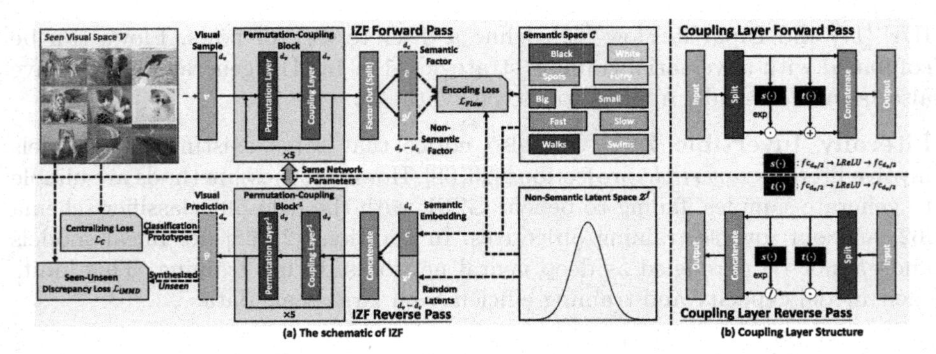

Fig. 2. **(a)** The architecture of the proposed IZF model. The forward pass and reverse pass are indeed sharing network parameters as invertible structures are used. Also note that only *seen* visual samples are accessible during training and IZF is an **inductive** ZSL model. **(b)** A typical illustration of the coupling layer [6] used in our model.

4 Formulation: Factorized Conditional Flow

ZSL aims at recognizing *unseen* data. The training set $\mathcal{D}^s = \{(\mathbf{v}^s, y^s, \mathbf{c}^s)\}$ of it is grounded on M^s *seen* classes, *i.e.*, $y^s \in \mathcal{Y}^s = \{1, 2, ..., M^s\}$. Let $\mathcal{V}^s \subseteq \mathbb{R}^{d_v}$ and $\mathcal{C}^s \subseteq \mathbb{R}^{d_c}$ respectively represent the visual space and the semantic space of *seen* data, of which $\mathbf{v}^s \in \mathcal{V}^s$ and $\mathbf{c}^s \in \mathcal{C}^s$ are the corresponding feature instances. The dimensions of these two spaces are denoted as d_v and d_c. Given an *unseen* label set $\mathcal{Y}^u = \{M^s + 1, M^s + 2, ..., M^s + M^u\}$ of M^u classes, the *unseen* data are denoted with the superscript of $\cdot^u$ as $\mathcal{D}^u = \{(\mathbf{v}^u, y^u, \mathbf{c}^u)\}$, where $\mathbf{v}^u \in \mathcal{V}^u$, $y^u \in \mathcal{Y}^u$ and $\mathbf{c}^u \in \mathcal{C}^u$. In this paper, the superscript are omitted when the referred sample can be both *seen* or *unseen*, *i.e.*, $\mathbf{v} \in \mathcal{V} = \mathcal{V}^s \cup \mathcal{V}^u, y \in \mathcal{Y} = \mathcal{Y}^s \cup \mathcal{Y}^u$ and $\mathbf{c} \in \mathcal{C} = \mathcal{C}^s \cup \mathcal{C}^u$.

The framework of IZF is demonstrated in Fig. 2(a). IZF factors out the high-level semantic information with its forward pass $f(\cdot)$, equivalently performing visual→semantic projection. The reverse pass handles conditional generation, *i.e.*, semantic→visual projection, with identical network parameters to the forward pass. To reflect label information in a flow, Eq. (1) is slightly extended to a conditional scheme with visual data $\mathbf{v}$ and their labels y:

$$p_\theta(\mathbf{v}|y) = p_{\mathcal{Z}}\left(f\left(\mathbf{v}\right)|y\right)\left|\det\frac{\partial f}{\partial \mathbf{v}}\right|. \tag{3}$$

Detailed proofs are given in the **supplementary material**. Next, we consider reflecting semantic knowledge in the encoder outputs for ZSL. To this end, a factorized model takes its shape.

4.1 Forward Pass: Factorizing the Semantics

High-dimensional image representations contain both high-level semantic-related information and non-semantic information such as low-level image details.

As factorizing image features has been proved effective for ZSL in [54], we adopt this spirit, but with different approach to fit the structure of flow. In [54], the factorization is basically only empirical, while IZF derives full likelihood model of a training sample.

As shown in Fig. 2(a), the proposed flow network learns factorized independent image representations $\hat{\mathbf{z}} = [\hat{\mathbf{c}}, \hat{\mathbf{z}}^f] = f(\mathbf{v})$ with its forward pass $f(\cdot)$, where $\hat{\mathbf{c}} \in \mathbb{R}^{d_c}$ denotes the predicted semantic factor of an arbitrary visual sample $\mathbf{v}$ and $\hat{\mathbf{z}}^f \in \mathbb{R}^{d_v - d_c}$ is the low-level non-semantic independent to $\hat{\mathbf{c}}$, *i.e.*, $\hat{\mathbf{z}}^f \perp\!\!\!\perp \hat{\mathbf{c}}$. We assume $\hat{\mathbf{z}}^f$ is not dependent on data label y, *i.e.*, $\hat{\mathbf{z}}^f \perp\!\!\!\perp y$ as it is designed to reflect no high-level semantic/category information. Therefore, we rewrite the conditional probability of Eq. (3) as

$$p_\theta(\mathbf{v}|y) = p_\mathcal{Z}\left([\hat{\mathbf{c}}, \hat{\mathbf{z}}^f] = f(\mathbf{v})|y\right) \left|\det \frac{\partial f}{\partial \mathbf{v}}\right| = p_{\mathcal{C}|\mathcal{Y}}(\hat{\mathbf{c}}|y) p_{\mathcal{Z}^f}(\hat{\mathbf{z}}^f) \left|\det \frac{\partial f}{\partial \mathbf{v}}\right|. \tag{4}$$

The conditional independence property gives $p_\mathcal{Z}(\hat{\mathbf{c}}, \hat{\mathbf{z}}^f|y) = p_{\mathcal{C}|\mathcal{Y}}(\hat{\mathbf{c}}|y) p_{\mathcal{Z}^f}(\hat{\mathbf{z}}^f)$. According to [16,57], this property is implicitly enforced by imposing fix-formed priors on each variable. In this work, the factored priors are

$$p_{\mathcal{C}|\mathcal{Y}}(\hat{\mathbf{c}}|y) = \mathcal{N}(\hat{\mathbf{c}}|\mathbf{c}(y), \mathbf{I}), \quad p_{\mathcal{Z}^f}(\hat{\mathbf{z}}^f) = \mathcal{N}(\hat{\mathbf{z}}^f|\mathbf{0}, \mathbf{I}), \tag{5}$$

where $\mathbf{c}(y)$ simply denotes the semantic embedding corresponding to y. Similar to the likelihood computation of VAEs [22], we empirically assign a uniformed Gaussian to $p_{\mathcal{C}|\mathcal{Y}}(\hat{\mathbf{c}}|y)$ centered at the corresponding semantic embedding $\mathbf{c}(y)$ of the visual sample so that it can be simply reduced to a $l2$ norm.

The Injected Semantic Knowledge. The benefits of the factorized $p_{\mathcal{C}|\mathcal{Y}}(\hat{\mathbf{c}}|y)$ are two-fold: **1)** it explicitly reflects the degree of similarity between different classes, ensuring smooth *seen-unseen* generalization for ZSL. This is also in line with the main motivation of several existing approaches [23,44]; **2)** a well-trained IZF model with $p_{\mathcal{C}|\mathcal{Y}}(\hat{\mathbf{c}}|y)$ factorizes the semantic meaning from non-semantic information of an image, making it possible to conditionally generate samples with $f^{-1}(\cdot)$ by directly feeding the semantic category embedding (see Eq. (6)).

4.2 Reverse Pass: Conditional Sample Generation

One advantage of deep generative ZSL models is the ability to observe synthesized *unseen* data. IZF fulfills this by

$$\mathbf{c} \in \mathcal{C}, \mathbf{z}^f \sim p_{\mathcal{Z}^f}, \hat{\mathbf{v}} = f^{-1}\left([\mathbf{c}, \mathbf{z}^f]\right). \tag{6}$$

The Use of Reverse Pass. Different from most generative ZSL approaches [38,62] where synthesized *unseen* samples simply feed a held-out classifier, IZF additionally uses these synthesized samples to measure the biased distributional overlap between *seen* and synthesized *unseen* data. We will elaborate the corresponding learning objectives and ideas in Sect. 5.3.

4.3 Network Structure

In the spirits of Eq. (4) and (6), we build the network of IZF as shown in Fig. 2(a). Concretely, IZF consists of 5 permutation-coupling blocks to shape a deep non-linear architecture. Inspired by [2,7], we combine the coupling layer with channel-wise permutation in each block. The permutation layer shuffles the elements of an input feature in a random but fixed manner so that the split of two successive coupling layers are different and the encoding/decoding performance is assured. We use identical structure for the built-in neural network $\mathbf{s}(\cdot)$ and $\mathbf{t}(\cdot)$ of the coupling layers in Eq. (2), *i.e.*, $\mathtt{fc}_{d_v/2} \rightarrow \mathtt{LReLU} \rightarrow \mathtt{fc}_{d_v/2}$, where LReLU is the leaky ReLU activation [33]. In the following, we show how the network is trained to enhance ZSL.

5 Training with the Merits of Generative Flow

To transfer knowledge from *seen* concepts to *unseen* ones, we employ the idea of bi-directional training of INNs [2] to optimize IZF. In principle, generative flows can be trained only with the forward pass (Sect. 5.1). However, considering the fact that the reverse pass of IZF is used for *zero-shot* classification, we impose additional learning objectives to its reverse pass to promote the ability of *seen-unseen* generalization (Sect. 5.2 and 5.3).

5.1 Learning to Decode by Encoding

The first learning objective of IZF comes from the definition of generative flow as depicted in Eq. (1). By analytic log-likelihood maximization of the forward pass, generative flows are ready to synthesize data samples. As only visual features of *seen* categories are observable to IZF, we construct this loss term upon $\mathcal{D}^s$ as

$$\mathcal{L}_{\text{Flow}} = \mathbb{E}_{(\mathbf{v}^s, y^s, \mathbf{c}^s)} \left[-\log p_\theta(\mathbf{v}^s | y^s) \right], \tag{7}$$

where $(\mathbf{v}^s, y^s, \mathbf{c}^s)$ are *seen* samples from the training set $\mathcal{D}^s$ and $p_\theta(\mathbf{v}^s|y^s)$ is computed according to Eq. (4). $\mathcal{L}_{\text{Flow}}$ is not only an encoding loss, but also can legitimate unconditional *seen* data generation due to the invertible nature of IZF. Compared with the training process of GAN/VAE-based ZSL models [38,62], IZF defines an explicit and simpler objective to fulfill the same functionality.

5.2 Centralizing Classification Prototypes

IZF supports naive Bayesian classification by projecting semantic embeddings back to the visual space with its reverse pass. For each class-wise semantic representation, we define a special generation procedure $\hat{\mathbf{v}}_c = f^{-1}([\mathbf{c}, \mathbf{0}])$ as the **classification prototype** of a class. As these prototypes are directly used to classify images by distance comparison, it would be harmful to the final accuracy when the prototypes are too close to unrelated visual samples. To address this issue, f^{-1} is expected to position them close to the centres $\bar{\mathbf{v}}_c$ of the respective classes they belong to. This idea is illustrated in Fig. 3, denoted as $\mathcal{L}_C$. In particular, this centralizing loss is imposed on the *seen* classes as

$$\mathcal{L}_{\mathrm{C}} = \mathbb{E}_{(\mathbf{c}^s, \bar{\mathbf{v}}_c^s)} \left[\| f^{-1}([\mathbf{c}^s, \mathbf{0}]) - \bar{\mathbf{v}}_c^s \|^2 \right], \tag{8}$$

where $\bar{\mathbf{v}}_c^s$ is the corresponding numerical mean of the visual samples that belong to the class with the semantic embedded $\mathbf{c}^s$. Similar to the semantic knowledge loss, we directly apply $l2$ norm to the model to regularize its behavior.

5.3 Measuring the *Seen-Unseen* Bias

Recalling the bias problem in ZSL with generative models, the synthesized *unseen* samples could be unexpectedly too close to the real *seen* ones. This would significantly decrease the classification performance for *unseen* classes, especially in the context of GZSL where *seen* and *unseen* data are both available. We propose to explicitly tackle the bias problem by preventing the **synthesized** ***unseen*** visual distribution $p_{\hat{\mathcal{V}}^u}$ from colliding with the **real** ***seen*** one $p_{\mathcal{V}^s}$. In other words, $p_{\mathcal{V}^s}$ is slightly pushed away from $p_{\hat{\mathcal{V}}^u}$.

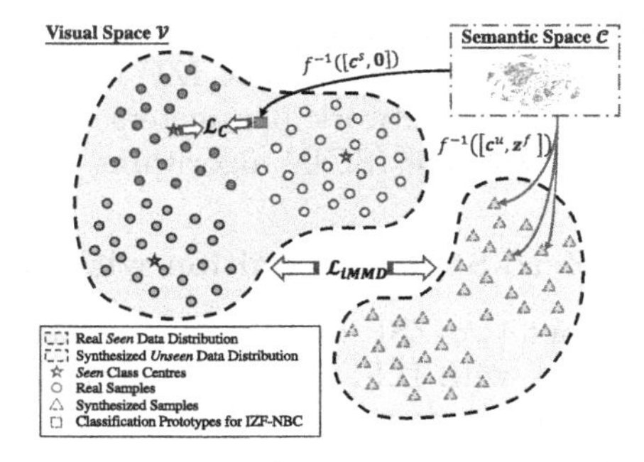

Fig. 3. Typical illustration of the IZF training losses *w.r.t.* the **reverse pass**, *i.e.*, $\mathcal{L}_{\mathrm{C}}$ in Sect. 5.2 and $\mathcal{L}_{\mathrm{iMMD}}$ in Sect. 5.3.

Our key idea is illustrated in Fig. 3, denoted as $\mathcal{L}_{\mathrm{iMMD}}$. With generative models, it is always possible to measure distributional discrepancy without acknowledging the true distribution parameters of $p_{\hat{\mathcal{V}}^u}$ and $p_{\mathcal{V}^s}$ by treating this as a negative two-sample-test problem. Hence, we resort to Maximum Mean Discrepancy (MMD) [2,53] as the measurement. Since we aim to increase the discrepancy, the last loss term of IZF is defined upon the **numerical negation** of MMD $(p_{\mathcal{V}^s} || p_{\hat{\mathcal{V}}^u})$ in a batch-wise fashion as

$$\begin{aligned}\mathcal{L}_{\mathrm{iMMD}} = &- \mathrm{MMD}\left(p_{\mathcal{V}^s} || p_{\hat{\mathcal{V}}^u}\right) = \tfrac{2}{n^2} \sum_{i,j} \kappa(\mathbf{v}_i^s, \hat{\mathbf{v}}_j^u) \\ &- \tfrac{1}{n(n-1)} \sum_{i \neq j} \left(\kappa(\mathbf{v}_i^s, \mathbf{v}_j^s) + \kappa(\hat{\mathbf{v}}_i^u, \hat{\mathbf{v}}_j^u) \right), \\ &\text{where } \mathbf{v}_i^s \in \mathcal{V}^s, \ \mathbf{c}_i^u \in \mathcal{C}^u, \ \mathbf{z}_i^f \sim p_{\mathcal{Z}^f}, \ \hat{\mathbf{v}}_i^u = f^{-1}([\mathbf{c}_i^u, \mathbf{z}_i^f]).\end{aligned} \tag{9}$$

Here n refers to the training batch size, and $\kappa(\cdot)$ is an arbitrary positive-definite reproducing kernel function. Importantly, as only *seen* visual samples $\mathbf{v}_i^s$ are directly used and $\hat{\mathbf{v}}_i^u$ are synthesized, $\mathcal{L}_{\mathrm{iMMD}}$ is indeed an **inductive** objective. The same setting has also been adopted in recent inductive ZSL methods [30,46, 51,62], *i.e.*, the names of the *unseen* classes are accessible during training while their visual samples remain inaccessible. We also note that replacing $\mathcal{L}_{\text{iMMD}}$ by simply tuning the values of *unseen* classification templates $f^{-1}([\mathbf{c}^u, \mathbf{0}])$ is infeasible in inductive ZSL since there exists no *unseen* visual reference sample for direct regularization.

5.4 Overall Objective and Training

By combining the above-discussed losses, the overall learning objective of IZF can be simply written as

$$\mathcal{L}_{\text{IZF}} = \lambda_1 \mathcal{L}_{\text{Flow}} + \lambda_2 \mathcal{L}_{\text{C}} + \lambda_3 \mathcal{L}_{\text{iMMD}}. \tag{10}$$

Three hyper-parameters λ_1, λ_2 and λ_3 are introduced to balance the contributions of different loss terms. IZF is fully differentiable *w.r.t.* $\mathcal{L}_{\text{IZF}}$. Hence, the corresponding network parameters can be directly optimized with Stochastic Gradient Descent (SGD) algorithms.

5.5 Zero-Shot Recognition with IZF

We adopt two ZSL classification strategies (*i.e.*, IZF-NBC and IZF-Softmax) that work with IZF. Specifically, IZF-NBC employs a naive Bayesian classifier to recognize a given test visual sample $\mathbf{v}_q$ by comparing the Euclidean distances between it and the classification prototypes introduced in Sect. 5.2. IZF-Softmax leverages a held-out classifier similar to the one used in [62]. The classification processes are performed as

$$\begin{aligned} &\text{IZF-NBC: } \hat{y}^q = \arg\min_y \| f^{-1}([\mathbf{c}(y), \mathbf{0}]) - \mathbf{v}^q \|, \\ &\text{IZF-Softmax: } \hat{y}^q = \arg\max_y \texttt{softmax}(\texttt{NN}(\mathbf{v}^q)). \end{aligned} \tag{11}$$

Here $\texttt{NN}(\cdot)$ is a single-layered fully-connected network trained with generated *unseen* data and the softmax cross-entropy loss on top of the softmax activation.

6 Experiments

6.1 Implementation Details

IZF is implemented with the popular deep learning toolbox PyTorch [39]. We build the INNs according to the framework of FrEIA [2,3]. The network architecture is elaborated in Sect. 4.3. The built-in networks $\mathbf{s}(\cdot)$ and $\mathbf{t}(\cdot)$ of all coupling layers of IZF are shaped by $\texttt{fc}_{d_v/2} \rightarrow \texttt{LReLU} \rightarrow \texttt{fc}_{d_v/2}$. Following [2,53], we employ the Inverse Multiquadratic (IM) kernel $\kappa(\mathbf{v}, \mathbf{v}') = 2d_v / \left(2d_v + \| \mathbf{v} - \mathbf{v}' \|^2\right)$ in Eq. (9) for best performance. We testify the choice of λ_1, λ_2 and λ_3 within $\{0.1, 0.5, 1, 1.5, 2\}$ and report the results of $\lambda_1 = 2, \lambda_2 = 1, \lambda_3 = 0.1$ for all comparisons. The Adam optimizer [20] is used to train IZF with a learning rate of 5×10^{-4} *w.r.t.* $\mathcal{L}_{\text{IZF}}$. The batch size is fixed to 256 for all experiments.

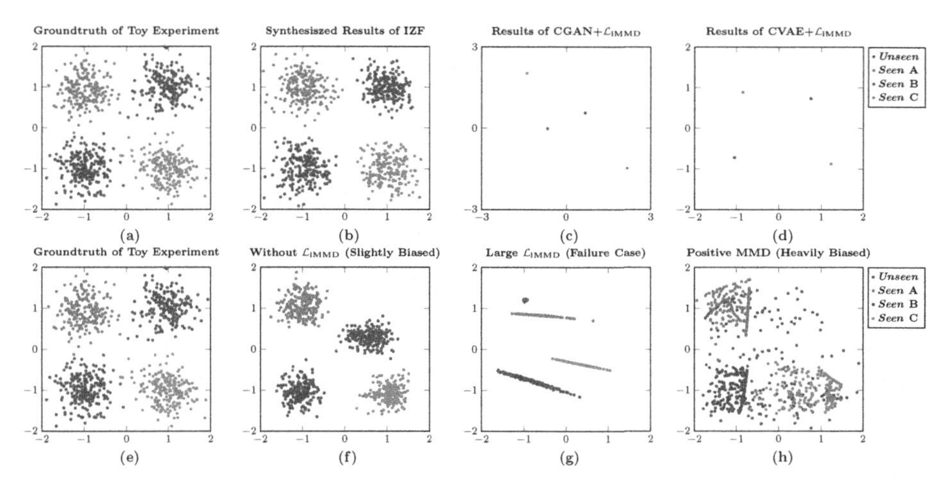

Fig. 4. Illustration of the 4-class toy experiment in Sect. 6.2. **(a, e)** 2-D Ground truth simulation data, with the top-right class being *unseen*. **(b)** Synthesized samples of IZF. **(c, d)** Synthesized results of conditional GAN and CVAE respectively with $\mathcal{L}_{\text{iMMD}}$. **(f)** Results without $\mathcal{L}_{\text{iMMD}}$ of IZF. **(g)** Failure results with extremely and unreasonably large $\mathcal{L}_{\text{iMMD}}$ ($\lambda_3 = 10$) of IZF. **(h)** Results with positive MMD of IZF.

6.2 Toy Experiments: Illustrative Analysis

Before evaluating IZF with real data, we firstly provide a toy ZSL experiment to justify our motivation. Particularly, the following themes are discussed:

1. Why Do We Resort to Flows Instead of GAN/VAE with $\mathcal{L}_{\text{iMMD}}$?
2. The effect of $\mathcal{L}_{\text{iMMD}}$ regarding the bias problem.

Setup. We consider a 4-class simulation dataset with 1 class being *unseen*. The class-wise attributes are defined as $\mathcal{C}^s = \{[0,1],[0,0],[1,0]\}$ for the *seen* classes **A**, **B** and **C** respectively, while the *unseen* class would have attribute of $\mathcal{C}^u = \{[1,1]\}$. The ground truth data are randomly sampled around a linear transformation of the attributes, *i.e.*, $\mathbf{v} := 2\mathbf{c} - 1 + \epsilon \in \mathbb{R}^2$, where $\epsilon \sim \mathcal{N}(\mathbf{0}, \frac{1}{3}\mathbf{I})$. To meet the dimensionality requirement, *i.e.*, $d_v > d_c$, we follow the convention of [2] to pad two zeros to data when feeding them to the network, *i.e.*, $\mathbf{v}' := [\mathbf{v}, 0, 0]$. The toy data are plotted in Fig. 4(a) and (e).

Why Do We Resort to Flows Instead of GAN/VAE? We firstly show the synthesized results of IZF in Fig. 4(b). It can be observed that IZF successfully interprets the relations of the *unseen* class to the *seen* ones, *i.e.*, being closer to **A** and **C** but further to **B**. To legit the use of generative flow, we accordingly build two baselines by combining Conditional GAN (CGAN) and CVAE with our $\mathcal{L}_{\text{iMMD}}$ loss (see our **supplementary document** for implementation details). The respective generated results are shown in Fig. 4(c) and (d). Aligning with our motivation, $\mathcal{L}_{\text{iMMD}}$ quickly fails the unstable training process of GAN in

Table 1. Inductive GZSL performance of IZF and the state-of-the-art methods with the PS setting [63].

Method	Reference	AwA1 [26]			AwA2 [26]			CUB [58]			SUN [40]			aPY [9]		
		A^s	A^u	H	A^s	A^u	H	A^s	A^u	H	A^s	A^u	H	A^s	A^u	H
DAP [26]	PAMI13	88.7	0.0	0.0	84.7	0.0	0.0	67.9	0.0	0.0	25.1	4.2	7.2	78.3	4.8	9.0
CMT [50]	NIPS13	86.9	8.4	15.3	89.0	8.7	15.9	60.1	4.7	8.7	28.0	8.7	13.3	74.2	10.9	19.0
DeViSE [10]	NIPS13	68.7	13.4	22.4	74.7	17.1	27.8	53.0	23.8	32.8	27.4	16.9	20.9	76.9	4.9	9.2
ALE [1]	CVPR15	16.8	76.1	27.5	81.8	14.0	23.9	62.8	23.7	34.4	33.1	21.8	26.3	73.7	4.6	8.7
SSE [70]	ICCV15	80.5	7.0	12.9	82.5	8.1	14.8	46.9	8.5	14.4	36.4	2.1	4.0	78.9	0.2	0.4
ESZSL [44]	ICML15	75.6	6.6	12.1	77.8	5.9	11.0	63.8	12.6	21.0	27.9	11.0	15.8	70.1	2.4	4.6
LATEM [60]	CVPR16	71.1	7.3	13.3	77.3	11.5	20.0	57.3	15.2	24.0	28.8	14.7	19.5	73.0	0.1	0.2
SAE [23]	CVPR17	77.1	1.8	3.5	82.2	1.1	2.2	54.0	7.8	13.6	18.0	8.8	11.8	**80.9**	0.4	0.9
DEM [69]	CVPR17	84.7	32.8	47.3	86.4	30.5	45.1	57.9	19.6	29.2	34.3	20.5	25.6	11.1	**75.1**	19.4
RelationNet [52]	CVPR18	**91.3**	31.4	46.7	**93.4**	30.0	45.3	61.1	38.1	47.0	–	–	–	–	–	–
DCN [30]	NIPS18	84.2	25.5	39.1	–	–	–	60.7	28.4	38.7	37.0	25.5	30.2	75.0	14.2	23.9
CRNet [67]	ICML19	74.7	58.1	65.4	78.8	52.6	63.1	56.8	45.5	50.5	36.5	34.1	35.3	68.4	32.4	44.0
LFGAA [31]	ICCV19	–	–	–	90.3	50.0	64.4	79.6	43.4	56.2	34.9	20.8	26.1	–	–	–
CVAE-ZSL [38]	ECCVW18	–	–	47.2	–	–	51.2	–	–	34.5	–	–	26.7	–	–	–
SE-GZSL [24]	CVPR18	67.8	56.3	61.5	68.1	58.3	62.8	53.3	41.5	46.7	30.5	40.9	34.9	–	–	–
f-CLSWGAN [62]	CVPR18	61.4	57.9	59.6	–	–	–	57.7	43.7	49.7	36.6	42.6	39.4	–	–	–
LisGAN [27]	CVPR19	76.3	52.6	62.3	–	–	–	57.9	46.5	51.6	37.8	42.9	40.2	–	–	–
SGAL [66]	NIPS19	75.7	52.7	62.2	81.2	55.1	65.6	44.7	47.1	45.9	31.2	42.9	36.1	–	–	–
CADA-VAE [46]	CVPR19	72.8	57.3	64.1	75.0	55.8	63.9	53.5	51.6	52.4	35.7	47.2	40.6	–	–	–
GDAN [18]	CVPR19	–	–	–	67.5	32.1	43.5	66.7	39.3	49.5	**89.9**	38.1	53.4	75.0	30.4	43.4
DLFZRL [54]	CVPR19	–	–	61.2	–	–	60.9	–	–	51.9	–	–	42.5	–	–	38.5
f-VAEGAN-D2 [64]	CVPR19	70.6	57.6	63.5	–	–	–	60.1	48.4	53.6	38.0	45.1	41.3	–	–	–
IZF-NBC	**Proposed**	75.2	57.8	65.4	76.0	58.1	65.9	56.3	44.2	49.5	50.6	44.5	47.4	58.3	39.8	47.3
IZF-Softmax	**Proposed**	80.5	**61.3**	**69.6**	77.5	**60.6**	**68.0**	68.0	**52.7**	**59.4**	57.0	**52.7**	**54.8**	60.5	42.3	**49.8**

ZSL. Besides, CVAE+$\mathcal{L}_{\text{iMMD}}$ isn't producing good-quality samples, undergoing the risk of obtaining biased classification hyper-planes of the held-out classifier. The side-effects of $\mathcal{L}_{\text{iMMD}}$ would slightly skew the generated data distributions from being realistic with its negative MMD, which aggravates the drawbacks of unstable training (GAN) and inaccurate ELBO (VAE) discussed in Sect. 1.

Towards the Bias Problem with $\mathcal{L}_{\text{iMMD}}$. We also illustrate the effects of $\mathcal{L}_{\text{iMMD}}$ with more baselines. It is shown in Fig. 4(f) that the model is biased by the *seen* classes without $\mathcal{L}_{\text{iMMD}}$ (also see `Baseline 4` of Sect. 6.5). The *unseen* generated samples are positioned closely to the *seen* ones. This would be harmful to the employed classifiers when there exist multiple *unseen* categories. Figure 4(g) is a failure case with large *seen-unseen* discrepancy loss, which dominates the optimization process and overfits the network to generate unreasonable samples. We also discuss this issue in hyper-parameter analysis (see Fig. 5(c)). Figure 4(h) describes an extreme situation when employing positive MMD to IZF (negative λ_3, `Baseline 5` of Sect. 6.5). The generated *unseen* samples are forced to fit the *seen* distribution and thus, the network is severely biased.

6.3 Real Data Experimental Settings

Benchmark Datasets. Five datasets are picked in our experiments. Animals with Attributes (AwA1) [26] contains 30,475 images of 50 classes and 85 attributes, of which AwA2 is a slightly extended version with 37,322 images. Caltech-UCSD Birds-200-20 (CUB) [58] carries 11,788 images from 200 kinds of birds with 312-attribute annotations. SUN Attribute (SUN) [40] consists of 14,340 images from 717 categories, annotated with 102 attributes. aPascal-aYahoo (aPY) [9] comes with 32 classes with 64 attributes, accounting 15,339 samples. We adopt the **PS** train-test setting [63] for both CZSL and GZSL.

Representations. All images $\mathbf{v}$ are represented using the 2048-D ResNet-101 [15] features and the semantic class embeddings $\mathbf{c}$ are category-wise attribute vectors from [61,63]. We pre-process the image features with min-max rescaling.

6.4 Comparison with the State-of-the-Arts

Baselines. IZF is compared with the state-of-the-art ZSL methods, including DAP [26], CMT [50], SSE [70], ESZSL [44], SAE [23], LATEM [60], ALE [1], DeViSE [10], DEM [69], RelationNet [52], DCN [30], CVAE-ZSL [38], SE-GZSL [24], f-CLSWGAN [62], CRNet [67], LisGAN [27], SGAL [66], CADA-VAE [46], GDAN [18], DLFZRL[54], f-VAEGAN-D2 [64] and LFGAA [31]. We report the official results of these methods from referenced articles with the identical experimental setting used in this paper for fair comparison.

Results. The GZSL comparison results are shown in Table 1. It can be observed that deep generative models obtains better on-average ZSL scores than the non-generative ones, while some simple semantic-visual projecting models hit comparable accuracy to them such as CRNet [67]. IZF-Softmax generally outperforms the compared methods, where the performance margins on AwA [26] are significant. IZF-NBC also works well on AwA [26] The proposed model produces balanced accuracy between *seen* and *unseen* data and obtains significant higher *unseen* accuracy. This shows the effectiveness of the discrepancy loss $\mathcal{L}_{\text{iMMD}}$ in solving the bias problem of ZSL. In addition to the GZSL results, we conduct CZSL experiments as well, which is shown

Table 2. CZSL per-class accuracy (%) comparison with the **PS** setting [63].

Method	AwA1	AwA2	CUB	SUN	aPY
DAP [26]	44.1	46.1	40.0	39.9	33.8
CMT [70]	39.5	37.9	34.6	39.9	28.0
SSE [70]	60.1	61.0	43.9	51.5	34.0
ESZSL [44]	58.2	58.6	53.9	54.5	38.3
SAE [23]	53.0	54.1	33.3	40.3	8.3
LATEM [60]	55.1	55.8	49.3	55.3	35.2
ALE [1]	59.9	62.5	54.9	58.1	39.7
DeViSE [10]	54.2	59.7	52.0	56.5	39.8
RelationNet [52]	68.2	64.2	55.6	–	–
DCN [30]	65.2	–	56.2	61.8	43.6
f-CLSWGAN [62]	68.2	–	57.3	60.8	–
LisGAN [27]	70.6	–	58.8	61.7	43.1
DLFZRL [54]	61.2	60.9	51.9	42.5	38.5
f-VAEGAN-D2 [64]	71.1	–	61.0	65.6	–
LFGAA [31]	–	68.1	**67.6**	62.0	–
IZF-NBC	72.7	71.9	59.6	63.0	**45.2**
IZF-Softmax	**74.3**	**74.5**	67.1	**68.4**	44.9

Table 3. Component analysis results on AwA1 [26] (Sect. 6.5). **NBC**: results with distance-based classifier. **Softmax**: results with a held-out trainable classifier.

Baseline	NBC			Softmax		
	A^s	A^u	H	A^s	A^u	H
1 CVAE + $\mathcal{L}_C$ + $\mathcal{L}_{\text{iMMD}}$	65.1	30.8	41.8	71.1	36.8	48.5
2 Without $\mathcal{L}_C$ and $\mathcal{L}_{\text{iMMD}}$	66.0	43.4	52.7	78.9	38.1	51.4
3 Without $\mathcal{L}_C$	67.0	41.7	51.4	79.2	60.9	68.8
4 Without $\mathcal{L}_{\text{iMMD}}$	79.6	49.0	60.7	81.3	53.2	64.3
5 Positive MMD	76.2	21.1	33.0	80.7	44.5	57.4
6 IM Kernel→Gaussian Kernel	73.6	54.9	62.9	79.6	61.7	69.5
IZF (full model)	75.2	57.8	65.4	80.5	61.3	69.6

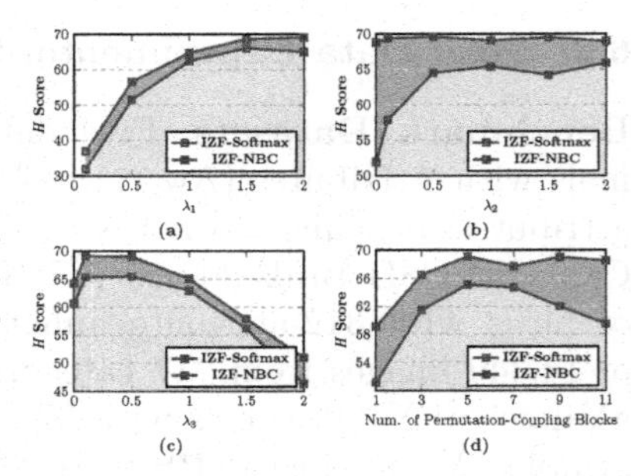

Fig. 5. **(a)**, **(b)** and **(c)** Hyper-parameter analysis for λ_1, λ_2 and λ_3. **(d)** Effect *w.r.t.* numbers of the permutation-coupling blocks.

in Table 2. As a relatively simpler setting, CZSL provides direct clues of the ability to transform knowledge from *seen* to *unseen*.

6.5 Component Analysis

We evaluate the effectiveness of each component of IZF to legitimate our design, including the loss terms and overall network structure. The following baselines are proposed. **(1) CVAE + $\mathcal{L}_C$ + $\mathcal{L}_{\text{iMMD}}$.** We firstly replace it with a simple CVAE [51] structure. This baseline uses the semantic representation as condition, and outputs synthesized visual features. $\mathcal{L}_C$ and $\mathcal{L}_{iMMD}$ are applied to this baseline. **(2) Without $\mathcal{L}_C$ & $\mathcal{L}_{\text{iMMD}}$.** All regularization on the reverse pass is omitted. **(3) Without $\mathcal{L}_C$.** The prototype centralizing loss is removed. **(4) Without $\mathcal{L}_{\text{iMMD}}$.** The discrepancy loss to control the *seen-unseen* bias problem of ZSL is deprecated. **(5) Positive MMD.** In Eq. (9), we employ negative MMD to tackle the bias problem. We propose a baseline with a positive MMD version of it to study its influence. This is realized by setting $\lambda_3 = -1$. **(6) IM Kernel→Gaussian Kernel.** Instead of the Inverse Multiquadratic kernel, another widely-used kernel function, *i.e.*, the Gaussian kernel, is tested in implementing Eq. (9).

Results. The above-mentioned baselines are compared in Table 3 on AwA1 [26]. The GZSL criteria are adopted here as they are more illustrative metrics for IZF, showing different performance aspects of the model. Through our test, `Baseline 1`, *i.e.*, CVAE+$\mathcal{L}_C$+$\mathcal{L}_{\text{iMMD}}$, is not working well with the distance-based classifier (Eq. (11)). With loss components omitted (`Baseline 2--4`), IZF does not work as expected. In `Baseline 4`, the classification results are significantly biased to the *seen* concepts. When imposing positive MMD to the loss function, the test accuracy of *seen* classes increases while the accuracy of *unseen* data drops quickly. This is because the bias problem gets severer and all generated samples, including the *unseen* classification prototypes, overfit to the *seen* domain. The choice of kernel is not a key factor in IZF, and `Baseline 7` obtains on-par

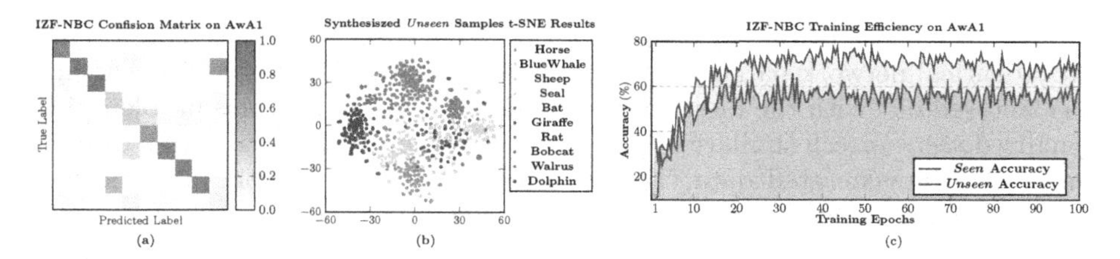

Fig. 6. **(a)** Confusion matrix of IZF on AwA1 with the CZSL setting. The order of labels is identical to the t-SNE legend. **(b)** t-SNE [34] results of the synthesized *unseen* samples on AwA1. **(c)** Training efficiency of IZF-NBC on AwA1.

accuracy to IZF. Similar to GAN/VAE-based models [27,38,62], IZF works with a held-out classifier, but it requires additional computational resources.

6.6 Hyper-Parameters

IZF involves 3 hyper-parameters in balancing the contribution of different loss items, shown in Eq. (10). The influences of the values of them on AwA1 are plotted in Fig. 5(a), (b) and (c) respectively. A large weight is imposed to the semantic knowledge loss $\mathcal{L}_{\text{Flow}}$, *i.e.*, $\lambda_1 = 2$, for best performance, as it plays an essential role in formulating the normalizing flow structure that ensures data generation with the sampled conditions and latents. A well-regressed visual-semantic projection necessitates conditional generation and, hence, bi-directional training. On the other hand, it is notable that a large value of λ_3 fails IZF overall. A heavy penalty to $\mathcal{L}_{\text{iMMD}}$ overfits the network to generate unreasonable samples to favour large *seen*-*unseen* distributional discrepancy, and further prevents the encoding loss $\mathcal{L}_{\text{Flow}}$ from functioning. We observe significant increase of $\mathcal{L}_{\text{Flow}}$ throughout the training steps with $\lambda_3 = 2$, though $\mathcal{L}_{\text{iMMD}}$ decreases quickly. We further report the training efficiency of IZF in Fig. 6(c), where IZF only requires $\sim$20 epochs to obtain best-performing parameters.

6.7 Discriminability on *Unseen* Classes

We intuitively analyze the discriminability and generation quality of IZF on *unseen* data by plotting the generated samples. The t-SNE [34] visualization of synthesized *unseen* data on AwA1 [26] is shown in Fig. 6(b). Although no direct regularization loss is applied to *unseen* classes, IZF manages to generate distinguishable samples according to their semantic meanings. In addition, the CZSL confusion matrix on AwA1 is reported in Fig. 6(a) as well.

7 Conclusion

In this paper, we proposed Invertible Zero-shot Flow (IZF), fully leveraging the merits of generative flows for ZSL. The invertible nature of flows enabled IZF to perform bi-directional mapping between the visual space and the semantic space with identical network parameters. The semantic information of a visual sample was factored-out with the forward pass of IZF. To handle the bias problem, IZF penalized *seen-unseen* similarity by computing kernel-based distribution discrepancy with the generated data. The proposed model consistently outperformed state-of-the-art baselines on benchmark datasets.

References

1. Akata, Z., Reed, S., Walter, D., Lee, H., Schiele, B.: Evaluation of output embeddings for fine-grained image classification. In: CVPR (2015)
2. Ardizzone, L., et al.: Analyzing inverse problems with invertible neural networks. In: ICLR (2019)
3. Ardizzone, L., Lüth, C., Kruse, J., Rother, C., Köthe, U.: Guided image generation with conditional invertible neural networks. arXiv preprint arXiv:1907.02392 (2019)
4. Cacheux, Y.L., Borgne, H.L., Crucianu, M.: Modeling inter and intra-class relations in the triplet loss for zero-shot learning. In: ICCV (2019)
5. Che, T., Li, Y., Jacob, A.P., Bengio, Y., Li, W.: Mode regularized generative adversarial networks. In: ICLR (2017)
6. Dinh, L., Krueger, D., Bengio, Y.: Nice: non-linear independent components estimation. In: ICLR Workshops (2014)
7. Dinh, L., Sohl-Dickstein, J., Bengio, S.: Density estimation using real NVP. In: ICLR (2017)
8. Elhoseiny, M., Elfeki, M.: Creativity inspired zero-shot learning. In: ICCV (2019)
9. Farhadi, A., Endres, I., Hoiem, D., Forsyth, D.A.: Describing objects by their attributes. In: CVPR (2009)
10. Frome, A., et al.: DeViSE: a deep visual-semantic embedding model. In: NeurIPS (2013)
11. Gao, R., et al.: Zero-VAE-GAN: generating unseen features for generalized and transductive zero-shot learning. IEEE Trans. Image Process. **29**, 3665–3680 (2020)
12. Gao, R., Hou, X., Qin, J., Liu, L., Zhu, F., Zhang, Z.: A joint generative model for zero-shot learning. In: Leal-Taixé, L., Roth, S. (eds.) ECCV 2018. LNCS, vol. 11132, pp. 631–646. Springer, Cham (2019). https://doi.org/10.1007/978-3-030-11018-5_50
13. Goodfellow, I., et al.: Generative adversarial nets. In: NeurIPS (2015)
14. Grover, A., Dhar, M., Ermon, S.: Flow-GAN: combining maximum likelihood and adversarial learning in generative models. In: AAAI (2018)
15. He, K., Zhang, X., Ren, S., Sun, J.: Deep residual learning for image recognition. In: CVPR (2016)
16. Higgins, I., et al.: Beta-VAE: learning basic visual concepts with a constrained variational framework. In: ICLR (2017)
17. Hoogeboom, E., Peters, J.W., van den Berg, R., Welling, M.: Integer discrete flows and lossless compression. In: NeurIPS (2019)
18. Huang, H., Wang, C., Yu, P.S., Wang, C.D.: Generative dual adversarial network for generalized zero-shot learning. In: CVPR (2019)
19. Jiang, H., Wang, R., Shan, S., Chen, X.: Transferable contrastive network for generalized zero-shot learning. In: ICCV (2019)
20. Kingma, D., Ba, J.: Adam: a method for stochastic optimization. In: ICLR (2015)

21. Kingma, D., Dhariwal, P.: Glow: Generative flow with invertible 1x1 convolutions. In: NeurIPS (2018)
22. Kingma, D., Welling, M.: Auto-encoding variational Bayes. In: ICLR (2014)
23. Kodirov, E., Xiang, T., Gong, S.: Semantic autoencoder for zero-shot learning. In: CVPR (2017)
24. Kumar Verma, V., Arora, G., Mishra, A., Rai, P.: Generalized zero-shot learning via synthesized examples. In: CVPR (2018)
25. Lampert, C.H., Nickisch, H., Harmeling, S.: Learning to detect unseen object classes by between-class attribute transfer. In: CVPR (2009)
26. Lampert, C.H., Nickisch, H., Harmeling, S.: Attribute-based classification for zero-shot visual object categorization. IEEE Trans. Pattern Anal. Mach. Intell. **36**(3), 453–465 (2013)
27. Li, J., Jing, M., Lu, K., Ding, Z., Zhu, L., Huang, Z.: Leveraging the invariant side of generative zero-shot learning. In: CVPR (2019)
28. Li, K., Min, M.R., Fu, Y.: Rethinking zero-shot learning: a conditional visual classification perspective. In: ICCV (2019)
29. Liu, R., Liu, Y., Gong, X., Wang, X., Li, H.: Conditional adversarial generative flow for controllable image synthesis. In: CVPR (2019)
30. Liu, S., Long, M., Wang, J., Jordan, M.I.: Generalized zero-shot learning with deep calibration network. In: NeurIPS (2018)
31. Liu, Y., Guo, J., Cai, D., He, X.: Attribute attention for semantic disambiguation in zero-shot learning. In: ICCV (2019)
32. Long, Y., Liu, L., Shen, Y., Shao, L.: Towards affordable semantic searching: zero-shot retrieval via dominant attributes. In: AAAI (2018)
33. Maas, A.L., Hannun, A.Y., Ng., A.Y.: Rectifier nonlinearities improve neural network acoustic models. In: ICML (2013)
34. van der Maaten, L., Hinton, G.: Visualizing data using t-SNE. J. Mach. Learn. Res. **9**(Nov), 2579–2605 (2008)
35. Mandal, D., et al.: Out-of-distribution detection for generalized zero-shot action recognition. In: CVPR (2019)
36. Mensink, T., Verbeek, J., Perronnin, F., Csurka, G.: Distance-based image classification: generalizing to new classes at near-zero cost. IEEE Trans. Pattern Anal. Mach. Intell. **35**(11), 2624–2637 (2013)
37. Mikolov, T., Sutskever, I., Chen, K., Corrado, G.S., Dean, J.: Distributed representations of words and phrases and their compositionality. In: NeurIPS (2013)
38. Mishra, A., Krishna Reddy, S., Mittal, A., Murthy, H.A.: A generative model for zero shot learning using conditional variational autoencoders. In: CVPR Workshops (2018)
39. Paszke, A., et al.: PyTorch: an imperative style, high-performance deep learning library. In: NeurIPS (2019)
40. Patterson, G., Hays, J.: Sun attribute database: discovering, annotating, and recognizing scene attributes. In: CVPR (2012)
41. Prenger, R., Valle, R., Catanzaro, B.: WaveGlow: a flow-based generative network for speech synthesis. In: ICASSP (2019)
42. Qin, J., et al.: Zero-shot action recognition with error-correcting output codes. In: CVPR (2017)
43. Radovanović, M., Nanopoulos, A., Ivanović, M.: Hubs in space: popular nearest neighbors in high-dimensional data. J. Mach. Learn. Res. **11**(Sep), 2487–2531 (2010)
44. Romera-Paredes, B., Torr, P.: An embarrassingly simple approach to zero-shot learning. In: ICML (2015)

45. Scheirer, W.J., de Rezende Rocha, A., Sapkota, A., Boult, T.E.: Toward open set recognition. IEEE Trans. Pattern Anal. Mach. Intell. **35**(7), 1757–1772 (2012)
46. Schonfeld, E., Ebrahimi, S., Sinha, S., Darrell, T., Akata, Z.: Generalized zero- and few-shot learning via aligned variational autoencoders. In: CVPR (2019)
47. Shen, Y., Liu, L., Shen, F., Shao, L.: Zero-shot sketch-image hashing. In: CVPR (2018)
48. Shen, Z., Lai, W.-S., Xu, T., Kautz, J., Yang, M.-H.: Exploiting semantics for face image deblurring. Int. J. Comput. Vis. **128**(7), 1829–1846 (2020). https://doi.org/10.1007/s11263-019-01288-9
49. Shen, Z., et al.: Human-aware motion deblurring. In: ICCV (2019)
50. Socher, R., Ganjoo, M., Sridhar, H., Bastani, O., Manning, C.D., Ng, A.Y.: Zero-shot learning through cross-modal transfer. In: NeurIPS (2013)
51. Sohn, K., Lee, H., Yan, X.: Learning structured output representation using deep conditional generative models. In: NeurIPS (2015)
52. Sung, F., Yang, Y., Zhang, L., Xiang, T., Torr, P.H., Hospedales, T.M.: Learning to compare: relation network for few-shot learning. In: CVPR (2018)
53. Tolstikhin, I., Bousquet, O., Gelly, S., Schoelkopf, B.: Wasserstein auto-encoders. In: ICLR (2018)
54. Tong, B., Wang, C., Klinkigt, M., Kobayashi, Y., Nonaka, Y.: Hierarchical disentanglement of discriminative latent features for zero-shot learning. In: CVPR (2019)
55. Tran, D., Vafa, K., Agrawal, K.K., Dinh, L., Poole, B.: Discrete flows: invertible generative models of discrete data. In: ICLR Workshops (2019)
56. Tsai, Y.H.H., Huang, L.K., Salakhutdinov, R.: Learning robust visual-semantic embeddings. In: ICCV (2017)
57. Tsai, Y.H.H., Liang, P.P., Zadeh, A., Morency, L.P., Salakhutdinov, R.: Learning factorized multimodal representations. In: ICLR (2019)
58. Wah, C., Branson, S., Welinder, P., Perona, P., Belongie, S.: The Caltech-UCSD Birds-200-2011 dataset. Technical report CNS-TR-2011-001, California Institute of Technology (2011)
59. Wang, Q., Chen, K.: Zero-shot visual recognition via bidirectional latent embedding. Int. J. Comput. Vis. **124**(3), 356–383 (2017). https://doi.org/10.1007/s11263-017-1027-5
60. Xian, Y., Akata, Z., Sharma, G., Nguyen, Q., Hein, M., Schiele, B.: Latent embeddings for zero-shot classification. In: CVPR (2016)
61. Xian, Y., Lampert, C.H., Schiele, B., Akata, Z.: Zero-shot learning-a comprehensive evaluation of the good, the bad and the ugly. IEEE Trans. Pattern Anal. Mach. Intell. **41**(9), 2251–2265 (2018)
62. Xian, Y., Lorenz, T., Schiele, B., Akata, Z.: Feature generating networks for zero-shot learning. In: CVPR (2018)
63. Xian, Y., Schiele, B., Akata, Z.: Zero-shot learning-the good, the bad and the ugly. In: CVPR (2017)
64. Xian, Y., Sharma, S., Schiele, B., Akata, Z.: f-VAEGAN-D2: a feature generating framework for any-shot learning. In: CVPR (2019)
65. Xie, G.S., et al.: Attentive region embedding network for zero-shot learning. In: CVPR (2019)
66. Yu, H., Lee, B.: Zero-shot learning via simultaneous generating and learning. In: NeurIPS (2019)
67. Zhang, F., Shi, G.: Co-representation network for generalized zero-shot learning. In: ICML (2019)

68. Zhang, H., Koniusz, P.: Zero-shot kernel learning. In: CVPR (2018)
69. Zhang, L., Xiang, T., Gong, S.: Learning a deep embedding model for zero-shot learning. In: CVPR (2017)
70. Zhang, Z., Saligrama, V.: Zero-shot learning via semantic similarity embedding. In: ICCV (2015)
71. Zhu, Y., Xie, J., Liu, B., Elgammal, A.: Learning feature-to-feature translator by alternating back-propagation for generative zero-shot learning. In: ICCV (2019)

GeoLayout: Geometry Driven Room Layout Estimation Based on Depth Maps of Planes

Weidong Zhang[1,2], Wei Zhang[1(✉)], and Yinda Zhang[3]

[1] School of Control Science and Engineering, Shandong University, Jinan, China
davidzhang@sdu.edu.cn
[2] School of Communications and Information Engineering, Xi'an University of Posts and Telecommunications, Xi'an, China
[3] Google Research, Mountain View, USA

Abstract. The task of room layout estimation is to locate the wall-floor, wall-ceiling, and wall-wall boundaries. Most recent methods solve this problem based on edge/keypoint detection or semantic segmentation. However, these approaches have shown limited attention on the geometry of the dominant planes and the intersection between them, which has significant impact on room layout. In this work, we propose to incorporate geometric reasoning to deep learning for layout estimation. Our approach learns to infer the depth maps of the dominant planes in the scene by predicting the pixel-level surface parameters, and the layout can be generated by the intersection of the depth maps. Moreover, we present a new dataset with pixel-level depth annotation of dominant planes. It is larger than the existing datasets and contains both cuboid and non-cuboid rooms. Experimental results show that our approach produces considerable performance gains on both 2D and 3D datasets.

Keywords: Room layout estimation · Plane segmentation · Dataset

1 Introduction

An indoor scene differs from the natural scenes in that it usually contains dominant planes such as floor, ceiling and walls. These planes are likely to be orthogonal to each other. Hence the spatial structure of an indoor scene tends to show some regularity and can be represented by the room layout. Currently, the task of room layout estimation is to locate the wall-floor, wall-ceiling, and wall-wall boundaries. It can provide a useful prior for a wide range of computer vision tasks, such as scene reconstruction [2,17,26] and augmented reality [18,24,35].

Electronic supplementary material The online version of this chapter (https://doi.org/10.1007/978-3-030-58517-4_37) contains supplementary material, which is available to authorized users.

A. Vedaldi et al. (Eds.): ECCV 2020, LNCS 12361, pp. 632–648, 2020.
https://doi.org/10.1007/978-3-030-58517-4_37

Recent methods achieve significant performance gains, which primarily focus on learning the feature maps with deep networks like fully convolutional networks (FCNs) [33]. One popular idea is to learn the wall-floor, wall-ceiling, and wall-wall edges [25,31,40]. Another is to learn the semantic surface labels such as floor, ceiling, front wall, left wall, and right wall [5,38]. Besides, there are also methods trying to infer the layout corners (keypoints) [20,43]. However, the gathered bottom-up information from edge/keypoint detection or semantic segmentation may not reflect the underlying geometry of room layout, e.g., orthogonal planes.

Essentially, the desired boundary between two surfaces appears because the two planes in 3D space intersect in a line. This motivates us to focus on the geometric model of the dominant surfaces (e.g., the floor, ceiling and wall) in the indoor scene. With this key insight, we propose to predict the depth maps of the dominant surfaces, and generate the layout by the intersection of the depth maps, as shown in Fig. 1. We first analyse the projection principle of a 3D plane into the depth map and obtain the representation without explicit camera intrinsics to parameterize the depth of a plane. Compared to the general 3D coordinate systems (e.g., the camera coordinate system), our parameterization can omit the need for the camera intrinsic parameters. It also makes the method applicable to the existing layout datasets whose intrinsic parameters are not provided like Hedau [13] and LSUN [39]. Then we train a deep network to predict the pixel-level surface parameters for each planar surface. The pixel-level parameters are further aggregated into an instance-level parameter to calculate the corresponding depth map, and the layout can be generated based on the predicted depth maps. Our method generally requires the depth map of the planar surfaces for learning. However, with our parameterization and geometric constraints, the model can also be trained with only 2D segmentation.

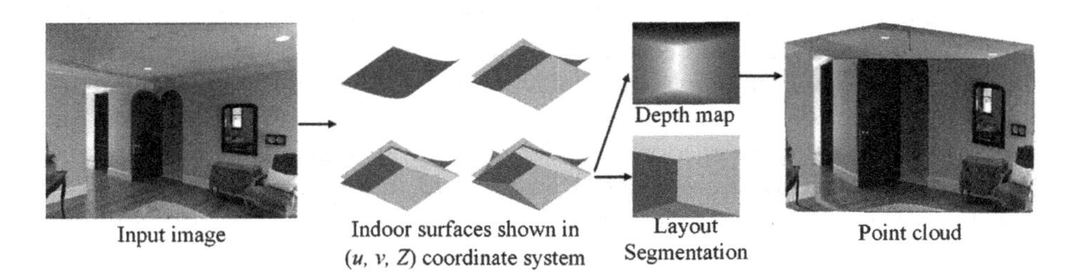

Fig. 1. Room layout estimation based on depth maps of the planar surfaces.

However, the existing datasets for layout estimation do not fully support the learning of the proposed 3D geometry-aware model as the 3D labels are not provided. Besides, all the images are exhibiting simple cuboid layout only. These shortcomings of the datasets severely limit the development of layout estimation algorithms and the practical applications. Therefore, we produce a new dataset for room layout estimation providing pixel-level depth annotation of the dominant planar surfaces. The ground truth is gathered semi-automatically with a combination of human annotation and plane fitting algorithm, and the dataset contains indoor scenes with complex non-cuboid layout.

The major contributions of this work are summarized as follows: (1) We propose to incorporate geometric reasoning to deep learning for the task of layout estimation, which is reformulated as predicting the depth maps of the dominant planes. (2) We demonstrate the proposed model can be effectively trained to predict the surface parameters, and it can also improve 2D layout performance with the learned 3D knowledge. (3) A dataset with 3D label for layout estimation is presented. The dataset is in large scale, complementary to previous datasets, and beneficial to the field of room layout estimation.

2 Related Work

The current work on layout estimation can be divided into two types according to whether the 3D spatial rules are exploited.

2D Based Layout Estimation. The layout estimation problem was first introduced by Hedau et al. [13], which consisted of two stages. First, a series of layout hypotheses were generated by ray sampling from the detected vanishing points. Second, a regressor was learned to rank the hypotheses. Later on, some methods were proposed to improve the framework [7,29,30,34], such as using different hand-crafted features and improving the hypotheses generation process.

Recently, the CNN and FCN-based methods were proposed for layout estimation and showed dramatic performance improvements on benchmark datasets. Mallya and Lazebnik [25] trained a FCN [33] to predict the edge maps, which are used for layout hypotheses generation and ranking. Ren et al. [31] used the edge map as a reference for generating layout hypotheses based on the vanishing lines, undetected lines, and occluded lines. Dasgupta et al. [5] instead predicted the pixel-level semantic labels with FCN, and the layout estimates are further optimized by vanishing lines. Zhang et al. [38] jointly learned the edge maps and semantic labels using an encoder-decoder network. The layout hypotheses were then generated and optimized based on the two information. Zhao et al. [40] transferred the semantic segmentation model to produce edge features and proposed a physics inspired optimization inference scheme. Lee et al. [20] adopted an end-to-end framework for layout estimation by predicting the locations of the layout keypoints. These methods generally predict the layout in 2D space, and the 3D knowledge of indoor scenes are usually ignored.

3D Based Layout Estimation and Related Work. Lee et al. [21] proposed the "Indoor World" model based on the Manhattan world assumption and symmetric floor and ceiling assumption. The model represented the scene layout in 2D space and could be translated into a 3D model by geometric reasoning on the configuration of edges. Choi et al. [4] trained several graph models that fused the room layout, scene type and the objects in 3D space. Zhao and Zhu [42] applied a hierarchical model that characterized a joint distribution of the layout and indoor objects. The layout hypotheses were evaluated by the 3D size and localization of the indoor objects. Guo et al. [11] trained five SVMs for the five layout categories using appearance, depth, and location features. These methods have exploited 3D spatial rules or considered the 3D relationship between the room layout and indoor objects, but none of these work has focused on the 3D geometry of the planar surfaces.

The 3D plane detection and reconstruction problem aims to segment plane instances and recover 3D plane parameters from an image, which is somewhat similar to layout estimation. The methods can be divided into two groups. The geometry-based methods [1,8,27] extract geometric cues such as vanishing points and line segments to recover 3D information. The appearance-based methods [10, 12,22,23,36,37] infer the 3D information based on the appearance and do not rely on the assumptions about the scenes. Specifically, Liu et al. [23] proposed a deep network that simultaneously learned a constant number of plane parameters and instance-level segmentation masks. Yang and Zhou [36] reformulated the problem and proved that the 3D plane fitting losses could be converted to depth prediction losses, and therefore did not require ground truth 3D planes. Yu et al. [37] presented a proposal-free instance segmentation approach that first learned pixel-level plane embeddings and then applied the mean shift clustering to generate plane instances. Since these methods are purely 3D based and require the camera intrinsic parameters to work, they cannot be applied to the current layout datasets like Hedau and LSUN.

3 Method

In this work, we intend to solve the layout estimation problem by predicting the depth maps of the dominant planes (e.g., floor, walls, ceiling) in a room. Then the layout can be obtained by the intersection of the depth maps of planar surfaces that intersect each other. In Sect. 3.1, we first analyze the depth map of a plane and give the general equation in the (u, v, Z) coordinate system, which can be used to parameterize the depth map of an arbitrary plane. Then we use a deep network to learn the surface parameters of the dominant planes and generate layout estimates. The illustration of our method is shown in Fig. 2.

3.1 Parameterizing Depth Maps of Planes

A 3D plane in the camera coordinate system can be represented with the equation: $aX + bY + cZ + d = 0$, where (a, b, c) is the normal vector and d is the distance to the camera center. A 3D point can be projected onto the image plane via perspective projection, i.e., $u = f_x \frac{X}{Z} + u_0, \quad v = f_y \frac{Y}{Z} + v_0$, where u, v are the pixel coordinates and f_x, f_y, u_0, v_0 are the camera intrinsic parameters, with f_x and f_y the focal lengths and u_0 and v_0 the coordinates of the principal point. Based on the perspective projection, The planar equation can be rewritten as: $\frac{1}{Z} = -\frac{a}{f_x d}u - \frac{b}{f_y d}v + \frac{1}{d}(\frac{a}{f_x}u_0 + \frac{b}{f_y}v_0 - c)$. Apparently, the inverse depth value $\frac{1}{Z}$ is proportional to the pixel coordinates u and v in the depth map. With the above observation, the depth map of a plane can be parameterized without explicit camera intrinsics by three new parameters $\hat{p}$, $\hat{q}$, $\hat{r}$ as shown in Eq. (1).

$$Z = \frac{1}{\hat{p}u + \hat{q}v + \hat{r}}. \tag{1}$$

In practice, the global scale of an indoor scene is ambiguous, which makes the three parameters involved with the scale. Hence, we introduce a scale factor $s = \sqrt{\hat{p}^2 + \hat{q}^2 + \hat{r}^2}$ and apply normalization to $\hat{p}$, $\hat{q}$, $\hat{r}$, i.e., $p = \hat{p}/s$, $q = \hat{q}/s$, $r = \hat{r}/s$. The normalized parameters p, q, r are therefore scale-invariant. Finally, the modified equation is given as below:

$$Z = \frac{1}{(pu + qv + r)s}. \tag{2}$$

3.2 Learning Depth Maps of Planes

We first introduce our method that learns the depth maps of planes with depth supervision in this section.

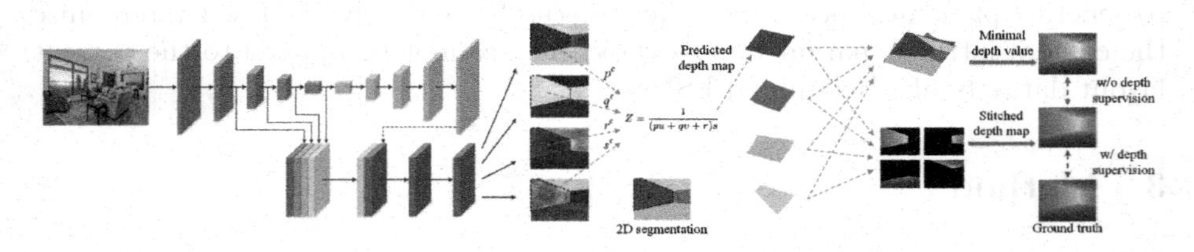

Fig. 2. An illustration of our method that can be trained w/wo depth supervision. Given an input image, the pixel-level surface parameters are predicted by the network. They are aggregated into several instance-level parameters to produce the depth maps of the planar surfaces. Based on the 2D segmentation, these depth maps can be combined into a stitched depth map, which is evaluated by either the ground truth (w/ depth supervision) or the minimal depth value of the predicted depth maps for each pixel localization (w/o depth supervision).

Pixel-Level Surface Parameter Estimation. As shown in Eq. (2), the depth map of a plane in the real world can be parameterized using p, q, r and s. Motivated by this, we train a deep network to predict the pixel-level surface parameters of the input image. We implement the surface parameter estimation network based on [16], which is originally designed for monocular depth estimation. The network consists of four modules: an encoder, a decoder, a multi-scale feature fusion module, and a refinement module. We replace the last layer of the refinement module by four output channels. For an input color image, the network outputs four heat maps representing the pixel-level surface parameters. It is worth noting that the network can be replaced to any architecture for pixel-wise prediction such as PSPNet [41] and FCN [33].

With the ground truth depth map of the dominant planes Z_i^*, we transform Eq. (1) to calculate the target parameters $\hat{p}_i^*$, $\hat{q}_i^*$, $\hat{r}_i^*$ at the i_{th} pixel:

$$\begin{aligned} \hat{p}_i^* &= \nabla_u(1/Z_i^*), \\ \hat{q}_i^* &= \nabla_v(1/Z_i^*), \\ \hat{r}_i^* &= 1/Z_i^* - \hat{p}_i^* u_i - \hat{q}_i^* v_i, \end{aligned} \tag{3}$$

where $\nabla_u(1/Z_i^*)$ represents the spatial derivative of $(1/Z_i^*)$ w.r.t. u computed at the i_{th} pixel, and so on. Following the same normalization operation, the scale factor s_i^* is calculated and the normalized parameters p_i^*, q_i^*, r_i^* are obtained. Let $P_i = (p_i, q_i, r_i, s_i)$ be the predicted parameters at the i_{th} pixel and P_i^* denotes the ground truth. We use L1 loss to supervise the regressed parameters for each pixel:

$$\mathcal{L}_p = \frac{1}{n} \sum_{i=1}^{n} \| P_i - P_i^* \| . \tag{4}$$

Besides, the surface parameters belonging to the same surface should be close together, while the parameters from different surfaces should be far apart. To this end, we employ the discriminative loss proposed in [6]. The loss function includes two terms: a variance term to penalize the parameter that is far from its corresponding instance center, and a distance term to penalize the pairs of different instance centers that are close to each other.

$$\mathcal{L}_d = \mathcal{L}_{var} + \mathcal{L}_{dist}, \tag{5}$$

$$\mathcal{L}_{var} = \frac{1}{C} \sum_{c=1}^{C} \frac{1}{n_c} \sum_{i=1}^{n_c} \max(\| P_i - P^c \| - \delta_v, 0), \tag{6}$$

$$\mathcal{L}_{dist} = \frac{1}{C(C-1)} \sum_{\substack{c_A=1 \\ c_A \neq c_B}}^{C} \sum_{c_B=1}^{C} \max(\delta_d - \| P^{c_A} - P^{c_B} \| , 0), \tag{7}$$

where C is the number of planar surfaces in the ground truth, n_c is the number of pixels in surface c, P^c is the mean of the pixel-level parameters belonging to c, δ_v and δ_d are the margins for the variance and distance loss, respectively. Here we employ the variance loss $\mathcal{L}_{var}$ to encourage the estimated parameters to be close within each surface. At last, we extract P^c as the instance-level parameters to generate the depth map of surface c.

Depth Map Generation. We found that exploiting the ground truth to supervise the predicted depth map makes the training more effective. For surface c with the predicted instance-level parameters $P^c = (p^c, q^c, r^c, s^c)$, its corresponding depth map Z^c can be produced using Eq. (2), i.e., $Z_i^c = 1/[(p^c u_i + q^c v_i +$

$r^c)s^c]$. In the training stage, the ground truth 2D segmentation l^* is used to combine the predicted depth maps into a stitched depth map, which is evaluated by the ground truth as below:

$$\mathcal{L}_z = \frac{1}{n}\sum_{i=1}^{n}\left\|\frac{1}{Z_i^{l_i^*}} - \frac{1}{Z_i^*}\right\|, \tag{8}$$

where $Z_i^{l_i^*}$ is the i_{th} pixel of the stitched depth map and Z^* is the ground truth. We use the inverse depth in the loss function as it is linear w.r.t. the pixel coordinates, which makes the training more stable and smooth. Finally, the overall objective is defined as follows:

$$\mathcal{L}_{3D} = \mathcal{L}_p + \alpha\mathcal{L}_{var} + \beta\mathcal{L}_z. \tag{9}$$

3.3 Training on 2D Layout Datasets

The current benchmark datasets Hedau [13] and LSUN [39] both use the ground truth 2D segmentation to represent the layout, and neither of them has ground truth depth. The supervised learning method in Sect. 3.2 is inapplicable for these datasets. In this section, we present a learning strategy that enables the model to be trained with only 2D segmentation.

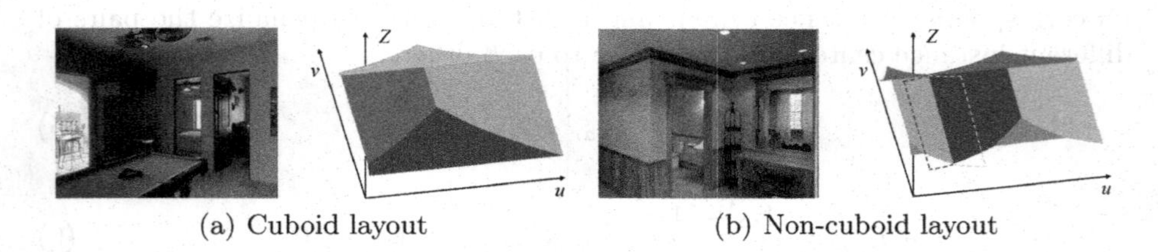

(a) Cuboid layout (b) Non-cuboid layout

Fig. 3. Depth representation of the dominant planes in the (u, v, Z) coordinate system. The layout of a cuboid room is determined by the nearest planar regions, but it is inapplicable for the non-cuboid room (see the red and yellow surfaces). (Color figure online)

First, we employ the same network structure as in Sect. 3.2 for surface parameter estimation. We use the full discriminative loss $\mathcal{L}_d$ to constrain the predicted surface parameters. Here we assume that the indoor image has cuboid layout, which means the room can be represented as a box. Such assumption is tenable for Hedau and LSUN datasets and has been widely adopted by many previous work. Based on this assumption, an important observation is that the layout is determined by the nearest planar regions in the indoor scene. As shown in Fig. 3(a), if representing the depth maps of all the dominant planes in the (u, v, Z) coordinate system, each surface will have a depth value at each (u, v) coordinate and the minimal value at each pixel will form the depth map that represents the layout. Also, the 2D segmentation map can be obtained according to which surface has the minimal depth value at each pixel. It is worth noting that the calculated depth map from Eq. (2) may have negative values, which should be excluded. We simply switch to inverse depth and extract the maximum of $\frac{1}{Z^c}$ at each pixel to produce the layout segmentation and corresponding depth map:

$$\frac{1}{Z_i} = \max_c \frac{1}{Z_i^c}, \quad l_i = \arg\max_c \frac{1}{Z_i^c}. \tag{10}$$

Here, l_i is the generated pixel-level layout segmentation and $\frac{1}{Z_i}$ is the corresponding depth map. Since arg max is not differentiable, it is unable to evaluate the generated layout estimates by the pixel error between l_i and ground truth l_i^*. Instead, we encourage the stitched depth map to be consistent with the minimal depth value. The loss function for the predicted depth map is defined as below:

$$\mathcal{L}_z = \frac{1}{n}\sum_{i=1}^{n}\left\|\frac{1}{Z_i^{l_i^*}} - \frac{1}{Z_i}\right\|. \tag{11}$$

However, we find that under the current objective the learned model tends to produce similar depth estimates for all the surfaces to reduce the loss. To deal with this problem, we propose a "stretch" loss to increase the mutual distance between the depth maps as follows:

$$\mathcal{L}_s = -\frac{1}{n}\sum_{i=1}^{n}\frac{e^{k/Z_i^{l_i^*}}}{\sum_{c=1}^{C} e^{k/Z_i^c}}, \tag{12}$$

where k is a scale factor in the softmax operation. The "stretch" loss encourages $1/Z_i^{l_i^*}$ to be much larger than the rest inverse depth values at i_{th} pixel, and therefore similar depth estimates will be punished. The overall objective is defined as follows:

$$\mathcal{L}_{2D} = \mathcal{L}_d + \eta\mathcal{L}_z + \theta\mathcal{L}_s. \tag{13}$$

It is worth noting that such learning strategy is inapplicable for the non-cuboid room, as shown in Fig. 3(b). Besides, the generated depth map can only infer the relative depth information, yet the precise depth value is unavailable.

3.4 Generating Layout Estimates

When the training stage is complete, a post-process step is employed to obtain the parameterized layout estimation results. Because of the discriminative loss (Eq. (5)–(7)), the predicted pixel-level surface parameters are likely to be piece-wise constant and can be easily grouped to produce a segmentation map representing the surface instances. We use standard mean-shift clustering as the number of clusters does not need to be pre-defined. After clustering, the small clusters with fewer than 1% of the overall pixels are abandoned. Next we extract the mean of the parameters within each cluster to obtain the instance-level parameters. Then the depth map for each planar surface can be generated.

To find the true layout among the depth maps that intersect each other, we evaluate the layout estimates based on its consistency with the clustered segmentation. Specifically, we sort the depth maps of different surfaces according to ascending order for each pixel, while the index indicating the surface instance will constitute multiple layers of segmentation maps. Starting from the first layer, we compare the current segmentation with the clustered segmentation. For each region of the current segmentation, if the label is inconsistent with the dominant label of the clustered segmentation, we use the labels from the next layer to replace the inconsistent label. This process continues until the current segmentation is consistent with the clustered segmentation. Then the predicted layout segmentation, depth map, and the corresponding surface parameters are all available. With the intrinsic camera parameters, the 3D point cloud representing the layout can also be generated based on the depth map. Finally, the layout corners can be computed based on the equations of the predicted depth maps, i.e., a layout corner is the point of intersection among three surfaces, or two surfaces and an image boundary.

4 Matterport3D-Layout Dataset

In this section, we introduce our large scale dataset with 3D layout ground truth for our training purpose, named Matterport3D-Layout. We use images from Matterport3D dataset [3] as the dataset contains real photos from complex scenes, which provides good layout diversity. It also provides depth image that can be used to recover 3D layout ground truth. We annotate the visible region of each plane and use Eq. (1) for parameter fitting in each surface. Then the depth maps of planar surfaces can be calculated using Eq. (1).

Annotation. We first filter out the images without recognizable layout. Then we draw 2D polygons using LabelMe [32] on the visible regions of the floor, ceiling and walls for each image. The polygons on different surfaces have different semantic categories. We also abandon the images with surfaces completely occluded by the indoor objects as the true depth of the surfaces are unavailable.

Layout Generation. Given the depth map and region annotation, we extract the depth value and pixel coordinates in each annotated region and employ RANSAC algorithm [9] for the curved surface fitting to obtain the instance-level surface parameters. Then the layout can be generated in a similar way as described in Sect. 3.4.

The original Matterport3D dataset includes 90 different buildings, so we randomly split the dataset into training, validation and testing set according to the building ID. The training set includes 64 buildings with a total of 4939 images. The validation set includes 6 buildings with 456 images. The testing set includes the remaining 20 buildings with a total of 1965 images. All images have the resolution of 1024×1280. The dataset contains the following fields: (1) Color image; (2) Depth map of the planar surfaces; (3) 2D segmentation of layout; (4) Original depth map containing indoor objects; (5) Visible region annotation; (6) Intrinsic matrix of the camera; (7) Surface parameters for each plane p, q, r; (8) The coordinates of the layout corners (u, v, Z); (9) Original surface normal.

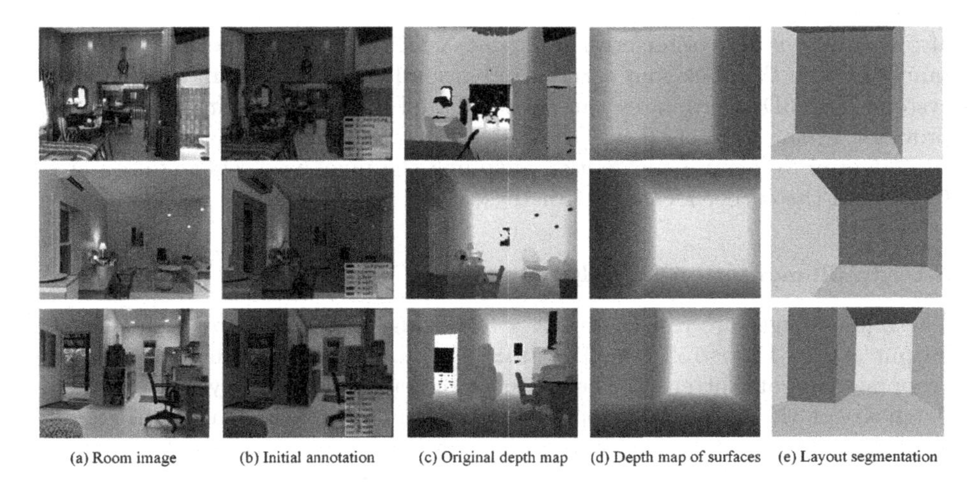

Fig. 4. Our Matterport3D-Layout dataset provides pixel-level depth label for the dominant planes.

Figure 4 shows some examples of our dataset. Prior to our dataset, there are two benchmark layout datasets: Hedau [13] and LSUN [39]. Statistics of the existing datasets are summarized in Table 1. As can be seen, the proposed dataset is the largest one and provides the richest kinds of ground truths. Besides, the proposed dataset contains non-cuboid layout samples which are absent in the other datasets. We hope this dataset can benefit the community and motivate the research about indoor layout estimation and related tasks.

Table 1. A brief summary of existing datasets in layout estimation.

Dataset	Train	Val	Test	Label	Layout type
Hedau [13]	209	–	105	seg	cuboid
LSUN [39]	4000	394	1000	seg. & corner	cuboid
Matterport3D-Layout	4939	456	1965	seg. & corner & depth	cuboid & non-cuboid

5 Experimental Results

In this section, we first evaluate our method on 3D room layout estimation. Next, we evaluate the effectiveness of transferring the knowledge to 2D room layout estimation. For 3D layout estimation, we use metrics for depth map evaluation, including root of the mean squared error (rms), mean relative error (rel), Mean log10 error (log10), and the percentage of pixels with the ratio between the prediction and the ground truth smaller than 1.25, 1.25^2, and 1.25^3. We also calculate a 3D corner error ($e_{3D\_cor.}$), which represents the Euclidean distance between the 3D layout corners and ground truth in the camera coordinate system. The 3D coordinates can be calculated using the intrinsic parameters provided in the dataset. For 2D layout estimation, we use two standard metrics adopted by many benchmarks, including the pixel-wise segmentation error ($e_{pix.}$) and the corner location error ($e_{cor.}$) [39].

5.1 Implementation Details

The input images are resized to 228×304 using bilinear interpolation and the output size is $114 \times 152 \times 4$. The training images are augmented by random cropping and color jittering. The model is implemented using PyTorch [28] with batch size of 32. We use Adam optimizer with an initial learning rate of 10^{-4} and a weight decay of 10^{-4}. The network is trained for 200 epochs and the learning rate is reduced to 10% for every 50 epochs. The values of the margins are set as $\delta_v = 0.1$, $\delta_d = 1.0$. The scale factor is set as $k = 20$. The weights in the final loss functions are set as $\alpha = 0.5$, $\beta = 1$, $\eta = 10$, $\theta = 0.03$.

5.2 Results on Matterport3D-Layout Dataset.

3D Layout Performance. The performance on the Matterport3D-Layout testing set is shown in Table 2. The existing layout estimation methods are mostly 2D based methods and cannot predict the 3D layout estimates. We compare to PlaneNet [23], which is the state-of-the-art method for 3D planar reconstruction. The major difference between our method and PlaneNet is that PlaneNet directly estimates a 2D segmentation with fixed number of regions together with the instance-level 3D planar parameters, while we estimate the pixel-level surface parameters first and infer segmentation geometrically. The results in Table 2 show that our method (GeoLayout-Ours) consistently outperforms PlaneNet on all the metrics. The reason might be that our method does not need to predict the error-prone 2D segmentation masks. In addition, the averaged instance-level surface parameters in GeoLayout are more robust against noise.

We also compare to a version of our method using plane parameterization (GeoLayout-Plane). Instead of the proposed surface representation, we estimate the 4 parameters of typical planar equation (i.e. 3 for surface normal and 1 for the offset to the origin). In the testing stage, the predicted plane parameters are converted to the surface parameters using intrinsic parameters and

Table 2. Layout estimation results on the Matterport3D-Layout dataset.

Method	$e_{pix.}$	$e_{cor.}$	$e_{3D\_cor.}$	rms	rel	log10	$\delta < 1.25$	$\delta < 1.25^2$	$\delta < 1.25^3$
PlaneNet [23]	6.89	5.29	14.00	0.520	0.134	0.057	0.846	0.954	0.984
GeoLayout-Plane	5.84	4.71	**12.05**	**0.448**	**0.109**	**0.046**	0.891	0.973	0.993
GeoLayout-Ours	**5.24**	**4.36**	12.82	0.456	0.111	0.047	**0.892**	**0.975**	**0.994**

the same layout generation process is performed to produce the layout estimates. GeoLayout-Plane shows comparable performance with GeoLayout-Ours. This indicates the network can successfully estimate surface parameters that already implicitly include camera intrinsics. However, as GeoLayout-Ours does not require the camera intrinsic parameters, it is more flexible in practice and can be easily run on images in the wild while GeoLayout-Plane cannot.

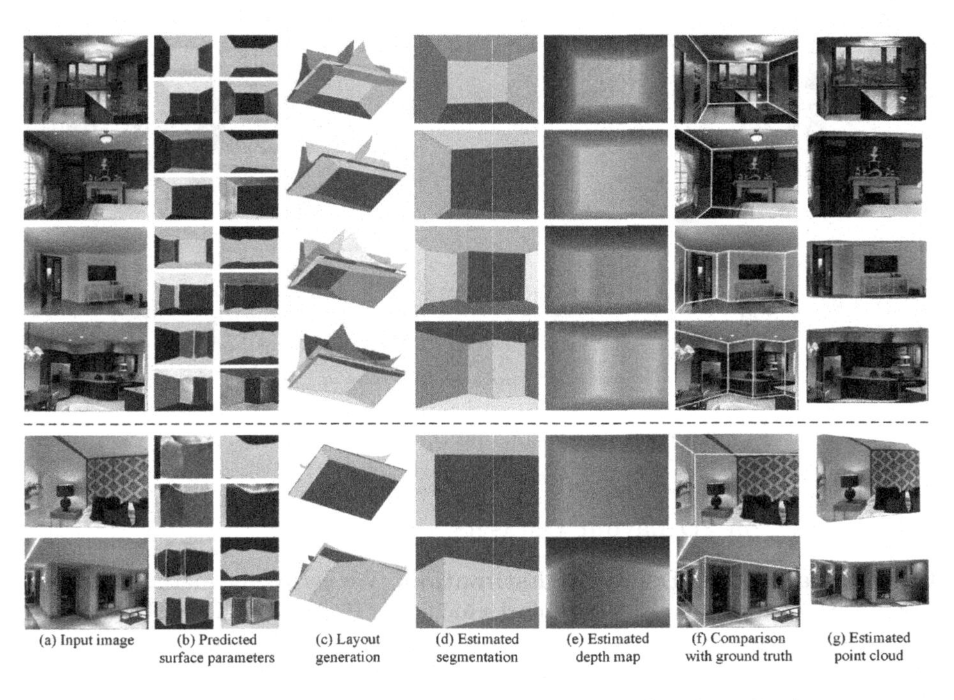

Fig. 5. Qualitative results on the Matterport3D-Layout dataset. The first two rows are cuboid rooms and the following two rows are non-cuboid rooms. Failure cases are shown in the last two rows.

Qualitative Results. The qualitative results are given in Fig. 5. The predicted pixel-level surface parameters are shown in (b), with p and q shown in the first row, r and s shown in the second row. Based on the surface parameters, the depth maps of the surfaces are calculated and displayed in the (u, v, Z) coordinate system as shown in (c). The estimated 2D segmentation and depth map are shown in (d) and (e), respectively. The comparison of the layout estimates (outlined by green) and the ground truth results (outlined by red) are shown in (f). We convert the estimated depth map into point cloud to better visualize the 3D layout estimates, as shown in (g). The first two rows are cuboid rooms and the following two rows are non-cuboid rooms. The results show that our method can reliably estimate the surface parameters and produces high quality layout estimates. Note that our method can handle the non-cuboid rooms with arbitrary number of walls. Two typical failure cases are shown in the last two rows. We found that most of the failure cases are either caused by the large prediction error of the surface parameters, or due to the error during clustering, especially for the non-cuboid rooms and those with more planar surfaces.

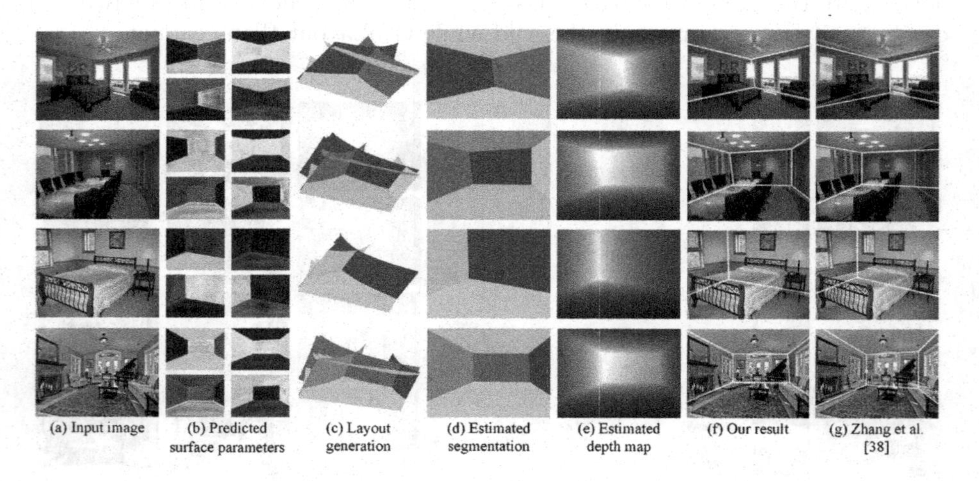

Fig. 6. Qualitative results on the LSUN validation set.

5.3 Results on 2D Layout Datasets

Generalization to 2D Layout Estimation. We verify our method on traditional 2D layout estimation benchmarks including Hedau [13] and LSUN [39]. We first directly run our model trained from Matterport3D-Layout dataset on the LSUN validation set without fine-tuning. The result is shown in Table 3 (w/o Fine-tune). The model still produces reasonable results, which indicates some generalization capability. We then fine-tune our model on LSUN as described in Sect. 3.3, and the performance is significantly improved (w/ Fine-tune). This indicates that the model can be effectively trained on 2D layout dataset with the proposed learning strategy.

Table 3. Comparison of the model w/wo fine-tuning on LSUN validation dataset.

Setting	$e_{pix.}$ (%)	$e_{cor.}$ (%)
w/o Fine-tune	12.67	8.12
w/ Fine-tune	**6.10**	**4.66**

2D Layout Performance. We compare our fine-tuned model on LSUN test set and Hedau dataset to other state-of-the-art methods in Table 4. The LSUN performance is reported by the dataset owner on withheld ground truth to prevent over-fitting. Our method achieves the best performance on LSUN dataset and the second best performance on Hedau dataset. Such result shows that incorporating 3D knowledge and geometric reasoning to layout estimation is beneficial and can significantly improve the 2D layout estimation performance.

Table 4. Layout estimation performance on LSUN [39] and Hedau [13] datasets.

Method	LSUN $e_{pix.}$ (%)	LSUN $e_{cor.}$ (%)	Hedau $e_{pix.}$ (%)
Hedau *et al.* (2009) [13]	24.23	15.48	21.20
Mallya *et al.* (2015) [25]	16.71	11.02	12.83
Dasgupta *et al.* (2016) [5]	10.63	8.20	9.73
Ren *et al.* (2016) [31]	9.31	7.95	8.67
Lee *et al.* (2017) [20]	9.86	6.30	8.36
Hirzer *et al.* (2020) [14]	7.79	5.84	7.44
Kruzhilov *et al.* (2019) [19]	6.72	5.11	7.85
Zhang *et al.* (2019) [38]	6.58	5.17	7.36
Hsiao *et al.* (2019) [15]	6.68	4.92	**5.01**
GeoLayout	**6.09**	**4.61**	7.16

Qualitative Results. Figure 6 shows the visual results on the LSUN validation set, with the results of Zhang et al. [38] for comparison. As can be seen, our method is less error-prone and generally produces more precise results than [38].

6 Conclusion

This paper proposed a novel geometry driven method for indoor layout estimation. The key idea is to learn the depth maps of planar surfaces and then generate the layout by applying geometric rules. We demonstrated that the model could be trained effectively using either 2D or 3D ground truths. The proposed method achieved state-of-the-art performance on benchmark datasets for both 2D and

3D layout. We also presented a new dataset with 3D layout ground truth, which we believe is beneficial to the field of room layout estimation.

Acknowledgements. This work was supported in part by the National Natural Science Foundation of China under Grant 61991411, and Grant U1913204, in part by the National Key Research and Development Plan of China under Grant 2017YFB1300205, and in part by the Shandong Major Scientific and Technological Innovation Project (MSTIP) under Grant 2018CXGC1503. We thank the LSUN organizer for the benchmarking service.

References

1. Barinova, O., Konushin, V., Yakubenko, A., Lee, K.C., Lim, H., Konushin, A.: Fast automatic single-view 3-D reconstruction of urban scenes. In: Forsyth, D., Torr, P., Zisserman, A. (eds.) ECCV 2008. LNCS, vol. 5303, pp. 100–113. Springer, Heidelberg (2008). https://doi.org/10.1007/978-3-540-88688-4_8
2. Camplani, M., Mantecon, T., Salgado, L.: Depth-color fusion strategy for 3-D scene modeling with kinect. IEEE Trans. Cybern. **43**(6), 1560–1571 (2013)
3. Chang, A., et al.: Matterport3D: learning from RGB-D data in indoor environments. In: International Conference on 3D Vision (3DV) (2017)
4. Choi, W., Chao, Y.W., Pantofaru, C., Savarese, S.: Understanding indoor scenes using 3D geometric phrases. In: Proceedings of the IEEE Conference on Computer Vision and Pattern Recognition, pp. 33–40 (2013)
5. Dasgupta, S., Fang, K., Chen, K., Savarese, S.: Delay: robust spatial layout estimation for cluttered indoor scenes. In: Proceedings of the IEEE Conference on Computer Vision and Pattern Recognition, pp. 616–624 (2016)
6. De Brabandere, B., Neven, D., Van Gool, L.: Semantic instance segmentation with a discriminative loss function. arXiv preprint arXiv:1708.02551 (2017)
7. Del Pero, L., Bowdish, J., Kermgard, B., Hartley, E., Barnard, K.: Understanding Bayesian rooms using composite 3D object models. In: Proceedings of the IEEE Conference on Computer Vision and Pattern Recognition, pp. 153–160 (2013)
8. Delage, E., Lee, H., Ng, A.Y.: Automatic single-image 3D reconstructions of indoor manhattan world scenes. In: Thrun, S., Brooks, R., Durrant-Whyte, H. (eds.) Robotics Research. STAR, vol. 28, pp. 305–321. Springer, Heidelberg (2007). https://doi.org/10.1007/978-3-540-48113-3_28
9. Fischler, M.A., Bolles, R.C.: Random sample consensus: a paradigm for model fitting with applications to image analysis and automated cartography. Commun. ACM **24**(6), 381–395 (1981)
10. Fouhey, D.F., Gupta, A., Hebert, M.: Unfolding an indoor origami world. In: Fleet, D., Pajdla, T., Schiele, B., Tuytelaars, T. (eds.) ECCV 2014. LNCS, vol. 8694, pp. 687–702. Springer, Cham (2014). https://doi.org/10.1007/978-3-319-10599-4_44
11. Guo, R., Zou, C., Hoiem, D.: Predicting complete 3D models of indoor scenes. arXiv preprint arXiv:1504.02437 (2015)
12. Haines, O., Calway, A.: Recognising planes in a single image. IEEE Trans. Pattern Anal. Mach. Intell. **37**(9), 1849–1861 (2014)
13. Hedau, V., Hoiem, D., Forsyth, D.: Recovering the spatial layout of cluttered rooms. In: Proceedings of the IEEE International Conference on Computer Vision, pp. 1849–1856. IEEE (2009)

14. Hirzer, M., Lepetit, V., Roth, P.: Smart hypothesis generation for efficient and robust room layout estimation. In: The IEEE Winter Conference on Applications of Computer Vision (WACV), March 2020
15. Hsiao, C.W., Sun, C., Sun, M., Chen, H.T.: Flat2layout: flat representation for estimating layout of general room types. arXiv preprint arXiv:1905.12571 (2019)
16. Hu, J., Ozay, M., Zhang, Y., Okatani, T.: Revisiting single image depth estimation: toward higher resolution maps with accurate object boundaries. In: 2019 IEEE Winter Conference on Applications of Computer Vision (WACV), pp. 1043–1051. IEEE (2019)
17. Izadinia, H., Shan, Q., Seitz, S.M.: Im2cad. In: Proceedings of the IEEE Conference on Computer Vision and Pattern Recognition, pp. 5134–5143 (2017)
18. Karsch, K., Hedau, V., Forsyth, D., Hoiem, D.: Rendering synthetic objects into legacy photographs. In: ACM Transactions on Graphics (TOG), vol. 30, p. 157. ACM (2011)
19. Kruzhilov, I., Romanov, M., Babichev, D., Konushin, A.: Double refinement network for room layout estimation. In: Palaiahnakote, S., Sanniti di Baja, G., Wang, L., Yan, W.Q. (eds.) ACPR 2019. LNCS, vol. 12046, pp. 557–568. Springer, Cham (2020). https://doi.org/10.1007/978-3-030-41404-7_39
20. Lee, C.Y., Badrinarayanan, V., Malisiewicz, T., Rabinovich, A.: Roomnet: end-to-end room layout estimation. In: Proceedings of the IEEE International Conference on Computer Vision, pp. 4865–4874 (2017)
21. Lee, D.C., Hebert, M., Kanade, T.: Geometric reasoning for single image structure recovery. In: Proceedings of the IEEE Conference on Computer Vision and Pattern Recognition, pp. 2136–2143. IEEE (2009)
22. Liu, C., Kim, K., Gu, J., Furukawa, Y., Kautz, J.: PlaneRCNN: 3D plane detection and reconstruction from a single image. In: Proceedings of the IEEE Conference on Computer Vision and Pattern Recognition, pp. 4450–4459 (2019)
23. Liu, C., Yang, J., Ceylan, D., Yumer, E., Furukawa, Y.: PlaneNet: piece-wise planar reconstruction from a single RGB image. In: Proceedings of the IEEE Conference on Computer Vision and Pattern Recognition, pp. 2579–2588 (2018)
24. Liu, C., Schwing, A.G., Kundu, K., Urtasun, R., Fidler, S.: Rent3D: floor-plan priors for monocular layout estimation. In: Proceedings of the IEEE Conference on Computer Vision and Pattern Recognition, pp. 3413–3421. IEEE (2015)
25. Mallya, A., Lazebnik, S.: Learning informative edge maps for indoor scene layout prediction. In: Proceedings of the IEEE International Conference on Computer Vision, pp. 936–944 (2015)
26. Martin-Brualla, R., He, Y., Russell, B.C., Seitz, S.M.: The 3D jigsaw puzzle: mapping large indoor spaces. In: Fleet, D., Pajdla, T., Schiele, B., Tuytelaars, T. (eds.) ECCV 2014. LNCS, vol. 8691, pp. 1–16. Springer, Cham (2014). https://doi.org/10.1007/978-3-319-10578-9_1
27. Micusk, B., Wildenauer, H., Vincze, M.: Towards detection of orthogonal planes in monocular images of indoor environments. In: 2008 IEEE International Conference on Robotics and Automation, pp. 999–1004. IEEE (2008)
28. Paszke, A., et al.: Automatic differentiation in pytorch (2017)
29. Pero, L.D., Bowdish, J., Fried, D., Kermgard, B., Hartley, E., Barnard, K.: Bayesian geometric modeling of indoor scenes. In: Proceedings of the IEEE Conference on Computer Vision and Pattern Recognition, pp. 2719–2726. IEEE (2012)
30. Ramalingam, S., Pillai, J., Jain, A., Taguchi, Y.: Manhattan junction catalogue for spatial reasoning of indoor scenes. In: Proceedings of the IEEE Conference on Computer Vision and Pattern Recognition, pp. 3065–3072 (2013)

31. Ren, Y., Li, S., Chen, C., Kuo, C.-C.J.: A coarse-to-fine indoor layout estimation (CFILE) method. In: Lai, S.-H., Lepetit, V., Nishino, K., Sato, Y. (eds.) ACCV 2016. LNCS, vol. 10115, pp. 36–51. Springer, Cham (2017). https://doi.org/10.1007/978-3-319-54193-8_3
32. Russell, B.C., Torralba, A., Murphy, K.P., Freeman, W.T.: LabelMe: a database and web-based tool for image annotation. Int. J. Comput. Vis. **77**(1–3), 157–173 (2008). https://doi.org/10.1007/s11263-007-0090-8
33. Shelhamer, E., Long, J., Darrell, T.: Fully convolutional networks for semantic segmentation. IEEE Trans. Pattern Anal. Mach. Intell. **39**(4), 640–651 (2017)
34. Wang, H., Gould, S., Roller, D.: Discriminative learning with latent variables for cluttered indoor scene understanding. Commun. ACM **56**(4), 92–99 (2013)
35. Xiao, J., Furukawa, Y.: Reconstructing the world's museums. Int. J. Comput. Vis. **110**(3), 243–258 (2014). https://doi.org/10.1007/s11263-014-0711-y
36. Yang, F., Zhou, Z.: Recovering 3D planes from a single image via convolutional neural networks. In: Ferrari, V., Hebert, M., Sminchisescu, C., Weiss, Y. (eds.) ECCV 2018. LNCS, vol. 11214, pp. 87–103. Springer, Cham (2018). https://doi.org/10.1007/978-3-030-01249-6_6
37. Yu, Z., Zheng, J., Lian, D., Zhou, Z., Gao, S.: Single-image piece-wise planar 3D reconstruction via associative embedding. arXiv preprint arXiv:1902.09777 (2019)
38. Zhang, W., Zhang, W., Gu, J.: Edge-semantic learning strategy for layout estimation in indoor environment. IEEE Trans. Cybern. **50**(6), 2730–2739 (2019)
39. Zhang, Y., Yu, F., Song, S., Xu, P., Seff, A., Xiao, J.: Largescale scene understanding challenge: room layout estimation. http://lsun.cs.princeton.edu/2016/
40. Zhao, H., Lu, M., Yao, A., Guo, Y., Chen, Y., Zhang, L.: Physics inspired optimization on semantic transfer features: an alternative method for room layout estimation. arXiv preprint arXiv:1707.00383 (2017)
41. Zhao, H., Shi, J., Qi, X., Wang, X., Jia, J.: Pyramid scene parsing network. In: Proceedings of the IEEE Conference on Computer Vision and Pattern Recognition, pp. 2881–2890 (2017)
42. Zhao, Y., Zhu, S.C.: Scene parsing by integrating function, geometry and appearance models. In: Proceedings of the IEEE Conference on Computer Vision and Pattern Recognition, pp. 3119–3126 (2013)
43. Zou, C., Colburn, A., Shan, Q., Hoiem, D.: LayoutNet: reconstructing the 3D room layout from a single RGB image. In: Proceedings of the IEEE Conference on Computer Vision and Pattern Recognition, pp. 2051–2059 (2018)

Location Sensitive Image Retrieval and Tagging

Raul Gomez[1,2(✉)], Jaume Gibert[1], Lluis Gomez[2], and Dimosthenis Karatzas[2]

[1] Eurecat, Centre Tecnològic de Catalunya, Unitat de Tecnologies Audiovisuals, Barcelona, Spain
{raul.gomez,jaume.gibert}@eurecat.org
[2] Computer Vision Center, Universitat Autònoma de Barcelona, Barcelona, Spain
{lgomez,dimos}@cvc.uab.es

Abstract. People from different parts of the globe describe objects and concepts in distinct manners. Visual appearance can thus vary across different geographic locations, which makes location a relevant contextual information when analysing visual data. In this work, we address the task of image retrieval related to a given tag conditioned on a certain location on Earth. We present LocSens, a model that learns to rank triplets of images, tags and coordinates by plausibility, and two training strategies to balance the location influence in the final ranking. LocSens learns to fuse textual and location information of multimodal queries to retrieve related images at different levels of location granularity, and successfully utilizes location information to improve image tagging.

1 Introduction

Image tagging is the task of assigning tags to images, referring to words that describe the image content or context. An image of a beach, for instance, could be tagged with the words *beach* or *sand*, but also with the words *swim*, *vacation* or *Hawaii*, which do not refer to objects in the scene. On the other hand, image-by-text retrieval is the task of searching for images related to a given textual query. Similarly to the tagging task, the query words can refer to explicit scene content or to other image semantics. In this work we address the specific retrieval case when the query text is a single word (a tag).

Besides text and images, location is a data modality widely present in contemporary data collections. Many cameras and mobile phones with built-in GPS systems store the location information in the corresponding *Exif* metadata header when a picture is taken. Moreover, most of the web and social media platforms add this information to generated content or use it in their offered services.

Electronic supplementary material The online version of this chapter (https://doi.org/10.1007/978-3-030-58517-4_38) contains supplementary material, which is available to authorized users.

A. Vedaldi et al. (Eds.): ECCV 2020, LNCS 12361, pp. 649–665, 2020.
https://doi.org/10.1007/978-3-030-58517-4_38

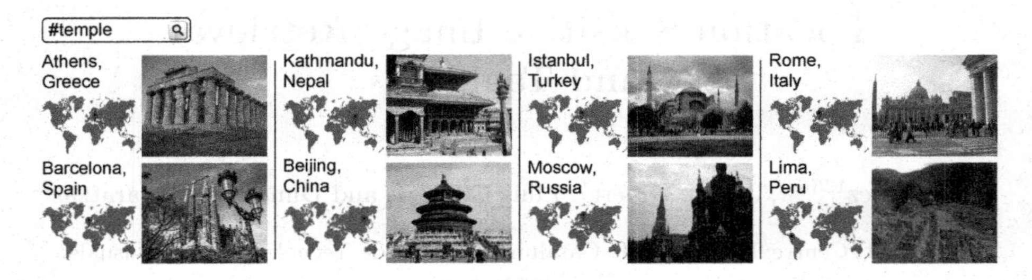

Fig. 1. Top retrieved image by *LocSens*, our location sensitive model, for the query hashtag "*temple*" at different locations.

In this work we leverage this third data modality: using location information can be useful in an image tagging task since location-related tagging can provide better contextual results. For instance, an image of a skier in France could have the tags *"ski, alps, les2alpes, neige"*, while an image of a skier in Canada could have the tags *"ski, montremblant, canada, snow"*. More importantly, location can also be very useful in an image retrieval setup where we want to find images related to a word in a specific location: the retrieved images related to the query tag *temple* in Italy should be different from those in China. In this sense, it could be interesting to explore which kind of scenes people from different countries and cultures relate with certain *broader* concepts. Location sensitive retrieval results produced by the proposed system are shown in Fig. 1.

In this paper we propose a new architecture for modeling the joint distribution of images, hashtags, and geographic locations and demonstrate its ability to retrieve relevant images given a query composed by a hashtag and a location. In this task, which we call location sensitive tag-based image retrieval, a retrieved image is considered relevant if the query hashtag is within its ground-truth hashtags and the distance between its location and the query location is smaller than a given threshold. Notice that distinct from previous work on GPS-aware landmark recognition or GPS-Constrained database search [13,14,24,27] in the proposed task the locations of the test set images are not available at inference time, thus simple location filtering is not an option.

A common approach to address these situations in both image by text retrieval and image tagging setups is to learn a joint embedding space for images and words [6,15,28,37]. In such a space, images are embedded near to the words with which they share semantics. Consequently, semantically *similar* images are also embedded together. Usually, word embedding models, such as Word2Vec [21] or GloVe [25] are employed to generate word representations, while a CNN is trained to embed images in the same space, learning optimal compact representations for them. Word models have an interesting and powerful feature: words with similar semantics have also similar representations and this is a feature that image tagging and retrieval models aim to incorporate, since learning a joint image and word embedding space with semantic structure provides a more flexible and less prone to drastic errors tagging or search engine.

Another approach to handle multiple modalities of data is by scoring tuples of multimodal samples aiming to get high scores on positive cases and low scores on negative ones [12,29,34,38]. This setup is convenient for learning from Web and Social Media data because, instead of strict similarities between modalities, the model learns more relaxed compatibility scores between them. Our work fits under this paradigm. Specifically, we train a model that produces scores for image-hashtag-coordinates triplets, and we use these scores in a ranking loss in order to learn parameters that discriminate between observed and unobserved triplets. Such scores are used to tag and retrieve images in a location aware configuration providing good quality results under the large-scale YFCC100M dataset [33]. Our summarized contributions are:

- We introduce the task of location sensitive tag-based image retrieval.
- We evaluate different baselines for learning image representations with hashtag supervision exploiting large-scale social media data that serve as initialization of the location sensitive model.
- We present the *LocSens* model to score images, tags and location triplets (Fig. 2), which allows to perform location sensitive image retrieval and outperforms location agnostic models in image tagging.
- We introduce novel training strategies to improve the location sensitive retrieval performance of *LocSens* and demonstrate that they are crucial in order to learn good representations of joint hashtag+location queries.

2 Related Work

Location-Aware Image Search and Tagging. O'Hare *et al.* [24] presented the need of conditioning image retrieval to location information, and targeted it by using location to filter out distant photos and then performing a visual search for ranking. Similar location-based filtering strategies have been also used for landmark identification [1] and to speed-up loop closure in visual SLAM [16]. The obvious limitation of such systems compared to *LocSens* is that they require geolocation annotations in the entire retrieval set. Kennedy *et al.* [13,14] and Rattenbury *et al.* [27] used location-based clustering to get the most representative tags and images for each cluster, and presented limited image retrieval results for a subset of tags associated to a given location (landmark tags). They did not learn, however, location-dependent visual representations for tags as we do here, and their system is limited to the use of landmark tags as queries. On the other hand, Zhang *et al.* [47] proposed a location-aware method for image tagging and tag-based retrieval that first identifies points of interest, clustering images by their locations, and then represents the image-tag relations in each of the clusters with an individual image-tag matrix [42]. Their study is limited to datasets on single city scale and small number of tags (1000). Their retrieval method is constrained to use location to improve results for tags with location semantics, and cannot retrieve location-dependent results (i.e. only the tag is used as query). Again, contrary to *LocSens*, this method requires geolocation annotations over the entire retrieval set. Other existing location-aware tagging methods [17,22] have also addressed constrained or small scale setups (e.g. a fixed number of cities) and small-size tag vocabularies, while in this paper we target a worldwide scale unconstrained scenario.

Location and Classification. The use of location information to improve image classification has also been previously explored, and has recently experienced a growing interest by the computer vision research community. Yuan *et al.* [46] combine GPS traces and hand-crafted visual features for events classification. Tang *et al.* [32] propose different ways to get additional image context from coordinates, such as temperature or elevation, and test the usefulness of such information in image classification. Herranz *et al.* [10,44] boost food dish classification using location information by jointly modeling dishes, restaurants and their menus and locations. Chu *et al.* [2] compare different methods to fuse visual and location information for fine-grained image classification. Mac *et al.* [18] also work on fine-grained classification by modeling the spatio-temporal distribution of a set of object categories and using it as a prior in the classification process. Location-aware classification methods that model the prior distribution of locations and object classes can also be used for tagging, but they can not perform location sensitive tag-based retrieval because the prior for a given query (tag+location) would be constant for the whole retrieval set.

Image Geolocalization. Hays *et al.* [8] introduced the task of image geolocalization, i.e. assigning a location to an image, and used hand-crafted features to retrieve nearest neighbors in a reference database of geotagged images. Gallagher *et al.* [4] exploited user tags in addition to visual search to refine geolocalization. Vo *et al.* [35] employed a similar setup but using a CNN to learn image representations from raw pixels. Weyand *et al.* [39] formulated geolocalization as a classification problem where the earth is subdivided into geographical cells, GPS coordinates are mapped to these regions, and a CNN is trained to predict them from images. Müller-Budack *et al.* [23] enhanced the previous setup using earth partitions with different levels of granularity and incorporating explicit scene classification to the model. Although these methods address a different task, they are related to *LocSens* in that we also learn geolocation-dependent visual representations. Furthermore, inspired by [35], we evaluate our models' performance at different levels of geolocation granularity.

Multimodal Learning. Multimodal joint image and text embeddings is a very active research area. DeViSE [3] proposes a pipeline that, instead of learning to predict ImageNet classes, learns to infer the Word2Vec [21] representations of their labels. This work inspired others that applied similar pipelines to learn from paired visual and textual data in a weakly-supervised manner [6,7,30]. More related to our work, Veit *et al.* [34] also exploit the YFCC100M dataset [33] to learn joint embeddings of images and hashtags for image tagging and retrieval. They work on user-specific modeling, learning embeddings conditioned to users to perform user-specific image tagging and tag-based retrieval. Apart from learning joint embeddings for images and text, other works have addressed tasks that need the joint interpretation of both modalities. Although some recent works have proposed more complex strategies to fuse different data modalities [5,20,26,36,45], their results show that their performance improvement compared to a simple feature concatenation followed by a Multi Layer Perceptron is marginal.

3 Methodology

Given a large set of images, tags and geographical coordinates, our objective is to train a model to score triplets of image-hashtag-coordinates and rank them to perform two tasks: (1) image retrieval querying with a hashtag and a location, and (2) image tagging when both the image and the location are available. We address the problem in two stages: first, we train a location-agnostic CNN to learn image representations using hashtags as weak supervision. We propose different training methodologies and evaluate their performance on image tagging and retrieval. These serve as benchmark and provide compact image representations to be later used within the location sensitive models. Second, using the learnt image and hashtags best performing representations and the locations, we train multimodal models to score triplets of these three modalities. We finally evaluate them on image retrieval and tagging and analyze how these models benefit from the location information.

3.1 Learning with Hashtag Supervision

Three procedures for training location-agnostic visual recognition models using hashtag supervision are considered: (1) multi-label classification, (2) softmax multi-class classification, and (3) hashtag embedding regression. In the following, let $\mathbb{H}$ be the set of H considered hashtags. $\mathbf{I_x}$ will stand for a training image and $\mathbf{H_x} \subseteq \mathbb{H}$ for the set of its groundtruth hashtags. The image model $f(\cdot;\theta)$ used is a ResNet-50 [9] with parameters θ. The three approaches eventually produce a vector representation for an image $\mathbf{I_x}$, which we denote by $\mathbf{r_x}$. For a given hashtag $h^i \in \mathbb{H}$, its representation—denoted $\mathbf{v_i}$—is either learnt externally or jointly with those of the images.

Multi-Label Classification (MLC). We set the problem in its most natural form: as a standard MLC setup over H classes corresponding to the hashtags in the vocabulary $\mathbb{H}$. The last ResNet-50 layer is replaced by a linear layer with H outputs, and each one of the H binary classification problems is addressed with a cross-entropy loss with sigmoid activation. Let $\mathbf{y_x} = (y_x^1, \ldots, y_x^H)$ be the multi-hot vector encoding the groundtruth hashtags of $\mathbf{I_x}$ and $\mathbf{f_x} = \sigma(f(\mathbf{I_x};\theta))$, where σ is the element-wise sigmoid function. The loss for image $\mathbf{I_x}$ is written as:

$$L = -\tfrac{1}{H} \sum_{h=1}^{H} [\, y_x^h \log f_x^h + (1 - y_x^h) \log(1 - f_x^h) \,]. \tag{1}$$

Multi-Class Classification (MCC). Despite being counter-intuitive, several prior studies [19,34] demonstrate the effectiveness of formulating multi-label problems with large numbers of classes as multi-class problems. At training time a random target class from the groundtruth set $\mathbf{H_x}$ is selected, and softmax activation with a cross-entropy loss is used. This setup is commonly known as softmax classification.

Let $h_x^i \in \mathbf{H_x}$ be a randomly selected class (hashtag) for $\mathbf{I_x}$. Let also f_x^i be the coordinate of $\mathbf{f_x} = f(\mathbf{I_x};\theta)$ corresponding to h_x^i. The loss for image $\mathbf{I_x}$ is set to be:

$$L = -\log\left(\frac{e^{f_x^i}}{\sum_{j=1}^{H} e^{f_x^j}}\right). \tag{2}$$

In this setup we redefine ResNet-50 by adding a linear layer with D outputs just before the last classification layer with H outputs. This allows getting compact image D-dimensional representations $\mathbf{r_x}$ as their activations in such layer. Since we are in a multi-class setup where the groundtruth is a one-hot vector, we are also implicitly learning hashtag embeddings: the weights of the last classification layer with input $\mathbf{r_x}$ and output $\mathbf{f_x}$ is an $H \times D$ matrix whose rows can be understood as D-dimensional representations of the hashtags in $\mathbb{H}$. Consequently, this approach learns at once D-dimensional embeddings for both images and hashtags. In our experiments, the dimensionality is set to $D = 300$ to match that of the word embeddings used in the next and last approach. This procedure does not apply to MLC for which groundtruth is multi-hot encoded.

Hashtag Embedding Regression (HER). We use pretrained GloVe [25] embeddings for hashtags, which are D-dimensional with $D = 300$. For each image $\mathbf{I_x}$, we sum the GloVe embeddings of its groundtruth hashtags $\mathbf{H_x}$, which we denote as $\mathbf{t_x}$. Then we replace the last layer of the ResNet-50 by a D-dimensional linear layer, and we learn the parameters of the image model by minimizing a cosine embedding loss. If, $\mathbf{f_x} = f(\mathbf{I_x};\theta)$ is the output of the vision model, the loss is defined by:

$$L = 1 - \left(\frac{\mathbf{t_x} \cdot \mathbf{f_x}}{\|\mathbf{t_x}\| \, \|\mathbf{f_x}\|}\right). \tag{3}$$

As already stated by [34], because of the nature of the GloVe semantic space, this methodology has the potential advantage of not penalizing predicting hashtags with close meanings to those in the groundtruth but that a user might not have used in the image description. Moreover, as shown in [3] and due to the semantics structure of the embedding space, the resulting image model will be less prone to drastic errors.

3.2 Location Sensitive Model (*LocSens*)

We design a location sensitive model that learns to score triplets formed by an image, a hashtag and a location. We use a siamese-like architecture and a ranking loss to optimize the model to score positive triplets (existing in the training set) higher than negative triplets (which we create). Given an image $\mathbf{I_x}$, we get its embedding $\mathbf{r_x}$ computed by the image model, the embedding $\mathbf{v_{x_i}}$ of a random hashtag $h^i_{\mathbf{x}}$ from its groundtruth set $\mathbf{H_x}$ and its groundtruth latitude and longitude $\mathbf{g_x} = [\varphi_{\mathbf{x}}, \lambda_{\mathbf{x}}]$, which constitute a positive triplet. Both $\mathbf{r_x}$ and $\mathbf{v_{x_i}}$ are L2 normalized and latitude and longitude are both normalized to range in $[0, 1]$. Note that 0 and 1 latitude fall on the poles while 0 and 1 represent the same longitude because of its circular nature and falls on the Pacific.

The three modalities are then mapped by linear layers with ReLU activations to 300 dimensions each, and L2 normalized again. This normalization guarantees that the magnitudes of the representations of the different modalities are equal when processed by subsequent layers in the multimodal network. Then the three vectors are concatenated. Although sophisticated multimodal data fusion strategies have been proposed, simple feature concatenation has also been proven to be an effective technique [34,36]. We opted for a simple concatenation as it streamlines the strategy. The concatenated representations are then forwarded through 5 linear layers with normalization and ReLU activations with $2048, 2048, 2048, 1024, 512$ neurons respectively. At the end, a linear layer with a single output calculates the score of the triplet. We have experimentally found that Batch Normalization [11] hampers learning, producing highly irregular gradients. We conjecture that all GPU-allowable batch size is in fact a small batch size for the problem at hand, since the number of triplets is potentially massive and the batch statistics estimation will always be erratic across batches. Group normalization [43] is used instead, which is independent of the batch size and permits learning of the models.

To create a negative triplet, we randomly replace the image or the tag of the positive triplet. The image is replaced by a random one not associated with the tag $h^i_{\mathbf{x}}$, and the tag by a random one not in $\mathbf{H_x}$. We have found that the performance in image retrieval is significantly better when all negative triplets are created replacing the image. This is because the frequency of tags is preserved in both the positive and negative triplets, while in the tagging configuration less common tags are more frequently seen in negative triplets.

We train with a Margin Ranking loss, with a margin set empirically to $m = 0.1$, use 6 negative triplets per positive triplet averaging the loss over them, and a batch size of 1024. If s_x is the score of the positive triplet and s_n the score of the negative triplet, the loss is written as:

$$L = max(0, s_n - s_x + m). \tag{4}$$

Figure 2 shows the model architecture and also the training strategies to balance location influence, which are explained next.

Balancing Location Influence on Ranking. One important challenge in multimodal learning is balancing the influence of the different data modalities. We started by introducing the raw location values into the *LocSens* model, but immediately observed that the learning tends to use the location information to

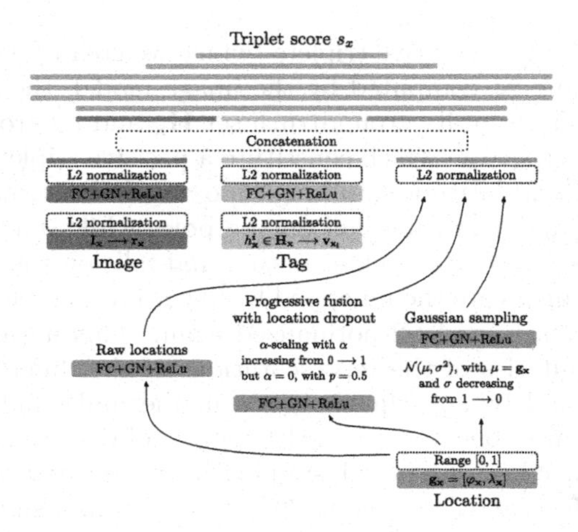

Fig. 2. The proposed *LocSens* multimodal scoring model trained by triplet ranking (bars after concatenation indicate fully connected + group normalization + ReLu activation layers). During training, location information is processed and inputted to the model with different strategies.

discriminate between triplets much more than the other two modalities, forgetting previously learnt relations between images and tags. This effect is especially severe in the image retrieval scenario, where the model ends up retrieving images close to the query locations but less related to the query tag. This suggests that the location information needs to be gradually incorporated into the scoring model for location sensitive image retrieval. For that, we propose the following two strategies, also depicted in Fig. 2.

Progressive Fusion with Location Dropout. We first train a model with *LocSens* architecture but silencing the location modality hence forcing it to learn to discriminate triplets without using location information. To do that, we multiply by $\alpha = 0$ the location representation before its concatenation. Once the training has converged we start introducing locations progressively, by slowly increasing α until $\alpha = 1$. This strategy avoids new gradients caused by locations to ruin the image-hashtags relations *LocSens* has learned in the first training phase. In order to force the model to sustain the capability to discriminate between triplets without using location information we permanently zero the location representations with a 0.5 probability. We call this *location dropout* in a clear abuse of notation but because of its resemblance to zeroing random neurons in the well-known regularization strategy [31]. For the sake of comparison, we report results for the *LocSens* model with zeroed locations, which is in fact a location agnostic model.

Location Sampling. Exact locations are particularly narrow with respect to global coordinates and such a fine-grained degree of granularity makes learning troublesome. We propose to progressively present locations from rough precision to more accurate values while training advances. For each triplet, we randomly sample the training location coordinates at each iteration from a $2D$ normal distribution with mean at the image real coordinates ($\mu = \mathbf{g_x}$) and with standard deviation σ decreasing progressively. We constrain the sampling between [0, 1] by taking modulo 1 on the sampled values.

We start training with $\sigma = 1$, which makes the training locations indeed random and so not informative at all. At this stage, the *LocSens* model will learn to rank triplets without using the location information. Then, we progressively decrease σ, which makes the sampled coordinates be more accurate and useful for triplet discrimination. Note that σ has a direct relation with geographical distance, so location data is introduced during the training to be first only useful to discriminate between very distant triplets, and progressively between more fine-grained distances. Therefore, this strategy allows training models sensitive to different location levels of detail.

4 Experiments

We conduct experiments on the YFCC100M dataset [33] which contains nearly 100 million photos from Flickr with associated hashtags and GPS coordinates among other metadata. We create the hashtag vocabulary following [34]: we remove numerical hashtags and the 10 most frequent hashtags since they are not informative. The hashtag set $\mathbb{H}$ is defined as the set of the next 100,000 most frequent hashtags. Then we select photos with at least one hashtag from $\mathbb{H}$ from which we filter out photos with more than 15 hashtags. Finally, we remove photos without location information. This results in a dataset of 24.8 M images, from which we separate a validation set of 250 K and a test set of 500 K. Images have an average of 4.25 hashtags.

4.1 Image by Tag Retrieval

We first study hashtag based image retrieval, which is the ability of our models to retrieve relevant images given a hashtag query. We define the set of querying hashtags $\mathbb{H}^q$ as the hashtags in $\mathbb{H}$ appearing at least 10 times in the testing set. The number of querying hashtags is 19,911. If R_h^k is the set of top k ranked images for the hashtag $h \in \mathbb{H}^q$ and G_h is the set of images labeled with the hashtag h, we define precision@k as:

$$P@k = \frac{1}{|\mathbb{H}^q|} \sum_{h \in \mathbb{H}^q} \frac{|R_h^k \cap G_h|}{k}. \quad (5)$$

We evaluate precision@10, which measures the percentage of the 10 highest scoring images that have the query hashtag in their groundtruth. Under these settings, precision@k is upper-bounded by 100. The precision@10 of the different location agnostic methods described in Sect. 3.1 is as follows: MLC: 1.01, MCC: **14.07**, HER (GloVe): 7.02. The Multi-Class Classification (MCC) model has the best performance in the hashtag based image retrieval task.

4.2 Location Sensitive Image by Tag Retrieval

In this experiment we evaluate the ability of the models to retrieve relevant images given a query composed by a hashtag and a location (Fig. 1). A retrieved image is considered relevant if the query hashtag is within its groundtruth hashtags and the distance between its location and the query location is smaller than a given threshold. Inspired by [35], we use different distance thresholds to evaluate the models' location precision at different levels of granularity. We define our query set of hashtag-location pairs by selecting the location and a random hashtag of 200, 000 images from the testing set. In this query set there will be repeated hashtags with different locations, and more frequent hashtags over all the dataset will also be more frequent in the query set (unlike in the location agnostic retrieval experiment of Sect. 4.1). This query set guarantees that the ability of the system to retrieve images related to the same hashtag but different locations is evaluated. To retrieve images for a given hashtag-location query with *LocSens*, we compute triplet plausibility scores with all test images and rank them.

Table 1 shows the performance of the different methods in location agnostic image retrieval and in different location sensitive levels of granularity. In location agnostic retrieval (first column) the geographic distance between the query and the results is not evaluated (infinite distance threshold). The evaluation in this scenario is the same as in Sect. 4.1, but the performances are higher because in this case the query sets contains more instances of the most frequent hashtags. The upper bound ranks the retrieval images containing the query hashtag by proximity to the query location, showcasing the optimal performance of any method in this evaluation. In location sensitive evaluations the optimal performance is less than 100% because we do not always have 10 or more relevant images in the test set.

Results show how the zeroed locations version of *LocSens* gets comparable results as MCC. By using raw locations in the *LocSens* model, we get the best results at fine level of location detail at the expense of a big drop in location agnostic retrieval. As introduced in Sect. 3.2, the reason is that it is relying heavily on locations to rank triplets decreasing its capability to predict relations between images and tags. As a result, it tends to retrieve images close to the query location, but less related to the query tag. The proposed dropout training strategy reduces the deterioration in location agnostic retrieval performance at a cost of a small drop in the fine levels of granularity. Also, it outperforms the former models in the coarse continent and country levels, due to its better balancing between using the query tag and location to retrieve related images. In its turn, the location sampling proposed approach with $\sigma = 1$ gets similar results as *LocSens* with zeroed locations because the locations are as irrelevant in both cases. When σ is decreased, the model improves its location sensitive retrieval performance while maintaining a high location agnostic performance. This is achieved because informative locations are introduced to the model in a progressive way, from coarse to fine, and always maintaining triplets where the

Table 1. Location sensitive hashtag based image retrieval: $P@10$. A retrieved image is considered correct if its groundtruth hashtags contain the queried hashtag and the distance between its location and the queried one is smaller than a given threshold

	$P@10$						
	Method	**Location Agnostic**	**Continent (2500 km)**	**Country (750 km)**	**Region (200 km)**	**City (25 km)**	**Street (1 km)**
	Upper Bound	100	96.08	90.51	80.31	64.52	42.46
Img + Tag	MLC	5.28	2.54	1.65	1.00	0.62	0.17
	MCC	**42.18**	29.23	24.2	18.34	13.25	4.66
	HER (GloVe)	37.36	25.03	20.27	15.51	11.23	3.65
	LocSens - Zeroed locations	40.05	28.32	24.34	18.44	12.79	3.74
Loc + Img + Tag	LocSens - Raw locations	32.74	28.42	25.52	**21.83**	**15.53**	**4.83**
	LocSens - Dropout	36.95	30.42	26.14	20.46	14.28	4.64
	LocSens - Sampling $\sigma = 1$	40.60	28.40	23.84	18.16	13.04	4.13
	LocSens - Sampling $\sigma = 0.1$	40.03	29.30	24.36	18.83	13.46	4.22
	LocSens - Sampling $\sigma = 0.05$	39.80	31.25	25.76	19.58	13.78	4.30
	LocSens - Sampling $\sigma = 0.01$	37.05	**31.27**	26.65	20.14	14.15	4.44
	LocSens - Sampling $\sigma = 0$	35.95	30.61	**27.00**	21.39	14.75	**4.83**

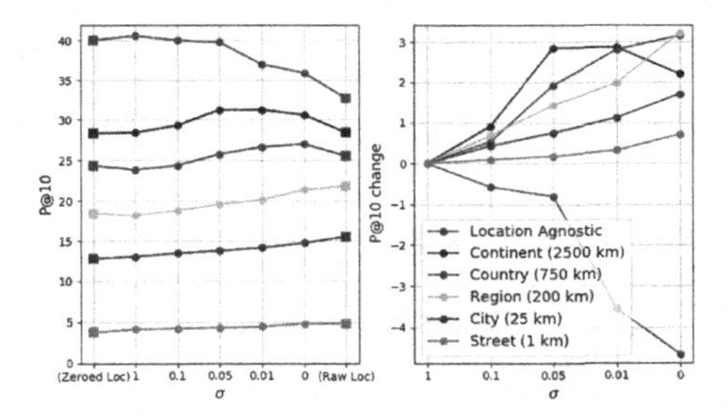

Fig. 3. Left: $P@10$ of the location sampling strategy for different σ and models with zeroed and raw locations. Right: $P@10$ difference respect to $\sigma = 1$.

location is not informative, forcing the network to retain its capacity to rank triplets using only the image and the tag.

Figure 3 shows the absolute and relative performances at different levels of granularity while σ is decreased. At $\sigma = 0.05$, it can be seen that the location sensitive performances at all granularities have improved with a marginal drop on location agnostic performance. When σ is further decreased, performances at finer locations keep increasing, while the location agnostic performance decreases. When $\sigma = 0$, the training scenario is the same as in the raw locations one, but the training schedule allows this model to reduce the drop in location agnostic performance and at coarse levels of location granularity.

The location sampling technique provides *LocSens* with a better balancing between retrieving images related to the query tag and their location. Furthermore, given that σ has a direct geographical distance interpretation, it permits to tune the granularity to which we want our model to be sensitive. Note that *LocSens* enables to retrieve images related to a tag and near to a given location, which location agnostic models cannot do. The performance improvements in Table 1 at the different levels of location granularity are indeed significant since for many triplets the geographic location is not informative at all.

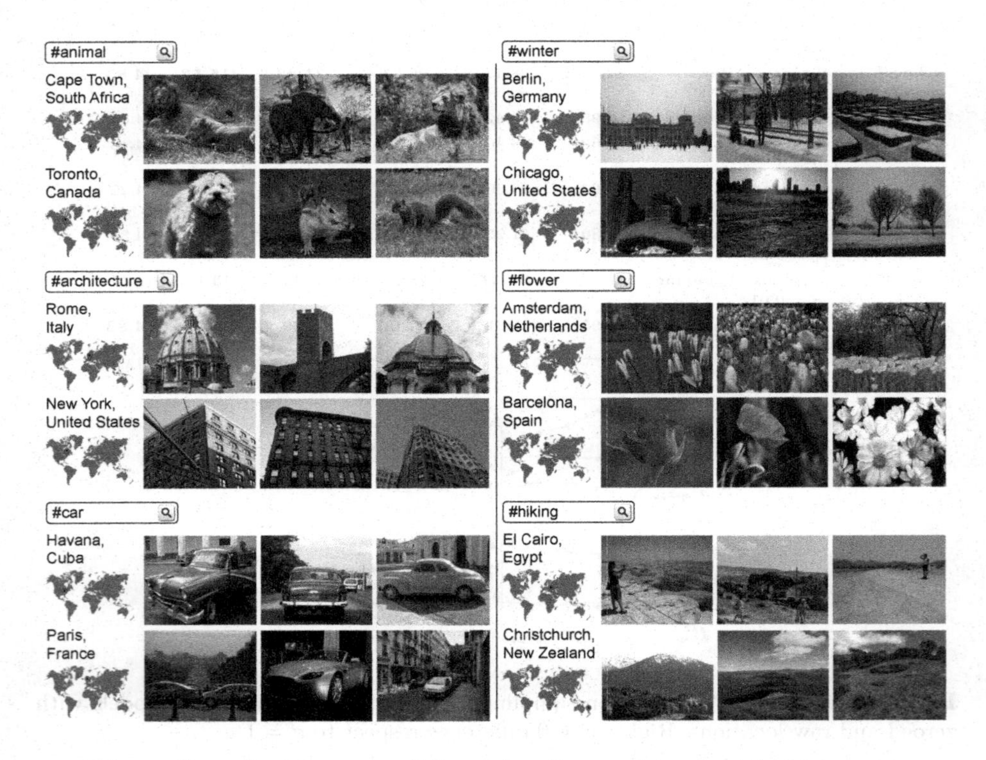

Fig. 4. Query hashtags with different locations and top 3 retrieved images.

Figures 1 and 4 show qualitative retrieval results of several hashtags at different locations. They demonstrate that the model successfully fuses textual and location information to retrieve images related to the joint interpretation of the two query modalities, being able to retrieve images related to the same concept across a wide range of locations with different geographical distances between them. *LocSens* goes beyond retrieving the most common images from each geographical location, as it is demonstrated by the *winter* results in Berlin or the *car* results in Paris.

4.3 Image Tagging

In this section we evaluate the ability of the models to predict hashtags for images in terms of $A@k$ (accuracy at k). If $\mathbf{H_x}$ is the set of groundtruth hashtags of $\mathbf{I_x}$, $\mathbf{R_x^k}$ denotes the k highest scoring hashtags for the image $\mathbf{I_x}$, and N is the number of testing images, $A@k$ is defined as:

$$A@k = \frac{1}{N} \sum_{n=1}^{N} \mathbb{1}\left[\mathbf{R_n^k} \cap \mathbf{H_n} \neq \emptyset\right], \tag{6}$$

where $\mathbb{1}[\cdot]$ is the indicator function having the value of 1 if the condition is fulfilled and 0 otherwise. We evaluate accuracy at $k = 1$ and $k = 10$, which measure how often the first ranked hashtag is in the groundtruth and how often at least one of the 10 highest ranked hashtags is in the groundtruth respectively.

A desired feature of a tagging system is the ability to infer diverse and distinct tags [40,41]. In order to measure the variety of tags predicted by the models, we measure the percentage of all the test tags predicted at least once in the whole test set (`%pred`) and the percentage of all the test tags correctly predicted at least once (`%cpred`), considering the top 10 tags predicted for each image.

Table 2 shows the performance of the different methods. Global Frequency ranks the tags according to the training dataset frequency. Among the location agnostic methods, MCC is the best one. This finding corroborates the experiments in [19,34] verifying that this simple training strategy outperforms others when having a large number of classes. To train the *LocSens* model we used the image and tag representations inferred by the MCC model, since it is the one providing the best results.

Table 2. Image tagging: $A@1$, $A@10$, `%pred` and `%cpred` of the frequency baseline, location agnostic prediction and the location sensitive model

Method	$A@1$	$A@10$	%pred	%cpred
Global Frequency	1.82	13.45	0.01	0.01
MLC	8.86	30.59	8.04	4.5
MCC	**20.32**	**47.64**	**29.11**	**15.15**
HER (GloVe)	15.83	31.24	18.63	8.74
LocSens - Zeroed locations	15.92	46.60	26.98	13.31
LocSens - Raw locations	**28.10**	**68.21**	**44.00**	**24.04**

LocSens - Raw locations stands for the model where the raw triplets locations are always inputted both at train and test time. It outperforms the location agnostic methods in accuracy, successfully using location information to improve the tagging results. Moreover, it produces more diverse tags than location agnostic models, demonstrating that using location is effective for augmenting the

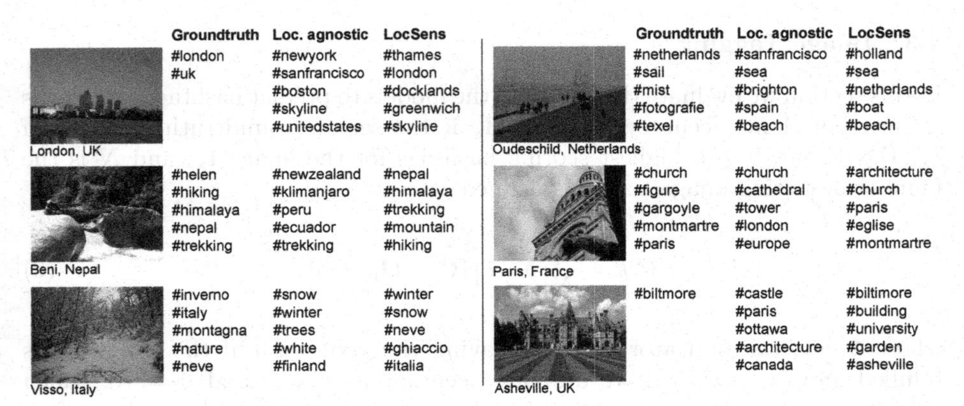

Groundtruth
Loc. agnostic
LocSens
#london #uk
#newyork #sanfrancisco #boston #skyline #unitedstates
#thames #london #docklands #greenwich #skyline
London, UK
#helen #hiking #himalaya #nepal #trekking
#newzealand #klimanjaro #peru #ecuador #trekking
#nepal #himalaya #trekking #mountain #hiking
Beni, Nepal
#inverno #italy #montagna #nature #neve
#snow #winter #trees #white #finland
#winter #snow #neve #ghiaccio #italia
Visso, Italy
Groundtruth
Loc. agnostic
LocSens
#netherlands #sail #mist #fotografie #texel
#sanfrancisco #sea #brighton #spain #beach
#holland #sea #netherlands #boat #beach
Oudeschild, Netherlands
#church #figure #gargoyle #montmartre #paris
#church #cathedral #tower #london #europe
#architecture #church #paris #eglise #montmartre
Paris, France
#biltmore
#castle #paris #ottawa #architecture #canada
#biltimore #building #university #garden #asheville
Asheville, UK

Fig. 5. Images with their locations and groundtruth hashtags and the corresponding top 5 hashtags predicted by the location agnostic MCC model and *LocSens*.

hashtag prediction diversity. Figure 5 shows some tagging examples of a location agnostic model (MCC) compared to *LocSens*, that demonstrate how the later successfully processes jointly visual and location information to assign tags referring to the concurrence of both data modalities. As seen in the first example, besides assigning tags directly related to the given location (*london*) and discarding tags related to locations far from the given one (*newyork*), *LocSens* predicts tags that need the joint interpretation of visual and location information (*thames*). Figure 6 shows *LocSens* tagging results on images with different faked locations, and demonstrates that *LocSens* jointly interprets the image and the location to assign better contextualized tags, such as *caribbean* if a sailing image is from Cuba, and *lake* if it is from Toronto. Note that *LocSens* infers tags generally related to the image content while clearly conditioned by the image location, benefiting from the context given by both modalities. Tagging methods based solely on location, however, can be very precise predicting tags directly referring to a location, like place names, but cannot predict tags related to the image semantics. We consider the later a requirement of an image tagging system, and we provide additional experimentation in the supplementary material.

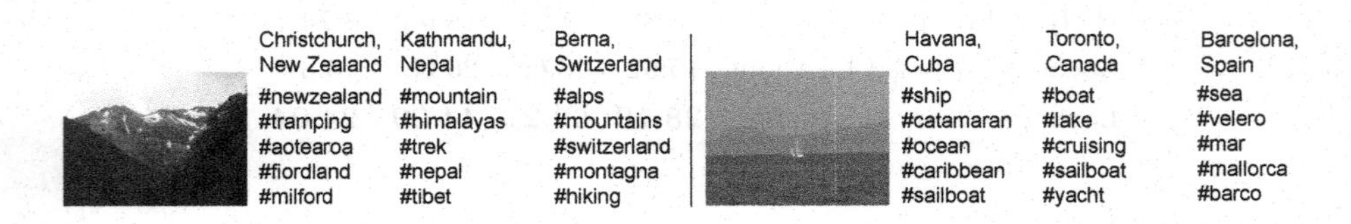

Fig. 6. *LocSens* top predicted hashtags for images with different faked locations.

5 Conclusions

We have confirmed that a multiclass classification setup is the best method to learn image and tag representations when a large number of classes is available. Using them, we have trained *LocSens* to rank image-tag-coordinates triplets by plausibility. We have shown how it is able to perform image by tag retrieval conditioned to a given location by learning location-dependent visual representations, and have demonstrated how it successfully utilizes location information for image tagging, providing better contextual results. We have identified a problem in the multimodal setup, especially acute in the retrieval scenario: *LocSens* heavily relies on location for triplet ranking and tends to return images close to the query location and less related to the query tag. To address this issue we have proposed two novel training strategies: progressive fusion with location dropout, which allows training with a better balance between the modalities influence on the ranking, and location sampling, which results in a better overall performance and enables to tune the model at different levels of distance granularity.

Acknowledgement. Work supported by project TIN2017-89779-P, the CERCA Programme/Generalitat de Catalunya and the PhD scholarship AGAUR 2016-DI-84.

References

1. Chen, D.M., et al.: City-scale landmark identification on mobile devices. In: CVPR (2011)
2. Chu, G., et al.: Geo-aware networks for fine-grained recognition. In: ICCVW (2019)
3. Frome, A., Corrado, G.S., Shlens, J., Bengio Jeffrey Dean, S., Ranzato, A., Mikolov, T.: DeViSE: a deep visual-semantic embedding model. In: NIPS (2013)
4. Gallagher, A., Joshi, D., Yu, J., Luo, J.: Geo-location inference from image content and user tags. In: CVPR (2009)
5. Gao, P., et al.: Question-guided hybrid convolution for visual question answering. In: Ferrari, V., Hebert, M., Sminchisescu, C., Weiss, Y. (eds.) ECCV 2018. LNCS, vol. 11205, pp. 485–501. Springer, Cham (2018). https://doi.org/10.1007/978-3-030-01246-5_29
6. Gomez, L., Patel, Y., Rusiñol, M., Karatzas, D., Jawahar, C.V.: Self-supervised learning of visual features through embedding images into text topic spaces. In: CVPR (2017)
7. Gordo, A., Larlus, D.: Beyond instance-level image retrieval: leveraging captions to learn a global visual representation for semantic retrieval. In: CVPR (2017)
8. Hays, J., Efros, A.A.: IM2GPS: estimating geographic information from a single image. In: CVPR (2008)
9. He, K., Zhang, X., Ren, S., Sun, J.: Deep residual learning for image recognition. In: CVPR (2016)
10. Herranz, L., Jiang, S., Xu, R.: Modeling restaurant context for food recognition. IEEE Trans. Multimedia **19**(2), 430–440 (2017)
11. Ioffe, S., Szegedy, C.: Batch normalization: accelerating deep network training by reducing internal covariate shift. arXiv (2015)
12. Jabri, A., Joulin, A., Van Der Maaten, L.: Revisiting visual question answering baselines. Technical report, Facebook AI Research (2016)

13. Kennedy, L., Naaman, M.: Generating diverse and representative image search results for landmarks. In: International Conference on World Wide Web (2008)
14. Kennedy, L., Naaman, M., Ahern, S., Nair, R., Rattenbury, T.: How flickr helps us make sense of the world: context and content in community-contributed media collections. In: ACM International Conference on Multimedia (2007)
15. Kiros, R., Salakhutdinov, R., Zemel, R.S.: Unifying visual-semantic embeddings with multimodal neural language models. arXiv (2014)
16. Kumar, A., Tardif, J.P., Anati, R., Daniilidis, K.: Experiments on visual loop closing using vocabulary trees. In: CVPR Workshops (2008)
17. Liu, J., Li, Z., Tang, J., Jiang, Y., Lu, H.: Personalized geo-specific tag recommendation for photos on social websites. IEEE Trans. Multimedia **16**(3), 588–600 (2014)
18. Mac Aodha, O., Cole, E., Perona, P.: Presence-only geographical priors for fine-grained image classification. In: ICCV (2019)
19. Mahajan, D., et al.: Exploring the limits of weakly supervised pretraining. In: Ferrari, V., Hebert, M., Sminchisescu, C., Weiss, Y. (eds.) ECCV 2018. LNCS, vol. 11206, pp. 185–201. Springer, Cham (2018). https://doi.org/10.1007/978-3-030-01216-8_12
20. Margffoy-Tuay, E., Pérez, J.C., Botero, E., Arbeláez, P.: Dynamic multimodal instance segmentation guided by natural language queries. In: Ferrari, V., Hebert, M., Sminchisescu, C., Weiss, Y. (eds.) ECCV 2018. LNCS, vol. 11215, pp. 656–672. Springer, Cham (2018). https://doi.org/10.1007/978-3-030-01252-6_39
21. Mikolov, T., Corrado, G., Chen, K., Dean, J.: Efficient estimation of word representations in vector space. In: ICLR (2013)
22. Moxley, E., Kleban, J., Manjunath, B.: SpiritTagger: a geo-aware tag suggestion tool mined from flickr. In: Proceedings of ACM ICMIR (2008)
23. Müller-Budack, E., Pustu-Iren, K., Ewerth, R.: Geolocation estimation of photos using a hierarchical model and scene classification. In: Ferrari, V., Hebert, M., Sminchisescu, C., Weiss, Y. (eds.) ECCV 2018. LNCS, vol. 11216, pp. 575–592. Springer, Cham (2018). https://doi.org/10.1007/978-3-030-01258-8_35
24. O'Hare, N., Gurrin, C., Jones, G.J., Smeaton, A.F.: Combination of content analysis and context features for digital photograph retrieval. In: European Workshop on the Integration of Knowledge, Semantics and Digital Media Technology (2005)
25. Pennington, J., Socher, R., Manning, C.: Glove: global vectors for word representation. In: EMNLP (2014)
26. Rajiv Jain, C.W.: Multimodal document image classification. In: ICDAR (2019)
27. Rattenbury, T., Good, N., Naaman, M.: Towards automatic extraction of event and place semantics from flickr tags. In: ACM SIGIR Conference on Research and Development in Information Retrieval (2007)
28. Ren, Z., Jin, H., Lin, Z., Fang, C., Yuille, A.: Joint image-text representation by Gaussian visual-semantic embedding. In: ACM Multimedia (2016)
29. Rohrbach, A., Rohrbach, M., Hu, R., Darrell, T., Schiele, B.: Grounding of textual phrases in images by reconstruction. Technical report, Max Planck Institute for Informatics (2015)
30. Salvador, A., et al.: Learning cross-modal embeddings for cooking recipes and food images. In: CVPR (2017)
31. Srivastava, N., Hinton, G., Krizhevsky, A., Salakhutdinov, R.: Dropout: a simple way to prevent neural networks from overfitting. J. Mach. Learn. Res. **15**(1), 1929–1958 (2014)
32. Tang, K., Paluri, M., Fei-Fei, L., Fergus, R., Bourdev, L.: Improving image classification with location context. In: ICCV (2015)

33. Thomee, B., et al.: YFCC100M: the new data in multimedia research. Commun. ACM **59**(2), 64–73 (2015)
34. Veit, A., Nickel, M., Belongie, S., Maaten, L.V.D.: Separating self-expression and visual content in hashtag supervision. In: CVPR (2018)
35. Vo, N., Jacobs, N., Hays, J.: Revisiting IM2GPS in the deep learning era. In: ICCV (2017)
36. Vo, N., et al.: Composing text and image for image retrieval - an empirical odyssey. In: CVPR (2019)
37. Wang, L., Li, Y., Huang, J., Lazebnik, S.: Learning two-branch neural networks for image-text matching tasks. In: CVPR (2017)
38. Wang, T., Wu, D.J., Coates, A., Ng, A.Y.: End-to-end text recognition with convolutional neural networks. In: ICPR (2012)
39. Weyand, T., Kostrikov, I., Philbin, J.: PlaNet - photo geolocation with convolutional neural networks. In: Leibe, B., Matas, J., Sebe, N., Welling, M. (eds.) ECCV 2016. LNCS, vol. 9912, pp. 37–55. Springer, Cham (2016). https://doi.org/10.1007/978-3-319-46484-8_3
40. Wu, B., Chen, W., Sun, P., Liu, W., Ghanem, B., Lyu, S.: Tagging like humans: diverse and distinct image annotation. In: CVPR (2018)
41. Wu, B., Jia, F., Liu, W., Ghanem, B.: Diverse image annotation. In: CVPR (2017)
42. Wu, L., Jin, R., Jain, A.K.: Tag completion for image retrieval. IEEE Trans. Pattern Anal. Mach. Intell. **35**(3), 716–727 (2012)
43. Wu, Y., He, K.: Group normalization. Int. J. Comput. Vis. **128**(3), 742–755 (2019). https://doi.org/10.1007/s11263-019-01198-w
44. Xu, R., Herranz, L., Jiang, S., Wang, S., Song, X., Jain, R.: Geolocalized modeling for dish recognition. IEEE Trans. Multimedia **17**(8), 1187–1199 (2015)
45. Yang, F., et al.: Exploring deep multimodal fusion of text and photo for hate speech classification. In: Workshop on Abusive Language Online (2019)
46. Yuan, J., Luo, J., Kautz, H., Wu, Y.: Mining GPS traces and visual words for event classification. Technical report, Northwestern University (2008)
47. Zhang, J., Wang, S., Huang, Q.: Location-based parallel tag completion for geo-tagged social image retrieval general terms. ACM Trans. Intell. Syst. Technol. **8**(3), 1–21 (2017)

Joint 3D Layout and Depth Prediction from a Single Indoor Panorama Image

Wei Zeng[1(✉)], Sezer Karaoglu[1,2], and Theo Gevers[1,2]

[1] Computer Vision Laboratory, University of Amsterdam, Amsterdam, The Netherlands
{w.zeng,th.gevers}@uva.nl
[2] 3DUniversum, Science Park 400, Amsterdam, The Netherlands
{s.karaoglu,theo.gevers}@3duniversum.com

Abstract. In this paper, we propose a method which jointly learns the layout prediction and depth estimation from a single indoor panorama image. Previous methods have considered layout prediction and depth estimation from a single panorama image separately. However, these two tasks are tightly intertwined. Leveraging the layout depth map as an intermediate representation, our proposed method outperforms existing methods for both panorama layout prediction and depth estimation. Experiments on the challenging real-world dataset of Stanford 2D–3D demonstrate that our approach obtains superior performance for both the layout prediction tasks (3D IoU: 85.81% v.s. 79.79%) and the depth estimation (Abs Rel: 0.068 v.s. 0.079).

Keywords: Indoor panorama image · Layout prediction · Depth estimation · Layout depth map

1 Introduction

Extracting 3D information from 2D indoor images is an important step towards the enabling of 3D understanding of indoor scenes and is beneficial for many applications such as robotics and virtual/augmented reality. Using the 3D information of indoor scenes, a computer vision system is able to understand the scene geometry, including both the apparent and hidden relationships between scene elements.

Although scene layout and depth can both be used for 3D scene understanding, previous methods focus on solving these two problems separately. For 3D layout prediction, methods mostly use 2D geometrical cues such as edges [20,25,35], corners [16,25,35], 2D floor-plans [19,30] or they make assumptions about the 3D scene geometry such that rooms are modelled by cuboids or by a

Electronic supplementary material The online version of this chapter (https://doi.org/10.1007/978-3-030-58517-4_39) contains supplementary material, which is available to authorized users.

A. Vedaldi et al. (Eds.): ECCV 2020, LNCS 12361, pp. 666–682, 2020.
https://doi.org/10.1007/978-3-030-58517-4_39

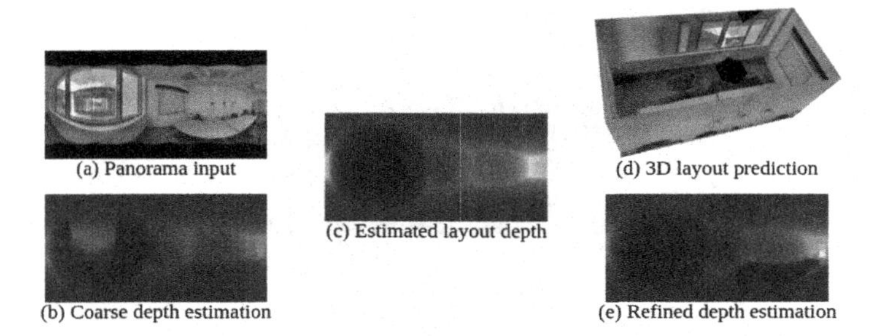

Fig. 1. Given (a) an indoor panorama as input, our proposed method utilizes the (b) coarse depth estimation to compute the (c) layout depth map. Leveraging the estimated layout depth map, our method improves the (d) 3D layout prediction and (e) refines the depth estimation (e.g. the ambiguous window depth is inferred correctly compared to the coarse depth estimation)

Manhattan World. For depth estimation, different features are used such as normals [17], planar surfaces [21] and semantic cues [22]. Hence, existing methods impose geometric assumptions but ignore to exploit the complementary characteristics of layout and depth information. In this paper, a different approach is taken. We propose a method that, from a single panorama, jointly exploits the 3D layout and depth cues via an intermediate layout depth map, as shown in Fig. 1. The intermediate layout depth map represents the distances from the camera to the room layout components (e.g. ceiling, floor and walls) and excludes all objects in the room (e.g. furniture), as illustrated in Fig. 2. Estimating the layout depth as an intermediate representation of the network encompasses the geometric information needed for both tasks. The use of depth information is beneficial to produce room layouts by reducing the complexity of object clutter and occlusion. Likewise, the use of room layout information diminishes the ambiguity of depth estimation and interposes planar information for the room layout parts (e.g. ceiling, floor and walls).

The proposed method estimates the 3D layout and detailed depth information from a single panorama image. To combine the depth and layout information, the proposed method predicts the layout depth map to relate these two tightly intertwined tasks. Previous methods on layout prediction provides proper reconstruction by predicting the layout edges and corners on the input panorama and by post-processing them to match the (Manhattan) 3D layout [16,25,35]. However, object clutter in the room poses a challenge to extract occluded edges and corners. In addition, estimating the 3D layout from 2D edge and corner maps is an ill-posed problem. Therefore, extra constraints are essential to perform 2D to 3D conversion in the optimization. In contrast, our method estimates the layout depth map by using more structural information to become less influenced by occlusions. Furthermore, the predicted layout depth map serves as a coarse 3D layout as it can be converted to the 3D point cloud of the scene layout. Thus the proposed method does not require extra constraints for the 2D to 3D

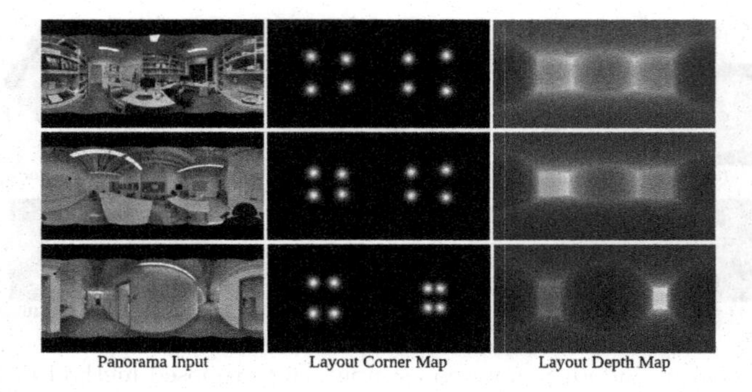

Fig. 2. Illustration of the layout depth maps. From left to right: the panorama input image, the original layout corner map and the layout depth map

conversion. This makes the proposed method more generic for parameterizing a 3D layout. After computing the estimated layout depth maps, the proposed method further enables the refinement of a detailed depth map. Monocular depth estimation methods usually have problems with planar room parts (ceiling, floor and walls) being rugged after the 3D reconstruction process. The layout depth map preserves the planar nature of the room layout components yielding robustness to these errors. Empirical results on the challenging Stanford 2D–3D indoor dataset show that jointly estimating 3D layout and depth outperforms previous methods for both tasks. The proposed method achieves state-of-the-art performance for both layout prediction and depth estimation from a single panorama image on the Stanford 2D–3D dataset. Our method also obtains state-of-the-art performance for 3D layout prediction on the PanoContext dataset.

In summary, our contributions are as follows:

- We propose a novel neural network pipeline which jointly learns layout prediction and depth estimation from a single indoor panorama image. We show that layout and depth estimation tasks are highly correlated and joint learning improves the performance for both tasks.
- We show that leveraging the layout depth map as an intermediate representation improves the layout prediction performance and refines the depth estimation.
- The proposed method outperforms the state-of-the-art methods for both layout prediction and depth estimation on the challenging real-world dataset Stanford 2D–3D and PanoContext dataset for layout prediction.

2 Related Work

Panorama Images: Operating directly on panorama input images is the primary difference between our method and most of the other layout prediction or depth estimation methods. Instead of perspective images, 360° panorama images are used as input by our proposed method because the field of view (FOV) of panoramas are larger and carry more scene information. However, the equirectangular projections may suffer from strong horizontal distortions. Su et al. [24] propose to learn a spherical convolutional network that translates a planar CNN to process 360° panorama images directly in its equirectangular projection. Tateno et al. [26] proposes a distortion-aware deformable convolution filter. Another approach is to use spherical convolutions as proposed by Cohen et al. [3]. Other recent papers [4,8,13] also focus on spherical CNNs and icosahedron representations for panorama processing. In this paper, standard convolutions with rectangular filter banks are applied on the input layers to account for the different distortion levels.

Layout Prediction: There are numerous papers that address the problem of predicting the 3D room layout from a single image taken from an indoor scene. Traditional methods treat this task as an optimization problem. Delage et al. [5] propose a dynamic Bayesian network model to recover the 3D model of the indoor scene. Hedau [10] models the room with a parametric 3D box by iteratively localizing clutter and refitting the box. Recently, neural network-based methods took stride in tackling this problem. Methods that train deep network to classify pixels into layout surfaces (e.g., walls, floor, ceiling) [12], boundaries [20], corners [16], or a combination [23]. Zou et al. [35] predict the layout boundary and corner map directly from the input panorama. Yang et al. [30] leverage both the equirectangular panorama-view and the perspective ceiling-view to learn different cues about the room layout. Sun et al. [25] encode the room layout as three 1D vectors and propose to recover the 3D room layouts from 1D predictions. Other work aims to leverage depth information for room reconstruction [18,32,36], but they all deal with perspective images and use the ground truth depth as input. In contrast, in our paper, we use the predicted depth and semantic content of the scene to predict the layout depth map as our intermediate representation to recover the 3D layout of the input panorama.

Depth Estimation: Single-view depth estimation refers to the problem of estimating depth from a single 2D image. Eigen et al. [9] show that it is possible to produce pixel depth estimations using a two scale deep network which is trained on images with their corresponding depth values. Several methods extend this approach by introducing new components such as CRFs to increase the accuracy [17], changing the loss from regression to classification [2], using other more robust loss functions [15], and by incorporating scene priors [29]. Zioulis et al. [34] propose a learning framework to estimate the depth of a scene from a single 360° panorama image. Eder et al. [7] present a method to train a plane-aware convolutional network for dense depth and surface normal estimation from panoramas. There are some other methods [6,27] to regress the layered depth image (LDI) to capture the occluded texture and depth. In our work, we demonstrate that the layout prediction and depth estimation are tightly coupled and can benefit from each other. Leveraging the estimated layout depth map, our method refines the depth estimation.

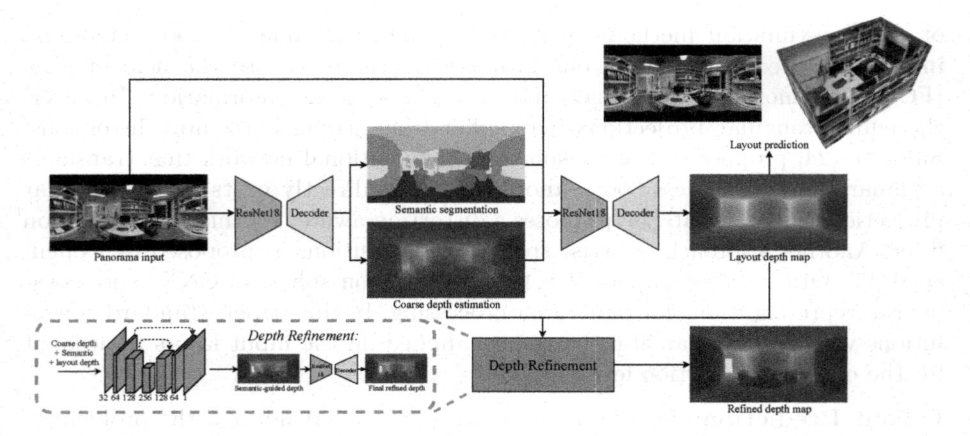

Fig. 3. Overview of the proposed pipeline. Our method first leverages the coarse depth and semantic prediction to enforce the layout depth prediction, and then uses the estimated layout depth map to recover the 3D layout and refine the depth estimation

3 Method

The goal of our approach is the joint learning of layout prediction and depth estimation from a single indoor panorama image. The proposed method leverages the layout depth map as an intermediate representation to relate the layout and depth estimation. Figure 3 shows an overview of our proposed pipeline.

Inferring high-quality 3D room layout from an indoor panorama image relies on the understanding of both the 3D geometry and the semantics of the indoor scene. Therefore, the proposed method uses the predicted coarse depth map and semantic segmentation of the input panorama to predict the layout depth map. The proposed method enables the refinement of depth estimation by integrating the coarse depth and layout depth with semantic information as a guidance.

3.1 Input and Pre-processing

Following [35], the first step of our method is to align the input panorama image to match the horizontal floor plane. The floor plane direction under equirectangular projection is estimated by first selecting the long line segments using the Line Segment Detector (LSD) [28] in overlapping perspective views and then vote for three mutually orthogonal vanishing directions [33]. This alignment ensures that wall-wall boundaries are vertical lines. The input of our network is the concatenation of the panorama image and the corresponding Manhattan line feature map provided by the alignment.

3.2 Coarse Depth and Semantics

Our approach receives the concatenation of a single *RGB* panorama and the Manhattan line feature map as input. The output of this module is the coarse depth estimation and semantic segmentation of the 2D panorama image.

An encoder-decoder architecture is used for the joint learning of the coarse depth information and semantic segmentation. The input panorama images suffer from horizontal distortions. To reduce the distortion effect, the encoder uses a modified input block in front of the ResNet-18 architecture. As shown by [34], the input block uses rectangle filters and varies the resolution to account for different distortion levels. The encoder is shared for both the depth estimation and semantic segmentation. The decoders restore the original input resolution by means of up-sampling operators followed by 3×3 convolutions. Skip connections are also added to link to the corresponding resolution in the encoder. The two decoders do not share weights and are trained to minimize the coarse depth estimation loss and semantic segmentation loss, respectively.

Loss Function: For coarse depth estimation, to account for both pixel-wise accuracy and spatially coherent results, this module incorporates the depth gradient and normals with the logarithm of the standard L1 loss, as done by [11]. So the loss function consists of three parts:

$$L_{coarse\_depth} = l_{depth} + \lambda l_{gradient} + \mu l_{normal} \tag{1}$$

where $\lambda, \mu \in R$ are hyper-parameters to balance the contribution of each component loss. The depth loss l_{depth}, the gradient loss $l_{gradient}$ and the surface normal loss l_{normal} are defined by:

$$l_{depth} = \frac{1}{n} \sum_{i=1}^{n} \ln(e_i + 1) \tag{2}$$

where $e_i = \|d_i - g_i\|_1$, d_i and g_i denote the predicted and ground truth depth maps respectively. n is the total number of pixels.

$$l_{gradient} = \frac{1}{n} \sum_{i=1}^{n} (\ln(|\nabla_x(e_i)| + 1) + \ln(|\nabla_y(e_i)| + 1)) \tag{3}$$

where $\nabla_x(e_i)$ is the spatial derivative of e_i computed at the ith pixel with respect to x, and so on.

$$l_{normal} = \frac{1}{n} \sum_{j=1}^{n} \left(1 - \frac{\langle n_j^d, n_j^g \rangle}{\sqrt{\langle n_j^d, n_j^d \rangle}\sqrt{\langle n_j^g, n_j^g \rangle}}\right) \tag{4}$$

where $n_i^d \equiv [-\nabla_x(d_i), -\nabla(d_i), 1]^\top$ and $n_i^g \equiv [-\nabla_x(g_i), -\nabla(g_i), 1]^\top$ denote the surface normal of the estimated depth map and the ground truth, respectively.

For semantic segmentation, the loss function is given by the per-pixel softmax cross-entropy between the predicted and ground-truth pixel-wise semantic labels:

$$L_{semantic} = -\sum_{i=1}^{n} p_i \log(\hat{p}_i) \tag{5}$$

where p and $\hat{p}$ are the ground truth and predicted semantic labels, respectively.

3.3 Layout Prediction

To obtain the global geometric structure of the scene, the proposed approach predicts the 3D layout of the scene. Instead of predicting 2D representations, our method directly predicts the layout depth maps of the input panoramas.

The input of this proposed module is a 8-channel feature map: the concatenation of *RGB* panorama, the corresponding Manhattan line feature map, and the predicted depth and semantics obtained by the previous modules of the pipeline. A ResNet-18 is used to build our encoder for the layout depth prediction network. The decoder architecture is similar to the previous ones for depth estimation and semantic segmentation, with nearest neighbor up-sampling operations followed by 3×3 convolutions. The skip connections are also added to prevent shifting of the prediction results during the up-sampling step. The output is the estimated layout depth map with the same resolution as the input panorama.

Loss Function: In addition to the pixel-wise depth supervision as described in Sect. 3.2, the virtual normal (VN) [31] is used as another geometric constraint to regulate the estimated layout depth map. The point cloud of the scene layout can be reconstructed from the estimated layout depth map based on the panoramic camera model. The virtual normal is the normal vector of a virtual plane formed by three randomly sampled non-colinear points in 3D space, which takes long-range relations into account from a global perspective. By minimizing the direction divergence between the ground-truth and predicted virtual normals, serving as a high-order 3D geometric constraint, the proposed method provides more accurate depth estimation and imposes the planar nature to the prediction of the layout depth map.

N group points are randomly sampled from the point cloud. In each group there are three points: $\Omega = \{P_i = (P_a, P_b, P_c)_i \mid i = 0, ..., N\}$. The three points in a group are restricted to be non-colinear as defined by condition C:

$$C = \{\alpha \geq \angle(\overrightarrow{P_aP_b}, \overrightarrow{P_aP_c}) \leq \beta, \alpha \geq \angle(\overrightarrow{P_bP_c}, \overrightarrow{P_bP_a}) \leq \beta \mid P_i \in \Omega\} \tag{6}$$

where $\alpha = 150°, \beta = 30°$ in our experiments.

Three points in each group establishes a virtual plane. The normal vector of the plane is computed by:

$$N = \{\boldsymbol{n}_i = \frac{\overrightarrow{P_aP_b} \times \overrightarrow{P_aP_c}}{\|\overrightarrow{P_aP_b} \times \overrightarrow{P_aP_c}\|} \mid P_i \in \Omega\} \tag{7}$$

where $\boldsymbol{n}_i$ is the normal vector of virtual plane P_i.

The virtual normal loss is computed by:

$$l_{vn} = \frac{1}{N} \sum_{i=1}^{N} \| \boldsymbol{n}_i^{pred} - \boldsymbol{n}_i^{gt} \|_1 \tag{8}$$

The overall loss for layout depth map estimation is defined by:

$$L_{layout\_depth} = l_{depth} + \lambda l_{gradient} + \mu l_{normal} + l_{vn} \tag{9}$$

The layout depth loss is based on both the local surface normal and the global virtual normal constraint. This ensures that the estimated layout depth map preserves the geometric structure of the scene layout accurately.

3D Layout Optimization: To constrain the layout shape so that the floor and ceiling are planar and the walls are perpendicular to each other (Manhattan world assumption), the proposed method recovers the parameterized 3D layout through optimization in 3D space. Previous methods [16,25,35] heavily rely on 2D image features (e.g. edge and corner maps). However, estimating the 3D layout from 2D edge and corner maps is an ill-posed problem and thus requires extra constraints. In contrast, our proposed method directly optimizes on the 3D layout point cloud and does not require extra constraints for the 2D to 3D layout conversion.

Using the point cloud of the scene layout converted from the predicted layout depth map, the floor/ceiling plan map is obtained by projecting the point cloud to the XZ plane. Similar to [30], a regression analysis is applied on the edges of the floor plan map and clustering them into sets of horizontal and vertical lines in 3D space. Then, the floor plan is recovered by using the straight, axis-aligned, wall-floor boundaries. The room height is efficiently computed by using the ceiling-floor distances along the Y axis.

3.4 Depth Refinement

After the coarse depth map and the layout depth map are obtained from the previous modules, a depth refinement step is taken.

A straight-forward way is to concatenate all the data representations as input and use an encoder-decoder network to predict the final depth estimation. This approach is denoted by *direct refinement*. The semantic approach is to use the semantic information as a guidance to dynamically fuse the two depth maps. This approach is denoted by *semantic-guided refinement*. The semantic-guided refinement step produces an attention map incorporating the coarse depth map and the layout depth map. For a structural background representing the scene layout components (ceiling, floor and wall), the network focuses more on the layout depth map. While for objects in the room (furniture), the network switches the attention to the coarse depth estimation. Therefore, in this paper, we combine these two concepts as shown in Fig. 3. First, an encoder-decoder network, taking the concatenation of the coarse depth, layout depth and semantic segmentation prediction as inputs, combines the previous depth maps with the semantic-guided attention map. This semantic-guided depth fusion maximizes the exploitation of the coarse depth and layout depth. Then, the depth refinement module takes the fused depth as input to predict the final refined depth. The encoder-decoder architecture of the depth refinement module is similar to the previous coarse depth estimation network.

Loss Function: The loss function for the depth refinement is the same as the layout depth estimation loss described in Sect. 3.3.

3.5 Training Details

Following the experimental setting of [35], the proposed method uses horizontal rotations, left-right flippings and luminance changes to augment the training samples. Our network uses the ADAM [14] optimizer with $\beta_1 = 0.9$ and $\beta_2 = 0.999$ to update the network parameters. To train the network, we first train the joint learning of coarse depth estimation and semantic segmentation, and then fix the weights of the depth and semantic network, and train the layout depth map prediction. Then, we set all the trained weights fixed to train the depth refinement module. Finally, we jointly train the whole network end-to-end.

4 Experiments

In this section, the performance of our proposed method is evaluated for both the layout prediction and depth estimation tasks.

Dataset: The dataset used for training is the Stanford 2D–3D dataset [1]. The Stanford 2D–3D dataset contains 1413 *RGB* panoramic images collected from 6 large-scale indoor environments, including offices, classrooms, and other open spaces like corridors, where 571 panoramas have layout annotations. Our experiments follow the official train-val-test split for evaluation. The PanoContext dataset is used to verify the generalizability of our approach for the task of layout prediction. The PanoContext [33] dataset contains 514 *RGB* panoramic images of two indoor environments, i.e., bedrooms and living rooms.

Evaluation Metrics: The following standard metrics are used to evaluate our approach:

3D IoU: $3D\ IoU = \frac{V_{pred} \cap V_{gt}}{V_{pred} \cup V_{gt}}$, where V_{pred} and V_{gt} stand for the volumetric occupancy of the predicted and ground truth 3D layout.

Corner error (CE): $CE = \frac{1}{\sqrt{H^2+W^2}} \sum_{i \in corners} \|c_i^{pred} - c_i^{gt}\|_2^2$, where H and W are the image height and width, c^{pred} and c^{gt} denote the predicted and ground truth corner positions.

Pixel error (PE): $PE = \frac{1}{|N|} \sum_{i=1}^{N} \mathbb{1}(s_i^{pred} \neq s_i^{gt})$, where s^{pred} and s^{gt} denotes the predicted and ground truth pixel-wise semantic (ceiling, floor and wall).

$\mathbb{1}(.)$ is an indicator function, setting to 1 when the pixel semantic prediction is incorrect.

Threshold: % of d_i that $max(\frac{d_i}{g_i}, \frac{g_i}{d_i}) = \delta < thr$

Absolute Relative Difference: $Abs\ Rel = \frac{1}{|N|}\sum_{i=1}^{N}||d_i - g_i||/g_i$

Squared Relative Difference: $Sq\ Rel = \frac{1}{|N|}\sum_{i=1}^{N}||d_i - g_i||^2/g_i$

RMSE (linear): $RMS = \sqrt{\frac{1}{|N|}\sum_{i=1}^{N}||d_i - g_i||^2}$

RMSE (log): $RMS(log) = \sqrt{\frac{1}{|N|}\sum_{i=1}^{N}||\log d_i - \log g_i||^2}$

where we use 3D IoU, corner error and pixel error to evaluate the layout prediction and the rest for depth estimation.

Table 1. Quantitative results of layout estimation on the Stanford 2D–3D dataset. Our method outperforms all existing methods

Method	3D IoU(%)	Corner error(%)	Pixel error(%)
LayoutNet [35]	76.33	1.04	2.70
DuLa-Net [30]	79.36	-	-
HorizonNet [25]	79.79	0.71	2.39
Ours	**85.81**	**0.67**	**2.20**

4.1 Layout Prediction

A quantitative comparison of different methods on the Stanford 2D–3D dataset is summarized in Table 1. LayoutNet [35] predicts the layout boundary and corner maps directly from the input panorama. DuLa-Net [30] leverages both the equirectangular panorama-view and the perspective ceiling-view to learn different cues for the room layout. HorizonNet [25] encodes the room layout as three 1D vectors and proposes to recover the 3D room layout from 1D predictions by a RNN. The proposed method shows state-of-the-art performance and outperforms other existing methods. By leveraging the layout depth map as an intermediate representation, the proposed network abstracts the geometric structure of the scene from both a local and global perspective. This results in more geometric cues for the scene layout prediction and is less affected by occlusions.

LayoutNet [35] and HorizonNet [25] also combine the Stanford 2D–3D [1] and PanoContext [33] training data to train their methods. Since the PanoContext dataset does not contain any depth or semantic ground truth, our model is first initialized with the Stanford 2D–3D dataset, and then the model is trained on the same mixed dataset with the weight-fixed coarse depth and semantic prediction modules. Table 2 shows the quantitative results trained on this mixed training data. Although the PanoContext dataset has different indoor configurations and no depth or semantic ground truth, our method still obtains competitive performance.

Table 2. Quantitative results on the (a) Stanford 2D–3D and (b) PanoContext for models trained with mixed PanoContext and Stanford 2D–3D training data. Our method outperforms other methods on both datasets

Method	3D IoU(%)	CE(%)	PE(%)	**Method**	3D IoU(%)	CE(%)	PE(%)
LayoutNet [35]	77.51	0.92	2.42	LayoutNet [35]	75.12	1.02	3.18
HorizonNet [25]	83.51	**0.62**	**1.97**	HorizonNet [25]	84.23	0.69	1.90
Ours	**86.21**	0.71	2.08	Ours	**84.40**	**0.61**	**1.74**

(a) Results for Stanford 2D-3D (b) Results for PanoContext

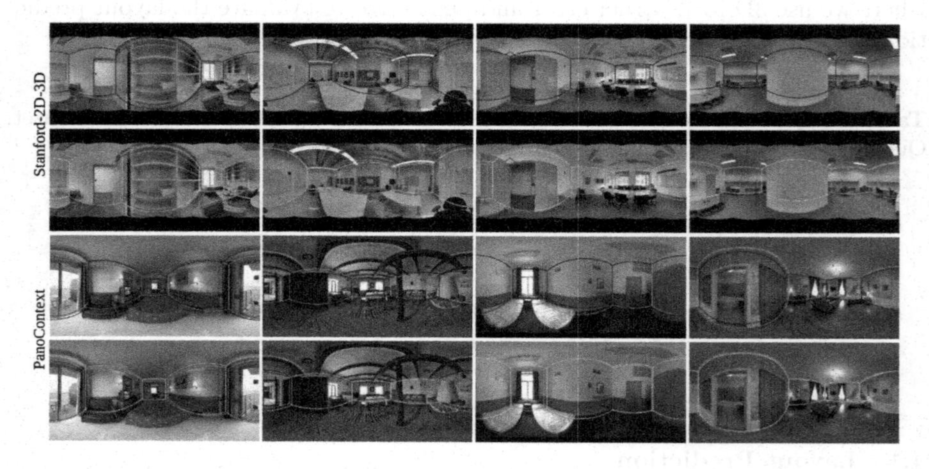

Fig. 4. Qualitative comparison on layout prediction. Results are shown of testing the baseline LayoutNet [35] (blue), our proposed method (green) and the ground truth (orange) on the Stanford 2D–3D dataset and PanoContext dataset. (Color figure online)

The qualitative results for the layout prediction are shown in Fig. 4. The first two rows demonstrate the results of the LayoutNet and our proposed method on the Stanford 2D–3D dataset. The last two rows are the results obtained for the PanoContext dataset. The proposed method outperforms the other methods on both datasets and shows robustness to occlusion. As presented by the second example for Stanford 2D–3D, since the proposed method explicitly incorporates the depth information, the corners are located more precisely (avoiding locations in the middle of the wall which has continuous depth). The semantic content ensures the detection of the occluded corners, as shown in the third example of Stanford 2D–3D (corners occluded by the door). The last example of the Stanford 2D–3D shows a failure case for both methods. For non-negligible occlusions in the scene, both methods fail to predict the corner positions accurately. Similar improvements are shown for the results obtained for the PanoContext dataset.

Ablation Study: The goal is to evaluate the performance of our layout prediction and layout depth estimation with different configurations: 1) *wo/ depth&semantic*: predicting the layout depth directly from the input; 2) *w/ pred.*

Table 3. Ablation study of layout prediction and layout depth map estimation on the Stanford 2D–3D dataset. We evaluate the influence of different modules and show that our final proposed approach performs the best

				Lower is better				Higher is better		
	3D IoU(%)	CE(%)	PE(%)	Abs Rel	Sq Rel	RMS	RMS(log)	$\delta < 1.25$	$\delta < 1.25^2$	$\delta < 1.25^3$
Wo/ depth & Semantic	77.28	1.21	3.31	0.089	0.044	0.327	0.056	0.914	0.987	0.996
W/ pred. depth	82.65	0.83	2.92	0.069	0.029	0.257	0.045	0.952	0.993	**0.998**
W/ pred. semantic	78.57	1.14	3.18	0.079	0.034	0.311	0.053	0.927	0.990	0.997
Wo/ VN	84.22	0.75	2.42	0.065	0.028	0.238	0.043	0.955	0.993	**0.998**
edg & Cor maps	82.03	1.05	2.61	-	-	-	-	-	-	-
layout depth -> edg & Cor maps	83.67	0.92	2.52	0.067	0.029	0.238	0.044	0.955	0.992	**0.998**
Proposed Final	**85.81**	**0.67**	**2.20**	**0.064**	**0.026**	**0.237**	**0.042**	**0.957**	**0.994**	**0.998**

depth: only with the predicted depth; 3) *w/ pred. semantic*: only with the predicted semantic; 4) *wo/ VN*: without the VN loss; 5) *edg&cor maps*: predicting the edge and corner maps from the concatenation of input panorama, predicted depth and semantic; 6) *layout depth -> edg&cor maps*: predicting the edge and corner maps from the layout depth map. As shown in Table 3, training with either predicted depth or semantic information increases the accuracy. The VN loss further regulates the estimated layout depth to preserve surface straightness, thus improving the recovered layout. In comparison with the edge and corner maps, the layout depth map contains both local and global information to recover the 3D layout of the scene.

Fig. 5. Qualitative results of non-cuboid layout prediction. It can be derived that our proposed method also works well for non-cuboid layouts

Non-cuboid Layout: To verify the generalization ability of our proposed method to non-cuboid layout, our model is fine-tuned on the non-cuboid rooms labeled by [25]. As shown in Fig. 5, our proposed method is able to handle non-cuboid layout rooms. Please see more results in the supplemental materials.

Table 4. Quantitative results and ablation study of depth estimation on the Stanford 2D–3D dataset. Our method outperforms all existing methods

	Lower is better				Higher is better		
	Abs Rel	Sq Rel	RMS	RMS(log)	$\delta < 1.25$	$\delta < 1.25^2$	$\delta < 1.25^3$
FCRN [15]	0.091	0.057	0.364	0.134	0.913	0.982	0.995
RectNet [34]	0.082	0.046	0.399	0.123	0.928	0.988	0.997
DistConv [26]	0.176	-	0.369	0.083	-	-	-
Plane-aware [7]	0.079	0.029	0.290	0.120	0.934	0.990	**0.998**
Proposed Coarse-depth	0.105	0.045	0.352	0.094	0.934	0.989	0.997
Proposed Direct-refinement	0.089	0.033	0.269	0.095	0.944	0.989	**0.998**
Proposed Semantic-guided	0.086	0.033	0.273	0.096	0.944	0.989	**0.998**
Proposed Final	**0.068**	**0.026**	**0.264**	**0.080**	**0.954**	**0.992**	**0.998**

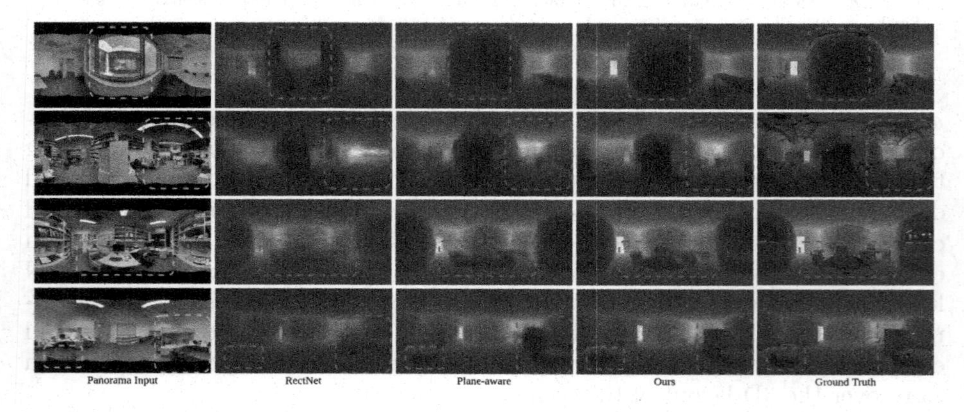

Fig. 6. Qualitative comparison on depth estimation. Results are shown for testing the baseline RectNet [34], Plane-aware network [7] and our proposed method on the Stanford 2D–3D dataset

4.2 Depth Estimation

Table 4 presents the quantitative results of different methods for depth estimation on the Stanford 2D–3D dataset. FCRN [15] designs a supervised fully convolutional residual network with up-projection blocks. RectNet [34] proposes a specific pipeline for depth estimation using panoramas as input. DistConv [26] trains on perspective images and then regress depth for panorama images by distortion-aware deformable convolution filters. Plane-aware [7] designs the plane-aware loss which leverages principal curvature as an indicator of planar boundaries. The results demonstrate that our proposed method obtains state-of-the-art depth estimation results from a single panorama image. The qualitative comparison is shown in Fig. 6. In the first image, the RectNet [34] is confused by the transparent window, which is a common failure case in depth estimation. The Plane-aware network [7] and our proposed network overcome this issue. Our result for the window region is smoother due to the constraints from the layout depth. In the second image, the distant regions are too ambiguous to predict the corresponding depth. Our proposed method predicts a proper depth map because of the explicit inter-positioning of the layout depth. Because of the proposed semantic-guided refinement, the proposed method also preserves better object details compared to the other two methods, as shown in the third and fourth image. Fig. 7 illustrates the derived surface normals from the estimated depth map. Constrained by the layout depth map, the surface normal results demonstrate that our proposed method preserves the planar property for depth estimation.

Table 5. Quantitative comparison of the proposed method for joint training. It is shown that joint training improves the performance for all the proposed modules

				Lower is better				Higher is better		
	3D IoU(%)	CE(%)	PE(%)	Abs Rel	Sq Rel	RMS	RMS(log)	$\delta < 1.25$	$\delta < 1.25^2$	$\delta < 1.25^3$
Coarse depth	-	-	-	0.112	0.049	0.379	0.116	0.930	0.988	**0.997**
Coarse depth (joint)	-	-	-	**0.105**	**0.045**	**0.352**	**0.094**	**0.934**	**0.989**	**0.997**
Depth refinement	-	-	-	0.084	0.032	0.273	0.088	0.950	0.989	**0.998**
Depth refinement (joint)	-	-	-	**0.068**	**0.026**	**0.264**	**0.080**	**0.954**	**0.992**	**0.998**
Layout depth	84.69	0.75	2.43	0.069	0.029	0.257	0.046	0.951	0.993	**0.998**
Layout depth (joint)	**85.81**	**0.67**	**2.20**	**0.064**	**0.026**	**0.237**	**0.042**	**0.957**	**0.994**	**0.998**

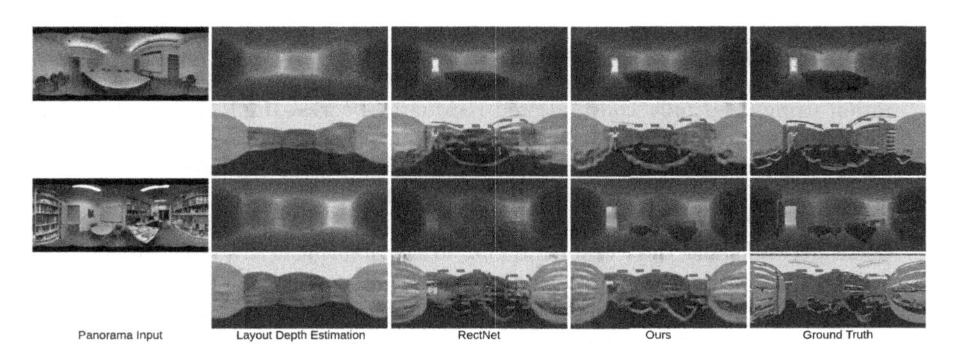

Fig. 7. Comparison of the derived surface normal from the depth estimation. Our proposed method produces smoother surfaces for planar regions

Ablation Study: An ablation study is conducted to evaluate the performance of the proposed method for different configurations, as shown in Table 4: 1) *Proposed Coarse-depth*: the depth estimation from the first decoder; 2) *Proposed Direct-refinement*: the depth refinement using all the data representation as input, as stated in Sect. 3.4; 3) *Proposed Semantic-guided*: the depth fusion using semantic-guided attention map, as state in Sect. 3.4. It is shown that the direct-refinement performs better than the coarse-depth. This indicates that the joint learning with layout prediction already improves the depth estimation. Semantic-guided refinement improves the performance which supports our argument to dynamically fuse the layout depth map and the coarse depth estimation based on background and foreground regions. Our proposed final method obtains the best overall performance for all variations.

Table 5 shows the quantitative comparison for each module of the proposed pipeline before and after joint training. It demonstrates that all the modules benefit from joint training.

5 Conclusion

We proposed a method to jointly learn the layout and depth from a single indoor panorama image. By leveraging the layout depth map as an intermediate representation, the optimization of 3D layout does not require extra constraints and the refined depth estimation preserves the planarity for the layout components. Experiment results on challenging indoor datasets show that, with the proposed method for joint learning, the performance of both the layout prediction and depth estimation from single panorama images is significantly improved and that our method outperforms the state-of-the-art.

References

1. Armeni, I., Sax, S., Zamir, A.R., Savarese, S.: Joint 2d–3d-semantic data for indoor scene understanding. arXiv preprint arXiv:1702.01105 (2017)
2. Cao, Y., Wu, Z., Shen, C.: Estimating depth from monocular images as classification using deep fully convolutional residual networks. IEEE Trans. Circ. Syst. Video Technol. **28**(11), 3174–3182 (2017)
3. Cohen, T., Geiger, M., Köhler, J., Welling, M.: Convolutional networks for spherical signals. arXiv preprint arXiv:1709.04893 (2017)
4. Cohen, T.S., Weiler, M., Kicanaoglu, B., Welling, M.: Gauge equivariant convolutional networks and the icosahedral cnn. arXiv preprint arXiv:1902.04615 (2019)
5. Delage, E., Lee, H., Ng, A.Y.: A dynamic bayesian network model for autonomous 3d reconstruction from a single indoor image. In: 2006 IEEE Computer Society Conference on Computer Vision and Pattern Recognition (CVPR'06). vol. 2, pp. 2418–2428. IEEE (2006)
6. Dhamo, H., Tateno, K., Laina, I., Navab, N., Tombari, F.: Peeking behind objects: layered depth prediction from a single image. Pattern Recogn. Lett. **125**, 333–340 (2019)
7. Eder, M., Moulon, P., Guan, L.: Pano popups: indoor 3d reconstruction with a plane-aware network. In: 2019 International Conference on 3D Vision (3DV), pp. 76–84. IEEE (2019)
8. Eder, M., Price, T., Vu, T., Bapat, A., Frahm, J.M.: Mapped convolutions. arXiv preprint arXiv:1906.11096 (2019)

9. Eigen, D., Puhrsch, C., Fergus, R.: Depth map prediction from a single image using a multi-scale deep network. In: Advances in Neural Information Processing Systems, pp. 2366–2374 (2014)
10. Hedau, V., Hoiem, D., Forsyth, D.: Recovering the spatial layout of cluttered rooms. In: 2009 IEEE 12th International Conference on Computer Vision, pp. 1849–1856. IEEE (2009)
11. Hu, J., Ozay, M., Zhang, Y., Okatani, T.: Revisiting single image depth estimation: toward higher resolution maps with accurate object boundaries. In: 2019 IEEE Winter Conference on Applications of Computer Vision (WACV), pp. 1043–1051. IEEE (2019)
12. Izadinia, H., Shan, Q., Seitz, S.M.: Im2cad. In: Proceedings of the IEEE Conference on Computer Vision and Pattern Recognition, pp. 5134–5143 (2017)
13. Jiang, C., Huang, J., Kashinath, K., Marcus, P., Niessner, M., et al.: Spherical cnns on unstructured grids. arXiv preprint arXiv:1901.02039 (2019)
14. Kingma, D.P., Ba, J.: Adam: a method for stochastic optimization. arXiv preprint arXiv:1412.6980 (2014)
15. Laina, I., Rupprecht, C., Belagiannis, V., Tombari, F., Navab, N.: Deeper depth prediction with fully convolutional residual networks. In: 3D Vision (3DV), 2016 Fourth International Conference on, pp. 239–248. IEEE (2016)
16. Lee, C.Y., Badrinarayanan, V., Malisiewicz, T., Rabinovich, A.: Roomnet: end-to-end room layout estimation. In: Proceedings of the IEEE International Conference on Computer Vision, pp. 4865–4874 (2017)
17. Li, B., Shen, C., Dai, Y., Van Den Hengel, A., He, M.: Depth and surface normal estimation from monocular images using regression on deep features and hierarchical crfs. In: Proceedings of the IEEE Conference on Computer Vision and Pattern Recognition, pp. 1119–1127 (2015)
18. Liu, C., Kohli, P., Furukawa, Y.: Layered scene decomposition via the occlusion-crf. In: Proceedings of the IEEE Conference on Computer Vision and Pattern Recognition, pp. 165–173 (2016)
19. Liu, C., Schwing, A.G., Kundu, K., Urtasun, R., Fidler, S.: Rent3d: floor-plan priors for monocular layout estimation. In: Proceedings of the IEEE Conference on Computer Vision and Pattern Recognition, pp. 3413–3421 (2015)
20. Mallya, A., Lazebnik, S.: Learning informative edge maps for indoor scene layout prediction. In: Proceedings of the IEEE International Conference on Computer Vision, pp. 936–944 (2015)
21. Micusik, B., Kosecka, J.: Piecewise planar city 3d modeling from street view panoramic sequences. In: 2009 IEEE Conference on Computer Vision and Pattern Recognition, pp. 2906–2912. IEEE (2009)
22. Zama Ramirez, P., Poggi, M., Tosi, F., Mattoccia, S., Di Stefano, Luigi: Geometry meets semantics for semi-supervised monocular depth estimation. In: Jawahar, C.V., Li, H., Mori, G., Schindler, K. (eds.) ACCV 2018. LNCS, vol. 11363, pp. 298–313. Springer, Cham (2019). https://doi.org/10.1007/978-3-030-20893-6_19
23. Ren, Y., Li, S., Chen, C., Kuo, C.C.J.: A coarse-to-fine indoor layout estimation (cfile) method. In: Asian Conference on Computer Vision, pp. 36–51. Springer (2016)
24. Su, Y.C., Grauman, K.: Learning spherical convolution for fast features from 360 imagery. In: Advances in Neural Information Processing Systems, pp. 529–539 (2017)
25. Sun, C., Hsiao, C.W., Sun, M., Chen, H.T.: Horizonnet: learning room layout with 1d representation and pano stretch data augmentation. In: Proceedings of the IEEE Conference on Computer Vision and Pattern Recognition, pp. 1047–1056 (2019)

26. Tateno, K., Navab, N., Tombari, F.: Distortion-aware convolutional filters for dense prediction in panoramic images. In: Proceedings of the European Conference on Computer Vision (ECCV), pp. 707–722 (2018)
27. Tulsiani, S., Tucker, R., Snavely, N.: Layer-structured 3d scene inference via view synthesis. In: Proceedings of the European Conference on Computer Vision (ECCV), pp. 302–317 (2018)
28. Von Gioi, R.G., Jakubowicz, J., Morel, J.M., Randall, G.: Lsd: a fast line segment detector with a false detection control. IEEE Trans. Pattern Anal. Mach. Intell. **32**(4), 722–732 (2008)
29. Wang, X., Fouhey, D., Gupta, A.: Designing deep networks for surface normal estimation. In: Proceedings of the IEEE Conference on Computer Vision and Pattern Recognition, pp. 539–547 (2015)
30. Yang, S.T., Wang, F.E., Peng, C.H., Wonka, P., Sun, M., Chu, H.K.: Dula-net: a dual-projection network for estimating room layouts from a single rgb panorama. In: Proceedings of the IEEE Conference on Computer Vision and Pattern Recognition, pp. 3363–3372 (2019)
31. Yin, W., Liu, Y., Shen, C., Yan, Y.: Enforcing geometric constraints of virtual normal for depth prediction. In: Proceedings of the IEEE International Conference on Computer Vision, pp. 5684–5693 (2019)
32. Zhang, J., Kan, C., Schwing, A.G., Urtasun, R.: Estimating the 3d layout of indoor scenes and its clutter from depth sensors. In: Proceedings of the IEEE International Conference on Computer Vision, pp. 1273–1280 (2013)
33. Zhang, Y., Song, S., Tan, P., Xiao, J.: Panocontext: a whole-room 3d context model for panoramic scene understanding. In: European Conference on Computer Vision, pp. 668–686. Springer (2014)
34. Zioulis, N., Karakottas, A., Zarpalas, D., Daras, P.: Omnidepth: dense depth estimation for indoors spherical panoramas. In: Proceedings of the European Conference on Computer Vision (ECCV), pp. 448–465 (2018)
35. Zou, C., Colburn, A., Shan, Q., Hoiem, D.: Layoutnet: reconstructing the 3d room layout from a single rgb image. In: Proceedings of the IEEE Conference on Computer Vision and Pattern Recognition, pp. 2051–2059 (2018)
36. Zou, C., Guo, R., Li, Z., Hoiem, D.: Complete 3d scene parsing from an rgbd image. Int. J. Comput. Vis. **127**(2), 143–162 (2019)

Guessing State Tracking for Visual Dialogue

Wei Pang and Xiaojie Wang(✉)

Center for Intelligence Science and Technology, School of Computer Science, Beijing University of Posts and Telecommunications, Beijing, China
{pangweitf,xjwang}@bupt.edu.cn

Abstract. The Guesser is a task of visual grounding in GuessWhat?! like visual dialogue. It locates the target object in an image supposed by an Oracle oneself over a question-answer based dialogue between a Questioner and the Oracle. Most existing guessers make one and only one guess after receiving all question-answer pairs in a dialogue with the predefined number of rounds. This paper proposes a guessing state for the Guesser, and regards guess as a process with change of guessing state through a dialogue. A guessing state tracking based guess model is therefore proposed. The guessing state is defined as a distribution on objects in the image. With that in hand, two loss functions are defined as supervisions to guide the guessing state in model training. Early supervision brings supervision to Guesser at early rounds, and incremental supervision brings monotonicity to the guessing state. Experimental results on GuessWhat?! dataset show that our model significantly outperforms previous models, achieves new state-of-the-art, especially the success rate of guessing 83.3% is approaching the human-level accuracy of 84.4%.

Keywords: Visual dialogue · Visual grounding · Guessing state tracking · GuessWhat?!

1 Introduction

Visual dialogue has received increasing attention in recent years. It involves both vision and language processing and interactions between them in a continuous conversation and brings some new challenging problems. Some different tasks of visual dialogue have been proposed, such as Visual Dialog [5], GuessWhat?! [20], GuessWhich [4], and MNIST Dialog [11,14,28]. Among them, GuessWhat?! is a typical object-guessing game played between a Questioner and an Oracle. Given an image including several objects, the goal of the Questioner is to locate the target object supposed by the Oracle oneself at the beginning of a game by asking a series of yes/no questions. The Questioner, therefore, has two sub-tasks: one is Question Generator (QGen) that asks questions to the Oracle, the other is Guesser that identifies the target object in the image based on the generated dialogue between the QGen and Oracle. The Oracle answers questions with yes

A. Vedaldi et al. (Eds.): ECCV 2020, LNCS 12361, pp. 683–698, 2020.
https://doi.org/10.1007/978-3-030-58517-4_40

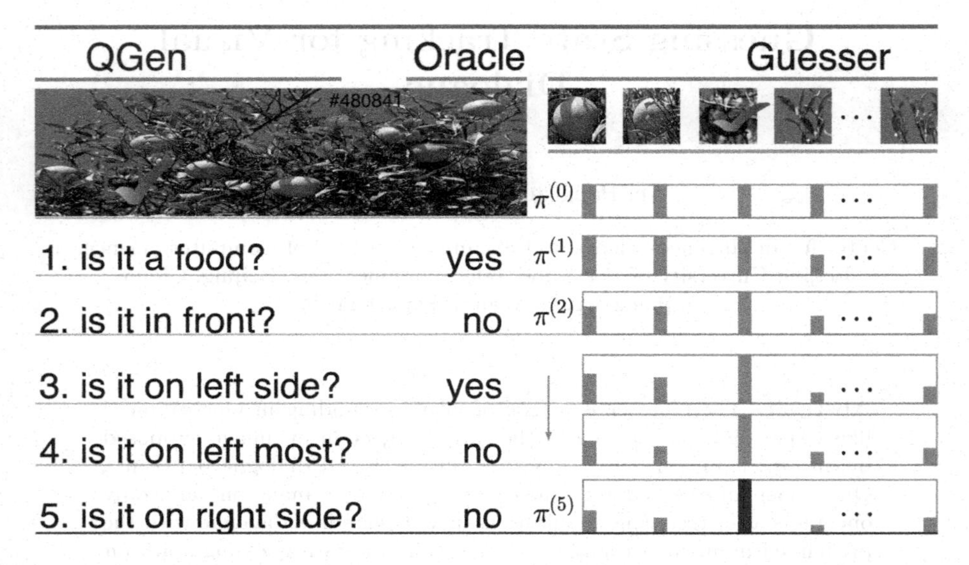

Fig. 1. The left part shows a game of GuessWhat?!. The right part illustrates the guess in Guesser as a process instead of a decision in a single point (the strips lineup denotes a probability distribution over objects, the arrowhead represents the tracking process).

or no. The left part of Fig. 1 shows a game played by the QGen, Oracle, and Guesser. The Guesser, which makes the final decision, is the focus of this paper.

Compared with QGen, relatively less work has been done on Guesser. It receives as input a sequence of question-answer (QA) pairs and a list of candidate objects in an image. The general architecture for Guesser introduced in [19,20] that encodes the QA pairs into a dialogue representation and encodes each object into an embedding. Then, it compares the dialogue representation with any object embedding via a dot product and outputs a distribution of probabilities over objects, the object with higher probability is selected as the target. Most current work focuses on encoding and fusing multiple types of information, such as QA pairs, images, and visual objects. For example, some models [3,13,16–20,26] convert the dialogue into a flat sequence of QA pair handled by a Long Short-Term Memory (LSTM)[8], some models [2,6,23,27] introduce attention and memory mechanism to obtain a multi-modal representation of the dialogue.

Most of the existing Guesser models make a guess after fixed rounds of QA pairs, and this does not fully utilize the information from the sequence of QA pairs, we refer to that way as single-step guessing. Different games might need different rounds of QA pairs. Some work [2,16] has therefore been done on choosing when to guess, i.e., make a guess after different rounds of question-answer for different games.

No matter the number of question-answer rounds is fixed or changed in different dialogues, existing Guesser models make one and only one guess after the final round of question-answer, i.e., Guesser is not activated until it reaches the final round of dialogue.

This paper models the Guesser in a different way. We think the Guesser to be active throughout the conversation of QGen and Oracle, rather than just only guessing at the end of the conversation. It keeps on updating a guess distribution after each question-answer pair from the beginning and does not make a final guess until the dialogue reaches a predefined round or it can make a confident guess. For example, as shown in Fig. 1, a guess distribution is initiated as uniform distribution, i.e., each object has the same probability as the target object at the beginning of the game. After receiving the first pair of QA, the guesser updates the guess distribution and continues to update the distribution in the following rounds of dialogue. It makes a final guess after predefined five rounds of dialogue.

We think that modeling the Guesser as a process instead of a decision in a single point provides more chances to not only make much more detailed use of dialogue history but also combine more information for making better guesses. One such information is monotonicity, i.e., a good enough guesser will never reduce the guessing probability on the target object by making proper use of each question-answer pair. A good guess either raises the probability of a target object in guess distribution when the pair contains new information about the target object or does not change the probability when the pair contains no new information.

This paper proposes a guessing state tracking (GST) based Guesser model for implementing the above idea. Guessing state (GS) is at first time introduced into the game. A GS is defined as a distribution on candidate objects. A GST mechanism, which includes three sub-modules, is proposed to update GS after each question-answer pair from the beginning. Update of Visual Representation (UoVR) module updates the representation of image objects according to the current guessing state, QAEncoder module encodes the QA pair, and Update of Guessing State (UoGS) module updates the guessing state by combining both information from the image and QA. GST brings a series of GS, i.e., let the Guesser make a series of guesses during the dialogue.

Two loss functions are designed on making better use of a series of GS, or the introduction of GS into visual dialogue makes the two new loss functions possible. One is called early supervision loss that tries to lead GS to the target object as early as possible, where ground-truth is used to guide the guesses after each round of QA, even the guess after the first round where a successful guess is impossible at that time. The other is called incremental supervision loss that tries to bring monotonicity mentioned above to the probability of target object in the series of GS.

Experimental results show that the proposed model achieves new state-of-the-art performances in all different settings on GuessWhat?!. To summarize, our contributions are mainly three-fold:

- We introduce guessing state into visual dialogue for the first time and propose a Guessing State Tracking (GST) based Guesser model, a novel mechanism that models the process of guessing state updating over question-answer pairs.

- We introduce two guessing states based supervision losses, early supervision loss, and incremental supervision loss, which are effective in model training.
- Our model performs significantly better than all previous models, and achieves new state-of-the-art in all different settings on GuessWhat?!. The guessing accuracy of 83.3% approaches the human's level of 84.4%.

2 Related Work

Visual Grounding is an essential language-to-vision problem of finding the most relevant object in an image by a natural language expression, which can be a phrase, a sentence, or a dialogue. It has attracted considerable attention in recent years [1,6,7,12,22,24,25], and has been studied in the Guesser task in the GuessWhat?! [6]. This paper focuses on grounding a series of language descriptions (QA pair) in an image gradually by dialoguing.

Most previous work views Guesser as making a single-step guess based on a sequence of QA pairs. In [3,11,13,16–20,23,26], all the multi-round QA pairs are considered as a flat sequence and encoded into a single representation using either an LSTM or an HRED [15] encoder, each object is represented as an embedding encoded from their object category embedding and 8-d spatial position embedding. A score is obtained by performing a dot product between the dialogue representation and each object embedding, then followed a softmax layer on the scores to output distribution of probabilities over objects, the object with higher probability is chosen as the most relevant object. As we can see, only one decision is made by the Guesser.

Most of the guesser models explored to encode the dialogue of multi-round QA pairs in an effective way. For example, in [2,27], they integrate Memory and Attention into the Guesser architecture used in [19]. Where the memory is consist of some facts that are separately encoded from each QA pairs, the image feature vector is used as a key to attend the memory. In [6], an accumulated attention (ATT) mechanism is proposed. It fuses three types of information, i.e., dialogue, image, and objects, by three attention models. Similarly, [23] proposed a history-aware co-attention network (HACAN) to encode the QA pairs.

As we can see, the models, as mentioned above, all make a single-step guess at the time that the dialogue ended, these might be counterintuitive. Different from them, we consider the guess as a process, and explicitly track the guessing states after every dialogue round. Compared with prior works, we refer to the proposed GST as multi-step guessing.

Our GST based Guesser model is related to the VDST [13] based QGen model. [13] proposed a well-defined questioning state for the QGen and implemented a suitable tracking mechanism through the dialogue. The crucial difference in tracking state is that the QGen requires to track changes on the representations of objects because it needs more detailed information concerning the attended objects for asking more questions, while the Guesser does not need it.

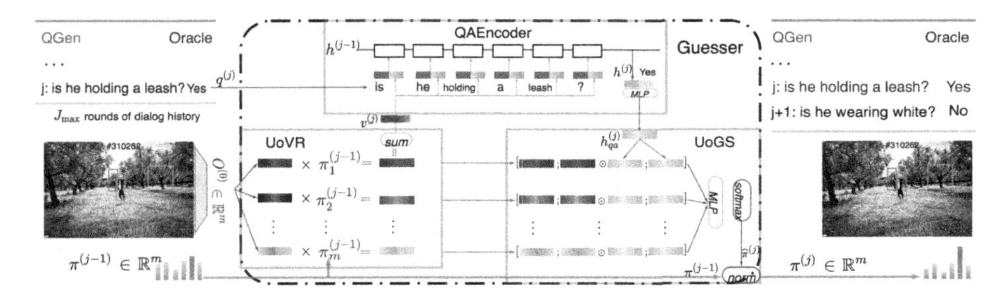

Fig. 2. Overview of the proposed Guesser model.

3 Model: Guessing State Tracking

The framework of our guessing state tracking (GST) model is illustrated in Fig. 2. Three modules are implemented in each round of guessing. There are Update of Visual Representation (UoVR), Question-Answer Encoder (QAEncoder), and Update of Guessing State (UoGS). Where, UoVR updates representation of an image for Guesser according to the previous round of guessing state, new visual representation is then combined into QAEncoder for synthesizing information from both visual and linguistic sides up to the current round of dialogue for the Guesser. Finally, UoGS is applied to update the guessing state of the guesser. We give details of each module in the following sub-sections.

3.1 Update of Visual Representation (UoVR)

Following previous work [19,20], candidate objects in an image are represented by their category and spatial features as in Eq. 1:

$$O^{(0)} = \{o_i^{(0)} | o_i^{(0)} = \text{MLP}([o_{cate}; o_{spat}])\}_{i=1}^{m}, \tag{1}$$

where $O^{(0)} \in \mathbb{R}^{m \times d}$ consists of m initial objects. For each object $o_i^{(0)}$, it is a concatenation of an 512-d category embedding o_{cate} and an 8-d vector o_{spat} of object location in an image. Where o_{cate} are learnable parameters, o_{spat} are coordinates $[x_{min}, y_{min}, x_{max}, y_{max}, x_{center}, y_{center}, w_{box}, h_{box}]$ as in [20], w_{box} and h_{box} denote width and height of an object, the coordinates range from -1 to 1 scaled by the image width and height. To map object and word embedding to the same dimension, the concatenation is passed through an MLP to obtain a d-dimensional vector.

Let $\pi^{(j)} \in \mathbb{R}^m$ be an accumulative probability distribution over m objects after jth round of dialogue. It is defined as the guessing state and will be updated with the guessing process. At the beginning of a game, $\pi^{(0)}$ is a uniform distribution. With the progress of guessing, the visual representation in guesser's mind would update accordingly. Two steps are designed. The first step is an update of representations of objects. Pang and Wang [13] use an effective representation update in the VDST model. We borrow it for our GST model, as written in Eq. 2:

$$O^{(j)} = (\pi^{(j-1)})^T O^{(0)}, \tag{2}$$

where $O^{(j)} \in \mathbb{R}^{m \times d}$ is a set of m updated objects at round j. Second, the summed embedding of all objects in $O^{(j)}$ is used as new visual representation as shown in Eq. 3,

$$v^{(j)} = \mathrm{sum}(O^{(j)}), \tag{3}$$

where $v^{(j)} \in \mathbb{R}^d$ denotes updated visual information for the guesser at round j.

3.2 Question-Answer Encoder (QAEncoder)

For encoding linguistic information in the current question with visual information in hand, we concatenate $v^{(j)}$ to each word embedding $w_i^{(j)}$ in jth turn question $q^{(j)}$, take the concatenation as input to a single-layer LSTM encoder one by one as shown in Eq. 4,

$$h^{(j)} = \mathrm{LSTM}([w_i^{(j)}; v^{(j)}]_{i=1}^{N^{(j)}}, h^{(j-1)}), \tag{4}$$

where $N^{(j)}$ is the length of question $q^{(j)}$. The last hidden state of the LSTM, $h^{(j)}$, is used as representation of $q^{(j)}$, and $h^{(j-1)}$ is used as initial input of the LSTM as shown in Fig. 2.

$h^{(j)}$ is then concatenated to $a^{(j)}$, which is the embedding of the answer to jth turn question, the result $[h^{(j)}; a^{(j)}]$ are passed through an MLP to obtain the representation of QA pair at round j, as written in Eq. 5,

$$h_{qa}^{(j)} = \mathrm{MLP}([h^{(j)}; a^{(j)}]), \tag{5}$$

where $h_{qa}^{(j)} \in \mathbb{R}^d$ synthesizes information from both questions and answers up to jth round dialogue for the guesser. It will be used to update the guessing state in the next module.

3.3 Update of Guessing State (UoGS)

When a new QA pair is received from the QGen and the Oracle, the Guesser needs to make a decision on which object would be ruled out, or which one would be gained more confidence, then renews its guessing state over objects in the image. Three steps are designed for updating the guessing state.

First, to fuse two types of information from QA and visual objects, we perform an element-wise product of $h_{qa}^{(j)}$ and each object embedding in $O^{(j)}$ to generate a joint feature for any object, as shown in Eq. 6,

$$O_{qa}^{(j)} = h_{qa}^{(j)} \odot O^{(j)}, \tag{6}$$

where $\odot$ denotes element-wise product, $O_{qa}^{(j)} \in \mathbb{R}^{m \times d}$ contains m joint feature objects.

Second, to measure how much the belief changes on ith object after jth dialog round, three feature vectors: the QA pair feature, joint feature of the ith object and updated representation of the ith object are concatenated and passed through a two-layer linear with a tanh activation, followed a softmax layer to produce change of belief as described in Eq. 7,

$$\hat{\pi}_i^{(j)} = \text{softmax}(W_2^T(tanh(W_1^T([h_{qa}^{(j)}; (O_{qa}^{(j)})_i; (O^{(j)})_i])))), \tag{7}$$

where $i \in \{1, 2, \ldots, m\}$, $W_1 \in \mathbb{R}^{1536 \times 128}$ and $W_2 \in \mathbb{R}^{128 \times 1}$ are learnable parameters, the bias b is omitted in the linear layer for simplicity. $\hat{\pi}_i^{(j)} \in [0, 1]$ means the belief changes on ith object after jth round. We find that this type of symmetric concatenation $[;;]$ in Eq. 7, where language and visual information are in a symmetrical position, is an effective way to handle multimodal information, which is also used in [9].

Finally, the previous rounds of guessing state $\pi^{(j-1)}$ are updated by multiplying $\hat{\pi}^{(j)} \in \mathbb{R}^m$ as follows: $\pi^{(j)} = norm(\pi^{(j-1)} \odot \hat{\pi}^{(j)})$. Where $\pi^{(j)} \in \mathbb{R}^m$ is the accumulated guessing state till round j, $norm$ is a sum-normalization method to make it a valid probability distribution, e.g., by dividing the sum of it.

3.4 Early and Incremental Supervision

The introduction of guessing states provides useful information for model training. Because the guessing states are tracked from the beginning of a dialogue, supervision of correct guess can be employed from an early stage, which is called early supervision. Because the guessing states are tracked at each round of dialogue, the change of guessing state can also be supervised to ensure that the guessing is alone in the right way. We call this kind of supervision, incremental supervision. Two supervision functions are introduced as follows.

Early Supervision. Early supervision (ES) tries to maximize the probability of the right object from the beginning of a dialogue and keeps on using up to the penultimate round of the dialogue. It is defined as the summary of a series of cross-entropy between the guessing state and the ground-truth. That is:

$$L_{ES} = \frac{1}{J_{max} - 1} \sum_{j=1}^{J_{max}-1} \text{CrossEntropy}(\pi^{(j)}, y^{GT}), \tag{8}$$

where y^{GT} is a one-hot vector with 1 in the position of the ground-truth object, J_{max} is the maximum number of rounds. The cross-entropy at the final round, i.e. $CrossEntropy(\pi^{(J_{max})}, y^{GT})$, we refer to as **plain supervision** loss (L_{PS} in briefly).

Incremental Supervision. Incremental supervision (IS) tries to keep the probability of the target object in guessing state increasing or nondecreasing as written in:

$$L_{IS} = -\sum_{j=1}^{J_{max}} \log(\pi_{target}^{(j)} - \pi_{target}^{(j-1)} + c), \tag{9}$$

where $\pi_{target}^{(j)}$ denotes the target probability at round j. IS is defined as the change in probability to the ground-truth object before and after a round of dialog. Besides the log function that served as an extra layer of smooth, IS is somewhat similar to the progressive reward used in [26] that is from Guesser but as a reward for training QGen model. c is a parameter that ensures the input to log be positive.

3.5 Training

Our model is trained in two stages, including supervised and reinforcement learning. For supervised learning, the guesser network is trained by minimizing the following objective, $L_{SL}(\theta) = \alpha(L_{ES} + L_{PS}) + (1 - \alpha)L_{IS}$, where α is a balancing parameter. For reinforcement learning, the guesser network is refined by maximizing the reward given in $L_{RL}(\theta) = -E_{\pi_\theta}[\alpha(L_{ES} + L_{PS}) + (1 - \alpha)L_{IS})]$, where π_θ denotes a policy parameterized by θ which associates guessing state over actions, e.g., an action corresponds to select an object over m candidate objects. Following [27], we use the REINFORCE algorithm [21] without baseline that updates policy parameters θ.

4 Experiments and Analysis

4.1 Experimental Setup

Dataset. GuessWhat?! dataset containing 66k images, about 800k question-answer pairs in 150 K games. It is split at random by 70%, 15%, 15% of the games into the training, validation, and test set [19,20].

Baseline Models. A GuessWhat?! game involves Oracle, QGen, and Guesser. Almost all existing work uses the same Oracle model [19,20], which will be used in all our experiments. Two different QGen models are used for validating our guesser model. One is the often used model in previous work [19], the other is a new QGen model which achieves new state-of-the-art [13]. Several different existing Guesser models are compared with our model. They are guesser [1,19, 20], guesser(MN) [2,27], GDSE [16,17], ATT [6] and HACAN [23]. The models are first trained in a supervised way on the training set, and then, one Guesser and one QGen model are jointly refined by reinforcement learning or cooperative learning from self-play with the Oracle model fixed.

Implementation Details. The maximum round J_{max} is set to 5 or 8 as in [13,18,26]. The balancing parameter in loss and reward objectives are set to 0.7, because we observe that our model achieves the minimum error rate on validation and test set when $\alpha = 0.7$. The parameter c in Eq. 9 is set as 1.1. The size of word embedding and LSTM hidden unit number are set to 512. Early stopping is used on the validation set.

We use success rate of guessing for evaluation. Following previous work [13, 19,20], both success rates on NewObject and NewGame are reported. Results by three inference methods described in [2], including Sampling (S), Greedy (G) and Beam-search (BS, beam size is set to 20) are used on both NewObject and NewGame. Following [2,27], during joint reinforcement learning of Guesser and QGen models, only the generated successful games are used to tune the Guesser model, while all the generated games are used to optimize the QGen.

Supervised Learning (SL). We separately train the Guesser and Oracle model for 20 epochs, the QGen for 50 epochs using Adam optimizer [10] with a learning rate of 3e−4 and a batch size of 64.

Reinforcement Learning (RL). We use momentum stochastic gradient descent with a batch size of 64 with 500 epochs and learning rate annealing. The base learning rate is 1e−3 and decayed every 25 epochs with exponential rate 0.99. The momentum parameter is set to 0.9.

4.2 Comparison with the State-of-the-Art

Task Success Rate. Table 1 reports the success rate of guessing with different combinations of QGen and Guesser models with the same Oracle model used in [19,20] for the GuessWhat?! game.

In the first part of Table 1, all models are trained in SL way. We can see that no matter which QGen models are used, qgen [19] or VDST [13], our guesser model GST significantly outperforms other guesser models in both 5 and 8 rounds dialogue at all different settings. Specifically, GST achieves a new state-of-the-art of 54.10% and 50.97% on NewObject and NewGame in Greedy way by SL.

In the second part of Table 1, two combinations trained in cooperative learning (CL) way are given. Our model is not trained in this way. So we do not have a comparison in CL case with the performance of these models are lower than those in the RL part.

In the third part of Table 1, all QGen and Guesser models are trained by RL. We can see that our GST Guesser model combined with the VSDT QGen model achieves the best performance in both 5 and 8 rounds dialogue at all different settings. It significantly outperforms other models. For example, it outperforms the best previous model at Sampling (S) setting on NewObject (i.e. guesser(MN)[27] + ISM [1] with 72.1%) by nearly 9%, outperforms the best previous model at Greedy (G) setting on NewObject (i.e. guesser(MN) [27] + TPG [27] with 74.3%) by more than 9%, outperforms the best previous model in NewObject at Beam-search (BS) setting on NewObject (i.e. guesser [19] + VDST [13] with 71.03%) by more than 12%. The same thing happens on NewGame case. That is to say, our model consistently outperforms previous models in all different settings on both NewObject and NewGame. Especially, GST achieves 83.32% success rate on NewObject in Greedy way, which is approaching human performance 84.4%. Fig. 3(a) shows the learning curve for joint training of GST Guesser and VDST QGen with 500 epochs in Sampling way, it shows superior accuracy compared to the Guesser model [19] trained with VDST QGen.

Table 1. Success rates of guessing (%) with same Oracle (higher is better).

	Questioner		Max Q's	New object			New game		
	Guesser	QGen		S	G	BS	S	G	BS
Pretrained in SL	Guesser [20]	qgen [20]	5	41.6	43.5	47.1	39.2	40.8	44.6
	Guesser (MN) [27]	TPG [27]	8	–	48.77	–	–	–	–
	Guesser [19]	qgen [19]	8	–	44.6	–	–	–	–
	GST (ours)		8	41.73	**44.89**	–	39.97	41.36	–
	Guesser [19]	VDST [13]	5	45.02	49.49	–	42.92	45.94	–
			8	46.70	48.01	–	44.24	45.03	–
	GST (ours)		5	**49.55**	**53.35**	**53.17**	**46.95**	**50.58**	**50.71**
			8	**52.71**	**54.10**	**54.32**	**50.19**	**50.97**	**50.99**
SL	GDSE-SL [17]	GDSE-SL [17]	5	–	–	–	–	47.8	–
			8	–	–	–	–	49.7	–
CL	GDSE-CL [17]	GDSE-CL [17]	5	–	–	–	–	53.7	–
			8	–	–	–	–	58.4	–
AQM	Guesser [11]	randQ [11]	5	–	–	–	–	42.48	–
		countQ [11]	5	–	–	–	–	61.64	–
Trained by RL	Guesser(MN) [27]	TPG[27]	5	62.6	–	–	–	–	–
			8	–	–	–	–	74.3	–
		ISM [1]	–	74.4	–	–	72.1	–	–
		TPG [27]	8	–	74.3	–	–	–	–
		ISD [2]	5	68.3	69.2	–	66.3	67.1	–
	Guesser [19]	VQG [26]	5	63.2	63.6	63.9	59.8	60.7	60.8
		ISM [1]	–	–	64.2	–	–	62.1	–
		ISD [2]	5	61.4	62.1	63.6	59.0	59.8	60.6
		RIG (rewards) [18]	8	65.20	63.00	63.08	64.06	59.0	60.21
		RIG (loss) [18]	8	67.19	63.19	62.57	65.79	61.18	59.79
	Guesser [19]	qgen [19]	5	58.5	60.3	60.2	56.5	58.4	58.4
			8	62.8	58.2	53.9	60.8	56.3	52.0
	Guesser (MN) [27]		5	59.41	60.78	60.28	56.49	58.84	58.10
			8	62.05	62.73	–	59.04	59.50	–
	GST (ours)		5	**64.78**	**67.06**	**67.01**	**61.77**	**64.13**	**64.26**
	Guesser [19]	VDST [13]	5	66.22	67.07	67.81	63.85	64.36	64.44
			8	69.51	70.55	71.03	66.76	67.73	67.52
	GST (ours)		5	**77.38**	**77.30**	**77.23**	**75.11**	**75.20**	**75.13**
			8	**83.22**	**83.32**	**83.46**	**81.50**	**81.55**	**81.62**
	Human [19]	–	–	–	84.4	–	–	84.4	–

Specifically, with same QGen (no matter which QGen models are used, qgen used in [19] or VDST used in [13]), our guesser model GST significantly outperforms other guesser models in both 5 and 8 rounds dialogue at all different settings. It demonstrates that GST is more able to ground a multi-round QA pairs dialogue in the image compared to previous single-step guessing models for the GuessWhat?! game.

Error Rate. For a fair comparison of Guesser models alone, we follow the previous work [6,23] by measuring error rate on training, validation, and test set. In Table 2, we can see that GST trained in SL is comparable to more complex attention algorithms, such as ATT [6] and HACAN [23]. After reinforcement learning, GST model achieves a lower error rate than the compared models in

Table 2. Error rate (%) on the GuessWhat?! dataset (lower is better).

Model	Train err	Val err	Test err	Max Q's
Random [20]	82.9	82.9	82.9	–
LSTM [20]	27.9	37.9	38.7	–
HRED [20]	32.6	38.2	39.0	–
Guesser [19]	–	–	36.2	–
LSTM+VGG [20]	26.1	38.5	39.5	–
HRED+VGG [20]	27.4	38.4	39.6	–
ATT-r2 [6]	29.3	35.7	36.5	–
ATT-r3 [6]	30.5	35.1	35.8	–
ATT-r4 [6]	29.8	35.3	36.3	–
ATT-r3(w2v) [6]	26.7	33.7	34.2	–
Guesser [18]	–	–	35.8	–
HACAN [23]	26.1	32.3	33.2	–
GST (ours, trained in SL)	24.7	33.7	34.3	–
GST (ours, trained in RL)	**22.7**	**23.1**	**24.7**	5
GST (ours, trained in RL)	**16.7**	**16.9**	**18.4**	8
Human [20]	9.0	9.2	9.2	

Table 3. Comparison of success rate with different supervisions in SL.

#	Model	New object	
		S	G
1	GST with ES&PS and IS (full)	52.71	54.10
2	−ES&PS	37.10	42.58
3	−IS	48.96	53.49
		New game	
1	GST with ES&PS and IS (full)	50.19	50.97
2	−ES&PS	34.41	39.48
3	−IS	46.10	50.33

both 5 and 8 rounds, especially at 8 rounds, it obtains error rate of 16.7%, 16.9%, and 18.4%, respectively.

4.3 Ablation Study

Effect of Individual Supervision. In this section, we conduct ablation studies to separate contribution of supervisions: Plain Supervision (PS), Early Supervision (ES) and Incremental Supervision (IS).

Table 3 reports the success rate of guessing after supervised learning. Removing ES& PS from the full model, the game success rate significantly drops 11.52

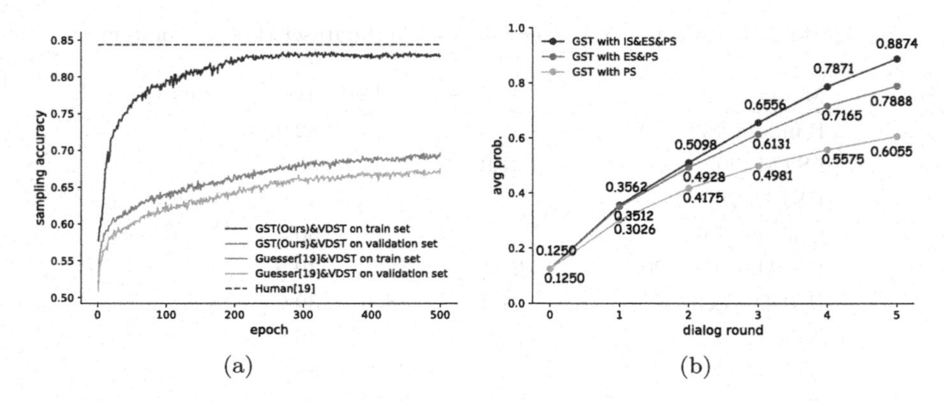

Fig. 3. **a**, Sampling accuracy of reinforcement learning on training and validation set, our GST outperforms guesser [19] by a large margin. **b**, Average belief of the ground-truth object at each round, changes with an increase in the number of dialogue rounds. (Color figure online)

(from 54.10% to 42.58%) and 11.49 (from 50.97% to 39.48%) points on NewObject and NewGame on Greedy case. Removing IS, the success rate drops 0.61 (from 54.10% to 53.49%) and 0.64 (from 50.97% to 50.33%), respectively. It shows that early supervision pair with ES&PS contributes more than incremental supervision.

We then analyze the impact of supervision losses to guessing state. We train three GST models with RL using three different loss functions, i.e. PS, PS&ES, and PS&ES&IS respectively and then count the averaged probability of the ground-truth object based on all the successful games in test set at each round. Fig. 3(b) shows three curves of averaged belief changing with rounds of dialogue. As is observed, we have three notes.

First, guess probability is progressively increasing in all three different losses. It demonstrates our core idea: thinking of the guess as a process instead of a single decision is an effective practical way to ground a multi-round dialogue in an image. Because GST based Guesser makes use of more detailed information in the series of guessing states (GS), i.e. the two losses.

Second, average probability in the blue line, trained with ES&PS, is higher than that in the gray line (trained in PS alone), it demonstrates the effectiveness of early supervision loss.

Third, average probability in the red line, trained with IS&ES&PS, is better than that in the blue line, it shows incremental supervision gives further improvement to guess.

Overall, these results demonstrate the effectiveness of early supervision and incremental supervision. It is the combination of these supervisions that train GST based guesser model efficiently.

Table 4. Error rate of different c in Eq. 9 during SL.

Concat	Train err	Val err	Test err
$[h_{qa}^{(j)}; h_{qa}^{(j)} \odot O^{(j)}; O^{(j)}]$	**24.8**	**33.7**	**34.4**
$[h_{qa}^{(j)} \odot O^{(j)}]$	26.3	35.7	36.7
$[h_{qa}^{(j)}; O^{(j)}]$	27.3	36.5	37.8

Table 5. Comparison of error rate (%) for three types of concatenation during SL.

C	Train	Val	Test
C=1.1	24.7	**33.7**	**34.3**
C=1.5	26.5	34.1	34.8
C=2.0	**23.3**	34.0	34.8

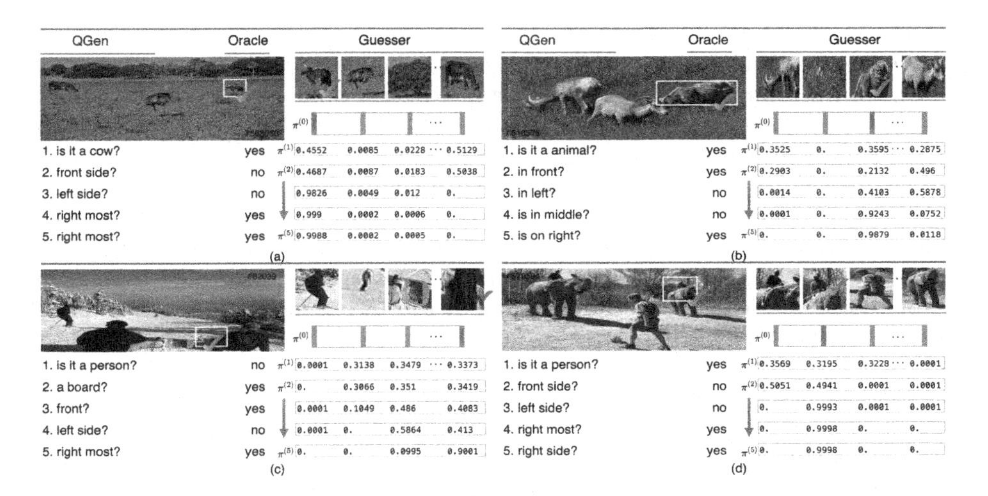

Fig. 4. Four successful games show the process of tracking guessing state.

Effect of Symmetric Concatenation. In Table 4, compared with symmetric concatenation appears in Eq. 7, average error rate increases 2.9 points on all three sets if $[h_{qa}^{(j)}; O^{(j)}]$ used and increases 1.9 points if $[h_{qa}^{(j)} \odot O^{(j)}]$ used. It indicates that symmetric concatenation serves as a valuable part in Eq. 7.

Effect of c in Eq. 9. Table 5 shows error rate of different c in Eq. 9 trained with SL on three dataset. As is observed, c is insensitive to the error rate. We set c to 1.1 as it obtains a lower error rate on Val err and Test err.

4.4 Qualitative Evaluation

In Fig. 4, we show four successful dialogues to visualize the process in guessing. We plot 4 candidate objects for simplicity, $\pi^{(0)}$ represents a uniform distribution of initial guessing state, $\pi^{(1)}$ to $\pi^{(5)}$ show the process of tracking GS. Taking Fig. 4(a) as an example. Guesser has an initial uniform guess on all candidates, i.e. $\pi^{(0)}$. QGen starts a dialogue by asking "is it a cow?", Oracle answer "yes", then Guesser renews its $\pi^{(0)}$ to $\pi^{(1)}$. Specifically, the probabilities on the ostrich and tree approaches go down to close to 0, the cow on both sides increases to

0.45 and 0.51 respectively. At last, all the probabilities are concentrated on the cow on the right with high confidence of 0.9988, which is the guessed object. In Figs. 4(b) to 4(d), three more success cases are shown.

4.5 Discussion on Stop Questioning

When to stop questioning is also a problem in GuessWhat?! like visual dialogue. Most of the previous work chooses a simple policy, i.e., a QGen model stops questioning after a predefined number of dialogue rounds, and the guessing model selects an object as the guess.

Our model can implement this policy by making use of $\pi^{(j)}$, the guessing state after the jth round dialogue. If K is the predefined number, the guesser model will keep on updating $\pi^{(j)}$ till $j = K$. The object with the highest probability in $\pi^{(K)}$ will be then selected as the guess.

A same number of questions are asked for any game under this policy, no matter how different the different games are. The problem of the policy is obvious. On the one hand, the guesser model does not select any object even if it is confident enough about a guess and make a QGen model keep on asking till K questions are asked. On the other hand, the QGen model cannot ask more questions when K questions are asked even if the guesser model is not confident about any guess at that time. The guesser model must give a guess.

Our model can provide a chance to adopt some other policies for stopping questioning. A simple way is to predefine a threshold of confidence. Once the biggest probability in a guessing state is equal to or bigger than the threshold, question answering is stopped, and the guesser model output the object with the biggest probability as the guess. Another way involves the gain of guessing state. Once the information gain from the jth state to the j + 1th state is less than a threshold, the guesser model outputs the object with the biggest probability as the guess.

5 Conclusion

The paper proposes a novel guessing state tracking (GST) based model for the Guesser, which models guess as a process with change of guessing state, instead of making one and only one guess, i.e. a single decision, over the dialogue history in all the previous work. To make full use of the guessing state, two losses, i.e., early supervision loss and incremental supervision loss, are introduced. Experiments show that our GST based guesser significantly outperforms all of the existing methods, and achieves new strong state-of-the-art accuracy that closes the gap to humans, the success rate of guessing 83.3% is approaching the human-level accuracy of 84.4%.

Acknowledgments. We thank the reviewers for their comments and suggestions. This paper is supported by NSFC (No. 61906018), Huawei Noah's Ark Lab and MoE-CMCC "Artificial Intelligence" Project (No. MCM20190701).

References

1. Abbasnejad, E., Wu, Q., Abbasnejad, I., Shi, J., van den Hengel, A.: An active information seeking model for goal-oriented vision-and-language tasks. arXiv preprint arXiv:1812.06398 (2018)
2. Abbasnejad, E., Wu, Q., Shi, J., van den Hengel, A.: What's to know? Uncertainty as a guide to asking goal-oriented questions. In: CVPR (2019)
3. Bani, G., et al.: Adding object detection skills to visual dialogue agents. In: ECCV (2018)
4. Chattopadhyay, P., et al.: Evaluating visual conversational agents via cooperative human-ai games. In: HCOMP (2017)
5. Das, A., et al.: Visual dialog. In: CVPR (2017)
6. Deng, C., Wu, Q., Wu, Q., Hu, F., Lyu, F., Tan, M.: Visual grounding via accumulated attention. In: CVPR (2018)
7. Fukui, A., Park, D.H., Yang, D., Rohrbach, A., Darrell, T., Rohrbach, M.: Multimodal compact bilinear pooling for visual question answering and visual grounding. arXiv preprint arXiv:1606.01847 (2016)
8. Hochreiter, S., Schmidhuber, J.: Long short-term memory. Neural Comput. **9**(8), 1735–1780 (1997)
9. Kim, H., Tan, H., Bansal, M.: Modality-balanced models for visual dialogue. In: AAAI (2020)
10. Kingma, D.P., Ba, J.: Adam: a method for stochastic optimization. In: ICLR (2015)
11. Lee, S.W., Heo, Y.J., Zhang, B.T.: Answerer in questioner's mind: information theoretic approach to goal-oriented visual dialog. In: NeurIPS (2018)
12. Mao, J., Huang, J., Toshev, A., Camburu, O., Yuille, A.L., Murphy, K.: Generation and comprehension of unambiguous object descriptions. In: CVPR, pp. 11–20 (2016)
13. Pang, W., Wang, X.: Visual dialogue state tracking for question generation. In: AAAI (2020)
14. Seo, P.H., Lehrmann, A., Han, B., Sigal, L.: Visual reference resolution using attention memory for visual dialog. In: NeurIPS (2017)
15. Serban, I., Sordoni, A., Bengio, Y., Courville, A., Pineau, J.: Hierarchical neural network generative models for movie dialogues. In: arXiv preprint arXiv:1507.04808 (2015)
16. Shekhar, R., Venkatesh, A., Baumgärtner, T., Bruni, E., Plank, B., Bernardi, R., Fernández, R.: Ask no more: deciding when to guess in referential visual dialogue. In: COLING (2018)
17. Shekhar, R., et al.: Beyond task success: a closer look at jointly learning to see, ask, and guesswhat. In: NAACL (2019)
18. Shukla, P., Elmadjian, C., Sharan, R., Kulkarni, V., Wang, W.Y., Turk, M.: What should I ask? Using conversationally informative rewards for goal-oriented visual dialogue. In: ACL (2019)
19. Strub, F., de Vries, H., Mary, J., Piot, B., Courville, A., Pietquin, O.: End-to-end optimization of goal-driven and visually grounded dialogue systems. In: IJCAI (2017)
20. de Vries, H., Strub, F., Chandar, S., Pietquin, O., Larochelle, H., Courville, A.C.: Guesswhat?! Visual object discovery through multi-modal dialogue. In: CVPR (2017)
21. Williams, R.J.: Simple statistical gradient-following algorithms for connectionist reinforcement learning. Mach. Learn. **8**(3–4), 229–256 (1992). https://doi.org/10.1007/978-1-4615-3618-5_2

22. Xiao, F., Sigal, L., Lee, Y.J.: Weakly-supervised visual grounding of phrases with linguistic structures. In: CVPR (2017)
23. Yang, T., Zha, Z.J., Zhang, H.: Making history matter: history-advantage sequence training for visual dialog. In: ICCV (2019)
24. Yu, L., Poirson, P., Yang, S., Berg, A.C., Berg, T.L.: Modeling context in referring expressions. In: Leibe, B., Matas, J., Sebe, N., Welling, M. (eds.) ECCV 2016. LNCS, vol. 9906, pp. 69–85. Springer, Cham (2016). https://doi.org/10.1007/978-3-319-46475-6_5
25. Yu, L., Tan, H., Bansal, M., Berg, T.L.: A joint speaker-listener-reinforcer model for referring expressions. In: CVPR (2017)
26. Zhang, J., Wu, Q., Shen, C., Zhang, J., Lu, J., van den Hengel, A.: Asking the difficult questions: goal-oriented visual question generation via intermediate rewards. In: ECCV (2018)
27. Zhao, R., Tresp, V.: Improving goal-oriented visual dialog agents via advanced recurrent nets with tempered policy gradient. In: IJCAI (2018)
28. Zhao, R., Tresp, V.: Efficient visual dialog policy learning via positive memory retention. In: NeurIPS (2018)

Memory-Efficient Incremental Learning Through Feature Adaptation

Ahmet Iscen[1(✉)], Jeffrey Zhang[2], Svetlana Lazebnik[2], and Cordelia Schmid[1]

[1] Google Research, Meylan, France
iscen@google.com
[2] University of Illinois at Urbana-Champaign, Champaign, USA

Abstract. We introduce an approach for incremental learning that preserves feature descriptors of training images from previously learned classes, instead of the images themselves, unlike most existing work. Keeping the much lower-dimensional feature embeddings of images reduces the memory footprint significantly. We assume that the model is updated incrementally for new classes as new data becomes available sequentially. This requires adapting the previously stored feature vectors to the updated feature space without having access to the corresponding original training images. Feature adaptation is learned with a multi-layer perceptron, which is trained on feature pairs corresponding to the outputs of the original and updated network on a training image. We validate experimentally that such a transformation generalizes well to the features of the previous set of classes, and maps features to a discriminative subspace in the feature space. As a result, the classifier is optimized jointly over new and old classes without requiring old class images. Experimental results show that our method achieves state-of-the-art classification accuracy in incremental learning benchmarks, while having at least an order of magnitude lower memory footprint compared to image-preserving strategies.

1 Introduction

Deep neural networks have shown excellent performance for many computer vision problems, such as image classification [15,22,37] and object detection [14, 31]. However, most common models require large amounts of labeled data for training, and assume that data from all possible classes is available for training at the same time.

By contrast, *class incremental learning* [30] addresses the setting where training data is received sequentially, and data from previous classes is discarded as data for new classes becomes available. Thus, classes are not learned all at once. Ideally, models should learn the knowledge from new classes while maintaining

Electronic supplementary material The online version of this chapter (https://doi.org/10.1007/978-3-030-58517-4_41) contains supplementary material, which is available to authorized users.

A. Vedaldi et al. (Eds.): ECCV 2020, LNCS 12361, pp. 699–715, 2020.
https://doi.org/10.1007/978-3-030-58517-4_41

the knowledge learned from previous classes. This poses a significant problem, as neural networks are known to quickly forget what is learned in the past – a phenomenon known as *catastrophic forgetting* [27]. Recent approaches alleviate catastrophic forgetting in neural networks by adding regularization terms that encourage the network to stay similar to its previous states [20,23] or by preserving a subset of previously seen data [30].

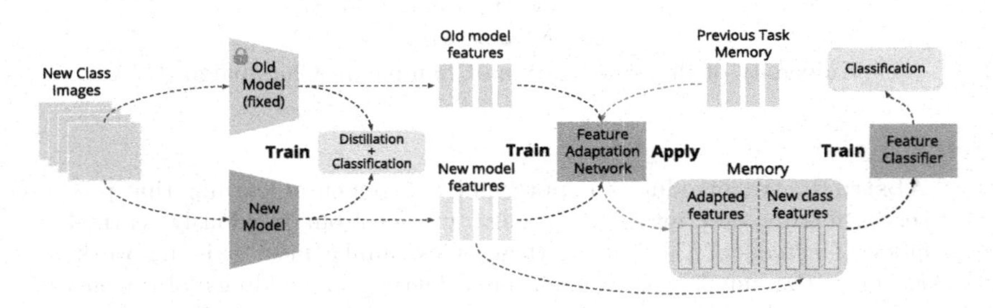

Fig. 1. An overview of our method. Given new class images, a new model is trained on the data with distillation and classification losses. Features are extracted using the old and new models from new class images to train a feature adaptation network. The learned feature adaptation network is applied to the preserved vectors to transform them into the new feature space. With features from all seen classes represented in the same feature space, we train a feature classifier.

One of the criteria stated by Rebuffi *et al.* [30] for a successful incremental learner is that "computational requirements and memory footprint should remain bounded, or at least grow very slowly, with respect to the number of classes seen so far". In our work, we significantly improve the memory footprint required by an incremental learning system. We propose to preserve a subset of *feature descriptors* rather than images. This enables us to compress information from previous classes in low-dimensional embeddings. For example, for ImageNet classification using ResNet-18, storing a 512-dimensional feature vector has $\sim$1% of the storage requirement compared to storing a $256 \times 256 \times 3$ image (Sect. 5.3). Our experiments show that we achieve better classification accuracy compared to state-of-the-art methods, with a memory footprint of at least an order of magnitude less.

Our strategy of preserving feature descriptors instead of images faces a serious potential problem: as the model is trained with more classes, the feature extractor changes, making the preserved feature descriptors from previous feature extractors obsolete. To overcome this difficulty, we propose a *feature adaptation* method that learns a mapping between two feature spaces. As shown in Fig. 1, our novel approach allows us to learn the changes in the feature space and adapt the preserved feature descriptors to the new feature space. With all image features in the same feature space, we can train a feature classifier to correctly classify features from all seen classes. To summarize, our contributions in this paper are as follows:

- We propose an incremental learning framework where previous feature descriptors, instead of previous images, are preserved.
- We present a *feature adaptation* approach which maps previous feature descriptors to their correct values as the model is updated.
- We apply our method on popular class-incremental learning benchmarks and show that we achieve top accuracy on ImageNet compared to other state-of-the-art methods while significantly reducing the memory footprint.

2 Related Work

The literature for *incremental learning* prior to the deep-learning era includes incrementally trained support vector machines [5], random forests [32], and metric-based methods that generalize to new classes [28]. We restrict our attention mostly to more recent deep-learning-based methods. Central to most of these methods is the concept of *rehearsal*, which is defined as preserving and replaying data from previous sets of classes when updating the model with new classes [33].

Non-rehearsal methods do not preserve any data from previously seen classes. Common approaches include increasing the network capacity for new sets of classes [35,38], or weight consolidation, which identifies the important weights for previous sets of classes and slows down their learning [20]. Chaudhry *et al.* [6] improve weight consolidation by adding KL-divergence-based regularization. Liu *et al.* [24] rotate the parameter space of the network and show that the weight consolidation is more effective in the rotated parameter space. Aljundi *et al.* [1] compute the importance of each parameter in an unsupervised manner without labeled data. Learning without Forgetting (LwF) [23] (discussed in more detail in Sect. 3) reduces catastrophic forgetting by adding a knowledge distillation [16] term in the loss function, which encourages the network output for new classes to be close to the original network output. Learning without Memorizing [8] extends LwF by adding a distillation term based on attention maps. Zhang *et al.* [45] argue that LwF produces models that are either biased towards old or new classes. They train a separate model for new classes, and consolidate the two models with unlabeled auxiliary data. Lastly, Yu *et al.* [44] updates previous class centroids for NME classification [30] by estimating the feature representation shift using new class centroids.

Rehearsal with Exemplars. Lopez-Paz and Ranzato [25] add constraints on the gradient update, and transfer information to previous sets of classes while learning new sets of classes. Incremental Classifier and Representation Learning (iCARL) by Rebuffi *et al.* [30] preserves a subset of images, called *exemplars*, and includes the selected subset when updating the network for new sets of classes. Exemplar selection is done with an efficient algorithm called herding [39]. The authors also show that the classification accuracy increases when the mean class vector [28] is used for classification instead of the learned classifier of the network. iCARL is one of the most effective existing methods in the literature, and will be considered as our main baseline. Castro *et al.* [4] extend iCARL by learning the network and classifier with an end-to-end approach. Similarly, Javed and Shafait [18] learn an end-to-end classifier by proposing a dynamic threshold moving algorithm. Other recent work extend iCARL by correcting the bias and introducing additional constraints in the loss function [2,17,41].

Rehearsal with Generated Images. These methods use generative models (GANs [10]) to generate *fake* images that mimic the past data, and use the generated images when learning the network for new classes [36,40]. He *et al.* [13] use multiple generators to increase capacity as new sets of classes become available. A major drawback of these methods is that they are either applied to less complex datasets with low-resolution images, or their success depends on combining the generated images with real images.

Feature-Based Methods. Earlier work on *feature generation*, rather than image generation, focuses on zero-shot learning [3,42]. Kemker *et al.* [19] use a dual-memory system which consists of fast-learning memory for new classes and long-term storage for old classes. Statistics of feature vectors, such as the mean vector and covariance matrix for a set of vectors, are stored in the memory. Xiang *et al.* [43] also store feature vector statistics, and learn a feature generator to generate vectors from old classes. The drawback of these methods [19,43] is that they depend on a pre-trained network. This is different than other methods (LwF, iCARL) where the network is learned from scratch.

In this paper, we propose a method which performs *rehearsal* with *features*. Unlike existing feature-based methods, we do not *generate* feature descriptors from class statistics. We preserve and adapt feature descriptors to new feature spaces as the network is trained incrementally. This allows training the network from scratch and does not depend on a pre-trained model as in [19,43]. Compared to existing rehearsal methods, our method has a significantly lower memory footprint by preserving features instead of images.

Our feature adaptation method is inspired by the feature hallucinator proposed by Hariharan and Girschick [12]. Their method learns intra-class feature transformations as a way of data augmentation in few-shot learning problem. Our method is quite different as we learn the transformations between feature pairs of the same image, extracted at two different increments of the network. Finally, whereas Yu *et al.* [44] uses interpolation to estimate changes for class centroids of features, our feature adaptation method learns a generalizable transformation function for all stored features.

3 Background on Incremental Learning

This section introduces the incremental learning task and summarizes popular strategies for training the network and handling *catastrophic forgetting*, namely, distillation and preservation of old data.

Problem formulation. We are given a set of images $\mathcal{X}$ with labels $\mathcal{Y}$ belonging to classes in $\mathcal{C}$. This defines the dataset $\mathcal{D} = \{(x, y) | x \in \mathcal{X}, y \in \mathcal{Y}\}$. In class-incremental learning, we want to expand an existing model to classify new classes. Given T tasks, we split $\mathcal{C}$ into T subsets $\mathcal{C}^1, \mathcal{C}^2, \ldots, \mathcal{C}^T$, where $\mathcal{C} = \mathcal{C}^1 \cup \mathcal{C}^2 \cup \cdots \cup \mathcal{C}^T$ and $\mathcal{C}^i \cap \mathcal{C}^j = \emptyset$ for $i \neq j$. We define task t as introducing new classes $\mathcal{C}^t$ using dataset $\mathcal{D}^t = \{(x, y) | y \in \mathcal{C}^t\}$. We denote $\mathcal{X}^t = \{x | (x, y) \in \mathcal{D}^t\}$ and $\mathcal{Y}^t = \{y | (x, y) \in \mathcal{D}^t\}$ as the training images and labels used at task t. The goal is to train a classifier which accurately classifies examples belonging to the new set of classes $\mathcal{C}^t$, while still being able to correctly classify examples belonging to classes $\mathcal{C}^i$, where $i < t$.

The Classifier. The learned classifier is typically a *convolutional neural network* (CNN) denoted by $f_{\theta,W} : \mathcal{X} \to \mathbb{R}^K$, where K is the number of classes. Learnable parameters θ and W correspond to two components of the network, the *feature extractor* h_θ and the *network classifier* g_W. The feature extractor $h_\theta : \mathcal{X} \to \mathbb{R}^d$ maps an image to a d-dimensional feature vector. The network classifier $g_W : \mathbb{R}^d \to \mathbb{R}^K$ is applied to the output of the feature extractor h_θ and outputs a K-dimensional vector for each class classification score. The network $f_{\theta,W}$ is the mapping from the input space directly to confidence scores, where $x \in \mathcal{X}$:

$$f_{\theta,W}(x) := g_W(h_\theta(x)). \tag{1}$$

Training the parameters θ and W of the network is typically achieved through a loss function, such as cross-entropy loss,

$$L_{\mathrm{CE}}(x, y) := -\sum_{k=1}^{K} y_k \log(\boldsymbol{\sigma}(f_{\theta,W}(x))_k), \tag{2}$$

where $y \in \mathbb{R}^K$ is the label vector and $\boldsymbol{\sigma}$ is either a *softmax* or *sigmoid* function.

In incremental learning, the number of classes our models output increases at each task. $K^t = \sum_i^t |\mathcal{C}^i|$ denotes the total number of classes at task t. Notice at task t, our model is expected to classify $|\mathcal{C}^t|$ more classes than task $t-1$. The network $f_{\theta,W}^t$ is only trained with $\mathcal{X}^t$, the data available in the current task. Nevertheless, the network is still expected to accurately classify any images belonging to the classes from the previous tasks.

Distillation. One of the main challenges in incremental learning is *catastrophic forgetting* [11,27]. At a given task t, we want to expand a previous model's capability to classify new classes $\mathcal{C}^t$. We train a new model $f_{\theta,W}^t$ initialized from $f_{\theta,W}^{t-1}$. Before the training of the task, we freeze a copy of $f_{\theta,W}^{t-1}$ to use as reference. We only have access to $\mathcal{X}^t$ and not to previously seen data $\mathcal{X}^i$, where $i < t$. As the network is updated with $\mathcal{X}^t$ in Eq. (2), its knowledge of previous tasks quickly disappears due to catastrophic forgetting. *Learning without Forgetting* (LwF) [23] alleviates this problem by introducing a *knowledge distillation loss* [16]. This loss is a modified cross-entropy loss, which encourages the network $f_{\theta,W}^t$ to mimic the output of the previous task model $f_{\theta,W}^{t-1}$:

$$L_{\mathrm{KD}}(x) := -\sum_{k=1}^{K^{t-1}} \boldsymbol{\sigma}(f_{\theta,W}^{t-1}(x))_k \log(\boldsymbol{\sigma}(f_{\theta,W}^t(x))_k), \tag{3}$$

where $x \in \mathcal{X}^t$. $L_{\mathrm{KD}}(x)$ encourages the network to make similar predictions to the previous model. The knowledge distillation loss term is added to the classification loss (2), resulting in the overall loss function:

$$L(x, y) := L_{\mathrm{CE}}(x, y) + \lambda L_{\mathrm{KD}}(x), \quad (4)$$

where λ is typically set to 1 [30]. Note that the network $f^t_{\theta,W}$ is continuously updated at task t, whereas the network $f^{t-1}_{\theta,W}$ remains frozen and will not be stored after the completion of task t.

Preserving Data of the Old Classes. A common approach is to preserve some images for the old classes and use them when training new tasks [30]. At task t, new class data refers to $\mathcal{X}^t$ and old class data refers to data seen in previous tasks, *i.e.* $\mathcal{X}^i$ where $i < t$. After each task t, a new *exemplar* set $\mathcal{P}^t$ is created from $\mathcal{X}^t$. *Exemplar* images in $\mathcal{P}^t$ are the selected subset of images used in training future tasks. Thus, training at task t uses images $\mathcal{X}^t$ and $\mathcal{P}^i$, where $i < t$.

Training on this additional old class data can help mitigate the effect of catastrophic forgetting for previously seen classes. In iCARL [30] the exemplar selection used to create $\mathcal{P}^t$ is done such that the selected set of exemplars should approximate the class mean vector well, using a *herding* algorithm [39]. Such an approach can bound the memory requirement for stored examples.

4 Memory-Efficient Incremental Learning

Our goal is to preserve compact feature descriptors, *i.e.* $\mathbf{v} := h_\theta(x)$, instead of images from old classes. This enables us to be significantly more memory-efficient, or to store more examples per class given the same memory requirement.

The major challenge of preserving only the feature descriptors is that it is not clear how they would evolve over time as the feature extractor h_θ is trained on new data. This introduces a problem for the new tasks, where we would like to use all preserved feature descriptors to learn a feature classifier $g_{\tilde{W}}$ on all classes jointly. Preserved feature descriptors are *not compatible* with feature descriptors from the new task because h_θ is different. Furthermore, we cannot re-extract feature descriptors from h_θ if we do not have access to old images.

We propose a *feature adaptation* process, which directly updates the feature descriptors as the network changes with a *feature adaptation network* ϕ_ψ. During training of each task, we first train the CNN using classification and distillation losses (Sect. 4.1). Then, we learn the feature adaptation network (Sect. 4.2) and use it to adapt stored features from previous tasks to the current feature space. Finally, a feature classifier $g_{\tilde{W}}$ is learned with features from both the current task and the adapted features from the previous tasks (Sect. 4.3). This feature classifier $g_{\tilde{W}}$ is used to classify the features extracted from test images and is independent from the network classifier g_W, which is used to train the network. Fig. 1 gives a visual overview of our approach (see also Algorithm 1 in Appendix A). We describe it in more detail in the following.

4.1 Network Training

This section describes the training of the backbone convolutional neural network $f_{\theta,W}$. Our implementation follows the same training setup as in Sect. 3 with two additional components: *cosine normalization* and *feature distillation.*

Cosine normalization was proposed in various learning tasks [23, 26], including incremental learning [17]. The prediction of the network (1) is based on cosine similarity, instead of simple dot product. This is equivalent to $\hat{W}^\top \hat{\mathbf{v}}$, where $\hat{W}$ is the column-wise ℓ_2-normalized counterpart of parameters W of the classifier, and $\hat{\mathbf{v}}$ is the ℓ_2-normalized counterpart of the feature $\mathbf{v}$.

Feature distillation is an additional distillation term based on feature descriptors instead of logits. Similar to (3), we add a constraint in the loss function which encourages the new feature extractor h_θ^t to mimic the old one h_θ^{t-1}:

$$L_{\text{FD}}(x) := 1 - \cos(h_\theta^t(x), h_\theta^{t-1}(x)), \tag{5}$$

where $x \in \mathcal{X}^t$ and h_θ^{t-1} is the frozen feature extractor from the previous task. The feature distillation loss term is minimized together with other loss terms,

$$L(x, y) := L_{\text{CE}}(x, y) + \lambda L_{\text{KD}}(x) + \gamma L_{\text{FD}}(x), \tag{6}$$

where γ is a tuned hyper-parameter. We study its impact in Sect. 5.4.

Feature distillation has already been applied in incremental learning as a replacement for the knowledge distillation loss (3), but only to the feature vectors of preserved images [17]. It is also similar in spirit to *attention distillation* [8], which adds a constraint on the attention maps produced by the two models.

Cosine normalization and feature distillation improve the accuracy of our method and the baselines. The practical impact of these components will be studied in more detail in Sect. 5.

4.2 Feature Adaptation

Overview. Feature adaptation is applied after CNN training at each task. We first describe feature adaptation for the initial two tasks and then extend it to subsequent tasks. At task $t = 1$, the network is trained with images $\mathcal{X}^1$ belonging to classes $\mathcal{C}^1$. After the training is complete, we extract feature descriptors $\mathcal{V}^1 = \{(h_\theta^1(x) | x \in \mathcal{X}^1\}$, where $h_\theta^1(x)$ refers to the feature extractor component of $f_{\theta,W}^1$. We store these features in memory $\mathcal{M}^1 = \mathcal{V}^1$ after the first task[1]. We also reduce the number of features stored in $\mathcal{M}^1$ to fit specific memory requirements, which is explained later in the section. At task $t = 2$, we have a new set of images $\mathcal{X}^2$ belonging to new classes $\mathcal{C}^2$. The network $f_{\theta,W}^2$ is initialized from $f_{\theta,W}^1$, where $f_{\theta,W}^1$ is fixed and kept as reference during training with distillation (6). After the training finishes, we extract features $\mathcal{V}^2 = \{(h_\theta^2(x) | x \in \mathcal{X}^2\}$.

We now have two sets of features, $\mathcal{M}^1$ and $\mathcal{V}^2$ extracted from two tasks that correspond to different sets of classes. Importantly, $\mathcal{M}^1$ and $\mathcal{V}^2$ are extracted

[1] We also store the corresponding label information.

with different feature extractors, h_θ^1 and h_θ^2, respectively. Hence, the two sets of vectors lie in different feature spaces and are not compatible with each other. Therefore, we must transform features $\mathcal{M}^1$ to the same feature space as $\mathcal{V}^2$. We train a feature adaptation network $\phi_\psi^{1\to 2}$ to map $\mathcal{M}^1$ to the same space as $\mathcal{V}^2$ (training procedure described below).

Once the feature adaptation network is trained, we create a new memory set $\mathcal{M}^2$ by transforming the existing features in the memory $\mathcal{M}^1$ to the same feature space as $\mathcal{V}^2$, *i.e.* $\mathcal{M}^2 = \mathcal{V}^2 \cup \phi_\psi^{1\to 2}(\mathcal{M}^1)$. The resulting $\mathcal{M}^2$ contains new features from the current task and adapted features from the previous task, and can be used to learn a discriminative feature classifier explained in Sect. 4.3. $\mathcal{M}^1$ and $f_{\theta,W}^1$ are no longer stored for future tasks.

We follow the same procedure for subsequent tasks $t > 2$. We have a new set of data with images $\mathcal{X}^t$ belonging to classes $\mathcal{C}^t$. Once the network training is complete after task t, we extract features descriptors $\mathcal{V}^t = \{(h_\theta^t(x)|x \in \mathcal{X}^t\}$. We train a feature adaptation network $\phi_\psi^{(t-1)\to t}$ and use it to create $\mathcal{M}^t = \mathcal{V}^t \cup \phi_\psi^{(t-1)\to t}(\mathcal{M}^{t-1})$. The memory set $\mathcal{M}^t$ will have features stored from *all classes* $\mathcal{C}^i$, $i \leq t$, transformed to the current feature space of h_θ^t. $\mathcal{M}^{t-1}$ and $f_{\theta,W}^{t-1}$ are no longer needed for future tasks.

Training the Feature Adaptation Network ϕ_ψ. At task t, we transform $\mathcal{V}^{t-1}$ to the same feature space as $\mathcal{V}^t$. We do this by learning a transformation function $\phi_\psi^{(t-1)\to t} : \mathbb{R}^d \to \mathbb{R}^d$, that maps output of the previous feature extractor h_θ^{t-1} to the current feature extractor h_θ^t using the current task images $\mathcal{X}^t$.

Let $\mathcal{V}' = \{(h_\theta^{t-1}(x), h_\theta^t(x))|x \in \mathcal{X}^t\}$ and $(\overline{\mathbf{v}}, \mathbf{v}) \in \mathcal{V}'$. In other words, given an image $x \in \mathcal{X}^t$, $\overline{\mathbf{v}}$ corresponds to its feature extracted with $h_\theta^{t-1}(x)$, the state of the feature extractor after task $t-1$. On the other hand, $\mathbf{v}$ corresponds to the feature representation of the same image x, but extracted with the model at the end of the current task, *i.e.* $h_\theta^t(x)$. Finding a mapping between $\overline{\mathbf{v}}$ and $\mathbf{v}$ allows us to map other features in $\mathcal{M}^{t-1}$ to the same feature space as $\mathcal{V}^t$.

When training the *feature adaptation network* $\phi_\psi^{(t-1)\to t}$, we use a similar loss function as the feature hallucinator [12]:

$$L_{\text{fa}}(\overline{\mathbf{v}}, \mathbf{v}, y) := \alpha L_{\text{sim}}(\mathbf{v}, \phi_\psi(\overline{\mathbf{v}})) + L_{\text{cls}}(g_W, \phi_\psi(\overline{\mathbf{v}}), y), \tag{7}$$

where y is the corresponding label to $\mathbf{v}$. The first term $L_{\text{sim}}(\mathbf{v}, \phi_\psi(\overline{\mathbf{v}})) = 1 - \cos(\mathbf{v}, \phi_\psi(\overline{\mathbf{v}}))$ encourages the adapted feature descriptor $\phi_\psi(\overline{\mathbf{v}})$ to be similar to $\mathbf{v}$, its counterpart extracted from the updated network. Note that this is the same loss function as feature distillation (5). The purpose of this method is transforming features between different feature spaces, whereas feature distillation is helpful by preventing features from drifting too much in the feature space. The practical impact of feature distillation will be presented in more detail in Sect. 5.4. The second loss term $L_{\text{cls}}(g_W, \phi_\psi(\overline{\mathbf{v}}), y)$ is the cross-entropy loss and g_W is the *fixed* network classifier of the network $f_{\theta,W}$. This term encourages adapted feature descriptors to belong to the correct class y.

Reducing the Size of $\mathcal{M}^t$. The number of stored vectors in memory $\mathcal{M}^t$ can be reduced to satisfy specified memory requirements. We reduce the number of features in the memory by *herding* [30, 39]. Herding is a greedy algorithm that chooses the subset of features that best approximates the class mean. When updating the memory after task t, we use herding to only keep a fixed number (L) of features per class, *i.e.* $\mathcal{M}^t$ has L vectors per class.

4.3 Training the Feature Classifier $g_{\tilde{W}}$

Our goal is to classify unseen test images belonging to $K^t = \sum_{i=1}^{t} |\mathcal{C}^i|$ classes, which includes classes from previously seen tasks. As explained in Sect. 3, the learned network $f^t_{\theta,W}$ is a mapping from images to K^t classes and can be used to classify test images. However, training only on $\mathcal{X}^t$ images results in sub-optimal performance, because the previous tasks are still forgotten to an extent, even when using distillation (5) during training. We leverage the preserved *adapted* feature descriptors from previous tasks to learn a more accurate feature classifier.

At the end of task t, a new feature classifier $g^t_{\tilde{W}}$ is trained with the memory $\mathcal{M}^t$, which contains the adapted feature descriptors from previous tasks as well as feature descriptors from the current task. This is different than the network classifier g^t_W, which is a part of the network $f^t_{\theta,W}$. When given a test image, we extract its feature representation with h^t_θ and classify it using the feature classifier $g^t_{\tilde{W}}$. In practice, $g^t_{\tilde{W}}$ is a linear classifier which can be trained in various ways, *e.g.* linear SVM, SGD *etc.* We use Linear SVMs in our experiments.

5 Experiments

We describe our experimental setup, then show our results on each dataset in terms of classification accuracy. We also measure the quality of our feature adaptation method, which is independent of the classification task. Finally, we study in detail the impact of key implementation choices and parameters.

5.1 Experimental Setup

Datasets. We use CIFAR-100 [21], ImageNet-100 and ImageNet-1000 in our experiments. ImageNet-100 [30] is a subset of the ImageNet-1000 dataset [34] containing 100 randomly sampled classes from the original 1000 classes. We follow the same setup as iCARL [30]. The network is trained in a class-incremental way, only considering the data available at each task. We denote the number of classes at each task by M, and total number of tasks by T. After each task, classification is performed on all classes seen so far. Every CIFAR-100 and ImageNet-100 experiment was performed 5 times with random class orderings. Reported results are averaged over all 5 runs.

Two evaluation metrics are reported. The first is a curve of classification accuracies on all classes that have been trained after each task. The second is the *average incremental accuracy*, which is the average of points in first metric. Top-1 and top-5 accuracy is computed for CIFAR-100 and ImageNet respectively.

Baselines. Our main baselines are given by the two methods in the literature that we extend. Learning Without Forgetting (LwF) [23] does not preserve any data from earlier tasks and is trained with classification and distillation loss terms (4). We use the multi-class version (LwF.MC) proposed by Rebuffi *et al.* [30]. iCARL [30] extends LwF.MC by preserving representative training images of previously seen classes. All experiments are reported with our implementation unless specified otherwise. Rebuffi *et al.* [30] fix the total number of exemplars stored at any point, and change the number of exemplars per class depending on the total number of classes. Unlike the original iCARL, we fix the number of exemplars per class as P in our implementation (as in [17]). We extend the original implementations of iCARL and LwF by applying cosine normalization and feature distillation loss (see Sect. 4.1), as these variants have shown to improve the accuracy. We refer to the resulting variants as γ-iCARL and γ-LwF respectively (γ is the parameter that controls the feature distillation in Eq. (5)).

Implementation Details. The feature extraction network h_θ is Resnet-32 [15] ($d = 64$) for CIFAR100 and Resnet-18 [15] ($d = 512$) for ImageNet-100 and ImageNet-1000. We use a Linear SVM [7,29] for our feature classifier $g_{\tilde{W}}$. The feature adaptation network ϕ_ψ is a 2-layer multilayer perceptron (MLP) with ReLU [9] activations and d input/output and $d' = 16d$ hidden dimensions. We use binary cross-entropy for the loss function (4), and λ for the knowledge distillation (4) is set to 1. Consequently, the activation function $\boldsymbol{\sigma}$ is sigmoid. We use the same hyper-parameters as Rebuffi *et al.* [30] when training the network, a batch size of 128, weight decay of 1e−5, and learning rate of 2.0. In CIFAR-100, we train the network for 70 epochs at each task, and reduce the learning rate by a factor of 5 at epochs 50 and 64. For ImageNet experiments, we train the network for 60 epochs at each task, and reduce the learning rate by a factor of 5 at epochs 20, 30, 40 and 50.

5.2 Impact of Memory Footprint

Our main goal is to improve the memory requirements of an incremental learning framework. We start by comparing our method against our baselines in terms of memory footprint. Figure 2 shows the memory required by each method and the corresponding average incremental accuracy. The memory footprint is all the preserved data (features or images) for all classes of the dataset. Memory footprint for γ-iCARL is varied by changing P, the fixed number of images preserved for each class. Memory footprint for our method is varied by changing L, the fixed number of feature descriptors per class (Sect. 4.2). We also present *Ours-hybrid*, a variant of our method where we keep P images and L feature descriptors. In this variant, we vary P to fit specified memory requirements.

Figure 2 shows average incremental accuracy for different memory usage on CIFAR-100, ImageNet-100 and ImageNet-1000. Note that while our method still achieves higher or comparable accuracy compared on CIFAR-100, the memory savings are less significant. That is due to the fact that images have lower resolution ($32 \times 32 \times 3$ uint8, 3.072KB) and preserving feature descriptors ($d = 64$

floats, 0.256KB) has less impact on the memory in that dataset. However, due to the lower computation complexity of training on CIFAR-100, we use CIFAR-100 to tune our hyperparameters (Sect. 5.4). The memory savings with our method are more significant for ImageNet. The resolution of each image in ImageNet is $256 \times 256 \times 3$, *i.e.*, storing a single uint8 image in the memory takes 192 KB. Keeping a feature descriptor of $d = 512$ floats is significantly cheaper; it only requires 2 KB. This is about ~1% of the memory required for an image. Note there are many compression techniques for both images and features (*e.g.* JPEG, HDF5, PCA). Our analysis will solely focus on uncompressed data.

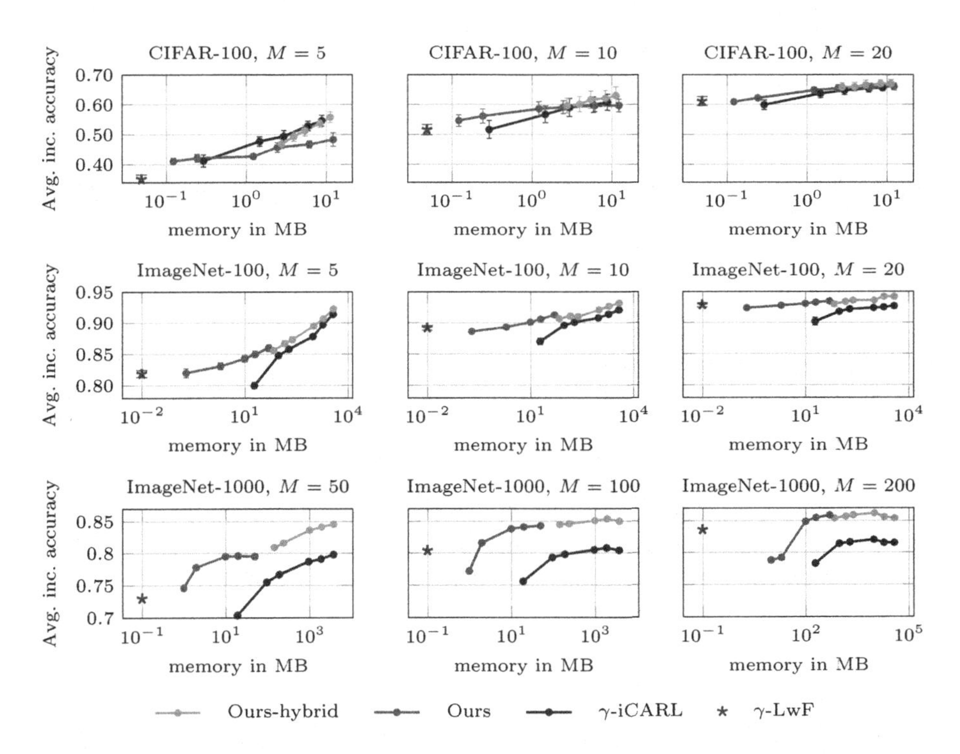

Fig. 2. Memory (in MB) vs average incremental accuracy on CIFAR-100, ImageNet-100 and ImageNet-1000 for different number of classes per task (M). We vary the memory requirement for our method and γ-iCARL by changing the number of preserved feature descriptors (L) and images (P) respectively. For Ours-hybrid, we set $L = 100$ (CIFAR) and $L = 250$ (ImageNet-100 and ImageNet-1000) and vary P.

We achieve the same accuracy with significantly less memory compared to γ-iCARL on ImageNet datasets. The accuracy of our method is superior to γ-iCARL when $M \geq 100$ on ImageNet-1000. Memory requirements are at least an order of magnitude less in most cases. Ours-hybrid shows that we can preserve features with smaller number of images and further improve accuracy. This results in higher accuracy compared to γ-iCARL for the same memory footprint.

Figure 3 shows the accuracy for different number of preserved *data points* on ImageNet-1000, where *data points* refer to features for our method and images for γ-iCARL. Our method outperforms γ-iCARL in most cases, even if ignoring the memory savings of features compared to images.

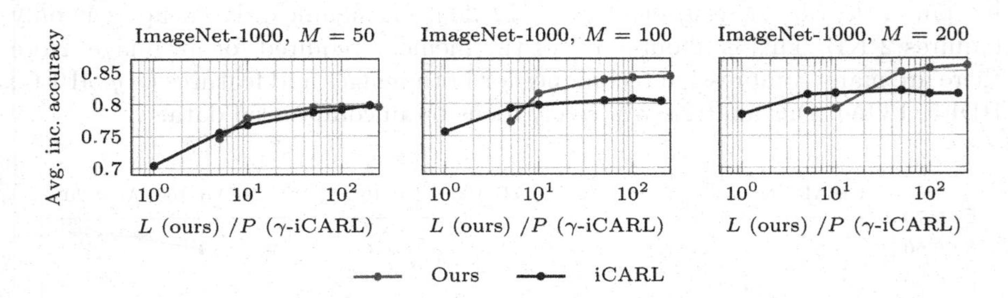

Fig. 3. Impact of L, the number of features stored per class for ours, and P, the number of images stored per class for γ-iCARL. M is the number of classes per task.

5.3 Comparison to State of the Art

Table 1 shows the total memory cost of the preserved data and average incremental accuracy of our method and existing works in the literature. Accuracy per task is shown in Fig. 4. We report the average incremental accuracy for Ours when preserving $L = 250$ features per class. Ours-hybrid preserves $L = 250$ features and $P = 10$ images per class. Baselines and state-of-the-art methods preserve $P = 20$ images per class. It is clear that our method, which is the first work storing and adapting feature descriptors, consumes significantly less memory than the other methods while improving the classification accuracy.

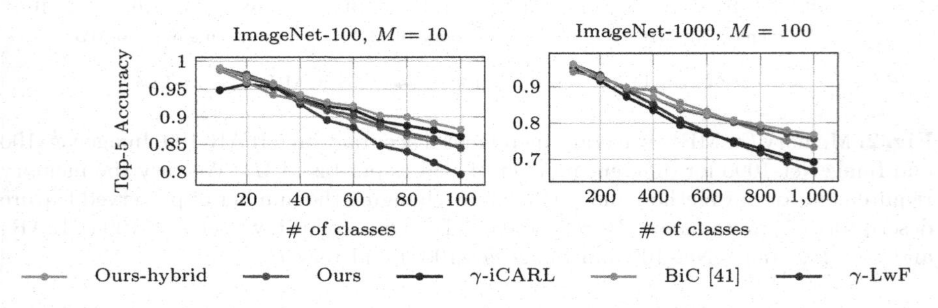

Fig. 4. Classification curves of our method and state-of-the-art methods on ImageNet-100 and ImageNet-1000. M is the number of classes per task.

Table 1. Average incremental accuracy on ImageNet-100 with $M = 10$ classes per task and ImageNet-1000 with $M = 100$ classes per task. Memory usage shows the cost of storing images or feature vectors for all classes. † indicates that the results were reported from the paper. ∗ indicates numbers were estimated from figures in the paper.

	ImageNet-100		ImageNet-1000	
	Mem. in MB	Accuracy	Mem. in MB	Accuracy
	STATE-OF-THE-ART METHODS			
Orig. LwF [23]†∗	-	0.642	-	0.566
Orig. iCARL [30]†∗	375	0.836	3750	0.637
EEIL [4]†	375	0.904	3750	0.694
Rebalancing [17]	375	0.681	3750	0.643
BiC w/o corr. [41]†	375	0.872	3750	0.776
BiC [41]†	375	0.906	3750	0.840
	BASELINES			
γ-LwF	-	0.891	-	0.804
γ-iCARL	375	0.914	3750	0.802
	OUR METHOD			
Ours	**48.8**	0.913	**488.3**	0.843
Ours-hybrid	236.3	**0.927**	2863.3	**0.846**

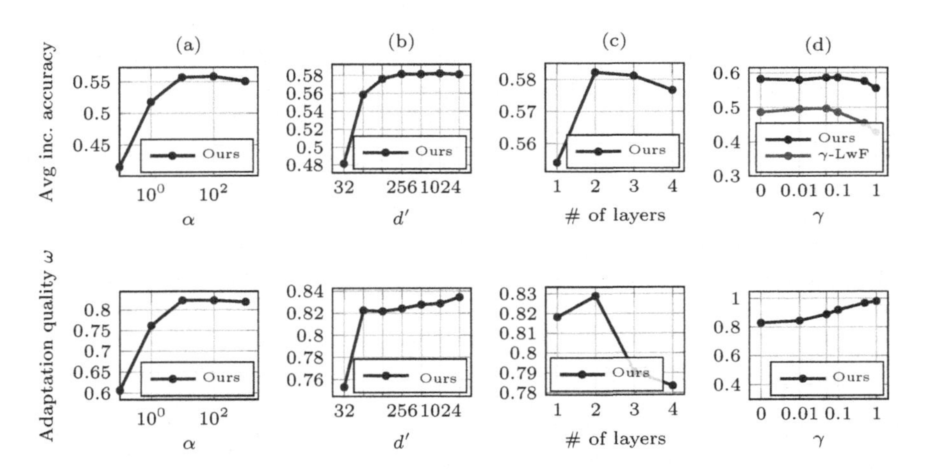

Fig. 5. Impact of different parameters in terms of classification accuracy (top) and adaptation quality ω as defined in Sect. 5.4 (bottom) on CIFAR-100: (a) similarity coefficient α (7) (b) size of hidden layers d' of the feature adaptation network (c) number of hidden layers in the feature adaptation network (d) feature distillation coefficient γ (5).

5.4 Impact of Parameters

We show the impact of the hyper-parameters of our method. All experiments in this section are performed on a validation set created by holding out 10% of the original CIFAR-100 training data.

Impact of cosine classifier is evaluated on the base network, *i.e.* LwF.MC. We achieve 48.7 and 45.2 accuracy with and without cosine classifier respectively. We include cosine classifier in all baselines and methods.

Impact of α. The parameter α controls the importance of the similarity term w.r.t. classification term when learning the feature adaptation network (7). Figure 5 top-(a) shows the accuracy with different α. The reconstruction constraint controlled by α requires a large value. We set $\alpha = 10^2$ in our experiments.

Impact of d'. We evaluate the impact of d', the dimensionality of the hidden layers of feature adaptation network ϕ_ψ in Fig. 5 top-(b). Projecting feature vectors to a higher dimensional space is beneficial, achieving the maximum validation accuracy with $d' = 1,024$. We set $d' = 16d$ in our experiments.

Impact of the Network Depth. We evaluate different number of hidden layers of the feature adaptation network ϕ_ψ in Fig. 5 top-(c). The accuracy reaches its peak with two hidden layers, and starts to decrease afterwards, probably because the networks starts to overfit. We use two hidden layers in our experiments.

Impact of Feature Distillation. We evaluate different γ for feature distillation (5) for γ-LwF and our method, see Fig. 5 top-(d). We set $\gamma = 0.05$ and include feature distillation in all baselines and methods in our experiments.

Quality of Feature Adaptation. We evaluate the quality of our feature adaptation process by measuring the average similarity between the adapted features and their *ground-truth value.* The ground-truth vector $h_\theta^t(x)$ for image x is its feature representation if we actually had access to that image in task t. We compare it against $\mathbf{v}$, the corresponding vector of x in the memory, that has been adapted over time. We compute the feature adaptation quality by dot product $\omega = \mathbf{v}^\top h_\theta^t(x)$. This measures how accurate our feature adaptation is compared to the real vector if we had access to image x[2].

We repeat the validation experiments, this time measuring average ω of all vectors instead of accuracy (Fig. 5 bottom row). Top and bottom rows of Fig. 5 shows that most trends are correlated meaning better feature adaptation results in better accuracy. One main exception is the behavior of γ in feature distillation (5). Higher γ results in higher ω but lower classification accuracy. This is expected, as high γ forces the network to make minimal changes to its feature extractor between different tasks, making feature adaptation more successful, but feature representations less discriminative.

Effect of balanced feature classifier. Class-imbalanced training is shown to lead to biased predictions [41]. We investigate this in Supplementary Section C. Our experiments show that balancing the number of instances per class leads to improvements in the accuracy when training the feature classifier $g_{\tilde{W}}$.

[2] x is normally not available in future tasks, we use it here for the ablation study.

6 Conclusions

We have presented a novel method for preserving feature descriptors instead of images in incremental learning. Our method introduces a *feature adaptation* function, which accurately updates the preserved feature descriptors as the network is updated with new classes. The proposed method is thoroughly evaluated in terms of classification accuracy and adaptation quality, showing that it is possible to achieve state-of-the-art accuracy with a significantly lower memory footprint. Our method is orthogonal to existing work [23,30] and can be combined to achieve even higher accuracy with low memory requirements.

Acknowledgements. This research was funded in part by NSF grants IIS 1563727 and IIS 1718221, Google Research Award, Amazon Research Award, and AWS Machine Learning Research Award.

References

1. Aljundi, R., Babiloni, F., Elhoseiny, M., Rohrbach, M., Tuytelaars, T.: Memory aware synapses: learning what (not) to forget. In: ECCV (2018)
2. Belouadah, E., Popescu, A.: Il2m: class incremental learning with dual memory. In: ICCV (2019)
3. Bucher, M., Herbin, S., Jurie, F.: Generating visual representations for zero-shot classification. In: ICCV (2017)
4. Castro, F.M., Marín-Jiménez, M.J., Guil, N., Schmid, C., Alahari, K.: End-to-end incremental learning. In: ECCV (2018)
5. Cauwenberghs, G., Poggio, T.: Incremental and decremental support vector machine learning. In: NeurIPS (2001)
6. Chaudhry, A., Dokania, P.K., Ajanthan, T., Torr, P.H.: Riemannian walk for incremental learning: understanding forgetting and intransigence. In: ECCV (2018)
7. Cortes, C., Vapnik, V.: Support-vector networks. Mach. Learn. **20**(3), 273–297 (1995). https://doi.org/10.1007/BF00994018
8. Dhar, P., Singh, R.V., Peng, K.C., Wu, Z., Chellappa, R.: Learning without memorizing. In: CVPR (2019)
9. Glorot, X., Bordes, A., Bengio, Y.: Deep sparse rectifier neural networks. In: AISTATS (2011)
10. Goodfellow, I., et al.: Generative adversarial nets. In: NeurIPS (2014)
11. Goodfellow, I.J., Mirza, M., Xiao, D., Courville, A., Bengio, Y.: An empirical investigation of catastrophic forgetting in gradient-based neural networks. arXiv preprint arXiv:1312.6211 (2013)
12. Hariharan, B., Girshick, R.: Low-shot visual recognition by shrinking and hallucinating features. In: CVPR (2017)
13. He, C., Wang, R., Shan, S., Chen, X.: Exemplar-supported generative reproduction for class incremental learning. In: BMVC (2018)
14. He, K., Gkioxari, G., Dollár, P., Girshick, R.: Mask r-cnn. In: CVPR (2017)
15. He, K., Zhang, X., Ren, S., Sun, J.: Deep residual learning for image recognition. In: CVPR (2016)
16. Hinton, G., Vinyals, O., Dean, J.: Distilling the knowledge in a neural network. In: NIPS Deep Learning and Representation Learning Workshop (2015)

17. Hou, S., Pan, X., Loy, C.C., Wang, Z., Lin, D.: Learning a unified classifier incrementally via rebalancing. In: CVPR (2019)
18. Javed, K., Shafait, F.: Revisiting distillation and incremental classifier learning. In: Jawahar, C.V., Li, H., Mori, G., Schindler, K. (eds.) ACCV 2018. LNCS, vol. 11366, pp. 3–17. Springer, Cham (2019). https://doi.org/10.1007/978-3-030-20876-9_1
19. Kemker, R., Kanan, C.: Fearnet: brain-inspired model for incremental learning. In: ICLR (2018)
20. Kirkpatrick, J., et al.: Overcoming catastrophic forgetting in neural networks. Proc. Nat. Acad. Sci. **114**(13), 3521–3526 (2017)
21. Krizhevsky, A., Hinton, G.: Learning multiple layers of features from tiny images. University of Toronto, Technical report (2009)
22. Krizhevsky, A., Sutskever, I., Hinton, G.E.: Imagenet classification with deep convolutional neural networks. In: NeurIPS (2012)
23. Li, Z., Hoiem, D.: Learning without forgetting. IEEE Trans. Pattern Anal. Mach. Intell. **40**(12), 2935–2947 (2017)
24. Liu, X., Masana, M., Herranz, L., Van de Weijer, J., Lopez, A.M., Bagdanov, A.D.: Rotate your networks: better weight consolidation and less catastrophic forgetting. In: ICPR (2018)
25. Lopez-Paz, D., Ranzato, M.: Gradient episodic memory for continual learning. In: NeurIPS (2017)
26. Luo, C., Zhan, J., Xue, X., Wang, L., Ren, R., Yang, Q.: Cosine normalization: using cosine similarity instead of dot product in neural networks. In: Kůrková, V., Manolopoulos, Y., Hammer, B., Iliadis, L., Maglogiannis, I. (eds.) ICANN 2018. LNCS, vol. 11139, pp. 382–391. Springer, Cham (2018). https://doi.org/10.1007/978-3-030-01418-6_38
27. McCloskey, M., Cohen, N.J.: Catastrophic interference in connectionist networks: the sequential learning problem. In: Bower, G.H. (ed.) Psychology of Learning and Motivation, vol. 24. Acadamic Press, New York (1989)
28. Mensink, T., Verbeek, J., Perronnin, F., Csurka, G.: Distance-based image classification: generalizing to new classes at near-zero cost. IEEE Trans. Pattern Anal. Mach. Intell. **35**(11), 2624–2637 (2013)
29. Pedregosa, F., et al.: Scikit-learn: machine learning in python. J. Mach. Learn. Res. **12**, 2825–2830 (2011)
30. Rebuffi, S.A., Kolesnikov, A., Sperl, G., Lampert, C.H.: Icarl: incremental classifier and representation learning. In: CVPR (2017)
31. Ren, S., He, K., Girshick, R., Sun, J.: Faster r-cnn: towards real-time object detection with region proposal networks. In: NeurIPS (2015)
32. Ristin, M., Guillaumin, M., Gall, J., Van Gool, L.: Incremental learning of random forests for large-scale image classification. IEEE Trans. Pattern Anal. Mach. Intell. **38**(3), 490–503 (2015)
33. Robins, A.: Catastrophic forgetting, rehearsal and pseudorehearsal. Connection Sci. **7**(2), 123–146 (1995)
34. Russakovsky, O., et al.: Imagenet large scale visualrecognition challenge. Int. J. Comput. Vis. **115**(3), 211–252 (2015). https://doi.org/10.1007/s11263-015-0816-y
35. Rusu, A.A., et al.: Progressive neural networks. arXiv preprint arXiv:1606.04671 (2016)
36. Shin, H., Lee, J.K., Kim, J., Kim, J.: Continual learning with deep generative replay. In: NeurIPS (2017)
37. Simonyan, K., Zisserman, A.: Very deep convolutional networks for large-scale image recognition. ICLR (2014)

38. Wang, Y.X., Ramanan, D., Hebert, M.: Growing a brain: fine-tuning by increasing model capacity. In: CVPR (2017)
39. Welling, M.: Herding dynamical weights to learn. In: ICML (2009)
40. Wu, C., et al.: Memory replay gans: learning to generate new categories without forgetting. In: NeurIPS (2018)
41. Wu, Y., et al.: Large scale incremental learning. In: CVPR (2019)
42. Xian, Y., Lorenz, T., Schiele, B., Akata, Z.: Feature generating networks for zero-shot learning. In: CVPR (2018)
43. Xiang, Y., Fu, Y., Ji, P., Huang, H.: Incremental learning using conditional adversarial networks. In: ICCV (2019)
44. Yu, L., et al.: Semantic drift compensation for class-incremental learning. In: CVPR (2020)
45. Zhang, J., et al.: Class-incremental learning via deep model consolidation. arXiv preprint arXiv:1903.07864 (2019)

Neural Voice Puppetry: Audio-Driven Facial Reenactment

Justus Thies[1(✉)], Mohamed Elgharib[2], Ayush Tewari[2], Christian Theobalt[2], and Matthias Nießner[1]

[1] Technical University of Munich, Munich, Germany
justus.thies@tum.de
[2] Max Planck Institute for Informatics, Saarland Informatics Campus, Saarbrücken, Germany

Abstract. We present *Neural Voice Puppetry*, a novel approach for audio-driven facial video synthesis (Video, Code and Demo: https://justusthies.github.io/posts/neural-voice-puppetry/). Given an audio sequence of a source person or digital assistant, we generate a photo-realistic output video of a target person that is in sync with the audio of the source input. This audio-driven facial reenactment is driven by a deep neural network that employs a latent 3D face model space. Through the underlying 3D representation, the model inherently learns temporal stability while we leverage neural rendering to generate photo-realistic output frames. Our approach generalizes across different people, allowing us to synthesize videos of a target actor with the voice of any unknown source actor or even synthetic voices that can be generated utilizing standard text-to-speech approaches. *Neural Voice Puppetry* has a variety of use-cases, including audio-driven video avatars, video dubbing, and text-driven video synthesis of a talking head. We demonstrate the capabilities of our method in a series of audio- and text-based puppetry examples, including comparisons to state-of-the-art techniques and a user study.

1 Introduction

In the recent years, speech-based interaction with computers made significant progress. Digital voice assistants are now ubiquitous due to their integration into many commodity devices such as smartphone, tvs, cars, etc.; even companies use more and more machine learning techniques to drive service bots that interact with their customers. These virtual agents aim for a user-friendly man-machine interface while keeping maintenance costs low. However, a significant challenge is to appeal to humans by delivering information through a medium that is most comfortable to them. While speech-based interaction is already very successful, such as shown in virtual assistants like Siri, Alexa, Google, etc., the visual counterpart is largely missing. This comes to no surprise given that a user would also

Electronic supplementary material The online version of this chapter (https://doi.org/10.1007/978-3-030-58517-4_42) contains supplementary material, which is available to authorized users.

A. Vedaldi et al. (Eds.): ECCV 2020, LNCS 12361, pp. 716–731, 2020.
https://doi.org/10.1007/978-3-030-58517-4_42

Fig. 1. *Neural Voice Puppetry* enables applications like facial animation for digital assistants or audio-driven facial reenactment.

like to associate the visuals of a face with the generated audio, similar to the ideas behind video conferencing. In fact, the level of engagement for audio-visual interactions is higher than for purely audio ones [9,24].

The aim of this work is to provide the missing visual channel by introducing *Neural Voice Puppetry*, a photo-realistic facial animation method that can be used in the scenario of a visual digital assistant (Fig. 1). To this end, we build on the recent advances in text-to-speech synthesis literature [14,21], which is able to provide a synthetic audio stream from a text that can be generated by a digital agent. As visual basis, we leverage a short target video of a real person. The key component of our method is to estimate lip motions that fit the input audio and to render the appearance of the target person in a convincing way. This mapping from audio to visual output is trained using the ground truth information that we can gather from a target video (aligned real audio and image data). We designed *Neural Voice Puppetry* to be an easy to use audio-to-video translation tool which does not require vast amount of video footage of a single target video or any manual user input. In our experiments, the target videos are comparably short (2–3 min), thus, allowing us to work on a large amount of video footage that can be downloaded from the Internet. To enable this easy applicability to new videos, we generalize specific parts of our pipeline. Specifically, we compute a latent expression space that is generalized among multiple persons (in our experiments 116). This also ensures the capability of being able to handle different audio inputs. Besides the generation of a visual appearance of a digital agent, our method can also be used as audio-based facial reenactment. Facial reenactment is the process of re-animating a target video in a photo-realistic manner with the expressions of a source actor [28,34]. It enables a variety of applications, ranging from consumer-level teleconferencing through photo-realistic virtual avatars [20, 26,27] to movie production applications such as video dubbing [11,17]. Recently, several authors started to exploit the audio signal for facial reenactment [5,23, 31]. This has the potential of avoiding failures of visual-based approaches, when the visual signal is not reliable, e.g., due to occluded face, noise, distorted views and so on. Many of these approaches, however, lack video-realism [5,31], since they work in a normalized space of facial imagery (cropped, frontal faces), to be agnostic to head movements. An exception is the work of Suwajanakorn et al. [23], where they have shown photo-realistic videos of President Obama

that can be synthesized just from the audio signal. This approach, however, requires very large quantities of data for training (17 h of President Obama weekly speeches) and, thus, limits its application and generalization to other identities. In contrast, our method only needs 2–3 min of a target video to learn the person-specific talking style and appearance. Our underlying latent 3D model space inherently learns 3D consistency and temporal stability that allows us to generate natural, full frame imagery. Especially, it enables the disentanglement of rigid head motions from facial expressions.
To enable photo-realistic renderings of digital assistants as well as audio-driven facial reenactment, we have the following contributions:

- A temporal network architecture called *Audio2ExpressionNet* is proposed to map an audio stream to a 3D blendshape basis that can represent person-specific talking styles. Exploiting features from a pre-trained speech-to-text network, we generalize the *Audio2ExpressionNet* on a dataset of news-speaker.
- Based on a short target video sequence (2–3 min), we extract a representation of *person-specific talking styles*, since our goal is to preserve the talking style of the target video during reenactment.
- A novel *light-weight neural rendering network* using neural textures is presented that allows us to generate photo-realistic video content reproducing the person-specific appearance. It surpasses the quality and speed of state-of-the-art neural rendering methods [10,29].

2 Related Work

Neural Voice Puppetry is a facial reenactment approach based only on audio input. In the literature, there are many video-based facial reenactment systems that enable dubbing and other general facial expression manipulation. Our focus in this related work section lies on audio-based methods. These methods can be organized in facial animation and facial reenactment. Facial animation concentrates on the prediction of expressions that can be applied to a predefined avatar. In contrast, audio-driven facial reenactment aims to generate photo-realistic videos of an existing person including all idiosyncrasies.

Video-Driven Facial Reenactment: The state-of-the-art report of Zollhöfer et al. [34] discusses several works for video-driven facial reenactment. Most methods, rely on a reconstruction of a source and target face using a parametric face model. The target face is reenacted by replacing its expression parameters with that of the source face. Thies et al. [28] uses a static skin texture and a data-driven approach to synthesize the mouth interior. In Deep Video Portraits [18], a generative adversarial network is used to produce photo-realistic skin texture that can handle skin deformations conditioned on synthetic renderings. Recently, Thies et al. [29] proposed the usage of neural textures in conjunction with a deferred neural renderer. Results show that neural textures can be used to generate high quality facial reenactments. For instance, it produces high fidelity mouth interiors with less artifacts. Kim et al. [17] analyzed the notion of style for facial expressions and showed its importance for dubbing. In contrast to Kim et al. [17], we directly estimate the expressions in the target talking-style domain, thus, we don't need to apply any transfer or adaption method.

Audio-Driven Facial Animation: These methods do not focus on photo-realistic results, but on the prediction of facial motions [6,16,22,25,30]. Karras et al. [16] drives a 3D facial animation using an LSTM that maps input waveforms to the 3D vertex coordinates of a face mesh, also considering the emotional state of the person. In contrast to our method, it needs high quality 3D reconstructions for supervised training and does not render photo-realistic output. Taylor et al. [25] use a neural network to map phonemes into the parameters of a reference face model. It is trained on data collected for only one person speaking for 8 hours. They show animations of different synthetic avatars using deformation retargeting. VOCA [6] is an end-to-end deep neural network for speech-to-animation translation trained on multiple subjects. Similar to our approach, a low-dimensional audio embedding based on features of the DeepSpeech network [13] is used. From this embedding, VOCA regresses 3D vertices on a FLAME face model [19] conditioned on a subject label. It requires high quality 4D scans recorded in a studio setup. Our approach works on 'in the wild' videos, with a focus on temporally coherent predictions and photo-realistic renderings.

Audio-Driven Facial Reenactment: Audio-driven facial reenactment has the goal to generate photo-realistic videos that are in sync with the input audio stream. There is a number of techniques for audio-driven facial reenactment [2,3,5,8,31,32] but only a few generate photo-realistic, natural, full frame images [23]. Suwajanakorn et al. [23] uses an audio stream from President Barack Obama to synthesize a high quality video of him speaking. A Recurrent Neural Network is trained on many hours of his speech to learn the mouth shape from the audio. The mouth is then composited with proper 3D matching to reanimate an original video in photo-realistic manner. Because of the huge amount of used training data (17h), it is not applicable to other target actors. In contrast, our approach only needs a 2–3 min long video of a target sequence.

Chung et al. [5] present a technique that animates the mouth of a still, normalized image to follow an audio speech. First, the image and audio is projected into a latent space through a deep encoder. A decoder then utilizes the joint embedding of the face and audio to synthesize the talking head. The technique is trained on tens of hours of data in an unsupervised manner. Another 2D image-based method has been presented by Vougioukas et al. [31]. They use a temporal GAN to produce a video of a talking face given a still image and an audio signal as input. The generator feeds the still image and the audio to an encoder-decoder architecture with a RNN to better capture temporal relations. It uses discriminators that work on per-frame and on a sequence level to improve temporal coherence. As conditioning, it also takes the audio signal as input to enforce the synthesized mouth to be in sync with the audio. In [32] a dedicated mouth-audio syn discriminator is used to improve the results. In contrast to our method, the 2D image-based methods are restricted to a normalized image space of cropped and frontalized images. They are not applicable to generate full frame images with 3D consistent motions.

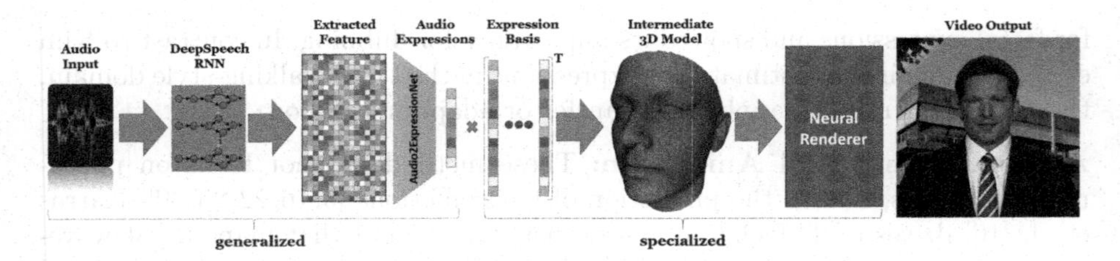

Fig. 2. Pipeline of *Neural Voice Puppetry.* Given an audio sequence we use the DeepSpeech RNN to predict a window of character logits that are fed into a small network. This generalized network predicts coefficients that drive a person-specific expression blendshape basis. We render the target face model with the new expressions using a novel light-weight neural rendering network.

Text-Based Video Editing: Fried et al. [10] presented a technique for text-based editing of videos. Their approach allows overwriting existing video segments with new texts in a seamless manner. A face model [12] is registered to the examined video and a viseme search finds video segments with similar mouth movements to the edited text. The corresponding face parameters of the matching video segment are blended with the original sequence parameters based on a heuristic, followed by a deep renderer to synthesize photo-realistic results. The method is person-specific and requires a one hour long training sequence of the target actor and, thus, is not applicable to short videos from the Internet. The viseme search is slow (∼5 min for three words) and does not allow for interactive results.

3 Overview

Neural Voice Puppetry consists of two main components (see Fig. 2): a generalized and a specialized part. A generalized network predicts a latent expression vector, thus, spanning an *audio-expression space*. This *audio-expression space* is shared among all persons and allows for reenactment, i.e., transferring the predicted motions from one person to another. To ensure generalizability w.r.t. the input audio, we use features extracted by a pretrained speech-to-text network [13] as input to estimate the audio-expressions. The audio-expressions are interpreted as blendshape coefficients of a 3D face model rig. This face model rig is person-specific and is optimized in the second part of our pipeline. This specialized stage captures the idiosyncrasies of a target person including the facial motion and appearance. It is trained on a short video sequence of $2-3$ minutes (in comparison to hours that are required by state-of-the-art methods). The 3D facial motions are represented as delta-blendshapes which we constrain to be in the subspace of a generic face template [1,28]. A neural texture in conjunction with a novel neural rendering network is used to store and to rerender the appearance of the face of an individual person.

Fig. 3. Samples of the training corpus used to optimize the *Audio2ExpressionNet*.

4 Data

In contrast to previous model-based methods, *Neural Voice Puppetry* is based on *'in-the-wild' videos that can be download from the internet.* The videos have to be synced with the audio stream, such that we can extract ground truth pairs of audio features and image content. In our experiments the videos have a resolution of 512×512 with $25fps$.

Training Corpus for the *Audio2ExpressionNet*: Figure 3 shows an overview of our video training corpus that is used for the training of the small network that predicts the 'audio expressions' from the input audio features (see Sect. 5.1). The dataset consists of 116 videos with an average length of 1.7 min (in total 302750 frames). We selected the training corpus, such that the persons are in a neutral mood (commentators of the German public TV).

Target Sequences: For a target sequence, we extract the person-specific talking style in the sequence. I.e., we compute a mapping from the generalized audio-expression space to the actual facial movements of the target actor (see Sect. 5.3). The sequences are 2–3 min long and, thus, easy to obtain from the Internet.

4.1 Preprocessing:

In an automatic preprocessing step, we extract face tracking information as well as audio features needed for training.

3D Face Tracking: Our method is using a statistical face model and delta-blendshapes [1,28] to represent a 3D latent space for modelling facial animation. The 3D face model space reduces the face space to only a few hundred parameters (100 for shape, 100 for albedo and 76 for expressions) and stays fixed in this work. Using the dense face tracking method of Thies et al. [28], we estimate the model parameters for every frame of a sequence. During tracking, we extract the per-frame expression parameters that are used to train the audio to expression network. To train our neural renderer, we also store the rasterized texture coordinates of the reconstructed face mesh.

Audio-Feature Extraction: The video contains a synced audio stream. We use the recurrent feature extractor of the pre-trained speech-to-text model DeepSpeech [13] (v0.1.0). Similar to Voca [6], we extract a window of character logits per video frame. Each window consists of 16 time intervals à 20ms, resulting in an audio feature of 16×29. The DeepSpeech model is generalized among thousands of different voices, trained on Mozilla's CommonVoice dataset.

5 Method

To enable photo-realistic facial reenactment based on audio signals, we employ a 3D face model as intermediate representation of facial motion. A key component of our pipeline is the audio-based expression estimation. Since every person has his own talking style and, thus, different expressions, we establish person-specific expression spaces that can be computed for every target sequence. To ensure generalization among multiple persons, we created a latent *audio-expression space* that is shared by all persons. From this audio-expression space, one can map to the person specific expression space, enabling reenactment. Given the estimated expression and the extracted audio features, we apply a novel light-weight neural rendering technique that generates the final output image.

5.1 Audio2ExpressionNet

Our method is designed to generate temporally smooth predictions of facial motions. To this end, we employ a deep neural network with two stages. First, we predict per-frame facial expression predictions. These expressions are potentially noisy, thus, we use an expression aware temporal filtering network. Given the noisy per-frame predictions as input the neural network predicts filter weights to compute smooth audio-expressions for a single frame. The per-frame as well as the filtering network can be trained jointly and outputs audio-expression coefficients. This audio-expression space is shared among multiple persons and is interpreted as blendshape coefficients. Per person, we compute a blendshape basis which is in the subspace of our generic face model [28]. The networks are trained with a loss that works on a vertex level of this face model.

Per-Frame Audio-Expression Estimation Network: Since our goal is a generalized audio-based expression estimation, we rely on generalized audio features. We use the RNN-part of the speech to text approach DeepSpeech [13] to extract these features. These features represent the logits of the DeepSpeech alphabet for 20 ms audio signal. For each video frame, we extract a time window of 16 features around the frame that consist of 29 logits (length of the DeepSpeech alphabet is 29). This, 16×29 tensor is input to our per-frame estimation network (see Fig. 4). To map from this feature space to the per-frame audio-expression space, we apply 4 convolutional layer and 3 fully connected layer. Specifically, we apply 1D convolutions with kernel dimensions (3) and stride (2), filtering in the time dimension. The convolutional layers have a bias and are followed by a leaky ReLU (slope 0.02). The feature dimensions are reduced successively from (16×29), (8×32), (4×32), (2×64) to (1×64). This reduced feature is input to the fully connected layers that have a bias and are also followed by a leaky ReLU (0.02), except the last layer. The fully connected layers map the 64 features from the convolutional network to 128, then to 64 and, finally, to the audio-expression space of dimension 32, where a TanH activation is applied.

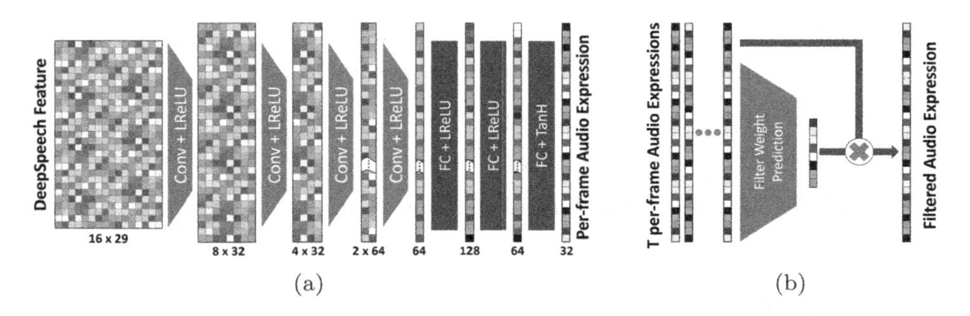

Fig. 4. Audio2ExpressionNet: (a) Per-frame audio-expression estimation network that gets DeepSpeech features as input, (b) to get smooth audio-expressions, we employ a content-aware filtering along the time dimension.

Temporally Stable Audio-Expression Estimation: To generate temporally stable audio-expression predictions, we jointly learn a filtering network that gets T per-frame estimates as input (see Fig. 4(b)). Specifically, we estimate the audio-expressions for frame t using a linear combination of the per-frame predictions of the timesteps $t - T/2$ to $t + T/2$. The weights for the linear combination are computed using a neural network that gets the audio-expressions as input (which results in an expression-aware filtering). The filter weight prediction network consists of five 1D convolutions followed by a linear layer with softmax activation (see supplemental material for detailed description). This content aware temporal filtering is also inspired by the self-attention mechanism [33].

Person-Specific Expressions: To retrieve the 3D model from this audio-expression space, we learn a person-specific audio-expression blendshape basis which we constrain by the generic blendshape basis of our statistical face model. I.e., the audio-expression blendshapes of a person are a linear combination of the generic blendshapes. This linear relation, results in a linear mapping from the audio-expression space which is output of the generalized network to the generic blendshape basis. This linear mapping is person specific, resulting in N matrices with dimension 76×32 during training (N being the number of training sequences and 76 being the number of generic blendshapes).

Loss: The network and the mapping matrices are learned end-to-end using the visually tracked training corpus and a vertex-based loss function, with a higher weight ($10\times$) on the mouth region of the face model. Specifically, we compute a vertex-to-vertex distance from the audio-based predicted and the visually tracked face model in terms of a root mean squared (RMS) distance:

$$L_{expr} = RMS(v_t - v_t^*) + \lambda \cdot L_{temp}$$

with v_t, the vertices based on the filtered expression estimation of frame t and v_t^* being the visual tracked face vertices. In addition to the absolute loss between predictions and the visual tracked face geometry, we use a temporal loss that considers the vertex displacements of consecutive frames:

$$L_{temp} = RMS((v_t - v_{t-1}) - (v_t^* - v_{t-1}^*)) + RMS((v_{t+1} - v_t) - (v_{t+1}^* - v_t^*)) \\ + RMS((v_{t+1} - v_{t-1}) - (v_{t+1}^* - v_{t-1}^*))$$

These forward, backward and central differences are weighted with λ (in our experiments $\lambda = 20$). The losses are measured in millimeters.

5.2 Neural Face Rendering

Based on the recent advances in neural rendering, we employ a novel light-weight neural rendering technique that is based on neural textures to store the appearance of a face. Our rendering pipeline synthesizes the lower face in the target video based on the audio-driven expression estimations. Specifically, we use two networks (see supplemental material for an overview figure). One network that focuses on the face interior, and another network that embeds this rendering into the original image. The estimated 3D face model is rendered using the rigid pose observed from the original target image using a neural texture [29]. The neural texture has a resolution of $256 \times 256 \times 16$. The network for the face interior translates these rendered feature descriptors to RGB colors. The network is using a similar structure as a classical U-Net with 5 layers. But instead of using strided convolutions that result in a downsampling in each layer, we are using dilated convolutions with increasing dilation factor and a stride of one. Instead of transposed convolutions we are using standard convolutions. All convolutions have kernel size 3×3. Note, dilated instead of strided convolutions do not increase the number of learnable parameters, but it increases the memory load during training and testing. Dilated convolutions reduce visual artifacts and result in smoother results (also temporally, see video). The second network that blends the face interior with the 'background image' has the same structure. To remove potential movements of the chin in the background image, we erode the background image around the rendered face. The second network inpaints these missing regions.

Loss: We use a per-frame loss function that is based on an ℓ_1 loss to measure absolute errors and a VGG style loss [15].

$$L_{rendering} = \ell_1(I, I^*) + \ell_1(\hat{I}, I^*) + VGG(I, I^*)$$

with I being the final synthetic image, I^* the ground truth image and $\hat{I}$ the intermediate result of the first network that focuses on the face interior (loss is masked to this region).

5.3 Training

Our training procedure has two stages – the generalization and the specialization phase. In the first phase, we train the *Audio2ExpressionNet* among all sequences from our dataset (see Sect. 4) in a supervised fashion. Given the visual face tracking information, we know the 3D face model of a specific person for every frame. In the training process, we reproduce these 3D reconstructions based on the audio input by optimizing the network parameters and the person-specific mappings from the audio-expression space to the 3D space. In the second phase, the rendering network for a specific target sequence is trained. Given the ground truth images and the visual tracking information, we train the neural renderer end-to-end including the neural texture.

New Target Video: Since the audio-based expression estimation network is generalized among multiple persons, we can apply it to unseen actors. The person specific mapping between the predicted audio-expression space coefficients and the expression space of the new person can be obtained by solving a linear system of equations. Specifically, we extract the audio-expression for all training images and compute the linear mapping to the expressions that are visually estimated. In addition to this step, the person-specific rendering network for the new target video is trained from scratch (see supplement for further information).

5.4 Inference

At test time, we only require a source audio sequence. Based on the target actor selection, we use the corresponding person-specific mapping. The mapping from the audio features to the person specific expression space takes less than 2 ms on an Nvidia 1080Ti. Generation of the 3D model and the rasterization using these predictions takes another 2 ms. The deferred neural rendering takes ~ 5 ms which results in a real-time capable pipeline.

Text-to-Video: Our pipeline is trained on real video sequences, where the audio is in sync with the visual content. Thus, we learned a mapping directly from audio to video that ensures synchronicity. Instead of going directly from text to video, where such a natural training corpus is not available, we synthesize voice from the text and feed this into our pipeline. For our experiments we used samples from the DNN-based text-to-speech demo of IBM Watson[1]. Which gives us state-of-the-art synthetic audio streams that are comparable to the synthetic voices of virtual assistants.

6 Results

*Neural Voice Puppetry*has several important use cases, i.e., audio-driven video avatars, video dubbing and text-driven video synthesis of a talking head, see supplemental video. In the following sections, we discuss these results including comparisons to state-of-the-art approaches.

[1] https://text-to-speech-demo.ng.bluemix.net/.

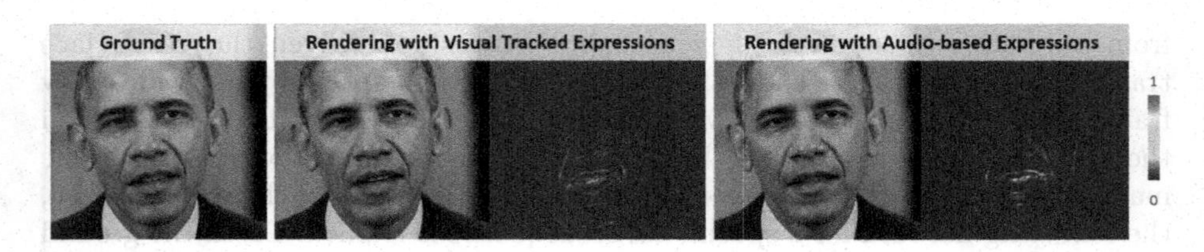

Fig. 5. Self-reenactment: Evaluation of our rendering network and the audio-prediction network. Error plot shows the euclidean photo-metric error.

6.1 Ablation Studies

Self-reenactment: We use self-reenactment to evaluate our pipeline (Fig. 5), since it gives us access to a ground truth video sequence where we can also retrieve visual face tracking. As a distance measurement, we use an ℓ_2 distance in color space (colors in [0,1]) and the corresponding PSNR values. Using this measure, we evaluate the rendering network and the entire reenactment pipeline. Specifically, we compare the results using visual tracked mouth movements to the results using audio-based predictions (see video). The mean color difference of the re-rendering on the test sequence of 645 frame is 0.003 for the visual and 0.005 for the audio-based expressions, which corresponds to a PSNR of 41.48 and 36.65 respectively. In addition to the photo-metric measurements, we computed the 2D mouth landmark distances relative to the eye distance using Dlib, resulting in 0.022 for visual tracking and 0.055 for the audio-based predictions.

In the supplemental video, we also show a side-by-side comparison of our rendering network using dilated convolutions and our network with strided convolutions (and a kernel size of 4 to reduce block artifacts in the upsampling). Both networks are trained with the same number of epochs (50). As can be seen, dilated convolutions lead to visually more pleasing results (smoother in spatial and temporal domain). As a comparison to the results using dilated convolutions reported above, strided convolutions result in a lower PSNR of 40.12 with visual tracking and 36.32 with audio-based predictions.

Temporal Smoothness: We also evaluated the benefits of using a temporal-based expression prediction network. Besides temporally smooth predictions shown in the supplemental video, it also improves the prediction accuracy of the mouth shape. The relative 2D mouth landmark error improves from 0.058 (per frame prediction) to 0.055 (temporal prediction).

Generalization/Transferability: Our results are covering different target persons which demonstrates the wide applicability of our method, including the reproduction of different person-specific talking styles and appearances. As can be seen in the supplemental video, the expression estimation network that is trained on multiple target sequences (302750 frames) results in more coherent predictions than the network solely trained on a sequence of Obama (3145 frames). The usage of more target videos increases the training corpus size and the variety of input voices and, thus, leads to more robustness. In the video,

Table 1. Analysis of generated videos with different source/target languages. Based on SyncNet [4], we measure the audio-visual sync (offset/confidence). As a reference, we list the sync measurements for the original target video (right).

Target	Source						
	Bengali (Male)	Chinese (Female)	German (Female)	Greek (Male)	Spanish (Male)	English (Male)	Reference (Original)
Obama (English)	(−3/5.137)	(−3/3.234)	(−3/5.544)	(−4/1.952)	(−3/4.179)	(−3/4.927)	(−3/7.865)
Macron (French)	(−3/3.994)	(−3/2.579)	(−2/3.012)	(−3/1.856)	(−3/3.752)	(−3/3.163)	(−1/3.017)
News-speaker (German)	(−2/5.361)	(−2/6.505)	(−2/5.734)	(−2/5.752)	(−2/6.408)	(−2/6.036)	(−1/9.190)
Woman (English)	(−1/6.265)	(−1/4.431)	(−1/3.841)	(−1/4.206)	(−1/3.684)	(−1/4.716)	(−1/6.036)

we also show a comparison of the transfer from different source languages to different target videos that are originally also in different languages. In Table 1, we show the corresponding quantitative measurements of the achieved lip sync using SyncNet [4]. SyncNet is trained on the BBC news program (English), nevertheless, the authors state that it works also good for other languages. As a reference for the measurements of the different videos, we list the values for the original target videos. Higher confidence values are better, while a value below 1 refers to uncorrelated audio video streams. The original video of Macron has the lowest measured confidence which propagates to the reenactment results.

6.2 Comparisons to State-of-the-art Methods

In the following, as well as in the supplemental document, we compare to model-based and pure image-based approaches for audio-driven facial reenactment.

Preliminary User Study: In a preliminary user study, we evaluated the visual quality and audio-visual sync of the state-of-the-art methods. The user study is based on videos taken from the supplemental materials of the respective publications (assuming the authors showing the best case scenario). Note that the videos of the different methods show (potentially) different persons (see supplemental material). In total, 56 attendees with a computer science background judged upon synchronicity and visual quality ('very bad', 'bad', 'neither bad nor good', 'good', 'very good') of 24 videos in randomized order (in total). In Fig. 6, we show the percentage of attendees that rated the specific approach with good or very good. As can be seen, the 2D image-based approaches achieve a high audio-visual sync (especially, Vougioukas [32]), but they lack visual quality and are not able to synthesize natural videos (outside of the normalized space). Our approach gives the best visual quality and also a high audio-visual sync, similar to state-of-the-art video-based reenactment approaches like Thies et al. [29].

Image-Based Methods: Our method aims for high quality output that is embedded in a real video, including the person-specific talking style, exploiting an explicit 3D model representation of the face to ensure 3D consistent movements. This is fundamentally different from image-based approaches that are

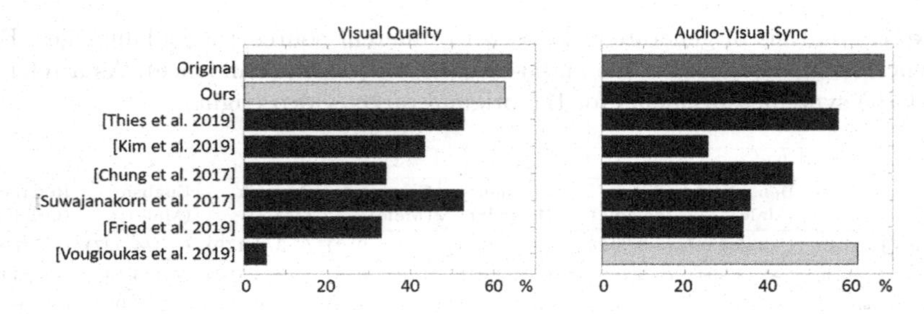

Fig. 6. User study: percentage of attendees (in total 56) that rated the visual and audio-visual quality good or very good.

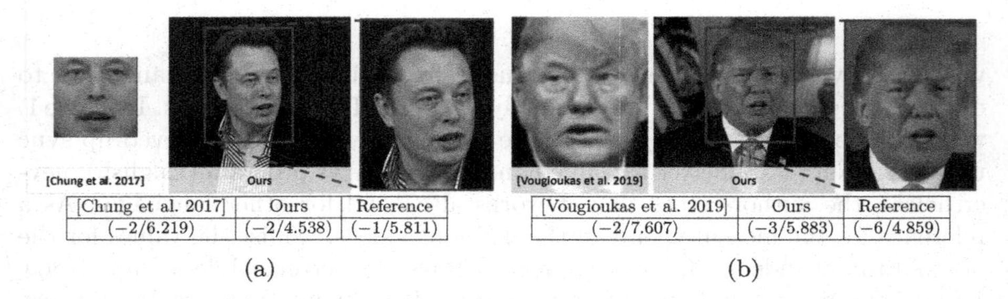

[Chung et al. 2017]	Ours	Reference
(−2/6.219)	(−2/4.538)	(−1/5.811)

(a)

[Vougioukas et al. 2019]	Ours	Reference
(−2/7.607)	(−3/5.883)	(−6/4.859)

(b)

Fig. 7. Visual quality comparison to the image-based methods (a) 'You said that?' [5] and (b) 'Realistic Speech-Driven Facial Animation with GANs' [32] (driven by the same input audio stream, respectively), including the synchronicity measurements using SyncNet [4] (offset/confidence).

operating in a normalized space of facial imagery (cropped, frontal faces) and do not capture person-specific talking styles, but, therefore, can be applied to single input images. In Fig. 7 as well as in the video, we show generated images of state-of-the-art image-based methods [5,32]. It illustrates the inherent visual quality differences that has also been quantified in our user study (see Fig. 6). The figure also includes the quantitative synchronicity measurements using SyncNet. Especially, Vougioukas et al. [32] achieves a high confidence score, while our method is in the range of the target video it has been trained on (compare to Fig. 6).

Model-based Audio-Driven Methods: In Fig. 8 we show a representative image from a comparison to Taylor et al. [25], Karras et al. [16] and Suwajanakorn et al. [23]. Only the method of Suwajanakorn et al. is able to produce photo-realistic output. The method is fitted to the scenario where a large video dataset of the target person is available and, thus, limited in its applicability. They demonstrate it on sequences of Obama, using 14 hours of training data and 3 hours for validation. In contrast, our method works on short $2-3$ min target video clips. Measuring the audio visual sync with SyncNet [4], the generated Obama videos in Fig. 8 (top row) result in $-2/5.9$ (offset/confidence) for the person-specific approach of Suwajanakorn et al., and $0/5.2$ for our generalized expression prediction network.

Fig. 8. Comparison to state-of-the-art audio-driven model-based video avatars using the same input audio stream. Our approach is applicable to multiple targets, especially, where only 2 – 3min of training data are available.

7 Limitations

As can be seen in the supplemental video, our approach works robustly on different audio sources and target videos. But it still has limitations. Especially, in the scenario of multiple voices in the audio stream our method fails. Recent work is solving this 'cocktail party' issue using visual cues [7]. As all other reenactment approaches, the target videos have to be occlusion free to allow good visual tracking. In addition, the audio-visual sync of the original videos has to be good since it transfers to the quality of the reenactment.

We assume that the target actor has a constant talking style during a target sequence. In follow-up work, we plan to estimate the talking style from the audio signal to adaptively control the expressiveness of the facial motions.

8 Conclusion

We presented a novel audio-driven facial reenactment approach that is generalized among different audio sources. This allows us not only to synthesize videos of a talking head from an audio sequence from another person, but also to generate a photo-realistic video based on a synthesized voice. I.e., text-driven video synthesis can be achieved that is in sync with artificial voice. We hope that our work is a stepping stone in the direction to photo-realistic audio-visual assistants.

Acknowledgments. We gratefully acknowledge the support by the AI Foundation, Google, Sony, a TUM-IAS Rudolf Mößbauer Fellowship, the ERC Starting Grant *Scan2CAD* (804724), the ERC Consolidator Grant *4DRepLy* (770784), and a Google Faculty Award.

References

1. Blanz, V., Vetter, T.: A morphable model for the synthesis of 3D faces. In: ACM Transactions on Graphics (Proceedings of SIGGRAPH), pp. 187–194 (1999)
2. Bregler, C., Covell, M., Slaney, M.: Video rewrite: driving visual speech with audio. In: Proceedings of the 24th Annual Conference on Computer Graphics and Interactive Techniques. SIGGRAPH 1997, pp. 353–360 (1997)
3. Chen, L., Maddox, R.K., Duan, Z., Xu, C.: Hierarchical cross-modal talking face generation with dynamic pixel-wise loss. In: Proceedings of the IEEE Conference on Computer Vision and Pattern Recognition, pp. 7832–7841 (2019)
4. Chung, J.S., Zisserman, A.: Out of time: automated lip sync in the wild. In: Chen, C.-S., Lu, J., Ma, K.-K. (eds.) ACCV 2016. LNCS, vol. 10117, pp. 251–263. Springer, Cham (2017). https://doi.org/10.1007/978-3-319-54427-4_19
5. Chung, J.S., Jamaludin, A., Zisserman, A.: You said that? In: British Machine Vision Conference (BMVC) (2017)
6. Cudeiro, D., Bolkart, T., Laidlaw, C., Ranjan, A., Black, M.: Capture, learning, and synthesis of 3D speaking styles. In: Computer Vision and Pattern Recognition (CVPR) (2019)
7. Ephrat, A., et al.: Looking to listen at the cocktail party: a speaker-independent audio-visual model for speech separation. ACM Trans. Graph. **37**(4), 112:1–112:11 (2018)
8. Ezzat, T., Geiger, G., Poggio, T.: Trainable videorealistic speech animation. ACM Trans. Graph. **21**, 388–398 (2002)
9. Finn, K.E.: Video-Mediated Communication. L. Erlbaum Associates Inc., Mahwah (1997)
10. Fried, O., et al.: Text-based editing of talking-head video. ACM Trans. Graph. (Proceedings of SIGGRAPH) **38**(4), 68:1–68:14 (2019)
11. Garrido, P., et al.: VDub - modifying face video of actors for plausible visual alignment to a dubbed audio track. In: Computer Graphics Forum (Proceedings of EUROGRAPHICS) (2015)
12. Garrido, P., et al.: Reconstruction of personalized 3D face rigs from monocular video. ACM Trans. Graph. (Proceedings of SIGGRAPH) **35**(3), 28 (2016)
13. Hannun, A., et al.: DeepSpeech: scaling up end-to-end speech recognition, December 2014
14. Jia, Y., et al.: Transfer learning from speaker verification to multispeaker text-to-speech synthesis. In: International Conference on Neural Information Processing Systems (NIPS), pp. 4485–4495 (2018)
15. Johnson, J., Alahi, A., Fei-Fei, L.: Perceptual losses for real-time style transfer and super-resolution. In: Leibe, B., Matas, J., Sebe, N., Welling, M. (eds.) ECCV 2016. LNCS, vol. 9906, pp. 694–711. Springer, Cham (2016). https://doi.org/10.1007/978-3-319-46475-6_43
16. Karras, T., Aila, T., Laine, S., Herva, A., Lehtinen, J.: Audio-driven facial animation by joint end-to-end learning of pose and emotion. ACM Trans. Graph. (Proceedings of SIGGRAPH) **36**(4), 1–12 (2017)
17. Kim, H., et al.: Neural style-preserving visual dubbing. ACM Trans. Graph. (SIGGRAPH Asia) **38**, 1–13 (2019)
18. Kim, H., et al.: Deep video portraits. ACM Trans. Graph. (Proceedings of SIGGRAPH) **37**(4), 163:1–163:14 (2018)
19. Li, T., Bolkart, T., Black, M.J., Li, H., Romero, J.: Learning a model of facial shape and expression from 4D scans. ACM Trans. Graph. **36**(6), 194:1–194:17 (2017). Two first authors contributed equally

20. Lombardi, S., Saragih, J., Simon, T., Sheikh, Y.: Deep appearance models for face rendering. ACM Trans. Graph. (Proceedings of SIGGRAPH) **37**(4), 68:1–68:13 (2018)
21. van den Oord, A., et al.: WaveNet: a generative model for raw audio. arXiv (2016). https://arxiv.org/abs/1609.03499
22. Pham, H.X., Cheung, S., Pavlovic, V.: Speech-driven 3D facial animation with implicit emotional awareness: a deep learning approach. In: 2017 IEEE Conference on Computer Vision and Pattern Recognition Workshops, CVPR Workshops 2017, Honolulu, HI, USA, 21–26 July 2017, pp. 2328–2336. IEEE Computer Society (2017)
23. Suwajanakorn, S., Seitz, S.M., Kemelmacher-Shlizerman, I.: Synthesizing Obama: learning lip sync from audio. ACM Trans. Graph. (Proceedings of SIGGRAPH) **36**(4), 1–13 (2017)
24. Tarasuik, J., Kaufman, J., Galligan, R.: Seeing is believing but is hearing? comparing audio and video communication for young children. Front. Psychol. **4**, 64 (2013)
25. Taylor, S., et al.: A deep learning approach for generalized speech animation. ACM Trans. Graph. **36**(4), 1–11 (2017)
26. Thies, J., Zollhöfer, M., Stamminger, M., Theobalt, C., Nießner, M.: FaceVR: real-time gaze-aware facial reenactment in virtual reality. ACM Trans. Graph. **37**(2), 1–15 (2018)
27. Thies, J., Zollhöfer, M., Stamminger, M., Theobalt, C., Nießner, M.: HeadOn: real-time reenactment of human portrait videos. ACM Trans. Graph. (Proceedings of SIGGRAPH) **37**, 1–13 (2018)
28. Thies, J., Zollhöfer, M., Stamminger, M., Theobalt, C., Nießner, M.: Face2Face: real-time face capture and reenactment of RGB videos. In: CVPR (2016)
29. Thies, J., Zollhöfer, M., Nießner, M.: Deferred neural rendering: image synthesis using neural textures. ACM Trans. Graph. (Proceedings of SIGGRAPH) **38**, 1–12 (2019)
30. Tzirakis, P., Papaioannou, A., Lattas, A., Tarasiou, M., Schuller, B.W., Zafeiriou, S.: Synthesising 3D facial motion from "in-the-wild" speech. CoRR abs/1904.07002 (2019)
31. Vougioukas, K., Petridis, S., Pantic, M.: End-to-end speech-driven facial animation with temporal GANs. In: BMVC (2018)
32. Vougioukas, K., Petridis, S., Pantic, M.: Realistic speech-driven facial animation with GANs. Int. J. Comput. Vis. **128**(5), 1398–1413 (2019)
33. Zhang, H., Goodfellow, I.J., Metaxas, D.N., Odena, A.: Self-attention generative adversarial networks. arXiv:1805.08318 (2018)
34. Zollhöfer, M., et al.: State of the art on monocular 3D face reconstruction, tracking, and applications. Comput. Graph. Forum (Eurographics State of the Art Reports) **37**, 523–550 (2018)

One-Shot Unsupervised Cross-Domain Detection

Antonio D'Innocente[1,3], Francesco Cappio Borlino[2], Silvia Bucci[2,3](✉), Barbara Caputo[2,3], and Tatiana Tommasi[2,3]

[1] Sapienza University of Rome, Rome, Italy
dinnocente@diag.uniroma1.it
[2] Politecnico di Torino, Turin, Italy
{francesco.cappio,silvia.bucci,barbara.caputo,tatiana.tommasi}@polito.it
[3] Italian Institute of Technology, Turin, Italy

Abstract. Despite impressive progress in object detection over the last years, it is still an open challenge to reliably detect objects across visual domains. All current approaches access a sizable amount of target data at training time. This is a heavy assumption, as often it is not possible to anticipate the domain where a detector will be used, nor to access it in advance for data acquisition. Consider for instance the task of monitoring image feeds from social media: as every image is uploaded by a different user it belongs to a different target domain that is impossible to foresee during training. Our work addresses this setting, presenting an object detection algorithm able to perform unsupervised adaptation across domains by using only one target sample, seen at test time. We introduce a multi-task architecture that one-shot adapts to any incoming sample by iteratively solving a self-supervised task on it. We further enhance this auxiliary adaptation with cross-task pseudo-labeling. A thorough benchmark analysis against the most recent cross-domain detection methods and a detailed ablation study show the advantage of our approach.

Keywords: Object detection · Cross-domain analysis · Self-supervision

1 Introduction

Social media feed us every day with an unprecedented amount of visual data. Images are uploaded by various actors, from corporations to political parties, institutions, entrepreneurs and private citizens, with roughly $10^2 M$ unique images shared everyday on Twitter, Facebook and Instagram. For the sake of freedom of expression, control over their content is limited, and their vast majority is uploaded without any textual description of their content. Their sheer magnitude makes it imperative to use algorithms to monitor and make sense of them, finding the right balance between protecting the privacy of citizens and their right of expression, and tracking fake news (often associated with malicious

Electronic supplementary material The online version of this chapter (https://doi.org/10.1007/978-3-030-58517-4_43) contains supplementary material, which is available to authorized users.

A. Vedaldi et al. (Eds.): ECCV 2020, LNCS 12361, pp. 732–748, 2020.
https://doi.org/10.1007/978-3-030-58517-4_43

intentions) while fighting illegal and hate content. This in most cases boils down to the ability to automatically associate as many tags as possible to images, which in turns means determining which objects are present in a scene.

Object detection has been largely investigated since the infancy of computer vision [11,47] and continues to attract a large attention in the current deep learning era [10,19,30,52]. Most of the algorithms assume that training and test data come from the same visual domain [18,19,40]. Recently, some authors have started to investigate the more challenging yet realistic scenario where the detector is trained on data from a visual source domain, and deployed at test time in a different target domain [32,33,44,46]. This setting is usually referred to as *cross-domain detection* and heavily relies on concepts and results from the domain adaptation literature [14,20,32]. Specifically, it inherits the standard tansductive logic, according to which unsupervised target data is available at training time together with annotated source data, and can be used to adapt across domains. This approach is not suitable, neither effective, for monitoring social media feeds. Consider for instance the scenario depicted in Fig. 1, where there is an incoming stream of images from various social media and the detector is asked to look for instances of the class bicycle. The images come continuously, but they are produced by different users that share them on different social platforms. Hence, even though they might contain the same object, each of them has been acquired by a different person, in a different context, under different viewpoints and illuminations. In other words, *each image comes from a different visual domain, distinct from the visual domain where the detector has been trained.* This poses two key challenges to current cross-domain detectors: (1) to adapt to the target data, these algorithms need first to gather feeds, and only after enough target data has been collected they can learn to adapt and start performing on the incoming images; (2) even if the algorithms have learned to adapt on target images from the feed up to time t, there is no guarantee that the images that will arrive from time $t+1$ will come from the same target domain.

This is the scenario we address. We focus on cross-domain detection when only one target sample is available for adaptation, without any form of supervision. We propose an object detection method able to adapt from one target image, hence suitable for the social media scenario described above. Specifically, we build a multi-task deep architecture that adapts across domains by leveraging over a pretext task. This auxiliary knowledge is further guided by a cross-task pseudo-labeling that injects the locality specific of object detection into self-supervised learning. The result is an architecture able to perform unsupervised adaptive object detection from a single image. Extensive experiments show the power of our method compared to previous state-of-the-art approaches. To summarize, the contributions of our paper are as follows:

(1) we introduce the One-Shot Unsupervised Cross-Domain Detection setting, a cross-domain detection scenario where the target domain changes from sample to sample, hence adaptation can be learned only from one image. This scenario is especially relevant for monitoring social media image feeds. We are not aware of previous works addressing it.

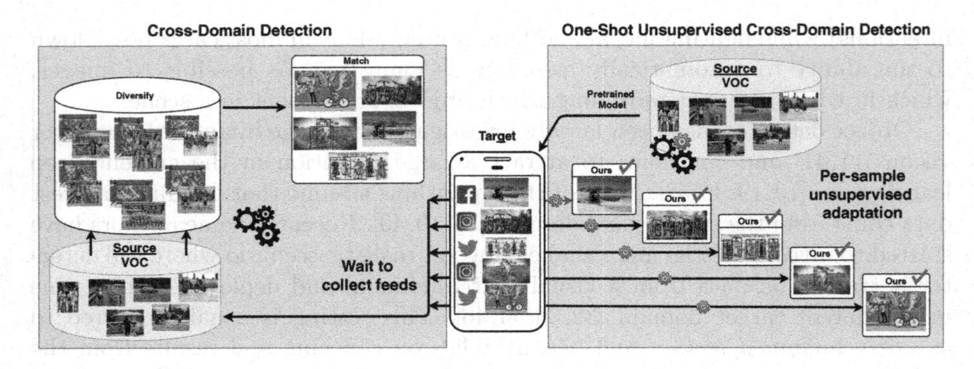

Fig. 1. Each social media image comes from a different domain. Existing Cross-Domain Detection algorithms (*e.g.*[28] in the left gray box) struggle to adapt in this setting. OSHOT (right) is able to adapt across domains from one single target image, thanks to the combined use of self-supervision and pseudo-labeling

(2) We propose OSHOT, the first cross-domain object detector able to perform one-shot unsupervised adaptation. Our approach leverages over self-supervised one-shot learning guided by a cross-task pseudo-labeling procedure, embedded into a multi-task architecture. A thorough ablation study showcases the importance of each component.
(3) We present a new experimental setup for studying one-shot unsupervised cross-domain adaptation, designed on three existing databases plus a new test set collected from social media feed. We compare against recent algorithms in cross-domain adaptive detection [28, 42] and one-shot unsupervised learning [8], achieving the state-of-the-art.

We make the code of our project available at https://github.com/VeloDC/oshot_detection.

2 Related Work

Object Detection. Many successful object detection approaches have been developed during the past several years, starting from the original sliding window methods based on handcrafted features, till the most recent deep-learning empowered solutions. Modern detectors can be divided into *one-stage* and *two-stage* techniques. In the former, classification and bounding box prediction is performed on the convolution feature map either solving a regression problem on grid cells [39], or exploiting anchor boxes at different scales and aspect ratios [31]. In the latter, an initial stage deals with the region proposal process and is followed by a refinement stage that adjusts the coarse region localization and classifies the box content. Existing variants of this strategy differ mainly in the region proposal algorithm [18, 19, 40]. Regardless of the specific implementation, the detector robustness across visual domains remains a major issue.

Cross-Domain Detection. When training and test data are drawn from two different distributions a model learned on the first is doomed to fail on the second. Unsupervised domain adaptation methods attempt to close the domain gap between the annotated source on which learning is performed, and the target samples on which the model is deployed. Most of the literature has focused on object classification with solutions based on feature alignment [2,32,33,44] or adversarial approaches [15,46]. GAN-based methods allow to directly update the visual style of the annotated source data and reduce the domain shift directly at pixel level [23,41]. Only in the last two years adaptive detection methods have been developed considering three main components: (i) including multiple and increasingly more accurate feature alignment modules at different internal stages, (ii) adding a preliminary pixel-level adaptation and (iii) pseudo-labeling. The last one is also known as *self-training* and consists in using the output of the source model detector as coarse annotation on the target. The importance of considering both global and local domain adaptation, together with a consistency regularizer to bridge the two, was first highlighted in [7]. The Strong-Weak (SW) method of [42] improves over the previous one pointing out the need of a better balanced alignment with strong global and weak local adaptation. It was also further extended by [49], where the adaptive steps are multiplied at different depth in the network. By generating new source images that look like those of the target, the Domain-Transfer (DT, [25]) method was the first to adopt pixel adaptation for object detection and combine it with pseudo-labeling. More recently the Div-Match approach [28] re-elaborated the idea of domain randomization [45]: multiple CycleGAN [53] applications with different constraints produce three extra source variants with which the target can be aligned at different extent through an adversarial multi-domain discriminator. A weak self-training procedure (WST) to reduce false negatives is combined with adversarial background score regularization (BSR) in [27]. Finally, [26] followed the pseudo-labeling strategy including an approach to deal with noisy annotations.

Adaptive Learning on a Budget. There is a wide literature on learning from a limited amount of data, both for classification and detection. However, in case of domain shift, learning on a target budget becomes extremely challenging. Indeed, the standard assumption for adaptive learning is that a large amount of unsupervised target samples are available at training time, so that a source model can capture the target domain style from them and adapt to it.

Only few attempts have been done to reduce the target cardinality. In [36] the considered setting is that of *few-shot supervised domain adaptation*: only a few target samples are available but they are fully labeled. In [3,8] the focus is on *one-shot unsupervised style transfer* with a large source dataset and a single unsupervised target image. These works propose time-costly autoencoder-based methods to generate a version of the target image that maintains its content, but visually resembles the source in its global appearance. Thus the goal is image generation with no discriminative purpose. A related setting is that of *online domain adaptation* where unsupervised target samples are initially scarce but accumulate in time [22,34,48]. In this case target samples belong to a continuous data stream with smooth domain changing, so the coherence among subsequent samples can be exploited for adaptation.

Self-supervised Learning. Despite not-being manually annotated, unsupervised data is rich of structural information that can be learned by self-supervision, *i.e.*hiding a subpart of the data information and then trying to recover it. This procedure is generally indicated as *pretext* task and possible examples are image completion [38], colorization [29,51], relative position of patches [12,37], rotation recognition [17] and many more. Self-supervised learning has been extensively used as an initialization step for scarcely annotated supervised learning settings and very recently [1] has shown with a thorough analysis the potential of self-supervised learning from a single image. Recent works also indicated that self-supervision supports adaptation and generalization when combined with supervised learning in a multi-task framework [4,6,50].

Our approach for cross-domain detection relates to the described scenario of learning on a budget and exploits self-supervised learning to perform one-shot unsupervised adaptation. Specifically with OSHOT we show how to recognize objects and their location on a single target image starting from a pre-trained source model, thus without the need of accessing the source data during testing.

3 Method

Problem Setting. We introduce the *one-shot unsupervised cross-domain detection scenario* where our goal is to predict on a single image x^t, with t being any target domain not available at training time, starting from N annotated samples of the source domain $S = \{x_i^s, y_i^s\}_{i=1}^N$. Here the structured labels $y^s = (c, b)$ describe class identity c and bounding box location b in each image x^s, and we aim to obtain y^t that precisely detects objects in x^t despite the domain shift.

OSHOT Strategy. To pursue the described goal, our strategy is to train the parameters of a detection learning model such that it can be ready to get the maximal performance on a single unsupervised sample from a new domain after few gradient update steps on it. Since we have no ground truth on the target sample, we implement this strategy by learning a representation that exploits inherent data information as that captured by a *self-supervised* task, and then finetune it on the target sample (see Fig. 2). Thus, we design our OSHOT to include (1) an initial *pretraining* phase where we extend a standard deep detection model adding an image rotation classifier, and (2) a following *adaptation* stage where the network features are updated on the single target sample by further optimization of the rotation objective. Moreover, we exploit *pseudo-labeling* to focus the auxiliary task on the local object context. A clear advantage of this solution is that we decouple source training from target testing, with no need to access the source data while adapting on the target sample.

Preliminaries. We leverage on Faster R-CNN [40] as our base detection model. It is a two-stage detector with three main components: an initial block of convolutional layers, a region proposal network (RPN) and a region-of-interest (ROI)

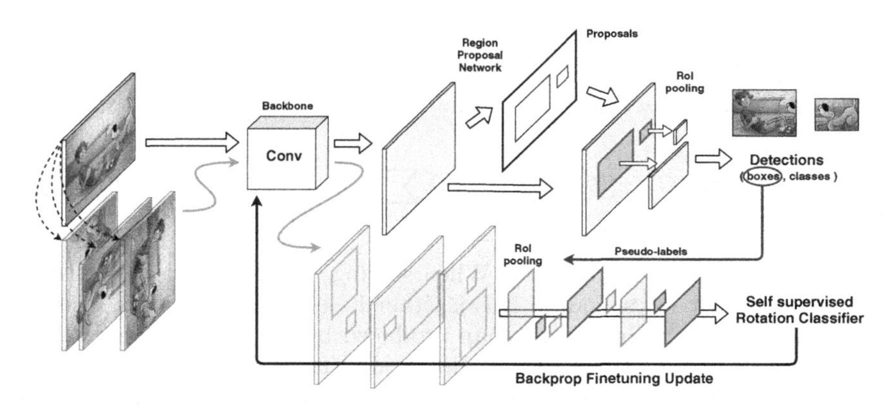

Fig. 2. Visualization of the adaptive phase of OSHOT with cross-task pseudo-labeling. The target image passes through the network and produces detections. While the class information is not used, the identified boxes are exploited to select object regions from the feature maps of the rotated image. The obtained region-specific feature vectors are finally sent to the rotation classifier. A number of subsequent finetuning iterations allows to adapt the convolutional backbone to the domain represented by the test image

based classifier. The bottom layers transform any input image x into its convolutional feature map $G_f(x|\theta_f)$ where θ_f is used to parametrize the feature extraction model. The feature map is then used by RPN to generate candidate object proposals. Finally the ROI-wise classifier predicts the category label from the feature vector obtained using ROI-pooling. The training objective combines the loss of both RPN and ROI, each of them composed by two terms:

$$\begin{aligned}\mathcal{L}_d(G_d(G_f(x|\theta_f)|\theta_d), y) = &\big(\mathcal{L}_{class}(c^*) + \mathcal{L}_{regr}(b)\big)_{RPN} + \\ &\big(\mathcal{L}_{class}(c) + \mathcal{L}_{regr}(b)\big)_{ROI}\ .\end{aligned} \tag{1}$$

Here $\mathcal{L}_{class}$ is a classification loss to evaluate the object recognition accuracy, while $\mathcal{L}_{regr}$ is a regression loss on the box coordinates for better localization. To maintain a simple notation we summarize the role of ROI and RPN with the function $G_d(G_f(x|\theta_f)|\theta_d)$ parametrized by θ_d. Moreover, we use c^* to highlight that RPN deals with a binary classification task to separate foreground and background objects, while ROI deals with the multi-class objective needed to discriminate among c foreground object categories. As mentioned above, ROI and RPN are applied in sequence: they both elaborate on the feature maps produced by the convolutional block, and then influence each other in the final optimization of the multi-task (classification, regression) objective function.

OSHOT Pretraining. As a first step, we extend Faster R-CNN to include image rotation recognition. Formally, to each source training image x^s we apply four geometric transformations $R(x, \alpha)$ where $\alpha = q \times 90°$ indicates rotations with $q \in \{1, \dots, 4\}$. In this way we obtain a new set of samples $\{R(x)_j, q_j\}_{j=1}^{M}$

where we dropped the α without loss of generality. We indicate the auxiliary rotation classifier and its parameters respectively with G_r and θ_r and we train our network to optimize the following multi-task objective

$$\underset{\theta_f,\theta_d,\theta_r}{\operatorname{argmin}} \sum_{i=1}^{N} \mathcal{L}_d(G_d(G_f(x_i^s|\theta_f)|\theta_d), y_i^s) + \lambda \sum_{j=1}^{M} \mathcal{L}_r(G_r(G_f(R(x^s)_j|\theta_f)|\theta_r), q_j^s), \tag{2}$$

where $\mathcal{L}_r$ is the cross-entropy loss. When solving this problem, we can design G_r in two different ways. Indeed it can either be a Fully Connected layer that naïvely takes as input the feature map produced by the whole (rotated) image $G_r(\cdot|\theta_r) = \text{FC}_{\theta_r}(\cdot)$, or it can exploit the ground truth location of each object with a subselection of the features only from its bounding box in the original map $G_r(\cdot|\theta_r) = \text{FC}_{\theta_r}(\mathit{boxcrop}(\cdot))$. The *boxcrop* operation includes pooling to rescale the feature dimension before entering the final FC layer. In this last case the network is encouraged to focus only on the object orientation without introducing noisy information from the background and provides better results with respect to the whole image option as we will discuss in Sect. 4.4. In practical terms, both in the case of image and box rotations, we randomly pick one rotation angle per instance, rather than considering all four of them: this avoids any troublesome unbalance between rotated and non-rotated data when solving the multi-task optimization problem.

OSHOT Adaptation. Given the single target image x^t, we finetune the backbone's parameters θ_f by iteratively solving a self-supervised task on it. This allows to adapt the original feature representation both to the content and to the style of the new sample. Specifically, we start from the rotated versions $R(x^t)$ of the provided sample and optimize the rotation classifier through

$$\underset{\theta_f,\theta_r}{\operatorname{argmin}} \mathcal{L}_r(G_r(G_f(R(x^t)|\theta_f)|\theta_r), q^t) \ . \tag{3}$$

This process involves only G_f and G_r, while the RPN and ROI detection components described by G_d remain unchanged. In the following we use γ to indicate the number of gradient steps (*i.e.*iterations), with $\gamma = 0$ corresponding to the OSHOT pretraining phase. At the end of the finetuning process, the inner feature model is described by θ_f^* and the detection prediction on x^t is obtained by $y^{t*} = G_d(G_f(x^t|\theta_f^*)|\theta_d)$.

Cross-Task Pseudo-labeling. As in the pretraining phase, also in the adaptation stage we have two possible choices to design G_r: either considering the whole feature map $G_r(\cdot|\theta_r) = \text{FC}_{\theta_r}(\cdot)$, or focusing on the object locations $G_r(\cdot|\theta_r) = \text{FC}_{\theta_r}(\mathit{pseudoboxcrop}(\cdot))$. For both variants we include dropout to prevent overfitting on the single target sample. With *pseudoboxcrop* we mean a localized feature extraction operation analogous to that discussed for pretraining, but obtained through a particular form of *cross-task self-training.* Specifically, we follow the self-training strategy used in [25,27] with a cross-task variant: instead of reusing the pseudo-labels produced by the source model on the target to update the detector, we exploit them for the self-supervised rotation classifier. In this way we keep the advantage of the self-training initialization, largely reducing the risks of error propagation due to wrong class pseudo-labels.

More practically, we start from the (θ_f, θ_d) model parameters of the pre-training stage and we get the feature maps from all the rotated versions of the target sample $G_f(\{R(x^t), q\}|\theta_f)$, $q = 1, \ldots, 4$. Only the feature map produced by the original image ($i.e. q = 4$) is provided as input to the RPN and ROI network components to get the predicted detection $y^t = (c, b) = G_d(G_f(x^t|\theta_f)|\theta_d)$. This pseudo-label is composed by the class label c and the bounding box location b. We discard the first and consider only the second to localize the region containing an object in all the four feature maps, also recalibrating the position to compensate for the orientation of each map. Once passed through this *pseudoboxcrop* operation, the obtained features are used to finetune the rotation classifier, updating the bottom convolutional network block.

4 Experiments

4.1 Datasets

Real-World (VOC). Pascal-VOC [13] is the standard real-world image dataset for object detection benchmarks. VOC2007 and VOC2012 both contain bounding boxes annotations of 20 common categories. VOC2007 has 5011 images in the train-val split and 4952 images in the test split, while VOC2012 contains 11540 images in the train-val split.
Artistic Media Datasets (AMD). Clipart1k, Comic2k and Watercolor2k [25] are three object detection datasets designed for benchmarking Domain Adaptation methods when the source domain is Pascal-VOC. Clipart1k shares its 20 categories with Pascal-VOC: it has 500 images in the training set and 500 images in the test set. Comic2k and Watercolor2k both have the same 6 classes (a subset of the 20 classes of Pascal-VOC), and 1000-1000 images in the training-test splits each.

Cityscapes [9] is an urban street scene dataset with pixel level annotations of 8 categories. It has 2975 and 500 images respectively in the training and validation splits. We use the instance level pixel annotations to generate bounding boxes of objects, as in [7].

Foggy Cityscapes [43] is obtained by adding different levels of synthetic fog to Cityscapes images. We only consider images with the highest amount of artificial fog, thus training-validation splits have 2975-500 images respectively.

KITTI [16] is a dataset of images depicting several driving urban scenarios. By following [7], we use the full 7481 images for both training (when used as source) and evaluation (when used as target).

Social Bikes is our new concept-dataset containing 30 images of scenes with persons/bicycles collected from Twitter, Instagram and Facebook by searching for *#bike* tags. Square crops of the full dataset are presented in Fig. 3: images

Fig. 3. The Social Bikes concept-dataset. A random data acquisition from multiple users/feeds leads to a target distribution with several, uneven domain shifts

acquired randomly from social feeds show diverse style properties and cannot be grouped under a single shared domain.

4.2 Performance Analysis

Experimental Setup. We evaluate OSHOT on several testbeds using the described datasets. In the following we will use an arrow *Source* → *Target* to indicate the experimental setting. Our base detector is Faster-RCNN [35] with a ResNet-50 [21] backbone pre-trained on ImageNet, RPN with 300 top proposals after non-maximum-supression, anchors at three scales (128, 256, 512) and three aspect ratios (1:1, 1:2, 2:1). For all our experiments we set the IoU threshold at 0.5 for the mAP results, and report the average of three independent runs.

OSHOT Pretraining. We always resize the image's shorter size to 600 pixels and apply random horizontal flipping. Unless differently specified, we train the base network for 70k iterations using SGD with momentum set at 0.9, the initial learning rate is 0.001 and decays after 50k iterations. We use a batch size of 1, keep batch normalization layers fixed for both pretraining and adaptation phases and freeze the first 2 blocks of ResNet50. The weight of the auxiliary task is set to $\lambda = 0.05$.

OSHOT Adaptation. We increase the weight of the auxiliary task to $\lambda = 0.2$ to speed up adaptation and keep all other training hyperparameters fixed. For *each* test instance, we finetune the *initial* model on the auxiliary task for 30 iterations before testing.

Benchmark Methods. We compare OSHOT with the following algorithms. *FRCNN*: baseline Faster-RCNN with ResNet50 backbone, trained on the source domain and deployed on the target without further adaptation. *DivMatch* [28]: cross-domain detection algorithm that, by exploiting target data, creates multiple randomized domains via CycleGAN and aligns their representations using an adversarial loss. *SW* [42]: adaptive detection algorithm that aligns source and target features based on global context similarity. For both DivMatch and SW, we use a ResNet-50 backbone pretrained on ImageNet for fair comparison. Since all cross-domain algorithms need target data in advance and are not designed to

Table 1. (left) VOC → Social Bikes mAP results; (right) visualization of DivMatch and OSHOT detections. The number associated with each bounding box indicates the model's confidence in localization. Examples show how OSHOT detection is accurate, while most DivMatch boxes are false positives

One-Shot *Target*			
Method	person	bicycle	mAP
FRCNN	67.7	56.6	62.1
OSHOT ($\gamma = 0$)	72.1	52.8	62.4
OSHOT ($\gamma = 30$)	69.4	59.4	**64.4**
Full *Target*			
DivMatch [28]	63.7	51.7	57.7
SW [42]	63.2	44.3	53.7

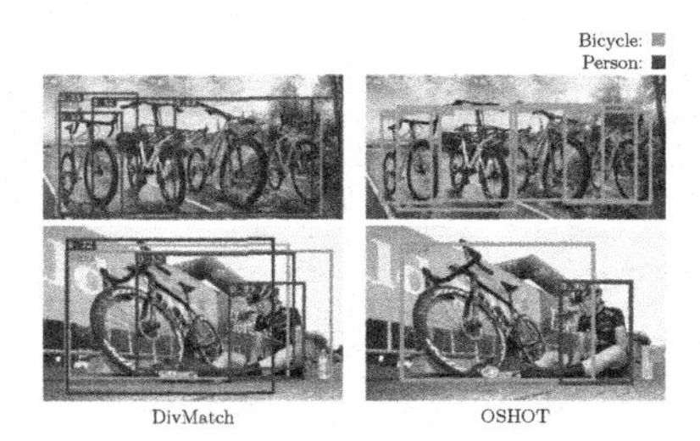

work in our one-shot unsupervised setting, we provide them with the advantage of 10 target images accessible during training and randomly selected at each run. We collect average precision statistics during inference under the favorable assumption that the target domain will not shift after deployment.

Adapting to Social Feeds. When data is collected from multiple sources, the assumption that all target images originate from the same underlying distribution does not hold and standard cross-domain detection methods are penalized regardless of the number of seen target samples. We pretrain the source detector on Pascal VOC, and deploy it on Social Bikes. We consider only the bicycle and person annotations for this target, since all other instances of VOC classes are scarce. We report results in Table 1. OSHOT outperforms all considered competitors, with a mAP score of 64.4. Despite granting them access to the full target, adaptive algorithms incur in negative transfer due to data scarcity and large variety of target styles.

Large Distribution Shifts. Artistic images are difficult benchmarks for cross-domain methods. Unpredictable perturbations in shape and color are challenging to detectors trained only on realistic images. We investigate this setting by training the source detector on Pascal VOC an deploying it on Clipart, Comic and Watercolor datasets. Table 2 summarizes results on the three adaptation splits. We can see how OSHOT with 30 finetuning iterations outperforms all competitors, with mAP gains ranging from 7.5 points on Clipart to 9.2 points on Watercolor. Cross-detection methods perform poorly in this setting, despite using 9 more samples in the adaptation phase compared to OSHOT that only uses the test sample. These results confirm that they are not designed to tackle data scarcity conditions and exhibit negligible improvements compared to the baseline.

Adverse Weather. Some peculiar environmental conditions, such as fog, may be disregarded in source data acquisition, yet adaptation to these circumstances is crucial for real world applications. We assess the performance of OSHOT on Cityscapes → FoggyCityscapes. We train our base detector on Cityscapes for

Table 2. mAP results for VOC → AMD

(a) VOC → Clipart

One-Shot Target																					
Method	aero	bike	bird	boat	bottle	bus	car	cat	chair	cow	table	dog	horse	mbike	person	plant	sheep	sofa	train	tv	mAP
FRCNN	18.5	43.3	20.4	13.3	21.0	47.8	29.0	16.9	28.8	12.5	19.5	17.1	23.8	40.6	34.9	34.7	9.1	18.3	40.2	38.0	26.4
OSHOT ($\gamma = 0$)	23.1	55.3	22.7	21.4	26.8	53.3	28.9	4.6	31.4	9.2	27.8	9.6	30.9	47.0	38.2	35.2	11.1	20.4	36.0	33.6	28.3
OSHOT ($\gamma = 10$)	25.4	61.6	23.8	21.1	31.3	55.1	31.6	5.3	34.0	10.1	28.8	7.3	33.1	59.9	44.2	38.8	15.9	19.1	39.5	33.9	31.0
OSHOT ($\gamma = 30$)	25.4	56.0	24.7	25.3	36.7	58.0	34.4	5.9	34.9	10.3	29.2	11.8	46.9	70.9	52.9	41.5	21.1	21.0	38.5	31.8	**33.9**
Ten-Shot Target																					
DivMatch [28]	19.5	57.2	17.0	23.8	14.4	25.4	29.4	2.7	35.0	8.4	22.9	14.2	30.0	55.6	50.8	30.2	1.9	12.3	37.8	37.2	26.3
SW [42]	21.5	39.9	21.7	20.5	32.7	34.1	25.1	8.5	33.2	10.9	15.2	3.4	32.2	56.9	46.5	35.4	14.7	15.2	29.2	32.0	26.4

(b) VOC → Comic

One-Shot Target							
Method	bike	bird	car	cat	dog	person	mAP
FRCNN	25.2	10.0	21.1	14.1	11.0	27.1	18.1
OSHOT ($\gamma = 0$)	26.9	11.6	22.7	9.1	14.2	28.3	18.8
OSHOT ($\gamma = 10$)	35.5	11.7	25.1	9.1	15.8	34.5	22.0
OSHOT ($\gamma = 30$)	35.2	14.4	30.0	14.8	20.0	46.7	**26.9**
Ten-Shot Target							
DivMatch [28]	27.1	12.3	26.2	11.5	13.8	34.0	20.8
SW [42]	21.2	14.8	18.7	12.4	14.9	43.9	21.0

(c) VOC → Watercolor

One-Shot Target							
Method	bike	bird	car	cat	dog	person	mAP
FRCNN	62.5	39.7	43.4	31.9	26.7	52.4	42.8
OSHOT ($\gamma = 0$)	70.2	46.7	45.5	31.2	27.2	55.7	46.1
OSHOT ($\gamma = 10$)	70.2	46.7	48.1	30.9	32.3	59.9	48.0
OSHOT ($\gamma = 30$)	77.1	44.7	52.4	37.3	37.0	63.3	**52.0**
Ten-Shot Target							
DivMatch [28]	64.6	44.1	44.6	34.1	24.9	60.0	45.4
SW [42]	66.3	41.1	41.1	30.5	20.5	52.3	42.0

Table 3. mAP results for Cityscapes → FoggyCityscapes

One-Shot Target									
Method	Person	Rider	Car	Truck	Bus	Train	Mcycle	Bicycle	mAP
FRCNN	30.4	36.3	41.4	18.5	32.8	9.1	20.3	25.9	26.8
OSHOT ($\gamma = 0$)	31.8	42.0	42.6	20.1	31.6	10.6	24.8	30.7	29.3
OSHOT ($\gamma = 10$)	31.9	41.9	43.0	19.7	38.0	10.4	25.5	30.2	30.1
OSHOT ($\gamma = 30$)	32.1	46.1	43.1	20.4	39.8	15.9	27.1	32.4	**31.9**
Ten-Shot Target									
DivMatch [28]	27.6	38.1	42.9	17.1	27.6	14.3	14.6	32.8	26.9
SW [42]	25.5	30.8	40.4	21.1	26.1	34.5	6.1	13.4	24.7
Full Target									
DivMatch [28]	32.3	43.5	47.6	23.9	38.0	23.1	27.6	37.2	34.2
SW [42]	31.3	32.1	47.4	19.6	28.8	41.0	9.8	20.1	28.8

30k iterations without stepdown, as in [5]. We select the best performing model on the Cityscapes validation split and deploy it to FoggyCityscapes. Experimental evaluation in Table 3 shows that OSHOT outperforms all compared approaches. Without finetuning iterations, performance using the auxiliary rotation task increases compared to the baseline. Subsequent finetuning iterations on the target sample improve these results, and 30 iterations yield models able to outperform the second-best method by 5 mAP. Cross-domain algorithms used in this setting struggle to surpass the baseline (DivMatch) or suffer negative transfer (SW).

Cross-Camera Transfer. Dataset bias between training and testing is unavoidable in practical applications, as for urban scene scenarios collected in different cities and with different cameras. We test adaptation between KITTI and Cityscapes in both directions. For cross-domain evaluation we consider only the

Table 4. mAP of car class in KITTI/Cityscapes detection experiments

	One-Shot *Target*	
Method	KITTI → Cityscapes	Cityscapes → KITTI
FRCNN	26.5	75.1
OSHOT $\gamma = 0$	26.2	**75.4**
OSHOT $\gamma = 10$	33.2	75.3
OSHOT $\gamma = 30$	**33.5**	75.0
	Ten-Shot *Target*	
DivMatch [28]	37.9	74.1
SW [42]	39.2	74.6

Table 5. Comparison between baseline, one-shot syle transfer and OSHOT in the one-shot unsupervised cross-domain detection setting

	FRCNN	BiOST [8]	OSHOT ($\gamma = 30$)
mAP on Clipart100	27.9	29.8	**30.7**
mAP on Social Bikes	62.1	51.1	**64.4**
Adaptation time (seconds per sample)	–	$\sim 2.4 * 10^4$	7.8

label car as standard practice. In Table 4, OSHOT improves by 7 mAP points on KITTI → Cityscapes compared to the FRCNN baseline. DivMatch and SW both show a gain in this split, with SW obtaining the highest mAP of 39.2 in the ten-shot setting. We argue that this is not surprising considering that, as shown in the visualization of Table 4, the Cityscapes images share all a uniform visual style. As a consequence, 10 target images may be enough for standard cross-domain detection methods. Despite visual style homogeneity, the diversity among car instances in Cityscapes is high enough for learning a good car detection model. This is highlighted by the results in Cityscapes → KITTI task, for which adaptation performance for all methods is similar, and OSHOT with $\gamma = 0$ obtains the highest mAP of 75.4. The FRCNN baseline on KITTI scores a high mAP of 75.1: in this favorable condition detection doesn't benefit from adaptation.

4.3 Comparison with One-Shot Style Transfer

Although not specifically designed for cross-domain detection, in principle it is possible to apply one-shot style transfer methods as an alternative solution for our setting. We use BiOST [8], the current state-of-the-art method for one-shot transfer, to modify the style of the target sample towards that of the source domain before performing inference. Due to the time-heavy requirements to perform BiOST on each test sample[1], we test it on Social Bikes and on a random subset of 100 Clipart images that we name Clipart100. We compare performance

[1] To get the style update, BiOST trains of a double-variational autoencoder using the entire source besides the single target sample. As advised by the authors through personal communications, we trained the model for 5 epochs.

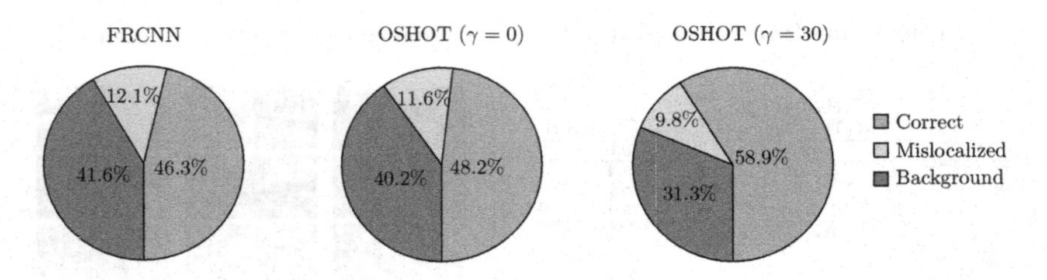

Fig. 4. Detection error analysis on the most confident detections on Clipart

and time requirements of OSHOT and BiOST on these two targets. Speed has been computed on an RTX2080Ti with full precision settings.

Table 5 shows summary mAP results using BiOST and OSHOT. On Clipart100, the baseline FRCNN detector obtains 27.9 mAP. We can see how BiOST is effective in the adaptation from one-sample, gaining 1.9 points over the baseline, however it is outperformed by OSHOT, which obtains 30.7 mAP. On Social Bikes, while OSHOT still outperforms the baseline, BiOST incurs in negative transfer, indicating that it was not able to effectively modify the source's style on the images we collected. Furthermore, BiOST is affected by two strong issues: (1) as already mentioned, it has an extremely high time complexity, with more than 6 hours needed to modify the style of a single source instance; (2) it works under the strict assumption of accessing at the same time the entire source training set and the target sample. Due to these weaknesses, and the fact that OSHOT still outperforms BiOST, we argue that existing one-shot translation methods are not suitable for one shot unsupervised cross-domain adaptation.

4.4 Ablation Study

Detection Error Analysis. Following [24], we provide detection error analysis for VOC $\rightarrow$ Clipart setting in Fig. 4. We select the 1000 most confident detections, and assign error classes based on IoU with ground truth (IoUgt). Errors are categorized as: correct (IoUgt $\geqslant 0.5$), mislocalized ($0.3 \leqslant$ IoUgt < 0.5) and background (IoUgt < 0.3). Results show that, compared to the baseline FRCNN model, the regularization effect of adding a self-supervised task at training time ($\gamma = 0$) marginally increases the quality of detections. Instead subsequent fine-tuning iterations on the test sample substantially improve the number of correct detections, while also decreasing both false positives and mislocalization errors.

Cross-Task Pseudo-labeling Ablation. As explained in Sect. 3 we have two options in the OSHOT adaptation phase: either considering the whole image, or focusing on pseudo-labeled bounding boxes obtained from the detector after the first OSHOT pretraining stage. For all the experiments presented above we focused on the second case. Indeed by solving the auxiliary task only on objects, we limit the use of background features which may mislead the network towards solutions of the rotation task not based on relevant semantic information

Table 6. Rotating image vs rotating objects via pseudo-labeling on OSHOT

	$G_r(image)$	$G_r(pseudoboxcrop)$
VOC → Clipart	31.0	**33.9**
VOC → Comic	21.0	**26.9**
VOC → Watercolor	48.2	**52.0**
Cityscapes → Foggy Cityscapes	27.7	**31.9**

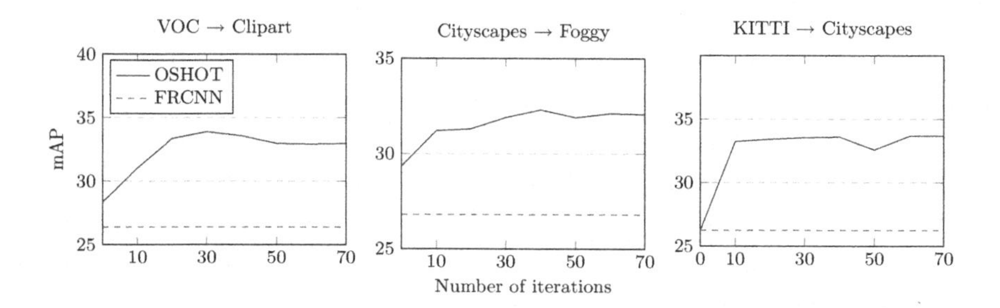

Fig. 5. Performance of OSHOT at different self-supervised iterations

(*e.g.*: finding fixed patterns in images, exploiting watermarks). We validate our choice by comparing it against using the rotation task on the entire image in both training and adaptation phases. Table 6 shows results for VOC → AMD and Cityscapes → Foggy Cityscapes using OSHOT. We observe that the choice of rotated regions is critical for the effectiveness of the algorithm. Solving the rotation task on objects using pseudo-annotations results in mAP improvements that range from 2.9 to 5.9 points, indicating that we learn better features for the main task.

Self-supervised Iterations. We study the effects of adaptating with up to $\gamma = 70$ iterations on VOC → Clipart, Cityscapes → FoggyCityscapes and KITTI → Cityscapes. Results are shown in Fig. 5. We observe a positive correlation between number of finetuning iterations and final mAP of the model in the earliest steps. This correlation is strong for the first 10 iterations and gets to a plateau after about 30 iterations: increasing γ beyond this point doesn't affect the final results.

5 Conclusions

This paper introduced the *one-shot unsupervised cross-domain detection* scenario, which is extremely relevant for monitoring image feeds on social media, where algorithms are called to adapt to a new visual domain from one single image. We showed that existing cross-domain detection methods suffer in this setting, as they are all explicitly designed to adapt from far larger quantities of target data. We presented OSHOT, the first deep architecture able to reduce the domain gap between source and target distribution by leveraging over one single target image. Our approach is based on a multi-task structure that exploits self-supervision and cross-task self-labeling. Extensive quantitative experiments and a qualitative analysis clearly demonstrate its effectiveness.

Acknowledgements. This work was partially founded by the ERC grant 637076 RoboExNovo (AD, FCB, SB, BC) and took advantage of the GPU donated by NVIDIA (Academic Hardware Grant, TT). We acknowledge the support provided by Tomer Cohen and Kim Taekyung on their code respectively of BiOST and DivMatch.

References

1. Asano, Y.M., Rupprecht, C., Vedaldi, A.: A critical analysis of self-supervision, or what we can learn from a single image. In: ICLR (2020)
2. Ben-David, S., Blitzer, J., Crammer, K., Kulesza, A., Pereira, F., Vaughan, J.: A theory of learning from different domains. Mach. Learn. **79**, 151–175 (2010). https://doi.org/10.1007/s10994-009-5152-4
3. Benaim, S., Wolf, L.: One-shot unsupervised cross domain translation. In: NIPS (2018)
4. Bucci, S., D'Innocente, A., Tommasi, T.: Tackling partial domain adaptation with self-supervision. In: Ricci, E., Rota Bulò, S., Snoek, C., Lanz, O., Messelodi, S., Sebe, N. (eds.) ICIAP 2019. LNCS, vol. 11752, pp. 70–81. Springer, Cham (2019). https://doi.org/10.1007/978-3-030-30645-8_7
5. Cai, Q., Pan, Y., Ngo, C.W., Tian, X., Duan, L., Yao, T.: Exploring object relation in mean teacher for cross-domain detection. In: CVPR (2019)
6. Carlucci, F.M., D'Innocente, A., Bucci, S., Caputo, B., Tommasi, T.: Domain generalization by solving jigsaw puzzles. In: CVPR (2019)
7. Chen, Y., Li, W., Sakaridis, C., Dai, D., Van Gool, L.: Domain adaptive faster R-CNN for object detection in the wild. In: CVPR (2018)
8. Cohen, T., Wolf, L.: Bidirectional one-shot unsupervised domain mapping. In: ICCV (2019)
9. Cordts, M., et al.: The cityscapes dataset for semantic urban scene understanding. In: CVPR (2016)
10. Dai, J., Li, Y., He, K., Sun, J.: R-FCN: object detection via region-based fully convolutional networks. In: NIPS (2016)
11. Dalal, N., Triggs, B.: Histograms of oriented gradients for human detection. In: CVPR (2005)
12. Doersch, C., Gupta, A., Efros, A.A.: Unsupervised visual representation learning by context prediction. In: ICCV (2015)
13. Everingham, M., Van Gool, L., Williams, C.K., Winn, J., Zisserman, A.: The pascal visual object classes (VOC) challenge. IJCV **88**(2), 303–338 (2010). https://doi.org/10.1007/s11263-009-0275-4
14. Ganin, Y., Lempitsky, V.: Unsupervised domain adaptation by backpropagation. In: ICML (2015)
15. Ganin, Y., et al.: Domain-adversarial training of neural networks. JMLR **17**(1), 2030–2096 (2016)
16. Geiger, A., Lenz, P., Stiller, C., Urtasun, R.: Vision meets robotics: the KITTI dataset. Int. J. Robot. Res. **32**(11), 1231–1237 (2013)

17. Gidaris, S., Singh, P., Komodakis, N.: Unsupervised representation learning by predicting image rotations. In: ICLR (2018)
18. Girshick, R.: Fast R-CNN. In: ICCV (2015)
19. Girshick, R., Donahue, J., Darrell, T., Malik, J.: Rich feature hierarchies for accurate object detection and semantic segmentation. In: CVPR (2014)
20. Goodfellow, I., et al.: Generative adversarial nets. In: NIPS (2014)
21. He, K., Zhang, X., Ren, S., Sun, J.: Deep residual learning for image recognition. In: CVPR (2016)
22. Hoffman, J., Darrell, T., Saenko, K.: Continuous manifold based adaptation for evolving visual domains. In: CVPR (2014)
23. Hoffman, J., et al.: CyCADA: cycle-consistent adversarial domain adaptation. In: ICML (2018)
24. Hoiem, D., Chodpathumwan, Y., Dai, Q.: Diagnosing error in object detectors. In: Fitzgibbon, A., Lazebnik, S., Perona, P., Sato, Y., Schmid, C. (eds.) ECCV 2012. LNCS, vol. 7574, pp. 340–353. Springer, Heidelberg (2012). https://doi.org/10.1007/978-3-642-33712-3_25
25. Inoue, N., Furuta, R., Yamasaki, T., Aizawa, K.: Cross-domain weakly-supervised object detection through progressive domain adaptation. In: CVPR (2018)
26. Khodabandeh, M., Vahdat, A., Ranjbar, M., Macready, W.G.: A robust learning approach to domain adaptive object detection. In: ICCV (2019)
27. Kim, S., Choi, J., Kim, T., Kim, C.: Self-training and adversarial background regularization for unsupervised domain adaptive one-stage object detection. In: ICCV (2019)
28. Kim, T., Jeong, M., Kim, S., Choi, S., Kim, C.: Diversify and match: a domain adaptive representation learning paradigm for object detection. In: CVPR (2019)
29. Larsson, G., Maire, M., Shakhnarovich, G.: Colorization as a proxy task for visual understanding. In: CVPR (2017)
30. Liu, S., Huang, D., Wang, Y.: Receptive field block net for accurate and fast object detection. In: Ferrari, V., Hebert, M., Sminchisescu, C., Weiss, Y. (eds.) ECCV 2018. LNCS, vol. 11215, pp. 404–419. Springer, Cham (2018). https://doi.org/10.1007/978-3-030-01252-6_24
31. Liu, W., et al.: SSD: single shot multibox detector. In: Leibe, B., Matas, J., Sebe, N., Welling, M. (eds.) ECCV 2016. LNCS, vol. 9905, pp. 21–37. Springer, Cham (2016). https://doi.org/10.1007/978-3-319-46448-0_2
32. Long, M., Cao, Y., Wang, J., Jordan, M.I.: Learning transferable features with deep adaptation networks. In: ICML (2015)
33. Long, M., Zhu, H., Wang, J., Jordan, M.I.: Deep transfer learning with joint adaptation networks. In: ICML (2017)
34. Mancini, M., Karaoguz, H., Ricci, E., Jensfelt, P., Caputo, B.: Kitting in the wild through online domain adaptation. In: IROS (2018)
35. Massa, F., Girshick, R.: maskrcnn-benchmark: Fast, modular reference implementation of Instance Segmentation and Object Detection algorithms in PyTorch. https://github.com/facebookresearch/maskrcnn-benchmark (2018). Accessed 22 Aug 2019
36. Motiian, S., Jones, Q., Iranmanesh, S., Doretto, G.: Few-shot adversarial domain adaptation. In: NIPS (2017)
37. Noroozi, M., Favaro, P.: Unsupervised learning of visual representations by solving jigsaw puzzles. In: Leibe, B., Matas, J., Sebe, N., Welling, M. (eds.) ECCV 2016. LNCS, vol. 9910, pp. 69–84. Springer, Cham (2016). https://doi.org/10.1007/978-3-319-46466-4_5

38. Pathak, D., Krähenbühl, P., Donahue, J., Darrell, T., Efros, A.: Context encoders: feature learning by inpainting. In: CVPR (2016)
39. Redmon, J., Divvala, S., Girshick, R., Farhadi, A.: You only look once: unified, real-time object detection. In: CVPR (2016)
40. Ren, S., He, K., Girshick, R., Sun, J.: Faster R-CNN: towards real-time object detection with region proposal networks. In: NIPS (2015)
41. Russo, P., Carlucci, F.M., Tommasi, T., Caputo, B.: From source to target and back: symmetric bi-directional adaptive GAN. In: CVPR (2018)
42. Saito, K., Ushiku, Y., Harada, T., Saenko, K.: Strong-weak distribution alignment for adaptive object detection. In: CVPR (2019)
43. Sakaridis, C., Dai, D., Van Gool, L.: Semantic foggy scene understanding with synthetic data. IJCV **126**(9), 973–992 (2018). https://doi.org/10.1007/s11263-018-1072-8
44. Sun, B., Saenko, K.: Deep CORAL: correlation alignment for deep domain adaptation. In: Hua, G., Jégou, H. (eds.) ECCV 2016. LNCS, vol. 9915, pp. 443–450. Springer, Cham (2016). https://doi.org/10.1007/978-3-319-49409-8_35
45. Tobin, J., Fong, R.H., Ray, A., Schneider, J., Zaremba, W., Abbeel, P.: Domain randomization for transferring deep neural networks from simulation to the real world. In: IROS (2017)
46. Tzeng, E., Hoffman, J., Darrell, T., Saenko, K.: Adversarial discriminative domain adaptation. In: CVPR (2017)
47. Viola, P., Jones, M.: Rapid object detection using a boosted cascade of simple features. In: CVPR (2001)
48. Wulfmeier, M., Bewley, A., Posner, I.: Incremental adversarial domain adaptation for continually changing environments. In: ICRA (2018)
49. Xie, R., Yu, F., Wang, J., Wang, Y., Zhang, L.: Multi-level domain adaptive learning for cross-domain detection. In: ICCV Workshops (2019)
50. Xu, J., Xiao, L., López, A.M.: Self-supervised domain adaptation for computer vision tasks. arXiv abs/1907.10915 (2019)
51. Zhang, R., Isola, P., Efros, A.A.: Colorful image colorization. In: Leibe, B., Matas, J., Sebe, N., Welling, M. (eds.) ECCV 2016. LNCS, vol. 9907, pp. 649–666. Springer, Cham (2016). https://doi.org/10.1007/978-3-319-46487-9_40
52. Zhang, S., Wen, L., Bian, X., Lei, Z., Li, S.Z.: Single-shot refinement neural network for object detection. In: CVPR (2018)
53. Zhu, J.Y., Park, T., Isola, P., Efros, A.A.: Unpaired image-to-image translation using cycle-consistent adversarial networks. In: ICCV (2017)

Stochastic Frequency Masking to Improve Super-Resolution and Denoising Networks

Majed El Helou(✉), Ruofan Zhou(✉), and Sabine Süsstrunk

School of Computer and Communicatison Sciences, EPFL, Lausanne, Switzerland
{majed.elhelou,ruofan.zhou,sabine.susstrunk}@epfl.ch

Abstract. Super-resolution and denoising are ill-posed yet fundamental image restoration tasks. In blind settings, the degradation kernel or the noise level are unknown. This makes restoration even more challenging, notably for learning-based methods, as they tend to overfit to the degradation seen during training. We present an analysis, in the frequency domain, of degradation-kernel overfitting in super-resolution and introduce a conditional learning perspective that extends to both super-resolution and denoising. Building on our formulation, we propose a stochastic frequency masking of images used in training to regularize the networks and address the overfitting problem. Our technique improves state-of-the-art methods on blind super-resolution with different synthetic kernels, real super-resolution, blind Gaussian denoising, and real-image denoising.

Keywords: Image restoration · Super-resolution · Denoising · Kernel overfitting

1 Introduction

Image super-resolution (SR) and denoising are fundamental restoration tasks widely applied in imaging pipelines. They are crucial in various applications, such as medical imaging [33,38,45], low-light imaging [12], astronomy [7], satellite imaging [8,50], or face detection [24]. However, both are challenging ill-posed inverse problems. Recent learning methods based on convolutional neural networks (CNNs) achieve better restoration performance than classical approaches, both in SR and denoising. CNNs are trained on large datasets, sometimes real [65] but often synthetically generated with either one kernel or a limited set [54,63]. They learn to predict the restored image or the residual between the restored target and the input [27,56]. However, to be useful in practice, the networks should perform well on test images with unknown degradation kernels

The first two authors have similar contributions.

Electronic supplementary material The online version of this chapter (https://doi.org/10.1007/978-3-030-58517-4_44) contains supplementary material, which is available to authorized users.

A. Vedaldi et al. (Eds.): ECCV 2020, LNCS 12361, pp. 749–766, 2020.
https://doi.org/10.1007/978-3-030-58517-4_44

for SR, and unknown noise levels for denoising. Currently, they tend to overfit to the set of degradation models seen during training [16].

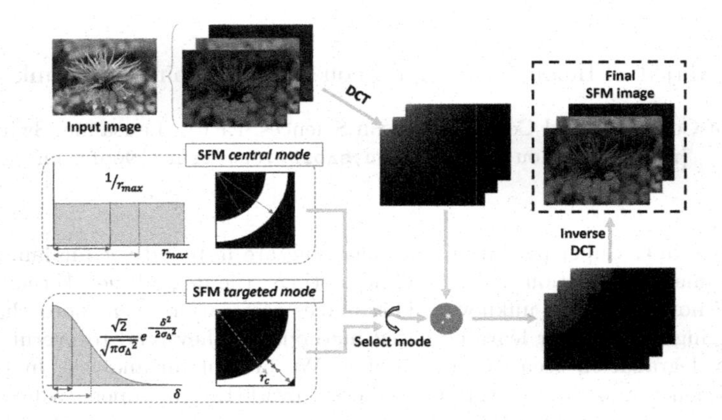

Fig. 1. Overview of Stochastic Frequency Masking (SFM). In the *central mode*, two radii values are sampled uniformly to delimit a masking area, and in the *targeted mode*, the sampled values delimit a quarter-annulus away from a target frequency. The obtained mask, shown with inverted color, is applied channel-wise to the discrete cosine transform of the image. We invert back to the spatial domain to obtain the SFM image that we use to train SR and denoising networks.

We investigate the SR degradation-kernel overfitting with an analysis in the frequency domain. Our analysis reveals that an implicit conditional learning is taking place in SR networks, namely, the learning of residual high-frequency content given low frequencies. We additionally show that this result extends to denoising as well. Building on our insights, we present **S**tochastic **F**requency **M**asking (SFM), which stochastically masks frequency components of the *images used in training*. Our SFM method (Fig. 1) is applied to a subset of the training images to regularize the network. It encourages the conditional learning to improve SR and denoising networks, notably when training under the challenging blind conditions. It can be applied during the training of *any* learning method, and has no additional cost at test time.[1]

Our experimental results show that SFM improves the performance of state-of-the-art networks on blind SR and blind denoising. For SR, we conduct experiments on synthetic bicubic and Gaussian degradation kernels, and on real degraded images. For denoising, we conduct experiments on additive white Gaussian denoising and on real microscopy Poisson-Gaussian image denoising. SFM improves the performance of state-of-the-art networks on each of these tasks.

Our contributions are summarized as follows. We present a frequency-domain analysis of the degradation-kernel overfitting of SR networks, and highlight the implicit conditional learning that, as we also show, extends to denoising. We

[1] Code available at: https://github.com/majedelhelou/SFM.

present a novel technique, SFM, that regularizes the learning of SR and denoising networks by only filtering the training data. It allows the networks to better restore frequency components and avoid overfitting. We empirically show that SFM improves the results of state-of-the-art learning methods on blind SR with different synthetic degradations, real-image SR, blind Gaussian denoising, and real-image denoising on high noise levels.

2 Related Work

Super-resolution. Depending on their image prior, SR algorithms can be divided into prediction models [43], edge-based models [11], gradient-profile pior methods [48] and example-based methods [21]. Deep example-based SR networks hold the state-of-the-art performance. Zhang *et al.* propose a very deep architecture based on residual channel attention to further improve these networks [63]. It is also possible to train in the wavelet domain to improve the memory and time efficiency of the networks [64]. Perceptual loss [25] and GANs [30,54] are leveraged to mitigate blur and push the SR networks to produce more visually-pleasing results. However, these networks are trained using a limited set of kernels, and studies have shown that they have poor generalization to unseen degradation kernels [22,46]. To address blind SR, which is degradation-agnostic, recent methods propose to incorporate the degradation parameters including the blur kernel into the network [46,57,59,60]. However, these methods rely on blur-kernel estimation algorithms and thus have a limited ability to handle arbitrary blur kernels. The most recent methods, namely IKC [22] and KMSR [65], propose kernel estimation and modeling in their SR pipeline. However, it is hard to gather enough training kernels to cover the real-kernel manifold, while also ensuring effective learning and avoiding that these networks overfit to the chosen kernels. Recently, real-image datasets were proposed [10,61] to enable SR networks to be trained and tested on high- and low-resolution (HR-LR) pairs, which capture the same scene but at different focal lengths. These datasets are also limited to the degradations of only a few cameras and cannot guarantee that SR models trained on them would generalize to unseen degradations. Our SFM method, which builds on our degradation-kernel overfitting analysis and our conditional learning perspective, can be used to improve the performance of *all* the SR networks we evaluate, including ones that estimate and model degradation kernels.

Denoising. Classical denoisers such as PURE-LET [34], which is specifically aimed at Poisson-Gaussian denoising, KSVD [2], WNNM [23], BM3D [14], and EPLL [66], which are designed for Gaussian denoising, have the limitation that the noise level needs to be known at test time, or at least estimated [20]. Recent learning-based denoisers outperform the classical ones on Gaussian denoising [4,39,56], but require the noise level [58], or pre-train multiple models for different noise levels [31,57], or more recently attempt to predict the noise level internally [18]. For a model to work under blind settings and adapt to any noise level, a common approach is to train the denoiser network while varying the training noise level [4,39,56]. Other recent methods, aimed at real-image denoising such as microscopy imaging [62], learn image statistics without requiring ground-truth samples on which noise is synthesized. This is practical because ground-truth data can be extremely difficult and costly to acquire in, for instance, medical applications. Noise2Noise [32] learns to denoise from pairs of noisy images. The noise is assumed to be zero in expectation and decorrelated from the signal. Therefore, unless the network memorizes it, the noise would not be predicted by it, and thus gets removed [32,53]. Noise2Self [6], which is a similar but more general version of Noise2Void [29], also assumes the noise to be decorrelated, conditioned on the signal. The network learns from single noisy images, by learning to predict an image subset from a separate subset, again with the assumption that the noise is zero in expectation. Although promising, these two methods do not yet reach the performance of Noise2Noise. By regularizing the conditional learning defined from our frequency-domain perspective, our SFM method improves the high noise level results of *all* tested denoising networks, notably under blind settings.

One example that uses frequency bands in restoration is the method in [5] that defines a prior based on a distance metric between a test image and a dataset of same-class images used for a deblurring optimization. The distance metric computes differences between image frequency bands. In contrast, we apply frequency masking on training images to regularize deep learning restoration networks, improving performance and generalization. Spectral dropout [26] regularizes network activations by dropping out components in the frequency domain to remove the least relevant, while SFM regularizes training by promoting the conditional prediction of different frequency components through masking the training images themselves. The most related work to ours is a recent method proposed in the field of speech recognition [37]. The authors augment speech data in three ways, one of which is in the frequency domain. It is a random separation of frequency bands, which splits different speech components to allow the network to learn them one by one. A clear distinction with our approach is that we do not aim to separate input components to be each individually learned. Rather, we mask targeted frequencies from the *training* input to strengthen the conditional frequency learning, and indirectly simulate the effect of a variety of kernels in SR and noise levels in denoising. The method we present is, to the best of our knowledge, the first frequency-based input *masking* method to regularize SR and denoising training.

3 Frequency Perspective on SR and Denoising

3.1 Super-Resolution

Preliminaries. Downsampling, a key element in modeling SR degradation, can be well explained in the frequency domain where it is represented by the sum of shifted and stretched versions of the frequency spectrum of a signal. Let x be a one-dimensional discrete signal, e.g., a pixel row in an image, and let z

be a downsampled version of x with a sampling interval T. In the discrete-time Fourier transform domain, with frequencies $\omega \in [-\pi, \pi]$, the relation between the transforms X and Z of the signals x and z, respectively, is given by $Z(\omega) = \frac{1}{T}\sum_{k=0}^{T-1} X((\omega+2\pi k)/T)$. The T replicas of X can overlap in the high frequencies and cause aliasing. Aside from complicating the inverse problem of restoring x from z, aliasing can create visual distortions. Before downsampling, low-pass filtering is therefore applied to attenuate if not completely remove the high-frequency components that would otherwise overlap.

These low-pass filtering blur kernels are applied through a spatial convolution over the image. The set of real kernels spans only a subspace of all mathematically-possible kernels. This subspace is, however, not well-defined analytically and, in the literature, is often limited to the non-comprehensive subspace spanned by 2D Gaussian kernels. Many SR methods thus model the anti-aliasing filter as a 2D Gaussian kernel, attempting to mimic the point spread function (PSF) of capturing devices [15,44,55]. In practice, even a single imaging device results in multiple kernels, depending on its settings [17]. For real images, the kernel can also be different from a Gaussian kernel [16,22]. The essential point is that the anti-aliasing filter causes the loss of high-frequency components, and that this filter can differ from image to image.

Frequency Visualization of SR Reconstructions. SR networks tend to overfit to the blur kernels used in the degradation for obtaining the training images [59]. To understand that phenomenon, we analyze in this section the relation between the frequency-domain effect of a blur kernel and the reconstruction of SR networks. We carry out the following experiment with a network trained with a unique and known blur kernel. We use the DIV2K [1] dataset to train a 20-block RRDB [54] x4 SR network with images filtered by a Gaussian blur kernel called F_1^{LP} (standard deviation $\sigma = 4.1$), shown in the top row of Fig. 2(a). We then run an inference on 100 test images filtered with a different Gaussian blur kernel called F_2^{LP} ($\sigma = 7.4$), shown in the bottom row of Fig. 2(a), to analyze the potential network overfitting.

We present a frequency-domain visualization in Fig. 2(b). The power spectral density (PSD) is the distribution of frequency content in an image. The typical PSD of an image (green curve) is modeled as $1/f^{\alpha}$, where f is the spatial frequency, with $\alpha \in [1, 2]$ and varying depending on the scene (natural vs. man-made) [9,19,51,52]. The $1/f^{\alpha}$ trend is visible in the PSD of HR images (green fill). The degraded LR test images are obtained with a low-pass filter on the HR image, before downsampling, and their frequency components are mostly low frequencies (pink fill). The SR network outputs contain high-frequency components restored by the network (red fill). However, these frequencies are mainly above 0.2π. This is only the range that was filtered out by the kernel used in creating the *training* LR images. The low-pass kernel used in creating the test LR images filters out a larger range of frequencies; it has a lower cutoff than the training kernel (the reverse case is also problematic and is illustrated in *Supplementary Material*). This causes a gap of missing frequency components not obtained in

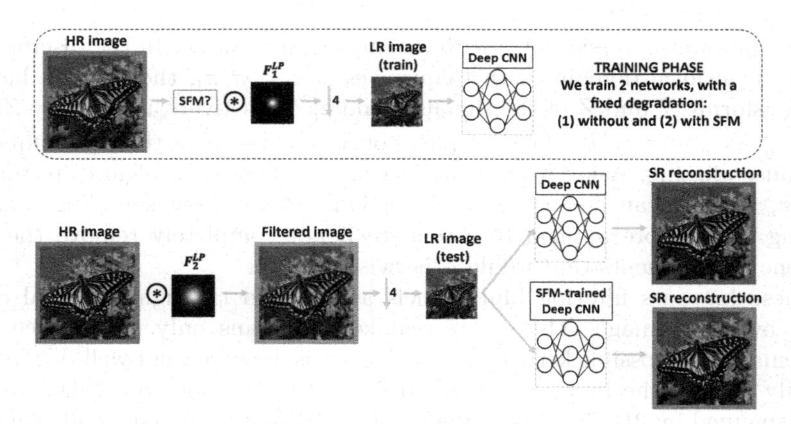

(a) Experimental setup

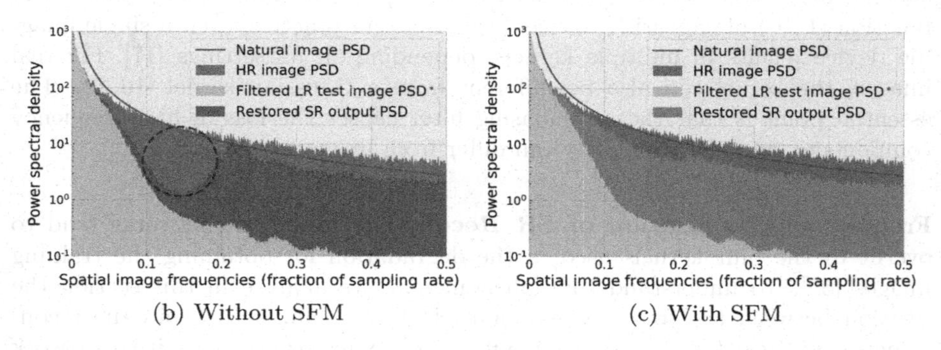

(b) Without SFM (c) With SFM

Fig. 2. (a) Overview of our experimental setup, with image border colors corresponding to the plot colors shown in (b,c). We train 2 versions of the same network on the same degradation kernel (F_1^{LP} anti-aliasing filter), one without and one with SFM, and test them using F_2^{LP}. (b) Average PSD (power spectral density) of HR images in green fill, with a green curve illustrating a typical natural-image PSD ($\alpha = 1.5$ [52]). The pink fill illustrates the average PSD of the low-pass filtered LR test images (*$shown$ *before downsampling for better visualization*). In red fill is the average PSD of the restored SR output image. The blue dashed circle highlights the learning gap due to degradation-kernel overfitting. (c) The same as (b), except that the output is that of the network trained with SFM. Results are averaged over 100 random samples.

the restored SR output, illustrated with a blue dashed circle in Fig. 2(b). The results suggest that an implicit conditional learning takes place in the SR network, on which we expand further in the following section. The results of the network trained with 50% SFM (masking applied to half of the training set) are shown in Fig. 2(c). A key observation is that the missing frequency components are predicted to a far better extent when the network is trained with SFM.

Implicit Conditional Learning. As we explain in the Preliminaries of Sect. 3.1, the high-frequency components of the original HR images are removed by the anti-aliasing filter. If that filter is *ideal*, it means that the low-frequency components are not affected and the high frequencies are perfectly removed. We propose that the SR networks in fact learn implicitly a conditional probability

$$P\left(I^{HR} \circledast F^{HP} \mid I^{HR} \circledast F^{LP}\right), \tag{1}$$

where F^{HP} and F^{LP} are ideal high-pass and low-pass filters, applied to the high-resolution image I^{HR}, and $\circledast$ is the convolution operator. The low and high frequency ranges are theoretically defined as $[0, \pi/T]$ and $[\pi/T, \pi]$, which is the minimum condition (largest possible cutoff) to avoid aliasing for a downsampling rate T. The components of I^{HR} that survive the low-pass filtering are the same frequencies contained in the LR image I^{LR}, when the filters F are ideal. In other words, the frequency components of $I^{HR} \circledast F^{LP}$ are those remaining in the LR image that is the network input.

The anti-aliasing filters are, in practice, not ideal, resulting in: (a) some low-frequency components of I^{HR} being attenuated, (b) some high frequencies surviving the filtering and causing aliasing. Typically, the main issue is the first issue (a), because filters are chosen in a way to remove the visually-disturbing aliasing at the expense of attenuating some low frequencies. We expand further on this in *Supplementary Material*, and derive that even with non-ideal filters, there is still conditional and residual learning components to predict a set of high-frequencies. These frequencies are, however, conditioned on a set of low-frequency components potentially attenuated by the non-ideal filter we call F_o^{LP}. This filter fully removes aliasing artifacts but can affect the low frequencies. The distribution can hence be defined by the components

$$P\left(I^{HR} \circledast F^{HP} \mid I^{HR} \circledast F_o^{LP}\right), \quad P\left(I^{HR} \circledast F^{LP} - I^{HR} \circledast F_0^{LP} \mid I^{HR} \circledast F_o^{LP}\right). \tag{2}$$

This is supported by our results in Fig. 2. The SR network trained with degradation kernel F_1^{LP} ($\sigma = 4.1$ in our experiment) restores the missing high frequencies of I^{HR} that would be erased by F_1^{LP}. However, that is the case even though the test image is degraded by $F_2^{LP} \neq F_1^{LP}$. As F_2^{LP} ($\sigma = 7.4$) removes a wider range of frequencies than F_1^{LP}, not predicted by the network, these frequencies remain missing. We observe a gap in the PSD of the output, highlighted by a blue dashed circle. This illustrates the degradation-kernel overfitting issue from a frequency-domain perspective. We also note that these missing frequency components are restored by the network trained with SFM.

3.2 Extension to Denoising

We highlight a connection between our conditional learning proposition and denoising. As discussed in Sect. 3.1, the average PSD of an image can be approximated by $1/f^{\alpha}$. The Gaussian noise samples added across pixels are independent and identically distributed. The PSD of the additive white Gaussian noise is uniform. Figure 3 shows the PSD of a natural image following a power law with $\alpha = 2$, that of white Gaussian noise (WGN), and the resulting signal-to-noise

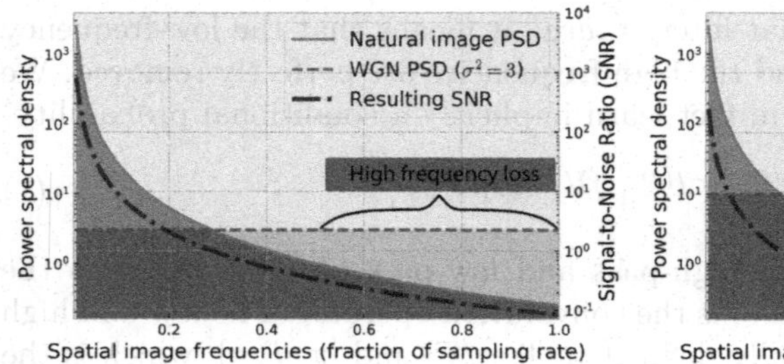

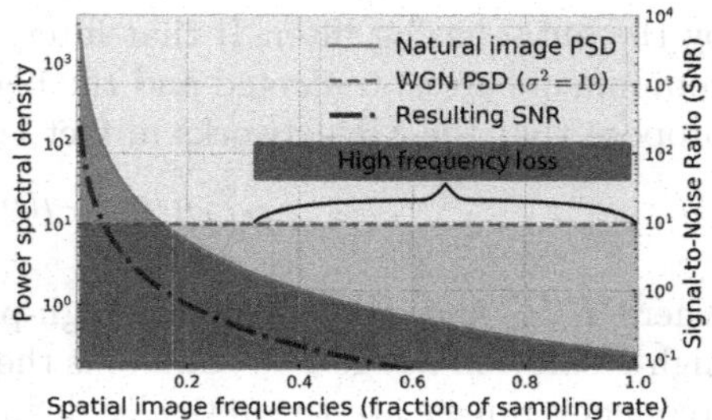

Fig. 3. Natural image PSD follows a power law as a function of spatial frequency. The plotted examples follow a power law with $\alpha = 2$ [52] and additive WGN ($\sigma^2 = 3$ on the left, and $\sigma^2 = 10$ on the right). The resulting SNR in the noisy image is exponentially smaller the higher the frequency, effectively causing a high frequency loss. The higher the noise level, the more frequency loss is incurred, and the more similar denoising becomes to our SR formulation.

ratio (SNR) when the WGN is added to the image. The resulting SNR decreases proportionally to $1/f^\alpha$.

The relation between SNR and frequency shows that with increasing frequency, the SNR becomes exponentially small. In other words, high frequencies are almost completely overtaken by the noise, while low frequencies are much less affected by it. And, *the higher the noise level, the lower the starting frequency beyond which the SNR is significantly small*, as illustrated by Fig. 3. This draws a direct connection to our SR analysis. Indeed, in both applications there exists an implicit conditional learning to predict lost high-frequency components given low-frequency ones that are less affected.

4 Stochastic Frequency Masking (SFM)

4.1 Motivation and Implementation

The objective of SFM is to improve the networks' prediction of high frequencies given lower ones, whether for SR or denoising. We achieve this by stochastically masking high-frequency bands from some of the training images in the learning phase, to encourage the conditional learning of the network. Our masking is carried out by transforming an image to the frequency domain using the Discrete Cosine Transform (DCT) type II [3,47], multiplying channel-wise by our stochastic mask, and lastly transforming the image back (Fig. 1). See *Supplementary Material* for the implementation details of the DCT type we use. We define frequency bands in the DCT domain over quarter-annulus areas, to cluster together similar-magnitude frequency content. Therefore, the SFM mask is delimited with a quarter-annulus area by setting the values of its inner and outer radii. We define two masking modes, the *central mode* and the *targeted mode*.

In the *central mode*, the inner and outer radius limits r_I and r_O of the quarter-annulus are selected uniformly at random from $[0, r_M]$, where $r_M = \sqrt{a^2 + b^2}$ is the maximum radius, with (a, b) being the dimensions of the image. We ensure that $r_I < r_O$ by permuting the values if $r_I > r_O$. With this mode, the resulting probability of a given frequency band r_ω to be masked is

$$P(r_\omega = 0) = P(r_I < r_\omega < r_O) = 2\left(\frac{r_\omega}{r_M} - \left(\frac{r_\omega}{r_M}\right)^2\right), \tag{3}$$

meaning the central bands are the more likely ones to be masked, with the likelihood *slowly* decreasing for higher or lower frequencies. In the *targeted mode*, a target frequency r_C is selected, with a parameter σ_Δ. The quarter-annulus is delimited by $[r_C - \delta_I, r_C + \delta_O]$, where δ_I and δ_O are independently sampled from the half-normal Δ distribution $f_\Delta(\delta) = \sqrt{2}/\sqrt{\pi\sigma_\Delta^2}e^{-\delta^2/(2\sigma_\Delta^2)}, \forall\delta \geq 0$. Therefore, with this mode, the frequency r_C is always masked, and the frequencies away from r_C are less and less likely to be masked, with a normal distribution decay.

We use the *central mode* for SR networks, and the *targeted mode* with a high target r_C for denoisers (Fig. 1). The former has a slow probability decay that covers wider bands, while the latter has an exponential decay adapted for targeting specific narrow bands. In both settings, the highest frequencies are most likely to be masked. The *central mode* masks the highest frequencies in SR, because central frequencies are the highest ones remaining after the anti-aliasing filter is applied. It is also worth noting that SFM thus simulates the effect of different blur kernels by stochastically masking different frequency bands.

4.2 Learning SR and Denoising with SFM

We apply SFM *only* on the input training data. For the simulated-degradation data, SFM is applied in the process of generating the LR inputs. We apply SFM on HR images before applying the degradation model to generate the LR inputs (blur kernel and downsampling). The target output of the network remains the original HR images. For real images where the LR inputs are given and the degradation model is unknown, we apply SFM on the LR inputs and keep the original HR images as ground-truth targets. Therefore, the networks trained with SFM do not use any additional data relative to the baselines. We apply *the same* SFM settings for all deep learning experiments. During training, we apply SFM on 50% of the training images, using the *central mode* of SFM, as presented in Sect. 4.1. Ablation studies with other rates are in our *Supplementary Material.* We add SFM to the training of the original methods with no other modification.

When training for additive white Gaussian noise (AWGN) removal, we apply SFM on the clean image before the synthetic noise is added. When the training images are real and the noise cannot be separated from the signal, we apply SFM on the noisy image. Hence, we ensure that networks trained with SFM do not utilize any additional training data relative to the baselines. In all denoising experiments, and for all of the compared methods, we use *the same* SFM settings. We apply SFM on 50% of training images, and use the *targeted mode* of our SFM (ablation studies including other rates are in our *Material*). We use a central band $r_C = 0.85\,r_M$ and $\sigma_\Delta = 0.15\,r_M$. As presented in Sect. 4.1, this means that the highest frequency bands are masked with high likelihood, and lower frequencies are exponentially less likely to be masked the smaller they are. We add SFM to the training of the original methods with no other modification.

5 Experiments

5.1 SR: Bicubic and Gaussian Degradations

Methods. We evaluate our proposed SFM method on state-of-the-art SR networks that can be divided into 3 categories. In the first category, we evaluate RCAN [63] and RRDB [54], which are networks that target pixel-wise distortion for a single degradation kernel. RCAN leverages a residual-in-residual structure and channel attention for efficient non-blind SR learning. RRDB [54] employs a residual-in-residual dense block as its basic architecture unit. The second category covers perception-optimized methods for a single degradation kernel, and includes ESRGAN [54]. It is a version of the RRDB network using a GAN for better SR perceptual quality and obtains the state-of-the-art results in this category. The last category includes algorithms for blind SR, we experiment on IKC [22], which incorporates into the training of the SR network a blur-kernel estimation and modeling to explicitly address blind SR. **Setup.** We train all the models using the DIV2K [1] dataset, which is a high-quality dataset that is commonly used for single-image SR evaluation. RCAN, RRDB, and ESRGAN are trained with the bicubic degradation, and IKC with Gaussian kernels ($\sigma \in [0.2, 4.0]$ [22]). For all models, 16 LR patches of size 48×48 are extracted per training batch. All models are trained using the Adam optimizer [28] for 50 epochs. The initial learning rate is set to 10^{-4} and decreases by half every 10 epochs. Data augmentation is performed on the training images, which are randomly rotated by 90°, 180°, 270°, and flipped horizontally.

Results. To generate test LR images, we apply bicubic and Gaussian blur kernels on the DIV2K [1] validation set. We also evaluate all methods trained with 50% SFM, following Sect. 4.2. Table 1 shows the PSNR results on x4 upscaling SR, with different blur kernels. Results show that the proposed SFM consistently improves the performance of the various SR networks on the different degradation kernels, even up to $0.27dB$ on an unseen test kernel for the recent IKC [22] that explicitly models kernels during training. We improve by up to $0.56\,dB$ for the other methods. With SFM, RRDB achieves comparable or better results than RCAN, which has double the parameters of RRDB. Sample visual results are shown in Fig. 4.

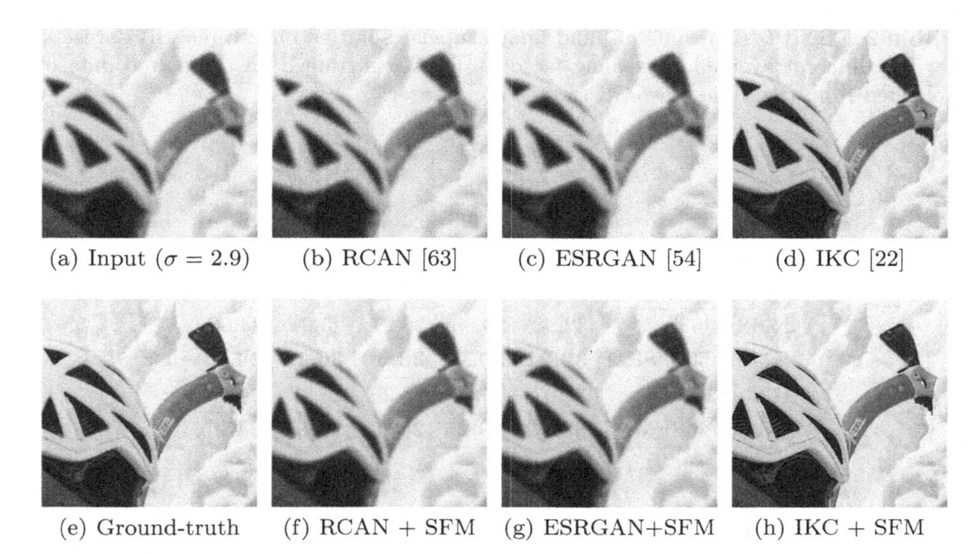

(a) Input ($\sigma = 2.9$) (b) RCAN [63] (c) ESRGAN [54] (d) IKC [22]

(e) Ground-truth (f) RCAN + SFM (g) ESRGAN+SFM (h) IKC + SFM

Fig. 4. Cropped SR results (x4) of different methods (top row), and with our SFM added (bottom row), for image 0844 of DIV2K. The visual quality improves for all methods when trained with SFM (images best viewed on screen).

Table 1. Single-image SR, with x4 upscaling factor, PSNR (dB) results on the DIV2K validation set. RCAN, RRDB and ESRGAN are trained using bicubic degradation, and IKC using Gaussian kernels ($\sigma \in [2.0, 4.0]$). Kernels seen in training are shaded gray. The training setups of the networks are presented in Sect. 5.1, and identical ones are used with SFM. We note that SFM improves the results of the various methods, even the IKC method that explicitly models kernels during its training improves by up to 0.27 dB with SFM on unseen kernels.

	Test blur kernel (g_σ is a Gaussian kernel, standard deviation σ)									
	bicubic	$g_{1.7}$	$g_{2.3}$	$g_{2.9}$	$g_{3.5}$	$g_{4.1}$	$g_{4.7}$	$g_{5.3}$	$g_{5.9}$	$g_{6.5}$
RCAN [63]	29.18	23.80	24.08	23.76	23.35	22.98	22.38	22.16	21.86	21.72
RCAN+SFM	**29.32**	**24.21**	**24.64**	**24.19**	**23.72**	**23.27**	**22.54**	**22.23**	**21.91**	**21.79**
IKC [22]	**27.81**	26.07	26.15	25.48	25.03	24.41	23.39	22.78	22.41	22.08
IKC+SFM	27.78	**26.09**	**26.18**	**25.52**	**25.11**	**24.52**	**23.54**	**22.97**	**22.62**	**22.35**
RRDB [54]	28.79	23.66	23.72	23.68	23.29	22.75	22.32	22.08	21.83	21.40
RRDB+SFM	**29.10**	**23.81**	**23.99**	**23.79**	**23.41**	**22.90**	**22.53**	**22.37**	**21.98**	**21.56**
ESRGAN [54]	25.43	21.22	22.49	22.03	21.87	21.63	21.21	20.99	20.05	19.42
ESRGAN+SFM	**25.50**	**21.37**	**22.78**	**22.26**	**22.08**	**21.80**	**21.33**	**21.10**	**20.13**	**19.77**

5.2 SR: Real-Image Degradations

Methods. We train and evaluate the same SR models as the networks we use in Sect. 5.1, except for ESRGAN and RRDB, as ESRGAN is a perceptual-quality-driven method and does not achieve high PSNR, and RCAN outperforms RRDB according to our experiments in 5.1. We also evaluate on KMSR [65] for the

Table 2. PSNR (dB) results of blind image super-resolution on two real SR datasets, for the different available upscaling factors. ‡RCAN is trained on the paired dataset collected from the same sensor as the testing dataset.

	Dataset and upscaling factor				
	RealSR [10]			SR-RAW [61]	
Method	x2	x3	x4	x4	x8
RCAN‡ [63]	33.24	30.24	28.65	26.29	24.18
RCAN 50% SFM	**33.32**	**30.29**	**28.75**	**26.42**	**24.50**
KMSR [65]	32.98	30.05	28.27	25.91	24.00
KMSR 50% SFM	**33.21**	**30.11**	**28.50**	**26.19**	**24.31**
IKC [22]	33.07	30.03	28.29	25.87	24.19
IKC 50% SFM	**33.12**	**30.25**	**28.42**	**25.93**	**24.25**

real SR experiments. KMSR collects real blur kernels from real LR images to improve the generalization of the SR network on unseen kernels.

Setup. We train and evaluate the SR networks on two digital zoom datasets: the SR-RAW dataset [61] and the RealSR dataset [10]. The training setup of the SR networks is the same as in Sect. 5.1. Note that we follow the same training procedures for each method as in the original papers. IKC is trained with Gaussian kernels ($\sigma \in [0.2, 4.0]$) and KMSR with the blur kernels estimated from LR images in the dataset. RCAN is trained *on the degradation of the test* data; a starting advantage over other methods.

Results. We evalute the SR methods on the corresponding datasets and present the results in Table 2. Each method is also trained with 50% SFM, following Sect. 4.2. SFM consistently improves all methods on all upscaling factors, pushing the state-of-the-art results by up to 0.23 dB on both of these challenging real-image SR datasets.

5.3 Denoising: AWGN

Methods. We evaluate different state-of-the-art AWGN denoisers. DnCNN-B [56] learns the noise residual rather than the final denoised image. Noise2Noise (N2N) [32] learns only from noisy image pairs, with no ground-truth data. N3Net [39] relies on learning nearest neighbors similarity, to make use of different similar patches in an image for denoising. MemNet [49] follows residual learning with memory transition blocks. Lastly, RIDNet [4] also does residual learning, but leverages feature attention blocks. **Setup**. We train all methods on the 400 Berkeley images [36], typically used to benchmark denoisers [13,42,56]. All methods use the Adam optimizer with a starting learning rate of 10^{-3}, except RIDNet that uses half that rate. We train for 50 epochs and synthesize noise instances per training batch. For blind denoising training, we follow the settings initially set in [56]: noise is sampled from a Gaussian distribution with standard

Table 3. PSNR (*dB*) results on BSD68 for different methods and noise levels. SFM improves the various methods, and the improvement increases with higher noise levels, supporting our hypothesis. We clamp test images to [0,255] as in camera pipelines. Denoisers are trained with levels up to 55 (shaded in gray), thus half the test range is not seen in training. ‡Re-trained under blind settings.

	Test noise level (standard deviation of the stationary AWGN)									
	10	20	30	40	50	60	70	80	90	100
DnCNN-B [56]	33.33	29.71	27.66	26.13	24.88	23.69	22.06	19.86	17.88	16.35
DnCNN-B + SFM	**33.35**	**29.78**	**27.73**	**26.27**	**25.09**	**24.02**	**22.80**	**21.24**	**19.46**	**17.87**
Noise2Noise [32]	**32.67**	28.84	26.61	25.02	23.76	22.69	21.74	20.88	20.11	19.41
Noise2Noise + SFM	32.55	**28.94**	**26.84**	**25.31**	**24.11**	**23.05**	**22.14**	**21.32**	**20.61**	**19.95**
Blind‡ N3Net [39]	**33.53**	**30.01**	**27.84**	26.30	25.04	23.93	22.87	21.84	20.87	19.98
N3Net + SFM	33.41	29.86	**27.84**	**26.38**	**25.19**	**24.15**	**23.20**	**22.32**	**21.51**	**20.78**
Blind‡ MemNet [49]	**33.51**	29.75	27.61	26.06	24.87	23.83	22.67	21.00	18.92	17.16
MemNet + SFM	33.36	**29.80**	**27.76**	**26.31**	**25.14**	**24.09**	**23.09**	**22.00**	**20.77**	**19.46**
RIDNet [4]	**33.65**	**29.87**	27.65	26.04	24.79	23.65	22.25	20.05	18.15	17.09
RIDNet + SFM	33.43	29.81	**27.76**	**26.30**	**25.12**	**24.08**	**23.11**	**22.08**	**20.74**	**19.17**

deviation chosen at random in $[0, 55]$. This splits the range of test noise levels into levels seen or not seen during training, which provides further insights on generalization. We also note that we use a U-Net [40] for the architecture of N2N as in the original paper. For N2N, we apply SFM on top of the added noise, to preserve the particularity that N2N can be trained without ground-truth data.

Results. We evaluate all methods on the BSD68 [41] test set. Each method is also trained with 50% SFM as explained in Sect. 4.2 and the results are in Table 3. SFM improves the performance of a variety of different state-of-the-art denoising methods on high noise levels (seen during training, such as 40 and 50, or not even seen), and the results support our hypothesis presented in Sect. 3.2 that *the higher the noise level* the more similar is denoising to SR and the more applicable is SFM. Indeed, the higher the noise level the larger the improvement of SFM, and this trend is true across all methods. Figure 5 presents sample results.

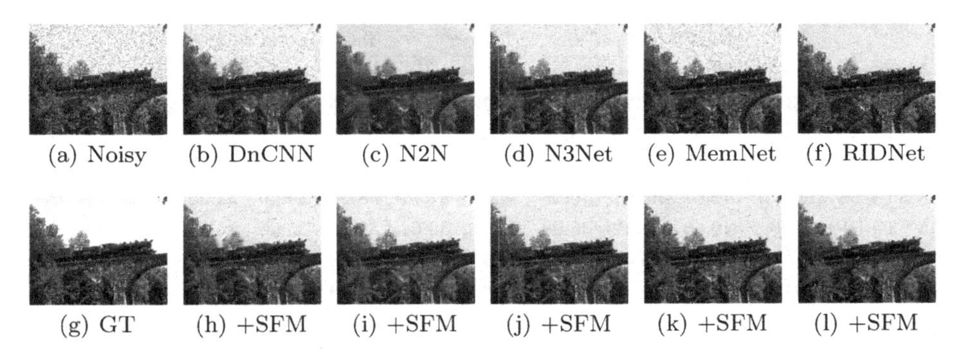

(a) Noisy (b) DnCNN (c) N2N (d) N3Net (e) MemNet (f) RIDNet

(g) GT (h) +SFM (i) +SFM (j) +SFM (k) +SFM (l) +SFM

Fig. 5. Denoising ($\sigma = 50$) results with different methods (top row), and with our SFM added (bottom row), for the last image (#67) of the BSD68 benchmark.

5.4 Denoising: Real Poisson-Gaussian Images

Methods. Classical methods are often a good choice for denoising in the absence of ground-truth datasets. PURE-LET [34] is specifically aimed at Poisson-Gaussian denoising, and KSVD [2], WNNM [23], BM3D [14], and EPLL [66] are designed for Gaussian denoising. Recently, learning methods were presented such as N2S [6] (and the similar, but less general, N2V [29]) that can learn from a dataset of only noisy images, and N2N [32] that can learn from a dataset of only noisy image pairs. We incorporate SFM into the learning-based methods.

Setup. We train the learning-based methods on the recent real fluorescence microscopy dataset [62]. The noise follows a Poisson-Gaussian distribution, and the image registration is of high quality due to the stability of the microscopes, thus yielding reliable ground-truth obtained by averaging 50 repeated captures. Noise parameters are estimated using the fitting approach in [20] for all classical denoisers. Additionally, the parameters are used for the variance-stabilization transform (VST) [35] for the Gaussian-oriented methods. In contrast, the learning methods can directly be applied under blind settings. We train N2S/N2N using a U-Net [40] architecture, for 100/400 epochs using the Adam optimizer with a starting learning rate of $10^{-5}/10^{-4}$ [62].

Results. We evaluate on the mixed and two-photon microscopy test sets [62]. We also train the learning methods with 50% SFM as explained in Sect. 4.2, and present the results in Table 4. A larger number of averaged raw images is equivalent to a lower noise level. N2N with SFM achieves the state-of-the-art performance on both benchmarks and for all noise levels, with an improvement of up to $0.42dB$. We also note that the improvements of SFM are larger on the more challenging two-photon test set where the noise levels are higher on average. SFM does not consistently improve N2S, however, this is expected.

Table 4. PSNR (dB) results on microscopy images with Poisson-Gaussian noise. We train under blind settings and apply SFM on noisy input images to preserve the fact that N2S and N2N can be trained without clean images.

	# raw images for averaging									
	Mixed test set [62]					Two-photon test set [62]				
Method	16	8	4	2	1	16	8	4	2	1
PURE-LET [34]	39.59	37.25	35.29	33.49	31.95	37.06	34.66	33.50	32.61	31.89
VST+KSVD [2]	40.36	37.79	35.84	33.69	32.02	38.01	35.31	34.02	32.95	31.91
VST+WNNM [23]	40.45	37.95	36.04	34.04	32.52	38.03	35.41	34.19	33.24	32.35
VST+BM3D [14]	40.61	38.01	36.05	34.09	32.71	38.24	35.49	34.25	33.33	32.48
VST+EPLL [66]	40.83	38.12	36.08	34.07	32.61	38.55	35.66	34.35	33.37	32.45
N2S [6]	36.67	35.47	34.66	33.15	31.87	34.88	33.48	32.66	31.81	30.51
N2S 50% SFM	36.60	35.62	34.59	33.44	32.40	34.39	33.14	32.48	31.84	30.92
N2N [32]	41.45	39.43	37.59	36.40	35.40	38.37	35.82	34.56	33.58	32.70
N2N 50% SFM	**41.48**	**39.46**	**37.78**	**36.43**	**35.50**	**38.78**	**36.10**	**34.85**	**33.90**	**33.05**

In fact, unlike other methods, N2S trains to predict a subset of an image given a surrounding subset. It applies spatial masking where the mask is made up of random pixels that interferes with the frequency components. For these reasons, N2S is not very compatible with SFM, which nonetheless improves results on the largest noise levels in both test sets.

6 Conclusion

We analyze the degradation-kernel overfitting of SR networks in the frequency domain. Our frequency-domain insights reveal an implicit conditional learning that also extends to denoising, especially on high noise levels. Building on our analysis, we present SFM, a technique to improve SR and denoising networks, without increasing the size of the training set or any cost at test time. We conduct extensive experiments on state-of-the-art networks for both restoration tasks. We evaluate SR with synthetic degradations, real-image SR, Gaussian denoising and real-image Poisson-Gaussian denoising, showing improved performance, notably on generalization, when using SFM.

References

1. Agustsson, E., Timofte, R.: NTIRE 2017 challenge on single image super-resolution: dataset and study. In: CVPR Workshops (2017)
2. Aharon, M., Elad, M., Bruckstein, A.: K-SVD: an algorithm for designing overcomplete dictionaries for sparse representation. IEEE Trans. Signal Process. **54**(11), 4311–4322 (2006)
3. Ahmed, N., Natarajan, T., Rao, K.R.: Discrete cosine transform. IEEE Trans. Comput. **100**(1), 90–93 (1974)
4. Anwar, S., Barnes, N.: Real image denoising with feature attention. In: ICCV (2019)
5. Anwar, S., Huynh, C.P., Porikli, F.: Image deblurring with a class-specific prior. IEEE Trans. Pattern Anal. Mach. Intell. **41**(9), 2112–2130 (2018)
6. Batson, J., Royer, L.: Noise2Self: blind denoising by self-supervision. In: ICML (2019)
7. Beckouche, S., Starck, J.L., Fadili, J.: Astronomical image denoising using dictionary learning. Astron. Astrophys. **556**, A132 (2013)
8. Benazza-Benyahia, A., Pesquet, J.C.: Building robust wavelet estimators for multicomponent images using Stein's principle. IEEE Trans. Image Process. **14**(11), 1814–1830 (2005)
9. Burton, G.J., Moorhead, I.R.: Color and spatial structure in natural scenes. Appl. Opt. **26**(1), 157–170 (1987)
10. Cai, J., Zeng, H., Yong, H., Cao, Z., Zhang, L.: Toward real-world single image super-resolution: a new benchmark and a new model. In: ICCV (2019)
11. Chan, T.M., Zhang, J., Pu, J., Huang, H.: Neighbor embedding based super-resolution algorithm through edge detection and feature selection. Pattern Recogn. Lett. **30**(5), 494–502 (2009)
12. Chatterjee, P., Joshi, N., Kang, S.B., Matsushita, Y.: Noise suppression in low-light images through joint denoising and demosaicing. In: CVPR (2011)

13. Chen, Y., Pock, T.: Trainable nonlinear reaction diffusion: a flexible framework for fast and effective image restoration. IEEE Trans. Pattern Anal. Mach. Intell. **39**(6), 1256–1272 (2016)
14. Dabov, K., Foi, A., Katkovnik, V., Egiazarian, K.: Image denoising by sparse 3-D transform-domain collaborative filtering. IEEE Trans. Image Process. **16**(8), 2080–2095 (2007)
15. Dong, W., Zhang, L., Shi, G., Li, X.: Nonlocally centralized sparse representation for image restoration. IEEE Trans. Image Process. **22**(4), 1620–1630 (2012)
16. Efrat, N., Glasner, D., Apartsin, A., Nadler, B., Levin, A.: Accurate blur models vs. image priors in single image super-resolution. In: ICCV (2013)
17. El Helou, M., Dümbgen, F., Süsstrunk, S.: AAM: an assessment metric of axial chromatic aberration. In: ICIP (2018)
18. El Helou, M., Süsstrunk, S.: Blind universal Bayesian image denoising with Gaussian noise level learning. IEEE Trans. Image Process. **29**, 4885–4897 (2020)
19. Field, D.J.: Relations between the statistics of natural images and the response properties of cortical cells. JOSA **4**(12), 2379–2394 (1987)
20. Foi, A., Trimeche, M., Katkovnik, V., Egiazarian, K.: Practical Poissonian-Gaussian noise modeling and fitting for single-image raw-data. IEEE Trans. Image Process. **17**(10), 1737–1754 (2008)
21. Freeman, W.T., Jones, T.R., Pasztor, E.C.: Example-based super-resolution. Comput. Graph. Appl. **22**(2), 56–65 (2002)
22. Gu, J., Lu, H., Zuo, W.Z., Dong, C.: Blind super-resolution with iterative kernel correction. In: CVPR (2019)
23. Gu, S., Zhang, L., Zuo, W., Feng, X.: Weighted nuclear norm minimization with application to image denoising. In: CVPR (2014)
24. Gunturk, B.K., Batur, A.U., Altunbasak, Y., Hayes, M.H., Mersereau, R.M.: Eigenface-domain super-resolution for face recognition. IEEE Trans. Image Process. **12**(5), 597–606 (2003)
25. Johnson, J., Alahi, A., Fei-Fei, L.: Perceptual losses for real-time style transfer and super-resolution. In: ECCV (2016)
26. Khan, S.H., Hayat, M., Porikli, F.: Regularization of deep neural networks with spectral dropout. Neural Netw. **110**, 82–90 (2019)
27. Kim, J., Lee, J.K., Lee, K.M.: Accurate image super-resolution using very deep convolutional networks. In: CVPR (2016)
28. Kingma, D.P., Ba, J.: Adam: a method for stochastic optimization. arXiv preprint arXiv:1412.6980 (2014)
29. Krull, A., Buchholz, T.O., Jug, F.: Noise2Void-learning denoising from single noisy images. In: CVPR (2019)
30. Ledig, C., et al.: Photo-realistic single image super-resolution using a generative adversarial network. In: CVPR (2017)
31. Lefkimmiatis, S.: Universal denoising networks: a novel CNN architecture for image denoising. In: CVPR (2018)
32. Lehtinen, J., et al.: Noise2Noise: learning image restoration without clean data. In: ICML (2018)
33. Li, S., Yin, H., Fang, L.: Group-sparse representation with dictionary learning for medical image denoising and fusion. IEEE Trans. Biomed. Eng. **59**(12), 3450–3459 (2012)
34. Luisier, F., Blu, T., Unser, M.: Image denoising in mixed Poisson-Gaussian noise. IEEE Trans. Image Process. **20**(3), 696–708 (2011)
35. Makitalo, M., Foi, A.: Optimal inversion of the generalized Anscombe transformation for Poisson-Gaussian noise. IEEE Trans. Image Process. **22**(1), 91–103 (2012)

36. Martin, D., Fowlkes, C., Tal, D., Malik, J., et al.: A database of human segmented natural images and its application to evaluating segmentation algorithms and measuring ecological statistics. In: ICCV (2001)
37. Park, D.S., et al.: SpecAugment: a simple data augmentation method for automatic speech recognition. arXiv preprint arXiv:1904.08779 (2019)
38. Peled, S., Yeshurun, Y.: Superresolution in MRI: application to human white matter fiber tract visualization by diffusion tensor imaging. Magn. Reson. Med. Official J. Int. Soc. Magn. Reson. Med. **45**(1), 29–35 (2001)
39. Plötz, T., Roth, S.: Neural nearest neighbors networks. In: NeurIPS (2018)
40. Ronneberger, O., Fischer, P., Brox, T.: U-Net: convolutional networks for biomedical image segmentation. In: International Conference on Medical Image Computing and Computer-Assisted Intervention (MICCAI), pp. 234–241 (2015)
41. Roth, S., Black, M.J.: Fields of experts. Int. J. Comput. Vision **82**(2), 205 (2009)
42. Schmidt, U., Roth, S.: Shrinkage fields for effective image restoration. In: CVPR (2014)
43. Schulter, S., Leistner, C., Bischof, H.: Fast and accurate image upscaling with super-resolution forests. In: CVPR (2015)
44. Shi, W., et al.: Real-time single image and video super-resolution using an efficient sub-pixel convolutional neural network. In: CVPR (2016)
45. Shi, W., et al.: Cardiac image super-resolution with global correspondence using multi-atlas patchMatch. In: Medical Image Computing and Computer-Assisted Intervention (MICCAI), pp. 9–16 (2013)
46. Shocher, A., Cohen, N., Irani, M.: Zero-shot super-resolution using deep internal learning. In: CVPR (2018)
47. Strang, G.: The discrete cosine transform. SIAM **41**(1), 135–147 (1999)
48. Sun, J., Xu, Z., Shum, H.Y.: Image super-resolution using gradient profile prior. In: CVPR (2008)
49. Tai, Y., Yang, J., Liu, X., Xu, C.: MemNet: a persistent memory network for image restoration. In: ICCV (2017)
50. Thornton, M.W., Atkinson, P.M., Holland, D.: Sub-pixel mapping of rural land cover objects from fine spatial resolution satellite sensor imagery using super-resolution pixel-swapping. Int. J. Remote Sens. **27**(3), 473–491 (2006)
51. Tolhurst, D., Tadmor, Y., Chao, T.: Amplitude spectra of natural images. Ophthalmic Physiol. Opt. **12**(2), 229–232 (1992)
52. Torralba, A., Oliva, A.: Statistics of natural image categories. Netw. Comput. Neural Syst. **14**(3), 391–412 (2003)
53. Ulyanov, D., Vedaldi, A., Lempitsky, V.: Deep image prior. In: CVPR (2018)
54. Wang, X., et al.: ESRGAN: enhanced super-resolution generative adversarial networks. In: ECCV Workshops (2018)
55. Yang, C.Y., Ma, C., Yang, M.H.: Single-image super-resolution: a benchmark. In: ECCV (2014)
56. Zhang, K., Zuo, W., Chen, Y., Meng, D., Zhang, L.: Beyond a Gaussian denoiser: residual learning of deep CNN for image denoising. IEEE Trans. Image Process. **26**(7), 3142–3155 (2017)
57. Zhang, K., Zuo, W., Gu, S., Zhang, L.: Learning deep CNN denoiser prior for image restoration. In: CVPR (2017)
58. Zhang, K., Zuo, W., Zhang, L.: FFDNet: toward a fast and flexible solution for CNN-based image denoising. IEEE Trans. Image Process. **27**(9), 4608–4622 (2018)
59. Zhang, K., Zuo, W., Zhang, L.: Learning a single convolutional super-resolution network for multiple degradations. In: CVPR (2018)

60. Zhang, K., Zuo, W., Zhang, L.: Deep plug-and-play super-resolution for arbitrary blur kernels. In: CVPR (2019)
61. Zhang, X., Chen, Q., Ng, R., Koltu, V.: Zoom to learn, learn to zoom. In: ICCV (2019)
62. Zhang, Y., et al.: A poisson-Gaussian denoising dataset with real fluorescence microscopy images. In: CVPR (2019)
63. Zhang, Y., Li, K., Li, K., Wang, L., Zhong, B., Fu, Y.: Image super-resolution using very deep residual channel attention networks. In: ECCV (2018)
64. Zhou, R., Lahoud, F., El Helou, M., Süsstrunk, S.: A comparative study on wavelets and residuals in deep super resolution. Electron. Imaging **13**, 135-1–135-6 (2019)
65. Zhou, R., Süsstrunk, S.: Kernel modeling super-resolution on real low-resolution images. In: ICCV (2019)
66. Zoran, D., Weiss, Y.: From learning models of natural image patches to whole image restoration. In: ICCV (2011)

Probabilistic Future Prediction for Video Scene Understanding

Anthony Hu[1,2(✉)], Fergal Cotter[1], Nikhil Mohan[1], Corina Gurau[1], and Alex Kendall[1,2]

[1] Wayve, London, UK
anthony@wayve.ai
[2] University of Cambridge, Cambridge, UK

Abstract. We present a novel deep learning architecture for probabilistic future prediction from video. We predict the future semantics, geometry and motion of complex real-world urban scenes and use this representation to control an autonomous vehicle. This work is the first to jointly predict ego-motion, static scene, and the motion of dynamic agents in a probabilistic manner, which allows sampling consistent, highly probable futures from a compact latent space. Our model learns a representation from RGB video with a spatio-temporal convolutional module. The learned representation can be explicitly decoded to future semantic segmentation, depth, and optical flow, in addition to being an input to a learnt driving policy. To model the stochasticity of the future, we introduce a conditional variational approach which minimises the divergence between the present distribution (what could happen given what we have seen) and the future distribution (what we observe actually happens). During inference, diverse futures are generated by sampling from the present distribution.

1 Introduction

Building predictive cognitive models of the world is often regarded as the essence of intelligence. It is one of the first skills that we develop as infants. We use these models to enhance our capability at learning more complex tasks, such as navigation or manipulating objects [50].

Unlike in humans, developing prediction models for autonomous vehicles to anticipate the future remains hugely challenging. Road agents have to make reliable decisions based on forward simulation to understand how relevant parts of the scene will evolve. There are various reasons why modelling the future is incredibly difficult: natural-scene data is rich in details, most of which are irrelevant for the driving task, dynamic agents have complex temporal dynamics, often controlled by unobservable variables, and the future is inherently uncertain, as multiple futures might arise from a unique and deterministic past.

Electronic supplementary material The online version of this chapter (https://doi.org/10.1007/978-3-030-58517-4_45) contains supplementary material, which is available to authorized users.

A. Vedaldi et al. (Eds.): ECCV 2020, LNCS 12361, pp. 767–785, 2020.
https://doi.org/10.1007/978-3-030-58517-4_45

Current approaches to autonomous driving individually model each dynamic agent by producing hand-crafted behaviours, such as trajectory forecasting, to feed into a decision making module [8]. This largely assumes independence between agents and fails to model multi-agent interaction. Most works that holistically reason about the temporal scene are limited to simple, often simulated environments or use low dimensional input images that do not have the visual complexity of real world driving scenes [49]. Some approaches tackle this problem by making simplifying assumptions to the motion model or the stochasticity of the world [8,42]. Others avoid explicitly predicting the future scene but rather rely on an implicit representation or Q-function (in the case of model-free reinforcement learning) in order to choose an action [28,34,37].

Real world future scenarios are difficult to model because of the stochasticity and the partial observability of the world. Our work addresses this by encoding the future state into a low-dimensional *future distribution*. We then allow the model to have a privileged view of the future through the future distribution at training time. As we cannot use the future at test time, we train a *present distribution* (using only the current state) to match the future distribution through a Kullback-Leibler (KL) divergence loss. We can then sample from the present distribution during inference, when we do not have access to the future. We observe that this paradigm allows the model to learn accurate and diverse probabilistic future prediction outputs.

In order to predict the future we need to first encode video into a motion representation. Unlike advances in 2D convolutional architectures [27,62], learning spatio-temporal features is more challenging due to the higher dimensionality of video data and the complexity of modelling dynamics. State-of-the-art architectures [63,66] decompose 3D filters into spatial and temporal convolutions in order to learn more efficiently. The model we propose further breaks down convolutions into many space-time combinations and context aggregation modules, stacking them together in a more complex hierarchical representation. We show that the learnt representation is able to jointly predict ego-motion and motion of other dynamic agents. By explicitly modelling these dynamics we can capture the essential features for representing causal effects for driving. Ultimately we use this motion-aware and future-aware representation to improve an autonomous vehicle control policy.

Our main contributions are threefold. Firstly, we present a novel deep learning framework for future video prediction. Secondly, we demonstrate that our probabilistic model is able to generate visually diverse and plausible futures. Thirdly, we show our future prediction representation substantially improves a learned autonomous driving policy.

2 Related Work

This work falls in the intersection of learning scene representation from video, probabilistic modelling of the ambiguity inherent in real-world driving data, and using the learnt representation for control.

Temporal Representations. Current state-of-the-art temporal representations from video use recurrent neural networks [55,56], separable 3D convolutions [26,30,61,63,65], or 3D Inception modules [7,66]. In particular, the separable 3D Inception (S3D) architecture [66], which improves on the Inception 3D module (I3D) introduced by Carreira *et al.* [7], shows the best trade-off between model complexity and speed, both at training and inference time. Adding optical flow as a complementary input modality has been consistently shown to improve performance [5,19,57,58], in particular using flow for representation warping to align features over time [22,68]. We propose a new spatio-temporal architecture that can learn hierarchically more complex features with a novel 3D convolutional structure incorporating both local and global space and time context.

Visual Prediction. Most works for learning dynamics from video fall under the framework of model-based reinforcement learning [17,21,33,43] or unsupervised feature learning [15,59], both regressing directly in pixel space [32,46,51] or in a learned feature space [20,31]. For the purpose of creating good representations for driving scenes, directly predicting in the high-dimensional space of image pixels is unnecessary, as some details about the appearance of the world are irrelevant for planning and control. Our approach is similar to that of Luc *et al.* [45] which trains a model to predict future semantic segmentation using pseudo-ground truth labels generated from a teacher model. However, our model predicts a more complete scene representation with segmentation, depth, and flow and is probabilistic in order to model the uncertainty of the future.

Multi-modality of Future Prediction. Modelling uncertainty is important given the stochastic nature of real-world data [35]. Lee *et al.* [41], Bhattacharyya *et al.* [4] and Rhinehart *et al.* [52] forecast the behaviour of other dynamic agents in the scene in a probabilistic multi-modal way. We distinguish ourselves from this line of work as their approach does not consider the task of video forecasting, but rather trajectory forecasting, and they do not study how useful the representations learnt are for robot control. Kurutach *et al.* [39] propose generating multi-modal futures with adversarial training, however spatio-temporal discriminator networks are known to suffer from mode collapse [23].

Our variational approach is similar to Kohl *et al.* [38], although their application domain does not involve modelling dynamics. Furthermore, while Kohl *et al.* [38] use multi-modal training data, i.e. multiple output labels are provided for a given input, we learn directly from real-world driving data, where we can only observe one future reality, and show that we generate diverse and plausible futures. Most importantly, previous variational video generation methods [16,40] were restricted to single-frame image generation, low resolution (64×64) datasets that are either simulated (Moving MNIST [59]) or with static scenes and limited dynamics (KTH actions [54], Robot Pushing dataset [18]). Our new framework for future prediction generates entire video sequences on complex real-world urban driving data with ego-motion and complex interactions.

Learning a Control Policy. The representation learned from dynamics models could be used to generate imagined experience to train a policy in a model-based reinforcement learning setting [24,25] or to run shooting methods for planning [11]. Instead we follow the approaches of Bojarski *et al.* [6], Codevilla *et al.* [13] and Amini *et al.* [1] and learn a policy which predicts longitudinal and lateral control of an autonomous vehicle using Conditional Imitation Learning, as this approach has been shown to be immediately transferable to the real world.

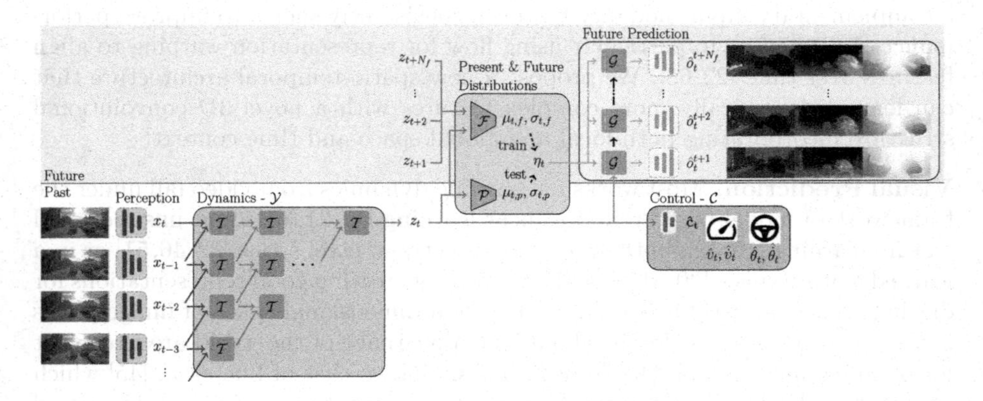

Fig. 1. Our architecture has five modules: *Perception*, *Dynamics*, *Present/Future Distributions*, *Future Prediction* and *Control*. The Perception module learns scene representation features, x_t, from input images. The Dynamics model builds on these scene features to produce a spatio-temporal representation, z_t, with our proposed *Temporal Block* module, $\mathcal{T}$. Together with a noise vector, η_t, sampled from a future distribution, $\mathcal{F}$, at training time, or the present distribution, $\mathcal{P}$, at inference time, this representation predicts future video scene representation (segmentation, depth and optical flow) with a convolutional recurrent model, $\mathcal{G}$, and decoders, $\mathcal{D}$. Lastly, we learn a Control policy, $\mathcal{C}$, from the spatio-temporal representation, z_t.

3 Model Architecture

Our model learns a spatio-temporal feature to jointly predict future scene representation (semantic segmentation, depth, optical flow) and train a driving policy. The architecture contains five components: *Perception*, an image scene understanding model, *Dynamics*, which learns a spatio-temporal representation, *Present/Future Distributions*, our probabilistic framework, *Future Prediction*, which predicts future video scene representation, and *Control*, which trains a driving policy using expert driving demonstrations. Figure 1 gives an overview of the model and further details are described in this section and Appendix A.

3.1 Perception

The perception component of our system contains two modules: the encoder of a scene understanding model that was trained on single image frames to reconstruct semantic segmentation and depth [36], and the encoder of a flow network [60], trained to predict optical flow. The combined perception features $x_t \in \mathbb{R}^{C\times H\times W}$ form the input to the dynamics model. These models can also be used as a teacher to distill the information from the future, giving pseudo-ground truth labels for segmentation, depth and flow $\{s_t, d_t, f_t\}$. See Subsect. 4.1 for more details on the teacher model.

3.2 Dynamics

Learning a temporal representation from video is extremely challenging because of the high dimensionality of the data, the stochasticity and complexity of natural scenes, and the partial observability of the environment. To train 3D convolutional filters from a sequence of raw RGB images, a large amount of data, memory and compute is required. We instead learn spatio-temporal features with a temporal model that operates on perception encodings, which constitute a more powerful and compact representation compared to RGB images.

The dynamics model $\mathcal{Y}$ takes a history of perception features $(x_{t-T+1} : x_t)$ with temporal context T and encodes it into a dynamics feature z_t:

$$z_t = \mathcal{Y}(x_{t-T+1} : x_t) \tag{1}$$

Temporal Block. We propose a spatio-temporal module, named *Temporal Block*, to learn hierarchically more complex temporal features as follows:

- **Decomposing the filters**: instead of systematically using full 3D filters (k_t, k_s, k_s), with k_t the time kernel dimension and k_s the spatial kernel dimension, we apply four parallel 3D convolutions with kernel sizes: $(1, k_s, k_s)$ (spatial features), $(k_t, 1, k_s)$ (horizontal motion), $(k_t, k_s, 1)$ (vertical motion), and (k_t, k_s, k_s) (complete motion). All convolutions are preceded by a $(1, 1, 1)$ convolution to compress the channel dimension.
- **Global spatio-temporal context**: in order to learn contextual features, we additionally use three spatio-temporal average pooling layers at: full spatial size (k_t, H, W) (H and W are respectively the height and width of the perception features x_t), half size $(k_t, \frac{H}{2}, \frac{W}{2})$ and quarter size $(k_t, \frac{H}{4}, \frac{W}{4})$, followed by bilinear upsampling to the original spatial dimension (H, W) and a $(1, 1, 1)$ convolution.

Figure 2 illustrates the architecture of the Temporal Block. By stacking multiple temporal blocks, the network learns a representation that incorporates increasingly more temporal, spatial and global context. We also increase the number of channels by a constant α after each temporal block, as after each block, the network has to represent the content of the k_t previous features.

3.3 Future Prediction

We train a future prediction model that unrolls the dynamics feature, which is a compact scene representation of the past context, into predictions about the

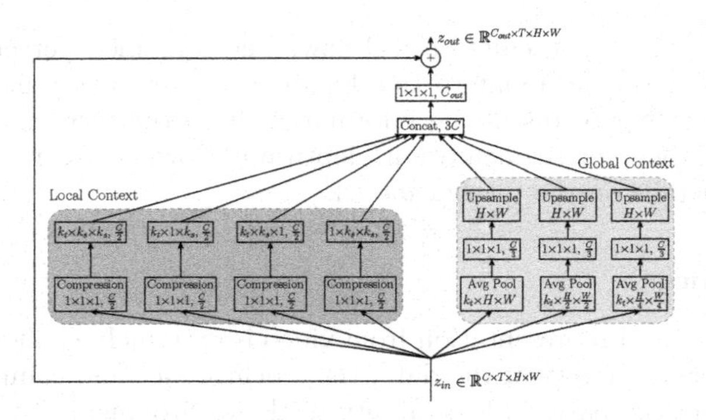

Fig. 2. A Temporal Block, our proposed spatio-temporal module. From a four-dimensional input $z_{in} \in \mathbb{R}^{C \times T \times H \times W}$, our module learns both local and global spatio-temporal features. The local head learns all possible configurations of 3D convolutions with filters: $(1, k_s, k_s)$ (spatial features), $(k_t, 1, k_s)$ (horizontal motion), $(k_t, k_s, 1)$ (vertical motion), and (k_t, k_s, k_s) (complete motion). The global head learns global spatio-temporal features with a 3D average pooling at full, half and quarter size, followed by a $(1, 1, 1)$ convolution and upsampling to the original spatial dimension $H \times W$. The local and global features are then concatenated and combined in a final $(1, 1, 1)$ 3D convolution.

state of the world in the future. The future prediction model is a convolutional recurrent network $\mathcal{G}$ which creates future features g_t^{t+i} that become the inputs of individual decoders $\mathcal{D}_s, \mathcal{D}_d, \mathcal{D}_f$ to decode these features to predicted segmentation $\hat{s}_t^{t+i}$, depth $\hat{d}_t^{t+i}$, and flow $\hat{f}_t^{t+i}$ values in the pixel space. We have introduced a second time superscript notation, *i.e.*g_t^{t+i}, represents the prediction about the world at time $t+i$ given the dynamics features at time t. Also note that $g_t^t \triangleq z_t$.

The structure of the convolutional recurrent network $\mathcal{G}$ is the following: a convolutional GRU [2] followed by three spatial residual layers, repeated D times, similarly to Clark *et al.* [12]. For deterministic inference, its input is $u_t^{t+i} = \mathbf{0}$, and its initial hidden state is z_t, the dynamics feature. The future prediction component of our network computes the following, for $i \in \{1, .., N_f\}$, with N_f the number of predicted future frames:

$$g_t^{t+i} = \mathcal{G}(u_t^{t+i}, g_t^{t+i-1}) \tag{2}$$

$$\hat{s}_t^{t+i} = \mathcal{D}_s(g_t^{t+i}) \tag{3}$$

$$\hat{d}_t^{t+i} = \mathcal{D}_d(g_t^{t+i}) \tag{4}$$

$$\hat{f}_t^{t+i} = \mathcal{D}_f(g_t^{t+i}) \tag{5}$$

3.4 Present and Future Distributions

From a unique past in the real-world, many futures are possible, but in reality we only observe one future. Consequently, modelling multi-modal futures from deterministic video training data is extremely challenging. We adopt a conditional variational approach and model two probability distributions: a *present distribution* P, that represents what could happen given the past context, and a *future distribution* F, that represents what actually happened in that particular observation. This allows us to learn a multi-modal distribution from the input data while conditioning the model to learn from the specific observed future from within this distribution.

The present and the future distributions are diagonal Gaussian, and can therefore be fully characterised by their mean and standard deviation. We parameterise both distributions with a neural network, respectively $\mathcal{P}$ and $\mathcal{F}$.

Present Distribution. The input of the network $\mathcal{P}$ is $z_t \in \mathbb{R}^{C_d \times H \times W}$, which represents the past context of the last T frames (T is the time receptive field of our dynamics module). The present network contains two downsampling convolutional layers, an average pooling layer and a fully connected layer to map the features to the desired latent dimension L. The output of the network is the parametrisation of the present distribution: $(\mu_{t,\text{present}}, \sigma_{t,\text{present}}) \in \mathbb{R}^L \times \mathbb{R}^L$.

Future Uistribution. $\mathcal{F}$ is not only conditioned by the past z_t, but also by the future corresponding to the training sequence. Since we are predicting N_f steps in the future, the input of $\mathcal{F}$ has to contain information about future frames $(t+1, ..., t+N_f)$. This is achieved using the learned dynamics features $\{z_{t+j}\}_{j \in J}$, with J the set of indices such that $\{z_{t+j}\}_{j \in J}$ covers all future frames $(t+1, ..., t+N_f)$, as well as z_t. Formally, if we want to cover N_f frames with features that have a receptive field of T, then:

$$J = \{nT \mid 0 \leq n \leq \lfloor N_f/T \rfloor\} \cup \{N_f\}.$$

The architecture of the future network is similar to the present network: for each input dynamics feature $z_{t+j} \in \mathbb{R}^{C_d \times H \times W}$, with $j \in F$, we apply two downsampling convolutional layers and an average pooling layer. The resulting features are concatenated, and a fully-connected layer outputs the parametrisation of the future distribution:

$$(\mu_{t,\text{future}}, \sigma_{t,\text{future}}) \in \mathbb{R}^L \times \mathbb{R}^L.$$

Probabilistic Future Prediction. During training, we sample from the future distribution a vector $\eta_t \sim \mathcal{N}(\mu_{t,\text{future}}, \sigma^2_{t,\text{future}})$ that conditions the predicted future perception outputs (semantic segmentation, depth, optical flow) on the observed future. As we want our prediction to be consistent in both space and time, we broadcast spatially $\eta_t \in \mathbb{R}^L$ to $\mathbb{R}^{L \times H \times W}$, and use the same sample throughout the future generation as an input to the GRU to condition the future: for $i \in \{1, .., N_f\}$, input $u_t^{t+i} = \eta_t$.

We encourage the present distribution P to match the future distribution F with a mode-covering KL loss:

$$L_{\text{probabilistic}} = D_{\text{KL}}(F(\cdot|Z_t, ..., Z_{t+N_f}) \mid\mid P(\cdot|Z_t)) \tag{6}$$

As the future is multimodal, different futures might arise from a unique past context z_t. Each of these futures will be captured by the future distribution F

that will pull the present distribution P towards it. Since our training data is extremely diverse, it naturally contains multimodalities. Even if the past context (sequence of images $(i_1, ..., i_t)$) from two different training sequences will never be the same, the dynamics network will have learned a more abstract spatio-temporal representation that ignores irrelevant details of the scene (such as vehicle colour, weather, road material etc.) to match similar past context to a similar z_t. In this process, the present distribution will learn to cover all the possible modes contained in the future.

During inference, we sample a vector η_t from the present distribution $\eta_t \sim \mathcal{N}(\mu_{t,\text{present}}, \sigma^2_{t,\text{present}})$, where each sample corresponds to a different future.

3.5 Control

From this rich spatio-temporal representation z_t explicitly trained to predict the future, we train a control model $\mathcal{C}$ to output a four dimensional vector consisting of estimated speed $\hat{v}$, acceleration $\hat{\dot{v}}$, steering angle $\hat{\theta}$ and angular velocity $\hat{\dot{\theta}}$:

$$\hat{c}_t = \{\hat{v}_t, \hat{\dot{v}}_t, \hat{\theta}_t, \hat{\dot{\theta}}_t\} = \mathcal{C}(z_t) \tag{7}$$

$\mathcal{C}$ compresses $z_t \in \mathbb{R}^{C_d \times H \times W}$ with strided convolutional layers, then stacks several fully connected layers, compressing at each stage, to regress the four dimensional output.

3.6 Losses

Future Prediction. The future prediction loss at timestep t is the weighted sum of future segmentation, depth and optical flow losses. Let the segmentation loss at the future timestep $t+i$ be L_s^{t+i}. We use a top-k cross-entropy loss [64] between the network output $\hat{s}_t^{t+i}$ and the pseudo-ground truth label s_{t+i}. L_s is computed by summing these individual terms over the future horizon N_f with a weighted discount term $0 < \gamma_f < 1$:

$$L_s = \sum_{i=0}^{N_f-1} \gamma_f^i L_s^{t+i} \tag{8}$$

For depth, L_d^{t+i} is the scale-invariant depth loss [44] between $\hat{d}_t^{t+i}$ and d_{t+i}, and similarly L_d is the discounted sum. For flow, we use a Huber loss betwen $\hat{f}_t^{t+i}$ and f_{t+i}. We weight the summed losses by factors $\lambda_s, \lambda_d, \lambda_f$ to get the future prediction loss $L_{\text{future-pred}}$.

$$L_{\text{future-pred}} = \lambda_s L_s + \lambda_d L_d + \lambda_f L_f \tag{9}$$

Control. We use imitation learning, regressing to the expert's true control actions $\{v, \theta\}$ to generate a *control loss* L_c. For both speed and steering, we have access to the expert actions.

We compare to the linear extrapolation of the generated policy's speed/steering for future time-steps up to N_c frames in the future:

$$L_c = \sum_{i=0}^{N_c-1} \gamma_c^i \left(\left(v_{t+i} - \left(\hat{v}_t + i\hat{\dot{v}}_t \right) \right)^2 + \left(\theta_{t+i} - \left(\hat{\theta}_t + i\hat{\dot{\theta}}_t \right) \right)^2 \right) \tag{10}$$

where $0 < \gamma_c < 1$ is the control discount factor penalizing less speed and steering errors further into the future.

Total Loss. The final loss L can be decomposed into the future prediction loss ($L_{\text{future-pred}}$), the probabilistic loss ($L_{\text{probabilistic}}$), and the control loss (L_c) .

$$L = \lambda_{fp} L_{\text{future-pred}} + \lambda_c L_c + \lambda_p L_{\text{probabilistic}} \tag{11}$$

In all experiments we use $\gamma_f = 0.6$, $\lambda_s = 1.0$, $\lambda_d = 1.0$, $\lambda_f = 0.5$, $\lambda_{fp} = 1$, $\lambda_p = 0.005$, $\gamma_c = 0.7$, $\lambda_c = 1.0$.

4 Experiments

We have collected driving data in a densely populated, urban environment, representative of most European cities using multiple drivers over the span of six months. For the purpose of this work, only the front-facing camera images i_t and the measurements of the speed and steering c_t have been used to train our model, all sampled at 5

4.1 Training Data

Perception. We first pretrain the scene understanding encoder on a number of heterogeneous datasets to predict semantic segmentation and depth: CityScapes [14], Mapillary Vistas [48], ApolloScape [29] and Berkeley Deep Drive [67]. The optical flow network is a pretrained PWC-Net from [60]. The decoders of these networks are used for generating pseudo-ground truth segmentation and depth labels to train our dynamics and future prediction modules.

Dynamics and Control. The dynamics and control modules are trained using 30 hours of driving data from the urban driving dataset we collected and described above. We address the inherent dataset bias by sampling data uniformly across lateral and longitudinal dimensions. First, the data is split into a histogram of bins by steering, and subsequently by speed. We found that weighting each data point proportionally to the width of the bin it belongs to avoids the need for alternative approaches such as data augmentation.

4.2 Metrics

We report standard metrics for measuring the quality of segmentation, depth and flow: respectively intersection-over-union, scale-invariant logarithmic error, and average end-point error. For ease of comparison, additionally to individual metrics, we report a unified perception metric $\mathcal{M}_{\text{perception}}$ defined as improvement of segmentation, depth and flow metrics with respect to the *Repeat Frame* baseline (repeats the perception outputs of the current frame):

$$\mathcal{M}_{\text{perception}} = \frac{1}{3}(\text{seg}_{\%\text{increase}} + \text{depth}_{\%\text{decrease}} + \text{flow}_{\%\text{decrease}}) \tag{12}$$

Inspired by the energy functions used in [3,53], we additionally report a *diversity distance metric* (DDM) between the ground truth future Y and samples from the predicted present distribution P:

$$\text{DDM}(Y, P) = \min_{S} \big[d(Y, S)\big] - \mathbb{E}\big[d(S, S')\big] \tag{13}$$

where d is an error metric and S, S', are independent samples from the present distribution P. This metric measures performance both in terms of accuracy, by looking at the minimum error of the samples, as well as the diversity of the predictions by taking the expectation of the distance between N samples. The distance d is the scale-invariant logarithmic error for depth, the average end-point error for flow, and for segmentation $d(x, y) = 1 - \text{IoU}(x, y)$.

To measure control performance, we report mean absolute error of speed and steering outputs, balanced by steering histogram bins.

5 Results

We first compare our proposed spatio-temporal module to previous state-of-the-art architectures and show that our module achieves the best performance on future prediction metrics. Then we demonstrate that modelling the future in a probabilistic manner further improves performance. And finally, we show that our probabilistic future prediction representation substantially improves a learned driving policy. All the reported results are evaluated on test routes with no overlap with the training data.

5.1 Spatio-Temporal Representation

We analyse the quality of the spatio-temporal representation our temporal model learns by evaluating future prediction of semantic segmentation, depth, and optical flow, two seconds in the future. Several architectures have been created to learn features from video, with the most successful modules being: the Convolutional GRU [2], the 3D Residual Convolution [26] and the Separable 3D Inception block [66].

We also compare our model to two baselines: *Repeat frame* (repeating the perception outputs of the current frame at time t for each future frame $t+i$ with

$i = 1, ..., N_f$), and *Static* (without a temporal model). As shown in Table 1, deterministic section, every temporal model architecture improves over the *Repeat frame* baseline, as opposed to the model without any temporal context (*Static*), that performs notably worse. This is because it is too difficult to forecast how the future is going to evolve with a single image.

Table 1. Perception performance metrics for two seconds future prediction on the collected urban driving data. We measure semantic segmentation with mean IoU, depth with scale-invariant logarithmic error, and depth with average end-point error. $\mathcal{M}_{\text{perception}}$ shows overall performance — we observe our model outperforms all baselines.

	Temporal Model	$\mathcal{M}_{\text{perception}}$ (↑)	Depth (↓)	Flow (↓)	Seg. (↑)
	Repeat frame	0.0%	1.467	5.707	0.356
	Static	-40.3%	1.980	8.573	0.229
Deterministic	Res. 3D Conv. [26]	6.9%	1.162	5.437	0.339
	Conv. GRU [2]	7.4%	1.097	5.714	0.346
	Sep. Inception [66]	9.6%	1.101	5.300	0.344
	Ours	**13.6%**	**1.090**	**5.029**	**0.367**
Probabilistic	Res. 3D Conv. [26]	8.1%	1.107	5.720	0.356
	Conv. GRU [2]	9.0%	1.101	5.645	0.359
	Sep. Inception [66]	13.8%	1.040	5.242	0.371
	Ours	**20.0%**	**0.970**	**4.857**	**0.396**

Further, we observe that our proposed temporal block module outperforms all preexisting spatio-temporal architectures, on all three future perception metrics: semantic segmentation, depth and flow. There are two reasons for this: the first one is that learning 3D filters is hard, and as demonstrated by the Separable 3D convolution [66] (i.e. the succession of a $(1, k_s, k_s)$ spatial filter and a $(k_t, 1, 1)$ time filter), decomposing into two subtasks helps the network learn more efficiently. In the same spirit, we decompose the spatio-temporal convolutions into all combinations of space-time convolutions: $(1, k_s, k_s)$, $(k_t, 1, k_s)$, $(k_t, k_s, 1)$, (k_t, k_s, k_s), and by stacking these temporal blocks together, the network can learn a hierarchically more complex representation of the scene. The second reason is that we incorporate global context in our features. By pooling the features spatially and temporally at different scales, each individual feature map also has information about the global scene context, which helps in ambiguous situations. Appendix A.3 contains an ablation study of the different component of the Temporal Block.

5.2 Probabilistic Future

Since the future is inherently uncertain, the deterministic model is training in a chaotic learning space because the predictions of the model are penalised with the ground truth future, which only represents a subset of all the possible outcomes. Therefore, if the network predicts a plausible future, but one that did not match the given training sequence, it will be heavily penalised. On the other hand, the probabilistic model has a very clean learning signal as the future distribution conditions the network to generate the correct future. The present distribution is encouraged to match the distribution of the future distribution during training, and therefore has to capture all the modes of the future.

During inference, samples $\eta_t \sim \mathcal{N}(\mu_{t,\text{present}}, \sigma^2_{t,\text{present}})$ from the present distribution should give a different outcome, with $p(\eta_t|\mu_{t,\text{present}}, \sigma^2_{t,\text{present}})$ indicating the relative likelihood of a given scenario. Our probabilistic model should be accurate, that is to say at least one of the generated future should match the ground truth future. It should also be diverse: the generated samples should capture the diversity of the possible futures with the correct probability. Next, we analyse quantitatively and qualitatively that our model generates diverse and accurate futures.

Table 2. Diversity Distance Metric for various temporal models evaluated on the urban driving data, demonstrating that our model produces the most accurate and diverse distribution.

Temporal Model	Depth (↓)	Flow (↓)	Seg. (↓)
Res. 3D Conv. [26]	0.823	2.695	0.474
Conv. GRU [2]	0.841	2.683	0.493
Sep. Inception [66]	0.799	2.914	0.469
Ours	**0.724**	**2.676**	**0.424**

Table 1 shows that every temporal architecture have superior performance when trained in a probabilistic way, with our model benefiting the most (from 13.6% to 20.0%) in future prediction metrics. Table 2 shows that our model outperforms other temporal representations also using the diversity distance metric (DDM) described in subsection 4.2. The DDM measures both accuracy and diversity of the distribution.

Perhaps the most striking result of the model is observing that our model can predict diverse and plausible futures from a single sequence of past frames at 5 Hz, corresponding to one second of past context and two seconds of future prediction. In Fig. 3 and Fig. 4 we show qualitative examples of our video scene understanding future prediction in real-world urban driving scenes. We sample from the present distribution, $\eta_{t,j} \sim \mathcal{N}(\mu_{t,\text{present}}, \sigma^2_{t,\text{present}})$, to demonstrate multi-modality.

Further, our framework can automatically infer which scenes are unusual or unexpected and where the model is uncertain of the future, by computing the differential entropy of the present distribution. Simple scenes (e.g. one-way streets) will tend to have a low entropy, corresponding to an almost deterministic future. Any latent code sampled from the present distribution will correspond to the

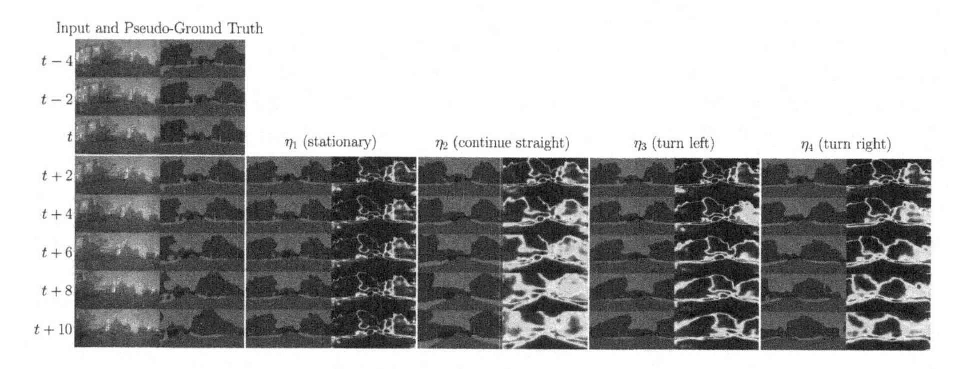

Fig. 3. Predicted futures from our model while driving through an urban intersection. From left, we show the actual past and future video sequence and labelled semantic segmentation. Using four different noise vectors, η, we observe the model imagining different driving manoeuvres at an intersection: being stationary, driving straight, taking a left or a right turn. We show both predicted semantic segmentation and entropy (uncertainty) for each future. This example demonstrates that our model is able to learn a probabilistic embedding, capable of predicting multi-modal and plausible futures.

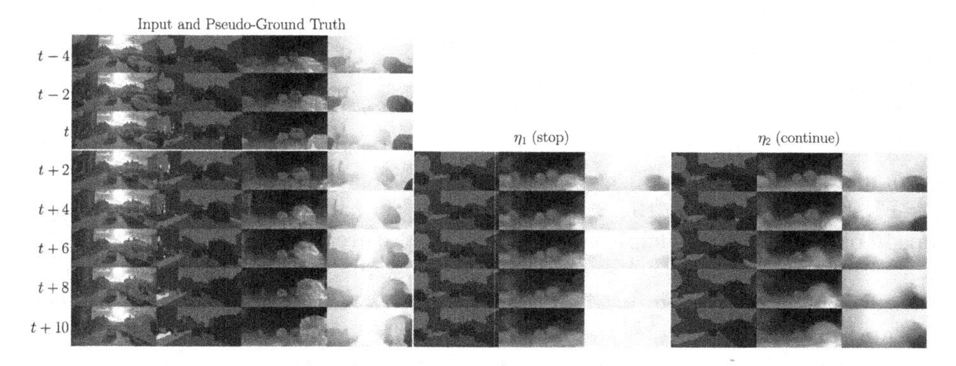

Fig. 4. Predicted futures from our model while driving through a busy urban scene. From left, we show actual past and future video sequence and labelled semantic segmentation, depth and optical flow. Using two different noise vectors, η, we observe the model imagining either stopping in traffic or continuing in motion. This illustrates our model's efficacy at jointly predicting holistic future behaviour of our own vehicle and other dynamic agents in the scene across all modalities.

same future. Conversely, complex scenes (e.g. intersections, roundabouts) will be associated with a high-entropy. Different samples from the present distribution will correspond to different futures, effectively modelling the stochasticity of the future.[1]

[1] In the accompanying blog post, we illustrate how diverse the predicted future becomes with varying levels of entropy in an intersection scenario and an urban traffic scenario.

Finally, to allow reproducibility, we evaluate our future prediction framework on Cityscapes [14] and report future semantic segmentation performance in Table 3. We compare our predictions, at resolution 256×512, to the ground truth segmentation at 5 and 10 frames in the future. Qualitative examples on Cityscapes can be found in Appendix C.

Table 3. Future semantic segmentation performance on Cityscapes at $i = 5$ and $i = 10$ frames in the future (corresponding to respectively 0.29 s and 0.59 s).

	Temporal Model	$\text{IoU}_{i=5}$ (↑)	$\text{IoU}_{i=10}$ (↑)
	Repeat frame	0.393	0.331
	Nabavi *et al.*[47]	-	0.274
	Chiu *et al.*[10]	-	0.408
Probabilistic	Res. 3D Conv. [26]	0.445	0.399
	Conv. GRU [2]	0.449	0.397
	Sep. Inception [66]	0.455	0.402
	Ours	**0.464**	**0.416**

5.3 Driving Policy

We study the influence of the learned temporal representation on driving performance. Our baseline is the control policy learned from a single frame.

First we compare to this baseline a model that was trained to directly optimise control, without being supervised with future scene prediction. It shows only a slight improvement over the static baseline, hinting that it is difficult to learn an effective temporal representation by only using control error as a learning signal.

All deterministic models trained with the future prediction loss outperform the baseline, and more interestingly the temporal representation's ability to better predict the future (shown by $\mathcal{M}_{\text{perception}}$) directly translate in a control performance gain, with our best deterministic model having, respectively, a 27% and 38% improvement over the baseline for steering and speed.

Finally, all probabilistic models perform better than their deterministic counterpart, further demonstrating that modelling the uncertainty of the future produces a more effective spatio-temporal representation. Our probabilistic model achieves the best performance with a 33% steering and 46% speed improvement over the baseline.

Table 4. Evaluation of the driving policy. The policy is learned from temporal features explicitly trained to predict the future. We observe a significant performance improvement over non-temporal and non-future-aware baselines.

	Temporal Model	$\mathcal{M}_{\text{perception}}(\uparrow)$	Steering ($\downarrow$)	Speed ($\downarrow$)
	Static	-	0.049	0.048
	Ours w/o future pred.	-	0.043	0.039
Deterministic	Res. 3D Conv. [26]	6.9%	0.039	0.031
	Conv. GRU [2]	7.4%	0.041	0.032
	Sep. Inception [66]	9.6%	0.040	0.031
	Ours	**13.6%**	**0.036**	**0.030**
Probabilistic	Res. 3D Conv. [26]	8.1%	0.040	0.028
	Conv. GRU [2]	9.0%	0.038	0.029
	Sep. Inception [66]	13.8%	0.036	0.029
	Ours	**20.0%**	**0.033**	**0.026**

6 Conclusions

This work is the first to propose a deep learning model capable of probabilistic future prediction of ego-motion, static scene and other dynamic agents. We observe large performance improvements due to our proposed temporal video encoding architecture and probabilistic modelling of present and future distributions. This initial work leaves a lot of future directions to explore: leveraging known priors and structure in the latent representation, conditioning the control policy on future prediction and applying our future prediction architecture to model-based reinforcement learning.

References

1. Amini, A., Rosman, G., Karaman, S., Rus, D.: Variational end-to-end navigation and localization. In: Proceedings of the International Conference on Robotics and Automation (ICRA). IEEE (2019)
2. Ballas, N., Yao, L., Pas, C., Courville, A.: Delving deeper into convolutional networks for learning video representations. In: Proceedings of the International Conference on Learning Representations (ICLR) (2016)
3. Bellemare, M.G., Danihelka, I., Dabney, W., Mohamed, S., Lakshminarayanan, B., Hoyer, S., Munos, R.: The cramer distance as a solution to biased wasserstein gradients. arXiv preprint (2017)
4. Bhattacharyya, A., Schiele, B., Fritz, M.: Accurate and diverse sampling of sequences based on a best of many sample objective. In: Proceedings of the IEEE Conference on Computer Vision and Pattern Recognition (CVPR) (2018)
5. Bilen, H., Fernando, B., Gavves, E., Vedaldi, A., Gould, S.: Dynamic image networks for action recognition. In: Proceedings of the IEEE Conference on Computer Vision and Pattern Recognition (CVPR) (2016)
6. Bojarski, M., et al.: End to end learning for self-driving cars. arXiv preprint (2016)

7. Carreira, J., Zisserman, A.: Quo Vadis, action recognition? A new model and the kinetics dataset. In: Proceedings of the IEEE Conference on Computer Vision and Pattern Recognition (CVPR) (2017)
8. Casas, S., Luo, W., Urtasun, R.: IntentNet: Learning to predict intention from raw sensor data. In: Proceedings of the Conference on Robot Learning (CoRL) (2018)
9. Chen, L.C., Papandreou, G., Schroff, F., Adam, H.: Rethinking Atrous convolution for semantic image segmentation. arXiv preprint (2017)
10. Chiu, H.-K., Adeli, E., Niebles, J.C.: Segmenting the future. arXiv preprint (2019)
11. Chua, K., Calandra, R., McAllister, R., Levine, S.: Deep reinforcement learning in a handful of trials using probabilistic dynamics models. In: Advances in Neural Information Processing Systems (NeurIPS) (2018)
12. Clark, A., Donahue, J., Simonyan, K.: Adversarial video generation on complex datasets. arXiv preprint (2019)
13. Codevilla, F., Miiller, M., López, A., Koltun, V., Dosovitskiy, A.: End-to-end driving via conditional imitation learning. In: Proceedings of the International Conference on Robotics and Automation (ICRA) (2018)
14. Cordts, M., et al.: The cityscapes dataset for semantic urban scene understanding. In: Proceedings of the IEEE Conference on Computer Vision and Pattern Recognition (CVPR) (2016)
15. Denton, E., Birodkar, V.: Unsupervised learning of disentangled representations from video. In: Advances in Neural Information Processing Systems (NeurIPS) (2017)
16. Denton, E., Fergus, R.: Stochastic video generation with a learned prior. In: Proceedings of the International Conference on Machine Learning (ICML), Proceedings of Machine Learning Research (2018)
17. Ebert, F., Finn, C., Dasari, S., Xie, A., Lee, A.X., Levine, S.: Visual foresight: model-based deep reinforcement learning for vision-based robotic control. arXiv preprint (2018)
18. Ebert, F., Finn, C., Lee, A., Levine, S.: Self-supervised visual planning with temporal skip connections. In: Proceedings of the Conference on Robot Learning (CoRL) (2017)
19. Feichtenhofer, C., Pinz, A., Wildes, R.P.: Spatiotemporal residual networks for video action recognition. In: Advances in Neural Information Processing Systems (NeurIPS) (2016)
20. Finn, C., Goodfellow, I., Levine, S.: Unsupervised learning for physical interaction through video prediction. In: Advances in Neural Information Processing Systems (NeurIPS) (2016)
21. Finn, C., Levine, S.: Deep visual foresight for planning robot motion. In: Proceedings of the International Conference on Robotics and Automation (ICRA) (2017)
22. Gadde, R., Jampani, V., Gehler, P.V.: Semantic video CNNs through representation warping. In: Proceedings of the IEEE Conference on Computer Vision and Pattern Recognition (CVPR) (2017)
23. Goodfellow, I.: NIPS 2016 tutorial: generative adversarial networks (2016)
24. Ha, D., Schmidhuber, J.: World models. In: Advances in Neural Information Processing Systems (NeurIPS) (2018)
25. Hafner, D., et al.: Learning latent dynamics for planning from pixels. In: Proceedings of the International Conference on Machine Learning (ICML) (2019)
26. Hara, K., Kataoka, H., Satoh, Y.: Learning Spatio-temporal features with 3D residual networks for action recognition. In: Proceedings of the International Conference on Computer Vision, Workshop (ICCVW) (2017)

27. He, K., Zhang, X., Ren, S., Sun, J.: Deep residual learning for image recognition. In: Proceedings of the IEEE Conference on Computer Vision and Pattern Recognition (CVPR) (2016)
28. Hessel, M., et al.: Rainbow: combining improvements in deep reinforcement learning. In: AAAI Conference on Artificial Intelligence (2018)
29. Huang, X., et al.: The ApolloScape dataset for autonomous driving. In: Proceedings of the IEEE Conference on Computer Vision and Pattern Recognition, Workshop (CVPRW) (2018)
30. Ioannou, Y., Robertson, D., Shotton, J., Cipolla, R., Criminisi, A.: Training CNNs with low-rank filters for efficient image classification. In: Proceedings of the International Conference on Learning Representations (ICLR) (2016)
31. Jaderberg, M., et al.: Reinforcement learning with unsupervised auxiliary tasks. In: Proceedings of the International Conference on Learning Representations (ICLR) (2017)
32. Jayaraman, D., Ebert, F., Efros, A., Levine, S.: Time-agnostic prediction: predicting predictable video frames. In: Proceedings of the International Conference on Learning Representations (ICLR) (2018)
33. Kaiser, L., et al.: Model-based reinforcement learning for Atari. In: Proceedings of the International Conference on Learning Representations (ICLR) (2020)
34. Kalashnikov, D., et al.: Qt-Opt: scalable deep reinforcement learning for vision-based robotic manipulation. In: Proceedings of the International Conference on Machine Learning (ICML)(2018)
35. Kendall, A., Gal, Y.: What uncertainties do we need in Bayesian deep learning for computer vision? In: Advances in Neural Information Processing Systems (NeurIPS) (2017)
36. Kendall, A., Gal, Y., Cipolla, R.: Multi-task learning using uncertainty to weigh losses for scene geometry and semantics. In: Proceedings of the IEEE Conference on Computer Vision and Pattern Recognition (CVPR) (2018)
37. Kendall, A., et al.: Learning to drive in a day. In: Proceedings of the International Conference on Robotics and Automation (ICRA) (2019)
38. Kohl, S., et al.: A probabilistic U-net for segmentation of ambiguous images. In: Advances in Neural Information Processing Systems (NeurIPS) (2018)
39. Kurutach, T., Tamar, A., Yang, G., Russell, S.J., Abbeel, P.: Learning plannable representations with causal InfoGAN. In: Advances in Neural Information Processing Systems (NeurIPS) (2018)
40. Lee, A.X., Zhang, R., Ebert, F., Abbeel, P., Finn, C., Levine, S.: Stochastic adversarial video prediction. arXiv preprint (2018)
41. Lee, N., Choi, W., Vernaza, P., Choy, C.B., Torr, P.H.S., Chandraker, M.K.: DESIRE: distant future prediction in dynamic scenes with interacting agents. In: Proceedings of the IEEE Conference on Computer Vision and Pattern Recognition (CVPR) (2017)
42. Levine, S., Abbeel, P.: Learning neural network policies with guided policy search under unknown dynamics. In: Advances in Neural Information Processing Systems (NeurIPS) (2014)
43. Levine, S., Finn, C., Darrell, T., Abbeel, P.: End-to-end training of deep visuomotor policies. J. Mach. Learn. Res. **17**, 1334–1373 (2016)
44. Li, Z., Snavely, N.: MegaDepth: learning single-view depth prediction from internet photos. In: Proceedings of the IEEE Conference on Computer Vision and Pattern Recognition (CVPR) (2018)

45. Luc, P., Neverova, N., Couprie, C., Verbeek, J., LeCun, Y.: Predicting deeper into the future of semantic segmentation. In: Proceedings of the International Conference on Computer Vision (ICCV) (2017)
46. Mathieu, M., Couprie, C., LeCun, Y.: Deep multi-scale video prediction beyond mean square error. In: Proceedings of the International Conference on Learning Representations (ICLR) (2016)
47. Nabavi, S.S., Rochan, M., Wang, Y.: Future semantic segmentation with convolutional LSTM. In: Proceedings of the British Machine Vision Conference (BMVC) (2018)
48. Neuhold, G., Ollmann, T., Bulo, S.R., Kontschieder, P.: The mapillary vistas dataset for semantic understanding of street scenes. In: Proceedings of the International Conference on Computer Vision (ICCV) (2017)
49. Oh, J., Guo, X., Lee, H., Lewis, R., Singh, S.: Action-conditional video prediction using deep networks in Atari games. In: Advances in Neural Information Processing Systems (NeurIPS) (2015)
50. Piaget, J.: The Origins of Intelligence in the Child. Routledge and Kegan Paul, London (1936)
51. Ranzato, M., Szlam, A., Bruna, J., Mathieu, M., Collobert, R., Chopra, S.: Video (language) modeling: a baseline for generative models of natural videos. arXiv preprint (2014)
52. Rhinehart, N., McAllister, R., Kitani, K.M., Levine, S.: PRECOG: prediction conditioned on goals in visual multi-agent settings. In: Proceedings of the International Conference on Computer Vision (ICCV) (2019)
53. Salimans, T., Zhang, H., Radford, A., Metaxas, D.N.: Improving GANs using optimal transport. In: Proceedings of the International Conference on Learning Representations (ICLR) (2018)
54. Schuldt, C., Laptev, I., Caputo, B.: Recognizing human actions: a local SVM approach. In: Proceedings of the International Conference on Pattern Recognition (2004)
55. Shi, X., Chen, Z., Wang, H., Yeung, D.Y., Wong, W.K., Woo, W.C.: Convolutional LSTM network: a machine learning approach for precipitation nowcasting. In: Advances in Neural Information Processing Systems (NeurIPS) (2015)
56. Siam, M., Valipour, S., Jagersand, M., Ray, N.: Convolutional gated recurrent networks for video segmentation. In: Proceedings of the International Conference on Image Processing (2017)
57. Simonyan, K., Zisserman, A.: Very deep convolutional networks for large-scale image recognition. In: Proceedings of the International Conference on Learning Representations (ICLR) (2015)
58. Simonyan, K., Zisserman, A.: Two-stream convolutional networks for action recognition in videos. In: NIPS (2014)
59. Srivastava, N., Mansimov, E., Salakhudinov, R.: Unsupervised learning of video representations using LSTMs. In: ICML (2015)
60. Sun, D., Yang, X., Liu, M.Y., Kautz, J.: PWC-Net: CNNs for optical flow using pyramid, warping, and cost volume. In: Proceedings of the IEEE Conference on Computer Vision and Pattern Recognition (CVPR) (2018)
61. Sun, L., Jia, K., Yeung, D., Shi, B.E.: Human action recognition using factorized Spatio-temporal convolutional networks. In: Proceedings of the International Conference on Computer Vision (ICCV) (2015)
62. Szegedy, C., et al.: Going deeper with convolutions. In: Proceedings of the IEEE Conference on Computer Vision and Pattern Recognition (CVPR) (2015)

63. Tran, D., Wang, H., Torresani, L., Ray, J., LeCun, Y., Paluri, M.: A closer look at spatiotemporal convolutions for action recognition. In: Proceedings of the IEEE Conference on Computer Vision and Pattern Recognition (CVPR) (2018)
64. Wu, Z., Shen, C., van den Hengel, A.: Bridging category-level and instance-level semantic image segmentation. arXiv preprint (2016)
65. Xie, S., Girshick, R.B., Dollár, P., Tu, Z., He, K.: Aggregated residual transformations for deep neural networks. In: Proceedings of the IEEE Conference on Computer Vision and Pattern Recognition (CVPR) (2017)
66. Xie, S., Sun, C., Huang, J., Tu, Z., Murphy, K.: Rethinking spatiotemporal feature learning for video understanding. In: Proceedings of the European Conference on Computer Vision (ECCV) (2018)
67. Yu, F., et al.: BDD100K: a diverse driving video database with scalable annotation tooling. In: Proceedings of the International Conference on Computer Vision, Workshop (ICCVW) (2018)
68. Zhu, X., Xiong, Y., Dai, J., Yuan, L., Wei, Y.: Deep feature flow for video recognition. In: Proceedings of the IEEE Conference on Computer Vision and Pattern Recognition (CVPR) (2017)

Suppressing Mislabeled Data via Grouping and Self-attention

Xiaojiang Peng[1,2], Kai Wang[1,2], Zhaoyang Zeng[3], Qing Li[4], Jianfei Yang[5], and Yu Qiao[1,2](✉)

[1] Guangdong-Hong Kong-Macao Joint Laboratory of Human-Machine Intelligence-Synergy Systems, Shenzhen Institutes of Advanced Technology, Chinese Academy of Sciences, Shenzhen 518055, China
yu.qiao@siat.ac.cn
[2] SIAT Branch, Shenzhen Institute of Artificial Intelligence and Robotics for Society, Shenzhen, China
[3] Sun Yat-sen University, Guangzhou, China
[4] Southwest Jiaotong University, Chengdu, China
[5] Nanyang Technological University, Singapore, Singapore

Abstract. Deep networks achieve excellent results on large-scale clean data but degrade significantly when learning from noisy labels. To suppressing the impact of mislabeled data, this paper proposes a conceptually simple yet efficient training block, termed as Attentive Feature Mixup (AFM), which allows paying more attention to clean samples and less to mislabeled ones via sample interactions in small groups. Specifically, this plug-and-play AFM first leverages a *group-to-attend* module to construct groups and assign attention weights for group-wise samples, and then uses a *mixup* module with the attention weights to interpolate massive noisy-suppressed samples. The AFM has several appealing benefits for noise-robust deep learning. (i) It does not rely on any assumptions and extra clean subset. (ii) With massive interpolations, the ratio of useless samples is reduced dramatically compared to the original noisy ratio. (iii) It jointly optimizes the interpolation weights with classifiers, suppressing the influence of mislabeled data via low attention weights. (iv) It partially inherits the vicinal risk minimization of mixup to alleviate over-fitting while improves it by sampling fewer feature-target vectors around mislabeled data from the mixup vicinal distribution. Extensive experiments demonstrate that AFM yields state-of-the-art results on two challenging real-world noisy datasets: Food101N and Clothing1M.

Keywords: Noisy-labeled data · Mixup · Noisy-robust learning

1 Introduction

In recent years, deep neural networks (DNNs) have achieved great success in various tasks, particularly in supervised learning tasks on large-scale image recognition challenges, such as ImageNet [6] and COCO [21]. One key factor that

X. Peng, K. Wang and Z. Zeng—Equally-contributed first authors.

A. Vedaldi et al. (Eds.): ECCV 2020, LNCS 12361, pp. 786–802, 2020.
https://doi.org/10.1007/978-3-030-58517-4_46

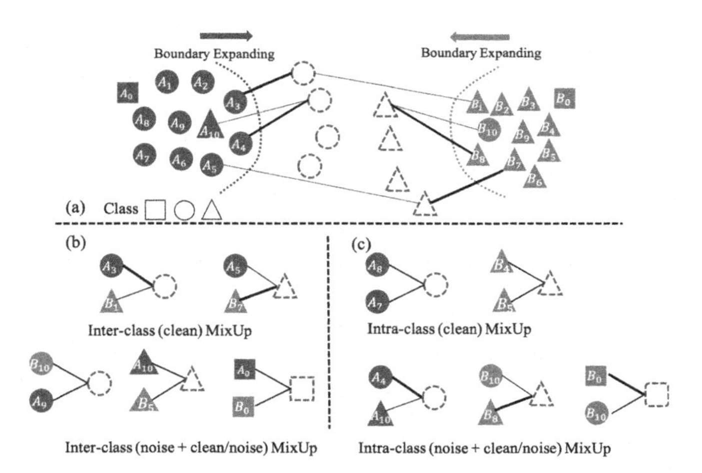

Fig. 1. Suppressing mislabeled samples by grouping and self-attention mixup. *Different colors and shapes denote given labels and ground truths. Thick and thin lines denote high and low attention weights, respectively.* A_0, A_{10}, B_0, and B_{10} are supposed to be mislabeled samples, and can be suppressed by assigning low interpolation weights in mixup operation.

drives impressive results is the large amount of well-labeled images. However, high-quality annotations are laborious and expensive, even not always available in some domains. To address this issue, an alternative solution is to crawl a large number of web images with tags or keywords as annotations [8,19]. These annotations provide weak supervision, which are noisy but easy to obtain.

In general, noisy labeled examples hurt generalization because DNNs easily overfit to noisy labels [2,7,30]. To address this problem, it is intuitive to develop noise-cleaning methods which aim to correct the mislabeled samples either by joint optimization of classification and relabeling [31] or by iterative self-learning [11]. However, the noise-cleaning methods often suffer from the main difficulty in distinguishing mislabeled samples from hard samples. Another solution is to develop noise-robust methods which aims to reduce the contributions of mislabeled samples for model optimization. Along this solution, some methods estimate a matrix for label noise modeling and use it to adapt output probabilities and loss values [26,30,35]. Some others resort to curriculum learning [4] by either designing a step-wise easy-to-hard strategy for training [10] or introducing an extra MentorNet [12] for sample weighting. However, these methods independently estimate the importance weights for individuals which ignore the comparisons among different samples while they have been proven to be the key of humans to perceive and learn novel concepts from noisy input images [29]. Some other solutions follow semi-supervised configuration where they assume a small manually-verified set can be used [15,17,20,32]. However, this assumption may be not supported in real-world applications. With the Vicinal Risk Minimization(VRM) principle, mixup [33,36] exploits a vicinal distribution for sampling virtual sample-target vectors, and proves its robustness for synthetic noisy data. But its effectiveness is limited in real-world noisy data [1].

In this paper, we propose a conceptually simple yet efficient training block, termed as Attentive Feature Mixup (AFM), to suppress mislabeled data thus to make training robust to noisy labels. The AFM is a plug-and-play block for training any networks and is comprised of two crucial parts: 1) a *Group-to-Attend* (GA) module that first randomly groups a minibatch images into small subsets and then estimates sample weights within those subsets by an attention mechanism, and 2) a *mixup* module that interpolates new features and soft labels according to self-attention weights. Particularly, for the GA module, we evaluate three feature interactions to estimate group-wise attention weights, namely concatenation, summary, and element-wise multiplication. The interpolated samples and original samples are respectively fed into an interpolation classifier and a normal classifier. Figure 1 illustrates how AFM suppress mislabeled data. Generally, there exists two main types of mixup: intra-class mixup (Fig. 1(c)) and inter-class mixup (Fig. 1(b)). For both types, the interpolations between mislabeled samples and clean samples may become useful for training with adaptive attention weights, *i.e.* low weights for the mislabeled samples and high weights for the clean samples. In other words, our AFM hallucinates numerous useful noisy-reduced samples to guide deep networks learn better representations from noisy labels. Overall, as a noisy-robust training method, our AFM is promising in the following aspects.

- It does not rely on any assumptions and extra clean subset.
- With AFM, the ratio of harmful noisy interpolations (*i.e.* between noisy samples) over all interpolations is largely less than the original noisy ratio.
- It jointly optimizes the mixup interpolation weights and classifier, suppressing the influence of mislabeled data via low attention weights.
- It partially inherits the vicinal risk minimization of mixup to alleviate overfitting while improves it by sampling less feature-target vectors around mislabeled data from the mixup vicinal distribution.

We validate our AFM on two popular real-world noisy-labeled datasets: Food101N [15] and Clothing1M [35]. Experiments show that our AFM outperforms recent state-of-the-art methods significantly with accuracies of **87.23%** on Food101N and **82.09%** on Clothing1M.

2 Related Work

2.1 Learning with Noisy Labeled Data

Learning with noisy data has been vastly studied on the literature of machine learning and computer vision. Methods on learning with label noise can be roughly grouped into three categories: noise-cleaning methods, semi-supervised methods and noise-robust methods.

First, noise-cleansing methods aim to identify and remove or relabel noisy samples with filter approaches [3,24]. Brodley *et al.* [5] propose to filter noisy samples using ensemble classifiers with majority and consensus voting. Sukhbaatar *et al.* [30] introduce an extra noise layer into a standard CNN which adapts the network outputs to match the noisy label distribution. Daiki *et al.* [31] propose a joint optimization framework to train deep CNNs with label noise, which updates the network parameters and labels alternatively. Based on the consistency of the noisy groundtruth and the current prediction of the model, Reed *et al.* [27] present a 'Soft' and a 'Hard' bootstrapping approach to relabel noisy data. Li *et al.* [20] relabel noisy data using the noisy groundtruth and the current prediction adjusted by a knowledge graph constructed from DBpedia-Wikipedia.

Second, semi-supervised methods aim to improve performance using a small manually-verified clean set. Lee *et al.* [15] train an auxiliary CleanNet to detect label noise and adjust the final sample loss weights. In the training process, the CleanNet needs to access both the original noisy labels and the manually-verified labels of the clean set. Veit *et al.* [32] use the clean set to train a label cleaning network but with a different architecture. These methods assume there exists such a label mapping from noisy labels to clean labels. Xiao *et al.* [35] mix the clean set and noisy set, and train an extra CNN and a classification CNN to estimate the posterior distribution of the true label. Li *et al.* [18] first train a teacher model on clean and noisy data, and then distill it into a student model trained on clean data.

Third, the noise-robust learning methods are assumed to be not too sensitive to the presence of label noise, which directly learn models from the noisy labeled data [13,14,25,26,34]. Manwani *et al.* [23] present a noise-tolerance algorithm under the assumption that the corrupted probability of an example is a function of the feature vector of the example. With synthetic noisy labeled data, Rolnick *et al.* [28] demonstrate that deep learning is robust to noise when training data is sufficiently large with large batch size and proper learning rate. Guo *et al.* [10] develop a curriculum training scheme to learn noisy data from easy to hard. Han *et al.* [11] propose a Self-Learning with Multi-Prototype (SMP) method to learn robust features via alternatively training and clustering which is time-consuming. Wang *et al.* [34] propose to suppress uncertain samples with self-attention, ranking loss, and relabeling. Our method is most related to Meta-Cleaner [37], which hallucinates a clean (precisely noise-reduced) representation by mixing samples (the ratio of the noisy images need to be small) from the same category. Our work differs from it in that i) we formulate the insight as attentive mixup, and ii) hallucinate noisy-reduced samples not only within class but also between classes which significantly increases the number of interpolations and expands the decision boundaries. Moreover, we introduce more sample interactions rather than the concatenation in [37], and find a better one.

2.2 Mixup and Variations

Mixup [36] regularizes the neural network to favor simple linear behavior in-between training examples. Manifold Mixup [33] leverages semantic interpolations in random layers as additional training signal to optimize neural networks.

The interpolation weights of those two methods are drawn from a β distribution. Meta-Mixup [22] introduces a meta-learning based online optimization approach to dynamically learn the interpolation policy from a reference set. AdaMixup [9] also learns the interpolation policy from dataset with an additional network to infer the policy and an intrusion discriminator. Our work differs from these variations in that i) we design a Group-to-Attend mechanism to learn attention weights for interpolating in a group-wise manner which is the key to reduce the influence of noises and ii) we address the noisy-robust problem on real-world noisy data and achieve state-of-the-art performance.

3 Attentive Feature Mixup

As proven in cognitive studies, we human mainly perceive and learn novel concepts from noisy input images by comparing and summary [29]. Based on this motivation, we propose a simple yet efficient model, called Attentive Feature Mixup (AFM), which aims to learn better features by making clean and noisy samples interact with each other in small groups.

3.1 Overview

Our AFM works on traditional CNN backbones and includes two modules: i) Group-to-Attend (GA) module and ii) mixup module, as shown in Fig. 2.

Let $\mathcal{B} = \{(\mathbf{I}_1, y_1), (\mathbf{I}_2, y_2), \cdots, (\mathbf{I}_n, y_n)\}$ be the mini-batch set of a noisy labeled dataset, which contains n samples, and $y_i \in \mathcal{R}^C$ is the noisy one-hot label vector of image $\mathbf{I}_i$. The AFM works as the following procedure. First, a CNN backbone $\phi(\cdot;\theta)$ with parameter θ is used to extract image features $\{x_1, x_2, \cdots, x_n\}$. Then, the Group-to-Attend (GA) module is used to divide the mini-batch images into small groups and learn attention weights for each samples within each group. Subsequently, with the group-wise attention weights, a mixup module is used to interpolate new samples and soft labels. Finally, these interpolations along with the original image features are fed into an interpolation classifier f_{c1} (*i.e.* FC layer) and a normal classifier f_{c2} (*i.e.* FC layer), respectively. Particularly, the interpolation classifier is supervised by the soft labels from the mixup module and the normal classifier by the original given labels which are noisy. Our AFM partially inherits the vicinal risk minimization of mixup to alleviate over-fitting with massive interpolations. Further, with jointly optimizing the mixup interpolation weights and classifier, AFM improves mixup by sampling less feature-target vectors around mislabeled data from the mixup vicinal distribution.

3.2 Group-to-Attend Module

In order to obtain meaningful attention weights, *i.e.* high weights for clean samples and low weights for mislabeled samples, we elaborately design a Group-to-Attend module, which consists of four crucial steps. First, we randomly and

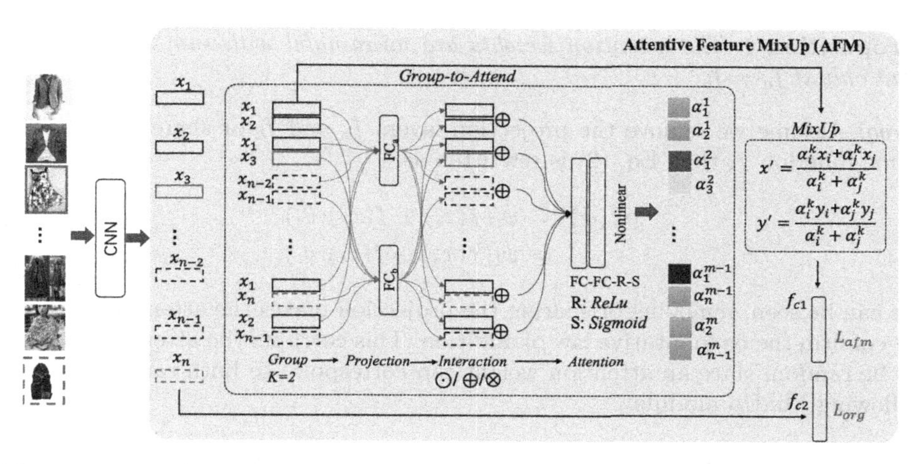

Fig. 2. The pipeline of Attentive Feature Mixup (AFM). Given a mini-batch of n images, a backbone CNN is first applied for feature extraction. Then, a Group-to-Attend (GA) module randomly composites massive groups with the group size K and linearly projects each element within a group with a separated FC layer, and then combines each group with an interaction (*i.e.* concatenation ($\odot$), sum ($\oplus$), and element-wise multiplication ($\otimes$)), and finally outputs K attention weights for each group. With the group-wise attention weights, a mixup module is used to interpolate new samples and soft labels.

repeatedly selecting K samples to construct groups as many as possible (the number of groups depends on the input batch size and the GPU memory). Second, we use K fully-connected (FC) layers to map the ordered samples of each group into new feature embeddings for sample interactions. As an example of $K = 2$ in Fig. 2, x_i and x_j are linearly projected as,

$$\tilde{x}_i = f_a(x_i; w_a), \quad \tilde{x}_j = f_b(x_j; w_b), \tag{1}$$

where w_a and w_b are the parameters of FC layer f_a and f_b, respectively. Third, we further make $\tilde{x}_i$ and $\tilde{x}_j$ interact for group-wise weight learning. Specifically, we experimentally explore three kinds of interactions: concatenation ($\odot$), sum ($\oplus$), and element-wise multiplication ($\otimes$). Last, we apply a light-weight self-attention network to estimate group-wise attention weights. Formally, for $K = 2$ and the sum interaction, this step can be defined as follows,

$$\begin{aligned}[\alpha_i^k, \alpha_j^k] &= \psi_t(\tilde{x}_i \oplus \tilde{x}_j; \theta_t) \\ &= \psi_t(f_a(x_i; w_a) \oplus f_b(x_j; w_b); \theta_t),\end{aligned} \tag{2}$$

where ψ_t is the attention network, θ_t denotes its parameters, and k denotes the $k-$th group. For the architecture of ψ_t, we follow the best one of [37], *i.e.* FC-FC-ReLu-Sigmoid. It is worth noting that feature interaction is crucial for learning meaningful attention weights since the relationship between noisy and clean samples within a group can be learned efficiently while not the case of non-interaction (*i.e.* learning weights for each other separately).

Proposition 1 *The attention weights are meaningful with sum interaction if and only if* $f_a \neq f_b$.

Proof Assume we remove the projection layers f_a and f_b or share them as the same function f, then Eq. (2) is rewritten as,

$$\begin{aligned} [\alpha_i^k, \alpha_j^k] &= \psi_t(f(x_i) \oplus f(x_j); \theta_t) \\ &= \psi_t(f(x_j) \oplus f(x_i); \theta_t). \end{aligned} \tag{3}$$

As can be seen, removing or sharing the projection makes the attention network ψ_t confirm the commutative law of addition. This corrupts the attention weights to be random since an attention weight can correspond to both samples for the following MixUp module.

The Effect of GA. An appealing benefit of our GA is that it reduces the impact of noisy-labeled samples significantly. Let N_{noisy} and N_{total} represent the number of the noisy images and total images in a noisy dataset, respectively. The noise ratio is $\frac{N_{noisy}}{N_{total}}$ in the image-wise case. Nevertheless, the number of pure noisy groups (*i.e.* all the images are mislabeled in these groups) in the group-wise case becomes $A_{N_{noisy}}^K$. With $K = 2$, we have,

$$\frac{N_{noisy}}{N_{total}} > \frac{A_{N_{noisy}}^2}{A_{N_{total}}^2} = \frac{N_{noisy}(N_{noisy}-1)}{N_{total}(N_{total}-1)} \approx \frac{N_{noisy}^2}{N_{total}^2}. \tag{4}$$

We argue that GA can reduce the pure noisy ratio dramatically and partial noisy groups (*i.e.* some images within these groups are corrected-labeled) may provides useful supervision by the well-trained attention network. However, though the ratio is smaller when K becomes larger, large K may lead to over-smooth features for the new interpolations which are harmful for discriminative feature learning.

3.3 Mixup Module

The mixup module interpolates virtual feature-target vectors for training. Specifically, following classic mixup vicinal distribution, we normalize the attention weights into range [0, 1]. Formally, for $K = 2$ and group members $\{x_i, x_j\}$, the mixup can be written as follows,

$$x' = \frac{1}{\sum_\alpha}(\alpha_i x_i + \alpha_j x_j), \tag{5}$$

$$y' = \frac{1}{\sum_\alpha}(\alpha_i y_i + \alpha_j y_j), \tag{6}$$

where x' and y' are the interpolated feature and soft label.

3.4 Training and Inference

Training. Our AFM along with the CNN backbone can be trained in an end-to-end manner. Specifically, we conduct a multi-task training scheme to separate the contributions of original training samples and new interpolations. Let f_{c1} and f_{c2} respectively denote the classifiers (include the Softmax or Sigmoid operations) of interpolations and original samples, we can formulate the training loss in a mini batch as follows,

$$\begin{aligned} \mathcal{L}_{total} &= \lambda \mathcal{L}_{afm} + (1-\lambda)\mathcal{L}_{org} \\ &= \frac{\lambda}{m}\sum_{i=1}^{m}\mathcal{L}(f_{c1}(x'_i), y'_i) + \frac{(1-\lambda)}{n}\sum_{i=1}^{n}\mathcal{L}(f_{c2}(x_i), y_i), \end{aligned} \tag{7}$$

where n is the batch size, m is the number of interpolations, and λ is a trade-off weight. We use the Cross-Entropy loss function for both L_{afm} and L_{org}. In this way, our AFM can be viewed as a regularizer over the training data by massive interpolations. As proven in [33,36], this regularizer can largely improve the generalization of deep networks. In addition, the parameters of f_{c1} and f_{c2} can be shared since both original features and interpolations are in same dimensions.

Inference. After training, both the GA module and mixup module can be simply removed since we do not need to compose new samples at test stage. We keep the classifiers f_{c1} and f_{c2} for inference. Particularly, they are identical and we can conduct inference as traditional CNNs if the parameters are shared.

4 Experiments

In this section, we first introduce datasets and implementation details, and then compare our AFM with the state-of-the-art methods. Finally, we conduct ablation studies with qualitative and quantitative results.

4.1 Datasets and Implementation Details

In this paper, we conduct experiments on two popular real-world noisy datasets: Food101N [16] and Clothing1M [35]. **Food101N** consists of 365k images that are crawled from Google, Bing, Yelp, and TripAdvisor using the Food-101 taxonomy. The annotation accuracy is about 80%. The clean dataset Food-101 is collected from *foodspotting.com* which contains 101 food categories with 101,000 real-world food images totally. For each class, 750 images are used for training, the other 250 images for testing. In our experiments, following the common setting, we use all images of Food-101N as the noisy dataset, and report the overall accuracy on the Food-101 test set. **Clothing1M** contains 1 million images of clothes with 14 categories. Since most of the labels are generated by the surrounding text of the images on the Web, a large amount of annotation noises exist, leading to a low annotation accuracy of 61.54% [35]. The human-annotated set of Clothing1M is used as the clean set which is officially divided into training, validation and

Table 1. Comparison with the state-of-the-art methods on Food101N dataset. VF(55k) is the noise-verification set used in CleanNet [16].

Method	Training data	Training time	Acc
Softmax [16]	Food101	–	81.67
Softmax [16]	Food101N	–	81.44
Weakly Supervised [38]	Food101N	–	83.43
CleanNet(w_{hard}) [16]	Food101N + VF(55K)	–	83.47
CleanNet(w_{soft}) [16]	Food101N + VF(55K)	–	83.95
MetaCleaner [37]	Food101N	–	85.05
SMP [11]	Food101N	–	85.11
ResNet50 (baseline)	Food101N	4 h 16 min 40 s	84.51
AFM (Ours)	Food101N	4 h17 min 4 s	**87.23**

testing sets, containing 50k, 14k and 10k images respectively. We report the overall accuracy on the clean test set of Clothing1M.

Implementation Details. As widely used in existing works, ResNet50 is used as our CNN backbone and initialized by the official ImageNet pre-trained model. For each image, we resize the image with a short edge of 256 and random crop 224 × 224 patch for training. We use SGD optimizer with a momentum of 0.9. The weight decay is 5×10^{-3}, and the batch size is 128. For Food101N, the initial learning rate is 0.001 and divided by 10 every 10 epochs. We stop training after 30 epochs. For Clothing1M, the initial learning rate is 0.001 and divided by 10 every 5 epochs. We stop training after 15 epochs. All the experiments are implemented by Pytorch with 4 NVIDIA V100 GPUs. The default λ and K are 0.75 and 2, respectively. By default, the classifiers f_{c1} and f_{c2} are shared.

4.2 Comparison on Food101N

We compare AFM to the baseline model and existing state-of-the-art methods in Table 1. AFM improves our strong baseline from 84.51% to 87.23%, and consistently outperforms recent state-of-the-art methods with large margins. Moreover, our AFM is almost free since it only increases training time by 24s. Specifically, AFM outperforms [38] by 3.80%, CleanNet(w_{soft}) by 3.28%, and SMP [11] by 2.12%. We notice that, CleanNet(w_{hard}) and CleanNet(w_{soft}) use extra 55k manually-verified images, while we do not use any extra images. In particular, MetaCleaner [37] uses a similar scheme but limited in intra-class mixup and its single feature interaction type, which leads to 2.18% worse than our AFM. An ablation study will further discuss these issues in the following section.

4.3 Comparison on Clothing1M

For the comparison on Clothing1M, we evaluate our AFM in three different settings following [11,16,26,37]: (1) only the noisy set are used for training, (2)

Table 2. Comparison with the state-of-the-art methods on Clothing1M. VF(25k) is the noise-verification set used in CleanNet [16].

Method	Training data	Acc. (%)
Softmax [16]	1M noisy	68.94
Weakly Supervised [38]	1M noisy	71.36
JointOptim [37]	1M noisy	72.23
MetaCleaner [37]	1M noisy	72.50
SMP (Final) [11]	1M noisy	**74.45**
SMP (Initial) [11]	1M noisy	72.09
AFM (Ours)	1M noisy	**74.12**
CleanNet(w_{hard}) [16]	1M noisy + VF(25K)	74.15
CleanNet(w_{soft}) [16]	1M noisy + VF(25K)	74.69
MetaCleaner [37]	1M noisy + VF(25K)	76.00
SMP [11]	1M noisy + VF(25K)	76.44
AFM (Ours)	1M noisy + VF(25K)	**77.21**
CleanNet(w_{soft}) [16]	1M noisy + Clean(50K)	79.90
MetaCleaner [37]	1M noisy + Clean(50K)	80.78
SMP [11]	1M noisy + Clean(50K)	81.16
CurriculumNet [10]	1M noisy + Clean(50K)	81.50
AFM (Ours)	1M noisy + Clean(50K)	**82.09**

the 25K extra manually-verified images [16] are added into the noisy set for training, and (3) the 50K clean training images are added into the noisy set.

The comparison results are shown in Table 2. For the first setting, our AFM improves the baseline method from 68.94% to 74.12%, and consistently outperforms MetaCleaner, JointOptim, and SMP (Initial) by about 2%. Although SMP (Final) performs on par with AFM in this setting, it needs several training-and-correction loops and careful parameter tuning. Compared to SMP (Final), our AFM is simpler and almost free in computational cost.

For the second setting, other methods except for MetaCleaner mainly apply the 25K verified images to train an accessorial network [16,26] or to select the class prototypes [11]. Following [37], we train our AFM on 1M noisy training set, and then fine-tune it on the 25K verified images. As shown in Table 2, AFM obtains 77.21% which sets new record in this setting. Specifically, our AFM is better than MetaCleaner and SMP by 1.21% and 0.77%, respectively.

For the third setting, all the methods first train models on the noisy set and then fine-tune them on the clean set. CurriculumNet [10] uses a deeper CNN backbone and obtains accuracy 81.5%, which is slightly better than SMP and other methods. Our AFM outperforms CurriculumNet by 0.59%, and is better than MetaCleaner by 1.31%. It is worth emphasizing that both CurriculumNet and SMP need to train repeatedly after model convergence which are complicated and time-consuming, while AFM is much simpler and almost free.

Table 3. Results of different feature interactions in Group-to-Attend module. *It removes FC_a and FC_b in GA module.

#	Interaction type	Training data	Acc. (%)
1	Concatenation	Food101N	86.95
2	Concatenation*	Food101N	86.51
3	Sum	Food101N	**87.23**
4	Sum*	Food101N	86.12
5	Multiplication	Food101N	86.64

Table 4. Evaluation of trade-off λ.

λ	0.00	0.25	0.50	0.75	1.00
Acc. (%)	84.51	86.75	86.97	**87.23**	86.47

Table 5. Evaluation of group size.

Size	2	3	4	5	6
Acc. (%)	**87.23**	86.46	86.01	85.92	85.46

4.4 Ablation Study

Evaluation of Feature Interaction Types. *Concatenation*, *sum* and *element-wise multiplication* are three popular feature fusion or interaction methods. MetaCleaner [37] simply takes the *concatenation*, and ignores the impact of the interaction types. We conduct an ablation study for them along with the projection in Group-to-Attend module. Specifically, the group size is set to 2 for this study. Table 3 presents the results on Food101N. Two observations can be concluded as following. First, with FC_a and FC_b, the *sum* interaction consistently performs better than the others. Second, for both *concatenation* and *sum*, it is better to use the projection process. As mentioned in Sect. 3.2, removing FC_a and FC_b leads to random attention weights for *sum* interaction, which may degrade our AFM to standard Manifold mixup [36]. Nevertheless, it still improves the baseline (*i.e.* 84.51%) slightly.

Evaluation of the Trade-Off Weight λ. In training phase, λ is used to trade-off the loss ratio between $\mathcal{L}_{afm}$ and $\mathcal{L}_{org}$. We evaluate it by increasing λ from 0 to 1 on Food101N, and present the results in Table 5. We achieve the best accuracy with default λ (*i.e.* 0.75). Decreasing λ means to use less interpolations from AFM, which gradually degrades the final performance. Particularly, $\lambda = 0$ is our baseline that only uses original noisy training data. In the other extreme case, using only the interpolations from AFM is better than the baseline but slight worse than the default one. This may be explained by that the massive interpolations are more or less smoothed by our AFM since the interpolation weights cannot be zeros due to the GA module. Hence, adding original features can be better since these features fill this gap naturally.

Evaluation of the Group Size. Our previous experiments fix the group size as 2 which construct pairwise samples for generating virtual feature-target vectors. Here we explore different group sizes for our attentive feature mixup, Specifically, we increase the group size from 2 to 6, and present the results in Table 4. As can be seen, enlarging the group size gradually degrades the final performance. This may be explained by that large group size interpolates over-smoothed features which are not discriminative for any classes.

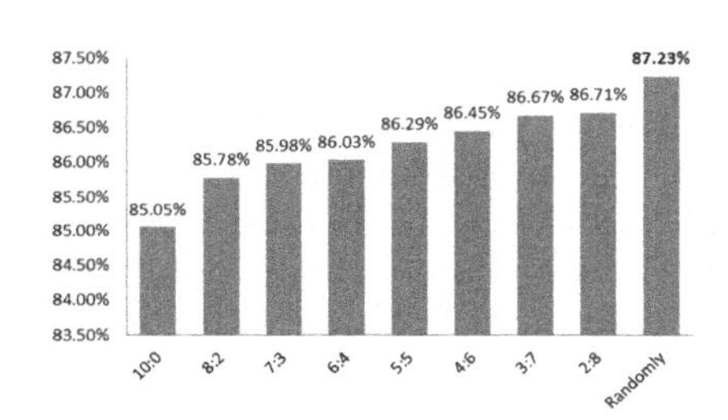

Fig. 3. Evaluation of the ratios of Intra- and Inter-class mixup.

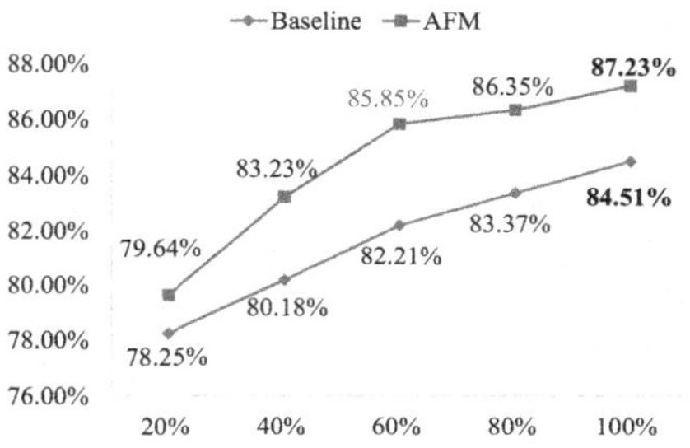

Fig. 4. Evaluation of AFM on synthetic small datasets.

Intra-class Mixup vs. Inter-class Mixup. To investigate the contributions of intra-class mixup and inter-class mixup, we conduct an evaluation by exploring different ratios between intra- and inter-class interpolations with group size 2. Specifically, we constrain the number of interpolations for both mixup types in each minibatch with 8 varied ratios from 10:0 to 2:8 on the Food101N dataset. The results are shown in Fig. 3. Several observations can be concluded as following. First, removing the inter-class mixup (*i.e.* 10:0) degrades the performance (it is similar with MetaCleaner [37]) while adding a small ratio (*e.g.* 8:2) of inter-class mixup significantly improves the final result. This indicates that the inter-class mixup is more useful for better feature learning. Second, increasing the ratio of inter-class mixup further boosts performance but the performance gaps are small. Third, we get the best result by random selecting group-wise samples. We argue that putting constraints on the ratio of mixup types may result in different data distribution compared to the original dataset while random choice avoids this problem.

AFM for Learning from Small Dataset. Since AFM can generate numerous of noisy-reduced interpolations in training stage, we intuitively check the power of AFM on small datasets. To this end, we construct sub-datasets from Food101N by randomly decreasing the size of Food101N to 80%, 60%, 40%, and 20%. The results of our default AFM on these synthetic datasets are shown in Fig. 4.

Table 6. Comparison of our AFM with mixup [36] and Manifold mixup [33]. We also evaluate f_{c1} and f_{c2} for them.

Method	f_{c1} + f_{c2}	f_{c1} + f_{c2} (Shared)
mixup [36]	85.36%	85.63%
Manifold mixup [33]	85.85%	86.12%
AFM (Ours)	**86.97%**	**87.23%**

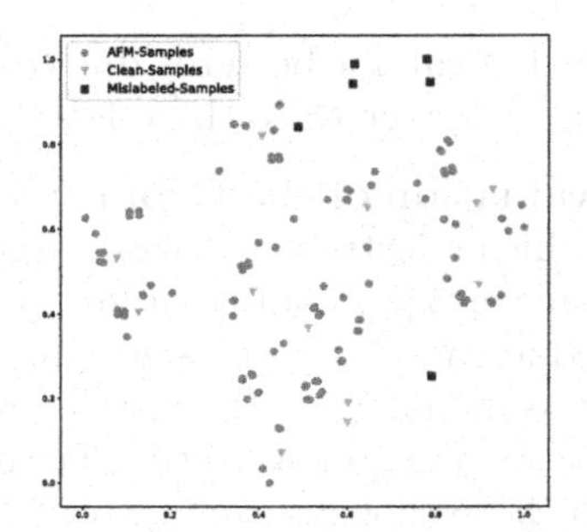

Fig. 5. AFM sample distribution.

Several observations can be concluded as following. First, our AFM consistently improves the baseline significantly. Second, the improvements from data size 40% to 100% are larger than that of 20%. This may be because that small dataset leads to less diverse interpolations. Third, we interestingly find that our AFM already obtains the state-of-the-art performance with only 60% data on Food101N.

AFM vs. Classic Mixup. Since our AFM is related to the mixup scheme, we compare it to the Standard mixup [36] and Manifold mixup [33]. The Standard mixup [36] interpolates samples in image level while Manifold mixup [33] in feature level. Both of them drawl the interpolation weights randomly from a β distribution. Our method introduce a Group-to-Attend (GA) module to generate meaningful weights for noise-robust training. As the new interpolations and the original samples can contribute differently, we separately apply classifiers for them, *i.e.* f_{c1} for interpolations and f_{c2} for original samples. Table 6 presents the comparison. Several observations are concluded as following. First, for both classifier setting, our AFM outperforms the others largely, *e.g.* AFM is better than standard mixup by 1.6% and the Manifold mixup by 1.11% in the shared classifier setting. Second, the shared classifiers are slightly better than the independent classifiers for all methods, which may be explained by that sharing parameters makes the classifier favor linear behavior over all samples thus reducing over-fitting and encouraging the model to discover useful features.

4.5 Visualizations

To better investigate the effectiveness of our AFM, we make two visualizations: i) attentive mixup sample distribution between clean and noisy samples in Fig. 5 and ii) the normalized attention weights in Fig. 6. For the former, we randomly select several noisy samples and clean samples on the VK(25) set of Food101N and apply our trained AFM model to generate virtual samples (*i.e.* AFM samples), and then use t-SNE to visualize all the real samples and attentive mixup samples. Figure 5 evidently shows that our AFM samples are mainly distributed

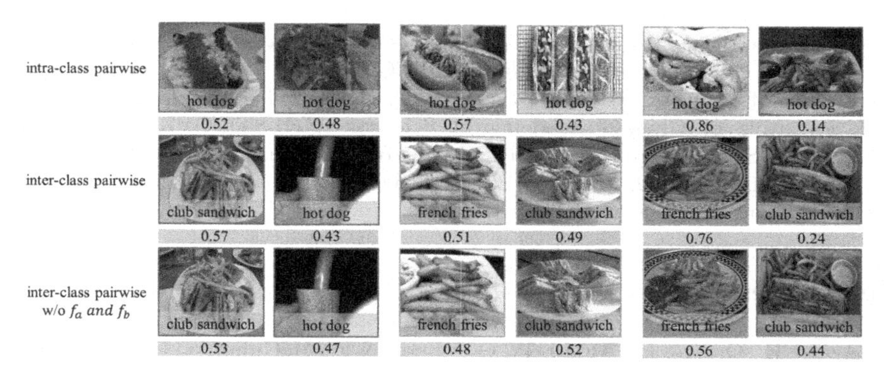

Fig. 6. Visualization of the attention weights in our AFM. The green and red boxes represent the clean and noisy samples.

around the clean samples, demonstrating our AFM suppresses noisy samples effectively. *It is worth noting that classical mixup samples are doomed to distribute around all the real samples rather than only clean samples.*

For the latter visualization, the first row of Fig. 6 shows three types of pairs for the intra-class case, the second row for the inter-class case, and the third row for the inter-class case without projection in the Group-to-Attend module. The first column denotes the "noisy+noisy" interpolations, the second column denotes "clean+clean", and the third column denotes "clean+noisy". Several finds can be observed as following. First, for both intra- and inter-class cases, the weights of "noisy+noisy" and "clean+clean" interpolations trend to be equal since these interpolations may lie in the decision boundaries which make the network hard to identify which is better for training. Second, for the "clean+noisy" interpolations on the first two rows, our AFM assigns evidently low weights to these noisy samples which demonstrates the effectiveness of AFM. Last, without projection in the Group-to-Attend module, our default AFM loses the ability to identify noisy samples as shown in the last image pair.

5 Conclusion

This paper proposed a conceptually simple yet efficient training block, termed as Attentive Feature Mixup (AFM), to address the problem of learning with noisy labeled data. Specifically, AFM is a plug-and-play training block, which mainly leverages grouping and self-attention to suppress mislabeled data and does not rely on any assumptions and extra clean subset. We conducted extensive experiments on two challenging real-world noisy datasets: Food101N and Clothing1M. Quantitative and qualitative results demonstrated that our AFM is superior to recent state-of-the-art methods. In addition, the grouping and self-attention is expected to extend in other topics, *e.g.* semi-supervised learning, where one may conduct this module for real annotations and pseudo labels to automatically suppress incorrect pseudo labels.

Acknowledge. This work is partially supported by National Key Research and Development Program of China (No. 2020YFC2004800), National Natural Science Foundation of China (U1813218, U1713208), Science and Technology Service Network Initiative of Chinese Academy of Sciences (KFJ-STS-QYZX-092), Guangdong Special Support Program (2016TX03X276), and Shenzhen Basic Research Program (JSGG20180507182100698, CXB201104220032A), Shenzhen Institute of Artificial Intelligence and Robotics for Society.

References

1. Arazo, E., Ortego, D., Albert, P., O'Connor, N.E., McGuinness, K.: Unsupervised label noise modeling and loss correction (2019)
2. Arpit, D., et al.: A closer look at memorization in deep networks. In: Proceedings of the 34th International Conference on Machine Learning, vol. 70, pp. 233–242. JMLR. org (2017)
3. Barandela, R., Gasca, E.: Decontamination of training samples for supervised pattern recognition methods. In: Ferri, F.J., Iñesta, J.M., Amin, A., Pudil, P. (eds.) SSPR /SPR 2000. LNCS, vol. 1876, pp. 621–630. Springer, Heidelberg (2000). https://doi.org/10.1007/3-540-44522-6_64
4. Bengio, Y., Louradour, J., Collobert, R., Weston, J.: Curriculum learning. In: ICML, pp. 41–48. ACM (2009)
5. Brodley, C.E., Friedl, M.A.: Identifying mislabeled training data. J. Artif. Intell. Res. **11**, 131–167 (1999)
6. Deng, J., Dong, W., Socher, R., Li, L.J., Li, K., Fei-Fei, L.: ImageNet: a large-scale hierarchical image database. In: CVPR, pp. 248–255. IEEE (2009)
7. Frénay, B., Verleysen, M.: Classification in the presence of label noise: a survey. TNNLS **25**(5), 845–869 (2014)
8. Gong, Y., Ke, Q., Isard, M., Lazebnik, S.: A multi-view embedding space for modeling internet images, tags, and their semantics. IJCV **106**(2), 210–233 (2014). https://doi.org/10.1007/s11263-013-0658-4
9. Guo, H., Mao, Y., Zhang, R.: Mixup as locally linear out-of-manifold regularization. In: AAAI, vol. 33, pp. 3714–3722 (2019)
10. Guo, S., et al.: CurriculumNet: weakly supervised learning from large-scale web images. In: Ferrari, V., Hebert, M., Sminchisescu, C., Weiss, Y. (eds.) ECCV 2018. LNCS, vol. 11214, pp. 139–154. Springer, Cham (2018). https://doi.org/10.1007/978-3-030-01249-6_9
11. Han, J., Luo, P., Wang, X.: Deep self-learning from noisy labels. In: ICCV (2019)
12. Jiang, L., Zhou, Z., Leung, T., Li, L.J., Fei-Fei, L.: MentorNet: regularizing very deep neural networks on corrupted labels. arXiv preprint arXiv:1712.05055 (2017)
13. Joulin, A., van der Maaten, L., Jabri, A., Vasilache, N.: Learning visual features from large weakly supervised data. In: Leibe, B., Matas, J., Sebe, N., Welling, M. (eds.) ECCV 2016. LNCS, vol. 9911, pp. 67–84. Springer, Cham (2016). https://doi.org/10.1007/978-3-319-46478-7_5
14. Krause, J., et al.: The unreasonable effectiveness of noisy data for fine-grained recognition. In: Leibe, B., Matas, J., Sebe, N., Welling, M. (eds.) ECCV 2016. LNCS, vol. 9907, pp. 301–320. Springer, Cham (2016). https://doi.org/10.1007/978-3-319-46487-9_19
15. Lee, K.H., He, X., Zhang, L., Yang, L.: CleanNet: transfer learning for scalable image classifier training with label noise. arXiv preprint arXiv:1711.07131 (2017)

16. Lee, K.H., He, X., Zhang, L., Yang, L.: CleanNet: transfer learning for scalable image classifier training with label noise. In: CVPR, pp. 5447–5456 (2018)
17. Li, J., Wong, Y., Zhao, Q., Kankanhalli, M.S.: Learning to learn from noisy labeled data. In: CVPR, June 2019
18. Li, Q., Peng, X., Cao, L., Du, W., Xing, H., Qiao, Y.: Product image recognition with guidance learning and noisy supervision. Comput. Vis. Image Underst. **196**, 102963 (2020)
19. Li, W., Wang, L., Li, W., Agustsson, E., Van Gool, L.: Webvision database: visual learning and understanding from web data. arXiv preprint arXiv:1708.02862 (2017)
20. Li, Y., Yang, J., Song, Y., Cao, L., Luo, J., Li, L.J.: Learning from noisy labels with distillation. In: ICCV, pp. 1928–1936 (2017)
21. Lin, T.-Y., et al.: Microsoft COCO: common objects in context. In: Fleet, D., Pajdla, T., Schiele, B., Tuytelaars, T. (eds.) ECCV 2014. LNCS, vol. 8693, pp. 740–755. Springer, Cham (2014). https://doi.org/10.1007/978-3-319-10602-1_48
22. Mai, Z., Hu, G., Chen, D., Shen, F., Shen, H.T.: MetaMixUp: learning adaptive interpolation policy of MixUp with meta-learning. arXiv preprint arXiv:1908.10059 (2019)
23. Manwani, N., Sastry, P.: Noise tolerance under risk minimization. IEEE Trans. Cybern. **43**(3), 1146–1151 (2013)
24. Miranda, A.L.B., Garcia, L.P.F., Carvalho, A.C.P.L.F., Lorena, A.C.: Use of classification algorithms in noise detection and elimination. In: Corchado, E., Wu, X., Oja, E., Herrero, Á., Baruque, B. (eds.) HAIS 2009. LNCS (LNAI), vol. 5572, pp. 417–424. Springer, Heidelberg (2009). https://doi.org/10.1007/978-3-642-02319-4_50
25. Misra, I., Lawrence Zitnick, C., Mitchell, M., Girshick, R.: Seeing through the human reporting bias: visual classifiers from noisy human-centric labels. In: CVPR, pp. 2930–2939 (2016)
26. Patrini, G., Rozza, A., Menon, A.K., Nock, R., Qu, L.: Making deep neural networks robust to label noise: a loss correction approach. In: CVPR, pp. 2233–2241 (2017)
27. Reed, S., Lee, H., Anguelov, D., Szegedy, C., Erhan, D., Rabinovich, A.: Training deep neural networks on noisy labels with bootstrapping. arXiv preprint arXiv:1412.6596 (2014)
28. Rolnick, D., Veit, A., Belongie, S., Shavit, N.: Deep learning is robust to massive label noise. arXiv preprint arXiv:1705.10694 (2017)
29. Schmidt, R.A., Bjork, R.A.: New conceptualizations of practice: common principles in three paradigms suggest new concepts for training. Psychol. Sci. **3**(4), 207–218 (1992)
30. Sukhbaatar, S., Bruna, J., Paluri, M., Bourdev, L., Fergus, R.: Training convolutional networks with noisy labels. arXiv preprint arXiv:1406.2080 (2014)
31. Tanaka, D., Ikami, D., Yamasaki, T., Aizawa, K.: Joint optimization framework for learning with noisy labels. arXiv preprint arXiv:1803.11364 (2018)
32. Veit, A., Alldrin, N., Chechik, G., Krasin, I., Gupta, A., Belongie, S.J.: Learning from noisy large-scale datasets with minimal supervision, In: CVPR. pp. 6575–6583 (2017)
33. Verma, V., et al.: Manifold mixup: better representations by interpolating hidden states, pp. 6438–6447 (2019)
34. Wang, K., Peng, X., Yang, J., Lu, S., Qiao, Y.: Suppressing uncertainties for large-scale facial expression recognition. In: CVPR, June 2020
35. Xiao, T., Xia, T., Yang, Y., Huang, C., Wang, X.: Learning from massive noisy labeled data for image classification. In: CVPR, pp. 2691–2699 (2015)

36. Zhang, H., Cisse, M., Dauphin, Y.N., Lopez-Paz, D.: mixup: beyond empirical risk minimization. arXiv preprint arXiv:1710.09412 (2017)
37. Zhang, W., Wang, Y., Qiao, Y.: MetaCleaner: learning to hallucinate clean representations for noisy-labeled visual recognition. In: CVPR, pp. 7373–7382 (2019)
38. Zhuang, B., Liu, L., Li, Y., Shen, C., Reid, I.: Attend in groups: a weakly-supervised deep learning framework for learning from web data. In: CVPR, pp. 1878–1887 (2017)

Author Index

Printed in the United States
By Bookmasters